United States Naval Observatory

Zones of Stars Observed at the United States Naval Observatory

With the Meridan Transit Instrument in the Years 1846, 1847, 1848, and 1849

United States Naval Observatory

Zones of Stars Observed at the United States Naval Observatory
With the Meridan Transit Instrument in the Years 1846, 1847, 1848, and 1849

ISBN/EAN: 9783337030704

Printed in Europe, USA, Canada, Australia, Japan

Cover: Foto ©berggeist007 / pixelio.de

WASHINGTON OBSERVATIONS FOR 1870.—APPENDIX IV.

ZONES OF STARS

OBSERVED AT

THE UNITED STATES NAVAL OBSERVATORY

WITH

THE MERIDIAN TRANSIT INSTRUMENT

IN

THE YEARS 1846, 1847, 1848, AND 1849.

BY

Professor REUEL KEITH, U. S. N.
Professor MARK H. BEECHER, U. S. N.
Professor JOSEPH S. HUBBARD, U. S. N.
Lieutenant JOHN J. ALMY, U. S. N.
Lieutenant WILLIAM A. PARKER, U. S. N.

PUBLISHED BY AUTHORITY OF THE
HON. SECRETARY OF THE NAVY.

REAR-ADMIRAL B. F. SANDS, U. S. N.,
SUPERINTENDENT.

WASHINGTON:
GOVERNMENT PRINTING OFFICE
1872.

TABLE OF CONTENTS.

INTRODUCTION.

The following zones of stars were observed with the Transit Instrument in the years 1846, 1847, 1848, and 1849, and form the complement of the zones observed with the Mural Circle in the same years; it having been the intention that the Transit Instrument should furnish for the reductions the standard right ascensions and the Mural Circle the standard declinations.

§ 1.

THE INSTRUMENT.

The instrument employed in making these observations was designated in those years as the West Transit Instrument. It was made by Ertel & Son, of Munich. The objective had a focal length of 7 feet 1 inch and a clear aperture of 5.3 inches. A description of this instrument and its mountings, together with a plate, may be found in the annual volume of the Observatory for 1845.

The system of wires used in the observations of right ascension consisted of seven vertical wires. These wires were lettered in order from the clamp side of the axis, and the equatorial reductions of each wire to the mean of the seven were as follows:

Wires.	1846.	1847.	1848.	1849, Jan. 23.	1849.
	s.	s.	s.	s.	s.
A	+ 37.686	+ 37.66	+ 37.66	+ 37.482	+ 37.536
B	+ 25.030	+ 24.97	+ 24.97	+ 25.086	+ 25.005
C	+ 12.389	+ 12.37	+ 12.37	+ 12.536	+ 12.535
D	− 0.138	− 0.16	− 0.16	− 0.077	0.000
E	− 12.515	− 12.58	− 12.58	− 12.570	− 12.499
F	− 24.922	− 24.72	− 24.72	− 25.002	− 25.020
G	− 37.532	− 37.54	− 37.54	− 37.457	− 37.548

In order to observe differences of declination, the diaphragm, which carried a wire movable by a micrometer-screw, and commonly used in determining the values of the collimation and level errors, was turned round 90° and provided with a system of ten wires. The value of one revolution of the micrometer-screw was nearly 24″.8 The books and papers containing the determinations of a value of one revolution of the micrometer-screw, and the values of the intervals of the wires, together with the tables of reduction prepared by Professor Keith, have not yet been found. The only way of recovering these values will be by a comparison of the stars common to the zones observed with the Transit Instrument and the Mural Circle.

§ 2.

In the following table are given, for the year 1849, the corrections of the clock and its hourly rates, and also the quantities *m*, *n*, *c*, which furnish the reduction of the observations of right ascension to the plane of the meridian. The similar quantities, for the years 1846, 1847, and 1848 are given in the observations near the bottom of the page. These quantities are given for sidereal time.

Date.		Corr.	Hourly rate.	*m*	*n*	*c*	Zone.
1849.	h.	s.	s.				
Jan. 23,	5	− 2.21	*l.* 0.019	. .	. .	.	210
27,	5	− 6.33	*l.* 0.025	. .	. .	. .	211
Feb. 10,	7	− 25.56	*l.* 0.019	− 0.572	+ 0.813	+ 0.031	212
13,	7	− 26.91	*l.* 0.004	− 0.589	+ 0.836	+ 0.031	213
13,	7	− 26.91	*l.* 0.004	− 0.589	+ 0.836	+ 0.031	214
19,	7	− 25.26	0.000	. .	. .	. .	215
19,	7	− 25.26	0.000	. .	.	. .	216
23,	7	+ 33.88	0.000	. .	. .	. .	217
Mar. 7,	7	+ 28.32	*g.* 0.055	. .	. .	. .	218
12,	9	+ 25.07	*g.* 0.020	.	. .	. .	219
16,	9	+ 24.74	*l.* 0.010	.	. .	. .	220
16,	9	+ 24.74	*l.* 0.010	.	. .	. .	221
19,	9	+ 23.66	*l.* 0.020	.	. .	.	222
22,	9	+ 23.63	*l.* 0.043	.	. .	.	223
22,	9	+ 23.63	*l.* 0.043	. .	. .	.	224
23,	9	+ 24.66	*l.* 0.043	. .	. .	. .	225
23,	9	+ 24.66	*l.* 0.043	. .	. .	. .	226
29,	9	+ 24.61	*g.* 0.035	. .	. .	. .	227
30,	9	+ 23.76	*g.* 0.035	. .	. .	. .	228
April 2,	10	+ 24.45	*l.* 0.032	− 0.665	+ 1.026	+ 0.021	229
5,	10	+ 26.88	*l.* 0.060	. .	. .	. .	230
5,	10	+ 26.88	*l.* 0.060	. .	. .	. .	231
10,	10	+ 34.53	*l.* 0.018	.	. .	. .	232
11,	10	+ 35.62	*l.* 0.080	.	. .	. .	233
11,	10	+ 35.62	*l.* 0.080	.	. .	. .	234
12,	10	+ 38.03	*l.* 0.077	.	. .	. .	235
14,	12	+ 42.02	*l.* 0.092	.	. .	. .	236
16,	12	+ 47.06	*l.* 0.095	.	. .	. .	237
20,	12	− 4.23	*l.* 0.084	.	. .	. .	238
May 2,	13	+ 10.03	*l.* 0.072	.	. .	.	239
11,	14	+ 20.16	*l.* 0.052	.	. .	. .	240
19,	15	+ 33.11	*l.* 0.022	. .	. .	. .	241
23,	15	+ 40.97	*l.* 0.071	. .	. .	. .	242
June 18,	16	+ 21.12	*l.* 0.082	− 0.674	+ 1.306	+ 0.075	243
20,	16	+ 25.56	*l.* 0.092	. .	. .	. .	244
22,	16	+ 30.01	*l.* 0.093	. .	. .	. .	245

The readings of the meteorological instruments employed in the reductions appear to have been taken, for the most part, from the observing-books of the Mural Circle.

§ 3.

The method of reduction was the same as that used for the Mural Zones, and which has been described on pages XIX and XX of the introduction to those zones.

The observers were Professor Reuel Keith, U. S. N., who observed from the beginning until the end of the observations; Professor Mark H. Beecher, U. S. N., who began observing in the early part of 1847 and continued until the end; and Lieutenants John J. Almy and William A. Parker, U. S. N. A few zones were observed in 1847 by Professor Joseph S. Hubbard, U. S. N.

The adjustments of the instrument were made by Professor Keith, who also made observations for determining the value of a revolution of the screw of the micrometer, and for determining the values of the intervals of the wires used in observing differences of declination.

The proofs have been compared with the observing-books and made to agree with them. The results deduced by Dr. Gould, and his notes to the observations, have been printed without change. In reading the proofs I have been assisted chiefly by Mr. A. N. Skinner. Professor Nourse assisted in reading a few of the signatures.

A. HALL,
Professor of Mathematics, U. S. N.

SEPTEMBER 14, 1872.

INDEX OF ZONES.

In the following index, D denotes the declination of the middle of the zone, and the following columns give the extent in right ascension, and the number of the zone, the page on which it will be found, and the number of stars it contains.

The whole number of observations of stars in these zones is 12,033; of which—

Professor Keith made 4,881
Professor Beecher made 3,109
Professor Hubbard made 752
Lieutenant Almy made 2,146
Lieutenant Parker made 1,145

D = −41° 15′.

Right Ascension.	Zone.	Page.	No. Stars.
h. m. h. m.			
8 2 to 11 0	222	235	93

D = −39° 25′.

Right Ascension.	Zone.	Page.	No. Stars.
9 16 to 9 58	1	3	8
14 17 to 15 0	2	3	12
15 47 to 17 5	3	3	10
9 37 to 11 12	4	4	15
9 39 to 10 27	5	4	21
11 10 to 12 49	6	4	43
16 10 to 16 55	34	36	23
16 54 to 18 5	43	43	43
18 59 to 20 0	60	61	31
18 28 to 19 0	65	64	22
20 7 to 20 58	78	80	15
21 3 to 22 58	94	92	35
23 8 to 23 31	96	94	11
23 30 to 0 29	97	94	10
0 14 to 3 0	103	99	36
11 57 to 14 17	228	246	75

D = −38° 30′.

Right Ascension.	Zone.	Page.	No. Stars.
11 4 to 12 59	7	5	52
14 30 to 19 3	21	20	37
9 38 to 10 14	216	229	10

D = −38° 15′.

Right Ascension.	Zone.	Page.	No. Stars.
h. m. h. m.			
14 1 to 15 29	8	7	28
15 15 to 15 39	23	22	12
15 59 to 16 43	24	22	14
15 27 to 18 12	29	28	80
17 44 to 19 0	47	46	35
19 47 to 0 2	71	70	62
8 15 to 11 15	223	237	96

D = −37° 30′.

Right Ascension.	Zone.	Page.	No. Stars.
11 30 to 14 1	232	252	98

D = −36° 55′.

Right Ascension.	Zone.	Page.	No. Stars.
9 39 to 11 47	9	7	13
14 32 to 15 33	10	8	9
14 48 to 18 15	22	21	50
16 40 to 16 57	38	40	12
18 5 to 21 0	56	55	46
20 58 to 23 58	74	74	70
0 31 to 1 39	92	91	23
7 34 to 9 42	220	233	70

D = −36° 15′.

Right Ascension.	Zone.	Page.	No. Stars.
11 43 to 13 4	214	227	21

II—Z

D = −35° 40′.

Right Ascension.	Zone.	Page.	No. Stars.
h. m. h. m.			
10 47 to 13 41	11	8	44
14 36 to 17 43	12	9	40
17 6 to 17 35	31	33	15
18 34 to 20 20	41	42	50
17 20 to 18 32	44	44	50
20 15 to 21 4	61	61	29
21 1 to 21 59	68	67	27
22 56 to 0 5	69	68	21
19 56 to 20 45	75	75	33
22 23 to 22 59	85	87	17
8 33 to 9 14	219	233	19
13 27 to 14 40	226	243	38

D = −35° 0′.

Right Ascension.	Zone.	Page.	No. Stars.
19 28 to 19 56	67	67	9
12 35 to 13 47	224	239	45

D = −34° 25′.

Right Ascension.	Zone.	Page.	No. Stars.
9 17 to 11 34	13	10	21
14 3 to 17 0	19	17	77
17 50 to 19 16	20	19	63
16 5 to 16 15	37	40	7
17 2 to 17 56	39	40	37
20 8 to 0 0	73	72	80
22 54 to 23 20	88	89	12
1 6 to 2 53	98	94	55
0 0 to 1 3	100	97	20
7 25 to 10 2	218	231	95
11 32 to 12 36	221	235	18

D = −33° 15′.

Right Ascension.	Zone.	Page.	No. Stars.
10 54 to 13 2	14	10	60
14 25 to 17 50	15	12	80
16 44 to 18 41	26	25	90
19 48 to 20 45	27	27	35
0 1 to 0 57	93	92	16
1 57 to 3 2	102	98	19
7 54 to 9 56	213	226	55

D = −31° 55′.

Right Ascension.	Zone.	Page.	No. Stars.
h. m. h. m.			
14 38 to 15 28	28	28	16
15 37 to 21 12	30	30	145
21 20 to 22 46	48	47	27
18 14 to 18 35	62	62	16
22 18 to 1 2	87	88	50
6 0 to 8 31	217	229	88
8 21 to 12 0	225	240	148

D = −31° 15′.

Right Ascension.	Zone.	Page.	No. Stars.
10 52 to 13 55	115	107	64
20 6 to 21 48	130	130	46
22 55 to 0 58	131	131	38

D = −30° 40′.

Right Ascension.	Zone.	Page.	No. Stars.
14 44 to 16 46	18	16	18
16 26 to 19 30	25	23	113
19 52 to 21 5	49	48	34
19 32 to 20 0	72	72	17
20 59 to 0 58	77	78	104
0 56 to 3 2	99	96	54
9 6 to 11 56	107	102	58
13 40 to 17 6	119	111	119
17 17 to 19 1	120	113	87
19 35 to 21 56	134	134	85
11 55 to 13 59	227	244	75
11 30 to 14 1	232	252	98

D = −30° 0′.

Right Ascension.	Zone.	Page.	No. Stars.
16 43 to 16 53	46	46	8
8 33 to 9 35	104	100	48
9 14 to 13 0	109	103	104
14 16 to 16 0	110	105	26
15 13 to 21 2	122	117	104
15 8 to 19 6	123	119	67
23 45 to 1 24	140	146	46

D = −29° 25′.

Right Ascension.	Zone.	Page.	No. Stars.
13 47 to 16 46	17	14	111
14 54 to 19 22	32	33	91
18 1 to 18 18	40	41	12
19 31 to 21 45	59	59	80
21 25 to 23 57	76	76	80

29° 25'—Continued.

Right Ascension.	Zone.	Page.	No. Stars.
h. m. h. m.			
21 25 to 23 57	79	80	75
0 39 to 1 28	80	82	19
0 0 to 0 30	89	89	12
1 32 to 2 53	101	97	40
10 22 to 10 25	108	103	2
10 28 to 10 51	114	107	9
17 29 to 20 42	125	121	78
18 41 to 20 48	126	123	66
12 6 to 14 1	241	265	67

D = −28° 45'.

Right Ascension.	Zone.	Page.	No. Stars.
9 48 to 11 0	111	106	21
11 1 to 11 3	112	106	3
14 46 to 15 0	113	106	4
11 51 to 14 49	116	108	51
15 53 to 17 25	117	109	36
19 36 to 21 0	128	126	58
22 29 to 1 1	138	141	126
2 1 to 4 0	139	144	97
6 58 to 9 1	215	227	52

D = −28° 10'.

Right Ascension.	Zone.	Page.	No. Stars.
11 52 to 13 57	16	13	33
15 12 to 16 11	33	35	34
16 22 to 16 49	42	43	25
21 42 to 23 9	50	49	25
17 10 to 18 44	51	49	50
18 45 to 21 17	52	50	75
21 11 to 23 3	81	83	54
23 16 to 0 54	82	84	50
1 0 to 3 0	95	93	46
13 41 to 15 56	118	110	33
19 59 to 21 48	132	132	57

D = −27° 30'.

Right Ascension.	Zone.	Page.	No. Stars.
20 9 to 1 56	127	124	74
12 29 to 13 1	234	256	19

D = −26° 55'.

Right Ascension.	Zone.	Page.	No. Stars.
16 2 to 19 3	36	37	137
19 0 to 19 33	63	62	25
19 31 to 22 8	70	68	100
22 11 to 23 56	84	85	58

D = −26° 55'—Continued.

Right Ascension.	Zone.	Page.	No. Stars.
h. m. h. m.			
23 56 to 2 27	90	89	67
8 59 to 10 36	105	101	24
9 50 to 10 59	106	101	20
15 12 to 17 20	121	115	56
20 10 to 23 24	133	133	61
23 24 to 2 2	135	136	75
2 57 to 3 58	157	167	24
17 25 to 17 50	177	189	15
17 53 to 19 52	178	189	60
6 10 to 7 37	210	224	37
4 4 to 4 49	211	224	16
10 23 to 11 3	236	259	23

D = −26° 15'.

Right Ascension.	Zone.	Page.	No. Stars.
22 19 to 2 2	136	138	144
2 37 to 3 59	155	163	44
6 22 to 8 14	156	164	100
16 4 to 18 28	165	176	58
4 15 to 5 14	212	225	30

D = −25° 40'.

Right Ascension.	Zone.	Page.	No. Stars.
17 31 to 19 50	53	52	89
20 47 to 22 30	54	54	54
22 15 to 23 59	85	87	22
20 7 to 3 0	129	127	127
23 52 to 0 20	148	158	13
4 17 to 4 54	152	162	9
5 44 to 6 11	160	170	17
7 14 to 9 4	161	171	58
8 36 to 10 59	162	172	70
14 4 to 15 47	163	174	25
15 5 to 19 49	164	174	106
11 48 to 15 1	235	257	100

D = −25° 0'.

Right Ascension.	Zone.	Page.	No. Stars.
0 29 to 3 58	149	159	68
7 8 to 7 35	150	160	17
7 17 to 7 38	153	162	8
7 41 to 8 28	159	170	25
15 5 to 16 27	166	178	36
16 52 to 16 59	172	184	5
16 37 to 19 38	174	185	100
17 34 to 18 24	175	187	31
9 14 to 11 0	230	250	70
13 7 to 15 5	231	251	46

D = −24° 25′.

Right Ascension.	Zone.	Page.	No. Stars.
h. m. h. m.			
18 15 to 21 42	57	56	108
21 50 to 23 25	64	63	45
21 20 to 21 50	83	85	17
19 55 to 20 12	91	91	13
21 57 to 0 0	137	141	37
1 52 to 3 53	151	161	33
15 11 to 17 27	168	180	62
10 10 to 14 0	229	247	115

D = −23° 45′.

Right Ascension.	Zone.	Page.	No. Stars.
22 58 to 1 26	141	147	160
2 27 to 5 0	142	151	177
6 7 to 8 4	154	162	61
5 54 to 7 42	158	167	125
19 15 to 21 41	186	202	79
10 24 to 12 56	238	261	83
14 55 to 16 17	239	263	55
13 42 to 14 59	240	264	48
16 16 to 16 34	242	267	9

D = −23° 10′.

Right Ascension.	Zone.	Page.	No. Stars.
22 7 to 0 1	143	154	29
1 29 to 2 32	144	155	21
15 51 to 17 20	243	267	41

D = −22° 30′.

Right Ascension.	Zone.	Page.	No. Stars.
22 16 to 0 19	145	155	76
3 8 to 4 23	146	157	37
20 59 to 22 1	176	188	31
20 13 to 22 31	179	190	70
18 55 to 20 35	192	208	50
10 12 to 14 31	237	259	109

D = −21° 55′.

Right Ascension.	Zone.	Page.	No. Stars.
22 27 to 23 26	147	158	19
18 26 to 22 16	182	194	95
22 31 to 0 32	189	205	44
15 55 to 18 32	245	270	69

D = −21° 15′.

Right Ascension.	Zone.	Page.	No. Stars.
h. m. h. m.			
19 18 to 21 47	180	192	65
22 38 to 23 27	187	204	19
22 35 to 23 56	188	204	40
15 45 to 17 51	244	268	74

D = −20° 40′.

Right Ascension.	Zone.	Page.	No. Stars.
17 16 to 18 12	171	184	26
16 34 to 17 6	170	184	9
18 1 to 20 31	184	197	86
20 4 to 21 49	190	206	69
23 6 to 0 32	191	208	37

D = −20° 0′.

Right Ascension.	Zone.	Page.	No. Stars.
15 42 to 20 29	169	181	124
22 36 to 0 18	198	215	36

D = −19° 25′.

Right Ascension.	Zone.	Page.	No. Stars.
20 29 to 22 1	181	193	42
18 54 to 20 38	183	196	54
22 42 to 0 9	200	217	35
3 10 to 4 4	203	218	22

D = −18° 45′.

Right Ascension.	Zone.	Page.	No. Stars.
17 57 to 21 45	185	199	134
20 13 to 22 0	193	210	59
23 33 to 23 53	206	221	6
1 49 to 2 18	207	221	11
21 35 to 22 18	208	222	10
0 0 to 2 57	209	222	74

D = −17° 30′.

Right Ascension.	Zone.	Page.	No. Stars.
20 32 to 21 14	194	211	14
22 38 to 23 59	199	216	37

D = −16° 55′.

Right Ascension.	Zone.	Page.	No. Stars.
h. m. h. m.			
19 48 to 20 54	195	211	23
20 5 to 22 0	196	212	70
23 6 to 2 1	197	213	79

D = −16° 15′.

Right Ascension.	Zone.	Page.	No. Stars.
23 0 to 23 20	201	217	9
1 44 to 2 56	202	218	22

D = −15° 40′.

Right Ascension.	Zone.	Page.	No. Stars.
h. m. h. m.			
21 33 to 22 27	204	219	21
20 46 to 0 12	205	219	89

D = −9° 20′.

Right Ascension.	Zone.	Page.	No. Stars.
18 57 to 20 57	167	178	53

ZONES OF STARS

OBSERVED AT THE NAVAL OBSERVATORY

WITH

THE MERIDIAN TRANSIT INSTRUMENT

IN THE

YEARS 1846-'47-'48-'49.

1—z

ZONES OF STARS

OBSERVED WITH THE

MERIDIAN TRANSIT INSTRUMENT AT THE NAVAL OBSERVATORY IN THE YEARS 1846–'47–'48–'49.

No.	Mag.	Seconds of Transit. I.	II.	III.	IV.	V.	VI.	VII.	T.		a_1	Micrometer.			$i + a_1'$		d_1	Mean Right Ascension, 1850.0.	Mean Declination, 1850.0.
									Zone 1. March 21. A. $D_0 = -39° 0' 0''$.										
									h. m. s.		s.			r.	′ ″		″	h. m. s.	° ′ ″
1	7	. .	53.5	9.5	25.7	42.0	58.5	14.6	9 16 25.88	−	16.86	VII.	7	4.41	−31 21.4		. .	9 16 9.02	
2	9	59.5	16.4	32.5	48.5	4.5	20.5	37.5	21 48.49		16.82	VII.	5	2.0	20 55.4		. .	21 31.67	
3	10	. .	. .	33.5	49.2	5.5	21.5	38.5	27 49.38		16.77	VII.	5	2.9	20 59.9		. .	27 32.61	
4	9	. .	. .	16.5	32.2	. .	. .	. .	33 32.12		16.72	IV.	4	2.10	15 59.3		. .	33 15.40	
5	9	18.0	34.5	50.5	6.5	22.5	38.6	55.4	38 6.57		16.69	VII.	6	3.42	25 49.5		. .	37 49.88	
6	8	21.2	37.2	53.2	9.2	25.3	42.2	58.5	44 9.53		16.63	VII.	5	8.41	24 18.8		. .	43 52.90	
7	8	55.8	11.8	28.2	44.0	0.3	16.3	33.0	48 44.11		15.59	VII.	4	0.40	15 13.6		. .	48 28.52	
8	7	. .	5.3	22.4	37.9	54.0	10.2	. .	9 57 37.74	−	16.50	VI.	7	3.18	−30 39.4		. .	9 57 21.24	
									Zone 2. April 6. K. $D_0 = -39° 1' 30''$.										
1	8	. .	15.2	31.7	. .	. .	. .	. .	14 17 47.33	−	14.80	IV.	2	8.25	− 9 7.7	−	55.2	14 17 32.53	− 39 11 32.9
2	8	. .	. .	3.7	20.0	. .	. .	. .	21 19.80		14.74	.	8	. .	. .		. .	21 5.06	
3	8	. .	. .	. .	. .	21.1	. .	53.2	22 4.91		14.73	VI.	7	13.48	35 59.2		54.6	21 50.18	38 23.8
4	7	. .	2.4	18.4	34.7	50.8	. .	. .	27 34.56		14.63	VI.	6	3.29	30 47.5		53.9	27 19.93	33 11.4
5	11	. .	. .	. .	. .	37.8	54.1	. .	29 21.77		14.60	VI.	3	8.22	14 7.1		53.7	29 7.17	16 30.8
6	8	43.9	0.1	. .	. .	. .	. .	. .	38 32.41		14.45	III.	3	10.33	15 13.6		52.2	38 17.96	17 35.8
7	8	. .	. .	25.1	41.6	. .	. .	. .	39 41.38		14.41	VI.	9	11.1	44 36.1		52.1	39 26.97	46 58.2
8	8	. .	56.7	13.2	. .	. .	. .	. .	44 28.76		14.35	III.	1	11.55	5 23.0		51.4	44 14.41	7 44.4
9	7	. .	44.2	0.7	16.8	. .	. .	. .	49 16.48		14.28	II.	3	11.8	15 31.4		50.6	49 2.20	17 52.0
10	9	. .	. .	. .	. .	59.2	15.3	. .	50 43.08		14.25	V.	2	9.58	9 24.5		50.4	50 28.83	11 44.9
11	9	. .	5.6	22.0	. .	. .	. .	. .	55 37.92		14.17	III.	5	13.54	26 57.6		49.6	55 23.75	29 17.2
12	11	. .	. .	. .	. .	30.4	. .	. .	14 59 14.34	−	14.11	V.	9	1.55	−39 58.9	−	48.9	14 59 0.23	− 39 42 17.8
									Zone 3. April 6. K. $D_0 = -39° 1' 30''$.										
1	8	29.8	46.1	. .	. .	. .	50.6	. .	15 47 18.45	−	13.36	IV.	5	6.0	−22 57.1	−	33.5	15 47 5.09	− 39 25 0.6
2	10	. .	43.1	59.4	15.7	. .	. .	. .	16 0 15.59		13.17	IV.	9	5.5	41 35.3		30.6	16 0 2.42	43 35.0
3	11	. .	31.3	47.9	3.8	. .	. .	. .	5 3.56		13.10	V.	3	5.2	12 25.7		29.5	4 50.46	14 25.2
4	10	. .	46.3	2.5	. .	. .	. .	. .	10 18.34		13.03	IV.	3	7.38	13 44.9		28.2	10 5.31	15 43.1
5	9	. .	8.2	24.6	. .	56.6	. .	. .	34 40.26		12.70	IV.	1	8.55	4 22.1		22.2	34 27.56	6 14.3
6	6	. .	. .	. .	45.6	. .	. .	. .	36 45.42		12.67	VI.	1	8.00	3 54.0		21.5	36 32.75	5 45.5
7	8	. .	. .	. .	. .	. .	. .	46.8	36 58.09		12.67	VII.	3	6.59	13 23.9		21.5	36 45.42	15 15.4
8	9	. .	0.0	16.2	32.3	. .	. .	. .	16 45 32.06		12.56	III.	3	7.13	13 32.1		19.3	16 45 19.50	15 21.4
9	7	18.6	34.8	51.3	7.3	. .	. .	. .	17 2 7.11		12.36	II.	4	4.5	16 57.4		14.8	17 1 54.75	18 42.2
10	8	. .	44.2	0.6	17.0	. .	. .	. .	17 4 16.71	−	12.34	IV.	7	8.22	−33 13.7	−	14.2	17 4 4.37	− 39 34 57.9

CORRECTIONS.

Date.		Corr. of Clock.	Hourly rate.	m	n	c
1846.	h.	s.	s.	s.	s.	s.
March 21,	6	− 23.15	− 0.001	− 0.203	+ 0.325	− 0.108
April 6,	6	− 25.68	− 0.007	− 0.207	+ 0.195	+ 0.116

INSTRUMENT READINGS.

	Date.	Barom.	Thermom. At.	Thermom. Ex.
	1846. h. m.	In.	°	°
Zone 1	March 21, 9 16	30.227	53.0	45.0
Zone 2	April 3,	32.156	55.0	

REMARKS.

(1) 1 to 8. Instrument evidently not firmly clamped. Declinations rejected.

ZONE 4. APRIL 9. A. $D_0 = -39° 2' 30''$.

No.	Mag.	Seconds of Transit. I.	II.	III.	IV.	V.	VI.	VII.	T.	a_1	Micrometer.			$i + d_2$	d_1	Mean Right Ascension, 1850.0.	Mean Declination, 1850.0.
									h. m. s.	s.			r.	' "	"	h. m. s.	° ' "
1	9	17.5	34.0	50.5	6.6	22.8	38.6	55.0	9 38 6.43	– 16.77	I.	4	11.25	–20 40.2	– 7.9	9 37 49.66	– 39 23 18.1
2	8	20.6	36.9	53.2	9.4	25.8	41.7	58.0	44 9.37	16.73	II.	5	11.16	25 36.6	9.4	43 52.64	28 16.0
3	9	. .	11.6	28.0	44.2	0.2	16.2	32.2	48 43.96	16.69	II.	6	9.25	27 42.5	10.6	48 27.27	30 23.1
4	10	. .	50.0	6.3	22.6	38.5	54.6	11.0	52 22.36	16.66	II.	5	9.52	24 54.5	11.4	52 5.70	27 35.0
5	7	. .	49.5	5.6	21.9	37.9	53.7	10.2	9 58 21.57	16.61	II.	3	4.52	12 20.4	12.7	9 58 4.96	15 3.1
6	8	. .	17.9	34.1	. .	6.5	22.3	38.8	10 5 50.07	16.55	II.	3	5.14	12 31.6	14.4	10 5 33.52	15 16.0
7	10	7.2	23.6	40.1	56.2	12.2	28.5	44.5	11 56.04	16.49	I.	6	5.14	26 36.3	15.8	11 39.55	29 22.1
8	10	54.5	9.8	26.3	42.5	58.6	14.9	31.1	18 42.53	16.43	I.	6	4.42	26 19.4	17.2	18 26.10	29 6.6
9	9	17.5	33.9	50.2	6.4	22.5	38.8	54.9	25 6.31	16.37	II.	6	6.3	27 0.9	18.6	24 49.94	29 49.5
10	7	. .	. .	. .	. .	26.3	42.5	58.8	27 10.24	16.35	V.	6	1.27	25 11.4	18.9	26 53.89	26 0.3
11	10	. .	7.5	24.0	39.5	56.5	12.2	28.8	38 39.96	16.22	II.	6	3.9	25 32.6	21.3	38 23.74	28 23.9
12	9	. .	59.0	15.2	31.8	47.9	3.8	20.0	46 31.55	16.14	II.	7	10.35	34 20.9	22.7	46 15.41	37 13.6
13	8	53.6	10.0	26.4	42.5	58.6	14.9	31.1	53 42.44	16.06	I.	5	7.45	23 49.7	23.9	53 26.38	26 13.6
14	7	. .	53.8	9.7	26.6	42.6	58.4	14.6	10 57 26.16	16.01	II.	6	6.23	27 11.1	24.6	10 57 10.15	30 5.7
15	8	. .	. .	. .	18.0	34.0	49.9	6.2	11 12 17.80	– 15.63	.	8	7.39	–37 52.4	– 27.0	11 12 1.97	– 39 40 49.4

ZONE 5. APRIL 13. K. $D_0 = -39° 2' 30''$.

No.	Mag.	I.	II.	III.	IV.	V.	VI.	VII.	T.	a_1	Micrometer.			$i + d_2$	d_1	Mean Right Ascension, 1850.0.	Mean Declination, 1850.0.
1	10	. .	0.4	. .	33.2	. .	. .	. .	9 39 32.88	– 16.37	V.	5	9.56	–24 56.7	– 7.2	9 39 16.51	– 39 27 33.9
2	10	. .	. .	. .	. .	. .	27.1	43.3	40 54.80	16.36	VII.	4	9.37	9 45.4	7.7	40 38.44	22 23.1
3	.	. .	. .	. .	. .	25.1	41.2	57.3	44 8.88	16.34	.	.	. .	. . .	. .	43 52.54	
4	8	. .	. .	36.0	52.0	. .	. .	. .	46 51.77	16.32	IV.	3	10.17	15 6.6	9.7	46 35.45	17 46.3
5	7	. .	. .	27.5	43.8	. .	. .	. .	48 43.54	16.31	IV.	6	7.19	27 39.8	10.3	48 27.23	30 20.1
6	8	. .	. .	. .	. .	51.	7.4	. .	50 35.01	16.30	VII.	9	3.45	40 54.0	10.9	50 18.71	43 34.9
7	10	. .	. .	. .	. .	. .	. .	10.3	52 21.77	16.28	VII.	6	2.03	24 58.8	11.5	52 5.49	27 40.3
8	7	48.7	5.2	21.3	37.1	. .	. .	. .	58 37.13	16.24	VI.	3	4.38	12 13.2	13.4	58 20.89	14 56.6
9	8	. .	. .	. .	. .	12.7	. .	45.1	9 59 56.48	16.23	VII.	4	5.40	17 45.1	13.9	9 59 40.25	20 29.0
10	9	. .	51.2	7.3	24.0	. .	. .	. .	10 3 23.69	16.21	VI.	9	5.11	41 38.1	14.9	10 3 7.48	44 23.0
11	8	. .	17.2	. .	49.6	. .	. .	. .	5 49.40	16.18	IV.	3	5.10	12 29.8	15.7	5 33.22	15 15.5
12	7	. .	. .	23.1	39.5	. .	. .	. .	7 39.24	16.17	III.	7	8.49	33 27.3	16.2	7 23.07	36 13.5
13	7	. .	. .	. .	. .	. .	. .	37.6	7 49.15	16.17	VII.	7	4.40	31 20.4	16.3	7 32.98	34 6.7
14	9	. .	. .	. .	. .	. .	. .	27.9	9 39.47	16.15	VII.	7	7.54	32 58.8	16.9	9 23.32	35 45.7
15	9	. .	. .	39.6	. .	. .	. .	. .	11 55.54	16.13	V.	6	5.10	26 34.2	17.6	11 39.41	29 21.8
16	9	. .	. .	22.2	. .	. .	. .	. .	14 38.28	16.11	III.	8	2.8	35 4.4	18.4	14 22.17	37 52.8
17	9	. .	. .	. .	. .	. .	38.1	. .	15 5.86	16.11	VII.	9	1.16	29 36.8	18.6	14 49.75	32 25.4
18	9	13.0	29.6	. .	. .	. .	. .	. .	19 1.88	16.08	III.	6	4.45	26 21.6	19.7	18 45.80	29 11.3
19	10	. .	. .	. .	34.7	. .	. .	. .	22 34.52	16.05	IV.	6	4.44	26 21.1	20.7	22 18.47	29 11.8
20	9	16.8	33.2	49.9	5.8	. .	. .	. .	25 5.66	16.02	V.	6	5.58	26 58.6	21.5	24 49.64	29 50.1
21	7	. .	37.0	53.6	9.3	. .	. .	. .	10 27 9.32	– 16.00	IV.	5	10.3	–25 0.1	– 22.2	10 26 53.32	– 39 27 52.3

ZONE 6. APRIL 13. K. $D_0 = -39° 2' 30''$.

No.	Mag.	I.	II.	III.	IV.	V.	VI.	VII.	T.	a_1	Micrometer.			$i + d_2$	d_1	Mean Right Ascension, 1850.0.	Mean Declination, 1850.0.
1	10	. .	. .	9.3	25.2	41.6	. .	. .	11 10 25.17	– 15.54	V.	3	8.13	–14 2.5	– 33.7	11 10 9.63	– 39 27 6.2
2	9	8.4	24.5	40.6	56.7	. .	. .	. .	12 56.79	15.51	IV.	5	9.4	24 30.4	34.3	12 41.28	27 34.7
3	9	. .	23.7	39.9	56.3	12.1	. .	. .	16 56.10	15.46	VI.	8	8.28	38 17.2	35.3	16 40.64	41 22.5
4	9	. .	. .	. .	. .	28.3	44.6	. .	18 12.27	15.45	VI.	7	10.28	34 17.3	35.6	17 56.82	37 22.9
5	9	. .	. .	. .	. .	. .	35.1	51.3	11 19 2.85	– 15.44	VII.	7	7.32	–27 45.6	– 35.8	11 18 47.41	– 39 30 51.4

CORRECTIONS.

Date.		Corr. of Clock.	Hourly rate.	m	n	c
1846.	h.	s.	s.	s.	s.	s.
April 9.	6	– 24.01	+ 0.019	– 0.207	+ 0.195	+ 0.116
13.	6	– 23.44	+ 0.008	– 0.260	+ 0.344	+ 0.120

INSTRUMENT READINGS.

	Date.		Barom.	Thermom. At.	Thermom. Ex.
	1846.	h. m.	in.	°	°
Zone 6	April 13.		29.814	49.0	

REMARKS.

(4) 10. Micrometer reading assumed as 6 2r.27 instead of 6 1r.27.
(6) 5. Micrometer reading assumed as 6 7r.32 instead of 7 7r.32.

ZONE 6. APRIL 13. K. $D_o = -39° 2' 30''$—Continued.

No.	Mag.	Seconds of Transit I.	II.	III.	IV.	V.	VI.	VII.	T.	a_1	Micrometer.			$i + d_2$	d_1	Mean Right Ascension, 1850.0.	Mean Declination, 1850.0.
									h. m. s.	s.			r.	′ ″	″	h. m. s.	° ′ ″
6	11	..	..	47.2	3.4	19.6	..	..	11 22 3.33	− 15.41	V.	7	9.13	−33 39.5	− 36.5	11 21 47.92	− 39 36 46.0
7	8	..	..	23.3	39.3	55.1	..	..	24 39.04	15.37	IV.	4	6.21	18 6.6	37.0	24 23.67	21 13.6
8	6	..	..	..	..	..	18.0	34.4	26 45.86	15.35	VII.	7	9.13	33 39.9	37.5	25 30.51	36 47.4
9	9	..	59.3	15.5	32.0	..	..	..	29 31.76	15.32	IV.	8	8.54	38 30.6	38.1	29 16.44	41 38.7
10	7	..	7.8	24.2	40.6	56.4	..	..	31 40.22	15.29	VI.	5	10.33	25 15.3	38.6	31 24.93	28 23.9
11	11	..	..	27.6	..	..	..	..	34 43.79	15.25	VI.	8	14.24	41 17.8	39.2	34 28.54	44 27.0
12	10	..	43.3	59.6	15.5	32.0	..	..	38 15.59	15.21	IV.	6	4.41	26 19.6	40.0	38 0.38	29 29.6
13	5	..	..	..	..	54.5	10.6	26.8	39 38.39	15.19	VII.	8	8.54	37 59.5	40.5	39 23.20	41 9.8
14	9	..	..	..	..	..	..	34.4	41 45.85	15.15	VII.	5	8.48	24 21.7	40.6	41 30.70	27 32.3
15	8	..	..	28.1	44.1	0.3	16.4	..	43 44.06	15.13	IV.	5	7.4	23 29.6	41.1	42 28.93	26 40.7
16	8	..	..	8.3	..	40.6	..	..	44 24.32	15.12	VI.	5	10.37	25 17.3	41.3	44 9.20	28 28.6
17	9	..	42.8	..	..	..	..	3.8	49 15.25	15.06	II.	6	8.57	28 29.2	42.3	49 0.19	31 41.5
18	9	..	26.8	43.3	59.8	..	..	..	51 59.47	15.02	IV.	8	10.43	39 25.9	42.8	51 44.45	42 38.7
19	11	..	..	15.6	..	48.4	..	..	55 32.05	14.97	V.	8	11.45	39 57.3	43.4	55 17.08	43 10.7
20	12	..	..	42.3	58.4	..	..	..	57 58.27	14.94	IV.	7	4.57	31 29.6	43.9	57 43.33	34 43.5
21	11	..	..	22.2	38.2	55.0	..	..	11 59 38.29	14.91	IV.	5	2.46	21 18.6	44.2	11 59 23.38	24 32.8
22	11	..	..	59.0	15.2	31.4	..	..	12 3 15.18	14.87	V.	9	3.46	40 55.2	44.9	12 3 0.31	44 10.1
23	12	..	..	35.6	..	..	..	..	5 51.26	14.84	IV.	3	3.16	11 31.9	45.3	5 36.42	14 47.2
24	12	..	..	40.2	..	..	..	..	7 55.89	14.81	IV.	3	6.54	13 22.5	45.7	7 41.08	16 38.2
25	7	..	..	42.6	58.5	14.8	31.0	..	10 58.49	14.77	VII.	2	4.27	7 6.4	46.2	10 43.72	10 22.6
26	7	..	..	..	..	31.6	..	..	11 15.45	14.76	VI.	5	8.51	24 23.5	46.2	11 0.69	27 39.7
27	12	..	..	..	15.5	..	..	..	14 15.32	14.72	VI.	2	7.1	8 24.9	46.7	14 0.60	11 41.6
28	11	..	..	44.5	0.8	16.9	..	..	16 0.63	14.70	V.	6	9.4	28 33.0	47.0	15 45.93	31 50.0
29	8	..	..	..	..	12.8	..	45.4	17 56.71	14.67	VII.	4	11.45	20 50.4	47.3	17 42.04	24 7.7
30	11	..	..	..	..	..	3.2	19.5	19 31.01	14.65	VII.	7	15.56	37 3.5	47.6	19 16.36	40 21.1
31	.	..	12.8	29.1	..	..	..	..	22 45.07	14.60	.	.	. .	. . .	48.1	22 30.47	
32	9	..	..	..	..	59.6	..	31.8	24 43.34	14.57	VII.	5	7.16	23 35.0	48.4	24 28.77	26 53.4
33	9	..	29.2	45.7	..	..	..	..	30 1.53	14.50	II.	5	5.32	22 43.0	49.2	29 47.03	26 2.2
34	.	..	..	..	..	..	54.8	11.0	32 22.50	14.46	.	.	. .	. . .	49.5	32 8.04	
35	9	..	..	..	..	..	..	3.9	33 15.06	14.45	V.	4	7.55	18 54.3	49.7	33 0.61	22 13.8
36	9	..	..	..	25.3	..	..	..	34? 25.12	14.43	VI.	1	7.8	3 27.9	49.8	34 10.69	6 47.6
37	5	..	..	..	..	46.8	2.5	19.3	36 30.64	14.40	V.	4	6.24	18 8.0	50.1	36 16.24	21 28.1
38	9	..	..	..	..	..	..	45.0	36 56.41	14.40	VII.	5	3.52	21 51.5	50.2	36 42.01	25 11.7
39	9	..	..	23.2	..	..	..	..	38 39.13	14.37	II.	6	5.54	26 56.4	50.4	38 24.76	30 16.8
40	8	..	..	..	..	..	26.6	..	39 54.39	14.35	V.	6	3.16	25 36.4	50.6	39 40.04	28 57.0
41	10	..	..	46.7	3.1	..	..	..	45 2.87	14.27	V.	8	8.48	38 27.5	51.2	44 48.60	41 48.7
42	9	..	..	..	32.7	48.2	4.6	..	47 32.33	14.24	VI.	4	6.30	18 11.0	51.6	47 18.09	21 32.6
43	9	..	5.8	22.5	38.6	54.6	..	..	12 48 38.34	− 14.22	VI.	5	7.14	−23 34.3	− 51.7	12 48 24.12	− 39 26 56.0

ZONE 7. APRIL 15. K. $D_o = -38° 3' 40''$.

No.	Mag.	I.	II.	III.	IV.	V.	VI.	VII.	T.	a_1	Micrometer.		r.	$i + d_2$	d_1	Mean Right Ascension, 1850.0.	Mean Declination, 1850.0.
1	12	..	..	19.5	..	51.5	..	..	11 4 35.36	− 16.00	V.	4	10.16	−20 6.1	− 19.5	11 4 19.36	− 38 24 5.6
2	6	..	45.8	1.8	17.8	33.9	..	..	8 17.78	15.96	V.	8	2.58	35 29.2	20.4	8 1.82	39 29.6
3	7	..	..	..	..	..	..	47.6	10 0.05	15.94	VII.	6	5.29	26 43.2	20.9	9 44.11	30 44.1
4	6	..	53.6	9.7	25.6	41.3	57.4	..	15 25.38	15.88	VI.	3	5.54	12 52.5	22.1	15 9.50	16 54.6
5	11	..	..	..	..	34.8	..	6.7	20 18.96	15.83	VI.	7	6.42	32 22.2	23.3	20 3.13	36 25.5
6	6	..	9.8	25.8	42.2	..	..	..	23 41.66	15.79	IV.	4	10.42	20 19.3	24.0	23 25.87	24 23.3
7	11	..	..	..	55.7	..	..	..	11 24 55.52	− 15.78	VI.	4	6.15	−18 3.7	− 24.3	11 24 39.74	− 38 22 8.0

CORRECTIONS.

Date.		Corr. of Clock.	Hourly rate.	m	n	c
1846.	h.	s.	s.	s.	s.	s.
April 15.	9	− 23.94	− 0.010	− 0.280	+ 0.344	+ 0.120

INSTRUMENT READINGS.

	Date.		Barom.	Thermom. At.	Thermom. Ex.
	1846.	h. m.	in.	°	°
Zone 7	April 15.	11 4	30.042	54.0	45.5

REMARKS.

(6) 8. Minutes assumed as 25 instead of 26.
(6) 13. Minutes assumed as 39 instead of 40.
(6) 14. Minutes of transit one larger than Mural Z., April 13.
(6) 15. Minutes assumed as 42 instead of 43.

ZONE 7. APRIL 15. K. $D_0 = -38° 3' 40''$—Continued.

No.	Mag.	SECONDS OF TRANSIT. I.	II.	III.	IV.	V.	VI.	VII.	T.	a_1	MICROMETER.			$i + d_2$	d_1	Mean Right Ascension, 1850.0.	Mean Declination, 1850.0.
									h. m. s.	s.			r.	' "	"	h. m. s.	° ' "
8	11	..	51.2	7.2	..	..	..	..	11 29 23.18	— 15.73	V.	8	9.28	—38 46.9	— 25.2	11 29 7.45	— 38 42 52.1
9	6	..	..	..	25.7	41.7	..	..	30 25.64	15.72	VI.	6	7.8	27 33.6	25.4	30 9.92	31 39.2
10	8	..	..	..	33.0	48.8	..	..	31 32.84	15.70	VI.	4	11.52	20 54.6	25.7	31 17.14	25 0.3
11	9	..	..	..	..	29.8	46.1	..	32 14.18	15.69	VII.	6	10.2	29 1.6	25.9	31 58.49	33 7.5
12	8	..	..	..	..	..	..	53.2	33 5.26	15.68	VII.	3	9.49	14 51.3	26.0	32 49.58	18 57.3
13	8	..	..	..	..	2.5	..	34.0	35 46.34	15.65	VII.	5	5.56	22 54.5	26.6	35 30.69	27 1.1
14	9	..	..	..	..	..	36.1	52.2	36 4.37	15.65	VII.	7	6.53	32 27.5	26.6	35 48.72	36 34.1
15	10	..	0.7	16.8	..	..	..	..	43 32.46	15.56	V.	5	3.23	21 37.5	28.2	43 16.90	25 45.7
16	11	..	..	..	37.0	53.1	..	..	44 37.01	15.55	VI.	8	6.31	37 17.0	28.4	44 21.46	41 25.4
17	11	..	..	..	..	..	0.3	..	46 28.56	15.53	VI.	5	8.32	24 14.0	28.7	46 13.03	28 22.7
18	11	..	..	..	..	34.0	50.0	6.0	47 18.21	15.52	VII.	6	7.15	27 37.0	28.9	47 2.69	31 45.9
19	5	..	..	..	34.2	50.0	..	..	48 34.09	15.51	V.	10	6.1	47 5.3	29.1	48 18.58	51 14.4
20	12	..	..	..	7.5	23.2	..	..	50 7.30	15.49	VI.	4	7.16	18 34.6	29.4	49 51.81	22 44.0
21	9	..	..	..	..	..	..	44.5	51 56.95	15.47	VII.	9	9.25	43 45.5	29.8	51 41.48	47 54.3
22	9	..	..	..	..	..	49.9	5.8	53 18.14	15.45	VII.	8	4.42	36 21.4	30.0	53 2.69	40 31.4
23	11	..	..	..	..	..	..	31.2	54 43.34	15.43	VII.	2	5.45	7 46.9	30.3	54 27.91	11 57.2
24	8	..	..	..	..	..	39.7	56.1	55 8.20	15.43	VII.	2	5.10	7 29.2	30.4	54 52.77	11 39.6
25	6	..	..	..	..	..	47.1	3.8	56 15.73	15.40	VII.	2	2.37	6 11.7	30.6	56 0.33	10 22.3
26	8	..	17.8	..	49.8	..	..	..	11 58 49.59	15.39	IV.	4	7.53	18 53.7	31.1	11 58 34.20	23 4.8
27	8	..	14.6	31.0	..	..	..	..	12 10 46.62	15.21	IV.	4	14.10	22 4.8	33.1	12 10 31.41	26 17.9
28	7	..	..	..	..	17.2	33.2	..	11 1.35	15.21	VI.	7	13.56	36 1.6	33.2	10 46.14	40 14.8
29	7	..	..	..	..	..	36.0	52.2	12 4.38	15.20	VII.	6	8.54	28 27.1	33.3	11 49.18	32 40.4
30	10	..	..	..	..	..	..	7.2	14 19.52	15.17	VII.	5	10.24	25 10.4	33.7	14 4.35	29 24.1
31	7	..	..	..	..	19.4	35.1	..	15 3.47	15.16	VI.	3	14.35	17 16.6	33.8	14 48 31	21 30.4
32	6	..	..	..	..	0.2	29.1	45.1	16 57.38	15.13	VII.	5	8.17	24 6.0	34.1	16 42.25	28 20.1
33	10	..	8.1	24.6	..	..	..	..	19 40.28	15.09	IV.	6	7.19	27 39.6	34.5	19 25.19	31 54.1
34	8	..	..	..	20.8	36.5	..	..	20 20.63	15.08	VI.	7	13.53	36 0.7	34.6	20 5.55	40 15.3
35	8	..	..	..	38.9	54.7	..	..	21 38.80	15.07	VI.	9	14.30	46 20.5	34.9	21 23.73	50 35.4
36	6	..	17.6	33.7	49.8	..	..	..	28 49.50	14.97	IV.	5	13.37	26 48.8	35.9	28 34.53	31 4.7
37	7	..	19.7	35.7	51.8	..	..	..	30 51.60	14.94	IV.	6	13.37	30 51.2	36.2	30 36.66	35 7.4
38	9	..	..	..	..	..	..	54.5	31 6.93	14.94	VII.	8	6.18	36 39.6	36.2	30 51.99	40 55.8
39	8	..	..	..	..	..	48.8	4.6	33 16.99	14.91	VII.	7	10.2	34 3.3	36.5	33 2.08	38 19.8
40	10	..	..	..	28.0	43.6	..	..	34 27.79	14.89	VI.	8	7.34	37 48.9	36.6	34 12.90	42 5.5
41	9	..	..	..	..	..	28.2	44.4	35 56.60	14.87	VII.	8	5.13	36 37.1	36.9	35 41.73	40 54.0
42	10	..	..	12.1	28.0	43.8	..	..	39 27.98	14.83	VI.	8	2.35	35 17.3	37.5	39 13.15	39 34.8
43	12	..	..	..	57.3	..	..	..	43 57.13	14.76	IV.	5	3.9	21 30.5	37.9	43 42.37	25 48.4
44	7	..	..	..	..	2.7	..	..	45 46.82	14.74	IV.	6	3.10	25 33.4	38.1	45 32.08	29 51.5
45	6	..	..	..	..	..	44.2	0.4	46 12.59	14.73	VII.	7	2.39	30 18.7	38.2	45 57.86	34 36.9
46	9	..	..	..	..	..	13.5	29.7	47 41.90	14.71	VII.	8	6.39	37 20.7	38.3	47 27.19	41 39.0
47	7	..	18.2	44.3	..	..	..	..	49 50.10	14.68	III.	5	6.14	23 5.4	38.6	49 35.42	27 24.0
48	7	..	..	..	49.0	5.2	..	..	51 49.06	14.65	VII.	3	7.6	13 28.6	38.8	51 34.41	17 47.4
49	10	..	..	23.8	39.8	..	..	..	52 39.65	14.64	V.	5	9.46	24 51.7	38.9	52 25.01	29 10.6
50	6	..	..	..	41.4	57.3	..	..	53 41.36	14.63	VI.	9	10.25	44 16.3	39.0	53 26.73	48 35.3
51	10	..	..	..	..	..	7.3	23.2	55 35.54	14.60	VII.	8	8.50	38 27.1	39.2	55 20.94	42 46.3
52	10	..	..	..	..	46.8	2.8	..	12 58 30.96	— 14.56	VI.	7	14.42	—36 25.5	— 39.5	12 58 16.40	— 38 40 45.0

CORRECTIONS.							INSTRUMENT READINGS.				
Date.		Corr. of Clock.	Hourly rate.	*m*	*n*	*c*	Date.		Barom.	THERMOM. At.	Ex.
1846.	h.	s.	s.	s.	s.	s.	1846.	h. m.	In.	°	°

REMARKS.

(7) 25. Minutes of transit assumed as 56 instead of 57; *vide* Lacaille 4993 and Gilliss' Santiago Observations.

ZONE 8. APRIL 15. K. $D_0 = -37° 55' 40''$.

No.	Mag.	Seconds of Transit. I.	II.	III.	IV.	V.	VI.	VII.	T.	a_1	Micrometer.			$i + d_s$	d_1	Mean Right Ascension, 1850.0.	Mean Declination, 1850.0.
									h. m. s.	s.			r.	′ ″	″	h. m. s.	° ′ ″
1	10	..	..	..	34.5	49.6	..	..	14 1 34.06	− 13.64	VI.	9	14.00	−46 5.3	− 10.4	14 1 20.42	− 38 41 55.7
2	11	..	..	52.5	8.7	..	..	..	6 8.53	13.57	VI.	7	8.11	33 7.4	10.8	5 54.96	28 58.2
3	11	..	..	..	..	29.2	..	0.8	8 13.31	13.54	VII.	8	10.16	39 10.8	11.0	7 59.77	35 1.8
4	10	..	..	..	..	..	..	2.8	9 15.02	13.52	VII.	3	10.36	15 15.1	11.1	9 1.50	11 6.2
5	9	..	..	35.9	51.3	7.7	..	..	13 51.42	13.44	V.	1	6.55	3 22.4	11.4	13 37.98	59 13.8
6	10	..	..	..	23.2	39.2	..	..	15 23.14	13.43	VII.	2	0.00	4 52.6	11.6	15 9.71	0 44.2
7	9	..	..	42.5	58.3	14.1	..	..	18 58.10	13.38	VI.	3	8.40	14 16.7	11.8	17 44.72	10 8.5
8	5.6	..	35.9	52.0	7.8	..	..	..	22 7.66	13.33	IV.	3	12.4	16 0.3	12.0	21 54.33	11 52.3
9	11	..	..	6.3	23.2	..	..	..	22.78	(13.29)	VI.	7	14.50	36 29.6	(12.2)	9.49	32 21.8
10	6	..	..	12.3	28.0	43.8	..	..	29 27.80	13.22	VI.	3	4.50	12 20.1	12.5	29 14.67	8 12.6
11	10	..	..	42.0	..	13.6	..	..	36 57.65	13.11	VI.	2	11.51	10 52.8	12.9	35 44.54	6 45.7
12	10	..	..	54.7	10.6	26.6	..	..	39 10.60	13.08	VI.	6	7.51	27 55.5	13.0	38 57.52	23 48.5
13	10	..	56.9	13.2	29.0	..	..	..	43 28.83	13.02	V.	4	7.8	18 30.5	13.2	43 15.81	14 23.7
14	9	..	..	..	19.9	36.1	..	..	45 20.00	12.99	VI.	10	3.6	45 36.3	13.2	45 7.01	41 29.5
15	9	..	..	0.6	16.2	32.1	..	..	49 16.24	12.93	V.	5	7.56	23 55.9	13.4	49 3.31	19 49.3
16	11	..	..	47.6	..	19.9	..	..	14 59 3.87	12.79	VI.	8	3.42	35 51.3	13.7	14 58 57.08	31 45.0
17	6	..	..	28.2	43.8	0.0	..	..	15 0 43.83	12.77	VI.	4	4.5	16 57.8	13.7	15 0 31.06	12 51.5
18	10	..	..	..	..	39.9	55.2	..	1 23.77	12.76	VII.	3	8.7	13 59.6	13.7	1 11.01	9 53.3
19	11	..	..	58.9	14.8	30.0	..	..	8 14.40	12.66	VII.	4	6.40	18 16.1	13.8	8 1.74	14 9.9
20	11	..	..	12.2	28.4	..	0.1	..	12 28.27	12.60	VII.	9	11.38	44 53.0	13.9	12 15.67	40 46.9
21	5	..	18.3	34.2	50.0	6.2	..	..	15 49.82	12.55	VI.	3	11.55	15 55.0	13.9	15 37.27	11 48.9
22	6	..	..	31.8	47.9	3.9	..	..	19 47.65	12.49	V.	3	1.20	10 33.8	13.9	19 35.16	6 27.7
23	9	..	..	..	..	51.3	..	23.3	20 35.29	12.48	VII.	2	5.55	7 52.0	14.0	20 22.81	3 46.0
24	7	..	..	..	..	..	3.1	19.5	22 31.52	12.46	VII.	2	10.44	10 18.5	13.9	22 19.06	6 12.4
25	10	..	37.2	..	9.4	..	..	..	25 9.24	12.42	V.	8	13.13	40 41.0	13.9	24 56.82	36 34.9
26	9	..	..	..	57.1	..	..	..	27 56.93	12.38	V.	8	10.45	39 26.0	13.9	27 44.55	35 19.9
27	6	..	..	..	..	10.3	..	..	28 54.42	12.36	VI.	9	9.34	43 50.3	13.9	28 42.06	39 44.2
28	6	..	..	..	..	..	..	51.4	15 29 3.72	− 12.36	VII.	9	5.57	−42 0.0	− 13.9	15 28 51.36	− 38 37 53.9

ZONE 9. APRIL 16. A. $D_0 = -36° 39' 10''$.

No.	Mag.	I.	II.	III.	IV.	V.	VI.	VII.	T.	a_1	Micrometer.			$i + d_s$	d_1	Mean Right Ascension, 1850.0.	Mean Declination, 1850.0.
1	9	53.0	8.9	24.6	40.4	55.6	11.5	27.5	9 39 40.25	− 16.45	V.	6	7.58	−27 59.2	− 10.5	9 39 23.78	− 37 7 19.7
2	11	..	..	59.0	15.1	30.5	46.2	1.7	9 45 14.87	16.42	VI.	9	4.15	41 8.0	11.6	9 44 58.45	20 29.6
3	8	..	..	28.0	44.3	59.8	15.1	30.8	10 22 43.96	16.18	IV.	8	8.6	38 4.8	18.8	10 22 27.78	17 33.6
4	8	..	35.0	50.8	6.9	22.2	38.0	53.3	27 6.60	16.14	III.	9	5.37	41 49.7	19.6	26 50.46	21 19.3
5	9	..	..	..	..	23.0	38.1	53.5	38 6.95	16.06	VI.	9	5.41	41 51.5	21.5	37 50.89	21 23.0
6	9	1.0	17.0	33.0	48.5	4.0	19.2	35.0	50 48.23	15.94	II.	5	6.0	22 57.0	23.3	50 32.29	2 30.3
7	8.9	29.0	45.0	1.1	16.5	32.2	47.6	3.4	10 55 16.39	15.90	II.	5	4.15	22 3.8	23.9	10 55 0.49	1 37.7
8	12	..	..	..	19.0	35.0	50.6	6.0	11 9 19.13	15.76	IV.	4	11.45	20 51.4	25.9	11 9 3.37	37 0 27.3
9	7	..	..	18.2	33.2	49.2	4.7	20.3	18 33.37	15.66	III.	3	11.29	15 42.9	27.0	18 17.71	36 55 19.9
10	8	..	..	..	..	..	21.0	36.8	32 49.81	15.53	VI.	8	6.	37 0.7	28.5	32 34.28	37 16 39.2
11	9	..	..	..	16.3	32.0	47.2	3.5	38 16.20	15.46	V.	2	5.32	7 41.7	29.1	38 0.74	36 47 20.8
12	9	..	..	42.8	58.2	13.5	29.0	44.9	44 57.92	15.39	III.	3	10.34	15 15.0	29.7	44 42.53	54 54.7
13	6	..	..	..	..	..	41.2	57.2	11 47 10.07	− 15.36	VI.	3	9.42	−15 18.7	− 29.9	11 46 54.71	− 36 54 58.6

CORRECTIONS.

Date.		Corr. of Clock.	Hourly rate.	m	n	c
1846.	h.	s.	s.	s.	s.	s.
April 15,	9	− 23.94	− 0.010	− 0.280	+ 0.344	+ 0.120
16,	9	− 23.83	− 0.012	− 0.280	+ 0.344	+ 0.120

INSTRUMENT READINGS.

Date.		Barom.	Thermom. At.	Thermom. Ex.
1846.	h. m.	in.	°	°

REMARKS.

(9) 13. Micrometer reading assumed as $10^r.42$ instead of $9^r.42$.

ZONE 10. APRIL 16. A. $D_0 = -36° 40' 0''$.

No.	Mag.	SECONDS OF TRANSIT. I.	II.	III.	IV.	V.	VI.	VII.	T.	a_1	MICROMETER.			$i + d_2$	d_1	Mean Right Ascension, 1850,0.	Mean Declination, 1850,0.
									h. m. s.	s.			r.	′ ″	″	h. m. s.	° ′ ″
1	4	5.2	21.2	37.0	52.8	8.3	23.8	39.3	14 32 52.53	− 13.16	II.	6	6.51	−28 25.6	− 17.6	14 32 39.37	− 37 8 43.2
2	5	53.0	8.8	24.6	40.2	56.0	11.6	27.1	43 40.22	13.00	II.	7	2.25	30 42.8	16.3	43 27.22	10 58.4
3	7	..	..	..	59.5	15.0	30.9	46.4	49 59.46	12.91	IV.	8	4.47	36 24.0	15.4	49 46.55	16 39.4
4	9	..	11.4	26.5	..	57.5	13.0	29.0	55 42.03	12.83	III.	4	9.4	24 29.0	14.6	55 29.20	4 43.6
5	9	..	..	8.0	23.0	39.3	54.5	10.5	14 59 23.33	12.78	III.	4	10.22	20 9.3	14.1	14 59 10.55	37 0 23.4
6	9	..	..	10.8	26.0	41.5	57.2	13.5	15 3 26.05	12.73	III.	4	3.52	16 51.8	13.5	15 3 13.32	36 57 5.3
7	7	59.2	15.0	31.5	47.0	2.5	18.1	33.8	12 46.73	12.60	I.	5	10.38	25 17.5	12.0	12 34.13	37 5 29.5
8	9	..	..	..	..	10.0	25.5	42.0	18 54.55	12.51	VII.	5	12.	25 59.0	10.9	18 42.04	37 6 9.9
9	8	..	..	50.5	6.0	22.2	37.6	53.1	15 33 6.13	− 12.32	IV.	3	12.	−15 58.7	− 8.4	15 32 53.81	− 36 56 7.1

ZONE 11. APRIL 17. K. $D_0 = -35° 22' 10''$.

No.	Mag.	I.	II.	III.	IV.	V.	VI.	VII.	T.	a_1	MICROMETER.			$i + d_2$	d_1	Mean Right Ascension, 1850,0.	Mean Declination, 1850,0.
1	8	..	..	..	..	..	48.3	3.7	10 47 17.51	− 16.54	VII.	4	3.53	−16 54.4	− 34.2	10 47 0.97	− 35 39 38.6
2	11	..	..	11.0	26.5	..	..	..	52 26.29	16.50	V.	4	8.38	19 17.0	34.9	52 9.79	42 1.9
3	8	..	59.3	14.7	..	..	..	..	10 57 30.14	16.45	III.	7	6.7	32 4.0	35.5	10 57 13.69	54 49.5
4	10	..	31.4	47.0	..	..	..	..	11 9 1.96	16.35	IV.	2	4.5	6 58.6	36.8	11 8 45.61	29 45.4
5	8	..	..	..	48.7	4.0	..	..	10 48.54	16.33	IV.	4	10.30	20 13.7	37.0	10 32.21	43 0.7
6	9	..	47.7	3.4	18.3	..	..	..	15 18.35	16.29	IV.	4	4.8	17 0.4	37.4	15 2.06	35 39 47.8
7	10	..	0.8	16.4	..	..	..	..	19 31.86	16.24	IV.	9	3.23	40 41.0	37.9	19 15.62	36 3 28.9
8	10	..	..	..	18.5	33.7	..	..	25 18.27	16.19	VI.	2	3.59	6 55.3	38.3	25 2.08	35 29 43.6
9	10	..	50.3	6.1	21.4	..	..	..	30 21.29	16.14	IV.	7	4.11	31 5.3	38.8	30 5.15	53 54.1
10	11	..	..	..	..	..	14.3	30.2	30 43.77	16.13	VII.	5	2.31	21 11.0	38.8	30 27.64	43 59.8
11	10	..	18.4	54.1	..	..	..	..	36 48.96	16.07	IV.	1	10.25	5 10.8	39.2	36 32.89	28 0.0
12	9	..	..	..	..	35.7	..	..	37 40.23	16.06	V.	2	13.40	11 49.5	39.3	37 24.17	34 38.8
13	9	..	..	..	..	..	..	52.1	38 5.64	16.05	VII.	2	12.41	11 19.2	39.4	37 49.59	35 34 8.6
14	9	..	..	39.1	54.9	..	..	..	43 54.69	15.99	V.	9	8.27	43 14.8	39.7	43 38.70	36 6 4.5
15	8	..	27.3	43.1	58.4	..	..	..	45 58.20	15.96	IV.	4	13.43	21 51.4	39.9	45 42.24	35 44 41.3
16	8	..	..	39.3	54.4	9.7	..	..	54.22	. .	V.	1	8.18	4 6.5	. .	. . .	. . .
17	10	..	..	5.7	21.3	..	..	..	11 55 21.22	15.86	VII.	10	4.58	46 30.6	40.4	11 55 5.36	36 9 21.0
18	10	..	..	29.5	..	..	1.5	..	12 1 44.57	15.80	VI.	6	9.17	28 38.8	40.7	12 1 28.77	35 51 29.5
19	8	48.2	..	..	..	..	50.2	..	5 34.53	15.75	VI.	1	13.29	6 43.7	40.9	5 18.78	29 34.6
20	9	..	5.0	..	..	..	..	21.5	5 35.44	15.75	VII.	4	10.58	20 27.3	40.9	5 19.69	43 18.2
21	9.10	..	..	58.6	13.2	..	..	..	10 12.96	15.70	IV.	1	14.24	7 11.8	41.1	9 57.26	30 2.9
22	10	..	..	..	..	56.2	11.5	27.0	12 40.78	15.67	VII.	5	10.11	25 3.9	41.1	12 25.11	47 55.0
23	12	..	..	29.9	45.3	0.7	..	..	17 45.19	15.61	V.	5	12.16	26 2.6	41.3	17 29.58	48 53.9
24	10	..	45.8	1.3	16.8	32.1	..	..	20 16.65	15.58	V.	6	11.18	29 40.3	41.3	20 1.07	52 31.6
25	11	..	..	38.5	53.6	8.8	..	..	25 53.42	15.51	V.	2	10.29	10 12.9	41.4	25 37.91	33 4.3
26	11	..	..	..	5.9	21.1	36.2	..	27 5.66	15.50	VII.	4	6.36	18 14.8	41.4	26 50.16	41 6.2
27	9	..	23.0	38.6	54.2	..	24.7	..	29 53.95	15.46	VI.	7	4.51	31 25.3	41.5	29 38.49	54 16.8
28	12	..	..	0.8	16.0	..	..	..	35 15.80	15.40	V.	3	11.00	15 28.8	41.5	35 0.40	38 20.3
29	7.8	..	..	..	8.0	23.2	38.3	54.4	36 7.80	15.39	VII.	2	7.30	8 41.8	41.5	35 52.41	31 33.3
30	12	..	23.7	39.3	54.6	..	..	..	40 54.47	15.33	V.	5	5.45	22 49.7	41.5	40 39.14	45 41.2
31	9	..	10.4	26.2	41.1	56.5	..	..	50 41.00	15.20	V.	1	9.57	4 56.6	41.4	50 25.80	27 48.0
32	12	..	..	40.8	56.1	11.5	..	..	52 56.00	15.17	VI.	4	15.06	22 33.1	41.3	52 40.83	45 24.4
33	11	..	..	40.7	55.7	..	..	..	12 58 55.56	15.09	V.	1	9.29	4 42.4	41.1	12 58 40.47	27 33.5
34	10	..	..	26.8	42.3	..	..	..	13 1 42.06	15.06	V.	4	12.0	20 59.2	41.1	13 1 27.00	43 50.3
35	8	..	29.0	44.6	0.0	15.1	..	..	13 5 59.68	− 15.00	VI.	3	3.23	−11 37.3	− 40.9	13 5 44 68	− 35 34 28.2

CORRECTIONS.

Date.	Corr. of Clock.	Hourly rate.	m	n	c
1846. h.	s.	s.	s.	s.	s.
April 17, 9	− 24.50	− 0.012	− 0.280	+ 0.344	+ 0.120

INSTRUMENT READINGS.

	Date.	Barom.	THERMOM. At.	THERMOM. Ex.
	1846. h. m.	in.	°	°
Zone 10	April 16,	30.337	54.	46.
Zone 11	17,	29.950	63.0	62.5

REMARKS.

(10) 1. Micrometer reading assumed as 8ʳ.51 instead of 6ʳ.51.
(10) 2. Micrometer reading assumed as 3ʳ 25 instead of 2ʳ.25.
(10) 4. Micrometer assumed as 5 instead of 4.

ZONE 11. APRIL 17. K. D₀ = −35° 22′ 10″—Continued.

No.	Mag.	I.	II.	III.	IV.	V.	VI.	VII.	T.	a_1	MICROMETER.		r	$i + d_2$	d_1	Mean Right Ascension, 1850.0.	Mean Declination, 1850.0.
									h. m. s.	s.			r	′ ″	″	h. m. s.	° ′ ″
36	4.5	. .	. .	10.2	26.0	41.3	56.4	. .	13 12 25.75	− 14.92	VI.	7	6.35	−32 18.0	− 40.6	13 12 10.83	− 35 55 8.6
37	12	. .	. .	. .	. .	. .	55.0	. .	13 24.37	14.91	VII.	5	5.43	22 48.2	40.6	13 9.46	35 45 38.8
38	10	. .	. .	. .	. .	. .	58.7	. .	24 27.94	14.76	VII.	9	6.13	42 6.5	40.0	24 13.18	36 4 56.5
39	10	. .	36.6	52.2	7.9	. .	. .	. .	28 7.49	14.72	VII.	5	10.56	25 26.6	39.8	27 52.77	35 48 16.4
40	10	. .	. .	46.1	. .	16.4	. .	. .	31 0.96	14.68	VI.	1	11.36	5 46.5	39.6	30 46.28	28 36.1
41	12	. .	. .	. .	47.2	. .	. .	. .	35 47.03	14.61	VI.	5	10.32	25 15.4	39.2	35 32.42	48 4.6
42	7	. .	. .	56.7	. .	27.6	. .	. .	37 11.88	14.59	V.	1	14.13	7 6.1	39.1	36 57.29	29 55.2
43	11	. .	. .	. .	. .	. .	44.0	59.2	39 13.18	14.56	VII.	6	8.52	28 25.9	39.0	38 58.62	51 14.9
44	10	. .	. .	. .	10.9	. .	41.5	. .	13 41 10.78	− 14.54	VII.	6	9.17	−28 38.5	− 38.8	13 40 56.24	− 35 51 27.3

ZONE 12. APRIL 17. K. D₀ = −35° 21′ 50″.

No.	Mag.	I.	II.	III.	IV.	V.	VI.	VII.	T.	a_1	MICROMETER.		r	$i + d_2$	d_1	Mean Right Ascension, 1850.0.	Mean Declination, 1850.0.
1	10	. .	. .	. .	20.2	35.8	. .	. .	14 36 20.17	− 13.84	VI.	2	5.41	− 7 46.9	− 48.7	14 36 6.33	− 35 30 25.6
2	7	47.4	3.3	18.8	34.4	49.5	. .	. .	38 31.19	13.60	VI.	8	6.50	37 25.4	48.4	38 20.39	36 0 3.8
3	9	. .	. .	33.2	. .	. .	. .	34.7	44 48.43	13.72	III.	6	4.20	26 8.7	47.2	44 34.71	35 48 45.9
4	10	. .	. .	. .	7.6	. .	38.1	. .	49 7.50	13.65	VI.	1	9.35	4 45.3	46.7	48 53.85	27 22.0
5	8	. .	. .	. .	53.0	. .	23.1	. .	49 52.71	13.64	VI.	1	6.1	2 56.9	46.6	49 39.07	25 33.5
6	11	. .	45.3	0.9	16.2	. .	. .	. .	54 16.04	13.59	VI.	4	7.57	18 56.0	45.9	54 2.45	41 31.9
7	7	. .	32.2	48.0	3.3	18.4	. .	. .	56 3.04	13.56	V.	4	6.6	18 0.1	45.6	55 49.48	40 35.7
8	11	. .	16.0	31.7	47.2	. .	. .	. .	14 59 46.96	13.51	VI.	6	9.1	28 30.7	44.9	14 59 33.45	51 5.6
9	7	8.2	23.7	39.3	54.2	. .	. .	. .	15 3 54.26	13.46	V.	2	7.27	8 40.8	44.2	15 3 40.80	31 15.0
10	4.5	44.9	0.7	16.0	31.5	46.7	2.0	. .	10 31.32	13.37	VI.	4	10.23	20 •9.9	43.0	12 17.95	42 42.9
11	11	. .	40.4	56.0	11.2	. .	. .	. .	15 11.08	13.31	VI.	3	14.49	17 24.4	42.1	14 57.77	35 39 56.5
12	8	. .	. .	. .	. .	. .	. .	42.0	16 55.87	13.29	VII.	8	14.24	41 15.9	41.7	16 42.58	36 3 47.6
13	9	. .	3.9	19.5	34.8	50.1	. .	. .	47 34.54	12.89	VI.	2	3.20	6 35.6	35.3	47 21.65	35 39 0.9
14	8	. .	. .	. .	. .	. .	. .	35.9	47 49.57	12.89	VII.	5	4.23	22 7 7	35.3	47 36.68	44 33.0
15	11	. .	42.2	57.7	13.0	. .	. .	. .	52 12.91	12.83	VI.	5	6.11	23 2.8	34.3	52 0.08	45 27.1
16	11	. .	25.7	41.2	56.3	. .	. .	. .	55 56.34	12.79	V.	5	5.24	22 39.1	33.4	55 43.55	45 2.5
17	8	42.4	58.1	13.8	29.0	. .	. .	. .	15 57 28.93	12.77	IV.	6	5.48	26 53.3	33.0	15 57 16.16	49 16.3
18	12	. .	52.2	8.0	23.2	. .	. .	. .	16 14 23.05	12.56	VI.	4	10.5	20 0.8	28.9	16 14 10.49	42 19.7
19	12	. .	. .	9.6	. .	. .	. .	. .	25 24.70	12.43	V.	3	12.56	16 27.4	26.2	25 12.27	38 43.6
20	8	. .	. .	43.3	58.5	13.8	. .	. .	25 58.35	12.43	VI.	3	7.57	13 56.0	26.0	25 45.92	35 36 12.0
21	8	. .	. .	40.4	56.4	10.2	. .	. .	39 56.10	12.28	VII.	9	13.51	45 58.3	22.4	39 43.82	36 8 10.7
22	10	. .	. .	. .	50.3	5.9	21.2	. .	43 50.39	12.24	VII.	8	4.11	36 4.6	21.3	43 38.15	35 58 15.9
23	7	. .	18.2	33.8	49.2	4.5	. .	. .	49 49.00	12.17	V.	4	9.31	19 43.8	19.7	49 36.83	41 53.5
24	8	4.5	20.2	35.7	51.2	. .	. .	. .	56 51.00	12.10	V.	5	11.38	25 48.4	17.6	56 38.90	47 56.0
25	9	. .	. .	. .	20.7	. .	. .	. .	16 58 20.53	12.09	V.	1	8.39	4 17.4	17.3	16 58 8.44	26 24.7
26	6.7	. .	. .	18.1	33.4	48.6	3.7	. .	17 5 33.16	12.02	VI.	2	12.54	11 26.1	15.1	17 5 21.14	33 31.2
27	10	. .	. .	31.1	46.7	2.0	. .	. .	8 46.54	11.98	VI.	7	7.56	32 59.0	14.3	8 34.56	55 3.3
28	7.8	. .	14.6	30.3	45.5	0.1	. .	. .	11 45.23	11.96	VI.	5	7.6	23 30.5	13.4	11 33.27	45 33.9
29	10	. .	. .	35.9	51.2	. .	. .	. .	15 51.00	11.92	V.	3	9.22	14 39.2	12.2	15 39.08	36 41.4
30	10	. .	52.9	8.2	. .	39.0	. .	. .	19 23.54	11.89	V.	5	3.44	21 48.5	11.2	19 11.65	43 49.7
31	10	. .	. .	. .	. .	. .	44.3	0.0	20 13.67	11.88	VII.	5	9.3	24 29.4	10.5	20 1.79	46 29.9
32	9	. .	51.5	7.0	22.3	. .	. .	. .	23 22.13	11.85	V.	3	10.33	15 15.1	10.0	23 10.28	37 15.1
33	11	. .	. .	15.2	30.7	46.0	. .	. .	27 30.45	11.81	V.	3	11.2	15 29.8	8.8	27 18.64	37 28.8
34	9	. .	13.7	29.2	44.5	. .	. .	. .	30 44.29	11.78	V.	2	12.47	11 22.7	7.9	30 32.51	33 20.6
35	9	. .	. .	9.9	25.1	40.3	. .	. .	32 24.96	11.77	VI.	4	12.18	21 8.1	7.4	32 13.19	43 5.5
36	8	. .	39.5	55.4	10.3	25.8	. .	. .	17 34 10.20	− 11.75	V.	1	10.57	− 5 26.9	− 6.8	17 33 58.45	− 35 27 23.7

CORRECTIONS.

Date.	Corr. of Clock.	Hourly rate.	m	n	c
1846.	h. s.	s.	s.	s.	s.

INSTRUMENT READINGS.

	Date.	Barom.	THERMOM. At.	THERMOM. Ex.
	1846. h. m.	In.	°	°
Zone 12	April 17.	29.960	61.0	56.5

REMARKS.

(12) 10. Minutes assumed as 12 instead of 10.

ZONE 12. APRIL 17. K. $D_0 = -35° 21' 50''$—Continued.

No.	Mag.	I.	II.	III.	IV.	V.	VI.	VII.	T.		a_1	MICROMETER.			$i + d_2$		d_1	Mean Right Ascension, 1850.0.	Mean Declination, 1850.0.
									h. m. s.		s.			r.	′ ″		″	h. m. s.	° ′ ″
37	8.9	. .	50.3	5.9	21.4	36.5	. .	. .	17 37 21.16	—	11.73	VI.	6	7.59	−27 59.4	—	5.9	17 37 9.43	− 35 49 55.3
38	10	. .	. .	. .	51.2	6.9	22.1	. .	39 51.34		11.70	VII.	2	3.46	6 48.4		5.1	39 39.64	28 43.5
39	7	. .	33.7	49.6	4.7	20.1	. .	. .	42 4.34		11.68	VI.	3	5.34	12 43.6		4.6	41 52.86	34 38.2
40	10	. .	. .	. .	. .	33.2	48.0	4.1	17 43 17.66	—	11.67	VII.	6	9.23	−28 41.5	—	4.1	17 43 5.99	− 35 50 35.6

ZONE 13. APRIL 18. A. $D_0 = -34° 8' 0'$.

No.	Mag.	I.	II.	III.	IV.	V.	VI.	VII.	T.		a_1	MICROMETER.			$i + d_2$		d_1	Mean Right Ascension, 1850.0.	Mean Declination, 1850.0.
1	8	. .	. .	39.5	54.5	9.5	25.0	40.0	9 17 54.51	—	16.03	VII.	6	8.34	−28 16.7	—	5.4	9 17 38.48	− 34 36 22.1
2	10	25.5	40.4	56.0	11.2	26.2	41.2	56.6	34 11.00		15.96	II.	5	3.55	21 54.0		10.6	33 55.04	30 4.6
3	12	49.0	4.5	19.5	34.8	50.0	. .	20.0	45 34.67		15.91	II.	5	8.40	24 18.1		14.2	45 18.76	32 32.3
4	9	. .	31.6	47.0	2.0	17.1	32.0	47.5	50 1.84		15.89	II.	3	8.9	14 2.4		15.6	49 45.95	22 18.0
5	9	15.0	30.0	45.5	1.0	15.9	31.0	46.0	54 0.65		15.87	I.	6	6.57	27 26.7		16.8	53 44.78	35 43.5
6	9	53.5	9.0	24.3	39.5	54.6	9.4	24.9	9 58 39.26		15.85	I.	4	3.29	16 40.3		18.3	9 58 23.41	24 58.6
7	10	. .	. .	. .	. .	. .	46.8	2.0	10 0 16.51		15.84	VI.	4	7.32	18 43.6		18.8	10 0 0.67	27 2.4
8	7	48.2	3.8	19.0	34.1	49.1	4.4	19.5	5 34.03		15.81	I.	6	5.39	26 48.2		20.4	5 18.22	35 8.6
9	9	. .	34.2	50.0	1.6	20.0	34.9	50.3	8 4.69		15.79	II.	5	3.35	21 43.9		21.1	7 48.90	30 5.0
10	8	. .	. .	59.5	14.4	29.5	45.5	. .	24 14.30		15.69	.	3	5.10	12 32.0		25.9	23 58.61	20 57.8
11	8	27.0	42.5	57.5	12.6	28.5	43.0	58.2	38 12.77		15.58	II.	6	2.8	25 1.7		29.6	37 57.19	33 31.3
12	7	33.0	48.0	3.2	18.9	33.8	48.8	4.0	46 18.61		15.51	II.	7	8.11	33 6.3		31.7	46 3.10	41 38.0
13	8	. .	5.0	20.2	35.5	51.0	6.0	21.0	48 35.56		15.49	II.	7	8.34	33 18.0		32.3	48 20.07	41 50.3
14	8	. .	. .	. .	26.6	42.5	56.5	11.0	10 51 26.60		15.47	IV.	9	9.4	43 33.0		33.0	10 51 11.13	52 6.0
15	9	7.0	23.0	38.5	53.2	8.9	23.6	39.0	11 0 53.37		15.40	II.	7	4.24	31 11.5		35.4	11 0 37.97	39 46.9
16	11	39.0	54.2	. .	25.2	40.0	55.0	10.2	3 24.78		15.37	I.	6	4.52	26 24.4		36.0	3 9.41	35 0.4
17	9	. .	. .	26.2	41.4	56.8	12.0	27.0	5 41.46		15.36	III.	6	5.24	26 41.0		36.5	5 26.10	35 17.5
18	7	24.1	39.2	54.8	10.0	25.0	40.1	55.5	14 9.89		15.28	I.	7	9.11	33 36.4		38.5	13 54.61	42 14.9
19	9	6.0	22.0	37.5	52.5	7.5	22.9	37.8	20 52.41		15.22	II.	7	11.34	34 49.0		40.0	20 37.19	43 29.0
20	8	20.2	36.0	51.0	6.1	21.0	36.2	51.4	28 6.10		15.16	III.	8	8.35	38 18.3		41.5	27 50.94	46 59.8
21	7	. .	. .	. .	11.0	26.2	41.2	56.2	11 34 10.92	—	15.10	IV.	8	7.17	−37 38.9	—	42.8	11 33 55.82	− 34 46 21.7

ZONE 14. APRIL 20. K. $D_0 = -32° 51' 20''$.

No.	Mag.	I.	II.	III.	IV.	V.	VI.	VII.	T.		a_1	MICROMETER.			$i + d_2$		d_1	Mean Right Ascension, 1850.0.	Mean Declination, 1850.0.
1	8	. .	25.8	40.7	55.6	. .	. .	. .	10 54 55.35	—	15.96	IV.	3	8.25	−14 10.9	—	46.9	10 54 39.39	− 33 6 17.8
2	10	. .	. .	. .	. .	. .	18.7	. .	57 19.00		15.94	VI.	3	11.46	15 52.4		47.6	57 3.06	6 0.0
3	10	. .	. .	. .	. .	. .	. .	40.8	10 57 55.92		15.93	. .	.	. .	. . .		. .	57 39.99	
4	10	. .	. .	. .	. .	16.1	. .	45.8	11 0 1.10		15.92	VII.	7	10.37	34 19.7		48.4	10 58 45.18	26 28.1
5	9	. .	. .	. .	. .	. .	. .	22.2	0 37.34		15.92	VII.	6	8.11	28 5.0		48.6	11 0 21.42	20 13.6
6	9	. .	. .	. .	. .	. .	. .	8.9	1 24.16		15.91	VII.	8	7.32	37 45.7		48.8	1 8.25	29 54.5
7	9	. .	24.7	. .	. .	. .	24.8	39.6	5 54.91		15.87	VII.	9	8.37	43 18.5		50.1	5 39.04	35 26.6
8	10	. .	. .	. .	39.4	. .	. .	. .	7 39.24		15.66	IV.	7	8.19	33 10.4		50.6	7 23.38	25 21.0
9	10	. .	. .	57.1	12.1	. .	. .	. .	9 11.98		15.85	IV.	7	8.14	33 7.9		51.0	8 56.13	25 18.9
10	9	. .	. .	. .	7.2	21.9	36.9	. .	10 7.04		15.84	VII.	9	2.45	40 20.4		51.2	9 51.20	32 31.6
11	8	. .	. .	. .	38.4	53.2	. .	. .	11 38.19		15.82	V.	1	7.55	3 55.9		51.6	11 22.37	56 7.5
12	11	. .	. .	36.1	51.3	. .	. .	. .	14 51.09		15.80	V.	7	9.41	33 51.8		52.5	14 35.29	26 4.3
13	11	. .	. .	. .	50.3	. .	. .	. .	15 50.14		15.79	IV.	7	7.34	32 47.8		52.8	15 34.35	25 0.6
14	8	. .	58.9	14.0	29.2	. .	. .	. .	17 28.98		15.78	I.	8	3.11	35 33.7		53.2	17 13.20	27 46.9
15	8	. .	57.2	. .	. .	. .	. .	. .	19 27.13		15.76	IV.	6	5.34	26 46.1		53.8	19 11.37	18 59.9
16	8	. .	. .	. .	. .	. .	53.6	8.6	11 19 23.73	—	15.76	VII.	6	9.27	−28 43.4	—	53.7	11 19 8.02	− 33 20 57.1

CORRECTIONS.

Date.		Corr. of Clock.	Hourly rate.	m	n	c
1846.	h.	s.	s.	s.	s.	s.
April 18,	9	− 23.60	+ 0.023	− 0.280	+ 0.344	+ 0.120
20,	10	− 24.25	− 0.019	− 0.280	+ 0.344	+ 0.120

INSTRUMENT READINGS.

	Date.	Barom.	THERMOM. At.	THERMOM. Ex.
	1846. h. m.	in.	°	°
Zone 13	April 18,	30.057	67.	64.5
Zone 14	20,	30.088	64.0	63.5

REMARKS.

(12) 37. Differs 1^m from observation July 14.
(14) 3. Minutes assumed as 57.
(14) 4. Hours and minutes assumed as 10^h 59^m instead of 11^h 0^m.
(14) 7. Minutes of transit 1 larger than Mural Z., April 20.

ZONE 14. APRIL 20. K. $D_0 = -32° 51' 20''$—Continued.

No.	Mag.	I.	II.	III.	IV.	V.	VI.	VII.	T.	a_1	Micrometer.			$i + d_2$	d_1	Mean Right Ascension, 1850.0.	Mean Declination, 1850.0.
									h. m. s.	s.			r.	′ ″	″	h. m. s.	° ′ ″
17	12	..	58.4	13.6	..	..	..	..	11 22 28.20 −	15.74	V.	3	8.39	−14 18.0 −	54.6	11 22 12.46	− 33 6 32.6
18	8	..	..	..	2.3	17.2	..	47.7	23 2.37	15.73	VII.	4	11.52	20 55.0	54.7	22 46.64	13 9.7
19	9	..	..	46.4	..	..	..	..	27 0.99	15.70	III.	2	12.41	11 20.4	55.7	26 45.29	3 36.1
20	9	..	..	..	..	32.9	..	2.8	27 17.98	15.70	VII.	6	14.4	31 3.5	55.8	27 2.28	23 19.3
21	8	..	..	..	..	35.0	..	4.9	32 20.53	15.65	VII.	7	5.0	31 29.2	57.1	32 4.88	23 46.3
22	9	..	..	..	48.3	3.2	..	..	33 48.17	15.63	VI.	3	7.3	13 29.2	57.5	33 32.54	5 46.7
23	8	..	..	1.3	16.0	30.8	..	..	36 15.83	15.62	VI.	3	1.2	10 26.7	58.1	36 0.21	2 44.8
24	8	..	43.8	58.8	13.8	..	..	..	40 13.52	15.58	IV.	3	3.27	11 40.2	59.0	39 57.94	3 59.2
25	11	..	..	..	..	..	10.6	25.7	40 40.82	15.57	VII.	5	5.26	22 39.8	59.2	40 25.25	14 59.0
26	12	..	..	..	..	55.1	9.9	..	43 40.16	15.55	VII.	7	11.39	34 51.1	59.9	43 24.61	27 11.0
27	5	..	..	21.2	36.0	50.7	5.6	..	45 35.82	15.53	V.	2	14.17	12 8.9	60.3	45 20.29	4 29.2
28	11	..	7.1	22.2	..	..	..	..	48 37.09	15.50	V.	7	5.53	31 56.5	61.1	48 21.59	23 17.6
29	12	1.8	..	..	..	..	..	..	52 46.72	15.47	IV.	2	8.32	9 14.6	62.0	52 31.25	1 36.6
30	7	15.9	31.0	46.1	1.0	..	..	..	54 0.87	15.45	IV.	4	11.16	20 37.2	62.3	53 45.42	12 59.5
31	11	..	..	..	..	..	..	56.3	54 11.41	15.45	VII.	5	12.12	26 5.1	62.3	53 55.96	18 27.4
32	7	15.7	30.7	45.5	..	..	..	..	58 0.65	15.42	V.	6	12.15	30 8.8	63.2	57 45.23	22 32.0
33	11	..	..	16.9	..	..	..	..	11 59 31.87	15.40	V.	7	12.4	35 4.1	63.5	11 59 16.47	27 27.6
34	12	..	..	38.3	..	..	..	..	12 1 53.13	15.38	IV.	6	5.4	26 30.9	64.0	12 1 37.75	18 54.9
35	9.10	..	..	..	53.1	7.8	..	..	4 52.92	15.35	VI.	7	11.1	34 32.1	64.7	4 37.57	26 56.8
36	5	..	..	..	..	57.1	11.9	..	5 42.15	15.34	VII.	5	10.27	25 12.0	64.9	5 26.81	17 36.9
37	8	..	..	..	..	..	16.6	32.1	6 46.95	15.33	VII.	1	8.52	4 24.3	65.1	6 31.62	56 49.4
38	10	..	..	..	45.6	0.2	..	..	8 45.36	15.32	VII.	7	9.17	33 39.2	65.5	8 30.04	26 4.7
39	9	..	..	..	54.3	..	24.3	39.1	13 54.33	15.26	VII.	8	9.40	38 50.4	66.5	13 39.07	31 16.9
40	9	..	..	..	..	30.2	..	..	15 15.21	15.25	VII.	4	5.3	17 28.1	66.8	14 59.96	9 54.9
41	10	..	50.4	5.3	..	..	..	..	17 20.38	15.23	IV.	8	7.47	37 53.8	67.2	17 5.15	30 21.0
42	10	49.2	4.4	19.7	..	..	..	..	19 34.40	15.20	IV.	5	11.27	25 42.8	67.6	19 19.20	18 10.4
43	10	24.1	39.2	54.1	..	..	..	..	27 8.79	15.12	IV.	2	9.46	9 52.0	69.1	26 53.67	33 2 21.1
44	7	..	..	..	24.4	39.3	..	..	28 24.24	15.11	IV.	1	4.52	2 23.5	69.3	28 9.13	32 54 52.8
45	10	..	..	54.2	..	..	..	..	31 9.04	15.08	IV.	6	6.30	27 14.4	69.8	30 53.96	33 19 44.2
46	10	..	..	..	..	55.3	..	..	32 40.33	15.06	VI.	5	5.55	22 54.7	70.1	32 25.27	15 24.8
47	9	23.6	38.8	..	9.0	..	..	..	35 8.85	15.04	V.	8	4.1	35 59.4	70.5	34 53.81	28 29.9
48	8	19.1	34.3	49.3	4.3	..	..	..	40 4.14	14.98	V.	5	4.18	22 5.9	71.3	39 49.16	14 37.2
49	9	6.8	21.9	37.2	..	..	..	..	41 51.91	14.96	IV.	5	9.18	24 37.6	71.6	41 36.95	17 9.2
50	5	..	..	33.7	48.7	3.4	..	..	42 48.46	14.95	V.	4	7.6	18 30.8	71.8	42 33.51	11 2.6
51	11	..	..	..	..	..	6.3	21.3	43 36.65	14.94	VII.	4	8.9	19 2.2	71.9	43 21.71	11 34.1
52	10	..	..	..	..	..	35.2	..	45 5.44	14.93	VII.	6	4.28	26 12.2	73.1	44 50.51	18 44.3
53	10	..	30.8	45.9	1.0	..	..	..	49 0.85	14.88	IV.	8	3.47	35 52.4	72.7	48 45.97	28 25.1
54	10	..	..	..	..	53.6	8.2	..	49 38.56	14.87	VII.	3	2.43	11 17.5	72.8	49 23.69	3 50.3
55	9	..	9.5	24.2	39.6	..	..	..	52 39.32	14.85	V.	6	10.53	29 27.4	73.3	52 24.47	22 0.7
56	8	..	..	46.5	1.7	16.5	..	..	55 1.55	14.82	V.	8	5.35	36 47.0	73.7	54 46.73	29 20.7
57	7	..	..	..	..	..	14.9	..	55 45.10	14.82	VII.	7	10.34	34 18.2	73.7	55 30.28	26 51.9
58	7	11.4	26.6	..	..	..	..	..	12 58 56.54	14.77	II.	6	5.3	26 30.2	74.2	12 58 41.77	19 4.4
59	8	..	50.6	5.8	..	..	..	..	13 0 20.58	14.76	III.	6	5.12	26 34.9	74.4	13 0 5.82	19 9.3
60	10	..	..	..	..	10.9	25.8	..	13 1 56.00 −	14.74	V.	9	2.58	−40 27.5 −	74.7	13 1 41.26	− 33 33 2.2

CORRECTIONS.

Date.	Corr. of Clock.	Hourly rate.	m	n	c
1846.	h. s.	s.	s.	s.	s.

INSTRUMENT READINGS.

Date.		Barom.	Thermom. At.	Thermom. Ex.
1846.	h. m.	in.	°	°

REMARKS.

ZONE 15. APRIL 20. K. $D_o = -32° 51' 20''$.

No.	Mag.	SECONDS OF TRANSIT. I.	II.	III.	IV.	V.	VI.	VII.	T.		a_1	MICROMETER.			$i + d_2$	d_1	Mean Right Ascension, 1850.0.	Mean Declination, 1850.0.
									h. m. s.		s.			r.	′ ″	′ ″	h. m. s.	° ′ ″
1	10	..	..	..	..	22.1	36.9	..	14 26 7.15	−	13.75	VII.	5	8.23	−24 9.3	− 45.2	14 25 53.40	− 33 16 14.5
2	10	..	2.9	18.1	..	..	..	..	32 32.87		13.67	IV.	5	12.5	26 2.0	44.6	32 19.20	18 6.6
3	8.9	..	..	..	25.3	40.3	..	..	33 25.20		13.66	V.	1	14.31	7 16.2	44.5	33 11.54	59 20.7
4	10	..	58.1	13.3	28.1	..	..	..	37 27.93		13.61	IV.	3	11.28	15 43.5	44.1	37 14.32	7 47.6
5	9	..	51.7	6.8	..	..	..	..	40 21.68		13.58	IV.	7	4.43	31 21.2	43.8	40 8.10	23 25.0
6	9	..	25.5	40.8	..	..	..	..	43 55.27		13.54	V.	2	6.47	8 21.4	43.3	43 41.73	0 24.7
7	9	..	..	36.3	..	..	..	..	44 50.89		13.52	IV.	3	4.52	12 23.2	43.2	44 37.37	4 26.4
8	8	..	..	..	..	..	26.8	42.1	44 57.07		13.52	VI.	2	8.50	9 23.4	43.2	44 43.55	1 26.6
9	7	..	..	..	46.7	1.1	..	..	46 46.48		13.50	VI.	5	5.9	22 31.5	43.0	46 32.98	14 34.5
10	11	..	58.3	13.5	..	..	..	..	49 28.07		13.47	IV.	3	4.50	12 22.2	42.7	49 14.60	4 24.9
11	11	..	..	..	57.7	..	..	..	51 57.54		13.44	VII.	4	4.42	17 17.5	42.4	51 44.10	9 19.9
12	9	..	..	..	4.0	..	33.2	48.9	52 3.76		13.44	VII.	5	4.41	22 17.0	42.4	51 50.32	14 19.4
13	11	..	..	..	..	51.1	5.9	..	53 36.15		13.42	VII.	5	4.39	22 16.0	42.2	53 22.73	14 18.2
14	11	..	..	27.2	42.1	..	..	..	55 41.84		13.39	V.	2	8.32	9 14.5	41.9	55 28.45	1 16.4
15	11	22.9	38.3	..	8.3	..	..	..	14 58 8.18		13.36	V.	7	5.19	31 39.3	41.6	14 57 54.82	23 40.9
16	8	..	11.7	26.8	41.4	..	..	..	15 2 41.36		13.31	IV.	3	4.4	11 59.0	41.0	15 2 28.05	33 4 0.0
17	11	..	..	..	47.0	1.8	..	..	4 46.79		13.28	V.	1	10.1	4 59.6	40.7	4 33.51	32 51 0.3
18	11	..	..	..	..	..	..	21.7	5 36.84		13.27	VI.	6	8.17	28 8.3	40.5	5 23.57	33 20 8.8
19	8	..	43.9	59.1	..	..	..	..	10 14.02		13.22	II.	8	5.31	36 44.8	39.9	10 0.80	28 44.7
20	8	..	48.5	3.6	18.7	..	..	..	13 18.64		13.18	V.	9	11.49	44 56.0	39.4	13 5.46	36 55.4
21	8	..	23.1	38.2	52.9	..	..	..	18 52.77		13.11	II.	2	8.2	8 59.2	38.6	18 39.66	0 57.8
22	8	..	57.4	12.6	27.5	..	..	..	21 27.22		13.08	IV.	3	3.22	11 37.7	38.2	21 14.14	3 35.9
23	7.8	..	..	..	..	..	35.0	50.0	22 5.18		13.08	VII.	6	4.15	26 5.6	38.1	21 52.10	33 18 3.7
24	8	..	..	56.2	10.9	25.9	..	..	34 10.74		12.94	V.	1	6.59	3 27.6	36.1	33 57.80	32 55 23.7
25	11	..	20.3	35.8	50.6	..	..	..	35 50.26		12.92	V.	2	6.45	8 20.4	35.8	35 37.34	33 0 16.2
26	11	..	58.2	..	..	..	..	..	38 27.90		12.89	IV.	2	5.49	7 52.1	35.3	38 15.01	32 59 47.4
27	11	..	..	..	..	..	17.2	..	38 47.54		12.88	VII.	2	9.20	9 38.3	35.3	38 34.66	33 1 33.6
28	11	..	56.4	11.4	..	..	..	..	41 26.19		12.85	IV.	4	10.18	20 7.9	34.8	41 13.34	12 2.7
29	5	..	..	..	..	..	..	24.3	41 39.31		12.85	VII.	4	6.10	18 2.0	34.8	41 26.46	9 56.8
30	7	46.3	1.2	16.2	..	..	..	..	47 31.38		12.79	IV.	8	10.42	39 22.3	33.7	47 18.59	31 16.0
31	7	..	..	..	..	..	2.0	..	47 32.16		12.79	VII.	8	10.36	39 18.8	33.7	47 19.37	33 31 12.5
32	7	..	5.1	20.4	35.0	..	..	..	50 34.83		12.75	III.	1	10.27	5 12.8	33.2	50 22.08	32 57 6.0
33	11	..	..	45.4	0.2	..	..	..	53 0.09		12.73	V.	5	3.45	21 49.2	32.7	52 47.36	33 13 41.9
34	11	..	33.2	48.3	..	..	..	..	55 3.21		12.71	V.	7	7.39	32 50.1	32.3	54 50.50	24 42.4
35	10	..	..	49.0	4.1	..	..	..	56 4.01		12.69	V.	9	8.46	43 23.5	32.1	55 51.32	35 15.6
36	10	..	23.4	38.6	..	..	..	..	15 57 53.37		12.67	V.	5	12.4	26 1.5	31.8	15 57 40.70	17 53.3
37	6.7	..	..	14.2	29.1	43.9	..	..	16 0 28.91		12.65	III.	4	3.22	16 37.5	31.3	16 0 16.26	8 28.8
38	11	..	..	9.2	..	..	..	..	2 23.85		12.63	IV.	3	11.28	15 43.5	30.9	2 11.22	33 7 34.4
39	7	..	..	45.7	0.3	15.2	..	..	12 0.15		12.53	VI.	1	8.42	4 19.6	29.0	11 47.62	32 56 8.6
40	9	31.9	46.9	2.2	17.0	..	..	..	16 16.88		12.48	IV.	4	12.20	21 9.6	28.0	16 4.40	33 12 57.6
41	10	..	29.9	45.1	59.9	..	..	..	19 59.65		12.45	III.	2	4.42	7 18.2	27.3	19 47.20	32 59 5.5
42	11	..	8.0	23.6	..	..	..	..	22 38.04		12.42	IV.	4	3.43	16 48.2	26.7	22 25.62	33 8 34.9
43	9.10	..	..	15.2	..	45.2	..	..	23 30.08		12.42	V.	4	11.9	20 33.6	26.5	23 17.66	12 20.1
44	11	..	21.3	..	51.5	..	..	..	25 51.35		12.39	V.	8	8.0	38 0.3	26.0	25 38.96	28 46.3
45	8	..	..	24.9	40.2	54.9	..	..	30 39.96		12.34	VII.	7	11.26	34 44.4	25.0	30 27.62	26 29.4
46	8.9	..	30.7	45.7	0.7	15.7	..	..	36 0.74		12.29	V.	9	6.14	42 6.6	23.8	35 48.45	32 50.4
47	8	..	32.1	47.3	2.3	..	..	..	38 2.16		12.27	V.	7	9.5	33 33.6	23.4	37 49.89	25 17.0
48	11	..	..	..	23.1	..	..	..	39 22.94		12.26	VII.	7	13.28	35 46.1	23.1	39 10.68	27 29.2
49	9	..	4.9	19.9	34.7	..	..	..	16 43 34.54	−	12.22	V.	2	9.51	− 9 54.4	− 22.1	16 43 22.32	− 33 1 36.5

CORRECTIONS.

Date.		Corr. of Clock.	Hourly rate.	m	n	c
1846.	h.	s.	s.	s.	s.	s.

INSTRUMENT READINGS.

Date.		Barom.	THERMOM. At.	Ex.
1846.	h. m.	in.	°	°

REMARKS.

ZONE 15. APRIL 20. K. $D_0 = -32° 51' 20''$—Continued.

No.	Mag.	Seconds of Transit I.	II.	III.	IV.	V.	VI.	VII.	T.	a_1	Micrometer			$i + d_0$	d_1	Mean Right Ascension, 1850.0.	Mean Declination, 1850.0.
									h. m. s.	s.			r.	′ ″	″	h. m. s.	° ′ ″
50	8	..	..	..	52.0	6.8	21.7	..	16 44 51.89	− 12.20	V.	4	13.18	−21 38.9	− 21.8	16 44 39.69	− 33 13 20.7
51	7	..	..	7.8	22.7	37.4	..	..	46 22.52	12.19	V.	5	7.55	23 55.6	21.4	46 10.33	15 37.0
52	7	..	..	..	37.3	52.1	..	..	47 37.11	12.18	VI.	2	8.35	9 15.8	21.2	47 24.93	0 57.0
53	8	..	..	..	43.9	..	13.6	..	48 43.79	12.16	VI.	6	12.59	30 30.9	20.9	48 31.63	22 11.8
54	8.9	..	..	..	..	..	56.6	11.6	49 26.79	12.15	VII.	5	11.31	25 44.4	20.8	49 14.64	17 25.2
55	8.9	..	38.3	53.6	..	..	..	..	51 8.16	12.14	II.	3	13.21	16 40.4	20.4	50 56.04	8 20.8
56	10	..	..	..	47.7	..	..	..	52 47.54	12.12	VI.	5	13.32	26 45.8	19.9	52 35.42	18 25.7
57	10	..	12.9	28.0	..	..	..	..	55 43.03	12.10	V.	9	4.44	41 21.1	19.3	55 30.93	33 0.4
58	11	..	14.3	29.4	..	..	..	..	16 59 44.22	12.06	V.	6	4.39	26 18.3	18.3	16 59 32.16	17 56.6
59	10	..	51.5	6.7	..	..	..	..	17 2 21.57	12.04	IV.	7	9.4	33 33.2	17.7	17 2 9.53	25 10.9
60	8.9	..	..	..	9.3	24.2	39.1	..	3 9.24	12.03	VI.	8	7.44	37 52.0	17.5	2 57.21	29 29.5
61	6	55.8	..	25.8	..	..	..	..	5 40.85	11.92	III.	6	13.1	30 32.1	16.9	5 28.93	22 9.0
62	7	..	..	..	..	27.2	41.2	..	6 11.85	11.92	VII.	8	15.54	41 59.6	16.8	5 59.93	33 36.4
63	9.10	..	36.2	..	..	..	..	..	9 6.10	11.98	II.	5	9.36	24 46.5	16.0	8 54.12	16 22.5
64	10	..	..	..	19.8	..	..	..	12 19.64	11.95	V.	5	7.59	23 57.6	15.3	12 7.69	15 32.9
65	11	..	..	..	20.9	..	..	..	13 20.74	11.94	VI.	2	13.23	11 41.5	15.0	13 8.80	3 16.5
66	9	..	..	..	44.7	..	..	..	14 44.54	11.93	IV.	3	6.34	13 14.8	14.6	14 32.61	4 49.4
67	11	..	..	..	0.0	..	..	..	18 59.84	11.89	VI.	7	12.58	35 31.3	13.6	18 47.95	27 4.9
68	8	..	..	9.4	..	..	..	..	20 24.52	11.88	IV.	9	9.42	43 51.8	13.2	20 12.64	35 25.0
69	9	..	..	..	..	7.6	..	..	21 52.72	11.86	VI.	9	2.34	40 15.2	12.9	21 40.86	33 31 48.1
70	9	..	..	13.4	..	..	..	..	23 27.87	11.85	III.	1	9.41	4 49.5	12.5	23 16.02	32 56 22.0
71	10	..	..	21.8	..	..	..	21.8	26 36.68	11.83	IV.	4	10.49	20 23.6	11.7	26 24.85	33 11 55.3
72	10	..	29.3	44.6	..	..	..	..	28 59.03	11.81	IV.	1	11.24	5 41.7	11.1	28 47.22	32 57 12.8
73	10	..	..	..	46.9	..	..	..	29 46.74	11.80	V.	8	13.14	40 39.1	10.9	29 34.94	33 32 10.0
74	10	..	..	..	..	..	47.3	..	30 17.57	11.80	VII.	2	12.26	11 12.4	10.7	30 5.77	2 43.1
75	7	..	..	3.0	18.2	..	..	..	32 17.99	11.78	IV.	7	9.5	33 33.7	10.2	32 6.21	33 25 3.9
76	7	..	..	13.5	28.3	..	..	..	33 28.07	11.78	V.	1	13.19	6 39.8	9.9	33 16.29	32 58 9.7
77	10	..	54.3	9.6	..	..	..	..	41 24.01	11.71	IV.	1	9.27	4 42.5	7.9	41 12.30	32 56 10.4
78	10	..	..	27.7	43.3	..	..	..	44 42.96	11.68	IV.	9	7.39	42 49.6	7.0	44 31.28	33 34 16.6
79	9	35.5	50.3	..	..	..	..	..	47 20.25	11.66	VII.	3	4.20	12 6.6	6.3	47 8.59	3 32.9
80	8	..	45.9	0.0	15.1	..	..	..	17 50 15.24	− 11.64	IV.	7	5.38	−37 49.0	− 5.6	17 50 3.60	− 33 23 14.6

ZONE 16. APRIL 27. K. $D_0 = -27° 52' 30''$.

No.	Mag.	I.	II.	III.	IV.	V.	VI.	VII.	T.	a_1	Micrometer			$i + d_0$	d_1	Mean Right Ascension, 1850.0.	Mean Declination, 1850.0.
1	8	..	..	..	21.9	..	50.0	..	11 52 21.77	− 15.54	V.	10	6.31	−47 15.6	− 11.5	11 52 6.23	− 28 39 57.1
2	9	2.9	17.2	31.3	..	..	..	..	55 45.57	15.52	III.	7	7.42	32 51.2	11.8	55 30.05	25 33.0
3	12	..	..	..	17.7	..	46.1	..	57 17.76	15.51	VI.	6	8.28	28 13.7	11.9	57 2.25	20 55.6
4	10	44.0	58.3	12.6	..	..	..	..	11 59 26.64	15.49	V.	5	9.21	24 39.2	12.1	59 11.15	17 21.3
5	10	..	..	..	..	24.0	37.9	52.4	12 0 9.81	15.49	VII.	4	11.50	20 54.3	12.2	11 59 54.32	13 36.5
6	10	..	49.4	..	17.8	..	..	..	4 17.77	15.46	IV.	9	4.5	41 0.4	12.5	12 4 2.31	33 42.9
7	9	52.9	7.2	21.5	35.8	49.9	4.0	18.1	7 35.65	15.44	IV.	7	4.58	31 28.4	12.7	7 20.21	24 11.1
8	10	44.6	58.9	13.2	..	..	..	..	12 27.17	15.40	IV.	4	8.33	19 15.2	13.1	12 11.77	11 58.3
9	11	..	..	..	..	..	26.0	40.1	12 57.71	15.40	VII.	7	9.47	33 54.0	13.1	12 42.31	26 37.1
10	12	..	36.5	50.9	..	..	..	..	17 4.0	15.36	IV.	3	8.1	13 59.4	13.4	16 49.34	6 42.8
11	11	27.7	42.0	56.2	10.4	..	..	..	22 10.39	15.32	V.	8	4.27	36 11.9	13.7	21 55.07	28 55.6
12	10	..	40.3	54.6	8.8	..	..	..	34 8.55	15.24	V.	4	4.38	17 16.4	14.2	33 53.31	10 0.6
13	10	..	..	..	34.3	48.3	..	..	36 34.15	15.22	VII.	10	7.47	47 53.6	14.3	36 18.93	28 40 37.9
14	9	..	..	..	..	..	59.9	14.3	12 37 31.69	− 15.21	VII.	1	7.56	− 3 57.4	− 14.4	12 37 16.48	− 27 56 41.8

CORRECTIONS.

Date.		Corr. of Clock.	Hourly rate.	m	n	c
1846.	h.	s.	s.	s.	s.	s.
April 27,	10	− 24.49	+ 0.002	− 0.280	+ 0.344	+ 0.120

INSTRUMENT READINGS.

Date.		Barom.	Thermom. At.	Thermom. Ex.
1846.	h. m.	in.	°	°

REMARKS.

ZONE 16. APRIL 27. K. $D_o = -27° 52' 30''$—Continued.

No.	Mag.	I.	II.	III.	IV.	V.	VI.	VII.	T.	a_1	MICROMETER.			$i + d_2$	d_1	Mean Right Ascension, 1850.0.	Mean Declination, 1850.0.
									h. m. s.	s.			r.	′ ″	″	h. m. s.	° ′ ″
15	10	..	..	..	24.0	..	52.3	..	12 41 24.05	− 15.18	VII.	2	8.56	− 9 27.1	− 14.5	12 41 8.87	− 28 2 11.6
16	9	..	..	4.7	19.0	33.0	46.9	..	43 16.82	15.16	VI.	6	9.44	28 52.1	14.6	43 3.66	21 36.7
17	8	..	35.0	49.2	..	..	..	..	46 3.40	15.14	IV.	8	5.57	36 57.4	14.6	45 48.26	29 42.0
18	10	..	..	..	..	37.2	51.0	..	46 22.95	15.14	VII.	9	7.16	42 36.6	14.6	46 7.81	33 21.2
19	8	47.1	1.6	15.8	30.0	..	..	..	49 29.71	15.11	IV.	2	11.5	10 32.8	14.7	49 14.60	3 17.5
20	8	..	..	19.0	33.3	..	..	..	51 33.21	15.10	VI.	10	6.55	47 27.6	14.7	51 18.11	40 12.3
21	7	49.1	3.7	17.8	32.0	..	..	..	54 31.95	15.07	V.	7	11.24	34 43.3	14.8	54 16.88	27 28.1
22	11	..	..	..	0.8	..	..	..	55 0.64	15.07	VII.	6	11.5	29 32.8	14.8	54 45.57	22 17.6
23	11	..	13.5	..	42.3	..	..	..	58 42.10	15.05	VI.	10	3.58	45 58.1	14.9	58 27.05	38 43.0
24	9	..	..	..	..	5.1	..	..	12 59 51.00	15.03	VI.	10	7.9	47 34.6	14.9	12 59 35.97	40 19.5
25	10	..	..	..	13.4	..	41.8	..	13 1 13.48	15.02	VII.	5	3.37	21 45.0	14.9	13 0 58.46	14 29.9
26	9	17.1	32.0	46.2	..	..	..	..	5 0.10	14.99	IV.	5	10.43	25 20.6	14.9	4 45.11	28 18 5.5
27	9	..	..	..	..	11.9	..	40.8	10 57.83	14.94	VII.	1	13.45	6 53.6	14.9	10 42.89	27 59 38.5
28	9	..	..	10.6	..	38.8	..	..	13 12.42	14.92	VI.	10	4.52	46 25.4	14.9	12 57.50	28 39 10.3
29	9	44.3	58.9	13.2	..	..	..	..	24 27.31	14.83	IV.	8	9.3	38 36.4	14.7	24 12.48	31 21.1
30	9	..	17.7	31.9	46.0	..	..	..	27 46.08	14.80	VI.	9	9.37	43 47.9	14.6	27 31.28	36 32.5
31	8	9.1	23.3	37.6	..	..	..	..	41 51.70	14.67	IV.	6	6.30	27 14.3	14.2	41 37.03	19 58.5
32	8	..	31.7	..	0.0	14.3	..	..	45 59.91	14.64	IV.	2	4.9	7 22.8	14.0	45 45.27	0 6.8
33	11	..	..	..	19.7	..	..	..	13 56 19.54	− 14.55	VII.	6	6.41	−27 19.4	− 13.4	13 56 4.99	− 28 20 2.8

ZONE 17. MAY 4. K. $D_o = -29° 0' 30''$.

No.	Mag.	I.	II.	III.	IV.	V.	VI.	VII.	T.	a_1	MICROMETER.			$i + d_2$	d_1	Mean Right Ascension, 1850.0.	Mean Declination, 1850.0.
1	8	..	11.4	25.8	..	..	..	..	13 47 39.92	− 14.59	IV.	2	4.13	− 7 4.4	− 64.8	13 47 28.33	− 29 8 39.2
2	8	..	..	39.4	..	..	..	..	49 53.57	14.58	IV.	4	7.38	18 47.3	64.5	49 41.99	20 21.8
3	8	..	..	..	34.2	..	2.7	..	50 34.01	14.57	VI.	5	7.25	23 40.3	64.2	50 22.44	25 14.5
4	8	14.4	28.8	..	..	..	..	..	55 57.58	14.52	III.	2	7.44	8 51.0	63.1	55 46.06	10 24.1
5	11	..	..	..	11.5	..	..	..	13 57 11.34	14.51	VII.	8	2.58	35 26.7	62.8	13 56 59.83	36 59.5
6	8	48.3	2.8	17.4	..	..	..	..	14 0 31.65	14.48	IV.	5	2.58	21 25.6	62.1	14 0 20.17	22 57.7
7	9	..	..	..	25.0	39.2	..	..	1 24.84	14.47	V.	9	6.16	42 6.7	61.9	1 13.37	43 38.6
8	7	..	..	..	..	..	22.4	37.0	1 53.66	14.47	VI.	5	1.33	20 42.5	61.8	1 42.19	22 14.3
9	8	..	..	..	..	..	18.6	..	2 49.99	14.45	VI.	1	5.37	2 47.1	61.6	2 38.54	4 18.7
10	9.10	..	..	..	26.9	..	..	..	4 26.74	14.44	IV.	4	2.26	16 9.7	61.2	4 15.30	17 40.9
11	8	..	..	34.6	49.0	..	..	..	5 48.87	14.43	IV.	6	7.6	27 32.5	60.9	5 37.44	29 3.4
12	9	..	..	10.6	..	..	..	..	7 24.56	14.41	III.	1	9.17	4 38.4	60.6	7 13.15	6 9.0
13	9	..	..	..	0.8	..	29.3	..	8 0.59	11.41	VI.	7	7.20	32 39.9	60.4	7 49.18	34 10.3
14	8	7.0	21.6	..	..	..	..	..	11 50.39	11.37	III.	4	7.52	18 54.3	59.5	11 39.02	20 23.8
15	9	27.3	41.8	..	..	..	..	..	14 10.69	11.35	III.	5	6.22	23 8.6	59.0	13 59.34	24 37.6
16	10	..	..	..	57.0	..	..	..	14 56.84	11.34	V.	2	2.25	6 9.8	58.8	14 45.50	7 38.6
17	10	..	..	..	..	12.6	..	..	15 58.16	11.34	VI.	5	4.21	22 7.4	58.5	15 46.62	23 35.9
18	8	..	..	9.8	24.3	38.5	..	..	17 24.16	11.32	V.	7	11.23	34 42.9	58.2	17 12.84	36 11.1
19	7.8	..	..	29.6	43.8	..	..	..	19 43.92	11.30	IV.	10	6.58	47 29.4	57.7	19 32.62	48 57.1
20	10	..	..	..	..	..	36.8	..	21 8.13	11.29	VI.	3	6.47	13 21.8	57.3	20 56.84	14 49.1
21	8	..	..	..	..	35.1	..	..	22 20.67	11.28	VI.	5	8.10	24 3.1	57.0	22 9.39	25 30.1
22	9.10	..	31.4	..	0.6	..	..	..	24 0.40	11.26	V.	8	8.37	38 18.4	56.6	23 49.14	39 45.0
23	11	..	..	..	..	23.2	37.3	..	27 8.68	11.24	VII.	7	11.56	34 59.2	55.8	26 57.44	36 25.0
24	7	..	..	..	..	46.2	0.5	15.0	28 31.79	11.22	VII.	7	11.17	34 39.5	55.4	28 20.57	36 4.9
25	7	..	..	..	4.7	..	..	..	33 4.54	11.18	IV.	1	3.9	1 32.5	54.3	32 53.36	2 56.8
26	8	..	..	10.6	25.1	..	..	..	14 34 25.02	− 11.17	IV.	9	4.49	−41 22.8	− 54.0	14 34 13.85	− 29 42 46.8

CORRECTIONS.

Date.		Corr. of Clock.	Hourly rate.	m	n	c
1846.	h.	s.	s.	s.	s.	s.
May 4.	10	− 21.53	+ 0.019	− 0.280	+ 0.344	+ 0.120

INSTRUMENT READINGS.

Date.		Barom.	THERMOM. At.	THERMOM. Ex.
1846.	h. m.	In.	°	°

REMARKS.

(16) 28. Transits over T.'s IV and VI assumed as recorded over T.'s III and V.
(16) 32. Micrometer reading assumed as 4ʳ.49 instead of 4ʳ.9.

ZONE 17. MAY 4. K. $D_0 = -29° 0' 30''$—Continued.

No.	Mag.	SECONDS OF TRANSIT. I.	II.	III.	IV.	V.	VI.	VII.	T.	a_1	MICROMETER.			$i + d_2$	d_1	Mean Right Ascension, 1850.0.	Mean Declination, 1850.0.
									h. m. s.	s.			r.	′ ″	″	h. m. s.	° ′ ″
27	8						13.0		14 34 44.25	− 11.17	VI.	7	6.53	−32 26.4	− 53.9	14 34 33.08	− 29 33 50.3
28	7							44.5	35 1.10	11.16	VII.	4	9.30	19 43.5	53.8	34 49.94	21 7.3
29	8		10.8	25.6					38 39.96	11.13	IV.	9	5.0	41 28.4	52.8	38 28.83	42 51.2
30	8			10.3	25.4				39 25.06	11.12	IV.	10	3.12	45 35.3	52.7	39 13.94	46 58.0
31	8						15.8		39 46.98	11.12	VI.	10	7.4	47 32.4	52.6	39 35.86	48 55.0
32	10						14.6		40 45.79	11.11	VII.	10	3.14	45 35.9	52.3	40 34.68	46 58.2
33	10				54.1				45 53.94	11.07	IV.	3	5.36	12 46.1	50.9	45 42.87	14 7.0
34	8	50.9	5.4	19.8					47 34.17	11.05	IV.	4	9.1	19 29.2	50.5	47 23.12	20 49.7
35	8				0.7				48 0.54	11.05	VI.	5	7.57	23 56.5	50.3	47 49.49	25 16.8
36	9.10		27.2	41.7					50 56.20	11.02	IV.	10	5.3	46 31.4	49.5	50 45.18	47 50.9
37	9						31.3		51 2.50	11.02	V.	9	9.56	43 57.9	49.5	50 51.48	45 17.4
38	9	6.1	20.6						54 49.38	10.99	III.	3	7.51	13 54.2	48.4	54 38.39	15 12.6
39	9			22.9					55 37.11	10.98	III.	4	11.52	20 55.5	48.2	55 26.13	22 13.7
40	7			12.9	27.1				14 56 27.03	10.97	V.	5	5.24	22 39.3	48.0	14 56 16.06	23 57.3
41	11			45.0					15 1 59.10	10.92	III.	3	8.37	14 17.5	46.4	15 0 48.18	15 33.9
42	11					6.1			1 57.71	10.92	V.	7	7.40	32 50.2	46.4	1 46.79	34 6.6
43	8	3.0	17.6						4 46.30	10.90	II.	2	10.18	10 8.7	45.6	4 35.40	11 24.3
44	9		46.7						8 15.77	10.87	III.	10	6.28	47 14.3	44.6	8 4.90	48 28.9
45	5				53.8	8.8	22.3		8 53.86	10.86	VI.	7	10.31	34 16.5	44.4	8 43.00	35 30.9
46	10						28.6		10 59.85	10.84	V.	7	6.8	32 3.8	43.8	10 49.01	33 17.6
47	10				9.0				16 8.84	10.80	V.	7	5.0	31 29.4	42.2	15 58.04	32 41.6
48	10					22.4			24 7.69	10.72	V.	2	5.13	7 34.7	39.8	23 57.17	8 44.5
49	9			38.5					26 52.43	10.70	III.	1	5.46	2 51.8	38.9	26 41.73	4 0.7
50	9		52.4						29 21.03	10.68	II.	1	9.12	4 35.7	38.1	29 10.35	5 43.8
51	5					52.0		21.0	29 37.89	10.68	VII.	3	11.2	15 30.4	38.1	29 27.21	16 38.5
52	9					7.0			30 52.53	10.66	IV.	3	8.4	14 0.8	37.6	30 41.87	15 8.4
53	9						47.3		31 18.55	10.66	VII.	7	6.38	27 18.4	37.5	31 7.89	28 25.9
54	8			36.0		4.7			32 50.40	10.64	V.	8	8.38	38 18.8	37.0	32 39.76	39 25.8
55	9		4.6						34 33.28	10.63	II.	2	9.15	9 36.8	36.4	34 22.65	10 47.2
56	8			2.3	16.9				35 16.78	10.62	IV.	9	6.1	41 59.2	36.2	35 6.16	43 5.4
57	7		13.3	27.9					36 42.24	10.61	III.	7	7.22	32 41.0	35.8	36 31.63	33 46.8
58	10						15.3		36 46.56	10.61	VI.	5	5.10	22 32.1	35.7	36 35.95	23 37.8
59	10	14.6		43.6					38 57.85	10.59	III.	3	11.6	15 32.8	35.0	38 47.26	16 37.8
60	9				32.2				39 32.04	10.58	V.	7	5.11	31 35.1	34.9	39 21.46	32 40.0
61	8						35.1		40 7.30	10.58	VI.	9	8.56	43 27.4	34.7	39 56.72	44 32.1
62	9							11.3	40 28.23	10.58	VII.	9	13.41	45 51.2	34.5	40 17.65	46 55.7
63	8						26.1		41 57.33	10.56	VII.	8	4.16	36 6.1	34.1	41 46.77	37 10.2
64	6					22.6			43 8.17	10.55	VI.	5	8.55	24 25.8	33.7	42 57.62	25 29.5
65	9							33.6	43 50.23	10.54	VII.	5	4.20	22 6.6	33.5	43 39.69	23 10.1
66	8	26.7	41.3						48 10.24	10.51	III.	7	6.4	32 1.7	32.0	47 59.73	33 3.7
67	9		36.5	50.9					50 5.33	10.49	IV.	7	4.51	31 24.9	31.4	49 54.84	32 26.3
68	7				30.7		59.3		50 30.53	10.49	VI.	8	7.42	37 50.4	31.2	50 20.04	38 51.6
69	9				23.3				54 37.56	10.45	IV.	9	5.38	41 47.6	29.9	54 27.11	42 47.5
70	9		59.1						58 27.70	10.42	IV.	1	5.19	2 38.2	28.5	58 17.28	3 36.7
71	8			6.6					15 59 21.02	10.41	VI.	7	11.43	34 52.9	28.2	15 59 10.61	35 51.2
72	11			31.4					16 2 45.71	10.39	III.	6	8.6	28 2.8	27.1	16 2 35.32	28 59.9
73	7				27.0				3 26.84	10.38	V.	9	11.2	44 31.2	26.8	3 16.46	45 28.0
74	9						27.2		3 58.53	10.38	VI.	3	8.42	14 19.9	26.6	3 48.15	15 16.5
75	8	57.0	11.3						16 7 40.26	− 10.34	III.	4	11.42	−20 50.5	− 25.4	16 7 29.92	− 29 21 45.9

CORRECTIONS.

Date.		Corr. of Clock.	Hourly rate.	m	n	c
1846.	h.	s.	s.	s.	s.	s.

INSTRUMENT READINGS.

Date.		Barom.	THERMOM. At.	Ex.
1846.	h. m.	in.	°	°

REMARKS.

(17) 53. Hor. thread assumed as 6 instead of 7.
(17) 69. Transit over T. III assumed as recorded over T. IV.

ZONE 17. MAY 4. K. $D_0 = -29° 0' 30''$—Continued.

No.	Mag.	I.	II.	III.	IV.	V.	VI.	VII.	T.	a_1	MICROMETER.			$i + d_2$	d_1	Mean Right Ascension, 1850.0.	Mean Declination, 1850.0.
									h. m. s.	s.			r.	′ ″	″	h. m. s.	° ′ ″
76	8		2.5	16.8					16 8 31.19	−10.33	III.	5	8.13	−24 4.9	−25.1	16 8 20.86	−29 25 0.0
77	8		50.6						10 19.53	10.32	IV.	7	9.6	33 33.8	24.4	10 9.21	34 28.2
78	9					46.0			10 31.47	10.32	V.	1	6.31	3 14.5	24.4	10 21.15	4 6.9
79	7					36.1			11 21.59	10.31	V.	2	5.45	7 50.9	24.1	11 11.28	8 45.0
80	8						29.6		12 0.80	10.30	VI.	9	8.27	43 12.8	23.8	11 50.50	44 6.6
81	9			28.9	43.4				13 43.34	10.29	V.	9	8.32	43 15.4	23.2	13 33.05	44 8.6
82	8			36.9					14 51.49	10.28	IV.	10	6.35	47 17.9	22.8	14 41.21	48 10.7
83	6				25.0	39.3			15 24.85	10.28	V.	4	9.47	19 52.4	22.6	15 14.57	20 45.0
84	7				33.2				16 33.04	10.27	VI.	7	8.42	33 21.4	22.2	16 22.77	34 13.6
85	9		36.5						18 5.14	10.26	III.	1	12.12	6 6.8	21.7	17 54.88	6 58.5
86	9		54.1		23.2				20 23.02	10.24	IV.	7	3.41	30 49.5	20.9	20 12.78	31 40.4
87	8			40.4					21 54.88	10.22	III.	8	9.50	38 55.2	20.3	21 44.66	39 45.5
88	8				26.3				22 26.14	10.22	IV.	7	7.5	32 32.7	20.1	22 15.92	33 22.8
89	9						15.4		22 46.60	10.22	VI.	9	7.22	42 39.9	20.0	22 36.38	43 29.9
90	9					20.4			24 6.03	10.21	V.	8	4.55	36 26.2	19.5	23 55.82	37 15.7
91	8.9					7.2			24 52.79	10.20	V.	6	11.15	29 38.3	19.2	24 42.59	30 27.5
92	8						38.0		26 9.34	10.19	VI.	3	3.51	11 52.8	18.8	25 59.15	12 41.6
93	9					39.9			27 25.42	10.18	V.	3	6.17	13 6.7	18.3	27 15.24	13 55.0
94	9			40.1					28 54.30	10.17	IV.	5	2.28	21 10.5	17.8	28 44.13	21 58.3
95	8				27.8				29 27.64	10.16	VII.	7	10.7	34 4.1	17.6	29 17.48	34 51.7
96	7.8						27.0		29 58.23	10.16	VII.	8	4.41	36 18.7	17.4	29 48.07	37 6.1
97	9	44.2							30 27.50	10.16	I.	2	6.30	8 13.2	17.2	30 17.34	9 0.4
98	10		16.0						33 44.66	10.13	IV.	2	7.24	8 41.0	16.0	33 34.53	9 27.0
99	8			24.0					34 38.25	10.13	IV.	4	4.13	17 3.7	15.7	34 28.12	17 49.4
100	10				16.3				35 16.14	10.12	IV.	4	5.24	17 39.6	15.4	35 6.02	18 25.0
101	9							2.4	35 19.01	10.12	VII.	4	10.6	20 1.6	15.4	35 8.89	20 47.0
102	9							58.8	36 15.27	10.11	VII.	2	7.5	8 30.9	15.0	36 5.16	9 15.9
103	8				46.6				37 46.44	10.10	IV.	4	6.8	18 1.8	14.5	37 36.34	18 46.3
104	9				26.2				38 26.04	10.10	VI.	8	9.46	38 53.1	14.2	38 15.94	39 37.3
105	9				27.4				39 27.24	10.08	IV.	9	5.00	41 28.4	13.9	39 17.16	42 12.3
106	10							27.0	39 43.86	10.08	VII.	9	1.20	39 36.8	13.8	39 33.78	40 20.6
107	8		47.2						42 15.86	10.06	III.	2	6.2	7 59.5	12.8	42 5.80	8 42.3
108	9		25.4						43 54.42	10.05	III.	9	7.18	42 38.0	12.2	43 44.37	43 20.2
109	8.9				28.3				44 28.14	10.05	IV.	2	2.30	6 12.4	12.0	44 18.09	6 54.4
110	8						20.0		44 51.37	10.04	VI.	2	4.29	7 12.4	11.8	44 41.33	7 54.2
111	8							23.4	16 45 40.19	−10.03	VII.	8	2.38	−35 16.6	−11.5	16 45 30.16	−29 35 58.1

ZONE 18. MAY 19. A. $D_0 = -30° 23' 50''$.

No.	Mag.	I.	II.	III.	IV.	V.	VI.	VII.	T.	a_1	MICROMETER.			$i + d_2$	d_1	Mean Right Ascension, 1850.0.	Mean Declination, 1850.0.
1	9	27.8	42.3	57.0	12.0	26.4	40.7	55.2	14 45 11.70	−12.60	II.	8	4.44	−36 20.7	−28.3	14 44 59.10	−31 0 39.0
2	10				37.5	51.4	6.5		51 37.24	12.55	IV.	2	4.20	7 7.7	27.7	51 24.69	30 31 25.4
3	9		9.5	24.2	38.5	53.4	7.5	22.2	14 54 38.55	12.53	II.	5	5.15	22 34.6	27.4	14 54 26.02	46 52.0
4	11				21.0	35.0	50.2	5.0	15 1 20.94	12.48	IV.	4	11.	20 29.3	26.6	15 1 8.46	44 45.9
5	4	55.2	10.0	24.6	39.4	53.7	8.2	22.8	5 39.18	12.45	I.	7	8.8	33 4.1	26.1	5 26.73	57 20.2
6	6		13.5	28.4	42.8	57.4	11.5	26.5	15 42.64	12.37	II.	3	10.2	15 0.1	24.9	9 30.27	39 15.0
7	8	41.8	56.2	10.9	25.5	40.0	54.2	9.1	17 25.37	12.36	I.	5	6.4	22 59.1	21.7	17 13.01	47 13.8
8	9			43.2	57.6	12.0	26.5		19 57.53	12.34	IV.	7	7.35	32 47.9	24.3	19 45.19	57 2.2
9	9				17.3	31.5	46.0	1.0	15 23 17.07	−12.32	IV.	3	7.10	−13 33.4	−23.9	15 23 4.75	−30 57 47.3

CORRECTIONS.

Date.	Corr. of Clock.	Hourly rate.	*m*	*n*	*c*
1846. h.	s.	s.	s.	s.	s.
May 19. 11	− 22.90	− 0.002	− 0.280	+ 0.344	+ 0.120

INSTRUMENT READINGS.

	Date.	Barom.	THERMOM. At.	THERMOM. Ex.
	1846. h. m.	In.	°	°
Zone 18	May 19.	30.057	64.5	53.

REMARKS.

(18) 6. Minutes assumed as 9 instead of 15.

ZONE 18. MAY 19. A. $D_0 = -30° 23' 50''$—Continued.

No.	Mag.	Seconds of Transit. I.	II.	III.	IV.	V.	VI.	VII.	T.	a_1	Micrometer.			$i + d_2$	d_1	Mean Right Ascension, 1850.0.	Mean Declination, 1850.0.
									h. m. s.	s.			r.	' "	"	h. m. s.	° ' "
10	9	2.8	17.2	32.5	46.9	1.5	15.6	30.4	15 28 46.68	− 12.28	I.	4	11.39	−20 48.5	− 23.1	15 28 34.40	− 30 45 1.6
11	7	. .	41.3	55.8	10.3	24.7	39.0	54.0	31 10.16	12.26	II.	4	8.4	19 0.2	22.8	30 57.90	30 43 13.0
12	9	. .	. .	. .	55.5	9.5	24.2	38.7	34 55.17	12.24	IV.	9	8.3	43 1.2	22.2	34 42.93	31 7 13.4
13	7	. .	. .	. .	19.9	35.0	49.0	3.8	46 20.04	12.15	IV.	3	8.13	14 5.2	20.5	46 7.89	30 38 15.7
14	7	. .	11.8	26.7	41.1	55.6	9.9	25.0	50 41.00	12.11	II.	4	9.36	19 46.6	19.8	50 28.89	30 43 56.4
15	11	. .	. .	. .	. .	. .	36.0	50.0	52 6.73	12.11	VI.	9	8.7	43 3.0	19.6	51 54.62	31 7 12.6
16	10	26.0	40.5	55.5	10.0	. .	. .	. .	15 56 9.94	12.08	III.	8	8.28	38 14.0	18.9	15 55 57.86	31 2 22.9
17	8	. .	. .	. .	14.0	28.3	42.9	57.5	16 0 13.78	12.05	V.	2	9.27	14 42.6	18.3	16 0 1.73	30 38 50.9
18	7	. .	. .	. .	22.5	37.2	51.5	6.5	16 45 22.56	− 12.74	V.	4	9.11	−19 34.2	− 10.0	16 45 10.82	− 30 43 34.2

ZONE 19. MAY 20. K. $D_0 = -34° 5' 50''$.

No.	Mag.	Seconds of Transit. I.	II.	III.	IV.	V.	VI.	VII.	T.	a_1	Micrometer.			$i + d_2$	d_1	Mean Right Ascension, 1850.0.	Mean Declination, 1850.0.
1	11	54.0	. .	26.1	. .	. .	. .	. .	14 3 40.60	− 12.94	IV.	7	12.3	−35 3.9	− 66.1	14 3 27.66	− 34 42 0.0
2	10	4.3	19.6	. .	. .	. .	. .	. .	5 49.90	12.93	III.	4	3.33	16 42.8	66.2	5 36.97	23 39.0
3	8	13.4	29.0	. .	. .	. .	. .	. .	7 59.08	12.91	IV.	3	3.4	11 28.3	66.2	7 46.17	18 24.5
4	9	45.0	0.6	. .	. .	. .	. .	. .	11 30.99	12.88	III.	7	11.41	34 52.8	66.2	11 18.11	41 49.0
5	10	. .	. .	13.6	. .	. .	. .	. .	12 28.96	12.87	III.	9	13.5	45 34.9	66.2	12 16.09	52 31.1
6	11	46.7	2.2	. .	. .	. .	. .	. .	16 32.61	12.83	III.	7	7.10	32 35.6	66.2	16 19.78	39 31.8
7	6	54.5	10.1	25.2	. .	. .	. .	. .	18 40.52	12.81	IV.	8	10.6	39 4.4	66.2	18 27.71	46 0.6
8	8.9	. .	. .	. .	. .	. .	26.4	41.7	18 56.19	12.81	VI.	6	7.10	27 34.5	66.2	18 43.38	34 30.7
9	10	. .	8.5	23.8	. .	. .	. .	. .	22 38.95	12.78	IV.	7	9.57	34 0.1	66.1	22 26.17	40 56.2
10	11	. .	. .	47.2	. .	. .	. .	. .	24 2.55	12.76	V.	9	11.19	44 41.3	66.1	23 49.79	51 37.4
11	8.9	. .	. .	. .	. .	. .	59.3	14.8	22 29.13	12.78	VII.	2	10.29	10 12.8	66.1	22 16.35	17 8.9
12	10	. .	. .	. .	. .	15.0	. .	. .	25 59.80	12.75	V.	4	12.35	21 17.0	66.1	25 47.05	28 13.1
13	8	. .	. .	. .	. .	28.8	43.8	. .	27 13.63	12.74	VII.	2	5.53	7 53.2	66.1	27 0.89	14 49.3
14	8	. .	. .	33.2	48.8	. .	. .	. .	28 48.59	12.72	V.	9	11.32	44 47.9	66.1	28 35.87	51 44.0
15	6	. .	. .	. .	. .	45.1	0.2	15.3	29 29.92	12.71	VII.	6	12.32	30 17.1	66.0	29 17.21	37 13.1
16	4	56.6	12.3	27.2	. .	. .	. .	. .	34 42.40	12.67	V.	5	8.54	24 25.4	65.9	34 29.73	31 21.3
17	5	. .	30.0	45.6	0.8	. .	. .	. .	36 0.54	12.66	V.	6	4.23	26 10.3	65.9	35 47.88	33 6.2
18	10	. .	. .	. .	. .	45.2	. .	15.5	41 29.94	12.60	VII.	4	12.9	21 3.4	65.7	41 17.34	27 59.1
19	10	. .	58.2	. .	28.7	. .	. .	. .	45 28.45	12.57	V.	2	12.31	11 15.0	65.6	45 15.68	18 10.6
20	9	. .	36.8	52.2	. .	. .	. .	. .	47 7.39	12.55	V.	8	11.3	39 33.2	65.6	46 54.84	46 28.8
21	8	. .	. .	. .	37.6	52.7	. .	. .	47 37.49	12.55	V.	6	13.18	30 40.8	65.5	47 24.94	37 36.3
22	7.8	. .	. .	. .	. .	43.4	58.6	. .	48 28.30	12.54	VI.	8	11.7	39 35.1	65.5	48 15.76	46 30.6
23	10	. .	. .	. .	53.2	. .	23.5	. .	49 53.24	12.52	VI.	1	12.9	6 3.8	65.5	49 40.72	12 59.3
24	10	. .	. .	. .	. .	44.1	. .	. .	51 28.90	12.51	VI.	4	10.20	20 8.6	65.4	51 16.39	27 4.0
25	10	30.1	45.3	0.6	. .	. .	. .	. .	54 15.51	12.48	VI.	2	11.13	10 36.4	65.3	54 3.03	17 30.7
26	9	. .	38.7	. .	9.3	. .	. .	. .	14 57 9.19	12.46	VI.	9	6.23	42 11.4	65.1	14 56 56.73	49 6.5
27	11	. .	. .	. .	32.7	. .	. .	. .	15 1 32.53	12.42	V.	5	3.24	21 38.5	64.9	15 1 20.11	28 33.4
28	7	. .	. .	. .	38.1	. .	. .	. .	2 37.93	12.41	IV.	10	7.33	47 49.0	64.9	2 25.52	54 43.9
29	8	. .	. .	. .	. .	. .	39.6	55.2	3 9.50	12.41	VII.	4	3.52	16 52.0	64.9	2 57.09	23 46.9
30	9	. .	7.7	. .	38.1	. .	. .	. .	6 37.88	12.38	V.	2	5.14	7 33.9	64.7	6 25.50	14 28.6
31	8.7	14.9	30.3	45.6	. .	. .	. .	. .	11 0.51	12.33	IV.	3	11.13	15 35.6	64.4	10 48.18	22 30.0
32	6.7	. .	41.9	57.0	11.9	. .	. .	. .	13 11.79	12.32	V.	1	6.1	2 57.8	64.3	12 59.47	9 52.1
33	8.9	. .	. .	5.8	20.7	35.8	. .	. .	14 20.52	12.31	VI.	1	10.19	5 8.2	64.2	14 8.21	12 2.4
34	8	. .	37.0	52.2	7.6	. .	. .	. .	20 7.53	12.26	VI.	9	13.28	45 46.4	63.9	19 55.27	52 40.3
35	10	. .	52.8	8.2	. .	. .	. .	. .	24 23.46	12.22	IV.	9	10.33	44 18.1	63.6	24 11.24	51 11.7
36	10	. .	2.2	17.7	. .	. .	. .	. .	15 27 32.33	− 12.19	V.	1	9.19	− 4 37.9	− 63.4	15 27 20.14	− 34 11 31.3

CORRECTIONS.

Date.		Corr. of Clock.	Hourly rate.	m	n	c
1846.	h.	s.	s.	s.	s.	s.
May 20,	11	− 23.07	+ 0.010	− 0.280	+ 0.344	+ 0.120

INSTRUMENT READINGS.

Date.		Barom.	Thermom. At.	Ex.
1846.	h. m.	in.	°	°

REMARKS.

(18) 17. Hor. thread assumed as 3 instead of 2.
(19) 9. Discordant from Mural Zone of same date.
(19) 20. Right Ascension discordant from Mural Zone of same date.

ZONE 19. MAY 20. K. $D_0 = -34° 5' 50''$—Continued.

No.	Mag.	SECONDS OF TRANSIT. I.	II.	III.	IV.	V.	VI.	VII.	T.	a_1	MICROMETER.			$i + d_1$	d_1	Mean Right Ascension, 1850.0.	Mean Declination, 1850.0.
									h. m. s.	s.			r.	′ ″	″	h. m. s.	° ′ ″
37	9	. .	35.0	50.3	. .	. .	. .	. .	15 30 5.12	— 12.17	V.	2	10.39	−10 18.3	— 63.2	15 29 52.95	— 34 17 11.5
38	8	. .	. .	. .	. .	48.9	. .	. .	30 33.62	12.17	V.	1	5.47	2 50.7	63.1	30 20.45	9 43.8
39	8	. .	1.9	17.2	. .	. .	. .	. .	32 32.11	12.15	IV.	4	3.20	16 36.3	62.7	32 19.96	23 29.0
40	5.6	. .	. .	5.2	20.5	35.4	. .	. .	33 20.15	12.14	V.	2	3.15	6 33.8	62.5	33 8.01	13 26.3
41	10	. .	. .	. .	. .	. .	24.3	. .	33 54.03	12.14	VII.	8	9.10	38 35.6	62.4	33 41.89	45 28.0
42	10	. .	. .	. .	. .	8.1	. .	. .	34 53.01	12.13	IV.	8	13.59	41 2.3	62.2	34 40.88	47 54.5
43	8	. .	. .	. .	. .	. .	13.2	28.6	35 43.07	12.13	VI.	6	13.1	30 32.1	62.0	35 30.94	37 24.1
44	6	. .	. .	7.4	22.3	. .	. .	. .	37 22.12	12.11	IV.	1	11.18	5 38.2	61.6	37 10.01	12 29.8
45	11	. .	. .	. .	. .	. .	41.3	. .	39 11.20	12.09	VI.	2	9.49	9 52.9	61.2	38 59.11	16 44.1
46	10	. .	59.5	15.1	. .	. .	. .	. .	42 30.03	12.07	II.	6	8.37	28 18.6	60.5	42 17.96	35 9.1
47	10	. .	. .	. .	35.2	. .	. .	. .	42 35.03	12.07	VI.	3	10.46	15 21.8	60.4	42 22.96	22 12.2
48	10	. .	. .	. .	. .	10.4	. .	. .	43 55.29	12.05	VI.	8	5.28	36 43.6	60.1	43 43.24	43 33.7
49	9	. .	34.8	46.8	. .	. .	. .	. .	46 5.34	12.04	VI.	10	4.7	46 4.5	59.6	45 53.30	52 54.1
50	11	. .	. .	. .	18.0	. .	. .	. .	48 18.83	12.02	IV.	8	6.52	37 26.3	59.1	48 6.81	44 15.4
51	8	. .	50.0	5.4	. .	. .	. .	. .	51 20.10	11.99	III.	1	10.22	5 9.8	58.4	51 8.11	11 58.2
52	9	. .	. .	48.7	. .	. .	. .	. .	53 4.03	11.98	IV.	9	9.28	43 45.2	58.0	52 52.05	50 33.2
53	9	. .	. .	. .	. .	. .	. .	53.2	53 7.89	11.98	VII.	9	7.44	42 52.1	58.0	52 55.91	49 40.1
54	7.8	. .	. .	. .	. .	. .	13.6	. .	53 43.47	11.98	.	.	. .	. . .	. .	53 31.49	
55	8	. .	. .	44.7	0.0	. .	. .	. .	58 59.89	11.92	IV.	8	9.5	38 33.6	56.6	58 47.97	45 20.2
56	7.8	. .	. .	38.8	. .	. .	. .	. .	15 59 53.85	11.92	IV.	6	6.13	27 5.9	56.3	59 41.93	33 52.2
57	9	. .	. .	. .	. .	. .	. .	45.8	16 0 0.42	11.92	VII.	8	8.25	38 12.8	56.3	15 59 48.52	44 59.1
58	8	7.9	23.2	38.8	. .	. .	. .	. .	7 53.61	11.86	IV.	4	9.52	19 54.6	54.4	16 7 41.75	26 39.0
59	7	. .	. .	. .	41.3	56.1	. .	. .	8 41.02	11.85	IV.	5	11.13	25 35.8	54.1	8 29.17	32 19.9
60	6.7	. .	. .	. .	. .	. .	7.0	22.2	13 36.74	11.81	VII.	6	6.18	27 7.9	52.9	13 24.93	33 50.8
61	6.7	. .	. .	. .	. .	. .	25.0	. .	14 54.70	11.80	VII.	9	6.58	42 28.6	52.6	14 42.90	49 11.2
62	6	. .	. .	. .	. .	. .	. .	54.2	15 8.66	11.80	.	.	. .	. . .	. .	14 56.86	
63	10	. .	. .	. .	. .	. .	3.6	18.8	16 33.36	11.79	VII.	7	4.16	31 7.2	52.2	16 21.57	37 49.4
64	8	. .	. .	. .	15.0	. .	45.0	. .	18 14.79	11.78	VII.	8	13.58	41 1.3	51.7	18 3.01	47 43.0
65	5	. .	16.6	31.8	. .	. .	. .	. .	21 16.74	11.74	III.	3	11.15	15 36.6	50.8	21 35.00	22 17.4
66	10	. .	21.9	. .	. .	. .	. .	. .	25 52.11	11.71	II.	3	8.16	14 5.9	49.7	25 40.40	20 45.6
67	9	. .	. .	16.3	31.3	. .	. .	. .	32 31.04	11.66	IV.	1	6.35	3 15.0	48.0	32 19.38	9 53.0
68	10	. .	18.3	. .	48.8	. .	. .	. .	35 48.56	11.64	IV.	3	6.23	13 9.0	47.1	35 36.92	19 46.1
69	9	. .	. .	53.9	. .	. .	. .	. .	46 8.65	11.56	IV.	2	6.8	8 1.3	44.3	45 57.09	14 35.6
70	9	. .	. .	. .	. .	28.7	. .	. .	48 13.42	11.55	V.	1	6.40	3 17.5	43.7	48 1.87	9 51.2
71	9	. .	. .	. .	. .	. .	. .	28.6	48 42.91	11.55	VII.	3	8.51	14 23.2	43.6	48 31.36	20 56.8
72	9	. .	. .	28.8	. .	. .	. .	. .	53 43.64	11.50	IV.	3	9.17	14 37.0	42.2	53 32.14	21 9.2
73	9	. .	10.9	26.0	. .	. .	. .	. .	55 41.42	11.49	IV.	9	12.34	45 19.3	41.6	55 29.93	51 50.9
74	9	56.1	14.4	. .	. .	. .	. .	. .	57 42.06	11.48	IV.	9	7.14	42 37.4	41.1	57 30.58	49 8.5
75	9	. .	. .	. .	. .	. .	19.5	. .	57 49.39	11.48	VI.	3	3.3	11 27.6	41.0	57 37.91	17 58.6
76	7	. .	. .	. .	. .	. .	. .	14.0	58 28.68	11.47	VII.	9	4.56	41 27.1	40.8	58 17.21	47 57.9
77	7	. .	. .	. .	. .	. .	8.9	24.0	16 59 38.59	— 11.46	VII.	6	5.27	−26 42.1	— 40.5	16 59 27.13	— 34 33 12.6

CORRECTIONS.

Date.		Corr. of Clock.	Hourly rate.	m	n	c
1846.	h.	s.	s.	s.	s.	s.

INSTRUMENT READINGS.

	Date.		Barom.	THERMOM. At.	Ex.
	1846.	h. m.	in.	°	°

REMARKS.

ZONE 20. MAY 20. K. $D_o = -34° 5' 50''$.

No.	Mag.	SECONDS OF TRANSIT. I.	II.	III.	IV.	V.	VI.	VII.	T.	a_1	MICROMETER.			$i + d_2$	d_1	Mean Right Ascension, 1850.0.	Mean Declination, 1850.0.
									h. m. s.	s.			r.	′ ″	″	h. m. s.	° ′ ″
1	9	. .	. .	. .	. .	. .	27.3	42.3	17 50 56.97	− 11.15	VII.	8	2.35	−35 15.7	− 28.2	17 50 45.82	− 34 41 33.9
2	9	. .	. .	. .	. .	. .	7.3	. .	52 37.23	11.14	VI.	1	9.9	4 32.7	27.8	52 26.09	10 50.5
3	9	. .	. .	. .	. .	. .	. .	23.1	53 37.81	11.13	VII.	9	10.19	44 10.5	27.6	53 26.68	50 28.1
4	9	. .	. .	. .	. .	. .	. .	8.2	54 22.89	11.13	VII.	9	7.34	42 47.0	27.4	54 11.76	49 4.4
5	9	. .	. .	. .	. .	. .	. .	45.0	54 59.69	11.13	VII.	9	6.58	42 28.8	27.3	54 48.56	48 46.1
6	9	. .	. .	. .	. .	. .	23.3	38.6	55 53.11	11.12	VII.	7	3.41	30 49.4	27.1	55 41.99	37 6.5
7	8	. .	. .	. .	. .	. .	. .	19.5	56 33.79	11.12	VII.	3	5.59	12 56.3	26.9	56 22.67	19 13.2
8	8	. .	. .	. .	. .	12.0	. .	. .	57 56.75	11.11	VI.	2	7.55	8 55.2	26.7	57 45.64	15 11.9
9	9	. .	. .	. .	. .	. .	6.0	. .	17 58 35.82	11.10	VI.	5	10.14	25 5.7	26.5	58 24.72	31 22.2
10	9	. .	. .	. .	. .	16.6	. .	. .	18 0 1.51	11.10	V.	8	13.19	39 41.3	26.2	17 59 50.41	45 57.5
11	9	18.4	32.8	. .	. .	. .	. .	. .	2 3.73	11.09	IV.	6	13.58	31 1.1	25.7	18 1 52.64	37 16.8
12	9	. .	. .	. .	. .	27.4	. .	. .	2 12.25	11.09	VI.	6	13.34	30 48.8	25.7	2 1.16	37 4.5
13	8	30.0	45.4	. .	. .	. .	. .	. .	4 15.60	11.08	III.	3	6.11	13 2.8	25.3	4 4.52	19 18.1
14	7.8	. .	. .	. .	17.7	32.8	. .	. .	4 17.59	11.08	VI.	7	5.11	31 35.3	25.3	4 6.51	37 50.6
15	7.8	. .	. .	. .	39.2	54.2	. .	. .	5 39.04	11.07	VI.	7	5.24	31 41.8	25.0	5 27.97	37 56.8
16	9	9.9	. .	. .	. .	. .	. .	. .	7 55.61	11.06	I.	4	10.59	20 28.0	24.5	7 44.55	26 42.5
17	8	. .	11.2	26.6	. .	. .	. .	. .	8 41.76	11.05	IV.	8	7.49	37 55.1	24.3	8 30.71	44 9.4
18	8.9	. .	. .	. .	. .	. .	17.9	. .	8 47.73	11.05	VII.	5	6.30	23 12.1	24.3	8 36.68	29 26.4
19	8	. .	. .	. .	. .	. .	. .	14.7	9 29.38	11.05	VII.	9	4.52	41 25.1	24.1	9 18.33	47 39.2
20	8	. .	. .	. .	. .	. .	. .	5.1	10 19.81	11.05	VII.	9	9.9	43 35.1	24.0	10 8.76	49 49.1
21	10	. .	. .	. .	11.4	. .	. .	. .	13 11.23	11.04	IV.	8	5.9	36 34.2	23.3	13 0.19	42 47.5
22	8	. .	. .	. .	. .	22.6	. .	. .	14 7.69	11.03	IV.	8	6.57	37 28.8	23.1	13 56.66	43 41.9
23	3.2	. .	. .	. .	. .	. .	. .	9.3	14 23.71	11.03	VII.	4	13.37	20 47.2	23.1	14 12.68	27 0.3
24	9	50.8	. .	. .	. .	. .	. .	. .	17 36.83	11.02	II.	9	8.4	43 2.5	22.4	17 25.81	49 14.9
25	9	. .	. .	. .	51.6	. .	. .	. .	17 51.43	11.02	IV.	2	7.50	8 52.9	22.3	17 40.41	15 5.2
26	9	. .	. .	. .	35.9	. .	. .	. .	19 35.73	11.01	IV.	7	4.37	31 18.3	22.0	19 24.72	37 30.3
27	8	. .	. .	. .	. .	. .	28.9	44.3	19 58.69	11.01	VII.	3	10.47	15 22.0	21.8	19 47.68	21 33.8
28	7.8	. .	. .	. .	34.9	49.9	. .	. .	21 34.77	11.01	VI.	9	11.1	44 32.0	21.5	21 23.76	50 43.5
29	10	. .	. .	42.0	. .	. .	. .	. .	24 56.95	10.99	IV.	5	3.46	21 49.6	20.8	24 45.96	28 0.4
30	9	. .	. .	. .	25.7	. .	. .	. .	26 25.53	10.99	IV.	2	5.8	7 31.0	20.4	26 14.54	13 41.4
31	9	. .	. .	. .	17.6	. .	. .	. .	27 17.43	10.99	IV.	7	6.44	32 22.6	20.2	27 6.44	38 32.8
32	7.8	. .	. .	. .	. .	11.5	. .	42.0	27 56.47	10.98	VI.	7	9.27	33 44.5	20.1	27 45.49	39 54.6
33	5	. .	. .	. .	32.9	. .	3.0	. .	29 32.83	10.97	V.	2	12.36	11 17.5	19.7	29 21.86	17 27.2
34	7	. .	. .	. .	. .	. .	46.3	1.5	32 16.07	10.96	VII.	6	11.34	29 47.8	19.1	32 5.11	35 56.9
35	7	. .	38.3	53.7	. .	. .	. .	. .	36 8.86	10.95	III.	8	6.33	37 16.6	18.3	35 57.91	43 24.9
36	6	. .	31.0	45.3	. .	. .	. .	. .	40 1.40	10.93	IV.	7	12.59	35 32.3	17.4	39 50.47	41 39.7
37	8	. .	. .	. .	. .	. .	37.6	. .	40 7.47	10.93	VII.	3	10.29	15 12.9	17.4	39 56.54	21 20.3
38	6.7	. .	. .	. .	. .	. .	32.3	47.5	41 2.03	10.93	VII.	5	4.27	22 9.8	17.2	40 51.10	28 17.0
39	9	. .	. .	. .	. .	. .	38.2	. .	43 7.94	10.92	VI.	8	7.20	37 40.2	16.7	42 57.02	43 46.9
40	9	. .	. .	. .	. .	. .	. .	22.7	43 37.38	10.92	VII.	9	4.51	41 24.5	16.6	43 26.46	47 31.1
41	9	. .	. .	. .	. .	50.1	. .	. .	45 35.02	10.91	VI.	9	9.13	43 37.5	16.2	45 24.11	49 43.7
42	9	. .	. .	. .	. .	. .	4.7	. .	46 34.53	10.91	VI.	5	7.45	23 50.3	15.9	46 23.62	29 56.2
43	9	. .	. .	14.7	. .	. .	. .	. .	48 29.65	10.90	IV.	5	3.26	21 39.5	15.5	48 18.75	27 45.0
44	7	. .	. .	. .	. .	59.4	14.3	30.1	48 44.27	10.90	VII.	4	6.59	18 26.6	15.5	48 33.37	24 32.1
45	7	. .	. .	. .	. .	14.8	. .	45.2	49 59.72	10.90	VII.	7	9.55	33 58.7	15.2	49 48.82	40 3.9
46	7	. .	. .	27.7	. .	. .	. .	. .	51 42.91	10.90	IV.	8	4.52	36 25.6	14.8	51 32.01	42 30.4
47	8	. .	. .	. .	. .	. .	. .	30.3	51 44.80	10.90	VII.	6	9.27	26 43.5	14.8	51 33.90	34 48.3
48	7	. .	. .	. .	. .	. .	18.1	. .	52 47.80	10.89	VII.	9	10.28	44 15.0	14.6	52 36.91	50 19.6
49	9	. .	. .	. .	. .	36.8	. .	. .	18 54 21.64	− 10.89	IV.	6	10.6	−29 3.8	− 14.2	18 54 10.75	− 34 35 8.0

CORRECTIONS.

Date.		Corr. of Clock.	Hourly rate.	m	n	c
1846.	h.	s.	s.	s.	s.	s.

INSTRUMENT READINGS.

	Date.		Barom.	THERMOM. At.	Ex.
,	1846.	h. m.	in.	°	°

REMARKS.

(20) 10. Micrometer reading assumed as 11ʳ.19 instead of 13ʳ.19.
(20) 23. Micrometer reading assumed as 11ʳ.37 instead of 13ʳ.37.

Zone 20. May 20. K. $D_o = -34° 5' 50''$—Continued.

No.	Mag.	Seconds of Transit. I.	II.	III.	IV.	V.	VI.	VII.	T.	σ_1	Micrometer.			$i + d_2$	d_1	Mean Right Ascension, 1850.0.	Mean Declination, 1850.0.
									h. m. s.	s.			r.	′ ″	″	h. m. s.	° ′ ″
50	10	..	..	..	..	..	37.8	..	18 55 7.60	− 10.88	VII.	6	7.30	−27 44.3	− 14.1	18 54 56.72	− 34 33 48.4
51	7	33.3	48.9	4.3	..	..	..	..	58 19.39	10.87	IV.	8	4.25	36 11.9	13.3	58 8.52	42 15.2
52	8	..	7.9	23.3	..	..	..	..	18 59 38.46	10.87	III.	8	6.29	37 14.6	13.1	18 59 27.59	43 17.7
53	9	..	25.6	..	..	..	..	..	19 1 55.91	10.87	III.	5	3.41	21 47.0	12.5	19 1 45.04	27 49.5
54	8	19.8	35.5	..	..	..	..	..	3 5.84	10.87	IV.	7	11.29	34 46.8	12.3	2 54.97	40 49.1
55	10	28.3	..	..	..	..	..	..	5 13.91	10.86	I.	3	7.13	13 33.5	11.8	5 3.05	19 35.3
56	8	..	..	20.8	36.0	..	..	..	5 35.83	10.86	IV.	6	4.46	26 21.9	11.7	5 24.97	32 23.6
57	10	..	..	..	..	..	..	49.8	6 4.26	10.85	VII.	5	11.01	25 29.2	11.6	5 53.41	31 30.8
58	9	..	..	..	..	..	55.0	..	7 24.87	10.85	VII.	4	13.5	21 31.7	11.3	7 14.02	27 33.0
59	9	..	..	..	..	..	44.1	..	8 13.82	10.84	VII.	9	3.59	40 58.2	11.1	8 2.98	46 59.3
60	8	..	..	..	..	..	43.2	..	9 13.04	10.84	VII.	5	5.35	22 44.3	10.9	9 2.20	28 45.2
61	10	..	..	..	..	..	9.0	..	10 38.81	10.84	VII.	6	4.56	26 26.4	10.6	10 27.97	32 27.0
62	10	11.9	..	42.9	..	..	..	..	15 57.84	10.83	IV.	6	8.46	28 23.3	9.4	15 47.01	34 22.7
63	9	..	..	..	..	23.8	..	54.2	19 16 8.74	− 10.83	VI.	8	3.34	−35 45.9	− 9.4	19 15 57.91	− 34 41 45.3

Zone 21. May 21. A. $D_o = -38° 24' 20''$.

No.	Mag.	I.	II.	III.	IV.	V.	VI.	VII.	T.	σ_1	Micrometer.			$i + d_2$	d_1	Mean Right Ascension, 1850.0.	Mean Declination, 1850.0.
1	9	..	..	..	..	41.5	57.6	13.8	14 30 25.58	− 11.72	VII.	6	10.51	−29 26.5	− 63.3	14 30 13.86	− 38 54 49.8
2	9	36.1	52.5	8.6	24.6	40.7	56.6	12.5	34 24.51	11.68	I.	5	8.	23 57.4	63.0	34 12.83	49 20.4
3	11	44.2	0.0	16.8	32.7	48.6	4.5	21.0	41 32.52	11.59	VII.	4	11.20	20 37.9	62.5	41 20.93	38 46 0.4
4	10	16.5	32.8	48.8	5.2	21.0	37.0	53.2	47 5.03	11.53	VII.	7	13.6	35 36.8	62.1	46 53.50	39 0 58.9
5	9	..	..	25.0	41.0	57.2	12.9	..	14 50 41.03	11.48	III.	10	4.42	46 25.6	61.8	14 50 29.55	11 47.4
6	9	..	..	45.0	1.2	17.0	33.0	49.2	15 7 1.07	11.33	VII.	9	1.30	34 44.2	60.3	15 6 49.74	0 4.5
7	7	..	..	..	10.0	26.0	41.6	57.5	15 9.72	11.25	VI.	9	11.45	44 57.2	59.5	14 58.47	39 10 16.7
8	6	..	..	26.8	42.9	58.8	14.5	30.9	24 42.69	11.16	VII.	5	9.56	24 56.1	58.4	24 31.53	38 50 14.5
9	9	..	24.5	40.5	56.0	..	..	..	27 56.10	11.14	III.	2	9.57	9 54.9	58.1	27 44.96	35 13.0
10	8	..	..	..	9.5	25.0	41.5	..	29 53.26	11.13	IV.	3	9.10	14 31.9	57.9	28 42.13	39 49.8
11	10	17.8	34.7	50.7	7.2	..	..	..	34 6.76	11.06	IV.	7	8.43	33 24.1	57.3	33 55.70	38 58 41.4
12	11	..	..	..	32.0	48.0	4.0	20.0	42 31.94	10.98	VII.	8	8.	38 1.0	56.3	42 20.96	39 3 17.3
13	8	..	..	..	..	14.5	30.5	46.8	50 58.53	10.95	VI.	3	8.15	14 3.8	55.1	50 47.58	38 39 18.9
14	9	..	..	37.5	53.5	9.1	25.2	41.2	54 53.26	10.88	VII.	8	3.30	35 45.1	54.6	54 42.38	39 0 59.7
15	10	..	..	..	..	58.0	14.0	30.0	56 41.94	10.87	VII.	6	3.25	25 40.4	54.4	56 31.07	38 50 54.8
16	8	..	..	..	..	59.5	15.0	31.5	15 58 43.27	10.85	VII.	4	3.30	16 39.5	54.1	15 58 32.42	41 53.6
17	10	1.5	17.5	34.0	49.5	6.0	..	..	16 4 49.67	10.80	V.	4	8.49	19 21.9	53.2	16 4 38.87	38 44 35.1
18	8	47.0	3.5	19.5	35.8	51.5	7.8	23.7	10 35.67	10.73	VII.	8	8.54	38 29.5	52.3	10 24.94	39 3 41.8
19	7	..	..	47.5	3.4	19.4	35.4	51.7	14 3.41	10.71	VII.	6	9.53	24 56.0	51.8	13 52.70	38 50 7.8
20	7	55.0	11.3	27.4	43.7	59.5	15.7	31.8	16 36 43.61	10.52	VII.	8	12.35	40 21.5	48.2	16 36 33.15	39 5 29.7
21	7	..	..	53.3	9.5	25.2	41.1	57.6	17 2 9.19	10.36	VII.	3	5.48	12 48.9	43.9	17 1 58.83	38 37 52.8
22	11	..	..	..	51.0	6.8	22.8	40.0	4 51.06	10.33	VII.	5	1.56	20 52.8	43.4	4 40.73	45 56.2
23	10	..	36.5	..	8.5	24.7	40.2	56.8	21 8.51	10.21	VII.	6	9.41	28 51.0	40.4	20 58.30	53 51.4
24	9	7.8	24.2	40.2	56.3	12.1	28.2	44.5	24 56.17	10.19	VI.	5	3.36	21 43.8	39.7	24 45.98	46 43.5
25	6	..	..	..	..	..	..	12.0	26 23.62	10.19	VII.	2	3.36	6 10.8	39.4	26 13.43	31 10.2
26	12	..	..	16.5	32.0	48.5	..	..	17 57 32.24	10.00	IV.	6	6.17	27 8.2	33.3	17 57 22.24	38 52 1.5
27	9	..	..	..	..	1.0	16.5	32.0	18 2 44.49	9.97	VII.	9	14.14	46 12.5	32.3	18 2 34.52	39 11 4.8
28	8	..	29.0	45.2	1.0	17.0	33.0	49.5	11 1.17	9.93	VII.	9	2.	40 0.2	30.6	10 51.24	39 4 50.8
29	8	..	..	8.5	24.5	40.5	56.5	12.6	13 24.49	9.92	VII.	7	3.7	30 33.0	30.1	13 14.57	38 55 23.1
30	10	33.0	50.2	6.0	22.0	..	..	..	18 21.94	9.89	IV.	7	6.40	32 21.6	29.1	18 12.05	57 10.7
31	8	..	..	..	5.0	21.0	37.5	53.0	18 19 5.04	− 9.88	VI.	7	5.19	−31 40.2	− 29.0	18 18 55.16	− 38 56 29.2

CORRECTIONS.

Date.		Corr. of Clock.	Hourly rate.	m	n	c
1846.	h.	s.	s.	s.	s.	s.
May 21,	11	− 22.36	+ 0.015	− 0.280	+ 0.344	+ 0.120

INSTRUMENT READINGS.

	Date.		Barom.	Thermom. At.	Ex.
	1846.	h. m.	in.	°	°
Zone 21	May 21,		30.120	67.	58.

REMARKS.

(21) 6. Micrometer thread assumed as 8 instead of 9.
(21) 10. Transits over T.'s V–VII assumed to have been recorded as over T.'s IV–VI.
(21) 19. Micrometer thread assumed as 5 instead of 6.
(21) 25. Micrometer reading assumed as $2^r.36$ instead of $3^r.36$.

Zone 21. May 21. A. $D_0 = -38° 24' 20''$—Continued.

No.	Mag.	Seconds of Transit. I.	II.	III.	IV.	V.	VI.	VII.	T.	a_1	Micrometer.			$i + d_2$	d_1	Mean Right Ascension, 1850.0.	Mean Declination, 1850.0.
									h. m. s.	s.			r.	′ ″	″	h. m. s.	° ′ ″
32	7	23.5	40.2	56.2	12.6	28.3	44.5	0.5	18 23 12.26	− 9.87	I.	5	9.18	−24 36.9	− 28.1	18 23 2.39	− 38 49 25.0
33	6	..	..	39.3	55.5	11.0	27.8	..	33 39.60	9.88	VII.	2	. .	5	26.0	33 29.72	
34	11	19.0	35.0	51.5	8.0	23.5	..	..	45 7.39	9.76	V.	4	11.50	20 53.6	23.7	44 57.63	38 45 37.3
35	9	..	..	..	55.0	11.0	27.0	43.0	46 54.93	9.75	VII.	8	4.51	41 26.9	23.3	46 45.18	39 6 10.2
36	6	..	..	..	13.0	29.2	44.8	1.0	18 53 12.89	9.73	VI.	1	6.41	3 14.7	22.0	18 53 3.16	38 27 56.7
37	11	..	..	..	2.0	18.5	34.0	50.5	19 3 2.17	− 9.67	VII.	7	3.55	−30 57.3	− 20.0	19 2 52.50	− 38 55 37.3

Zone 22. May 25. A. $D_0 = -36° 31' 40''$.

No.	Mag.	I.	II.	III.	IV.	V.	VI.	VII.	T.	a_1	Micrometer.			$i + d_2$	d_1	Mean Right Ascension, 1850.0.	Mean Declination, 1850.0.
1	9	..	..	..	..	..	..	36.5	14 48 49.44	− 6.26	VII.	4	5.45	−17 48.7	. .	14 48 43.18	
2	9	..	..	..	40.2	56.8	11.8	27.5	15 12 40.61	6.12	VII.	7	9.24	33 43.5	. .	15 12 34.49	
3	11	..	..	..	..	56.0	12.0	28.0	18 40.74	6.09	VII.	4	6.25	18 8.9	. .	18 34.65	
4	7	..	28.9	44.8	0.5	16.0	..	..	33 0.28	5.97	V.	5	9.2	24 29.4	. .	32 54.31	− 36 56
5	9	5.7	21.9	37.6	53.2	..	..	..	41 53.01	5.90	IV.	5	6.1	22 57.8	. .	41 47.11	
6	9	..	..	54.0	9.6	25.5	40.8	56.5	43 9.59	5.89	VII.	5	9.36	24 46.1	. .	43 3.70	
7	11	..	..	..	..	..	..	57.5	45 10.59	5.86	VII.	7	6.52	32 26.5	. .	45 4.73	
8	7	58.6	14.5	30.3	46.1	..	..	..	48 45.89	5.83	IV.	7	6.23	32 12.4	. .	48 40.06	
9	10	..	..	..	..	52.5	8.2	23.8	49 36.94	5.82	VII.	7	9.35	33 49.0	. .	49 31.12	
10	9	25.2	41.2	56.9	12.7	..	..	..	52 12.54	5.81	IV.	7	11.22	34 43.8	. .	52 6.73	
11	8	..	..	22.5	37.9	53.6	..	..	53 37.79	5.83	V.	2	11.39	10 47.8	. .	53 31.96	
12	11	..	..	52.3	7.6	23.5	..	..	56 7.62	5.80	V.	3	11.39	15 48.2	. .	56 1.82	
13	11	..	..	28.0	43.5	0.0	..	..	15 59 43.31	5.77	V.	3	9.41	14 48.4	. .	15 59 37.54	
14	7	28.2	44.1	0.2	15.6	..	..	..	16 13 15.48	5.66	III.	6	6.24	32 11.5	. .	16 13 9.82	
15	6	..	..	22.3	38.0	53.6	..	..	14 37.90	5.63	V.	8	3.55	40 56.4	. .	14 32.27	
16	7	34.0	50.0	6.0	21.2	..	..	..	18 21.16	5.63	IV.	4	7.14	18 34.3	. .	18 15.53	
17	10	44.5	0.5	16.2	..	..	..	..	23 31.68	5.59	III.	5	7.24	23 39.8	. .	23 26.09	
18	8	..	..	..	..	1.0	16.5	32.2	23 45.34	5.57	VII.	7	5.46	31 53.1	. .	23 39.77	
19	11	..	29.3	44.8	0.5	..	..	..	26 0.45	5.56	IV.	7	4.34	31 17.2	. .	25 54.89	
20	9	..	..	..	58.5	14.0	29.8	45.5	26 58.46	5.57	VII.	4	6.12	18 2.4	. .	26 52.89	
21	6	..	..	..	9.5	25.0	40.2	56.1	29 9.22	5.55	VII.	5	6.1	22 57.2	. .	29 3.67	
22	8	..	..	..	50.3	5.9	21.5	37.3	30 50.28	5.55	VII.	4	9.16	19 35.5	. .	30 44.73	
23	8	..	..	..	33.0	48.8	4.0	20.0	32 32.95	5.53	VII.	3	10.33	15 14.3	. .	32 27.42	
24	8	52.2	8.0	23.9	39.6	..	..	..	38 39.38	5.48	IV.	6	5.44	26 51.3	. .	38 33.90	
25	7	..	..	..	59.5	15.0	30.5	46.0	40 59.31	5.45	VII.	9	8.35	43 19.1	. .	40 53.86	
26	7	..	..	..	..	..	49.0	4.8	16 56 17.87	5.35	VII.	6	10.42	29 21.7	. .	16 56 12.52	
27	9	..	..	..	..	..	15.5	30.5	17 2 44.01	5.28	VII.	9	11.24	44 44.6	. .	17 2 38.73	
28	10	..	..	15.5	31.2	46.8	2.5	..	7 31.16	5.26	VI.	7	9.53	33 58.5	. .	7 25.90	
29	9	..	..	..	..	26.0	41.7	57.2	9 10.44	5.25	VII.	8	12.	40 2.5	. .	9 5.19	
30	9	37.2	53.1	..	..	..	..	..	16 24.39	5.22	II.	6	5.51	26 54.6	. .	16 19.17	
31	10	..	..	20.5	..	..	..	..	17 36.09	5.20	III.	7	5.18	31 39.4	. .	17 30.89	
32	8	..	..	..	53.0	8.5	23.8	39.5	18 52.69	5.24	VII.	2	4.58	7 24.3	. .	18 47.45	
33	5	..	..	..	..	55.2	10.7	26.5	20 39.59	5.18	VII.	8	9.30	38 46.5	. .	20 34.41	
34	3	43.5	59.5	..	..	..	..	..	23 30.75	5.18	II.	6	7.54	27 56.9	. .	23 25.57	
35	9	..	..	..	..	59.5	..	31.0	39 44.00	5.09	VII.	7	8.30	33 16.1	. .	38 38.91	
36	8	20.8	36.8	52.5	8.0	..	..	..	43 8.03	5.08	IV.	7	4.17	31 8.6	. .	43 2.95	
37	8	2.0	17.8	33.5	..	..	..	..	45 49.15	5.07	III.	7	3.6	30 32.6	. .	45 44.08	
38	7	2.2	18.0	34.0	..	..	..	..	48 49.29	5.09	III.	4	8.12	19 3.6	. .	48 44.20	
39	9	..	..	..	..	22.5	38.2	54.0	17 49 7.00	− 5.08	VII.	5	8.12	−24 3.6	. .	17 49 1.92	

CORRECTIONS.

Date.		Corr. of Clock.	Hourly rate.	m	n	c
1846.	h.	s.	s.	s.	s.	s.
May 25,	11	− 16.80	− 0.042	− 0.280	+ 0.344	+ 0.120

INSTRUMENT READINGS.

Date.		Barom.	Thermom. At.	Thermom. Ex.
1846.	h. m.	in.	°	°

REMARKS.

(21) 33. Assumed to be λ Coronæ, and the transits to have been over T.s IV–VII.
(21) 35. Micrometer thread assumed as 9 instead of 8.
(22) 14. Micrometer assumed as 7 $6^r.24$ instead of 6 $6^r.24$.
(22) 15. Micrometer assumed as 9 $3^r.55$ instead of 8 $3^r.55$.

ZONE 22. MAY 25. A. $D_0 = -36° 31' 40''$—Continued.

No.	Mag.	Seconds of Transit. I.	II.	III.	IV.	V.	VI.	VII.	T.	a_1	Micrometer.			$i + d_2$	d_1	Mean Right Ascension, 1850.0.	Mean Declination, 1850.0.
									h. m. s.	s.			r.	′ ″	′ ″	h. m. s.	° ′ ″
40	9	..	..	..	..	..	4.0	19.5	17 50 32.72	− 5.07	VII.	5	12.	−25 59.0	. .	17 50 27.65	
41	10	13.8	29.6	45.5	1.0	..	..	..	55 0.91	5.05	IV.	5	9.18	24 37.6	. .	54 55.86	
42	9	..	28.0	44.0	59.5	15.0	..	..	56 59.19	5.07	V.	1	11.47	5 51.7	. .	56 54.12	
43	10	..	..	54.8	10.3	26.0	..	..	17 59 10.27	5.02	V.	6	7.58	27 59.2	. .	17 59 5.25	
44	9	..	..	..	..	36.2	52.0	7.5	18 0 20.71	5.00	VII.	9	8.59	43 31.2	. .	18 0 15.71	
45	11	..	50.0	5.5	21.0	37.0	..	..	3 21.02	5.03	V.	3	9.26	14 56.2	. .	3 15.99	
46	9	..	..	..	32.5	48.2	3.5	19.1	4 32.35	5.01	VII.	5	10.	24 58.3	. .	4 27.34	
47	3	..	..	18.2	33.9	49.2	4.5	20.5	7 33.51	5.01	VII.	3	4.35	16 27.9	. .	7 28.50	
48	9	47.0	3.0	18.7	34.5	..	..	..	10 34.20	5.00	IV.	4	7.54	18 54.6	. .	10 29.20	
49	7	..	..	33.0	48.4	4.0	19.3	35.2	12 48.25	4.99	VII.	3	6.15	13 3.6	. .	12 43.26	
50	8	..	..	..	52.5	8.0	24.0	39.5	18 14 52.53	− 4.96	VII.	6	6.45	−27 21.7	. .	18 14 47.57	
									ZONE 23. MAY 27. K. $D_0 = -37° 50' 27''$.								
1	10	..	7.1	..	..	..	..	..	15 15 38.99	9.68	IV.	7	5.24	−31 42.8	. .	15 15 29.31	
2	6	..	..	..	..	..	18.8	35.0	15 47.14	9.69	VII.	5	2.56	21 23.3	. .	15 37.45	
3	10	..	53.8	9.7	..	..	..	..	18 25.37	9.68	III.	4	3.32	16 41.4	. .	18 15.69	
4	10	..	..	..	..	..	19.5	..	18 47.69	9.64	V.	9	2.42	40 21.5	. .	18 38.05	
5	6	..	..	..	..	0.8	16.3	..	19 44.76	9.66	V.	4	2.14	16 1.9	. .	19 35.10	
6	9	..	..	..	..	..	4.1	..	20 32.52	9.65	V.	3	6.50	13 21.2	. .	20 22.87	
7	7	..	..	13.1	..	44.7	..	..	22 28.71	9.64	IV.	3	11.37	15 46.8	. .	22 19.07	
8	10	..	..	..	..	6.3	22.0	..	24 50.28	9.59	VII.	10	4.9	46 7.5	. .	24 40.69	
9	10	..	..	..	36.4	52.3	..	..	26 36.30	9.59	VI.	7	2.5	30 1.7	. .	26 26.71	
10	9	..	..	..	..	10.1	26.2	..	27 54.28	9.57	VII.	10	1.44	44 54.0	. .	27 44.71	
11	9	..	58.5	14.6	..	..	..	..	30 30.27	9.57	III.	5	5.9	22 31.3	. .	30 20.70	
12	10	..	41.1	..	13.1	..	..	..	15 39 12.84	− 9.51	IV.	4	6.48	−18 20.8	. .	15 39 3.33	
									ZONE 24. MAY 27. K. $D_0 = -37° 47' 40''$.								
1	5	..	..	..	..	40.0	55.7	..	15 59 24.07	− 9.35	VII.	4	11.6	−20 31.0	. .	15 59 14.72	
2	11	..	..	6.3	..	..	..	..	16 1 22.37	9.30	III.	9	5.0	41 31.4	. .	16 1 13.07	
3	9	57.8	13.9	29.9	..	..	..	..	7 46.05	9.46	IV.	9	10.6	44 6.6	. .	7 36.59	
4	11	29.5	..	..	..	..	..	..	10 17.53	9.26	III.	5	10.58	25 28.2	. .	10 8.27	
5	10	..	..	..	..	..	53.2	9.3	10 21.00	9.25	VII.	8	8.47	38 25.4	. .	10 11.75	
6	11	..	..	51.9	..	..	..	..	14 7.82	9.22	III.	7	9.00	33 32.2	. .	13 58.60	
7	11	..	50.3	..	..	..	..	..	16 22.41	9.20	III.	9	9.59	44 3.1	. .	16 13.20	
8	10	..	..	..	..	6.5	..	..	16 50.50	9.23	VI.	2	7.38	8 44.7	. .	16 41.27	
9	10	1.0	17.4	33.5	..	..	..	..	20 49.38	9.17	IV.	8	8.34	38 19.4	. .	20 40.21	
10	10	39.6	55.7	..	..	..	..	..	23 27.29	9.19	III.	2	8.45	9 19.0	. .	23 18.10	
11	10	..	..	..	22.3	..	..	..	24 22.13	9.18	IV.	1	7.32	3 41.6	. .	24 12.95	
12	11	..	..	58.9	..	..	..	..	30 14.95	9.10	V.	9	3.49	40 55.4	. .	30 5.85	
13	9	..	12.7	28.6	44.3	..	..	..	33 44.10	9.12	IV.	2	2.9	5 58.4	. .	33 34.98	
14	10	..	..	40.0	55.7	..	..	..	16 42 55.55	− 9.05	IV.	3	10.33	−15 14.2	. .	16 42 46.50	

CORRECTIONS.

Date.		Corr. of Clock.	Hourly rate.	m	n	c
1846.	h.	s.	s.	s.	s.	s.
May 27,	12	− 20.36	− 0.054	− 0.366	+ 0.292	+ 0.120

INSTRUMENT READINGS.

	Date.		Barom.	Thermom. At.	Thermom. Ex.
	1846.	h. m.	in.	°	°
Zone 23	May 27,		29.482	77.8	75.0

REMARKS.

(22) 47. Micrometer probably 4 3r.35.

ZONE 25. JUNE 3. K. $D_0 = -30° 22' 30''$.

No.	Mag.	SECONDS OF TRANSIT. I.	II.	III.	IV.	V.	VI.	VII.	T.	a_1	MICROMETER.			$i + d_2$	d_1	Mean Right Ascension, 1850.0.	Mean Declination, 1850.0.
									h. m. s.	s.			r.	′ ″	″	h. m. s.	° ′ ″
1	10	..	..	54.3	8.8	..	..	..	16 27 8.66	− 20.41	V.	5	5.55	−22 55.0	− 54.8	16 26 48.25	− 30 46 19.8
2	10	..	..	..	..	..	..	7.0	27 23.30	20.41	VII.	6	5.38	26 47.6	54.7	27 2.89	50 12.3
3	10	..	..	..	28.5	..	57.3	..	29 28.	20.41	.	.	. .	. . .	. .	29 8.	
4	8	..	..	..	31.4	46.1	..	..	31 31.35	20.41	V.	2	6.59	8 28.1	53.7	31 10.94	30 31 51.8
5	8	10.2	24.8	..	..	..	..	..	33 54.16	20.36	III.	8	11.32	39 46.9	53.1	33 33.80	31 3 10.0
6	8	..	..	..	19.9	34.5	..	..	34 19.84	20.37	V.	7	7.39	32 49.8	53.0	33 59.47	30 56 12.8
7	8	..	..	..	..	..	25.8	..	34 56.88	20.39	VI.	4	9.38	19 47.7	52.9	34 36.49	30 43 10.6
8	8	..	..	..	..	..	19.3	..	35 50.26	20.35	VI.	9	6.41	42 19.4	52.7	35 29.91	31 5 42.1
9	7	..	..	22.3	..	51.7	..	..	37 37.13	20.34	IV.	10	6.14	47 7.6	52.2	37 16.79	31 10 29.8
10	7	..	55.3	9.8	24.3	..	..	..	39 9.70	20.35	IV.	7	6.49	32 24.6	51.8	38 49.35	30 55 46.4
11	10	4.3	19.1	..	..	..	..	..	42 48.28	20.34	III.	7	7.24	32 42.2	50.9	42 27.94	30 56 3.1
12	9	..	..	..	6.0	..	..	..	43 5.84	20.32	V.	10	1.35	44 46.5	50.8	42 45.52	31 8 7.3
13	9	..	23.5	38.3	..	..	..	..	44 52.40	20.36	III.	2	3.11	6 32.8	50.4	44 32.04	30 29 53.2
14	8	..	..	31.5	..	..	..	..	45 45.73	20.35	V.	4	0.37	15 14.4	50.1	45 25.38	38 34.5
15	9	..	..	..	..	..	..	35.7	45 52.10	20.32	V.	8	2.4	34 59.9	50.1	45 31.78	30 58 20.0
16	8	..	..	..	..	..	..	44.1	47 0.56	20.30	V.	9	2.33	40 14.2	49.8	46 40.26	31 3 34.0
17	9	..	51.0	5.8	..	..	..	..	49 19.99	20.34	III.	3	6.12	13 4.1	49.3	48 59.65	30 36 23.4
18	9	..	45.3	..	14.4	..	..	..	50 14.37	20.31	IV.	7	10.3	34 2.6	49.0	49 54.06	57 21.6
19	10	34.1	..	3.6	..	..	..	..	52 17.75	20.33	III.	2	5.24	7 40.1	48.5	51 57.42	30 58.6
20	8	..	..	..	..	48.3	..	17.3	52 33.58	20.33	VII.	2	7.36	9 46.9	48.4	52 13.25	33 5.3
21	9	..	..	..	..	..	..	16.9	53 33.20	20.31	VII.	6	5.48	26 52.7	48.2	53 12.89	30 50 10.9
22	8	..	..	..	28.1	..	..	..	55 27.94	20.27	V.	10	3.4	45 31.5	47.7	55 7.67	31 3 49.2
23	10	..	29.3	..	..	..	..	..	57 58.47	20.28	IV.	7	6.3	32 1.4	47.1	57 38.19	30 55 18.5
24	7	..	..	..	26.4	..	55.4	..	16 58 26.33	20.29	VI.	6	9.51	28 55.7	46.9	16 58 6.04	52 12.6
25	8	5.3	20.0	..	..	..	..	..	17 0 49.01	20.30	III.	3	10.14	15 6.4	46.3	17 0 28.71	38 22.7
26	9	..	..	..	..	..	..	36.2	2 52.33	20.29	VII.	3	3.30	11 41.8	45.8	2 32.04	30 34 57.6
27	8	..	14.7	44.1	..	..	..	..	43.96	20.29	IV.	9	13.3	45 32.7	45.0	5 23.67	31 8 47.7
28	9	5.2	20.0	..	..	..	..	..	7 48.95	20.28	IV.	3	4.44	12 19.7	44.5	7 28.67	30 35 34.2
29	10	..	..	..	26.4	..	..	9.8	8 26.09	20.28	VII.	3	5.23	12 38.9	44.3	8 5.81	35 53.2
30	8	..	..	..	..	55.4	..	..	9 40.85	20.24	VII.	7	8.55	33 27.8	44.0	9 20.61	56 41.8
31	8	..	..	..	..	..	..	24.7	40.7	20.							
32	10	..	..	39.2	..	..	..	..	11 53.75	20.23	IV.	7	11.7	34 35.0	43.4	11 33.52	30 57 48.4
33	8	..	32.3	..	2.0	..	..	..	13 1.73	20.21	IV.	9	12.36	45 19.0	43.1	12 41.52	31 8 32.1
34	10	..	..	..	20.6	..	..	..	14 20.44	20.22	V.	7	5.14	31 36.5	42.8	14 0.22	30 54 49.3
35	8	51.9	6.6	..	..	..	..	..	17 35.85	20.21	III.	7	10.0	34 1.1	41.9	17 15.64	30 57 13.0
36	7	7.4	..	36.7	..	..	..	..	18 51.41	20.19	III.	9	4.32	41 14.4	41.6	18 31.22	31 4 26.0
37	7	..	..	..	26.3	..	55.1	..	19 26.14	20.21	VII.	5	9.42	24 49.3	41.4	19 5.93	30 48 0.7
38	11	53.7	..	..	..	..	..	..	23 37.73	20.17	I.	8	5.6	36 31.5	40.3	23 17.56	59 41.8
39	11	..	..	46.9	..	..	..	..	24 1.47	20.17	IV.	8	3.21	35 38.9	40.2	23 41.30	58 49.1
40	10	..	..	32.3	..	1.5	..	..	24 46.90	20.17	IV.	7	10.10	34 6.2	40.0	24 26.73	57 16.2
41	9	..	..	40.6	..	..	..	..	26 54.66	20.22	IV.	1	6.9	3 3.1	39.4	26 34.44	26 12.5
42	8	47.2	..	16.8	..	..	..	..	28 31.14	20.18	IV.	5	10.56	25 27.2	39.0	28 10.96	48 36.2
43	10	18.3	33.3	..	..	..	..	..	29 2.23	20.19	III.	4	10.47	20 22.7	38.9	28 42.04	30 43 31.6
44	10	..	50.0	..	14.3	..	..	..	19.05	20.16	IV.	3	10.6	15 2.4	37.8	32 58.89	31 38 10.2
45	10	..	29.5	..	..	..	..	..	35 58.75	20.14	III.	8	10.50	39 25.7	37.0	35 38.61	30 2 32.7
46	9	..	..	..	..	..	37.7	..	36 8.69	20.14	VII.	8	4.7	36 1.7	37.0	35 48.55	30 59 8.7
47	8	30.4	45.3	..	..	..	..	..	38 14.58	20.13	IV.	10	1.5	44 31.4	36.4	37 54.45	31 7 37.8
48	10	..	..	..	..	50.0	..	..	38 35.43	20.15	VII.	6	6.13	27 5.3	36.3	38 15.28	30 50 11.6
49	6	..	..	43.2	..	12.0	..	..	17 39 57.36	− 20.17	IV.	2	8.29	− 9 13.6	− 35.9	17 39 37.19	− 30 32 19.5

CORRECTIONS.

Date.		Corr. of Clock.	Hourly rate.	m	n	c
1846.	h.	s.	s.	s.	s.	s.
June 3,	12	− 31.03	− 0.054	− 0.366	+ 0.272	+ 0.120

INSTRUMENT READINGS.

Date.	Barom.	THERMOM. At.	Ex.
1846. h. m.	In.	°	°

REMARKS.

(25) 10. Transits over T.'s III–V assumed as recorded over T.'s II–IV.
(25) 20. Micrometer reading assumed as $9^r.36$ instead of $7^r.36$.
(25) 27. Transit over T. IV assumed as recorded over T. III, and minutes to have been 5.
(25) 44. Transit over T. IV assumed as $19^s.3$, not $14^s.3$, and minutes as 33.

ZONE 25. JUNE 3. K. $D_0 = -30° 22' 30''$—Continued.

No.	Mag.	SECONDS OF TRANSIT. I.	II.	III.	IV.	V.	VI.	VII.	T.	a_1	MICROMETER.			$i + d_2$	d_1	Mean Right Ascension, 1850.0.	Mean Declination, 1850.0.
									h. m. s.	s.			r.	′ ″	″	h. m. s.	° ′ ″
50	8	. .	. .	. .	. .	2.6	. .	. .	17 40 48.05	– 20.14	VI.	7	4.40	–31 19.2	– 35.7	17 40 27.91	– 30 54 24.9
51	9	. .	. .	. .	. .	. .	58.7	. .	41 29.72	20.14	VII.	7	4.16	31 6.7	35.5	41 9.58	54 12.2
52	9	. .	. .	. .	. .	. .	. .	27.4	41 43.76	20.14	VII.	7	6.14	32 6.5	33.5	41 23.62	55 12.0
53	10	3.8	. .	. .	. .	. .	. .	. .	44 47.70	20.14	III.	6	5.16	26 36.9	34.6	44 27.57	49 41.5
54	9	. .	. .	. .	. .	5.8	. .	. .	44 51.23	20.14	VI.	6	4.34	26 15.5	34.6	44 31.09	49 20.1
55	8	. .	. .	. .	. .	11.0	. .	. .	45 56.37	20.16	V.	3	1.56	10 54.7	34.3	45 36.21	33 59.0
56	6	. .	. .	. .	. .	. .	. .	9.6	46 25.70	20.16	VII.	2	8.31	9 14.2	34.2	46 5.54	32 18.4
57	10	. .	. .	. .	. .	10.0	. .	. .	47 55.39	20.15	IV.	3	9.12	14 34.1	33.8	47 35.24	30 37 37.9
58	10	. .	. .	. .	23.6	. .	. .	. .	49 23.44	20.10	IV.	9	9.6	43 32.9	33.4	49 3.34	31 6 36.3
59	10	. .	. .	. .	. .	17.8	. .	. .	50 3.18	20.15	IV.	3	8.37	14 24.5	33.2	49 43.03	30 37 27.7
60	9.10	. .	. .	. .	14.7	. .	. .	. .	51 14.54	20.10	IV.	9	5.11	41 34.1	32.9	50 54.44	31 4 37.0
61	8	41.2	. .	. .	. .	. .	. .	. .	53 25.21	20.11	III.	7	11.17	34 40.0	32.3	53 5.10	30 57 42.3
62	8	. .	. .	. .	37.6	. .	. .	. .	53 37.44	20.11	V.	7	11.13	34 38.0	32.2	53 17.33	57 40.2
63	8	. .	. .	. .	. .	. .	. .	29.3	53 45.56	20.12	VII.	5	7.13	23 34.0	32.2	53 25.44	46 36.2
64	9	. .	. .	. .	. .	44.7	. .	. .	55 30.07	20.14	VI.	2	11.27	10 43.4	31.7	55 9.97	33 45.1
65	9	. .	. .	. .	. .	. .	41.0	. .	56 12.03	20.11	VII.	6	6.57	27 27.5	31.5	55 51.92	50 29.0
66	9	. .	. .	. .	. .	. .	0.8	. .	57 31.90	20.12	VI.	4	12.36.	21 17.6	31.2	57 11.78	30 44 18.8
67	8	. .	. .	. .	. .	. .	10.1	. .	17 58 41.08	20.08	VII.	8	7.40	37 49.3	30.8	17 58 21.00	31 0 50.1
68	5	. .	16.3	. .	45.7	. .	. .	. .	18 0 45.45	20.10	IV.	5	3.37	21 45.3	30.3	18 0 25.35	30 44 45.6
69	9	. .	. .	28.7	. .	. .	. .	. .	3 43.26	20.07	VII.	7	13.15	35 39.2	29.5	3 23.19	58 38.7
70	8	. .	. .	58.1	. .	26.9	. .	. .	7 12.20	20.12	IV.	1	7.48	3 53.2	28.5	6 52.08	26 51.7
71	11	. .	33.5	. .	. .	. .	. .	. .	9 2.57	20.08	III.	5	5.22	22 38.3	28.0	8 42.49	45 36.3
72	10	. .	19.3	. .	. .	. .	. .	. .	11 48.33	20.09	II.	4	8.8	19 2.2	27.3	11 28.24	30 41 59.5
73	5	. .	. .	. .	. .	9.9	24.6	. .	11 55.47	20.06	VII.	8	6.14	37 5.9	27.2	11 35.41	31 0 3.1
74	7	. .	35.4	. .	. .	. .	. .	. .	15 4.47	20.07	III.	5	5.49	22 52.0	26.4	14 44.40	30 45 48.4
75	4.5	. .	. .	. .	43.0	. .	11.8	. .	15 42.84	20.06	V.	6	5.25	26 41.4	26.2	15 22.78	49 37.6
76	7	. .	. .	8.1	. .	. .	. .	. .	17 22.18	20.10	V.	1	10.38	5 19.1	25.7	17 2.08	30 28 14.8
77	9	14.8	. .	. .	. .	. .	. .	. .	20 58.87	20.04	II.	8	12.37	40 19.7	24.8	20 38.83	31 3 14.5
78	9	. .	. .	0.2	14.6	. .	. .	. .	21 14.58	20.04	V.	7	6.37	32 18.5	24.7	20 54.54	30 55 13.2
79	10	. .	. .	. .	21.9	. .	. .	. .	22 21.74	20.04	IV.	8	6.58	37 28.5	24.4	22 1.70	31 0 22.9
80	9	. .	. .	19.3	. .	. .	. .	. .	23 33.98	20.02	III.	9	7.44	42 51.4	24.1	23 13.96	31 5 45.5
81	7	. .	26.8	. .	. .	. .	. .	. .	24 56.02	20.03	III.	8	5.00	36 28.8	23.7	24 35.99	30 59 22.5
82	9	. .	. .	20.9	. .	. .	. .	. .	25 35.35	20.04	IV.	6	7.7	27 33.0	23.5	25 15.31	50 26.5
83	8	22.3	. .	. .	. .	. .	. .	. .	28 6.00	20.03	II.	6	12.45	30 23.6	22.8	27 45.97	53 16.4
84	10	. .	43.8	. .	. .	. .	. .	. .	29 13.02	20.02	III.	8	5.00	36 28.8	22.6	28 53.00	59 21.4
85	8	. .	. .	. .	34.0	. .	. .	. .	29 33.84	20.05	IV.	4	0.52	15 22.1	22.4	29 13.79	38 14.5
86	8	. .	13.3	. .	42.6	. .	. .	. .	31 42.37	20.05	IV.	4	3.30	16 41.9	22.9	31 22.32	30 39 34.8
87	10	. .	29.6	. .	. .	. .	. .	. .	37 58.83	20.00	III.	8	8.8	38 3.9	21.2	37 38.83	31 0 55.1
88	9	. .	24.8	. .	. .	. .	. .	. .	38 53.71	20.04	II.	2	6.8	8 2.2	20.9	38 33.67	30 30 53.1
89	10	. .	. .	26.9	. .	. .	. .	. .	39 41.30	20.02	V.	5	9.4	24 30.5	20.7	39 21.28	47 21.2
90	9	. .	. .	. .	. .	. .	33.3	. .	40 4.39	20.04	VII.	3	10.23	15 10.5	19.6	39 44.35	30 38 0.1
91	8	. .	. .	. .	38.8	. .	. .	. .	41 38.64	19.99	IV.	9	11.47	44 54.3	19.2	41 18.65	31 7 43.5
92	10	. .	. .	. .	54.9	. .	. .	. .	42 54.74	20.00	IV.	7	12.20	35 11.9	18.9	42 34.74	30 58 0.8
93	6	. .	. .	. .	. .	38.0	52.5	. .	43 23.48	20.00	VII.	7	5.20	31 39.2	18.7	43 3.48	30 54 27.9
94	10	. .	42.0	. .	. .	. .	. .	. .	47 11.28	19.98	III.	9	6.21	42 9.5	17.7	46 51.30	31 4 57.2
95	6	4.4	19.3	. .	. .	. .	. .	. .	48 48.49	19.99	IV.	8	8.30	38 15.0	17.3	48 28.50	31 1 2.3
96	9	. .	. .	. .	. .	36.1	. .	. .	49 21.57	19.99	VI.	8	4.57	36 27.2	17.1	49 1.58	30 59 14.3
97	9	20.4	35.0	. .	. .	. .	. .	. .	53 4.43	19.97	IV.	10	4.00	45 59.8	16.2	52 44.46	31 8 46.0
98	9	. .	. .	. .	. .	55.4	. .	. .	18 52 40.77	– 20.03	V.	3	1.17	–10 35.0	– 16.3	18 53 20.74	– 30 33 21 3

CORRECTIONS.

Date.		Corr. of Clock.	Hourly rate.	m	n	c
1846.	h.	s.	s.	s.	s.	s.

INSTRUMENT READINGS.

Date.		Barom.	THERMOM. At.	Ex.
1846.	h. m.	in.	°	°

REMARKS.

(25) 84. Differs from Transit, 1847, May 28, 14ˢ.05 (one transit T.) in right ascension.
(25) 98. Minutes assumed as 53 instead of 52.

Zone 25. June 3. K. $D_o = -30° 22' 30''$—Continued.

No.	Mag.	Seconds of Transit. I.	II.	III.	IV.	V.	VI.	VII.	T.		a_1	Micrometer.			$i + d_1$		d_1	Mean Right Ascension, 1850.0.		Mean Declination, 1850.0.
									h. m. s.		s.			r.	′ ″		″	h. m. s.		° ′ ″
99	9	8.1							18 59 52.02	−	19.99	II.	6	9.27	−28 43.6	−	14.4	18 59 32.03	−	30 51 28.0
100	10				8.7				19 0 8.54		19.99	IV.	6	11.47	29 54.6		14.3	18 59 48.55		30 52 38.9
101	8	50.1	4.9						5 34.19		19.96	III.	9	7.20	42 39.3		12.9	19 5 14.23		31 5 22.2
102	9		23.8						6 52.79		20.01	III.	3	9.0	14 29.0		12.6	6 32.78		30 37 11.6
103	7				59.6		28.2		6 59.35		20.00	VI.	4	10.35	20 16.5		12.5	6 39.35		42 59.0
104	9				29.7				8 29.54		20.01	IV.	3	8.56	14 27.0		12.1	8 9.53		37 9.1
105	8		11.2						10 40.41		19.97	III.	7	12.23	35 13.4		11.6	10 20.44		57 55.0
106	9						10.8		10 41.84		20.00	VII.	4	12.30	21 14.3		11.6	10 21.84		43 55.9
107	9	40.0							14 23.68		20.01	II.	2	7.2	8 29.4		10.8	14 3.67		30 31 10.0
108	7	35.7	50.7						15 19.88		19.96	III.	9	6.40	42 19.1		10.4	14 59.92		31 4 59.5
109	9		51.3						16 20.26		20.01	III.	3	4.49	12 22.1		10.1	16 0.25		30 35 2.2
110	8	43.4	58.3						20 27.54		19.96	III.	9	7.31	42 44.8		9.1	20 7.58		31 5 23.9
111	6			38.0	52.6				23 52.36		20.00	IV.	4	5.59	17 57.2		8.2	23 32.36		30 40 35.4
112	8	58.5	13.4						28 42.32		20.01	III.	3	9.57	14 57.8		7.0	28 22.31		37 34.8
113	8		13.9		43.0				19 29 42.86	−	20.01	V.	3	8.41	14 19.4	−	6.7	19 29 22.85	−	30 36 56.1

Zone 26. June 15. K. $D_o = -32° 55' 50''$.

No.	Mag.	I.	II.	III.	IV.	V.	VI.	VII.	T.		a_1	Micrometer.			$i + d_1$		d_1	Mean Right Ascension, 1850.0.		Mean Declination, 1850.0.
1	9		7.3						16 43 37.11	+	33.93	III.	4	4.47	−17 19.5	−	37.6	16 44 11.04	−	33 13 47.1
2	8					21.0			44 6.01		33.93	V.	4	3.43	16 48.1		37.5	44 39.94		13 15.6
3	7			21.8					45 36.51		33.94	IV.	4	8.13	19 4.7		37.1	46 10.45		15 31.8
4	6			36.6					46 51.07		33.92	III.	1	9.1	4 29.4		36.8	47 24.99		0 56.2
5	7					12.7			47 57.74		33.95	VI.	5	11.28	25 43.1		36.5	48 31.69		22 9.6
6	8						10.3		48 40.55		33.95	VII.	4	11.55	20 56.4		36.3	49 14.50		17 22.7
7	10				41.9				50 41.74		33.99	IV.	9	6.40	42 19.7		35.8	51 15.73		38 45.5
8	8						58.2		51 28.34		33.99	VI.	9	9.58	43 59.6		35.6	52 2.33		40 25.2
9	8							46.2	52 1.24		33.96	VII.	4	12.2	21 0.0		35.5	52 35.20		17 25.5
10	10				43.8				56 43.64		34.01	V.	8	3.35	35 46.3		34.3	57 17.65		32 10.6
11	9			41.0					57 56.09		34.02	V.	9	8.28	43 14.3		34.0	58 30.11		39 38.3
12	8			43.3					16 58 57.12		34.00	V.	5	13.00	26 29.8		33.7	16 59 31.12		22 53.5
13	9			48.3		18.1			17 0 3.30		34.03	V.	9	6.40	42 19.6		33.4	17 0 37.33		38 43.0
14	8				35.6				1 35.44		34.01	VI.	6	9.26	28 43.2		33.0	2 9.45		25 6.2
15	8					53.2			2 39.29		34.02	VI.	7	8.4	33 2.6		32.7	3 12.31		29 25.3
16	8						34.5		3 4.82		33.99	VII.	3	5.48	12 51.1		32.6	3 38.81		9 13.7
17	6			39.9					4 54.71		34.02	IV.	5	11.30	25 44.4		32.2	5 28.73		22 6.6
18	7					40.2			5 25.31		34.04	VII.	8	6.22	37 10.3		32.0	5 59.35		33 32.3
19	9			18.2					7 32.95		34.03	IV.	5	4.16	22 4.9		31.5	8 6.98		18 26.4
20	.			5.1					20.		. .	.	.	. .	. . .		. .	8 54.		
21	8					34.9			8 [illegible].92		34.02	V.	4	10.3	20 0.3		31.3	8 53.94		16 21.6
22	8		3.8						11 33.64		34.03	III.	4	8.21	19 9.7		30.4	12 7.67		15 30.1
23	8		20.9						12 51.03		34.08	III.	9	9.30	43 45.6		30.1	13 25.11		40 5.7
24	8				58.8				13 58.64		34.03	IV.	2	7.00	8 28.0		29.8	14 32.67		4 47.8
25	10		43.8						18 13.77		34.07	IV.	7	3.30	30 44.3		28.7	18 47.84		27 3.0
26	8				58.1				18 57.94		34.09	VI.	9	4.51	41 24.4		28.5	19 32.03		37 42.9
27	6					53.6			19 38.72		34.09	VI.	8	10.6	33 3.8		28.3	20 12.81		35 22.1
28	7					36.7			20 21.78		34.08	VI.	7	5.40	31 49.8		28.1	20 55.86		28 7.9
29	6					35.1			21 20.22		34.10	VI.	8	9.8	35 34.5		27.8	21 54.32		34 52.3
30	8					24.9			17 22 20.04	+	34.10	VI.	9	10.51	−44 26.4	−	27.6	17 22 54.14	−	33 40 44.0

CORRECTIONS.

Date.		Corr. of Clock.		Hourly rate.		m		n		c
1846.	h.		s.		s.		s.		s.	s.
June 15.	12	−	22.90	+	0.003	−	0.366	+	0.272	+ 0.120

INSTRUMENT READINGS.

	Date.		Barom.	Thermom. At.	Thermom. Ex.
	1846.	h. m.	in.	°	°
Zone 26	June 15.		29.655	77.0	83.7

REMARKS.

(26) 20. Probably identical with following star.

ZONE 26. JUNE 15. K. $D_o = -32° 55' 50''$—Continued.

No.	Mag.	Seconds of Transit: I.	II.	III.	IV.	V.	VI.	VII.	T.		a_1	Micrometer.	Wire No.	r.	$i + d_2$		d_1	Mean Right Ascension, 1850.0.	Mean Declination, 1850.0.
									h. m. s.		s.			r.	′ ″		″	h. m. s.	° ′ ″
31	9					26.6			17 24 11.71	+	34.10	VI.	8	5.34	−36 46.3	−	27.1	17 24 45.81	− 33 33 3.4
32	8					19.0			25 4.13		34.11	VI.	9	5.23	41 40.6		26.8	25 38.24	37 57.4
33	9						20.3		25 50.60		34.08	VII.	4	1.49	15 50.0		26.7	26 24.68	12 6.7
34	10				40.0				27 40.24		34.09	IV.	6	6.59	27 29.1		26.1	28 14.33	23 45.2
35	6			46.1					29 1.10		34.12	III.	8	4.6	36 2.0		25.8	29 35.22	32 17.8
36	10						33.2		29 3.42		34.11	VI.	7	4.44	31 21.5		25.6	29 37.53	27 37.3
37	9			36.4	36.4				30 51.45		34.13	IV.	9	1.59	39 57.6		25.3	31 25.58	36 12.9
38	6				32.3				31 32.14		34.11	VI.	6	9.52	28 56.4		25.1	32 6.25	25 11.5
39	11				48.8				32 48.64		34.12	IV.	7	13.24	35 41.6		24.8	33 22.76	31 59.4
40	10			30.2					38 44.84		34.10	IV.	3	10.40	15 19.2		23.2	39 18.94	11 32.4
41	5.6		53.7	8.8					40 23.86		34.16	IV.	9	8.3	43 1.7		22.7	40 58.02	39 14.4
42	11						30.7		41 0.97		34.12	VI.	4	10.39	20 18.3		22.7	41 35.09	16 31.0
43	10		5.6		35.8				43 35.30		34.13	IV.	6	11.25	29 43.6		21.8	44 9.43	25 55.4
44	10						27.1		43 57.27		34.15	VI.	8	8.29	38 14.8		21.7	44 31.42	34 26.5
45	10				33.3				45 33.14		34.16	IV.	8	10.50	39 26.3		21.3	46 7.30	35 37.6
46	7			19.8	34.6				46 34.54		34.14	VI.	6	5.9	26 33.3		21.0	47 8.68	22 44.3
47	11		46.3						48 16.17		34.15	III.	5	3.59	21 56.2		20.6	48 50.32	18 6.8
48	7	43.7		14.3					49 29.00		34.15	IV.	6	6.23	27 10.9		20.2	50 3.15	23 21.1
49	10		15.3						52 45.18		34.16	IV.	5	8.55	24 26.0		19.3	53 19.34	20 35.3
50	9				32.9				53 32.74		34.15	IV.	4	4.8	17 0.8		19.1	54 6.89	13 9.9
51	11			54.6					55 9.36		34.17	IV.	5	5.17	22 35.7		18.7	55 43.53	18 44.4
52	8	54.0	9.3						58 39.33		34.19	III.	8	8.50	38 25.6		17.7	59 13.52	34 33.3
53	9					56.3			58 41.34		34.17	VI.	5	7.31	23 43.3		17.7	59 15.51	19 51.0
54	8							45.0	17 59 0.01		34.17	VII.	4	7.50	18 52.6		17.6	17 59 34.18	15 0.2
55	7			41.8					18 0 56.59		34.18	IV.	5	9.24	24 40.6		17.1	18 1 30.77	20 47.7
56	8		17.8						2 47.86		34.20	II.	8	8.27	38 13.8		16.6	3 22.06	34 20.4
57	7				7.1				3 6.94		34.20	V.	7	12.8	35 6.0		16.5	3 41.14	31 12.5
58	8						5.8		4 28.27		34.18	VI.	4	9.49	19 53.0		16.1	5 2.45	15 59.1
59	8						45.8		4 16.10		34.17	VI.	3	9.58	14 57.8		16.2	4 50.27	11 4.0
60	7						44.1		5 14.42		34.17	VI.	3	13.18	16 38.9		15.9	5 48.59	12 44.8
61	10			15.0					6 59.75		34.19	IV.	4	13.32	21 46.0		15.4	7 33.94	17 51.4
62	7						53.5		7 23.75		34.19	VI.	6	12.54	30 28.4		15.3	7 57.94	26 33.7
63	8		41.4		11.6				9 11.27		34.18	V.	2	6.53	8 25.6		14.8	9 45.45	4 30.4
64	6.7			13.9					10 28.74		34.20	IV.	6	6.45	27 22.0		14.4	11 2.94	23 26.4
65	8				6.7				11 6.54		34.22	VI.	8	10.40	39 21.0		14.3	11 40.76	35 25.9
66	8		39.2						13 8.91		34.19	III.	2	7.42	8 49.2		13.7	13 43.10	4 52.9
67	8		32.8						14 2.71		34.21	II.	5	10.2	24 59.6		13.5	14 36.92	21 3.1
68	9				29.4				14 29.24		34.23	III.	8	9.15	38 38.2		13.3	15 3.47	34 41.5
69	9					18.8			15 3.82		34.20	VI.	4	7.25	18 40.2		13.2	15 38.02	14 43.4
70	9					16.8			16 1.82		34.20	VI.	4	9.28	19 42.4		12.9	16 36.02	15 45.3
71	6				4.8				17 4.64		34.22	V.	7	9.3	33 32.6		12.6	17 38.86	29 35.2
72	8				45.7				17 45.54		34.21	VI.	5	11.54	25 56.3		12.4	18 19.75	21 58.7
73	8						52.7		18 22.87		34.24	VII.	8	8.55	38 27.6		12.3	18 57.11	34 29.9
74	7				36.4				19 36.24		34.20	VI.	3	4.42	12 18.0		11.9	20 10.44	8 19.9
75	8							34.2	19 49.19		34.21	VII.	4	3.2	16 26.9		11.9	20 23.40	12 28.8
76	6							25.3	20 40.15		34.20	VII.	2	7.55	8 55.6		11.6	21 14.35	4 57.2
77	7						23.8		21 59.19		34.19	VI.	1	12.38	6 18.9		11.2	22 33.38	2 20.1
78	5				33.0				23 32.84		34.20	IV.	2	12.34	11 16.9		10.8	24 7.04	17 17.7
79	6							25.6	18 23 40.47	+	34.19	VII.	1	13.59	− 6 59.6	−	10.8	18 24 14.66	− 33 3 0.4

CORRECTIONS.

Date.		Corr. of Clock.	Hourly rate.	m	n	c
1846.	h.	s.	s.	s.	s.	s.

INSTRUMENT READINGS.

Date.		Barom.	Thermom. At.	Thermom. Ex.
1846.	h. m.	In.	°	°

REMARKS.

(26) 37. Transit over T. III assumed as correct.

ZONE 26. JUNE 15. K. $D_o = -32° 55' 50''$—Continued.

No.	Mag.	SECONDS OF TRANSIT. I.	II.	III.	IV.	V.	VI.	VII.	T.	a_1	MICROMETER.			$i + d_2$	d_1	Mean Right Ascension, 1850.0.	Mean Declination, 1850.0.
									h. m. s.	s.			r.	' "	"	h. m. s.	° ' "
80	9	. .	. .	. .	. .	. .	57.6	. .	18 25 27.90	+ 34.22	VI.	4	12.27	−21 12.9	− 10.3	18 26 2.12	− 33 17 13.2
81	7	. .	. .	49.3	. .	. .	. .	. .	28 4.36	34.25	IV.	8	12.2	40 2.7	9.5	28 38.61	36 2.2
82	7	. .	. .	51.6	. .	. .	. .	. .	29 6.19	34.21	V.	2	12.21	11 9.3	9.3	29 40.40	7 8.6
83	10	. .	34.9	. .	. .	. .	. .	. .	31 4.87	34.23	III.	6	11.12	29 36.8	8.7	31 39.10	25 35.5
84	9	. .	. .	. .	40.5	. .	. .	. .	32 40.34	34.26	VI.	9	5.56	41 57.3	8.3	33 14.60	37 55.6
85	9	. .	. .	. .	. .	15.8	. .	. .	34 0.93	34.26	V.	9	3.16	40 36.5	7.9	34 35.19	36 34.4
86	7	. .	. .	. .	. .	. .	54.5	. .	34 24.68	34.25	VI.	8	3.19	35 38.0	7.8	34 58.93	31 35.8
87	9	. .	. .	. .	57.1	. .	. .	. .	35 56.94	34.24	V.	5	3.19	21 36.0	7.4	36 31.18	17 33.4
88	9	. .	. .	. .	. .	4.7	. .	. .	36 49.68	34.22	VI.	3	2.28	11 10.2	7.4	37 23.90	7 7.3
89	9	. .	. .	. .	. .	. .	19.6	. .	37 49.92	34.22	VI.	3	3.35	11 44.1	6.8	38 24.14	7 40.9
90	8	. .	. .	. .	46.3	. .	. .	. .	18 39 46.14	+ 34.27	IV.	10	2.40	−45 20.1	− 6.3	18 40 20.41	− 33 41 16.4

ZONE 27. JUNE 15. K. $D_o = -32° 56' 0''$.

No.	Mag.	I.	II.	III.	IV.	V.	VI.	VII.	T.	a_1	MICROMETER.			$i + d_2$	d_1	Mean Right Ascension, 1850.0.	Mean Declination, 1850.0.
1	7	. .	. .	. .	. .	. .	54.8	. .	19? 48 25.19	+ 34.22	VI.	1	11.57	− 5 58.2	. .	19 48 59.41	
2	8	. .	. .	. .	. .	. .	47.5	. .	49 17.84	34.23	VI.	2	10.53	10 25.6	. .	49 52.07	
3	9	. .	. .	. .	51.0	. .	. .	. .	50 50.84	34.24	III.	4	7.49	18 52.5	. .	51 25.08	
4	8	. .	. .	. .	38.8	. .	. .	. .	51 38.64	34.25	IV.	5	10.59	25 28.7	. .	52 12.89	
5	5	37.9	. .	8.3	. .	. .	. .	. .	55 23.18	34.26	IV.	6	11.2	29 32.0	. .	55 57.44	
6	8	53.7	9.0	24.0	. .	. .	. .	. .	57 38.80	34.25	IV.	4	12.10	21 4.5	. .	58 13.05	
7	9	. .	. .	. .	. .	. .	. .	22.7	57 37.55	34.23	VII.	1	12.49	6 24.2	. .	58 11.78	
8	8	. .	. .	. .	. .	28.0	. .	57.9	19? 59 12.92	34.24	VII.	3	6.58	13 26.5	. .	19 59 47.16	
9	7	. .	25.1	40.2	. .	. .	. .	. .	20? 2 54.87	34.24	IV.	3	10.59	15 28.8	. .	20 3 29.11	
10	8	. .	. .	. .	. .	. .	25.4	. .	5 55.56	34.30	IV.	9	15.37	46 51.3	. .	6 29.86	
11	10	. .	. .	. .	. .	. .	37.3	. .	8 7.55	34.27	VII.	6	11.41	29 51.2	. .	8 41.82	
12	7	. .	. .	. .	. .	22.5	. .	. .	9 7.51	34.25	VI.	3	12.9	16 4.0	. .	9 41.76	
13	8	. .	. .	. .	. .	. .	11.9	. .	9 42.15	34.27	VI.	5	9.4	24 30.3	. .	10 16.42	
14	7	. .	. .	. .	. .	. .	. .	40.8	9 55.69	34.25	VII.	2	8.34	9 14.3	. .	10 29.94	
15	6	. .	. .	. .	. .	24.3	. .	. .	11 9.31	34.26	VI.	4	3.36	16 44.4	. .	11 43.57	
16	10	. .	. .	. .	. .	17.0	. .	. .	12 2.00	34.25	VI.	3	10.21	15 9.1	. .	12 36.25	
17	6.7	. .	. .	. .	. .	16.6	. .	. .	13 1.70	34.28	V.	7	13.23	35 44.0	. .	13 35.98	
18	9.10	. .	. .	16.8	. .	. .	. .	. .	14 31.76	34.28	III.	7	11.23	34 43.4	. .	15 6.04	
19	8	. .	8.2	. .	. .	. .	. .	. .	15 37.88	34.24	II.	1	11.5	5 31.9	. .	16 12.12	
20	8	. .	. .	. .	. .	. .	. .	44.8	18 0.13	34.30	VII.	9	6.35	42 16.7	. .	18 34.43	
21	7	. .	. .	. .	. .	. .	. .	25.1	19 40.34	34.29	VII.	8	3.19	35 37.7	. .	20 14.63	
22	8	. .	. .	. .	. .	. .	. .	12.1	20 27.45	34.30	VII.	9	9.1	43 30.5	. .	21 1.75	
23	8	. .	. .	. .	22.6	. .	. .	. .	22 22.44	34.26	IV.	4	8.5	19 0.6	. .	22 56.70	
24	5	. .	40.4	. .	10.8	. .	. .	. .	29 10.60	34.30	IV.	10	4.38	46 19.8	. .	29 44.90	
25	8	. .	. .	14.2	. .	. .	. .	. .	30 29.04	34.27	IV.	6	6.15	27 6.8	. .	31 3.31	
26	9	11.8	. .	. .	. .	. .	. .	. .	33 56.91	34.27	II.	5	6.11	23 2.8	. .	34 31.18	
27	11	. .	. .	. .	. .	2.9	. .	. .	33 47.99	34.28	VI.	7	7.33	32 46.9	. .	34 22.27	
28	10	. .	25.3	. .	. .	. .	. .	. .	35 55.14	34.26	III.	4	8.50	19 23.3	. .	36 29.40	
29	7	49.3	. .	19.7	. .	. .	. .	. .	37 34.63	34.29	III.	8	3.32	35 44.8	. .	38 8.92	
30	7	. .	21.7	. .	. .	. .	. .	. .	38 51.85	34.30	III.	10	1.33	44 46.2	. .	39 26.15	
31	8	. .	. .	. .	. .	24.3	. .	. .	39 9.42	34.29	VI.	8	8.38	38 19.3	. .	39 44.71	
32	8	. .	. .	. .	13.2	. .	. .	. .	40 13.04	34.29	IV.	8	8.53	38 27.1	. .	40 47.33	
33	9	. .	27.6	. .	. .	. .	. .	. .	41 57.38	34.25	II.	3	9.10	14 33.5	. .	42 31.63	
34	10	. .	32.7	. .	. .	. .	. .	. .	43 2.59	34.27	IV.	5	7.3	23 29.3	. .	43 36.86	
35	9	. .	. .	5.0	. .	. .	. .	. .	20? 44 19.87	+ 34.27	IV.	6	10.36	−29 18.8	. .	20 44 54.14	

CORRECTIONS.

Date.		Corr. of Clock.	Hourly rate.	m	n	c
1846.	h.	s.	s.	s.	s.	s.

INSTRUMENT READINGS.

Date.		Barom.	THERMOM. At.	Ex.
1846.	h. m.	in.	°	°

REMARKS.

Zone 28. June 16. A. $D_o = -31° 34' 10''$.

No.	Mag.	I.	II.	III.	IV.	V.	VI.	VII.	T.	n_1	Micrometer.			$i + d_3$	d_1	Mean Right Ascension, 1850.0.	Mean Declination, 1850.0.
									h. m. s.	s.			r.	′ ″	′ ″	h. m. s.	° ′ ″
1	8	..	..	..	17.5	32.0	46.6	..	14 38 17.31	+ 33.65	.	1	8.	− 3 59.1	− 57.2	14 38 50.96	− 31 39 6.3
2	7	..	..	..	11.0	25.6	..	..	40 10.51	33.67	IV.	3	5.46	12 50.8	56.5	40 44.18	31 47 57.3
3	8	5.2	20.2	35.3	49.6	..	..	..	49.67	33.72	I.	6	4.50	26 23.4	54.1	47 23.39	32 1 27.5
4	9	..	..	..	..	53.0	7.5	22.2	48 38.18	33.73	VII.	8	8.48	38 23.8	53.4	49 11.91	13 27.2
5	10	..	48.0	3.1	..	..	..	..	50 17.76	33.74	II.	8	9.50	38 55.4	52.8	50 51.50	13 58.2
6	8	31.0	46.5	1.0	15.9	..	..	..	53 15.70	33.75	II.	6	7.33	27 46.0	51.7	53 49.45	2 47.7
7	12	3.5	18.5	..	..	..	..	..	56 47.96	33.76	I.	6	5.38	26 47.6	50.4	57 21.72	32 1 48.0
8	10	..	..	..	..	18.5	32.3	47.5	57 3.31	33.76	V.	4	4.10	17 2.0	50.3	14 57 37.07	31 52 2.3
9	11	..	..	13.2	28.2	..	..	..	14 59 28.04	33.80	IV.	9	2.52	40 24.1	49.4	15 0 1.84	32 15 23.5
10	11	..	..	..	..	20.8	35.0	50.0	15 1 5.86	33.79	VII.	7	5.24	31 41.3	48.8	1 39.65	6 40.1
11	12	..	8.5	23.5	38.0	..	..	..	6 38.01	33.81	IV.	7	3.25	30 41.6	46.5	7 11.82	5 38.1
12	10	..	..	..	21.5	..	..	..	10 21.34	33.83	IV.	8	9.34	38 47.6	45.2	10 55.17	32 13 42.8
13	9	..	..	5.5	19.8	..	..	..	12 19.69	33.81	III.	1	7.35	3 46.4	44.4	12 53.50	31 38 40.8
14	10	..	..	..	28.0	42.5	57.0	12.5	13 27.86	33.83	VI.	3	2.50	11 21.8	44.0	14 1.69	46 15.8
15	8	..	..	..	..	..	28.2	42.5	14 58.50	33.82	VII.	1	4.33	2 13.9	43.4	15 32.82	31 37 7.3
16	11	6.2	21.1	36.0	..	..	..	..	21 50.62	33.89	III.	7	5.3	31 31.1	40.7	22 24.51	32 6 21.8
17	8	..	..	..	..	..	..	38.5	21 54.49	33.90	VII.	10	5.46	46 53.2	40.7	22 28.39	21 43.9
18	10	3.0	..	32.5	..	..	..	..	15 26 47.37	+ 33.91	I.	7	8.	−33 0.1	− 38.7	15 27 21.28	− 32 7 43.8

Zone 29. June 17. K. $D_o = -37° 52' 0''$.

No.	Mag.	I.	II.	III.	IV.	V.	VI.	VII.	T.	n_1	Micrometer.			$i + d_3$	d_1	Mean Right Ascension, 1850.0.	Mean Declination, 1850.0.
1	8	..	..	..	11.8	..	..	..	15 27 11.64	+ 33.36	III.	9	7.3	−42 33.9	− 46.5	15 27 45.00	− 38 35 20.4
2	9	..	..	..	..	..	9.1	..	27 37.34	33.36	VI.	7	10.37	34 21.2	46.4	28 10.70	27 7.6
3	5.6	..	..	..	..	..	49.4	..	28 17.63	33.37	VI.	9	12.7	45 7.7	46.3	28 51.00	37 54.0
4	9	..	..	..	..	3.2	..	..	29 47.28	33.34	V.	4	10.27	20 11.7	46.0	30 20.62	12 57.7
5	8	..	..	14.5	..	46.2	..	..	38 30.16	33.50	VI.	3	12.7	16 1.7	44.3	39 3.66	8 46.0
6	9	..	55.0	..	27.2	..	..	..	44 26.98	33.47	IV.	8	6.11	37 6.9	43.1	45 0.45	29 50.0
7	10	..	..	..	42.9	..	..	..	45 42.74	33.47	IV.	7	4.30	31 15.5	42.8	46 16.21	23 58.3
8	7	..	..	..	..	..	..	8.3	48 20.73	33.50	VII.	9	13.4	40 33.7	42.3	48 54.23	33 16.0
9	6	..	..	..	..	6.2	..	..	49 50.34	33.51	VI.	8	13.45	40 56.8	42.0	50 23.85	33 38.8
10	5	..	..	..	..	..	45.7	..	50 13.90	33.51	VII.	9	15.7	46 38.6	41.9	50 47.41	39 20.5
11	4.5	4.8	21.3	37.1	..	..	..	..	52 52.75	33.51	III.	4	6.10	18 1.5	41.3	15 53 26.26	38 10 42.8
12	9	..	..	..	..	2.3	..	..	15 59 46.31	33.51	V.	1	5.54	2 51.9	39.9	16 0 19.82	37 55 31.8
13	10.11	..	..	46.2	..	..	..	..	16 8 1.92	33.58	IV.	5	8.8	24 2.1	38.0	8 35.50	38 16 40.1
14	8	..	..	0.8	16.6	..	..	..	14 16.37	33.60	IV.	3	6.11	13 1.6	36.6	14 49.97	5 38.2
15	4	..	..	..	52.9	..	24.7	..	19 52.80	33.68	VI.	10	6.53	47 31.0	35.3	20 26.48	40 6.3
16	9	..	..	..	..	41.9	..	..	23 25.96	33.65	VII.	3	10.55	15 24.9	34.5	23 59.61	7 59.4
17	7.8	..	..	..	..	..	1.9	..	23 30.34	33.64	VII.	2	13.1	11 28.1	34.5	24 3.98	4 2.6
18	7.8	..	48.4	4.3	20.3	..	..	..	26 20.16	33.68	III.	6	10.19	29 10.7	34.0	26 53.84	21 44.7
19	8	..	..	0.7	..	32.3	..	..	32 16.48	33.71	IV.	6	12.17	30 10.6	32.4	32 50.19	22 43.0
20	8	..	..	..	..	29.9	..	..	33 13.94	33.69	VII.	2	10.18	10 5.5	32.2	33 47.63	38 2 37.7
21	6	..	..	..	2.0	..	33.2	..	40 1.76	33.72	VI.	2	3.55	6 51.8	30.6	40 35.48	37 59 22.4
22	5	..	..	..	47.1	..	18.6	..	41 46.95	33.74	VI.	4	6.17	18 4.8	30.1	42 20.69	38 10 34.9
23	9	..	15.0	31.2	..	..	..	..	43 46.86	33.77	III.	6	4.22	26 9.8	29.7	44 20.63	18 39.5
24	5	..	..	4.9	20.6	..	..	..	44 20.46	33.75	IV.	4	4.36	17 13.9	29.5	44 54.21	9 43.4
25	6	25.2	41.3	..	..	..	..	..	46 12.98	33.76	III.	4	7.44	18 49.1	29.1	46 46.74	11 18.2
26	7	..	..	..	..	35.3	51.3	..	48 19.51	33.79	VI.	6	6.32	27 15.5	28.6	48 53.30	38 19 44.1
27	7	..	55.4	11.6	..	..	..	..	16 50 26.90	+ 33.76	III.	1	11.58	− 5 56.4	− 28.1	16 51 0.66	− 37 58 24.5

CORRECTIONS.

Date.	Corr. of Clock.	Hourly rate.	m	n	c
1846.	h. s.	s.	s.	s.	s.
June 16,	12 − 23.39	− 0.022	− 0.366	+ 0.272	+ 0.120

INSTRUMENT READINGS.

	Date.	h. m.	Barom.	Thermom. At.	Thermom. Ex.
	1846.	h. m.	in.	°	°
Zone 28	June 16,	14 38	29.97	77.8	73.3
		15 27	29.98	75.5	71.0
Zone 29	June 17,	15 27	30.05	75.0	67.0
		18 11	29.94	74.0	67.0

REMARKS.

(28) 3. Minutes assumed as 46.
(29) 8. Micrometer assumed as 3r.4 instead of 13r.4.

ZONE 29. JUNE 17. K. $D_0 = -37° 52' 0''$—Continued.

No.	Mag.	I.	II.	III.	IV.	V.	VI.	VII.	T.	a_1	MICROMETER. (wire)		(r)	$i + d_2$	d_1	Mean Right Ascension, 1850.0.	Mean Declination, 1850.0.
									h. m. s.	s.			r	' "	"	h. m. s.	° ' "
28	6	..	..	..	..	51.3	..	23.2	16 50 35.33	+ 33.78	VII.	3	5.39	−12 44.8	− 28.0	16 51 9.11	− 38 5 12.8
29	5	..	..	..	..	22.0	..	53.9	52 6.09	33.80	VII.	5	3.40	21 35.6	27.7	52 39.89	38 14 3.3
30	4.5	..	..	..	..	..	..	51.4	53 3.42	33.77	VII.	1	7.2	3 25.8	27.4	53 37.19	37 55 53.2
31	9	..	51.0	7.0	..	..	..	..	16 55 22.86	33.83	III.	7	6.12	32 7.1	26.8	16 55 56.69	38 24 33.9
32	4	..	..	8.8	24.0	..	..	..	17 1 24.67	33.87	IV.	9	12.37	45 23.2	25.4	17 1 58.54	37 48.6
33	9	..	..	..	..	..	..	15.5	1 27.69	33.83	VII.	4	11.14	20 35.0	25.3	2 1.52	13 0.9
34	5	..	..	..	4.9	20.8	..	..	3 4.79	33.83	IV.	3	8.27	14 10.5	24.9	3 38.62	6 35.4
35	10	..	..	..	..	..	2.0	..	3 30.32	33.85	VI.	6	7.15	27 37.3	24.8	4 4.17	20 2.1
36	4	..	..	..	47.8	3.7	19.2	..	5 47.65	33.87	VII.	7	4.59	31 29.5	24.3	6 21.52	23 53.8
37	9	..	..	..	..	..	..	9.3	5 21.46	33.84	VII.	4	4.34	17 12.3	24.4	5 55.30	38 9 36.7
38	5	..	..	1.3	17.0	..	..	..	7 16.77	33.84	V.	2	4.34	7 11.8	23.9	7 50.61	37 59 35.7
39	8	..	..	..	..	..	50.3	..	8 18.76	33.83	VI.	1	7.17	3 33.7	23.6	8 52.65	37 55 57.3
40	9	..	..	..	..	49.8	..	..	9 33.94	33.90	IV.	8	10.17	39 11.6	23.4	10 7.84	38 31 35.0
41	9	..	..	..	..	..	..	16.5	9 58.86	33.90	VII.	8	6.16	37 8.8	23.2	10 32.76	29 32.0
42	7	..	..	..	52.6	8.2	..	..	12 52.33	33.86	VII.	2	11.12	10 32.9	22.5	13 26.19	2 55.4
43	6	56.1	12.2	..	..	..	..	..	15 43.88	33.88	II.	4	7.56	18 55.0	21.8	16 17.76	38 11 16.8
44	6	..	..	19.2	35.0	..	..	..	16 34.68	33.87	IV.	1	6.23	3 6.6	21.6	17 8.55	37 55 28.2
45	9	..	..	..	28.3	44.5	..	..	17 28.38	33.93	IV.	9	4.50	41 26.4	21.3	18 2.31	38 33 47.7
46	4.5	..	..	..	33.3	48.8	..	..	18 33.02	33.92	II.	7	3.49	30 54.4	21.1	19 6.94	23 15.5
47	10	..	..	..	..	..	15.3	..	19 43.60	33.93	VI.	7	4.33	31 16.7	20.8	20 17.53	23 37.5
48	8	..	..	..	..	..	12.9	28.8	20 41.16	33.94	VII.	8	10.13	39 9.0	20.5	21 15.10	38 31 29.5
49	8	..	26.3	42.4	..	..	..	..	22 57.76	33.69	III.	1	13.0	6 27.7	20.0	23 31.75	37 58 47.7
50	7	..	..	..	..	3.8	..	..	24 47.93	33.95	VI.	8	5.55	36 58.5	19.5	25 21.88	38 29 18.0
51	3.4	..	..	..	..	55.0	11.0	..	25 39.19	33.96	VI.	8	10.00	39 2.7	19.3	26 13.15	38 31 22.0
52	4.5	..	..	..	..	52.4	8.0	24.4	26 36.47	33.90	VII.	1	11.22	5 37.5	19.0	27 10.37	37 57 56.5
53	9	..	..	..	..	10.9	..	..	28 55.00	33.94	VI.	5	8.29	24 12.5	18.5	29 28.94	38 16 31.0
54	9	..	21.0	37.0	..	..	..	..	31 52.84	33.97	III.	7	3.43	30 51.6	17.7	32 26.81	23 9.3
55	8.9	..	..	35.7	21.3	..	..	..	32 51.19	33.94	IV.	3	9.39	14 47.0	17.5	33 25.13	7 4.5
56	6	28.0	44.1	..	..	..	..	..	36 15.65	33.94	II.	2	10.50	10 22.1	16.6	36 49.59	2 38.7
57	9	..	..	..	44.6	..	..	..	36 44.63	33.96	III.	5	10.50	25 24.1	16.5	37 18.59	17 40.6
58	7	..	..	..	..	52.1	..	..	37 36.15	33.94	V.	2	14.5	12 1.1	16.3	38 10.09	4 17.4
59	7	..	..	..	..	..	37.9	54.1	38 6.30	33.94	VII.	2	12.28	11 11.4	16.1	38 40.24	3 27.5
60	9	..	..	..	54.8	..	..	..	39 54.63	33.99	VI.	7	8.4	33 3.7	15.7	40 28.62	25 19.4
61	9	..	..	..	..	..	..	42.0	39 54.20	33.96	VII.	4	12.19	21 7.9	15.7	40 28.16	13 23.6
62	7	..	..	..	48.9	..	..	..	41 48.73	34.00	V.	9	8.56	43 31.1	15.2	42 22.73	35 46.3
63	6	41.0	57.4	..	..	..	..	..	43 29.32	34.01	III.	9	7.6	42 35.3	14.8	44 3.33	34 50.1
64	8	..	14.7	30.7	..	..	..	..	44 46.25	33.97	IV.	3	9.11	14 32.8	14.5	45 20.22	6 47.3
65	8	..	..	..	..	..	..	37.2	44 49.65	34.02	VII.	10	4.31	46 18.7	14.5	45 23.67	38 33.2
66	7	..	..	..	42.6	58.7	..	..	46 42.58	33.98	VI.	3	2.56	11 22.5	14.0	47 16.56	3 36.5
67	6	..	..	..	16.0	31.9	..	..	48 15.93	34.02	VI.	9	3.30	40 45.6	13.6	48 49.95	32 59.2
68	5	..	..	21.8	37.8	..	..	..	49 37.62	34.00	IV.	6	8.13	28 7.0	13.3	50 11.62	20 20.3
69	6	..	..	..	39.6	55.6	..	..	50 39.54	33.97	V.	2	13.41	11 49.0	13.0	51 13.51	4 2.0
70	7	..	0.4	16.6	..	..	..	..	52 32.03	33.99	IV.	3	7.17	13 35.0	12.5	53 6.02	5 47.5
71	9	..	..	15.2	..	..	..	..	54 30.98	34.01	IV.	6	3.29	25 43.0	12.0	55 4.99	17 55.0
72	8	..	..	..	38.9	..	..	..	55 38.73	34.05	IV.	10	1.10	44 37.4	11.8	56 12.78	36 49.2
73	7	..	..	..	59.8	..	..	..	56 59.63	34.04	IV.	9	6.27	42 15.6	11.4	57 33.67	31 27.0
74	7	..	49.9	..	..	..	..	..	58 21.96	34.04	II.	9	12.23	45 15.8	11.1	58 56.00	37 26.9
75	10	..	..	58.0	..	..	..	..	17 58 14.07	34.03	IV.	9	8.21	43 13.4	10.8	17 59 48.10	35 24.2
76	8	..	..	5.2	21.2	..	..	..	18 1 20.94	+ 34.01	V.	4	10.45	−20 20.9	− 10.3	18 1 54.95	− 38 12 31.2

CORRECTIONS.

Date.		Corr. of Clock.	Hourly rate.	m	n	c
1846.	h.	s.	s.	s.	s.	s.

INSTRUMENT READINGS.

Date.		Barom.	THERMOM. At.	THERMOM. Ex.
1846.	h. m.	in.	°	°

REMARKS.

(29) 75. Minutes assumed as 59 instead of 58.

ZONE 29. JUNE 17. K. $D_0 = -37° 52' 0''$—Continued.

No.	Mag.	Seconds of Transit I.	II.	III.	IV.	V.	VI.	VII.	T.	a_1	Micrometer			$i + \alpha_2$	d_1	Mean Right Ascension, 1850.0.	Mean Declination, 1850.0.
									h. m. s.	s.			r.	′ ″	″	h. m. s.	° ′ ″
77	7	24.9	40.9						18 3 12.89	+ 34.04	III.	7	10.45	−34 25.5	− 9.9	18 3 46.93	− 38 26 35.4
78	5			9.1		41.0			6 24.77	34.02	IV.	4	12.12	21 5.0	9.8	6 58.79	13 14.1
79	7				27.9		54.6		7 22.84	34.03	V.	5	8.53	24 24.8	8.8	7 56.87	16 33.6
80	7	35.9	52.2						18 11 23.80	+ 34.06	IV.	9	10.11	−44 9.2	− 7.8	18 11 57.86	− 38 36 17.0

ZONE 30. JUNE 18. A. $D_0 = -31° 34' 30''$.

No.	Mag.	I.	II.	III.	IV.	V.	VI.	VII.	T.	a_1	Micrometer			$i + \alpha_2$	d_1	Mean Right Ascension, 1850.0.	Mean Declination, 1850.0.
1	10	32.5	47.5	2.5					15 37 17.13	+ 33.06	III.	7	7.2	−32 31.2	− 91.6	15 37 50.19	− 32 8 32.8
2	9			28.0	43.0				40 42.87	33.09	IV.	9	11.1	44 31.3	90.9	41 15.96	20 32.2
3	9		2.0	17.0					41 31.70	33.09	III.	8	7.20	37 39.8	90.7	42 4.79	13 40.5
4	10			33.6	48.5				49 48.37	33.12	IV.	8	7.19	37 39.3	89.1	50 21.49	13 38.4
5	10				0.0	14.5			50 59.83	33.12	V.	8	10.40	39 20.9	88.9	51 32.95	32 15 19.8
6	9			39.6	53.5				15 58 53.10	33.12	II.	1	9.57	4 58.0	87.2	15 59 26.22	31 40 55.2
7	11						12.0	27.0	16 1 42.71	33.14	VII.	3	11.25	15 41.7	86.7	16 2 15.85	51 38.4
8	11					32.5	48.5		3 16.76	33.14	V.	2	6.57	8 26.8	86.3	3 49.90	44 23.1
9	11				11.5				23 11.34	33.21	IV.	2	10.48	10 23.7	81.9	23 44.55	31 46 15.6
10	11	5.0							25 49.51	33.25	I.	7	7.25	32 42.4	81.3	26 22.76	32 8 33.7
11	10				58.0	13.0	27.5		26 58.11	33.23	IV.	3	8.	13 58.6	81.1	27 31.34	31 49 49.7
12	10					24.0	38.8	53.5	28 9.35	33.25	VII.	5	11.15	26 36.3	80.8	28 42.60	32 2 27.1
13	11	35.2	50.6						31 19.83	33.26	I.	5	10.34	25 15.5	80.0	31 53.09	32 1 5.5
14	10		49.2	4.0					32 18.60	33.26	II.	5	9.04	24 30.3	79.8	32 51.66	31 0 20.1
15	6					42.0	56.8	11.5	34 27.29	33.24	V.	3	6.16	13 5.9	79.3	34 0.53	31 48 55.2
16	10				2.0	17.0			35 2.07	33.28	V.	8	6.37	37 18.0	79.2	35 35.35	32 13 7.1
17	9				38.5				38 38.34	33.24	IV.	1	4.40	2 18.0	78.3	39 11.58	31 38 6.3
18	9					0.0	14.5		39 45.23	33.28	V.	8	4.38	36 17.9	78.0	40 18.51	32 12 5.9
19	9			26.0	40.5				41 40.41	33.26	III.	4	8.43	19 19.9	77.5	42 13.67	31 55 7.4
20	10		50.5						43 20.01	33.28	II.	6	10.59	29 30.1	77.1	43 53.29	32 5 17.2
21	8	41.1	56.5						44 25.68	33.27	I.	4	10.59	20 28.4	76.3	44 58.95	31 56 14.7
22	8				12.5				45 12.34	33.30	IV.	7	10.59	39 29.8	76.7	45 45.64	32 15 16.5
23	9		13.0	28.0					46 42.59	33.29	II.	6	11.26	29 43.8	76.3	47 15.88	5 30.1
24	11	31.5	46.5						48 16.01	33.30	II.	7	5.32	31 45.6	75.9	48 49.31	32 7 31.5
25	11		19.5						49 48.91	33.30	II.	5	3.52	21 52.6	75.6	50 22.21	31 57 38.2
26	7		8.4	23.4					51 37.83	33.30	II.	4	8.15	19 5.6	75.1	52 11.13	31 54 50.7
27	9						41.5	56.0	52 11.98	33.31	VI.	6	4.32	26 14.5	75.0	52 45.29	32 1 59.5
28	10			51.2	6.0				54 5.86	33.32	IV.	7	2.	29 58.6	74.5	54 39.18	32 5 43.1
29	11					10.0	24.5		54 55.23	33.31	VI.	5	8.22	24 9.1	74.3	55 28.54	31 59 53.4
30	10							32.0	55 47.57	33.30	VII.	2	7.33	8 44.6	74.1	56 20.87	44 28.7
31	11					59.0	12.5	28.0	16 58 43.70	33.30	VII.	2	11.2	10 30.2	73.3	16 59 17.00	31 46 13.5
32	8		30.2	45.0					17 0 59.72	33.34	II.	7	6.21	32 10.4	72.8	17 1 33.06	32 7 53.2
33	8		10.8	25.5					2 40.37	33.35	III.	8	10.41	39 21.4	72.3	3 13.72	15 3.7
34	9				40.5				3 10.34	33.34	IV.	6	7.26	27 42.7	72.2	3 43.68	32 3 24.9
35	9		0.0	15.0					5 29.39	33.34	III.	4	2.39	16 16.0	71.6	6 2.73	31 51 57.6
36	9			46.5	1.5				7 1.15	33.34	IV.	4	5.22	17 38.4	71.2	7 34.49	31 53 19.6
37	7			1.5		30.7			8 16.25	33.38	V.	10	6.58	47 30.0	70.9	8 49.63	32 23 10.9
38	11						5.0	20.0	10 35.72	33.36	VII.	5	8.31	24 13.4	70.3	11 9.08	31 59 53.7
39	11		25.6	40.5					12 55.06	33.36	III.	5	10.48	25 23.1	69.7	13 28.42	32 1 2.8
40	10			45.6	0.0				13 59.92	33.35	IV.	3	9.19	14 38.5	69.4	14 33.27	31 50 17.9
41	8	39.0	54.0						17 16 23.33	+ 33.36	II.	4	4.34	−17 13.9	− 68.8	17 16 56.69	− 31 52 52.7

CORRECTIONS.

Date.	Corr. of Clock.	Hourly rate.	m	n	c
1846. h.	s.	s.	s.	s.	s.
June 18, 12	− 22.51	− 0.013	− 0.366	+ 0.272	+ 0.120
	+ 22.53	− 0.013	− 0.366	+ 0.272	+ 0.120

INSTRUMENT READINGS.

	Date.	Barom.	Thermom. At.	Ex.
	1846. h. m.	In.	°	°
Zone 30	June 18, 15 37	29.97	76.0	68.0
	18, 21 11	29.92	75.7	66.5

REMARKS.

(29) 79. Transit over T. IV assumed as 22^s.9 instead of 27^s.9.
(30) 15. Minutes assumed as 33 instead of 34.
(30) 22. Micrometer thread assumed as 8 instead of 7

Zone 30. June 18. A. $D_0 = -31° 34' 30''$—Continued.

No.	Mag.	Seconds of Transit. I.	II.	III.	IV.	V.	VI.	VII.	T_0	a_1	Micrometer.			$i + d_2$	d_1	Mean Right Ascension, 1850.0.	Mean Declination, 1850.0.
									h. m. s.	s.			r.	′ ″	″	h. m. s.	° ′ ″
42	11						15.5	30.5	17 17 46.22	+ 33.37	VI.	5	6.15	−23 4.9	− 68.4	17 18 19.59	− 31 58 43.3
43	9			42.5	57.5				19 57.20	33.38	IV.	5	8.18	24 7.3	67.9	20 30.58	59 45.2
44	8					21.0	35.5		21 6.24	33.37	VI.	3	10.27	15 12.6	67.6	21 39.61	50 50.2
45	10					2.5	16.5	31.7	22 47.43	33.38	VII.	4	1.42	15 46.7	67.1	23 20.81	51 23.8
46	8						50.5	5.5	24 21.17	33.37	VII.	3	0.00	9 55.6	66.7	24 54.54	31 45 32.3
47	7			25.0	39.8				26 39.62	33.39	IV.	6	3.50	25 53.5	66.1	27 13.01	32 1 29.6
48	11			50.5	5.0				28 4.91	33.39	III.	4	7.54	18 55.2	65.8	28 38.30	31 54 31.0
49	8		12.0	27.0					29 41.60	33.42	III.	7	4.	30 59.2	65.3	30 15.02	32 6 34.5
50	8						53.6	8.0	31 23.99	33.41	VII.	6	9.1	28 30.2	64.9	31 57.40	4 5.1
51	9	24.3	39.1						33 8.74	33.43	II.	7	9.18	33 39.8	64.4	33 42.17	9 14.2
52	10				2.5	16.5			34 2.08	33.44	IV.	9	6.58	42 28.3	64.2	34 35.52	32 18 2.5
53	12		52.5	8.0					36 22.16	33.41	II.	4	6.44	18 19.6	63.6	36 55.57	31 53 53.2
54	11				1.0				37 0.84	33.43	IV.	7	7.46	32 53.5	63.4	37 34.27	32 8 26.9
55	11		40.5	55.5					39 9.96	33.43	III.	5	2.41	21 16.9	62.8	39 43.39	31 56 49.7
56	8							20.0	39 35.50	33.40	VII.	1	6.35	3 15.6	62.7	39	38 48.3
57	10					31.5	46.2		41 16.83	33.43	VI.	5	7.22	23 48.8	62.3	41 50.26	59 21.1
58	7			38.0	53.0				42 52.70	33.43	IV.	5	7.40	23 48.1	61.8	43 26.13	31 59 19.9
59	11			13.5	28.5				44 28.25	33.44	IV.	6	10.	29 0.6	61.4	45 1.69	32 4 32.0
60	8				15.5	29.8	44.5		45 15.21	33.43	VI.	4	9.43	19 50.1	61.2	45 48.64	31 55 21.3
61	9			55.0	9.8				47 9.61	33.44	IV.	5	11.46	25 52.4	60.7	47 43.05	32 1 23.1
62	10		20.0	34.8	49.5				48 49.34	33.44	IV.	4	9.43	19 50.3	60.3	49 22.78	31 55 20.6
63	10		25.5	40.0					50 54.72	33.45	III.	5	4.14	22 3.9	59.7	51 28.17	31 57 33.6
64	9				42.9	57.5			51 42.75	33.45	V.	6	6.41	27 19.9	59.5	52 16.20	32 2 49.4
65	10					10.0	24.5		52 55.24	33.44	VI.	4	4.40	17 17.0	59.2	53 28.68	31 52 46.2
66	8				52.8	7.5			54 52.73	33.47	V.	8	12.	40 1.3	58.6	55 26.20	32 15 29.9
67	8						50.0		56 20.64	33.47	VI.	8	8.8	38 3.9	58.3	56 54.11	13 32.2
68	12			6.5					58 21.13	33.48	III.	6	6.53	27 26.0	57.7	17 58 54.61	2 53.7
69	8			27.5		57.0			17 59 42.27	33.49	V.	7	11.32	34 47.7	57.4	18 0 15.76	10 15.1
70	9		22.5	37.5					18 1 52.27	33.50	III.	9	8.5	43 2.3	56.8	2 25.77	32 18 29.1
71	8			3.2	17.8				3 17.70	33.47	IV.	5	9.	24 28.5	56.4	3 51.17	31 59 54.9
72	9					42.5	57.0		4 27.72	33.50	VI.	8	10.43	39 22.2	56.0	5 1.22	32 14 48.2
73	9			16.0	31.0				7 30.85	33.51	IV.	9	3.9	40 32.7	55.2	8 4.36	15 57.9
74	10					23.5			9 8.79	33.50	V.	7	5.	31 29.6	54.8	9 42.29	6 54.4
75	8						39.0	44.0	10 9.74	33.50	VII.	8	10.24	39 12.4	54.2	11 43.24	32 15 36.6
76	9				14.5	29.0	44.5		33 14.60	33.49	VI.	3	2.47	11 20.1	48.3	33 48.09	31 46 38.4
77	7	33.0	48.0	2.9					35 17.50	33.52	III.	6	8.15	28 7.4	47.7	35 51.02	32 3 25.1
78	7				55.2	10.0			36 55.14	33.50	IV.	4	6.44	18 19.8	47.2	37 28.64	31 53 37.0
79	8				5.0	20.0			38 5.05	33.52	V.	6	4.9	26 3.1	46.9	38 38.57	32 1 20.0
80	11					43.0	58.0	13.0	40 28.51	33.49	VII.	2	4.4	6 58.9	46.3	41 2.00	31 42 15.2
81	10			33.0	47.5				41 47.49	33.52	IV.	6	9.39	28 49.9	45.9	42 21.01	32 4 5.8
82	7		8.8	23.5	38.2				43 23.40	33.53	IV.	7	11.36	34 49.8	45.4	43 56.93	10 5.2
83	9			0.0	15.5				45 15.06	33.55	IV.	8	6.	36 59.4	45.0	45 48.61	32 12 14.4
84	6				11.6	26.1	41.0		46 11.50	33.51	VI.	4	4.37	17 15.5	44.7	46 45.01	31 52 30 2
85	8			35.0	50.0				47 49.72	33.53	IV.	6	2.4	24 59.9	44.3	48 23.25	32 0 14.2
86	9			58.5	13.8				50 13.52	33.55	IV.	9	10.35	44 18.2	43.7	50 47.07	19 31.9
87	9			36.0	51.0				51 50.90	33.56	IV.	10	8.7	48 5.0	43.2	52 24.46	32 23 18.2
88	10			17.3	32.0				53 31.78	33.51	IV.	4	1.41	15 46.7	42.8	54 5.29	31 50 59.5
89	9			19.0	33.8				55 33.54	33.51	IV.	4	4.42	17 18.2	42.3	56 7.05	52 30.5
90	7						24.5	39.5	18 19 55.21	+ 33.51	VII.	4	9.31	−19 43.8	− 42.2	18 56 28.72	− 31 54 56.0

CORRECTIONS.

Date.	Corr. of Clock.	Hourly rate.	m	n	c
1846. h.	s.	s.	s.	s.	s.

INSTRUMENT READINGS.

Date.	Barom.	Thermom. At.	Thermom. Ex.
1846. h. m.	in.	°	°

REMARKS.

(30) 56. Transit rejected. (See Zone of 1846, August 12.)
(30) 75. Transit over T. VII assumed as 54s instead of 44s, and minutes as 11 instead of 10.
(30) 82. Transits over T.'s III–V assumed as recorded over T.'s II–IV.

ZONE 30. JUNE 18. A. $D_0 = -31° 34' 30''$—Continued.

No.	Mag.	SECONDS OF TRANSIT. I.	II.	III.	IV.	V.	VI.	VII.	T.	a_1	MICROMETER.			$i + d_2$	d_1	Mean Right Ascension, 1850.0.	Mean Declination, 1850.0.
									h. m. s.	s.			r.	' "	"	h. m. s.	° ' "
91	8	..	..	..	..	37.5	52.0	7.2	18 57 22.78	+ 33.51	VII.	4	7 24	−18 39.6	− 41.8	18 57 56.29	− 31 53 51.4
92	10	..	..	..	..	42.5	57.0	..	18 59 27.73	33.54	VI.	7	9.40	33 51.0	41.2	19 0 1.27	32 9 2.2
93	10	..	..	..	..	..	56.0	10.5	19 1 26.55	33.55	VII.	9	12.	45 0.6	40.7	2 0.10	32 20 11.3
94	7	..	..	..	..	26.5	41.0	56.0	3 11.74	33.51	VII.	4	12.	21 0.2	40.2	3 45.25	31 56 10.4
95	6	42.6	57.5	12.0	..	..	..	..	6 26.97	33.54	III.	7	8.39	33 20.3	39.4	7 0.51	32 8 29.7
96	8	..	..	42.0	56.5	..	..	..	8 56.52	33.54	IV.	7	5.38	31 48.8	38.8	9 30.06	6 57.6
97	6	..	..	59.0	13.8	28.5	..	..	10 13.70	33.54	III.	7	2.26	30 11.7	38.4	10 47.24	5 20.1
98	6	..	43.0	58.0	12.8	..	..	..	12 12.68	33.55	IV.	8	7.23	37 41.3	37.9	12 46.23	32 12 49.2
99	11	37.5	53.0	..	..	..	..	..	13 22.09	33.57	II.	4	6.39	18 17.1	37.6	13 55.66	31 53 24.7
100	9	..	..	..	..	57.9	12.5	27.5	15 43.25	33.61	VII.	8	5.5	36 31.1	37.0	16 16.86	32 11 38.1
101	8	..	..	..	53.0	7.8	22.5	..	17 53.08	33.57	IV.	4	7.26	18 41.1	36.4	18 26.60	31 53 47.5
102	10	35.4	50.5	..	..	..	..	..	18 19.88	33.58	I.	5	9.4	24 30.1	36.3	18 53.46	31 59 36.4
103	6	..	..	46.5	1.5	16.0	..	..	20 1.39	33.62	III.	10	9.4	48 33.7	35.9	20 35.01	32 23 39.6
104	9	..	..	..	..	18.5	33.0	48.0	21 3.75	33.59	VII.	6	7.16	27 37.2	35.6	21 37.34	2 42.8
105	11	..	..	29.0	44.0	..	..	..	23 43.78	33.60	IV.	7	9.34	33 48.1	34.9	24 17.38	8 53.0
106	7	..	..	..	56.2	11.0	..	..	25 56.15	33.58	V.	5	11.1	25 29.6	34.3	26 29.73	0 33.9
107	8	..	..	..	..	24.5	39.5	53.9	26 9.89	33.60	VII.	7	10.48	34 25.1	34.2	26 43.49	32 9 29.3
108	8	..	..	30.0	44.5	59.3	..	..	28 44.42	33.57	V.	3	9.29	14 43.5	33.5	29 17.99	31 49 47.0
109	9	..	..	..	6.0	20.8	..	..	31 5.95	33.59	V.	6	6.0	26 59.2	32.9	31 39.54	32 2 2.1
110	9	24.8	39.5	..	..	..	..	..	33 9.07	33.57	II.	5	8.0	23 58.0	32.4	33 42.64	31 59 0.4
111	5	..	..	..	53.3	7.7	..	..	36 53.08	33.60	V.	9	3.44	40 50.3	31.4	36 26.68	32 15 51.7
112	8	..	..	..	52.0	6.5	..	..	40 51.85	33.61	V.	10	8.29	48 16.0	30.3	41 25.46	23 16.3
113	8	..	8.0	22.5	37.6	..	..	..	42 37.52	33.61	IV.	10	4.35	46 17.8	29.8	43 11.13	32 21 17.6
114	8	..	43.2	58.0	..	..	..	..	46 12.60	33.56	III.	5	9.30	24 43.6	28.9	46 46.16	31 59 42.5
115	10	..	..	..	..	..	38.5	53.0	50 8.96	33.56	VII.	5	4.5	21 58.9	27.9	50 42.52	56 56.8
116	8	41.7	56.5	11.5	..	..	..	..	56 26.03	33.55	III.	5	8.27	24 11.8	26.3	56 59.58	59 8.1
117	11	..	..	3.5	18.0	33.0	..	..	19 58 17.95	33.53	V.	2	6.28	8 12.3	25.8	19 58 51.48	43 8.1
118	9	..	49.3	4.0	18.5	..	..	..	20 0 18.42	33.55	IV.	5	4.56	22 25.2	25.2	20 0 51.97	31 57 20.4
119	8	..	..	..	..	49.2	..	18.5	2 34.48	33.58	VII.	9	6.50	42 23.9	24.7	3 8.06	32 17 18.6
120	9	..	46.0	0.5	15.0	..	..	..	4 15.07	33.54	IV.	4	10.45	20 21.7	24.4	4 48.61	31 55 16.1
121	9	..	..	53.0	7.5	22.0	..	..	6 7.27	33.52	V.	2	3.23	6 38.7	23.8	6 40.79	31 41 32.5
122	6	..	9.5	24.5	39.5	..	..	..	9 39.23	33.57	II.	8	5.18	36 37.9	22.9	10 12.80	32 11 30.8
123	8	19.5	34.7	49.0	..	..	..	..	18 3.66	33.50	III.	2	8.18	9 7.8	20.8	18 37.16	31 43 58.6
124	9	..	9.5	24.3	39.2	..	..	..	20 39.05	33.54	IV.	7	10.41	34 22.0	20.1	21 12.59	32 9 12.1
125	8	..	..	..	..	5.0	19.3	34.2	22 50.05	33.50	VII.	4	6.28	18 11.3	19.6	23 23.55	31 53 0.9
126	10	..	..	..	18.0	32.8	..	..	27 17.92	33.48	V.	2	10.15	10 6.9	18.5	27 51.40	44 55.4
127	9	..	..	..	..	..	36.6	51.8	28 7.41	33.49	VII.	4	7.5	18 30.0	18.3	28 40.90	31 53 18.3
128	5	46.8	1.8	16.6	31.5	..	..	..	31 31.31	33.52	I.	7	7.6	32 32.8	17.5	32 4.83	32 7 20.3
129	10	48.5	3.5	..	..	..	..	..	33 32.99	33.50	I.	6	12.0	30 0.7	17.0	34 6.49	32 4 47.7
130	9	..	48.0	3.0	17.5	..	..	..	35 17.37	33.48	IV.	4	3.38	16 45.8	16.6	35 50.85	31 51 32.4
131	8	..	..	..	..	44.2	59.0	13.5	37 29.49	33.49	VII.	6	10.4	29 0.9	16.1	38 2.98	32 3 47.0
132	8	..	..	..	40.0	54.6	9.3	..	39 39.90	33.48	VI.	5	8.11	24 3.6	15.5	40 13.38	31 58 49.1
133	7	..	21.0	35.5	50.5	..	..	..	40 50.44	33.51	IV.	9	5.39	41 48.5	15.3	41 23.95	32 16 33.9
134	9	..	2.5	17.2	31.8	..	..	..	42 31.76	33.47	IV.	5	4.39	22 16.6	14.9	43 5.23	31 57 1.5
135	7	..	..	..	..	38.2	53.0	8.0	44 23.67	33.48	VII.	6	11.12	29 36.4	14.4	44 57.15	32 4 20.8
136	8	19.2	34.1	49.1	..	..	..	..	45 3.71	33.49	III.	7	6.46	32 23.2	14.3	45 37.20	7 7.5
137	10	..	..	..	..	..	27.0	41.2	46 57.37	33.49	VII.	9	5.15	41 35.9	13.9	47 30.86	16 19.8
138	8	..	..	..	44.0	58.5	..	..	43.83	33.49	IV.	9	5.50	41 54.1	13.4	49 17.32	32 16 37.5
139	8	..	..	..	..	47.0	2.0	16.5	20 57 32.33	+ 33.41	VII.	1	6.20	− 3 8.0	− 11.4	20 58 5.74	− 31 37 49.4

CORRECTIONS.

Date.	Corr. of Clock.	Hourly rate.	m	n	c
1846. h.	s.	s.	s.	s.	s.

INSTRUMENT READINGS.

Date.	Barom.	THERMOM. At.	Ex.
1846. h. m.	In.	°	°

REMARKS.

(30) 111. Minutes assumed as 35 instead of 36.
(30) 136. Minutes assumed as 48.

ZONE 30. JUNE 18. A. $D_0 = -31° 34' 30''$ —Continued.

No.	Mag.	Seconds of Transit. I.	II.	III.	IV.	V.	VI.	VII.	T.		a_1	Micrometer.			$i + d_2$		d_1	Mean Right Ascension, 1850.0.	Mean Declination, 1850.0.
									h. m. s.		s.			r.	′ ″		″	h. m. s.	° ′ ″
140	8	. .	. .	. .	. .	. .	31.8	47.0	20 58 2.58	+	33.40	VII.	2	9.45	− 4 52.3	−	11.3	20 58 35.98	− 31 39 33.6
141	9	. .	1.0	15.9	30.5	. .	. .	. .	21 0 30.48		33.45	IV.	7	4.8	31 3.3		10.8	21 1 3.93	32 5 44.1
142	10	. .	46.5	1.5	. .	. .	. .	. .	5 15.85		33.41	II.	3	6.27	13 11.3		9.7	5 49.26	31 47 51.0
143	10	. .	. .	22.0	36.9	51.4	. .	. .	7 36.52		33.38	V.	1	10.11	5 5.2		9.2	8 9.90	39 44.4
144	11	. .	. .	. .	44.5	59.2	13.5	. .	10 44.34		33.40	IV.	3	7.19	13 37.8		8.6	11 17.74	48 16.4
145	9	. .	. .	. .	25.0	40.0	. .	. .	21 11 25.02	+	33.38	IV.	2	11.17	−10 38.3	−	8.4	21 11 58.40	− 31 45 16.7

ZONE 31. JUNE 22. K. $D_0 = -35° 14' 40''$.

No.	Mag.	I.	II.	III.	IV.	V.	VI.	VII.	T.		a_1	Micrometer.			$i + d_2$		d_1	Mean Right Ascension, 1850.0.	Mean Declination, 1850.0.
1	7	. .	. .	. .	. .	. .	21.6	. .	17 5 51.17	+	30.51	VII.	4	6.58	−18 25.9	−	30.2	17 6 21.68	− 35 33 36.1
2	9	. .	. .	. .	. .	. .	. .	42.1	6 55.87		30.50	VII.	2	3.34	6 42.5		29.8	7 26.37	21 52.3
3	6	. .	. .	48.0	3.4	. .	. .	49.2	12 3.24		30.55	IV.	7	3.3	30 30.9		28.2	12 33.79	45 39.1
4	9	22.4	. .	53.7	. .	. .	. .	. .	14 8.68		30.52	III.	5	3.23	21 37.9		27.5	14 39.20	36 45.4
5	9	. .	10.8	. .	41.8	. .	. .	. .	19 41.56		30.56	IV.	6	9.41	28 51.2		25.7	20 12.12	43 56.9
6	9	. .	. .	16.2	31.6	. .	. .	. .	20 31.45		30.57	V.	7	4.58	31 29.0		25.4	21 2.02	46 34.5
7	8	53.8	9.2	. .	. .	. .	. .	. .	23 39.88		30.57	II.	5	4.27	22 9.9		24.4	24 10.45	37 14.3
8	10	. .	. .	. .	48.9	. .	. .	. .	24 48.73		30.55	IV.	2	10.53	10 25.2		24.0	25 19.28	25 29.2
9	9	2.1	. .	33.2	. .	. .	. .	. .	26 48.29		30.58	III.	5	4.55	22 24.4		23.4	27 18.87	37 27.8
10	9	36.7	. .	8.0	. .	. .	. .	. .	28 23.25		30.60	III.	8	9.21	38 41.9		22.9	28 53.85	53 44.8
11	8	15.8	31.5	47.3	. .	. .	. .	. .	30 2.10		30.58	III.	4	6.51	18 22.9		22.4	30 32.68	33 25.3
12	7	56.5	12.1	. .	. .	. .	. .	. .	31 42.76		30.57	III.	6	8.4	28 2.1		21.8	32 12.74	43 3.9
13	8	. .	. .	. .	. .	3.9	18.9	34.9	31 48.58		30.57	VII.	2	9.53	9 54.2		21.8	32 19.15	24 56.0
14	10	. .	. .	. .	. .	32.8	. .	. .	33 17.41		30.58	VI.	2	10.42	10 19.4		21.3	33 47.99	25 20.7
15	7	. .	. .	. .	28.2	. .	58.8	. .	17 34 28.21	+	30.58	VI.	3	4.58	−18 25.4	−	20.9	17 34 58.79	− 35 33 26.3

ZONE 32. JUNE 24. A. $D_0 = -29° 4' 50''$.

No.	Mag.	I.	II.	III.	IV.	V.	VI.	VII.	T.		a_1	Micrometer.			$i + d_2$		d_1	Mean Right Ascension, 1850.0.	Mean Declination, 1850.0.
1	8	. .	. .	. .	. .	23.8	37.5	52.5	14 54 9.09	+	29.39	VI.	2	8.40	− 9 22.2	−	70.6	14 54 38.48	− 29 15 22.8
2	7	. .	. .	. .	. .	1.3	15.5	. .	14 56 46.89		29.40	VI.	4	6.21	18 8.2		70.1	14 56 16.29	24 8.3
3	10	. .	. .	. .	6.0	20.5	34.5	49.5	15 4 6.01		29.42	VII.	1	11.12	5 36.2		68.6	15 4 35.43	11 34.8
4	5	30.3	45.0	59.5	13.8	. .	. .	. .	8 13.67		29.45	IV.	6	11.14	29 37.8		67.8	8 43.12	35 35.6
5	11	. .	. .	. .	20.0	. .	. .	3.0	11 19.83		29.46	VII.	6	8.54	27 26.0		67.2	10 49.29	33 23.2
6	10	. .	. .	52.0	6.0	. .	. .	. .	13 51.71		29.48	IV.	9	10.14	44 7.0		66.8	13 21.19	50 3.8
7	5	. .	. .	. .	58.0	12.6	26.8	41.0	28 57.96		29.49	VII.	2	11.45	10 52.4		63.5	29 27.45	16 45.9
8	11	. .	. .	. .	. .	53.5	8.0	. .	30 39.24		29.52	VI.	5	6.11	22 32.8		63.1	31 8.76	28 25.9
9	8	. .	. .	. .	. .	. .	. .	53.8	32 10.68		29.53	VII.	7	9.10	33 35.3		62.8	32 40.21	39 28.1
10	7	. .	33.5	48.0	2.4	. .	. .	. .	36 2.18		29.54	IV.	6	8.3	28 1.3		61.9	36 31.72	33 53.2
11	8	45.2	0.2	14.2	. .	. .	. .	. .	42 28.50		29.55	III.	4	9.26	19 41.8		60.4	42 58.05	25 32.2
12	9	. .	1.2	16.3	. .	. .	. .	. .	47 30.13		29.57	III.	6	6.54	27 26.4		59.3	47 59.70	33 15.7
13	8	. .	. .	. .	. .	. .	. .	34.0	49 50.87		29.58	VII.	7	8.5	33 2.5		58.7	50 20.45	38 51.2
14	9	. .	. .	. .	. .	26.5	. .	. .	53 12.19		29.60	V.	9	11.6	44 33.2		57.9	53 41.79	50 21.1
15	10	. .	. .	13.5	. .	. .	. .	. .	15 59 28.04	+	29.62	III.	9	11.45	−44 52.9	−	56.4	15 59 57.66	− 29 50 39.3

CORRECTIONS.

Date.			Corr. of Clock.	Hourly rate.	m	n	c
1846.		h.	s.	s.	s.	s.	s.
June	22	12	+ 19.45	− 0.013	− 0.447	+ 0.372	+ 0.120
	24	12	+ 19.17	− 0.013	− 0.447	+ 0.372	+ 0.120

INSTRUMENT READINGS.

	Date.			Barom.	Thermom. At.	Thermom. Ex.
	1846.	h.	m.	in.	°	°
Zone 31	June 22,	17	6	30.10	69.0	51.7
		17	34	30.09	69.0	51.5
Zone 32	June 24,	14	54	30.09	71.0	63.1
		19	21	30.07	70.5	61.5

REMARKS.

(30) 140. Micrometer assumed as thread 1 instead of thread 2.
(32) 2. Minutes assumed as 55 instead of 56.
(32) 5. Minutes assumed as 10 instead of 11, and micrometer reading as 6r.54 instead of 8r.54.
(32) 6. Transits over T.'s IV and V assumed as recorded over T.'s III and IV, and minutes as 12 instead of 13.
(32) 8. Micrometer reading assumed as 5r.11 instead of 6r.11.
(32) 13. Transit over T. VII assumed as 34s.0 instead of 3s.4.

ZONE 32. JUNE 24. A. $D_o = -29° 4' 50''$—Continued.

No.	Mag.	SECONDS OF TRANSIT. I. II. III. IV. V. VI. VII.	T.	a_1	MICROMETER.			$i + d_o$	d_1	Mean Right Ascension, 1850.0.	Mean Declination, 1850.0.
			h. m. s.	s.			r.	′ ″	″	h. m. s.	° ′ ″
16	10	. . . . 32.5 47.0	16 2 46.90	+ 29.63	IV.	8	11.11	−39 26.2	− 55.6	16 3 16.53	− 29 45 21.8
17	9	. . . . . 58.0 12.5	4 58.01	29.63	V.	9	8.39	43 29.1	55.1	4 27.64	49 14.2
18	9	. . . . 5.5 20.0 35.0	7 51.39	29.62	VII.	4	8.44	19 20.2	54.4	8 21.01	25 4.6
19	8	. . . . . 42.0	10 41.84	29.62	IV.	1	6.55	2 56.6	53.7	11 11.46	8 40.3
20	7	. . . . 0.0 14.2 29.0	14 45.61	29.64	VII.	3	10.12	15 5.1	52.6	15 15.25	20 47.7
21	8	. . . . 37.0	15 53.82	29.65	VII.	6	9.25	28 42.3	52.3	16 23.47	34 24.6
22	10	. . . . 12.0 27.0	19 43.58	29.66	VII.	6	4.23	26 9.7	51.4	20 13.24	31 51.1
23	11	. . . . . . 29.5 45.0 . .	21 15.74	29.66	V.	7	9.55	33 58.4	51.0	21 45.40	39 39.4
24	9	30.5	24 13.88	29.67	I.	5	9.35	24 45.8	50.2	24 43.55	30 26.0
25	9	. . . . 16.0 30.5	25 30.16	29.66	IV.	2	4.21	7 8.5	49.9	25 59.82	12 48.4
26	9	5.0 19.5	28 48.22	29.68	II.	6	10.10	29 5.3	49.0	29 17.90	34 44.3
27	7	. . . . . 19.0 33.5	29 18.98	29.69	IV.	7	5.10	31 34.5	48.9	29 48.67	37 13.4
28	10	. . . . . 59.0 13.2	34 58.79	29.69	IV.	3	4.55	12 25.4	47.4	34 28.48	18 2.8
29	9	. . . . . 36.0	35 35.84	29.68	IV.	1	7.38	3 48.5	47.2	36 5.52	9 25.7
30	9	. . 38.2 52.5 7.0	37 6.64	29.70	IV.	3	6.42	13 19.5	46.8	37 36.34	18 56.3
31	19	. . . . 16.0 31.0	37 47.61	29.71	VII.	7	10.32	34 16.8	46.6	38 17.32	39 53.4
32	11	. . . . 33.0 48.0	39 4.63	29.71	VII.	7	11.30	34 46.1	46.3	39 34.34	40 22.4
33	9	. . 51.5 6.1	41 20.39	29.72	III.	9	8.42	43 20.5	45.7	41 50.11	48 56.2
34	7	. . . . 20.5	41 37.01	29.69	VII.	1	6.42	3 19.8	45.6	42 6.70	8 55.4
35	11	. . . . . . 29.5	43 15.19	29.73	V.	8	8.33	38 16.3	45.2	43 44.90	43 51.5
36	7	. . . . 46.5 1.0	45 0.83	29.72	IV.	7	2.10	30 3.6	44.7	45 30.55	35 38.3
37	11	22.5 37.0	48 5.74	29.73	II.	7	5.15	31 36.8	43.8	48 35.47	37 10.6
38	11	. . . . . 48.5	48 34.07	29.75	IV.	10	6.50	47 25.4	43.7	49 3.82	52 59.1
39	7	. . 33.0	51 1.36	29.73	II.	4	11.39	20 48.9	43.1	51 31.09	26 22.0
40	10	. . . . 55.5	51 12.19	29.73	VII.	4	3.42	21 50.8	43.0	51 41.90	27 23.8
41	10	. . . . 56.2 12.0	54 11.31	29.76	IV.	10	5.10	46 34.9	42.2	54 41.07	52 7.1
42	11	. . . . 3.5 17.0 32.5	55 48.91	29.73	VII.	4	11.41	20 49.8	41.7	56 18.64	26 21.5
43	8	. . . . 0.5 15.0	58 14.94	29.78	IV.	10	1.36	45 17.0	41.1	58 44.72	50 48.1
44	9	. . . . . 19.0 32.5 47.0 . .	16 59 18.44	29.77	VI.	8	4.48	36 22.5	40.8	16 59 48.21	41 53.3
45	11	. . . . 15.5	17 8 19.64	29.75	III.	4	5.32	36 44.9	38.2	17 8 49.39	42 13.1
46	7	. . 58.0 13.0	10 26.61	29.75	III.	2	3.27	6 41.2	37.7	10 56.36	12 8.9
47	9	. . . . . 50.5 5.0 19.2 . .	13 50.51	29.78	VI.	7	1.36	34 44.9	36.7	14 20.29	40 11.6
48	7	. . . . . 27.0 41.5	15 26.97	29.78	V.	5	12.00	25 59.4	36.3	15 56.75	31 25.7
49	5	. . 48.5 3.0	17 17.27	29.80	III.	8	8.00	37 59.6	35.7	17 47.07	43 25.3
50	7	. . . . 3.0 17.5	17 34.36	29.78	VII.	6	11.37	29 49.0	35.6	18 4.14	35 14.6
51	11	. . . . 31.5	18 48.24	29.78	VII.	5	4.35	22 14.2	35.3	19 18.02	27 39.5
52	8	16.8 31.5 45.8	23 0.08	29.79	III.	6	5.14	26 35.9	34.1	23 29.87	32 0.0
53	9	. . . . . 38.5	24 38.34	29.76	IV.	1	6.48	3 23.2	33.6	25 8.10	8 46.8
54	7	34.0 48.5 3.0	29 17.14	29.79	III.	4	11.33	20 46.0	32.3	29 46.93	26 8.3
55	11	. . 20.5	30 48.97	29.80	II.	5	10.27	25 12.3	31.8	31 18.77	30 34.1
56	8	. . 55.2 10.0	32 23.84	29.79	III.	4	2.7	16 0.0	31.4	32 53.63	21 21.4
57	12	. . . . 43.0	33 57.18	29.79	III.	4	10.19	20 8.6	30.9	34 26.97	25 29.5
58	9	. . . . . 11.8	35 11.64	29.84	IV.	9	4.48	41 22.3	30.6	35 41.48	46 42.9
59	8	. . . . 19.0 33.0	36 33.08	29.82	IV.	7	3.14	30 35.9	30.2	37 2.90	35 56.1
60	10	. . 54.0 9.0	39 22.76	29.80	III.	4	5.	17 27.4	29.3	39 52.56	22 46.7
61	8	. . . . 24.5 39.0	40 38.66	29.79	IV.	2	6.30	8 13.7	28.9	41 8.45	13 32.6
62	9	. . . . 9.0 23.0	42 23.04	29.81	IV.	5	10.2	24 59.9	28.5	42 52.85	30 18.4
63	12	. . . . 31.8	43 45.90	29.79	III.	3	10.	14 59.4	28.1	44 15.69	20 17.5
64	8	. . . . . 3.5	17 45 3.34	+ 29.85	IV.	9	4.42	−41 19.2	− 27.7	17 45 33.19	− 29 46 36.9

CORRECTIONS.

Date.		Corr. of Clock.	Hourly rate.	m	n	c
1846.	h.	s.	s.	s.	s.	s.

INSTRUMENT READINGS.

Date.		Barom.	THERMOM. At.	THERMOM. Ex.
1846.	h. m.	in.	°	°

REMARKS.

(32) 17. Minutes of transit assumed as 3 instead of 4.
(32) 19. Micrometer reading assumed as 1 $5^r.55$ instead of 1 $6^r.55$.
(32) 28. Minutes assumed as 33 instead of 34.
(32) 38. Transit over T. III assumed as recorded over T. IV.
(32) 39. Micrometer reading assumed as $11^r.39$ instead of $1^r.39$.
(32) 40. Micrometer reading assumed as $13^r.42$ instead of $3^r.4$[illegible]
(32) 43. Micrometer reading assumed as 10 $2^r.36$ instead of 10 $1^r.36$.
(32) 45. T. III assumed as $5^s.5$ instead of $15^s.5$, and horizontal thread 8 instead of 4.
(32) 47. Micrometer reading assumed as 8 $1^r.36$ instead of 7 $1^r.36$.

ZONE 32. JUNE 24. A. $D_0 = -29° 4' 50''$—Continued.

No.	Mag.	Seconds of Transit. I.	II.	III.	IV.	V.	VI.	VII.	T.		a_1	Micrometer.			$i + d_4$		d_1	Mean Right Ascension, 1850.0.	Mean Declination, 1850.0.
									h. m. s.		s.			r.	′ ″		″	h. m. s.	° ′ ″
65	10				15.0	30.0			17 46 15.22	+	29.82	V.	5	8.40	−24 18.4	−	27.3	17 46 45.04	− 29 29 35.7
66	10					56.0			47 41.62		29.82	V.	6	7.14	27 36.5		26.9	48 11.44	32 53.4
67	7			57.0	11.5	26.0			49 11.34		29.81	V.	4	3.37	16 45.5		26.5	49 41.15	22 2.0
68	8		39.5						51 7.92		29.81	II.	4	10.42	20 20.1		26.0	51 37.73	25 36.1
69	11		10.5						52 43.97		29.82	II.	6	4.43	16 18.1		25.5	53 13.79	21 33.6
70	8				45.0	59.0			53 30.42		29.85	V.	10	4.12	46 5.5		25.2	54 0.27	51 20.7
71	5				56.5	11.0	25.3		55 56.55		29.83	VI.	6	11.2	29 31.5		24.6	55 26.38	34 46.1
72	9					13.5	28.0		57 59.25		29.82	VI.	4	11.35	20 45.9		23.9	58 29.07	25 59.8
73	9					36.5			57 22.10		29.83	V.	5	7.49	23 52.6		24.1	57 51.93	29 6.7
74	9							20.0	17 57 36.90		29.83	VII.	5	3.40	21 46.4		24.0	17 58 6.73	27 0.4
75	9		46.8						18 0 15.09		29.81	II.	2	7.23	8 40.3		23.3	18 0 44.90	13 53.6
76	7			55.0	9.5				19 9.22		26.82	IV.	3	11.5	15 32.3		17.7	19 39.04	20 40.0
77	8			57.5	11.8				22 11.59		29.82	IV.	3	4.33	12 14.3		16.8	22 41.41	17 21.1
78	7				55.5	9.8	24.0		26 55.37		29.87	VI.	9	9.5	43 31.9		15.5	26 25.24	48 37.4
79	9					29.0	43.0		27 14.49		29.85	VI.	7	1.44	30 20.5		15.4	27 44.34	35 25.9
80	10			23.5	38.2				32 37.94		29.85	IV.	7	5.41	31 50.1		13.8	33 7.79	36 53.9
81	9		51.0	5.0					34 19.41		29.84	III.	6	10.6	29 3.4		13.3	34 49.25	34 6.7
82	8					10.0	24.2	38.5	34 55.52		29.84	VII.	7	8.6	33 33.3		13.1	35 25.36	38 36.4
83	10					55.8	9.5	24.5	36 41.22		29.86	VII.	7	9.36	33 48.5		12.6	37 11.08	38 51.1
84	6	52.2	7.0	21.2	35.5				42 35.48		29.85	III.	6	7.55	27 57.2		10.8	43 5.33	32 58.0
85	11				19.5				52 19.34		29.85	IV.	6	7.51	27 55.2		8.0	52 49.19	32 53.2
86	8				48.5	2.5			18 55 48.19		29.83	V.	3	6.8	13 2.2		7.0	18 56 18.02	17 59.2
87	7				18.6	33.0	47.5		19 1 18.71		29.86	VI.	7	10.50	39 24.7		5.4	19 1 48.57	44 20.1
88	9	58.3	13.0	27.3					4 41.56		29.85	III.	5	9.6	24 31.5		4.4	5 11.41	29 25.9
89	8				4.0	18.5			12 4.01		29.87	V.	9	7.53	42 55.7		2.3	12 33.88	47 48.0
90	8			53.0	7.5				15 7.33		29.85	IV.	7	3.28	30 42.9	−	1.4	15 37.18	35 34.3
91	9							52.5	19 21 23.88	+	29.87	VII.	9	8.8	−43 2.9	+	0.4	19 21 39.37	− 29 47 52.5

ZONE 33. JULY 6. A. $D_0 = -27° 51' 0''$.

No.	Mag.	Seconds of Transit. I.	II.	III.	IV.	V.	VI.	VII.	T.		a_1	Micrometer.			$i + d_4$		d_1	Mean Right Ascension, 1850.0.	Mean Declination, 1850.0.
1	8					17.5	31.3	46.0	15 12 3.25	+	25.73	VII.	4	1.51	−16 21.9	−	44.9	15 12 28.98	− 28 8 6.8
2	9		17.0	31.2					17 45.21		25.74	III.	4	9.56	19 57.0		44.5	18 10.95	11 41.5
3	9						37.0	51.0	18 8.63		25.76	VII.	5	8.45	24 20.6		44.5	18 34.39	16 5.1
4	6					41.5	55.5		19 27.33		25.76	VI.	6	7.18	28 38.8		44.4	19 53.09	20 23.2
5	8							38.0	19 55.43		25.76	VII.	6	7.15	27 36.6		44.4	20 21.19	19 21.0
6	9		38.5	53.0					25 7.01		25.78	III.	7	5.36	31 47.5		44.1	25 32.79	23 31.6
7	7				46.5	0.5			25 46.30		25.77	VII.	5	3.33	21 43.0		44.0	26 12.07	32 25.8
8	6						6.5	20.5	26 38.18		25.78	VII.	8	7.54	37 56.0		44.0	27 3.96	29 40.0
9	10	46.0	2.5						29 30.81		25.77	II.	4	8.27	19 12.0		43.8	29 56.58	10 55.8
10	8			44.5	59.0				30 58.80		25.80	IV.	9	8.42	43 20.2		43.7	31 24.60	35 3.9
11	7						28.0	42.0	31 59.74		25.80	VII.	10	12.	50 1.2		43.7	32 25.54	41 44.9
12	9			18.5	32.6				34 32.49		25.79	IV.	6	10.37	29 19.1		43.4	34 58.28	21 2.5
13	7				38.0	52.0	5.8		35 37.77		25.78	VI.	3	3.13	11 33.9		43.3	36 3.55	3 17.2
14	10						51.5	6.0	37 23.31		25.78	VII.	2	7.48	8 52.9		43.2	37 49.09	0 36.1
15	7						14.0	28.2	38 45.74		25.80	VII.	6	7.15	27 36.6		43.1	39 11.54	19 19.7
16	9							17.0	39 34.43		25.80	VII.	6	6.15	27 6.3		43.1	40 0.23	18 49.4
17	8							16.5	40 33.87		25.80	VII.	5	5.57	22 55.7		43.0	40 59.67	14 38.7
18	9				11.5	25.5			15 42 11.32	+	25.80	V.	6	2.56	−25 26.2	−	42.8	15 42 37.12	− 28 17 9.0

CORRECTIONS.

Date.		Corr. of Clock.		Hourly rate.		m		n	c
1846.	h.		s.		s.		s.	s.	s.
July 6,	12	+	15.75	−	0.013	−	0.463	+ 0.300	0.000
6,	12	−	15.73	−	0.013	−	0.463	+ 0.300	0.000

INSTRUMENT READINGS.

	Date.		Barom.	Thermom. At.	Ex.
	1846	h. m.	in.	°	°
Zone 33	July 6,	15 12	29.80	60.0	77.5
		16 10	29.81	80.0	74.0

REMARKS.

(32) 69. T. II assumed as 15ˢ.5, not 10ˢ.5; and micrometer as 4 2ʳ.43 instead of 6 4ʳ.43.
(32) 70. Transits over T.'s V and VI assumed as recorded over T.'s IV and V.
(32) 71. Minutes assumed as 54 instead of 55.
(32) 74. Transit over T. VII assumed as recorded over T. VI.
(32) 78. Minutes assumed as 25 instead of 26.
(32) 79. Micrometer reading assumed as 2ʳ.44 instead of 1ʳ.44.
(32) 82. Right ascension differs 10ˢ from Arg. Z. 221, 119; micrometer reading assumed as 9ʳ.6 instead of 8ʳ.6.
(32) 87. Micrometer reading assumed as 8 10ʳ.50 instead of 7 10ʳ.50
(33) 1. Micrometer reading assumed as 4 2ʳ.51 instead of 4 1ʳ.51.
(33) 4. Micrometer reading assumed as 9ʳ.18 instead of 7ʳ.18.
(33) 7. Micrometer thread assumed as 9 instead of 5.

ZONE 33. JULY 6. A. D_0 = −27° 51′ 0″—Continued.

No.	Mag.	Seconds of Transit. I.	II.	III.	IV.	V.	VI.	VII.	T.		a_1	Micrometer.			$i + d_2$		d_1	Mean Right Ascension, 1850.0.	Mean Declination, 1850.0.
									h. m. s.		s.			r.	′ ″		″	h. m. s.	° ′ ″
19	8	..	..	..	..	..	15.3	29.5	15 43 47.10	+	25.82	VII.	9	2.25	−40 10.9	−	42.7	15 44 12.92	− 28 31 53.6
20	10	..	..	13.5	27.5	..	..	..	46 27.48		25.82	IV.	7	10.5	34 3.4		42.5	46 53.30	25 45.9
21	7	..	..	..	28.0	42.0	..	..	47 27.82		25.81	V.	5	4.49	22 21.7		42.4	47 53.63	14 4.1
22	9	..	..	38.5	53.0	..	..	..	49 52.79		25.83	IV.	9	5.16	41 36.2		42.2	50 18.62	28 33 18.4
23	6	..	..	..	32.5	46.5	..	..	51 32.27		25.80	V.	1	5.17	2 37.5		42.1	51 58.07	27 54 19.6
24	9	..	..	..	..	..	..	1.0	52 18.25		25.81	VII.	3	7.24	13 40.4		42.0	52 44.06	28 5 22.4
25	8	..	..	..	..	..	..	57.2	53 14.42		25.81	VII.	3	1.42	11 17.9		41.9	53 40.23	2 59.8
26	9	..	..	..	23.0	37.0	..	..	55 22.79		25.81	V.	3	7.7	13 32.2		41.7	55 48.60	5 13.9
27	8	..	..	..	..	..	23.0	37.8	55 55.00		25.81	VII.	4	5.56	17 55.5		41.7	56 20.81	9 37.2
28	8	..	..	..	..	..	9.8	23.7	57 41.43		25.84	VII.	8	3.40	35 47.7		41.5	58 7.27	27 29.2
29	8	..	..	..	..	..	26.0	40.0	15 58 57.68		25.84	VII.	8	1.35	34 44.6		41.4	15 59 23.52	28 26 26.0
30	9	..	..	..	..	..	23.0	37.5	16 0 54.80		25.81	VII.	1	8.24	4 11.6		41.2	16 1 20.61	27 55 52.8
31	5	..	..	..	..	48.8	2.5	17.0	2 34.39		25.81	VII.	2	9.16	9 37.4		41.1	3 0.20	28 1 18.5
32	8	..	..	..	..	17.0	30.9	..	6 2.78		25.86	VI.	9	12.0	46 0.7		40.8	6 28.64	37 41.5
33	4	51.5	5.5	20.2	34.2	..	..	..	8 34.07		25.84	IV.	5	5.3	22 28.8		40.5	8 59.91	14 9.3
34	9	..	..	..	31.0	45.0	..	..	16 10 30.82	+	25.84	V.	6	10.30	−29 15.5	−	40.4	16 10 56.66	− 28 20 55.9

ZONE 34. JULY 7. K. D_0 = −39° 9′ 40″.

No.	Mag.	Seconds of Transit. I.	II.	III.	IV.	V.	VI.	VII.	T.		a_1	Micrometer.			$i + d_2$		d_1	Mean Right Ascension, 1850.0.	Mean Declination, 1850.0.
1	7	..	..	..	39.9	..	..	..	16 9 39.72	+	25.54	IV.	1	11.40	− 5 46.2	−	17.7	16 10 5 26	− 39 15 43.9
2	8	..	..	..	56.3	..	..	..	11 56.12		25.60	IV.	8	13.50	41 0.5		17.1	12 21.72	50 57.6
3	8	48.0	..	..	..	..	..	..	16 36.66		25.58	II.	4	7.14	18 33.4		15.8	17 2.24	28 29.2
4	9	9.8	..	..	..	..	..	..	20 58.86		25.63	I.	9	9.15	43 41.2		14.5	21 24.49	53 35.7
5	7	..	14.4	..	47.0	..	..	..	27 46.69		25.60	IV.	2	11.26	10 39.8		12.6	28 12.29	20 32.4
6	7	27.6	..	..	..	..	..	..	29 16.27		25.62	II.	4	8.27	19 10.4		12.2	29 41.89	29 2.6
7	9	..	..	..	..	..	25.3	..	28 53.25		25.61	VI.	2	6.49	8 19.1		12.3	29 18.86	18 11.4
8	8	..	..	..	..	..	13.6	..	29 41.43		25.63	VII.	6	4.33	26 14.9		12.0	30 7.06	36 6.9
9	8	..	..	..	..	..	32.2	..	30 0.06		25.63	VII.	5	10.9	25 2.8		11.9	30 25.69	34 54.7
10	8	..	..	..	..	42.0	58.1	..	31 25.91		25.63	VI.	4	11.51	20 53.8		11.5	31 51.54	30 45.3
11	8	..	..	..	..	37.8	..	..	32 21.54		25.61	V.	1	6.36	3 11.9		11.2	32 47.15	13 3.1
12	8	22.7	39.1	..	..	..	..	..	35 11.47		25.65	II.	5	11.26	25 42.2		10.4	35 37.12	35 32.6
13	7	..	..	..	30.3	..	..	..	35 30.12		25.64	VII.	4	10.36	20 15.5		10.3	35 55.76	30 5.8
14	6	..	..	..	..	..	..	8.6	36 19.78		25.62	VII.	1	11.2	5 36.3		10.1	36 45.40	15 26.4
15	8	..	..	..	..	..	..	2.1	37 13.65		25.66	VII.	7	4.55	31 27.9		9.8	37 39.31	41 17.7
16	10	26.5	..	..	..	..	..	..	40 15.42		25.67	I.	7	10.19	34 7.2		9.0	40 41.09	43 56.2
17	9	..	..	..	32.0	..	..	..	40 31.82		25.66	IV.	5	8.53	24 24.8		8.9	40 57.48	34 13.7
18	10	..	14.0	..	..	..	..	..	43 46.19		25.65	II.	3	8.15	14 3.5		7.9	44 11.84	23 51.4
19	6	..	..	37.9	..	..	..	..	44 53.46		25.64	IV.	1	11.13	5 32.3		7.6	45 19.10	15 19.9
20	9	30.6	..	..	..	..	..	..	46 19.50		25.68	II.	7	9.51	33 58.4		7.2	46 45.18	43 45.6
21	9	16.8	..	49.6	..	..	..	..	48 5.29		25.66	III.	2	10.5	9 58.8		6.7	48 30.95	19 45.5
22	7	..	0.8	..	33.5	..	..	..	52 33.38		25.71	IV.	9	6.3	42 4.5		5.3	52 59.09	51 49.8
23	9	..	..	54.2	..	..	..	..	16 54 10.37	+	25.71	IV.	8	11.32	−39 50.5	−	4.8	16 54 36.08	− 39 49 35.3

CORRECTIONS.

Date.		Corr. of Clock.	Hourly rate.	m	n	c
1846.	h.	s.	s.	s.	s.	s.
July 7,	12	− 14.66	− 0.030	− 0.463	+ 0.300	0.000

INSTRUMENT READINGS.

	Date.		Barom.	Thermom. At.	Thermom. Ex.
	1846.	h. m.	in.	°	°
Zone 34	July 7,	16 10	29.93	78.0	68.0
		16 54	29.93	77.0	67.0

REMARKS.

(33) 25. Micrometer reading assumed as 2r.42 instead of 1r.42.
(33) 32. Micrometer reading assumed as 14r.0 instead of 12r.0.
(34) 14. Micrometer reading assumed as 11r.22 instead of 11r.2.

Zone 35. July 7. K. $D_s = -34° 6'$.

No.	Mag.	Seconds of Transit I.	II.	III.	IV.	V.	VI.	VII.	T.	a_1	Micrometer.			$i + d_1$	d_1	Mean Right Ascension, 1850.0.	Mean Declination, 1850.0.
									h. m. s.	s.			r.	′ ″	″	h. m. s.	° ′ ″
1	9	..	..	..	..	..	..	18.1	19 15 32.76	+ 25.23	VII.	8	3.30	−35 43.5	. .	19 15 57.99	
2	8	30.9	..	..	..	..	..	..	20 16.36	25.17	II.	2	4.12	7 2.5	. .	20 41.53	
3	11	..	19.3	..	..	..	..	..	22 49.73	25.21	III.	8	5.41	36 50.2	. .	23 14.94	
4	5.6	..	..	..	29.0	44.1	..	..	23 28.88	25.18	V.	5	9.43	24 50.2	. .	23 54.06	
5	8	46.3	..	..	..	..	..	..	25 32.27	25.21	III.	9	10.10	44 6.3	. .	25 57.48	
6	8	15.1	..	..	..	..	..	..	27 0.89	25.19	II.	7	4.29	31 14.0	. .	27 26.08	
7	9	..	..	..	..	51.5	..	..	27 36.25	25.14	VI.	1	8.19	4 7.5	. .	28 1.39	
8	7	..	..	..	..	..	..	5.8	28 20.61	25.22	VII.	10	5.19	46 40.5	. .	28 45.83	
9	9	..	..	..	..	9.4	..	..	29 54.24	25.16	V.	5	4.28	22 10.8	. .	30 19.40	
10	10	..	..	..	..	..	..	27.3	30 41.62	25.14	VII.	2	9.52	9 54.2	. .	31 6.76	
11	10	..	6.0	..	..	..	..	..	39 36.37	25.15	III.	7	5.34	31 47.0	. .	40 1.52	
12	9	56.5	..	..	..	..	..	..	47 42.29	25.12	II.	7	3.50	30 54.3	. .	48 7.41	
13	8	..	54.5	..	25.0	..	..	..	51 24.82	25.09	II.	5	11.57	25 57.8	. .	51 49.91	
14	8	..	..	..	27.8	..	..	..	52 27.62	25.06	IV.	2	11.57	10 57.9	. .	52 52.68	
15	9	..	..	..	56.7	..	..	..	54 56.52	25.05	IV.	1	8.53	4 25.0	. .	55 21.57	
16	6.7	36.6	52.4	..	..	..	..	..	19 59 22.33	+ 25.05	II.	3	5.14	−12 33.9	. .	19 59 47.38	

Zone 36. July 9. K. $D_s = -26° 35' 40''$.

No.	Mag.	I.	II.	III.	IV.	V.	VI.	VII.	T.	a_1	Micrometer.			$i + d_1$	d_1	Mean Right Ascension, 1850.0.	Mean Declination, 1850.0.
1	7	17.4	..	..	..	..	..	..	16 1 59.52	+ 23.38	II.	2	7.29	− 8 43.7	− 52.0	16 2 22.90	− 26 45 15.7
2	8	..	..	..	..	..	16.1	..	1 48.29	23.37	VI.	1	8.57	4 28.7	52.1	2 11.66	26 41 0.8
3	8	..	..	..	..	..	..	57.1	2 15.08	23.41	VII.	7	7.48	32 53.7	52.0	2 38.49	27 9 25.7
4	9	..	..	..	52.8	..	..	..	3 52.65	23.39	V.	5	8.36	24 16.4	51.6	4 16.04	27 0 48.0
5	7	2.3	16.4	..	..	58.8	..	..	5 44.49	23.38	II.	3	5.30	12 43.2	51.2	6 7.87	26 49 14.4
6	10	59.8	..	..	..	..	..	..	11 41.99	23.38	I.	3	11.0	15 29.6	49.8	12 5.37	51 59.4
7	7	11.4	25.4	..	..	..	..	..	14 53.42	23.38	III.	2	12.31	11 15.4	49.0	15 16.80	26 47 44.4
8	8	..	..	..	32.8	..	..	..	15 32.65	23.43	VI.	9	6.51	42 23.8	48.8	15 56.08	27 18 52.6
9	8	..	..	..	50.8	..	..	..	16 50.65	23.43	IV.	9	9.58	43 58.4	48.5	17 14.08	27 20 26.9
10	7	..	..	..	35.0	..	..	..	17 34.85	23.39	V.	3	8.24	14 11.2	48.3	17 58.24	26 50 39.5
11	10	..	..	..	..	..	..	19.0	19 36.90	23.40	VII.	5	11.17	25 37.3	47.9	20 0.30	27 2 5.2
12	6	..	..	..	..	11.1	..	39.3	20 57.03	23.39	VII.	3	8.15	14 6.3	47.5	21 20.42	26 50 33.8
13	7	..	..	..	..	21.0	..	..	25 6.90	23.39	V.	2	12.0	24 4.9	46.5	25 30.29	27 0 31.4
14	10	..	..	25.7	..	..	..	..	26 39.72	23.42	III.	7	9.27	33 44.1	46.1	27 3.14	27 10 10.2
15	10	..	..	..	..	31.1	..	..	27 17.00	23.39	V.	2	6.29	9 14.2	46.0	27 40.39	26 45 40.2
16	7	..	..	44.0	57.6	..	..	..	28 57.51	23.38	IV.	1	9.0	4 30.4	45.6	29 20.89	40 56.0
17	7	..	..	37.8	..	..	..	..	29 51.65	23.40	III.	5	4.58	22 26.3	45.3	30 15.05	58 51.6
18	8	..	28.8	..	..	..	..	..	31 56.71	23.39	III.	3	4.14	12 4.9	44.8	32 20.10	48 29.7
19	8	..	..	..	..	24.1	..	..	32 10.01	23.39	VI.	3	4.26	12 10.9	44.8	32 33.40	48 35.7
20	6.7	..	..	..	..	20.9	..	..	33 6.83	23.40	VI.	4	4.35	17 14.9	44.5	33 30.23	26 53 39.4
21	5.6	..	..	21.9	..	..	..	..	34 35.92	23.42	III.	7	9.23	33 42.0	44.1	34 59.34	27 10 6.1
22	8	..	..	..	8.7	..	..	..	35 8.55	23.42	V.	7	9.36	33 48.6	44.0	35 31.97	10 12.6
23	10	..	..	26.6	..	..	..	..	36 40.47	23.40	III.	5	8.52	25 25.0	43.6	37 3.87	1 48.6
24	7	..	..	..	..	8.0	..	..	36 54.00	23.42	VII.	7	13.25	35 43.9	43.6	37 17.42	12 7.5
25	9	..	57.1	..	..	..	..	..	39 25.30	23.43	III.	9	5.20	41 38.0	42.9	39 48.73	27 18 0.9
26	9	..	..	..	6.3	..	..	..	40 6.15	23.38	V.	1	11.0	5 31.0	42.8	40 29.53	26 41 53.8
27	8	..	20.3	..	..	..	..	..	41 48.23	23.39	II.	3	9.37	14 47.9	42.3	42 11.62	51 10.2
28	8	..	..	19.1	..	..	..	..	42 32.65	23.38	III.	1	6.17	3 8.0	42.2	42 56.03	39 30.2
29	8	..	..	..	..	..	..	19.4	16 42 37.08	+ 23.39	VII.	2	7.53	− 8 55.6	− 42.1	16 43 9.47	− 26 45 17.7

CORRECTIONS.

Date.		Corr. of Clock.	Hourly rate.	m	n	c
1846.	h.	s.	s.	s.	s.	s.
July 9.	12	− 13.46	− 0.034	− 0.576	+ 0.428	0.000

INSTRUMENT READINGS.

Date.				Barom.	Thermom. At.	Thermom. Ex.
		1846.	h. m.	in.	°	°
Zone	35	July 7.	19 15	29.93	75.0	65.0
			19 59	29.94	74.5	64.8
Zone	36	July 9.	16 2	29.97	81.0	73.5
			19 2	29.99	78.0	71.0

REMARKS.

(36) 23. Micrometer reading assumed as 10ʳ.52 instead of 8ʳ.52.

Zone 36. July 9. K. $D_c = -26° 35' 40''$—Continued.

No.	Mag.	Seconds of Transit. I.	II.	III.	IV.	V.	VI.	VII.	T.		a_1	Micrometer.			$i + d_2$		d_1	Mean Right Ascension, 1850.0.	Mean Declination, 1850.0.
									h. m. s.		s.			t.	' "		"	h. m. s.	° ' "
30	10			59.3					16 44 13.48	+	23.43	III.	9	11.37	−44 48.4	−	41.7	16 44 36.91	− 27 21 10.1
31	8					57.1			44 43.12		23.43	V.	9	7.20	42 38.6		41.6	45 6.55	19 0.2
32	10				9.1				47 8.95		23.42	V.	8	12.1	40 1.0		41.0	47 32.37	27 16 22.0
33	8							58.8	47 16.45		23.38	VII.	1	12.19	6 10.5		40.9	47 39.83	26 42 31.4
34	7							51.4	48 9.55		23.43	VII.	9	13.45	45 52.7		40.7	48 32.98	27 22 13.4
35	7				41.5				49 41.35		23.40	V.	4	10.31	20 14.8		40.3	50 4.75	26 56 35.1
36	8	53.5							51 35.78		23.40	II.	5	10.0	24 58.6		39.8	51 59.18	27 1 18.4
37	9			30.2	44.2				51 44.08		23.41	V.	6	4.51	26 24.2		39.8	52 7.49	27 2 44.0
38	8							44.4	52 2.06		23.39	VII.	2	4.23	7 9.6		39.7	52 25.45	26 43 29.3
39	7					35.5			53 21.42		23.40	IV.	4	2.37	16 15.5		39.4	53 44.82	52 34.9
40	8					32.5			54 18.45		23.40	V.	5	6.51	23 23.4		39.1	54 41.85	59 42.5
41	7							15.8	54 33.45		23.39	VII.	2	2.27	6 11.0		39.0	54 56.84	42 30.0
42	9			13.0					56 26.63		23.39	IV.	2	5.16	7 36.7		38.5	56 50.02	43 55.2
43	10				16.8				57 16.65		23.40	IV.	4	3.22	16 38.2		38.3	57 40.05	26 52 56.5
44	9		52.8						16 59 20.94		23.42	IV.	8	3 5	35 30.4		37.8	16 59 44.36	27 11 48.2
45	8	12.0		40.4					17 0 54.12		23.39	IV.	3	4.32	12 14.1		37.4	17 1 17.51	26 48 31.5
46	6.7	45.8							2 27.98		23.39	II.	3	9.17	14 37.8		36.9	2 51.37	50 54.7
47	8			39.1					2 52.82		23.39	IV.	3	7.9	13 33.4		36.8	3 16.21	49 50.2
48	6.7	48.3	2.2				58.5		4 30.27		23.39	II.	3	3.22	11 38.5		36.4	4 53.66	26 47 54.9
49	8		16.0	30.0					6 44.22		23.42	III.	10	5.49	46 54.0		35.8	7 7.64	27 23 9.8
50	8	26.3							11 8.70		23.41	I.	7	11.30	34 45.9		34.6	11 32.11	11 0.5
51	7				19.4				11 19.25		23.41	IV.	8	10.46	39 23.2		34.6	11 42.66	15 37.8
52	9				2.9				14 21.75		23.41	IV.	8	12.59	40 30.4		33.7	14 45.16	27 16 44.1
53	10						18.2		14 50.37		23.38	VII.	2	11.1	10 30.6		33.6	15 13.75	26 46 44.2
54	8					12.9			15 58.82		23.38	V.	3	9.35	14 47.0		33.3	16 22.20	26 51 0.3
55	8		19.9						20 47.98		23.40	III.	6	9.1	28 30.5		32.0	21 11.38	27 4 42.5
56	9				33.5				23 33.35		23.42	IV.	9	12.9	45 4.6		31.2	23 56.77	27 21 15.8
57	6				2.1				25 1.95		23.37	IV.	1	3.24	1 40.7		30.8	25 25.32	26 37 51.5
58	10					19.9			26 5.85		23.38	V.	5	3.20	21 36.8		30.5	26 29.23	57 47.3
59	9				18.1				27 17.95		23.38	IV.	5	7.28	23 42.1		30.2	27 41.33	59 52.3
60	7	38.4							29 20.57		23.37	II.	3	8.35	14 16.6		29.6	29 43.94	26 50 26.2
61	9					36.4			29 22.43		23.41	V.	9	11.33	44 46.4		29.6	29 45.84	27 20 56.0
62	7		36.9	50.9					31 5.07		23.41	IV.	9	6.45	42 21.0		29.1	31 28.48	27 18 30.1
63	6.7		52.1						32 19.99		23.37	II.	2	9.8	9 33.7		28.8	32 43.36	26 45 42.5
64	9	51.4							33 33.79		23.40	II.	7	9.18	33 39.4		28.4	33 57.19	27 9 47.8
65	8		58.0						34 26.24		23.41	III.	10	5.45	46 52.0		28.2	34 49.65	23 0.2
66	7			44.8					34 58.81		23.40	III.	7	8.28	33 14.3		28.0	35 22.21	9 22.3
67	9					33.3			35 19.34		23.40	IV.	10	6.28	47 13.8		28.0	35 42.74	23 21.8
68	9						31.2		36 3.26		23.39	VII.	7	7.33	32 46.1		27.7	36 26.65	8 53.8
69	8							24.7	36 42.85		23.40	VII.	10	4.38	46 17.8		27.6	37 6.25	27 22 25.4
70	6			28.0		56.0			38 41.85		23.38	V.	4	7.39	18 48.0		27.0	39 5.23	26 51 55.0
71	9				50.1				39 49.95		23.40	IV.	9	10.53	44 26.2		26.7	40 13.35	27 20 32.9
72	7					34.0			40 19.90		23.36	V.	2	8.40	9 19.7		26.6	40 43.26	26 45 26.3
73	6							17.0	40 34.88		23.37	VII.	5	8.48	24 22.1		26.5	.) 58.25	27 0 28.6
74	8					9.8			41 55.72		23.36	V.	3	7.9	13 33.3		26.1	42 19.08	26 49 39.4
75	8							1.0	42 18.76		23.36	VII.	3	9.43	14 50.7		26.0	42 42.12	26 50 56.7
76	6					0.4			43 46.41		23.39	V.	8	8.57	38 28.1		25.6	44 10.80	27 14 33.7
77	9						8.2		44 40.23		23.39	VII.	8	10.38	39 18.8		25.3	45 3.62	27 15 24.1
78	7	56.9							17 46 39.01	+	23.35	II.	2	6.33	− 8 15.5	−	24.8	17 47 2.36	26 44 20.3

CORRECTIONS.

Date.		Corr. of Clock.	Hourly rate.	m	n	c
1846.	h.	s.	s.	s.	s.	s.

INSTRUMENT READINGS.

Date.		Barom.	Thermom. At.	Thermom. Ex.
1846.	h. m.	in.	°	°

REMARKS.

ZONE 36. JULY 9. K. $D_0 = -26° 35' 40''$—Continued.

No.	Mag.	Seconds of Transit. I.	II.	III.	IV.	V.	VI.	VII.	T.	a_1	Micrometer. Thread	Rev.	r.	$i + d_1$	d_1	Mean Right Ascension, 1850.0.	Mean Declination, 1850.0.
									h. m. s.	s.			r.	′ ″	″	h. m. s.	° ′ ″
79	8			56.0					17 47 9.96	+ 23.38	IV.	7	2.45	−30 21.1	− 24.6	17 47 33.34	− 27 6 25.7
80	10			49.9					48 3.75	23.36	IV.	5	6.21	23 8.2	24.4	48 27.11	26 59 12.6
81	9			55.1					49 9.29	23.39	IV.	10	3.19	45 38.3	24.1	49 32.68	27 21 42.4
82	9						59.2		49 31.29	23.36	VII.	5	.32	20 11.6	24.0	49 54.65	26 56 15.6
83	8							31.8	49 49.72	23.37	VII.	6	6.8	27 2.8	23.9	50 13.09	27 3 6.7
84	9							19.6	51 37.45	23.36	VII.	4	13.22	21 40.8	23.4	52 0.81	26 57 44.2
85	7					12.8			52 58.77	23.37	VI.	6	8.32	28 15.7	23.0	53 22.14	27 4 18.7
86	7					8.3			53 54.22	23.35	V.	3	11.43	15 51.7	22.7	54 17.57	26 51 54.4
87	7				44.4				54 44.25	23.35	IV.	3	9.44	14 51.6	22.5	55 7.60	26 50 54.1
88	8					30.0			55 15.97	23.36	VI.	6	10.22	29 10.5	22.3	55 39.33	27 5 12.8
89	8					25.7			56 11.65	23.36	VI.	5	5.34	22 44.4	22.1	56 35.01	26 58 46.5
90	8						59.6		56 31.75	23.34	VII.	3	3.27	11 40.8	22.0	56 55.09	26 47 42.8
91	9						42.5		57 14.57	23.37	VI.	7	3.11	30 34.0	21.8	57 37.94	27 6 35.8
92	8							59.6	17 58 17.73	23.39	VII.	9	9.11	43 34.3	21.5	17 58 41.12	27 19 35.8
93	9		4.8						18 1 32.63	23.33	II.	1	8.11	4 5.5	20.6	18 1 55.96	26 40 6.1
94	8					45.4			3 31.41	23.38	VI.	8	9.34	38 46.7	20.0	3 54.79	27 14 46.7
95	10			23.0					6 37.10	23.37	IV.	8	11.35	39 48.0	19.1	7 0.47	15 47.1
96	4.5		48.3	2.4					8 16.42	23.36	III.	7	10.55	29 28.1	18.6	8 39.78	5 26.7
97	9							20.5	8 38.51	23.36	VII.	7	12.32	35 17.2	18.6	9 1.87	27 11 15.8
98	8		12.1						12 40.11	23.35	III.	5	5.4	22 29.3	17.4	13 3.46	26 58 26.7
99	8					58.1			12 44.05	23.35	V.	5	5.16	22 35.4	17.4	13 7.40	58 32.8
100	9					43.2			13 29.12	23.33	V.	3	8.59	14 29.0	17.2	13 52.45	26 50 26.2
101	9							21.5	13 39.39	23.35	VII.	5	10.42	25 19.7	17.1	14 2.74	27 1 16.8
102	8			57.7					16 11.94	23.37	IV.	10	9.20	48 40.6	16.4	16 35.21	27 24 37.0
103	6.5				45.4				17 50.27	23.31	V.	1	14.7	7 5.4	16.0	18 22.58	26 43 1.4
104	5.6						49.0		18 21.14	23.32	VII.	3	9.8	14 33.1	15.8	18 44.46	50 28.9
105	7					27.0			19 12.88	23.31	V.	1	8.53	4 26.8	15.6	19 36.19	26 40 22.4
106	10			28.1					20 41.98	23.34	III.	5	10.31	25 14.5	15.1	21 5.32	27 1 9.6
107	10					28.7			21 14.67	23.34	VI.	6	8.20	28 9.6	15.0	21 38.01	4 4.6
108	7				28.8				22 28.65	23.37	IV.	9	8.45	43 21.6	14.6	22 52.02	19 16.2
109	9					5.4			22 51.42	23.37	VI.	9	5.9	41 32.3	14.5	23 14.79	27 17 26.8
110	9					0.9			23 46.85	23.33	V.	5	5.5	22 29.8	14.2	24 10.18	26 58 24.0
111	8						42.0		24 14.12	23.32	VII.	4	6.30	18 12.7	14.1	24 37.44	26 54 6.8
112	10	38.8							25 21.27	23.36	II.	9	6.40	42 18.3	13.8	26 44.63	27 18 12.1
113	9		15.0						26 43.11	23.33	II.	7	7.26	32 42.8	13.4	27 6.44	27 8 36.2
114	9						53.2		26 25.38	23.29	VI.	1	10.19	5 10.1	13.5	26 48.67	26 41 3.6
115	8			8.2					31 21.95	23.30	IV.	3	13.4	16 32.6	12.1	31 45.25	26 52 24.7
116	8			46.1					32 0.09	23.33	IV.	7	6.47	32 23.3	11.9	32 23.42	27 8 15.2
117	8			37.6					32 51.24	23.30	IV.	2	7.15	8 36.8	11.7	33 14.54	26 44 28.5
118	4	10.9	25.5						35 53.44	23.32	II.	7	7.2	32 30.7	10.8	36 16.76	27 8 21.5
119	8	27.0							38 9.44	23.33	I.	8	9.25	38 41.9	10.1	38 32.77	14 32.0
120	9	16.2							38 58.48	23.31	II.	5	9.38	24 47.6	9.9	39 21.79	0 37.5
121	8		23.0						39 51.20	23.34	IV.	9	4.58	41 26.9	9.7	40 14.54	27 17 16.6
122	7	11.5	25.8						40 53.79	23.30	III.	4	10.41	20 19.8	9.4	41 17.09	26 56 9.2
123	7			12.8					41 40.98	23.34	III.	9	10.12	44 5.5	9.3	42 4.32	27 19 54.8
124	6.7	3.0		31.3					42 45.08	23.29	IV.	3	6.57	13 27.3	8.9	43 8.37	26 49 16.2
125	7	11.8							44 54.12	23.30	II.	6	8.49	28 23.8	8.3	45 17.42	27 4 12.1
126	10	41.8							46 24.12	23.30	II.	6	7.7	27 32.8	7.8	46 47.42	27 3 20.6
127	9				33.9				18 46 33.75	+ 23.29	V.	4	5.22	−17 38.7	− 7.8	18 46 57.04	− 26 53 26.5

CORRECTIONS.

Date.		Corr. of Clock.	Hourly rate.	m	n	c
1846.	h.	s.	s.	s.	s.	s.

INSTRUMENT READINGS.

Date.		Barom.	Thermom. At.	Ex.
1846.	h. m.	In.	°	°

REMARKS.

(36) 96. Hor. thread assumed as 6 instead of 7.
(36) 103. Transit over T. III assumed as recorded over T. IV.
(36) 112. Minutes assumed as 26 instead of 25.
(36) 123. Transit over T. II assumed as recorded over T. III.

ZONE 36. JULY 9. K. $D_0 = -26° 35' 40''$—Continued.

No.	Mag.	I.	II.	III.	IV.	V.	VI.	VII.	T.	a_1	MICROMETER.			$i + d_1$	d_1	Mean Right Ascension, 1850.0.	Mean Declination, 1850.0.
									h. m. s.	s.			r.	′ ″	″	h. m. s.	° ′ ″
128	9	. .	. .	19.3	. .	. .	. .	. .	18 48 33.36	+ 23.31	IV.	8	6.45	−37 21.5	− 7.2	18 48 56.67	− 27 13 8.7
129	8	. .	16.4	. .	. .	. .	. .	. .	49 44.24	23.26	III.	1	11.44	5 53.2	6.9	50 7.50	26 41 40.1
130	9	. .	. .	. .	. .	14.6	. .	. .	51 0.53	23.27	VI.	3	13.7	16 34.0	6.5	51 23.80	52 20.5
131	9	. .	. .	49.3	. .	. .	. .	. .	56 3.04	23.26	IV.	3	11.39	15 49.7	5.1	56 26.30	26 51 34.8
132	7	. .	57.4	. .	. .	. .	. .	. .	58 25.48	23.28	III.	6	8.6	28 2.7	4.5	58 48.76	27 3 47.2
133	9	. .	. .	. .	. .	42.6	. .	. .	58 28.57	23.28	V.	6	8.59	28 29.5	4.4	18 58 51.85	4 13.9
134	6	. .	32.0	. .	. .	. .	. .	. .	18 59 2.42	23.30	III.	9	12.13	45 6.6	4.1	19 0 25.72	27 20 50.7
135	8	. .	. .	. .	. .	22.1	. .	. .	19 0 8.00	23.25	V.	2	11.44	10 52.6	4.0	0 31.25	26 46 36.6
136	7	. .	. .	. .	. .	. .	. .	53.9	0 11.67	23.25	VII.	3	14.34	17 17.7	4.0	0 34.92	26 53 1.7
137	8	. .	. .	. .	. .	. .	17.0	. .	19 1 49.02	+ 23.30	IV.	9	8.35	−43 16.5	− 3.5	19 2 12.32	− 27 19 0.0
									ZONE 37. JULY 10. A. $D_0 = -34° 7' 30''$.								
1	11	. .	. .	. .	. .	. .	36.0	51.5	16 5 6.03	+ 24.47	V.	9	11.20	−44 41.6	. .	16 5 30.50	
2	8	. .	. .	2.5	17.5	. .	. .	. .	7 17.36	24.45	IV.	4	7.32	18 43.9	. .	7 41.81	
3	7	. .	. .	. .	. .	20.2	35.4	50.6	8 5.14	24.46	VII.	3	8.46	24 20.9	. .	8 29.60	
4	10	. .	. .	2.5	17.5	. .	. .	. .	10 17.44	24.47	IV.	6	9.5	28 32.9	. .	10 41.91	
5	11	. .	. .	25.0	40.2	. .	. .	. .	11 40.00	24.48	IV.	5	8.5	24 0.7	. .	12 4.48	
6	8	. .	. .	45.8	1.0	16.2	. .	. .	13 0.89	24.48	V.	6	3.43	25 49.9	. .	13 25.37	
7	8	. .	. .	. .	32.8	47.9	. .	. .	16 14 32.68	+ 24.49	V.	5	8.40	−24 18.3	. .	16 14 57.17	
									ZONE 38. JULY 10. A. $D_0 = -36° 37' 30''$.								
1	7	. .	. .	13.8	29.2	. .	. .	. .	16 40 29.26	+ 24.59	IV.	8	6.43	−37 22.4	. .	16 40 53.85	
2	11	. .	. .	. .	37.0	52.6	. .	. .	41 36.88	24.58	V.	4	8.11	19 3.2	. .	42 1.46	
3	11	. .	. .	53.0	8.5	. .	. .	. .	43 8.52	24.60	IV.	8	8.23	38 13.1	. .	43 33.12	
4	10	. .	. .	. .	. .	. .	. .	22.5	43 35.70	24.61	VII.	9	5.10	41 34.0	. .	44 0.31	
5	8	. .	. .	. .	29.0	44.5	. .	. .	45 28.87	24.60	V.	8	8.15	38 9.0	. .	45 53.47	
6	9	. .	. .	. .	. .	. .	51.2	7.5	47 20.30	24.58	VII.	2	12.	10 58.0	. .	47 44.88	
7	9	. .	. .	. .	. .	10.3	25.6	41.5	47 54.56	24.61	VII.	7	6.41	32 20.9	. .	48 19.17	
8	10	. .	. .	. .	. .	. .	41.0	57.0	49 9.97	24.62	VII.	7	10.55	34 29.5	. .	49 34.59	
9	8	. .	52.0	7.8	. .	. .	. .	. .	51 23.30	24.61	III.	6	5.40	26 49.2	. .	51 47.91	
10	8	19.3	35.5	51.1	. .	. .	. .	. .	53 6.52	24.63	III.	4	7.45	18 50.0	. .	53 31.15	
11	9	. .	. .	58.8	14.8	. .	. .	. .	54 14.55	24.64	IV.	8	3.55	35 57.4	. .	54 39.19	
12	5	. .	16.5	32.3	48.0	. .	. .	. .	16 55 47.79	+ 24.65	IV.	5	6.44	−23 19.6	. .	16 56 12.44	
									ZONE 39. JULY 10. A. $D_0 = -34° (8)'$.								
1	10	. .	. .	36.0	51.2	. .	. .	. .	17 1 51.11	+ 24.66	IV.	8	4.26	−36 12.3	. .	17 2 15.77	
2	8	23.9	39.2	54.4	. .	. .	. .	. .	4 9.57	24.66	III.	6	11.31	29 46.6	. .	4 34.23	
3	8	. .	54.5	10.0	25.0	. .	. .	. .	5 25.00	24.67	IV.	8	7.49	37 55.0	. .	5 49.67	
4	9	15.5	30.8	. .	. .	. .	. .	. .	7 1.26	24.65	II.	7	7.33	32 47.0	. .	7 25.91	
5	10	. .	. .	. .	. .	32.5	47.5	3.0	7 17.38	24.65	VII.	4	3.52	16 52.0	. .	7 42.03	
6	8	. .	. .	49.0	4.0	19.0	. .	. .	3.75	24.66	V.	1	8.27	4 11.8	. .	(10) 28.41	
7	10	. .	38.5	53.5	. .	. .	. .	. .	13 8.46	24.66	III.	3	5.42	12 52.7	. .	13 33.12	
8	7	. .	. .	. .	. .	1.5	16.5	. .	17 13 46.35	+ 24.68	VI.	6	3.37	−25 46.7	. .	17 14 11.03	

CORRECTIONS.

Date.		Corr. of Clock.	Hourly rate.	m	n	c
1846.	h.	s.	s.	s.	s.	s.
July 10,	13	− 13.48	+ 0.098	− 0.463	+ 0.300	0.000

INSTRUMENT READINGS.

	Date.		Barom.	THERMOM. At.	THERMOM. Ex.
	1846.	h. m.	In.	°	°
Zone 37	July 10,	16 5	30.5	86.0	81.5
		16 14	30.5	85.8	80.7
Zone 38	July 10,	16 40	30.5	85.3	79.2
		16 56	30.5	85.0	78.0
Zone 39	July 10,	17 2	30.5	85.0	78.0

REMARKS.

(36) 134. Transit over T. I assumed as recorded over T. II.
(39) Telescope not firmly clamped.

ZONE 39. JULY 10. A. $D_0 = -34°$ (8)′—Continued.

No.	Mag.	Seconds of Transit. I.	II.	III.	IV.	V.	VI.	VII.	T.	a_1	Micrometer.			$i + d_2$	d_1	Mean Right Ascension, 1850.0.	Mean Declination, 1850.0.
									h. m. s.	s.			r.	′ ″	″	h. m. s.	° ′ ″
9	6	. .	. .	. .	. .	. .	11.0	. .	17 14 40.87	+ 24.68	VI.	6	1.40	−24 47.6	. .	17 15 5.55	
10	9	7.2	23.0	. .	. .	. .	. .	. .	17 53.05	24.69	II.	5	2.	20 55.8	. .	18 17.74	
11	10	. .	. .	. .	30.5	. .	. .	. .	18 30.33	24.69	IV.	6	11.36	29 49.2	. .	18 55.02	
12	8	. .	. .	. .	58.5	13.5	. .	. .	19 58.31	24.68	V.	3	7.10	13 32.8	. .	20 22.99	
13	8	. .	. .	. .	. .	. .	. .	54.5	20 8.92	24.69	VII.	4	4.18	17 5.2	. .	20 33.61	
14	9	. .	. .	. .	23.0	38.0	53.0	. .	22 22.85	24.71	VI.	7	4.53	31 26.1	. .	22 47.56	
15	7	. .	. .	9.5	25.0	40.0	. .	. .	24 24.81	24.72	V.	8	1.55	34 55.9	. .	24 49.53	
16	9	. .	. .	51.0	6.0	. .	. .	. .	26 5.88	24.71	IV.	5	4.22	22 7.9	. .	26 30.59	
17	8	. .	. .	. .	10.5	24.5	. .	. .	27 9.83	24.71	V.	5	10.	24 58.8	. .	27 34.54	
18	10	. .	. .	. .	. .	. .	. .	16.5	27 31.07	24.71	VII.	6	10.	29 0.2	. .	27 55.78	
19	8	. .	. .	. .	. .	. .	7.0	22.5	28 36.93	24.71	VII.	4	10.55	20 26.0	. .	29 1.64	
20	8	. .	. .	. .	40.5	55.5	10.8	. .	30 40.45	24.73	VI.	8	3.27	35 42.2	. .	31 5.18	
21	7	. .	54.0	9.5	. .	. .	. .	. .	33 24.24	24.71	III.	3	9.25	14 41.0	. .	33 48.95	
22	7	. .	. .	. .	. .	6.2	21.3	36.3	33 51.07	24.74	VII.	7	10.	34 1.1	. .	34 15.81	
23	6	. .	33.0	48.2	3.5	. .	. .	. .	36 3.38	24.74	IV.	8	4.50	36 24.4	. .	36 28.12	
24	7	. .	. .	. .	. .	18.5	34.0	. .	37 3.60	24.74	VII.	8	10.46	39 24.0	. .	37 28.34	
25	6	. .	. .	. .	. .	18.0	33.2	. .	38 2.96	24.74	VII.	5	7.15	23 34.9	. .	38 27.70	
26	7	. .	. .	39.5	. .	. .	. .	. .	39 54.38	24.73	III.	4	5.51	17 52.7	. .	40 19.11	
27	5	3.0	18.5	. .	. .	. .	. .	. .	41 48.92	24.75	II.	8	7.9	37 34.4	. .	42 13.67	
28	5	. .	24.5	39.8	. .	. .	. .	. .	42 54.91	24.76	III.	7	9.10	33 36.3	. .	43 19.67	
29	5	. .	. .	. .	. .	. .	0.5	15.5	43 30.25	24.76	VII.	7	12.	35 1.8	. .	43 55.01	
30	6	. .	. .	. .	. .	. .	. .	10.0	44 24.49	24.75	VII.	5	5.26	22 39.7	. .	44 49.24	
31	7	. .	. .	2.0	17.3	. .	. .	. .	46 17.03	24.75	IV.	5	4.54	22 24.1	. .	46 41.78	
32	10	. .	. .	. .	. .	28.8	44.0	. .	48 13.75	24.77	VI.	8	9.33	38 47.4	. .	48 38.52	
33	8	. .	. .	. .	44.3	59.0	. .	. .	49 44.01	24.77	V.	7	7.25	32 43.2	. .	50 8.78	
34	8	. .	. .	. .	. .	. .	51.8	7.0	51 21.63	24.77	VII.	7	9.50	33 56.0	. .	51 46.40	
35	11	. .	. .	. .	. .	11.5	26.5	. .	51 56.35	24.76	VI.	5	10.10	25 3.7	. .	52 21.11	
36	8	. .	. .	. .	. .	. .	32.0	47.5	53 1.98	24.79	VI.	8	7.40	37 49.9	. .	53 26.77	
37	8	. .	47.2	2.0	17.6	. .	. .	. .	17 55 17.35	+ 24.77	IV.	6	10.44	−29 22.9	. .	17 55 42.12	

ZONE 40. JULY 10. A. $D_0 = -29° 9′ 10″$.

No.	Mag.	I.	II.	III.	IV.	V.	VI.	VII.	T.	a_1	Micrometer.			$i + d_2$	d_1	Mean Right Ascension, 1850.0.	Mean Declination, 1850.0.
1	7	. .	. .	. .	3.5	18.0	. .	. .	18 1 3.50	+ 24.35	V.	8	6.13	−37 5.5	. .	18 1 27.88	
2	7	. .	. .	. .	. .	16.0	30.2	. .	2 1.59	24.37	VI.	7	5.7	31 32.5	. .	2 25.96	
3	10	. .	. .	. .	. .	36.5	51.0	6.0	3 22.40	24.35	VII.	4	7.31	18 43.4	. .	3 46.75	
4	8	. .	. .	. .	. .	. .	6.0	20.5	4 37.35	24.37	VII.	7	2.24	30 10.2	. .	5 1.72	
5	9	. .	. .	. .	. .	. .	14.5	28.5	6 45.60	24.38	VII.	7	3.55	30 56.1	. .	7 9.98	
6	8	. .	. .	. .	. .	. .	37.5	51.5	8 8.64	24.40	VII.	9	5.	36 28.8	. .	8 33.00	
7	7	. .	. .	. .	. .	57.3	11.4	25.5	9 42.68	24.38	VII.	7	4.20	31 8.6	. .	10 7.06	
8	7	. .	. .	. .	28.2	42.4	56.6	. .	11 28.01	24.38	VI.	7	5.	31 29.2	. .	11 52.39	
9	9	. .	. .	. .	. .	. .	55.9	10.0	12 27.06	24.39	VII.	8	3.30	35 42.8	. .	12 51.45	
10	8	. .	13.7	28.0	. .	. .	. .	. .	14 42.28	24.38	III.	7	4.42	31 20.3	. .	15 6.66	
11	11	. .	. .	2.5	17.4	. .	. .	. .	16 17.13	24.40	IV.	9	9.20	43 39.6	. .	16 41.53	
12	7	. .	. .	. .	. .	21.2	35.3	49.8	18 17 6.71	+ 24.38	VII.	7	3.25	−30 41.0	. .	18 17 31.09	

CORRECTIONS.

Date.		Corr. of Clock.	Hourly rate.	m	n	c
1846.	h.	s.	s.	s.	s.	s.

INSTRUMENT READINGS.

	Date.	Barom.	Thermom. At.	Thermom. Ex.
	1846. h. m.	in.	°	°
Zone 40	July 10, 18 17	30.05	84.0	77.7

REMARKS.

(40) 6. Micrometer thread assumed as 8 instead of 9.
(40) 9. Discordant from Mural, 1846.

ZONE 41. JULY 10. A. $D_0 = -35° 21' 50''$—Continued.

No.	Mag.	Seconds of Transit I.	II.	III.	IV.	V.	VI.	VII.	T.		u_1	Micrometer.			$i + d_0$		d_1	Mean Right Ascension, 1850.0.		Mean Declination, 1850.0.
									h. m. s.		s.			r.	' "		"	h. m. s.		° ' "
1	5	. .	. .	. .	. .	8.0	22.1	37.7	18 33 51.49	+	25.00	VII.	5	9.40	−24 48.2	−	23.7	18 34 16.49	−	35 47 1.9
2	6	. .	. .	. .	. .	. .	20.0	35.0	34 49.11		25.00	VII.	8	7.58	37 59.4		23.5	35 14.11		36 0 12.9
3	11	. .	. .	. .	. .	. .	11.0	. .	36 40.36		24.99	VI.	7	3.48	30 53.4		23.1	37 5.35		35 53 6.5
4	11	. .	. .	. .	. .	35.5	50.8	. .	38 20.13		24.97	VI.	5	10.2	24 59.6		22.8	38 45.10		47 12.4
5	11	. .	. .	. .	. .	. .	19.5	35.2	39 48.94		24.98	VI.	6	7.25	27 42.1		22.5	40 13.92		49 54.6
6	8	. .	. .	35.5	50.9	. .	. .	. .	41 50.73		24.97	IV.	5	10.43	25 20.6		22.1	42 15.70		47 32.7
7	9	56.5	12.3	. .	. .	. .	. .	. .	43 43.08		24.99	II.	7	7.28	32 44.7		21.7	44 8.07		54 56.4
8	9	. .	. .	. .	29.0	44.2	59.5	. .	44 28.84		24.99	VI.	7	6.46	32 23.5		21.6	44 53.83		54 35.1
9	8	. .	. .	. .	21.2	36.8	. .	. .	46 21.16		24.95	V.	2	4.17	7 4.8		21.2	46 46.11		35 29 16.0
10	6	. .	. .	. .	26.0	41.5	. .	. .	48 26.01		25.01	V.	10	6.46	47 25.6		20.8	48 51.02		36 9 36.4
11	10	. .	. .	. .	29.5	44.5	. .	. .	50 29.23		24.99	V.	7	9.40	33 51.7		20.4	50 54.22		35 56 2.1
12	7	. .	. .	. .	. .	. .	50.0	5.2	51 19.22		25.00	VII.	9	6.30	42 15.0		20.2	51 44.22		36 4 25.2
13	10	. .	. .	36.0	51.0	. .	. .	. .	53 50.92		24.96	IV.	3	5.16	12 34.8		19.7	54 15.88		35 34 44.5
14	8	. .	. .	58.2	13.2	. .	. .	. .	55 13.13		24.96	IV.	3	6.10	13 2.1		19.5	55 38.09		35 11.6
15	9	. .	. .	. .	. .	. .	. .	19.5	55 33.29		24.98	VII.	6	5.38	26 47.7		19.4	55 58.27		48 57.1
16	11	. .	. .	. .	. .	. .	24.5	40.0	56 53.84		24.98	VII.	6	8.52	28 25.9		19.1	57 18.82		50 35.0
17	11	. .	. .	. .	. .	. .	. .	40.0	18 57 53.76		24.97	VII.	5	7.32	23 43.4		18.9	18 58 18.73		45 52.3
18	8	. .	36.0	51.0	. .	. .	. .	. .	19 5 6.16		24.94	III.	1	5.01	2 27.0		17.5	19 5 31.10		24 34.5
19	8	. .	. .	. .	. .	50.5	7.0	23.2	6 36.77		24.94	VII.	1	8.38	4 16.3		17.3	7 1.71		26 23.6
20	5	. .	47.5	3.2	18.5	. .	. .	. .	9 18.29		24.95	IV.	4	8.22	19 9.0		16.7	9 43.24		41 15.7
21	11	. .	. .	. .	. .	. .	. .	27.8	10 41.51		24.95	IV.	7	10.42	20 19.3		16.4	11 6.46		42 25.7
22	9	. .	. .	. .	. .	. .	22.0	38.0	12 51.57		24.95	VII.	4	5.40	17 46.5		16.0	13 16.52		39 52.5
23	11	. .	. .	. .	. .	52.0	7.0	23.0	15 36.59		24.95	VII.	5	7.38	23 46.4		15.5	16 1.54		45 51.9
24	7	. .	. .	. .	2.5	17.5	32.5	. .	18 2.12		24.92	VI.	1	6.47	3 20.4		15.0	18 27.04		25 25.4
25	8	. .	. .	. .	. .	35.0	. .	. .	19 19.54		24.92	V.	1	2.5	0 57.9		14.8	19 44.46		23 2.7
26	9	. .	44.0	59.5	15.0	. .	. .	. .	22 14.76		24.95	IV.	5	8.11	24 3.7		14.2	22 39.71		46 7.9
27	11	. .	. .	. .	31.0	46.5	. .	. .	24 30.96		24.95	V.	5	10.36	25 17.0		13.8	24 55.91		47 20.8
28	8	. .	39.8	55.2	. .	. .	. .	. .	27 10.30		24.93	III.	3	4.18	12 5.4		13.3	27 35.23		34 8.7
29	9	. .	. .	1.2	17.0	. .	. .	. .	31 16.76		24.97	IV.	9	4.50	41 24.9		12.6	31 41.73		35 3 27.5
30	8	. .	. .	20.2	35.3	. .	. .	. .	33 35.18		24.93	IV.	3	7.5	13 30.0		12.1	34 0.11		36 35 32.1
31	11	. .	. .	. .	. .	. .	23.0	39.0	34 52.55		24.93	VII.	3	3.25	11 38.1		11.9	35 17.48		36 33 40.0
32	7	. .	43.2	59.0	14.4	. .	. .	. .	37 13.99		24.91	IV.	2	10.17	10 7.0		11.5	37 38.90		35 32 8.5
33	11	. .	. .	. .	. .	. .	28.0	43.5	37 57.37		24.96	VII.	9	6.53	42 25.1		11.3	38 22.33		36 4 26.4
34	7	17.5	33.3	48.9	. .	. .	. .	. .	41 3.95		24.92	III.	4	8.14	19 3.9		10.8	41 28.87		36 41 4.7
35	6	. .	. .	. .	. .	. .	29.8	45.4	47 59.15		24.90	VII.	3	6.5	12 59.1		9.6	48 24.05		35 34 58.7
36	5	. .	. .	. .	. .	48.5	3.3	19.2	49 32.90		24.90	VII.	4	7.20	18 37.0		9.3	49 57.80		40 36.3
37	11	27.5	43.0	58.5	. .	. .	. .	. .	52 13.72		24.90	III.	4	8.	18 57.8		8.8	52 38.62		40 56.6
38	8	. .	. .	39.5	54.8	10.0	. .	. .	53 54.63		24.90	V.	5	3.59	21 56.1		8.5	54 19.53		43 54.6
39	9	. .	. .	. .	7.5	22.8	38.0	. .	54 7.37		24.90	VI.	5	6.47	23 20.9		8.5	54 32.27		45 19.4
40	7	. .	39.0	54.8	10.2	. .	. .	. .	57 9.80?		24.93	IV.	8	1.47	34 22.0		8.0	57 34.73?		56 20.0
41	8	. .	47.2	2.9	18.0	. .	. .	. .	59 17.64		24.88	IV.	3	8.8	14 1.8		7.6	19 59 42.52		35 59.4
42	7	. .	. .	. .	. .	7.0	22.5	37.9	19 59 52.21		24.89	VII.	5	8.28	24 11.7		7.5	20 0 17.10		46 9.2
43	11	. .	. .	2.5	17.8	. .	. .	. .	20 3 17.58		24.89	IV.	6	6.50	27 24.7		7.0	3 42.47		49 21.7
44	6	34.3	50.2	5.8	. .	. .	. .	. .	6 20.32		24.87	III.	4	4.39	17 16.1		6.5	6 45.19		39 12.6
45	11	. .	. .	. .	. .	19.5	35.0	50.5	7 4.76		24.88	VII.	6	10.	29 0.3		6.3	7 29.64		35 50 56.6
46	6	. .	. .	29.5	45.2	0.5	15.6	. .	11 45.17		24.90	VI.	10	5.9	46 36.3		5.5	12 10.07		36 8 31.8
47	7	. .	24.8	40.5	56.0	. .	. .	. .	12 55.51		24.87	IV.	6	10.10	29 5.9		5.4	13 20.38		35 51 1.3
48	7	. .	. .	. .	. .	57.5	12.7	28.2	13 42.52		24.85	VII.	3	8.49	14 22.0		5.2	14 7.37		36 17.2
49	10	. .	. .	. .	. .	9.0	24.5	40.0	16 54.28		24.87	VII.	7	5.32	31 45.7		4.7	17 19.15		53 40.4
50	8	. .	. .	. .	48.0	2.8	18.5	33.9	20 18 47.92	+	24.85	VII.	6	10.35	−29 18.0	−	4.4	20 19 12.77	−	35 51 12.4

CORRECTIONS.

Date.	Corr. of Clock.	Hourly rate.	m	n	c
1846.	h. s.	s.	s.	s.	s.
July 10,	13 − 13.48	+ 0.098	− 0.463	+ 0.300	0.000

INSTRUMENT READINGS.

	Date.		Barom.	Thermom. At.	Thermom. Ex.
	1846.	h. m.	in.	°	°
Zone 41	July 10,	18 34	30.05	83.7	77.5
	10,	20 19	30.05	83.0	76.3

REMARKS.

(41) 40. Micrometer reading assumed as 0r.47 instead of 1r.47

ZONE 42. JULY 11. K. $D_o = -27° 50' 40''$—Continued.

No.	Mag.	I.	II.	III.	IV.	V.	VI.	VII.	T.	a_1	MICROMETER.			$i + d_1$	d_1	Mean Right Ascension, 1850.0.	Mean Declination, 1850.0.
									h. m. s.	s.			r.	′ ″	″	h. m. s.	° ′ ″
1	10		12.0						16 21 40.41	+ 26.06	II.	7	4.12	−31 4.9 −	37.4	16 22 6.47	− 28 22 22.3
2	10					6.4			21 52.15	26.03	VI.	3	6.17	13 6.8	37.3	22 18.18	4 24.1
3	10			9.4					25 23.37	26.06	IV.	5	7.44	23 50.2	36.9	25 49.43	15 7.1
4	4				7.2	21.4			26 7.07	26.04	IV.	1	5.29	2 43.7	36.8	26 33.11	27 54 0.5
5	8							58.9	26 16.14	26.05	VII.	3	2.47	11 20.5	36.7	26 42.19	28 2 37.2
6	9					4.2			27 50.00	26.07	V.	6	6.12	27 5.2	36.5	28 16.07	18 21.7
7	8			3.3					29 17.58	26.10	IV.	9	11.14	44 37.0	36.3	29 43.68	35 53.3
8	7			6.8					30 21.11	26.10	V.	10	6.1	47 0.2	36.2	30 47.21	38 16.4
9	8	4.3							31 47.11	26.09	II.	6	4.13	26 4.9	36.0	32 13.20	17 20.9
10	9		48.3						32 16.84	26.11	II.	9	5.25	41 40.5	35.9	32 42.95	32 56.4
11	9						46.3		32 18.19	26.08	VI.	5	3.9	21 31.1	35.9	32 44.27	12 47.0
12	7				39.0				33 38.84	26.07	VI.	2	9.57	9 58.3	35.7	34 4.91	1 14.0
13	9						36.2		34 8.04	26.10	VI.	7	4.55	31 26.6	35.7	34 34.14	22 42.3
14	7					26.0			35 11.79	26.08	VI.	5	4.36	22 15.0	35.5	35 37.87	13 30.5
15	9							26.3	35 43.91	26.12	VII.	9	6.33	42 14.6	35.4	36 10.03	33 30.0
16	8						21.0		36 52.77	26.12	VII.	10	5.15	46 36.6	35.3	37 18.89	37 51.9
17	8					23.0			38 8.76	26.08	V.	3	10.6	15 2.6	35.1	38 34.84	6 17.7
18	10					19.0			39 4.77	26.10	VI.	4	7.34	18 45.2	34.9	39 30.87	10 0.1
19	10	55.0							41 37.73	26.10	II.	4	11.13	20 35.8	34.6	42 3.83	11 50.4
20	10					51.0			42 36.84	26.12	VI	8	5.0	36 28.4	34.4	43 2.96	27 42.8
21	8					37.2			43 22.98	26.11	VI.	4	10.33	20 15.6	34.3	43 49.09	11 29.9
22	10			26.1					44 40.11	26.12	III.	6	6.46	27 22.3	34.1	45 6.23	18 36.4
23	10		16.8						45 45.03	26.10	VI.	3	7.25	13 41.3	34.0	46 11.13	4 55.3
24	10			17.8					46 31.99	26.13	IV.	8	9.19	38 39.4	33.9	46 58.12	29 53.3
25	9	18.6							16 48 1.50	+ 26.13	II.	8	5.28	−36 42.5 −	33.6	16 48 27.63	− 28 27 56.1

ZONE 43. JULY 11. K. $D_o = -39° 6' 50''$.

No.	Mag.	I.	II.	III.	IV.	V.	VI.	VII.	T.	a_1	MICROMETER.			$i + d_1$	d_1	Mean Right Ascension, 1850.0.	Mean Declination, 1850.0.
1	10				9.6				16 54 9.32	+ 26.16	VI.	9	7.0	−42 32.9 −	13.7	16 54 35.58	− 39 49 36.6
2	7			12.3		44.3			17 1 28.03	27.02	V.	3	3.36	11 42.3	12.8	17 1 55.05	18 45.1
3	6.7	48.5		21.4					3 37.33	27.05	III.	6	7.50	27 55.3	12.5	4 4.38	34 57.8
4	10		56.5						5 28.76	27.04	III.	4	8.4	18 59.0	12.3	5 55.80	26 1.3
5	7.8		44.2						6 16.32	27.03	III.	2	6.42	8 15.8	12.2	6 43.35	15 18.0
6	7		49.0						7 11.33	27.06	III.	5	8.23	24 9.6	12.0	7 38.39	31 11.6
7	10			27.8					7 43.53	2..05	IV.	3	10.45	15 20.0	12.0	8 10.58	22 22.0
8	9						32.8		8 0.71	27.05	VII.	3	7.47	13 49.0	11.9	8 27.76	20 50.9
9	10			8.7					16 24.58	27.10	IV.	6	9.6	28 34.0	10.8	16 51.68	35 34.8
10	10		27.0						18 59.39	27.10	III.	6	7.18	27 39.1	10.5	19 26.49	34 39.6
11	10				38.8				20 38.62	27.14	VI.	9	10.0	44 4.2	10.3	21 5.76	51 4.5
12	10	25.6							25 14.08	27.09	II.	1	13.14	6 33.5	9.6	25 41.17	13 33.1
13	9		32.5						26 4.87	27.13	III.	6	5.51	26 55.0	9.5	26 32.00	33 54.5
14	8	5.1							28 53.61	27.10	III.	2	6.16	8 2.6	9.1	29 20.71	15 1.7
15	8						10.0		28 37.44	27.14	VII.	7	10.16	34 10.6	9.2	29 4.58	41 9.8
16	8						47.2		29 14.59	27.17	VII.	9	12.0	45 4.8	9.1	29 41.76	52 3.9
17	10				40.0				30 39.82	27.10	VI.	1	11.49	5 50.6	8.9	31 6.92	12 49.5
18	10			43.6					31 59.89	27.14	IV.	6	13.0	30 32.7	8.7	32 27.03	37 31.4
19	9			21.9					32 38.31	27.16	IV.	8	7.11	37 38.0	8.6	33 5.47	44 36.6
20	9						19.2		17 32 46.62	+ 27.16	VI.	8	5.10	−36 36.3 −	8.6	17 33 13.78	− 39 43 34.9

CORRECTIONS.

Date.		Corr. of Clock.	Hourly rate.	m	n	c	
1846.	h.	s.	s.	s.	s.	s.	
July 11,	13	− 15.39	+ 0.112	− 0.463	+ 0.300	0.000	Zone 42
							Zone 43

INSTRUMENT READINGS.

Date.		Barom.	THERMOM. At.	THERMOM. Ex.
1846.	h. m.	in.	°	°
July 11,	16 22	30.00	87.0	85.6
11,	16 48	29.99	86.0	84.9
11,	16 54	29.99	86.0	84.9
11,	18 4	29.98	85.0	83.5

REMARKS.

ZONE 43. JULY 11. K. $D_0 = -39° 6' 50''$—Continued.

No.	Mag.	Seconds of Transit. I.	II.	III.	IV.	V.	VI.	VII.	T.	a_1	Micrometer.			$i + d_3$	d_1	Mean Right Ascension, 1850.0.	Mean Declination, 1850.0.
									h. m. s.	s.			r.	' "	"	h. m. s.	° ' "
21	9						23.3		17 33 53.70	+ 27.17	VII.	8	10.42	−39 24.4	− 8.4	17 34 17.87	− 39 46 22.8
22	7		7.4		39.7				35 39.40	27.11	II.	1	9.45	4 47.8	8.2	36 6.51	11 46.0
23	8			11.3					36 27.28	27.13	IV.	3	8.44	14 18.6	8.1	36 54.41	21 16.7
24	7	58.7	15.1						37 47.64	27.18	II.	9	8.27	43 17.1	7.9	38 14.82	50 15.0
25	9					15.8			37 59.76	27.13	VI.	3	10.25	15 9.5	7.9	38 26.89	22 7.4
26	10					39.3			39 23.34	27.18	VI.	8	6.44	37 24.0	7.7	39 50.52	44 21.7
27	10		27.3						41 59.49	27.16	III.	6	4.27	26 12.4	7.3	42 26.65	33 9.7
28	9				44.3				42 44.12	27.20	VI.	9	8.20	43 13.5	7.2	43 11.32	50 10.7
29	10				53.8				43 53.62	27.15	IV.	3	10.0	14 47.1	7.0	44 20.77	21 44.1
30	9			40.9					44 56.84	27.15	IV.	3	3.47	11 48.0	6.9	45 23.99	18 44.9
31	7			38.8					45 55.41	27.22	IV.	10	8.17	48 15.0	6.8	46 22.63	55 11.8
32	7			11.2					49 27.05	27.16	III.	2	3.30	6 38.5	6.2	49 54.21	13 31.7
33	8						10.9		49 38.35	27.20	VI.	7	5.59	32 0.6	6.2	50 5.55	38 56.8
34	10		30.9						52 2.89	27.17	III.	3	5.39	12 44.7	5.9	52 30.06	19 40.6
35	9			30.2					52 46.25	27.17	IV.	4	4.4	16 57.3	5.8	53 13.42	23 53.1
36	9			8.8					53 24.82	27.17	IV.	4	1.26	15 37.2	5.7	53 51.99	22 32.9
37	9	58.0							56 46.77	27.19	II.	5	9.21	24 38.8	5.2	57 13.96	31 34.0
38	8	11.1							17 57 59.86	27.19	II.	5	7.0	23 27.3	5.0	17 58 27.05	30 22.3
39	9		15.0						18 1 17.03	27.18	II.	3	10.28	15 11.0	4.6	18 1 44.21	22 5.6
40	8				37.0				1 37.82	27.18	IV.	3	10.24	15 9.3	4.5	2 5.00	22 3.8
41	8						39.4		2 7.06	27.16	VI.	1	8.25	4 7.1	4.4	2 34.22	11 1.5
42	8							22.3	2 34.00	27.22	VII.	8	6.24	37 13.4	4.4	3 1.24	44 7.8
43	10						40.9		18 4 8.53	+ 27.17	VII.	2	4.18	− 7 2.2	− 4.2	18 4 35.70	− 39 13 56.4

ZONE 44. JULY 14. K. $D_0 = -35° 16' 0''$.

No.	Mag.	I.	II.	III.	IV.	V.	VI.	VII.	T.	a_1	Micrometer.			$i + d_3$	d_1	Mean Right Ascension, 1850.0.	Mean Declination, 1850.0.
1	9	24.0							17 20 10.44	+ 1.74	II.	6	7.4	−27 31.5	− 22.8	17 20 12.18	− 35 43 54.3
2	9	13.8							21 0.28	1.74	II.	6	12.16	30 9.4	22.7	21 2.02	46 32.1
3	10					43.3			21 27.98	1.76	V.	8	13.43	40 54.8	22.6	21 29.74	57 17.2
4	8				9.1				24 8.93	1.72	IV.	4	11.44	20 51.2	22.2	24 10.65	37 13.4
5	10				17.5				25 17.33	1.72	V.	2	8.13	9 4.1	22.0	25 19.05	25 26.1
6	9				17.3				27 17.13	1.73	IV.	4	12.17	21 7.9	21.7	27 18.86	37 29.6
7	9		21.0						28 51.91	1.77	III.	8	6.47	37 24.0	21.4	28 53.68	33 45.4
8	8				31.0				30 30.83	1.73	V.	4	4.17	17 4.9	21.2	30 32.56	33 26.1
9	8		40.9						32 11.68	1.75	III.	6	5.26	26 42.1	20.9	32 13.43	43 3.0
10	8					32.9			32 17.44	1.72	VI.	2	7.17	8 35.6	20.9	32 19.16	24 56.5
11	9			31.3					33 46.27	1.72	III.	2	8.1	8 58.0	20.7	33 47.99	25 18.7
12	7.8		26.4						34 56.98	1.72	III.	2	12.14	11 6.0	20.5	34 58.70	27 26.5
13	8		51.4						37 22.12	1.75	IV.	5	3.54	21 53.6	20.1	37 23.87	38 13.7
14	9					18.0			37 32.64	1.77	VI.	7	8.36	33 19.2	20.1	37 34.41	49 39.3
15	7						38.7		38 8.04	1.77	VII.	7	9.17	33 39.6	20.0	38 9.81	49 59.6
16	8					40.0			39 24.51	1.72	V.	1	6.2	2 57.8	19.8	39 26.23	19 17.6
17	9					33.1			40 17.69	1.75	VI.	5	6.22	23 8.3	19.7	40 19.44	39 28.0
18	9							24.7	40 38.34	1.74	VII.	3	5.2	12 27.1	19.6	40 40.08	28 46.7
19	7						17.6		41 47.14	1.72	VI.	1	5.32	2 42.4	19.4	41 48.86	19 1.8
20	7.6					7.0			42 51.57	1.74	VI.	4	6.49	18 21.8	19.3	42 53.31	34 41.1
21	9				5.0				44 4.83	1.77	V.	7	10.34	34 19.1	19.1	44 6.60	50 38.2
22	9						1.1		17 44 30.55	+ 1.73	VI.	2	12.45	−11 21.6	− 19.0	17 44 32.28	− 35 27 40.6

CORRECTIONS.

Date.		Corr. of Clock.	Hourly rate.	m	n	c
1846.	h.	s.	s.	s.	s.	s.
July 14,	13	− 23.27	+ 0.104	+ 0.272	− 0.039	+ 0.130
	15	− 10.53	+ 0.051	+ 0.272	− 0.031	+ 0.130

INSTRUMENT READINGS.

	Date.	Barom.	Thermom. At.	Thermom. Ex.
	1846. h. m.	in.	°	°
Zone 44	July 14, 17 20	30.01	76.0	65.5
	14, 18 31	30.02	74.0	61.2

REMARKS.

(44) 14. Differs 5s and 20" from No. 6 of Mural Zone of same date.
(44) 15. Differs 1m from observation April 17.

ZONE 44. JULY 14. K. D$_0$ = −35° 16′ 0″—Continued.

No.	Mag.	Seconds of Transit. I.	II.	III.	IV.	V.	VI.	VII.	T.	a_1	Micrometer.		r.	$i + d_0$	d_1	Mean Right Ascension, 1850.0.	Mean Declination, 1850.0.
									h. m. s.	s.			r.	′ ″	″	h. m. s.	° ′ ″
23	6.7	1.9							17 47 48.57 +	1.78	II.	9	7.7	−42 34.1 −	18.5	17 47 50.35	− 35 58 52.6
24	8	14.4							49 1.09	1.78	II.	9	10.38	44 20.9	18.3	49 2.87	36 0 39.2
25	8			4.3					49 19.85	1.78	III.	9	9.25	43 44.1	18.3	49 21.63	36 0 2.4
26	8	12.4							51 58.79	1.75	II.	5	6.29	23 11.8	17.9	52 0.54	35 39 29.7
27	9	20.8							53 7.15	1.74	II.	4	10.10	30 3.4	17.7	53 8.89	36 21.1
28	6	56.0	11.6						54 42.65	1.78	III.	8	7.21	37 41.2	17.4	54 44.43	53 58.6
29	10					11.0			54 55.54	1.73	VI.	2	7.12	8 33.1	17.4	54 57.27	24 50.5
30	10						5.9		55 35.29	1.75	VII.	5	9.21	24 38.5	17.3	55 37.04	35 40 55.8
31	6							59.7	56 13.69	1.78	VII.	9	12.29	45 16.8	17.2	56 15.47	36 1 34.0
32	10					9.9			57 54.45	1.74	VI.	3	3.44	11 48.0	16.9	57 56.19	35 28 4.9
33	8			3.0					59 18.20	1.76	III.	6	6.13	27 5.9	16.7	59 20.02	43 22.6
34	10				55.0				17 59 54.83	1.74	IV.	3	9.51	14 53.9	16.6	17 59 56.57	31 10.5
35	10	23.3							18 4 9.52	1.73	II.	2	11.4	10 30.5	16.0	18 4 11.25	26 46.5
36	10	18.8							5 5.36	1.77	II.	7	11.58	35 1.4	15.8	5 7.13	51 17.2
37	7.8		13.0						5 43.75	1.75	III.	5	8.40	24 18.3	15.7	5 45.50	40 34.0
38	9		24.2						9 55.14	1.78	III.	8	12.47	40 26.2	15.1	9 56.92	35 56 41.3
39	9						22.1		9 51.36	1.79	VI.	10	7.3	47 34.1	15.1	9 53.15	36 3 49.2
40	6			58.6					15 14.22	1.79	III.	10	7.30	47 37.8	14.2	15 16.01	36 3 52.0
41	9			29.8					16 44.70	1.72	III.	1	11.36	5 46.8	14.0	16 46.42	35 22 0.8
42	9						23.7		16 53.20	1.73	VI.	2	7.8	8 31.0	13.9	16 54.93	35 24 44.9
43	8					15.3			17 59.99	1.79	VI.	10	6.15	47 9.8	13.8	18 1.78	36 3 23.6
44	10					18.9			19 3.50	1.75	VI.	5	8.1	23 58.4	13.6	19 5.25	35 40 12.0
45	9			35.6				38.5	. . . .	. .	IV.	2	7.43	8 49.0			
46	10					1.0			20 45.68	1.78	VII.	9	5.9	41 34.0	13.3	20 47.46	57 47.3
47	6.7				29.4	44.6			22 29.18	1.73	VII.	2	11.36	10 46.3	13.0	22 30.91	26 59.3
48	7	35.3							26 21.60	1.74	II.	3	14.48	17 24.0	12.4	26 23.34	33 36.4
49	9		6.1						29 36.65	1.73	II.	2	7.53	8 53.8	11.9	29 38.38	25 5.7
50	9						3.2		18 31 32.58 +	1.76	VI.	6	6.55	−27 27.0 −	11.6	18 31 34.34	− 35 43 38.6

ZONE 45. JULY 14. K. D$_0$ = −33° 51′ 30″.

No.	Mag.	I.	II.	III.	IV.	V.	VI.	VII.	T.	a_1	Micrometer.		r.	$i + d_0$	d_1	Mean Right Ascension, 1850.0.	Mean Declination, 1850.0.
1	10				16.3				18 41 16.14 +	1.43	IV.	2	11.39	−10 49.2 −	27.9	18 41 17.57	− 34 2 47.1
2	7.8		37.5						46 7.48	1.45	III.	7	11.6	34 34.8	26.7	46 8.93	26 31.5
3	7				28.2				46 28.04	1.45	V.	7	12.26	35 15.2	26.6	46 29.49	27 11.8
4	7					20.3			47 5.44	1.46	VI.	8	10.5	39 3.3	26.5	47 6.90	30 59.8
5	9		10.0						48 39.70	1.42	III.	2	12.38	11 19.9	26.1	48 41.12	3 16.0
6	9			15.0					40 29.58	1.42	IV.	3	5.10	12 32.4	25.6	50 31.00	4 28.0
7	7.8				9.6				51 9.44	1.45	VI.	7	10.11	34 6.8	25.5	51 10.89	34 26 2.3
8	8		45.0						53 14.63	1.41	III.	1	10.44	5 21.5	24.9	53 16.04	33 57 16.4
9	9			49.7					55 4.58	1.45	IV.	7	4.33	31 16.1	24.5	55 6.03	34 23 10.6
10	9					44.8			54 29.85	1.43	VI.	5	4.24	22 8.7	24.4	55 31.28	14 3.1
11	9					34.5			56 19.55	1.43	VI.	5	3.31	21 41.9	24.2	56 20.98	13 36.1
12	9			44.1					18 58 59.18	1.46	III.	9	8.6	43 3.2	23.5	18 59 0.64	34 56.7
13	9			41.7					19 1 56.79	1.46	IV.	9	10.20	44 11.0	22.8	19 1 58.25	36 3.8
14	8				37.9				2 37.74	1.43	VI.	6	4.58	26 27.7	22.6	2 39.17	18 20.3
15	9	52.5							6 37.60	1.42	II.	6	6.36	27 17.3	21.6	6 39.02	19 8.9
16	9					58.9			19 6 44.05 +	1.45	V.	9	5.13	−41 35.7 −	21.6	19 6 45.50	− 31 33 27.3

CORRECTIONS.

Date.		Corr. of Clock.	Hourly rate.	m	n	c
1846.	h.	s.	s.	s.	s.	s.

INSTRUMENT READINGS.

	Date.	Barom.	Thermom. At.	Thermom. Ex.
	1846. h. m.	in.	°	°
Zone 45	July 14, 18 41	30.02	74.0	61.0
	14, 19 49	30.3	73.0	58.8

REMARKS.

ZONE 45. JULY 14. K. $D_0 = -33° 51' 30''$—Continued.

No.	Mag.	I.	II.	III.	IV.	V.	VI.	VII.	T.	a_1	MICROMETER.			$i + d_1$	d_1	Mean Right Ascension, 1850.0.	Mean Declination, 1850.0.
									h. m. s.	s.			r.	' ''	''	h. m. s.	° ' ''
17	8	. .	3.9	. .	. .	. .	. .	. .	19 8 33.86	+ 1.42	III.	7	8.2	–33 1.7	– 21.1	19 8 35.28	– 34 24 52.8
18	7	. .	. .	56.5	. .	. .	. .	. .	9 11.53	1.44	III.	9	2.50	40 23.4	20.9	9 12.97	32 14.3
19	8	. .	. .	. .	. .	. .	48.6	. .	9 18.64	1.42	VII.	7	9.9	33 35.2	20.9	9 20.26	25 26.1
20	9	. .	. .	. .	7.2	. .	. .	. .	11 7.04	1.41	V.	6	6.29	27 13.9	20.5	11 8.45	19 4.4
21	8	. .	. .	51.9	. .	. .	. .	. .	13 6.59	1.39	IV.	4	7.56	18 56.1	20.0	13 7.98	10 46.1
22	7.8	. .	. .	. .	14.2	. .	. .	. .	21 14.04	1.37	V.	2	6.45	8 20.4	18.0	21 15.41	0 8.4
23	9	. .	. .	. .	. .	. .	19.0	. .	22 49.28	1.38	VII.	4	12.3	21 0.5	17.6	22 50.66	12 48.1
24	9	. .	. .	. .	. .	29.7	. .	. .	24 14.83	1.40	VII.	8	4.15	36 6.0	17.1	24 16.23	27 53.1
25	8	. .	. .	. .	. .	. .	. .	53.9	25 8.93	1.36	VII.	3	9.49	14 52.9	17.0	25 10.29	6 39.9
26	8	. .	. .	. .	. .	. .	. .	46.8	26 1.87	1.36	VII.	3	13.48	16 53.8	16.8	26 3.23	8 40.6
27	7	. .	. .	. .	. .	. .	. .	39.5	26 54.46	1.36	VII.	2	8.18	9 7.0	16.6	26 55.82	0 53.6
28	7	. .	. .	. .	. .	32.7	. .	. .	28 17.75	1.37	VI.	5	5.28	22 41.1	16.2	28 19.12	14 27.3
29	9	48.1	. .	. .	. .	. .	. .	. .	31 33.11	1.36	II.	4	11.28	20 43.1	15.4	31 34.47	12 28.5
30	9	50.3	. .	. .	. .	. .	. .	. .	40 35.41	1.35	III.	6	8.47	28 23.6	13.2	40 36.76	34 20 6.8
31	8	. .	. .	. .	. .	. .	37.1	. .	41 7.40	1.33	VII.	2	4.29	7 11.2	13.1	41 8.73	33 58 54.3
32	9	. .	. .	50.0	. .	. .	. .	. .	43 4.65	1.33	IV.	4	3.45	16 49.2	12.6	43 5.98	34 8 31.8
33	9	. .	. .	47.0	. .	. .	. .	. .	44 1.56	1.32	IV.	3	1.56	10 54.3	12.4	44 2.88	2 36.7
34	6.7	41.3	. .	11.6	. .	. .	. .	. .	45? 26.52	. .	IV.	7	10.21	34 12.1	12.1	. . .	25 54.2
35	9	13.0	. .	. .	. .	. .	. .	. .	48 57.87	1.30	II.	2	9.41	9 49.3	11.2	48 59.17	1 30.5
36	9	. .	20.9	. .	. .	. .	. .	. .	19 49 50.63	+ 1.30	III.	3	8.38	–14 17.5	– 11.0	19 49 51.93	– 34 5 58.5

ZONE 46. JULY 15. K. $D_0 = -29° 42' 20''$.

No.	Mag.	I.	II.	III.	IV.	V.	VI.	VII.	T.	a_1	MICROMETER.			$i + d_1$	d_1	Mean Right Ascension, 1850.0.	Mean Declination, 1850.0.
1	7	. .	. .	33.4	. .	. .	. .	. .	16 43 47.74	– 12.83	III.	6	7.12	–27 35.5	– 22.7	16 43 34.91	– 30 10 18.2
2	9	. .	. .	. .	23.6	. .	. .	. .	44 23.44	12.83	IV.	6	4.43	26 20.3	22.3	44 10.61	9 2.6
3	6.7	. .	. .	. .	15.2	. .	. .	. .	45 15.04	12.82	IV.	8	6.51	37 24.9	21.6	45 2.22	20 6.5
4	8.9	33.4	48.1	. .	. .	. .	. .	. .	47 17.10	12.82	III.	8	6.52	37 25.4	19.9	47 4.28	30 20 5.3
5	9	. .	18.8	33.4	. .	. .	. .	. .	48 47.56	12.84	IV.	3	11.39	15 49.4	18.6	48 34.72	29 58 28.0
6	9	. .	. .	. .	. .	. .	45.7	. .	49 17.03	12.84	VII.	2	10.22	10 10.3	18.2	49 4.19	29 52 48.5
7	9	. .	. .	. .	. .	18.1	. .	. .	51 3.61	12.83	V.	4	5.29	17 42.0	16.7	50 50.78	30 0 18.7
8	9	. .	17.1	. .	. .	. .	. .	. .	16 52 46.17	– 12.80	IV.	9	7.54	–42 56.4	– 15.3	16 52 33.37	– 30 25 31.7

ZONE 47. JULY 15. K. $D_0 = -37° 51' 30''$.

No.	Mag.	I.	II.	III.	IV.	V.	VI.	VII.	T.	a_1	MICROMETER.			$i + d_1$	d_1	Mean Right Ascension, 1850.0.	Mean Declination, 1850.0.
1	8	. .	43.5	. .	. .	. .	. .	. .	17 44 15.54	– 11.99	III.	9	8.3	–43 4.3	. .	17 44 3.55	
2	9	. .	. .	16.6	. .	. .	. .	. .	45 32.36	12.03	IV.	3	10.4	14 59.6	. .	45 20.33	
3	8	41.0	. .	. .	. .	. .	. .	. .	47 28.72	12.03	II.	3	3.43	11 46.3	. .	47 16.69	
4	10	. .	. .	41.4	. .	. .	. .	. .	47 57.07	12.01	IV.	5	2.41	21 17.9	. .	47 45.06	
5	7	. .	29.9	. .	. .	. .	. .	. .	49 1.90	11.98	III.	9	4.21	41 11.0	. .	48 49.92	
6	8	35.9	. .	. .	. .	. .	. .	. .	50 23.89	12.01	II.	6	9.4	28 32.5	. .	50 11.88	
7	10	. .	21.0	. .	. .	. .	. .	. .	50 52.62	12.02	III.	4	7.42	18 48.1	. .	50 40.60	
8	7	. .	. .	10.0	. .	. .	. .	. .	51 25.52	12.02	IV.	3	4.52	12 21.5	. .	51 13.49	
9	8	30.8	. .	. .	. .	. .	. .	. .	53 18.56	12.03	II.	3	8.13	14 3.1	. .	53 6.53	
10	9	. .	46.8	. .	. .	. .	. .	. .	57 18.47	12.01	III.	4	12.40	21 19.1	. .	57 6.46	
11	8	. .	. .	. .	45.9	. .	. .	. .	57 45.73	11.97	V.	9	7.28	42 46.6	. .	57 33.76	
12	10	. .	. .	. .	. .	. .	53.0	. .	58 21.38	12.00	VI.	5	3.54	21 53.1	. .	58 9.38	
13	8	. .	. .	. .	. .	. .	39.9	. .	59 8.09	11.97	VI.	10	3.29	45 47.7	. .	58 56.12	
14	9	. .	. .	. .	. .	. .	31.7	. .	17 59 59.90	– 11.97	VI.	9	9.21	–43 43.6	. .	17 59 47.93	

CORRECTIONS.

Date.	Corr. of Clock.	Hourly rate.	m	n	c
1846. h.	s.	s.	s.	s.	s.
July 15. 15	– 24.52	+ 0.063	+ 0.241	– 0.008	+ 0.130

INSTRUMENT READINGS.

	Date.		Barom.	THERMOM. At.	THERMOM. Ex.
	1846.	h. m.	In.	°	°
Zone 46	July 15.	16 43	30.23	69.6	60.3
		16 52	30.23	69.0	60.3
Zone 47	July 15.	17 44	30.23	68.4	60.3
		18 59	30 23	66.0	58.0

REMARKS.

ZONE 47. JULY 15. K. D₀ = −37° 51′ 30″—Continued.

No.	Mag.	I.	II.	III.	IV.	V.	VI.	VII.	T.	a_1	Micrometer.			$i + d_2$	d_1	Mean Right Ascension, 1850.0.	Mean Declination, 1850.0.
									h. m. s.	s.			r.	′ ″	″	h. m. s.	° ′ ″
15	8		35.3						18 2 6.96	− 12.01	III.	4	11.44	−20 50.8	. .	18 1 54.95	
16	7.8	10.7		43.0					3 58.85	11.99	IV.	7	11.50	34 58.5	. .	3 46.86	
17	6.7		39.3						7 10.98	12.00	III.	4	13.14	21 36.4	. .	6 58.98	
18	8		36.9						8 8.63	11.99	III.	5	9.50	24 53.7	. .	7 56.64	
19	7.8	22.0	38.3						12 10.31	11.96	III.	9	11.27	44 47.8	. .	11 58.35	
20	7			32.8					14 48.94	11.96	IV.	10	6.17	47 13.2	. .	14 36.98	
21	8		26.7						17 58.60	11.97	III.	7	11.23	34 44.8	. .	17 46.63	
22	6			25.1					18 40.66	12.00	IV.	3	10.27	15 11.2	. .	18 28.66	
23	8				9.1				22 8.93	11.99	IV.	4	10.40	20 18.4	. .	21 56.94	
24	9			3.8					25 19.26	12.02	IV.	2	9.1	9 27.0	. .	25 7.24	
25	7			57.1					26 12.81	11.97	IV.	5	7.33	23 44.3	. .	26 0.84	
26	9	4.3							28 51.97	12.01	III.	2	6.29	8 10.0	. .	28 39.96	
27	5			25.3	41.3				33 41.19	11.98	IV.	8	4.20	36 10.7	. .	33 29.21	
28	9				36.0				34 35.83	12.00	V.	5	9.37	24 47.1	. .	34 23.83	
29	9		42.4						42 13.97	12.03	III.	3	11.25	15 40.6	. .	42 1.94	
30	9	7.8							47 55.81	12.02	II.	6	11.16	29 39.4	. .	47 43.79	
31	6.5	26.9		59.0					52 14.98	12.01	III.	7	14.41	36 25.1	. .	52 2.97	
32	9.8	11.0							54 58.84	12.04	II.	4	8.23	19 8.7	. .	54 46.80	
33	9	42.6							56 30.60	12.03	II.	6	10.42	29 22.2	. .	56 18.57	
34	6.7			40.0					56 55.46	12.05	IV.	2	9.20	9 36.7	. .	56 43.41	
35	5.4		55.9		28.0				18 59 27.66	− 12.05	IV.	4	3.4	−16 27.3	. .	18 59 15.61	

ZONE 48. JULY 24. K. D₀ = −31° 34′ 30″.

No.	Mag.	I.	II.	III.	IV.	V.	VI.	VII.	T.	a_1	Micrometer.			$i + d_2$	d_1	Mean Right Ascension, 1850.0.	Mean Declination, 1850.0.
1	6.7						34.0		21 20 4.74	+ 0.58	VI.	4	7.00	−18 27.7	− 19.5	21 20 5.32	− 31 53 17.2
2	8					5.4			22 50.71	0.59	VI.	8	6.58	37 28.5	19.1	22 51.30	32 12 17.6
3	10			53.4					25 8.31	0.58	IV.	9	11.25	44 43.4	18.6	25 8.89	32 19 32.2
4	8		37.2		6.8				33 6.62	0.52	IV.	5	3.18	21 35.6	17.7	33 7.14	31 56 23.3
5	8			24.6			8.3		35 39.09	0.51	VII.	4	12.5	21 1.6	17.3	35 39.60	55 48.9
6	8			59.3		28.3			47 13.50	0.43	VI.	1	6.25	3 10.8	15.8	47 13.93	31 37 56.6
7	8						20.8		47 51.48	0.46	VII.	7	4.4	31 0.8	15.7	47 51.94	32 5 46.5
8	9	39.7	54.6						50 24.28	0.46	III.	9	3.53	40 54.9	15.4	50 24.74	15 40.3
9	7			54.8	9.7				51 9.58	0.45	V.	8	11.16	39 39.1	15.3	51 10.03	14 24.4
10	10				19.3				52 19.14	0.45	VI.	8	9.28	38 44.3	15.2	52 19.59	13 29.5
11	8	46.0		18.0					54 32.65	0.43	IV.	7	12.20	35 12.0	14.9	54 33.08	9 56.9
12	9						11.3		54 41.88	0.44	VII.	9	11.38	44 49.5	14.9	54 42.37	19 34.4
13	8				52.3	7.0			21 57 52.24	0.43	V.	9	8.10	43 4.8	14.5	21 57 52.67	17 49.3
14	11			2.6					22 3 17.32	0.39	III.	7	7.11	32 35.8	13.9	22 3 17.71	7 19.7
15	9			59.9					4 14.73	0.36	IV.	8	11.33	39 47.7	13.7	5 15.11	14 31.4
16	10			22.1					7 36.82	0.37	IV.	7	8.47	33 24.4	13.5	7 37.19	32 8 7.9
17	9			5.4	20.1				9 19.91	0.34	VI.	4	8.3	18 59.6	13.3	9 20.25	31 53 42.0
18	10		7.2						13 36.56	0.32	III.	4	6.28	18 11.7	12.9	13 36.88	52 54.6
19	8			12.8	27.7				15 27.58	0.34	V.	8	10.46	39 23.9	12.7	15 27.92	31 14 6.6
20	9	14.0	29.0						18 58.53	0.31	III.	7	8.27	33 14.2	12.4	18 58.84	32 7 56.6
21	7					13.0			18 58.34	0.32	VII.	9	10.22	44 11.1	12.4	18 58.66	18 53.5
22	9		48.8						22 18.46	0.30	III.	9	9.12	43 36.1	12.0	22 18.76	32 18 18.0
23	7					45.5			23 30.71	0.27	VII.	3	5.37	12 45.8	12.0	22 30.98	31 47 27.8
24	6					49.8			22 23 35.05	+ 0.27	VII.	5	2.56	−21 24.0	− 11.9	22 23 35.32	− 31 56 5.9

CORRECTIONS.

Date.		Corr. of Clock.	Hourly rate.	m	n	c
1846.	h.	s.	s.	s.	s.	s.
July 24.	15	− 10.85	+ 0.063	+ 0.572	− 0.411	0.000

INSTRUMENT READINGS.

	Date.		Barom.	Thermom. At.	Thermom. Ex.
	1846.	h. m.	in.	°	°
Zone 48	July 24,	21 20	30.07	63.9	73.0
		22 45	29.92	75.8	71.5

REMARKS.

(48) 15. Minutes assumed as 5 instead of 4.

ZONE 48. JULY 24. K. $D_o = -31°\ 34'\ 30''$—Continued.

No.	Mag.	Seconds of Transit. I.	II.	III.	IV.	V.	VI.	VII.	T.		a_1	Micrometer.			$i+d_1$		d_1	Mean Right Ascension, 1850.0.	Mean Declination, 1850.0.
									h. m. s.		s.			r.	′ ″		″	h. m. s.	° ′ ″
25	10	..	35.2	..	5.0	..	..	..	22 36 4.71	+	0.26	IV.	4	10.14	−20 6.0	−	10.9	22 36 4.97	−31 54 46.9
26	8	11.7	..	..	..	..	..	..	40 56.12		0.19	V.	5	9.53	24 55.2		10.6	40 56.31	59 35.8
27	11	..	20.8	..	..	..	..	..	22 45 50.05	+	0.16	IV.	2	5.41	−7 48.5	−	10.3	22 45 50.21	−31 42 28.8

ZONE 49. JULY 29. A. $D_o = -30°\ 19'\ 0''$.

No.	Mag.	I.	II.	III.	IV.	V.	VI.	VII.	T.		a_1	Micrometer.			$i+d_1$		d_1	Mean Right Ascension, 1850.0.	Mean Declination, 1850.0.
1	11	..	..	..	2.0	16.0	31.0	..	19 52 1.79	+	5.54	VI.	6	5.56	−26 57.0	−	44.1	19 52 7.33	−30 46 41.1
2	10	..	..	..	..	..	33.0	47.5	53 3.98		5.56	V.	7	6.	31 59.8		43.6	53 9.54	51 43.4
3	7	..	..	..	56.2	10.5	..	..	54 55.97		5.52	V.	3	9.23	15 10.8		42.8	55 1.49	34 53.6
4	11	..	..	..	31.5	..	..	..	56 31.34		5.52	IV.	4	6.38	18 16.9		42.1	56 36.86	37 59.0
5	11	..	..	..	7.0	21.5	..	..	58 6.87		5.51	V.	4	2.50	16 21.7		41.4	58 12.38	36 3.1
6	7	..	..	..	28.5	43.0	..	..	19 59 28.40		5.52	V.	7	5.56	31 57.8		40.8	19 59 33.92	51 38.6
7	10	..	..	..	..	1.5	15.8	..	20 0 46.92		5.50	VI.	4	10.	19 58.8		40.2	20 0 52.42	39 39.0
8	10	..	4.5	19.2	..	..	..	..	3 33.77		5.50	III.	8	11.25	39 43.4		39.0	3 39.27	59 22.4
9	9	..	..	..	..	21.2	36.0	..	4 6.88		5.48	VII.	4	12.	20 59.2		38.8	4 12.36	40 38.0
10	6	..	56.5	11.5	..	..	..	..	6 25.50		5.44	III.	2	6.10	8 5.4		37.8	6 30.94	27 43.2
11	11	..	..	20.5	..	..	..	..	7 35.11		5.48	III.	8	9.30	38 45.3		37.3	7 40.59	58 22.6
12	8	..	..	..	47.8	2.0	..	..	8 47.56		5.47	V.	7	10.2	34 2.1		36.8	8 53.03	53 38.9
13	7	..	..	..	..	44.8	59.0	14.0	12 30.25		5.45	VII.	7	4.56	31 27.0		35.2	12 35.70	51 2.2
14	8	..	..	..	46.5	0.9	15.3	..	14 46.35		5.44	VI.	8	8.45	38 22.4		34.2	14 51.79	57 56.6
15	11	..	..	58.5	..	..	..	..	17 12.70		5.40	III.	3	6.30	13 13.2		33.2	17 18.10	32 46.4
16	8	..	..	..	..	..	..	19.0	17 35.31		5.41	VII.	5	6.54	23 24.4		33.0	17 40.72	42 57.4
17	8	..	38.5	53.0	..	..	..	..	20 7.59		5.41	III.	7	9.24	33 42.9		31.9	20 13.00	53 14.8
18	9	..	5.2	20.0	..	..	..	..	22 34.27		5.38	III.	5	2.19	21 5.8		30.9	22 39.65	40 36.7
19	8	33.2	48.2	..	..	..	..	..	24 17.09		5.37	II.	5	4.2	21 57.8		30.1	24 22.46	41 27.9
20	8	5.0	19.8	..	..	..	..	..	25 48.68		5.35	II.	3	1.30	10 41.4		29.5	25 54.03	30 30 10.9
21	8	..	..	..	..	50.0	4.5	..	27 35.50		5.38	VI.	9	10.	43 59.9		28.7	26 40.88	31 3 28.6
22	8	..	..	..	49.0	..	..	..	28 48.84		5.38	IV.	9	12.	45 0.8		28.2	28 54.22	31 4 29.0
23	9	..	35.2	50.0	..	..	..	..	32 4.45		5.35	III.	7	11.5	34 33.9		26.8	32 9.80	30 54 0.7
24	9	..	..	..	4.0	18.0	..	..	33 3.67		5.35	V.	9	8.5	43 2.0		26.4	33 9.02	31 2 28.4
25	7	..	..	52.0	6.7	..	..	..	36 6.59		5.33	IV.	9	4.50	41 23.5		25.2	36 11.92	31 0 48.7
26	8	18.1	33.0	47.7	..	..	..	..	40 2.02		5.28	III.	5	9.57	24 57.3		23.5	40 7.30	30 44 20.8
27	10	..	..	47.0	1.5	..	..	..	41 1.47		5.29	IV.	8	9.53	38 56.9		23.1	41 6.76	58 20.0
28	8	..	..	..	..	15.0	29.8	44.2	46 0.61		5.24	VII.	5	6.17	23 5.7		21.1	46 5.85	30 42 26.8
29	7	..	..	..	..	..	41.5	56.0	50 12.53		5.25	VII.	10	3.55	45 56.8		19.3	50 17.78	31 5 16.1
30	9	..	15.8	30.2	..	..	..	..	52 44.82		5.23	III.	7	7.40	32 50.3		18.3	52 49.05	30 52 8.6
31	9	34.0	49.0	..	..	..	..	..	56 18.05		5.21	II.	7	11.15	34 38.8		16.9	56 23.26	53 55.7
32	5	..	..	..	52.0	6.2	20.3	..	56 51.62		5.19	VI.	5	7.32	23 43.8		16.6	56 56.81	43 0.4
33	9	..	17.5	32.0	..	..	..	..	20 59 46.39		5.17	III.	4	8.40	19 18.5		15.5	20 59 51.56	38 34.0
34	11	..	..	28.0	42.5	..	..	..	21 4 42.40	+	5.15	IV.	6	10.34	−29 17.7	−	13.5	21 4 47.55	−30 48 31.2

CORRECTIONS.

Date.	Corr. of Clock.	Hourly rate.	m	n	c
1846. h.	s.	s.	s.	s.	s.
July 29, 15	− 5.82	− 0.015	+ 0.572	− 0.411	0.000

INSTRUMENT READINGS.

Date.		Barom.	Thermom. At.	Thermom. Ex.
1846.	h. m.	in.	°	°
Zone 49 July 29,	19 52	29.92	77.0	69.5
	21 4	29.90	76.8	68.8

REMARKS.

(49) 3. Micrometer reading assumed as 10r.23 instead of 9r.23.
(49) 21. Minutes assumed as 26 instead of 27.

ZONE 50. AUGUST 3. A. $D_0 = -27° 50' 40''$.

No.	Mag.	Seconds of Transit. I.	II.	III.	IV.	V.	VI.	VII.	T.		a_1	Micrometer.			$i + d_3$		d_1	Mean Right Ascension, 1850.0.	Mean Declination, 1850.0.
									h. m. s.		s.			r.	′ ″		″	h. m. s.	° ′ ″
1	7	..	..	2.5	16.0	30.0	..	..	21 42 1.99	+	3.72	V.	10	6.35	−47 17.6	−	12.3	21 42 5.71	− 28 38 9.9
2	8	..	..	..	..	..	15.7	30.0	43 47.42		3.67	VII.	2	10.22	10 10.6		12.2	43 51.09	1 2.8
3	7	11.5	26.0	40.2	..	..	..	..	50 54.33		3.64	III.	6	11.38	29 49.8		11.4	50 57.97	20 41.2
4	9	..	..	..	..	36.5	..	4.5	50 22.54		3.63	VII.	4	6.27	18 11.1		11.4	51 25.67	9 2.5
5	6	..	35.4	49.9	3.8	..	..	..	21 54 3.67		3.61	IV.	3	8.45	14 21.7		11.1	21 54 7.28	5 12.8
6	9	..	..	19.5	33.5	..	..	..	22 1 33.47		3.58	IV.	7	8.36	33 18.5		10.3	22 1 37.05	28 24 8.8
7	4	..	..	..	53.5	7.2	21.2	..	2 53.14		3.55	VI.	1	4.53	2 25.2		10.2	2 56.69	27 53 15.4
8	9	..	..	..	5.0	18.5	..	..	5 4.57		3.55	V.	6	12.	20 59.7		10.0	5 8.12	28 11 49.7
9	5	..	..	..	..	..	12.5	26.8	5 44.37		3.56	VII.	8	11.7	39 33.6		9.9	5 47.93	30 23.5
10	7	2.0	16.5	30.8	..	..	..	..	8 44.71		3.52	III.	4	4.29	17 11.9		9.7	8 48.23	8 1.6
11	7	34.2	49.0	3.3	..	..	..	..	11 17.12		3.50	III.	4	5.8	17 31.6		9.4	11 20.62	8 21.0
12	9	41.0	55.5	..	..	..	..	..	14 23.85		3.50	II.	6	7.35	27 47.0		9.2	14 27.35	18 36.2
13	6	..	..	..	..	14.5	..	..	15 0.38		3.53	V.	10	7.28	47 44.3		9.1	15 3.91	38 33.4
14	9	..	13.0	27.0	..	..	..	..	17 41.04		3.48	III.	3	10.33	15 16.2		8.9	17 44.52	28 6 5.1
15	8	..	..	30.0	44.5	..	..	..	19 44.03		3.46	IV.	2	4.57	7 26.9		8.7	19 47.49	27 58 15.6
16	8	..	..	51.0	5.0	..	..	..	33 4.81		3.39	IV.	3	3.8	11 30.5		7.8	33 8.20	28 2 18.3
17	7	..	..	..	..	11.5	25.0	39.6	33 57.03		3.38	VII.	4	8.17	19 6.7		7.7	34 0.41	28 9 54.4
18	9	..	..	..	37.0	51.0	..	..	35 36.78		3.38	V.	2	6.57	8 27.5		7.6	35 40.16	27 59 15.1
19	8	21.8	36.0	..	..	..	..	..	39 4.41		3.36	II.	4	8.47	19 22.0		7.4	39 7.77	28 10 9.4
20	6	..	41.0	55.3	9.5	..	..	..	40 9.38		3.36	IV.	6	12.	30 1.0		7.3	40 12.74	28 20 48.3
21	6	..	..	48.0	2.0	..	..	..	51 1.74		3.27	IV.	1	6.10	3 4.3		6.8	51 5.01	27 53 51.1
22	5	..	..	25.6	39.7	53.6	..	..	22 56 39.38		3.25	V.	1	11.44	5 52.9		6.5	22 56 42.63	27 56 39.4
23	9	..	..	52.0	6.0	..	..	..	23 2 5.83		3.23	IV.	3	10.30	15 14.7		6.3	23 2 9.06	28 6 1.0
24	9	..	8.5	23.0	37.0	..	..	..	7 36.77		3.20	IV.	2	8.	8 59.3		6.2	7 39.97	59 45.5
25	6	..	..	..	..	39.0	53.0	..	23 8 24.84	+	3.18	VI.	1	6.32	− 3 15.2	−	6.1	23 8 28.02	− 27 54 1.3

ZONE 51. AUGUST 5. A. $D_0 = -27° 48' 0''$.

No.	Mag.	I.	II.	III.	IV.	V.	VI.	VII.	T.		a_1	Micrometer.			$i + d_3$		d_1	Mean Right Ascension, 1850.0.	Mean Declination, 1850.0.
1	8	..	..	..	..	..	9.5	24.5	17 10 41.56	+	4.75	VII.	2	5.6	−12 33.7	−	33.0	17 10 46.31	− 28 1 6.7
2	7	..	..	..	47.0	1.3	..	..	12 46.92		4.74	V.	1	4.50	2 23.9		32.2	12 51.66	27 50 56.1
3	5	..	..	..	..	5.0	19.1	33.5	13 50.83		4.74	VII.	3	2.5	10 59.3		31.9	13 55.57	27 59 31.2
4	8	..	..	..	..	..	..	32.5	14 50.11		4.79	VII.	9	5.55	41 55.4		31.5	14 54.90	28 30 26.9
5	7	..	..	29.5	43.5	..	..	..	16 43.44		4.76	IV.	6	8.4	28 1.8		30.9	16 48.20	16 32.7
6	8	31.0	45.4	..	..	..	..	..	18 13.77		4.75	II.	5	9.52	24 54.6		30.4	18 18.52	13 25.0
7	9	..	..	..	..	..	..	52.0	18 9.26		4.73	VII.	3	8.29	14 13.2		30.4	18 13.99	2 43.6
8	9	..	..	..	9.0	23.0	..	..	20 8.85		4.78	VII.	9	6.3	41 59.5		29.7	20 13.63	30 29.2
9	8	55.5	10.2	..	..	..	..	..	25 38.34		4.73	II.	4	4.42	17 18.5		27.8	25 43.07	5 46.3
10	7	28.7	43.0	..	..	..	..	..	28 11.49		4.74	II.	7	5.48	31 53.4		26.9	28 16.23	28 20 20.3
11	6	..	..	..	..	55.2	8.8	..	29 40.83		4.71	VI.	2	7.6	8 32.0		26.4	28 45.54	27 56 58.4
12	7	..	..	..	..	..	56.5	11.0	30 28.41		4.74	VII.	7	3.5	30 30.8		26.2	29 33.15	28 18 57.0
13	8	17.2	31.8	..	..	..	..	..	33 0.07		4.72	II.	5	10.58	25 27.0		25.3	33 4.79	13 52.3
14	8	..	36.2	50.6	..	..	..	..	35 4.48		4.71	III.	4	5.50	17 52.8		24.6	35 9.19	6 17.4
15	7	..	..	49.5	4.0	..	..	..	36 3.71		4.73	IV.	7	5.28	31 43.4		24.2	36 8.44	20 7.6
16	8	..	38.2	53.0	..	..	..	..	38 6.91		4.73	III.	8	3.42	35 49.1		23.5	[illegible] 11.64	24 12.6
17	8	..	49.5	4.0	..	..	..	..	39 17.89		4.73	III.	5	6.11	23 3.1		23.1	39 22.62	11 26.2
18	10	..	..	..	..	..	40.0	..	40 11.84		4.72	VI.	7	7.12	32 35.8		22.8	40 16.56	20 58.6
19	8	54.0	8.5	22.5	..	..	..	..	42 36.85		4.72	III.	8	3.55	35 55.7		22.0	42 41.57	24 17.7
20	4	26.5	41.0	55.5	9.2	..	..	..	17 47 9.18	+	4.67	IV.	3	7.27	−13 42.4	−	20.5	17 47 13.85	− 28 2 2.9

CORRECTIONS.

Date.		Corr. of Clock.	Hourly rate.	m	n	c
1846.	h.	s.	s.	s.	s.	s.
Aug. 3,	15	− 6.67	− 0.005	+ 0.372	− 0.163	0.000
Aug. 5,	18	− 6.86	− 0.002	+ 0.372	− 0.163	0.000

INSTRUMENT READINGS.

	Date.		Barom.	Thermom. At.	Thermom. Ex.
	1846.	h. m.	in.	°	°
Zone 50	Aug. 3,	21 42	30.23	74.2	60.5
		23 8	30.22	74.5	60.0
Zone 51	Aug. 5,	17 10	30.05	82.0	77.0
		18 43	30.04	80.0	75.5

REMARKS.

(50) 1. Transits over T.'s IV–VI assumed to have been recorded as over T.'s III–V.
(50) 4. Minutes assumed as 51 instead of 50.
(50) 8. Hor. thread assumed as 4 instead of 6.
(51) 1. Micrometer assumed as $15^r.6$ instead of $5^r.6$.
(51) 11. Minutes assumed as 28 instead of 29.
(51) 12. Minutes assumed as 29 instead of 30.
(51) 13. Micrometer assumed as $6^r.58$ instead of $10^r.58$.

ZONE 51. AUGUST 5. A. $D_0 = -27° 48' 0''$—Continued.

No.	Mag.	Seconds of Transit I.	II.	III.	IV.	V.	VI.	VII.	T.	a_1	Micrometer.			$i + d_2$	d_1	Mean Right Ascension, 1850.0.	Mean Declination, 1850.0.
									h. m. s.	s.			r.	′ ″	″	h. m. s.	° ′ ″
21	9	. .	8.0	22.8	. .	. .	. .	. .	17 48 36.49	+ 4.68	III.	4	7.50	−18 53.4	− 19.9	17 48 41.17	− 28 7 13.3
22	9	. .	. .	19.2	33.5	. .	. .	. .	52 33.33	4.68	IV.	7	11.25	34 43.8	18.6	52 38.01	28 23 2.4
23	6	. .	. .	. .	. .	37.2	51.5	. .	53 23.19	4.64	VI.	1	0.30	0 12.4	18.3	53 27.83	27 48 30.7
24	7	. .	. .	35.8	49.8	. .	. .	. .	55 49.53	4.64	IV.	1	3.35	1 46.1	17.5	55 54.17	27 50 3.6
25	6	. .	. .	. .	12.2	26.3	40.4	. .	57 12.14	4.67	VI.	7	9.50	33 55.6	17.0	57 16.81	28 22 12.6
26	4	. .	. .	. .	. .	. .	58.5	12.4	17 58 30.16	4.68	VII.	8	11.32	39 46.2	19.8	17 58 34.84	28 6.0
27	9	. .	. .	45.2	59.5	. .	. .	. .	18 0 59.23	4.65	IV.	5	4.46	22 20.2	19.0	18 1 3.88	10 39.2
28	9	. .	. .	. .	. .	. .	. .	8.2	1 25.69	4.66	VII.	7	7.4	32 31.6	18.8	1 30.35	20 50.4
29	7	. .	. .	50.0	4.4	. .	. .	. .	3 4.13	4.64	IV.	6	7.9	27 40.0	18.3	3 8.77	15 52.3
30	10	. .	. .	. .	. .	. .	. .	7.5	3 24.99	4.65	VII.	7	7.5	32 32.1	18.2	3 29.64	20 50.3
31	8	. .	. .	. .	4.5	18.8	. .	. .	5 4.44	4.62	V.	3	5.29	12 42.7	17.6	5 9.06	1 0.3
32	8	. .	. .	. .	. .	20.5	34.5	49.2	6 6.39	4.62	VII.	4	7.44	18 50.0	17.3	6 11.01	7 7.3
33	5	. .	. .	. .	49.0	3.0	. .	. .	7 48.83	4.64	V.	7	4.42	31 20.2	16.7	7 53.47	19 36.9
34	9	. .	. .	16.0	30.5	. .	. .	. .	9 30.18	4.62	IV.	6	7.49	27 54.2	16.2	9 34.80	16 10.4
35	8	40.4	55.0	. .	. .	. .	. .	. .	11 23.25	4.62	II.	5	5.7	22 30.7	15.6	11 27.87	10 46.3
36	5	. .	57.5	12.0	26.2	. .	. .	. .	12 26.10	4.64	IV.	9	4.38	41 17.0	15.2	12 30.74	29 32.2
37	9	. .	. .	. .	. .	. .	. .	30.5	12 47.93	4.61	VII.	6	6.2	26 59.7	15.1	12 52.54	15 14.8
38	8	. .	. .	. .	31.6	45.5	. .	. .	18 31.39	4.62	V.	8	11.21	39 41.0	13.2	18 36.01	27 54.2
39	8	. .	54.0	8.4	22.5	. .	. .	. .	21 22.27	4.57	IV.	3	9.3	14 30.8	12.4	21 26.84	2 43.2
40	7	. .	. .	. .	19.0	32.5	. .	. .	22 18.61	4.61	V.	10	8.38	48 17.2	12.0	22 23.22	36 29.2
41	9	. .	. .	. .	58.0	. .	26.0	. .	23 57.86	4.58	VI.	5	8.59	24 27.9	11.5	24 2.44	12 39.4
42	7	. .	. .	. .	. .	. .	. .	22.0	24 39.57	4.59	VII.	8	9.58	38 58.7	11.3	24 44.16	27 10.0
43	7	. .	. .	. .	. .	18.8	33.2	47.2	26 4.81	4.59	VII.	8	8.14	38 6.1	10.8	26 9.40	26 16.9
44	5	. .	. .	. .	. .	. .	. .	11.5	27 29.19	4.59	VII.	10	10.38	49 19.9	10.4	27 33.78	37 30.3
45	7	. .	. .	27.5	41.8	. .	. .	. .	29 41.59	4.55	IV.	6	12.	30 1.0	9.7	29 46.14	18 10.7
46	8	. .	4.5	18.8	33.1	. .	. .	. .	31 32.95	4.56	IV.	8	2.25	35 10.3	9.1	31 37.51	23 19.4
47	8	. .	18.0	32.7	. .	. .	. .	. .	33 46.80	4.57	III.	10	6.21	47 10.4	8.3	33 51.37	35 18.7
48	8	. .	8.8	23.3	37.6	. .	. .	. .	35 37.37	4.54	IV.	7	10.5	34 3.4	7.7	35 41.91	22 11.1
49	7	22.2	36.8	51.2	. .	. .	. .	. .	38 5.34	4.55	III.	8	8.13	38 6.0	7.0	38 9.89	26 13.0
50	9	. .	44.6	59.6	13.8	. .	. .	. .	18 43 27.77	+ 4.52	IV.	7	4.15	−31 6.6	− 5.3	18 43 32.29	− 28 19 11.9

ZONE 52. AUGUST 8. A. $D_0 = -27° 47' 0''$.

No.	Mag.	I.	II.	III.	IV.	V.	VI.	VII.	T.	a_1	Micrometer.			$i + d_2$	d_1	Mean Right Ascension, 1850.0.	Mean Declination, 1850.0.
1	8	. .	. .	. .	. .	. .	25.5	40.2	18 44 57.43	+ 4.65	VII.	3	6.21	−13 8.6	− 43.7	18 45 2.08	− 28 0 52.3
2	7	. .	. .	. .	. .	42.5	56.5	10.8	46 28.31	4.68	VII.	7	4.28	31 11.5	43.4	46 32.99	18 54.9
3	9	. .	. .	. .	. .	. .	4.2	18.7	47 36.11	4.67	VII.	7	3.18	30 37.4	43.2	47 40.78	28 18 20.6
4	7	. .	. .	17.5	31.0	45.5	. .	. .	49 31.06	4.62	V.	1	6.20	3 9.4	42.8	49 35.68	27 50 52.2
5	6	. .	. .	. .	. .	44.5	58.3	12.9	50 30.26	4.65	VII.	6	6.37	27 17.4	42.6	50 34.91	28 15 0.0
6	8	. .	51.0	5.5	. .	. .	. .	. .	54 19.20	4.61	III.	2	6.28	8 12.4	41.9	54 23.81	27 55 54.3
7	9	. .	17.8	32.2	. .	. .	. .	. .	55 46.26	4.64	III.	7	6.16	32 7.7	41.6	55 50.90	28 19 49.3
8	10	. .	. .	. .	. .	. .	34.5	48.8	56 6.32	4.64	VII.	7	10.46	34 23.7	41.5	56 10.96	28 22 5.2
9	10	. .	46.0	0.5	14.6	. .	. .	. .	18 59 14.29	4.59	IV.	2	7.39	8 48.8	40.9	18 59 18.88	27 56 29.7
10	8	42.6	57.0	. .	. .	. .	. .	. .	19 1 25.43	4.62	II.	7	5.33	31 45.8	40.5	19 1 30.05	28 19 26.3
11	9	. .	24.6	38.[illegible]	. .	. .	. .	. .	2 53.01	4.62	III.	8	3.3	35 29.4	40.2	2 57.63	23 9.6
12	10	40.5	54.5	. .	. .	. .	. .	. .	5 23.11	4.59	II.	6	10.15	29 7.7	39.7	5 27.70	16 47.4
13	10	. .	44.5	59.2	. .	. .	. .	. .	6 12.95	4.58	III.	4	8.22	19 9.6	39.6	6 17.53	6 49.2
14	9	. .	. .	50.6	4.5	. .	. .	. .	9 4.65	4.56	IV.	4	3.58	16 56.3	38.9	9 9.21	4 35.2
15	11	. .	. .	. .	. .	59.0	13.5	. .	19 9 45.08	+ 4.56	VI.	4	7.56	−18 56.3	− 38.9	19 9 49.64	− 28 6 35.2

CORRECTIONS.

Date.		Corr. of Clock.	Hourly rate.	m	n	c
1846.	h.	s.	s.	s.	s.	s.
Aug. 8,	18	− 6.72	+ 0.009	+ 0.372	− 0.163	0.000

INSTRUMENT READINGS.

	Date.		Barom.	Thermom. At.	Thermom. Ex.
	1846.	h. m.	In.	°	°
Zone 52	Aug. 8,	18 45	30.01	83.2	79.2
		21 16	29.9	82.0	75.5

REMARKS.

(51) 26. The instrument apparently moved a little in declination.
(51) 50. Transits over T.'s I–III assumed as recorded over T.'s II–V.
(52) 14. Transits over T.'s III and IV assumed as 50s.6 and 5s.0 instead of 30s.6 and 45s.0, to agree with Arg. Z. 231, 17; 241, 13; and 394, 2.

ZONE 52. AUGUST 8. A. $D_0 = -27° 47' 0''$—Continued.

No.	Mag.	Seconds of Transit. I.	II.	III.	IV.	V.	VI.	VII.	T.	a_1	Micrometer.			$i + d_1$	d_1	Mean Right Ascension, 1850.0.	Mean Declination, 1850.0.
									h. m. s.	s.			r.	′ ″	″	h. m. s.	° ′ ″
16	8	..	..	..	..	24.5	38.5	52.6	19 11 10.24	+ 4.56	VII.	6	9.15	−28 37.2	− 38.6	19 11 14.80	− 28 16 15.8
17	9	..	37.5	51.6	..	..	..	..	14 5.59	4.53	III.	3	11.28	15 44.0	38.1	14 10.12	3 22.1
18	4	..	..	50.5	4.5	18.8	32.5	47.1	15 4.44	4.55	VII.	5	3.5	21 28.8	37.9	15 8.99	9 6.7
19	11	..	..	..	32.0	..	..	..	21 31.84	4.50	IV.	4	9.8	19 32.9	36.7	21 36.34	7 9.6
20	10	..	..	..	..	..	..	37.0	21 54.44	4.51	VII.	6	8.17	28 7.9	36.6	21 58.95	15 44.5
21	6	..	..	..	..	29.0	43.0	57.0	23 14.71	4.52	VII.	7	3.7	30 31.8	36.4	23 19.23	18 8.2
22	8	..	..	..	..	..	..	48.8	24 6.30	4.51	VII.	7	8.20	33 10.0	36.2	24 10.81	20 46.2
23	9	..	..	..	..	50.0	4.0	..	25 35.83	4.50	VI.	7	8.2	33 1.1	35.9	25 40.33	28 20 37.0
24	6	..	..	..	35.5	49.5	..	..	27 35.27	4.46	V.	1	4.12	2 4.7	35.5	27 39.73	27 49 40.2
25	9	..	..	..	1.5	15.8	29.8	..	29 1.54	4.46	VI.	4	9.8	19 32.7	35.3	29 6.00	28 7 8.0
26	8	..	..	24.0	37.8	..	..	..	30 37.65	4.44	IV.	1	9.32	4 46.4	35.0	30 42.09	27 52 21.4
27	5	..	..	..	26.8	41.0	54.9	..	31 26.74	4.44	VI.	3	9.4	14 31.2	34.8	31 31.18	28 2 6.0
28	8	..	..	..	..	..	34.2	48.8	36 6.10	4.43	VII.	4	3.36	16 44.8	34.0	36 10.53	4 18.8
29	8	..	3.2	17.5	31.5	..	..	..	38 31.40	4.42	IV.	4	6.23	18 9.5	33.5	38 35.82	5 43.0
30	8	21.0	35.6	49.8	..	..	..	..	42 3.95	4.42	III.	7	10.31	34 16.5	32.9	42 8.37	21 49.4
31	8	..	..	8.8	23.5	..	..	..	43 23.20	4.44	IV.	9	9.29	43 43.9	32.7	43 27.64	28 31 16.6
32	8	..	..	..	..	56.2	10.0	..	44 41.94	4.37	VI.	1	9.41	4 50.7	32.4	44 46.31	27 52 23.1
33	8	..	..	3.0	17.2	31.3	..	..	47 16.96	4.37	V.	3	6.59	13 28.2	32.0	47 21.33	28 1 0.2
34	9	..	50.0	4.2	..	..	..	..	49 18.21	4.36	III.	4	10.18	20 8.2	31.6	49 22.57	7 39.8
35	8	..	..	41.0	55.2	9.5	..	..	50 55.04	4.35	V.	3	9.35	14 47.0	31.3	50 59.39	2 18.3
36	7	..	18.5	33.0	..	..	..	..	52 47.02	4.38	III.	8	5.43	36 50.2	31.0	52 51.40	24 21.2
37	3	..	..	..	..	35.8	49.7	4.1	53 21.53	4.35	VII.	4	9.46	19 51.7	30.9	53 25.88	7 22.6
38	7	..	..	..	8.5	22.6	36.5	..	55 8.37	4.35	VI.	5	4.43	22 18.5	30.6	55 12.72	9 49.1
39	6	..	..	..	..	..	28.2	42.5	55 59.99	4.34	VII.	6	5.0	26 28.4	30.5	56 4.33	13 58.9
40	8	4.5	19.5	..	..	..	..	..	19 59 47.68	4.34	II.	8	2.51	35 23.2	29.8	19 59 52.02	22 53.0
41	6	..	..	..	..	17.5	31.6	45.9	20 0 3.40	4.34	VII.	8	3.30	35 42.7	29.8	20 0 7.74	23 12.5
42	9	..	..	..	27.5	41.5	..	..	3 27.34	4.32	V.	7	10.14	34 7.9	29.2	3 31.66	21 37.1
43	8	..	55.0	9.1	23.8	..	..	..	6 23.53	4.32	IV.	9	10.29	44 14.4	28.7	6 27.85	31 43.1
44	8	..	..	..	..	34.0	48.0	2.8	7 19.92	4.28	VII.	4	2.22	16 7.4	28.5	7 24.20	3 35.9
45	9	..	..	..	57.8	11.5	25.5	40.0	10 57.44	4.28	VII.	7	4.8	31 2.7	28.0	11 1.72	18 30.7
46	8	..	..	..	..	18.5	32.0	46.5	15 4.08	4.27	VII.	9	4.22	41 8.5	27.3	15 8.35	28 35.8
47	9	..	..	..	..	23.0	36.8	51.5	18 8.74	4.20	VII.	3	7.10	13 33.3	26.8	18 12.94	28 1 0.1
48	5	..	..	..	29.3	43.5	57.0	..	20 29.11	4.18	VII.	1	1.39	0 51.5	26.4	20 33.29	27 48 17.9
49	8	..	..	0.0	14.0	28.2	..	..	22 13.95	4.20	V.	6	6.48	27 23.3	26.1	22 18.15	28 14 49.4
50	11	..	..	50.5	..	..	..	..	24 4.44	4.19	III.	5	6.52	23 23.8	25.8	24 8.63	10 49.6
51	9	..	54.5	..	23.0	..	..	..	26 22.94	4.21	IV.	9	6.34	42 15.6	25.5	26 27.15	29 41.1
52	11	..	..	41.0	..	..	..	..	28 55.09	4.18	III.	7	5.55	31 57.1	25.1	28 59.27	19 22.2
53	7	4.6	19.2	..	..	..	..	..	30 47.52	4.16	II.	6	6.00	26 59.0	24.8	30 51.68	14 23.8
54	7	..	59.8	14.2	28.5	..	..	..	31 28.19	4.14	IV.	4	8.37	19 17.2	24.7	31 32.33	6 41.9
55	6	..	..	..	..	38.8	52.8	6.9	32 24.59	4.18	VII.	9	10.19	44 8.8	24.5	32 28.77	31 33.3
56	7	..	18.7	33.0	17.5	..	..	..	34 47.25	4.16	IV.	8	11.15	39 38.0	24.2	34 51.41	28 27 2.2
57	8	..	..	..	..	..	..	13.8	35 31.02	4.11	VII.	3	2.23	11 8.4	24.1	35 35.13	27 58 32.5
58	7	28.5	42.9	57.2	..	..	..	..	38 11.27	4.12	III.	6	6.15	27 6.6	23.7	38 15.39	28 14 30.3
59	5	..	..	..	47.0	1.0	15.2	..	38 46.90	4.12	VI.	7	3.	30 28.6	23.6	38 51.02	28 17 52.2
60	6	..	33.0	47.5	..	..	..	..	41 1.20	4.07	III.	2	5.35	7 46.1	23.3	41 5.27	27 55 9.4
61	5	..	..	..	30.5	44.5	..	..	42 30.36	4.12	V.	10	3.35	45 46.5	23.0	42 34.48	28 33 9.5
62	7	..	..	..	14.5	28.5	..	..	44 14.36	4.11	V.	10	7.28	47 44.2	22.8	44 18.47	35 7.0
63	5	..	..	..	..	..	..	44.5	45 2.10	4.10	VII.	9	5.45	41 50.4	22.7	45 6.20	29 13.1
64	8	55.0	9.2	23.8	..	..	..	..	20 48 37.79	+ 4.06	III.	7	5.26	−31 42.4	− 22.2	20 48 41.85	− 28 19 4.6

CORRECTIONS.

Date.		Corr. of Clock.	Hourly rate.	m	n	c
1846.	h.	s.	s.	s.	s.	s.

INSTRUMENT READINGS.

Date.		Barom.	Thermom. At.	Thermom. Ex.
1846.	h. m.	in.	°	°

REMARKS.

(52) 45. Declination apparently 15″ too large.

ZONE 52. AUGUST 8. A. $D_o = -27° 47' 0''$—Continued.

No.	Mag.	SECONDS OF TRANSIT. I. II. III. IV. V. VI. VII.	T.	a_1	MICROMETER.			$i + d_1$	d_1	Mean Right Ascension, 1850.0.	Mean Declination, 1850.0.
			h. m. s.	s.			r.	' "	"	h. m. s.	° ' "
65	11	. . . . . . 41.5	20 49 41.34	+ 4.02	IV.	2	4.45	− 7 20.9	− 22.0	20 49 45.36	− 27 54 42.9
66	7	. . 49.6 4.4 18.3	51 18.08	4.01	IV.	2	5.46	7 51.7	21.8	51 22.09	27 55 13.5
67	7	. . . . . . 31.2 45.0 59.3 . .	52 31.00	4.05	VI.	8	3.37	35 46.4	21.6	52 35.05	28 23 8.0
68	5	. . . . . . 9.2 23.2 37.1 . .	54 9.00	4.03	VI.	7	5.27	31 42.8	21.4	54 13.03	19 4.2
69	8	. . . . 3.5	55 20.83	4.00	VII.	4	10.00	19 58.7	21.3	55 24.83	7 20.0
70	8	. . 23.0 37.2 51.8	20 57 51.42	4.00	IV.	6	8.20	28 9.8	20.9	20 57 55.42	15 30.7
71	7	. . 50.8 4.8 19.4	21 3 19.22	3.99	IV.	9	8.51	43 24.8	20.2	21 3 23.21	30 45.0
72	4	. . . . . . 19.5 33.5 47.2 1.5	4 19.15	3.96	VII.	6	4.46	26 21.4	20.1	4 23.11	28 13 41.5
73	5	. . 29.5 44.0 58.0	10 57.71	3.89	IV.	1	6.5	3 1.8	19.3	11 1.60	27 50 21.1
74	8	. . . . . . 14.5 28.0 42.5 56.5	12 14.13	3.93	VII.	8	1.25	34 39.6	19.1	12 18.06	28 21 58.7
75	8	43.8 58.3 12.5	21 16 26.66	+ 3.90	III.	7	6.26	−32 12.7	− 18.6	21 16 30.56	− 28 19 31.3

ZONE 53. AUGUST 11. A. $D_o = -25° 17' 0''$.

No.	Mag.	SECONDS OF TRANSIT. I. II. III. IV. V. VI. VII.	T.	a_1	MICROMETER.			$i + d_1$	d_1	Mean Right Ascension, 1850.0.	Mean Declination, 1850.0.
1	8	. . . . 56.0	17 31 14.25	+ 4.04	VII.	4	6.56	−13 27.0	− 111.0	17 31 18.29	− 25 32 18.0
2	9	. . . . 21.5 34.5 . .	33 7.25	4.04	VI.	5	6.21	23 7.9	110.4	33 11.29	41 58.3
3	10	49.0 3.5	35 30.97	4.03	II.	4	11.10	20 34.4	109.6	35 35.00	39 24.0
4	8	39.0 53.8	37 21.14	4.03	II.	5	5.21	22 37.8	109.0	37 25.17	41 26.8
5	11	. . . . 11.5	37 29.82	4.03	VII.	5	7.26	23 40.7	108.9	37 33.85	42 29.6
6	7	. . 11.2 25.0	39 38.80	4.02	III.	5	5.43	22 49.0	108.2	39 42.82	41 37.2
7	8	. . . . . . 15.0 28.8 42.0 . .	40 14.72	4.04	VI.	8	5.45	36 50.9	108.0	40 18.76	55 38.9
8	7	51.5 5.5	42 33.27	4.01	II.	5	9.43	24 50.1	107.3	42 37.28	43 37.4
9	10	. . . . 7.0 21.5	42 39.63	4.01	VII.	6	4.12	26 4.2	107.2	42 43.64	44 51.4
10	8	. . . . 25.2 39.2	43 57.63	4.03	VII.	8	10.00	38 59.5	106.8	44 1.66	57 46.3
11	9	. . . . 23.0 37.0	44 55.42	4.03	VII.	8	5.36	36 46.2	106.5	44 59.45	55 32.7
12	6	. . 26.2 40.0 54.0	46 53.79	3.99	IV.	4	9.30	19 44.1	105.8	46 57.78	38 29.9
13	6	. . . . . . 7.0 21.5	48 7.18	3.97	V.	1	6.00	2 59.6	105.4	48 11.15	25 21 45.0
14	7	. . . . 17.0 31.0 . .	49 3.26	4.02	VI.	10	3.00	45 28.4	105.1	49 7.28	26 4 13.5
15	6	. . . . 35.8	49 54.19	3.99	VII.	6	10.12	29 6.0	104.8	49 58.16	25 47 50.8
16	11	. . 51.0 5.5	52 18.88	3.97	III.	4	4.52	17 23.6	104.0	52 22.85	25 36 7.6
17	8	. . . . 21.0	53 7.18	4.01	V.	10	7.53	47 26.2	103.8	53 11.19	26 6 10.0
18	7	. . . . 13.0 27.0 41.0	53 59.33	4.00	VII.	8	3.48	35 51.6	103.5	54 3.33	25 54 35.1
19	6	. . . . 41.2 54.8 9.2	55 27.32	3.96	VII.	4	5.30	17 42.5	103.0	55 31.28	36 25.5
20	7	. . . . . . 12.5 26.0 40.0 . .	57 12.29	3.95	VI.	2	10.20	10 10.2	102.4	57 16.24	28 52.6
21	7	. . . . 41.0 55.2	58 13.44	3.95	VII.	3	12.	16 0.0	102.1	58 17.39	34 42.1
22	7	. . . . 42.0	17 59 0.34	3.96	VII.	5	10.16	25 6.6	101.8	17 59 4.30	43 48.4
23	5	1.5 15.5 29.5	18 1 43.30	3.96	III.	6	8.56	28 27.9	100.9	18 1 47.26	25 47 8.8
24	8	. . . . 29.5 43.3 56.8	2 15.57	3.97	VII.	9	8.56	43 26.6	100.7	2 19.54	26 2 7.3
25	7	7.8 21.8 35.8	5 49.64	3.95	III.	7	4.5	31 1.4	99.6	5 53.59	25 49 41.0
26	8	. . . . . . 36.0 50.0	6 35.98	3.94	V.	6	4.00	26 28.7	99.3	6 39.92	45 8.0
27	8	. . . . 25.5 39.8	6 57.99	3.93	VII.	4	3.48	16 51.0	99.2	7 1.92	35 30.2
28	5	. . . . 7.0 20.8 34.7	9 20.70	3.91	V.	4	11.15	20 37.0	98.4	9 24.61	39 15.4
29	7	. . . . 30.0	9 48.22	3.90	VII.	3	11.37	15 48.4	98.2	9 52.12	34 26.6
30	9	. . . . 41.5 55.0 . .	11 27.51	3.92	VI.	6	10.58	29 29.4	97.7	11 31.43	48 7.1
31	9	. . . . 18.0	12 36.40	3.92	VII.	7	2.37	30 16.8	97.3	12 40.32	25 48 54.1
32	8	. . . . 53.0 6.5 . .	15 39.00	3.93	VI.	9	6.39	42 17.6	96.3	15 42.93	26 0 53.9
33	8	. . . . . . 17.2 31.0	17 17.10	3.92	V.	9	4.29	41 12.1	95.8	17 21.02	25 59 47.9
34	10	. . . . 52.5 6.5 . .	18 18 38.75	+ 3.89	VI.	6	6.46	−27 22.2	− 95.3	18 18 42.64	− 25 45 57.5

CORRECTIONS.

Date.	Corr. of Clock.	Hourly rate.	m	n	c
1846. h.	s.	s.	s.	s.	s.
Aug. 11, 18	− 7.22	− 0.008	+ 0.372	− 0.163	0.000

INSTRUMENT READINGS.

Date.	Barom.	Thermom. At.	Thermom. Ex.
1846. h. m.	In.	°	°
Zone 53 Aug. 11, 17 31	30.06	78.0	71.5

REMARKS.

(53) 1. Micrometer assumed as wire 3 instead of wire 4.
(53) 17. Micrometer reading assumed as 6r.53 instead of 7r.53.
(53) 26. Micrometer reading assumed as 5r.00 instead of 4r.00.

ZONE 53. AUGUST 11. A. $D_0 = -25° 17' 0''$—Continued.

No.	Mag.	Seconds of Transit. I.	II.	III.	IV.	V.	VI.	VII.	T.	a_1	Micrometer.			$i + d_0$	d_1.	Mean Right Ascension, 1850.0.	Mean Declination, 1850.0.
									h. m. s.	s.			r.	′ ″	″	h. m. s.	° ′ ″
35	7	. .	55.0	9.5	. .	. .	. .	. .	18 20 22.91	+ 3.87	III.	4	9.3	−20 0.6	− 94.7	18 20 26.78	− 25 38 35.3
36	7	. .	. .	. .	59.2	. .	27.0	. .	20 59.24	3.87	VI.	4	9.54	19 56.0	94.5	21 3.11	38 30.5
37	7	. .	. .	. .	19.5	33.5	. .	. .	22 19.48	3.87	V.	6	9.28	28 44.1	94.1	22 23.35	47 18.2
38	11	. .	. .	1.5	16.0	. .	. .	. .	24 15.58	3.86	IV.	6	11.45	29 53.3	93.4	24 19.44	48 26.7
39	8	. .	. .	. .	. .	. .	15.5	29.8	24 47.96	3.84	VII.	2	11.6	10 33.2	93.3	24 51.80	29 6.5
40	10	. .	. .	. .	. .	. .	47.0	1.0	27 19.37	3.85	VII.	5	10.5	25 1.0	92.4	26 23.22	43 33.4
41	6	. .	. .	. .	50.5	4.0	18.0	. .	27 50.29	3.85	VI.	6	8.19	28 9.1	92.3	27 54.44	46 41.4
42	6	. .	. .	. .	. .	. .	23.3	37.5	28 55.73	3.83	VII.	3	8.16	14 6.9	91.9	28 59.56	32 38.8
43	8	. .	. .	. .	32.0	46.0	0.0	13.5	30 32.04	3.83	VII.	4	8.59	19 28.0	91.4	30 35.87	37 59.4
44	9	. .	. .	. .	. .	. .	. .	41.0	31 59.21	3.81	VII.	3	10.21	15 10.0	90.9	32 3.02	33 40.9
45	9	. .	. .	. .	. .	47.0	0.8	14.6	34 33.07	3.81	VII.	4	9.28	19 42.7	90.0	34 36.88	38 12.7
46	9	. .	. .	. .	. .	. .	19.2	33.0	34 51.44	3.81	VII.	4	7.30	18 43.1	89.9	34 55.25	37 13.0
47	8	. .	. .	46.5	. .	13.5	. .	. .	37 59.72	3.78	V.	1	7.35	3 47.6	88.9	38 3.50	22 16.5
48	8	. .	. .	. .	20.5	. .	48.0	. .	40 20.38	3.80	VI.	6	8.8	28 3.6	88.1	40 24.18	46 31.7
49	8	. .	. .	50.0	3.5	. .	. .	. .	42 3.38	3.76	IV.	1	6.27	3 13.3	87.5	42 7.14	21 40.8
50	7	9.5	23.3	. .	. .	. .	. .	. .	47 51.12	3.75	II.	4	11.41	20 50.0	85.6	47 54.87	39 15.6
51	11	. .	. .	. .	. .	50.5	4.0	18.5	48 36.59	3.75	VII.	4	4.44	17 19.3	85.3	48 40.34	35 41.6
52	8	. .	. .	6.0	20.0	. .	. .	. .	50 19.75	3.74	IV.	4	8.32	19 14.8	84.8	50 23.49	37 39.6
53	8	. .	. .	. .	40.5	54.0	. .	. .	51 40.24	3.76	V.	8	6.33	37 15.3	84.3	51 44.00	55 39.6
54	7	. .	. .	51.5	5.5	. .	. .	. .	53 5.37	3.76	IV.	8	3.3	35 29.3	83.8	53 9.13	53 53.1
55	8	. .	. .	. .	26.5	. .	. .	. .	54 26.35	3.70	IV.	1	8.27	4 13.9	83.4	54 30.05	22 37.3
56	5	. .	. .	. .	29.0	43.0	. .	. .	55 28.94	3.71	V.	2	7.0	8 29.3	83.0	55 32.65	25 26 52.3
57	8	. .	. .	. .	. .	25.5	39.5	. .	57 11.75	3.75	VI.	9	10.12	44 5.2	82.5	57 15.50	26 2 27.7
58	6	. .	. .	. .	36.2	50.0	3.7	. .	58 36.09	3.74	VI.	8	6.57	37 27.3	82.0	58 40.83	25 55 49.3
59	9	. .	. .	. .	. .	. .	17.0	31.2	18 59 49.48	3.71	VII.	5	9.55	24 56.0	81.6	18 59 53.19	43 17.6
60	8	. .	39.0	53.0	. .	. .	. .	. .	19 2 6.76	3.70	III.	6	4.26	26 11.6	80.9	19 2 10.46	44 32.5
61	8	. .	57.8	11.5	. .	. .	. .	. .	3 25.52	3.69	III.	8	5.00	36 28.3	80.5	3 29.21	54 48.8
62	8	. .	. .	. .	. .	18.0	31.5	46.0	4 4.13	3.69	VII.	6	8.55	28 27.1	80.2	4 7.82	46 47.3
63	10	. .	. .	. .	. .	. .	. .	56.0	5 14.24	3.68	VII.	4	4.50	17 22.3	79.9	5 17.92	35 42.2
64	5	. .	. .	40.0	54.5	. .	. .	. .	6 54.13	3.69	IV.	8	6.7	37 2.2	79.3	6 57.82	55 21.5
65	8	. .	. .	. .	. .	. .	35.0	50.0	7 7.84	3.67	VII.	4	5.20	17 37.4	79.3	7 11.51	35 56.7
66	9	. .	. .	. .	. .	55.5	9.2	. .	10 41.61	3.64	VI.	3	7.35	13 46.4	78.1	10 45.25	32 4.5
67	7	. .	. .	28.5	42.0	. .	. .	. .	11 42.08	3.65	IV.	6	12.0	30 0.9	77.8	11 45.73	25 48 18.7
68	8	. .	. .	. .	. .	53.0	6.5	. .	12 39.00	3.67	VI.	9	5.57	41 56.4	77.5	12 42.67	26 0 13.9
69	6	. .	. .	. .	. .	8.0	22.0	36.0	13 54.29	3.64	VII.	6	5.45	26 51.2	77.0	13 57.93	25 45 8.2
70	9	. .	. .	. .	. .	. .	. .	33.5	14 51.77	3.63	VII.	4	9.8	19 32.6	76.7	14 55.40	37 49.3
71	8	47.5	1.5	. .	. .	. .	. .	. .	17 29.07	3.63	II.	6	10.25	34 13.3	75.9	17 32.70	52 29.2
72	8	19.2	33.3	. .	. .	. .	. .	. .	20 1.05	3.61	II.	6	7.40	27 49.4	75.1	20 4.66	46 4.5
73	7	. .	30.5	44.5	58.5	. .	. .	. .	20 58.29	3.61	IV.	6	4.38	26 17.7	74.7	21 1.90	25 44 32.4
74	5	. .	. .	. .	1.5	15.0	29.0	. .	22 1.29	3.62	VI.	9	10.52	44 25.4	74.4	22 4.91	26 2 39.8
75	7	. .	28.0	42.2	. .	. .	. .	. .	23 55.87	3.59	III.	6	7.45	27 52.1	73.8	23 59.46	25 46 5.9
76	8	. .	. .	. .	15.5	. .	. .	. .	24 45.35	3.61	IV.	9	8.21	43 9.4	73.5	24 48.96	26 1 22.9
77	10	. .	. .	. .	. .	48.5	. .	. .	26 4.62	3.59	V.	7	2.44	30 20.5	73.1	26 8.21	25 48 33.6
78	6	. .	. .	. .	. .	4.5	18.0	32.4	27 50.50	3.55	VII.	2	2.17	6 6.1	72.5	27 54.05	24 18.6
79	9	. .	. .	. .	. .	. .	4.5	18.0	29 36.67	3.57	VII.	8	4.32	36 13.9	72.0	29 40.24	54 25.9
80	9	26.0	40.0	. .	. .	. .	. .	. .	33 7.74	3.54	II.	5	4.24	22 9.0	70.8	33 11.28	40 19.8
81	8	. .	. .	31.5	46.0	. .	. .	. .	33 45.61	3.56	IV.	7	11.22	34 42.1	70.6	33 49.17	52 52.7
82	10	57.5	11.5	25.5	. .	. .	. .	. .	35 39.18	3.53	III.	5	3.22	21 37.8	70.0	35 42.71	39 47.8
83	9	. .	. .	. .	. .	. .	27.7	41.5	19 35 59.98	+ 3.53	VII.	6	3.31	−25 43.5	− 69.9	19 36 3.51	− 25 43 53.4

CORRECTIONS.

Date.		Corr. of Clock.	Hourly rate.	m	n	c
1846.	h.	s.	s.	s.	s.	s.

INSTRUMENT READINGS.

Date.		Barom.	Thermom. At.	Ex.
1846.	h. m.	in.	°	°

REMARKS.

(53) 35. Micrometer reading assumed as 10r.3 instead of 9r.3.
(53) 40. Minutes assumed as 26 instead of 27.
(53) 71. Micrometer assumed as wire 7 instead of wire 6.

ZONE 53. AUGUST 11. A. $D_o = -25° 17' 0''$—Continued.

No.	Mag.	Seconds of Transit. I.	II.	III.	IV.	V.	VI.	VII.	T.	a_1	Micrometer.			$i + d_2$	d_1	Mean Right Ascension, 1850.0.	Mean Declination, 1850.0.
									h. m. s.	s.			r.	′ ″	″	h. m. s.	° ′ ″
84	7	45.6	0.0	14.1	. .	. .	. .	. .	19 38 27.85	+ 3.54	III.	9	4.40	−41 17.7	− 69.1	19 38 31.39	− 25 59 26.8
85	8	47.4	2.0	15.7	. .	. .	. .	. .	41 29.43	3.49	III.	5	5.19	22 36.9	68.2	41 32.92	40 45.1
86	7	15.5	30.1	44.0	. .	. .	. .	. .	43 57.71	3.50	III.	7	6.50	32 24.8	67.4	44 1.21	50 32.2
87	11	. .	. .	. .	. .	. .	29.5	. .	46 1.87	3.48	VI.	8	4.56	36 26.2	66.8	46 5.35	54 33.0
88	8	. .	51.8	5.5	19.7	. .	. .	. .	48 19.36	3.45	IV.	4	2.39	16 16.5	66.0	48 22.81	34 22.5
89	9	. .	. .	. .	. .	. .	22.5	37.0	19 49 55.07	+ 3.43	VII.	3	3.4	−11 29.3	− 65.5	19 49 58.50	− 25 29 34.8

ZONE 54. AUGUST 11. A. $D_o = -25° 17' 0''$.

No.	Mag.	I.	II.	III.	IV.	V.	VI.	VII.	T.	a_1	Micrometer.			$i + d_2$	d_1	Mean Right Ascension, 1850.0.	Mean Declination, 1850.0.
1	8	32.8	46.8	. .	. .	. .	. .	. .	20 47 14.60	+ 3.14	II.	6	7.38	−27 48.4	− 66.4	20 47 17.74	− 25 45 54.8
2	7	. .	. .	. .	. .	3.2	17.0	31.0	48 49.29	3.11	VII.	2	8.58	9 28.6	65.9	48 52.40	27 34.5
3	7	. .	. .	. .	. .	. .	37.0	50.8	50 9.34	3.13	VII.	8	9.28	38 43.3	65.5	50 12.47	56 48.8
4	7	. .	11.5	26.0	39.5	. .	. .	. .	53 39.32	3.06	IV.	2	9.10	9 35.0	64.5	53 42.33	27 39.5
5	7	. .	. .	. .	. .	39.5	53.0	7.4	55 25.58	3.08	VII.	5	3.30	21 41.6	64.0	54 28.66	39 45.6
6	4	. .	50.0	3.3	17.5	31.6	. .	. .	20 58 17.40	3.06	V.	4	6.10	18 3.0	63.1	20 58 20.46	36 6.1
7	8	. .	53.0	7.0	21.0	. .	. .	. .	21 0 20.61	3.03	IV.	1	10.13	5 7.4	62.6	21 0 23.64	23 10.0
8	8	. .	. .	. .	. .	31.0	44.2	58.5	1 16.78	3.02	IV.	1	12.	7 32.2	62.3	1 19.80	25 34.5
9	11	. .	. .	. .	. .	. .	41.0	54.5	3 13.20	3.06	VII.	9	5.4	41 29.5	61.7	3 16.26	59 31.2
10	9	. .	. .	. .	. .	. .	17.0	31.0	4 49.39	3.04	VII.	7	1.18	29 36.0	61.3	4 52.43	47 37.3
11	9	5.0	19.0	33.0	. .	. .	. .	. .	9 46.71	2.99	III.	5	1.29	20 40.8	59.8	9 49.70	38 40.6
12	7	. .	. .	. .	. .	58.0	11.0	. .	12 43.76	2.95	VI.	1	5.4	2 31.2	59.0	12 46.71	20 30.2
13	6	27.3	41.7	55.7	. .	. .	. .	. .	15 9.41	2.98	III.	7	6.48	32 23.7	58.4	15 12.39	50 22.1
14	9	56.0	10.8	. .	. .	. .	. .	. .	16 38.07	2.94	II.	3	9.16	14 37.4	57.9	16 41.01	32 35.3
15	7	. .	. .	. .	. .	16.0	30.3	44.0	17 2.41	2.97	VII.	7	11.45	34 53.4	57.8	17 5.38	52 51.2
16	8	. .	55.5	9.5	. .	. .	. .	. .	19 23.39	2.96	III.	8	8.7	38 2.8	57.2	19 26.35	56 0.0
17	9	. .	. .	. .	. .	52.0	6.0	20.0	19 38.29	2.94	VII.	6	5.24	26 40.6	57.1	19 41.23	44 37.7
18	6	. .	14.8	28.8	. .	. .	. .	. .	21 42.62	2.94	III.	7	7.35	32 47.4	56.5	21 45.56	50 43.9
19	7	. .	. .	. .	. .	. .	. .	37.5	22 56.02	2.93	VII.	8	10.49	39 24.2	56.2	22 58.95	57 20.4
20	9	. .	. .	47.0	1.0	15.0	. .	. .	25 0.97	2.92	V.	8	4.6	36 1.0	55.6	25 3.89	53 56.6
21	9	. .	. .	. .	. .	. .	15.5	29.5	25 47.85	2.89	VII.	4	8.11	19 3.8	55.4	25 50.74	36 59.2
22	8	16.0	30.0	44.2	. .	. .	. .	. .	28 57.80	2.88	III.	5	6.35	23 15.3	54.6	29 0.68	41 9.9
23	10	. .	56.5	. .	. .	. .	. .	. .	30 24.28	2.88	II.	6	9.50	28 55.1	54.2	30 27.16	46 49.3
24	8	. .	. .	0.5	14.5	28.2	. .	. .	31 14.36	2.88	V.	8	3.46	35 51.0	54.0	31 17.24	53 45.0
25	6	25.0	39.1	53.2	. .	. .	. .	. .	33 6.91	2.86	III.	6	9.47	28 53.7	53.5	33 9.77	46 47.2
26	8	. .	. .	. .	. .	. .	35.0	49.0	34 7.28	2.83	VII.	1	10.8	5 4.5	53.2	34 10.11	22 57.7
27	6	. .	. .	. .	. .	8.0	21.8	36.0	35 54.19	2.83	VII.	4	5.19	17 37.0	52.7	35 57.02	35 29.7
28	8	. .	. .	. .	36.2	50.0	3.5	. .	37 36.02	2.81	VI.	3	5.2	12 29.2	52.3	37 38.83	30 21.5
29	7	. .	. .	. .	. .	. .	49.5	3.0	38 21.63	2.83	VII.	6	7.23	27 40.6	52.1	38 24.46	45 32.7
30	8	34.2	48.5	2.2	. .	. .	. .	. .	41 16.15	2.82	III.	7	7.55	32 57.6	51.3	41 18.97	50 48.9
31	10	18.3	32.5	. .	. .	. .	. .	. .	43 0.19	2.81	II.	6	5.51	26 54.4	50.9	43 3.00	25 44 45.3
32	8	. .	. .	. .	. .	1.0	14.5	. .	43 47.01	2.82	VI.	9	12.	46 30.5	50.7	43 49.83	26 4 21.2
33	9	. .	. .	. .	. .	31.0	44.5	59.0	45 17.15	2.80	VII.	7	6.39	32 18.9	50.3	45 19.95	25 50 9.2
34	8	. .	. .	. .	. .	9.0	23.0	36.8	46 55.22	2.78	VII.	5	10.50	25 23.7	49.9	46 58.00	25 43 13.6
35	7	. .	4.5	18.5	32.8	. .	. .	. .	50 32.54	2.78	IV.	10	4.31	46 14.5	49.0	50 35.32	26 4 3.5
36	8	58.3	12.2	26.4	. .	. .	. .	. .	52 40.07	2.75	III.	6	3.36	25 46.3	48.5	52 42.82	25 43 34.8
37	8	. .	. .	. .	. .	. .	36.0	50.3	53 8.51	2.74	VII.	5	4.22	17 7.1	48.4	53 11.25	34 55.5
38	8	. .	. .	. .	19.0	33.0	46.5	. .	55 18.96	2.72	VI.	3	8.58	14 28.3	47.8	55 21.68	32 16.1
39	9	. .	50.0	4.0	18.0	. .	. .	. .	21 59 17.63	+ 2.70	IV.	2	5.45	− 7 51.5	− 46.9	21 59 20.33	− 25 25 38.4

CORRECTIONS.

Date.	Corr. of Clock.	Hourly rate.	m	n	c
1846. h.	s.	s.	s.	s.	s.

INSTRUMENT READINGS.

	Date.	Barom.	Thermom. At.	Thermom. Ex.
	1846. h. m.	in.	°	°
Zone 54	Aug. 11, 22 29	30.08	76.0	68.0

REMARKS.

(54) 5. Minutes assumed as 54 instead of 55.
(54) 8. Micrometer reading assumed as 15r.00 instead of 12r.00.
(54) 32. Micrometer reading assumed as 15r.00 instead of 12r.00.
(54) 37. Hor. thread assumed as 4 instead of 5.

ZONE 54. AUGUST 11. A. $D_0 = -25° 17' 0''$—Continued.

No.	Mag.	Seconds of Transit I.	II.	III.	IV.	V.	VI.	VII.	T.	a_1	Micrometer.			$i + d_s$	d_1	Mean Right Ascension, 1850.0.	Mean Declination, 1850.0.
									h. m. s.	s.			r.	′ ″	″	h. m. s.	° ′ ″
40	10		12.0						22 1 39.78	+ 2.70	II.	6	11.16	−29 38.5	− 46.3	22 1 42.48	− 25 47 24.8
41	9		27.5	41.4					2 55.14	2.68	III.	5	4.30	22 12.2	46.0	2 57.82	39 58.2
42	10				58.5				3 58.35	2.70	IV.	8	11.32	39 46.3	45.8	4 1.05	57 32.1
43	5					29.4	43.0	56.8	5 15.40	2.69	VII.	8	7.10	37 33.6	45.5	5 18.09	55 19.1
44	9				51.5	5.5	19.0		6 51.45	2.67	VI.	6	4.20	26 8.4	45.1	6 54.12	43 53.5
45	8				22.0	35.5			12 21.74	2.65	V.	7	10.35	34 18.4	43.8	12 24.39	52 2.2
46	9		50.0	3.8					14 17.43	2.61	III.	2	7.58	8 58.6	43.4	14 20.04	26 42.0
47	4			52.5	6.0	19.8			15 5.92	2.60	V.	3	6.48	13 22.8	43.2	15 8.52	31 6.0
48	9		49.5	3.5					17 17.07	2.59	III.	2	12.	11 31.1	42.7	17 19.66	29 13.8
49	10			39.0	52.8				21 52.58	2.56	IV.	2	10.42	10 21.4	41.7	21 55.14	28 3.1
50	9				24.5				23 24.35	2.55	IV.	2	10.30	10 15.4	41.4	23 26.90	27 56.8
51	8				21.5	35.0			24 21.19	2.54	V.	1	10.49	5 25.6	41.2	24 23.73	23 6.8
52	10		10.0						27 37.75	2.54	II.	6	4.29	26 13.0	40.5	27 40.29	43 53.5
53	9				54.0	8.0			27 53.99	2.55	V.	8	3.53	35 54.5	40.4	27 56.54	53 34.9
54	10		13.0	26.8					22 29 40.69	+ 2.53	III.	6	10.15	−29 7.8	− 40.0	22 29 43.22	− 25 46 47.8

ZONE 55. AUGUST 12. K. $D_0 = -30° 56' 10''$.

No.	Mag.	I.	II.	III.	IV.	V.	VI.	VII.	T.	a_1	Micrometer.			$i + d_s$	d_1	Mean Right Ascension, 1850.0.	Mean Declination, 1850.0.
1	9		49.7	4.3	19.0				17 34 18.78	+ 4.55	V.	3	11.22	−15 40.7	− 6.0	17 34 23.33	− 31 11 56.7
2	9	6.5		36.2					37 50.45	4.54	IV.	3	2.45	11 19.5	5.2	37 54.99	7 34.7
3	6	37.7	52.7	7.2	22.2				39 22.05	4.58	IV.	9	7.00	42 29.3	4.9	39 26.63	38 44.2
4	8	4.6	19.4	34.0					41 48.63	4.53	IV.	5	11.17	25 37.8	4.4	41 53.16	21 52.2
5	7			16.0	30.8	45.3			42 30.56	4.52	VI.	4	11.22	20 40.2	4.3	42 35.08	16 54.5
6	9.10			40.6	55.3	9.6			46 55.11	4.54	VII.	7	8.36	33 18.2	3.4	46 59.65	29 31.6
7	9.10		58.8		29.0				17 49 28.47	+ 4.52	VI.	6	10.15	−29 7.9	− 2.9	17 49 32 99	− 31 25 20.8

ZONE 56. AUGUST 12. K. $D_0 = -36° 31' 20''$.

No.	Mag.	I.	II.	III.	IV.	V.	VI.	VII.	T.	a_1	Micrometer.			$i + d_s$	d_1	Mean Right Ascension, 1850.0.	Mean Declination, 1850.0.
1	7						7.9		18 5 36.93	+ 4.42	VI.	1	9.17	− 4 35.6	− 51.7	18 5 41.35	− 36 36 47.3
2	4.5			8.8	24.3				7 24.12	4.44	IV.	3	11.43	15 50.3	51.2	7 28.56	48 1.5
3	10				10.8				17 10.63	4.41	VI.	5	11.51	25 54.8	48.4	17 15.04	36 58 3.2
4	10			15.8					18 31.42	4.43	VI.	7	7.41	32 51.6	48.0	18 35.85	37 4 59.6
5	10			35.7					18 51.33	4.42	VI.	7	9.28	33 45.8	47.9	18 55.75	37 5 53.7
6	9	9.1							20 56.22	4.39	III.	4	14.15	22 7.4	47.3	21 0.61	36 54 14.7
7	10	38.1							23 25.24	4.39	II.	5	4.55	22 24.1	46.6	23 29.63	54 30.7
8	11				38.1				25 37.93	4.38	VII.	5	6.33	23 13.4	45.9	25 42.31	55 19.3
9	7.8		23.3	39.1					33 54.50	4.33	IV.	4	8.55	19 25.5	43.5	33 58.83	36 51 29.0
10	7			51.1	7.0				38 6.82	4.35	V.	8	8.14	38 8.5	42.2	38 11.17	37 10 10.7
11	9			2.0	17.6				40 17.46	4.32	V.	6	5.29	26 43.7	41.6	40 21.78	36 58 45.3
12	8				27.2				41 27.03	4.28	V.	2	1.45	5 47.1	41.3	41 31.31	36 37 48.4
13	7			15.9		47.3			48 31.74	4.30	V.	9	13.33	45 50.4	39.2	48 36.04	37 17 49.6
14	7		19.0		50.6				50 50.48	4.28	VI.	9	9.26	43 45.1	38.5	50 54.76	15 43.6
15	6.7			35.8					50 51.60	4.28	VI.	9	9.32	43 48.2	38.5	50 55.88	15 46.7
16	9							14.0	51 27.10	4.26	VII.	7	9.12	33 37.4	38.3	51 31.36	5 35.7
17	8			17.8					54 33.42	4.24	VI.	7	7.6	32 33.9	37.4	54 37.66	4 31.3
18	5		40.7	56.2					18 56 12.11	4.25	IV.	9	10.41	44 23.4	36.9	18 56 16.36	16 20.3
19	7		54.6		26.0				19 5 25.94	+ 4.19	IV.	8	12.14	−40 10.1	− 34.2	19 5 30.13	− 37 12 4.3

CORRECTIONS.

Date.		Corr. of Clock.	Hourly rate.	m	n	c
1846.	h.	s.	s.	s.	s.	s.
Aug. 12,	18	− 7.78	− 0.008	+ 0.372	− 0.163	0.000

INSTRUMENT READINGS.

	Date.		Barom.	Thermom. At.	Thermom. Ex.
	1846.	h. m.	in.	°	°
Zone 55	Aug. 12,	17 34	30.07	80.0	74.0
		20 59	30.06	82.0	77.0

REMARKS.

(54) 48. Micrometer reading assumed as 13r.00 instead of 12r.00.

ZONE 56. AUGUST 12. A. $D_0 = -36° 31' 20''$—Continued.

No.	Mag.	SECONDS OF TRANSIT. I.	II.	III.	IV.	V.	VI.	VII.	T.	a_1	MICROMETER.			$i + d_4$	d_1	Mean Right Ascension, 1850.0.	Mean Declination, 1850.0.
									h. m. s.	s.			r.	′ ″	″	h. m. s.	° ′ ″
20	8	..	..	..	14.9	..	..	..	19 9 14.73	+ 4.17	V.	8	7.18	−37 40.1	− 33.1	19 9 18.90	− 37 9 33.2
21	9	5.6	..	37.2	..	..	..	..	25 52.91	4.06	IV.	8	3.48	35 53.9	28.2	25 56.97	7 42.1
22	8	58.7	14.8	..	..	..	..	..	31 46.22	4.02	III.	9	7.59	42 0.7	26.5	31 50.24	37 13 47.2
23	8	..	..	..	..	26.9	..	..	32 11.26	4.00	VI.	6	7.27	27 43.2	26.4	32 15.26	36 59 29.6
24	7	..	..	..	6.5	22.1	..	..	45 6.34	3.86	VI.	1	7.14	3 33.3	22.7	45 10.20	36 35 16.0
25	7	..	43.8	59.8	..	..	..	..	48 15.29	3.89	IV.	7	8.16	33 9.6	21.9	48 19.18	37 4 51.5
26	8	..	..	..	37.4	..	..	..	51 37.23	3.87	V.	7	8.53	33 28.3	20.9	51 41.10	5 9.2
27	8	49.0	..	..	..	..	..	..	53 36.22	3.85	II.	6	9.37	28 49.2	20.4	53 40.07	37 0 29.6
28	9	..	..	7.1	..	..	..	..	55 22.43	3.82	IV.	3	12.16	16 7.0	19.9	55 26.25	36 47 46.9
29	9	..	..	..	..	7.8	..	..	55 52.14	3.82	VI.	5	7.4	23 29.5	19.7	55 55.96	55 9.2
30	9	..	..	14.8	..	..	..	..	19 59 30.26	3.81	V.	5	8.26	24 11.2	18.7	19 59 34.07	55 49.9
31	8	46.2	2.1	..	..	..	..	..	20 6 33.34	3.74	III.	5	5.33	22 43.6	16.8	20 6 37.08	54 20.4
32	11	..	..	..	6.4	..	..	..	10 6.33	3.72	VI.	6	5.2	26 29.8	15.9	10 9.95	36 58 5.7
33	9	5.9	21.9	..	..	..	..	..	11 52.24	3.71	III.	7	8.7	33 5.0	15.4	11 55.95	37 4 40.4
34	10	..	57.8	13.5	29.4	..	..	..	13 29.22	3.72	V.	8	8.20	38 11.6	15.0	13 32.94	37 9 46.6
35	8.9	..	16.6	32.2	47.8	..	..	..	18 47.58	3.62	V.	2	9.0	9 27.4	13.6	18 51.20	36 41 1.0
36	9	..	31.9	47.8	..	..	..	..	23 3.02	3.56	IV.	3	4.47	12 19.6	12.4	23 6.58	43 52.0
37	10	..	..	..	23.6	..	..	..	29 23.43	3.55	V.	4	12.7	21 2.6	10.8	29 26.98	52 33.4
38	7	..	..	..	..	28.3	..	..	30 12.56	3.52	V.	1	3.42	1 46.1	10.6	30 16.08	33 16.7
39	8	..	10.0	..	..	..	..	..	36 41.05	3.47	III.	2	6.14	8 3.3	9.0	36 44.52	36 39 32.3
40	9	..	6.8	22.6	..	..	..	..	43 38.19	3.45	IV.	7	7.43	32 52.9	7.3	43 41.64	37 4 20.2
41	9	..	..	..	..	21.3	36.7	..	44 5.64	3.42	VII.	3	7.55	13 54.3	7.2	44 9.06	36 45 21.5
42	7	..	56.1	12.1	27.6	..	..	..	50 27.32	3.36	V.	2	11.55	10 56.0	5.7	50 30.68	42 21.7
43	8	..	..	..	22.5	..	53.4	..	51 22.33	3.35	VII.	2	3.51	6 50.4	5.4	51 25.68	36 38 15.8
44	8	..	..	..	..	..	..	5.5	52 18.66	3.27	VII.	8	8.9	38 5.5	5.2	52 21.93	37 9 30.7
45	10	..	59.8	..	..	..	..	..	56 31.00	3.28	III.	4	9.36	19 46.2	4.2	56 34.28	36 51 10.4
46	8.7	..	21.5	37.3	..	..	..	..	20 59 53.11	+ 3.34	IV.	10	5.36	−46 51.1	− 3.4	20 59 56.45	− 37 18 14.5

ZONE 57. AUGUST 13. A. $D_0 = -24° 1' 0''$.

No.	Mag.	I.	II.	III.	IV.	V.	VI.	VII.	T.	a_1	MICROMETER.			$i + d_4$	d_1	Mean Right Ascension, 1850.0.	Mean Declination, 1850.0.
1	8	..	..	36.3	49.8	..	..	..	18 15 49.76	+ 3.53	IV.	6	7.58	−27 58.7	− 44.0	18 15 53.29	− 24 29 42.7
2	8	..	..	..	49.0	2.2	..	..	18 48.70	3.54	V.	8	6.32	37 14.7	43.2	18 52.24	38 57.9
3	8	..	..	..	1.2	14.8	..	..	20 1.08	3.53	V.	7	4.15	31 6.4	43.0	20 4.61	32.49.4
4	8	..	..	..	10.5	24.2	..	..	21 10.38	3.49	V.	2	5.43	7 50.6	42.7	21 13.87	9 33.3
5	8	..	..	37.5	51.2	..	..	..	22 51.20	3.54	IV.	10	8.18	48 8.9	42.3	22 54.74	49 51.2
6	5	..	..	..	..	14.0	28.0	..	24 0.53	3.48	VI.	3	1.31	11 13.0	42.0	24 4.01	12 55.0
7	6	..	..	..	..	32.5	46.0	0.0	26 18.79	3.48	VII.	4	6.40	18 17.9	41.5	25 22.27	19 59.4
8	8	..	..	..	53.0	6.5	..	..	27 52.83	3.49	V.	7	5.1	31 29.6	41.1	26 56.32	33 10.7
9	10	..	..	..	7.5	21.5	..	..	28 7.59	3.49	V.	7	10.45	34 23.3	41.1	28 11.08	36 4.4
10	11	..	..	..	53.5	..	..	..	30 53.35	3.48	IV.	7	9.11	33 35.9	40.4	30 56.83	35 16.3
11	6	54.0	8.0	..	..	..	..	..	32 35.35	3.46	II.	6	7.22	27 40.3	40.0	32 38.81	29 20.3
12	10	..	..	..	..	11.5	24.5	..	32 57.53	3.44	VI.	2	6.52	8 25.3	39.9	33 0.97	10 5.2
13	7	..	..	..	..	..	..	26.5	33 45.12	3.43	VII.	1	9.50	4 55.6	39.8	33 48.55	6 35.4
14	9	..	30.5	44.5	..	..	..	..	35 57.96	3.45	III.	5	9.9	24 33.0	39.2	36 1.41	26 12.2
15	7	..	50.3	4.0	..	..	..	..	38 17.62	3.44	III.	6	6.3	27 0.6	38.7	38 21.06	28 39.3
16	11	..	..	..	..	..	14.0	..	38 46.77	3.43	VI.	6	8.19	28 9.1	38.6	38 50.20	29 47.7
17	8	..	..	..	16.0	29.5	..	..	40 15.86	3.46	V.	9	3.38	40 46.2	38.2	40 19.32	42 24.4
18	7	..	..	..	..	..	23.5	37.3	18 40 56.30	+ 3.46	VII.	10	5.30	−46 43.8	− 38.1	18 40 59.76	− 24 48 21.9

CORRECTIONS. INSTRUMENT READINGS.

Date.		Corr. of Clock.	Hourly rate.	m	n	c		Date.	Barom.	THERMOM. At.	THERMOM. Ex.
1846.	h.	s.	s.	s.	s.	s.		1846. h. m.	in.	°	°
Aug. 13.	18	− 7.55	− 0.002	+ 0.372	− 0.163	0.000	Zone 57	Aug. 13, 18 15	29.95	82.0	77.0
								13, 21 41	29.95	80.5	73.0

REMARKS.

(56) 22. Micrometer reading assumed as $5^r.59$ instead of $7^r.59$.
(57) 6. Micrometer reading assumed as $2^r.31$ instead of $1^r.31$.
(57) 7. Minutes assumed as 25 instead of 26.
(57) 8. Minutes assumed as 26 instead of 27.

ZONE 57. AUGUST 13. A. $D_0 = -24° 1' 0''$—Continued.

No.	Mag.	Seconds of Transit. I.	II.	III.	IV.	V.	VI.	VII.	T.	a_1	Micrometer.			$i + d_2$	d_1	Mean Right Ascension, 1850.0.	Mean Declination, 1850.0.
									h. m. s.	s.			r.	′ ″	″	h. m. s.	° ′ ″
19	8					34.0	47.5	1.5	18 42 20.36	+3.43	VII.	7	12.	−35 0.9	−37.8	18 42 23.79	−24 36 38.7
20	8						41.5	56.0	44 14.50	3.39	VII.	2	7.3	8 30.7	37.4	44 17.89	10 8.1
21	7			14.5	28.5				47 28.34	3.43	IV.	10	6.10	47 4.3	36.6	47 31.77	48 40.9
22	8						57.5		53 30.26	3.38	VI.	7	7.35	32 47.3	35.2	53 33.64	34 22.5
23	10				25.5				56 25.35	3.36	IV.	6	6.47	27 22.8	34.5	56 28.71	28 57.3
24	9		13.0	27.0					57 40.43	3.35	III.	5	6.43	23 19.4	34.2	57 43.78	24 53.6
25	9				20.0				58 19.85	3.34	IV.	4	7.55	18 56.1	34.1	58 23.19	20 30.2
26	8				34.5	48.0			18 59 34.35	3.36	V.	8	6.0	36 58.5	33.8	18 59 37.71	38 32.3
27	9					54.0	7.5		19 0 40.29	3.34	VI.	5	5.58	22 56.5	33.6	19 0 43.63	24 30.1
28	7					2.5	16.0		1 48.79	3.33	VI.	5	7 25	23 40.4	33.3	1 52.12	25 13.7
29	9				15.0	28.5			3 14.83	3.34	V.	7	5.39	31 48.8	33.0	3 18.17	33 21.8
30	5	40.0	54.0	7.8					6 21.33	3.32	III.	5	8.11	24 3 8	32.3	6 24.65	25 36.1
31	9					59.0	12.5	27.0	6 45.45	3.30	VII.	4	7.42	18 49.2	32.2	6 48.75	20 21.4
32	9					25.0	38.0		8 11.03	3.29	VI.	2	8.53	9 26.4	31.9	8 14.32	10 58.3
33	8						2.5	16.8	9 35.39	3.28	VII.	2	4.48	7 22.5	31.5	9 38.67	8 54.0
34	6	17.0	31.0						11 58.34	3.29	II.	6	5.36	26 46.8	31.0	12 1.63	28 17.8
35	8			28.0	41.5	55.0			12 41.48	3.32	V.	9	2.19	40 6.3	30.8	12 44.80	41 37.1
36	9			0.5	14.0				14 13.96	3.28	IV.	6	8.39	28 19.4	30.5	14 17.24	29 49.9
37	4		37.5	51.5	5.2				16 5.15	3.30	IV.	10	3.57	45 57.2	30.1	16 8.45	47 27.3
38	8							53.0	16 12.03	3.30	VII.	9	2.48	40 20.7	30.1	16 15.33	41 50.8
39	8					2.8	16.3		17 49.09	3.25	VI.	4	7.20	18 38.3	29.7	17 52.34	20 8.0
40	10						15.6	29.7	18 48.44	3.25	VII.	4	6.8	18 1.8	29.5	18 51.69	19 31.3
41	8				22.5	36.5			20 22.58	3.25	V.	6	10.32	29 16.4	29.1	20 25.83	30 45.5
42	6			33.3	47.0				21 46.76	3.22	IV.	3	7.51	13 54.7	28.8	21 49.98	15 23.5
43	7						30.3	44.3	22 3.10	3.24	VII.	5	5.10	22 32.1	28.8	22 6.34	24 0.9
44	11			45.0	58.8				23 58.65	3.24	IV.	7	8.5	33 2.6	28.4	24 1.89	34 31.0
45	11					59.0	12.5		24 45.29	3.24	VI.	7	9.19	33 39.8	28.2	24 48.53	35 8.0
46	9						5.5	18.5	25 37.90	3.25	VII.	10	6.42	47 20.1	28.0	25 41.15	48 48.1
47	9						13.5	27.5	26 46.38	3.25	VII.	9	2.53	40 23.2	27.7	26 49.63	41 50.9
48	9	42.8	56.9						29 24.18	3.20	II.	5	9.49	24 53.1	27.1	29 27.38	26 20.2
49	9		40.0	53.5					30 7.25	3.20	III.	6	10.16	29 8.3	27.00	30 10.45	30 35.3
50	9					43.0	56.8	10.7	30 29.49	3.20	VII.	6	6.2	26 59.8	26.9	30 32.69	28 26.7
51	9							22.0	31 41.00	3.21	VII.	8	9.	38 29.1	26.6	31 44.21	39 55.7
52	6			15.5	29.6				33 29.35	3.21	IV.	9	5.55	41 55.4	26.2	33 32.56	43 21.6
53	10						56.5	10.0	35 29.14	3.20	VII.	9	9.55	43 56.3	25.8	35 32.34	45 22.1
54	8			49.8	3.0				38 2.94	3.13	IV.	1	8.21	4 11.0	25.2	38 6.07	5 36.2
55	11		29.0						39 56.42	3.15	II.	6	11.32	29 46.6	24.8	39 59.57	31 11.4
56	8		25.0	39.0					40 52.21	3.12	III.	1	7.48	3 54.3	24.6	40 55.33	5 18.9
57	8		30.8	44.5	58.2				42 58.26	3.17	IV.	10	7.	47 29.6	24.2	43 1.43	48 53.8
58	5	34.0	47.0						45 14.75	3.11	II.	4	3.51	16 52.8	23.7	45 17.86	18 16.5
59	8		39.4	53.5					46 6.80	3.11	III.	4	2.34	16 14.0	23.5	46 9.91	17 37.5
60	8	15.2	29.2						47 56.69	3.14	II.	9	6.6	42 0.8	23.1	47 59.83	43 23.9
61	8				53.5	7.0			48 53.37	3.14	V.	10	8.24	48 11.9	22.9	48 56.51	49 34.8
62	8					31.0	44.8	58.3	51 17.33	3.09	VII.	4	10.3	20 0.4	22.4	51 20.42	21 22.8
63	9		31.0	45.0					54 58.68	3.11	III.	9	10.58	44 28.4	21.6	55 1.79	45 50.0
64	8						52.2	6.0	19 55 24.90	3.08	VII.	5	5.17	22 35.6	21.5	19 55 27.98	23 57.1
65	9								20 12	3.00	VI.	9	5.15	41 35.1	18.1	20 12 3.00	42 53.2
66	8		27.0	41.0					14 54.59	2.98	III.	8	6.10	37 3.6	17.6	14 57.57	38 21.2
67	8				17.5				20 18 17.35	+2.97	IV.	10	9.59	−48 59.9	−16.9	20 18 20.32	−24 50 16.8

CORRECTIONS.

Date.	Corr. of Clock.	Hourly rate.	m	n	c
1846. h.	s.	s.	s.	s.	s.

INSTRUMENT READINGS.

Date.		Barom.	Thermom. At.	Thermom. Ex.
1846.	h. m.	in.	°	°

REMARKS.

ZONE 57. AUGUST 13. A. $D_0 = -24° 1' 0''$—Continued.

No.	Mag.	SECONDS OF TRANSIT. I.	II.	III.	IV.	V.	VI.	VII.	T.	a_1	MICROMETER.			$i + d_2$	d_1	Mean Right Ascension, 1850.0.	Mean Declination, 1850.0.
									h. m. s.	s.			r.	′ ″	″	h. m. s.	° ′ ″
68	7	. .	. .	17.5	31.5	. .	. .	. .	20 19 31.28	+ 2.95	IV.	8	7.13	−37 35.4	− 16.7	20 19 34.23	− 24 38 52.1
69	6	. .	. .	. .	. .	. .	14.5	28.4	19 47.27	2.94	VII.	6	5.56	26 56.7	16.6	19 50.21	28 13.3
70	9	. .	. .	39.0	12.5	26.5	. .	. .	22 12.44	2.91	V.	2	2.30	6 13.2	16.1	22 15.35	7 29.3
71	9	. .	. .	. .	. .	45.5	59.2	13.0	23 31.88	2.92	VII.	6	5.31	26 44.1	15.9	23 34.80	28 0.0
72	10	58.5	12.2	. .	. .	. .	. .	. .	26 39.67	2.90	II.	5	6.31	23 13.2	15.3	26 42.57	24 28.5
73	9	. .	. .	. .	. .	51.5	5.2	. .	28 37.89	2.92	VI.	9	4.43	41 18.9	14.9	28 40.81	42 33.8
74	8	24.0	38.0	. .	. .	. .	. .	. .	30 5.41	2.90	II.	8	1.20	34 37.0	14.6	30 7.31	35 51.6
75	6	44.0	58.2	. .	. .	. .	. .	. .	32 25.53	2.89	II.	8	5.16	36 36.2	14.2	31 28.42	37 50.4
76	5	. .	. .	. .	. .	. .	. .	54.0	31 12.76	2.88	VII.	4	5.34	17 44.6	14.4	31 15.64	18 59.0
77	8	44.8	58.5	. .	. .	. .	. .	. .	36 25.97	2.85	II.	5	7.14	23 34.9	13.4	36 28.82	24 48.3
78	9	. .	53.5	7.0	21.0	. .	. .	. .	37 20.71	2.84	IV.	4	7.57	18 57.2	13.3	37 23.55	20 10.5
79	7	. .	. .	. .	. .	. .	. .	6.5	37 25.24	2.83	VII.	3	9.12	14 35.3	13.3	37 28.07	15 48.6
80	9	58.0	12.0	. .	. .	. .	. .	. .	40 39.36	2.83	II.	6	8.48	28 23.8	12.7	40 42.19	29 36.5
81	9	. .	. .	. .	. .	. .	7.0	20.8	40 39.72	2.83	VII.	6	7.5	27 31.6	12.6	40 42.55	28 44.2
82	9	. .	. .	12.5	26.2	. .	. .	. .	43 26.15	2.83	IV.	9	5.32	41 43.8	12.2	43 28.98	42 56.0
83	4	. .	. .	. .	. .	23.8	37.3	51.3	44 10.09	2.80	VII.	4	8.25	19 10.9	12.0	44 12.89	20 22.9
84	9	47.2	1.0	. .	. .	. .	. .	. .	46 28.47	2.80	II.	6	11.11	29 35.9	11.6	47 31.27	30 47.5
85	8	0.1	14.0	28.0	. .	. .	. .	. .	48 27.68	2.77	III.	5	7.44	23 50.1	11.2	48 30.45	25 1.3
86	9	13.0	27.0	. .	. .	. .	. .	. .	50 54.34	2.77	II.	6	4.14	26 5.4	10.8	50 57.11	27 16.2
87	6	. .	27.0	41.0	. .	. .	. .	. .	52 54.41	2.75	III.	5	3.36	21 44.9	10.5	52 57.16	22 55.4
88	7	. .	. .	31.0	44.6	. .	. .	. .	20 59 44.64	2.74	IV.	10	6.23	47 10.9	9.3	20 59 47.38	48 20.2
89	6	. .	. .	. .	. .	. .	54.0	8.1	21 1 26.81	2.70	VII.	3	5.22	12 39.1	9.0	21 0 29.51	13 48.1
90	10	. .	. .	. .	. .	57.5	11.5	25.0	1 43.96	2.71	VII.	5	2.13	21 2.7	9.0	1 46.67	22 11.7
91	7	. .	. .	. .	. .	30.2	43.7	57.2	3 16.41	2.72	VII.	9	7.33	42 44.6	8.7	3 19.13	43 53.3
92	11	. .	. .	. .	. .	0.0	13.5	. .	5 46.28	2.67	VI.	2	5.36	7 46.9	8.3	5 48.95	8 55.2
93	9	32.0	46.0	. .	. .	. .	. .	. .	8 13.38	2.68	II.	7	3.34	30 45.6	7.9	8 16.06	31 53.5
94	8	. .	. .	. .	49.5	3.0	16.8	. .	8 49.41	2.68	VI.	8	7.54	37 56.0	7.8	8 52.09	39 3.8
95	8	. .	. .	. .	. .	15.5	29.0	42.8	10 1.74	2.65	VII.	5	9.48	24 52.4	7.6	10 4.39	26 0.0
96	8	. .	. .	. .	. .	. .	. .	30.0	10 48.81	2.65	VII.	5	5.46	22 50.2	7.5	10 51.46	23 57.7
97	7	. .	. .	20.0	33.5	. .	. .	. .	13 33.27	2.61	IV.	1	3.53	1 55.7	7.1	13 35.88	3 2.8
98	5	52.5	6.7	20.5	. .	. .	. .	. .	15 33.86	2.61	III.	3	10.13	15 6.4	6.8	15 36.47	16 13.2
99	8	. .	22.5	36.2	. .	. .	. .	. .	16 49.64	2.61	III.	3	3.39	11 47.5	6.6	16 52.25	12 54.1
100	4	. .	. .	. .	. .	21.0	34.7	48.4	17 7.35	2.63	VII.	6	5.25	26 41.1	6.6	17 9.98	27 47.7
101	8	. .	. .	. .	. .	18.8	32.2	46.2	29 5.03	2.55	VII.	4	11.39	20 48.9	4.8	29 7.58	21 53.7
102	7	. .	37.3	51.2	. .	. .	. .	. .	32 4.80	2.55	III.	7	9.40	33 50.5	4.4	32 7.35	34 54.9
103	11	. .	. .	. .	. .	. .	45.0	59.0	32 17.82	2.54	VII.	6	7.55	27 56.8	4.3	32 20.36	29 1.1
104	6	. .	. .	. .	. .	. .	11.5	25.0	34 44.15	2.55	VII.	10	8.28	48 13.6	4.0	34 46.70	49 17.6
105	9	14.0	27.8	. .	. .	. .	. .	. .	38 55.19	2.50	II.	4	11.13	20 36.0	3.4	38 57.69	21 39.4
106	9	. .	. .	. .	21.0	34.0	. .	. .	39 20.58	2.51	V.	6	7.5	27 31.9	3.4	39 23.09	28 35.3
107	9	. .	. .	. .	21.5	35.0	48.7	. .	40 21.38	2.49	VI.	4	8.9	19 3.0	3.2	40 23.87	20 6.2
108	9	. .	. .	. .	. .	44.5	58.0	. .	21 41 30.78	+ 2.51	VI.	7	2.51	−30 23.9	− 3.1	21 41 33.29	− 24 31 27.0

CORRECTIONS.

Date.		Corr. of Clock.	Hourly rate.	m	n	c
1846.	h.	s.	s.	s.	s.	s.

INSTRUMENT READINGS.

Date.		Barom.	THERMOM. At.	Ex.
1846.	h. m.	in.	°	°

REMARKS.

(57) 75. Minutes assumed as 31 instead of 32.
(57) 84. Minutes assumed as 47 instead of 46.
(57) 85. Transits over T.'s II–IV assumed as recorded over T.'s I–III.
(57) 89. Minutes assumed as 0 instead of 1.

ZONE 58. AUGUST 18. A. $D_0 = -28°$ (3)′.

No.	Mag.	I.	II.	III.	IV.	V.	VI.	VII.	T.	a_1	Micrometer		r.	$i + d_0$	d_1	Mean Right Ascension, 1850.0.	Mean Declination, 1850.0.
									h. m. s.	s.			r.	′ ″	″	h. m. s.	° ′ ″
1	8	6.5	21.0	. .	. .	. .	. .	. .	18 45 49.60	+ 2.67	II.	8	12.	−40 0.6	. .	18 45 52.27	
2	6	. .	. .	16.0	30.0	. .	. .	. .	46 29.87	2.62	IV.	3	10.21	15 10.2	. .	46 32.49	
3	7	. .	. .	23.6	37.8	. .	. .	. .	47 37.56	2.62	IV.	3	8.57	14 27.8	. .	47 40.18	
4	11	. .	. .	. .	. .	. .	33.5	48.0	48 5.43	2.64	VII.	8	5.40	36 48.4	. .	48 8.07	
5	6	. .	. .	. .	. .	51.0	5.5	. .	49 37.07	2.64	VI.	9	7.19	42 39.2	. .	49 39.71	
6	5	. .	. .	. .	. .	. .	. .	14.3	50 31.51	2.60	VII.	2	10.29	10 14.2	. .	50 34.11	
7	10	. .	. .	. .	. .	39.0	. .	. .	52 24.80	2.61	V.	6	7.12	27 35.4	. .	52 27.41	
8	7	. .	. .	. .	. .	. .	30.0	44.5	53 1.94	2.63	VII.	10	3.23	45 40.2	. .	53 4.57	
9	6	5.0	19.5	. .	. .	. .	. .	. .	55 47.68	2.57	II.	3	8.35	14 16.5	. .	55 50.25	
10	8	. .	. .	49.0	3.2	. .	. .	. .	57 3.16	2.61	IV.	10	3.12	45 35.0	. .	57 5.77	
11	4	. .	. .	. .	. .	15.0	29.2	43.0	58 0.82	2.60	VII.	10	3.55	45 56.3	. .	18 58 3.42	
12	8	. .	. .	. .	. .	12.5	26.5	40.5	18 59 58.24	2.59	VII.	9	3.38	40 46.3	. .	19 0 0.83	
13	7	. .	. .	. .	. .	. .	57.2	11.3	19 2 28.93	2.57	VII.	8	11.57	39 55.6	. .	2 31.50	
14	8	. .	. .	. .	. .	. .	6.5	20.5	2 38.16	2.56	VII.	8	6.59	37 28.3	. .	2 40.72	
15	7	51.5	5.5	. .	. .	. .	. .	. .	19 5 34.04	+ 2.53	II.	5	3.35	−21 44.2	. .	19 5 36.57	

ZONE 59. AUGUST 18. A. $D_0 = -29°$ (3–5)′.

No.	Mag.	I.	II.	III.	IV.	V.	VI.	VII.	T.	a_1	Micrometer		r.	$i + d_0$	d_1	Mean Right Ascension, 1850.0.	Mean Declination, 1850.0.
1	11	. .	24.5	. .	. .	. .	. .	. .	19 31 52.99	+ 2.57	II.	6	7.23	−27 40.9	. .	19 31 55.56	
2	8	. .	. .	17.8	32.8	. .	. .	. .	32 32.41	2.57	IV.	7	11.33	34 48.0	. .	32 34.98	
3	9	4.0	19.0	. .	. .	. .	. .	. .	35 47.41	2.54	II.	5	8.44	24 20.2	. .	35 49.95	
4	7	. .	. .	21.0	35.0	. .	. .	. .	37 34.91	2.51	IV.	2	5.54	7 55.5	. .	37 37.42	
5	5	. .	. .	. .	27.0	41.2	. .	. .	38 26.83	2.52	V.	6	7.9	27 34.0	. .	38 29.35	
6	6	. .	. .	. .	. .	30.0	44.3	. .	39 15.64	2.50	VI.	3	4.4	11 59.4	. .	39 18.14	
7	11	. .	. .	46.8	1.5	. .	. .	. .	43 1.22	2.48	IV.	6	8.59	28 29.6	. .	43 3.70	
8	5	24.0	38.6	. .	. .	. .	. .	. .	46 7.23	2.47	II.	5	10.41	25 19.4	. .	46 9.70	
9	6	. .	. .	. .	. .	35.5	. .	. .	46 21.13	2.48	V.	7	6.21	32 10.3	. .	46 23.61	
10	8	. .	. .	. .	. .	33.0	47.0	. .	48 18.48	2.45	VI.	5	10.	24 58.7	. .	48 20.93	
11	7	. .	. .	. .	50.8	5.0	. .	. .	49 50.59	2.42	VI.	3	2.45	11 19.5	. .	49 53.01	
12	9	. .	29.5	44.5	. .	. .	. .	. .	51 58.26	2.42	III.	4	5.44	17 49.7	. .	52 0.68	
13	5	14.5	29.2	44.0	. .	. .	. .	. .	55 57.94	2.40	III.	6	3.53	25 54.9	. .	56 0.34	
14	7	. .	. .	42.8	57.0	. .	. .	. .	56 56.98	2.40	IV.	7	2.46	30 21.7	. .	56 59.38	
15	8	. .	36.7	51.5	. .	. .	. .	. .	19 59 5.33	2.36	III.	3	11.9	15 31.3	. .	19 59 7.69	
16	11	. .	. .	48.5	2.8	. .	. .	. .	20 0 2.68	2.37	IV.	5	8.14	24 5.3	. .	20 0 5.05	
17	11	. .	. .	. .	. .	. .	53.5	8.0	0 24.82	2.37	VII.	5	9.43	24 49.8	. .	0 27.19	
18	8	. .	8.0	22.5	. .	. .	. .	. .	2 36.51	2.35	III.	4	6.31	18 13.4	. .	2 38.86	
19	9	. .	. .	. .	3.0	. .	. .	. .	5 2.84	2.35	IV.	6	5.31	36 44.4	. .	5 5.19	
20	9	. .	. .	. .	. .	5.5	19.5	34.5	5 51.06	2.32	VII.	4	5.54	17 54.4	. .	5 53.38	
21	9	. .	. .	. .	. .	33.5	47.5	. .	7 18.99	2.32	VI.	6	8.50	28 24.9	. .	7 21.31	
22	6	. .	. .	. .	. .	. .	46.0	59.5	8 16.92	2.35	VII.	10	7.52	47 26.1	. .	8 19.27	
23	6	. .	29.0	. .	. .	. .	. .	. .	10 57.53	2.31	II.	7	6.20	32 9.7	. .	10 59.84	
24	6	. .	. .	. .	10.5	24.8	39.0	53.0	11 10.26	2.31	VII.	8	4.6	36 1.0	. .	11 12.57	
25	7	. .	. .	. .	. .	54.0	8.7	23.0	13 39.85	2.28	VII.	7	9.42	33 51.5	. .	13 42.13	
26	4	. .	. .	. .	27.0	41.2	55.3	. .	15 26.78	2.27	VI.	6	11.11	29 36.1	. .	15 29.05	
27	5	. .	. .	3.3	17.8	. .	. .	. .	17 17.63	2.25	IV.	6	11.21	29 41.3	. .	17 19.88	
28	6	. .	. .	. .	14.5	29.0	. .	. .	18 14.45	2.23	V.	3	9.8	14 33.2	. .	18 16.68	
29	7	45.5	0.0	14.4	. .	. .	. .	. .	20 28.69	2.23	III.	6	6.57	27 27.9	. .	20 30.92	
30	5	. .	13.5	28.2	. .	. .	. .	. .	20 21 42.30	+ 2.23	III.	7	7.50	−32 55.3	. .	20 21 44.53	

CORRECTIONS.

Date.		Corr. of Clock.	Hourly rate.	m	n	c
1846.	h.	s.	s.	s.	s.	s.
Aug. 18,	18	− 8.62	− 0.009	+ 0.372	− 0.163	0.000

INSTRUMENT READINGS.

	Date.		Barom.	Thermom. At.	Thermom. Ex.
	1846.	h. m.	in.	°	°
Zone 58	Aug. 18,	18 45	30.11	75.7	70.0
	18,	19 0	30.13	74.1	68.4
Zone 59	18,	20 0	30.13	74.1	68.0
	18,	21 44	30.09	75.0	64.0

REMARKS.

(58) Instrument not clamped properly.
(59) Instrument not clamped properly.
(59) 22. Micrometer reading assumed as 6r.52 instead of 7r.52.

ZONE 59. AUGUST 18. A. $D_0 = -29°$ (3-5)'—Continued.

No.	Mag.	SECONDS OF TRANSIT. I.	II.	III.	IV.	V.	VI.	VII.	T.		a_1	MICROMETER.			$i + d_2$	d_1	Mean Right Ascension, 1850.0.	Mean Declination, 1850.0.
									h. m. s.		s.			r.	′ ″	″	h. m. s.	° ′ ″
31	6	. .	. .	55.5	9.8	. .	. .	. .	20 23 9.52	+	2.18	IV.	1	5.15	− 2 36.3	. .	20 23 11.70	
32	4	. .	. .	. .	. .	. .	58.8	13.0	23 29.89		2.17	VII.	1	4.49	2 22.7	. .	23 32.06	
33	5	. .	. .	. .	. .	. .	. .	17.0	24 36.02		2.23	VII.	10	0.37	44 16.5	. .	24 38.25	
34	7	. .	36.0	50.9	. .	. .	. .	. .	28 4.62		2.16	III.	2	9.56	9 57.7	. .	28 6.78	
35	8	. .	. .	. .	. .	. .	5.0	19.5	38 36.29		2.17	VII.	4	3.50	16 51.7	. .	28 38.46	
36	7	. .	29.4	44.2	. .	. .	. .	. .	30 58.10		2.15	III.	4	10.10	20 4.1	. .	31 0.25	
37	10	. .	39.0	53.5	. .	. .	. .	. .	32 7.58		2.15	III.	5	7.11	23 33.4	. .	32 9.73	
38	5	. .	. .	. .	. .	51.5	5.0	19.8	32 36.64		2.11	VII.	3	8.21	13 38.8	. .	32 38.75	
39	5	. .	. .	. .	. .	. .	21.8	36.3	33 53.08		2.11	VII.	3	6.59	14 28.2	. .	33 55.19	
40	9	. .	. .	. .	52.5	6.5	21.0	. .	35 52.28		2.13	VI.	7	5.16	31 37.3	. .	35 54.41	
41	6	. .	. .	13.0	27.0	. .	. .	. .	37 26.99		2.10	IV.	4	5.51	17 53.2	. .	37 29.09	
42	9	59.0	13.7	. .	. .	. .	. .	. .	39 42.28		2.09	II.	5	10.42	25 19.9	. .	39 44.37	
43	8	. .	. .	. .	. .	7.5	21.5	36.0	39 52.94		2.10	VII.	7	4.45	31 21.4	. .	39 55.04	
44	7	. .	. .	. .	24.0	37.5	. .	. .	41 23.48		2.09	V.	7	3.54	30 25.8	. .	41 25.57	
45	9	. .	. .	1.0	15.5	. .	. .	. .	43 15.28		2.07	IV.	5	8.3	23 59.7	. .	43 17.35	
46	10	. .	. .	. .	. .	. .	. .	44.0	45 0.56		2.03	VII.	2	5.45	7 50.6	. .	45 2.59	
47	9	. .	. .	. .	. .	. .	44.0	58.5	46 15.31		2.04	VII.	5	1.43	20 47.3	. .	46 17.35	
48	10	5.5	20.0	. .	. .	. .	. .	. .	50 48.72		2.00	II.	6	10.32	29 16.4	. .	50 50.72	
49	8	. .	. .	24.0	38.0	. .	. .	. .	51 37.99		1.99	IV.	4	6.37	18 16.5	. .	51 39.98	
50	5	. .	23.8	38.4	. .	. .	. .	. .	53 52.62		1.99	III.	8	8.22	38 10.7	. .	53 54.61	
51	10	. .	. .	. .	. .	. .	9.5	23.5	56 40.62		1.97	VII.	8	9.42	38 50.8	. .	56 42.59	
52	9	29.5	41.8	. .	. .	. .	. .	. .	59 13.13		1.96	II.	7	4.18	31 8.0	. .	20 59 15.09	
53	11	. .	. .	50.0	5.0	. .	. .	. .	20 0 4.63		1.95	IV.	8	5.31	36 44.4	. .	21 0 6.58	
54	5	. .	. .	15.7	30.0	. .	. .	. .	21 1 29.71		1.91	IV.	1	3.27	1 41.7	. .	1 31.62	
55	8	. .	. .	. .	. .	. .	14.0	58.5	2 15.33		1.93	IV.	5	10.37	25 17.6	. .	2 17.26	
56	7	. .	. .	9.0	23.0	. .	. .	. .	4 23.00		1.90	IV.	4	9.	19 28.8	. .	4 24.90	
57	10	. .	. .	41.0	55.5	. .	. .	. .	5 55.39		1.90	IV.	8	8.54	38 27.0	. .	5 57.29	
58	10	. .	. .	. .	59.0	. .	. .	. .	6 58.84		1.87	IV.	3	6.	12 58.2	. .	7 0.71	
59	7	. .	. .	. .	. .	. .	. .	13.0	7 31.52		1.85	VII.	1	7.42	3 50.1	. .	7 33.37	
60	6	. .	. .	. .	9.5	23.6	. .	. .	9 9.25		1.85	V.	3	9.57	14 57.9	. .	9 11.10	
61	4	. .	. .	. .	. .	14.3	28.5	. .	9 59.89		1.85	VI.	4	8.33	19 14.9	. .	10 1.74	
62	6	. .	23.0	37.7	. .	. .	. .	. .	12 51.88		1.85	III.	8	9.32	43 45.7	. .	12 53.73	
63	9	35.0	50.0	. .	. .	. .	. .	. .	14 18.56		1.84	II.	8	8.20	38 9.6	. .	14 20.40	
64	8	. .	. .	. .	. .	. .	. .	1.5	14 18.43		1.84	VII.	8	8.12	38 5.3	. .	14 20.27	
65	9	. .	. .	. .	. .	. .	. .	0.5	15 17.52		1.84	VII.	9	11.4	44 31.8	. .	15 19.36	
66	7	. .	19.5	34.0	. .	. .	. .	. .	17 48.23		1.81	III.	7	12.	35 1.6	. .	17 50.04	
67	11	. .	. .	. .	. .	56.5	. .	. .	18 42.05		1.77	V.	3	3.5	11 29.6	. .	18 43.82	
68	9	. .	9.5	23.5	. .	. .	. .	. .	20 37.82		1.77	III.	5	5.11	22 32.8	. .	20 39.59	
69	8	. .	. .	16.5	30.8	. .	. .	. .	21 30.69		1.77	IV.	6	2.44	25 20.1	. .	21 32.46	
70	9	. .	. .	. .	. .	23.0	37.5	52.0	22 8.75		1.77	VII.	5	7.52	18 54.0	. .	22 10.52	
71	11	. .	. .	. .	. .	. .	1.0	15.5	23 32.32		1.75	VII.	5	6.39	23 16.9	. .	23 34 07	
72	6	5.0	19.5	. .	. .	. .	. .	. .	25 48.21		1.74	II.	6	9.45	28 52.6	. .	25 49.95	
73	9	28.0	43.0	. .	. .	. .	. .	. .	28 11.33		1.71	II.	4	4.9	17 1.5	. .	28 13.04	
74	9	. .	. .	. .	. .	. .	18.5	33.0	31 49.77		1.67	VII.	3	1.39	10 45.9	. .	31 51.44	
75	9	. .	. .	. .	. .	40.0	. .	. .	35 25.51		1.65	V.	2	2.18	6 6.4	. .	35 27.16	
76	6	. .	. .	. .	. .	0.2	14.2	. .	36 45.69		1.66	VI.	7	1.27	20 39.5	. .	36 47.35	
77	9	. .	. .	. .	. .	34.0	48.5	3.5	41 19.86		1.60	VII.	2	10.28	10 13.5	. .	41 21.46	
78	8	14.0	28.5	. .	. .	. .	. .	. .	43 57.24		1.61	II.	7	5.24	31 41.4	. .	43 58.85	
79	8	. .	. .	. .	13.0	27.0	. .	. .	44 12.72		1.60	V.	5	7.54	23 55.1	. .	44 14.32	
80	6	. .	. .	. .	. .	. .	26.0	40.3	21 44 57.20	+	1.57	VII.	4	5.53	−17 53.8	. .	21 44 58.77	

CORRECTIONS.

Date.		Corr. of Clock.	Hourly rate.	m	n	c
1846.	h.	s.	s.	s.	s.	s.

INSTRUMENT READINGS.

Date.		Barom.	THERMOM. At.	Ex.
1846.	h. m.	In.	°	°

REMARKS.

(59) 33. Transit over T. VII assumed as 19ˢ instead of 17ˢ.
(59) 38. Micrometer reading assumed as 7ʳ.21 instead of 3ʳ.21.
(59) 44. Micrometer reading assumed as 2ʳ.54 instead of 3ʳ.54.
(59) 59. Transit over T. VII assumed as 15ˢ.0 instead of 13ˢ.0.
(59) 62. Hor. thread assumed as 9 instead of 8.
(59) 70. Hor. thread assumed as 4 instead of 5.
(59) 78. Hor. thread assumed as 7 instead of [illegible].

ZONE 60. AUGUST 29. A. $D_o = -39° 4' 40''$.

No.	Mag.	SECONDS OF TRANSIT. I.	II.	III.	IV.	V.	VI.	VII.	T.	a_1	MICROMETER.			$i + d_0$	d_1	Mean Right Ascension, 1850.0.	Mean Declination, 1850.0.
									h. m. s.	s.			r.	' "	"	h. m. s.	° ' "
1	5						10.0	26.5	18 59 37.89	+ 4.77	VII.	6	8.56	−28 59.3	− 43.23	18 59 42.66	− 39 34 22.5
2	8			20.2	36.4				19 1 36.15	4.75	IV.	5	10.14	25 5.9	42.18	19 1 40.90	30 28.1
3	8		6.2	22.5	39.0				3 38.50	4.71	IV.	3	8.34	14 13.4	41.07	3 43.21	19 34.5
4	8		48.2	4.8					5 20.54	4.73	III.	5	2.40	21 15.6	40.18	5 25.27	26 35.6
5	7		46.5	3.0	19.5				8 19.20	4.73	IV.	9	6.57	42 31.8	38.57	8 23.93	47 50.4
6	11			44.5	1.0				15 0.65	4.64	IV.	6	11.5	29 34.4	35.02	15 5.29	34 49.4
7	10	50.1	6.8						18 39.00	4.60	II.	5	10.22	25 9.7	33.07	18 43.60	30 22.8
8	10				59.5	15.5			18 59.34	4.59	V.	4	10.18	20 6.9	32.89	19 3.93	25 19.8
9	7						16.0	32.3	19 43.80	4.60	VII.	7	4.50	31 25.3	32.49	19 48.40	36 37.6
10	11					36.0	52.2		21 19.92	4.58	VI.	6	6.51	27 25.2	31.66	21 24.50	32 36.9
11	10							58.5	22 9.91	4.57	VII.	5	3.50	21 50.5	31.22	22 14.48	27 1.7
12	11			52.0	9.0				25 8.29	4.54	IV.	4	5.35	17 43.4	29.67	25 12.83	22 53.1
13	6	49.5	6.0						28 38.05	4.48	II.	2	5.36	7 42.0	27.79	28 42.53	12 49.8
14	6			21.5	38.0				29 37.74	4.53	IV.	9	1.55	39 58.6	27.26	29 42.27	45 5.9
15	8		59.0	15.5					32 31.66	4.50	III.	9	9.51	44 0.1	25.77	32 36.16	49 5.9
16	7				23.5	40.0			33 23.54	4.44	V.	2	4.47	7 17.4	25.31	33 27.98	12 22.7
17	10					45.5	1.5		34 29.35	4.43	VI.	2	9.	9 25.5	24.75	34 33.78	14 30.3
18	6				16.3	32.4			36 16.23	4.47	V.	9	4.15	41 9.6	23.79	36 20.70	46 13.4
19	11				22.5	39.5			37 22.31	4.42	V.	4	3.4	16 26.8	23.22	37 26.73	21 30.0
20	11				48.5	4.5			38 48.32	4.40	V.	4	7.27	18 40.1	22.47	38 52.72	23 42.6
21	9		46.0	2.5	19.0				42 18.72	4.41	IV.	9	10.40	44 25.0	20.65	42 23.13	49 25.7
22	10						46.5	3.0	43 14.34	4.35	VII.	3	2.36	21 14.8	20.17	43 18.69	26 15.0
23	11			42.0	58.2				45 58.09	4.35	IV.	7	7.30	32 47.2	18.74	46 2.44	37 45.9
24	6				9.0	25.0			47 8.88	4.36	V.	9	3.6	40 34.6	18.13	47 13.24	45 32.7
25	8						24.8	40.8	47 52.46	4.33	VII.	7	4.9	31 4.5	17.75	47 56.79	36 2.2
26	10		53.0	9.5					51 25.18	4.27	III.	3	5.38	12 44.1	15.95	51 29.45	17 40.0
27	8					44.0	0.5		53 27.08	4.29	VI.	8	9.54	39 0.5	14.86	53 31.37	43 55.4
28	5							17.0	53 28.24	4.24	VII.	2	9.53	9 52.0	14.87	53 32.48	14 46.9
29	9		27.0	43.0					55 59.40	4.27	III.	9	7.24	42 45.5	13.57	56 3.67	47 39.1
30	8		36.5	53.5					57 9.38	4.27	III.	9	3.39	40 51.3	13.00	57 13.65	45 44.3
31	7					31.5	47.5	4.5	19 59 15.48	+ 4.18	VII.	2	6.	− 7 53.8	− 11.90	19 59 19.66	− 39 12 45.7

ZONE 61. AUGUST 29. A. $D_o = -35° (17)'$.

No.	Mag.	I.	II.	III.	IV.	V.	VI.	VII.	T.	a_1	MICROMETER.			$i + d_0$	d_1	Mean Right Ascension, 1850.0.	Mean Declination, 1850.0.
1	11			30.5	16.0				20 15 45.81	+ 3.89	IV.	6	8.56	−28 28.4	. .	20 15 49.69	
2	11					43.0	59.0		16 27.99	3.86	VI.	5	8.8	24 1.9	. .	16 31.85	
3	9						46.0	1.0	18 15.11	3.86	VII.	8	3.38	35 47.8	. .	18 18.97	
4	9						45.5	1.0	19 14.84	3.84	VII.	7	2.4	30 0.5	. .	19 18.68	
5	8						39.5	55.0	19 8.85	3.84	VII.	7	8.40	33 20.9	. .	19 12.69	
6	9					53.0	8.0	23.5	21 37.45	3.63	VII.	7	8.29	33 15.3	. .	21 41.08	
7	8					16.0	31.5	47.0	22 0.81	3.65	VII.	9	8.28	43 14.7	. .	22 4.46	
8	9		18.5	34.0	49.6				28 49.36	3.55	IV.	7	4.49	31 24.5	. .	28 52.91	
9	7					45.0	0.6	16.0	30 29.83	3.55	VII.	8	9.10	38 35.8	. .	30 33.38	
10	9			0.0	15.5				33 15.40	3.62	IV.	8	10.6	39 4.8	. .	33 19.02	
11	8					13.5	29.0	44.5	33 58.32	3.52	VII.	10	4.44	46 8.2	. .	34 2.84	
12	4				39.0	54.0			35 38.71	3.47	V.	5	8.44	24 20.3	. .	35 42.18	
13	8						32.0	47.2	36 1.21	3.49	VII.	8	5.45	36 52.1	. .	36 4.70	
14	7				42.2	57.5			20 37 42.06	+ 3.45	V.	5	7.42	−23 48.9	. .	20 37 45.51	

CORRECTIONS.

Date.		Corr. of Clock.	Hourly rate.	m	n	c
1846.	h.	s.	s.	s.	s.	s.

INSTRUMENT READINGS.

	Date.		Barom.	THERMOM. At.	THERMOM. Ex.
	1846.	h. m.	in.	°	°
Zone 60	Aug. 29,	18 59	30.06	77.5	73.5
		19 59	30.06	77.5	75.5
Zone 61	Aug. 29,	20 15	30.06	77.0	73.0
		21 3	30.07	77.0	71.3

REMARKS.

(60) 1. Micrometer reading assumed as $9^r.56$ instead of $8^r.56$.

ZONE 61. AUGUST 29. A. $D_0 = -35° (17)'$—Continued.

No.	Mag.	I.	II.	III.	IV.	V.	VI.	VII.	T.	a_1	MICROMETER.			$i + d_2$	d_1	Mean Right Ascension, 1850.0.	Mean Declination, 1850.0.
									h. m. s.	s.			r.	° ′ ″	″	h. m. s.	° ′ ″
15	8	. .	. .	. .	. .	. .	58.0	13.0	20 38 27.11	+ 3.47	VII.	8	8.17	−38 9.0	. .	20 38 50.58	
16	11	. .	. .	. .	16.5	32.0	. .	. .	40 16.47	3.43	V.	6	7.24	27 41.8	. .	40 19.90	
17	11	. .	. .	. .	3.5	19.0	. .	. .	42 3.47	3.41	V.	6	6.25	27 11.9	. .	42 6.88	
18	9	. .	. .	. .	10.0	25.2	. .	. .	43 9.82	3.40	V.	6	4.53	26 25.4	. .	43 13.22	
19	8	. .	. .	15.0	30.0	. .	. .	. .	44 29.97	3.38	IV.	4	4.55	17 23.1	. .	44 33.35	
20	8	. .	. .	. .	. .	31.8	47.0	2.5	46 16.31	3.35	VII.	2	8.	8 57.1	. .	46 19.66	
21	8	. .	. .	. .	26.5	41.5	57.0	. .	50 26.29	3.30	VI.	2	3.52	6 51.9	. .	50 29.59	
22	6	. .	. .	. .	. .	4.5	19.3	35.0	51 48.87	3.30	VII.	6	7.55	27 57.0	. .	51 52.17	
23	8	. .	. .	. .	. .	30.0	45.5	1.0	53 14.79	3.31	VII.	8	4.38	36 18.2	. .	53 18.10	
24	9	. .	. .	. .	. .	4.0	19.5	. .	54 48.77	3.25	VI.	2	2.50	6 20.5	. .	54 52.02	
25	9	50.0	6.0	21.5	. .	. .	. .	. .	57 36.58	3.23	III.	5	3.4	21 29.3	. .	57 39.81	
26	10	. .	. .	. .	. .	9.5	25.2	. .	20 58 54.35	3.23	VI.	7	11.	34 32.0	. .	20 58 57.58	
27	9	30.0	45.2	. .	. .	. .	. .	. .	21 1 15.95	3.18	II.	2	4.30	7 11.1	. .	21 1 19.13	
28	9	. .	. .	. .	47.5	2.8	. .	. .	1 47.35	3.18	V.	4	6.9	18 1.6	. .	1 50.53	
29	8	6.4	22.5	. .	. .	. .	. .	. .	21 3 53.00	+ 3.17	II.	5	5.47	−22 50.6	. .	21 3 56.17	

ZONE 62. SEPTEMBER 2. A. $D_0 = -31° 34' 10''$.

No.	Mag.	I.	II.	III.	IV.	V.	VI.	VII.	T.	a_1	MICROMETER.			$i + d_2$	d_1	Mean Right Ascension, 1850.0.	Mean Declination, 1850.0.
1	5	. .	. .	. .	48.0	2.5	17.0	. .	18 14 47.76	+ 1.60	VI.	10	7.37	−47 49.6	. .	18 14 49.36	
2	6	. .	. .	. .	8.2	23.2	. .	. .	16 8.23	1.54	V.	4	1.50	15 51.2	. .	16 9.77	
3	9	. .	. .	. .	. .	. .	21.0	36.1	16 51.76	1.54	VII.	4	3.50	16 51.4	. .	16 53.30	
4	9	. .	. .	. .	. .	. .	. .	20.5	17 36.23	1.54	VII.	5	7.14	23 34.5	. .	17 37.77	
5	8	. .	. .	. .	. .	5.5	20.5	35.5	18 51.07	1.54	VII.	5	8.20	24 7.8	. .	18 52.61	
6	10	. .	. .	. .	. .	. .	18.5	33.5	19 49.24	1.54	VII.	6	9.12	28 35.8	. .	19 50.78	
7	7	. .	. .	. .	21.5	36.0	. .	. .	21 21.35	1.56	V.	10	9.36	48 49.9	. .	21 22.91	
8	10	. .	. .	. .	. .	38.0	52.5	. .	22 23.22	1.54	VI.	9	3.8	40 31.9	. .	22 24.76	
9	9	. .	. .	6.0	21.5	. .	. .	. .	24 21.14	1.54	IV.	10	6.13	47 7.3	. .	24 22.68	
10	10	. .	. .	45.2	59.5	. .	. .	. .	26 59.48	1.47	IV.	3	10.26	15 12.4	. .	27 0.95	
11	10	. .	. .	2.0	17.0	. .	. .	. .	28 16.67	1.46	IV.	4	11.11	20 34.8	. .	28 18.13	
12	7	. .	8.5	23.4	. .	. .	. .	. .	29 38.12	1.49	III.	8	3.14	35 35.4	. .	29 39.61	
13	8	. .	34.0	48.8	. .	. .	. .	. .	31 3.53	1.47	III.	7	10.52	34 27.5	. .	31 5.00	
14	6	. .	43.5	58.5	13.0	. .	. .	. .	32 12.74	1.42	IV.	1	4.45	2 20.6	. .	32 14.16	
15	7	. .	2.2	17.3	. .	. .	. .	. .	33 31.59	1.43	III.	3	4.57	12 26.0	. .	33 33.02	
16	10	. .	. .	. .	10.0	24.2	. .	. .	18 34 9.67	+ 1.47	V.	8	5.42	−36 50.2	. .	18 34 11.14	

ZONE 63. SEPTEMBER 2. A. $D_0 = -26° 32' 40''$.

No.	Mag.	I.	II.	III.	IV.	V.	VI.	VII.	T.	a_1	MICROMETER.			$i + d_2$	d_1	Mean Right Ascension, 1850.0.	Mean Declination, 1850.0.
1	7	. .	2.3	16.5	. .	. .	. .	. .	19 0 30.23	+ 0.78	III.	3	8.58	−14 28.4	. .	19 0 31.01	
2	7	. .	. .	. .	. .	. .	2.0	16.5	0 34.23	0.78	VII.	4	11.58	20 58.4	. .	0 35.01	
3	8	. .	. .	. .	. .	25.5	39.2	. .	2 11.37	0.82	VI.	10	6.11	47 5.0	. .	2 12.19	
4	10	. .	. .	. .	. .	. .	37.8	52.0	3 9.91	0.79	VII.	7	2.59	30 27.8	. .	3 10.70	
5	5	. .	. .	. .	. .	51.0	4.9	18.9	4 36.96	0.78	VII.	7	12.	35 1.0	. .	4 37.74	
6	8	1.4	15.7	29.8	. .	. .	. .	. .	6 43.64	0.74	III.	4	10.19	20 8.7	. .	6 44.38	
7	11	. .	. .	. .	15.0	29.0	. .	. .	7 14.90	0.75	V.	5	6.55	23 25.4	. .	7 15.65	
8	10	. .	. .	. .	. .	25.5	39.3	. .	8 11.42	0.72	VI.	3	5.30	12 43.2	. .	8 12.14	
9	8	. .	. .	. .	40.0	53.5	. .	. .	9 39.67	0.76	V.	10	7.1	47 30.4	. .	9 40 45	
10	8	. .	. .	46.3	0.6	14.1	. .	. .	19 11 0.28	+ 0.73	V.	7	7.35	−32 47.5	. .	19 11 1.01	

CORRECTIONS.

Date.		Corr. of Clock.	Hourly rate.	m	n	c
1846.	h.	s.	s.	s.	s.	s.
Sept. 2,	18	− 10.53	− 0.029	+ 0.372	− 0.163	0.000

INSTRUMENT READINGS.

	Date.		Barom.	THERMOM. At.	THERMOM. Ex.
	1846.	h. m.	In.	°	°
Zone 62	Sept. 2,	18 14	30.06	82.5	80.5
		18 34	30.07	82.2	79.5
Zone 63	Sept. 2,	19 0	30.07	82.0	78.5
		19 32	30.07	82.0	78.0

REMARKS.

ZONE 63. SEPTEMBER 2. A. $D_o = -26° 32' 40''$—Continued.

No.	Mag.	SECONDS OF TRANSIT. I.	II.	III.	IV.	V.	VI.	VII.	T.		a_1	MICROMETER.			$i + d_1$		d_1	Mean Right Ascension, 1850.0.	Mean Declination, 1850.0.
									h. m. s.		s.			r.	′ ″		″	h. m. s.	° ′ ″
11	8	..	..	..	..	4.5	18.2	..	19 11 50.37	+	0.75	VI.	10	7.42	−47 51.0		..	19 11 51.12	
12	8	2.6	16.8	..	..	..	..	..	13 44.89		0.71	II.	6	6.26	27 12.1		..	13 45.60	
13	9	..	..	..	..	20.5	35.0	..	14 6.77		0.74	VI.	9	8.40	43 18.9		..	14 7.51	
14	8	..	..	..	..	..	41.9	55.8	15 13.92		0.73	VII.	10	2.49	45 22.8		..	15 14.65	
15	8	59.5	14.0	..	..	..	..	..	17 41.99		0.69	II.	7	6.1	31 59.9		..	17 42.68	
16	9	..	..	..	..	16.0	30.0	44.0	18 2.07		0.71	VII.	9	4.42	41 18.5		..	18 2.78	
17	5	..	..	..	34.0	48.2	..	..	20 34.04		0.69	V.	10	1.22	44 39.2		..	20 34.73	
18	9	..	..	..	..	5.5	20.0	..	22 51.77		0.63	VI.	2	5.56	7 56.8		..	22 52.40	
19	10	..	..	32.0	46.0	..	..	..	24 45.95		0.65	IV.	8	6.34	37 15.9		..	24 46.60	
20	11	..	43.0	58.0	..	..	..	..	26 11.55		0.63	III.	7	7.43	32 51.6		..	26 12.18	
21	7	..	..	..	40.0	53.9	..	..	26 39.81		0.58	V.	1	9.32	4 46.5		..	26 40.39	
22	9	..	..	..	..	..	48.0	2.5	28 20.21		0.60	VII.	4	6.47	18 21.3		..	28 20.81	
23	9	..	..	..	..	42.2	56.0	..	29 18.12		0.70	VI.	6	5.19	26 38.2		..	29 18.82	
24	10	..	..	..	12.0	26.0	..	..	31 11.92		0.59	V.	7	8.5	33 2.6		..	31 12.51	
25	10	..	..	22.0	36.0	..	..	..	19 32 35.99	+	0.60	IV.	9	5.59	−41 57.7		..	19 32 36.59	

ZONE 64. SEPTEMBER 7. A. $D_o = -24° 2' 0''$.

No.	Mag.	I.	II.	III.	IV.	V.	VI.	VII.	T.		a_1	MICROMETER.			$i + d_1$		d_1	Mean Right Ascension, 1850.0.	Mean Declination, 1850.0.
1	5	..	..	41.5	54.0	8.0	..	..	21 50 54.42	−	4.75	V.	6	11.55	−29 58.3	−	55.6	21 50 49.67	− 24 32 53.9
2	10	..	..	..	..	..	4.0	18.0	51 36.85		4.76	VII.	6	12.	30 0.5		55.3	51 32.09	32 55.8
3	7	9.6	23.6	..	..	..	..	..	53 50.86		4.78	II.	4	5.5	17 30.1		54.4	53 46.08	20 24.5
4	9	47.5	1.3	..	..	..	..	..	56 28.73		4.80	II.	5	9.14	24 35.4		53.3	56 23.93	27 28.7
5	8	..	..	45.5	59.0	..	..	..	56 58.81		4.82	IV.	2	4.1	6 59.1		53.1	56 53.99	9 52.2
6	9	..	..	31.5	44.8	..	..	..	58 44.80		4.82	IV.	4	8.	18 58.7		52.3	58 39.98	21 51.0
7	7	..	27.5	41.5	..	..	..	..	21 59 54.96		4.83	III.	5	9.57	24 57.3		51.8	21 59 50.13	27 49.1
8	6	28.0	41.8	..	..	..	..	..	22 1 9.19		4.84	II.	4	11.20	20 39.5		51.4	22 1 4.35	23 30.9
9	7	..	45.0	59.0	..	..	..	..	2 12.33		4.85	III.	3	8.35	14 16.9		50.9	2 7.48	17 7.8
10	9	..	..	..	..	..	47.5	1.0	2 20.13		4.83	VII.	9	8.11	43 3.8		50.9	2 15.30	45 54.7
11	8	54.6	8.5	..	..	..	..	..	5 35.86		4.87	II.	5	4.39	22 16.6		49.5	5 30.99	25 6.1
12	9	..	..	..	..	59.0	12.8	..	5 45.44		4.87	VI.	8	6.	36 58.4		49.5	5 40.57	39 47.9
13	6	..	..	..	..	..	55.0	8.7	6 27.73		4.86	VII.	9	6.3	41 59.2		49.1	6 22.87	44 48.3
14	9	32.0	16.0	..	..	..	..	..	9 13.31		4.91	II.	5	9.24	24 40.5		48.0	9 8.40	27 28.5
15	5	..	..	..	13.0	26.5	..	..	10 12.77		4.93	V.	1	3.42	1 50.1		47.6	10 7.84	4 37.7
16	8	..	..	..	..	25.0	38.6	..	11 11.33		4.92	VI.	7	3.2	30 29.4		47.2	11 6.41	33 16.6
17	5	..	..	7.0	20.7	..	..	..	16 20.36		4.98	IV.	1	9.23	4 42.3		45.3	16 15.38	7 27.6
18	4	15.6	29.6	..	..	..	..	..	17 56.92		4.98	II.	5	7.48	23 52.0		44.6	17 51.94	26 36.6
19	9	25.0	37.5	..	..	..	..	..	18		4.97	II.	8	3.32	35 43.7		44.5	18	38 28.2
20	7	59.0	13.0	..	..	..	..	..	21 40.41		5.01	II.	7	9.50	34 26.7		43.1	21 35.40	37 9.8
21	9	..	..	10.0	23.8	..	..	..	22 23.56		5.02	IV.	5	3.10	21 31.8		42.9	22 18.54	24 14.7
22	4	44.0	51.8	..	..	..	..	..	26		5.03	II.	9	8.13	43 5.0		41.5	26	45 46.5
23	8	..	..	..	..	58.0	11.2	..	27 57.54		5.05	VI.	9	9.41	43 49.4		40.8	27 52.49	46 30.2
24	9	50.2	4.3	..	..	..	..	..	31 31.56		5.10	II.	5	10.4	25 0.7		39.5	31 26.46	27 40.2
25	10	..	..	..	..	..	48.5	2.0	35 21.07		5.12	VII.	6	6.26	27 11.9		37.9	35 15.75	29 49.8
26	8	..	..	4.5	..	32.0	..	..	36 18.21		5.13	V.	7	3.14	30 35.6		37.5	36 13.08	33 13.1
27	7	32.3	46.0	..	..	..	..	..	43 13.54		5.18	II.	7	3.59	30 58.2		34.9	43 8.36	33 33.1
28	7	..	52.0	6.2	..	..	..	..	45 19.73		5.19	III.	9	3.47	40 50.8		34.1	45 14.54	43 24.9
29	8	25.8	39.8	..	..	..	..	..	47 7.30		5.21	II.	9	8.38	43 17.6		33.5	47 2.09	45 51.1
30	8	..	..	59.2	13.0	..	..	..	22 48 12.87	−	5.22	IV.	8	4.54	−36 25.3	−	33.1	22 48 7.65	− 24 38 58.4

CORRECTIONS.

Date.		Corr. of Clock.	Hourly rate.	m	n	c
1846.	h.	s.	s.	s.	s.	s.
Sept. 7.	18	− 14.57	− 0.027	+ 0.497	− 0.319	0.000

INSTRUMENT READINGS.

	Date.		Barom.	THERMOM. At.	Ex.
	1846	h. m.	in.	°	°
Zone 64	Sept. 7.	21 50	30.09	81.0	72.0
		23 24	30.08	79.0	49.0

REMARKS.

(63) 23. Transits over T.'s V and VI assumed as 32s.2 and 46s.0 instead of 42s.2 and 56s.0.
(64) 20. Micrometer assumed as 10r.50 instead of 9r.50.
(64) 23. Transits over T.'s IV and V assumed as recorded over T.'s V and VI.
(64) 24. Transit over T. VI rejected.

ZONE 64. SEPTEMBER 7. A. $D_o = -24° 2' 0''$—Continued.

No.	Mag.	I.	II.	III.	IV.	V.	VI.	VII.	T.	a_1	Micrometer			$i + d_1$	d_1	Mean Right Ascension, 1850.0.	Mean Declination, 1850.0.
									h. m. s.	s.			r.	′ ″	″	h. m. s.	° ′ ″
31	9	..	..	3.2	17.0	..	..	..	22 49 16.86	− 5.23	IV.	8	3.35	−35 45.4	− 32.6	22 49 11.63	− 24 38 18.0
32	8	..	7.5	21.7	..	..	..	..	50 35.07	5.25	III.	6	6.0	26 59.0	32.2	50 29.82	29 31.2
33	5	..	..	..	..	..	13.5	27.5	50 46.26	5.26	VII.	3	4.46	12 21.0	32.1	50 41.00	14 53.1
34	8	47.2	1.2	..	..	..	..	..	54 28.49	5.28	II.	4	10.57	20 28.6	30.8	54 23.21	22 59.4
35	9	..	42.0	55.5	..	..	..	..	56 9.30	5.29	III.	7	8.26	33 13.2	30.2	56 4.01	35 43.4
36	8	..	..	48.0	2.0	..	..	..	57 1.77	5.30	IV.	8	4.58	36 27.3	29.9	56 56.47	38 57.2
37	4	0.3	14.5	28.0	..	..	..	..	22 58 41.72	5.31	III.	7	3.3	30 30.1	29.3	22 58 36.41	32 59.4
38	7	51.5	5.9	19.5	..	..	..	..	23 4 33.14	5.36	III.	8	3.0	35 27.7	27.2	23 4 27.78	37 54.9
39	8	..	24.0	37.5	51.3	..	..	..	6 51.27	5.38	IV.	8	5.0	36 28.3	26.4	6 45.89	38 54.7
40	8	..	..	57.0	11.5	..	..	..	9 11.08	5.39	IV.	10	2.40	45 18.3	25.6	9 5.69	47 43.9
41	6	16.5	30.5	..	..	..	..	..	14 57.72	5.46	II.	3	7.39	13 48.5	23.6	14 52.26	16 12.1
42	8	..	..	..	..	27.0	40.4	..	15 13.23	5.47	VI.	2	10.15	10 7.8	23.5	15 7.76	12 31.3
43	8	..	51.0	5.0	..	..	..	..	17 18.31	5.48	III.	3	4.20	12 8.2	22.8	17 12.83	14 31.0
44	6	..	..	22.5	36.0	..	..	..	21 36.10	5.49	IV.	10	9.43	49 22.2	21.4	21 30.61	51 43.6
45	10	..	..	..	28.5	42.0	..	..	23 24 28.33	− 5.52	V.	6	5.55	−26 56.5	− 20.5	23 24 22.81	− 24 29 17.0

ZONE 65. SEPTEMBER 9. A. $D_o = -39° 4' 40''$.

No.	Mag.	I.	II.	III.	IV.	V.	VI.	VII.	T.	a_1	Micrometer			$i + d_1$	d_1	Mean Right Ascension, 1850.0.	Mean Declination, 1850.0.
1	8	44.0	1.5	..	..	..	..	..	18 28 33.42	− 2.78	II.	7	8.11	−33 7.7	− 53.15	18 28 30.64	− 39 38 40.9
2	8	..	50.0	6.2	..	..	..	..	29 22.23	2.80	III.	5	11.25	25 41.9	52.74	29 19.43	31 14.6
3	9	..	..	..	..	20.8	36.8	..	31 4.64	2.83	VI.	4	5.24	17 37.5	51.85	31 1.81	23 9.4
4	6	..	..	..	32.5	48.5	..	..	32 32.32	2.83	V.	4	9.47	19 51.1	51.07	32 29.49	25 22.2
5	5	..	..	..	34.5	50.5	..	..	34 34.39	2.81	V.	9	10.30	44 20.0	50.01	34 31.58	49 50.0
6	9	..	19.5	35.5	..	..	..	..	36 51.82	2.84	III.	8	7.23	37 44.2	48.84	36 48.98	43 13.0
7	7	..	23.0	39.5	..	..	..	..	37 55.61	2.83	III.	9	2.18	40 10.3	48.24	37 52.78	45 38.5
8	9	..	..	..	..	..	43.5	59.5	38 11.17	2.85	VII.	7	8.27	33 15.5	48.11	38 8.32	38 43.6
9	8	..	..	..	..	56.5	12.0	..	40 40.11	2.90	VI.	5	10.6	25 1.6	46.81	40 37.21	30 28.4
10	6	..	..	..	..	5.5	22.0	..	40 49.62	2.92	VI.	3	10.9	15 1.2	46.73	40 46.70	20 27.9
11	9	21.5	38.0	..	..	..	..	..	43 10.30	2.91	II.	5	9.36	24 46.4	45.50	43 7.39	30 11.9
12	9	..	45.0	1.0	..	..	..	..	44 17.34	2.89	III.	8	9.35	38 51.2	44.91	44 14.45	44 16.1
13	10	..	..	..	..	5.0	21.5	..	44 49.11	2.94	VI.	4	10.57	20 26.4	44.64	44 46.17	25 51.0
14	7	..	..	..	27.2	43.2	..	..	46 27.07	2.91	V.	8	4.35	36 19.0	43.75	46 24.16	41 42.8
15	6	2.0	18.8	..	..	..	..	..	48 51.14	2.93	II.	8	8.19	38 12.4	42.56	48 48.21	43 35.0
16	8	..	55.5	11.7	..	..	..	..	50 27.98	2.93	III.	9	5.11	41 38.1	41.71	50 25.05	46 59.8
17	9	..	15.5	32.5	..	..	..	..	51 48.10	2.99	III.	5	8.3	23 59.4	41.01	51 45.11	29 20.4
18	8	29.0	45.0	..	..	..	..	..	53 17.69	2.98	II.	7	10.5	34 5.6	40.23	53 14.71	39 25.8
19	8	..	..	..	..	57.0	13.5	..	54 40.10	2.97	VI.	9	10.55	44 32.6	39.53	54 37.13	49 52.1
20	8	..	..	..	..	28.8	44.0	..	55	.97	VI.	9	8.19	43 13.3			
21	9	..	..	..	..	9.0	25.3	..	56 52.98	3.04	VI.	4	2.24	16 6.2	38.37	56 49.94	21 24.6
22	4	56.0	12.5	..	..	..	..	..	18 59 44.85	− 3.05	II.	6	9.20	−29 11.4	− 36.92	18 59 41.80	− 39 34 28.3

CORRECTIONS.

Date.		Corr. of Clock.	Hourly rate.	m	n	c
1846. Sept. 9,	h. 18	s. − 15.75	s. − 0.052	s. + 0.497	s. − 0.319	s. 0.000

INSTRUMENT READINGS.

	Date.		Barom.	Thermom. At.	Thermom. Ex.
Zone 65	1846. Sept. 9,	h. m. 18 28	in. 30.16	° 76.0	° 66.1

REMARKS.

(64) 44. Micrometer assumed as 10r.43 instead of 9r.43.
(65) 22. Micrometer reading assumed as 10r.20 instead of 9r.20.

ZONE 66. SEPTEMBER 9. A. $D_0 = -32° 49' 30''$.

No.	Mag.	SECONDS OF TRANSIT. I.	II.	III.	IV.	V.	VI.	VII.	T.	a_1	MICROMETER.			$i + d_2$	d_1	Mean Right Ascension, 1850.0.	Mean Declination, 1850.0.
									h. m. s.	s.			r.	' "	"	h. m. s.	° ' "
1	9	. .	. .	. .	. .	6.5	21.5	. .	20 40 51.67	— 4.70	VI.	9	10.27	−44 14.3	— 17.4	20 40 46.97	— 33 34 1.7
2	9	. .	. .	. .	36.0	51.0	. .	. .	42 35.94	4.76	V.	4	10.28	20 12.9	17.2	42 31.18	10 0.1
3	9	. .	28.5	43.5	. .	. .	. .	. .	44 58.50	4.75	III.	8	2.38	35 17.5	16.8	44 53.75	25 4.3
4	10	. .	. .	. .	. .	. .	38.0	53.0	45 8.23	4.77	VII.	6	5.1	26 28.9	16.7*	45 3.46	16 15.6
5	8	. .	36.0	50.9	. .	. .	. .	. .	47 5.85	4.80	III.	7	4.15	31 7.0	16.4	47 1.05	20 53.4
6	10	9.5	24.8	. .	. .	. .	. .	. .	51 54.62	4.85	II.	6	4.30	26 13.5	15.6	51 49.77	15 59.1
7	5	. .	. .	. .	. .	. .	45.4	0.6	52 15.79	4.82	VII.	9	9.47	38 54.0	15.5	52 10.97	28 39.5
8	7	. .	. .	. .	49.0	4.0	. .	. .	53 48.94	4.89	V.	4	11.23	20 40.7	15.3	53 44.05	10 26.0
9	7	11.5	26.8	. .	. .	. .	. .	. .	55 56.74	4.88	II.	8	2.57	35 26.9	15.0	55 51.86	25 11.9
10	8	. .	20.0	35.0	. .	. .	. .	. .	56 49.99	4.89	III.	8	5.13	36 30.8	14.8	56 45.10	33 25 15.6
11	4	. .	. .	. .	. .	33.5	48.1	. .	20 57 18.49	4.93	VI.	2	3.32	6 27.5	14.7	20 57 13.56	32 56 12.2
12	7	. .	44.1	59.4	. .	. .	. .	. .	21 1 14.25	4.93	III.	8	7.20	37 40.1	14.1	21 1 9.32	33 27 24.2
13	9	. .	35.5	51.0	. .	. .	. .	. .	5 5.59	5.00	III.	6	4.57	26 24.8	13.5	5 0.59	16 8.3
14	7	. .	. .	. .	. .	34.5	49.3	. .	5 19.58	5.01	VI.	4	8.56	19 26.2	13.5	5 14.57	9 9.7
15	8	. .	. .	42.6	57.8	. .	. .	. .	6 57.56	5.02	IV.	7	4.0	30 59.4	13.3	6 52.54	20 42.7
16	8	. .	. .	45.5	1.0	. .	. .	. .	9 0.68	5.03	IV.	8	10.58	39 30.4	12.9	8 55.65	33 29 13.3
17	5	. .	. .	. .	2.5	17.4	. .	. .	10 2.35	5.08	V.	2	8.35	9 16.0	12.8	9 57.27	32 58 58.8
18	9	. .	. .	. .	. .	. .	7.0	22.0	10 37.18	5.09	VII.	3	8.28	14 10.4	12.7	10 32.09	33 3 53.1
19	6	. .	. .	. .	. .	. .	45.0	0.2	11 15.30	5.08	VII.	4	6.2	18 58.6	12.6	11 10.22	8 41.2
20	9	. .	58.5	13.5	. .	. .	. .	. .	14 28.47	5.09	III.	8	3.48	33 52.0	12.2	14 23.38	25 35.1
21	8	48.8	4.0	. .	. .	. .	. .	. .	18 34.08	5.14	II.	9	5.21	41 39.6	11.6	18 28.94	31 21.2
22	10	. .	. .	. .	. .	. .	5.0	20.2	18 35.29	5.18	VII.	4	8.21	19 8.2	11.6	18 30.11	8 49.8
23	9	5.0	20.5	. .	. .	. .	. .	. .	25 50.25	5.25	II.	6	7.15	27 37.0	10.6	25 45.00	17 17.6
24	10	. .	. .	. .	53.0	7.5	. .	. .	27 52.68	5.29	V.	4	6.33	18 14.1	10.3	27 47.39	7 54.4
25	8	. .	17.5	32.6	. .	. .	. .	. .	30 47.53	5.29	III.	8	4.40	36 19.2	9.9	30 42.24	25 59.1
26	6	2.4	17.2	. .	. .	. .	. .	. .	32 47.24	5.31	II.	5	5.50	22 52.2	9.6	32 41.90	12 31.8
27	7	. .	. .	7.0	21.7	. .	. .	. .	33 21.63	5.31	IV.	5	3.16	21 34.5	9.6	33 16.29	11 14.1
28	7	. .	. .	. .	. .	. .	57.0	12.5	34 27.51	5.33	VII.	8	7.2	37 30.5	9.4	34 22.18	27 9.9
29	8	. .	. .	. .	59.0	13.6	. .	. .	36 58.77	5.37	V.	7	7.43	32 52.1	9.1	36 53.40	22 31.2
30	6	. .	4.0	19.3	. .	. .	. .	. .	37 34.10	5.39	III.	7	10.40	34 21.7	9.1	37 28.71	24 0.8
31	10	. .	. .	. .	. .	29.0	43.5	. .	38 13.93	5.41	VI.	4	3.20	16 36.3	9.0	38 8.52	6 15.3
32	6	. .	. .	. .	. .	31.2	46.0	. .	39 16.28	5.43	VI.	3	4.58	12 26.2	8.8	39 10.85	2 5.0
33	10	. .	. .	. .	. .	. .	35.5	51.0	40 5.95	5.44	VII.	3	11.50	15 54.1	8.7	40 0.51	5 32.8
34	10	. .	. .	. .	. .	. .	54.0	9.0	41 24.23	5.43	VII.	6	6.43	27 20.5	8.6	41 18.80	16 59.1
35	8	. .	. .	. .	44.5	59.3	. .	. .	42 44.35	5.46	V.	5	8.37	24 16.8	8.4	42 38.89	13 55.2
36	10	. .	. .	. .	. .	. .	59.5	15.0	43 29.98	5.46	VII.	6	10.13	29 6.7	8.3	43 24.52	18 45.0
37	8	. .	31.2	46.5	. .	. .	. .	. .	46 1.43	5.46	III.	9	9.49	43 55.3	8.0	45 55.97	33 33.3
38	9	. .	. .	53.0	8.0	. .	. .	. .	47 7.78	5.50	IV.	5	4.20	22 6.9	7.9	47 2.28	11 44.8
39	7	. .	58.0	12.8	. .	. .	. .	. .	48 27.67	5.53	III.	4	11.14	20 36.2	7.8	48 22.14	10 14.0
40	7	. .	. .	. .	57.0	12.5	. .	. .	48 57.21	5.53	V.	6	7.35	27 47.2	7.7	48 51.68	17 24.9
41	7	. .	. .	. .	55.5	10.5	. .	. .	49 55.42	5.56	V.	3	8.49	14 23.0	7.6	49 49.86	33 4 0.6
42	5	. .	. .	. .	. .	8.2	22.5	. .	53 53.04	5.61	VI.	1	3.32	1 42.8	7.2	53 47.43	32 51 20.0
43	8	. .	. .	. .	. .	21.2	35.7	50.8	55 6.03	5.61	VII.	3	10.11	15 4.0	7.1	55 0.42	33 4 41.1
44	8	. .	. .	. .	. .	. .	36.9	52.1	56 7.23	5.61	VII.	6	5.57	26 57.2	7.0	56 1.62	16 34.2
45	8	. .	52.5	7.5	. .	. .	. .	. .	21 59 22.21	5.66	III.	4	4.19	17 6.3	6.6	21 59 16.55	6 42.9
46	7	. .	. .	1.3	16.1	. .	. .	. .	22 0 16.00	5.66	IV.	5	7.46	23 51.1	6.5	22 0 10.34	13 27.6
47	4	. .	. .	11.2	26.4	. .	. .	. .	1 26.13	5.67	IV.	6	6.40	27 19.5	6.4	1 20.46	16 55.9
48	8	48.4	3.4	. .	. .	. .	. .	. .	3 33.34	5.71	II.	5	5.33	22 43.6	6.2	3 27.63	12 19.8
49	10	. .	. .	. .	. .	54.0	8.4	. .	22 3 38.88	— 5.72	VI.	4	8.30	−19 13.1	— 6.2	22 3 33.16	— 33 8 49.3

CORRECTIONS.

Date.		Corr. of Clock.	Hourly rate.	m	n	c
1846.	h.	s.	s.	s.	s.	s.
Sept. 9.	18	— 15.75	— 0.052	+ 0.497	— 0.319	0.000

INSTRUMENT READINGS.

	Date.		Barom.	THERMOM. At.	Ex.
	1846.	h. m.	in.	°	°
Zone 66	Sept. 9.	20 40	30.21	75.0	62.6

REMARKS.

(66) 7. Micrometer assumed as wire 8 instead of wire 9.
(66) 11. Micrometer assumed as 2 3r.2 instead of 2 3r.32.

ZONE 66. SEPTEMBER 9. A. $D_o = -32° 49' 30''$—Continued.

No.	Mag.	I.	II.	III.	IV.	V.	VI.	VII.	T.	a_1	MICROMETER.			$i + d_4$	d_1	Mean Right Ascension, 1850.0.	Mean Declination, 1850.0.
									h. m. s.	s.			r.	′ ″	″	h. m. s.	° ′ ″
50	10		30.5	45.5					22 5 0.15	− 5.73	III.	3	6.40	−13 17.8	− 6.1	22 4 54.42	− 33 2 53.9
51	10			32.0	47.3				7 46.96	5.76	IV.	5	10.26	25 12.0	5.8	7 41.20	14 47.8
52	8		49.5	4.7					9 19.40	5.78	III.	5	7.9	23 32.3	5.7	9 13.62	13 8.0
53	9			1.0	16.0				10 15.93	5.78	IV.	8	10.41	39 21.8	5.6	10 10.15	28 57.4
54	10			1.5	16.2				11 16.14	5.80	IV.	5	5.25	22 39.8	5.5	11 10.34	12 15.3
55	9			42.5	57.8				12 57.56	5.82	IV.	8	5.7	36 32.9	5.4	12 51.74	33 26 8.3
56	8			53.0	8.0				18 7.63	5.91	IV.	1	6.4	2 59.9	5.0	18 1.72	32 52 34.9
57	9				28.0	44.0			19 27.87	5.91	V.	3	4.54	12 24.2	4.9	19 21.96	33 1 59.1
58	4	18.8	34.0	49.1					23 3.79	5.96	III.	4	7.51	17 22.5	4.6	22 57.83	6 57.1
59	9	47.5	53.0						24 32.73	5.95	II.	8	5.20	36 39.2	4.5	24 26.78	33 26 13.7
60	6							26.0	25 40.90	6.00	VII.	1	10.29	5 13.4	4.4	25 34.90	32 54 47.8
61	11					34.5	49.0		27 19.43	5.99	VI.	5	7.2	23 28.6	4.3	27 13.44	33 13 2.9
62	9			37.0	51.8				28 51.73	6.02	IV.	6	6.43	27 21.0	4.2	28 45.71	33 16 55.2
63	9		39.5	54.7					30 9.21	6.05	III.	2	9.46	9 51.9	4.1	30 3.16	32 59 26.0
64	8	15.4	31.0						32 0.83	6.04	II.	8	3.16	35 36.5	4.0	31 54.79	33 25 10.5
65	9				22.1	37.0			32 22.00	6.05	V.	6	6.21	27 9.8	4.0	32 15.95	16 43.8
66	10			36.5	51.5				34 51.45	6.08	IV.	9	8.4	43 2.2	3.9	34 45.37	33 32 36.1
67	8		46.0	1.0					37 15.53	6.13	III.	1	8.7	4 2.0	3.8	37 9.40	32 53 35.8
68	7	3.2	18.4						38 48.43	6.13	II.	8	8.7	38 3.7	3.7	38 42.30	33 27 37.4
69	9	9.0	24.5						40 54.45	6.15	II.	9	8.6	43 3.0	3.6	40 48.30	32 36.6
70	10				11.5	26.5			41 11.47	6.16	V.	7	5.20	31 39.8	3.6	41 5.31	21 13.4
71	10					31.5			42 16.56	6.17	V.	5	7.11	23 33.3	3.5	42 10.39	13 6.8
72	9						8.0	23.0	42 38.23	6.19	VII.	6	6.58	27 28.1	3.5	42 32.04	17 1.6
73	10				51.0	5.5			44 50.70	6.21	V.	5	9.39	24 48.2	3.4	44 44.49	14 21.6
74	4	59.7	14.0						47 43.89	6.25	II.	7	3.54	30 56.2	3.3	47 37.64	20 29.5
75	9						0.4	15.4	52 30.65	6.30	VII.	7	6.20	32 9.7	3.1	52 24.35	21 42.8
76	10		36.0	51.0					55 5.96	6.33	III.	7	11.12	34 37.8	3.0	54 59.63	33 24 10.8
77	7					58.8	13.2		55 43.68	6.37	VI.	1	10.48	5 23.3	3.0	55 37.31	32 54 56.3
78	9					42.0	56.5		57 26.92	6.34	VI.	8	6.55	37 27.3	2.9	57 20.58	33 27 0.2
79	8			0.5	15.4				22 59 15.17	6.39	IV.	3	8.43	14 20.0	2.9	22 59 8.78	33 3 52.9
80	7			59.2	13.8				23 1 13.68	6.42	IV.	2	7.37	8 46.8	2.8	23 1 7.26	32 58 19.6
81	7			21.2	36.2				2 35.89	6.45	IV.	2	9.33	9 45.4	2.8	2 29.44	32 59 18.2
82	6	10.9	26.0						4 55.83	6.46	II.	4	6.31	18 12.9	2.8	4 49.37	33 7 45.7
83	11				9.0	24.0			9 8.96	6.50	V.	6	9.8	28 34.3	2.7	9 2.46	18 7.0
84	10				4.0	19.0			10 3.95	6.51	V.	5	7.24	23 39.9	2.7	9 57.44	13 12.6
85	4					4.2	19.0	33.9	10 49.25	6.52	VII.	7	4.55	31 26.8	2.7	10 42.73	20 59.5
86	9					34.0	49.2		12 19.27	6.52	VI.	8	4.22	36 9.9	2.7	12 12.75	25 42.6
87	11			42.0	57.0				16 56.84	6.59	IV.	6	7.1	27 30.1	2.7	16 50.25	17 2.8
88	9						21.5	37.0	17 51.93	6.62	VII.	3	9.54	14 55.4	2.7	17 45.31	4 28.1
89	10		9.6	25.0					21 39.75	6.65	III.	7	10.44	34 23.7	2.7	21 33.10	23 56.4
90	8		8.0	23.0					24 37.67	6.70	III.	3	8.34	14 15.4	2.7	24 30.97	3 48.1
91	9			58.0	13.0				28 12.92	6.72	IV.	8	7.25	37 42.7	2.8	28 6.20	27 15.5
92	10						30.0	45.0	29 0.26	6.74	VII.	6	3.41	35 48.9	2.8	28 53.52	25 21.7
93	9			59.5	14.0				31 14.01	6.77	IV.	4	6.34	18 14.6	2.8	31 7.24	7 47.4
94	5					16.5	31.4		32 1.62	6.76	VI.	9	9.57	43 59.2	2.8	31 54.86	33 32.0
95	9		52.5	7.7					34 22.61	6.79	III.	8	9.30	38 45.8	2.9	34 15.82	28 18.7
96	9			7.5	22.3				35 22.28	6.81	IV.	7	9.8	33 35.2	2.9	35 15.47	23 8.1
97	10	48.0	3.5						39 33.21	6.87	II.	5	9.43	24 50.0	3.0	39 26.34	33 14 23.0
98	9			41.2	56.0				23 40 55.79	− 6.90	IV.	2	10.23	−10 10.7	− 3.0	23 40 48.89	− 32 59 43.7

CORRECTIONS.

Date.	Corr. of Clock.	Hourly rate.	m	n	c
1846. h.	s.	s.	s.	s.	s.

INSTRUMENT READINGS.

Date.	Barom.	THERMOM. At.	THERMOM. Ex.
1846. h. m.	in.	°	°

REMARKS.

(66) 56. Micrometer reading assumed as 4r.51 instead of 7r.51.

ZONE 66. SEPTEMBER 9. A. $D_0 = -32° 49' 30''$—Continued.

No.	Mag.	Seconds of Transit I.	II.	III.	IV.	V.	VI.	VII.	T.	d_1	Micrometer		r.	$i + d_2$	d_1	Mean Right Ascension, 1850.0.	Mean Declination, 1850.0.
									h. m. s.	s.			r.	′ ″	″	h. m. s.	° ′ ″
99	10	..	..	17.2	32.5	..	..	..	23 42 32.20	– 6.91	IV.	7	3.17	−30 37.7	– 3.1	23 42 25.29	– 33 20 10.8
100	7	..	..	..	47.0	1.6	..	..	43 46.75	6.91	V.	6	4.29	26 13.2	3.2	43 39.84	15 46.4
101	9	..	..	..	..	..	..	1.0	44 16.22	6.91	VII.	7	2.19	30 7.9	3.2	44 9.31	19 41.1
102	8	13.5	19.0	..	..	..	..	..	46 58.62	7.94	II.	6	9.17	28 38.7	3.3	46 51.68	18 12.0
103	6	..	38.5	53.6	..	..	..	..	52 8.20	7.01	III.	3	6.21	13 8.2	3.5	52 1.19	2 41.7
104	11	..	..	51.5	7.0	..	..	..	53 6.63	7.01	IV.	7	9.41	33 51.9	3.6	52 59.62	23 29.5
105	9	26.8	42.0	..	..	..	..	..	57 11.84	7.06	II.	5	4.31	22 12.3	3.8	57 4.78	11 46.1
106	6	..	..	23.0	37.9	..	..	..	57 37.78	7.07	IV.	6	8.42	28 21.2	3.8	57 30.71	17 55.0
107	8	..	33.2	48.2	..	..	..	..	59 3.21	7.07	III.	8	4.14	36 6.0	3.9	58 56.14	33 25 39.9
108	7	..	..	..	48.0	3.0	..	..	23 59 7.90	– 7.11	V.	1	7.37	− 3 46.9	– 3.9	23 59 40.79	– 32 53 20.8

ZONE 67. SEPTEMBER 14. K. $D_0 = -34° 41' 10''$.

No.	Mag.	I.	II.	III.	IV.	V.	VI.	VII.	T.	d_1	Micrometer		r.	$i + d_2$	d_1	Mean Right Ascension, 1850.0.	Mean Declination, 1850.0.
1	8	..	..	..	..	..	23.4	39.2	19 28 53.16	– 7.36	VII.	2	12.10	−11 2.8	– 20.2	19 28 45.80	– 34 52 33.0
2	9	..	45.3	0.6	15.8	..	..	..	34 15.68	7.39	IV.	4	11.2	20 30.0	18.5	34 8.29	35 1 58.5
3	9	..	..	..	41.2	56.4	11.7	..	39 41.17	7.44	VII.	3	14.59	17 29.4	16.8	39 33.73	34 58 56.2
4	9	..	..	..	..	..	8.6	..	40 38.42	7.44	VII.	4	11.21	20 39.1	16.5	40 30.98	35 2 5.6
5	8	..	..	17.7	32.9	..	..	..	47 32.69	7.52	V.	3	11.56	15 57.3	14.4	47 25.17	34 57 21.7
6	6.7	27.9	43.2	58.8	14.2	..	..	..	50 13.90	7.52	V.	5	8.57	24 26.4	13.6	50 6.38	35 5 50.0
7	9	18.2	33.8	..	..	..	..	..	52 4.20	7.54	III.	4	6.22	18 8.5	13.0	51 56.66	34 59 31.5
8	9	..	..	..	23.0	38.2	..	..	52 22.88	7.54	VII.	5	10.21	25 8.9	12.9	52 15.34	35 6 31.8
9	8	5.1	20.8	36.2	..	..	..	..	19 55 51.24	– 7.58	IV.	5	7.36	−23 46.0	– 11.9	19 55 43.66	– 35 5 7.9

ZONE 68. SEPTEMBER 14. K. $D_0 = -35° 16' 40''$.

No.	Mag.	I.	II.	III.	IV.	V.	VI.	VII.	T.	d_1	Micrometer		r.	$i + d_2$	d_1	Mean Right Ascension, 1850.0.	Mean Declination, 1850.0.
1	10	12.2	..	..	..	..	..	..	21 1 58.52	– 8.32	III.	4	7.37	−18 45.2	– 24.8	21 1 50.20	– 35 35 50.0
2	10	..	..	..	4.3	..	..	..	4 4.13	8.33	VI.	6	7.20	27 39.6	24.3	3 55.80	44 43.9
3	9	..	..	..	4.6	19.8	..	..	5 4.45	8.33	VI.	9	6.39	42 19.8	24.1	4 56.12	59 23.9
4	9	..	..	..	..	16.1	..	..	6 0.68	8.37	VI.	4	10.50	20 23.6	23.9	5 52.31	37 27.5
5	9	..	5.2	20.8	..	..	..	..	13 35.85	8.48	V.	3	11.45	15 51.6	22.1	13 27.37	32 53.7
6	7.8	49.8	5.3	..	..	..	..	..	15 36.06	8.49	III.	4	8.45	19 20.6	21.7	15 27.57	36 22.3
7	7	57.5	..	28.0	..	..	..	..	16 42.37	8.53	VI.	3	5.30	12 41.7	21.4	16 34.84	29 43.1
8	9	..	..	33.8	..	..	..	..	17 48.98	8.52	IV.	5	5.43	22 48.8	21.3	17 40.46	39 50.1
9	9	..	..	..	..	28.9	..	59.9	18 13.51	8.53	VII.	3	11.47	15 52.1	21.2	18 4.98	32 53.3
10	9	..	7.8	23.3	..	..	..	..	22 38.58	8.58	IV.	6	6.58	27 28.7	20.2	22 30.00	44 28.9
11	8.9	..	8.0	23.5	..	..	..	..	24 38.66	8.61	IV.	4	7.59	18 57.4	19.8	24 30.05	35 57.2
12	8.9	34.8	50.5	6.2	..	..	..	..	27 21.43	8.62	IV.	7	9.28	33 45.7	19.3	27 12.81	50 45.0
13	8	..	..	48.3	..	18.9	..	..	29 3.30	8.69	VI.	1	8.4	3 59.4	18.9	28 54.61	20 58.3
14	9	..	..	..	..	38.3	..	9.0	30 22.83	8.68	VII.	5	7.26	23 40.4	18.6	30 14.15	40 39.0
15	9.10	..	..	..	10.3	..	40.8	..	38 10.24	8.80	VII.	1	11.7	5 31.7	17.1	38 1.44	22 28.8
16	10	..	..	..	50.7	..	..	..	39 50.53	8.78	VI.	8	5.26	36 42.8	16.8	39 41.75	53 39.6
17	9	..	31.1	..	2.0	..	..	..	43 1.86	8.83	V.	6	6.8	27 3.4	16.2	42 53.03	43 59.6
18	9	..	..	4.3	19.6	..	..	..	44 19.41	8.86	VI.	4	4.19	17 5.8	16.0	44 10.55	34 1.8
19	8	..	..	..	..	48.3	..	..	46 32.82	8.92	VII.	1	8.25	4 9.7	15.6	46 23.90	21 5.3
20	7.8	..	..	51.6	7.2	..	..	..	48 6.72	8.93	.	.	..	..	15.3	47 57.79	
21	9	..	..	..	..	30.9	..	..	49 15.52	8.91	VI.	6	10.24	29 12.7	15.1	49 6.61	46 7.8
22	10	..	..	..	..	..	19.0	..	21 49 48.33	– 8.92	VII.	8	5.24	−36 41.5	– 15.0	21 49 39.41	– 35 53 36.5

CORRECTIONS.

Date.		Corr. of Clock.	Hourly rate.	m	n	c
1846.	h.	s.	s.	s.	s.	s.
Sept. 14.	9	– 19.65	– 0.017	+ 0.497	– 0.319	0.000

INSTRUMENT READINGS.

	Date.		Barom.	Thermom. At.	Thermom. Ex.
	1846.	h. m.	in.	°	°
Zone 67	Sept. 14.	19 28	29.92	81.0	75.0
		19 55	29.91	81.0	75.0
Zone 68	Sept. 14.	21 1	29.91	80.5	75.0

REMARKS.

[(67.) The observing-book has the right ascension 18h.]

ZONE 68. SEPTEMBER 14. K. $D_o = -35° 16' 40''$—Continued.

No.	Mag.	Seconds of Transit (I. II. III. IV. V. VI. VII.)	T.	a_1	Micrometer.			$i + d_2$	d_1	Mean Right Ascension, 1850.0.	Mean Declination, 1850.0.
			h. m. s.	s.			r.	′ ″	″	h. m. s.	° ′ ″
23	9	22.5	21 51 7.13	− 8.93	VI.	7	6.4	−32 2.2	− 14.8	21 50 58.20	− 35 48 57.0
24	10	19.1	54 34.13	9.01	V.	3	6.20	13 7.1	14.1	54 25.12	30 1.2
25	9	47.4	56 18.31	9.00	IV.	8	5.59	36 59.7	13.8	56 9.31	53 53.5
26	9	9.4	56 38.73	9.01	VII.	8	4.54	36 26.3	13.8	56 29.72	53 20.1
27	9	6.9	21 58 22.37	− 9.02	V.	8	10.12	−39 7.7	− 13.5	21 58 13.35	− 35 56 1.2

ZONE 69. SEPTEMBER 14. K. $D_o = -35° 16' 40''$.

No.	Mag.	Seconds of Transit	T.	a_1	Micrometer.			$i + d_2$	d_1	Mean Right Ascension, 1850.0.	Mean Declination, 1850.0.
1	9	54.0	22 56 23.44	− 9.81	VII.	4	7.52	−18 53.2	− 1.4	22 56 13.63	− 35 35 34.6
2	8.9	40.9 56.7	22 59 11.77	9.85	IV.	5	7.41	23 48.5	1.3	22 59 1.92	40 29.8
3	9	10.2 25.8	23 1 40.94	9.89	IV.	5	4.49	22 21.4	1.2	23 1 31.05	39 2.6
4	10	30.3 45.6	6 0.95	9.93	IV.	6	4.47	26 22.4	1.0	5 51.02	35 43 3.4
5	9	1.8 17.0 32.8	7 32.67	9.93	V.	10	1.22	44 41.6	0.9	7 22.74	36 1 22.5
6	10	17.0	12 16.83	10.02	IV.	2	8.3	8 59.2	0.8	12 6.81	35 25 40.0
7	9	28.8 14.3	15 14.88	10.07	IV.	1	12.11	6 4.6	0.7	15 4.81	22 45.3
8	9	6.1	15 50.64	10.06	V.	2	7.41	8 48.0	0.7	15 40.58	25 28.7
9	9	34.3 49.9	23 5.30	10.14	IV.	8	10.28	39 15.9	0.5	22 55.16	55 56.4
10	8	47.5	26 2.37	10.22	IV.	1	7.23	3 38.9	0.5	25 52.15	20 19.4
11	8.7	49.6 4.8	27 4.84	10.19	IV.	8	7.52	37 56.9	0.5	26 54.65	54 37.4
12	9	41.5	32 56.56	10.30	IV.	3	9.38	14 47.4	0.5	32 46.26	35 31 27.9
13	9	2.3	34 2.13	10.27	V.	10	4.20	46 11.7	0.5	33 51.86	36 2 52.2
14	9	12.6	37 43.45	10.33	IV.	7	6.8	32 0.4	0.5	37 33.12	35 48 40.9
15	8	29.7 0.6	44 0.38	10.44	IV.	3	9.30	14 43.3	0.6	43 49.94	31 23.9
16	9	8.7	52 8.53	10.51	V.	7	8.22	33 12.2	0.7	51 58.02	49 52.9
17	9.8	58.5	23 55 29.18	10.57	III.	4	6.14	18 4.2	0.8	23 55 18.61	34 45.0
18	8.9	3.6 18.8	0 1 18.69	10.64	V.	5	2.14	21 3.0	1.0	0 1 8.05	37 44.0
19	8	21.0	1 34.92	10.63	VII.	8	9.39	38 50.5	1.0	1 24.29	55 31.5
20	6.7	35.3	3 19.90	10.67	V.	5	10.1	24 59.3	1.0	3 9.23	41 40.3
21	6	31.9 47.0	0 4 16.42	− 10.65	VII.	9	5.21	−41 40.0	− 1.1	0 4 5.77	− 35 58 21.1

ZONE 70. SEPTEMBER 15. A. $D_o = -26° 33' 20''$.

No.	Mag.	Seconds of Transit	T.	a_1	Micrometer.			$i + d_2$	d_1	Mean Right Ascension, 1850.0.	Mean Declination, 1850.0.
1	11	53.5 7.0	19 31 21.29	− 8.89	III.	7	5.33	−31 45.9	− 50.5	19 31 12.40	− 27 5 56.4
2	11	20.5 35.5	33 3.11	8.89	II.	5	1.55	20 53.7	50.0	32 54.22	26 55 3.7
3	7	33.0 47.5	33 19.27	8.91	VI.	3	7.2	13 29.6	49.9	33 10.36	47 39.5
4	8	41.0 55.2	34 54.92	8.90	IV.	4	9.36	19 47.0	49.4	34 46.02	53 56.4
5	8	50.0 4.0	35 21.95	8.91	VII.	3	9.33	14 45.7	49.3	35 13.04	26 48 55.0
6	9	57.0 11.3	36 43.18	8.88	VI.	6	11.49	29 55.2	48.9	36 34.30	27 4 4.1
7	7	10.6 24.5	37 56.58	8.89	VI.	4	4.26	17 10.3	48.5	37 47.69	26 51 18.8
8	8	16.0 30.0	39 15.90	8.87	V.	6	6.50	27 24.3	48.1	39 7.03	27 1 32.4
9	6	35.6 49.9	41 18.08	8.86	II.	7	4.38	31 18.0	47.5	41 9.22	27 5 25.5
10	9	39.0 53.3	41 53.01	8.86	IV.	6	. .	. . .	. .	41 44.15	26 58 4.7
11	10	54.5 9.0	43 22.76	8.86	III.	6	9.39	28 49.7	46.9	43 13.90	27 2 56.6
12	7	8.0 21.9	43 39.99	8.84	VII.	8	10.10	39 4.6	46.8	43 31.15	13 11.4
13	6	7.0 21.0 35.0	45 6.97	8.85	VI.	7	3.45	30 51.2	46.4	44 58.12	27 4 57.6
14	5	22.0 36.6	46 50.23	8.85	III.	5	5.8	22 31.3	45.9	46 41.38	26 56 37.2
15	8	29.3 13.5	47 15.42	8.84	VI.	7	5.30	31 44.3	45.8	47 6.58	27 5 50.1
16	10	42.0 56.0	19 48 14.03	− 8.84	VII.	7	9.13	−33 36.7	− 45.5	19 48 5.19	− 27 7 42.2

CORRECTIONS.

Date.	Corr. of Clock.	Hourly rate.	m	n	c
1846.	h. s.	s.	s.	s.	s.
Sept. 13,	9 − 19.67	− 0.037	+ 0.497	− 0.319	0.000

INSTRUMENT READINGS.

	Date.		Barom.	Thermom. At.	Thermom. Ex.
	1846.	h. m.	In.	°	°
Zone 69	Sept. 14,	0 4	29.89	78.5	72.0
Zone 70	Sept. 15,	19 31	30.00	80.0	69.5
		20 7	30.07	74.7	56.5

REMARKS.

ZONE 70. SEPTEMBER 15. A. D₀ = −26° 33′ 20″—Continued.

No.	Mag.	Seconds of Transit. I.	II.	III.	IV.	V.	VI.	VII.	T.	a_1	Micrometer.			$i + d_4$	d_1	Mean Right Ascension, 1850.0.	Mean Declination, 1850.0.
									h. m. s.	s.			r.	′ ″	″	h. m. s.	° ′ ″
17	4				57.0		24.8		19 49 56.92	− 8.87	VI.	1	3.56	− 1 56.7	− 45.0	19 49 48.05	− 26 36 1.7
18	10			40.5	54.5				51 54.43	8.82	IV.	7	9.21	33 41.1	44.4	51 45.61	27 7 45.5
19	9	32.5	46.4						54 14.65	8.83	II.	6	9.27	28 43.5	43.7	54 5.82	27 2 47.2
20	7	36.5	50.5						55 18.51	8.85	II.	2	11.7	10 33.8	43.4	55 9.66	26 44 37.2
21	6			57.0	11.0				56 10.97	8.80	IV.	8	11.45	39 53.0	43.2	56 2.17	27 13 56.2
22	9			58.0	11.5				57 11.57	8.83	IV.	4	10.25	20 11.8	42.9	57 2.74	26 54 14.7
23	7					1.5	15.3		57 47.42	8.83	VI.	4	5.31	17 43.3	42.7	57 38.59	51 46.0
24	9						16.0	29.5	19 58 47.70	8.83	VII.	3	9.40	14 49.2	42.4	58 38.87	48 51.6
25	5					21.2	34.8		20 0 7.04	8.84	VI.	1	10.21	5 11.1	42.0	19 59 58.20	39 13.1
26	9				6.0	20.5			2 6.14	8.81	V.	4	6.35	18 15.6	41.4	20 1 57.33	52 17.0
27	9				5.0	19.0			3 4.88	8.82	V.	3	7.47	13 52.5	41.2	2 56.06	47 53.7
28	9	15.5	30.0						4 57.89	8.80	II.	5	7.31	23 43.4	40.6	4 49.09	57 44.0
29	10					55.0	8.5		6 40.77	8.81	VI.	3	2.16	11 5.2	40.1	6 31.96	26 45 5.3
30	8						24.0	38.5	6 56.29	8.76	VII.	9	2.35	40 14.3	40.1	6 47.53	27 14 14.4
31	10					49.5	4.0		9 35.77	8.78	VI.	6	3.17	25 36.6	39.3	9 26.99	26 59 35.9
32	9				37.0	51.0			10 36.88	8.80	V.	3	8.26	14 12.2	30.0	10 28.08	26 48 11.2
33	9					51.0	5.0		12 37.02	8.75	VI.	8	10.30	39 15.0	38.5	12 28.27	27 13 13.5
34	8				4.0	18.0			13 3.86	8.80	V.	1	11.3	5 32.4	38.3	12 55.06	26 39 30.7
35	8						4.5	18.5	13 36.40	8.80	VII.	1	7.11	3 34.9	38.2	13 27.60	26 37 33.1
36	6							19.0	14 37.06	8.75	VII.	8	9.8	38 33.3	37.9	14 28.31	27 12 31.2
37	7	11.3	25.4						16 53.55	8.76	II.	6	8.36	28 17.8	37.3	16 44.79	2 15.1
38	9		7.5	21.5					17 35.55	8.75	III.	7	7.23	32 41.5	37.1	17 26.80	27 6 38.6
39	11						10.5	25.0	17 42.73	8.76	VII.	5	7.57	23 56.3	37.1	17 33.97	26 57 53.4
40	8							23.5	18 41.17	8.77	VII.	2	5.11	7 33.8	36.8	18 32.40	26 41 30.6
41	8			13.0	27.2				20 27.00	8.74	IV.	7	2.12	30 4.4	36.3	20 18.26	27 4 0.7
42	9						37.5	52.0	21 9.71	8.76	VII.	4	4.58	17 26.2	36.1	21 0.95	26 51 22.3
43	10		34.5	48.5					25 2.60	8.72	III.	8	5.38	36 47.6	35.1	24 53.88	27 10 42.7
44	8							3.0	25 20.70	8.76	VII.	2	9.21	9 40.0	35.0	25 11.94	26 43 35.0
45	8	56.0	11.0						27 38.71	8.73	II.	6	10.9	29 4.7	34.4	27 29.98	27 2 59.1
46	8						7.0	21.0	27 39.00	8.73	VII.	6	10.17	29 8.5	34.4	27 30.27	27 3 2.9
47	9					10.0	24.0		28 56.02	8.73	VI.	5	5.35	22 44.8	34.0	28 47.29	26 56 38.8
48	9		34.5	49.0					31 2.79	8.72	III.	7	4.5	31 1.5	33.4	30 54.07	27 4 54.9
49	6			30.3	44.5				31 44.35	8.70	IV.	8	4.12	36 4.2	33.2	31 35.65	9 57.4
50	9		50.0	4.0					33 17.99	8.72	III.	6	5.2	26 29.8	32.8	33 9.27	27 0 22.6
51	8	10.0	24.5						34 52.29	8.74	II.	3	6.15	13 5.4	32.4	34 43.55	26 46 57.8
52	8	26.0	40.0						36 8.03	8.73	II.	3	5.50	12 53.3	32.1	35 59.30	46 45.4
53	8		4.0	18.5					37 32.06	8.73	III.	3	7.57	13 57.5	31.7	37 23.33	47 49.2
54	6		3.0	17.2					38 31.04	8.70	III.	5	7.11	23 33.4	31.4	38 22.34	57 24.8
55	9					2.5	16.5		38 46.52	8.71	VI.	4	8.18	19 7.5	31.4	38 39.81	52 58.9
56	7		6.5	20.3					40 34.28	8.71	III.	4	6.40	19 18.7	30.9	40 25.57	53 9.6
57	9	32.5	47.0						42 14.91	8.72	II.	3	5.48	12 52.3	30.5	42 6.19	46 42.8
58	9						42.0	56.5	42 14.18	8.72	VII.	3	5.46	12 51.0	30.5	42 5.46	46 41.5
59	6	27.2	41.7						45 9.54	8.70	II.	4	7.35	18 45.8	29.7	45 0.84	52 35.5
60	8		7.5	22.0					46 35.48	8.71	III.	2	3.25	6 40.6	29.3	46 26.77	40 29.9
61	5	18.7	32.5						48 0.68	8.69	II.	4	5.30	17 42.6	29.0	47 51.99	51 31.6
62	8	19.5	34.0						49 1.80	8.70	II.	3	9.4	14 31.2	28.7	48 53.10	48 19.9
63	9			18.5	33.0				49 32.59	8.68	IV.	5	3.7	21 30.3	28.6	49 23.91	55 18.9
64	9						45.0	2.5	57 20.16	8.69	VII.	3	3.24	11 39.3	26.6	57 11.47	26 45 25.9
65	9	3.5	18.0						20 59 45.99	− 8.65	II.	7	7.2	−32 30.8	− 26.0	20 59 36.34	− 27 6 16.8

CORRECTIONS.

Date.		Corr. of Clock.	Hourly rate.	m	n	c
1846.	h.	s.	s.	s.	s.	s.

INSTRUMENT READINGS.

Date.		Barom.	Thermom. At.	Thermom. Ex.
1846.	h. m.	In.	°	°

REMARKS.

ZONE 70. SEPTEMBER 15. A. $D_0 = -26° 33' 20''$—Continued.

No.	Mag.	I.	II.	III.	IV.	V.	VI.	VII.	T.	a_1	Micrometer.			$i + d_1$	d_1	Mean Right Ascension, 1850.0.	Mean Declination, 1850.0.
									h. m. s.	s.			r.	' "	"	h. m. s.	° ' "
66	8	..	..	..	..	1.0	15.0	..	20 59 47.02	− 8.64	VI.	8	9.37	−38 48.2	− 26.0	20 59 38.38	− 27 12 34.2
67	7	..	..	..	..	30.0	44.0	..	21 1 16.03	8.68	VI.	3	4.33	12 14.4	25.7	21 1 7.35	26 46 0.1
68	8	..	..	..	..	..	57.5	11.5	3 29.51	8.64	VII.	7	5.17	31 37.5	25.1	2 20.87	27 5 22.6
69	6	..	..	..	..	20.5	34.5	..	4 6.53	8.62	VI.	9	12.	44 59.9	25.0	3 57.91	27 16 44.9
70	9	..	..	51.5	5.0	..	..	..	6 5.02	8.66	IV.	3	4.20	12 8.0	24.5	5 56.36	26 45 52.5
71	9	..	..	..	..	8.0	22.0	..	7 54.03	8.64	VI.	5	9.41	24 50.6	24.1	7 45.39	26 58 34.7
72	7	..	..	..	..	13.0	27.5	..	8 59.27	8.62	VI.	7	7.36	32 47.9	23.8	8 50.65	27 6 31.7
73	7	..	..	..	..	22.5	36.5	..	10 8.53	8.60	VI.	9	3.39	40 46.9	23.5	9 59.93	27 14 30.4
74	6	..	..	..	..	28.0	42.0	..	11 14.03	8.63	VI.	5	8.39	24 17.8	23.3	11 5.40	26 58 1.1
75	8	..	..	17.0	31.0	..	..	..	12 30.89	8.63	IV.	6	8.39	28 19.4	23.0	12 22.26	27 2 2.4
76	8	..	..	22.0	36.0	..	..	..	16 35.76	8.64	IV.	3	3.45	11 20.0	22.0	16 27.12	26 45 2.0
77	7	..	15.4	29.5	..	..	..	..	19 43.57	8.60	III.	8	7.16	37 37.1	21.3	19 34.97	27 11 18.4
78	11	25.7	40.0	..	..	..	..	..	23 8.49	8.62	II.	5	7.49	23 52.3	20.5	22 59.87	26 57 32.8
79	9	..	..	..	..	..	28.5	43.0	23 0.68	8.64	VII.	3	3.7	11 30.7	20.6	22 52.04	26 45 11.3
80	9	..	..	..	..	..	54.0	8.5	25 26.28	8.60	VII.	8	5.20	36 38.2	20.0	25 17.68	27 10 18.2
81	9	..	..	8.0	22.0	..	..	..	26 21.77	8.63	IV.	3	5.30	12 43.3	19.8	26 13.14	26 46 23.1
82	5	55.0	10.0	..	..	..	..	..	27 37.57	8.62	II.	4	3.28	16 41.0	19.5	27 28.95	26 50 20.5
83	10	..	..	..	55.0	9.0	..	..	28 54.92	8.59	V.	6	5.44	36 50.7	19.2	28 46.33	27 10 29.9
84	9	..	..	..	..	30.0	44.0	..	30 16.03	8.62	VI.	4	9.3	19 30.2	18.9	30 7.41	26 53 9.1
85	8	..	52.5	7.0	..	..	..	..	34 20.49	8.62	III.	2	4.18	7 7.4	18.1	34 11.87	40 45.5
86	8	..	..	..	..	6.4	20.0	..	34 52.22	8.62	VI.	2	6.5	8 1.3	18.0	34 43.60	41 39.3
87	8	..	..	..	20.5	34.5	..	..	36 20.40	8.60	V.	6	2.50	25 23.1	17.7	36 11.80	26 59 0.8
88	8	..	..	..	16.0	30.0	..	..	38 15.91	8.56	V.	9	8.31	43 14.5	17.2	38 7.38	27 16 51.7
89	7	..	33.5	47.0	..	..	..	..	40 1.17	8.59	III.	5	4.17	22 5.6	16.9	39 52.58	26 55 42.5
90	8	..	12.5	27.0	..	..	..	..	41 40.64	8.60	III.	4	9.29	19 43.5	16.5	41 32.04	53 20.0
91	8	..	..	55.5	9.5	23.5	..	..	44 9.30	8.61	V.	2	9.9	9 34.3	16.0	44 0.69	26 43 10.3
92	9	..	..	..	..	53.0	7.0	..	45 39.03	8.55	VI.	9	3.37	40 45.9	15.7	45 30.48	27 14 21.6
93	7	..	9.0	23.0	..	..	..	..	48 37.11	8.56	III.	8	7.18	37 38.1	15.1	48 28.55	27 11 13.2
94	9	..	..	20.0	34.0	..	..	..	50 33.78	8.60	IV.	3	7.35	13 46.5	14.7	50 25.18	26 47 21.2
95	11	..	..	6.0	20.5	..	..	..	57 20.17	8.56	IV.	7	5.23	31 39.6	13.4	57 11.61	27 5 13.0
96	11	..	..	..	..	..	4.5	19.0	21 58 36.79	8.55	VII.	8	7.25	37 41.3	13.2	21 58 28.24	11 14.5
97	10	..	..	..	35.5	50.0	..	..	22 0 35.67	8.55	V.	8	4.31	36 13.8	12.8	22 0 27.12	9 46.6
98	7	56.6	11.4	..	..	..	..	..	4 39.22	8.56	II.	7	2.45	30 20.9	12.1	4 30.66	27 3 53.0
99	7	26.5	..	10.5	54.5	..	..	..	6 54.30	8.58	II.	2	7.58	8 58.5	11.6	6 45.72	26 42 30.1
100	8	..	..	..	55.0	8.9	..	..	22 7 54.86	− 8.56	V.	6	4.59	−26 28.3	− 11.5	22 7 46.30	− 26 59 59.8

ZONE 71. SEPTEMBER 16. K. $D_0 = -37° 48' 10''$.

No.	Mag.	I.	II.	III.	IV.	V.	VI.	VII.	T.	a_1	Micrometer.			$i + d_1$	d_1	Mean Right Ascension, 1850.0.	Mean Declination, 1850.0.
1	8	..	16.6	..	..	..	..	..	19 47 48.56	− 8.68	IV.	8	8.7	−38 5.8	− 42.4	19 47 39.88	− 38 26 58.2
2	7.8	..	..	11.0	27.0	..	..	..	50 26.71	8.73	IV.	4	4.52	17 22.0	41.7	50 17.98	6 13.7
3	7	55.4	11.6	..	..	..	..	..	53 43.45	8.75	III.	7	6.15	32 8.8	40.8	53 34.70	38 29 59.6
4	8.9	..	..	..	..	4.8	..	..	53 48.83	8.79	V.	2	4.58	7 23.9	40.8	53 40.04	37 56 14.7
5	8	..	..	..	..	..	..	0.9	54 13.16	8.76	VII.	6	7.18	27 38.4	40.7	54 4.40	38 16 29.1
6	10	..	..	..	..	10.1	..	..	19 55 54.18	8.80	VI.	4	8.30	19 12.2	40.2	19 55 45.38	8 2.4
7	11	..	..	..	..	20.1	..	..	20 7 4.15	8.93	VI.	3	3.51	11 50.3	37.3	20 6 55.22	0 37.6
8	9.10	..	..	..	..	34.9	..	7.0	9 19.12	8.94	VII.	6	4.24	26 10.2	36.8	9 10.18	14 57.0
9	11	..	..	57.0	..	..	..	..	11 12.69	8.97	IV.	4	13.59	21 59.2	36.3	11 3.72	10 45.5
10	10	..	..	..	37.8	..	..	..	20 12 37.63	− 8.99	IV.	6	6.58	−27 28.9	− 36.0	20 12 28.64	− 38 16 14.9

CORRECTIONS.

Date.		Corr. of Clock.	Hourly rate.	m	n	c	
1846.	h.	s.	s.	s.	s.	s.	
Sept. 16,	9	− 20.97	− 0.024	+ 0.394	− 0.190	0.000	Zone 71

INSTRUMENT READINGS.

Date.		Barom.	Thermom. At.	Thermom. Ex.
1846.	h. m.	in.	°	°
Sept. 16,	19 47	30.10	71.5	61.5
	20 38	30.10	71.5	56.0
	0 2	30.09	70.5	51.0

REMARKS.

(70) 68. Minutes assumed as 2 instead of 3.
(70) 76. Micrometer reading assumed as $2^r.45$ instead of $3^r.45$.

ZONE 71. SEPTEMBER 16. K. D_0 = −37° 48′ 10″—Continued.

No.	Mag.	SECONDS OF TRANSIT. I.	II.	III.	IV.	V.	VI.	VII.	T.	a_1	MICROMETER.			$i + d_0$	d_1	Mean Right Ascension, 1850.0.	Mean Declination, 1850.0.
									h. m. s.	s.			r.	′ ″	″	h. m. s.	° ′ ″
11	11	. .	. .	44.9	. .	. .	. .	. .	20 14 0.35	− 9.03	IV.	2	8.13	− 9 2.7	− 35.6	20 13 51.32	− 37 57 48.3
12	10	. .	. .	. .	. .	. .	42.4	. .	14 10.87	9.03	VI.	2	9.40	9 46.6	35.6	14 1.84	58 32.2
13	7	. .	. .	3.0	18.7	. .	. .	. .	17 18.44	9.07	IV.	1	8.44	4 18.0	34.8	17 9.37	37 53 2.8
14	10	. .	. .	. .	9.4	. .	. .	. .	21 9.23	9.08	V.	5	10.6	25 1.8	33.8	21 0.15	38 13 45.6
15	10	. .	. .	. .	. .	20.0	. .	. .	22 4.13	9.08	V.	8	5.20	36 41.0	33.6	21 55.05	38 25 24.6
16	9	. .	. .	. .	. .	. .	53.3	. .	27 21.81	9.20	VI.	1	10.53	5 23.1	32.3	27 12.61	37 54 5.4
17	10	. .	8.9	25.0	. .	. .	. .	. .	30 40.55	9.22	IV.	4	5.21	17 36.7	31.5	30 31.33	38 6 18.2
18	11	. .	. .	. .	3.4	. .	. .	. .	32 3.23	9.21	IV.	7	3.12	30 36.0	31.2	31 54.02	38 19 17.2
19	10	. .	. .	. .	55.0	. .	. .	. .	38 54.83	9.35	VI.	1	11.56	5 55.0	29.6	38 45.48	37 54 34.6
20	5.6	. .	. .	15.0	31.1	47.2	. .	. .	41 31.09	9.33	V.	8	10.36	39 21.2	29.0	41 21.76	38 28 0.2
21	10	. .	. .	. .	. .	19.1	. .	. .	47 3.12	9.46	IV.	1	11.11	5 32.5	27.8	46 53.66	37 54 10.3
22	10	. .	. .	. .	. .	37.0	. .	. .	55 21.07	9.53	VI.	4	7.4	18 28.6	26.0	55 11.54	38 7 4.6
23	9	12.8	29.1	. .	. .	. .	. .	. .	57 0.73	9.55	III.	5	4.48	22 20.6	25.7	56 51.18	10 56.3
24	10	. .	. .	. .	. .	. .	50.4	. .	57 18.79	9.55	VII.	5	2.29	21 9.6	25.6	57 9.24	9 45.2
25	7.8	. .	. .	41.3	57.8	. .	. .	. .	20 59 57.51	9.56	VI.	9	9.7	43 36.5	25.0	20 59 47.95	32 11.5
26	8.9	. .	49.1	5.2	. .	. .	. .	. .	21 14 21.02	9.79	IV.	7	8.40	33 22.2	22.2	21 14 11.24	21 54.4
27	5	. .	36.9	. .	. .	. .	. .	. .	17 8.88	9.82	IV.	8	11.36	39 51.7	21.6	16 59.06	28 23.3
28	9.10	. .	12.7	28.8	. .	. .	. .	. .	19 44.30	9.89	IV.	3	9.18	14 36.3	21.1	19 34.41	3 7.4
29	9	. .	. .	25.7	41.3	. .	. .	. .	20 41.35	9.87	VI.	7	5.56	31 58.8	20.9	20 31.48	20 29.7
30	11	. .	. .	. .	. .	. .	52.2	. .	30 20.48	10.01	VII.	6	11.12	29 37.1	19.1	30 10.47	18 6.2
31	10	. .	. .	46.1	. .	. .	. .	. .	33 2.00	10.04	IV.	7	9.40	33 52.6	18.6	32 51.96	22 21.2
32	2.3	. .	. .	44.2	0.3	16.2	. .	. .	45 0.05	10.23	IV.	3	11.19	15 37.6	16.5	44 49.82	38 4 4.1
33	4.5	. .	58.9	14.9	. .	. .	. .	. .	47 30.35	10.27	IV.	2	8.41	9 16.9	16.0	47 20.08	37 57 42.9
34	8.9	. .	. .	. .	. .	24.0	. .	. .	48 8.14	10.23	VI.	8	9.52	38 58.7	15.9	47 57.91	38 27 24.6
35	8.9	. .	. .	. .	. .	. .	. .	55.9	48 8.22	10.24	VII.	7	9.23	33 43.4	15.9	47 57.98	38 22 9.3
36	10	. .	. .	26.6	. .	. .	. .	. .	50 42.08	10.32	V.	2	11.18	10 36.4	15.5	50 31.76	37 59 1.9
37	9	. .	. .	. .	. .	24.2	. .	. .	21 53 8.22	10.36	VI.	1	10.37	5 15.0	15.1	52 57.86	37 53 40.1
38	9	. .	31.2	47.3	. .	. .	. .	. .	22 0 3.28	10.43	IV.	9	6.22	42 13.2	14.0	21 59 52.85	38 30 37.2
39	8	. .	. .	32.7	48.8	. .	. .	. .	1 48.67	10.47	V.	9	2.52	40 26.6	13.7	22 1 38.20	28 50.3
40	9	. .	. .	24.3	40.3	. .	. .	. .	4 40.08	10.51	V.	5	9.54	24 55.7	13.3	4 29.57	13 19.0
41	11	. .	. .	. .	. .	2.2	. .	. .	13 46.29	10.65	VI.	5	5.55	22 51.4	11.9	13 35.64	11 16.3
42	8	. .	. .	6.2	22.0	. .	. .	. .	34 21.83	10.95	V.	4	8.38	19 16.5	9.2	34 10.88	7 35.7
43	8	. .	. .	. .	. .	. .	10.4	. .	34 38.60	10.94	VI.	8	9.38	38 51.6	9.2	34 27.66	27 10.8
44	7.8	. .	. .	27.4	. .	59.7	. .	. .	39 43.67	11.02	V.	9	10.57	44 32.5	8.6	39 32.65	32 51.1
45	11	. .	. .	53.2	. .	. .	. .	. .	42 8.99	11.06	VI.	6	4.30	26 13.7	8.3	41 57.93	14 32.0
46	11	. .	. .	35.0	. .	. .	. .	. .	50 50.59	11.20	VI.	3	13.17	16 37.1	7.4	50 39.39	4 54.5
47	10	. .	16.9	. .	. .	. .	. .	. .	52 48.58	11.23	IV.	4	12.56	21 27.3	7.2	52 37.35	9 44.5
48	10	. .	. .	. .	. .	. .	35.6	. .	53 3.80	11.21	VI.	8	9.29	38 47.1	7.2	52 52.59	38 27 4.3
49	9	. .	. .	. .	51.2	. .	. .	. .	54 51.03	11.27	IV.	1	11.45	5 49.7	7.0	54 39.76	37 54 6.7
50	10	. .	. .	. .	. .	38.2	. .	. .	22 59 22.22	11.33	V.	1	8.46	4 18.9	6.6	22 59 10.89	52 35.5
51	8	. .	. .	51.0	. .	. .	. .	. .	23 8 6.43	11.46	IV.	2	5.51	7 50.8	5.8	23 7 54.97	37 56 6.6
52	11	. .	. .	. .	59.0	. .	. .	. .	12 58.83	11.50	VI.	9	6.20	42 11.9	5.4	12 47.33	38 30 27.3
53	7.8	. .	59.1	. .	. .	. .	. .	. .	16 30.62	11.57	III.	3	6.20	13 6.0	5.2	16 19.05	1 21.2
54	10	. .	. .	. .	. .	30.3	. .	. .	33 14.36	11.82	VI.	3	11.58	15 57.1	4.1	33 2.54	4 11.2
55	9	. .	. .	. .	. .	. .	. .	43.1	33 55.36	11.82	VII.	6	7.13	27 35.9	4.0	33 43.54	15 49.9
56	8.9	. .	. .	10.3	. .	. .	. .	. .	42 25.98	11.95	IV.	4	12.55	21 26.8	3.6	42 14.03	9 40.4
57	9	. .	. .	14.4	. .	46.3	. .	. .	47 30.25	12.01	V.	5	7.20	23 37.7	3.4	47 18.24	38 11 51.1
58	9	. .	. .	. .	21.5	. .	. .	. .	50 21.33	12.07	V.	2	5.45	7 47.7	3.2	50 9.26	37 56 0.9
59	8.7	37.5	. .	9.7	. .	. .	. .	. .	23 54 25.28	− 12.12	V.	3	11.10	−15 33.0	− 3.1	23 54 13.16	− 38 3 46.1

CORRECTIONS.

Date.		Corr. of Clock.	Hourly rate.	m	n	c
1846.	h.	s.	s.	s.	s.	s.

INSTRUMENT READINGS.

Date.		Barom.	THERMOM. At.	Ex.
1846.	h. m.	in.	°	°

REMARKS.

Zone 71. September 16. K. $D_0 = -37° 48' 10''$—Continued.

No.	Mag.	I.	II.	III.	IV.	V.	VI.	VII.	T.	a_i	Micrometer.			$i + d_s$	d_i	Mean Right Ascension, 1850.0.	Mean Declination, 1850.0.
									h. m. s.	s.			r.	′ ″	″	h. m. s.	° ′ ″
60	10	..	..	..	..	..	15.1	..	23 56 43.52	− 12.15	VI.	4	4.48	−17 19.7	− 3.0	23 56 31.37	− 38 5 32.7
61	9	..	..	..	..	15.6	..	47.2	23 57 59.53	12.17	VII.	4	10.5	20 0.0	3.0	23 57 47.36	8 13.0
62	9.10	..	..	..	17.6	33.3	..	..	0 2 17.40	− 12.23	VII.	4	11.27	−20 41.6	− 2.9	0 2 5.17	− 38 8 54.5

Zone 72. September 19. A. $D_0 = -30° 18' 30''$.

No.	Mag.	I.	II.	III.	IV.	V.	VI.	VII.	T.	a_i	Micrometer.			$i + d_s$	d_i	Mean Right Ascension, 1850.0.	Mean Declination, 1850.0.
1	9	..	..	8.5	23.5	..	..	..	19 32 23.22	− 10.39	IV.	8	8.15	−38 7.5	− 11.6	19 32 12.83	− 30 56 49.1
2	9	..	..	..	..	..	..	27.5	32 43.80	10.43	VII.	5	3.9	21 30.7	11.5	32 33.37	40 12.2
3	8	..	..	..	49.5	4.8	..	..	35 49.77	10.46	V.	4	2.48	16 20.6	10.8	35 39.31	30 35 1.4
4	7	..	..	9.0	23.8	..	..	..	37 23.66	10.43	IV.	9	3.4	43 5.1	10.4	37 13.23	31 1 45.5
5	7	2.7	17.9	..	..	..	..	..	39 46.76	10.48	II.	6	7.25	27 41.9	9.9	39 36.28	30 46 21.8
6	7	..	57.2	11.5	..	..	..	..	40 26.00	10.50	III.	4	9.26	19 41.8	9.8	40 15.50	38 21.6
7	9	..	24.0	39.0	..	..	..	..	43 53.18	10.52	III.	5	3.39	21 46.2	9.0	43 42.66	40 25.2
8	8	..	..	..	..	25.5	40.0	..	44 11.02	10.54	VI.	4	10.16	20 6.9	8.9	44 0.48	38 45.8
9	6	..	3.0	17.7	..	..	..	..	47 32.27	10.53	III.	8	10.9	39 5.0	8.2	47 21.74	57 43.2
10	6	..	..	..	..	15.0	29.5	..	48 0.52	10.57	VI.	4	4.8	17 0.9	8.1	47 49.95	35 39.0
11	5	..	..	..	39.7	54.0	..	..	49 39.51	10.55	V.	8	7.6	37 32.5	7.7	49 28.96	56 10.2
12	9	..	..	4.0	18.5	..	..	..	51 18.43	10.57	IV.	7	9.28	33 45.0	7.4	51 7.86	52 22.4
13	11	..	..	17.5	..	..	..	..	52 31.94	10.59	III.	6	7.41	27 50.2	7.1	52 21.35	46 27.3
14	9	..	..	..	20.0	35.0	..	..	53 20.14	10.61	V.	5	8.0	23 58.2	6.9	53 9.53	42 35.1
15	8	..	..	..	12.5	27.5	..	..	55 12.62	10.64	V.	4	2.40	16 16.6	6.5	55 1.98	30 34 53.1
16	7	..	43.5	58.0	..	..	..	..	59 12.69	10.64	III.	9	4.57	41 27.0	5.7	59 2.05	− 31 0 2.7
17	8	..	..	..	..	59.0	13.5	..	19 59 44.50	− 10.66	VI.	7	..	−29	− 5.6	19 59 33.84	

Zone 73. September 19. A. $D_0 = -34° 4' 0''$.

No.	Mag.	I.	II.	III.	IV.	V.	VI.	VII.	T.	a_i	Micrometer.			$i + d_s$	d_i	Mean Right Ascension, 1850.0.	Mean Declination, 1850.0.
1	9	..	51.0	6.5	..	..	..	..	20 8 21.34	− 10.43	III.	5	2.58	−21 25.3	− 42.9	20 8 10.91	− 34 26 8.2
2	11	..	..	..	..	14.5	29.5	..	8 59.35	10.43	VI.	8	5.35	36 47.1	42.7	8 48.92	41 29.8
3	10	..	..	..	31.5	46.5	..	..	13 31 37	10.46	V.	8	7.23	37 44.4	41.5	13 20.91	42 25.9
4	9	..	..	..	..	46.5	1.5	..	14 31.36	10.48	VI.	7	8.29	33 15.5	41.2	14 20.88	37 56.7
5	8	..	..	..	..	5.0	20.0	..	15 49.85	10.48	VI.	9	7.47	42 53.9	40.9	15 39.37	47 34.8
6	9	14.0	29.5	..	..	..	..	..	19 59.82	10.55	II.	6	5.30	26 43.9	39.8	19 49.27	31 23.7
7	8	..	..	..	7.5	23.0	..	..	21 7.61	10.55	V.	7	9.50	33 56.6	39.5	20 57.06	38 36.1
8	9	..	48.5	3.5	..	..	..	..	25 19.75	10.61	III.	6	10.38	29 19.9	38.4	25 9.14	33 58.3
9	8	..	..	..	..	..	58.8	14.5	25 28.84	10.63	VII.	4	12.0	20 58.8	38.4	25 18.21	25 37.2
10	7	..	..	..	..	..	..	5.2	26 19.84	10.61	VII.	7	9.13	33 37.4	38.2	26 9.23	38 15.6
11	8	..	..	..	58.5	13.5	..	..	27 58.37	10.63	V.	8	3.41	35 49.6	37.8	27 47.74	40 27.4
12	8	..	..	..	..	32.0	47.0	..	29 16.87	10.68	VI.	2	4.41	7 17.1	37.4	29 6.19	11 54.5
13	8	..	..	..	52.5	7.5	..	..	31 52.34	10.68	V.	6	11.17	29 39.6	36.8	31 41.66	34 16.4
14	8	..	..	56.0	11.5	..	..	..	32 11.23	10.67	IV.	7	8.24	33 13.2	36.7	32 0.56	37 49.9
15	11	..	..	..	..	..	3.5	21.0	32 33.40	10.67	VII.	4	10.26	20 11.3	36.6	32 22.73	24 47.9
16	8	..	..	..	..	..	23.0	38.0	33 52.74	10.69	VII.	8	9.45	38 53.3	36.3	33 42.05	43 29.6
17	9	..	..	13.5	28.5	..	..	..	37 28.24	10.78	IV.	1	8.16	4 6.2	35.4	37 17.46	8 41.6
18	6	..	..	..	48.0	3.0	..	..	39 47.78	10.82	V.	1	6.30	3 12.5	34.8	39 36.96	7 47.3
19	5	1.0	16.5	..	..	..	..	..	40 46.61	10.81	II.	3	10.33	15 15.2	34.6	40 35.80	19 49.8
20	9	..	..	..	..	..	6.5	21.5	40 36.19	10.80	VII.	5	7.59	23 57.1	34.6	40 25.39	28 31.7
21	8	8.2	23.5	..	..	..	..	..	20 42 53.81	− 10.82	II.	5	3.52	−21 52.4	− 34.0	20 42 42.99	− 34 26 26.4

CORRECTIONS.

Date.			Corr. of Clock.	Hourly rate.	m	n	c
1846.		h.	s.	s.	s.	s.	s.
Sapt	19,	9	− 22.16	− 0.021	+ 0.394	− 0.190	0.000

INSTRUMENT READINGS.

	Date.		Barom.	Thermom. At.	Thermom. Ex.
	1846.	h. m.	in.	°	°
Zone 72	Sept. 19,	19 32	30.17	71.5	64.2
		19 59	30.17	71.0	61.8
Zone 73	Sept. 19,	20 8	30.22	70.0	62.0
		23 59	30.22	65.0	57.0

REMARKS.

(73) 12. Differs $13^s.6$ from Mural Zone of July 14.
(73) 32. Transit over T. VI assumed as recorded over T. VII.

ZONE 73. SEPTEMBER 19. A. $D_c = -34° 4' 0''$—Continued.

No.	Mag.	SECONDS OF TRANSIT. I.	II.	III.	IV.	V.	VI.	VII.	T.	a_1	MICROMETER.			$i + d_2$	d_1	Mean Right Ascension, 1850.0.	Mean Declination, 1850.0.
									h. m. s.	s.			r.	′ ″	″	h. m. s.	° ′ ″
22	10	1.8	17.5	. .	. .	. .	. .	. .	20 46 47.70	− 10.86	II.	6	10.13	−29 7.1	− 33.1	20 46 36.84	− 34 33 40.2
23	6	56.5	12.0	. .	. .	. .	. .	. .	47 42.31	10.87	II.	6	11.15	29 38.4	32.9	47 31.44	34 11.3
24	8	. .	. .	. .	49.0	4.0	. .	. .	49 48.81	10.93	V.	2	7.41	8 48.3	32.4	49 37.88	13 20.7
25	6	. .	. .	7.0	22.0	. .	. .	. .	50 21.76	10.94	IV.	2	3.42	6 47.5	32.2	50 10.82	11 19.7
26	9	. .	. .	24.0	39.0	. .	. .	. .	51 38.92	10.92	IV.	6	4.17	26 7.2	32.0	51 28.00	30 39.2
27	9	. .	. .	44.5	59.6	. .	. .	. .	52 59.53	10.93	IV.	7	7.0	32 30.7	31.7	52 48.60	37 2.4
28	8	. .	11.5	26.8	. .	. .	. .	. .	58 42.06	11.00	III.	9	8.18	43 9.7	30.8	58 31.06	47 40.0
29	8	. .	. .	. .	30.5	45.5	. .	. .	20 59 30.29	11.06	V.	1	6.59	3 27.2	30.1	20 59 19.23	7 57.3
30	9	54.0	10.0	. .	. .	. .	. .	. .	21 8 40.10	11.12	II.	7	7.49	32 55.2	28.0	21 8 28.98	37 23.2
31	10	. .	. .	. .	. .	. .	50.0	5.0	8 19.73	11.12	VII.	8	7.25	37 42.4	28.1	8 8.61	42 10.5
32	6	. .	. .	. .	. .	. .	. .	46.0	9 15.80	11.13	VII.	7	7.25	32 42.8	28.0	9 4.67	37 10.8
33	6	. .	. .	. .	. .	34.0	49.0	. .	10 18.85	11.15	VI.	6	4.19	26 8.00	27.7	10 7.70	30 35.7
34	10	. .	27.5	43.0	. .	. .	. .	. .	17 57.94	11.25	III.	6	9.44	28 52.6	26.0	17 46.69	33 18.6
35	10	. .	58.0	13.5	. .	. .	. .	. .	22 28.44	11.31	III.	6	8.32	28 16.2	25.0	22 17.13	32 41.2
36	5	. .	. .	5.5	20.8	. .	. .	. .	23 20.62	11.31	IV.	7	5.31	31 45.6	24.9	23 9.31	36 10.5
37	11	51.0	6.0	. .	. .	. .	. .	. .	25 36.48	11.36	II.	5	7.48	23 51.8	24.4	25 25.12	28 16.2
38	5	30.2	45.5	. .	. .	. .	. .	. .	30 15.74	11.43	II.	4	3.30	16 41.2	23.4	30 4.31	21 4.6
39	9	. .	. .	. .	. .	39.0	54.0	. .	30 23.86	11.43	VI.	4	10.11	20 4.0	23.4	30 12.43	24 27.4
40	11	13.0	28.0	. .	. .	. .	. .	. .	39 58.54	11.54	II.	6	8.4	28 1.8	21.4	39 47.00	32 23.2
41	8	. .	. .	45.0	0.0	. .	. .	. .	41 59.79	11.59	IV.	2	9.20	9 38.5	21.1	41 48.20	13 59.6
42	9	. .	27.5	43.0	. .	. .	. .	. .	42 58.00	11.57	III.	7	5.9	31 34.4	20.9	42 46.43	35 55.3
43	8	. .	. .	. .	. .	. .	18.5	33.5	45 48.12	11.64	VII.	2	11.35	10 46.2	20.3	45 36.48	15 6.5
44	9	. .	. .	51.0	6.0	. .	. .	. .	52 5.79	11.73	IV.	2	8.10	9 3.0	19.2	51 54.06	13 22.2
45	6	. .	. .	. .	. .	3.5	18.5	. .	52 48.35	11.71	VI.	6	9.9	28 34.7	19.0	52 36.64	32 53.7
46	8	7.5	23.0	. .	. .	. .	. .	. .	55 53.24	11.76	II.	5	9.6	24 31.3	18.4	55 41.48	28 49.7
47	9	. .	. .	56.5	11.5	. .	. .	. .	57 11.34	11.79	IV.	4	5.15	17 34.5	18.2	56 59.55	21 52.7
48	10	. .	. .	. .	. .	. .	59.5	. .	21 58 29.29	11.76	VI.	8	8.16	38 8.6	18.0	21 58 17.53	42 26.6
49	5	. .	. .	. .	. .	49.0	4.5	. .	22 0 34.09	11.80	VI.	9	6.11	42 5.3	17.6	22 0 22.29	46 22.9
50	5	. .	. .	. .	20.2	35.5	. .	. .	1 20.23	11.80	V.	9	3.25	40 41.5	17.5	2 8.43	44 59.0
51	8	0.0	15.5	. .	. .	. .	. .	. .	8 45.79	11.92	II.	6	8.1	28 0.3	16.2	8 33.87	32 16.5
52	9	29.5	45.0	. .	. .	. .	. .	. .	10 15.17	11.96	II.	4	9.5	19 30.6	16.0	10 3.21	23 46.6
53	7	50.2	5.5	. .	. .	. .	. .	. .	11 35.83	11.97	II.	5	4.20	22 6.6	15.7	11 23.86	26 22.3
54	8	55.0	11.0	. .	. .	. .	. .	. .	15 40.99	12.03	II.	5	9.6	24 31.3	15.1	15 28.96	28 46.4
55	11	. .	. .	. .	57.5	12.5	. .	. .	17 57.32	12.07	V.	4	7.58	18 56.9	14.7	17 45.25	23 11.6
56	9	. .	. .	. .	14.5	30.0	. .	. .	19 14.55	12.09	V.	3	2.40	11 16.1	14.5	19 2.46	15 30.6
57	7	. .	. .	. .	. .	33.5	48.5	. .	20 18.35	12.10	VI.	6	7.25	27 42.1	14.4	20 6.25	31 56.5
58	7	. .	. .	. .	. .	23.5	38.5	. .	21 8.36	12.11	VI.	7	7.25	32 43.1	14.2	20 56.25	36 57.3
59	9	. .	42.0	57.0	. .	. .	. .	. .	23 12.25	12.14	III.	7	7.31	32 46.3	13.9	23 0.11	37 0.2
60	8	38.2	54.0	. .	. .	. .	. .	. .	25 24.23	12.17	II.	7	11.0	34 31.8	13.6	25 12.06	38 45.4
61	11	. .	. .	. .	. .	46.5	1.5	. .	26 31.35	12.18	VI.	6	6.41	27 19.8	13.4	26 19.17	31 33.2
62	7	. .	14.5	29.5	. .	. .	. .	. .	29 44.44	12.25	III.	2	10.52	10 24.9	12.9	29 32.19	14 37.8
63	9	. .	. .	. .	49.5	5.0	. .	. .	34 49.57	12.30	V.	4	4.35	17 14.2	12.2	34 37.27	21 26.4
64	8	. .	. .	49.0	4.0	. .	. .	. .	48 4.01	12.45	IV.	8	4.57	36 28.1	10.5	47 51.56	40 38.6
65	11	. .	. .	. .	. .	26.0	41.0	. .	50 10.86	12.50	VI.	4	9.6	19 31.1	10.2	49 58.36	23 41.3
66	8	. .	51.5	7.0	. .	. .	. .	. .	52 21.94	12.52	III.	6	8.53	28 26.8	9.9	52 9.42	32 36.7
67	8	. .	. .	48.5	3.0	. .	. .	. .	22 54 3.07	12.56	IV.	3	8.5	14 0.6	9.8	22 53 50.51	18 10.4
68	7	5.2	20.7	. .	. .	. .	. .	. .	23 19 50.93	12.89	II.	5	5.52	22 53.2	7.1	23 19 38.04	27 0.3
69	8	45.5	1.0	. .	. .	. .	. .	. .	24 31.26	12.95	II.	6	4.0	25 58.4	6.7	24 18.31	30 5.1
70	8	23.0	38.7	. .	. .	. .	. .	. .	23 26 8.85	− 12.97	II.	5	10.28	−25 12.8	− 6.6	23 25 55.68	− 34 29 19.4

CORRECTIONS.

Date.		Corr. of Clock.	Hourly rate.	m	n	c
1846.	h.	s.	s.	s.	s.	s.

INSTRUMENT READINGS.

Date.		Barom.	THERMOM. At.	Ex.
1846.	h. m.	in.	°	°

REMARKS.

(73) 32. Transit over T. IV assumed as recorded over T. VII.
(73) 50. Minutes of transit assumed as 2 instead of 1.

ZONE 73. SEPTEMBER 19. A. $D_0 = -34° 4' 0''$—Continued.

No.	Mag.	Seconds of Transit I.	II.	III.	IV.	V.	VI.	VII.	T.	a_1	Micrometer.			$i + d_2$	d_1	Mean Right Ascension, 1850.0.	Mean Declination, 1850.0.
									h. m. s.	s.			r.	' "	"	h. m. s.	° ' "
71	9				38.5	54.0			23 29 38.59	− 13.02	V.	6	3.40	−25 48.4	− 6.3	23 29 25.57	− 34 29 54.7
72	7		11.8	27.0					31 41.90	13.05	III.	3	10.23	15 10.3	6.1	31 28.85	19 16.4
73	10				39.0				33 38.83	13.08	IV.	4	9.43	19 50.1	6.0	33 25.75	23 56.1
74	8		43.0	58.5					36 13.20	13.11	III.	3	3.1	11 26.7	5.8	36 0.09	15 32.5
75	9	36.0	51.5						43 21.67	13.20	II.	4	8.40	19 18.0	5.3	43 8.47	23 23.3
76	7					30.0	44.5		44 14.61	13.21	VI.	3	6.32	13 13.3	5.3	44 1.40	17 18.6
77	9	26.0	41.0						47 11.49	13.23	II.	5	8.22	24 9.0	5.1	46 58.26	28 14.1
78	8		30.5	45.5					54 0.76	13.32	III.	7	8.55	33 28.8	4.8	53 47.44	37 33.6
79	9				56.5	11.5			57 56.31	13.38	V.	4	2.5	15 58.3	4.6	57 42.93	20 2.9
80	9						37.0	52.0	23 59 6.14	− 13.39	VII.	5	9.13	−24 34.5	− 4.5	23 59 52.75	− 34 28 39.0

ZONE 74. SEPTEMBER 21. K. $D_0 = -36° 28' 30''$.

No.	Mag.	I.	II.	III.	IV.	V.	VI.	VII.	T.	a_1	Micrometer.			$i + d_2$	d_1	Mean Right Ascension, 1850.0.	Mean Declination, 1850.0.
1	7.8					22.6			20 59 7.03	− 12.16	V.	10	10.32	−49 21.3	− 25.4	20 58 54.87	− 37 18 16.7
2	5			52.0	7.8				21 4 7.64	12.22	V.	7	8.28	33 15.7	24.6	21 3 55.42	37 2 10.3
3	4.5					54.0			6 38.31	12.28	VI.	4	11.31	20 44.2	24.4	6 26.03	36 49 38.6
4	8.9					45.4			7 29.70	12.28	VI.	4	7.25	18 39.6	24.3	7 17.42	36 47 33.9
5	12					2.3			9 46.71	12.29	VI.	9	10.32	44 18.8	24.0	9 34.42	37 13 12.8
6	11				43.7				11 43.53	12.31	VI.	10	8.42	48 25.3	23.7	11 31.22	17 19.0
7	11				41.2				12 41.03	12.34	V.	7	10.12	34 8.4	23.5	12 28.69	3 1.9
8	11						31.7		13 0.49	12.34	VII.	7	9.34	33 48.6	23.5	12 48.15	2 42.1
9	10		51.8						18 23.26	12.41	V.	8	4.36	36 18.2	22.7	18 10.85	37 5 10.9
10	10						22.2		17 51.18	12.44	VII.	2	4.45	7 17.5	22.8	17 38.74	36 36 10.3
11	10		59.9						21 31.02	12.49	VII.	2	10.22	10 8.2	22.3	21 18.53	36 39 0.5
12	8			22.6					25 38.54	12.51	IV.	10	11.22	49 46.6	21.8	25 26.03	37 18 38.4
13	10			8.3					27 23.59	12.57	IV.	2	14.12	12 5.2	21.5	27 11.02	36 40 56.7
14	6.5						32.8		28 1.80	12.59	VII.	1	8.33	4 12.7	21.4	27 49.21	33 4.1
15	9				24.3				39 24.13	12.75	IV.	1	6.15	3 3.4	20.0	39 11.38	31 53.4
16	9		28.6			15.1			42 59.54	12.80	VI.	2	10.22	10 8.5	19.6	42 46.74	38 58.1
17	8						26.4		43 55.33	12.82	VII.	3	6.30	13 11.1	19.5	43 42.51	42 0.6
18	8						22.0		44 51.00	12.83	VII.	1	6.41	3 16.0	19.4	44 38.17	32 5.4
19	7						2.6		45 31.51	12.82	VII.	4	4.11	17 1.0	19.3	45 18.69	45 50.3
20	10					35.0			47 19.31	12.85	IV.	4	12.43	21 20.9	19.1	47 6.46	50 10.0
21	11			46.8					50 2.24	12.89	III.	4	11.33	20 45.4	18.8	49 49.35	36 49 34.2
22	12					26.4			51 10.77	12.89	VI.	7	9.21	33 42.3	18.6	50 57.88	37 2 30.9
23	7					27.6			52 12.02	12.89	VI.	10	6.55	47 31.1	18.5	51 59.13	37 16 19.6
24	9						18.4		52 47.31	12.93	VII.	4	5.38	17 45.0	18.5	52 34.38	36 46 33.5
25	8						15.8		21 53 44.80	12.96	VII.	1	8.49	4 20.8	18.4	21 53 31.84	33 9.2
26	7	34.2	50.3						22 0 21.45	13.03	III.	4	7.7	18 30.6	17.7	22 0 8.42	47 18.3
27	10	51.0	6.8						1 38.02	13.07	III.	3	7.53	13 53.6	17.5	1 24.95	42 41.1
28	11		16.2	32.2					2 47.50	13.07	IV.	4	5.52	17 52.7	17.4	2 34.43	46 40.1
29	10					53.6			3 37.93	13.09	VI.	5	11.8	25 33.0	17.4	3 24.84	54 20.4
30	10			59.2					6 14.74	13.11	VI.	6	6.8	27 3.2	17.1	6 1.63	55 50.3
31	10		15.0	30.9					8 46.20	13.17	IV.	3	8.7	14 0.8	16.8	8 33.03	42 47.6
32	10				26.7				9 26.53	13.16	VI.	6	11.59	30 1.0	16.8	9 13.37	36 58 47.8
33	9						9.0		9 37.81	13.16	VII.	7	5.25	31 42.5	16.8	9 24.65	37 0 29.3
34	9		33.9						13 5.05	13.23	IV.	3	5.25	12 38.7	16.5	12 51.82	36 41 25.2
35	10			28.3					22 18 44.11	− 13.27	IV.	9	7.53	−42 58.5	− 16.0	22 18 30.84	− 37 11 44.5

CORRECTIONS.

Date.		Corr. of Clock.	Hourly rate.	m	n	c
1846.	h.	s.	s.	s.	s.	s.
Sept. 21,	18	− 23.49	− 0.019	+ 0.215	+ 0.014	0.000

INSTRUMENT READINGS.

	Date.		Barom.	Thermom. At.	Thermom. Ex.
	1846.	h. m.	In.	°	°
Zone 74	Sept. 21,	20 59	30.08	72.0	62.0
		22 0	30.09	71.0	64.0

REMARKS.

ZONE 74. SEPTEMBER 21. K. $D_o = -36° 28' 30''$—Continued.

No.	Mag.	Seconds of Transit. I.	II.	III.	IV.	V.	VI.	VII.	T.	a_1	Micrometer.			$i + d_1$	d_1	Mean Right Ascension, 1850.0.	Mean Declination, 1850.0.
									h. m. s.	s.			r.	′ ″	″	h. m. s.	° ′ ″
36	10	. . .	. . .	36.7	52.6	. . .	. . .	. . .	22 20 52.44	− 13.31	V.	8	6.36	−38 19.8	− 15.8	22 20 39.13	− 37 7 5.6
37	10	. . .	. . .	. . .	. . .	55.7	. . .	. . .	21 40.02	13.35	VI.	5	3.39	21 45.6	15.7	21 26.67	36 50 31.3
38	9	4.5	20.5	. . .	. . .	. . .	. . .	. . .	23 51.60	13.39	III.	3	5.43	12 47.8	15.6	23 38.21	41 33.4
39	9	. . .	. . .	46.6	. . .	. . .	. . .	. . .	27 1.80	13.43	IV.	2	3.20	6 35.1	15.4	26 48.37	36 35 20.5
40	10	. . .	. . .	. . .	10.2	. . .	. . .	. . .	29 10.03	13.42	VI.	10	8.39	48 23.8	15.2	28 56.61	37 17 9.0
41	10	. . .	. . .	8.1	. . .	. . .	. . .	. . .	34 23.30	13.53	V.	2	3.29	6 39.6	14.8	34 9.77	36 35 24.4
42	10	. . .	. . .	. . .	8.9	. . .	. . .	. . .	35 8.73	13.56	VI.	3	3.14	11 32.2	14.8	34 55.17	40 17.0
43	10	20.8	36.7	. . .	. . .	. . .	. . .	. . .	38 8.06	13.88	IV.	6	6.9	27 4.0	14.6	37 54.48	36 55 48.6
44	10	. . .	31.5	47.2	. . .	. . .	. . .	. . .	42 2.94	13.63	V.	8	5.8	36 34.4	14.4	41 49.31	37 5 18.8
45	9	. . .	. . .	. . .	. . .	. . .	. . .	58.7	42 11.56	13.65	VII.	4	6.41	18 17.0	14.4	41 57.91	36 47 1.4
46	8.7	. . .	. . .	. . .	. . .	48.3	4.0	. . .	43 32.76	13.67	VII.	3	4.30	12 10.3	14.3	43 19.09	40 54.6
47	10	. . .	. . .	. . .	. . .	1.0	16.1	. . .	44 45.14	13.69	VII.	5	9.27	24 41.5	14.2	44 31.45	53 25.7
48	10	. . .	. . .	. . .	. . .	5.4	. . .	. . .	45 49.70	13.71	VII.	4	6.36	18 14.4	14.2	45 35.99	36 46 58.6
49	5.6	. . .	. . .	. . .	1.7	17.1	32.8	. . .	47 1.52	13.70	VII.	9	6.59	42 30.6	14.1	46 47.82	37 11 14.7
50	10	. . .	2.4	. . .	. . .	. . .	. . .	. . .	53 33.70	13.82	IV.	5	8.30	24 13.3	13.8	53 19.88	36 52 57.1
51	5	. . .	. . .	. . .	25.3	. . .	. . .	. . .	54 25.13	13.80	VI.	9	11.35	44 50.7	13.8	54 11.33	37 13 34.5
52	5	. . .	57.0	. . .	. . .	. . .	53.3	9.2	56 22.14	13.86	III.	3	8.55	14 25.0	13.7	56 8.28	36 43 8.7
53	8	. . .	0.9	16.8	. . .	. . .	. . .	. . .	22 59 32.24	13.91	IV.	5	8.37	24 16.8	13.6	22 59 18.33	36 53 0.4
54	10	. . .	. . .	. . .	. . .	. . .	27.3	. . .	23 6 56.10	13.99	VI.	7	8.00	33 1.3	13.4	23 6 42.11	37 1 44.7
55	9	. . .	. . .	. . .	38.3	. . .	. . .	. . .	7 38.13	14.05	VI.	1	6.3	2 57.1	13.4	7 24.08	36 31 40.5
56	9	. . .	. . .	. . .	. . .	. . .	33.7	. . .	8 2.70	14.05	VI.	1	7.3	3 27.5	13.3	7 48.65	32 10.8
57	11	. . .	. . .	. . .	48.2	. . .	. . .	. . .	9 48.03	14.06	VI.	5	6.11	23 2.6	13.3	9 33.97	51 45.9
58	8.9	. . .	. . .	. . .	28.7	. . .	59.8	. . .	11 28.54	14.07	VII.	6	8.21	28 10.3	13.2	11 14.47	56 53.5
59	10	. . .	6.8	22.6	. . .	. . .	. . .	. . .	13 38.03	14.11	V.	4	9.16	19 36.0	13.2	13 23.92	48 19.2
60	10	. . .	16.3	32.3	. . .	. . .	. . .	. . .	14 47.17	14.13	IV.	5	6.42	23 18.5	13.2	14 33.04	36 52 1.7
61	7.8	34.5	. . .	6.2	21.8	. . .	. . .	. . .	17 21.72	14.14	VI.	7	5.39	31 49.9	13.2	17 7.58	37 0 33.1
62	8.9	. . .	. . .	30.3	46.4	. . .	. . .	. . .	23 46.13	14.24	V.	8	8.19	38 11.2	13.1	23 31.89	6 54.3
63	8.7	9.8	25.7	. . .	. . .	. . .	. . .	. . .	25 57.20	14.26	III.	8	3.57	36 59.2	13.1	25 42.94	37 5 42.3
64	9.10	. . .	. . .	. . .	. . .	32.0	. . .	. . .	30 16.28	14.36	VI.	3	7.42	13 47.9	13.1	30 1.92	36 42 31.0
65	9	. . .	15.5	. . .	. . .	. . .	. . .	. . .	42 46.75	14.53	IV.	4	9.57	19 56.8	13.2	42 32.22	48 40.0
66	8	. . .	. . .	33.7	49.2	. . .	. . .	. . .	43 49.09	14.54	VII.	5	6.14	23 3.8	13.3	43 34.55	51 47.1
67	11	. . .	10.3	. . .	. . .	. . .	. . .	. . .	52 41.54	14.67	III.	4	8.16	19 5.6	13.4	52 26.87	36 47 49.0
68	7.8	10.8	26.7	42.4	. . .	. . .	. . .	. . .	55 58.16	14.69	IV.	8	5.7	36 34.0	13.5	55 43.47	37 5 17.5
69	8	47.8	3.4	19.1	. . .	. . .	. . .	. . .	57 34.75	14.74	IV.	5	5.13	22 33.5	13.6	57 20.01	36 51 17.1
70	8.9	27.2	42.9	58.7	. . .	. . .	. . .	. . .	23 58 14.31	− 14.73	IV.	6	4.58	−26 28.1	− 13.6	23 57 59.58	− 36 55 11.7

ZONE 75. SEPTEMBER 22. A. $D_o = -35° 8' 40''$.

No.	Mag.	I.	II.	III.	IV.	V.	VI.	VII.	T.	a_1	Micrometer.			$i + d_1$	d_1	Mean Right Ascension, 1850.0.	Mean Declination, 1850.0.
1	8	. . .	37.6	53.3	. . .	. . .	. . .	. . .	19 57 8.22	− 12.52	III.	2	10.38	. . .	. .	19 56 55.70	
2	9	. . .	16.5	32.0	. . .	. . .	. . .	. . .	57 47.05	12.52	III.	3	4.30	. . .	. .	57 34.53	
3	6	. . .	. . .	. . .	39.0	54.5	. . .	. . .	58 38.92	12.54	V.	1	3.55	. . .	. .	58 26.38	
4	9	. . .	. . .	. . .	54.0	10.0	. . .	. . .	19 59 54.22	12.53	V.	6	4.23	. . .	. .	19 59 41.69	
5	7	. . .	. . .	. . .	. . .	42.8	58.2	. . .	20 0 27.49	12.52	VI.	8	4.53	. . .	. .	20 0 14.97	
6	9	. . .	22.3	38.0	. . .	. . .	. . .	. . .	3 53.36	12.56	III.	8	11.16	. . .	. .	3 40.80	
7	10	3.5	19.2	. . .	. . .	. . .	. . .	. . .	6 50.05	12.59	II.	7	9.54	. . .	. .	6 37.46	
8	6	. . .	. . .	. . .	57.0	12.5	. . .	. . .	6 56.97	12.59	V.	6	11.10	. . .	. .	6 44.38	
9	8	. . .	. . .	. . .	. . .	. . .	. . .	57.0	7 10.71	12.60	VII.	4	10.41	. . .	. .	6 58.11	
10	10	. . .	. . .	. . .	3.0	18.5	. . .	. . .	20 9 2.95	− 12.62	V.	4	9.23	. . .	. .	20 8 50.33	

CORRECTIONS.

Date.		Corr. of Clock.	Hourly rate.	m	n	c
1846.	h.	s.	s.	s.	s.	s.
Sept. 22,	18	− 23.39	− 0.023	+ 0.215	+ 0.014	0.000

INSTRUMENT READINGS.

Date.		Barom.	Thermom. At.	Thermom. Ex.
1846.	h. m.	in.	°	°

REMARKS.

(74) 55. Differs by 3ˢ.9 from Mural Zone, October 10.
(74) 60. Transit over T. II assumed as 15ˢ.3 instead of 13ˢ.3.

ZONE 75. SEPTEMBER 22. A. $D_0 = -35° 8' 40''$—Continued.

No.	Mag.	Seconds of Transit. I.	II.	III.	IV.	V.	VI.	VII.	T.	a_1	Micrometer.			$i + d_2$	d_1	Mean Right Ascension, 1850.0.	Mean Declination, 1850.0.
									h. m. s.	s.			r.	' "	"	h. m. s.	° ' "
11	8	. .	. .	. .	. .	. .	. .	6.0	20 9 19.60	− 12.63	VII.	2	11.31	. . .	. .	20 9 6.97	
12	10	. .	. .	. .	12.0	27.5	. .	. .	11 11.97	12.63	V.	6	5.53	. . .	. .	10 59.34	
13	9	. .	51.5	7.0	. .	. .	. .	. .	13 21.95	12.67	III.	1	9.40	. . .	. .	13 9.28	
14	6	. .	. .	. .	18.5	34.0	. .	. .	14 18.47	12.66	V.	6	5.19	. . .	. .	14 5.81	
15	8	. .	. .	. .	. .	. .	. .	10.0	14 23.68	12.67	VII.	4	4.28	. . .	. .	14 11.01	
16	10	. .	. .	. .	18.5	34.0	. .	. .	16 18.45	12.69	V.	4	11.55	. . .	. .	16 5.76	
17	8	. .	. .	. .	. .	44.5	59.5	. .	17 28.98	12.68	VI.	9	9.42	. . .	. .	17 16.30	
18	8	. .	. .	8.0	23.0	. .	. .	. .	19 23.17	12.70	IV.	9	5.36	. . .	. .	19 10.47	
19	9	. .	. .	36.0	51.7	. .	. .	. .	20 51.52	12.72	IV.	9	5.18	. . .	. .	20 38.80	
20	8	. .	11.5	27.5	. .	. .	. .	. .	25 42.20	12.79	III.	1	9.40	. . .	. .	25 29.41	
21	12	. .	. .	30.0	45.5	. .	. .	. .	27 45.38	12.78	IV.	8	5.20	. . .	. .	27 32.60	
22	9	. .	. .	. .	14.0	29.0	. .	. .	29 13.75	12.79	V.	8	11.11	. . .	. .	29 0.96	
23	8	. .	. .	3.5	19.0	. .	. .	. .	30 18.88	12.80	IV.	8	6.6	. . .	. .	30 6.08	
24	8	. .	. .	. .	. .	. .	24.0	40.0	30 53.53	12.84	VII.	1	9.32	. . .	. .	30 40.69	
25	8	. .	. .	. .	. .	58.0	13.0	. .	32 42.52	12.86	VIII.	1	4.55	. . .	. .	32 29.66	
26	5	7.0	22.5	. .	. .	. .	. .	. .	35 53.43	12.87	II.	7	7.42	. . .	. .	35 40.56	
27	9	. .	. .	. .	. .	10.5	26.0	. .	35 55.24	12.86	VI.	8	9.39	. . .	. .	35 42.38	
28	7	10.5	26.4	. .	. .	. .	. .	. .	37 57.13	12.89	II.	7	6.48	. . .	. .	37 44.24	
29	8	. .	20.5	36.2	. .	. .	. .	. .	38 51.10	12.92	III.	2	6.52	. . .	. .	38 38.18	
30	12	. .	. .	. .	31.5	47.0	. .	. .	40 31.49	12.91	V.	8	3.21	. . .	. .	40 18.58	
31	9	. .	. .	. .	23.5	38.5	. .	. .	42 23.20	12.94	V.	3	11.17	. . .	. .	42 10.26	
32	8	. .	. .	. .	. .	. .	55.0	10.2	43 24.21	12.94	VII.	7	11.12	. . .	. .	43 11.27	
33	8	. .	. .	. .	. .	1.0	15.8	. .	20 41 45.41	− 12.96	VI.	5	11.19	. . .	. .	20 41 32.45	

ZONE 76. SEPTEMBER 22. A. $D_0 = -29° 2' 10''$.

No.	Mag.	Seconds of Transit. I.	II.	III.	IV.	V.	VI.	VII.	T.	a_1	Micrometer.			$i + d_2$	d_1	Mean Right Ascension, 1850.0.	Mean Declination, 1850.0.
1	7	. .	34.5	49.0	. .	. .	. .	. .	21 26 3.17	− 13.12	III.	7	3.22	−30 39.9	− 21.4	21 25 50.05	− 29 33 11.3
2	9	. .	. .	. .	. .	. .	33.0	48.0	26 4.58	13.13	VII.	6	5.48	26 52.7	21.4	25 51.45	29 24.1
3	6	43.2	58.0	. .	. .	. .	. .	. .	28 26.40	13.17	II.	3	8.14	14 5.7	21.0	28 13.23	16 36.7
4	10	. .	. .	. .	. .	11.0	55.0	. .	28 26.49	13.16	VI.	4	7.33	18 44.6	21.0	28 13.33	21 15.6
5	11	. .	. .	. .	. .	. .	0.5	15.0	34 31.80	13.24	VII.	4	8.4	19 0.0	19.9	34 18.56	21 29.9
6	7	18.1	32.8	. .	. .	. .	. .	. .	37 1.34	13.26	II.	5	4.28	22 10.0	19.5	36 48.08	24 40.4
7	9	24.0	38.5	. .	. .	. .	. .	. .	38 7.16	13.28	II.	5	8.0	23 58.0	19.3	37 53.88	26 27.3
8	9	. .	17.0	31.5	. .	. .	. .	. .	38 45.63	13.28	III.	6	7.39	27 49.1	19.2	38 32.35	30 18.3
9	8	. .	. .	11.5	25.5	. .	. .	. .	40 25.41	13.31	IV.	2	6.45	8 21.2	18.9	40 12.10	10 50.1
10	8	. .	. .	22.0	36.2	. .	. .	. .	41 36.04	13.33	IV.	3	3.25	11 39.9	18.7	41 22.71	14 8.6
11	9	28.8	43.2	. .	. .	. .	. .	. .	44 12.01	13.32	II.	7	8.17	33 8.8	18.2	43 58.69	35 37.0
12	8	. .	. .	. .	28.0	42.5	. .	. .	44 27.97	13.34	V.	5	10.37	25 17.5	18.2	44 14.63	27 45.7
13	10	. .	. .	. .	. .	. .	35.5	50.0	45 6.86	13.33	VII.	7	7.31	32 45.3	18.1	44 53.53	35 13.4
14	8	. .	. .	. .	. .	21.0	35.5	. .	46 6.74	13.35	VI.	7	2.0	29 58.3	17.9	45 53.39	32 26.2
15	7	. .	52.0	6.4	. .	. .	. .	. .	48 20.61	13.38	III.	6	10.7	29 3.9	17.6	48 7.23	31 31.5
16	6	40.8	55.2	. .	. .	. .	. .	. .	51 23.84	13.42	II.	4	5.16	17 35.4	17.1	51 10.42	20 2.5
17	8	. .	. .	. .	. .	. .	5.5	20.0	52 36.87	13.43	VII.	8	8.58	38 28.6	16.9	51 23.44	40 55.5
18	9	. .	. .	. .	. .	17.5	32.5	. .	53 3.49	13.44	VI.	6	7.10	27 34.3	16.8	52 50.05	30 1.1
19	9	. .	. .	25.5	40.0	. .	. .	. .	54 39.92	13.44	IV.	9	6.5	42 1.2	16.6	54 26.48	44 27.8
20	6	. .	49.2	4.0	. .	. .	. .	. .	56 17.72	13.49	III.	2	4.25	7 10.5	16.3	56 4.23	9 36.8
21	7	8.5	23.4	. .	. .	. .	. .	. .	57 51.85	13.51	II.	5	6.55	23 25.2	16.1	57 38.34	25 51.3
22	9	. .	. .	2.5	17.0	. .	. .	. .	21 58 16.79	− 13.50	IV.	6	3.58	−25 57.5	− 16.0	21 58 3.29	− 29 28 23.5

CORRECTIONS.

Date.	Corr. of Clock.	Hourly rate.	m	n	c
1846. h.	s.	s.	s.	s.	s.

INSTRUMENT READINGS.

Date.	Barom.	Thermom. At.	Thermom. Ex.
1846. h. m.	in.	°	°

REMARKS.

(76) 7. Differs from Mural Zone, 1847, September 16, by $2^s.17$ in right ascension and $1' 32''.5$ in declination.
(76) 17. Minutes assumed as 51 instead of 52.

ZONE 76. SEPTEMBER 22. A. $D_o = -29° 2' 10''$—Continued.

No.	Mag.	Seconds of Transit. I.	II.	III.	IV.	V.	VI.	VII.	T.	a_1	Micrometer.			$i + d_1$	d_1	Mean Right Ascension, 1850.0.	Mean Declination, 1850.0.
									h. m. s.	s.			r.	′ ″	″	h. m. s.	° ′ ″
23	9	46.5	1.0						22 5 29.64	− 13.59	II.	5	4.45	−22 19.5	− 15.0	22 5 16.05	− 29 24 44.5
24	10	38.0	53.0						6 21.52	13.58	II.	7	10.48	34 25.1	14.8	6 7.94	36 49.9
25	8	38.3	53.0						13 21.47	13.67	II.	6	10.19	29 9.8	13.9	13 7.80	31 33.7
26	8	14.5	29.4						15 57.85	13.72	II.	5	7.16	23 35.8	13.5	15 44.13	25 59.3
27	8	58.0	12.5						17 41.18	13.73	II.	6	4.46	26 21.6	13.3	17 27.45	28 41.9
28	7				14.0	28.3			18 13.82	13.76	V.	1	8.8	4 3.5	13.2	18 0.06	6 26.7
29	11			17.0	31.5				20 31.31	13.76	IV.	6	7.28	27 43.6	12.9	20 17.55	30 6.5
30	6				13.8	28.3			21 13.76	13.77	V.	5	6.12	23 3.6	12.8	20 59.99	25 26.4
31	9		43.5	58.0					24 12.05	13.81	III.	5	3.18	21 35.7	12.5	23 58.24	23 58.2
32	7					2.4	16.6		24 47.97	13.79	VI.	9	7.33	42 45.5	12.4	24 34.18	45 7.9
33	9		18.0	32.8					26 46.84	13.82	III.	7	5.32	31 45.6	12.2	26 33.02	34 7.8
34	8				36.0	50.5			27 35.94	13.87	V.	3	5.36	12 46.0	12.1	27 22.07	15 8.1
35	9	39.5	54.5						29 22.97	13.85	II.	7	3.30	30 43.8	11.8	29 9.12	33 5.6
36	6	53.6	8.0						30 36.76	13.88	II.	6	9.58	28 59.2	11.7	30 22.88	31 20.9
37	7	57.5	12.5						31 41.03	13.88	II.	8	2.35	35 15.3	11.6	31 27.15	37 36.9
38	5							20.5	31 37.02	13.92	VII.	1	7.28	3 43.0	11.6	31 23.10	6 4.6
39	9			41.0	55.5				34 55.27	13.94	IV.	5	5.32	22 43.4	11.2	34 41.33	25 4.6
40	8			14.5	58.0				35 58.41	13.92	IV.	9	3.11	40 33.3	11.1	35 44.49	42 54.4
41	8		29.8	44.2					39 58.24	14.00	III.	4	2.29	16 11.1	10.7	39 44.24	18 31.8
42	9				46.5	1.0			41 46.47	14.02	V.	5	8.36	24 16.4	10.5	41 32.45	26 36.9
43	7	44.0	58.5						45 27.01	14.06	II.	2	10.31	10 15.2	10.1	45 12.95	12 35.3
44	8	45.0	0.0						48 28.39	14.09	II.	5	4.18	22 5.8	9.8	48 14.30	24 25.6
45	8		14.0	29.0					50 42.59	14.14	III.	1	8.9	4 4.1	9.6	50 28.45	6 23.7
46	5	37.0	52.0						53 20.55	14.13	II.	6	6.22	37 10.0	9.4	53 6.42	39 29.4
47	9						38.5	52.5	57 9.53	14.20	VII.	3	9.20	14 38.8	9.0	56 55.33	16 57.8
48	9	58.2	13.0						22 59 41.48	14.23	II.	5	3.9	21 31.0	8.8	22 59 27.25	23 49.8
49	5		58.2	12.8					23 0 26.98	14.21	III.	8	3.14	35 35.0	8.8	23 0 12.77	37 53.8
50	6			59.0	13.5				2 13.43	14.22	IV.	9	8.15	43 6.9	8.6	1 59.21	45 25.5
51	8	44.0	58.8						5 27.16	14.29	II.	3	3.1	11 27.6	8.4	5 12.87	13 46.0
52	8			28.0	42.5				6 42.21	14.32	IV.	3	8.3	14 0.3	8.3	6 27.89	16 18.6
53	7	6.4	20.8						8 49.54	14.33	II.	6	7.20	27 39.4	8.1	8 35.21	29 57.5
54	5			6.0	20.2				9 20.05	14.34	IV.	3	5.47	12 51.6	8.1	9 5.71	15 9.7
55	8			27.0	41.2				10 41.07	14.36	IV.	3	10.34	15 16.6	8.0	10 26.71	17 34.6
56	6						32.5	47.1	11 3.90	14.33	VII.	7	4.29	31 13.4	7.9	10 48.57	33 31.3
57	9							33.0	11 49.84	14.34	VII.	7	3.51	30 57.2	7.9	11 35.50	33 15.1
58	10					15.0	29.0		13 0.49	14.36	VI.	6	3.19	25 37.6	7.8	12 46.13	27 55.4
59	7	58.4	13.0						15 41.64	14.40	II.	6	6.47	27 22.7	7.6	15 27.24	29 40.3
60	8				57.0	11.5			16 56.96	14.42	V.	5	4.7	22 47.9	7.6	16 42.54	25 5.5
61	9			27.8	42.3				21 42.21	14.44	IV.	9	3.12	40 33.8	7.3	21 27.77	42 51.1
62	9				47.5	1.5			24 47.26	14.48	V.	9	9.27	43 43.2	7.1	23 32.78	46 0.3
63	9		4.5	19.5					25 33.38	14.51	III.	6	7.40	27 49.6	7.1	25 18.87	30 6.7
64	8	57.5	11.5						27 40.49	14.53	II.	7	5.39	31 49.0	6.9	27 25.96	34 5.9
65	8					8.0	22.5		27 53.74	14.53	VI.	8	9.8	38 33.9	6.9	27 39.21	40 50.8
66	8		54.8	9.4					30 23.56	14.55	III.	7	8.8	33 4.4	6.8	30 9.01	35 21.2
67	7			58.8	12.5				31 12.53	14.60	IV.	1	7.20	3 39.3	6.8	30 57.93	5 56.1
68	9	19.5	34.0						37 2.75	14.63	II.	7	7.20	32 40.0	6.5	36 48.12	34 56.5
69	10				57.0	12.0			38 57.23	14.66	V.	6	7.49	27 54.2	6.4	38 42.57	30 10.6
70	9				12.0	26.5			39 11.99	14.65	V.	8	1.59	34 57.3	6.4	38 57.34	37 13.7
71	8			58.0	12.5				23 42 12.16	− 14.71	IV.	2	6.45	− 8 21.2	− 6.3	23 41 57.45	− 29 10 37.5

CORRECTIONS.

Date.	Corr. of Clock.	Hourly rate.	m	n	c
1846. h.	s.	s.	s.	s.	s.

INSTRUMENT READINGS.

Date.		Barom.	Thermom. At.	Ex.
1846.	h. m.	in.	°	°

REMARKS.

(76) 29. Differs in right ascension by 1ˢ.88 from Mural Zone, 1847, September 16.
(76) 62. Minutes assumed as 23 instead of 24.

ZONE 76. SEPTEMBER 22. A. $D_0 = -29° 2' 10''$—Continued.

No.	Mag.	Seconds of Transit I.	II.	III.	IV.	V.	VI.	VII.	T.	a_1	Micrometer.			$i + d_2$	d_1	Mean Right Ascension, 1850.0.	Mean Declination, 1850.0.
									h. m. s.	s.			r.	′ ″	″	h. m. s.	° ′ ″
72	7				23.0	37.5			23 44 22.93	− 14.73	V.	2	4.52	− 7 24.1	− 6.3	23 44 8.20	− 29 9 40.4
73	8					43.2	58.7		. . . .	. . .	VI.	2	7.54	8 55.9			
74	8						1.5	16.0	46 32.80	14.76	VII.	4	4.23	17 8.4	6.2	46 18.04	19 24.6
75	9	50.0	4.3						51 33.09	14.80	II.	6	6.	26 59.0	6.1	51 18.29	29 15.1
76	10		51.0	6.0					52 19.76	14.81	III.	4	6.4	17 59.8	6.0	52 4.95	20 15.8
77	6	10.6	25.0						54 53.79	14.81	II.	7	5.0	31 29.3	6.0	54 8.98	33 45.3
78	8	13.0	27.5						54 56.14	14.84	II.	5	4.12	22 2.8	6.0	54 41.30	24 18.8
79	7			57.0	11.5				56 11.20	14.86	IV.	3	6.4	13 0.2	6.0	55 56.34	15 16.2
80	7						55.0	10.0	23 56 26.52	− 14.86	VII.	3	3.0	− 11 26.8	− 6.0	23 56 11.66	− 29 13 42.8

ZONE 77. SEPTEMBER 23. K. $D_0 = -30° 16' 30''$.

No.	Mag.	Seconds of Transit I.	II.	III.	IV.	V.	VI.	VII.	T.	a_1	Micrometer.			$i + d_2$	d_1	Mean Right Ascension, 1850.0.	Mean Declination, 1850.0.
1	11			51.2	5.8				21 0 5.58	− 13.66	IV.	4	14.0	− 15 46.3	− 14.0	20 59 51.92	− 30 32 30.3
2	5.6				18.5		47.5		1 18.52	13.69	IV.	1	5.41	2 49.1	13.8	21 1 4.83	19 32.9
3	10			47.2	1.8				5 1.67	13.71	V.	7	5.43	31 51.2	13.2	4 47.96	48 34.4
4	10			54.1					6 8.53	13.73	IV.	6	5.48	26 53.1	13.0	5 54.80	30 43 36.1
5	10					57.0			7 42.52	13.73	VI.	9	10.23	44 11.6	12.8	7 28.79	31 0 54.4
6	10							24.6	11 40.85	13.82	VII.	3	14.11	17 5.7	12.1	11 27.03	33 47.8
7	9						28.7		12 59.73	13.82	VII.	6	12.23	30 12.3	12.0	12 45.91	46 54.3
8	11						42.7		14 13.80	13.84	VI.	5	5.12	22 33.1	11.8	13 59.96	39 14.9
9	10		33.3						17 2.24	13.89	III.	3	7.47	13 52.1	11.3	16 48.35	31 30 33.4
10	9				22.1				17 21.94	13.88	IV.	4	5.14	17 34.4	11.3	17 8.06	30 34 15.7
11	5.6			27.8					20 42.54	13.90	IV.	10	6.5	47 3.0	10.8	20 28.64	31 3 43.8
12	10					38.0			21 23.46	13.92	VI.	6	11.9	29 35.2	10.7	21 9.54	30 46 15.9
13	9	36.4	51.3	6.2	20.8				24 20.50	13.94	V.	7	4.31	31 14.8	10.3	24 6.56	47 55.1
14	9			12.0	26.8	40.9			25 26.49	13.96	VI.	6	9.3	28 31.5	10.1	25 12.53	45 11.6
15	11					16.2			26 1.65	13.98	VII.	5	11.35	25 46.4	10.0	25 47.67	42 26.4
16	6.7			28.5	43.0	57.3			30 42.94	14.00	VI.	9	5.59	41 58.2	9.4	30 28.94	58 37.6
17	9.10		44.8	59.7	14.0	28.5			32 13.81	14.05	IV.	2	8.45	9 21.7	9.2	31 59.76	26 0.9
18	9.10				28.5				33 28.34	14.07	IV.	1	8.49	4 24.1	9.0	33 14.27	21 3.1
19	10			40.9		9.8			37 55.05	14.13	IV.	1	6.51	3 24.4	8.4	37 40.92	20 2.8
20	11		43.1		12.7	27.1			41 12.45	14.13	IV.	7	6.12	32 5.9	8.0	40 58.32	48 43.9
21	10					15.5	29.9	44.6	42 0.95	14.15	VII.	5	8.10	24 2.8	7.9	41 46.80	30 40 40.7
22	7.8					41.8	56.0	10.3	44 27.06	14.15	VII.	9	10.53	44 26.5	7.6	43 12.91	31 1 4.1
23	10				54.1	8.6	23.0	37.7	45 54.08	14.18	VII.	8	13.6	40 34.0	7.4	45 39.90	30 57 11.4
24	8.9						21.5	36.3	46 52.62	14.20	VII.	6	7.41	27 49.8	7.3	46 38.42	44 27.1
25	11			52.3	6.4				49 6.37	14.25	IV.	3	8.37	14 17.4	7.0	48 52.12	30 54.4
26	9					16.0		45.0	50 1.28	14.26	VII.	2	10.52.	10 25.4	6.9	49 47.02	27 2.3
27	8.9					31.3	45.8	0.5	51 16.82	14.26	VII.	5	10.20	25 8.5	6.7	51 2.56	41 45.2
28	7	6.5	21.3	36.0	50.2	4.3			54 50.13	14.31	VI.	4	11.49	20 53.9	6.3	54 35.82	37 30.2
29	6			44.3		13.0	27.7	42.2	55 58.56	14.33	VII.	3	13.38	21 49.2	6.2	55 44.23	38 24.4
30	6.7				29.2	43.7	58.0		58 29.10	14.37	VI.	1	8.21	4 9.7	5.9	58 14.73	20 45.6
31	10				47.0				21 59 46.84	14.34	VII.	9	6.6	42 1.5	5.8	21 59 32.50	58 37.3
32	9					42.9		11.7	22 1 28.28	14.37	VII.	8	3.19	35 37.4	5.6	22 1 13.91	52 13.0
33	10		31.0	45.7	0.0	14.5			6 0.03	14.42	VI.	7	5.30	34 44.5	5.1	5 45.61	48 19.6
34	9			55.2		23.9			8 9.26	14.49	IV.	1	5.57	2 57.2	4.9	7 51.77	19 32.1
35	9						15.7	30.3	8 46.75	14.45	VII.	8	6.6	37 1.8	4.8	8 32.30	53 36.6
36	9.10	19.1	33.8	48.9	3.2				22 12 2.90	− 14.52	IV.	3	3.25	− 11 39.7	− 4.5	22 11 48.38	− 30 28 14.2

CORRECTIONS.

Date.		Corr. of Clock.	Hourly rate.	m	n	c
1846.	h.	s.	s.	s.	s.	s.
Sept. 23.	18	− 24.45	− 0.010	+ 0.215	+ 0.014	0.000

INSTRUMENT READINGS.

Date.		Barom.	Thermom. At.	Thermom. Ex.
1846	h. m.	in.	°	°

REMARKS.

(77) 22. Minutes assumed as 43 instead of 44.
(77) 29. Hor. thread assumed as 4 instead of 3.

Zone 77. September 23. K. $D_o = -30° 16'' 30'$—Continued.

No.	Mag.	Seconds of Transit. I.	II.	III.	IV.	V.	VI.	VII.	T.		a_1	Micrometer.			$i + d_2$		d_1	Mean Right Ascension, 1850.0.	Mean Declination, 1850.0.
									h. m. s.		s.			r.	′ ″		″	h. m. s.	° ′ ″
37	8.9	..	57.2	11.9	26.3	40.7	..	..	22 14 26.06	—	14.56	VI.	1	11.12	— 5 36.1	—	4.2	22 14 11.50	— 30 22 10.3
38	9	..	..	..	..	..	45.8	..	15 16.95		14.56	VII.	2	10.46	10 22.4		4.1	15 2.39	26 56.5
39	9	..	..	..	47.4	1.7	15.9	..	16 47.13		14.58	VII.	2	8.51	9 27.3		4.0	16 32.55	26 1.3
40	10	..	44.6	59.3	..	..	..	..	19 13.91		14.57	IV.	9	8.2	43 0.6		3.9	18 59.34	59 34.4
41	7.8	14.9	29.8	44.6	59.3	13.6	28.3	42.7	20 59.05		14.61	VII.	6	8.42	28 20.6		3.6	20 44.44	44 54.2
42	9	..	..	..	24.1	38.5	..	7.3	22 23.90		14.61	VII.	8	6.34	37 16.0		3.5	22 9.29	53 49.5
43	11	..	47.7	2.3	..	..	..	..	25 16.57		14.68	IV.	3	7.2	13 29.4		3.2	25 1.89	30 2.6
44	7.8	..	..	..	..	..	51.1	6.1	25 22.26		14.68	VII.	2	9.11	9 34.4		3.2	25 7.58	30 26 7.6
45	8	..	..	21.7	..	..	..	..	27 36.46		14.67	III.	10	12.10	50 7.4		3.0	27 21.79	31 6 40.4
46	9	..	48.9	3.8	18.3	32.4	..	..	29 18.05		14.71	IV.	6	7.39	27 49.2		2.9	29 3.34	30 44 22.1
47	10	..	..	..	42.2	..	..	..	31 42.04		14.76	VI.	3	12.45	16 7.4		2.7	31 27.28	32 40.1
48	8	..	..	..	47.0	1.2	16.0	..	33 46.85		14.75	VII.	7	6.12	32 5.5		2.5	33 32.10	30 48 38.0
49	9	..	..	..	..	5.8	20.5	34.8	34 51.40		14.76	VII.	10	6.44	47 22.3		2.4	34 36.64	31 3 54.7
50	9	..	31.1	45.8	..	..	..	..	41 0.15		14.86	IV.	5	6.51	23 23.3		2.0	40 45.29	30 39 55.3
51	10	..	..	..	..	33.7	48.0	2.8	41 19.12		14.86	VII.	5	5.29	22 41.4		2.0	41 4.26	39 13.4
52	5	..	50.1	4.3	19.2	..	..	..	43 18.77		14.90	VI.	1	6.31	3 14.1		1.9	43 3.87	19 46.0
53	9	..	..	..	30.8	..	..	..	44 30.64		14.90	IV.	5	9.43	24 50.2		1.8	44 15.74	30 41 22.0
54	7	..	..	..	..	8.8	23.3	37.8	44 54.34		14.87	VII.	9	11.5?	44 32.6		1.8	44 39.47	31 1 3.4
55	7.8	..	..	5.0	19.1	33.6	48.0	..	47 19.05		14.94	VI.	2	6.4	8 0.1		1.6	47 4.11	30 24 31.7
56	1	52.2	6.8	21.8	36.0	50.4	..	..	49 35.81		14.97	V.	2	6.57	8 24.0		1.5	49 20.84	24 55.5
57	10	..	..	24.0	38.7	53.2	..	..	53 38.48		15.00	VI.	4	7.55	18 55.6		1.3	53 23.48	35 26.9
58	10	..	..	..	57.3	11.8	26.2	..	54 57.22		15.02	VII.	5	6.4	22 59.1		1.2	54 42.20	30 39 30.3
59	8	..	..	..	23.9	..	..	..	56 23.74		15.00	IV.	10	10.3	49 3.3		1.1	56 8.74	31 5 34.4
60	5.6	18.2	33.1	47.8	2.7	17.0	31.3	..	59 2.36		15.05	VII.	7	11.58	35 0.3		1.0	58 47.31	30 51 31.3
61	9	..	..	..	..	..	..	31.0	22 59 47.59		15.05	VII.	9	14.5	46 3.5		1.0	22 59 32.54	31 2 34.5
62	8	..	..	..	11.7	26.1	40.7	55.0	23 1 11.58		15.10	VII.	8	12.25	40 13.3		0.9	23 0 56.48	30 56 44.2
63	5	..	..	40.4	..	9.3	..	..	2 54.55		15.13	V.	1	7.14	3 36.0		0.8	2 39.42	20 6.8
64	9	..	..	..	..	..	47.2	1.4	4 18.03		15.13	VII.	7	6.54	32 26.7		0.8	4 2.90	48 57.5
65	7.8	..	57.2	..	26.5	41.0	55.2	9.8	5 26.34		15.15	VII.	8	3.25	35 40.4		0.7	5 11.19	52 11.1
66	5.6	8.8	23.6	38.3	52.8	..	..	..	7 52.63		15.17	IV.	5	6.38	23 16.8		0.6	7 37.46	39 47.4
67	9	..	..	45.7	0.0	14.4	28.8	..	8 59.92		15.19	VI.	6	5.59	26 58.5		0.5	8 44.73	43 29.1
68	10	..	..	22.9	37.6	52.0	..	..	12 37.47		15.23	V.	8	7.00	37 29.5		0.5	12 22.24	54 0.0
69	7	57.4	12.3	26.8	41.4	55.8	10.3	24.8	17 41.31		15.29	VII.	7	8.39	33 19.8		0.3	17 26.02	49 50.1
70	9	..	..	46.8	0.9	15.6	..	..	20 0.86		15.33	IV.	1	8.15	4 6.9		0.2	19 45.53	20 37.1
71	11	46.0	0.9	15.6	30.3	..	..	..	22 30.18		15.35	IV.	8	10.33	39 18.2		0.2	22 14.83	55 48.4
72	11	..	..	..	..	23.2	..	52.2	28 8.62		15.42	VII.	6	10.21	29 10.6		0.1	27 53.20	45 40.7
73	10	..	..	..	..	5.9	..	..	30 44.49		15.45	VI.	8	8.49	38 24.4		0.1	30 29.04	54 54.5
74	9	..	..	55.8	..	24.6	..	..	32 9.92		15.47	V.	1	10.16	5 8.0		0.1	31 54.45	21 38.1
75	8	41.8	56.8	11.7	25.7	40.0	..	..	34 25.59		15.50	V.	2	11.35	10 47.6		0.1	34 10.09	27 17.7
76	10	..	..	..	..	38.2	..	..	39 23.65		15.54	VII.	5	11.25	25 41.3		0.0	39 8.11	42 11.3
77	10	..	..	15.8	..	45.0	59.2	..	42 30.30		15.57	VII.	4	14.17	22 8.4		0.0	42 14.73	38 38.4
78	9	..	..	..	..	..	16.8	..	43 47.85		15.59	VII.	7	5.8	31 33.1		0.0	43 32.26	48 3.1
79	7	..	..	48.1	2.6	17.0	31.3	..	49 2.36		15.66	VI.	1	8.20	4 9.2		0.1	48 46.70	20 39.3
80	9	..	..	59.0	..	..	..	..	51 13.05		15.68	IV.	1	7.23	3 40.6		0.1	50 57.37	20 10.7
81	4.5	..	..	..	0.7	15.3	..	..	52 0.60		15.69	IV.	1	5.30	2 43.5		0.1	51 44.91	19 13.6
82	9	..	59.3	13.8	28.6	..	..	..	53 28.44		15.69	IV.	8	7.00	37 29.5		0.1	53 12.75	53 59.6
83	4.5	9.9	24.8	39.4	53.8	..	..	..	54 53.67		15.71	IV.	4	3.49	16 51.5		0.2	54 37.96	33 21.7
84	9	..	..	..	..	43.8	..	..	55 29.31		15.71	VII.	9	7.21	42 39.4		0.2	55 13.60	59 9.6
85	8	..	..	32.0	46.8	..	..	..	23 56 46.65	—	15.72	V.	9	5.30	—41 43.7	—	0.2	23 56 30.93	— 30 58 13.9

CORRECTIONS.

Date.		Corr. of Clock.	Hourly rate.	m	n	c
1846.	h.	s.	s.	s.	s.	s.

INSTRUMENT READINGS.

Date.		Barom.	Thermom. At.	Ex.
1846.	h. m.	in.	°	°

REMARKS.

(77) 47. Micrometer reading assumed as 12ʳ.15 instead of 12ʳ.45.

ZONE 77. SEPTEMBER 23. K. $D_0 = -30° 16' 30''$—Continued.

No.	Mag.	I.	II.	III.	IV.	V.	VI.	VII.	T.	a_1	MICROMETER.			$i + d_s$	d_i	Mean Right Ascension, 1850.0.	Mean Declination, 1850.0.
									h. m. s.	s.			r.	′ ″	″	h. m. s.	° ′ ″
86	9.8	. .	. .	. .	. .	. .	54.8	. .	23 57 25.95	– 15.75	VII.	3	3.20	–11 36.7	– 0.2	23 57 10.20	– 30 28 6.9
87	10	. .	. .	41.4	55.2	. .	24.9	. .	0 3 55.56	15.82	VII.	3	4.12	12 3.0	0.4	0 3 39.74	28 33.4
88	9	. .	9.2	23.7	38.4	52.8	. .	. .	5 38.30	15.83	V.	8	6.14	37 6.2	0.4	5 22.47	53 36.6
89	9	. .	. .	. .	. .	. .	43.0	58.0	5 14.15	15.83	VII.	2	6.10	8 2.9	0.4	5 58.32	30 24 33.3
90	10	. .	. .	51.7	6.6	20.8	. .	. .	8 6.38	15.85	V.	9	9.22	43 40.9	0.5	7 50.53	31 0 11.4
91	9	39.6	54.3	8.8	23.7	. .	. .	. .	11 23.43	15.89	IV.	7	3.51	30 54.6	0.6	11 7.51	30 47 25.2
92	8.9	. .	10.0	. .	39.2	53.7	. .	. .	13 39.03	15.93	V.	3	9.22	14 40.1	0.7	13 23.10	31 10.8
93	7	22.9	37.7	52.4	7.0	21.3	35.5	50.4	16 6.73	15.95	VII.	5	8.25	24 0.4	0.8	15 50.78	40 31.2
94	9.10	16.5	. .	46.0	. .	15.4	. .	. .	20 0.34	16.00	VI.	2	6.30	8 13.3	1.0	19 44.34	24 44.3
95	4	. .	2.8	17.1	31.9	46.2	0.5	. .	26 31.57	16.07	VII.	2	3.17	6 35.5	1.4	26 15.50	23 6.9
96	10	. .	. .	37.8	. .	. .	20.8	. .	29 52.00	16.10	VII.	3	11.46	15 52.5	1.5	29 35.90	32 24.0
97	9	. .	. .	. .	. .	. .	22.0	. .	30 53.20	16.11	VI.	1	7.30	3 44.0	1.6	30 37.09	20 15.6
98	9	. .	. .	. .	26.3	40.5	54.7	9.8	33 25.99	16.12	VII.	5	4.28	22 10.5	1.8	32 9.87	38 42.3
99	9	. .	. .	. .	. .	. .	53.7	8.2	34 24.66	16.13	VII.	5	10.42	25 19.6	1.8	33 8.53	30 41 51.4
100	7	50.8	5.8	20.4	35.0	49.3	. .	. .	39 34.93	16.18	VI.	9	10.24	44 12.1	2.2	39 18.75	31 0 44.3
101	10	. .	19.9	. .	49.3	. .	. .	. .	46 49.14	16.25	V.	9	7.58	42 58.5	2.7	46 32.89	30 59 31.2
102	10	. .	. .	. .	. .	18.9	. .	. .	55 4.37	16.34	VII.	7	4.33	31 15.4	3.4	54 48.03	47 48.8
103	4.5	. .	. .	. .	24.3	38.6	53.0	. .	56 24.10	16.35	VII.	1	6.49	3 23.0	3.6	56 7.75	19 56.6
104	9	. .	. .	. .	36.0	. .	. .	19.5	0 57 35.80	– 16.36	VII.	4	9.19	–19 37.8	– 3.7	0 57 19.44	– 30 36 11.5

ZONE 78. SEPTEMBER 24. A. $D_0 = -39° 4' 10''$.

No.	Mag.	I.	II.	III.	IV.	V.	VI.	VII.	T.	a_1	MICROMETER.			$i + d_s$	d_i	Mean Right Ascension, 1850.0.	Mean Declination, 1850.0.
1	7	34.2	50.4	. .	. .	. .	. .	. .	20 7 22.75	– 12.07	II.	5	7.55	–23 55.1	. .	20 7 10.68	
2	8	40.0	56.5	. .	. .	. .	. .	. .	8 28.66	12.09	II.	5	2.42	21 16.3	. .	8 16.57	
3	9	. .	. .	9.2	25.5	. .	. .	. .	10 25.55	12.07	IV.	10	6.46	47 29.2	. .	10 13.48	
4	9	. .	. .	33.5	49.6	. .	. .	. .	11 49.62	12.13	IV.	7	8.37	33 21.2	. .	11 37.49	
5	10	. .	. .	. .	. .	. .	25.0	41.0	13 52.56	12.13	VII.	9	3.35	40 48.8	. .	13 40.43	
6	9	37.0	53.5	. .	. .	. .	. .	. .	17 25.72	12.21	II.	5	10.12	25 4.6	. .	17 13.51	
7	10	. .	. .	. .	29.5	45.5	. .	. .	22 29.41	12.28	V.	6	7.19	27 39.7	. .	22 17.13	
8	6	. .	12.4	28.7	. .	. .	. .	. .	36 44.99	12.43	III.	9	3.51	40 57.5	. .	36 32.56	
9	10	18.5	35.0	. .	. .	. .	. .	. .	43 7.21	12.54	II.	5	8.45	24 20.5	. .	42 54.67	
10	7	. .	. .	. .	12.5	28.6	. .	. .	43 12.48	12.64	V.	8	5.21	36 42.2	. .	42 59.84	
11	8	. .	7.2	23.5	. .	. .	. .	. .	51 39.37	12.69	III.	3	11.4	15 29.4	. .	51 26.68	
12	7	43.2	59.8	. .	. .	. .	. .	. .	53 32.15	12.57	II.	8	6.23	37 13.5	. .	53 19.58	
13	5	. .	. .	. .	. .	51.0	7.0	. .	54 34.78	12.74	VI.	2	8.47	9 18.8	. .	54 22.04	
14	10	3.0	19.5	. .	. .	. .	. .	. .	57 51.75	12.75	II.	6	5.26	26 42.1	. .	57 39.00	
15	7	. .	. .	. .	. .	25.4	41.2	. .	20 58 9.07	– 12.77	VI.	3	8.4	–13 57.9	. .	20 57 56.30	

ZONE 79. SEPTEMBER 24. A. $D_0 = -29° 2' 30''$.

No.	Mag.	I.	II.	III.	IV.	V.	VI.	VII.	T.	a_1	MICROMETER.			$i + d_s$	d_i	Mean Right Ascension, 1850.0.	Mean Declination, 1850.0.
1	8	21.0	35.4	. .	. .	. .	. .	. .	21 26 4.17	– 13.87	II.	7	3.0	–30 28.6	– 20.0	21 25 50.30	– 29 33 18.6
2	10	. .	. .	. .	. .	19.0	33.5	. .	26 4.74	13.87	VI.	6	5.32	26 44.8	20.0	25 50.87	29 34.8
3	7	. .	58.0	12.2	. .	. .	. .	. .	28 26.31	13.90	III.	3	7.57	13 59.8	19.6	28 12.41	16 49.4
4	10	. .	. .	. .	. .	. .	. .	10.0	28 26.70	13.90	VII.	4	7.25	18 40.0	19.6	28 12.80	21 29.6
5	11	. .	. .	. .	6.0	20.5	. .	. .	32 5.95	13.96	V.	4	5.37	17 46.1	19.0	31 51.99	20 35.1
6	7	18.5	33.2	. .	. .	. .	. .	. .	37 1.74	14.00	II.	5	4.12	22 2.8	18.2	36 47.74	24 51.0
7	10	. .	18.0	32.5	. .	. .	. .	. .	21 38 46.63	– 14.02	III.	6	7.23	–27 41.0	– 18.0	21 38 32.61	– 29 30 29.0

CORRECTIONS.

Date.		Corr. of Clock.	Hourly rate.	m	n	c
1846.	h.	s.	s.	s.	s.	s.
Sept. 24,	18	– 24.45	+ 0.004	+ 0.394	– 0.190	0.000

INSTRUMENT READINGS.

Date.		Barom.	THERMOM. At.	THERMOM. Ex.
1846.	h. m.	in.	°	°

REMARKS.

(77) 89. Minutes assumed as 6 instead of 5.
(77) 98. Minutes assumed as 32 instead of 33.
(77) 99. Minutes assumed as 33 instead of 34.

ZONE 79. SEPTEMBER 24. A. $D_o = -29° 2' 30''$—Continued.

No.	Mag.	SECONDS OF TRANSIT. I.	II.	III.	IV.	V.	VI	VII.	T.	a_1	MICROMETER.			$i + d_i$	d_i	Mean Right Ascension, 1850.0.	Mean Declination, 1850.0.
									h. m. s.	s.			r.	′ ″	″	h. m. s.	° ′ ″
8	8	. .	56.5	11.1	. .	. .	. .	. .	21 40 24.93	− 14.04	III.	2	6.32	− 8 14.7	− 17.7	21 40 10.89	− 29 11 2.4
9	9	29.8	44.0	. .	. .	. .	. .	. .	44 12.91	14.08	II.	7	8.28	33 14.3	17.1	43 58.83	36 1.4
10	9	. .	. .	. .	28.5	42.8	. .	. .	44 28.38	14.08	V.	6	10.43	29 22.1	17.1	44 14.30	32 9.2
11	6	. .	. .	. .	. .	27.5	41.7	. .	45 13.09	14.10	VI.	4	8.45	19 21.0	17.0	44 58.99	22 8.0
12	8	. .	. .	. .	. .	22.5	36.7	. .	46 8.10	14.10	VI.	6	11.40	29 50.7	16.9	45 54.00	32 37.6
13	7	38.4	53.0	. .	. .	. .	. .	. .	47	14.12	II.	6	9.48	28 54.2	16.7	47	31 40.9
14	6	. .	36.8	51.2	. .	. .	. .	. .	51 5.59	14.16	III.	9	9.5	43 32.1	16.1	50 51.43	46 18.2
15	8	. .	. .	. .	. .	. .	7.5	21.5	51 38.62	14.17	VII.	8	8.53	38 26.0	16.1	51 24.45	41 12.1
16	10	. .	. .	. .	. .	19.0	33.2	. .	53 4.59	14.19	VI	6	6.59	27 28.8	15.9	52 50.40	30 14.7
17	10	. .	. .	. .	. .	. .	32.0	46.5	54 3.30	14.18	VII.	4	8.22	19 9.2	15.8	53 49.12	21 55.0
18	10	. .	. .	. .	. .	. .	. .	41.0	54 57.66	14.20	VII.	3	11.46	15 52.6	15.6	54 43.46	18 38.2
19	5	. .	. .	. .	. .	. .	46.0	1.0	57 17.51	14.22	VII.	2	4.16	7 5.6	15.3	57 3.29	9 50.9
20	7	. .	. .	. .	52.9	7.0	. .	. .	21 58 52.66	14.23	V.	5	6.44	23 19.8	15.1	21 58 38.43	26 4.9
21	11	. .	. .	16.0	. .	. .	. .	. .	22 5 30.20	14.30	III.	5	4.8	22 1.0	14.2	22 5 15.90	24 45.2
22	9	. .	. .	. .	. .	37.0	51.2	. .	6 22.58	14.29	VI.	8	0.51	34 22.7	14.1	6 8.29	37 6.8
23	8	39.4	54.0	. .	. .	. .	. .	. .	13 22.66	14.39	II.	6	9.46	28 53.2	13.3	13 8.27	31 36.5
24	7	. .	. .	. .	59.0	13.0	. .	. .	16 58.71	14.43	V.	5	6.53	23 24.3	12.9	15 44.24	26 7.2
25	8	58.5	13.0	. .	. .	. .	. .	. .	17 41.68	14.45	II.	6	4.11	26 3.9	12.8	17 27.23	28 46.7
26	7	. .	. .	. .	. .	. .	24.0	38.6	17 55.47	14.42	VII.	10	8.2	48 1.4	12.8	17 41.05	50 41.2
27	6	31.2	46.0	. .	. .	. .	. .	. .	21 14.50	14.48	II.	5	5.59	22 56.9	12.4	21 0.02	25 39.3
28	9	. .	. .	. .	. .	27.2	41.5	. .	21 12.84	14.48	VI.	6	5.33	26 45.3	12.4	20 58.36	29 27.7
29	9	. .	. .	. .	. .	2.5	16.5	. .	26 47.99	14.54	VI.	7	5.15	31 36.8	11.8	26 33.45	34 18.6
30	8	. .	. .	. .	. .	. .	. .	19.2	27 35.82	14.57	VII.	3	5.26	12 40.6	11.8	27 21.25	15 22.4
31	10	. .	. .	. .	24.2	38.5	. .	. .	29 24.08	14.56	V.	7	3.20	30 38.8	11.6	29 9.52	33 20.4
32	5	. .	. .	. .	38.2	52.5	. .	. .	30 38.08	14.59	V.	6	10.1	29 0.9	11.5	30 23.49	31 42.4
33	7	. .	. .	. .	42.5	56.7	. .	. .	31 42.34	14.59	V.	7	11.54	34 58.6	11.4	31 27.75	37 40.0
34	7	12.0	27.0	. .	. .	. .	. .	. .	34 55.40	14.64	II.	5	5.9	22 31.6	11.1	34 40.76	25 12.7
35	9	. .	31.0	45.5	. .	. .	. .	. .	35 59.79	14.62	III.	9	2.59	40 27.1	11.0	35 45.17	43 8.1
36	8	. .	30.0	44.5	. .	. .	. .	. .	39 58.49	14.71	III.	4	2.21	16 7.1	10.6	39 43.78	18 47.7
37	9	. .	. .	33.0	47.5	. .	. .	. .	40 47.28	14.71	IV.	5	8.29	24 12.8	10.6	40 32.57	26 53.4
38	7	43.5	58.5	. .	. .	. .	. .	. .	45 26.76	14.78	II.	2	10.2	10 0.6	10.2	45 11.98	12 40.8
39	7	15.9	0.3	. .	. .	. .	. .	. .	48 28.98	14.78	II.	5	3.51	21 52.2	10.0	48 14.20	24 32.2
40	8	. .	14.0	28.2	. .	. .	. .	. .	50 42.18	14.85	III.	1	7.34	3 46.4	9.8	50 27.33	6 26.2
41	4	39.0	53.5	. .	. .	. .	. .	. .	53 22.30	14.83	II.	8	6.0	36 58.8	9.6	53 7.47	39 38.4
42	9	. .	13.2	27.5	. .	. .	. .	. .	22 59 41.65	14.93	III.	5	2.32	21 12.5	9.2	22 59 26.72	23 51.7
43	6	. .	. .	14.2	28.5	. .	. .	. .	23 0 28.47	14.91	IV.	8	2.57	35 26.6	9.2	23 0 13.56	38 5.8
44	8	. .	. .	0.5	15.0	. .	. .	. .	2 14.93	14.92	IV.	9	8.9	43 3.8	9.1	2 0.01	45 42.9
45	8	. .	. .	. .	. .	. .	. .	22.0	2 38.62	14.98	VII.	3	4.14	12 4.3	9.1	2 23.64	14 43.4
46	7	44.0	58.4	. .	. .	. .	. .	. .	5 26.96	15.01	II.	3	2.31	11 12.5	8.9	5 11.95	13 51.4
47	8	59.2	13.8	. .	. .	. .	. .	. .	6 42.29	15.02	II.	3	7.40	13 48.6	8.8	6 27.27	16 27.4
48	6	. .	21.8	36.0	. .	. .	11.0	25.5	50.27	15.04	III.	6	6.57	27 27.9	8.7	8 35.23	30 6.6
49	6	. .	. .	. .	. .	34.5	48.5	. .	9 19.99	15.05	VI.	3	5.7	12 31.3	8.7	9 4.94	15 10.0
50	10	. .	. .	. .	. .	39.0	53.5	. .	10 24.74	15.05	VI.	4	2.51	16 22.1	8.6	10 9.69	19 0.7
51	6	. .	. .	. .	. .	. .	44.4	59.0	11 15.71	15.07	VII.	3	0.0	9 55.9	6.6	11 0.64	12 34.5
52	9	. .	. .	. .	. .	15.0	30.5	. .	1.20	15.08	VI.	6	3.33	25 44.7	8.5	12 46.12	28 23.2
53	8	59.1	13.9	. .	. .	. .	. .	. .	15 42.44	15.10	II.	6	6.5	27 1.5	8.4	15 27.34	29 39.9
54	8	. .	. .	. .	51.8	12.0	. .	. .	57.62	15.10	V.	5	3.57	21 55.4	8.4	15 42.52	24 33.8
55	7	. .	. .	10.2	24.2	. .	. .	. .	21 24.07	15.19	IV.	1	5.35	2 46.4	8.2	21 8.88	5 24.6
56	9	. .	. .	. .	. .	3.2	17.5	. .	23 23 48.84	− 15.16	VI.	9	9.23	−43 41.0	− 8.1	23 23 33.68	− 29 46 19.1

CORRECTIONS.

Date.		Corr. of Clock.	Hourly rate.	m	n	c
1846.	h.	s.	s.	s.	s.	s.

INSTRUMENT READINGS.

Date.		Barom.	THERMOM. At.	Ex.
1846.	h. m.	in.	°	°

REMARKS.

(79) 24. Minutes assumed as 15 instead of 16.
(79) 48. Transit over T. VI rejected. Minutes assumed as 8.
(79) 52. Minutes of transit assumed as 13.
(79) 54. Transit over T. IV assumed as 57ˢ instead of 51ˢ, and minutes as 15.

ZONE 79. SEPTEMBER 24. A. $D_0 = -29° 2' 30''$—Continued.

No.	Mag.	Seconds of Transit: I.	II.	III.	IV.	V.	VI.	VII.	T.	a_1	Micrometer.			$i + d_1$	d_1	Mean Right Ascension, 1850.0.	Mean Declination, 1850.0.
									h. m. s.	s.			r.	′ ″	″	h. m. s.	° ′ ″
57	9	. .	. .	20.0	34.5	. .	. .	. .	23 25 34.31	−15.21	IV.	6	7.38	−27 48.7	−8.1	23 25 19.10	−29 30 26.8
58	8	58.5	13.0	. .	. .	. .	. .	. .	27 41.74	15.22	II.	7	5.9	31 33.8	8.0	27 26.52	34 11.8
59	10	. .	. .	47.5	2.0	. .	. .	. .	28 1.80	15.23	IV.	6	4.32	26 14.7	8.0	27 46.57	28 52.7
60	9	. .	. .	. .	24.8	39.0	. .	. .	30 24.64	15.25	V.	7	8.1	33 0.8	7.9	30 9.39	35 38.7
61	8	. .	. .	. .	. .	26.5	40.5	. .	31 11.99	15.30	VI.	1	7.5	3 31.6	7.9	30 56.69	6 9.5
62	7	. .	. .	. .	37.6	52.0	. .	. .	32 37.48	15.31	V.	2	7.18	8 37.9	7.9	32 22.17	11 15.8
63	7	. .	. .	. .	26.0	40.5	. .	. .	35 26.02	15.29	V.	10	11.52	49 58.0	7.8	35 10.73	52 35.8
64	8	. .	. .	49.5	4.0	. .	. .	. .	37 3.84	15.32	IV.	7	6.48	32 24.0	7.8	36 48.52	35 1.8
65	9	. .	44.5	59.2	. .	. .	. .	. .	39 13.32	15.35	III.	7	11.10	34 36.3	7.7	38 57.97	37 14.0
66	8	29.0	43.5	. .	. .	. .	. .	. .	42 11.98	15.41	II.	2	6.11	8 7.8	7.7	41 56.57	10 45.5
67	7	. .	. .	8.5	22.8	. .	. .	. .	44 22.56	15.43	IV.	2	4.35	7 15.6	7.6	44 7.13	9 53.2
68	9	. .	. .	. .	29.0	43.0	. .	. .	45 28.68	15.44	V.	2	7.31	8 44.5	7.6	45 13.24	11 22.1
69	9	. .	. .	. .	33.0	47.5	. .	. .	46 32.95	15.44	V.	4	4.3	16 58.7	7.6	46 17.51	19 36.3
70	9	. .	5.4	20.0	. .	. .	. .	. .	51 34.07	15.49	III.	6	5.20	26 38.9	7.6	51 18.58	29 16.5
71	10	. .	. .	. .	20.0	34.5	. .	. .	52 19.95	15.51	V.	4	6.18	18 6.9	7.6	52 4.44	20 44.5
72	6	. .	26.0	40.8	. .	. .	. .	. .	53 54.83	15.50	III.	7	4.25	31 11.7	7.6	53 39.33	33 49.3
73	9	. .	28.5	43.0	. .	. .	. .	. .	54 57.06	15.52	III.	5	3.50	21 51.9	7.6	54 41.54	24 29.5
74	7	. .	42.5	57.3	. .	. .	. .	. .	56 11.09	15.56	III.	3	5.11	12 33.4	7.6	55 55.53	15 11.0
75	7	. .	. .	. .	. .	. .	55.0	10.0	23 57 26.50	−15.57	VII.	3	2.31	−11 12.2	−7.6	23 56 10.93	−29 13 49.8

ZONE 80. SEPTEMBER 24. A. $D_0 = -29° 2' 30''$.

No.	Mag.	I.	II.	III.	IV.	V.	VI.	VII.	T.	a_1	Micrometer.			$i + d_1$	d_1	Mean Right Ascension, 1850.0.	Mean Declination, 1850.0.
1	9	44.5	59.0	. .	. .	. .	. .	. .	0 39 27.61	−15.58	II.	4	9.20	−19 38.7	−2.3	0 39 12.03	−29 22 11.0
2	6	53.5	6.2	. .	. .	. .	. .	. .	41 36.67	16.00	II.	4	2.49	16 21.1	2.5	41 20.67	18 53.6
3	8	2.0	17.0	. .	. .	. .	. .	. .	44 45.43	16.02	II.	5	11.0	25 29.0	2.8	44 29.41	28 1.8
4	8	5.2	20.0	. .	. .	. .	. .	. .	46 48.45	16.05	II.	4	7.52	18 54.2	3.0	46 32.40	21 27.2
5	9	. .	. .	. .	. .	9.2	23.3	. .	46 54.75	16.03	VI.	7	11.11	34 36.7	3.0	46 38.72	37 9.7
6	10	. .	. .	. .	. .	. .	46.5	1.5	55 18.12	16.10	VII.	8	9.53	38 56.3	4.0	55 2.02	41 30.3
7	7	. .	. .	. .	30.5	45.0	. .	. .	0 58 30.42	16.18	V.	1	9.53	4 56.6	4.4	0 58 14.24	7 31.0
8	9	. .	. .	27.5	42.0	. .	. .	. .	1 0 41.81	16.17	IV.	6	7.30	27 44.6	4.6	1 0 25.64	30 19.0
9	10	14.0	28.3	. .	. .	. .	. .	. .	2 57.14	16.17	II.	7	5.50	31 54.5	4.9	2 40.97	34 29.4
10	7	59.0	13.5	. .	. .	. .	. .	. .	11 42.21	16.26	II.	6	9.45	28 52.6	6.1	11 25.95	31 28.7
11	7	. .	. .	. .	34.8	49.0	. .	. .	13 34.66	16.24	V.	9	9.52	43 55.8	6.4	13 18.42	46 32.2
12	9	. .	. .	. .	36.2	51.0	. .	. .	14 36.34	16.27	V.	7	7.5	32 32.5	6.6	14 20.07	35 9.1
13	9	. .	. .	. .	. .	12.2	26.2	. .	15 57.59	16.30	VI.	3	10.46	15 22.6	6.7	15 41.29	17 59.3
14	7	. .	. .	23.5	37.5	. .	. .	. .	17 37.46	16.32	IV.	3	8.40	14 19.1	6.9	17 21.14	16 56.0
15	8	. .	. .	27.2	41.7	. .	. .	. .	18 41.54	16.30	IV.	7	4.48	31 23.4	7.1	18 25.24	34 0.5
16	9	. .	. .	49.5	4.0	. .	. .	. .	21 3.89	16.31	IV.	8	7.58	37 58.6	7.4	20 47.58	40 36.0
17	8	46.8	1.5	. .	. .	. .	. .	. .	23 30.18	16.32	II.	8	3.54	35 55.2	7.8	23 13.86	38 33.0
18	8	40.6	55.0	. .	. .	. .	. .	. .	27 23.77	16.37	II.	7	3.32	30 44.8	8.4	27 7.40	33 23.2
19	9	. .	. .	38.5	52.8	. .	. .	. .	1 27 52.66	−16.38	IV.	5	3.35	−21 44.3	−8.5	1 27 36.28	−29 24 22.8

CORRECTIONS.

Date.		Corr. of Clock.	Hourly rate.	m	n	c
1846.	h.	s.	s.	s.	s.	s.

INSTRUMENT READINGS.

Date.		Barom.	Thermom. At.	Thermom. Ex.
1846.	h. m.	in.	°	°

REMARKS.

(79) 75. Minutes assumed as 56 instead of 57.

ZONE 81. SEPTEMBER 25. K. $D_o = -27° 46' 40''$.

No.	Mag.	SECONDS OF TRANSIT. I.	II.	III.	IV.	V.	VI.	VII.	T.	a_1	MICROMETER.			$i + d_1$	d_1	Mean Right Ascension, 1850.0.	Mean Declination, 1850.0.
									h. m. s.	s.			r.	' "		h. m. s.	° ' "
1	10	. .	. .	. .	10.2	37.9	. .	. .	21 9	. .	V.	1	8.21	− 4 10.4			
2	8	. .	. .	. .	. .	30.0	. .	. .	11 15.71	− 14.03	V.	1	6.49	3 24.0	− 13.6	21 11 1.68	− 27 50 17.6
3	10	. .	. .	. .	. .	. .	. .	8.3	11 25.50	14.04	VII.	2	8.12	9 5.0	13.6	11 11.46	27 55 58.6
4	9	. .	. .	. .	. .	0.5	. .	. .	12 46.33	14.08	V.	7	11.55	34 58.9	13.3	12 32.25	28 21 42.2
5	10	1.8	16.2	. .	. .	. .	. .	. .	16 44.65	14.07	V.	7	6.58	32 28.9	12.7	16 30.58	19 21.6
6	8	. .	29.5	44.0	58.2	12.2	. .	. .	19 58.05	14.10	V.	8	3.10	35 33.0	12.2	19 43.95	22 25.2
7	11	. .	. .	. .	14.0	. .	. .	. .	20 13.84	14.10	VI.	8	4.45	36 20.8	12.2	19 59.74	23 13.0
8	8	. .	. .	41.8	56.1	10.1	. .	. .	25 56.00	14.15	VI.	10	4.14	46 6.1	11.4	25 41.85	32 57.5
9	10	. .	. .	. .	. .	. .	4.3	. .	25 36.07	14.15	VI.	10	5.39	46 49.1	11.4	25 21.92	33 40.5
10	11	. .	. .	. .	. .	. .	. .	15.8	26 33.26	14.18	VII.	6	11.29	29 44.9	11.3	26 19.08	16 36.2
11	10	9.0	23.2	37.4	. .	. .	. .	. .	28 51.65	14.19	IV.	7	3.52	30 55.0	10.9	28 37.46	17 45.9
12	7.8	. .	32.1	46.2	0.8	14.7	. .	. .	30 0.60	14.20	V.	10	6.17	47 8.4	10.8	29 46.40	28 33 59.2
13	7	. .	. .	. .	. .	. .	5.0	19.5	30 36.83	14.24	VII.	3	3.34	11 44.2	10.7	30 22.59	27 58 34.9
14	8	53.0	7.7	21.8	36.1	. .	. .	. .	33 35.99	14.25	IV.	8	2.17	35 6.2	10.2	33 21.74	28 21 56.4
15	9	. .	. .	. .	. .	. .	11.7	26.2	33 43.57	14.26	VII.	5	4.29	22 11.2	10.2	33 29.31	9 1.4
16	11	. .	. .	. .	. .	. .	. .	58.4	34 15.77	14.27	VII.	5	6.50	23 22.4	10.2	34 1.50	10 12.6
17	8	31.8	46.2	0.5	14.7	. .	. .	. .	39 14.67	14.31	V.	8	11.19	39 40.0	9.5	39 0.36	28 26 29.5
18	10	. .	. .	. .	. .	. .	2.8	17.6	39 34.77	14.35	VI.	1	13.31	6 46.9	9.4	39 20.42	27 53 36.3
19	7	. .	. .	14.5	28.6	42.8	. .	. .	41 28.47	14.36	VI.	4	8.21	19 8.9	9.2	41 14.11	28 5 58.1
20	9	. .	. .	. .	10.3	. .	38.3	. .	42 10.17	14.36	VI.	3	9.58	14 58.4	9.1	41 55.81	1 47.5
21	8	. .	. .	. .	. .	21.2	35.1	49.7	43 6.99	14.37	VII.	4	8.10	19 3.2	9.0	42 52.62	5 52.2
22	8	. .	. .	. .	. .	. .	33.6	48.0	44 5.39	14.38	VII.	3	8.22	14 9.7	8.8	43 51.01	0 58.5
23	10	. .	. .	. .	6.8	. .	. .	. .	46 6.64	14.38	IV.	9	9.15	43 36.9	8.6	45 52.26	30 25.5
24	10	. .	. .	. .	. .	17.0	31.1	45.4	47 2.82	14.42	VII.	4	3.46	16 49.8	8.4	46 48.40	3 38.2
25	11	. .	. .	6.0	. .	34.2	. .	. .	49 20.15	14.41	V.	9	7.28	42 42.8	8.2	49 5.74	29 31.0
26	7	. .	44.3	58.6	12.8	. .	. .	. .	51 12.71	14.44	IV.	7	9.48	33 54.8	7.9	50 58.27	20 42.7
27	9	. .	. .	. .	. .	54.3	. .	22.8	51 40.16	14.47	VII.	4	14.21	22 10.5	7.8	51 25.69	8 58.3
28	6.7	. .	53.4	. .	22.0	36.0	. .	. .	21 54 21.77	14.50	VI.	4	6.52	18 24.0	7.5	21 54 7.27	28 5 11.5
29	11	. .	. .	. .	40.9	. .	. .	. .	22 0 39.74	14.59	IV.	2	7.52	8 55.4	6.8	22 0 25.15	27 55 42.2
30	9.10	. .	. .	. .	. .	6.0	20.0	34.2	1 51.80	14.57	VII.	8	6.55	37 26.3	6.6	1 37.23	28 24 12.9
31	7	. .	. .	. .	. .	25.3	39.3	54.2	3 11.21	14.61	VII.	2	1.56	6 25.3	6.5	2 56.60	27 53 11.8
32	8	. .	. .	. .	. .	36.1	. .	. .	4 21.81	14.62	V.	1	5.10	2 34.0	6.3	4 7.19	27 49 20.3
33	10	. .	. .	. .	. .	. .	51.9	. .	5 23.77	14.62	VII.	5	10.10	25 3.5	6.2	5 9.15	28 11 49.7
34	4.5	. .	. .	. .	. .	. .	. .	44.8	6 2.42	14.61	VII.	9	9.30	43 44.1	6.2	5 47.81	30 30.3
35	7.8	. .	34.6	49.0	3.2	17.2	. .	. .	9 2.94	14.67	VI.	4	12.27	21 13.2	5.7	8 48.27	7 58.9
36	11	. .	. .	. .	11.2	. .	. .	. .	10 11.04	14.68	V.	3	11.30	15 45.0	5.7	9 56.36	2 30.7
37	8	. .	7.4	21.6	35.5	. .	. .	. .	11 35.54	14.69	V.	4	13.8	21 34.0	5.6	11 20.85	8 19.6
38	10	. .	. .	. .	42.4	56.2	10.4	. .	14 42.17	14.72	VI.	7	5.19	31 38.8	5.2	14 27.45	18 24.0
39	9	. .	31.2	45.5	. .	. .	. .	. .	17 59.45	14.78	IV.	4	8.35	19 16.2	4.9	17 44.67	28 6 1.1
40	9	19.8	. .	48.4	2.3	16.5	. .	. .	20 2.19	14.80	V.	3	3.2	11 28.4	4.7	19 47.39	27 58 13.1
41	9	. .	. .	. .	15.2	29.2	. .	. .	21 14.98	14.82	VI.	2	7.13	8 35.5	4.6	21 0.16	55 20.1
42	4.5	. .	. .	. .	. .	. .	4.8	19.6	21 36.77	14.83	VII.	1	11.15	5 38.0	4.6	21 21.94	27 52 22.6
43	11	. .	53.8	7.8	22.3	. .	. .	. .	26 22.01	14.86	IV.	5	6.9	23 2.2	4.1	25 7.15	28 9 46.3
44	11	. .	. .	. .	59.0	. .	. .	. .	26 58.84	14.86	IV.	6	7.49	27 54.2	4.1	26 43.98	28 14 38.3
45	11	. .	. .	. .	12.1	. .	. .	. .	28 20.84	14.90	IV.	2	7.52	8 55.4	4.0	28 5.94	27 55 39.4
46	10	18.3	32.7	47.0	. .	. .	. .	. .	30 1.07	14.90	IV.	6	5.9	26 33.4	3.8	29 46.17	28 13 17.2
47	2.3	. .	. .	22.0	36.0	50.2	3.9	. .	32 35.81	14.95	V.	1	5.34	2 46.1	3.6	32 20.86	27 49 29.7
48	9	. .	. .	. .	. .	. .	51.3	5.9	33 23.20	14.94	VII.	4	1.18	15 35.0	3.6	33 8.26	28 2 18.6
49	11	. .	. .	41.4	55.5	. .	. .	. .	22 35 55.26	− 14.97	IV.	3	4.56	−12 26.1	− 3.4	22 35 40.29	− 27 59 9.5

CORRECTIONS.

Date.		Corr. of Clock.	Hourly rate.	m	n	c
1846.	h.	s.	s.	s.	s.	s.
Sept. 25,	18	− 24.67	+ 0.017	+ 0.394	− 0.190	0.000

INSTRUMENT READINGS.

Date.		Barom.	THERMOM. At.	Ex.
1846.	h. m.	in.	°	°

REMARKS.

(81) 31. Micrometer reading assumed as 2r.56 instead of 1r.56.
(81) 43. Minutes assumed as 25 instead of 26.

ZONE 81. SEPTEMBER 25. K. $D_0 = -27° 46' 40''$—Continued.

No.	Mag.	Seconds of Transit. I.	II.	III.	IV.	V.	VI.	VII.	T.	a_1	Micrometer.			$i + d_1$	d_1	Mean Right Ascension, 1850.0.	Mean Declination, 1850.0.
									h. m. s.	s.			r.	′ ″	″	h. m. s.	° ′ ″
50	9		54.6	9.0	23.2				22 39 22.98	− 15.00	IV.	5	7.15	−23 35.5	− 3.1	22 39 7.98	− 28 10 18.6
51	7			13.8	28.0	42.0	56.2		40 27.90	15.00	VII.	7	10.35	34 18.2	3.0	40 12.90	28 21 1.2
52	10			6.4		34.7			52 20.27	15.17	IV.	2	4.34	7 15.3	2.2	52 5.10	27 53 57.5
53	6.7		29.8	44.0	58.2	12.2			22 56 57.93	15.22	VI.	2	9.50	9 54.8	2.0	22 56 42.71	27 56 36.8
54	11		56.4		24.8				23 2 24.67	− 15.27	IV.	4	8.41	−19 19.2	− 1.7	23 2 9.40	− 28 6 0.9

ZONE 82. SEPTEMBER 30. A. $D_0 = -27° 48' 20''$.

No.	Mag.	Seconds of Transit. I.	II.	III.	IV.	V.	VI.	VII.	T.	a_1	Micrometer.			$i + d_1$	d_1	Mean Right Ascension, 1850.0.	Mean Declination, 1850.0.
1	5	37.0	51.4						23 16 19.70	− 17.60	II.	4	6.5	−18 0.2	− 2.7	23 16 2.10	− 28 6 22.9
2	6		33.0	47.2					17 1.23	17.61	III.	5	4.28	22 11.1	2.6	16 43.62	10 33.7
3	6	47.8	2.1						18 30.54	17.63	II.	6	5.15	26 36.2	2.6	18 12.91	14 58.8
4	7	9.5	23.9						19 52.20	17.65	II.	4	6.1	17 58.2	2.5	19 34.55	6 20.7
5	8	54.5	8.8						21 37.35	17.67	II.	8	6.13	37 5.3	2.5	21 19.68	28 25 27.8
6	9					50.5	4.5		21 36.34	17.68	VI.	2	4.15	7 5.5	2.5	21 18.66	27 55 28.0
7	9			55.5	10.0				23 9.61	17.69	IV.	4	7.37	18 46.9	2.4	22 51.92	28 7 9.3
8	10					48.5	2.2		23 34.18	17.69	VI.	5	7.37	23 46.4	2.4	23 16.49	28 12 8.8
9	9							35.5	23 52.70	17.71	VII.	2	9.18	9 38.3	2.4	23 34.99	27 58 0.7
10	10			12.0	26.5				26 26.06	17.73	IV.	3	4.15	12 5.3	2.4	26 8.33	28 0 27.7
11	10					22.0	35.5		27 7.58	17.73	VI.	6	4.39	26 18.0	2.4	26 49.85	14 40.4
12	5					16.5	30.8		28 2.48	17.75	VI.	7	3.15	30 36.2	2.4	27 44.73	28 18 58.6
13	8	5.0	19.5						31 47.66	17.80	II.	2	11.36	10 48.3	2.3	31 29.86	27 59 10.6
14	10	59.5	13.8						33 42.22	17.81	II.	5	10.56	25 27.0	2.3	33 24.41	28 13 49.3
15	7		14.7	29.0					34 43.20	17.82	III.	8	11.35	39 48.2	2.3	34 25.38	28 10.5
16	11				35.0	48.5			36 34.54	17.86	V.	3	7.21	13 24.1	2.3	36 16.68	1 46.4
17	8		38.5	52.5					38 6.77	17.87	III.	7	7.15	32 37.5	2.3	37 48.90	20 59.8
18	10	42.0	56.0						39 24.32	17.87	II.	4	11.17	20 37.8	2.3	39 6.45	28 9 0.1
19	11			26.0	40.0				42 39.79	17.92	IV.	2	8.55	9 27.1	2.3	42 21.87	27 57 49.4
20	10		24.0	38.5					44 52.17	17.94	III.	1	10.46	5 23.6	2.3	44 34.23	53 45.9
21	7				53.5	8.0	21.8		46 53.61	17.97	VI.	1	8.38	4 18.9	2.4	46 35.64	27 52 41.3
22	9		56.5	10.8					51 25.12	17.99	III.	10	8.14	48 7.6	2.4	51 7.13	28 36 30.0
23	10		3.5	18.0					52 32.17	18.01	III.	9	9.52	43 55.6	2.5	52 14.16	28 32 18.1
24	8		23.0	37.5	52.0				54 51.43	18.05	IV.	2	9.27	9 46.5	2.6	54 33.38	27 58 9.1
25	9		53.0	7.5					57 21.32	18.07	III.	4	4.55	17 25.0	2.6	57 3.25	28 5 47.6
26	9					6.0	19.8		57 51.74	18.08	VI.	4	5.51	17 53.1	2.6	57 33.66	6 15.7
27	8		27.3	41.4					23 59 55.52	18.10	III.	5	11.3	25 30.6	2.7	23 59 37.42	13 53.3
28	6			0.8	15.0				0 4 14.99	18.13	IV.	10	10.51	49 42.1	2.9	0 3 56.86	38 5.0
29	7	19.6	34.4						7 2.66	18.18	II.	7	8.20	33 10.2	3.0	6 44.48	21 33.2
30	8					22.0	36.0		7 7.84	18.19	VI.	3	3.40	11 47.4	3.0	6 49.65	0 10.4
31	9							27.5	7 44.98	18.20	VII.	7	5.32	31 45.1	3.0	7 26.78	28 20 8.1
32	10						33.0	47.0	9 4.57	18.21	VII.	2	10.34	10 16.7	3.1	8 46.36	27 58 39.8
33	11		9.3	23.5					12 37.79	18.23	III.	9	4.30	41 13.0	3.3	12 19.56	28 29 36.3
34	7			53.0	7.0				15 6.74	18.27	IV.	1	5.33	2 45.5	3.4	14 48.47	27 51 8.9
35	10	2.0	17.5						16 45.33	18.28	II.	6	4.31	26 14.0	3.5	16 27.05	28 14 37.5
36	7					19.2	33.0		17 4.94	18.29	VI.	1	6.54	3 26.3	3.6	16 46.65	27 51 49.9
37	5					47.0	0.9	15.0	18 32.73	18.29	VII.	9	9.57	43 57.8	3.6	18 14.44	28 32 21.4
38	8		21.0	36.0					21 49.67	18.33	III.	5	9.51	24 54.2	3.8	21 31.34	13 18.0
39	9				45.5	0.0			22 45.55	18.34	V.	4	2.44	16 18.8	3.9	22 27.21	4 42.7
40	7	50.2	5.0						0 29 33.09	− 18.41	II.	4	5.46	−17 50.6	− 4.4	0 29 14.68	− 28 6 15.0

CORRECTIONS.

Date.		Corr. of Clock.	Hourly rate.	m	n	c
1846.	h.	s.	s.	s.	s.	s.
Sept. 30,	19	− 26.59	− 0.020	+ 0.265	− 0.106	− 0.086

INSTRUMENT READINGS.

Date.		Barom.	Thermom. At.	Thermom. Ex.
1846.	h. m.	in.	°	°

REMARKS.

(82) 16. Micrometer reading assumed as 6r.51 instead of 7r.21; *vide* M. C. Zone 60, 123.
(82) 28. Micrometer reading assumed as 11r.21 instead of 10r.51.

ZONE 82. SEPTEMBER 30. A. $D_0 = -27° 48' 20''$—Continued.

No.	Mag.	I.	II.	III.	IV.	V.	VI.	VII.	T.	a_1	MICROMETER.			$i + d_2$	d_1	Mean Right Ascension, 1850.0.	Mean Declination, 1850.0.
									h. m. s.	s.			r.	′ ″	″	h. m. s.	° ′ ″
41	6	..	..	..	..	49.8	3.5	..	0 29 35.48	− 18.41	VI.	6	4.49	−26 23.1	− 4.4	0 29 17.07	− 28 14 47.5
42	7	..	..	..	0.5	14.5	..	..	31 0.34	18.42	V.	8	8.49	38 24.3	4.5	30 41.92	26 48.8
43	8	43.0	58.0	..	..	..	..	..	33 25.99	18.44	II.	4	4.39	17 16.8	4.7	33 7.55	5 41.5
44	7	..	..	..	42.0	56.0	..	..	33 41.82	18.45	V.	5	6.45	23 20.3	4.7	33 23.37	11 45.0
45	8	38.5	52.5	..	..	..	..	..	42 21.20	18.54	II.	8	6.2	36 59.8	5.4	42 2.66	25 25.2
46	9	52.5	7.0	..	..	..	..	..	43 35.34	18.56	II.	6	5.8	26 32.8	5.5	43 16.78	14 58.3
47	9	..	..	44.0	58.0	..	..	..	44 58.01	18.56	IV.	8	7.43	37 51.0	5.6	44 39.45	26 16.6
48	5	..	..	42.9	57.5	..	..	..	48 57.27	18.59	IV.	10	6.10	47 5.0	5.9	48 38.68	28 35 30.9
49	7	..	50.5	5.0	..	..	..	..	51 18.67	18.63	III.	1	11.28	5 44.8	6.2	51 0.04	27 54 11.0
50	11	..	..	..	51.0	5.0	..	..	0 53 50.79	− 18.66	V.	3	6.58	−12 57.3	− 6.4	0 53 32.13	− 28 1 23.7

ZONE 83. OCTOBER 3. A. $D_0 = -24° 5' 10''$.

No.	Mag.	I.	II.	III.	IV.	V.	VI.	VII.	T.	a_1	MICROMETER.			$i + d_2$	d_1	Mean Right Ascension, 1850.0.	Mean Declination, 1850.0.
1	8	..	33.0	47.0	..	..	..	..	21 21 0.42	− 16.41	III.	5	4.13	−21 33.3	− 11.5	21 20 44.01	− 24 26 54.8
2	7	..	..	..	..	..	43.5	57.2	24 16.23	16.42	VII.	9	8.19	43 7.9	11.2	23 59.81	48 29.1
3	5	..	..	43.4	57.2	..	..	..	26 56.83	16.49	IV.	1	5.35	1 46.6	11.0	26 40.34	7 7.6
4	8	42.5	56.8	..	..	..	..	..	29 23.91	16.50	II.	4	3.53	16 53.8	10.7	29 7.41	22 14.5
5	8	42.3	56.2	..	..	..	..	..	32 23.62	16.52	II.	6	11.10	29 35.5	10.6	32 7.10	34 56.1
6	9	..	..	..	37.2	50.8	..	..	32 37.07	16.53	V.	5	7.54	23 55.2	10.6	32 20.51	29 15.8
7	6	21.8	36.0	..	..	..	..	..	35 3.38	16.53	II.	9	10.10	44 4.1	10.4	34 46.85	49 24.5
8	10	..	39.2	53.2	..	..	..	..	36 6.45	16.58	III.	2	4.26	7 11.6	10.4	35 49.87	12 32.0
9	9	..	46.8	0.8	..	..	..	..	39 14.16	16.60	III.	4	3.32	16 43.3	10.1	38 57.56	22 3.4
10	9	..	..	..	..	53.0	6.5	..	39 39.29	16.60	VI.	5	6.57	23 26.3	10.1	39 22.69	28 46.4
11	8	..	..	..	..	54.2	7.8	..	41 40.53	16.63	VI.	3	9.30	14 44.5	10.0	40 23.90	20 4.5
12	11	..	..	..	..	3.5	16.5	..	41 49.54	16.62	VI.	6	4.33	26 15.0	10.0	41 32.92	31 35.0
13	8	..	..	53.5	7.5	..	..	..	45 7.29	16.63	IV.	9	3.16	40 20.1	9.8	44 50.66	45 39.9
14	9	29.5	43.2	..	..	..	..	..	47 10.65	16.68	II.	5	3.5	21 29.1	9.7	46 53.97	26 48.8
15	9	..	29.2	43.0	..	..	..	..	47 56.56	16.69	III.	6	4.35	26 16.1	9.7	47 39.87	31 35.8
16	8	..	..	..	..	44.0	57.8	..	48 30.44	16.68	VI.	8	7.9	37 33.3	9.6	48 13.76	42 52.9
17	9	..	..	48.5	2.5	..	..	..	21 50 2.27	− 16.70	IV.	8	5.58	−36 57.6	− 9.6	21 49 44.57	− 24 42 17.2

ZONE 84. OCTOBER 3. A. $D_0 = -26° 32' 40''$.

No.	Mag.	I.	II.	III.	IV.	V.	VI.	VII.	T.	a_1	MICROMETER.			$i + d_2$	d_1	Mean Right Ascension, 1850.0.	Mean Declination, 1850.0.
1	11	..	..	..	..	..	27.0	41.5	22 11 59.24	− 16.82	VII.	6	4.35	−26 15.8	− 32.2	22 11 42.42	− 26 59 29.0
2	6	..	..	50.0	3.9	..	..	..	15 3.65	16.87	IV.	1	5.36	2 47.3	32.3	14 46.78	26 35 59.6
3	6	..	..	..	59.0	12.8	..	..	15 58.82	16.86	V.	7	11.0	34 31.1	32.0	15 41.96	27 7 43.1
4	5	..	5.6	19.9	..	..	..	..	19 33.86	16.90	III.	8	7.36	37 47.3	31.0	19 16.96	10 58.3
5	8	..	11.6	26.0	..	..	..	..	21 40.02	16.91	III.	10	4.24	46 11.2	30.5	21 23.11	19 21.7
6	9	..	..	47.5	2.0	..	..	..	23 1.68	16.95	IV.	7	9.58	33 59.9	30.1	22 44.73	6 10.0
7	8	56.5	11.0	..	..	..	..	..	24 39.03	16.96	II.	8	4.0	41 0.6	29.7	24 22.07	27 14 10.3
8	7	..	..	..	..	..	..	25.2	24 42.84	16.99	VII.	1	8.6	4 2.6	29.7	24 25.85	26 37 12.3
9	9	..	37.2	51.5	..	..	..	..	28 5.41	17.01	III.	7	8.32	33 16.4	28.8	27 48.40	27 6 25.2
10	7	..	..	43.0	57.2	..	..	..	28 56.90	17.03	IV.	4	3.40	16 47.3	28.5	28 39.87	26 49 55.8
11	9	..	..	..	..	..	3.0	17.5	29 35.16	17.05	VII.	1	9.59	4 59.7	28.3	29 18.11	38 8.0
12	9	23.2	37.5	..	..	..	..	..	32 5.50	17.05	II.	5	8.9	24 2.6	27.7	31 48.45	57 10.3
13	8	..	..	32.0	45.8	..	..	..	32 45.66	17.08	IV.	2	11.12	10 36.4	27.5	32 28.58	26 43 43.9
14	10	..	..	23.0	37.0	..	..	..	22 34 36.89	− 17.08	IV.	6	9.48	−28 54.3	− 27.1	22 34 19.81	− 27 2 1.4

CORRECTIONS.

Date.		Corr. of Clock.	Hourly rate.	m	n	c
1846.	h.	s.	s.	s.	s.	s.
Oct. 3.	20	− 26.61	+ 0.005	+ 0.265	− 0.106	− 0.086

INSTRUMENT READINGS.

Date.		Barom.	THERMOM.	
			At.	Ex.
1846.	h. m.	in.	°	°

REMARKS.

(82) 50. Micrometer reading assumed as 5r.58 instead of 6r.58.
(83) 1. Micrometer reading assumed as 3r.13 instead of 4r.13.
(83) 3. Micrometer reading assumed as 3r.35 instead of 5r.35.
(83) 11. Minutes assumed as 40 instead of 41.
(83) 13. Micrometer reading assumed as 2r.46 instead of 3r.16.
(84) 7. Micrometer reading assumed as 14r.00 instead of 4r.00.

ZONE 84. OCTOBER 3. A. $D_0 = -26° 32' 40''$—Continued.

No.	Mag.	SECONDS OF TRANSIT. I.	II.	III.	IV.	V.	VI.	VII.	T.	a_i	MICROMETER.			$i + d_i$	d_i	Mean Right Ascension, 1850.0.	Mean Declination, 1850.0.
									h. m. s.	s.			r.	′ ″	″	h. m. s.	° ′ ″
15	8	48.5	3.3	. .	. .	. .	. .	. .	22 36 31.13	— 17.10	II.	7	5.19	—31 38.7	— 26.6	22 36 14.03	— 27 4 45.3
16	9	. .	. .	. .	. .	. .	57.2	11.5	36 29.28	17.11	VII.	3	3.35	11 44.8	26.6	36 12.17	26 44 51.4
17	10	. .	21.5	35.8	. .	. .	. .	. .	38 49.62	17.13	III.	6	2.3	24 59.4	26.0	38 32.49	26 58 5.4
18	7	. .	. .	. .	. .	. .	25.5	39.7	38 57.62	17.13	VII.	7	6.20	32 9.3	26.0	38 40.49	27 5 15.3
19	9	. .	28.0	42.0	. .	. .	. .	. .	40 56.22	17.13	III.	10	5.45	46 52.1	25.4	40 39.09	19 57.5
20	9	. .	. .	. .	59.0	13.0	. .	. .	41 58.95	17.14	V.	10	1.12	34 33.3	25.2	41 41.81	7 38.5
21	8	. .	. .	. .	. .	17.5	31.0	. .	43 3.27	17.17	VI.	9	4.49	41 22.3	24.9	42 46.10	27 14 27.2
22	11	. .	. .	. .	42.5	. .	. .	. .	44 42.35	17.20	IV.	5	7.35	23 45.6	24.5	44 25.15	26 56 50.1
23	9	6.0	20.5	. .	. .	. .	. .	. .	48 48.47	17.25	II.	7	2.49	30 23.0	23.5	48 31.22	27 3 26.5
24	7	. .	. .	. .	18.2	32.2	. .	. .	49 18.10	17.25	V.	6	7.12	27 35.4	23.4	49 0.85	27 0 38.8
25	7	. .	. .	. .	. .	. .	21.8	36.4	49 54.10	17.26	VII.	5	11.37	20 47.7	23.2	49 36.84	26 53 50.9
26	6	. .	. .	. .	27.0	40.5	. .	. .	52 26.65	17.28	V.	5	8.17	24 6.8	22.7	52 9.37	57 9.5
27	8	5.5	19.9	. .	. .	. .	. .	. .	56 47.88	17.34	II.	6	4.47	26 22.1	21.6	56 30.54	59 23.7
28	10	. .	. .	29.5	43.8	. .	. .	. .	58 43.52	17.36	IV.	5	11.10	25 34.2	21.2	58 26.16	26 58 35.4
29	10	. .	. .	. .	39.5	53.5	. .	. .	22 59 39.41	17.37	V.	6	7.7	27 32.9	21.0	22 59 22.04	27 0 33.9
30	4	. .	. .	52.5	6.4	. .	. .	. .	23 1 6.17	17.40	IV.	1	10.57	5 29.4	20.7	23 0 48.77	26 38 20.1
31	8	. .	. .	14.5	29.0	. .	. .	. .	5 28.72	17.41	IV.	8	11.41	39 51.0	19.7	5 11.31	27 12 50.7
32	7	. .	. .	. .	. .	. .	13.0	27.0	5 45.09	17.41	VII.	10	8.45	48 22.7	19.6	5 27.68	21 22.3
33	9	24.0	38.2	. .	. .	. .	. .	. .	11 6.32	17.49	II.	6	11.6	29 33.5	18.5	10 48.83	2 32.0
34	8	. .	. .	33.7	47.5	. .	. .	. .	11 47.60	17.49	IV.	9	8.22	43 10.0	18.3	11 30.11	27 16 8.3
35	10	. .	. .	3.5	17.0	. .	. .	. .	13 16.96	17.53	IV.	1	8.57	4 28.8	18.0	12 59.43	26 37 26.8
36	9	. .	23.5	37.5	. .	. .	. .	. .	14 51.71	17.52	III.	9	12.0	45 0.1	17.7	14 34.19	27 17 57.8
37	9	. .	. .	26.4	40.3	. .	. .	. .	15 40.32	17.53	IV.	8	11.33	39 47.0	17.5	15 22.79	27 12 44.5
38	10	. .	. .	. .	. .	. .	0.0	14.8	16 32.32	17.57	VII.	2	7.32	8 45.0	17.3	16 14.75	26 41 42.3
39	9	. .	. .	16.5	30.5	. .	. .	. .	18 30.53	17.55	IV.	10	5.9	46 34.0	16.9	18 12.98	27 19 30.9
40	10	. .	. .	. .	55.0	8.8	. .	. .	19 54.78	17.60	V.	3	8.30	14 14.2	16.6	19 37.18	26 47 10.8
41	9	. .	49.5	4.2	. .	. .	. .	. .	24 17.58	17.65	III.	2	3.15	6 35.5	15.7	23 59.93	39 31.2
42	5	. .	. .	. .	. .	. .	46.5	. .	24 18.70	17.65	VI.	1	3.15	1 35.9	15.7	24 1.05	34 31.6
43	7	. .	. .	. .	16.5	30.9	. .	. .	26 16.56	17.67	V.	1	6.21	3 9.9	15.3	25 58.89	36 5.2
44	8	52.5	6.8	. .	. .	. .	. .	. .	29 34.64	17.72	II.	2	7.36	8 47.2	14.7	29 16.92	26 41 41.9
45	6	9.1	23.5	. .	. .	. .	. .	. .	32 51.50	17.75	II.	6	9.15	28 37.4	14.1	32 33.73	27 1 31.5
46	5	15.3	29.7	. .	. .	. .	. .	. .	36 57.73	17.77	II.	7	5.46	31 52.4	13.3	36 39.96	4 45.7
47	8	. .	. .	. .	. .	15.5	29.0	. .	37 1.27	17.75	VI.?	10	3.4	45 30.7	13.3	36 43.52	18 24.0
48	10	31.8	46.0	. .	. .	. .	. .	. .	40 14.10	17.81	II.	6	8.18	28 8.6	12.7	39 55.45	1 1.3
49	9	. .	. .	. .	50.0	3.8	. .	. .	41 49.83	17.82	V.	9	5.0	41 28.0	12.4	41 32.19	27 14 20.4
50	8	. .	. .	59.7	13.8	. .	. .	. .	43 13.51	17.85	IV.	3	3.26	11 40.7	12.2	42 55.49	26 44 32.9
51	8	. .	15.2	29.4	. .	. .	. .	. .	45 43.20	17.88	III.	4	11.11	20 35.0	11.8	45 24.83	53 26.8
52	8	. .	. .	26.5	40.0	. .	. .	. .	48 40.01	17.92	IV.	3	3.43	11 49.3	11.2	48 21.92	44 40.5
53	8	9.0	23.0	. .	. .	. .	. .	. .	51 51.11	17.94	II.	4	10.0	19 59.0	10.7	51 32.35	26 52 49.7
54	9	. .	13.5	27.5	. .	. .	. .	. .	52 41.50	17.94	III.	6	7.40	27 49.6	10.6	52 23.04	27 0 40.2
55	10	. .	43.5	57.5	. .	. .	. .	. .	54 11.71	17.93	III.	10	3.3	45 30.3	10.3	53 53.26	27 18 20.6
56	7	. .	. .	. .	. .	. .	. .	9.2	54 26.83	17.97	VII.	1	8.39	4 19.3	10.3	54 9.84	26 37 9.6
57	9	. .	. .	. .	2.0	15.8	. .	. .	56 1.83	— 17.97	V.	9	6.28	42 12.4	— 10.0	23 55 44.04	— 27 15 2.4
58	8	. .	. .	. .	. .	. .	11.2	36.0	22 57?	. .	VII.	2	11.15	—10 37.6			

CORRECTIONS.

Date.		Corr. of Clock.	Hourly rate.	m	n	c
1846.	h.	s.	s.	s.	s.	s.

INSTRUMENT READINGS.

Date.		Barom.	THERMOM. At.	Ex.
1846.	h. m.	in.	°	°

REMARKS.

(84) 20. Hor. thread assumed as 8 instead of 10.
(84) 25. Hor. thread assumed as 4 instead of 5.

ZONE 85. OCTOBER 5. K. $D_0 = -25° 14' 20''$.

No.	Mag.	Seconds of Transit. I.	II.	III.	IV.	V.	VI.	VII.	T.	n_1	Micrometer.			$i + d_0$	d_1	Mean Right Ascension, 1850.0.	Mean Declination, 1850.0.
									h. m. s.	s.			r.	′ ″	″	h. m. s.	° ′ ″
1	6	..	..	12.3	25.9	39.9	..	..	22 15 25.88	– 17.03	VI.	4	3.18	–16 36.0	– 1.8	22 15 8.85	– 25 30 57.8
2	10	..	..	..	41.3	54.9	8.3	..	24 40.95	17.16	VII.	2	7.26	8 42.1	1.2	24 23.79	25 23 3.3
3	7	..	7.7	21.5	35.6	49.1	..	..	37 35.48	17.23	V.	10	6.11	47 5.1	0.6	37 18.25	26 1 25.7
4	10	24.9	38.9	52.9	6.8	..	..	..	41 6.69	17.31	V.	6	8.30	28 14.8	0.5	40 49.38	25 42 35.3
5	10	..	..	..	59.0	..	..	..	44 58.85	17.33	V.	10	3.34	45 45.8	0.4	44 41.52	26 0 6.2
6	9	..	..	..	..	..	..	55.6	45 14.16	17.35	VII.	9	6.25	42 10.4	0.4	44 56.81	25 56 30.8
7	8	..	..	..	..	40.3	53.8	8.1	46 26.38	17.37	VII.	7	6.42	32 20.4	0.4	46 9.01	46 40.8
8	7	..	47.7	1.6	15.6	..	..	..	22 52 15.55	17.42	V.	9	8.52	43 25.0	0.3	22 51 58.13	57 45.3
9	8.9	..	..	..	..	..	3.3	17.6	23 7 35.84	17.61	VII.	5	11.27	25 42.4	0.1	23 7 18.23	40 2.5
10	10	..	..	..	41.8	..	..	..	10 41.65	17.64	VI.	7	11.50	34 56.1	0.0	10 24.01	49 16.1
11	10	0.0	13.8	28.2	42.0	..	..	..	13 41.81	17.67	IV.	6	6.37	27 17.8	0.0	13 24.14	41 37.8
12	9	52.1	6.3	20.2	..	..	..	..	15 33.80	17.70	IV.	3	3.35	11 45.3	0.0	15 16.10	26 5.3
13	9	..	..	..	54.0	..	..	..	19 53.85	17.76	V.	1	6.54	3 26.8	0.1	19 36.09	17 46.9
14	7	..	56.3	10.2	24.1	37.8	..	..	26 23.97	17.81	VI.	6	6.00	26 59.0	0.2	26 6.16	41 19.2
15	8	..	..	..	17.6	31.3	44.8	..	30 17.35	17.85	VII.	6	7.55	27 56.8	0.3	29 59.50	42 17.1
16	10	45.0	59.0	13.0	..	..	..	..	36 26.71	17.91	IV.	5	2.51	21 22.2	0.5	36 8.80	35 42.7
17	5.6	31.3	45.6	59.5	13.3	..	..	..	45 13.29	18.01	IV.	7	11.41	34 51.8	1.0	44 55.28	49 12.8
18	5.6	22.1	36.3	50.2	4.0	..	..	..	49 3.89	18.06	IV.	4	9.56	19 57.2	1.2	48 45.83	34 18.4
19	7	37.8	51.2	5.8	..	33.4	..	..	52 19.29	18.09	V.	3	8.51	14 24.9	1.5	52 1.20	28 46.4
20	9	5.2	19.4	33.4	47.1	1.0	..	..	55 46.96	18.14	VI.	2	12.7	11 4.2	1.8	55 28.82	25 26.0
21	10	..	..	..	53.5	..	..	..	58 53.35	18.17	IV.	1	11.50	5 56.1	2.0	58 35.18	20 18.1
22	10	..	..	..	..	..	..	59.9	23 59 18.08	– 18.16	VII.	3	5.43	–12 49.6	– 2.0	23 58 59.92	– 25 27 11.6

ZONE 86. OCTOBER 6. A. $D_0 = -35° 20' 30''$.

No.	Mag.	Seconds of Transit. I.	II.	III.	IV.	V.	VI.	VII.	T.	n_1	Micrometer.			$i + d_0$	d_1	Mean Right Ascension, 1850.0.	Mean Declination, 1850.0.
1	7	..	33.0	48.7	..	..	..	..	22 24 3.90	– 16.33	III.	6	10.46	–29 24.1	..	22 23 47.57	
2	6	..	..	41.6	57.2	..	..	..	24 57.01	16.34	IV.	7	11.44	34 54.6	..	24 40.67	
3	8	..	..	45.5	1.2	..	..	..	26 0.90	16.36	IV.	6	8.24	28 12.3	..	25 44.54	
4	5	..	..	..	..	..	..	55.5	26 9.07	16.36	VII.	2	4.35	7 13.2	..	25 52.71	
5	9	20.8	36.5	..	..	..	..	..	29 7.22	16.40	II.	5	7.7	23 31.0	..	28 50.82	
6	7	22.0	37.8	..	..	..	..	..	30 8.29	16.42	II.	2	9.8	9 31.6	..	29 51.87	
7	6	54.9	10.5	..	..	..	..	..	31 41.45	16.45	II.	8	7.32	37 46.7	..	31 25.00	
8	7	31.2	46.6	..	..	..	..	..	39 17.39	16.55	II.	4	4.55	17 23.9	..	39 0.84	
9	8	44.6	0.1	..	..	..	..	..	40 30.78	16.58	II.	3	8.26	14 10.6	..	40 14.20	
10	8	..	47.4	2.9	..	..	..	..	41 18.16	16.59	III.	6	5.46	26 52.2	..	41 1.57	
11	7	..	31.8	47.5	..	..	..	..	44 2.37	16.63	III.	2	2.38	6 14.5	..	43 45.74	
12	9	..	..	30.0	45.8	..	..	..	49 45.45	16.73	IV.	6	8.33	28 16.8	..	49 28.72	
13	9	0.8	16.5	..	..	..	..	..	51 47.32	16.76	II.	7	4.30	31 14.7	..	51 30.56	
14	9	..	..	..	..	59.0	13.8	..	53 43.40	16.79	VI.	3	10.41	15 18.9	..	53 26.61	
15	4	41.5	57.0	..	..	..	..	..	55 27.69	16.81	II.	3	8.6	14 0.5	..	55 10.88	
16	9	..	59.5	15.0	..	..	..	..	56 30.12	16.83	III.	4	2.8	15 59.6	..	56 13.29	
17	8	32.2	48.0	..	..	..	..	..	22 59 18.63	– 16.86	II.	5	2.30	–21 10.8	..	22 59 1.77	

CORRECTIONS.

Date.		Corr. of Clock.	Hourly rate.	m	n	c
1846.	h.	s.	s.	s.	s.	s.
Oct. 5.	20	– 26.74	+ 0.002	+ 0.265	– 0.106	– 0.086
Oct. 6,	20	– 26.33	+ 0.001	+ 0.265	– 0.106	– 0.086

INSTRUMENT READINGS.

Date.		Barom.	Thermom. At.	Thermom. Ex.
1846.	h. m.	in.	°	°

REMARKS.

Zone 87. October 7. K. $D_0 = -31° 35' 0''$.

No.	Mag.	Seconds of Transit. I.	II.	III.	IV.	V.	VI.	VII.	T.	a_1	Micrometer.			$i + d_2$	d_1	Mean Right Ascension, 1850.0.	Mean Declination, 1850.0.
									h. m. s.	s.			r.	′ ″	″	h. m. s.	° ′ ″
1	9	31.1	45.8	0.7	. .	. .	. .	. .	22 19 0.57	− 16.54	III.	7	7.36	−32 48.5	− 5.9	22 18 44.03	− 32 7 54.4
2	8	. .	. .	. .	. .	. .	. .	59.6	19 15.56	16.57	VII.	9	9.27	43 43.4	5.9	18 58.99	32 18 49.3
3	11	. .	. .	. .	37.3	. .	6.3	. .	21 37.11	16.62	VI.	3	3.29	11 41.3	5.8	21 20.49	31 46 47.1
4	8	. .	. .	. .	48.1	2.7	17.1	. .	22 47.91	16.66	VI.	3	4.51	12 22.8	5.7	22 31.25	47 28.5
5	9	. .	. .	. .	. .	6.9	. .	36.4	23 52.12	16.69	VII.	5	2.13	21 2.3	5.6	23 35.43	31 56 7.9
6	9	. .	. .	. .	28.8	. .	. .	. .	28 28.64	16.77	V.	10	11.53	49 59.3	5.3	28 11.87	32 25 4.6
7	8	. .	52.6	7.3	22.1	36.6	. .	. .	36 21.89	16.92	VI.	4	9.30	19 43.5	4.8	36 4.97	31 54 48.3
8	8	. .	43.8	58.7	13.6	28.2	. .	. .	41 13.34	17.00	VI.	5	9.8	24 32.3	4.5	40 56.34	59 36.8
9	11	. .	. .	13.8	. .	. .	. .	. .	45 28.08	17.07	IV.	1	11.56	5 58.2	4.3	45 11.01	41 2.5
10	10	. .	. .	. .	7.6	22.4	36.7	. .	46 7.51	17.08	VII.	2	5.1	7 27.7	4.3	45 50.43	31 42 32.0
11	6	. .	50.0	4.9	19.9	34.5	. .	. .	48 19.78	17.09	V.	10	4.57	46 29.0	4.2	48 2.69	32 21 33.2
12	10	50.5	5.6	20.3	. .	. .	. .	. .	51 34.75	17.17	IV.	2	10.24	10 11.4	4.0	51 17.58	31 45 15.4
13	9	. .	. .	. .	. .	. .	17.2	. .	51 47.90	17.17	VI.	4	11.20	20 39.1	4.0	51 30.73	55 43.1
14	10	. .	. .	. .	. .	. .	. .	3.2	52 18.89	17.16	VII.	4	9.29	19 42.7	4.0	52 1.73	54 46.7
15	9	30.4	45.6	0.3	. .	. .	. .	. .	55 14.87	17.20	IV.	5	3.47	21 50.3	3.8	54 57.67	56 54.1
16	8.9	. .	. .	19.3	33.8	. .	. .	. .	22 57 33.57	17.25	V.	1	4.10	2 2.6	3.8	22 57 16.32	37 6.4
17	11	. .	51.3	. .	20.8	. .	. .	. .	23 0 20.58	17.28	IV.	1	12.4	6 2.3	3.6	23 0 3.30	41 5.9
18	11	. .	. .	33.3	. .	. .	. .	. .	1 47.61	17.30	IV.	2	6.20	8 8.1	3.6	1 30.31	43 11.7
19	11	. .	. .	. .	. .	. .	19.0	34.3	6 49.86	17.36	VII.	5	4.39	22 16.1	3.4	6 32.50	31 57 19.5
20	10	10.5	. .	40.3	55.1	. .	. .	. .	11 55.01	17.43	V.	7	12.36	35 20.1	3.3	11 37.58	32 10 23.4
21	11	. .	. .	46.0	. .	15.4	. .	. .	14 0.78	17.44	VII.	9	3.36	40 45.9	3.3	13 43.34	32 15 49.2
22	7.8	. .	30.0	44.8	59.6	14.3	. .	. .	15 59.42	17.48	VI.	5	1.58	20 55.0	3.2	15 41.94	31 55 58.2
23	8	. .	. .	10.0	. .	39.3	53.5	. .	20 24.35	17.55	VI.	1	9.14	4 36.1	3.2	20 6.50	31 39 39.3
24	8	. .	. .	. .	. .	. .	. .	52.8	21 8.69	17.56	VII.	8	7.22	37 40.4	3.2	20 51.13	32 12 43.6
25	8	. .	3.2	. .	. .	. .	. .	. .	23 32.86	17.58	III.	9	9.11	43 35.7	3.1	23 15.28	18 38.8
26	8	. .	. .	. .	0.3	15.3	. .	. .	24 0.37	17.59	VI.	7	11.22	34 42.6	3.1	23 42.78	9 45.7
27	6.7	. .	16.5	31.6	46.4	. .	. .	. .	25 46.19	17.62	VI.	7	5.58	31 58.8	3.1	25 28.57	7 1.9
28	8	26.8	. .	56.9	11.7	. .	. .	. .	35 11.45	17.74	V.	6	9.3	28 31.7	3.1	34 53.71	3 34.8
29	8	17.0	32.1	47.0	. .	. .	. .	. .	42 1.71	17.84	IV.	8	10.12	39 6.8	3.2	41 43.87	32 14 10.0
30	9	51.8	6.8	21.4	36.3	. .	. .	. .	44 36.09	17.87	V.	4	9.3	19 30.0	3.2	44 18.22	31 54 33.2
31	8	. .	. .	8.1	23.1	37.8	52.3	. .	51 22.97	17.94	VI.	9	9.31	43 45.7	3.4	51 5.03	32 18 49.1
32	9	. .	. .	14.6	29.3	44.0	. .	. .	55 29.28	18.00	VI.	8	4.58	36 27.9	3.5	55 11.28	32 11 31.4
33	8	. .	. .	. .	. .	. .	42.1	57.0	56 12.78	18.02	VII.	4	11.8	20 32.8	3.5	55 54.76	31 55 36.3
34	9	. .	. .	. .	. .	. .	. .	58.2	23 57 13.95	18.03	VII.	5	10.27	25 12.0	3.5	23 56 55.92	32 0 15.5
35	9	. .	45.6	. .	15.0	. .	. .	. .	0 5 14.84	18.14	IV.	1	12.0	6 0.2	3.8	0 4 56.70	31 41 4.0
36	5	. .	21.8	36.5	51.4	6.0	20.3	. .	8 51.26	18.17	VI.	9	5.23	41 40.3	4.0	8 33.09	32 16 44.3
37	7	15.6	30.7	45.6	0.3	. .	. .	. .	16 0.04	18.27	V.	4	4.3	16 58.4	4.4	15 41.77	31 52 2.8
38	9.10	. .	. .	. .	. .	58.2	12.4	27.6	16 43.33	18.28	VII.	7	3.57	30 57.3	4.4	16 25.05	32 6 1.7
39	9	19.9	35.0	49.7	. .	. .	. .	. .	20 4.23	18.32	IV.	4	3.44	16 48.8	4.6	19 45.91	31 51 53.4
40	10	. .	. .	56.8	. .	. .	. .	. .	21 11.21	18.34	IV.	3	9.42	14 50.0	4.6	20 52.87	49 54.6
41	8	43.7	58.7	13.7	28.4	42.8	57.3	12.3	33 28.12	18.48	VII.	5	7.2	23 28.4	5.5	33 9.64	58 33.9
42	9	. .	. .	. .	19.2	. .	. .	. .	35 19.04	18.51	V.	1	5.22	2 39.0	5.6	35 0.53	31 37 44.6
43	7	. .	15.6	. .	45.2	59.8	. .	. .	37 45.12	18.54	V.	8	7.8	37 33.8	5.8	37 26.58	32 12 39.6
44	8	45.8	0.8	15.7	30.3	. .	. .	. .	40 30.18	18.57	V.	4	12.49	21 24.3	6.0	40 11.61	31 56 30.3
45	7	. .	. .	. .	. .	. .	11.5	26.4	40 42.22	18.58	VII.	7	12.31	35 17.2	6.1	40 23.64	32 10 23.3
46	9.10	. .	9.9	24.8	. .	. .	. .	. .	44 39.20	18.63	IV.	3	6.3	12 59.3	6.4	44 20.57	31 48 5.7
47	9.10	. .	. .	. .	. .	12.1	. .	41.8	44 57.35	18.63	VII.	3	3.58	11 55.7	6.4	44 38.72	31 47 2.1
48	6	. .	. .	37.1	52.2	. .	. .	. .	52 51.93	18.71	VII.	7	9.14	33 37.6	7.2	52 33.22	32 8 44.8
49	4.5	. .	5.1	20.2	35.1	. .	. .	. .	0 55 34.96	18.74	V.	10	5.3	46 32.0	7.4	0 55 16.22	32 21 39.4
50	8	. .	. .	. .	. .	22.3	. .	. .	1 2 7.49	− 18.83	VII.	2	6.48	− 8 21.8	− 8.1	1 1 48.66	− 31 43 29.9

CORRECTIONS.

Date.		Corr. of Clock.	Hourly rate.	m	n	c
1846.	h.	s.	s.	s.	s.	s.
Oct. 7,	20	− 26.65	+ 0.001	+ 0.265	− 0.106	− 0.086

INSTRUMENT READINGS.

Date.		Barom.	Thermom. At.	Thermom. Ex.
1846.	h. m.	In.	°	°

REMARKS.

ZONE 88. OCTOBER 8. A. $D_0 = -34° 4' 0''$.

No.	Mag.	Seconds of Transit. I.	II.	III.	IV.	V.	VI.	VII.	T.	a_1	Micrometer.			$i + d_4$	d_1	Mean Right Ascension, 1850.0.	Mean Declination, 1850.0.
									h. m. s.	s.			r.	′ ″	″	h. m. s.	° ′ ″
1	10	26.6	42.0	..	..	..	..	..	22 55 12.40	− 16.69	II.	7	7.37	−32 49.1	..	22 54 55.71	
2	7	25.5	41.0	..	..	..	..	..	23 1 11.43	16.78	II.	8	9.57	38 59.7	..	23 0 54.65	
3	7	..	..	29.6	44.5	..	..	..	2 59.65	16.81	III.	5	4.28	22 10.8	..	2 42.84	
4	10	..	..	0.0	15.2	..	..	..	8 30.24	16.89	III.	5	10.30	25 14.0	..	8 13.35	
5	9	..	..	27.5	42.5	..	..	..	9 57.40	16.93	III.	3	2.43	11 17.6	..	9 40.47	
6	5	6.5	22.0	..	..	..	..	..	11 52.28	16.95	II.	6	6.53	27 25.9	..	11 35.33	
7	8	10.5	26.0	..	..	..	..	..	12 56.43	16.95	II.	8	10.1	39 1.7	..	12 39.45	
8	8	2.0	17.8	..	..	..	..	..	13 47.89	16.98	II.	5	9.26	24 41.4	..	13 30.91	
9	9	42.5	58.0	..	..	..	..	..	15 28.17	17.00	II.	4	7.57	18 56.2	..	15 11.17	
10	8	..	..	46.0	1.0	..	..	..	16 0.78	17.03	IV.	2	6.0	7 57.3	..	15 43.75	
11	9	..	..	..	51.5	6.5	..	..	17 51.30	17.05	V.	3	3.51	11 52.0	..	17 34.25	
12	6	9.0	24.5	..	..	..	..	..	23 19 54.73	− 17.07	II.	5	5.57	−22 55.7	..	23 19 37.66	

ZONE 89. OCTOBER 8. A. $D_0 = -29° 3' 0''$.

No.	Mag.	I.	II.	III.	IV.	V.	VI.	VII.	T.	a_1	Micrometer.			$i + d_4$	d_1	Mean Right Ascension, 1850.0.	Mean Declination, 1850.0.
1	8	..	..	..	..	44.0	57.5	..	0 0 29.24	− 17.64	VI.	2	3.4	− 6 29.4	− 4.5	0 0 11.60	− 29 9 33.9
2	9	..	..	..	..	56.5	10.5	..	2 41.99	17.68	VI.	2	3.35	6 45.1	5.0	2 24.31	9 50.1
3	8	..	..	..	19.5	4.0	..	..	8 49.51	17.74	V.	9	6.49	47 24.3	6.2	7 31.77	50 30.5
4	5	..	47.3	1.9	..	..	..	..	14 16.21	17.81	III.	10	3.1	45 29.7	7.2	13 58.40	48 36.9
5	7	..	56.0	10.5	..	..	..	..	16 24.61	17.83	III.	6	4.20	26 8.6	7.7	16 6.78	29 16.5
6	6	..	..	..	5.2	19.2	..	..	17 4.96	17.84	V.	10	2.55	45 26.7	7.8	16 47.12	48 34.5
7	7	..	..	30.0	44.0	..	..	..	18 43.87	17.86	IV.	1	6.30	3 14.1	8.1	18 26.01	6 22.2
8	6	..	..	..	..	32.5	47.0	..	19 18.24	17.87	VI.	3	3.0	11 27.1	8.3	19 0.37	14 35.4
9	5	..	..	..	..	..	38.5	..	20 9.99	17.88	VI.	1	2.52	1 23.8	8.4	19 52.11	4 32.2
10	9	..	..	..	..	46.5	1.0	..	21 32.25	17.89	VI.	8	12.0	40 0.8	8.7	21 14.36	43 9.5
11	7	..	52.5	7.0	..	..	..	..	27 21.27	17.95	III.	8	8.2	38 0.7	10.0	27 3.32	41 10.7
12	8	42.8	57.2	..	..	..	..	..	0 29 26.05	− 17.98	II.	8	7.1	−37 29.7	− 10.4	0 29 8.07	− 29 40 40.1

ZONE 90. OCTOBER 9. K. $D_0 = -26° 24' 40''$.

No.	Mag.	I.	II.	III.	IV.	V.	VI.	VII.	T.	a_1	Micrometer.			$i + d_4$	d_1	Mean Right Ascension, 1850.0.	Mean Declination, 1850.0.
1	9	..	..	..	..	..	..	25.8	23 56 43.61	− 18.10	VII.	4	7.38	−18 47.0	− 7.1	23 56 25.51	− 26 43 34.1
2	9	25.9	40.3	54.3	8.3	..	..	..	0 1 6.21	18.15	IV.	5	8.7	24 1.8	7.0	0 0 50.06	26 48 48.8
3	9	..	..	..	..	..	2.1	16.4	1 34.30	18.16	VII.	8	9.30	38 44.5	7.0	1 16.14	27 3 31.5
4	8.9	..	..	..	..	..	..	20.4	2 38.20	18.17	VII.	4	5.48	17 51.5	7.0	2 20.03	26 42 38.5
5	10	..	..	..	9.2	..	36.8	..	4 8.94	18.17	VII.	9	2.40	40 16.9	6.9	4 50.77	27 5 3.8
6	8	..	..	42.6	..	..	..	..	5 56.78	18.20	IV.	10	1.46	44 51.4	6.9	5 38.58	9 38.2
7	7	..	..	..	..	39.9	..	..	6 25.92	18.19	VI.	9	6.55	42 25.9	6.9	6 7.73	27 7 12.8
8	10	..	..	..	..	..	49.0	..	7 21.12	18.21	VII.	4	6.45	18 20.3	6.9	7 2.91	26 43 7.2
9	8	57.7	11.9	26.0	40.0	..	..	..	10 39.87	18.26	IV.	4	10.22	20 10.3	6.9	10 21.61	26 44 57.2
10	9	43.0	57.3	11.4	..	..	..	..	13 25.52	18.27	III.	9	8.10	43 3.9	6.9	13 7.25	27 7 50.8
11	9	..	..	..	50.4	4.4	18.1	..	13 50.27	18.28	VI.	9	3.56	40 55.5	6.9	13 31.99	27 5 42.4
12	11	..	..	28.5	42.3	..	..	..	24 42.14	18.42	V.	2	5.15	7 36.1	6.9	24 23.72	26 32 23.0
13	7	44.6	58.7	12.9	27.1	..	..	..	26 26.89	18.42	IV.	7	2.52	30 24.7	7.0	26 8.47	55 11.7
14	7	..	25.0	39.2	53.3	7.2	..	..	27 53.16	18.45	IV.	7	6.49	32 24.4	7.0	27 34.71	57 11.4
15	8	39.9	54.1	8.2	22.1	..	..	..	30 21.98	18.48	IV.	2	9.27	9 43.5	7.0	30 3.50	34 30.5
16	9	35.2	49.2	3.5	..	..	..	..	32 17.36	18.49	III.	5	7.58	23 57.2	7.1	31 58.87	48 44.3
17	10	..	..	..	47.5	..	..	..	0 32 47.35	− 18.50	IV.	5	8.26	−24 11.4	− 7.1	0 32 28.85	− 26 48 58.5

CORRECTIONS.

Date.		Corr. of Clock.	Hourly rate.	m	n	c
1846.	h.	s.	s.	s.	s.	s.
Oct. 8,	20	− 26.23	+ 0.002	+ 0.265	− 0.106	− 0.086
Oct. 9,	20	− 26.67	− 0.012	+ 0.265	− 0.106	− 0.086

INSTRUMENT READINGS.

Date.		Barom.	Thermom. At.	Thermom. Ex.
1846.	h. m.	in.	°	°

REMARKS.

(89) 3. Minutes assumed as 7 instead of 8. Micrometer wire assumed as 10 instead of 9.

ZONE 90. OCTOBER 9. K. $D_0 = -26° 24' 40''$—Continued.

No.	Mag.	Seconds of Transit. I.	II.	III.	IV.	V.	VI.	VII.	T.	a_1	Micrometer.			$i + d_0$	d_1	Mean Right Ascension, 1850.0.	Mean Declination, 1850.0.
									h. m. s.	s.			r.	′ ″	″	h. m. s.	° ′ ″
18	8	. .	. .	. .	. .	. .	. .	44.2	0 33 2.12	− 18.50	VII.	6	7.24	−27 41.2	− 7.1	0 32 43.62	− 26 52 28.3
19	10	. .	. .	. .	45.3	. .	. .	. .	34 45.15	18.53	IV.	1	8.5	4 2.6	7.1	34 26.62	28 49.7
20	8	. .	51.9	5.8	19.9	. .	. .	. .	36 19.75	18.54	III.	4	9.59	19 58.6	7.2	36 1.21	26 44 45.8
21	9	. .	41.8	. .	9.9	. .	. .	. .	41 9.89	18.58	IV.	10	5.13	46 35.9	7.3	40 51.31	27 11 23.2
22	7	. .	. .	. .	. .	. .	56.2	10.3	42 28.33	18.59	VII.	10	5.29?	46 43.6	7.4	42 9.74	11 31.0
23	7.8	. .	. .	3.0	17.3	31.0	. .	. .	45 17.16	18.61	V.	10	8.21	48 10.9	7.5	44 58.55	27 12 58.4
24	8.9	0.5	14.7	28.8	42.8	. .	. .	. .	48 42.69	18.66	III.	5	5.50	22 52.6	7.6	48 24.03	26 47 40.2
25	9	. .	. .	. .	. .	. .	33.7	48.2	49 6.03	18.65	VII.	10	2.6	45 1.1	7.6	48 47.38	27 9 48.7
26	7	. .	. .	59.1	13.2	27.0	. .	. .	51 12.95	18.67	V.	4	3.7	16 30.5	7.7	50 54.28	26 41 18.2
27	8	. .	23.7	. .	51.9	. .	. .	. .	52 51.69	18.71	V.	3	11.17	15 38.5	7.8	52 32.98	40 26.3
28	7	. .	. .	. .	. .	48.8	2.4	16.8	53 34.58	18.71	VII.	2	7.19	8 38.4	7.8	53 15.87	33 26.2
29	8	36.0	50.2	4.4	18.4	. .	. .	. .	56 18.35	18.72	IV.	7	10.34	34 18.0	8.0	55 59.63	26 59 6.0
30	8	. .	24.6	38.6	53.0	6.6	. .	. .	0 58 52.75	18.74	VI.	9	5.41	41 48.5	8.1	0 58 34.01	27 6 36.6
31	8	8.2	22.6	36.7	50.3	. .	. .	. .	1 1 50.38	18.79	VI.	3	10.9	15 4.1	8.2	1 1 31.59	26 39 52.3
32	6.7	. .	. .	56.1	10.1	24.0	37.4	. .	3 9.88	18.79	VII.	7	11.45	34 53.5	8.3	2 51.09	26 59 41.8
33	8	37.0	51.0	. .	19.6	. .	. .	. .	5 19.37	18.79	IV.	9	5.12	41 34.0	8.4	5 0.58	27 6 22.4
34	8	. .	. .	. .	47.6	1.6	15.3	. .	9 47.47	18.86	VI.	1	6.28	3 13.4	8.7	9 28.61	26 28 2.1
35	8	4.3	18.6	32.8	. .	. .	. .	. .	11 46.53	18.87	IV.	4	3.23	16 38.6	8.8	11 27.66	41 27.4
36	10	11.8	26.0	40.3	54.4	. .	. .	. .	16 54.22	18.92	IV.	7	9.30	33 45.7	9.2	16 35.30	58 34.9
37	8	. .	12.9	26.9	41.1	. .	. .	. .	20 40.83	18.96	VI.	4	5.46	17 50.7	9.4	20 21.87	42 40.1
38	9	. .	. .	. .	. .	. .	24.9	39.6	21 57.16	18.97	VII.	2	4.44	7 20.1	9.5	21 38.19	32 9.6
39	4.5	. .	. .	. .	37.6	51.2	5.1	19.3	23 37.26	18.98	VII.	6	10.34	34 17.6	9.7	23 18.28	59 7.3
40	8	. .	. .	. .	. .	44.3	. .	. .	26 30.17	19.01	V.	1	3.0	1 28.5	9.9	26 11.16	26 18.4
41	9.10	. .	59.7	13.5	27.6	. .	. .	. .	29 27.45	19.03	IV.	3	13.10	16 35.6	10.1	29 8.42	41 25.7
42	10	. .	. .	. .	. .	29.6	. .	. .	30 15.58	19.03	VII.	7	7.34	32 46.7	10.2	29 56.55	57 36.9
43	10	. .	10.0	24.0	. .	. .	. .	. .	34 37.99	19.07	IV.	6	6.22	27 10.2	10.6	34 18.92	26 52 0.8
44	10	. .	. .	. .	. .	23.3	. .	. .	35 9.33	19.06	VII.	9	10.19	44 8.7	10.6	34 50.27	27 8 59.3
45	9	52.8	7.1	. .	. .	. .	. .	. .	38 35.29	19.10	III.	9	6.8	42 2.3	10.9	38 16.19	6 55.2
46	7	. .	. .	. .	55.0	. .	. .	. .	38 54.85	19.10	IV.	9	10.30	44 14.7	10.9	38 35.75	9 5.6
47	7	. .	. .	. .	. .	. .	33.6	. .	39 5.64	19.11	VII.	8	9.41	38 50.0	11.0	38 46.53	27 3 41.0
48	9	. .	. .	. .	. .	45.4	. .	. .	40 31.27	19.13	V.	1	3.49	1 53.2	11.1	40 12.14	26 26 44.3
49	7	. .	. .	44.7	59.0	. .	. .	. .	41 58.79	19.13	IV.	8	2.44	35 19.8	11.2	41 39.66	27 0 11.0
50	7	. .	. .	. .	. .	. .	51.3	5.7	43 23.48	19.14	VII.	5	4.10	22 1.7	11.4	42 4.34	26 46 53.1
51	8	. .	. .	0.5	14.2	28.0	. .	. .	46 14.03	19.17	V.	2	8.30	9 14.6	11.7	45 54.86	26 34 6.3
52	9	. .	23.9	37.9	52.3	. .	. .	. .	47 52.08	19.17	IV.	8	12.20	40 10.7	11.9	47 33.91	27 5 2.6
53	8	. .	17.8	31.8	46.0	. .	. .	. .	50 45.85	19.20	VI.	7	6.13	32 6.0	12.2	50 26.65	26 56 58.2
54	6	3.4	17.8	31.8	46.1	. .	. .	. .	52 45.97	19.20	V.	9	11.55	44 57.5	12.4	52 26.77	27 9 49.9
55	9	. .	. .	. .	54.8	. .	. .	. .	1 53 54.65	19.22	V.	7	6.52	32 25.8	12.5	1 53 35.43	26 57 18.3
56	9	. .	. .	18.4	. .	. .	. .	. .	2 0 32.24	19.27	V.	5	4.15	22 4.6	13.3	2 0 12.97	26 46 57.9
57	9	24.9	. .	. .	. .	. .	. .	. .	2 7.31	19.28	IV.	8	3.40	35 48.1	13.5	1 58.03	27 0 41.6
58	9	. .	. .	37.5	51.8	5.7	. .	. .	2 51.62	19.30	VI.	8	1.47	34 50.8	13.6	2 32.32	26 59 44.4
59	7.8	24.4	38.7	52.7	. .	. .	. .	. .	6 6.81	19.32	IV.	8	5.30	36 43.6	14.0	5 47.49	27 1 37.6
60	10	10.3	. .	. .	. .	. .	. .	. .	8 52.45	19.34	III.	3	3.53	11 54.3	14.4	8 33.11	26 36 48.7
61	7	. .	. .	. .	12.0	. .	. .	. .	9 11.85	19.32	VI.	10	9.4	48 32.4	14.4	8 52.53	27 13 26.8
62	7	. .	54.5	8.6	. .	. .	. .	. .	12 22.74	19.34	IV.	9	8.16	43 7.0	14.9	12 3.40	27 8 1.9
63	6.5	. .	. .	. .	. .	48.8	2.1	16.1	13 34.27	19.38	VII.	3	9.10	14 34.0	15.0	12 14.89	26 39 29.0
64	5	. .	26.8	41.0	55.1	9.0	22.7	. .	17 54.79	19.40	V.	2	3.51	6 53.7	15.6	17 35.39	26 31 49.3
65	7.8	10.9	25.6	39.3	53.6	. .	. .	. .	21 53.51	19.41	V.	9	4.54	41 21.7	16.2	21 34.10	27 6 17.9
66	10	. .	. .	. .	. .	43.3	. .	. .	23 29.25	19.43	VII.	5	3.3	21 27.8	16.5	23 9.82	26 46 24.3
67	7.8	14.0	28.3	42.7	. .	. .	. .	. .	2 26 56.49	− 19.45	IV.	7	6.31	−32 15.3	− 17.0	2 26 37.04	− 26 57 12.3

CORRECTIONS.

Date.		Corr. of Clock.	Hourly rate.	m	n	c
1846.	h.	s.	s.	s.	s.	s.

INSTRUMENT READINGS.

Date.		Barom.	Thermom. At.	Ex.
1846.	h. m.	in.	°	°

REMARKS.

(90) 39. Hor. thread assumed as 7 instead of 6.
(90) 50. Minutes assumed as 42 instead of 43.
(90) 63. Minutes assumed as 12 instead of 13.

ZONE 91. OCTOBER 10. A. $D_0 = -23° 58' 35''$.

No.	Mag.	SECONDS OF TRANSIT. I.	II.	III.	IV.	V.	VI.	VII.	T.	a_1	MICROMETER.			$i + d_1$	d_1	Mean Right Ascension, 1850.0.	Mean Declination, 1850.0.
									h. m. s.	s.			r.	′ ″	″	h. m. s	° ′ ″
1	8	. .	15.8	29.5	. .	. .	. .	. .	19 55 43.10	− 15.45	II.	6	3.43	−25 49.8	. .	19 55 27.65	
2	7	. .	30.0	44.0	. .	. .	. .	. .	58 57.40	15.49	II.	4	10.48	20 23.3	. .	58 41.91	
3	9	. .	. .	29.0	42.8	. .	. .	. .	19 59 42.63	15.48	IV.	7	3.49	30 53.4	. .	19 59 27.15	
4	9	. .	. .	. .	36.0	19.9	. .	. .	20 0 36.03	15.49	V.	6	10.31	29 15.9	. .	20 0 20.54	
5	11	. .	. .	. .	. .	. .	. .	31.0	0 49.80	15.51	VII.	4	10.29	20 13.5	. .	0 34.29	
6	8	. .	. .	. .	10.8	24.5	. .	. .	2 10.73	15.51	V.	6	10.41	29 21.0	. .	1 55.22	
7	7	. .	. .	. .	. .	. .	. .	26.2	2 45.25	15.49	VII.	9	6.00	41 57.7	. .	2 29.76	
8	8	. .	. .	39.5	53.0	. .	. .	. .	6 53.05	15.52	IV.	9	4.10	41 2.5	. .	6 37.53	
9	7	. .	. .	. .	. .	44.5	57.5	. .	7 30.53	15.56	VI.	4	. .	. . .	. .	7 14.97	
10	10	. .	. .	46.5	0.8	. .	. .	. .	9 0.40	15.56	IV.	7	8.32	33 16.3	. .	8 44.84	
11	6	. .	. .	0.9	14.5	. .	. .	. .	10 14.38	15.58	IV.	5	7.9	23 32.5	. .	9 58.80	
12	11	. .	. .	50.5	4.0	. .	. .	. .	11 3.88	15.60	IV.	4	5.35	17 45.4	. .	10 48.28	
13	11	. .	. .	. .	5.0	18.5	. .	. .	20 12 4.83	− 15.59	V.	7	7.20	−32 39.8	. .	20 11 49.24	

ZONE 92. OCTOBER 10. A. $D_0 = -36° 29' 30''$.

No.	Mag.	I.	II.	III.	IV.	V.	VI.	VII.	T.	a_1	MICROMETER.			$i + d_1$	d_1	Mean Right Ascension, 1850.0.	Mean Declination, 1850.0.
1	8	. .	21.0	37.0	. .	. .	. .	. .	0 31 52.25	. .	III.	4	3.26	−16 38.7	− 3.5	. . .	− 36 46 12.2
2	7	52.5	8.8	. .	. .	. .	. .	. .	35 39.81	. .	II.	5	2.35	21 13.2	3.8	. . .	50 47.0
3	8	. .	. .	. .	44.0	59.5	. .	. .	39 43.84	. .	V.	5	2.59	21 25.5	4.2	. . .	36 50 59.7
4	7	20.5	36.5	. .	. .	. .	. .	. .	42 7.85	. .	II.	7	11.50	34 57.9	4.4	. . .	37 4 32.3
5	8	. .	43.0	59.0	. .	. .	. .	. .	45 14.06	. .	III.	1	8.9	4 16.1	4.7	. . .	36 33 50.8
6	8	. .	. .	. .	. .	40.5	56.0	. .	50 24.84	. .	VI.	10	2.41	45 24.1	5.2	. . .	37 14 59.3
7	6	. .	18.0	33.8	. .	. .	. .	. .	0 53 49.39	. .	III.	7	8.0	33 1.6	5.5	. . .	37 2 37.1
8	9	. .	57.0	13.0	. .	. .	. .	. .	1 2 28.40	− 19.05	III.	6	6.11	27 5.0	6.5	1 2 9.35	36 56 41.5
9	7	. .	45.0	2.0	. .	. .	. .	. .	6 16.54	19.09	III.	1	5.58	2 54.6	6.9	5 57.45	32 31.5
10	8	. .	44.5	59.5	. .	. .	. .	. .	7 15.43	19.12	III.	7	0.40	29 18.7	7.0	6 56.31	36 58 55.7
11	8	. .	27.0	42.8	. .	. .	. .	. .	11 58.39	19.19	III.	7	8.23	33 13.2	7.6	11 39.20	37 2 50.8
12	7	. .	. .	27.0	43.0	. .	. .	. .	13 42.71	19.21	IV.	7	6.28	32 15.0	7.8	13 23.50	1 52.8
13	9	. .	. .	. .	19.5	35.0	. .	. .	20 19.37	19.30	V.	8	11.38	39 52.1	8.7	20 0.07	37 9 30.8
14	9	. .	. .	36.0	52.0	. .	. .	. .	21 51.62	19.31	IV.	5	3.40	21 46.3	8.9	21 32.31	36 51 25.2
15	8	. .	. .	20.0	35.5	. .	. .	. .	23 35.55	19.35	IV.	9	7.1	42 32.2	9.2	23 16.20	37 12 11.4
16	10	. .	. .	. .	. .	. .	31.0	46.0	23 59.47	19.35	VII.	7	10.30	34 17.1	9.2	23 40.12	37 3 56.3
17	9	. .	3.0	18.5	. .	. .	. .	. .	29 34.00	19.42	III.	4	3.31	16 41.2	10.0	29 14.58	36 46 21.2
18	6	. .	. .	. .	. .	. .	. .	23.5	29 36.66	19.43	VII.	8	8.26	38 15.4	10.0	29 17.23	37 7 55.4
19	5	. .	35.0	50.5	. .	. .	. .	. .	32 6.47	19.47	III.	10	7.1	47 34.5	10.4	31 47.00	37 17 14.9
20	9	. .	. .	55.0	11.0	. .	. .	. .	33 10.61	19.46	IV.	4	8.0	18 57.5	10.5	32 51.15	36 48 38.0
21	8	. .	. .	59.7	15.0	. .	. .	. .	34 15.17	19.48	IV.	9	10.35	44 20.7	10.7	33 55.69	37 14 1.4
22	8	. .	. .	. .	. .	. .	11.5	. .	36 40.54	19.50	VI.	1	8.37	4 15.0	11.0	36 21.04	36 33 56.0
23	8	21.5	37.5	. .	. .	. .	. .	. .	1 39 8.68	− 19.54	II.	5	4.9	−22 0.8	− 11.4	1 38 48.64	− 36 51 42.2

CORRECTIONS.

Date.		Corr. of Clock.	Hourly rate.	m	n	c
1846.	h.	s.	s.	s.	s.	s.
Oct. 10,	20	− 26.60	− 0.019	+ 0.265	− 0.106	− 0.086

INSTRUMENT READINGS.

Date.		Barom.	THERMOM. At.	Ex.
1846.	h. m.	in.	°	°

REMARKS.

(91) 7. Transit over T. VII assumed as 26ˢ.2 instead of 36ˢ.2.
(92) 5. Micrometer reading assumed as 8ʳ.39 instead of 8ʳ.9.

ZONE 93. OCTOBER 15. A. $D_0 = -32° 45' 0''$.

No.	Mag.	Seconds of Transit I.	II.	III.	IV.	V.	VI.	VII.	T.	a_1	Micrometer.			$i + d_2$	d_1	Mean Right Ascension, 1850.0.	Mean Declination, 1850.0.
									h. m. s.	s.	°		r.	′ ″	″	h. m. s.	° ′ ″
1	9	..	..	..	59.0	14.5	..	..	0 1 59.26	− 20.82	V.	10	8.0	− 1 11.3	..	0 1 38.44	
2	9	18.5	32.5	..	..	..	..	..	4 3.04	20.85	II.	7	3.9	30 33.5	..	3 42.29	
3	10	53.5	9.0	..	..	..	..	..	7 38.59	20.91	II.	3	9.58	14 57.7	..	7 17.68	
4	6	..	59.0	14.5	..	..	..	..	9 29.39	20.92	III.	10	6.46	47 24.7	..	9 8.47	
5	9	17.5	33.0	..	..	..	..	..	20 2.80	21.08	II.	7	5.19	31 39.2	..	19 41.72	
6	10	..	..	17.0	32.5	..	..	..	20 32.09	21.09	IV.	6	9.20	28 40.4	..	20 11.00	
7	7	..	58.2	15.0	..	..	..	..	22? 29.14?	..	III.	9	2.56	40 26.5	..	(22)	
8	8	..	..	..	..	..	17.5	32.0	22 47.46	21.12	VII.	5	3.45	21 48.7	..	22 26.34	
9	7	36.0	52.0	..	..	..	..	..	25 21.27	21.15	II.	2	7.57	8 56.5	..	25 0.12	
10	8	..	32.0	47.0	..	..	..	..	27 1.78	21.18	III.	5	5.59	22 56.9	..	26 40.60	
11	8	..	49.5	5.2	..	..	..	..	31 19.44	21.23	III.	2	5.35	7 44.9	..	30 58.21	
12	10	6.0	21.0	..	..	..	..	..	39 40.80	21.36	II.	4	9.4	19 30.2	..	39 19.44	
13	10	..	14.0	29.5	..	..	..	..	40 44.18	21.37	III.	7	8.2	33 1.8	..	40 22.81	
14	9	..	..	..	..	18.0	32.5	..	41 2.92	21.37	VI.	5	10.14	23 5.4	..	40 41.55	
15	6	..	42.0	57.0	..	..	..	..	45 11.81	21.43	III.	5	9.25	24 41.1	..	44 50.38	
16	6	52.5	7.5	..	..	..	..	..	0 56 37.26	− 21.59	II.	2	7.57	− 8 56.5	..	0 56 15.67	

ZONE 94. OCTOBER 16. K. $D_0 = -38° 58' 0''$.

No.	Mag.	I.	II.	III.	IV.	V.	VI.	VII.	T.	a_1	Micrometer.			$i + d_2$	d_1	Mean Right Ascension, 1850.0.	Mean Declination, 1850.0.
1	10	..	..	10.2	26.4	..	..	..	21 3 26.25	− 18.28	VII.	4	8.23	− 19 7.9	− 26.8	21 3 7.97	− 39 17 34.7
2	10	..	..	..	4.6	20.7	..	..	5 4.53	18.32	VI.	3	6.23	13 6.5	26.4	4 46.21	11 32.9
3	8	15.0	31.6	47.8	..	..	..	..	8 4.12	18.32	IV.	9	5.54	42 0.1	25.7	7 45.80	40 25.8
4	7.8	..	..	..	47.2	3.0	18.8	..	9 46.77	18.38	VI.	6	8.14	23 7.4	25.4	9 28.39	26 32.8
5	9	..	5.1	21.6	37.9	..	..	..	11 37.61	18.41	V.	6	4.55	26 26.6	25.0	11 19.20	24 51.6
6	10	..	..	..	..	57.1	12.8	..	19 40.68	18.54	VII.	7	7.4	32 33.4	23.2	19 22.14	30 56.6
7	10	..	..	..	14.5	30.4	46.2	..	20 14.19	18.55	VI.	6	12.41	30 22.8	23.1	19 55.59	28 45.9
8	8.9	..	..	..	..	55.5	11.2	27.8	22 39.19	18.61	VII.	4	8.10	19 1.3	22.6	22 20.58	17 23.9
9	6	..	1.1	17.4	..	..	..	..	35 33.22	18.84	IV.	3	11.2	15 28.4	20.1	35 14.43	13 48.5
10	9	..	..	..	2.8	19.2	..	..	36 2.80	18.84	VII.	2	8.39	9 14.3	20.0	35 41.04	7 34.3
11	10	..	..	..	..	52.8	..	..	37 36.66	18.85	VII.	6	4.42	26 19.5	19.7	37 17.95	24 39.2
12	9	..	..	..	..	..	26.2	..	39 53.84	18.86	VII.	10	8.7	48 9.8	19.3	39 34.66	46 29.1
13	7	..	..	..	..	27.8	43.3	0.1	42 11.42	18.93	VII.	4	9.52	19 53.0	18.9	41 52.49	18 11.9
14	10	..	..	..	..	46.6	..	..	45 30.39	19.00	VI.	3	4.18	12 3.2	18.3	45 11.55	10 21.5
15	8.7	..	..	..	..	..	56.8	13.3	47 24.76	18.98	VII.	9	9.23	43 45.5	18.0	47 5.63	42 3.5
16	5.4	..	..	16.0	32.1	48.0	3.7	..	50 31.72	19.09	VI.	2	6.40	8 17.3	17.4	50 12.67	6 34.7
17	8	48.1	4.8	21.0	..	53.5	..	..	55 37.23	19.16	VI.	8	7.13	37 38.9	16.6	55 18.13	35 55.5
18	10	..	..	..	31.3	..	..	..	21 58 31.12	19.18	VI.	9	4.55	41 29.8	16.2	21 58 11.94	39 46.0
19	7	..	53.2	9.5	25.3	..	..	..	22 3 25.14	19.32	IV.	1	8.34	4 11.6	15.4	22 3 5.82	2 27.0
20	9	29.5	45.8	2.4	..	..	..	..	10 18.04	19.44	IV.	3	5.30	12 40.0	14.5	9 58.60	10 54.5
21	8	..	0.4	16.6	33.0	..	..	..	11 32.77	19.41	V.	7	7.16	32 40.1	14.3	11 13.33	30 54.4
22	9	..	41.3	57.3	..	..	..	..	14 13.53	19.48	IV.	7	6.38	32 20.9	14.0	13 54.05	30 34.9
23	9	..	..	..	..	48.8	..	..	14 32.75	19.48	V.	9	4.44	41 24.5	13.9	14 13.27	39 38.4
24	10	..	17.7	33.9	..	..	..	..	24 50.03	19.68	VI.	7	6.37	32 20.1	12.7	24 30.35	30 32.8
25	10	..	33.2	49.8	..	..	..	..	27 5.45	19.70	V.	3	8.25	14 8.7	12.5	25 45.75	12 21.2
26	8	..	..	..	..	..	..	0.8	27 12.35	19.72	VII.	7	5.33	31 47.2	12.4	26 52.63	29 59.6
27	6	46.9	3.6	20.1	..	..	..	..	30 35.74	19.81	III.	4	4.15	17 2.7	12.0	30 15.93	15 14.7
28	10	..	35.6	51.8	7.7	..	..	..	36 7.69	19.90	V.	5	4.36	22 14.4	11.5	35 47.79	20 25.9
29	7	..	..	18.9	34.8	50.8	..	..	22 39 34.51	− 20.00	V.	1	4.48	− 2 16.9	− 11.2	22 39 14.51	− 39 0 28.1

CORRECTIONS.

Date.		Corr. of Clock.	Hourly rate.	m	n	c
1846.	h.	s.	s.	s.	s.	s.
Oct. 15,	20	− 29.05	− 0.014	+ 0.265	− 0.106	− 0.086
Oct. 16,	20	− 29.98	− 0.014	+ 0.265	− 0.106	− 0.086

INSTRUMENT READINGS.

Date.		Barom.	Thermom. At.	Thermom. Ex.
1846.	h. m.	In.	°	°

REMARKS.

(93) 1. Minutes assumed as 0 instead of 1, and micr. reading as 12r.00 instead of 8r.00, to agree with Mural Z., Sept. 19 and Dec. 4.
(94) 23. Differs 2s.36 in right ascension from Mural Z., 1846, October 17.
(94) 25. Minutes assumed as 26.

ZONE 94. OCTOBER 16. K. $D_0 = -38° 58' 0''$—Continued.

No.	Mag.	SECONDS OF TRANSIT. I.	II.	III.	IV.	V.	VI.	VII.	T.	n_1	MICROMETER.			$i + d_1$	d_1	Mean Right Ascension, 1850.0.	Mean Declination, 1850.0.
									h. m. s.	s.	°		r.	' "	"	h. m. s.	° ' "
30	8			40.6	57.0	12.8			22 43 56.83	— 20.01	VI.	10	5.13	—46 41.8	— 10.8	22 43 36.82	— 39 44 52.6
31	9	4.7	20.8	37.3					47 53.35	20.11	IV.	6	11.11	29 37.5	10.5	47 33.24	27 48.0
32	10		38.7	54.8	11.2				53 11.09	20.19	V.	8	13.16	40 43.4	10.1	52 50.90	38 53.5
33	6	38.3	54.6	11.0	27.1				55 26.85	20.27	V.	3	12.34	16 15.0	9.9	55 6.58	14 24.9
34	9		38.6	55.0					22 58 11.14	— 20.28	IV.	8	9.49	38 58.4	— 9.7	22 57 50.86	— 39 37 8.1
35	6				52.0	8.0	23.8		. . . .	. . .	VII.	9	9.53	—44 0.7			

ZONE 95. OCTOBER 16. K. $D_1 = -27° 43' 0''$.

No.	Mag.	I.	II.	III.	IV.	V.	VI.	VII.	T.	n_1	MICROMETER.			$i + d_1$	d_1	Mean Right Ascension, 1850.0.	Mean Declination, 1850.0.
1	9					11.2	25.0		1 0 56.94	— 22.13	VI.	1	8.45	— 4 22.4	— 3.8	1 0 34.81	— 27 47 26.2
2	10		11.3	26.3					4 39.80	22.17	V.	3	4.43	12 19.4	3.9	4 17.63	55 23.3
3	9				50.8		18.9		10 50.75	22.23	VI.	1	10.32	5 16.4	4.1	10 28.52	27 48 20.5
4	9	36.3	50.7	5.3	19.3				15 19.06	22.29	V.	4	6.28	18 12.0	4.3	14 56.77	28 1 16.3
5	8		52.0	6.1	20.1	34.6			21 20.06	22.34	VI.	2	10.52	10 26.0	4.6	20 57.72	27 53 30.6
6	8				21.3				23 21.14	22.36	V.	1	3.38	1 47.4	4.7	22 58.78	27 44 52.1
7	7		46.8	1.0	15.2				25 15.24	22.40	V.	10	3.18	45 38.0	4.8	24 52.84	28 28 42.8
8	8	57.7	12.2	26.2	40.7				27 40.44	22.42	VI.	5	10.1	24 59.2	4.9	27 18.02	28 8 4.1
9	9			8.7	22.8	36.7			31 22.54	22.44	VII.	3	12.22	16 10.9	5.2	31 0.10	27 59 16.1
10	10						46.7		32 18.60	22.46	VII.	4	5.57	17 56.0	5.2	31 56.14	28 1 1.2
11	10			46.1					39 0.11	22.52	IV.	6	5.30	26 44.0	5.6	38 37.59	9 49.6
12	5					41.4	55.1	9.9	39 27.15	22.52	VII.	5	5.55	22 54.7	5.7	39 4.63	6 0.4
13	7			37.2		5.3			42 51.35	22.57	VI.	10	8.40	48 20.6	5.9	42 28.78	31 26.5
14	8		15.2	29.3	43.8				45 43.68	22.59	VI.	10	5.54	46 56.7	6.2	45 21.09	28 30 2.9
15	9						40.7	55.4	46 12.64	22.58	VII.	4	2.17	16 4.8	6.2	45 50.06	27 59 11.0
16	8		36.3	50.5	5.2				49 4.89	22.62	VI.	9	9.48	43 53.5	6.4	48 42.27	28 26 59.9
17	9			55.5	10.0	23.9			50 9.57	22.62	VI.	7	11.38	34 50.2	6.5	49 46.95	28 17 56.7
18	11				33.5	47.4			52 33.25	22.64	VII.	3	11.43	15 51.2	6.7	52 10.61	27 58 57.9
19	9						48.3		53 20.15	22.65	VII.	6	9.59	28 59.5	6.8	52 57.50	28 12 6.3
20	9				39.6	53.9	7.4		1 54 39.47	22.66	VII.	8	6.12	37 4.6	6.9	1 54 16.81	20 11.5
21	5		16.2	30.7	45.1				2 1 44.81	22.73	V.	7	10.8	34 4.9	7.6	2 1 22.08	17 12.5
22	10				8.1				3 7.94	22.74	V.	10	5.39	46 49.3	7.7	2 45.20	28 29 57.0
23	8			22.4	36.5	50.6			4 36.29	22.74	VI.	3	5.28	12 42.0	7.9	4 13.55	27 55 49.9
24	7.8			34.9		3.0			5 48.89	22.76	VI.	7	2.7	30 1.8	8.0	5 26.13	28 13 9.8
25	8	20.3	34.6	49.0					14 2.88	22.82	IV.	3	9.19	14 38.9	8.8	13 40.06	27 57 47.7
26	9			11.0	25.0	39.2			19 24.81	22.86	VI.	1	12.19	6 10.4	9.4	19 1.95	27 49 19.8
27	10		45.2	59.9					21 13.84	22.89	IV.	7	11.49	34 56.0	9.6	20 50.95	28 18 5.6
28	9	8.0	22.6	36.2					22 50.54	22.89	IV.	3	12.59	21 29.0	9.8	22 27.65	28 4 38.8
29	8							41.1	22 58.34	22.89	VII.	3	5.24	12 39.7	9.8	22 35.45	27 55 49.5
30	7		50.6	4.8	19.0				25 18.88	22.92	IV.	6	6.17	27 7.7	10.1	24 55.96	28 10 17.8
31	7					14.0	27.8	42.3	25 59.79	22.93	VII.	9	10.22	44 10.4	10.2	25 36.86	28 27 20.6
32	10					0.6	14.6		29 46.44	22.94	VII.	2	3.6	6 30.4	10.7	29 23.50	27 49 41.1
33	8			20.9		49.2			32 34.95	22.97	VI.	6	5.11	26 34.2	11.0	32 11.98	28 9 45.2
34	6						22.4	36.5	32 54.14	22.98	VII.	9	2.6	45 0.8	11.1	32 31.16	28 11.9
35	8		53.1	7.4					35 21.44	22.99	IV.	6	4.49	26 23.3	11.4	34 58.45	28 9 34.7
36	8			44.5	58.6				35 58.36	22.98	IV.	3	4.7	12 1.3	11.5	35 35.38	27 55 12.8
37	11				16.3				41 16.14	23.02	IV.	3	5.22	12 39.1	12.2	40 53.12	27 55 51.3
38	9			8.3	22.6				45 22.44	23.05	V.	7	6.2	32 0.6	12.8	44 59.39	28 15 13.4
39	9		0.2		28.9				2 47 28.75	— 23.07	V.	9	11.58	—44 59.3	— 13.0	2 47 5.68	— 28 28 12.3

CORRECTIONS.

Date.		Corr. of Clock.	Hourly rate.	m	n	c
1846.	h.	s.	s.	s.	s.	s.

INSTRUMENT READINGS.

Date.		Barom.	THERMOM. At.	Ex.
1846.	h. m.	in.	°	°

REMARKS.

(95) 28. Micrometer thread assumed as 4 instead of 3.
(95) 34. Micrometer thread assumed as 10 instead of 9.

ZONE 95. OCTOBER 16. K. $D_0 = -27° 43' 0''$—Continued.

No.	Mag.	I.	II.	III.	IV.	V.	VI.	VII.	T.	a_1	Micrometer			$i + d_2$	d_1	Mean Right Ascension, 1850.0.	Mean Declination, 1850.0.
									h. m. s.	s.			r.	' "	"	h. m. s.	° ' "
41	8	..	..	30.7	..	59.0	..	..	2 48 44.94	− 23.08	V.	10	6.18	−47 9.0	− 13.2	2 48 21.86	− 28 30 22.2
42	10	..	7.3	21.6	..	..	..	..	51 35.95	23.09	IV.	7	8.5	33 2.8	13.6	51 12.66	16 16.4
43	10	..	..	..	..	..	16.0	30.3	51 47.83	23.09	VII.	7	12.42	35 22.3	13.6	51 24.74	18 35.9
44	6	57.8	12.0	26.3	40.8	..	..	..	54 40.52	23.11	V.	7	3.37	30 47.4	14.1	54 17.41	14 1.5
45	9	..	..	..	..	28.1	42.0	..	55 13.89	23.12	VII.	5	4.11	22 2.1	14.2	54 50.77	5 16.3
46	10	..	2.0	16.2	..	..	..	..	2 59 30.35	− 23.14	IV.	7	4.51	−31 24.8	− 14.8	2 59 7.21	− 28 14 39.6

ZONE 96. OCTOBER 17. A. $D_0 = -39° 0' 0''$.

No.	Mag.	I.	II.	III.	IV.	V.	VI.	VII.	T.	a_1	Micrometer			$i + d_2$	d_1	Mean Right Ascension, 1850.0.	Mean Declination, 1850.0.
1	9	30.0	46.5	..	..	..	..	..	23 9 18.83	− 20.56	II.	6	5.47	−26 52.8	− 61.4	23 8 58.27	− 39 27 54.2
2	7	..	25.0	41.3	..	..	..	..	11 57.12	20.63	II.	3	9.28	14 40.5	63.0	11 56.49	15 3.5
3	7	..	..	..	46.5	2.5	..	..	12 46.32	20.63	V.	4	6.1	17 56.5	63.4	12 25.69	18 59.9
4	10	..	..	..	..	32.2	48.0	..	16 15.93	20.70	VI.	3	4.4	11 56.1	65.5	15 55.23	13 1.6
5	7	..	6.8	23.0	..	..	..	..	20 39.05	20.77	III.	6	6.17	27 8.2	68.0	20 18.28	28 16.2
6	8	..	..	..	..	..	21.0	37.5	20 48.81	20.78	VII.	2	7.47	8 48.0	68.1	20 28.03	9 56.1
7	8	..	..	39.5	55.3	..	..	..	22 55.11	20.82	IV.	2	6.30	8 9.6	69.4	22 34.29	9 19.0
8	8	10.0	26.4	..	..	..	..	..	26 58.47	20.90	II.	1	9.8	4 28.7	71.8	26 37.57	5 40.5
9	10	..	..	21.5	38.5	..	..	..	27 37.79	20.91	IV.	4	6.30	18 11.3	72.2	27 16.88	19 23.5
10	7	..	..	..	..	..	39.0	55.5	28 6.96	20.90	VII.	10	4.0	46 4.3	72.5	27 46.06	47 16.8
11	7	57.8	14.2	..	..	..	..	..	23 30 46.45	− 20.96	II.	4	6.38	−18 15.0	− 74.1	23 30 25.49	− 39 19 29.1

ZONE 97. OCTOBER 19. K. $D_0 = -39° 1' 0''$.

No.	Mag.	I.	II.	III.	IV.	V.	VI.	VII.	T.	a_1	Micrometer			$i + d_2$	d_1	Mean Right Ascension, 1850.0.	Mean Declination, 1850.0.
1	10	59.1	..	32.0	..	..	..	..	23 30 47.75	− 22.20	IV.	4	4.59	−17 24.9	− 41.6	23 30 25.55	− 39 19 6.5
2	11	..	..	20.2	36.8	..	..	..	35 36.40	22.29	VI.	6	12.53	30 29.1	41.8	35 14.11	32 10.9
3	8	..	..	4.8	20.7	37.0	..	..	37 20.56	22.32	V.	2	6.38	8 13.2	41.9	36 58.24	9 55.1
4	9	..	52.9	..	25.5	..	..	..	42 25.43	22.40	IV.	10	5.16	46 44.1	42.2	42 3.03	48 26.3
5	8	..	..	19.8	35.8	52.7	..	..	50 36.13	22.55	V.	10	3.21	45 45.6	42.7	13.58	47 28.3
6	7	12.6	29.0	..	..	17.8	..	..	58 1.61	22.71	VI.	8	11.36	39 52.7	43.3	57 38.90	41 36.0
7	7	..	..	21.0	37.0	..	9.2	..	23 59 36.96	22.74	VII.	9	4.3	41 3.4	43.4	23 59 14.22	42 46.8
8	11	..	57.7	..	..	..	..	..	0 25 29.78	23.21	IV.	1	9.26	4 37.6	45.5	0 25 6.57	6 23.1
9	7	..	..	24.2	40.9	56.8	..	..	26 40.53	23.23	VI.	6	9.54	28 58.2	45.6	26 17.30	30 43.8
10	9	..	..	..	19.6	..	..	..	0 29 19.42	− 23.28	V.	1	10.57	− 5 23.7	− 45.8	0 28 56.14	− 39 7 9.5

ZONE 98. OCTOBER 24. A. $D_0 = -33° 58' 30''$.

No.	Mag.	I.	II.	III.	IV.	V.	VI.	VII.	T.	a_1	Micrometer			$i + d_2$	d_1	Mean Right Ascension, 1850.0.	Mean Declination, 1850.0.
1	9	..	..	..	..	..	38.0	53.0	1 7 7.72	− 27.38	VII.	7	8.13	−33 7.1	− 4.8	1 6 40.34	− 34 31 41.9
2	10	27.5	43.2	..	..	..	..	..	13 13.44	27.46	II.	7	6.14	32 7.2	5.3	12 45.98	30 42.5
3	10	..	..	50.0	5.0	..	..	..	14 4.81	27.47	IV.	3	6.3	12 58.8	5.4	13 37.34	11 34.2
4	5	..	..	54.8	9.7	..	..	..	15 9.48	27.48	IV.	1	5.25	2 39.5	5.5	14 42.00	1 15.0
5	6	3.5	19.0	..	..	..	..	..	16 49.21	27.52	II.	5	3.15	21 33.7	5.6	16 21.69	20 9.3
6	11	..	..	..	..	5.0	19.5	..	16 49.61	27.52	VI.	4	4.52	17 22.6	5.6	16 22.09	15 58.2
7	10	..	..	..	..	13.5	28.0	..	19 58.10	27.57	VI.	9	7.34	42 47.4	5.9	19 30.53	41 23.3
8	5	33.8	49.4	..	..	..	..	..	22 19.71	27.59	II.	7	9.29	33 45.9	6.1	21 52.12	32 22.0
9	9	..	..	..	..	51.5	7.0	..	23 36.62	27.60	VI.	2	8.19	9 7.2	6.2	23 9.02	7 43.4
10	9	..	..	19.6	34.8	..	..	..	1 25 34.52	− 27.63	IV.	3	6.26	−13 10.4	− 6.4	1 25 6.89	− 34 11 46.8

CORRECTIONS.

Date.		Corr. of Clock.	Hourly rate.	m	n	c
1846.	h.	s.	s.	s.	s.	s.
Oct. 17,	21	− 30.05	− 0.017	+ 0.265	− 0.106	− 0.086
Oct. 19,	21	− 31.25	− 0.037	+ 0.265	− 0.106	− 0.086
Oct. 24,	21	− 34.90	− 0.030	+ 0.265	− 0.106	− 0.086

INSTRUMENT READINGS.

Date.		Barom.	Thermom. At.	Thermom. Ex.
1846	h. m.	in.	°	°

REMARKS.

ZONE 98. OCTOBER 24. A. $D_o = -33° 58' 30''$—Continued.

No.	Mag.	SECONDS OF TRANSIT. I.	II.	III.	IV.	V.	VI.	VII.	T.	a_i	MICROMETER.			$i + d_2$	d_1	Mean Right Ascension, 1850.0.	Mean Declination, 1850.0.
									h. m. s.	s.			r.	′ ″	″	h. m. s.	° ′ ″
11	7	. .	29.8	45.0	. .	. .	. .	. .	1 29 59.84	− 27.69	III.	2	11.5	−10 31.4	− 6.8	1 29 32.15	− 34 9 8.2
12	10	. .	. .	. .	50.0	5.0	. .	. .	30 49.82	27.71	V.	4	6.7	18 0.7	6.9	30 22.11	16 37.6
13	9	. .	13.0	28.0	. .	. .	. .	. .	32 43.00	27.72	III.	3	9.54	14 55.6	7.0	31 15.28	13 32.6
14	5	. .	. .	. .	. .	18.5	33.0	. .	33 3.12	27.74	VI.	4	12.0	20 59.1	7.2	32 35.38	19 36.3
15	6	. .	23.5	39.0	. .	. .	. .	. .	34 53.69	27.75	III.	2	11.36	10 47.1	7.3	34 25.94	9 24.4
16	9	27.0	42.5	. .	. .	. .	. .	. .	38 12.91	27.80	II.	8	6.28	37 14.0	7.7	37 45.11	35 51.7
17	10	. .	15.0	30.0	. .	. .	. .	. .	39 45.02	27.83	III.	4	2.59	16 25.6	7.9	39 17.19	15 3.5
18	8	. .	1.0	16.5	. .	. .	. .	. .	42 31.37	27.86	III.	5	6.53	23 24.2	8.2	42 3.51	22 2.4
19	10	2.0	17.5	. .	. .	. .	. .	. .	44 47.74	27.89	II.	5	8.22	24 9.0	8.5	44 19.85	22 47.5
20	7	30.9	46.0	. .	. .	. .	. .	. .	47 16.37	27.92	II.	4	9.27	19 41.7	8.8	46 48.45	18 20.5
21	9	. .	. .	39.0	54.5	. .	. .	. .	47 54.23	27.92	IV.	7	7.35	32 48.4	8.9	47 26.31	31 27.3
22	10	. .	. .	. .	. .	54.0	9.0	. .	47 38.86	27.92	VI.	4	6.24	18 9.1	8.9	47 10.94	16 48.0
23	7	. .	. .	58.0	13.4	. .	. .	. .	50 13.14	27.96	IV.	6	10.45	29 23.5	9.2	49 45.18	28 2.7
24	10	. .	. .	. .	. .	9.0	24.5	. .	50 54.10	27.97	VI.	6	5.11	26 34.3	9.3	50 26.13	25 13.6
25	10	. .	. .	. .	. .	34.5	49.5	. .	52 19.38	27.98	VI.	2	11.0	10 28.7	9.5	51 51.40	9 8.2
26	10	. .	. .	. .	59.0	15.0	. .	. .	54 59.29	28.01	V.	1	8.30	4 13.0	9.8	54 31.28	2 52.8
27	9	28.2	43.6	. .	. .	. .	. .	. .	56 13.87	28.04	II.	5	5.1	22 27.3	10.0	55 45.83	21 7.3
28	10	. .	. .	. .	. .	. .	41.5	57.5	1 56 11.62	28.03	VII.	1	9.37	4 46.4	10.0	55 43.59	3 26.4
29	9	. .	. .	. .	5.0	19.8	. .	. .	2 0 4.77	28.09	V.	8	8.11	38 6.3	10.5	1 59 36.68	36 46.8
30	9	. .	7.0	22.0	. .	. .	. .	. .	1 36.95	28.10	III.	3	3.37	11 44.8	10.7	2 1 8.85	10 25.5
31	9	45.8	1.0	. .	. .	. .	. .	. .	5 31.60	28.17	II.	9	2.52	40 24.7	11.3	5 3.43	39 6.0
32	6	. .	59.0	14.7	. .	. .	. .	. .	7 29.19	28.16	III.	1	7.36	3 45.7	11.6	7 1.03	2 27.3
33	9	. .	. .	. .	. .	47.0	2.0	. .	10 31.85	28.22	VI.	6	10.0	29 0.5	12.0	10 3.63	27 42.5
34	9	. .	. .	. .	. .	58.5	13.0	. .	12 43.13	28.23	VI.	1	11.37	5 47.4	12.3	12 14.90	4 29.7
35	8	57.8	13.0	. .	. .	. .	. .	. .	13 43.32	28.25	II.	4	9.39	19 47.8	12.5	13 15.07	18 30.3
36	10	. .	35.2	50.5	. .	. .	. .	. .	15 5.65	28.27	III.	8	4.15	36 6.9	12.7	14 37.38	34 49.6
37	9	. .	. .	14.0	29.5	. .	. .	. .	17 29.35	28.30	IV.	10	6.51	47 27.8	13.1	17 1.05	46 10.9
38	9	. .	23.5	38.5	. .	. .	. .	. .	18 53.44	28.30	III.	3	2.38	11 15.0	13.3	18 25.14	9 58.3
39	10	48.5	3.5	. .	. .	. .	. .	. .	20 33.92	28.33	II.	4	7.6	18 30.4	13.5	20 5.59	17 13.9
40	10	. .	. .	. .	. .	3.5	18.5	. .	20 48.35	28.33	VI.	6	3.52	25 54.3	13.6	20 20.02	24 37.9
41	4	. .	. .	. .	. .	24.3	39.2	. .	22 9.11	28.34	VI.	7	3.7	30 32.6	13.8	21 40.77	29 16.4
42	9	51.0	6.5	. .	. .	. .	. .	. .	28 36.84	28.42	II.	7	5.51	31 55.5	14.9	28 8.42	30 40.4
43	9	. .	. .	18.0	32.5	. .	. .	. .	31 32.62	28.45	IV.	4	8.42	19 19.2	15.3	31 4.17	18 4.5
44	6	. .	. .	. .	. .	20.5	35.2	. .	32 5.22	28.45	VI.	4	12.0	20 59.1	15.4	31 36.77	19 44.5
45	8	. .	30.0	45.2	. .	. .	. .	. .	34 0.36	28.47	III.	7	8.9	33 5.6	15.8	33 31.89	31 51.4
46	8	3.4	18.9	. .	. .	. .	. .	. .	36 49.18	28.50	II.	6	7.16	27 37.5	16.3	36 20.68	26 23.8
47	6	. .	9.5	25.0	. .	. .	. .	. .	37 40.18	28.52	III.	9	11.1	44 32.3	16.4	37 11.66	43 18.7
48	8	6.5	22.3	. .	. .	. .	. .	. .	39 52.48	28.54	II.	7	4.36	31 17.6	16.8	39 23.94	30 4.4
49	8	. .	. .	. .	. .	12.0	27.0	. .	39 56.86	28.54	VI.	7	5.10	31 34.8	16.8	39 28.32	30 21.6
50	9	. .	. .	. .	. .	. .	13.8	29.2	40 43.73	28.55	VII.	8	5.32	36 45.4	17.0	40 15.18	35 32.4
51	6	. .	18.0	33.0	. .	. .	. .	. .	42 48.15	28.57	III.	5	11.21	25 39.8	17.3	42 19.58	24 27.1
52	8	46.8	2.0	. .	. .	. .	. .	. .	46 32.14	28.59	II.	1	10.33	5 15.0	18.1	46 3.55	4 3.1
53	6	. .	. .	53.9	8.7	. .	. .	. .	49 8.59	28.61	IV.	2	8.42	9 19.1	18.5	48 39.98	8 7.6
54	10	. .	58.0	14.0	. .	. .	. .	. .	28.69	28.64	III.	6	8.33	28 16.7	19.0	51 0.05	27 5.7
55	5	. .	. .	9.0	25.0	. .	. .	. .	2 53 24.61	− 28.67	IV.	10	8.39	−48 22.5	− 19.4	2 52 55.94	− 34 47 11.9

CORRECTIONS.

Date.	Corr. of Clock.	Hourly rate.	m	n	c
1846. h.	s.	s.	s.	s.	s.

INSTRUMENT READINGS.

Date.	Barom.	THERMOM. At.	Ex.
1846. h. m.	In.	°	°

REMARKS.

(98) 11. Differs 1[m] in right ascension from Mural Z., 1846, December 23.
(98) 13. Transits discordant; T. III assumed as 28ˢ.0; minutes perhaps 32.
(98) 54. Minutes of right ascension perhaps 50 or 52.

ZONE 99. OCTOBER 26. K. $D_o = -30° 13' 50''$.

No.	Mag.	Seconds of Transit. I.	II.	III.	IV.	V.	VI.	VII.	T.	a_1	Micrometer.			$i + d_s$	d_1	Mean Right Ascension, 1850.0.	Mean Declination, 1850.0.
									h. m. s.	s.			r.	′ ″	″	h. m. s.	° ′ ″
1	7	..	..	..	..	51.0	5.3	20.2	0 56 36.40	− 28.74	V.	1	12.00	− 6 0.4	− 5.3	0 56 7.66	− 30 19 55.7
2	10	..	..	..	..	..	16.9	31.8	0 57 48.06	28.75	VII.	4	14.27	22 13.4	5.3	57 19.31	36 8.7
3	10	..	51.3	6.3	..	..	..	..	1 0 20.33	28.78	V.	2	12.24	11 12.3	5.2	0 59 51.55	25 7.5
4	8	28.3	43.1	57.8	12.1	..	..	..	2 11.96	28.81	IV.	2	12.52	11 26.5	5.2	1 1 43.15	25 21.7
5	9.10	..	..	..	7.5	21.7	36.2	..	5 7.26	28.85	VII.	6	12.9	30 5.2	5.2	4 38.41	44 0.4
6	9.10	..	..	..	20.8	..	49.5	..	6 20.65	28.86	VII.	3	11.16	15 37.3	5.2	5 51.79	29 32.5
7	9	..	..	19.5	33.8	48.6	..	..	10 33.91	28.92	V.	7	7.23	32 41.8	5.2	10 4.99	46 37.0
8	9	..	..	44.7	59.3	..	..	..	11 59.27	28.94	IV.	9	11.5	44 33.1	5.2	11 30.33	58 28.3
9	9	..	..	..	..	35.8	..	4.0	13 21.27	28.95	VII.	7	4.34	31 16.0	5.2	12 52.32	45 11.2
10	8.9	..	..	..	..	5.2	19.4	..	13 50.57	28.96	VI.	2	8.49	9 23.5	5.2	13 21.61	30 23 18.7
11	8.7	..	..	..	59.6	13.7	..	..	19 59.33	29.04	VI.	10	11.9	49 36.5	5.2	19 30.29	31 3 31.7
12	6	..	..	..	..	50.2	4.2	..	20 35.45	29.05	VII.	10	5.52	46 56.1	5.2	20 6.40	31 0 51.3
13	8	..	..	..	14.4	28.9	42.9	..	22 14.19	29.06	VI.	6	4.36	26 16.6	5.3	21 45.13	30 40 11.9
14	7.8	..	..	..	12.1	26.4	40.6	..	23 11.82	29.08	VI.	8	2.52	35 24.0	5.3	22 42.74	49 19.3
15	6.7	..	..	46.7	1.3	15.7	29.8	..	25 1.09	29.10	VII.	7	5.29	31 43.8	5.3	24 32.99	45 39.1
16	9	..	..	..	..	16.0	30.1	..	26 1.33	29.11	VII.	3	11.55	15 57.0	5.3	25 32.22	29 52.3
17	7	..	..	..	..	12.2	..	41.3	26 57.67	29.12	VII.	6	9.12	28 35.8	5.4	26 28.55	42 31.2
18	9	..	..	..	..	..	13.2	..	27 44.21	29.13	VII.	7	11.18	34 40.2	5.4	27 15.08	48 35.6
19	7	..	..	..	..	..	1.9	..	28 32.97	29.14	VI.	6	6.14	27 6.1	5.4	28 3.83	41 1.5
20	5.6	..	..	..	41.7	56.2	10.3	..	29 41.52	29.16	VI.	6	5.23	26 40.3	5.4	29 12.36	40 35.7
21	9	43.0	57.8	..	26.9	..	..	..	37 26.76	29.24	V.	4	10.13	20 5.5	5.6	36 57.52	34 1.1
22	9	..	50.2	5.1	19.8	..	..	..	41 19.36	29.29	V.	3	8.36	14 16.8	5.8	40 50.07	30 28 12.6
23	8.9	..	..	1.2	16.1	30.2	..	..	46 15.87	29.35	VI.	10	7.56	47 59.0	5.9	45 46.52	31 1 54.9
24	9	13.9	28.7	43.3	..	..	..	..	53 57.61	29.44	IV.	3	13.26	16 43.4	6.3	53 28.17	30 30 39.7
25	5	..	33.4	48.0	2.3	..	..	..	55 2.39	29.45	V.	6	11.00	29 30.8	6.3	54 32.94	43 27.1
26	5	..	..	0.2	..	29.3	..	58.3	1 56 14.50	29.46	VII.	2	9.5	9 31.3	6.4	1 55 45.04	23 27.7
27	9	..	25.5	40.2	54.7	..	..	..	2 3 54.59	29.55	V.	6	8.42	28 21.0	6.8	2 3 25.04	42 17.8
28	11	..	..	..	..	12.2	..	..	4 57.71	29.56	VII.	9	6.40	42 18.7	6.8	4 28.15	56 15.5
29	9	..	..	..	54.7	..	23.0	..	11 54.32	29.63	VI.	4	4.32	17 13.0	7.2	11 24.69	31 10.2
30	9	..	..	..	..	..	56.0	..	12 37.17	29.63	VII.	2	5.46	7 50.7	7.3	12 7.54	21 48.0
31	9	..	..	10.1	..	..	..	..	14 24.69	29.65	IV.	8	7.00	37 29.6	7.4	13 55.04	51 27.0
32	8	..	2.7	17.3	31.7	..	..	..	15 31.47	29.67	V.	1	7.50	3 54.1	7.5	15 1.80	17 51.6
33	9	..	18.9	33.6	..	..	..	..	16 47.94	29.68	IV.	5	3.58	21 55.9	7.6	16 18.26	35 53.5
34	6	..	..	..	11.2	25.8	40.1	..	17 11.16	29.68	VI.	4	8.2	18 59.1	7.6	16 41.48	32 56.7
35	7	..	..	30.2	44.5	59.0	..	..	18 44.41	29.70	VI.	4	7.21	18 38.4	7.7	18 14.71	32 36.1
36	9	4.7	19.5	34.3	..	..	..	..	21 48.38	29.73	IV.	2	8.19	9 8.5	7.9	21 18.65	23 6.4
37	9	..	..	..	..	37.8	..	..	22 23.17	29.73	V.	1	10.22	5 10.9	8.0	21 53.44	19 8.9
38	11	..	..	44.9	..	..	..	..	23 59.06	29.75	IV.	2	10.32	10 46.0	8.1	23 29.31	30 24 44.1
39	6.7	..	..	..	..	36.0	50.0	..	24 21.25	29.75	VI.	10	6.35	47 18.1	8.1	23 51.50	31 1 16.2
40	7	5.2	20.0	34.8	49.2	..	..	..	27 49.07	29.79	IV.	5	3.51	21 52.4	8.4	27 19.28	30 35 50.8
41	8	..	..	..	32.8	47.1	1.3	..	28 32.53	29.80	VII.	5	5.17	22 35.4	8.4	28 2.73	36 33.8
42	5	..	41.9	56.8	11.2	..	..	..	30 11.10	29.81	V.	6	8.7	28 3.3	8.6	29 41.29	42 1.9
43	5.4	36.5	51.2	5.8	20.7	..	..	..	32 20.45	29.83	IV.	8	5.6	36 32.0	8.8	31 50.62	50 30.8
44	7	51.9	6.7	21.3	35.9	..	..	..	34 35.80	29.85	V.	7	8.12	33 6.5	9.0	34 5.92	47 5.5
45	5.6	..	14.0	28.9	43.5	..	..	..	35 43.26	29.86	V.	6	8.28	28 13.9	9.0	35 13.40	42 12.9
46	9	..	9.9	24.7	..	..	..	..	39 38.91	29.90	IV.	4	3.50	16 52.0	9.4	39 9.01	30 30 51.4
47	7.8	..	..	..	30.9	..	59.2	..	45 30.45	29.95	VI.	10	10.49	49 26.4	10.0	45 0.50	31 3 26.4
48	8.9	..	52.5	7.2	..	36.2	..	..	47 21.48	29.96	VI.	3	6.4	12 59.8	10.2	46 51.52	30 27 0.0
49	7.8	..	..	31.0	45.2	0.0	14.1	..	2 48 45.22	− 29.98	VI.	3	6.42	− 13 19.0	− 10.3	2 48 15.24	− 30 27 19.3

CORRECTIONS.

Date.		Corr. of Clock.	Hourly rate.	m	n	c
1846.	h.	s.	s.	s.	s.	s.
Oct. 26,	21	− 36.53	− 0.026	+ 0.265	− 0.106	− 0.086

INSTRUMENT READINGS.

Date.		Barom.	Thermom. At.	Thermom. Ex.
1846.	h. m.	in.	°	°

REMARKS.

(99) 38. Micrometer assumed as 11ʳ.32 instead of 10ʳ.32.

ZONE 99. OCTOBER 26. K. $D_o = -30° 13' 50''$—Continued.

No.	Mag.	SECONDS OF TRANSIT. I.	II.	III.	IV.	V.	VI.	VII.	T.	a_1	MICROMETER.			$i + d_3$	d_1	Mean Right Ascension, 1850.0.	Mean Declination, 1850.0.
									h. m. s.	s.			r.	′ ″	″	h. m. s.	° ′ ″
50	5.4	38.7	53.4	8.1	..	..	..	..	2 51 22.34	− 30.00	IV.	3	7.18	−13 37.4	− 10.6	2 50 52.34	− 30 27 38.0
51	8	15.3	30.2	44.7	..	..	..	..	53 59.25	30.02	IV.	7	6.19	32 9.5	10.8	53 29.23	46 10.3
52	7	..	40.7	55.1	10.0	..	..	..	2 55 9.86	30.03	V.	9	8.41	43 20.3	11.0	2 54 39.83	57 21.3
53	7	9.2	24.0	38.7	53.2	..	..	..	3 0 53.00	30.07	V.	4	9.59	19 58.4	11.6	3 0 22.93	34 0.0
54	7	..	42.3	56.8	11.3	..	..	..	3 2 11.36	− 30.08	V.	8	10.54	−39 27.8	− 11.8	3 1 41.28	− 30 53 29.6

ZONE 100. OCTOBER 28. A. $D_o = -33° 57' 40''$.

No.	Mag.	I.	II.	III.	IV.	V.	VI.	VII.	T.	a_1	MICROMETER.			$i + d_3$	d_1	Mean Right Ascension, 1850.0.	Mean Declination, 1850.0.
1	5	7.3	22.4	..	..	..	..	..	0 0 52.84	− 27.38	II.	5	8.7	−24 1.4	− 6.8	0 0 25.46	− 34 21 48.2
2	11	..	10.5	25.5	..	..	..	..	2 40.85	27.42	III.	8	10.4	39 3.5	7.0	2 13.43	36 50.5
3	9	..	..	16.5	31.2	..	..	..	5 31.18	27.46	IV.	2	8.39	9 17.6	7.3	5 3.68	7 4.9
4	10	..	25.0	40.3	..	..	..	..	7 55.32	27.49	III.	6	5.42	26 50.2	7.6	7 27.83	24 37.8
5	9	..	40.0	55.4	..	..	..	..	12 10.66	27.56	III.	10	5.41	46 52.4	8.1	11 43.10	44 40.5
6	8	..	45.0	59.5	..	..	..	..	16 14.57	27.63	III.	1	5.00	2 26.8	8.6	15 46.94	0 15.4
7	8	..	..	2.8	18.2	..	..	..	20 18.08	27.70	IV.	9	11.16	44 40.0	9.1	19 50.38	42 29.1
8	10	7.8	23.0	..	..	..	..	..	22 53.29	27.74	II.	4	4.42	17 17.5	9.4	21 25.55	15 6.9
9	8	11.8	27.0	..	..	..	..	..	26 57.29	27.80	II.	4	2.58	16 24.9	10.0	26 29.49	14 14.9
10	9	..	..	..	..	34.5	49.0	..	29 19.09	27.83	VI.	10	2.56	45 28.7	10.3	28 51.26	43 19.0
11	9	..	6.5	21.8	..	..	..	..	31 36.71	27.88	III.	4	9.30	19 43.4	10.7	31 8.83	17 34.1
12	6	..	..	..	..	17.0	31.4	..	33 1.82	27.90	VI.	10	9.50	48 58.2	10.9	32 33.62	46 49.1
13	10	..	10.5	26.0	..	..	..	..	37 40.97	27.97	III.	7	3.49	30 54.0	11.6	37 13.00	28 45.6
14	11	..	..	52.2	..	..	..	..	41 7.10	28.02	III.	4	7.29	18 42.2	12.1	40 39.08	16 34.3
15	9	..	..	5.0	20.2	..	..	..	42 20.15	28.04	IV.	9	3.54	40 46.3	12.3	41 52.11	38 48.6
16	9	..	..	40.0	55.8	..	..	..	47 55.51	28.13	IV.	10	8.45	48 25.5	13.2	47 27.38	46 18.7
17	10	..	..	9.8	25.0	..	..	..	50 24.96	28.16	IV.	9	7.34	42 47.6	13.6	49 56.80	40 41.2
18	7	..	12.6	27.9	..	..	..	..	54 42.70	28.24	III.	3	4.34	12 13.7	14.4	54 14.46	10 8.1
19	6	10.2	25.8	..	..	..	..	..	0 57 55.96	28.28	II.	5	4.35	22 14.2	15.0	0 57 27.68	20 9.2
20	10	44.9	0.5	15.7	..	..	..	..	1 2 30.71	− 28.36	III.	6	4.58	−26 27.9	− 15.8	1 2 2.35	− 34 24 23.7

ZONE 101. OCTOBER 28. A. $D_o = -28° 59' 50''$.

No.	Mag.	I.	II.	III.	IV.	V.	VI.	VII.	T.	a_1	MICROMETER.			$i + d_3$	d_1	Mean Right Ascension, 1850.0.	Mean Declination, 1850.0.
1	8	..	20.9	35.0	..	..	..	..	1 32 49.58	− 28.59	III.	10	6.43	−47 22.0	− 8.8	1 32 20.99	− 29 47 20.8
2	9	18.0	33.0	..	..	..	..	..	36 1.37	28.61	II.	4	10.4	20 0.9	9.6	35 32.76	20 0.5
3	7	50.0	4.8	..	..	..	..	..	38 33.36	28.65	II.	6	8.51	28 25.4	10.2	38 4.71	28 25.6
4	8	..	..	..	..	53.8	7.9	..	38 39.34	28.65	VI.	4	5.00	17 27.3	10.2	38 10.69	17 27.5
5	8	19.5	34.5	..	..	..	..	..	41 2.94	28.67	II.	5	12.00	25 59.3	10.8	40 34.27	26 0.1
6	10	..	..	..	..	..	28.0	42.0	40 59.11	28.67	VII.	7	7.13	32 36.2	10.8	40 30.44	32 37.0
7	7	..	6.5	21.0	..	..	..	..	43 35.38	28.71	III.	10	6.30	47 15.4	11.4	43 6.67	47 16.8
8	9	..	..	..	11.0	25.0	..	..	44 10.72	28.70	V.	5	7.25	23 40.5	11.6	43 41.02	23 42.1
9	11	..	..	..	..	..	20.5	35.0	44 51.80	28.71	VII.	4	8.7	19 1.5	11.7	44 23.09	19 3.2
10	8	31.0	45.9	..	..	..	..	..	50 14.49	28.77	II.	8	4.39	36 18.0	13.0	49 45.72	36 21.0
11	7	49.0	3.5	..	..	..	..	..	51 32.27	28.79	II.	7	11.17	34 39.8	13.4	51 3.48	34 43.2
12	8	..	..	..	45.0	59.2	..	..	51 44.84	28.79	V.	7	7.30	32 45.2	13.4	51 16.05	32 48.6
13	11	..	..	28.9	44.0	..	..	..	53 43.66	28.82	IV.	10	7.7	47 34.2	13.9	53 14.84	47 38.1
14	8	..	..	50.0	4.0	..	..	..	56 3.88	28.82	IV.	1	7.33	3 45.8	14.5	55 35.06	3 50.3
15	8	55.5	10.0	..	..	..	..	..	1 58 38.49	− 28.85	II.	1	9.11	− 4 35.2	− 15.2	1 58 9.60	− 29 4 40.4

CORRECTIONS.

Date.		Corr. of Clock.	Hourly rate.	m	n	c
1846.	h.	s.	s.	s.	s.	s.
Oct. 28,	21	− 36.04	− 0.009	+ 0.265	− 0.106	− 0.086

INSTRUMENT READINGS.

Date.		Barom.	THERMOM. At.	THERMOM. Ex.
1846.	h. m.	in.	°	°

REMARKS.

ZONE 101. OCTOBER 28. A. $D_0 = -28° 59' 50''$—Continued.

No.	Mag.	I.	II.	III.	IV.	V.	VI.	VII.	T.	a_1	MICROMETER.			$i + d_2$	d_1	Mean Right Ascension, 1850.0.	Mean Declination, 1850.0.
									h. m. s.	s.			r.	′ ″	″	h. m. s.	° ′ ″
16	7	. .	. .	. .	. .	14.5	28.5	. .	1 59 0.00	— 28.85	VI.	1	11.25	— 5 42.9	— 15.2	1 58 31.15	— 29 5 48.1
17	7	. .	. .	. .	. .	. .	39.5	54.5	2 1 11.12	28.88	VII.	8	5.17	36 37.0	15.8	2 0 42.24	36 42.8
18	7	. .	1.0	15.0	. .	. .	. .	. .	3 29 59	28.91	III.	9	8.5	43 1.9	16.4	3 0.68	43 8.3
19	9	. .	21.0	. .	. .	. .	. .	. .	5 49.62	28.93	II.	8	11.9	39 35.1	17.0	5 20.69	39 42.1
20	10	29.0	44.0	. .	. .	. .	. .	. .	7 12.45	28.94	II.	5	7.12	23 33.8	17.4	6 43.51	23 41.2
21	11	. .	. .	. .	. .	. .	41.2	. .	7 12.53	28.94	VI.	8	3.5	40 29.5	17.4	6 43.59	40 36.9
22	9	12.5	27.0	. .	. .	. .	. .	. .	9 55.58	28.97	II.	4	4.8	17 1.0	18.1	9 26.61	17 9.1
23	7	. .	. .	34.3	48.2	. .	. .	. .	11 48.13	28.98	IV.	1	7.48	3 53.4	18.6	11 19.15	4 2.0
24	9	. .	. .	. .	53.5	7.5	. .	. .	13 53.25	29.01	V.	8	5.53	36 55.5	19.2	13 24.24	37 4.7
25	8	. .	33.2	47.5	. .	. .	. .	. .	17 1.94	29.05	III.	9	8.20	43 9.4	20.0	16 32.89	43 19.4
26	8	53.0	8.0	. .	. .	. .	. .	. .	18 36.36	29.52	II.	4	9.6	19 31.6	20.4	18 6.84	19 42.0
27	9	. .	. .	. .	. .	28.0	42.0	. .	20 13.49	29.55	VI.	10	9.53	48 57.8	20.9	19 43.94	49 8.7
28	8	. .	21.0	35.5	. .	. .	. .	. .	21 49.64	29.55	III.	6	8.13	28 6.3	21.3	21 20.09	28 17.6
29	10	42.5	57.3	. .	. .	. .	. .	. .	24 25.75	29.55	II.	4	6.50	18 22.8	22.0	23 56.20	18 34.8
30	8	. .	. .	46.2	0.5	14.8	. .	. .	25 0.40	29.57	V.	6	5.54	26 56.1	22.2	24 31.83	27 8.3
31	7	24.2	38.8	. .	. .	. .	. .	. .	29 7.38	29.59	II.	5	2.31	21 11.8	23.4	28 37.79	21 25.2
32	7	19.0	34.0	. .	. .	. .	. .	. .	31 2.56	29.60	II.	8	8.40	38 50.0	23.9	30 32.96	39 3.9
33	9	. .	. .	6.0	20.2	. .	. .	. .	36 20.05	29.62	IV.	3	5.59	12 57.6	25.4	35 50.43	13 13.0
34	8	. .	19.5	34.0	. .	. .	. .	. .	37 48.13	29.64	III.	6	4.51	26 24.2	25.8	37 18.49	26 40.0
35	9	. .	. .	. .	. .	. .	24.0	40.0	37 56.00	29.63	VII.	1	8.20	4 9.2	25.9	37 26.37	4 25.1
36	9	. .	25.5	39.5	. .	. .	. .	. .	39 53.76	29.65	III.	4	5.32	17 43.6	26.5	39 24.11	18 0.1
37	9	. .	35.0	49.2	. .	. .	. .	. .	45 3.60	29.67	III.	8	6.58	36 59.6	28.0	44 33.93	37 17.6
38	10	. .	. .	12.0	26.5	. .	. .	. .	48 26.31	29.69	IV.	6	6.50	27 24.4	29.0	47 56.02	27 43.4
39	8	42.0	56.6	. .	. .	. .	. .	. .	50 25.31	29.70	II.	7	8.25	33 12.9	29.6	49 55.61	33 32.5
40	5	29.2	43.6	. .	. .	. .	. .	. .	2 53 12.37	— 29.71	II.	6	11.40	—29 50.8	— 30.5	2 52 42.66	— 29 30 21.3

ZONE 102. NOVEMBER 16. K. $D_0 = -32° 43' 40''$.

No.	Mag.	I.	II.	III.	IV.	V.	VI.	VII.	T.	a_1	MICROMETER.			$i + d_2$	d_1	Mean Right Ascension, 1850.0.	Mean Declination, 1850.0.
1	7	. .	. .	. .	. .	19.1	. .	48.8	1 58 4.12	— 40.10	VII.	7	3.14	—30 35.7	— 8.9	1 57 24.02	— 33 14 24.6
2	10	. .	. .	. .	. .	40.8	. .	. .	2 0 25.88	40.13	VII.	6	10.22	29 11.3	9.3	1 59 45.75	33 13 0.6
3	6.7	. .	. .	46.8	1.8	. .	. .	. .	5 1.43	40.18	VI.	1	4.52	2 23.2	10.1	2 4 21.55	32 46 13.3
4	9	. .	. .	. .	8.9	. .	. .	. .	6 8.74	40.21	V.	4	6.46	18 20.6	10.3	5 28.53	33 2 10.9
5	8	. .	. .	. .	. .	. .	. .	54.0	6 9.04	40.20	VII.	3	11.50	15 54.0	10.3	5 28.84	32 59 44.3
6	8.9	41.2	56.2	. .	. .	. .	. .	. .	9 26.18	40.25	III.	6	4.16	26 6.6	10.9	8 45.93	33 9 57.5
7	8	. .	7.1	22.4	. .	. .	. .	. .	14 37.06	40.33	IV.	5	8.30	24 13.3	11.9	13 56.73	8 5.2
8	5.6	. .	59.0	13.9	28.7	. .	. .	. .	26 28.76	39.48	VI.	7	6.41	32 20.7	14.2	25 49.28	16 14.9
9	7.8	. .	. .	41.4	56.5	. .	. .	. .	27 56.43	40.51	V.	10	2.42	45 21.3	14.5	27 15.92	33 29 15.8
10	8	. .	. .	. .	. .	. .	. .	57.3	28 12.17	40.49	VII.	1	5.35	2 44.6	14.5	27 31.68	32 40 39.1
11	9	. .	. .	. .	. .	. .	. .	44.9	29 0.04	40.52	VII.	5	7.26	23 40.5	14.7	28 19.52	33 7 35.2
12	9	. .	48.4	. .	. .	. .	. .	. .	31 18.38	40.54	V.	8	4.39	36 18.7	15.2	30 37.84	20 13.9
13	5	58.5	13.8	28.8	. .	. .	. .	. .	38 43.59	40.65	IV.	5	11.26	25 42.3	16.8	38 2.94	9 39.1
14	8	14.3	29.7	. .	. .	. .	. .	. .	40 59.48	40.67	IV.	5	11.50	25 54.5	17.3	40 18.81	9 51.8
15	3	44.6	59.3	. .	29.8	. .	. .	. .	44 29.33	40.71	V.	4	6.40	18 17.6	18.0	42 48.62	2 15.6
16	8	. .	. .	57.1	. .	27.2	. .	. .	49 12.32	40.78	IV.	10	9.42	48 53.7	19.1	48 31.54	32 52.8
17	9	. .	. .	. .	41.8	. .	. .	. .	53 41.64	40.84	V.	10	5.36	46 49.3	20.2	53 0.80	30 49.5
18	5.4	. .	. .	. .	. .	. .	. .	52.7	2 54 7.83	40.83	VII.	5	4.55	22 24.1	20.3	2 53 27.00	33 6 24.4
19	5.6	0.3	15.3	. .	. .	. .	. .	. .	3 1 45.10	— 40.91	III.	2	13.37	—11 48.6	— 22.0	3 1 4.19	— 32 55 50.6

CORRECTIONS.

Date.		Corr. of Clock.	Hourly rate.	m	n	c
1846.	h.	s.	s.	s.	s.	s.
Nov. 16,	21	— 46.59	— 0.038	— 0.020	+ 0.056	— 0.086

INSTRUMENT READINGS.

Date.	Barom.	THERMOM. At.	THERMOM. Ex.
1846. h. m.	In.	°	°

REMARKS.

(101) 21. Micrometer thread assumed as 9 instead of 8.
(101) 32. Micrometer reading assumed as $9^{r}.40$ instead of $6^{r}.40$.
(101) 37. Micrometer reading assumed as $6^{r}.08$ instead of $6^{r}.58$.
(102) 15. Minutes assumed as 43 instead of 44.

ZONE 103. NOVEMBER 20. K. $D_0 = -38° 58' 0''$.

No.	Mag.	SECONDS OF TRANSIT. I.	II.	III.	IV.	V.	VI.	VII.	T.	a_1	MICROMETER.			$i + d_1$	d_1	Mean Right Ascension, 1850.0.	Mean Declination, 1850.0.
									h. m. s.	s.			r	′ ″	″	h. m. s.	° ′ ″
1	9						54.9		0 15 22.77	− 42.14	VI.	3	8.23	−14 7.3	− 55.5	0 14 40.63	− 39 13 2.8
2	8				0.2	16.0			26 59.95	42.34	VI.	7	5.36	31 49.2	56.6	26 17.61	30 45.8
3	7.8			13.3	29.9	46.2			33 29.83	42.52	IV.	10	6.22	47 17.5	56.9	32 47.31	46 14.4
4	7	23.4	40.2	56.5	12.3				36 12.22	42.58	IV.	4	6.27	18 9.6	57.2	35 29.64	17 6.8
5	6		7.0	23.4	39.3				37 39.16	42.63	IV.	3	11.55	15 55.2	57.3	36 56.53	14 52.5
6	6	8.5	25.0	41.3	57.6				44 57.27	42.77	IV.	5	3.36	21 44.0	58.1	44 14.50	20 42.1
7	9.10		17.2	34.0	49.8				46 49.77	42.79	IV.	7	7.48	32 56.5	58.2	46 6.98	31 54.7
8	10					49.9			50 33.88	42.90	VI.	10	3.17	45 43.3	58.7	49 50.98	44 42.0
9	4.5			44.1		16.5	32.4	49.0	55 0.40	42.98	VI.	9	11.6	44 38.4	59.2	54 17.42	43 37.6
10	7				54.9				0 59 54.72	43.08	IV.	1	6.35	3 11.0	59.8	0 59 11.64	2 10.8
11	9		24.2	40.2					1 14 56.24	43.38	VI.	4	9.46	19 50.3	61.9	1 14 12.86	18 52.2
12	9				20.2				25 20.02	43.58	VI.	3	6.15	13 2.3	63.4	24 36.44	12 5.7
13	8				22.2				27 22.02	43.60	IV.	1	3.43	1 43.7	63.7	26 38.42	0 47.4
14	9			37.5					29 53.68	43.68	III.	5	8.27	24 11.6	64.2	29 10.00	23 15.8
15	9						27.4		29 55.22	43.68	V.	4	10.40	20 18.0	64.2	29 11.54	19 22.2
16	9						15.7	32.3	37 43.61	43.83	V.	4	2.2	15 55.1	65.5	36 59.78	15 0.6
17	9	8.1	24.4	40.4					42 56.63	43.92	III.	5	9.43	24 50.2	66.5	42 12.71	23 56.7
18	5.6		32.1	48.4	4.4				44 4.17	43.94	IV.	2	11.6	10 29.3	66.6	43 20.23	9 35.9
19	9						51.0	7.8	45 19.09	43.96	VII.	7	12.28	35 17.9	66.9	44 35.13	34 24.8
20	9				12.3	28.2			46 12.13	43.99	VI.	8	8.44	38 25.3	67.1	45 28.14	37 32.4
21	9						15.2	32.0	43.24	44.00	VII.	6	12.12	30 7.8	67.1	(45) 59.24	29 14.9
22	5.6						11.7	28.2	47 39.56	44.01	VI.	4	12.4	21 0.3	67.2	46 55.55	20 7.5
23	9					32.2			1 56 16.10	44.18	VI.	7	6.19	32 11.0	69.1	1 55 31.92	31 20.1
24	8			52.8	8.6	24.7			2 5 8.41	44.31	V.	1	10.45	5 17.8	70.8	2 4 24.10	4 28.6
25	9	11.2		44.3			34.4		0.25	(44.42)	VI.	8	13.43	40 57.1	71.8	15.83	40 8.9
26	9						48.8		15 16.45	44.53	VI.	9	10.4	44 7.0	73.0	14 31.92	43 20.0
27	9					45.0	0.6	17.1	17 28.60	44.54	VII.	5	8.46	24 20.6	73.5	16 44.06	23 34.1
28	9			8.2					23 24.35	44.66	V.	8	9.15	38 41.3	74.9	22 39.69	37 56.2
29	9	14.9	31.2	47.8					31 3.61	44.77	III.	5	7.42	23 48.7	76.7	30 18.84	23 5.4
30	9						22.5		31 50.23	44.79	VI.	7	11.40	34 53.9	76.9	31 5.44	34 10.8
31	9	56.2		29.0					36 45.03	44.88	II.	7	4.25	31 13.1	78.2	36 0.15	30 31.3
32	4.5						25.2		36 53.15	44.85	VI.	1	4.44	2 14.4	78.2	36 8.30	1 32.6
33	2.3				26.8	43.0	58.9		48 26.75	45.02	IV.	1	7.48	3 48.0	81.4	47 41.73	3 9.4
34	5					1.8	17.3	34.3	50 45.46	45.07	VII.	3	12.32	16 13.3	82.1	50 0.39	15 35.4
35	9			49.7	6.0				2 59 5.89	45.23	VI.	10	4.16	46 13.2	84.2	58 20.66	45 37.4
36	9			29.3		1.3			3 0 44.94	− 45.21	VI.	1	9.37	− 4 43.0	− 85.1	2 59 59.73	− 39 4 8.1

CORRECTIONS.

Date.		Corr. of Clock.	Hourly rate.	m	n	c
1846.	h.	s.	s.	s.	s.	s.
Nov. 20,	21	− 50.29	− 0.049	− 0.020	+ 0.056	− 0.086

INSTRUMENT READINGS.

Date.		Barom.	THERMOM. At.	THERMOM. Ex.
1846.	h. m.	in.	°	°

REMARKS.

(103) 25. Transit over T. VI assumed as 32ˢ.4 instead of 34ˢ.4; minutes unknown.

ZONE 104. FEBRUARY 5. II. BELT, −30° 1'. D_o = −29° 36' 50".

No.	Mag.	Seconds of Transit. I.	II.	III.	IV.	V.	VI.	VII.	T.	a_1	a_2	Micrometer.			i	d_1	d_2	Mean Right Ascension, 1850.0.	Mean Declination, 1850.0.
									h. m. s.	s.	s.			r.	' "	"	"	h. m. s.	° ' "
1	9	..	..	..	37.	..	5.3	..	8 33 36.69	−12.37	−1.00	IV.	5	10.150	−25 6.41	−29.86	−3.01	8 33 23.32	−30 4 29.3
2	9	..	..	6.2	21.	35.2	..	..	36 20.66	12.34	1.00	I.	5	5.040	22 29.15	30.23	2.67	36 7.32	30 1 52.0
3	9	..	..	..	3.3	17.5	..	..	38 3.04	12.32	0.97	II.	1	5.223	2 42.32	30.46	0.01	37 49.75	29 42 2.8
4	9	..	..	..	..	19.3	33.4	..	39 4.74	12.31	0.99	II.	3	9.333	14 46.99	30.60	1.61	38 51.44	29 54 9.2
5	9	..	..	25.2	39.1	..	..	..	40 39.20	12.29	1.00	IV.	5	4.365	22 15.73	30.81	2.63	40 25.91	30 1 39.2
6	9	..	..	..	13.3	28.	..	..	41 13.32	12.29	0.99	IV.	4	2.300	16 12.74	30.87	1.82	41 0.04	29 55 35.4
7	9	..	..	..	19.2	33.8	..	..	42 19.23	12.28	1.02	III.	9	8.360	43 15.25	31.02	5.50	42 5.93	30 22 41.8
8	9	..	..	..	29.5	44.	..	..	43 29.48	12.26	1.02	IV.	9	5.336	41 43.31	31.17	5.28	43 16.20	30 21 9.8
9	8	..	..	..	..	29.	43.1	..	44 14.46	12.25	0.98	IV.	3	4.286	12 13.53	31.26	1.27	44 1.23	29 51 36.1
10	9	..	..	19.2	33.5	..	2.2	..	47 33.39	12.22	0.99	IV.	4	3.143	16 35.07	31.69	1.87	47 20.18	55 58.6
11	8	..	..	21.2	35.2	50.	..	..	49 35.32	12.19	0.99	IV.	4	5.343	17 45.67	31.96	2.02	49 22.14	29 57 9.6
12	8.9	..	..	32.2	46.5	1.	..	..	50 46.49	12.18	1.00	IV.	6	8.506	28 24.95	32.11	3.46	50 33.31	30 7 50.5
13	8.9	..	..	57.0	11.6	..	..	..	52 11.53	12.16	1.02	IV.	9	10.092	44 2.29	32.28	5.60	51 58.35	23 30.2
14	9	..	..	..	2.2	17.	31.2	..	53 2.34	12.15	1.00	IV.	5	10.485	25 23.30	32.38	3.05	52 49.19	4 48.7
15	9	..	..	..	2.8	16.2	30.6	..	54 2.08	12.14	1.01	IV.	6	10.056	29 2.77	32.51	3.54	53 48.93	8 28.8
16	9	..	..	..	21.0	34.8	49.3	..	55 20.58	12.13	1.00	IV.	6	9.316	28 45.62	32.67	3.51	55 7.45	8 11.8
17	8	..	..	..	31.8	45.8	0.	..	56 31.40	12.12	1.02	IV.	10	4.324	46 13.36	32.81	5.90	56 18.26	30 25 42.1
18	9	..	..	..	44.8	59.2	..	..	57 44.64	12.10	0.98	IV.	1	5.270	2 44.88	32.96	0.01	57 31.56	29 42 7.9
19	9	..	..	..	..	12.	25.8	..	8 59 57.29	12.07	1.01	IV.	8	5.314	37 13.50	33.24	4.66	8 59 44.21	30 16 41.4
20	9	..	..	15.3	29.6	..	..	..	9 2 29.43	12.04	0.98	III.	3	4.013	11 59.72	33.53	1.24	9 2 16.41	29 51 24.5
21	9	..	..	..	6.	..	..	..	3 5.84	12.03	0.98	IV.	2	6.213	8 11.26	33.60	0.74	2 52.83	47 35.6
22	7	..	..	..	48.2	2.5	16.6	31.5	3 47.94	12.03	0.98	IV.	1	11.442	5 55.0	33.65	0.44	3 34.93	29 45 19.2
23	9	..	..	40.2	54.7	..	..	..	6 54.63	11.99	1.02	IV.	8	7.262	37 41.09	34.05	4.73	6 41.62	30 17 9.9
24	9	..	..	39.3	54.0	8.2	..	..	7 53.68	11.98	0.99	IV.	4	5.420	17 49.55	34.16	2.03	7 40.71	29 57 15.7
25	8	..	..	..	53.8	8.2	22.5	..	8 53.71	11.97	0.98	IV.	1	10.363	5 20.86	34.28	0.35	8 40.76	29 44 45.5
26	9	..	..	..	..	..	8.3	..	9 39.50	11.96	1.02	IV.	9	8.402	43 17.41	34.37	5.50	9 26.52	30 22 47.3
27	9	..	..	..	..	57.2	11.2	26.2	10 42.60	11.94	0.99	III.	3	7.156	13 37.69	34.49	1.46	10 29.67	29 53 3.6
28	9	..	..	..	26.5	..	..	..	13 26.34	11.91	1.01	VI.	6	10.486	29 24.26	34.79	3.59	13 13.42	30 8 52.6
29	9	..	..	20.2	35.0	..	..	..	14 34.68	11.89	1.00	IV.	5	10.572	25 27.69	34.92	3.06	14 21.79	4 55.7
30	8	..	..	0.5	15.	29.2	43.2	..	15 14.84	11.88	1.01	IV.	7	4.302	30 43.39	34.99	3.59	15 1.95	30 10 12.0
31	9	..	..	21.	35.	49.2	..	..	16 34.89	11.87	0.99	IV.	3	7.153	13 37.59	35.14	1.46	16 22.03	29 53 4.2
32	9	..	..	46.2	..	..	29.5	..	17 0.56	11.86	0.98	IV.	3	2.520	11 24.83	35.19	1.16	16 47.72	50 51.2
33	9	..	..	..	3.8	..	..	..	18 3.64	11.85	0.98	IV.	2	14.520	12 28.77	35.30	1.31	17 50.81	51 55.4
34	9	..	..	..	..	31.5	45.2	..	19 16.75	11.83	0.99	V.	3	7.36	13 47.99	35.44	1.48	19 3.93	29 53 14.9
35	9	..	..	..	..	..	51.	6.	19 22.41	11.83	1.00	VI.	5	11.545	25 56.39	35.45	3.12	19 9.58	30 5 25.0
36	9	..	..	..	..	..	..	26.3	20 43.05	11.81	1.02	V.	8	12.430	40 20.78	35.59	5.09	20 30.22	30 19 51.5
37	9	..	..	..	..	34.2	48.4	..	24 19.70	11.76	0.98	III.	1	7.589	4 1.42	35.97	0.18	24 6.96	29 43 27.6
38	9	..	..	..	..	..	40.5	..	25 11.81	11.75	0.98	VI.	3	4.142	12 6.09	36.07	1.26	24 59.08	51 33.4
39	9	..	..	..	..	7.2	..	36.	25 52.59	11.73	0.99	V.	4	4.343	17 15.37	36.14	1.94	25 39.87	29 56 43.5
40	9	..	..	..	42.5	57.0	10.2	..	28 42.11	11.71	1.01	III.	7	4.590	31 28.12	36.43	3.88	28 29.39	30 10 58.4
41	9	..	..	..	..	..	29.5	..	29 0.70	11.70	1.01	VI.	8	1.462	34 49.46	36.47	4.33	28 47.99	14 20.3
42	9	..	..	..	..	..	..	56.	29 12.58	11.70	1.00	VII.	6	4.35	26 15.63	36.49	3.17	28 59.88	5 45.3
43	7.8	..	..	45.	59.2	13.8	..	..	30 59.32	11.68	1.02	IV.	8	9.468	38 21.74	36.67	4.82	30 46.62	17 53.2
44	8.9	..	..	..	..	42.8	57.	11.5	31 28.20	11.67	0.99	V.	4	11.428	20 51.43	36.72	2.42	31 15.54	0 20.6
45	9	..	..	..	..	..	..	44.8	31 1.51	11.68	1.01	VII.	7	15.220	36 41.85	36.67	4.59	30 48.82	30 16 13.1
46	9	..	..	..	42.8	..	..	..	33 42.64	11.64	0.98	IV.	2	10.39	9 50.96	36.94	0.96	33 30.02	29 49 17.9
47	9	..	..	..	58.2	2.5	..	..	33 58.06	11.63	1.01	IV.	7	12.044	35 2.67	36.97	4.36	33 45.42	30 14 34.0
48	9	..	..	..	..	..	1.0	..	9 34 32.30	−11.62	−0.99	VI.	3	6.516	−13 25.45	−37.02	−1.43	9 34 19.69	−29 52 53.9

CORRECTIONS.

Date.		Corr. of Clock.	Hourly rate.	m	n	c
1847.	h.	s.	s.	s.	s.	s.

INSTRUMENT READINGS.

	Date.	Barom.	Thermom. At.	Thermom. Ex.
	1847. h. m.	in.	°	°
Zone 104	Feb. 5. 9 5	29.944	79.0	27.5

REMARKS.

Feb. 5, 8h. No observations were made for determining the instrumental corrections.
(104) 19. Micrometer reading assumed as 6r.314 instead of 5r.314.
(104) 30. Micrometer reading assumed as 3r.302 instead of 4r.302.
(104) 43. Micrometer reading assumed as 8r.468 instead of 9r.468.
(104) 46. Micrometer reading assumed as 9r.39 instead of 10r.39.
(104) 47. Transit over T. V assumed to have been at 12s.5 instead of 2s.5.

ZONE 105. APRIL 3. K. BELT, −26° 48′. D_0 = −26° 27′ 40″.

No.	Mag.	Seconds of Transit I.	II.	III.	IV.	V.	VI.	VII.	T.	a_1	a_2	Micrometer.			i	d_1	d_2	Mean Right Ascension, 1850.0.	Mean Declination, 1850.0.
									h. m. s.	s.	s.			r.	′ ″	″	″	h. m. s.	° ′ ″
1	9	. .	13.5	. .	41.4	. .	. .	. .	8 59 41.36	−19.31	+0.01	IV.	4	2.1	−15 58.11	−23.82	−2.13	8 59 22.06	−26 44 4.06
2	9	. .	24.2	38.7	52.4	. .	. .	. .	9 1 52.27	19.31	+0.02	V.	3	7.56	13 58.08	24.45	1.94	9 1 32.98	42 4.47
3	11	. .	26.6	. .	. .	. .	. .	. .	4 54.61	19.31	0.00	III.	5	3.20	21 37.10	25.33	2.67	4 35.30	49 45.10
4	8	. .	17.2	31.3	45.0	. .	. .	. .	9 44.94	19.31	+0.02	IV.	2	2.23	6 11.10	26.71	1.23	9 25.65	34 19.04
5	9	. .	26.3	. .	54.2	. .	. .	. .	11 54.13	19.31	+0.02	IV.	2	11.59	11 1.55	27.34	1.67	11 34.64	26 39 10.56
6	10	. .	. .	31.2	. .	59.2	. .	. .	13 45.27	19.31	−0.02	V.	8	12.54	40 26.34	27.87	4.46	13 25.94	27 8 38.69
7	9	. .	. .	. .	17.4	31.2	. .	. .	15 17.25	19.31	−0.02	VI.	9	6.14	42 3.52	28.31	4.64	14 57.92	27 10 16.47
8	9	. .	. .	30.9	44.9	. .	. .	. .	16 44.61	19.31	+0.02	VI.	1	8.1	4 2.37	28.73	1.04	16 25.32	26 32 12.14
9	9	. .	. .	. .	19.2	. .	46.8	. .	23 18.96	19.31	−0.01	VI.	7	13.54	35 57.76	30.60	4.05	22 59.64	27 4 12.41
10	7.8	. .	54.8	8.9	. .	. .	. .	. .	26 22.70	19.31	+0.02	IV.	4	3.45	16 50.55	31.46	2.21	26 3.41	26 45 4.22
11	8	. .	. .	. .	. .	51.2	4.8	. .	26 37.04	19.31	−0.01	VI.	6	7.6	32 32.04	31.52	3.72	26 17.72	27 0 47.28
12	8	. .	15.7	30.1	44.1	57.8	. .	. .	29 43.64	19.31	0.00	V.	5	5.47	22 51.22	32.40	2.79	29 24.53	26 51 6.41
13	4	8.4	22.3	36.4	50.7	. .	. .	. .	37 50.57	19.30	−0.01	III.	8	5.24	36 39.43	34.65	4.12	37 31.26	27 4 58.20
14	8	. .	. .	. .	. .	. .	53.0	7.2	38 25.11	19.30	0.00	VI.	6	8.23	28 10.86	34.81	3.30	38 5.81	26 56 28.97
15	8	. .	. .	. .	58.2	. .	26.1	. .	39 58.08	19.30	+0.02	VI.	1	13.15	6 40.69	35.24	1.26	39 38.80	34 57.19
16	9	. .	. .	. .	. .	41.4	. .	9.3	41 27.29	19.29	0.00	VII.	6	4.4	26 0.05	35.64	3.09	41 8.00	54 18.78
17	8	. .	53.2	. .	21.4	. .	. .	. .	9 59 21.15	19.26	+0.02	IV.	1	8.56	4 30.28	40.44	1.07	9 59 1.91	32 51.79
18	9	. .	. .	27.0	. .	55.0	. .	. .	10 9 40.97	19.24	0.00	VI.	6	10.39	29 19.45	43.11	3.41	10 9 21.73	57 45.97
19	9	. .	. .	3.8	. .	31.8	. .	. .	11 17.66	19.24	+0.01	V.	4	9.42	19 50.53	43.52	2.50	10 58.43	48 16.55
20	9	. .	. .	. .	. .	49.5	. .	17.6	17 35.36	19.22	0.02	VI.	2	13.11	11 37.67	45.11	1.73	17 16.16	40 4.51
21	7.8	. .	. .	35.5	. .	23.1	36.8	. .	27 9.02	19.19	+0.02	V.	1	12.8	6 7.04	47.45	1.22	26 49.85	34 35.71
22	7	17.4	31.8	45.8	. .	. .	. .	. .	29 59.72	19.18	0.00	IV.	5	8.38	24 17.50	48.15	2.93	29 40.54	52 48.58
23	5	. .	. .	. .	. .	45.3	58.8	13.3	30 31.05	19.18	+0.02	V.	2	9.29	9 45.87	48.27	1.56	30 11.89	38 15.70
24	9	. .	. .	. .	34.0	. .	1.8	. .	10 35 33.91	−19.16	+0.01	VI.	3	4.42	−12 20.13	−49.45	−1.79	10 35 14.76	−26 40 51.37

ZONE 106. APRIL 6. K. D_0 = −26° 27′ 10″.

No.	Mag.	Seconds of Transit I.	II.	III.	IV.	V.	VI.	VII.	T.	a_1	a_2	Micrometer.			i	d_1	d_2	Mean Right Ascension, 1850.0.	Mean Declination, 1850.0.
1	8	45.0	. .	13.5	27.3	41.4	. .	. .	9 50 27.27	−19.66	+0.01	III.	4	4.27	−18 12.20	−25.99	−2.20	9 50 7.62	−26 45 50.4
2	9	. .	. .	. .	. .	. .	. .	37.0	50 54.64	19.66	0.02	VII.	2	3.45	7 52.54	26.06	1.00	50 35.00	35 29.60
3	8	. .	53.3	7.5	21.3	. .	. .	. .	9 59 21.13	19.65	+0.02	IV.	1	10.30	5 17.68	27.41	0.83	9 59 1.50	26 32 55.92
4	10	. .	. .	. .	. .	1.4	. .	. .	10 2 47.44	19.64	−0.02	VII.	9	6.39	42 15.88	27.95	4.94	10 2 27.78	27 9 58.77
5	9	. .	12.9	. .	40.8	. .	8.4	. .	9 40.71	19.63	−0.01	VII.	7	2.22	30 8.59	29.01	3.57	9 21.07	26 57 51.16
6	9	. .	. .	. .	17.7	. .	45.4	. .	11 17.54	19.62	0.00	VII.	5	1.22	20 37.23	29.25	2.50	10 57.92	48 18.98
7	10	. .	. .	. .	30.2	. .	. .	. .	13 30.05	19.62	0.00	V.	4	10.8	20 3.63	29.57	2.44	13 10.43	47 45.64
8	9	. .	. .	21.4	. .	49.2	. .	. .	17 35.10	19.61	+0.01	VI.	3	4.51	12 24.64	30.17	1.60	17 15.50	40 6.41
9	9	. .	. .	. .	. .	35.3	48.8	. .	18 21.09	19.61	0.00	VI.	5	8.4	24 0.17	30.29	2.88	18 1.48	51 43.34
10	9	. .	. .	44.8	58.7	. .	. .	. .	21 58.48	19.60	+0.02	IV.	2	5.24	7 42.37	30.80	1.09	21 38.90	35 24.26
11	9.10	. .	. .	20.2	. .	47.9	. .	. .	35 33.85	19.55	0.01	IV.	3	6.11	13 5.17	32.66	1.67	35 14.31	40 49.50
12	9	. .	. .	. .	. .	. .	1.8	. .	36 34.00	19.55	0.02	VI.	1	6.11	3 6.86	32.79	0.59	36 14.47	30 50.26
13	8.9	. .	30.0	43.8	57.8	. .	. .	. .	38 57.71	19.54	+0.01	V.	3	11.43	15 52.53	33.11	1.98	38 38.18	43 37.62
14	10	. .	. .	13.5	. .	. .	. .	. .	40 27.46	19.54	−0.01	VII.	7	3.58	30 56.98	33.31	3.66	40 7.91	26 58 43.95
15	5.6	. .	41.8	55.8	. .	24.0	. .	. .	42 9.97	19.53	−0.01	V.	9	1.47	39 49.00	33.52	4.67	41 50.43	27 7 37.19
16	8	. .	. .	57.3	11.3	25.3	. .	. .	45 11.15	19.52	+0.01	VI.	4	7.50	18 53.90	33.89	2.31	44 51.64	26 46 40.10
17	9	. .	32.3	. .	. .	14.4	. .	. .	49 0.41	19.50	−0.01	VI.	7	9.1	33 30.00	34.37	3.95	48 40.90	27 1 18.32
18	10	. .	. .	24.2	. .	. .	. .	. .	52 37.90	19.49	+0.02	VII.	3	6.1	12 59.71	34.79	1.67	52 18.43	26 40 46.17
19	8	. .	54.3	8.8	. .	36.5	. .	. .	56 22.40	19.47	0.01	V.	3	9.48	14 54.54	35.23	1.87	56 2.91	42 41.64
20	3	. .	. .	. .	26.1	. .	. .	. .	10 58 25.95	−19.46	+0.02	III.	1	2.31	−1 16.09	−35.47	−0.37	10 58 6.51	−26 29 1.93

CORRECTIONS.

Date.	Corr. of Clock.	Hourly rate.	m	n	c
1847. h.	s.	s.	s.	s.	s.
April 3, 9	ƒ 25.21	/ 0.016	+ 0.083	− 0.013	− 0.121
April 6, 9	ƒ 25.70	g 0.002	+ 0.277	− 0.023	− 0.121

INSTRUMENT READINGS.

	Date.	Barom.	Thermom. At.	Thermom. Ex.
	1847. h. m.	in.	°	°
Zone 105	April 3. 8 59	29.64	57.0	52.7
Zone 106	April 6, 9 50	29.524	55.4	47.6
	10 58	29.524	55.0	46.4

REMARKS.

(105) 11. Micrometer thread assumed as 7 instead of 6.
(106) 1. Micrometer reading assumed as 6r.27 instead of 4r.27.
(106) 2. Micrometer reading assumed as 5r.45 instead of 3r.45.

ZONE 107. APRIL 9. K. BELT, −30° 38′. D₀=−30° 18′ 30″.

No.	Mag.	SECONDS OF TRANSIT. I.	II.	III.	IV.	V.	VI.	VII.	T.	a_1	a_2	MICROMETER.			i	d_1	d_2	Mean Right Ascension, 1850.0.	Mean Declination, 1850.0.
									h. m. s.	s.	s.			r.	′ ″	″	″	h. m. s.	° ′ ″
1	9	55.8	10.5	. .	. .	. .	. .	. .	9 6 39.77	−19.22	−0.01	III.	8	3.58	−35 56.05	−18.46	−4.45	9 6 20.54	−30 54 48.96
2	10	. .	. .	. .	. .	. .	. .	31.7	7 47.80	19.22	+0.02	VII.	2	9.51	9 56.55	18.81	1.02	7 28.60	28 46.38
3	9	15.1	30.0	44.7	59.2	. .	. .	. .	11 59.08	19.22	0.00	IV.	6	7.14	27 36.24	20.11	3.34	11 39.86	46 29.69
4	10	. .	. .	. .	. .	40.0	. .	. .	12 25.44	19.22	0.00	V.	5	8.4	24 0.31	20.24	2.86	12 6.22	30 42 53.41
5	10	. .	. .	. .	52.3	. .	. .	. .	13 52.14	19.22	−0.02	IV.	9	11.28	44 42.01	20.69	5.63	13 32.90	31 3 38.33
6	3	. .	. .	. .	43.2	. .	. .	. .	14 43.04	19.22	−0.03	IV.	10	9.14	48 35.35	20.95	6.16	14 23.79	31 7 32.46
7	8	. .	25.2	40.0	54.6	. .	. .	. .	19 54.37	19.22	0.00	III.	5	8.42	24 19.47	22.54	2.90	19 35.15	30 43 14.91
8	8	. .	58.9	13.6	28.1	. .	. .	. .	21 28.05	19.22	−0.01	V.	7	6.40	32 19.05	23.01	3.96	21 8.82	51 16.02
9	10	. .	. .	. .	. .	. .	. .	55.3	22 11.40	19.22	+0.01	VII.	2	8.20	9 10.67	23.24	0.93	21 52.19	28 4.84
10	7	. .	47.2	2.0	16.6	. .	. .	. .	24 16.47	19.22	−0.01	IV.	7	11.17	34 38.77	23.87	4.27	23 57.24	30 53 36.91
11	9	. .	. .	. .	. .	56.0	. .	. .	25 41.55	19.21	−0.03	III.	10	10.40	49 18.67	24.30	6.26	25 22.31	31 8 19.23
12	9	18.1	32.7	. .	1.9	. .	. .	. .	29 1.85	19.21	0.00	IV.	6	4.45	26 21.11	25.31	3.17	28 42.64	30 45 19.59
13	8	. .	. .	. .	54.0	8.7	22.7	. .	29 53.92	19.21	+0.01	VII.	3	9.39	14 49.61	25.58	1.65	29 34.72	30 33 46.84
14	9	. .	. .	20.1	34.7	. .	. .	. .	36 34.66	19.20	−0.02	IV.	9	7.43	42 48.56	27.57	5.38	36 15.44	31 1 51.51
15	9	. .	. .	. .	. .	21.1	35.5	50.3	37 6.60	19.20	−0.01	VII.	7	10.12	34 5.55	27.73	4.20	36 47.39	30 53 7.48
16	9	. .	. .	. .	. .	27.8	. .	57.2	38 13.35	19.20	0.00	VII.	5	7.14	23 34.69	28.07	2.81	37 54.15	42 35.57
17	8	. .	. .	2.8	17.7	. .	. .	. .	40 17.30	19.20	+0.01	IV.	3	11.56	15 59.14	28.69	1.82	40 58.11	34 59.65
18	7	. .	. .	. .	50.4	. .	19.3	34.0	40 50.27	19.20	0.01	VI.	3	9.57	15 29.19	28.85	1.75	40 31.08	34 29.79
19	8	13.7	28.5	43.4	. .	. .	. .	. .	43 57.60	19.20	+0.01	IV.	4	11.43	20 51.58	29.77	2.44	43 38.41	39 53.79
20	8	. .	. .	33.7	48.2	. .	17.0	. .	44 48.08	19.20	0.00	VI.	6	10.52	29 25.97	30.02	3.58	44 28.88	48 29.57
21	9	. .	. .	. .	. .	. .	53.8	. .	46 24.92	19.19	+0.01	VII.	3	6.39	13 18.85	30.49	1.46	46 5.74	32 20.80
22	7	. .	. .	. .	. .	. .	. .	40.5	49 56.53	19.19	0.02	VII.	1	7.35	3 48.98	31.53	0.25	49 37.36	22 50.76
23	7	. .	. .	. .	37.5	. .	6.0	21.2	9 53 37.28	19.18	0.01	VII.	4	7.41	18 49.10	32.60	2.18	9 53 18.11	37 53.88
24	8.9	. .	. .	. .	. .	27.8	. .	. .	10 2 13.15	19.17	+0.02	VI.	1	6.2	3 2.34	35.08	0.15	10 2 54.00	22 7.57
25	8	. .	. .	50.3	. .	19.6	. .	. .	7 5.01	19.15	0.00	V.	6	9.15	28 37.20	36.48	3.47	6 45.86	47 47.15
26	8	. .	. .	1.7	16.1	30.7	. .	. .	9 16.06	19.14	0.00	V.	5	9.57	24 57.28	37.10	2.98	8 56.92	44 7.36
27	8	. .	8.4	23.4	38.0	. .	. .	. .	13 37.98	19.13	−0.01	IV.	7	7.11	32 34.72	38.32	4.00	13 18.84	51 47.14
28	7	. .	. .	. .	. .	. .	12.7	27.6	13 43.84	19.13	−0.01	VII.	7	3.25	30 40.31	38.35	3.75	13 24.70	49 52.41
29	9	. .	50.2	. .	. .	. .	. .	. .	17 19.34	19.12	0.00	II.	6	9.18	28 38.56	39.35	3.47	17 0.22	47 51.38
30	10	. .	. .	. .	. .	. .	59.3	14.3	17 30.48	19.12	0.00	VII.	6	6.10	27 3.52	39.40	3.26	17 11.36	30 46 16.18
31	8	. .	. .	. .	. .	48.9	2.7	. .	25 34.05	19.09	−0.02	VII.	10	2.16	45 4.13	41.61	5.68	25 14.94	31 4 21.42
32	7	. .	. .	57.5	12.2	26.4	. .	. .	28 11.86	19.08	+0.01	V.	3	9.51	14 56.05	42.33	1.67	27 52.79	30 34 10.05
33	9	. .	. .	. .	49.3	3.7	. .	. .	29 49.15	19.08	0.00	VI.	6	6.30	27 13.86	42.70	3.29	29 30.07	46 29.85
34	5	. .	. .	. .	. .	51.1	5.2	20.3	30 36.52	19.08	−0.01	VII.	8	11.47	39 52.15	42.98	4.98	30 17.43	30 59 10.11
35	8	. .	. .	. .	41.3	. .	10.0	. .	34 41.05	19.07	0.02	VI.	9	10.6	44 0.48	44.05	5.54	34 21.96	31 3 20.07
36	6	. .	. .	. .	38.2	52.3	6.7	. .	36 37.84	19.06	−0.01	VI.	8	10.16	39 6.52	44.57	4.87	36 18.77	30 58 25.96
37	9	. .	56.8	. .	26.2	40.6	. .	. .	38 25.93	19.05	+0.01	VI.	3	7.28	13 43.79	44.93	1.52	38 6.89	33 0.24
38	7	. .	. .	37.1	51.6	6.3	. .	. .	40 51.59	19.05	0.00	IV.	6	10.58	29 29.19	45.68	3.59	40 32.54	48 48.46
39	6	. .	. .	. .	. .	46.9	1.0	15.7	41 32.17	19.04	−0.01	VII.	7	11.3	34 31.26	45.85	4.26	41 13.12	53 51.37
40	9	. .	. .	. .	. .	49.1	. .	18.4	42 34.60	19.04	0.00	VI.	5	7.8	23 31.91	46.11	2.80	42 15.56	42 50.82
41	7	. .	36.3	50.3	5.7	. .	. .	. .	49 5.35	19.00	−0.01	V.	8	11.42	39 50.03	47.77	4.97	48 46.34	59 12.77
42	7	. .	. .	. .	5.3	20.3	34.0	. .	50 5.32	19.00	+0.02	VI.	1	9.29	4 46.71	48.02	0.37	49 46.34	30 24 5.10
43	5.6	9.0	24.2	38.7	. .	. .	. .	. .	10 53 53.32	18.98	−0.01	IV.	9	7.50	42 52.09	48.93	5.39	10 53 34.33	31 2 16.41
44	11	. .	. .	50.3	. .	. .	. .	. .	11 2 4.48	18.95	+0.01	VI.	3	2.56	11 26.65	50.87	1.22	11 1 45.54	30 30 48.74
45	8	27.6	42.6	57.3	. .	. .	. .	. .	11 11.77	18.90	−0.01	III.	8	4.57	36 25.81	52.92	4.51	10 52.86	55 53.24
46	9	. .	. .	. .	. .	2.7	. .	32.	11 48.12	18.90	+0.01	VII.	3	5.43	12 50.60	53.05	1.40	11 29.23	32 15.05
47	9	51.4	6.3	21.3	. .	. .	. .	. .	14 35.42	18.89	0.00	IV.	5	5.17	22 36.14	53.65	2.68	14 16.53	30 42 2.47
48	8	. .	. .	12.7	. .	41.4	55.4	. .	21 26.89	18.85	−0.01	VI.	9	6.49	42 21.14	55.10	5.31	21 8.03	31 1 51.55
49	2.3	. .	27.9	42.7	57.3	11.3	. .	. .	11 25 57.13	−18.82	−0.01	VI.	9	6.29	−42 11.05	−56.01	−5.29	11 25 38.30	−31 1 42.35

CORRECTIONS.

Date.		Corr. of Clock.	Hourly rate.	m	n	c
1847.	h.	s.	s.	s.	s.	s.
April 9,	9	ƒ25.67	ℓ0.000	+ 0.260	− 0.055	− 0.121

INSTRUMENT READINGS.

	Date.		Barom.	THERMOM. At.	THERMOM. Ex.
	1847.	h. m.	in.	°	°
Zone 107	April 9,	9 6	29.736	63.4	55.7

REMARKS.

(107) 18. Micrometer reading assumed as 10ʳ.57 instead of 9ʳ.57.

ZONE 107. APRIL 9. K. BELT, −30° 38′. D_0 = −30° 18′ 30″—Continued.

No.	Mag.	Seconds of Transit I.	II.	III.	IV.	V.	VI.	VII.	T.	a_1	a_2	Micrometer.			i	d_1	d_2	Mean Right Ascension, 1850.0.	Mean Declination, 1850.0.
									h. m. s.	s.	s.			r.	′ ″	″	″	h. m. s.	° ′ ″
50	4.5	..	10.3	24.8	39.3	..	..	..	11 30 39.06	−18.80	+0.02	III.	1	7.24	− 3 43.83	−56.95	−0.20	11 30 20.28	−30 23 10.98
51	9	..	39.3	54.0	..	..	..	..	37 8.44	18.76	0.00	IV.	6	8.7	28 2.97	58.21	3.40	36 49.68	47 34.58
52	9	..	..	..	52.0	..	20.6	..	37 51.72	18.75	−0.01	VI.	8	6.17	37 6.00	58.35	4.61	37 32.96	56 38.96
53	8	..	..	..	..	27.8	41.8	56.8	40 12.99	18.74	+0.01	VI.	2	1.38	5 48.21	58.80	0.49	39 54.26	25 17.50
54	9	..	..	9.4	..	..	..	..	46 23.86	18.70	0.00	IV.	6	8.44	28 21.63	59.94	3.44	46 5.16	47 55.01
55	8	..	..	15.7	30.1	44.6	..	..	47 30.07	18.69	0.00	IV.	6	12.48	30 24.65	60.14	3.72	47 11.38	49 58.51
56	8	..	..	31.3	..	0.4	..	..	50 45.82	18.67	0.00	IV.	6	10.34	29 17.10	60.73	3.57	50 27.15	48 51.40
57	7	56.2	10.9	25.7	..	..	..	..	55 40.15	18.64	0.00	III.	7	5.42	31 49.80	61.59	3.90	55 21.51	51 25.29
58	8	..	..	..	..	18.4	32.7	47.8	11 56 3.88	−18.64	0.00	VII.	4	12.2	−21 0.71	−61.66	−2.47	11 55 45.24	−30 40 34.84

ZONE 108. APRIL 13. K. BELT, −29° 23′. D_0 = −29° 6′ 10″.

No.	Mag.	I.	II.	III.	IV.	V.	VI.	VII.	T.	a_1	a_2	Micrometer.			i	d_1	d_2	Mean Right Ascension, 1850.0.	Mean Declination, 1850.0.
1	9	..	0.1	14.0	..	..	..	..	10 22 43.22	−19.87	−0.01	V.	8	11.30	−39 43.98	−31.46	−4.77	10 22 23.34	−29 46 30.21
2	5	27.5	41.7	56.0	10.2	..	..	..	10 25 10.37	−19.86	0.00	IV.	5	10.00	−24 58.85	−31.72	−3.00	10 24 50.51	−29 31 43.57

ZONE 109. APRIL 16. K. BELT, −29° 46′. D_0 = −29° 21′ 40″.

No.	Mag.	I.	II.	III.	IV.	V.	VI.	VII.	T.	a_1	a_2	Micrometer.			i	d_1	d_2	Mean Right Ascension, 1850.0.	Mean Declination, 1850.0.
1	9	..	..	..	..	..	..	24.8	9 14 41.68	−20.10	−0.02	VII.	9	7.59	−42 56.20	−18.24	−5.25	9 14 21.56	−30 4 59.69
2	8.9	..	..	..	42.4	56.5	..	..	16 42.17	20.11	−0.01	V.	7	4.16	31 6.44	18.67	3.77	16 22.05	29 53 8.88
3	9	..	..	..	..	..	36.6	51.3	17 7.96	20.11	0.00	VII.	6	9.46	28 52.46	18.70	3.49	16 47.85	50 54.71
4	7	..	32.1	46.8	0.7	..	..	..	24 0.70	20.11	+0.02	V.	2	4.34	7 17.12	20.22	0.83	23 40.61	29 18.17
5	9	..	..	..	..	..	55.7	10.4	24 27.03	20.11	0.00	VI.	5	3.2	21 27.88	20.32	2.57	24 6.92	43 30.77
6	10	..	..	..	..	..	..	43.3	26 0.10	20.11	−0.01	VII.	7	11.19	34 39.35	20.69	4.22	25 39.98	56 44.22
7	9	..	..	..	34.8	..	..	..	28 34.64	20.11	+0.02	IV.	1	7.19	3 41.36	21.19	0.40	28 14.55	29 25 42.95
8	9	..	..	..	..	..	49.3	..	29 20.51	20.11	−0.02	VI.	9	9.38	43 46.36	21.39	5.35	29 0.38	30 5 53.06
9	10	..	..	..	..	..	55.1	..	30 26.40	20.11	0.00	VI.	6	5.58	26 57.72	21.58	3.26	30 6.29	29 49 2.56
10	6	..	..	..	..	..	..	45.6	31 2.09	20.11	+0.02	VII.	2	10.31	10 16.75	21.70	1.20	30 42.00	32 19.65
11	7	6.8	21.4	36.0	..	..	..	..	33 50.30	20.11	0.00	III.	6	6.35	27 16.53	22.28	3.29	33 30.19	29 49 22.10
12	10	..	..	..	19.2	..	..	..	34 19.04	20.11	−0.02	IV.	9	4.55	41 23.85	22.37	5.01	33 58.91	30 3 31.26
13	5	..	10.2	..	39.2	..	..	..	35 38.99	20.10	+0.01	IV.	3	10.28	15 14.76	22.65	1.80	35 18.90	29 37 19.21
14	9	..	18.2	32.7	..	..	..	..	37 46.89	20.10	0.01	III.	3	11.17	15 39.42	23.10	1.85	37 26.80	37 44.37
15	6	..	38.2	52.6	7.0	..	..	..	39 6.78	20.10	+0.02	IV.	2	7.25	8 43.38	23.37	1.01	38 46.70	30 47.76
16	10	..	..	..	0.0	..	28.2	..	39 59.66	20.10	−0.01	VII.	7	4.49	31 22.70	23.55	3.80	39 39.55	53 30.05
17	9	..	..	..	..	..	23.4	..	40 54.78	20.10	+0.01	VII.	3	5.33	12 45.58	23.73	1.50	40 34.69	34 30.81
18	7	..	..	..	16.9	31.1	..	..	42 16.69	20.10	+0.01	VI.	3	11.1	15 31.21	24.01	1.83	41 56.59	37 37.05
19	9	..	..	..	..	..	33.9	45.5	43 5.18	20.10	0.00	VI.	5	4.49	22 21.84	24.17	2.66	42 45.08	29 44 28.67
20	10	..	..	..	..	..	59.8	..	45 31.04	20.09	−0.01	VI.	8	11.11	39 34.25	24.65	4.83	45 10.94	30 1 43.73
21	9	..	..	..	19.3	..	47.6	..	47 19.00	20.09	−0.01	VI.	8	6.39	37 17.11	25.01	4.54	46 58.90	29 59 26.66
22	7	..	..	..	..	30.6	..	..	50 16.07	20.09	+0.02	V.	1	5.15	2 38.78	25.60	0.27	49 56.00	24 44.65
23	9	..	..	53.2	..	21.8	..	..	52 7.26	20.09	+0.02	VI.	2	6.18	8 9.41	25.96	0.92	51 47.19	30 16.29
24	8	..	..	..	..	11.3	25.3	40.2	52 56.83	20.09	−0.01	VI.	8	4.39	36 16.60	26.11	4.40	52 36.73	29 58 27.11
25	10	..	..	..	..	40.7	..	..	54 26.34	20.09	−0.01	V.	9	5.16	41 34.39	26.40	5.08	54 6.24	30 3 45.87
26	5.6	..	58.8	..	27.8	..	..	..	56 27.67	20.08	0.00	IV.	6	10.21	29 10.53	26.80	3.52	56 7.59	29 51 20.85
27	9	..	..	..	..	22.2	35.8	..	9 57 7.44	20.08	+0.01	VI.	2	8.51	9 26.56	26.92	1.10	9 56 47.37	31 34.58
28	9	..	..	..	39.8	..	..	..	10 6 39.64	20.07	0.02	IV.	1	10.30	5 17.68	28.73	0.58	10 6 19.59	27 26.99
29	8	..	..	..	..	..	29.1	43.8	7 0.48	20.07	+0.01	VII.	4	8.14	19 5.77	28.79	2.25	6 40.42	29 41 16.81
30	6	..	..	..	..	30.7	14.6	..	10 8 16.09	−20.06	−0.01	VI.	9	6.28	−42 10.55	−29.03	−5.16	10 7 56.02	−30 4 24.74

CORRECTIONS.

Date.		Corr. of Clock.	Hourly rate.	m	n	c
1847.	h.	s.	s.	s.	s.	s.
April 13,	11	ƒ25.95	ʃ 0.012	+ 0.210	+ 0.016	− 0.121
April 16,	11	ƒ25.99	ℓ 0.002	+ 0.092	+ 0.050	− 0.121

INSTRUMENT READINGS.

	Date.	Barom.	Thermom. At.	Thermom. Ex.
	1847. h. m.	in.	°	°
Zone 108	April 13, 10 22	29.89	58.7	48.0
Zone 109	April 16, 9 15	29.78	54.	46.8

REMARKS.

(108) 1. Transits over T.'s I and II assumed as recorded over T.'s II and III.

ZONE 109. APRIL 16. K. BELT, −29° 46′. D_0 = −29° 21′ 40″—Continued.

No.	Mag.	SECONDS OF TRANSIT. I.	II.	III.	IV.	V.	VI.	VII.	T.	a_1	a_2	MICROMETER.			i	d_1	d_2	Mean Declination, 1850.0.	Mean Right Ascension, 1850.0.
									h. m. s.	s.	s.			r.	′ ″	″	″	° ′ ″	h. m. s.
31	10	. .	. .	47.0	. .	15.7	. .	. .	10 10 1.24	−20.06	+0.01	V.	4	5.35	−17 45.97	−29.35	−2.10	10 9 41.19	−29 39 57.42
32	9	. .	. .	46.7	. .	15.4	. .	. .	14 0.95	20.05	0.00	V.	5	6.37	23 16.44	30.07	2.77	13 40.90	45 29.28
33	8	. .	. .	13.0	27.4	. .	. .	. .	15 27.18	20.05	−0.01	III.	7	9.2	33 30.65	30.32	4.06	15 7.12	55 45.03
34	7.8	. .	. .	. .	3.9	18.2	. .	. .	16 3.77	20.05	−0.01	VI.	7	5.43	31 15.16	30.43	3.79	15 43.71	53 29.38
35	6.7	. .	. .	. .	17.3	31.8	. .	. .	17 17.22	20.05	+0.01	IV.	2	8.30	8 15.66	30.76	1.04	16 57.18	29 30 27.46
36	9	. .	. .	. .	. .	. .	42.7	. .	18 13.92	20.04	−0.01	VI.	9	8.7	43 0.48	30.82	5.27	17 53.87	30 5 16.57
37	10	. .	. .	. .	. .	. .	. .	47.9	19 4.67	20.04	−0.01	VII.	7	7.14	32 35.61	30.97	3.96	18 44.62	29 54 30.74
38	9	. .	. .	. .	. .	. .	. .	32.3	19 49.04	20.04	0.00	VII.	6	11.47	29 53.47	31.11	3.62	19 29.00	52 8.20
39	9	. .	16.8	31.4	. .	. .	. .	. .	22 45.64	20.03	0.00	III.	5	8.20	24 8.37	31.61	2.88	22 25.61	46 22.86
40	4.5	. .	. .	. .	. .	16.2	30.3	. .	23 1.69	20.03	0.00	VI.	6	8.24	28 11.35	31.66	3.41	22 41.66	50 26.46
41	10	. .	. .	. .	. .	22.4	. .	. .	27 7.91	20.03	+0.01	V.	2	11.7	10 35.28	32.35	1.22	26 47.89	32 28.85
42	9	. .	. .	. .	33.7	. .	. .	. .	28 33.54	20.02	+0.01	IV.	2	13.1	11 32.80	32.60	1.31	28 13.53	29 33 46.74
43	9	. .	. .	32.1	46.6	. .	. .	. .	29 46.52	20.02	−0.01	IV.	8	12.18	40 8.22	32.81	4.89	29 26.49	30 2 25.92
44	7	. .	. .	. .	46.7	. .	14.8	. .	34 46.33	20.01	0.00	VI.	4	15.32	22 46.85	33.63	2.71	34 26.32	29 45 3.19
45	6.7	. .	. .	43.3	58.1	. .	. .	. .	36 57.89	20.00	−0.01	IV.	9	7.10	42 31.92	33.98	5.21	36 37.88	30 4 51.11
46	6.7	29.5	. .	. .	. .	27.3	. .	. .	39 12.94	19.99	−0.01	IV.	9	5.34	41 43.52	34.35	5.10	38 52.94	30 4 2.97
47	10	. .	. .	. .	44.5	. .	. .	. .	40 44.34	19.99	0.00	IV.	6	9.3	28 31.20	34.59	3.44	40 24.35	29 50 49.23
48	8	37.9	52.7	7.0	. .	. .	. .	. .	42 21.53	19.98	−0.01	III.	7	12.17	35 8.97	34.85	4.28	42 1.54	57 28.10
49	7	. .	. .	. .	. .	43.9	. .	. .	42 29.49	19.98	0.00	VI.	6	9.14	28 36.56	34.87	3.45	42 9.51	50 54.88
50	7	. .	. .	. .	. .	. .	. .	25.6	42 40.25	19.98	0.00	VII.	5	6.35	23 15.05	34.90	2.78	42 20.27	45 32.73
51	9	. .	. .	56.9	. .	25.4	. .	. .	46 10.93	19.97	+0.01	IV.	2	11.13	10 38.35	35.45	1.23	45 50.95	32 55.03
52	8	21.3	. .	50.5	4.8	. .	. .	. .	48 4.59	19.97	+0.01	III.	2	9.57	9 59.98	35.74	1.16	47 44.63	32 16.88
53	9	. .	. .	. .	. .	46.6	0.8	. .	48 32.14	19.96	0.00	VII.	6	5.50	26 53.45	35.81	3.26	48 12.18	49 12.52
54	8	. .	22.7	37.2	. .	. .	. .	. .	50 51.25	19.96	+0.02	IV.	2	0.28	5 13.11	36.17	0.56	50 31.31	29 27 29.84
55	9	. .	43.7	48.1	. .	. .	. .	. .	53 12.50	19.95	−0.01	III.	9	10.00	43 57.60	36.52	5.39	52 52.54	30 6 19.51
56	8	. .	. .	. .	44.9	58.9	. .	. .	56 44.57	19.93	+0.02	V.	1	10.20	5 12.58	37.04	0.57	56 24.66	29 27 30.19
57	6.7	. .	. .	. .	. .	56.2	10.4	. .	57 41.74	19.93	−0.01	VI.	8	11.22	39 39.80	37.19	4.85	57 21.80	30 2 1.84
58	6.7	. .	. .	. .	. .	. .	49.9	4.4	10 58 21.11	19.93	+0.01	VII.	3	10.40	15 20.39	37.28	1.80	58 1.19	29 37 39.47
59	5.6	. .	. .	46.2	0.5	. .	28.9	. .	11 0 0.33	19.92	0.00	VII.	5	4.29	22 11.51	37.53	2.65	10 59 40.41	44 31.69
60	9	. .	22.3	36.6	. .	. .	. .	. .	1 51.14	19.91	−0.01	IV.	8	2.28	35 10.72	37.79	4.28	11 1 31.22	57 32.79
61	9	. .	. .	26.3	40.8	. .	. .	. .	3 40.54	19.91	+0.01	IV.	3	12.17	16 9.72	38.05	1.89	3 20.64	38 29.66
62	8	. .	. .	. .	. .	. .	59.9	. .	7 31.22	19.90	0.00	VII.	5	8.51	24 23.62	38.60	2.92	7 11.32	29 46 45.14
63	7	. .	. .	. .	57.0	. .	. .	. .	8 56.84	19.89	−0.01	V.	8	10.55	39 26.33	38.80	4.83	8 36.94	30 1 49.96
64	6	. .	. .	13.3	27.4	. .	. .	. .	12 27.28	19.88	+0.01	IV.	1	15.26	7 46.91	39.28	0.88	12 7.41	29 30 7 07
65	6	. .	. .	. .	. .	. .	26.5	41.8	12 58.11	19.88	+0.01	VII.	3	10.27	15 13.83	39.35	1.78	12 38.24	29 37 34.96
66	7	. .	3.4	17.8	. .	. .	. .	. .	15 32.38	19.86	−0.01	III.	9	6.47	42 20.28	39.69	5.19	15 12.51	30 4 45.16
67	8	. .	. .	. .	0.7	. .	28.8	. .	16 0.28	19.86	−0.01	VI.	9	5.55	41 53.91	39.75	5.14	15 40.41	30 4 18.80
68	9	. .	. .	. .	. .	. .	. .	22.0	16 38.58	19.86	+0.01	VII.	4	5.17	17 36.51	39.84	2.06	16 18.73	29 39 58.41
69	9	. .	. .	33.8	. .	. .	. .	. .	20 47.96	19.84	0.01	IV.	4	5.50	17 33.58	40.39	2.11	20 28.13	40 16.08
70	9	. .	59.2	13.7	. .	. .	. .	. .	23 27.86	19.82	+0.01	III.	3	8.31	14 15.72	40.72	1.68	23 8.05	36 38.12
71	10	. .	. .	. .	. .	. .	18.0	. .	23 49.35	19.82	0.00	VI.	4	7.59	18 58.44	40.77	2.25	23 29.53	29 41 21.46
72	8	. .	. .	. .	24.4	. .	52.5	. .	25 23.97	19.82	−0.01	VI.	10	4.34	46 13.98	40.96	5.69	25 4.14	30 8 40.63
73	7.8	. .	. .	. .	41.9	56.2	. .	. .	26 41.76	19.81	0.00	VI.	6	6.24	27 10.84	41.13	3.28	26 21.95	29 49 35.25
74	7	. .	. .	. .	43.5	58.0	. .	. .	27 57.94	19.81	0.00	VI.	7	4.45	31 20.92	41.25	3.80	27 38.13	29 53 45.97
75	9	13.2	27.4	. .	. .	. .	. .	. .	31 56.64	19.79	−0.01	II.	9	2.51	40 21.13	41.76	4.94	31 36.84	30 2 47.83
76	8	. .	. .	. .	53.2	7.5	. .	. .	32 53.04	19.79	+0.01	VI.	4	3.2	16 28.68	41.87	1.93	32 33.26	29 38 52.48
77	8	. .	. .	. .	. .	. .	. .	11.4	38 28.31	19.76	−0.01	VII.	10	2.1	44 56.58	42.51	5.54	38 8.54	30 7 24.63
78	7	. .	. .	. .	. .	. .	. .	5.1	39 22.01	19.75	−0.01	VII.	10	1.13	44 32.37	42.61	5.48	39 2.25	30 7 0.46
79	9	. .	. .	. .	. .	. .	30.1	. .	11 42 1.40	−19.74	0.00	VII.	6	7.10	−27 33.79	−42.89	−3.33	11 41 41.66	−29 50 0.01

CORRECTIONS.

Date.	Corr. of Clock.	Hourly rate.	m	n	c
1846. h.	s.	s.	s.	s.	s.

INSTRUMENT READINGS.

Date.	Barom.	THERMOM. At.	Ex.
1846. h. m.	in.	°	°

REMARKS.

(109) 35. Micrometer reading assumed as 6 instead of 8.
(109) 50. T. VII assumed as 23ˢ.6 instead of 25ˢ.6.
(109) 52. Minutes of transit assumed as 48 instead of 47.
(109) 55. T. III assumed as 58ˢ.1 instead of 48ˢ.1.
(109) 74. Transits over T.'s III and IV assumed as recorded over T.'s IV and V.

ZONE 109. APRIL 16. K. BELT, −29° 46'. D_o = −29° 21' 40"—Continued.

No.	Mag.	I.	II.	III.	IV.	V.	VI.	VII.	T.	a_1	a_2	MICROMETER.			i	d_1	d_2	Mean Right Ascension, 1850.0.	Mean Declination, 1850.0.
									h. m. s.	s.	s.			r.	' "	"	"	h. m. s.	° ' "
80	4.5	. .	58.1	. .	27.3	. .	. .	. .	11 44 27.11	−19.73	−0.01	IV.	8	5.45	−36 50.06	−43.15	−4.49	11 44 7.37	−29 59 17.70
81	8	22.0	36.6	51.0	. .	. .	. .	. .	53 5.38	19.68	0.00	III.	5	7.12	23 34.08	44.04	2.81	52 45.70	46 0.93
82	8	. .	. .	. .	. .	. .	. .	49.0	53 5.46	19.68	+0.01	VII.	2	5.15	7 37.40	44.04	0.84	52 45.79	30 2.28
83	7	. .	. .	39.1	. .	7.3	. .	. .	54 53.11	19.67	0.00	V.	5	7.35	23 45.68	44.22	2.83	54 33.44	46 12.73
84	9	. .	23.2	. .	52.2	. .	. .	. .	11 59 52.04	19.64	0.00	IV.	6	11.50	29 55.41	44.71	3.63	11 59 32.40	52 23.75
85	7.8	. .	. .	. .	. .	. .	45.8	. .	12 4 17.12	19.61	0.00	VII.	5	7.18	23 36.72	45.10	2.63	12 3 57.51	29 46 4.65
86	9	. .	. .	54.2	. .	. .	. .	. .	11 8.74	19.57	−0.01	IV.	9	6.46	42 19.83	45.69	5.20	10 49.16	30 4 50.72
87	9	. .	. .	. .	3.8	. .	. .	. .	13 3.64	19.56	+0.01	IV.	3	7.58	13 59.12	45.84	1.63	12 44.09	29 36 26.59
88	9	. .	. .	. .	. .	. .	47.4	2.0	13 18.68	19.56	0.00	VII.	5	4.23	22 8.48	45.86	2.61	12 59.12	44 36.95
89	9	. .	. .	. .	. .	. .	. .	43.4	13 59.97	19.55	+0.01	VII.	4	3.23	16 39.02	45.92	1.95	13 40.43	39 6.89
90	5.6	. .	48.2	. .	16.9	. .	. .	. .	16 16.80	19.54	+0.01	IV.	2	5.26	7 43.38	46.10	0.86	15 57.27	30 10.34
91	9	. .	. .	. .	. .	. .	. .	14.3	16 30.88	19.54	0.00	VII.	4	5.50	17 53.15	46.12	2.12	16 11.34	40 21.39
92	9	. .	. .	27.5	. .	. .	. .	. .	21 41.96	19.51	−0.01	IV.	8	6.39	37 17.30	46.51	4.56	21 22.44	59 48.37
93	7.8	. .	. .	. .	37.0	51.0	. .	. .	22 36.70	19.50	0.00	V.	4	11.45	20 52.54	46.57	2.48	22 17.20	43 21.59
94	5.6	. .	. .	. .	. .	35.1	. .	3.7	23 20.56	19.50	0.00	VI.	6	8.47	28 22.95	46.62	3.43	23 1.06	50 53.00
95	9	. .	. .	. .	. .	. .	7.8	. .	30 39.07	19.45	0.00	VI.	7	8.49	33 23.96	47.13	4.06	30 19.62	55 55.15
96	4	. .	. .	. .	. .	. .	. .	8.5	31 25.03	19.44	+0.01	VII.	3	6.41	13 19.87	47.18	1.53	31 5.60	29 35 48.58
97	7	. .	. .	. .	6.8	20.8	. .	. .	35 6.55	19.42	−0.01	V.	9	11.40	44 48.02	47.40	5.51	34 47.12	30 7 20.93
98	6	. .	19.8	. .	48.8	. .	. .	. .	41 48.72	19.38	0.00	IV.	8	6.33	37 14.27	47.79	4.56	41 29.34	29 59 46.62
99	6	. .	. .	. .	. .	55.3	. .	24.0	42 40.76	19.37	0.00	VI.	5	6.12	23 3.69	47.84	2.75	42 21.39	45 34.28
100	8	. .	. .	. .	22.3	36.2	. .	. .	44 21.96	19.36	0.00	IV.	6	6.20	27 9.01	47.92	3.26	44 2.60	49 40.19
101	9	. .	51.8	6.2	. .	. .	. .	. .	52 20.34	19.30	0.00	IV.	5	9.59	24 58.34	48.31	3.00	52 1.04	29 47 29.65
102	8	. .	. .	. .	58.0	12.0	. .	. .	53 57.74	19.29	0.00	IV.	8	9.48	38 52.59	48.37	4.76	53 38.45	30 1 25.72
103	9	. .	. .	. .	57.0	. .	25.6	. .	54 56.92	19.29	0.00	IV.	2	8.37	9 19.70	48.41	1.06	54 37.63	29 31 49.17
104	9	. .	17.2	. .	. .	. .	. .	. .	12 59 45.93	−19.25	0.00	III.	3	10.39	−15 20.27	−48.61	−1.79	12 59 26.68	−29 37 50.67

ZONE 110. APRIL 16. K. BELT, −29° 46'. D_o = −29° 21' 50".

No.	Mag.	I.	II.	III.	IV.	V.	VI.	VII.	T.	a_1	a_2	MICROMETER.			i	d_1	d_2	Mean Right Ascension, 1850.0.	Mean Declination, 1850.0.
1	6	48.9	3.6	. .	. .	. .	. .	. .	14 16 32.63	−18.63	0.00	III.	9	10.53	−44 24.32	−40.27	−5.49	14 16 14.00	−30 7 0.08
2	5.4	52.7	7.1	21.8	. .	. .	. .	. .	24 36.26	18.56	0.00	III.	8	12.39	40 18.75	39.33	4.96	24 17.70	30 2 53.04
3	7	. .	17.2	25.7	. .	. .	. .	. .	28 39.80	18.52	0.00	III.	2	7.19	8 40.31	38.84	0.96	28 21.28	29 31 10.11
4	7	. .	20.6	35.0	49.3	. .	. .	. .	31 49.34	18.49	0.00	IV.	7	3.30	30 43.29	38.45	3.73	31 30.85	53 15.47
5	9	49.9	4.2	. .	. .	. .	. .	. .	34 33.16	18.47	0.00	II.	4	10.45	20 22.15	38.09	2.41	34 14.69	42 52.65
6	7	. .	. .	. .	. .	6.8	. .	. .	34 52.31	18.47	0.00	VI.	2	12.42	11 23.04	38.04	1.28	34 33.84	33 52.36
7	9	. .	. .	. .	. .	1.6	. .	. .	39 32.91	18.42	0.00	VI.	5	9.7	24 31.93	37.41	2.94	39 14.49	47 2.28
8	9	. .	. .	. .	. .	. .	. .	38.3	39 54.99	18.42	0.00	VII.	6	4.57	26 26.73	37.39	3.19	39 36.57	48 57.31
9	5	13.0	27.8	. .	. .	. .	. .	. .	43 56.68	18.38	0.00	II.	8	1.45	34 48.85	36.83	4.24	43 38.30	57 19.92
10	6	. .	34.7	49.2	. .	. .	. .	. .	55 3.58	18.28	0.00	IV.	6	12.42	30 21.63	35.25	3.68	54 45.30	52 50.56
11	5	. .	19.4	34.2	48.6	. .	. .	. .	56 48.48	18.27	0.00	IV.	7	13.00	35 30.70	34.99	4.35	56 30.21	58 0.04
12	8	. .	. .	. .	. .	40.6	54.6	. .	14 57 26.04	18.26	0.00	VII.	6	7.4	27 30.77	34.90	3.33	14 57 7.78	29 49 59.00
13	9	. .	50.2	4.6	. .	. .	. .	. .	15 0 19.20	18.24	+0.01	IV.	9	9.19	43 36.97	34.48	5.38	15 0 0.97	30 6 6.83
14	3.4	. .	32.8	47.4	1.8	. .	. .	. .	10 1.55	18.15	0.00	III.	3	6.23	13 11.17	32.96	1.50	8 43.40	29 35 35.63
15	9	. .	. .	25.6	. .	54.3	. .	. .	13 39.90	18.12	0.00	IV.	6	7.30	27 44.31	32.36	3.35	13 21.78	50 10.02
16	9	. .	. .	. .	. .	. .	55.3	. .	31 26.71	17.96	−0.02	VI.	1	12.16	6 10.93	29.34	0.64	31 8.73	28 30.91
17	8	. .	. .	. .	58.6	. .	27.8	. .	32 58.80	17.94	0.00	V.	4	4.20	17 8.15	29.05	1.96	32 40.86	29 39 29.16
18	7	. .	. .	. .	28.3	. .	56.6	. .	34 27.99	17.93	+0.01	VI.	9	3.38	40 44.83	28.76	5.03	34 10.07	30 3 8.64
19	6	. .	. .	. .	50.2	4.6	. .	. .	36 50.08	17.91	0.00	IV.	3	3.12	11 34.91	28.36	1.31	36 32.17	29 33 54.58
20	9	. .	. .	1.2	15.6	. .	. .	. .	15 40 15.43	−17.88	0.00	IV.	5	4.32	−22 13.46	−27.73	−2.63	15 39 57.55	−29 44 33.82

CORRECTIONS.

Date.		Corr. of Clock.	Hourly rate.	m	n	c
1847.	h.	s.	s.	s.	s.	s.

INSTRUMENT READINGS.

Date.		Barom.	THERMOM.	
			At.	Ex.
1847.	h. m.	in.	°	°

REMARKS.

(109) 101. Transit over T. II assumed as at 51ˢ.8.
(110) 7. Transit over T. VI assumed as recorded over T. V.
(110) 14. Minutes assumed as 9 instead of 10.

ZONE 110. APRIL 16. K. BELT, −29° 46′. D₀ = −29° 21′ 50″—Continued.

No.	Mag.	SECONDS OF TRANSIT. I.	II.	III.	IV.	V.	VI.	VII.	T.	a_1	a_2	MICROMETER.			i	d_1	d_2	Mean Right Ascension, 1850.0.	Mean Declination, 1850.0.
									h. m. s.	s.	s.			r.	′ ″	″	″	h. m. s.	° ′ ″
21	4	..	..	2.3	16.2	..	..	..	15 43 16.14	−17.85	−0.02	IV.	1	6.30	− 3 16.66	−27.16	−0.30	15 42 58.27	−29 25 34.12
22	9	42.8	57.2	..	..	..	..	..	45 26.30	17.83	+0.01	III.	8	4.5	35 59.59	26.76	4.42	45 8.48	58 20.77
23	6	..	..	..	..	..	7.4	22.1	50 38.71	17.78	+0.01	VII.	4	3.16	16 35.50	25.77	1.90	50 20.94	38 53.17
24	9	..	31.3	46.0	..	..	..	..	54 0.25	17.75	−0.01	IV.	6	8.4	28 1.46	25.10	3.37	53 42.49	50 19.93
25	9	..	..	..	31.1	45.5	..	..	54 31.00	17.77	0.00	IV.	4	11.1	20 30.40	25.00	2.43	54 13.25	42 47.83
26	9	15.8	30.5	45.0	..	..	..	..	15 59 59.35	−17.70	0.00	III.	6	9.51	−28 55.35	−27.91	−3.50	15 59 41.65	−29 51 12.76

ZONE 111. APRIL 21. K. BELT, −28° 46′. D₀ = −28° 23′ 10″.

No.	Mag.	I.	II.	III.	IV.	V.	VI.	VII.	T.	a_1	a_2	MICROMETER.			i	d_1	d_2	Mean Right Ascension, 1850.0.	Mean Declination, 1850.0.
1	9	..	..	..	53.4	..	21.8	..	9 48 53.36	−17.84	+0.02	IV.	2	10.35	−10 19.20	−22.88	−1.45	9 48 35.54	−28 33 53.53
2	9	..	..	..	22.8	36.9	..	..	50 22.60	17.84	0.02	V.	2	9.29	9 45.86	23.18	1.39	50 4.78	33 20.43
3	7	..	..	4.2	18.1	..	..	..	9 52 18.e3	17.84	+0.02	V.	2	13.30	11 47.38	23.55	1.59	9 52 0.21	35 22.52
4	9	..	..	..	28.3	..	56.6	..	10 10 28.15	17.81	−0.01	VI.	7	5.24	31 40.58	27.08	3.71	10 10 10.33	28 55 21.37
5	7	..	..	44.3	58.4	..	..	..	14 58.51	17.80	−0.02	III.	10	6.50	47 22.69	27.92	5.43	14 40.69	29 11 6.04
6	8	..	..	..	46.9	1.0	..	..	15 46.73	17.80	0.00	IV.	5	8.55	24 26.07	28.07	2.94	15 28.93	28 48 7.08
7	7	..	51.2	5.7	20.0	..	..	..	17 19.85	17.80	0.00	IV.	6	11.35	29 47.85	28.36	3.50	17 2.05	53 29.71
8	5	..	..	33.1	47.2	..	..	..	18 47.08	17.79	+0.01	I.	4	4.38	17 16.86	28.64	2.19	18 29.30	40 57.69
9	10	..	..	..	25.4	..	53.4	..	21 25.10	17.79	−0.01	V.	7	6.49	32 23.59	29.12	3.79	21 7.30	56 6.50
10	5	..	..	..	..	..	19.3	33.2	22 50.62	17.78	−0.01	VI.	7	2.24	30 9.82	29.38	3.55	22 32.83	53 52.75
11	10	..	..	..	..	6.6	..	..	25 52.29	17.78	+0.01	V.	4	9.55	19 57.05	29.93	2.47	25 34.52	43 39.45
12	10	..	..	32.8	..	1.5	..	..	33 47.17	17.76	−0.01	V.	7	10.28	34 14.01	31.32	4.00	33 29.40	57 59.33
13	10	..	..	..	..	..	..	19.8	34 7.01	17.76	−0.01	VII.	7	10.16	34 7.60	31.38	3.98	33 49.24	57 52.96
14	10	..	..	..	37.3	..	..	..	36 37.14	17.75	0.00	IV.	5	4.31	22 12.95	31.81	2.71	36 19.39	45 51.47
15	8	36.2	50.4	5.0	19.4	33.4	47.3	1.8	38 19.11	17.75	−0.01	I.	7	2.11	30 3.03	32.10	3.53	38 1.35	28 53 48.65
16	6	..	..	..	..	3.9	..	..	42 49.72	17.73	−0.02	V.	10	8.43	48 19.67	32.87	5.53	42 31.97	29 12 8.07
17	8	16.0	30.3	44.8	59.2	13.0	..	41.8	44 58.87	17.73	0.00	VII.	6	7.12	27 34.81	33.22	3.28	44 41.14	28 51 21.31
18	10	..	..	..	..	8.9	..	..	46 8.74	17.72	−0.02	V.	9	8.21	43 7.67	33.40	4.97	45 51.00	29 6 56.04
19	9	..	3.9	18.4	..	..	..	..	50 32.24	17.71	+0.02	III.	1	10.1	5 3.00	34.13	0.88	50 14.55	28 28 48.01
20	7	..	..	..	18.9	..	..	..	51 18.74	17.70	+0.02	IV.	1	4.55	2 28.75	34.26	0.63	51 1.06	28 26 13.64
21	8	..	..	12.3	26.8	40.7	..	..	10 59 26.63	−17.67	−0.02	IV.	10	3.43	−45 48.44	−35.54	−5.26	10 59 8.94	−29 9 39.24

ZONE 112. APRIL 24. K. BELT, −28° 46′. D₀ = −28° 24′ 30″.

No.	Mag.	I.	II.	III.	IV.	V.	VI.	VII.	T.	a_1	a_2	MICROMETER.			i	d_1	d_2	Mean Right Ascension, 1850.0.	Mean Declination, 1850.0.
1	6	33.8	48.3	..	..	..	..	..	11 1 17.07	−14.95	−0.00	III.	9	11.4	−44 29.87	−32.00	−5.23	11 1 2.11	−29 9 37.10
2	7	..	..	..	40.8	..	..	23.3	1 40.56	14.95	0.00	VII.	6	14.1	31 1.03	31.94	3.70	1 25.61	28 56 6.67
3	7	..	..	..	..	..	27.6	..	11 2 59.15	−14.95	−0.01	VII.	7	9.33	−33 45.91	−31.78	−4.01	11 2 44.19	−28 58 51.70

ZONE 113. APRIL 24. K. BELT, −28° 48′. D₀ = −28° 22′ 30″.

No.	Mag.	I.	II.	III.	IV.	V.	VI.	VII.	T.	a_1	a_2	MICROMETER.			i	d_1	d_2	Mean Right Ascension, 1850.0.	Mean Declination, 1850.0.
1	9	..	..	..	35.3	..	..	..	14 46 35.14	−13.62	0.00	IV.	7	5.57	−31 57.41	−42.36	−3.82	14 46 21.52	−28 55 13.59
2	7	47.8	2.6	16.7	31.1	..	..	..	48 30.81	13.60	+0.01	II.	2	9.12	9 37.29	42.52	1.24	48 17.22	32 51.05
3	9	..	..	..	27.6	..	..	..	58 27.44	13.52	0.00	IV.	7	5.51	31 54.38	43.39	3.81	58 13.92	55 11.58
4	9	..	..	41.0	55.0	..	..	..	14 59 54.98	−13.51	0.00	IV.	5	7.44	−23 50.27	−43.51	−2.87	14 59 41.47	−28 47 6.65

CORRECTIONS.

Date.		Corr. of Clock.	Hourly rate.	m	n	c
1847.	h.	s.	s.	s.	s.	s.
April 21,	11	ƒ 23.95	/ 0.043	+ 0.202	+ 0.050	− 0.121
April 24,	13	ƒ 21.32	/ 0.013	+ 0.179	+ 0.119	0.000

INSTRUMENT READINGS.

	Date.	Barom.	THERMOM. At.	THERMOM. Ex.
	1847. h. m.	in.	°	°
Zone 111	April 21, 9 48	29.82	69.2	71.2
Zone 113	April 24, 11 2	30.060	59.5	48.5

REMARKS.

(112) 1. Minutes of transit assumed as 1 instead of 2.

ZONE 114. APRIL 26. K. BELT, −29° 23′. D_o = −28° 58′ 20″.

No.	Mag.	I.	II.	III.	IV.	V.	VI.	VII.	T.	a_1	a_2	MICROMETER.			i	d_1	d_2	Mean Right Ascension, 1850.0.	Mean Declination, 1850.0.
									h. m. s.	s.	s.			r.	′ ″	″	″	h. m. s.	° ′ ″
1	10	. .	. .	. .	. .	. .	55.4	. .	10 28 26.78	−13.68	−0.01	VI.	8	1.38	−34 45.32	−36.45	−4.14	10 28 13.09	−29 33 45.91
2	10	. .	. .	. .	. .	55.2	. .	. .	30 40.78	13.68	+0.01	V.	3	6.40	13 19.75	37.00	1.67	30 27.11	12 18.42
3	10	. .	. .	25.6	. .	. .	. .	. .	34 40.14	13.67	−0.02	III.	10	4.14	46 4.03	37.93	5.48	34 26.45	45 7.44
4	10	30.2	. .	59.2	. .	. .	. .	. .	42 13.69	13.65	0.01	III.	9	5.58	41 55.56	39.69	5.00	42 0.03	41 0.25
5	9.10	. .	. .	. .	. .	48.3	2.1	. .	42 33.72	13.64	−0.02	VI.	10	5.9	46 31.62	39.74	5.54	42 20.06	45 36.90
6	9	. .	9.6	. .	38.4	. .	. .	. .	45 38.28	13.64	0.00	IV.	5	8.13	24 4.89	40.41	2.88	45 24.64	23 8.18
7	9	. .	. .	. .	23.4	. .	51.7	. .	46 23.20	13.64	+0.01	VI.	4	7.39	18 48.36	40.60	2.27	46 9.57	17 51.24
8	9	. .	. .	44.0	58.5	12.8	. .	. .	47 58.38	13.63	−0.01	VI.	7	8.23	33 10.84	40.92	3.96	47 44.74	32 15.72
9	10	. .	16.3	. .	. .	. .	. .	. .	10 50 45.06	−13.63	0.00	II.	6	8.47	−28 22.95	−41.55	−3.40	10 50 31.43	−29 27 27.90

ZONE 115. MAY 4. K. BELT, −31° 16′. D_o = −30° 54′ 0″.

No.	Mag.	I.	II.	III.	IV.	V.	VI.	VII.	T.	a_1	a_2	MICROMETER.			i	d_1	d_2	Mean Right Ascension, 1850.0.	Mean Declination, 1850.0.
1	9	. .	19.1	33.8	. .	. .	. .	. .	10 52 48.45	−12.39	−0.01	III.	7	6.10	−32 3.92	−10 91	−4.98	10 52 36.05	−31 26 19.81
2	9	. .	. .	. .	. .	. .	. .	50.7	53 6.85	12.39	−0.01	VII.	8	6.44	37 19.35	10.96	4.73	52 54.45	31 35.04
3	10	. .	. .	. .	. .	47.9	. .	. .	54 33.23	12.39	0.00	VII.	5	2.10	21 1.38	11.19	3.45	54 20.84	15 16.02
4	5	. .	. .	5.6	20.3	. .	. .	. .	56 20.04	12.38	+0.01	IV.	3	9.57	14 59.13	11.45	2.63	56 7.67	9 13.21
5	10	. .	. .	. .	23.7	. .	. .	. .	57 23.54	12.38	+0.01	IV.	3	12.00	16 1.15	11.62	2.77	57 11.17	10 15.54
6	10	. .	. .	. .	. .	26.3	40.5	. .	10 58 11.51	12.38	0.00	VI.	6	7.42	27 50.15	11.74	4.39	10 57 59.13	22 6.28
7	9	32.6	47.6	2.4	. .	. .	. .	. .	11 1 16.80	12.37	0.00	IV.	5	4.38	22 16.48	12.21	3.64	11 1 4.43	16 32.33
8	11	. .	. .	2.3	. .	. .	. .	. .	2 16.91	12.36	−0.01	III.	7	5.31	31 44.25	12.36	4.94	2 4.54	26 1.55
9	9	. .	. .	. .	44.6	. .	. .	. .	2 44.44	12.36	0.01	IV.	8	5.11	36 32.91	12.42	5.62	2 32.07	30 50.95
10	5	. .	. .	. .	. .	. .	22.5	37.4	2 53.44	12.36	0.01	VII.	8	10.3	38 59.69	12.45	5.99	2 41.07	33 18.13
11	6	29.6	. .	59.6	. .	. .	. .	. .	5 14.18	12.35	−0.02	IV.	9	7.52	42 53.10	12.79	6.53	5 1.81	37 12.42
12	10	. .	1.8	. .	. .	. .	. .	. .	8 30.89	12.34	+0.02	IV.	2	3.34	6 46.91	13.28	1.50	8 18.57	1 1.69
13	9	. .	. .	. .	. .	11.7	. .	40.8	8 57.04	12.34	−0.01	VII.	9	0.5	38 57.15	13.34	5.98	8 44.69	31 33 16.47
14	5	. .	. .	. .	. .	19.5	. .	. .	11 4.75	12.34	+0.02	V.	1	3.20	1 40.79	13.66	0.81	10 52.43	30 55 55.26
15	7	. .	. .	. .	21.7	. .	50.6	. .	12 21.45	12.33	−0.02	IV.	10	5.59	46 57.02	13.83	7.10	12 9.10	31 41 17.95
16	6	. .	. .	. .	. .	40.7	. .	10.3	13 26.04	12.33	0.00	VII.	5	5.23	22 38.68	13.99	3.66	13 13.71	16 56.33
17	9	. .	. .	. .	. .	. .	24.6	. .	15 55.38	12.32	−0.02	VI.	9	6.34	42 13.57	14.33	6.42	15 43.04	36 34.32
18	8	. .	. .	37.6	51.9	. .	. .	. .	17 51.85	12.32	+0.01	IV.	3	12.31	16 16.78	14.61	2.81	17 39.54	10 34.20
19	8	47.6	2.6	. .	. .	. .	. .	. .	20 31.83	12.31	0.00	IV.	5	10.19	25 8.43	14.98	4.00	20 19.52	19 27.41
20	7	. .	. .	. .	20.5	35.4	. .	. .	21 20.51	12.30	+0.02	IV.	2	5.15	7 37.83	15.08	1.63	21 8.23	1 54.54
21	3.4	. .	. .	36.4	50.3	. .	. .	. .	25 50.38	12.29	+0.02	IV.	2	4.51	7 25.73	15.69	1.61	25 38.11	1 43.03
22	9	. .	. .	. .	. .	50.9	. .	. .	27 36.32	12.28	−0.02	VII.	9	7.23	42 38.00	15.93	6.47	27 24.02	37 0.40
23	9	. .	. .	5.3	. .	. .	. .	. .	29 19.66	12.27	+0.01	III.	4	3.29	16 42.43	16.15	2.87	29 7.40	11 1.45
24	9	. .	. .	. .	. .	57.8	. .	. .	29 43.23	12.27	−0.02	VI.	10	1.28	44 40.15	16.21	6.79	29 30.94	39 3.15
25	4	. .	. .	. .	. .	. .	56.1	. .	34 26.86	12.25	+0.02	VI.	10	3.31	45 42.18	16.80	6.94	34 14.63	40 5.92
26	9	. .	13.4	28.3	. .	. .	. .	. .	36 42.62	12.24	0.01	III.	4	1.58	15 56.54	17.08	2.77	36 30.39	31 10 16.39
27	7	. .	. .	. .	. .	0.2	. .	. .	37 45.45	12.24	+0.02	V.	1	4.41	2 21.64	17.21	0.90	37 33.23	30 56 39.75
28	9	. .	. .	. .	. .	. .	. .	8.3	38 24.38	12.24	−0.01	VII.	7	5.7	31 31.73	17.29	4.91	38 12.13	31 25 53.93
29	9	. .	. .	. .	. .	. .	. .	58.2	39 14.39	12.24	0.02	VII.	9	3.5	40 27.91	17.39	6.20	39 2.13	34 51.50
30	9	. .	. .	8.5	. .	. .	. .	. .	42 23.27	12.23	−0.02	IV.	9	6.39	42 16.30	17.77	6.45	42 11.02	36 40.52
31	11	30.5	. .	. .	. .	. .	. .	. .	47 14.64	12.21	0.00	I.	5	6.17	23 5.93	18.32	3.74	47 2.43	17 27.99
32	8	. .	50.0	. .	19.7	. .	. .	. .	49 19.46	12.20	−0.01	IV.	7	5.18	31 37.74	18.56	4.91	49 7.25	26 1.21
33	9	46.7	. .	15.8	. .	. .	. .	. .	53 30.46	12.19	+0.01	IV.	4	3.51	16 53.57	19.01	2.89	53 18.28	11 15.47
34	9	. .	. .	. .	. .	32.4	. .	. .	54 17.76	12.18	0.00	VI.	6	7.57	27 57.71	19.09	4.42	54 5.58	22 21.22
35	8	. .	. .	. .	4.8	. .	33.8	. .	11 56 4.64	12.18	−0.01	VI.	7	4.8	31 2.24	19.29	4.84	11 55 52.45	25 26.37
36	7	. .	. .	. .	. .	. .	43.5	. .	12 6 14.30	−12.12	−0.01	VI.	8	9.26	−38 41.29	−20.26	−5.94	12 6 2.17	−31 33 7.49

CORRECTIONS.

Date.		Corr. of Clock.	Hourly rate.	m	n	c
1847.	h.	s.	s.	s.	s.	s.
April 26,	13	ƒ 19.98	/ 0.028	+ 0.211	+ 0.083	0.000
May 4,	13	ƒ 18.74	/ 0.010	+ 0.210	+ 0.200	0.000

INSTRUMENT READINGS.

	Date.		Barom.	THERMOM. At.	THERMOM. Ex.
	1847.	h. m.	in.	°	°
Zone 114	April 26,	9 5	29.798	74.	64.5
Zone 115	May 4,	10 52	29.96	62.0	57.0

REMARKS.

Zone 115. May 4. K. Belt, −31° 16′. D_0 = −30° 54′ 0″—Continued.

No.	Mag.	SECONDS OF TRANSIT. I.	II.	III.	IV.	V.	VI	VII.	T.	a_1	a_2	MICROMETER.			i	d_1	d_2	Mean Right Ascension, 1850.0.	Mean Declination, 1850.0.
									h. m. s.	s.	s.			r.	′ ″	″	″	h. m. s.	° ′ ″
37	9	. .	50.2	4.9	. .	. .	. .	. .	12 9 19.60	−12.10	−0.01	IV.	8	5.11	−36 32.91	−20.55	−5.62	12 9 7.49	−31 30 59.08
38	10	. .	57.4	. .	. .	. .	. .	. .	14 26.54	12.09	+0.01	IV.	3	4.9	12 3.66	20.97	2.21	14 14.46	6 26.84
39	9	. .	. .	21.0	. .	. .	. .	. .	16 35.66	12.08	−0.01	IV.	8	2.52	35 22.83	21.15	5.47	16 23.57	29 49.45
40	9	. .	. .	21.2	. .	. .	. .	. .	18 35.68	12.07	0.00	IV.	5	9.42	24 49.77	21.32	3.98	18 23.61	19 15.07
41	9	. .	51.7	. .	. .	. .	. .	. .	22 20.92	12.05	+0.01	IV.	4	8.40	19 19.31	21.59	3.21	22 8.88	13 44.11
42	9	. .	36.2	. .	. .	. .	. .	. .	24 5.33	12.04	+0.02	IV.	2	11.43	10 53.46	21.72	2.06	23 53.31	5 17.26
43	5	. .	. .	. .	. .	. .	49.2	. .	24 19.66	12.04	−0.01	VI.	10	7.45	47 50.26	21.74	7.26	24 7.61	42 19.26
44	8	. .	. .	. .	. .	. .	. .	34.5?	24 50.72	12.04	0.01	VII.	9	9.12	43 32.97	21.77	6.64	24 38.67	38 1.38
45	8	. .	. .	. .	. .	. .	49.1	. .	26 19.87	12.03	−0.01	VI.	9	12.9	45 2.47	21.88	6.85	26 7.83	39 31.20
46	9	. .	. .	52.5	. .	. .	. .	. .	29 7.03	12.01	0.00	V.	6	5.1	26 29.12	22.08	4.21	28 55.02	31 20 55.41
47	8	. .	. .	. .	13.2	. .	. .	. .	38 13.04	11.97	+0.01	IV.	1	3.59	2 0.51	22.65	0.85	38 1.08	30 56 24.01
48	9	. .	56.1	. .	25.6	. .	. .	. .	40 25.48	11.95	0.00	V.	8	3.39	35 46.48	22.78	5.52	40 13.53	31 30 14.78
49	10	. .	. .	. .	48.9	. .	. .	. .	43 48.74	11.93	0.00	IV.	4	8.22	19 10.23	22.96	3.20	43 36.81	13 36.39
50	9	. .	. .	24.6	. .	. .	. .	. .	49 39.26	11.90	0.00	IV.	7	11.21	34 40.78	23.27	5.37	49 27.36	29 9.42
51	6	. .	43.4	58.0	. .	. .	. .	. .	52 12.51	11.89	0.00	IV.	4	8.55	19 26.87	23.39	3.24	52 0.62	13 53.50
52	9	. .	. .	. .	. .	. .	59.0	. .	12 52 29.93	11.89	+0.01	VI.	3	5.15	12 36.72	23.39	2.26	12 52 18.05	7 2.37
53	9	. .	. .	. .	16.0	30.9	. .	. .	13 1 16.03	11.82	0.00	IV.	4	7.15	18 36.44	23.78	3.10	13 1 4.21	13 3.32
54	8	. .	. .	. .	. .	25.6	. .	54.5	2 10.70	11.82	0.00	VI.	5	6.23	23 9.21	23.81	3.72	1 58.88	17 36.74
55	6	57.5	11.4	27.2	. .	. .	. .	. .	4 41.14	11.80	0.00	IV.	2	8.31	9 16.67	23.89	1.82	4 29.34	3 42.38
56	6	. .	50.5	5.3	. .	. .	. .	. .	23 19.83	11.69	0.00	IV.	6	6.45	27 21.62	24.40	4.34	23 8.14	21 50.36
57	8	. .	. .	54.7	. .	24.3	. .	. .	24 9.40	11.69	0.00	V.	5	5.4	22 29.54	24.42	3.66	23 57.71	16 57.62
58	7	11.7	26.6	. .	. .	. .	. .	. .	26 55.90	11.67	0.00	III.	7	5.55	31 56.35	24.46	4.98	26 44.29	26 25.79
59	8	. .	. .	14.0	28.6	. .	. .	. .	27 28.60	11.67	0.00	IV.	9	5.29	41 40.99	24.48	6.38	27 16.93	36 11.85
60	8	. .	. .	. .	28.6	42.8	. .	. .	28 28.33	11.66	0.00	V.	9	9.16	43 35.41	24.49	6.66	28 16.67	38 6.56
61	3	2.6	. .	32.4	. .	. .	. .	. .	44 46.71	11.55	0.00	IV.	4	3.15	16 35.42	24.60	2.82	44 35.16	11 2.84
62	7	. .	. .	. .	54.2	. .	22.5	. .	45 53.65	11.54	0.00	VI.	10	7.30	47 42.70	24.59	7.25	45 42.11	42 14.54
63	7	. .	. .	. .	. .	18.2	. .	. .	47 3.56	11.54	0.00	VI.	6	6.14	27 5.78	24.59	4.28	46 52.02	31 21 34.65
64	6	. .	. .	. .	. .	46.4	. .	. .	13 54 31.65	−11.49	0.00	V.	1	6.32	−3 17.62	−24.55	−0.98	13 54 20.16	−30 57 43.15

Zone 116. May 6. K. Belt, −28° 46′. D_0 = −28° 24′ 50″.

No.	Mag.	I.	II.	III.	IV.	V.	VI	VII.	T.	a_1	a_2	MICROMETER.			i	d_1	d_2	Mean Right Ascension, 1850.0.	Mean Declination, 1850.0.
1	7	. .	. .	40.2	. .	. .	. .	. .	11 51 54.69	−10.01	−0.01	III.	10	9.30	−48 33.29	− 3.06	−5.69	11 51 44.67	−29 13 32.0
2	9	12.6	26.8	41.3	. .	. .	. .	. .	12 2 55.44	9.98	0.00	IV.	4	14.15	22 8.21	4.22	2.69	12 2 45.46	28 47 5.1
3	9	. .	. .	. .	13.2	. .	41.3	. .	4 13.01	9.97	+0.01	V.	2	7.38	8 49.89	4.35	1.25	4 3.05	33 45.5
4	9	. .	. .	. .	35.5	. .	. .	. .	5 35.34	9.97	0.00	V.	4	9.27	19 42.95	4.49	2.47	5 25.37	44 39.9
5	8	. .	58.7	13.2	. .	. .	. .	. .	8 27.25	9.96	0.00	IV.	5	4.56	22 25.56	4.77	2.70	8 17.29	47 23.0
6	9	. .	. .	4.4	. .	. .	. .	. .	10 18.48	9.95	0.00	V.	5	4.39	22 16.94	4.96	2.69	10 8.53	28 47 14.6
7	9	. .	. .	. .	10.6	. .	38.3	. .	11 10.12	9.95	−0.01	V.	9	7.14	42 33.89	5.03	4.96	11 0.16	29 7 33.9
8	9	. .	. .	. .	52.8	. .	. .	. .	12 52.64	9.94	+0.01	IV.	1	3.20	1 40.84	5.19	0.45	12 42.71	28 26 36.5
9	8	. .	. .	. .	. .	59.2	. .	. .	13 44.94	9.94	0.00	V.	6	9.1	28 30.14	5.27	3.38	13 35.00	28 53 28.8
10	9	. .	. .	30.6	. .	. .	. .	. .	17 45.07	9.92	−0.01	IV.	10	5.33	46 43.91	5.63	5.46	17 35.14	29 11 45.0
11	9	. .	. .	. .	. .	25.5	. .	. .	18 11.22	9.92	0.00	V.	5	10.26	25 11.91	5.67	3.02	18 1.30	28 50 10.6
12	9	. .	. .	31.9	. .	0.7	14.3	. .	19 46.15	9.91	0.00	VI.	7	6.32	32 14.88	5.82	3.83	19 36.24	57 14.5
13	9	. .	. .	. .	42.0	. .	. .	. .	21 41.84	9.90	+0.01	IV.	3	8.53	11 55.58	5.98	1.59	21 31.85	36 53.2
14	8	. .	. .	. .	. .	. .	. .	48.9	22 5.73	9.90	+0.01	VII.	1	7.59	4 1.11	6.01	0.71	21 55.84	28 28 57.8
15	8	. .	36.8	51.0	. .	. .	. .	48.3	25 5.59	9.88	−0.01	III.	10	8.15	48 5.55	6.26	5.62	24 55.70	29 13 7.4
16	10	. .	. .	. .	. .	44.7	. .	. .	28 29.69	9.97	0.00	VII.	4	11.36	20 47.64	6.54	2.55	28 19.72	28 45 46.7
17	7	54.0	8.3	23.3	37.3	. .	. .	. .	12 31 37.06	− 9.85	0.00	IV.	3	15.15	−17 39.46	− 6.78	−2.20	12 31 27.21	−28 42 38.4

CORRECTIONS.

Date.		Corr. of Clock.	Hourly rate.	m	n	c
1847.	h.	s.	s.	s.	s.	s.
May 6,	13	ƒ16.51	*l* 0.026			

INSTRUMENT READINGS.

	Date.		Barom.	Thermom. At.	Thermom. Ex.
	1847.	h. m.	in.	°	°
Zone 116	May 6,	11 51	29.81	62.5	57.5

REMARKS.

(116) 1. Micrometer reading assumed as 9r.10 instead of 9r.30.

ZONE 116. MAY 6. K. BELT, −28° 46′. D₀=−28° 24′ 50″—Continued.

No.	Mag.	SECONDS OF TRANSIT. I.	II.	III.	IV.	V.	VI.	VII.	T.	a_1	a_2	MICROMETER.			i	d_1	d_2	Mean Right Ascension, 1850.0.	Mean Declination, 1850.0.
									h. m. s.	s.	s.			r.	′ ″	″	″	h. m. s.	° ′ ″
18	8	. .	. .	. .	. .	. .	. .	43.5	12 32 0.58	− 9.85	0.00	VII.	5	9.33	−24 44.81	− 6.81	−2.97	12 31 50.73	−28 49 44.6
19	9	. .	. .	. .	. .	. .	. .	14.3	32 31.35	9.85	0.00	VII.	5	4.7	22 22.11	6.84	2.68	32 21.50	47 21.6
20	9	. .	. .	. .	. .	44.1	. .	. .	36 29.77	9.83	0.00	VI.	3	11.4	15 32.73	7.14	1.95	36 19.94	40 31.8
21	8	. .	. .	. .	. .	44.4	. .	. .	37 30.15	9.83	0.00	VI.	7	7.23	32 40.59	7.22	3.87	37 20.32	57 41.7
22	9	. .	21.8	36.3	. .	. .	. .	. .	39 50.50	9.82	0.00	IV.	7	8.59	33 29.19	7.39	3.99	39 40.68	58 30.6
23	9	. .	. .	28.8	. .	. .	. .	. .	44 42.79	9.80	0.00	IV.	3	11.20	15 40.98	7.70	1.98	44 32.99	28 40 40.7
24	7	. .	. .	. .	. .	36.8	. .	5.2	45 22.59	9.80	−0.01	VII.	10	5.54	46 54.08	7.74	5.50	45 12.78	29 11 57.3
25	9	. .	. .	. .	. .	. .	. .	1.3	46 24.21	9.79	+0.01	VII.	2	10.46	10 24.33	7.81	1.41	46 14.43	28 35 23.6
26	6.7	. .	. .	. .	. .	14.2	28.0	42.6	50 59.82	9.77	−0.01	VII.	9	14.23	46 13.82	8.11	5.41	50 50.04	29 11 17.3
27	6	. .	. .	. .	41.8	. .	. .	. .	54 41.64	9.75	+0.01	IV.	1	4.49	2 25.73	8.32	0.53	54 31.90	28 27 24.6
28	7	. .	. .	. .	24.6	38.8	. .	. .	12 55 24.49	9.75	0.00	IV.	6	4.55	26 26.15	8.36	3.16	12 55 14.74	51 27.7
29	6	. .	54.5	. .	23.0	. .	. .	. .	13 24 22.93	9.58	0.00	IV.	5	4.58	22 26.56	9.60	2.70	13 24 13.35	47 28.9
30	10	. .	. .	. .	. .	. .	. .	5.5	24 22.30	9.58	0.00	VII.	1	2.31	6 14.72	9.60	0.91	24 12.72	31 15.2
31	9	. .	. .	13.4	. .	41.8	. .	. .	27 27.39	9.57	0.00	IV.	3	3.6	11 31.89	9.70	1.46	27 17.82	28 36 33.1
32	7	24.6	39.3	. .	. .	. .	. .	. .	30 7.91	9.55	0.00	III.	8	10.51	39 24.31	9.76	4.68	29 58.36	29 4 28.7
33	5.6	. .	. .	. .	27.8	. .	55.8	. .	30 27.52	9.55	0.00	IV.	5	5.4	22 29.59	9.77	2.70	30 17.97	28 47 32.1
34	9	1.8	16.0	. .	. .	. .	. .	. .	32 44.80	9.53	0.00	III.	7	8.11	33 4.93	9.81	3.90	32 35.27	58 8.6
35	9	. .	4.2	. .	. .	. .	. .	. .	40 32.81	9.49	0.00	III.	6	11.2	29 31.16	9.96	3.52	40 23.32	28 54 34.6
36	9	. .	. .	. .	. .	36.8	50.4	. .	42 22.25	9.48	0.00	VII.	9	7.25	42 39.06	9.97	5.06	42 12.77	29 7 44.1
37	9	. .	. .	. .	39.0	. .	. .	. .	47 38.84	9.44	0.00	IV.	9	9.33	43 44.03	10.03	5.19	47 29.40	8 49.2
38	7	. .	17.6	31.8	. .	. .	. .	. .	49 46.18	9.43	0.00	IV.	7	13.1	35 31.20	10.05	4.20	49 36.75	0 35.4
39	9	. .	. .	42.3	. .	10.8	. .	. .	13 55 56.68	9.39	0.00	IV.	9	13.10	45 33.44	10.06	5.35	13 55 47.29	10 38.9
40	9	. .	20.5	. .	. .	. .	. .	. .	14 2 49.20	9.35	0.00	II.	8	10.44	39 20.65	10.06	4.65	14 2 39.85	29 4 25.4
41	5	48.3	2.6	17.0	. .	. .	. .	. .	6 31.03	9.33	0.00	II.	2	9.15	9 38.66	10.03	1.25	6 21.70	28 34 39.9
42	5.6	36.8	51.5	6.0	. .	. .	. .	. .	15 20.13	9.27	0.00	IV.	7	10.54	34 27.18	9.91	4.10	15 10.86	59 31.2
43	4.5	50.7	5.7	20.0	34.2	. .	. .	. .	19 34.03	9.24	0.00	V.	5	7.44	23 50.22	9.84	2.87	19 24.79	48 52.9
44	8	. .	. .	. .	. .	48.2	. .	. .	20 33.81	9.24	−0.01	V.	1	2.52	1 26.68	9.83	0.36	20 24.56	26 26.9
45	9	. .	. .	. .	. .	. .	27.8	. .	22 59.45	9.22	0.00	VII.	3	6.48	13 23.41	9.77	1.69	21 50.23	28 38 24.9
46	10	. .	. .	. .	48.7	. .	16.8	. .	28 48.44	9.19	0.00	VI.	8	5.15	36 34.75	9.63	4.32	28 39.25	29 1 38.7
47	9	2.6	16.9	31.6	. .	. .	. .	. .	31 45.82	9.17	0.00	IV.	3	8.13	14 6.69	9.54	1.76	31 36.65	28 39 8.0
48	9	20.3	34.9	49.5	. .	. .	. .	. .	33 3.65	9.15	0.00	IV.	8	8.13	38 4.69	9.50	4.51	32 54.50	29 3 8.7
49	10	. .	. .	. .	. .	. .	41.3	. .	33 12.93	9.16	0.00	VI.	4	5.9	17 32.72	9.50	2.14	33 3.77	28 42 34.4
50	9	48.3	. .	. .	. .	. .	. .	. .	46 31.43	9.06	0.00	II.	7	2.34	30 14.87	9.00	3.60	46 22.37	55 17.5
51	5.6	44.2	. .	13.2	27.3	. .	. .	. .	14 48 27.09	− 9.05	−0.01	V.	2	5.48	− 7 54.42	− 8.91	−1.09	14 48 18.03	−28 32 54.4

ZONE 117. MAY 6. K. D₀=−28° 24′ 20″.

No.	Mag.	I.	II.	III.	IV.	V.	VI.	VII.	T.	a_1	a_2	MICROMETER.			i	d_1	d_2	Mean Right Ascension, 1850.0.	Mean Declination, 1850.0.
1	4	. .	. .	. .	. .	. .	. .	15.2	15 53 32.20	− 8.60	−0.01	VII.	4	5.42	−17 49.13	−28.91	−2.19	15 53 23.59	−28 42 40.23
2	8	. .	. .	. .	24.3	38.2	. .	. .	55 23.99	8.58	0.01	VI.	2	1.53	5 55.78	28.69	0.85	55 15.40	30 45.32
3	9	. .	. .	. .	. .	. .	50.1	. .	15 59 21.78	8.55	−0.01	VI.	2	7.55	8 58.32	28.24	1.19	15 59 13.22	33 47.75
4	8	. .	45.5	. .	. .	. .	. .	. .	16 1 14.09	8.54	0.00	II.	6	8.15	28 6.81	28.01	3.35	16 1 5.55	28 52 58.17
5	3.4	. .	. .	38.7	. .	7.2	. .	. .	1 52.99	8.54	+0.01	IV.	8	4.00	35 57.12	27.93	4.26	1 44.46	29 0 49.31
6	7	46.7	1.3	. .	. .	. .	. .	. .	4 29.71	8.52	−0.01	III.	4	0.30	15 12.18	27.60	1.89	4 21.18	28 40 1.67
7	9	. .	52.0	. .	. .	. .	. .	. .	11 20.75	8.47	+0.01	III.	9	10.5	44 0.12	26.73	5.21	11 12.29	29 8 52.06
8	9	. .	. .	. .	26.0	40.1	. .	. .	16 25.86	8.44	+0.01	IV.	8	8.24	38 10.24	26.04	4.50	16 17.43	29 3 0.78
9	8	. .	47.8	2.2	. .	. .	. .	. .	18 16.44	8.42	0.00	IV.	7	5.38	31 47.83	25.79	3.75	18 8.02	28 56 37.37
10	9	. .	59.0	. .	27.6	. .	. .	. .	21 27.60	8.40	+0.01	IV.	9	10.22	44 8.72	25.35	5.22	21 19.21	29 8 59.29
11	9	28.5	. .	57.3	. .	. .	. .	. .	16 23 11.41	− 8.39	0.00	IV.	4	6.16	−18 6.70	−25.10	−2.20	16 23 3.02	−28 42 54.00

CORRECTIONS.

Date.		Corr. of Clock.	Hourly rate.	m	n	c
1847.	h.	s.	s.	s.	s.	s.

INSTRUMENT READINGS.

Date.	Barom.	THERMOM. At.	Ex.
1847. h. m.	in.	°	°

REMARKS.

(116) 19. Micrometer reading assumed as 4ʳ.50.
(116) 30. Hor. thread assumed as 2 instead of 1.
(116) 45. Minutes assumed as 21 instead of 22.
(117) The time of this zone seems uncertain, and the large stars marked (3.4) and (6) are not found.

ZONE 117. MAY 8. K. $D_o = -28° 24' 20''$—Continued.

No.	Mag.	Seconds of Transit. I.	II.	III.	IV.	V.	VI.	VII.	T.	a_1	a_2	Micrometer.			i	d_1	d_2	Mean Right Ascension, 1850.0.	Mean Declination, 1850.0.
									h. m. s.	s.	s.			r.	′ ″	″	″	h. m. s.	° ′ ″
12	9	..	..	..	9.2	..	..	..	16 26 9.04	− 8.37	+0.01	IV.	10	8.8	−48 2.07	−24.66	−5.70	16 26 0.68	−29 12 52.43
13	6	..	28.2	42.3	..	..	..	..	30 56.45	8.34	−0.01	IV.	3	7.3	13 31.39	23.96	1.69	30 48.10	28 38 17.04
14	9	8.4	..	..	..	..	..	..	32 51.46	8.32	0.00	II.	5	9.41	24 49.07	23.66	2.98	32 43.14	49 35.71
15	9	30.4	..	..	..	..	..	..	36 13.45	8.30	0.00	III.	5	6.24	23 9.88	23.15	2.78	36 5.15	47 55.81
16	8	..	..	..	38.9	..	..	..	36 38.74	8.29	0.00	IV.	7	3.22	30 39.25	23.09	3.65	36 30.45	55 25.99
17	8	..	..	..	..	42.3	..	..	37 27.96	8.29	−0.01	VI.	3	6.12	13 5.49	22.96	1.63	37 19.66	37 50.08
18	7	..	..	..	40.2	..	..	..	38 40.04	8.28	−0.01	IV.	4	2.28	16 11.72	22.77	1.98	38 31.75	40 56.47
19	8	..	30.7	..	..	..	..	..	39 59.27	8.27	0.00	II.	6	4.42	26 19.41	22.57	3.15	39 51.00	28 51 5.13
20	9	..	..	1.0	..	..	..	..	42 15.43	8.25	+0.01	IV.	9	10.17	44 6.22	22.22	5.22	42 7.19	29 8 53.66
21	8	44.4	..	..	..	..	..	..	44 27.65	8.24	0.01	II.	9	6.38	42 15.60	21.87	5.01	44 19.42	7 2.48
22	9	..	..	36.4	..	..	..	..	44 50.81	8.24	0.01	IV.	9	8.43	43 18.82	21.81	5.15	44 42.58	8 5.78
23	9	..	..	..	..	28.0	..	..	46 13.77	8.23	0.01	VI.	8	5.52	36 53.40	21.59	4.38	46 5.55	1 39.37
24	9	..	..	..	8.3	..	..	..	49 8.14	8.21	+0.02	IV.	9	12.42	45 19.33	21.13	5.39	48 59.95	29 10 5.85
25	9	..	..	..	9.5	..	..	..	53 9.34	8.18	−0.01	IV.	4	11.3	20 31.41	20.47	2.48	53 1.15	28 45 14.36
26	9	..	..	24.6	..	..	..	..	56 38.57	8.15	−0.01	IV.	3	9.50	14 55.60	19.90	1.84	56 30.41	28 39 37.34
27	9	..	31.2	..	..	..	..	..	16 58 59.96	8.14	+0.02	III.	9	11.17	44 36.42	19.51	5.35	16 58 51.84	29 9 21.28
28	10	..	53.4	..	..	..	..	..	17 8 21.91	8.07	−0.01	IV.	4	12.29	21 14.77	17.95	2.58	17 8 13.83	28 45 55.30
29	7	..	..	..	5.5	..	..	..	11 5.34	8.05	+0.02	IV.	10	7.13	47 34.33	17.48	5.66	10 57.31	29 12 17.47
30	9	..	..	..	..	..	..	53.2	11 10.51	8.05	+0.02	VII.	9	7.44	42 48.65	17.46	5.09	11 2.48	29 7 31.20
31	9	..	..	49.8	..	..	..	..	15 3.62	8.03	−0.02	III.	1	11.25	5 45.35	16.67	0.80	14 55.57	28 30 22.82
32	8	..	27.7	..	..	..	..	..	17 56.13	8.01	−0.01	III.	3	6.45	13 22.27	16.30	1.67	17 48.11	38 0.24
33	8	..	37.3	..	..	..	..	..	19 5.92	8.00	+0.01	III.	7	3.55	30 55.84	16.10	3.69	18 57.93	55 35.63
34	9	..	..	..	..	28.2	..	..	19 13.96	8.00	+0.01	VI.	8	2.21	35 7.00	16.07	4.18	19 5.97	59 47.25
35	8	..	..	..	45.0	..	..	..	22 44.84	7.98	−0.01	IV.	4	6.26	18 11.74	15.45	2.22	22 36.85	28 42 49.41
36	9	..	..	17.0	..	..	..	..	17 24 31.40	− 7.97	+0.02	IV.	9	6.11	−42 2.17	−15.14	−5.01	17 24 23.45	−29 6 42.32

ZONE 118. MAY 18. K. BELT, −28° 8′. $D_o = -27° 46' 10''$.

No.	Mag.	I.	II.	III.	IV.	V.	VI.	VII.	T.	a_1	a_2	Micrometer.			i	d_1	d_2	Mean Right Ascension, 1850.0.	Mean Declination, 1850.0.
1	8	..	..	27.2	41.6	55.5	..	..	13 41 41.38	− 4.02	0.00	V.	7	8.3	−33 0.90	−40.20	−3.86	13 41 37.36	−28 19 54.96
2	9	..	..	..	..	..	31.8	..	45 3.55	4.00	0.00	VI.	6	10.47	29 23.47	40.24	3.46	45 59.55	16 17.17
3	8	..	..	..	..	..	..	32.2	45 49.41	4.00	0.00	VII.	3	6.25	13 11.82	40.25	1.75	45 45.41	0 3.82
4	9	..	..	6.2	..	34.6	..	..	49 20.40	3.97	0.00	V.	7	7.17	32 37.70	40.30	3.83	49 16.43	19 31.83
5	10	..	56.8	..	..	..	..	..	52 25.29	3.96	0.00	II.	8	2.31	35 12.06	40.31	4.10	52 21.33	22 6.47
6	8	..	26.5	40.8	..	..	..	..	55 54.97	3.94	0.00	IV.	7	8.55	33 27.17	40.32	3.92	55 51.03	20 21.41
7	9	..	..	..	..	33.5	..	1.7	56 19.28	3.94	0.00	VII.	8	7.15	37 35.03	40.32	4.38	56 15.34	24 29.73
8	9	..	..	12.9	..	41.6	..	..	13 58 27.33	3.93	0.00	V.	9	4.31	41 11.70	40.33	4.75	13 58 23.40	28 6.78
9	8	19.3	33.4	47.9	..	..	..	..	14 3 1.86	3.90	0.00	II.	4	6.2	17 59.46	40.33	2.25	14 2 57.96	4 52.04
10	9	..	..	..	..	43.5	..	..	3 29.30	3.90	0.00	V.	5	8.9	24 2.83	40.33	2.89	3 25.40	10 56.05
11	6.7	..	..	..	..	39.0	..	..	6 24.89	3.89	+0.01	V.	10	7.46	47 50.93	40.32	5.47	6 21.01	34 46.72
12	10	..	..	..	..	..	43.0	..	9 14.71	3.87	0.00	VII.	8	6.37	37 15.88	40.30	4.33	9 10.84	24 10.51
13	9	..	..	..	..	53.3	..	21.4	10 39.02	3.87	0.00	VII.	7	11.33	34 46.43	40.30	4.05	10 35.15	21 40.78
14	8	..	..	48.3	..	16.6	..	..	13 2.30	3.85	0.00	IV.	3	13.31	16 47.03	40.27	2.11	12 58.45	3 39.41
15	9	..	..	17.8	..	..	..	..	14 31.99	3.85	0.00	IV.	7	10.34	34 17.10	40.25	4.00	14 28.14	21 11.35
16	7	..	..	32.8	47.3	..	..	..	16 47.00	3.84	0.00	V.	5	11.29	25 43.67	40.23	3.08	16 43.16	12 36.98
17	7	44.8	59.0	13.6	..	..	..	..	20 27.72	3.82	0.00	IV.	8	11.8	39 32.93	40.18	4.57	20 23.90	28 26 27.68
18	10	..	..	38.3	..	..	..	..	22 52.13	3.81	0.00	IV.	2	11.0	10 31.80	40.15	1.47	22 48.32	27 57 23.42
19	10	..	34.5	3.0	..	..	..	..	28 2.94	3.78	0.00	IV.	8	11.11	39 34.44	40.05	4.57	27 59.16	28 26 29.06
20	9	..	..	..	..	..	..	0.0	14 28 17.25	− 3.78	0.00	VII.	3	14.3	−17 2.75	−40.05	−2.14	14 28 13.47	−28 3 54.94

CORRECTIONS.

Date.		Corr. of Clock.	Hourly rate.	m	n	c
1847.	h.	s.	s.	s.	s.	s.
May 18,	14	ƒ 10.97	ƒ 0.035			

INSTRUMENT READINGS.

	Date.	Barom.	Thermom. At.	Thermom. Ex.
	1847. h. m.	In.	°	°
Zone 118	May 18, 13 41	29.59	65.5	56.1

REMARKS.

(118) 17. Transit over T. III assumed as at 13ˢ.6 instead of 18ˢ.6.

ZONE 118. MAY 18. K. BELT, −28° 8'. D_0 = −27° 46' 10"—Continued.

No.	Mag.	SECONDS OF TRANSIT. I.	II.	III.	IV.	V.	VI.	VII.	T.	a_1	a_2	MICROMETER.			i	d_1	d_2	Mean Right Ascension, 1850.0.	Mean Declination, 1850.0.
									h. m. s.	s.	s.			r.	' "	"	"	h. m. s.	° ' "
21	9	. .	44.8	. .	14.4	. .	. .	. .	14 33 13.71	− 3.75	0.00	IV.	5	7.4	−23 30.10	−39.93	−2.84	14 33 9.96	−28 10 22.87
22	7	. .	. .	6.7	. .	35.2	. .	. .	48 21.08	3.67	+0.01	IV.	10	4.6	46 0.05	39.47	5.27	48 17.42	32 54.79
23	8	. .	. .	. .	40.3	. .	8.0	. .	49 39.97	3.66	0.00	VI.	4	8.14	19 6.02	39.43	2.36	49 36.31	5 57.81
24	9	. .	. .	. .	. .	. .	. .	54.3	14 58 11.69	3.62	0.00	VII.	6	6.54	27 25.75	39.09	3.27	14 58 8.07	14 18.11
25	9	. .	. .	29.3	. .	57.5	. .	. .	15 5 43.51	3.58	+0.01	V.	9	9.34	43 44.49	38.75	5.01	15 5 39.94	30 38.25
26	9	. .	51.3	. .	20.2	. .	. .	. .	10 19.86	3.55	+0.01	V.	8	9.1	38 28.84	38.53	4.46	10 16.32	25 21.83
27	9	13.7	28.0	42.3	. .	. .	. .	. .	19 56.35	3.49	0.00	IV.	7	9.7	33 33.22	38.01	3.92	19 52.86	20 25.15
28	10	. .	. .	. .	. .	39.3	. .	. .	20 25.09	3.49	0.00	V.	7	7.3	32 30.64	37.99	3.81	20 21.60	28 19 22.44
29	9	. .	. .	. .	. .	56.8	10.7	. .	22 42.54	* 3.48	−0.01	VI.	2	3.10	6 30.08	37.85	1.05	22 39.05	27 53 18.98
30	9	. .	. .	56.8	. .	24.9	. .	. .	26 10.78	3.46	−0.01	V.	2	12.4	11 4.02	37.64	1.52	26 7.31	27 57 53.18
31	8	. .	. .	. .	. .	22.0	35.8	. .	27 7.68	3.45	+0.01	VI.	9	7.48	42 50.90	37.58	4.95	27 4.24	28 29 43.43
32	10	32.3	. .	0.8	. .	. .	. .	. .	39 15.02	3.39	0.00	VI.	7	6.57	32 27.49	36.81	3.81	39 11.63	19 18.11
33	9	. .	49.2	. .	18.1	. .	. .	. .	15 55 17.74	− 3.29	+0.01	IV.	9	10.4	−43 59.67	−35.68	−5.08	15 55 14.46	−28 30 50.43

ZONE 119. MAY 28. K. BELT, −30° 38'. D_0 = −30° 15' 40".

No.	Mag.	I.	II.	III.	IV.	V.	VI.	VII.	T.	a_1	a_2	MICROMETER.			i	d_1	d_2	Mean Right Ascension, 1850.0.	Mean Declination, 1850.0.
1	9	. .	. .	. .	. .	. .	. .	4.9	13 40 21.13	+ 5.78	0.00	VII.	4	12.2	−21 0.71	−47.81	−3.49	13 40 26.91	−30 37 32.01
2	8	. .	. .	. .	. .	15.6	. .	. .	42 0.98	5.79	0.00	VI.	2	11.58	11 0.84	47.79	2.22	42 6.77	27 30.85
3	9	. .	. .	. .	. .	. .	. .	56.5	42 12.64	5.79	0.00	VII.	2	15.0	12.32.35	47.79	2.42	42 18.43	29 2.56
4	8	. .	. .	. .	. .	. .	. .	45.8	43 2.00	5.79	0.00	VII.	3	15.38	17 50.61	47.78	3.08	43 7.79	34 21.47
5	9	. .	. .	. .	. .	. .	55.3	10.8	44 26.78	5.80	0.00	VII.	9	6.31	42 11.81	47.77	6.24	44 32.58	58 45.82
6	7	19.0	33.9	. .	. .	. .	. .	. .	47 2.83	5.81	0.00	III.	4	2.40	16 17.73	47.74	2.89	47 8.64	32 48.36
7	9	. .	. .	. .	. .	26.2	. .	. .	47 11.70	5.81	0.00	VI.	8	3.33	35 43.30	47.74	5.40	47 17.51	52 16.44
8	10	. .	. .	. .	51.3	. .	. .	. .	48 51.14	5.82	0.00	IV.	8	9.00	38 28.39	47.72	5.76	48 56.96	55 1.87
9	10	. .	. .	. .	56.8	. .	. .	. .	49 56.64	5.82	0.00	IV.	1	6.58	3 30.77	47.71	1.25	50 2.46	30 19 59.73
10	9	. .	. .	. .	. .	52.9	. .	. .	50 38.43	5.82	0.00	VI.	9	10.22	44 8.54	47.70	6.51	50 44.25	31 0 42.75
11	11	. .	. .	56.8	. .	. .	. .	. .	52 11.11	5.83	0.00	III.	4	9.15	19 36.90	47.67	3.31	52 16.94	30 36 7.88
12	8	. .	. .	. .	38.2	. .	. .	. .	52 38.04	5.63	0.00	IV.	4	10.47	20 23.35	47.66	3.38	52 43.87	36 54.39
13	10	. .	. .	. .	. .	. .	3.2	. .	53 34.22	5.84	0.00	VI.	7	5.30	31 43.60	47.65	4.86	53 40.06	48 16.11
14	6	. .	. .	. .	. .	. .	. .	58.5	54 14.98	5.84	0.00	VII.	9	4.32	41 11.81	47.63	6.12	54 20.82	57 45.56
15	8	36.8	51.1	. .	. .	. .	. .	. .	58 20.42	5.86	0.00	IV.	5	8.52	24 24.56	47.56	3.92	58 26.28	40 56.04
16	9	. .	. .	. .	. .	49.3	. .	. .	13 58 34.80	5.86	0.00	V.	8	3.36	35 44.97	47.56	5.40	13 58 40.66	52 17.93
17	9	. .	. .	4.9	. .	. .	. .	. .	14 0 19.00	5.87	−0.01	III.	2	1.25?	5 41.80	47.52	1.54	14 0 24.86	22 10.86
18	10	. .	24.4	. .	. .	. .	. .	. .	1 53.34	5.88	0.00	III.	3	2.8	11 2.59	47.48	2.20	1 59.22	27 32.27
19	7	. .	42.5	. .	. .	. .	. .	. .	4 11.48	5.89	0.00	III.	4	0.40	15 17.22	47.43	2.75	4 17.37	31 47.40
20	8	. .	. .	. .	. .	7.9	. .	. .	5 53.25	5.90	−0.01	V.	1	8.34	4 19.14	47.38	1.36	5 59.13	20 47.88
21	9	. .	. .	. .	. .	. .	. .	43.9	6 0.02	5.90	0.00	VII.	2	11.47	10 55.05	47.38	2.21	6 5.92	27 24.64
22	11	. .	9.2	. .	. .	. .	. .	. .	8 38.25	5.91	0.00	III.	5	4.4	21 59.29	47.31	3.60	8 44.16	38 30.20
23	9	. .	. .	. .	. .	13.5	. .	. .	9 58.85	5.92	−0.01	V.	1	7.15	3 39.29	47.28	1.28	10 4.76	30 20 7.85
24	9	. .	. .	. .	. .	14.4	. .	. .	10 59.93	5.92	+0.01	VI.	9	10.18	44 6.52	47.25	6.51	11 5.86	31 0 40.28
25	9	. .	8.8	. .	. .	. .	. .	. .	12 38.13	5.93	0.01	II.	9	12.53	45 24.67	47.20	6.69	12 44.07	1 58.56
26	8	. .	. .	21.7	. .	. .	. .	. .	13 36.41	5.94	0.01	III.	9	11.55	44 55.58	47.17	6.62	13 42.36	1 29.37
27	10	. .	. .	. .	. .	25.0	. .	. .	14 10.53	5.94	+0.01	V.	9	11.24	44 39.95	47.15	6.58	14 16.48	31 1 13.68
28	10	. .	. .	36.0	. .	. .	. .	. .	15 50.26	5.95	0.00	IV.	3	12.34	16 18.30	47.09	2.88	15 56.21	30 32 48.27
29	8	. .	. .	38.3	. .	. .	. .	. .	16 52.83	5.95	0.00	III.	7	7.13	32 35.68	47.06	4.99	16 58.78	49 7.73
30	10	. .	. .	. .	. .	32.3	. .	. .	19 17.69	5.97	0.00	V.	3	6.46	13 22.78	46.98	2.50	19 23.66	29 52.26
31	9	. .	. .	. .	. .	. .	. .	25.8	19 41.93	5.97	0.00	VII.	2	13.56	12 0.09	46.97	2.35	19 47.90	28 29.41
32	10	. .	. .	. .	. .	8.9	. .	. .	14 22 54.40	+ 5.98	0.00	VI.	8	4.52	−36 23.14	−46.84	−5.49	14 23 0.38	−30 52 55.47

CORRECTIONS.

Date.		Corr. of Clock.	Hourly rate.	m	n	c
1847. May 28,	h. 14	s. f 1.32	s. l 0.042	s.	s.	s.

INSTRUMENT READINGS.

	Date.	Barom.	THERMOM. At.	THERMOM. Ex.
	h. m.	in.	°	°
Zone 119	1847. May 28, 13 40	29.88	75.2	71.7

REMARKS.

(119) 23. Minutes assumed as 8 instead of 9.

ZONE 119. MAY 28. K. BELT, −30° 38′. D_o=−30° 15′ 40″—Continued.

No.	Mag.	SECONDS OF TRANSIT. I.	II.	III.	IV.	V.	VI.	VII.	T.	a_1	a_2	MICROMETER.			i	d_1	d_2	Mean Right Ascension, 1850.0.	Mean Declination, 1850.0.
									h. m. s.	s.	s.			r.	′ ″	″	″	h. m. s.	° ′ ″
33	9	. .	. .	7.8	. .	. .	. .	. .	14 24 21.83	+ 5.99	−0.01	IV.	1	3.51	− 1 56.47	−46.77	−1.06	14 24 27.81	−30 18 24.30
34	8	. .	10.2	. .	. .	. .	. .	. .	27 39.09	6.01	−0.01	III.	1	12.23	6 14.59	46.63	1.61	27 45.09	30 22 42.83
35	9.10	. .	. .	. .	. .	40.0	. .	. .	30 25.53	6.02	+0.01	VI.	9	10.5	43 59.97	46.51	6.50	30 31.56	31 0 32.98
36	7	. .	. .	. .	. .	4.2	. .	. .	32 49.53	6.03	−0.01	V.	1	1.30	0 45.33	46.40	0.89	32 55.55	30 17 12.62
37	10	. .	. .	. .	. .	. .	. .	3.2	33 19.55	6.04	0.00	VII.	7	2.42	30 18.64	46.37	4.69	33 25.59	46 49.70
38	10	. .	. .	. .	. .	14.9	. .	. .	35 0.38	6.05	0.00	V.	7	5.16	31 36.69	46.28	4.86	35 6.43	30 48 7.83
39	9	. .	. .	7.3	. .	. .	. .	. .	37 22.03	6.06	+0.01	III.	10	4.53	46 23.69	46.17	6.82	37 28.10	31 2 56.68
40	9	. .	47.6	. .	16.7	. .	. .	. .	39 16.65	6.07	0.00	IV.	6	11.25	29 42.82	46.08	4.62	39 22.72	30 46 13.52
41	10	. .	12.3	. .	. .	. .	. .	. .	40 41.43	6.07	0.00	III.	6	7.12	27 35.19	46.00	4.33	40 47.50	44 5.52
42	9	. .	15.6	. .	. .	. .	. .	. .	43 44.52	6.09	−0.01	III.	2	7.41	8 51.40	45.83	1.93	43 50.60	30 25 19.16
43	8	. .	. .	39.0	. .	. .	. .	. .	44 53.70	6.09	0.00	III.	9	10.29	44 12.22	45.77	6.55	44 59.79	31 0 44.54
44	9	34.4	49.3	. .	. .	. .	. .	. .	46 18.42	6.10	0.00	III.	7	5.43	31 50.30	45.69	4.87	46 24.52	30 48 20.86
45	10	. .	. .	. .	. .	. .	49.3	. .	46 20.29	6.10	0.00	VII.	8	9.6	38 30.97	45.69	5.78	46 26.39	55 2.44
46	9	. .	32.8	. .	. .	. .	. .	. .	49 2.02	6.12	0.00	II.	8	4.8	36 0.95	45.53	5.45	49 8.14	52 31.93
47	9	21.1	. .	. .	. .	. .	. .	. .	51 4.91	6.13	0.00	II.	4	9.29	19 43.81	45.41	3.32	51 11.04	36 12.54
48	8	. .	. .	. .	. .	53.5	. .	. .	51 39.01	6.13	0.00	II.	8	12.23	40 10.54	45.38	6.00	52 45.14	56 41.92
49	8	. .	. .	. .	3.4	. .	. .	. .	53 3.24	6.14	0.00	V.	9	8.26	43 10.20	45.29	6.39	53 9.38	59 41.88
50	7	. .	51.4	. .	. .	. .	. .	. .	54 20.55	6.14	0.00	IV.	6	12.54	30 27.68	45.21	4.72	54 26.69	46 57.61
51	9	. .	. .	. .	41.0	55.4	. .	. .	55 40.84	6.15	0.00	VII.	5	9.45	24 50.83	45.12	3.98	55 46.99	41 19.93
52	6	. .	. .	8.2	. .	. .	. .	. .	14 57 22.26	6.16	0.00	IV.	1	6.52	3 27.75	45.02	1.23	14 57 28.42	19 54.00
53	10	. .	. .	20.3	. .	. .	. .	. .	15 2 34.88	6.19	+0.01	IV.	8	4.55	36 21.85	44.67	5.52	15 2 41.08	52 55.04
54	7	. .	. .	. .	. .	. .	24.6	. .	2 55.78	6.19	−0.01	VI.	1	4.30	2 15.95	44.64	1.07	3 1.96	18 41.66
55	4.5	. .	52.0	6.8	. .	. .	. .	. .	5 21.36	6.20	+0.01	III.	9	3.42	40 46.99	44.47	6.07	5 27.57	57 17.53
56	11	. .	. .	. .	. .	14.5	. .	. .	5 59.87	6.21	−0.01	V.	2	6.49	8 25.19	44.35	1.86	6 6.07	24 51.40
57	9	37.6	. .	. .	. .	. .	. .	. .	8 21.63	6.22	+0.01	II.	8	7.26	37 40.79	44.26	5.68	8 27.86	54 10.73
58	7.6	41.0	55.6	. .	. .	. .	. .	. .	9 24.75	6.23	0.00	III.	5	5.49	22 52.23	44.19	3.72	9 30.98	39 20.14
59	10	. .	. .	. .	44 4	. .	. .	. .	12 44.24	6.24	+0.01	IV.	9	8.2	42 58.15	43.93	6.39	12 50.49	59 28.45
60	7	. .	38.3	53.1	. .	. .	. .	. .	17 7.53	6.27	0.00	III.	7	3.35	30 45.76	43.59	4.75	17 13.80	47 14.10
61	6	41.0	55.7	. .	. .	. .	. .	. .	19 25.02	6.28	+0.01	III.	9	3.11	40 31.35	43.41	6.06	19 31.31	57 0.82
62	10	. .	. .	. .	. .	. .	52.0	. .	19 23.04	6.28	0.00	VI.	7	1.42	29 48.63	43.41	4.64	19 29.32	46 16.68
63	9	. .	30.2	. .	59.6	. .	. .	. .	22 59.34	6.29	0.00	IV.	5	2.33	21 13.45	43.11	3.49	23 5.63	37 40.05
64	9	. .	5.9	20.6	. .	. .	. .	. .	24 34.85	6.30	−0.01	III.	3	7.59	13 59.58	42.99	2.56	24 41.14	30 25.13
65	8	. .	. .	. .	23.6	. .	. .	. .	53 23.44	6.46	−0.01	IV.	3	9.21	14 40.97	40.29	2.65	53 29.89	30 31 3.91
66	9	. .	11.3	. .	. .	. .	. .	. .	55 40.64	6.47	+0.02	III.	10	4.5	45 59.49	40.05	6.80	55 47.13	31 2 26.34
67	8	. .	. .	. .	. .	6.3	. .	. .	55 51.84	6.47	0.02	VII.	10	4.10	46 1.61	40.03	6.80	55 58.33	31 2 28.44
68	10	. .	. .	. .	. .	10.6	. .	. .	56 56.11	6.47	+0.01	VI	9	0.50	39 20.11	39.91	5.92	57 2.59	30 55 45.94
69	9	. .	. .	. .	. .	35.3	. .	. .	58 20.64	6.48	−0.02	V.	1	3.57	1 59.45	39.76	1.02	15 58 27.10	18 20.23
70	7	. .	26.8	41.5	. .	. .	. .	. .	59 55.85	6.49	0.00	IV.	5	4.58	22 26.56	39.61	3.66	16 0 2.34	38 49.83
71	10	. .	. .	. .	. .	. .	. .	4.8	15 59 21.20	6.49	+0.01	VII.	8	0.3	33 57.16	39.66	5.20	15 59 27.70	50 22.02
72	9	. .	. .	. .	. .	. .	. .	37.6	16 1 53.64	6.50	−0.02	VII.	1	8.18	4 10.66	39.38	1.30	16 2 0.12	20 31.34
73	9	. .	. .	. .	. .	. .	. .	41.7	2 57.90	6.51	0.01	VII.	4	6.51	18 23.89	39.15	3.14	3 4.40	34 46.18
74	9	. .	43.8	. .	. .	. .	. .	. .	7 12.74	6.53	−0.01	III.	3	7.35	13 47.48	38.78	2.52	7 19.26	30 8.78
75	10	. .	. .	46.6	. .	. .	. .	. .	9 3.06	6.54	0.00	IV.	6	9.31	28 45.32	38.58	4.50	9 9.60	45 8.40
76	10	. .	. .	. .	. .	. .	50.4	. .	9 21.50	6.55	−0.01	VI.	4	7.29	18 43.30	38.54	3.18	9 28.04	35 5.02
77	7	. .	. .	. .	. .	. .	. .	40.8	10 57.03	6.55	0.00	VII.	4	11.17	20 38.02	38.36	3.42	11 3.58	36 59.80
78	8	. .	. .	31.2	. .	. .	. .	. .	12 45.79	6.56	+0.01	IV.	8	6.7	37 1.16	38.15	5.59	12 52.36	53 24.90
79	8	. .	. .	. .	. .	6.3	. .	. .	12 51.78	6.56	0.01	VI.	7	7.58	32 58.22	38.14	5.07	12 58.35	30 49 21 43
80	7	. .	58.3	. .	. .	. .	. .	. .	15 27.66	6.58	0.02	III.	10	7.43	47 49.41	37.83	7.04	15 34.26	31 4 14.28
81	8	. .	54.9	. .	. .	. .	. .	. .	16 16 24.13	+ 6.58	+0.01	III.	8	5.7	−36 30.85	−37.72	−5.53	16 16 30.72	−30 52 54.10

CORRECTIONS.

Date.	Corr. of Clock.	Hourly rate.	m	n	c
1847. h.	s.	s.	s.	s.	s.

INSTRUMENT READINGS.

Date.	Barom.	THERMOM. At.	Ex.
1847. h. m.	in.	°	°

REMARKS.

(119) 48. **Minutes assumed as** 52 instead of 51.

ZONE 119. MAY 28. K. BELT, −30° 38′. D_o = −30° 15′ 40″—Continued.

No.	Mag.	SECONDS OF TRANSIT. I.	II.	III.	IV.	V.	VI.	VII.	T.	a_1	a_2	MICROMETER.			i	d_1	d_2	Mean Right Ascension, 1850.0.	Mean Declination, 1850.0.
									h. m. s.	s.	s.			r.	′ ″	″	″	h. m. s.	° ′ ″
82	10	. .	. .	59.0	. .	. .	. .	. .	16 17 13.59	+ 6.59	+0.01	IV.	8	5.40	−36 47.54	−37.62	−5.56	16 17 20.19	−30 53 10.72
83	9	. .	16.0	. .	. .	. .	. .	. .	18 45.16	6.59	0.00	III.	7	3.20	30 38.19	37.52	4.74	18 51.75	47 0.45
84	10	. .	. .	6.5	. .	. .	. .	. .	20 20.88	6.60	0.00	III.	5	8.45	24 20.98	37.25	3.92	20 27.48	40 42.15
85	9	. .	32.8	. .	2.3	. .	. .	. .	24 2.22	6.62	+0.01	III.	9	5.7	41 29.85	36.78	6.19	24 8.85	57 52.82
86	9	. .	. .	. .	. .	46.8	. .	. .	24 32.32	6.62	+0.01	V.	9	4.53	41 22.79	36.72	6.18	24 38.95	57 55.69
87	10	. .	13.5	. .	. .	. .	. .	. .	26 42.65	6.63	0.00	II.	7	2.5	30 0.23	36.45	4.66	26 49.28	30 46 21.34
88	11	. .	. .	. .	2.3	. .	. .	. .	29 2.14	6.65	+0.02	VI.	10	3.23	45 38.15	36.15	6.75	29 8.81	31 2 1.05
89	9	21.3	. .	. .	. .	. .	. .	. .	31 5.07	6.66	−0.01	II.	4	1.8	15 31.18	35.89	2.76	31 11.72	30 31 49.83
90	9	. .	. .	. .	28.2	. .	. .	. .	33 28.04	6.67	+0.02	IV.	10	5.54	46 54.50	35.58	6.92	33 34.73	31 3 17.00
91	8	. .	. .	. .	. .	21.9	. .	. .	34 52.89	6.67	+0.01	VII.	9	1.53	39 51.62	35.49	5.98	33 59.56	30 56 13.09
92	7	. .	. .	. .	. .	. .	. .	14.6	34 30.91	6.68	0.00	VII.	6	5.48	26 52.42	35.44	4.24	34 37.59	43 12.10
93	6	. .	. .	. .	. .	54.9	. .	24.0	35 40.24	6.68	−0.01	VII.	3	10.12	15 6.25	35.29	2.70	35 46.91	31 24.24
94	8	. .	. .	. .	. .	33.0	. .	. .	37 18.46	6.69	0.00	VI.	6	5.37	26 47.13	35.07	4.22	37 25.15	43 6.42
95	7	. .	. .	29.0	. .	. .	. .	. .	38 43.61	6.70	+0.01	IV.	8	9.53	39 25.26	34.89	5.93	38 50.32	55 46.18
96	10	. .	. .	. .	31.9	. .	. .	. .	39 31.74	6.71	−0.01	IV.	3	8.36	14 18.30	34.78	2.60	39 38.44	30 35.68
97	8	48.3	. .	. .	. .	. .	. .	. .	41 31.98	6.72	−0.01	II.	2	6.24	8 12.43	34.51	1.82	41 38.69	24 28.76
98	9	. .	. .	7.4	. .	. .	. .	. .	42 22.03	6.72	+0.01	IV.	8	11.36	39 47.06	34.40	5.97	42 28.76	56 7.43
99	9	. .	. .	. .	4.6	. .	. .	. .	43 4.44	6.73	−0.01	IV.	2	9.44	9 53.48	34.30	2.04	43 11.16	26 9.82
100	10	. .	57.2	. .	. .	. .	. .	. .	44 26.17	6.73	0.01	III.	3	7.19	13 39.41	34.11	2.53	44 32.89	29 56.05
101	9	. .	. .	. .	. .	1.5	. .	. .	44 46.89	6.73	−0.01	V.	3	3.34	11 45.96	34.07	2.27	44 53.61	28 2.30
102	8	. .	. .	. .	. .	. .	. .	49.1	45 5.42	6.74	0.00	VII.	6	6.45	27 21.17	34.02	1.30	45 12.16	30 43 39.49
103	7	. .	. .	. .	34.4	. .	. .	. .	46 34.24	6.74	+0.02	VI.	10	6.42	47 18.51	33.82	6.97	46 41.00	31 3 39.30
104	9	. .	. .	. .	. .	. .	. .	41.9	47 57.93	6.75	−0.02	VII.	1	7.53	3 38.05	33.63	1.27	47 4.66	30 20 12.95
105	8	. .	. .	39.6	. .	. .	. .	. .	48 54.12	6.75	0.00	IV.	4	10.15	20 7.21	33.50	3.35	49 0.87	36 24 06
106	9	. .	. .	. .	33.6	. .	. .	. .	49 33.44	6.76	+0.02	IV.	9	4.15	41 3.68	33.41	6.15	49 40.22	57 23.24
107	9	7.8	. .	. .	. .	. .	. .	. .	51 51.55	6.77	−0.01	II.	3	9.31	14 45.82	33.08	2.66	51 58.31	31 1.56
108	7	. .	. .	53.2	. .	. .	. .	. .	52 7.46	6.77	0.01	III.	4	3.43	16 49.49	33.05	2.93	52 14.22	33 5.47
109	8	. .	. .	. .	. .	42.0	. .	. .	52 27.38	6.77	−0.01	VI.	2	8.41	9 21.51	33.00	1.94	52 34.14	25 36.45
110	9	. .	. .	. .	. .	. .	35.6	. .	53 6.61	6.78	+0.01	VI.	7	9.54	33 56.72	32.91	5.20	53 13.40	50 14.83
111	11	. .	. .	. .	. .	. .	. .	23.0	53 39.09	6.78	−0.01	VII.	2	7.58	8 59.57	32.83	1.90	53 45.86	25 14.30
112	10	. .	. .	. .	. .	. .	28.3	. .	55 59.28	6.78	+0.01	VII.	9	4.14	41 2.73	32.51	6.15	55 6.07	57 21.39
113	9	2.5	. .	. .	. .	. .	. .	. .	57 46.15	6.80	−0.02	II.	2	1.17	5 37.62	32.26	1.46	57 52.93	21 51.34
114	6	. .	. .	. .	0.2	. .	. .	. .	16 58 0.04	6.80	+0.01	V.	8	4.1	35 57.57	32.22	5.46	16 58 6.85	52 15.25
115	8	39.0	. .	. .	. .	. .	. .	. .	17 0 22.83	6.81	0.00	II.	5	4.18	22 6.19	31.89	3.60	17 0 29.64	38 21.68
116	10	. .	. .	23.8	. .	. .	. .	. .	0 38.29	6.81	+0.01	IV.	7	2.20	30 7.97	31.85	4.67	0 45.11	46 24.49
117	9	42.3	. .	. .	. .	. .	. .	. .	2 26.10	6.82	−0.01	III.	4	7.23	18 40.42	31.59	3.17	2 32.91	34 55.18
118	10	. .	. .	. .	48.3	. .	. .	. .	2 48.14	6.83	+0.01	V.	7	12.0	35 0.40	31.53	5.33	2 54.98	51 17.26
119	9	. .	. .	. .	17.8	. .	. .	. .	17 5 17.64	+ 6.84	+0.01	IV.	7	12.37	−35 19.11	−31.17	−5.37	17 5 24.40	−30 51 35.65

ZONE 120.* MAY 28. K. BELT, −30° 38′. D_o = −30° 33′ 0″.

No.	Mag.	I.	II.	III.	IV.	V.	VI.	VII.	T.	a_1	a_2	MICROMETER.			i	d_1	d_2	Mean Right Ascension, 1850.0.	Mean Declination, 1850.0.
1	8	25.8	. .	. .	. .	. .	. .	. .	17 17 9.65	+ 6.91	0.00	II.	5	7.24	−23 39.98	−28.37	−3.82	17 17 16.56	−30 57 12.18
2	7	40.9	55.8	. .	. .	. .	. .	. .	18 24.91	6.91	+0.01	III.	7	3.51	30 53.82	28.19	4.78	18 31.83	31 4 26.79
3	10	. .	. .	. .	. .	48.5	. .	. .	18 33.87	6.91	−0.02	VII.	2	4.48	7 23.77	28.18	1.70	18 40.76	30 40 53.65
4	8	. .	. .	. .	. .	. .	. .	43.8	18 59.96	6.91	−0.01	VII.	3	9.5	14 32.46	28.11	2.62	19 6.86	48 3.19
5	10	. .	. .	56.8	. .	. .	. .	. .	23 11.23	6.94	0.00	IV.	6	4.30	26 13.55	27.54	4.17	23 18.17	59 45.26
6	8	. .	. .	. .	. .	5.0	. .	. .	22 50.45	6.93	0.00	V.	5	10.41	25 19.47	27.59	4.04	22 57.39	58 51.10
7	9	. .	. .	. .	. .	. .	49.5	. .	17 24 20.57	+ 6.94	0.00	VII.	5	7.33	−23 41.27	−27.39	−3.84	17 24 27.51	−30 57 15.50

CORRECTIONS. / INSTRUMENT READINGS.

Date.	Corr. of Clock.	Hourly rate.	m	n.	c		Date.	Barom.	THERMOM. At.	Ex.
1847. h.	s.	s.	s.	s.	s.		1847. h. m.	in.	°	°

* The stars during the first and last parts of the night's work evidently differ by about 16′ in declination. A jar or change of some kind evidently occurred between $17^h 0^m$ and $17^h 18^m$, and the new zone is assumed to begin previous to the observation of $17^h 17^m$.

REMARKS.

(119) 91. Transit over T. VI assumed as recorded over T. V, and minutes as 33.
(119) 95. Micrometer reading assumed as $10^r.53$ instead of $9^r.53$.
[Found that the instrument had been moved 20′ by accident at $17^h 5^m$.—Observer.]

ZONE 120. MAY 28. K. BELT, −30° 38′. D_o = −30° 33′ 0″—Continued.

No.	Mag.	Seconds of Transit. I.	II.	III.	IV.	V.	VI.	VII.	T.	a_1	a_2	Micrometer.			i	d_1	d_2	Mean Right Ascension, 1850.0.	Mean Declination, 1850.0.
									h. m. s.	s.	s.			r.	′ ″	″	″	h. m. s.	° ′ ″
8	9	20.8	...	...	...	...	...	...	17 28 4.56	+ 6.96	−0.01	II.	3	10.17	−15 9.02	−26.89	−2.70	17 28 11.51	−30 48 38.61
9	9	...	23.5	...	...	...	...	...	32 52.37	6.98	−0.02	III.	1	9.22	4 43.33	26.22	1.36	32 59.33	30 38 10.91
10	8	32.2	...	...	...	...	...	...	34 16.23	6.99	+0.01	II.	8	8.58	35 27.18	26.03	5.79	34 23.23	31 11 59.00
11	10	...	...	...	...	...	32.3	...	34 3.25	6.99	+0.02	VI.	10	5.21	46 37.65	26.06	6.90	34 10.26	20 10.61
12	9	...	...	...	32.5	...	...	...	35 32.34	7.00	0.00	IV.	6	9.49	28 54.40	25.85	4.52	35 39.34	31 2 24.77
13	9	...	...	...	...	...	11.2	...	35 42.26	7.00	0.00	VI.	5	11.22	25 39.99	25.83	4.09	35 49.26	30 59 9.91
14	9	...	18.9	33.4	...	...	...	...	37 48.03	7.01	+0.01	IV.	7	10.17	34 8.52	25.54	5.21	37 55.05	31 7 39.27
15	9	...	...	...	...	23.8	...	...	38 9.21	7.01	−0.01	V.	3	13.23	16 42.94	25.49	2.90	38 16.21	30 50 11.33
16	8	...	...	...	28.2	...	...	...	39 28.04	7.02	+0.01	V.	8	8.51	38 23.80	25.31	5.80	39 35.07	31 11 54.91
17	9	...	...	...	...	17.4	...	...	40 2.80	7.02	−0.01	VI.	3	12.29	16 15.57	25.24	2.84	40 9.81	30 49 43.65
18	8	...	...	...	...	...	50.4	...	40 21.49	7.02	0.01	VI.	4	11.53	20 56.42	25.18	3.46	40 28.50	54 25.66
19	9	...	...	...	...	...	32.0	...	41 3.09	7.03	0.01	VI.	4	11.29	20 44.32	25.09	3.43	41 10.11	54 12.84
20	8	...	...	...	...	...	...	1.1	41 17.34	7.03	−0.01	VII.	4	13.32	21 46.09	25.06	3.55	41 24.36	30 55 14.70
21	6	...	...	...	...	...	...	11.6	42 28.11	7.03	+0.02	VII.	9	9.0	43 26.94	24.89	6.49	42 35.16	31 16 58.32
22	10	...	...	...	...	...	...	0.3	43 16.82	7.04	0.02	VII.	9	10.40	44 17.37	24.78	6.62	43 23.88	17 48.77
23	10	...	...	...	...	41.3	...	...	44 26.81	7.04	+0.01	VI.	8	11.49	39 53.41	24.63	6.02	44 33.86	31 13 24.06
24	10	...	...	32.8	...	...	...	...	45 46.89	7.05	−0.02	III.	1	11.4	5 34.77	24.43	1.44	45 53.92	30 39 0.64
25	9	...	...	...	17.8	...	...	...	46 17.64	7.05	+0.01	IV.	8	11.17	39 37.47	24.36	5.98	46 24.70	31 13 7.81
26	8	...	15.0	...	...	...	...	...	47 44.10	7.06	0.00	III.	5	11.40	25 49.22	24.17	4.11	47 51.16	30 59 17.50
27	9	...	27.5	...	...	...	...	...	48 56.69	7.06	+0.01	III.	7	8.16	33 7.46	24.00	5.09	49 3.76	31 6 36.55
28	9	...	...	...	0.3	...	...	...	49 0.14	7.07	−0.01	VI.	3	2.29	11 13.03	23.99	2.18	49 7.20	30 44 39.20
29	9	...	...	33.5	...	...	...	...	50 48.00	7.07	+0.01	III.	7	4.22	31 9.46	23.74	4.82	50 55.08	31 4 38.02
30	10	...	...	...	25.6	...	...	...	51 25.44	7.08	0.00	IV.	5	8.56	24 26.58	23.65	3.93	51 32.52	30 57 54.16
31	9	14.4	...	...	...	...	...	...	52 58.26	7.08	0.00	III.	5	8.33	24 14.93	23.43	3.91	53 5.34	57 42.27
32	10	...	...	...	11.0	...	...	...	53 10.84	7.09	0.00	IV.	5	8.26	24 11.45	23.40	3.91	53 17.93	57 38.76
33	10	...	...	...	...	...	48.0	...	53 19.12	7.09	−0.01	VI.	3	6.22	13 10.52	23.39	2.42	53 26.20	30 46 36.33
34	10	...	...	...	30.8	...	...	...	55 30.64	7.10	0.00	IV.	6	11.40	29 50.37	23.08	4.64	55 37.74	31 3 18.09
35	9	...	...	...	20.0	...	...	...	56 19.84	7.10	+0.02	IV.	8	14.41	41 20.32	22.97	6.19	56 26.96	31 14 49.48
36	10	...	...	...	5.0	...	...	...	57 4.84	7.11	−0.02	V.	2	11.41	10 52.42	22.86	2.12	57 11.93	30 44 17.40
37	8	...	45.6	...	...	...	...	...	58 14.75	7.11	0.00	III.	6	6.48	27 23.08	22.70	4.31	58 21.86	31 0 50.09
38	9	...	...	...	31.0	...	...	...	58 30.84	7.11	0.00	V.	6	7.26	27 42.24	22.66	4.35	58 37.95	1 9.25
39	9	...	...	...	...	24.8	...	...	17 59 10.30	7.12	+0.01	VI.	8	6.13	37 3.98	22.56	5.62	17 59 17.43	31 10 32.16
40	5	...	...	...	19.1	33.7	...	...	18 0 19.02	7.12	−0.01	VI.	3	2.52	11 24.63	22.41	1.21	18 0 26.13	30 44 49.25
41	9	...	...	...	2.3	...	...	...	3 2.14	7.13	0.00	IV.	5	10.22	25 9.94	22.02	4.02	3 9.27	30 58 35.98
42	8	37.8	52.6	...	...	...	...	...	6 21.85	7.15	+0.01	III.	8	9.11	38 33.88	21.53	5.81	6 29.01	31 12 1.22
43	8	...	...	...	29.1	43.3	...	...	6 28.89	7.15	0.02	V.	10	8.47	48 21.69	21.52	7.15	6 36.06	21 50.36
44	8	...	...	...	26.8	...	...	...	7 26.64	7.15	+0.02	IV.	10	6.24	47 9.63	21.39	6.99	7 33.81	31 20 38.01
45	10	...	...	...	...	12.6	...	...	7 57.95	7.15	−0.02	V.	1	7.3	3 33.24	21.31	1.18	8 5.08	30 36 55.73
46	10	...	...	...	...	50.8	...	...	8 36.19	7.15	−0.01	VI.	3	4.35	12 16.57	21.22	2.31	8 43.33	30 45 40.10
47	9	...	7.7	...	...	...	...	...	10 37.04	7.16	+0.02	II.	10	10.11	49 3.89	20.93	7.26	10 44.22	31 22 32.08
48	6.7	44.8	59.3	...	...	...	...	...	11 28.59	7.17	0.00	III.	6	5.16	26 36.69	20.80	4.20	11 35.76	31 0 1.69
49	10	...	...	...	...	...	50.3	...	11 21.45	7.17	−0.02	VII.	2	7.19	8 39.91	20.81	1.84	11 28.60	30 42 2.57
50	9	36.5	...	...	...	...	...	...	14 20.58	7.18	+0.02	II.	9	7.57	42 55.42	20.39	6.44	14 27.78	31 16 22.25
51	8	...	...	23.9	...	...	...	...	14 38.09	7.18	−0.01	IV.	3	4.28	12 13.23	20.36	2.30	14 45.26	30 45 35.89
52	5	...	...	2.1	...	31.0	...	...	15 16.33	7.19	−0.02	V.	3	2.35	16 15.21	20.26	2.85	15 23.50	30 49 38.32
53	10	...	...	...	...	...	...	12.4	15 28.71	7.19	0.00	VII.	6	5.40	26 48.39	20.23	4.22	15 35.90	31 0 12.84
54	8	53.2	...	...	...	...	...	...	18 37.25	7.20	+0.02	II.	8	11.8	39 32.73	19.78	5.97	18 44.47	12 58.48
55	9	48.2	...	...	...	...	...	...	20 32.14	7.21	0.00	II.	6	11.41	29 50.67	19.51	4.64	20 39.35	31 3 14.82
56	9	...	...	33.6	...	...	...	...	18 20 47.94	+ 7.21	0.00	III.	4	13.30	−21 45.48	−19.46	−3.56	18 20 55.15	−30 55 8.50

CORRECTIONS.

Date.		Corr. of Clock.	Hourly rate.	m	n	c
1847.	h.	s.	s.	s.	s.	s.

INSTRUMENT READINGS.

Date.		Barom.	Thermom. At.	Thermom. Ex.
1847.	h. m.	in.	°	°

REMARKS.

(120) 52. Hor. thread assumed as 4 instead of 5.

ZONE 120. MAY 28. K. BELT, −30° 38′. D₀ = −30° 33′ 0″—Continued.

No.	Mag.	SECONDS OF TRANSIT. I.	II.	III.	IV.	V.	VI.	VII.	T.	a_1	a_2	MICROMETER.			i	d_1	d_2	Mean Right Ascension, 1850.0.	Mean Declination, 1850.0.
									h. m. s.	s.	s.			r.	′ ″	″	″	h. m. s.	° ′ ″
57	10	. .	. .	40.6	. .	. .	. .	. .	18 21 55.04	+ 7.22	0.00	IV.	6	5.59	−26 58.42	−19.30	−4.26	18 22 2.26	−31 0 21.98
58	9	. .	. .	. .	28.5	. .	. .	. .	22 28.34	7.22	+0.02	IV.	9	3.9	40 30.40	19.21	6.10	22 35.58	31 13 55.71
59	9	. .	. .	. .	. .	. .	. .	52.8	25 8.99	7.23	−0.01	VII.	4	4.17	17 6.24	18.84	2.93	25 16.21	30 50 28.01
60	8	55.7	. .	. .	. .	. .	. .	. .	27 39.51	7.24	−0.01	II.	4	9.51	19 54.90	18.46	3.30	27 46.74	53 16.66
61	9	. .	. .	17.3	. .	. .	. .	. .	28 31.71	7.24	0.00	IV.	5	11.58	25 58.34	18.34	4.13	28 38.95	59 20.81
62	7	. .	. .	. .	. .	21.8	. .	. .	29 7.16	7.25	−0.02	VI.	1	10.24	5 14.45	18.25	1.39	29 14.39	30 38 34.09
63	10	13.9	. .	. .	. .	. .	. .	. .	30 57.84	7.25	0.00	II.	6	11.14	29 37.06	17.97	4.62	31 5.09	31 2 59.65
64	8	. .	. .	1.7	. .	. .	. .	. .	31 15.80	7.26	−0.02	IV.	1	12.18	6 12.12	17.93	1.51	31 23.04	30 39 31.56
65	9	. .	. .	. .	59.1	. .	. .	. .	31 58.94	7.26	+0.02	IV.	9	11.39	44 47.57	17.81	6.69	32 6.22	31 18 12.07
66	9	. .	. .	. .	. .	. .	58.2	. .	32 29.32	7.26	−0.02	VII.	3	5.1	12 29.42	17.75	2.33	32 36.56	30 45 49.50
67	8	47.8	. .	. .	. .	. .	. .	. .	37 31.71	7.29	0.00	II.	6	7.10	27 34.02	17.02	4.33	37 39.00	31 0 55.37
68	8	. .	. .	0.3	. .	. .	. .	. .	39 14.52	7.30	−0.01	III.	3	8.1	14 0.59	16.78	2.52	39 21.81	30 47 19.89
69	8	. .	. .	. .	. .	. .	52.6	. .	39 37.96	7.30	−0.02	V.	1	9.21	4 42.82	16.75	1.32	39 45.24	30 38 0.89
70	8	. .	. .	. .	. .	57.2	. .	. .	40 42.69	7.30	+0.01	V.	7	10.49	34 24.61	16.56	5.28	40 50.00	31 7 46.45
71	9	. .	. .	. .	. .	. .	42.8	. .	41 13.86	7.30	0.00	VI.	5	9.2	24 29.40	16.48	3.94	41 21.16	30 57 49.82
72	8	. .	. .	. .	39.8	. .	. .	. .	42 39.64	7.31	+0.01	VI.	8	4.34	36 14.07	16.26	5.52	42 46.96	31 9 35.85
73	6.7	. .	. .	. .	. .	. .	25.8	. .	42 56.88	7.31	0.00	VI.	5	2.19	21 6.19	16.24	3.49	43 4.19	30 54 25.92
74	9	. .	24.3	. .	. .	. .	. .	. .	44 53.63	7.32	+0.02	II.	10	2.32	45 12.45	15.92	6.74	45 0.97	31 18 35.11
75	9	. .	15.7	. .	. .	. .	. .	. .	46 44.87	7.33	0.01	III.	7	5.21	31 39.20	15.68	4.90	46 52.21	4 59.78
76	10	3.0	. .	. .	. .	. .	. .	. .	47 46.96	7.33	+0.01	II.	7	5.41	31 49.14	15.52	4.92	47 54.30	5 9.59
77	8	. .	52.8	. .	. .	. .	. .	. .	48 21.93	7.33	0.00	III.	6	7.25	27 41.73	15.43	4.35	48 29.26	1 1.51
78	7.8	. .	. .	. .	36.5	. .	. .	. .	48 36.34	7.33	+0.02	V.	8	12.41	40 19.77	15.40	6.09	48 43.69	31 13 41.26
79	9	. .	. .	. .	. .	. .	23.9	. .	48 54.96	7.34	0.00	VI.	5	11.53	25 55.62	15.36	4.12	49 2.30	30 59 15.10
80	8	53.5	. .	. .	. .	. .	. .	. .	52 37.50	7.35	+0.01	II.	7	12.46	35 23.45	14.82	5.40	52 44.86	31 8 43.67
81	9	2.6	. .	. .	. .	. .	. .	. .	53 46.46	7.36	0.00	II.	5	8.57	24 26.88	14.65	3.93	53 53.82	30 57 45.46
82	6.5	55.8	10.8	. .	. .	. .	. .	. .	54 39.99	7.36	+0.02	III.	9	6.49	42 21.29	14.51	6.37	54 47.37	31 15 42.17
83	9	. .	. .	. .	0.9	. .	. .	. .	55 0.74	7.36	+0.01	IV.	8	5.15	36 34.93	14.47	5.57	55 8.11	9 54.97
84	9	. .	. .	. .	45.3	. .	. .	. .	57 45.14	7.37	0.00	IV.	6	6.47	27 22.63	14.06	4.31	57 52.51	0 41.00
85	8	. .	. .	32.8	. .	. .	. .	. .	58 47.41	7.38	+0.02	IV.	8	9.25	38 40.99	13.92	5.87	58 54.81	31 12 0.78
88	8	. .	. .	. .	25.8	. .	. .	. .	59 25.64	7.38	−0.01	IV.	4	6.26	18 11.74	13.87	3.08	18 59 33.01	30 51 28.69
87	9	. .	. .	. .	. .	. .	10.8	. .	18 59 56.22	+ 7.38	−0.01	VII.	4	8.45	−19 21.38	−13.75	−3.22	19 0 3.59	−30 52 38.35

ZONE 121. JUNE 11. K. BELT, −26° 53′. D₀ = −26° 30′ 20″.

No.	Mag.	I.	II.	III.	IV.	V.	VI.	VII.	T.	a_1	a_2	MICROMETER.			i	d_1	d_2	Mean Right Ascension, 1850.0.	Mean Declination, 1850.0.
1	9	42.9	57.4	. .	. .	. .	. .	. .	15 12 25.44	+15.46	0.00	III.	7	10.43	−34 21.59	−33.67	−3.88	15 12 40.90	−27 5 19.14
2	9	. .	. .	19.8	. .	47.6	. .	. .	13 33.59	15.46	0.00	V.	5	3.47	21 50.72	33.59	2.70	13 49.05	26 52 47.01
3	10	. .	. .	. .	. .	. .	50.2	. .	14 22.36	15.46	−0.01	VI.	2	4.7	12 5.93	33.52	1.86	14 37.81	43 1.31
4	8	. .	. .	. .	. .	. .	. .	30.1	14 47.81	15.46	−0.01	VII.	4	0.5	14 59.23	33.49	2.09	15 3.26	45 54.81
5	9	. .	25.8	. .	54.4	. .	. .	. .	18 54.02	15.47	0.00	IV.	4	3.44	16 50.05	33.17	2.26	19 9.49	26 47 45.48
6	9	. .	. .	. .	. .	. .	. .	36.2	18 54.17	15.47	+0.01	VII.	8	2.9	35 0.76	33.17	3.93	19 9.65	27 5 57.86
7	9	. .	. .	22.7	. .	. .	. .	. .	25 36.23	15.49	−0.01	IV.	2	5.31	7 45.90	32.63	1.43	25 51.81	26 38 39.96
8	8	. .	47.8	. .	. .	. .	. .	. .	30 15.98	15.50	+0.01	III.	8	8.7	38 1.63	32.23	4.23	30 31.49	27 8 58.09
9	9	. .	42.0	. .	0.2	. .	. .	. .	37 59.97	15.52	−0.01	IV.	1	12.5	6 5.57	31.52	1.27	38 15.45	26 36 58.36
10	9	. .	52.0	6.6	. .	. .	. .	. .	39 20.41	15.53	+0.01	IV.	8	4.12	36 3.17	31.40	4.02	39 35.95	27 6 58.59
11	9	. .	. .	. .	. .	0.0	. .	. .	39 45.98	15.53	0.00	V.	7	1.52	29 53.83	31.36	3.46	40 1.51	27 0 48.65
12	9	. .	53.7	7.8	. .	. .	. .	. .	43 21.51	15.54	−0.01	IV.	2	4 28	7 14.13	31.00	1.36	43 37.04	26 38 6.49
13	9	. .	. .	. .	. .	18.2	. .	. .	44 4.24	15.54	+0.01	V.	9	8.54	43 24.33	30.93	4.74	44 19.79	27 14 20.00
14	7.8	. .	. .	. .	7.0	. .	34.8	. .	15 45 6.88	+15.54	0.00	V.	5	4.58	−22 26.52	−30.82	−2.75	15 45 22.42	−26 53 20.09

CORRECTIONS.

Date.		Corr. of Clock.	Hourly rate.	m	n	c
1847.	h.	s.	s.	s.	s.	s.
June 11,	15	18.24	10.002			

INSTRUMENT READINGS.

	Date.		Barom.	THERMOM. At.	THERMOM. Ex.
	1847.	h. m.	in.	°	°
Zone 121	June 11,	9 5	29.718	77.5	71.

REMARKS.

(120) 61. Differs from Transit, 1846, June 3, 14ˢ.05 (1 transit T.) in right ascension.
(120) 69. Transit over T. VII assumed as recorded over T. VI.
(121) 3. Micrometer reading assumed as 14ʳ.7 instead of 4ʳ.7.
(121) 9. Transit over T. II assumed to have been at 32ˢ.0 instead of 42ˢ.0.

ZONE 121. JUNE 11. K. BELT, −26° 53'. D₀=−26° 30' 20"—Continued.

No.	Mag.	Seconds of Transit I.	II.	III.	IV.	V.	VI.	VII.	T.	a_1	a	Micrometer			i	d_1	d_2	Mean Right Ascension, 1850.0.	Mean Declination, 1850.0.
									h. m. s.	s.	s.			r.	' "	"	"	h. m. s.	° ' "
15	8	6.8	..	35.3	..	..	..	..	15 47 49.36	+15.55	+0.01	III.	9	4.12	−41 2.13	−30.55	−4.52	15 48 4.92	−27 11 57.20
16	7	..	..	51.2	4.8	..	..	..	50 4.72	15.55	−0.01	IV.	1	8.5	4 4.56	30.32	1.06	50 20.26	26 34 55.94
17	10	..	..	40.8	..	..	..	..	50 54.67	15.55	0.00	IV.	5	8.20	24 8.42	30.23	2.92	51 10.22	55 1.57
18	8	..	..	..	..	55.5	..	23.7	51 41.34	15.56	−0.01	VII.	1	12.18	6 11.73	30.15	1.26	51 56.89	37 3.14
19	9	..	16.3	..	..	..	..	..	55 44.17	15.57	0.01	III.	1	10.53	5 29.23	29.70	1.19	55 59.73	36 20.12
20	9	..	..	..	5.2	..	..	..	56 5.05	15.57	0.01	IV.	1	3.32	1 46.90	29.67	0.86	56 20.61	32 37.43
21	10	5.5	..	34.3	..	..	..	..	15 57 47.82	15.57	0.01	IV.	3	1.3	10 29.86	29.48	1.65	15 58 3.38	41 20.99
22	7.8	25.2	..	54.0	..	..	..	..	16 2 7.57	15.58	0.01	IV.	3	8.50	14 25.35	28.99	2.02	16 2 23.14	45 16.36
23	9	..	..	..	..	..	37.8	..	2 9.93	15.58	−0.01	VI.	3	10.20	15 10.56	28.99	2.08	2 25.50	26 46 1.63
24	10	..	..	..	0.4	..	..	..	4 0.25	15.59	0.00	IV.	6	11.51	29 55.91	28.77	3.46	4 15.84	27 0 48.14
25	8	10.2	..	38.9	..	..	..	..	5 52.51	15.59	−0.01	IV.	3	6.47	18 22.33	28.56	2.37	6 8.09	26 49 13.26
26	8	..	33.8	47.7	..	..	..	..	15 1.62	15.62	0.01	IV.	4	3.53	16 54.58	27.43	2.24	15 17.23	47 44.25
27	9	..	14.8	..	..	..	..	..	17 42.81	15.62	0.01	IV.	4	9.30	19 44.52	27.10	2.51	17 58.42	50 34.13
28	9	32.8	..	1.5	..	..	..	..	25 15.15	15.64	0.01	IV.	3	13.6	16 34.43	26.11	2.21	25 30.78	47 22.75
29	10	23.2	..	..	..	..	..	..	29 5.37	15.65	−0.01	III.	2	10.13	10 8.06	25.59	1.62	29 21.01	40 55.27
30	10	..	30.8	..	58.	..	..	..	29 58.37?	15.65	0.00	II.	6	8.9	28 3.81	25.47	3.27	30 14.02	58 52.55
31	8	..	46.8	..	14.8	..	..	..	33 14.74	15.66	0.00	IV.	5	5.42	22 48.75	25.03	2.80	33 30.40	26 53 36.58
32	6	..	15.4	..	43.6	..	..	..	34 43.52	15.66	+0.01	IV.	8	10.44	39 20.84	24.83	4.35	34 59.19	27 10 10.02
33	8	..	..	..	..	..	43.8	..	35 15.83	15.67	0.01	VI.	8	10.57	39 27.22	24.76	4.36	35 31.51	10 16.34
34	9	..	..	..	..	..	29.4	..	37 1.42	15.67	+0.01	VI.	9	4.47	41 19.65	24.52	4.55	37 17.10	27 12 8.72
35	8	..	..	4.9	..	..	..	..	42 18.50	15.68	−0.01	III.	2	1.56	5 57.45	23.79	1.22	42 34.17	26 36 42.46
36	9	..	..	..	..	54.8	..	..	42 40.69	15.69	−0.01	V.	2	7.30	8 45.87	23.74	1.49	42 56.37	26 39 31.10
37	9	..	..	37.0	..	..	..	..	44 51.24	15.69	+0.01	IV.	10	8.34	48 15.19	23.43	5.24	45 6.94	27 19 3.86
38	10	..	..	..	24.7	..	..	..	47 24.55	15.69	−0.01	IV.	3	3.31	11 44.49	23.07	1.77	47 40.23	26 42 29.33
39	9	..	42.7	56.7	..	..	..	..	49 10.63	15.70	0.00	III.	5	3.12	21 33.07	22.83	2.67	49 26.33	52 18.57
40	9	..	..	..	..	34.0	..	..	49 19.92	15.70	−0.01	V.	4	1.58	15 56.55	22.81	2.15	49 35.61	46 41.51
41	8	..	..	..	..	16.8	..	..	50 48.91	15.70	0.00	V.	6	3.47	25 51.82	22.70	3.07	50 4.61	26 56 37.59
42	8	..	..	29.8	..	57.8	..	..	51 43.78	15.71	0.00	V.	7	3.16	30 36.19	22.47	3.52	51 59.49	27 1 22.18
43	7	..	..	15.4	29.6	..	..	..	53 29.34	15.71	0.00	IV.	5	3.41	21 47.73	22.22	2.68	53 45.05	26 52 32.63
44	7.8	..	..	12.2	..	40.3	..	..	54 26.21	15.71	0.00	VI.	6	9.56	28 57.76	22.09	3.38	54 41.92	26 59 43.23
45	9	..	..	..	..	..	44.7	..	55 16.73	15.72	+0.01	VI.	8	10.36	39 16.64	21.97	4.37	55 32.46	27 10 2.98
46	9	..	..	..	..	48.9	..	..	56 34.81	15.72	−0.01	IV.	3	6.29	13 14.25	21.79	1.88	56 50.52	26 43 57.92
47	9	..	34.1	..	..	..	..	..	16 59 1.99	15.73	0.01	IV.	2	4.36	7 18.18	21.45	1.33	16 59 17.71	38 0.96
48	9	..	34.0	48.2	..	..	..	..	17 2 1.99	15.73	0.01	IV.	4	5.38	17 47.53	21.03	2.31	17 1 17.71	48 30.87
49	6	53.2	..	21.8	..	..	..	..	2 35.54	15.74	0.01	IV.	4	10.22	20 10.74	20.95	2.54	2 51.27	50 54.23
50	8	..	..	..	..	..	28.5	..	3 0.61	15.74	0.01	VI.	4	8.17	19 7.54	20.89	2.44	3 16.34	49 50.87
51	7	..	..	..	..	52.4	..	..	4 38.33	15.74	−0.01	IV.	4	4.32	17 14.26	20.66	2.36	4 54.06	26 47 57.28
52	9	..	48.3	2.7	..	..	..	..	11 16.66	15.75	+0.02	IV.	9	2.44	40 17.80	19.73	4.47	11 32.43	27 11 2.00
53	6	..	..	..	..	..	54.3	..	11 26.32	15.75	+0.02	VI.	10	2.2	44 57.35	19.71	4.93	11 42.09	27 15 41.99
54	9	..	..	..	..	12.4	..	..	14 58.32	15.76	−0.01	VI.	3	12.4	16 3.01	19.21	2.14	15 14.07	26 46 44.36
55	9	..	..	..	6.8	..	..	..	16 6.64	15.77	0.01	VI.	4	10.41	20 20.15	19.05	2.56	16 22.40	51 1.76
56	9	..	45.8	59.9	..	..	..	..	17 19 13.58	+15.78	−0.02	III.	1	10.15	− 5 10.07	−18.61	−1.13	17 19 29.34	−26 35 49.81

CORRECTIONS.

Date.		Corr. of Clock.	Hourly rate.	m	n	c
1847.	h.	s.	s.	s.	s.	s.

INSTRUMENT READINGS.

Date.		Barom.	Thermom. At.	Thermom. Ex.
1847.	h. m.	In.	°	°

REMARKS.

(121) 25. Hor. thread assumed as 4 instead of 5.
(121) 41. Minutes assumed as 49 instead of 50.
(121) 48. Minutes assumed as 1 instead of 2.

ZONE 122. JUNE 17. K. BELT, −30° 1'. D_o = −29° 38' 0".

No.	Mag.	SECONDS OF TRANSIT. I.	II.	III.	IV.	V.	VI.	VII.	T.	a_1	a'	MICROMETER.			i	d_1	d_2	Mean Right Ascension, 1850.0.	Mean Declination, 1850.0.
									h. m. s.	s.	s.			r.	' "	"	"	h. m. s.	° ' "
1	9		36.8						15 13 5.56	+15.45	−0.01	II.	2	12.36	−11 20.01	−47.57	−2.34	15 13 21.00	29 50 9.92
2	10						49.5		13 20.80	15.45	−0.01	VI.	3	5.46	12 52.37	47.56	2.55	13 36.24	29 51 42.47
3	9			58.3					30 12.96	15.48	+0.01	III.	10	7.33	47 44.37	46.35	6.84	30 28.45	30 26 37.56
4	10					53.9			30 39.49	15.48	+0.01	V.	7	13.21	35 41.23	46.31	5.36	30 54.98	30 14 32.90
5	6.7					38.8			32 24.23	15.48	−0.01	V.	1	1.28	0 44.31	46.18	1.08	32 39.70	29 39 31.57
6	6					8.8	22.8		33 54.19	15.49	0.00	VI.	5	8.43	24 19.82	46.06	3.91	34 9.68	30 3 9.79
7	7						2.8		34 34.00	15.49	+0.01	VI.	7	11.58	34 59.24	46.00	5.24	34 49.50	13 50.48
8	7							56.6	35 13.47	15.49	0.01	VII.	7	9.17	33 37.81	45.95	5.07	35 28.97	12 25.83
9	9			54.2					37 8.84	15.49	+0.01	IV.	10	4.44	45 48.95	44.80	6.66	37 24.34	30 24 40.41
10	9		12.7						39 41.41	15.50	−0.01	IV.	1	11.36	5 50.96	44.59	1.69	39 56.90	29 44 37.24
11	10		39.9						41 8.86	15.50	0.00	IV.	6	12.54	30 27.68	45.47	4.66	41 24.36	30 9 17.81
12	9				58.2				41 58.04	15.50	+0.01	IV.	8	3.46	35 50.06	45.40	5.35	42 13.55	14 40.81
13	8.7	53.7	8.3						43 37.42	15.51	+0.01	II.	9	3.19	40 35.24	45.26	5.94	43 52.94	30 19 26.44
14	9		23.5						44 52.35	15.51	0.00	III.	4	9.7	19 32.87	45.14	3.31	45 7.86	29 58 21.32
15	9		12.3						47 41.38	15.52	+0.01	III.	9	5.47	41 50.02	44.90	6.11	47 56.91	30 20 41.03
16	8	19.8	34.3						49 3.34	15.52	0.00	II.	6	8.48	28 23.44	44.78	4.43	49 18.86	7 12.65
17	9			27.8					49 42.10	15.52	0.00	IV.	5	9.8	24 32.62	44.73	3.93	49 57.62	30 3 21.28
18	9							9.8	53 26.20	15.52	−0.01	VII.	2	13.15	11 39.42	44.28	2.37	53 41.71	29 50 26.17
19	8	16.5	30.9						57 0.01	15.53	0.00	III.	6	12 51	30 26.11	44.04	4.67	57 15.54	30 9 14.82
20	8				10.2				58 10.04	15.53	+0.01	IV.	7	8.40	33 19.61	43.93	5.03	58 25.56	30 12 8.57
21	9			11.0					15 59 25.12	15.53	−0.01	IV.	3	5.4	12 31.39	43.82	2.47	15 59 40.64	29 51 17.68
22	10		15.1						16 0 44.05	15.53	0.00	IV.	6	9.58	28 58.93	43.69	4.50	16 0 59.58	30 7 47.12
23	9		15.1						1 44.18	15.53	+0.01	III.	9	5.24	41 38.42	43.59	6.08	1 59.72	30 20 28.09
24	8.9	17.5		46.6					3 0.89	15.54	−0.01	IV.	2	3.35	6 47.40	43.45	1.79	3 16.42	29 45 32.64
25	7.8		43.0	57.3					4 11.58	15.54	−0.01	IV.	3	1.10	10 33.39	43.33	2.25	4 27.11	29 49 18.97
26	8	38.6	53.2						6 22.28	15.54	+0.01	II.	8	3.7	35 30.19	43.11	5.31	6 37.83	30 14 18.61
27	8				9.8				14 9.64	15.56	−0.01	IV.	2	11.31	10 47.43	42.29	2.27	14 25.19	29 49 31.99
28	9							9.2	14 25.57	15.56	−0.01	VII.	2	8.56	9 28.84	42.27	2.12	14 41.12	29 48 13.23
29	7	51.2		20.8					29 35.04	15.59	0.00	IV.	7	3.55	30 55.89	40.59	4.74	29 50.63	30 9 41.22
30	8		49.8						32 18.81	15.59	+0.01	IV.	8	2.35	35 14.26	40.27	5.28	32 34.41	13 59.81
31	9						50.3		41 21.44	15.61	+0.01	VII.	9	13.42	45 49.14	39.21	6.63	41 37.06	24 34.98
32	7			4.5	18.8				43 18.78	15.62	0.00	IV.	7	5.17	31 37.24	38.98	4.81	43 34.40	10 21.03
33	8				54.8				44 54.64	15.62	0.00	IV.	7	2.45	30 20.60	. .	4.65	45 10.26	
34	5					0.4	14.6		44 45.89	15.62	+0.01	VI.	9	4.54	41 23.15	38.80	6.05	45 1.52	30 20 8.00
35	9		49.8						55 18.55	15.64	−0.01	IV.	2	10.5	10 4.07	37.51	2.17	55 34.18	29 48 43.75
36	8		11.5		40.9				56 40.54	15.64	0.01	IV.	4	5.21	17 38.95	37.33	3.11	56 56.17	56 19.39
37	7		59.7	14.1					58 28.34	15.64	0.01	III.	3	4.25	12 11.67	37.11	2.41	58 43.97	50 51.19
38	8						10.5		58 41.83	15.64	−0.01	VI.	2	6.31	8 15.96	37.08	1.95	58 57.46	29 46 54.99
39	5							40.1	16 58 56.77	15.64	+0.01	VII.	7	8.27	33 12.61	37.06	5.02	16 59 12.42	30 11 54.69
40	9.10			49.3					17 2 3.78	15.64	0.01	IV.	8	3.39	35 46.53	36.65	5.35	17 1 19.43	14 28.53
41	10.11					43.2			2 28.80	15.64	0.01	V.	9	0.52	39 21.27	36.60	5.80	2 44.45	18 3.67
42	8	39.6	54.4						5 23.45	15.65	+0.02	II.	9	7.34	42 43.83	36.23	6.25	5 39.12	21 26.31
43	9				38.2				5 38.04	15.65	0.00	IV.	6	5.11	26 34.21	36.20	4.20	5 53.69	5 14.61
44	5					15.2			6 0.73	15.65	0.00	V.	5	6.34	23 14.93	36.15	3.77	6 16.38	1 54.85
45	7				3.2				7 3.04	13.65	+0.01	IV.	7	6.3	32 0.44	36.01	4.89	7 18.70	30 10 41.34
46	7					44.7			7 30.20	15.66	0.00	V.	4	1.51	20 52.21	35.96	3.48	7 45.86	29 59 31.65
47	8					48.2			8 33.65	15.66	−0.02	V:	1	7.24	3 43.83	35.82	1.40	8 49.29	29 42 21.05
48	7	32.6	47.6						12 16.54	15.66	+0.02	II.	9	6.16	42 4.50	35.34	6.15	12 32.22	30 20 45.99
49	10.9				33.2				17 12 33.04	+15.67	0.00	IV.	5	8.15	−24 5.90	−35.30	−3.88	17 12 48.71	30 2 45.08

CORRECTIONS.

Date.		Corr. of Clock.	Hourly rate.	m	n	c
1847.	h.	s.	s.	s.	s.	s.
June 17,	15	s 8.17	g 0.034			

INSTRUMENT READINGS.

	Date.		Barom.	THERMOM. At.	Ex.
	1847.	h. m.	in.	°	°
Zone 122	June 17,	15 13	29.95	73.5	65.3

REMARKS.

(122) 9. Micrometer reading assumed as 3r.44 instead of 4r.44.
(122) 29. One of the transit observations erroneous by 5s, or transit over T. I should be 51s.2 instead of 57s.2. The latter assumed as most probable.
(122) 40. Minute of transit assumed as 1 instead of 2.
(122) 46. Micrometer thread 5 assumed to have been used instead of 4.

ZONE 122. JUNE 17. K. BELT, 30° 1′. D₀ = −29° 38′ 0″—Continued.

No.	Mag.	SECONDS OF TRANSIT. I.	II.	III.	IV.	V.	VI.	VII.	T.	a_1	a_2	MICROMETER.			i	d_1	d_2	Mean Right Ascension, 1850.0.	Mean Declination, 1850.0.
									h. m. s.	s.	s.			r.	′ ″	′	″	h. m. s	° ′ ″
50	9	...	...	50.2	...	...	...	...	17 14 4.15	+15.67	−0.02	IV.	1	3.12	− 1 36.81	−35.10	−1.14	17 14 19.80	−29 40 13.05
51	8	...	...	...	...	...	...	13.6	14 29.92	15.67	0.02	VII.	2	−0.27	4 45.47	35.05	1.49	14 45.57	43 22.01
52	2	...	...	16.8	30.8	45.3	...	...	17 30.73	15.68	−0.02	IV.	1	9.41	4 52.96	34.66	1.53	17 46.39	29 43 29.15
53	9	...	...	...	...	...	35.3	...	23 6.49	15.69	+0.01	VI.	8	6.7	37 0.96	33.91	5.53	23 22.19	30 15 40.40
54	8	...	...	...	...	22.6	...	51.8	24 8.37	15.69	+0.01	VII.	8	10.40	39 18.38	33.78	5.82	24 24.07	30 17 57.98
55	8	...	6.5	...	35.0	...	...	...	29 35.06	15.70	−0.01	IV.	3	6.43	13 21.31	33.03	2.55	29 50.75	29 51 56.89
56	10	...	...	...	28.8	...	...	...	30 28.64	15.70	0.00	IV.	6	10.28	29 14.06	32.91	4.53	30 44.34	30 7 51.50
57	8	...	...	...	...	...	...	20.7	33 37.30	15.70	0.00	VII.	6	6.41	27 19.16	32.45	4.30	33 53.00	30 5 55.94
58	8	...	...	...	...	40.2	...	...	35 25.67	15.71	−0.01	VI.	2	6.24	8 12.43	32.23	1.90	35 41.37	29 46 46.55
59	9	49.8	...	18.6	...	...	...	...	38 33.30	15.71	−0.01	IV.	8	5.38	35 46.02	31.78	5.36	38 49.02	30 14 23.16
60	10	...	...	...	...	...	11.5	...	38 42.64	15.71	+0.02	VI.	10	5.9	46 31.61	31.77	6.75	38 58.37	25 10.13
61	10	...	49.9	...	...	...	...	...	42 18.84	15.71	0.00	IV.	6	7.32	27 45.32	31.26	4.35	42 34.55	6 20.93
62	9	...	...	...	...	49.3	...	...	42 34.87	15.72	+0.01	V.	7	5.44	31 50.81	31.22	4.67	42 50.60	30 10 26.90
63	9	...	...	3.3	...	...	...	...	45 17.35	15.72	−0.02	IV.	2	6.3	8 2.04	30.84	1.89	45 33.05	29 46 34.77
64	8	...	...	...	...	...	4.5	...	45 35.77	15.72	0.00	VI.	5	1.35	20 44.00	30.79	3.46	45 51.49	29 59 18.25
65	8.7	...	...	...	...	47.2	...	...	46 32.77	15.72	+0.01	V.	7	4.57	31 27.11	30.67	4.82	46 48.50	30 10 2.60
66	8	...	...	53.3	...	...	...	...	48 7.59	15.73	0.00	IV.	5	7.44	23 50.27	30.44	3.85	48 23.32	2 24.56
67	4.5	...	...	57.0	11.3	...	...	...	49 11.30	15.73	+0.01	V.	7	12.40	35 20.57	30.28	5.31	49 27.04	13 56.16
68	9	...	...	...	56.0	...	...	...	49 55.84	15.73	+0.01	IV.	8	11.26	39 17.06	30.18	5.88	50 11.58	30 18 23.12
69	8	...	...	...	...	...	46.3	...	50 17.62	15.73	−0.01	VI.	2	11.9	10 36.14	30.13	2.20	50 33.34	29 49 8.47
70	8	...	...	...	...	...	...	19.3	50 35.67	15.73	0.01	VII.	2	9.56	9 59.09	30.08	2.13	50 51.39	48 31.30
71	7.8	0.7	15.4	...	...	...	...	...	53 44.17	15.73	−0.01	III.	3	5.42	12 50.50	29.63	2.49	53 59.89	29 51 22.62
72	2.3	...	...	39.8	...	8.6	...	...	54 54.34	15.73	+0.02	IV.	10	5.29	46 41.89	29.45	6.78	56 10.09	30 25 18.12
73	8	...	33.2	...	...	...	...	...	17 58 2.07	15.74	0.00	IV.	5	3.49	21 51.77	29.00	3.59	17 59 17.81	30 0 24.36
74	9	...	...	...	11.8	...	...	...	18 1 11.64	15.74	−0.02	IV.	2	5.40	7 50.44	28.53	1.87	18 1 27.36	29 46 20.84
75	9	50.0	...	...	...	...	...	...	18 2 33.59	15.75	0.00	III.	5	8.3	23 59.80	28.37	3.87	18 2 49.34	30 2 32.04
76	8	24.6	9.2	...	...	...	...	...	19 0 38.28	15.82	+0.01	III.	8	4.5	35 59.59	19.31	5.41	19 0 54.11	14 24.31
77	9	...	...	...	55.5	...	...	...	0 55.34	15.82	0.00	IV.	5	8.50	24 23.55	19.26	3.92	1 11.16	30 2 46.73
78	5.6	...	...	...	...	46.8	...	16.0	2 32.30	15.82	−0.02	VII.	1	12.9	6 7.14	19.00	1.62	1 48.10	29 44 27.76
79	8	4.8	20.0	...	...	...	...	...	4 48.66	15.82	0.00	IV.	5	12.59	26 29.10	18.64	4.19	5 4.48	30 4 51.93
80	8	9.2	23.6	...	...	...	...	...	11 52.64	15.82	−0.01	III.	3	8.55	14 27.82	17.52	2.67	12 8.45	29 52 48.01
81	7.8	...	...	3.3	17.5	...	...	...	12 17.36	15.82	0.02	IV.	2	9.3	9 33.81	17.46	2.05	12 33.16	47 53.32
82	9	...	...	53.6	...	...	...	...	21 7.58	15.83	0.02	IV.	1	6.0	3 1.53	16.09	1.25	21 23.39	41 18.87
83	8	...	...	...	...	38.0	52.1	...	21 23.45	15.83	0.02	VI.	2	9.26	9 44.20	16.04	2.08	21 39.26	48 2.32
84	8	...	...	...	19.2	...	...	...	57 19.04	15.86	−0.01	IV.	2	14.4	12 4.57	10.57	2.33	57 34.89	29 50 17.47
85	7	...	...	...	...	...	15.8	30.3	19 57 46.98	15.86	+0.01	VII.	6	13.21	30 40.84	10.50	4.73	19 58 2.85	30 8 56.07
86	9.10	...	...	23.5	...	...	...	...	20 2 37.75	15.86	0.00	III.	5	2.51	21 22.47	9.79	3.52	20 2 53.61	29 59 35.78
87	4.5	...	...	0.0	...	...	...	...	6 14.68	15.86	+0.02	III.	10	10.48	49 22.70	9.27	7.08	6 30.56	30 27 39.05
88	10	...	...	...	4.0	...	...	...	8 3.84	15.86	−0.01	IV.	3	6.2	13 0.64	9.00	2.48	8 19.69	29 51 12.12
89	8	6.8	...	36.0	...	...	...	...	9 50.39	15.86	0.00	IV.	6	6.13	27 5.48	8.75	4.27	10 6.25	30 5 18.50
90	8	...	...	42.8	...	...	...	...	10 56.75	15.86	−0.02	IV.	1	3.25	1 43.36	8.59	1.07	11 12.59	29 39 53.02
91	9	...	...	...	...	29.2	...	...	21 14.78	15.87	+0.01	V.	7	8.15	33 6.95	7.12	5.05	21 30.66	30 11 19.12
92	7.8	...	22.7	7.2	...	...	...	...	24 21.36	15.87	−0.02	IV.	2	9.33	9 47.93	6.67	2.06	24 37.21	29 47 56.66
93	9	...	31.3	...	0.2	...	...	...	31 0.02	15.87	0.02	IV.	1	10.41	5 23.22	5.75	1.50	31 15.87	43 30.47
94	5.6	...	...	36.8	51.3	...	...	...	33 51.08	15.87	0.01	IV.	4	7.30	18 44.01	5.37	3.22	34 6.94	56 52.60
95	9	...	...	...	...	38.0	...	...	37 23.50	15.87	−0.01	V.	3	10.50	15 25.81	4.88	2.78	37 39.36	53 33.47
96	7	34.0	...	3.2	...	...	...	...	42 17.51	15.87	0.00	IV.	4	13.5	21 32.92	4.21	3.55	42 33.38	29 59 40.68
97	7	...	...	...	...	...	...	59.7	42 16.45	15.87	+0.02	VII.	9	6.19	42 5.77	4.21	6.23	42 32.37	30 20 16.21
98	9	39.9	...	...	...	...	...	...	20 45 23.44	+15.87	−0.01	III.	4	10.40	−20 19.77	− 3.80	−3.39	20 45 39.30	−29 58 26.96

CORRECTIONS.

Date.		Corr. of Clock.	Hourly rate.	m	n	c
1847.	h.	s.	s.	s.	s.	s.

INSTRUMENT READINGS.

	Date.		Barom.	THERMOM. At.	Ex.
	1847.	h. m.	In.	°	°

REMARKS.

(122) 51. Hor. thread assumed as 1 instead of 2, and micrometer readings as 9r.27 instead of 0r.27.
(122) 72. Minutes assumed as 55 instead of 54.
(122) 73. Minutes assumed as 59 instead of 58.
(122) 76. One of the transit observations, supposed the former, erroneous by 30s.
(122) 78. Minutes assumed as 1 instead of 2.
(122) 92. Transit over. II assumed as 52s.7 instead of 22s.7.

ZONE 122. JUNE 17. K. BELT, −30° 1′. D_o=−29° 38′ 0″—Continued.

No.	Mag.	SECONDS OF TRANSIT. I.	II.	III.	IV.	V.	VI.	VII.	T.	a_1	a_2	MICROMETER.			i	d_1	d_4	Mean Declination, 1850.0.	Mean Right Ascension, 1850.0.
									h. m. s.	s.	s.			r.	′ ″	″	″	° ′ ″	h. m. s.
99	10	..	..	..	..	58.3	..	..	20 50 43.84	+15.87	0.00	VI.	5	10.31	−25 14.28	−3.09	−4.03	20 50 59.71	−30 3 21.40
100	7	..	..	..	..	..	11.0	..	51 42.17	15.87	+0.02	V.	8	12.53	40 25.82	2.96	6.02	51 58.06	30 18 34.80
101	6.7	..	..	24.4	..	..	..	..	53 38.39	15.87	−0.02	V.	1	7.28	3 45.84	2.71	1.30	53 54.24	29 41 49.85
102	9	..	..	..	..	..	23.2	..	54 54.54	15.87	−0.02	VI.	1	11.58	6 1.84	2.54	1.58	55 10.39	29 44 5.96
103	10	..	11.8	..	..	..	..	..	20 57 40.79	15.87	+0.01	III.	7	7.40	32 49.30	2.17	5.01	20 57 56.67	30 10 56.48
104	4.5	4.8	..	..	..	..	..	..	21 0 48.59	+15.87	+0.02	II.	9	5.3	−41 27.68	−1.77	−6.15	21 1 4.48	−30 19 35.60

ZONE 123. JUNE 21. B. BELT, −30° 1′. D_o=−29° 38′ 40″.

No.	Mag.	I.	II.	III.	IV.	V.	VI.	VII.	T.	a_1	a_2	MICROMETER.			i	d_1	d_4	Mean Declination, 1850.0.	Mean Right Ascension, 1850.0.
1	8	..	..	..	32.	..	0.0	..	15 8 31.55	+16.44	−0.01	VII.	3	12.8	−16 4.73	−33.75	−1.91	15 8 47.98	−29 55 20.39
2	9	..	..	..	..	11.5	25.5	..	11 56.64	16.44	+0.01	VII.	7	11.32	34 45.90	33.47	4.21	12 13.09	30 14 3.58
3	10	..	..	..	..	..	..	50.5	14 7.06	16.45	0.00	VII.	5	12.22	26 10.00	33.31	3.14	14 23.51	30 5 26.45
4	10	..	..	..	..	49.	..	18.	16 34.49	16.45	−0.01	VII.	3	13.15	16 38.52	33.11	1.98	16 50.93	29 55 53.61
5	10	..	..	..	..	..	17.	..	24 48.19	16.47	+0.01	VII.	8	6.12	37 3.24	32.41	4.49	25 4.67	30 16 20.14
6	8	..	..	..	12.	..	..	..	30 11.64	16.48	−0.01	V.	2	6.41	8 21.15	31.94	0.97	30 28.31	29 47 34.06
7	10	..	..	..	..	41.	..	9.	?32 26.14	16.48	+0.01	VII.	7	12.3	35 1.52	31.74	4.23	?32 42.63	30 14 17.49
8	8	..	..	..	..	48.	2.	..	34 33.39	16.49	+0.01	VI.	7	11.6	34 33.03	31.54	4.17	34 49.89	13 48.74
9	10	..	..	..	..	..	9.5	24.	?40.58	16.	−0.01	VII.	2	0.4	5 0.58	..	0.59	57.1	
10	7	..	..	..	..	11.	25.4	..	?56.59	16.5	+0.01	VII.	8	3.28	35 40.53	..	4.33	13.1	
11	7	..	..	..	..	..	..	19.	48 35.75	16.53	+0.01	VII.	9	2.32	40 11.31	30.18	4.88	48 52.29	19 26.37
12	10	..	..	..	..	..	..	41.5	46 58.15	16.52	0.00	VII.	7	5.55	31 55.96	30.34	3.87	47 14.67	11 10.17
13	8	..	..	..	..	55.5	..	24.5	49 41.09	16.53	0.00	VII.	6	8.21	28 9.58	30.07	3.40	49 57.62	7 23.05
14	8	..	..	..	..	40.5	..	9.0	51 25.84	16.53	0.00	VII.	6	10.58	29 28.75	29.89	3.57	51 42.37	8 42.21
15	9	..	..	..	..	..	33.5	..	15 53 4.70	16.54	+0.01	VII.	7	12.34	35 17.16	29.71	4.27	15 53 21.25	14 31.14
16	9	..	..	..	..	..	..	27.	16 0 43.61	16.55	0.00	VII.	6	9.19	28 38.83	28.90	3.46	16 1 0.16	30 7 51.19
17	8	..	..	..	..	14.	28.4	..	2 59.60	16.56	−0.01	VII.	2	11.38	10 50.52	28.64	1.28	3 16.15	29 50 0.44
18	8	37.5	52.8	..	..	..	..	..	6 21.52	16.56	+0.01	III.	7	12.15	35 7.96	28.27	4.26	6 38.09	30 14 20.49
19	8	28.5	43.	..	..	..	..	..	9 12.04	16.57	0.00	III.	6	8.35	28 17.04	27.94	3.41	9 28.61	7 28.39
20	8	..	..	..	..	..	51.	5.5	9 22.15	16.57	0.00	VII.	5	8.2	23 58.91	27.93	2.88	9 38.72	30 3 9.72
21	10	..	..	..	..	..	3.	17.	11 33.85	16.57	−0.01	VII.	2	10.01	5 2.61	27.68	0.54	11 50.41	29 44 10.83
22	9	..	..	..	..	23.5	..	..	14 9.00	16.58	−0.01	VI.	3	10.42	10 22.53	27.37	1.18	14 25.57	29 49 31.08
23	8	..	..	..	..	..	..	53.5	16 10.08	16.58	0.00	VII.	6	4.54	26 25.21	27.13	3.17	16 26.66	30 5 35.51
24	10	8.	..	37.5	..	..	..	..	22 51.54	16.60	−0.01	IV.	3	5.8	12 33.40	26.32	1.46	23 8.13	29 51 41.18
25	.	0.5	15.	..	..	..	..	..	25 43.93	16.60	−0.01	IV.	4	5.38	17 47.53	25.96	2.13	26 0.52	29 56 55.62
26	10	50.5	5.	19.3	..	..	..	..	29 33.95	16.61	0.00	IV.	7	3.5	30 30.68	25.49	3.70	29 50.56	30 9 39.87
27	10	44.4	..	13.2	..	..	..	..	34 13.14	16.62	0.00	IV.	4	8.14	19 6.20	24.85	2.26	34 29.76	29 58 13.31
28	9	56.4	11.	..	..	..	..	..	38 39.97	16.63	0.00	III.	6	4.51	26 24.08	24.30	3.17	38 56.60	30 5 31.55
29	11	..	..	28.	43.	..	..	..	40 42.54	16.63	0.00	V.	4	10.22	20 10.69	24.03	2.40	40 59.17	29 59 17.12
30	8	34.4	49.	..	..	..	..	..	43 18.02	16.64	0.00	III.	7	4.33	31 15.01	23.68	3.79	43 34.66	30 10 22.48
31	8	..	..	..	..	8.	22.	37.	43 53.47	16.64	0.00	VII.	7	2.04	29 59.49	23.60	3.63	44 10.11	9 6.72
32	10	..	..	..	..	..	47.5	2.	46 18.67	16.64	0.00	VII.	6	10.26	29 12.62	23.27	3.53	46 35.31	30 8 19.42
33	9	..	..	18.	32.	..	..	..	48 32.03	16.65	0.00	V.	4	8.50	19 24.30	22.97	2.31	48 48.68	29 58 29.58
34	9	..	..	19.2	34.	..	..	..	50 33.65	16.65	0.00	V.	4	12.30	21 15.23	22.69	2.53	50 50.30	30 0 20.45
35	10	..	11.	25.	..	..	..	..	53 39.49	16.66	−0.01	IV.	4	1.53	15 54.07	22.26	1.87	53 56.14	29 54 58.20
36	10	..	..	..	..	..	46.	1.	55 17.36	16.66	−0.01	VII.	3	9.32	9 46.99	22.03	1.11	55 34.01	29 48 50.13
37	7	12.4	27.	..	..	..	..	..	16 58 56.04	16.67	+0.01	IV.	7	7.45	32 51.87	21.52	3.98	16 59 12.72	30 11 57.37
38	10	..	33.	..	2.2	..	..	..	17 3 2.00	16.67	0.01	IV.	7	12.50	35 25.66	20.95	4.32	17 1 18.68	14 30.93
39	9	39.	53.5	8.	..	..	..	..	17 5 22.65	+16.68	+0.02	IV.	9	6.55	−42 24.36	−20.62	−5.22	17 5 39.35	−30 21 30.20

CORRECTIONS.

Date.		Corr. of Clock.	Hourly rate.	m	n	c
1847.	h.	s.	s.	s.	s.	s.
June 21,	15	*s* 9.15	*l* 0.004			

INSTRUMENT READINGS.

	Date.		Barom.	THERMOM. At.	THERMOM. Ex.
	1847.	h. m.	In.	°	°
Zone 123	June 21,	9 5	29.982	75.	68.

REMARKS.

(123) 21. Hor. thread assumed as 1 instead of 2.
(123) 22. Hor. thread assumed as 2 instead of 3.
(123) 27. Transits over T.'s II and IV assumed as recorded over T.'s I and III.
(123) 36. Hor. thread assumed as 2 instead of 3.
(123) 38. Minutes assumed as 1 instead of 3.

ZONE 123. JUNE 21. B. BELT, −30° 1′. D₀=−29° 38′ 40″—Continued.

No.	Mag.	Seconds of Transit. I.	II.	III.	IV.	V.	VI.	VII.	T.	a_1	a_4	MICROMETER.			i	d_1	d_2	Mean Right Ascension, 1850.0.	Mean Declination, 1850.0.
									h. m. s.	s.	s.			r.	′ ″	″	″	h. m. s.	° ′ ″
40	7	..	..	..	..	..	29.2	43.	17 6 0.00	+16.68	0.00	VII.	5	5.50	−22 52.34	−20.5	−2.74	17 6 16.68	−30 1 55.61
41	7	..	..	23.	37.	..	..	..	8 36.99	16.69	−0.01	IV.	3	7.32	18 46.92	20.16	0.99	8 53.67	29 47 48.07
42	9	48.	..	17.	..	..	..	..	12 31.44	16.70	0.00	IV.	5	7.37	23 46.74	19.59	2.85	12 48.14	30 2 49.18
43	9	..	5.	19.4	34.	..	..	..	14 33.98	16.70	+0.02	IV.	9	9.45	43 50.08	19.29	5.39	14 50.70	30 22 54.76
44	9	..	34.	48.5	..	..	..	..	17 2.77	16.70	−0.01	IV.	4	6.25	18 11.23	18.93	2.16	17 19.46	29 57 12.32
45	8	..	..	51.4	..	..	..	..	23 5.90	16.72	+0.01	IV.	8	5.20	36 37.45	18.05	4.47	23 22.63	30 15 39.97
46	8	..	..	..	33.	..	..	..	25 32.84	16.72	0.00	VI.	6	5.02	26 29.48	17.70	3.18	25 49.56	30 5 30.36
47	10	..	..	49.	..	..	..	..	29 3.20	16.73	−0.01	IV.	4	5.38	17 47.53	17.18	2.11	29 19.92	29 56 46.82
48	10	..	..	..	49.	3.	..	..	29 34.40	16.73	−0.01	V.	3	5.59	12 59.07	17.07	1.52	29 51.12	29 51 57.66
49	9	53.4	..	22.	..	..	..	..	? 36.69	..	..	IV.	6	5.52	26 54.89	..	3.23		
50	10	44.	..	59.	..	..	..	..	17 25 27.84	16.72	+0.01	IV.	7	7.22	32 40.27	17.71	3.99	25 44.57	30 11 41.97
51	11	28.6	43.	..	..	..	..	..	12 11.	..	0.00	IV.	6	12.12	30 6.50	..	3.64		
52	8	..	..	..	..	46.5	..	15.2	38 32.07	16.75	+0.02	VII.	9	11.49	35 20.88	15.76	4.30	38 48.84	14 20.94
53	10	..	10.	..	..	..	..	..	41 39.06	16.75	0.01	III.	8	12.46	40 22.30	15.29	4.94	41 55.82	19 22.53
54	9	..	..	..	..	48.	2.	..	42 33.40	16.75	+0.01	VI.	7	4.50	31 23.43	15.15	3.81	42 50.16	30 10 22.39
55	8	52.	..	..	..	..	..	..	45 35.54	16.76	0.00	II.	4	10.32	20 15.59	14.68	2.41	45 52.30	29 59 12.68
56	8	..	..	..	..	..	12.	27.	17 45 43.38	16.76	−0.01	VII.	4	5.15	17 35.49	14.67	2.08	17 46 0.13	29 56 32.24
57	8	..	..	34.5	49.	..	..	..	18 28 48.87	16.83	+0.01	IV.	6	12.17	30 9.02	7.91	3.65	18 29 5.71	30 9 0.58
58	8	6.5	21.	..	50.	..	..	..	34 49.98	16.83	0.00	IV.	6	9.16	28 37.76	6.94	3.45	35 6.81	30 7 28.15
59	7	..	..	43.	57.	..	..	..	36 56.95	16.84	−0.02	IV.	2	6.13	8 7.08	6.62	0.89	37 13.77	29 46 54.59
60	9	..	2.	..	30.	..	..	..	41 30.44	16.84	+0.01	IV.	8	8.42	38 19.32	5.89	4.71	41 47.29	30 17 9.92
61	10	..	26.	40.5	..	..	..	..	46 54.98	16.85	0.01	IV.	7	10.14	34 7.01	5.03	4.16	47 11.84	12 56.20
62	3	3.	17.4	32.	46.5	..	..	..	52 46.45	16.86	+0.01	IV.	7	5.07	?26 32.20	4.08	3.19	53 3.32	30 5 19.47
63	10	..	..	..	..	..	30.	44.5	55 1.12	16.86	−0.01	VII.	4	5.16	17 36.00	3.72	2.06	55 17.97	29 56 21.78
64	8	..	..	..	..	..	50.5	5.8	18 57 22.08	16.86	+0.01	VII.	7	5.00	31 28.23	3.33	3.83	18 57 38.95	30 10 15.39
65	8	..	8.	22.	..	..	..	..	19 0 36.75	16.87	+0.01	IV.	8	3.30	35 41.99	2.81	4.37	19 0 53.63	30 14 29.17
66	9	..	..	5.2	..	..	..	..	3 19.34	16.87	−0.01	IV.	3	8.12	14 6.19	2.41	1.63	3 36.20	29 52 50.23
67	8	..	..	..	..	..	16.2	31.3	19 4 47.76	+16.87	0.00	VII.	6	4.26	−21 11.09	− 2.14	−2.52	19 5 4.63	−29 59 55.75

ZONE 124. JUNE 24. K. BELT, −30° 23′. D₀=−30° 1′ 40″.

No.	Mag.	I.	II.	III.	IV.	V.	VI.	VII.	T.	a_1	a_4	MICROMETER.			i	d_1	d_2	Mean Right Ascension, 1850.0.	Mean Declination, 1850.0.
1	7	..	..	..	..	4.5	..	..	15 45 50.18	+17.90	+0.01	VI.	8	4.21	−36 7.51	−29.46	−4.28	15 46 8.09	−30 38 21.25
2	10	..	..	..	56.8	..	..	..	46 56.64	17.90	−0.01	IV.	2	7.51	8 56.49	29.43	1.19	47 14.53	11 7.11
3	8	..	..	..	..	53.4	..	..	47 39.00	17.90	0.01	V.	4	6.57	18 27.32	29.42	2.28	47 56.89	20 39.02
4	8	..	32.0	46.4	..	..	..	..	?48 0.55	17.90	−0.01	IV.	3	10.4	15 2.67	29.41	1.87	49 18.44	17 13.95
5	7.8	..	41.8	..	..	..	..	..	50 10.70	17.90	+0.01	II.	9	5.40	41 46.35	29.37	4.95	50 28.61	44 0.67
6	10	..	..	..	..	39.3	..	..	50 24.88	17.90	−0.01	V.	3	3.54	11 56.04	29.36	1.52	50 42.77	14 6.92
7	9	..	..	..	..	39.0	..	..	51 24.56	17.90	0.01	V.	2	3.5	6 32.23	29.34	0.90	51 42.45	8 42.47
8	10	..	..	..	..	..	48.2	..	52 19.67	17.90	−0.01	V.	3	9.13	14 36.89	29.31	1.82	52 37.56	16 48.02
9	7	..	..	..	..	25.8	39.8	..	53 11.33	17.91	0.00	VI.	6	9.42	28 50.68	29.29	3.45	53 29.24	31 3.42
10	9	..	..	..	..	12.3	..	..	57 57.86	17.92	0.00	V.	2	4.15	7 7.53	29.17	0.96	58 15.77	9 17.66
11	9	..	..	..	..	..	..	51.2	58 7.92	17.92	−0.01	VII.	4	0.0	14 56.67	29.17	1 87	15 58 25.83	17 7.71
12	7.8	..	..	..	..	..	12.3	..	15 59 43.67	17.92	+0.01	VI.	8	5.27	36 40.79	29.13	35	16 0 1.60	38 54.27
13	9	..	..	..	..	..	10.7	..	16 0 42.22	17.92	−0.02	VI.	1	11.16	5 40.68	29.10	0.82	1 0.12	7 50.60
14	8	..	..	..	..	56.3	..	..	1 41.90	17.92	0.00	VI.	4	6.39	18 18.11	29.07	2.26	1 59.82	20 29.44
15	10	..	..	..	..	..	14.4	..	2 45.82	17.93	0.00	VI.	6	4.8	26 2.26	29.04	3.11	2 3.75	28 14.41
16	9	..	..	..	..	..	28.3	..	4 59.83	17.93	−0.02	VII.	1	6.33	3 17.74	28.97	0.55	4 17.74	5 27.26
17	7	36.3	50.8	..	..	..	..	..	16 6 19.43	+17.93	−0.01	III.	3	4.29	−12 13.69	−28.92	−1.55	16 6 37.35	−30 14 24.16

CORRECTIONS.

Date.		Corr. of Clock.	Hourly rate.	m	n	c
1847.	h.	s.	s.	s.	s.	s.
June 24,	15	*x* 10.54	*l* 0.013			

INSTRUMENT READINGS.

	Date.		Barom.	Thermom. At.	Thermom. Ex.
	1847.	h. m.	in.	°	°
Zone 124	June 24,	15 46	29.94	78.0	70.7

REMARKS.

(123) 41. Hor. thread assumed as 2 instead of 3.
(123) 48. Transits over T.'s V and VI assumed as recorded over T.'s IV and V.
(123) 50. Transit over T. II assumed to have been recorded as over T. III.
(123) 52. Hor. thread assumed as 8 instead of 9, and micrometer reading as 2r.49 instead of 11r.49, to agree with Arg. Z. 221, 35.
(123) 62. Micrometer thread assumed as 6 instead of 7.
(124) 4. Minutes assumed as 49 instead of 48.
(124) 10. Minutes assumed as 3 instead of 4.

ZONE 124. JUNE 24. K. BELT, −30° 23′. D_0=−30° 1′ 40″.—Continued.

No.	Mag.	Seconds of Transit. I.	II.	III.	IV.	V.	VI.	VII.	T.	a_1	a_2	Micrometer.			i	d_1	d_2	Mean Right Ascension, 1850.0.	Mean Declination, 1850.0.
									h. m. s.	s.	s.			r.	′ ″	″	″	h. m. s.	° ′ ″
18	9		31.4						16 6 0.17	+17.93	0.00	IV.	6	9.56	−28 57.93	−28.93	−3.46	16 6 18.10	−30 31 10.32
19	9		40.4						9 9.21	17.94	+0.01	IV.	7	7.50	32 54.39	28.84	3.90	9 27.16	35 7.13
20	8				44.9				9 44.74	17.94	0.00	IV.	6	11.31	29 55.91	28.82	3.57	10 2.68	32 9.30
21	8				40.1				12 39.94	17.95	+0.02	IV.	10	6.26	47 10.64	28.72	5.58	12 57.91	49 24.94
22	9						34.2		13 5.64	17.95	0.00	VI.	5	4.12	22 3.18	28.70	2.66	13 23.59	24 14.54
23	10					40.8			14 26.40	17.95	−0.01	VI.	4	7.52	18 54.91	28.65	2.32	14 44.34	21 5.88
24	9						52.3		15 23.73	17.95	0.00	VI.	5	9.15	24 35.96	28.64	2.94	15 41.68	26 47.54
25	8						37.5		16 9.03	17.95	−0.02	VI.	1	6.56	3 29.58	28.62	0.58	16 26.06	5 38.78
26	9			54.0					20 8.42	17.96	+0.01	IV.	8	9.2	38 29.40	28.45	4.56	20 26.39	40 42.41
27	9			15.5					26 30.01	17.97	+0.02	IV.	9	10.25	44 10.25	28.20	5.22	26 47.00	46 23.67
28	7		3.8		32.7				29 32.44	17.98	−0.01	IV.	1	14.55	7 31.28	28.08	1.01	29 50.41	9 40.37
29	8				53.2				30 53.04	17.98	0.00	V.	6	11.21	29 40.73	28.02	3.53	31 11.02	31 52.28
30	7			2.3					32 16.31	17.98	−0.01	IV.	2	13.42	11 53.48	27.96	1.51	32 34.28	14 2.95
31	9		49.3						34 18.19	17.99	+0.01	IV.	8	13.55	40 57.13	27.87	4.86	34 36.19	43 9.86
32	7		58.9						35 27.67	17.99	0.00	III.	6	10.25	29 12.50	27.81	3.48	35 45.66	31 23.79
33	7				16.7				43 16.54	18.01	−0.01	IV.	2	6.18	8 9.60	27.45	1.08	43 34.54	10 18.13
34	9						21.0		43 52.51	18.01	−0.01	VI.	2	3.51	6 55.28	27.42	0.96	44 10.51	9 3.66
35	6						12.7		44 44.12	18.01	0.00	IV.	6	5.54	17 55.87	27.38	2.23	45 2.13	20 5.48
36	8		12.8						48 41.62	18.02	+0.01	III.	7	10.26	34 13.01	27.19	4.06	48 59.65	36 24.26
37	9			39.4					50 53.46	18.02	−0.01	III.	3	9.9	14 34.88	27.07	1.82	51 11.47	16 43.77
38	8				55.5				51 55.34	18.02	0.00	IV.	7	4.0	30 58.42	27.03	3.68	52 13.36	33 9.12
39	9						43.8		52 15.24	18.02	0.00	VI.	5	6.59	23 27.39	27.00	2.84	52 33.26	25 37.23
40	9						2.7		57 34.13	18.03	0.00	VI.	5	9.31	24 44.03	26.72	2.97	57 52.16	26 53.72
41	4.5				54.7	9.3			16 58 54.70	18.03	−0.01	IV.	2	9.36	9 49.45	26.65	1.28	16 59 12.72	11 57.38
42	8				10.3				17 0 10.14	+18.03	+0.01	V.	8	4.33	−36 13.71	−26.58	−4.29	17 0 28.18	−30 38 24.58

ZONE 125. JUNE 25. B. BELT, −29° 23½′. D_0=−29° 10′ 40″.

No.	Mag.	I.	II.	III.	IV.	V.	VI.	VII.	T.	a_1	a_2	Micrometer.			i	d_1	d_2	Mean Right Ascension, 1850.0.	Mean Declination, 1850.0.
1	7					47.	0.	15.	17 29 32.03	+19.15	+0.01	VII.	9	3.21	−40 36.01	−40.93	−4.83	17 29 51.19	−29 52 1.77
2	8						3.	18.	32 34.58	19.16	−0.01	VII.	2	10.4	10 3.14	40.47	1.27	32 53.73	21 24.68
3	8		53.5	8.	22.5				35 36.64	19.16	−0.01	III.	3	11.40	15 51.02	39.99	1.94	35 55.79	26 12.95
4	9				44.	58.	12.		36 43.63	19.16	0.00	VII.	5	9.7	24 31.69	39.81	2.94	37 2.79	35 54.44
5	8		4.5	19.					39 33.05	19.17	−0.01	IV.	3	3.2	11 29.87	39.37	1.45	39 52.21	22 50.69
6	9	29.	44.						42 12.51	19.17	0.00	IV.	5	7.27	23 41.69	38.95	2.95	42 31.68	35 3.49
7	8						2.	17.	42 33.61	19.17	0.00	VII.	4	8.14	19 5.77	38.90	2.30	42 52.78	30 26.97
8	8		0.5	14.5					45 28.92	19.18	0.00	VII.	4	12.22	21 10.81	38.43	2.56	45 48.10	32 31.80
9	8	9.5	24.						47 52.73	19.18	0.00	III.	4	12.54	21 27.33	38.05	2.59	48 11.91	32 47.97
10	7			8.6	22.				49 22.22	19.18	−0.01	VII.	2	11.48	10 55.57	37.82	1.38	49 41.39	22 14.77
11	7			18.					51 32.18	19.18	0.00	V.	5	6.18	23 6.85	37.48	2.79	51 51.36	34 27.12
12	8							33.	52 49.78	19.19	0.00	VII.	4	10.40	20 19.39	37.27	2.46	53 8.97	31 39.12
13	5.6			53.			36.	51.	55 7.48	19.19	0.00	VII.	5	7.1	23 28.15	36.90	2.82	55 26.67	34 47.87
14	8		4.	19.					57 32.88	19.20	−0.01	IV.	4	5.50	17 53.58	36.51	2.17	57 52.07	29 12.26
15	7						16.	31.	17 57 47.62	19.20	0.00	VII.	5	1.09	20 30.66	36.47	2.48	17 58 6.82	31 49.61
16	7		41.0				27.	42.	18 2 9.05	19.20	+0.01	VII.	7	12.12	35 6.07	35.80	4.19	18 1 28.26	46 26.06
17	9					57.	11.	26.	4 42.62	19.20	0.00	VII.	5	5.57	22 55.88	35.36	2.76	5 1.82	34 14.00
18	8						14.	29.	6 45.66	19.21	+0.01	VII.	8	5.25	36 39.54	35.03	4.38	7 4.90	47 58.95
19	8	4.5	19.						9 47.82	19.21	0.00	III.	6	10.7	29 3.43	34.53	3.47	10 7.03	40 21.43
20	9	49.	4.						18 11 32.51	+19.21	0.00	IV.	5	9.6	−24 31.62	−34.25	−2.94	18 11 51.72	−29 35 48.81

CORRECTIONS.

Date.		Corr. of Clock.	Hourly rate.	m	n	c
1847.	h.	s.	s.	s.	s.	s.
June 25,	15	+ 11.46	+ 0.024	+ 0.103	+ 0.271	0.000

INSTRUMENT READINGS.

	Date.	Barom.	Thermom. At.	Thermom. Ex.
	1847. h. m.	in.	°	°
Zone 125	June 25, 9 5	30.122	78.	73.5

REMARKS.

(124) 20. Micrometer reading assumed as 11r.51 instead of 11r.31.
(124) 35. Hor. thread assumed as 4 instead of 6.
(125) 16. Minutes assumed as 1 instead of 2, and transits over T.'s VI and VII as recorded 10s too early.

ZONE 125. JUNE 25. B. BELT, −29° 23½′. D_0 = −29° 10′ 40″—Continued.

No.	Mag.	Seconds of Transit. I.	II.	III.	IV.	V.	VI.	VII.	T.	a_1	a_2	Micrometer.			i	d_1	d_2	Mean Right Ascension, 1850.0.	Mean Declination, 1850.0.
									h. m. s.	s.	s.			r.	′ ″	″	″	h. m. s.	° ′ ″
21	8		2.	17.					18 14 30.97	+19.22	0.00	IV.	5	10.35	−25 16.50	−33.77	−3.03	18 14 50.19	−29 36 33.30
22	8	28.5	43.					14.	14 11.76	19.22	0.00	VII.	5	8.45	24 20.60	33.82	2.92	14	35 37.34
23	8	28.5	43.						17 11.76	19.22	0.00	III.	5	8.5	24 0.81	33.33	2.88	17 30.98	35 17.02
24	7						23.	28.	17 54.70	19.23	+0.02	VII.	9	7.44	42 48.64	33.21	5.12	18 13.95	54 6.97
25	9		45.	59.5					21 13.83	19.23	0.01	V.	7	4.8	31 2.40	32.67	3.71	21 33.07	42 18.78
26	10		1.5	16.					23 30.31	19.23	0.01	IV.	7	5.43	31 50.35	32.29	3.81	23 49.55	43 6.45
27	6		37.	51.					26 5.63	19.24	+0.01	IV.	8	6.58	37 26.87	31.87	4.48	26 24.88	48 43.22
28	7						53.4	8.	27 24.83	19.24	0.00	V.	5	8.37	24 16.95	31.65	2.91	27 44.07	35 31.51
29	7							38.	28 54.97	19.24	+0.01	VII.	7	12.24	35 12.12	31.40	4.21	29 14.22	46 27.73
30	9			56.			38.4		31 9.97	19.24	−0.01	VII.	3	10.3	15 31.98	31.03	1.84	31 29.20	26 44.85
31	8						17.	32.	32 48.64	19.25	0.00	VII.	5	11.26	25 41.78	30.76	3.08	33 7.89	36 55.62
32	8			42.	56.3				34 56.20	19.25	0.00	V.	6	6.40	27 19.05	30.41	3.28	35 15.45	38 32.74
33	7		26.	41.					36 55.11	19.25	+0.01	IV.	8	3.30	35 41.99	30.08	4.26	37 14.37	46 56.33
34	7				18.		46.		49 17.63	19.26	0.00	VII.	6	9.4	28 31.28	28.03	3.41	46 36.89	39 42.72
35	10	13.	27.5						48 56.34	19.26	+0.01	III.	6	13.31	30 46.28	28.09	3.67	49 15.61	41 58.04
36	7	47.	1.	15.4					? 29.84	19.27	0.00	IV.	5	3.38	21 46.22	. .	2.63	49.11	
37	8					57.		25.5	53 42.60	19.27	+0.01	VII.	8	8.45	38 20.40	27.30	4.58	54 1.88	49 32.28
38	7				59.		27.		18 55 58.68	19.28	−0.02	VII.	2	3.44	6 51.52	26.92	0.89	18 56 17.94	17 59.33
39	7	44.5	00.	14.7	29.				19 1 28.63	19.28	+0.01	V.	7	8.34	33 16.54	26.02	3.98	19 1 47.94	44 26.54
40	7			3.		31.5	46.		3 17.34	19.28	+0.02	VII.	9	5.40	41 46.11	25.72	5.00	3 36.64	52 56.83
41	8							35.5	4 52.24	19.28	−0.01	VII.	4	4.55	18 25.92	25.46	2.14	5 11.51	29 33.52
42	9					32.	46.		7 17.53	19.29	0.00	VII.	5	9.29	24 42.78	25.07	2.97	7 36.82	35 50.82
43	10		51.	6.					10 19.90	19.29	−0.01	IV.	4	9.3	19 30.90	24.56	2.36	10 39.18	30 37.82
44	7				50.	4.			11 49.77	19.29	+0.02	VI.	9	5.21	40 36.24	24.31	4.85	12 9.08	51 45.40
45	9							54.	13 10.65	19.29	−0.02	VII.	2	8.58	9 29.85	24.09	1.19	13 29.92	20 35.13
46	8					32.	46.	1.	15 17.63	19.30	0.00	VII.	5	9.3	24 29.67	23.74	2.94	15 36.93	35 36.35
47	9			21.	35.				17 20.74	19.30	0.00	VI.	5	9.45	24 51.09	23.37	2.99	17 40.04	35 57.45
48	8						38.	53.	19 9.56	19.30	−0.02	VII.	1	10.2	5 3.13	23.11	0.67	19 28.84	16 6.91
49	8						49.	4.	21 20.70	19.30	+0.02	VII.	9	5.47	41 53.63	22.75	5.00	21 40.02	53 1.38
50	9			9.	23.				24 22.98	19.30	−0.01	V.	4	5.12	17 34.37	22.25	2.14	24 42.27	28 36.76
51	9	33.	47.						32 16.05	19.31	0.00	III.	6	6.59	27 28.63	20.97	3.30	32 35.36	38 32.00
52	8	38.			11.		39.	54.	38 10.83	19.32	−0.01	VII.	4	10.40	20 19.39	20.01	2.46	38 30.14	31 21.86
53	9	7.4	22.						45 50.65	19.33	−0.01	III.	4	5.56	17 53.53	18.79	2.18	46 9.97	28 54.50
54	9							47.5	46 4.33	19.33	0.00	VII.	5	9.35	24 45.81	18.75	2.97	46 23.66	35 47.53
55	9					16.	30.	45.	48 1.60	19.33	−0.01	VII.	4	5.12	17 33.99	18.44	2.14	48 20.92	28 34.57
56	9						3.	18.3	49 34.71	19.33	0.02	VII.	1	8.28	4 15.72	18.19	0.56	49 54.02	15 14.47
57	8		13.		42.	56.	10.	25.	55 41.66	19.34	−0.01	VII.	4	7.38	18 47.61	17.23	2.27	56 0.99	29 47.11
58	8							24.	56 40.82	19.34	0.00	VII.	5	6.6	?23 0.43	17.07	2.78	57 0.16	34 0.28
59	9					3.	17.	32.	19 58 48.57	19.34	−0.02	VII.	2	7.2	8 31.36	16.74	1.05	19 59 7.89	19 29.15
60	10						49.2	4.2	20 2 20.58	19.34	−0.01	VII.	3	2.14	11 5.24	16.19	1.36	20 2 39.91	22 2.79
61	10					1.		29.5	4 46.53	19.34	0.00	VII.	6	10.42	29 20.70	15.81	3.51	5 5.87	40 20.02
62	9			46.3	0.5		29.	43.3	6 0.44	19.35	+0.01	VII.	8	11.22	40 9.81	15.31	4.80	8 19.60	51 9.92
63	8	58.	12.2						10 41.11	19.35	0.00	II.	5	9.49	24 53.25	14.90	2.99	11 0.46	35 51.14
64	8					8.2	22.	37.	10 53.72	19.35	0.00	VII.	6	9.42	28 50.44	14.86	3.45	11 13.07	39 48.75
65	9				23.4	38.		6.2	13 23.32	19.35	0.00	VII.	6	5.14	26 35.30	14.48	3.19	13 42.67	37 32.97
66	7					24.4	38.4	53.	15 9.89	19.36	0.00	VII.	5	4.53	22 23.61	14.21	2.70	15 29.25	33 20.52
67	7				1.2	15.	29.2	44.1	17 0.80	19.36	0.00	VII.	5	5.4	22 29.16	13.93	2.71	17 20.16	33 25.80
68	8						1.4	16.1	18 32.95	19.36	+0.02	VII.	9	3.44	40 47.62	14.01	4.88	18 52.33	51 46.51
69	8						40.2		20 20 11.65	+19.36	−0.01	VII.	4	10.26	−20 9.30	−13.44	−2.43	20 20 31.00	−29 31 5.17

CORRECTIONS.

Date.	Corr. of Clock.	Hourly rate.	m	n	c
1847.	h. s.	s.	s.	s.	s.

INSTRUMENT READINGS.

Date.	Barom.	Thermom. At.	Thermom. Ex.
1847. h. m.	in.	°	°

REMARKS.

(125) 22. Transits discordant, and apparently belong to the preceding and following stars.
(125) 30. Micrometer reading assumed as 11r.3 instead of 10r.3.
(125) 34. Minutes assumed as 46 instead of 49.
(125) 41. Micrometer reading assumed as 6r.55 instead of 4r.55.
(125) 44. Micrometer reading assumed as 3r.21 instead of 5r.21.
(125) 62. Micrometer reading assumed as 12r.22 instead of 11r.22.
(125) 67. Minutes assumed as 17, to correspond with ——.
(125) 69. Minutes of transit assumed as 18 instead of 16.

ZONE 125. JUNE 25. B. BELT, −29° 23½′. D₀=−29° 10′ 40″—Continued.

No.	Mag.	I.	II.	III.	IV.	V.	VI.	VII.	T.	a_1	a_2	MICROMETER.			i	d_1	d_2	Mean Right Ascension, 1850.0.	Mean Declination, 1850.0.
									h. m. s.	s.	s.			r.	′ ″	″	″	h. m. s.	° ′ ″
70	7						54.2	9.1	20 21 25.79	+19.36	0.00	VII.	5	11.24	−25 40.77	−13.26	−3.09	20 21 45.15	−29 36 37.12
71	7				18.	32.	46.4	1.2	24 17.87	19.36	+0.01	VII.	8	6.9	37 1.74	12.83	4.44	24 37.24	47 59.01
72	10			2.	16.2				29 16.22	19.36	0.01	VI.	8	4.17	36 5.50	12.08	4.32	29 35.59	47 1.90
73	9						25.3	40.	31 56.82	19.36	+0.01	VII.	7	7.3	32 30.26	11.68	3.89	32 16.19	43 25.83
74	9				21.	35.2	49.	4.	33 20.75	19.36	0.00	VII.	6	9.40	28 49.43	11.48	3.47	33 40.11	39 44.38
75	10				35.2		4.		35 35.24	19.37	0.00	VII.	5	8.47	24 21.61	11.15	2.92	35 54.61	35 15.68
76	10					34.5		3.3	37 20.28	19.37	+0.02	VII.	9	7.22	42 37.54	10.89	5.06	37 39.67	53 33.49
77	10							19.3	39 36.13	19.37	0.00	VII.	5	8.6	24 0.94	10.56	2.88	39 55.50	34 54.38
78	10						35.	50.	18 41 6.63	+19.37	0.00	VII.	5	6.47	−23 21.10	−10.34	−2.80	20 41 26.00	−29 34 14.24

ZONE 126. AUGUST 30. K. BELT, −29° 23′. D₀=−29° 0′ 0″.

No.	Mag.	I.	II.	III.	IV.	V.	VI.	VII.	T.	a_1	a_2	MICROMETER.			i	d_1	d_2	Mean Right Ascension, 1850.0.	Mean Declination, 1850.0.
1	10					39.4			18 41 25.07	+18.26	+0.01	V.	7	11.33	−34 46.79	−28.61	−4.10	18 41 43.34	−29 35 19.50
2	5				17.0	1.2			42 46.85	18.26	+0.01	V.	7	7.6	32 32.16	28.42	3.84	43 5.12	33 4.42
3	8			4.7		33.3			44 18.83	18.25	−0.01	V.	3	10.20	15 10.68	28.21	1.90	44 37.07	15 40.79
4	6			22.7	36.6				45 36.66	18.24	0.00	IV.	5	6.33	23 14.47	28.03	2.79	45 54.90	23 45.29
5	6					32.5			46 18.19	18.24	+0.01	V.	8	10.33	39 15.24	27.94	4.62	46 36.44	39 47.80
6	10							30.8	46 47.77	18.24	0.01	IX.	7	11.15	34 36.57	27.87	4.08	47 6.02	35 8.52
7	10		28.5						48 57.40	18.23	+0.02	II.	9	5.2	41 27.19	27.57	4.88	49 15.65	41 59.64
8	10					56.2			51 41.81	18.21	−0.01	V.	4	8.52	19 25.31	27.20	2.38	52 0.01	19 54.89
9	7					45.1		13.8	52 30.75	18.20	+0.01	IX.	7	6.51	32 23.45	27.09	3.82	52 48.96	32 54.36
10	10		15.0						54 43.78	18.19	0.00	III.	6	11.32	29 46.29	26.79	3.53	55 1.97	30 16.61
11	10		6.0	20.7					55 34.82	18.19	0.00	III.	5	10.13	25 5.35	26.67	3.01	55 53.01	25 35.03
12	7						25.0	42.5	55 59.36	18.18	−0.01	VI.	4	5.11	17 33.73	26.61	2.16	56 17.53	18 2.50
13	9					33.4			18 57 19.02	18.18	0.00	V.	5	5.18	22 36.59	26.43	2.73	18 57 37.20	23 5.75
14	9		2.0						19 0 30.53	18.16	−0.02	II.	2	1.58	5 58.30	26.00	0.89	19 0 48.67	6 25.19
15	5				30.4		58.5		1 30.04	18.15	+0.02	IV.	9	9.56	43 55.63	25.87	5.16	1 48.21	44 26.66
16	9			24.9					4 39.20	18.14	+0.01	III.	7	3.58	30 57.35	25.44	3.67	4 57.35	31 26.46
17	7					7.3			4 52.95	18.14	0.00	VI.	6	10.10	29 4.80	25.42	3.45	5 11.09	29 33.67
18	8	58.8	13.3	27.8					9 41.95	18.11	−0.01	IV.	4	3.45	16 50.55	24.76	2.08	10 0.05	17 17.39
19	10	28.2							11 11.51	18.10	0.00	II.	5	9.27	24 42.01	24.57	2.97	11 29.61	25 9.55
20	6.7			57.4	11.7				13 11.54	18.09	−0.01	IV.	4	10.21	20 10.23	24.30	2.46	13 29.62	20 36.99
21	5.6	35.2	49.8	4.4	18.8				15 18.67	18.08	+0.01	IV.	7	12.21	35 11.03	24.02	4.15	15 36.76	35 29.20
22	9	38.3	53.2						17 21.90	18.07	0.01	IV.	7	13.10	35 35.74	23.75	4.20	17 39.98	36 3.69
23	8		7.9	22.3					18 36.77	18.06	+0.01	IV.	8	13.22	40 40.49	23.58	4.80	18 54.84	41 8.87
24	7.8					24.8			19 10.39	18.05	−0.01	VI.	3	11.33	15 47.35	23.51	1.96	19 28.43	16 12.82
25	9		37.0	51.6					21 5.98	18.04	+0.02	IV.	8	14.35	40 47.05	23.26	4.86	21 24.04	41 15.17
26	9						23.1		28 54.62	18.00	−0.02	VI.	1	8.0	4 1.85	22.23	0.67	29 12.60	4 24.75
27	8	27.4		56.5					32 10.54	17.98	−0.01	IV.	3	2.45	11 21.30	21.81	1.48	32 28.51	11 44.59
28	9							0.0	32 17.01	17.98	+0.01	IX.	8	8.37	38 15.61	21.80	4.51	32 35.00	38 41.92
29	10		3.3						35 32.06	17.96	0.00	II.	6	7.24	27 41.09	21.38	3.31	35 50.02	28 5.78
30	9				19.9				37 19.74	17.95	−0.01	IV.	2	11.59	11 1.55	21.15	1.44	37 37.68	11 24.14
31	7				11.8	26.1			38 11.70	17.94	+0.01	IV.	7	4.6	31 1.45	21.04	3.67	38 29.65	31 26.16
32	8						20.2		38 51.73	17.94	−0.02	VI.	1	6.36	3 19.50	20.96	0.60	39 9.65	3 41.06
33	5						0.0	14.8	39 31.48	17.93	−0.02	VI.	2	7.34	8 47.74	20.87	1.19	39 49.39	9 9.80
34	8			37.9					45 52.17	17.89	0.00	III.	6	9.24	28 41.75	20.06	3.42	46 10.06	29 5.23
35	8					19.8			46 5.48	17.89	+0.01	V.	7	13.4	35 32.67	20.04	4.19	46 23.38	35 56.90
36	6.7							0.8	19 46 17.76	+17.89	+0.01	VII.	7	10.52	−34 25.74	−20.01	−4.08	19 46 35.66	−29 34 49.83

CORRECTIONS.

Date.		Corr. of Clock.	Hourly rate.	m	n	c
1847.	h.	s.	s.	s.	s.	s.
Aug. 30,	16	r 10.40	g 0.010			

INSTRUMENT READINGS.

	Date.	Barom.	THERMOM. At.	THERMOM. Ex.
	1847. h. m.	In.	°	°
Zone 126	Aug. 30, 18 41	29.88	78.5	73.5

REMARKS.

(125) 73. Minutes assumed as 30 instead of 31.
(126) 25. Micrometer reading assumed as 13ʳ.35 instead of 14ʳ.35

ZONE 126. AUGUST 30. K. BELT, −29° 23'. D_0 = −29° 0' 0"—Continued.

No.	Mag.	Seconds of Transit. I.	II.	III.	IV.	V.	VI.	VII.	T.	a_1	a_2	Micrometer.			i	d_1	d_2	Mean Right Ascension, 1850.0.	Mean Declination, 1850.0.
									h. m. s.	s.	s.			r.	' "	"	"	h. m. s.	° ' "
37	10			48.9					19 48 3.17	+17.88	0.00	IV.	6	8.35	−28 17.09	−19.79	−3.37	19 48 21.05	−29 28 40.25
38	7		44.4						49 12.90	17.87	−0.02	II.	1	5.27	2 44.69	19.04	0.53	49 30.75	3 4.86
39	8				35.8	50.0			49 35.62	17.87	−0.01	V.	3	9.48	14 54.55	19.59	1.87	49 53.48	15 16.01
40	5.6			28.8	43.1				55 43.01	17.84	0.00	IV.	6	10.37	29 18.61	18.84	3.48	56 0.85	29 40.93
41	8			27.8	42.3				19 56 42.15	17.83	+0.01	IV.	7	9.23	33 41.28	18.72	3.99	19 56 59.99	34 3.99
42	7			39.8	53.8				20 1 53.70	17.80	−0.02	IV.	2	5.9	7 34.81	18.10	1.07	20 2 11.48	7 53.98
43	6				42.3				10 42.14	17.74	+0.01	IV.	8	3.4	35 28.88	17.07	4.18	10 59.89	35 50.13
44	6						22.0		10 53.36	17.74	0.01	VII.	8	10.25	39 10.62	17.05	4.61	11 11.11	39 32.48
45	6							38.2	10 55.23	17.74	0.01	VII.	8	11.30	39 43.00	17.05	4.68	11 12.98	39 51.73
46	8							7.5	13 24.50	17.72	0.01	VII.	8	6.37	37 15.86	16.77	4.38	13 42.23	37 37.01
47	6				11.7	25.8			15 11.51	17.71	+0.01	IV.	7	8.00	32 59.44	16.57	3.89	15 29.23	33 19.90
48	8							21.3	15 38.08	17.71	−0.01	VII.	4	10.39	20 18.89	16.52	2.48	15 55.78	20 37.89
49	9							15.5	16 32.42	17.71	+0.01	VII.	7	4.13	31 4.54	16.42	3.68	16 50.14	31 24.64
50	6							45.2	17 2.15	17.70	+0.01	VII.	7	8.17	33 7.58	16.36	3.92	17 19.86	33 27.86
51	7.6							42.4	17 59.15	17.69	−0.01	VII.	4	5.40	17 48.11	16.26	3.20	18 16.83	18 7.57
52	9		44.8						20 13.59	17.68	+0.01	II.	7	3.50	30 53.18	16.01	3.65	20 31.28	31 12.84
53	5		58.5	12.9					21 27.32	17.67	+0.01	IV.	8	4.43	36 18.60	15.88	4.28	21 45.00	36 38.96
54	9			40.9					22 54.84	17.66	−0.02	III.	2	2.0	5 59.46	15.72	0.87	22 12.48	6 16.05
55	7					29.6			23 15.15	17.66	0.02	V.	2	1.19	5 38.78	15.69	0.83	23 32.79	5 55.30
56	10				27.8				25 27.64	17.65	0.02	IV.	2	1.37	5 47.91	15.45	0.85	25 45.27	6 4.21
57	8				49.8		17.8		27 49.47	17.63	−0.01	V.	3	6.56	13 27.62	15.20	1.72	28 7.09	13 44.74
58	9		14.5						30 43.20	17.62	0.00	III.	5	6.56	23 26.02	14.90	1.84	31 0.82	23 42.76
59	7				58.8	12.9			30 58.62	17.62	+0.02	V.	9	8.25	43 9.69	14.87	5.09	31 16.26	43 29.65
60	5						56.6		31 28.12	17.61	−0.02	VI.	1	8.35	4 19.50	14.82	0.69	31 45.71	4 35.01
61	6					36.0			32 21.60	17.60	0.01	VII	4	4.38	17 16.85	14.73	2.13	32 39.19	17 33.71
62	7						6.2	21.1	33 37.76	17.59	0.01	VII.	4	5.50	17 53.15	14.61	2.20	33 55.34	18 9.96
63	10		48.4						36 16.96	17.58	−0.01	II.	2	8.40	9 21.02	14.34	1.23	36 34.53	9 36.59
64	9	54.6	9.0						39 37.94	17.55	+0.01	II.	7	11.13	34 36.56	14.01	4.09	39 55.50	34 54.66
65	9	25.2	39.4						41 8.42	17.54	0.01	III.	7	9.52	33 55.86	13.86	3.97	41 25.97	34 13.69
66	10			37.2					20 46 51.62	+17.50	+0.01	IV.	8	8.8	−38 2.17	−13.32	−4.48	20 47 9.13	−29 38 19.97

ZONE 127. SEPTEMBER 3. B. BELT, −27° 31'. D_0 = −27° 8' 40".

No.	Mag.	I.	II.	III.	IV.	V.	VI.	VII.	T.	a_1	a_2	Micrometer.			i	d_1	d_2	Mean Right Ascension, 1850.0.	Mean Declination, 1850.0.
1	9	40.	54.4	9.					20 9 22.69	+17.10	0.00	IV.	5	6.20	−23 7.91	−18.84	−2.80	20 9 39.79	−27 32 9.55
2	10		5.	19.					11 33.04	17.08	0.00	III.	4	11.41	20 50.53	18.57	2.60	11 50.12	29 51.70
3	9	27.	41.	55.2					15 9.19	17.07	−0.01	IV.	3	7.13	13 36.43	18.15	1.92	15 26.25	22 36.50
4	9	20.2	34.3	49.					18 2.67	17.05	−0.01	IV.	4	5.30	17 43.50	17.81	2.33	18 19.71	26 43.64
5	7	34.	48.	2.5					20 16.65	17.04	+0.02	IV.	9	10.40	44 17.82	17.55	4.87	20 33.71	53 20.24
6	10	49.	3.						24 31.47	17.01	+0.01	III.	7	4.43	31 20.06	17.07	3.62	24 48.49	40 20.75
7	8				24.3	38.5			25 24.29	17.00	+0.01	VI.	8	3.43	38 48.36	16.97	4.33	25 41.30	47 49.66
8	6				18.4	32.			31 18.01	16.96	−0.02	VI.	1	2.35	1 17.98	16.31	0.74	31 34.95	10 15.03
9	10	28.	42.2						35 10.52	16.94	0.00	III.	6	4.28	26 12.49	15.89	3.12	35 27.46	35 11.50
10	7							51.	35 8.74	16.94	+0.01	VII.	8	3.49	35 51.17	15.89	4.04	35 25.69	44 51.10
11	7					25.	39.	51.4	37 10.90	16.93	−0.01	VII.	3	10.57	15 28.99	15.67	2.09	37 27.82	24 26.75
12	9			55.5	9.4				40 9.31	16.91	0.01	VII.	4	9.50	19 54.20	15.35	2.52	40 26.21	28 52.07
13	4	52.	6.2	21.	35.				42 34.65	16.89	−0.01	V.	4	9.8	19 33.38	15.10	2.48	42 51.53	28 30.96
14	9						14.	28.4	43 46.03	16.89	+0.01	VII.	8	11.38	34 45.10	14.98	4.10	44 2.93	43 44.18
15	10	15.		44.					20 47 57.92	+16.86	+0.01	IV.	8	4.00	−35 57.12	−14.53	−4.08	20 48 14.79	−27 44 55.73

CORRECTIONS.

Date.		Corr. of Clock.	Hourly rate.	m	n	c
1847.	h.	s.	s.	s.	s.	s.
Sept. 3.	18	s 9.83	g 0.010	+ 0.114	+ 0.102	0.000

INSTRUMENT READINGS.

	Date.			Barom.	Thermom. At.	Thermom. Ex.
	1847.	h.	m.	In.	°	°
Zone 127	Sept. 3,	9	5	30.030	81.	77.

REMARKS.

(126) 45. Micrometer reading assumed as 11r.3 instead of 11r.30.
(127) 14. Micrometer reading assumed as 1r.38 instead of 11r.38.

ZONE 127. SEPTEMBER 3. B. BELT, −27° 31′. D₀ = −27° 8′ 40″—Continued.

No.	Mag.	I.	II.	III.	IV.	V.	VI.	VII.	T.	a_1	a_2	MICROMETER.			i	d_1	d_2	Mean Right Ascension, 1850.0.	Mean Declination, 1850.0.
									h. m. s.	s.	s.			r.	′ ″	″	″	h. m. s.	° ′ ″
16	9	37.	51.	..	..	..	..	..	20 50 19.93	16.84	0.00	VII.	6	5.11	−26 33.81	−14.29	−3.15	20 50 36.77	−27 35 31.25
17	8	..	..	20.	..	48.	..	..	20 52 33.66	16.82	−0.01	VI.	4	7.51	18 54.41	14.06	2.40	20 52 50.67	27 50.87
18	8	59.	13.2	27.5	..	..	..	..	21 9 41.25	16.71	0.02	IV.	1	10.56	5 30.79	12.35	1.15	21 9 57.94	14 24.29
19	8	..	..	..	..	..	1.	15.5	11 32.95	16.70	−0.01	VII.	2	7.32	8 46.52	12.18	1.46	11 49.64	17 40.16
20	10	..	..	..	8.	..	..	50.	14 7.85	16.68	+0.02	VII.	9	12.56	45 25.99	11.94	5.00	14 24.55	54 22.93
21	10	..	..	..	..	27.2	..	55.	17 12.97	16.66	−0.01	VII.	8	11.28	39 42.61	11.65	4.44	17 29.62	48 38.70
22	7	..	..	..	..	31.	45.	..	19 16.90	16.65	−0.02	VII.	1	4.00	2 15.75	11.46	0.69	19 33.53	11 7.90
23	9	..	30.6	44.5	58.6	..	..	..	21 58.69	16.63	+0.01	IV.	8	9.24	38 40.49	11.22	4.33	22 15.33	47 36.04
24	9	20.	34.3	48.5	..	..	..	..	24 2.57	16.62	0.00	IV.	6	10.24	29 12.05	11.03	3.41	24 19.19	38 6.49
25	8	..	20.3	34.2	48.5	..	..	..	34 34.07	16.55	−0.01	V.	3	10.00	15 0.61	10.11	2.04	34 50.61	23 52.76
26	9	..	..	59.	13.	..	..	..	36 12.90	16.54	0.00	V.	5	10.37	25 17.47	9.99	3.03	36 29.44	34 10.49
27	7	..	..	..	..	..	..	2.	36 19.66	10.54	0.00	VII.	6	10.00	28 59.55	9.98	3.39	36 36.20	37 52.92
28	10	..	..	..	..	47.	1.	..	38 32.91	16.52	0.00	VII.	5	7.39	23 47.35	9.80	2.87	38 49.43	32 40.02
29	9	..	..	28.6	42.5	56.7	..	..	43 42.52	16.49	+0.01	VI.	8	10.46	39 21.67	8.64	4.41	43 59.02	48 14.72
30	10	..	25.2	40.2	54.5	..	..	..	21 47 8.33	16.46	0.01	VI.	7	11.14	34 37.08	8.32	3.95	21 48 24.80	43 29.35
31	8	..	..	56.	10.	24.	..	..	22 37 9.97	16.15	+0.01	VI.	8	5.30	36 42.32	29.70	4.16	22 37 26.11	45 56.18
32	10	..	..	9.	23.2	37.4	..	..	41 23.00	16.12	0.00	VI.	2	11.16	10 39.69	29.63	1.61	41 39.11	19 50.93
33	10	..	..	..	30.	44.	57.8	..	42 29.72	16.12	−0.01	VII.	2	10.00	10 1.15	29.62	1.54	42 45.83	18 12.31
34	9	28.5	42.7	57.	..	..	..	..	46 11.11	16.09	0.00	IV.	5	2.10	35 1.65	29.58	3.99	46 27.20	44 15.22
35	8	..	..	..	..	38.	52.	6.5	48 23.92	16.08	0.00	VII.	3	5.9	12 33.51	29.55	1.79	48 40.00	21 44.85
36	10	3.4	17.4	32.	..	..	..	..	51 45.88	16.06	0.00	IV.	6	6.50	27 24.14	29.52	3.24	52 1.94	36 36.90
37	7	31.2	45.5	0.0	..	..	..	..	56 13.86	16.03	0.00	IV.	6	7.55	27 56.91	29.51	3.30	55 29.89	37 9.72
38	6	..	..	..	26.3	40.5	..	..	22 56 26.30	16.03	+0.01	VI.	10	6.53	47 24.07	29.51	5.23	22 56 42.34	56 38.81
39	7	9.	23.2	37.5	..	..	..	..	23 2 51.63	16.00	0.00	VI.	7	10.42	34 20.95	29.51	3.92	23 3 7.63	43 34.38
40	9	..	45.	58.	12.	26.	..	..	5 11.81	15.98	0.00	VII.	3	4.21	12 9.30	29.52	1.74	5 27.79	21 20.56
41	11	..	..	..	..	8.	22.	36.4	8 53.91	15.97	0.00	VII.	4	7.22	18 39.57	29.53	2.39	9 9.88	27 51.49
42	10	..	46.5	..	14.5	..	..	..	11 14.45	15.95	0.00	V.	2	4.15	7 7.54	29.56	1.27	11 30.40	16 18.37
43	5	..	..	..	..	13.7	27.	41.5	12 59.93	15.94	0.00	VII.	8	10.26	39 11.36	29.58	4.41	13 15.87	48 25.35
44	10	..	..	..	..	..	35.	49.5	15 6.93	15.93	−0.01	VII.	1	7.23	3 42.97	29.61	0.93	15 21.95	12 53.51
45	9	49.	3.	17.2	..	..	..	..	19 31.27	15.90	0.00	IV.	4	11.52	20 56.12	29.66	2.60	19 47.17	30 8.38
46	10	..	..	..	3.	..	30.5	45.	21 2.68	15.89	0.00	VII.	10	4.13	46 3.17	29.69	5.10	21 18.57	35 17.96
47	10	..	14.3	28.5	..	..	..	..	26 42.57	15.87	0.00	IV.	7	3.36	30 46.32	29.80	3.57	26 58.44	39 59.69
48	7	..	..	..	..	15.	..	42.5	28 0.56	15.86	0.00	VII.	7	8.22	33 10.13	29.82	3.81	28 16.42	42 23.76
49	9	..	44.	58.	..	26.	..	..	32 12.11	15.84	0.00	VII.	7	9.59	33 59.04	29.92	3.89	32 27.95	43 12.85
50	9	..	..	..	57.7	12.	..	39.5	37 57.54	15.80	0.00	VII.	6	11.35	29 47.45	30.08	3.46	38 13.34	39 0.99
51	7	..	51.	5.1	19.2	33.5	..	..	46 19.31	15.76	0.00	VI.	9	8.52	43 23.18	30.37	4.84	46 35.07	52 38.39
52	6	..	..	54.	8.	22.	..	..	49 7.85	15.74	0.00	VI.	4	6.44	18 20.64	30.47	2.34	49 23.59	27 33.45
53	9	..	2.3	17.	31.	..	..	..	51 30.67	15.73	0.00	V.	3	4.45	12 21.77	30.56	1.76	51 46.40	21 34.09
54	6	..	..	..	17.	..	..	..	23 54 16.84	15.72	0.00	VI.	10	10.2	48 59.38	30.68	5.41	23 54 32.56	58 15.47
55	10	..	..	..	28.8	..	56.4	..	0 3 28.45	15.68	0.00	VII.	9	5.25	41 38.57	31.11	4.67	0 3 44.13	50 54.35
56	8	..	..	..	..	..	..	25.4	4 43.10	15.67	0.00	VII.	7	6.41	32 19.20	31.17	3.73	4 58.77	41 34.10
57	10	..	..	..	..	..	39.	53.5	14 11.01	15.63	0.00	VII.	5	7.4	23 29.70	31.69	2.95	14 26.64	32 44.24
58	9	..	..	17.	..	..	..	..	16 31.23	15.62	0.00	IV.	9	6.2	42 27.89	31.82	4.71	16 46.85	51 44.42
59	9	..	..	..	..	..	..	24.	16 41.74	15.62	0.00	VII.	8	3.28	35 40.57	31.83	4.06	16 57.36	44 56.46
60	8	..	..	..	22.3	36.3	50.	..	21 22.08	15.59	0.00	VII.	9	4.28	41 9.83	32.12	4.61	21 37.67	50 26.56
61	8	..	..	..	30.3	44.5	..	..	26 30.27	15.57	0.00	VI.	6	4.2	25 59.25	32.47	3.10	28 45.84	35 14.82
62	7	..	..	..	..	..	12.5	27.	36 44.45	15.53	0.00	VII.	2	12.45	11 24.34	33.22	1.66	36 59.98	20 39.22
63	10	..	..	..	..	16.	..	45.	40 2.36	15.52	0.00	VII.	8	8.59	38 27.49	33.46	4.36	40 17.88	47 45.31
64	8	..	..	..	..	..	..	25.3	0 44 42.63	+15.50	0.00	VII.	1	7.19	− 3 40.96	−33.85	−0.90	0 44 58.13	−27 12 55.71

CORRECTIONS.

Date.		Corr. of Clock.	Hourly rate.	m	n	c
1847.	h.	s.	s.	s.	s.	s.

INSTRUMENT READINGS.

Date.		Barom.	Thermom. At.	Thermom. Ex.
1847.	h. m.	in.	°	°

REMARKS.

(127) 22. Micrometer assumed as 4r.30 instead of 4r.00.
(127) 25. Transits over T.'s III, IV, V assumed as recorded over T.'s II, III, IV.
(127) 29. Transits over T.'s I, II, III assumed as recorded over T.'s III, IV, V, and minutes as 48 instead of 47.
(127) 34. Minutes of transit assumed as 46 instead of 43.
(127) 37. Minutes of transit assumed as 55 instead of 56.
(127) 39. Transit over T. II assumed as 23s.2 instead of 25s.2.
(127) 43. Transit observations very discordant.
(127) 58. Micrometer reading assumed as 7r.2 instead of 6r.2.
(127) 61. Minutes assumed as 28 instead of 26.

ZONE 127. SEPTEMBER 3. B. BELT, −27° 31′. D_c = −27 8° ′ 40″.

No.	Mag.	Seconds of Transit (I. II. III. IV. V. VI. VII.)	T.	a_1	a_2	Micrometer			i	d_1	d_2	Mean Right Ascension, 1850.0.	Mean Declination, 1850.0.
			h. m. s.	s.	s.			r.	′ ″	″	″	h. m. s.	° ′ ″
65	8	. . 59. 13.5 . . 41.	0 58 27.17	+15.45	0.00	VI.	5	5.35	−22 45.04	−35.06	−2.78	0 58 42.62	−27 32 2.88
66	10	. . 19. 33.2 47.4	1 10 47.25	15.41	0.00	V.	7	1.43	29 49.29	36.27	3.47	1 11 2.66	39 9.03
67	10	. . . . . . 10. 24.3	13 10.01	15.40	0.00	VI.	5	5.46	22 50.59	36.53	2.78	13 25.41	32 9.90
68	10	. . . . 49. 3.5	14 20.98	15.40	0.00	VII.	4	5.53	17 54.69	36.65	2.27	14 36.38	27 13.61
69	7	. . . . 28. 42.5	20 42.30	15.38	0.00	VII.	9	10.17	44 5.82	37.33	4.93	20 57.68	53 28.08
70	8	. . . . 25.4	22 43.14	15.37	0.00	VII.	7	13.12	35 36.35	37.62	4.05	22 58.51	44 58.02
71	8	. . . . . . 6.5 . . 34. 46.3	42 6.11	15.32	0.00	VII.	8	11.3	39 30.01	39.87	4.45	42 21.43	48 54.33
72	9	. . . . 23. . . 51.5	44 51.36	15.31	0.00	VI.	9	6.10	42 1.49	40.20	4.72	45 6.67	51 26.41
73	8	23. 37.2 52. 15.	51 5.51	15.30	0.00	V.	3	6.54	13 26.62	41.04	1.85	51 20.81	22 49.71
74	8	. . 47.2 1.5 16.	1 55 29.71	+15.29	0.00	V.	4	7.27	−18 42.45	−41.60	−2.38	1 55 45.00	−27 28 6.43

ZONE 128. SEPTEMBER 4. K. BELT, −28° 46′. D_c = −28° 22′ 0″.

No.	Mag.	Seconds of Transit (I. II. III. IV. V. VI. VII.)	T.	a_1	a_2	Micrometer			i	d_1	d_2	Mean Right Ascension, 1850.0.	Mean Declination, 1850.0.
1	9	. . . . 50.7	19 36 4.97	+16.70	+0.01	III.	7	9.44	−33 51.83	−13.20	−3.93	19 36 21.68	−28 56 8.96
2	8	28.7 43.2 57.4	38 11.74	16.68	0.00	III.	6	10.2	29 0.91	12.90	3.42	38 28.42	28 51 17.23
3	9	. . . . 38.8 . . 7.3	38 53.14	16.68	+0.01	V.	9	5.2	41 27.33	12.80	4.73	39 9.83	29 3 44.86
4	8	. . . . 0.9 . .	39 32.39	16.67	+0.02	VI.	10	6.0	46 57.34	12.70	5.34	39 49.08	29 9 15.38
5	9	7.2 21.8	41 50.35	16.66	0.00	II.	6	7.41	27 49.66	12.39	3.34	42 7.01	28 50 5.39
6	9	. . . . 0.4 14.8	42 14.66	16.66	+0.01	IV.	7	9.31	33 45.32	12.33	3.90	42 31.33	56 1.55
7	9	. . . . 5.4	42 22.56	16.66	0.00	VII.	6	12.33	30 16.67	12.31	3.55	42 39.22	52 32.53
8	9	. . 52.5 . . 21.2	48 21.03	16.62	0.00	IV.	4	11.59	20 59.65	11.49	2.58	48 37.65	28 43 13.72
9	8	. . . . 56.8	49 14.09	16.61	+0.01	VII.	8	13.48	40 53.18	11.37	4.68	49 30.71	29 3 9.23
10	8	. . . . 44.7	51 1.89	16.60	+0.01	VII.	7	6.39	37 16.89	11.12	4.25	51 18.50	28 59 32.26
11	9	. . . . . . 36.2 50.4	52 36.05	16.59	−0.01	VI.	3	8.50	14 25.16	10.92	1.90	52 52.63	36 37.98
12	9	. . . . 39.8 . .	53 11.40	16.59	0.00	VI.	5	10.13	25 5.21	10.83	3.01	53 27.99	47 19.05
13	9	. . . . 23.7 . . 52.3	54 37.72	16.58	−0.02	IV.	1	7.45	3 54.47	10.64	0.83	54 54.28	26 5.94
14	9	. . . . 45.9 . .	55 17.52	16.57	0.00	VI.	4	8.52	19 25.17	10.55	2.42	55 34.09	41 38.14
15	8	. . 39.1 53.4	57 7.59	16.56	0.00	IV.	5	10.57	25 27.59	10.31	3.05	19 57 24.15	47 40.95
16	7.8	. . . . 32.6 . . 1.0	19 59 46.78	16.54	0.00	IV.	6	12.0	30 0.45	9.95	3.52	20 0 3.32	52 13.92
17	9	. . . . 51.8 . .	20 0 23.42	16.54	0.00	VI.	4	11.54	20 56.94	9.87	2.58	0 39.96	28 43 9.39
18	8	. . . . 40.6 . . 8.9	1 54.88	16.53	+0.02	IV.	9	13.27	45 42.01	9.67	5.19	2 11.43	29 7 56.87
19	10	. . . . 39.9	4 54.31	16.51	+0.01	III.	9	7.37	42 45.49	9.28	4.89	5 10.83	4 59.66
20	10	. . . . 30.9	5 16.68	16.50	+0.01	V.	8	9.57	38 57.08	9.23	4.47	5 33.19	29 1 10.78
21	9	. . . . 24.2 38.7	5 55.83	16.50	+0.01	VI.	7	9.25	33 42.10	9.14	3.90	6 12.34	28 55 55.14
22	9	. . 2.2 16.3	8 30.44	16.48	−0.01	III.	3	6.38	13 18.74	8.81	1.79	8 46.91	35 29.34
23	10	. . . . . . 5.3	9 5.14	16.48	−0.01	IV.	3	7.33	13 46.52	8.73	1.84	9 21.61	35 57.09
24	10	. . 30.1	13 58.74	16.45	+0.01	II.	7	7.53	32 55.71	8.11	3.81	14 15.20	28 55 7.63
25	9	. . 12.4 . . 41.2	14 41.03	16.44	0.00	IV.	6	11.5	29 32.72	8.02	3.47	14 57.47	29 51 44.21
26	7	. . . . 44.3 . . 13.0	15 58.79	16.43	+0.02	VI.	10	5.15	46 34.64	7.86	5.28	16 15.24	29 8 47.78
27	9	. . . . 24.2	17 9.97	16.43	+0.01	V.	8	3.15	35 34.37	7.71	4.11	17 26.41	28 57 46.19
28	7	. . . . . . 29.8 43.8	18 29.57	16.42	0.00	V.	5	5.48	22 51.72	7.51	2.78	18 45.99	45 2.04
29	9	8.5 22.8	21 51.55	16.40	+0.01	II.	7	8.30	33 14.38	7.12	3.86	22 7.96	55 25.36
30	10	. . . . 6.8	22 20.90	16.39	0.00	III.	5	6.45	23 20.33	7.06	2.84	22 37.29	45 30.23
31	9	. . . . 5.0	22 21.93	16.39	+0.01	VII.	3	4.32	12 14.84	7.06	1.69	22 38.31	28 34 23.59
32	7	. . . . 58.9	23 16.17	16.39	0.01	IX.	9	−0.15	44 2.29	6.95	5.02	23 32.57	29 6 14.36
33	8	. . . . 11.5	25 28.82	16.37	0.01	VII.	9	9.56	43 55.21	6.68	5.00	25 45.20	29 6 6.89
34	7	. . . . 24.6 38.8	29 38.76	16.34	0.01	IV.	7	11.23	34 41.79	6.9	4.01	29 55.11	28 56 51.99
35	7	46.2 . . 14.9	20 31 29.38	+16.33	+0.01	IV.	9	6.51	−42 22.34	− 5.95	−4.84	20 31 45.72	−29 4 33.13

CORRECTIONS.

Date.	Corr. of Clock.	Hourly rate.	m	n	c
1847.	h. s.	s.	s.	s.	s.
Sept. 4.	18 *s* 9.13	*g* 0.012			

INSTRUMENT READINGS.

	Date.	Barom.	Thermom. At.	Thermom. Ex.
	1847. h. m.	In.	°	°
Zone 128	Sept. 4. 9 5	30.038	82.8	78.

REMARKS.

(127) 71. Transit observations discordant. T. VII rejected.
(127) 72. Transits over T.'s II and IV assumed as recorded over T.'s III and IV.
(127) 74. Transits over T.'s I, II, III assumed as recorded over T.'s II, III, IV.
(128) 10. Hor. thread assumed as 8 instead of 7.
(128) 32. Hor. thread assumed as 10 instead of 9.

ZONE 128. SEPTEMBER 4. K. BELT, −28° 46′. D_0=−28° 22′ 0″—Continued.

No.	Mag.	I.	II.	III.	IV.	V.	VI.	VII.	T.	a_1	a_2	MICROMETER.	No.	r.	i	d_1	d_2	Mean Right Ascension, 1850.0.	Mean Declination, 1850.0.
									h. m. s.	s.	s.			r.	′ ″	″	″	h. m. s.	° ′ ″
36	7.8	..	..	..	12.7	26.8	..	..	20 32 12.49	+16.32	−0.01	V.	2	8.31	− 9 16.62	− 5.87	−1.39	20 32 28.80	−28 31 23.88
37	9	..	6.8	20.9	..	..	..	..	34 34.93	16.31	−0.02	III.	1	9.43	4 53.92	5.59	0.93	34 51.22	27 0.44
38	7	..	..	..	14.2	28.5	..	..	35 14.12	16.30	0.00	IV.	5	4.43	22 19.00	5.51	2.72	35 30.42	28 44 27.23
39	9	..	..	..	..	32.3	..	..	36 18.11	16.30	+0.02	V.	10	7.3	47 29.24	5.39	5.40	36 34.43	29 9 40.03
40	9	..	..	..	..	..	..	37.3	36 54.54	16.29	+0.01	VII.	8	5.34	36 44.10	5.32	4.25	37 10.84	28 58 53.67
41	9	..	51.8	..	..	..	..	..	39 20.25	16.27	−0.01	III.	3	9.53	14 57.06	5.03	1.96	39 36.51	37 4.05
42	9	..	39.8	..	8.0	..	..	..	40 8.14	16.27	0.00	IV.	7	5.33	31 45.31	4.93	3.70	40 24.41	53 53.94
43	9	..	..	55.1	..	23.3	..	..	41 8.99	16.26	−0.01	IV.	3	3.14	11 35.92	4.82	1.62	41 25.24	33 42.36
44	7	..	..	..	18.1	32.7	..	..	42 18.15	16.25	−0.01	VI.	2	11.53	10 58.33	4.69	1.56	42 34.39	33 4.58
45	9	..	..	..	..	..	..	46.1	43 3.33	16.25	+0.01	VII.	7	13.45	35 52.97	4.60	4.14	43 19.59	58 1.71
46	9	..	..	..	..	..	..	45.1	44 1.98	16.24	−0.01	VII.	2	5.57	12 57.69	4.49	1.76	44 18.21	35 3.94
47	6.7	..	..	..	..	..	..	33.2	44 50.07	16.24	−0.01	VII.	2	4.20	7 9.68	4.39	1.14	45 6.30	28 29 15.21
48	8	..	..	..	..	..	18.2	..	45 49.72	16.23	+0.01	VII.	9	4.32	41 11.84	4.28	4.71	46 5.96	29 3 20.83
49	9	20.8	35.2	..	..	..	..	..	48 3.90	16.21	+0.01	II.	7	7.19	32 38.57	4.02	3.80	48 20.12	28 54 46.39
50	10	43.1	..	..	..	..	..	..	49 26.20	16.20	0.00	I.	6	5.41	26 48.92	3.87	3.17	49 42.40	48 55.96
51	9	47.c	1.6	..	..	..	..	..	51 30.19	16.19	0.00	II.	7	5.54	31 55.71	3.63	3.72	51 46.38	54 3.06
52	9	..	..	..	43.5	..	..	..	51 43.34	16.19	0.00	V.	6	4.54	26 25.60	3.61	3.15	51 59.53	48 32.36
53	10	..	..	..	..	..	38.7	..	52 10.31	16.18	0.00	VII.	5	6.15	23 4.97	3.56	2.79	52 26.49	45 11.32
54	9	37.9	52.5	..	..	..	..	..	54 20.83	16.17	−0.01	II.	2	5.5	7 32.60	3.31	1.18	54 36.99	28 29 37.09
55	10	..	..	34.2	..	..	..	..	55 48.61	16.16	+0.01	IV.	9	8.7	43 0.67	3.13	4.91	56 4.78	29 5 8.71
56	10	..	36.9	51.6	..	..	..	..	57 5.45	16.15	−0.01	IV.	3	8.40	14 20.31	3.01	1.88	57 21.59	28 36 25.20
57	9	..	..	..	..	55.3	..	..	57 41.09	16.14	+0.01	V.	9	3.27	40 39.42	2.94	4.67	20 57 57.24	29 2 47.03
58	6.7	..	..	..	44.8	58.7	..	..	20 59 44.56	+16.13	+0.01	VI.	9	6.51	−42 22.15	− 2.71	−4.85	21 0 0.70	−29 4 29.71

ZONE 129. SEPTEMBER 6. H. BELT, −25° 39′. D_0=−25° 15′ 0″.

No.	Mag.	I.	II.	III.	IV.	V.	VI.	VII.	T.	a_1	a_2	MICROMETER.	No.	r.	i	d_1	d_2	Mean Right Ascension, 1850.0.	Mean Declination, 1850.0.
1	10	..	25.5	39.	53.	..	..	..	20 6 52.92	+15.95	0.00	V.	5	4.2	−23 59.30	−24.79	−2.91	20 7 8.87	−25 39 27.00
2	11	..	..	..	..	6.	19.7	34.	8 52.10	15.94	−0.01	VII.	3	5.56	12 57.25	24.48	2.02	9 8.03	28 23.75
3	8	..	..	..	..	..	57.5	11.	10 29.60	15.93	0.00	VII.	6	3.32	25 43.96	24.22	3.06	10 45.53	41 11.24
4	9	..	14.	28.2	..	..	..	..	13 41.66	15.91	−0.01	IV.	2	7.28	8 44.89	23.73	1.68	13 57.56	24 10.30
5	9	..	..	..	..	..	..	43.	17 1.10	15.89	−0.01	VII.	2	10.50	10 26.40	23.22	1.81	17 16.98	25 51.43
6	11	..	..	..	..	..	21.	35.4	18 53.50	15.88	0.00	VII.	4	8.58	19 28.02	22.93	2.57	19 9.38	34 53.52
7	10	..	..	..	..	..	..	12.3	20 30.36	15.87	−0.02	VII.	2	3.54	6 56.63	22.68	1.52	20 46.21	22 20.83
8	9	..	..	..	..	..	40.5	55.	22 13.04	15.85	−0.01	VII.	4	2.00	15 57.25	22.43	2.22	22 28.88	31 21.90
9	6	..	..	..	..	54.	7.5	22.	23 40.03	15.84	−0.01	VII.	3	2.55	11 25.98	22.21	1.87	23 55.86	26 50.06
10	10	..	..	..	..	..	..	38.5	25 57.03	15.83	+0.02	VII.	9	10.24	44 9.39	21.87	4.59	26 12.88	59 35.85
11	6	58.	12.2	26.2	..	..	..	..	28 39.89	15.81	0.00	IV.	5	4.31	22 12.95	21.47	2.76	28 55.70	37 37.18
12	8	..	..	23.	..	51.	..	..	29 36.95	15.80	0.00	VI.	6	7.31	27 44.65	21.33	3.26	29 52.75	43 9.24
13	7	57.	11.2	..	..	..	..	..	31 38.86	15.79	−0.01	IV.	4	8.20	19 9.22	21.03	2.47	31 54.64	34 32.72
14	9	9.	..	..	..	..	..	..	36 50.85	15.76	0.00	II.	5	6.42	23 18.85	20.28	2.87	37 6.61	38 42.00
15	4	..	..	..	..	10.2	24.	38.	36 56.37	15.76	+0.01	VII.	7	8.5	33 1.60	20.27	3.70	37 12.14	48 25.57
16	6	41.5	55.	9.	..	..	..	..	41 22.97	15.73	−0.01	IV.	3	13.5	16 33.92	19.63	2.31	41 38.69	31 55.86
17	10	..	..	48.	2.2	16.2	..	..	47 2.06	15.69	0.00	VI.	7	2.40	30 17.92	18.84	3.46	47 17.75	45 40.22
18	9	..	..	..	..	50.5	4.	18.2	48 36.43	15.68	−0.01	VII.	3	4.21	12 9.34	18.62	2.00	48 52.10	27 29.96
19	8	..	..	13.3	27.3	40.5	..	..	53 26.85	15.65	−0.01	VI.	3	6.40	12 18.94	17.95	2.01	53 42.49	27 38.90
20	4	23.3	37.5	51.2	..	..	..	..	57 5.07	15.62	0.00	IV.	4	11.32	20 46.04	17.46	2.64	20 58 20.69	36 6.14
21	9	..	..	54.5	..	21.5	34.	..	21 0 7.32	15.60	−0.01	VII.	2	5.49	7 54.62	17.04	1.59	21 0 22.91	23 13.25
22	8	..	..	..	20.	34.	..	..	21 1 59.94	+15.59	−0.01	VII.	2	11.37	−10 50.10	−16.75	−1.81	21 2 15.52	−25 26 8.66

CORRECTIONS.

Date.	Corr. of Clock.	Hourly rate.	m	n	c
1847. h.	s.	s.	s.	s.	s.
Sept. 6, 18	[illegible] 8.71	[illegible] 0.015	+ 0.113	+ 0.053	0.000

INSTRUMENT READINGS.

	Date.	Barom.	THERMOM. At.	THERMOM. Ex.
	1847. h. m.	In.	°	°
Zone 129	Sept. 6, 9 5	30.100	76.5	69.

REMARKS.

(128) 46. Hor. thread assumed as 3 instead of 2.
(129) 1. Micrometer reading assumed as 8r.2 instead of 4r.2, to agree with Arg. 239, 105.
(129) 19. Micrometer reading assumed as 4r.40 instead of 6r.40.
(129) 20. Minutes assumed as 58 instead of 59.
(129) 22. Transits over T.'s IV and V assumed as 0s and 14s instead of 20s and 34s, and minutes as 1 instead of 2.

ZONE 129. SEPTEMBER 6. B. BELT, −25° 39′. D_0 = −25° 15′ 0″—Continued.

No.	Mag.	SECONDS OF TRANSIT. I.	II.	III.	IV.	V.	VI.	VII.	T.	a_1	a_2	MICROMETER.			i	d_1	d_2	Mean Right Ascension, 1850.0.	Mean Declination, 1850.0.
									h. m. s.	s.	s.			r.	′ ″	″	″	h. m. s.	° ′ ″
23	8	..	..	4.0	18.	..	..	..	21 4 17.70	+15.57	−0.01	VI.	3	4.30	−12 14.09	−16.50	−1.92	21 4 33.26	−25 27 32.51
24	10	52.	6.5	..	..	..	..	..	9 34.04	15.54	0.00	IV.	5	5.40	23 48.24	15.81	2.81	9 49.58	39 6.86
25	9	..	..	18.	31.5	45.	..	..	12 31.27	15.52	−0.01	VI.	1	10.18	5 11.46	15.43	1.35	12 46.78	20 28.24
26	7	6.	29.	43.2	57.	..	..	..	14 56.93	15.51	+0.01	V.	7	12.12	35 6.46	15.13	3.89	15 12.45	50 25.48
27	8	8.	22.	36.	..	..	..	..	16 49.94	15.50	+0.01	IV.	8	7.6	37 30.91	14.88	4.98	17 5.45	52 49.87
28	10	..	..	12.	26.	..	..	..	19 25.83	15.47	0.00	V.	6	10.48	29 24.11	14.55	3.39	19 41.30	44 42.05
29	8	..	2.	16.2	30.	..	..	..	21 29.94	15.46	+0.01	V.	8	3.4	35 28.83	14.30	3.92	21 45.41	50 47.05
30	8	..	..	..	44.	..	..	..	22 43.85	15.45	0.01	V.	9	6.18	42 5.66	14.14	4.45	22 59.31	57 24.25
31	11	..	..	34.5	48.7	..	..	..	24 48.50	15.44	0.01	V.	8	9.15	38 35.91	13.89	4.15	25 3.95	53 53.95
32	9	..	..	..	..	36.	49.2	3.	26 21.72	15.43	+0.01	VII.	8	9.35	43 44.65	13.70	4.65	26 37.16	59 3.01
33	11	..	..	32.	46.	..	..	..	28 45.80	15.42	0.00	V.	5	11.34	25 46.21	13.40	3.07	29 1.22	41 2.68
34	10	..	34.	48.	..	..	..	..	31 1.91	15.40	+0.01	IV.	8	8.58	38 27.38	13.13	4.16	31 17.32	53 44.67
35	7	..	26.3	40.5	54.5	..	..	..	32 54.27	15.39	0.00	V.	7	5.13	31 35.18	12.91	3.55	33 9.66	46 51.64
36	8	59.7	..	28.	42.	..	..	..	35 41.68	15.37	0.00	V.	4	10.26	20 12.72	12.55	2.59	35 57.05	35 27.89
37	8	41.3	55.5	..	23.3	..	..	..	37 23.13	15.36	−0.01	VI.	3	10.5	15 3.17	12.39	2.16	37 38.48	30 17.72
38	9	..	..	..	..	..	36.5	50.5	38 8.86	15.35	0.00	VII.	7	1.40	29 47.46	12.30	3.42	38 24.21	45 33.42
39	10	..	35.5	49.5	4.	..	..	..	41 3.53	15.33	+0.01	VI.	8	2.00	35 26.70	11.96	3.86	41 18.87	25 50 52.52
40	10	..	..	..	..	48.	0.5	..	43 34.50	15.32	+0.02	VI.	10	9.58	48 57.37	11.68	5.07	43 49.84	26 4 14.12
41	10	..	15.	29.	..	..	..	..	46 42.79	15.30	0.00	IV.	6	8.5	28 1.96	11.32	3.27	46 58.09	25 43 16.55
42	9	45.5	..	14.	..	..	..	..	52 27.60	15.26	0.00	IV.	6	8.42	28 20.62	10.69	3.30	52 42.86	43 34.61
43	9	..	..	..	..	..	23.5	37.5	52 55.81	15.26	0.00	VII.	4	9.45	19 51.72	10.64	2.56	53 11.07	35 4.92
44	8	..	..	..	6.2	..	34.	48.	55 6.22	15.24	0.00	VII.	4	7.54	18 55.75	10.40	2.48	55 21.46	34 8.63
45	11	..	..	..	5.	19.2	33.2	..	21 59 5.24	15.22	−0.01	VII.	2	10.42	10 22.37	9.99	1.77	21 59 20.45	25 34.13
46	9	32.	46.	..	..	..	..	..	22 2 13.68	15.20	−0.01	IV.	2	11.15	11 39.86	9.67	1.84	22 2 28.87	26 51.37
47	6	..	35.	49.	3.	..	..	..	5 2.91	15.18	+0.01	VI.	8	12.8	40 3.02	9.39	4.28	5 18.10	55 16.69
48	10	..	..	25.2	39.	53.	..	..	6 38.99	15.17	0.00	VI.	6	9.18	28 30.60	9.23	3.23	6 54.16	43 51.06
49	12	..	..	..	3.6	17.5	30.5	45.	11 3.25	15.14	−0.01	VII.	2	6.21	8 10.75	8.79	1.58	11 18.38	23 21.12
50	12	..	..	..	..	..	..	18.7	12 37.15	15.13	+0.01	VII.	8	7.25	37 40.12	8.65	4.09	12 52.29	52 52.86
51	5	..	..	..	..	7.5	21.	35.2	14 53.45	15.12	0.00	VII.	4	2.4	15 59.27	8.44	2.25	15 8.57	31 9.96
52	12	..	..	..	10.2	..	37.5	51.5	29 9.96	15.03	+0.01	VII.	9	4.57	41 24.50	7.17	4.42	29 25.00	56 36.09
53	6	..	..	..	3.	16.5	30.	44.	37 2.60	14.98	+0.01	VII.	10	4.41	40 17.33	6.54	4.29	37 17.59	55 28.16
54	9	52.	6.2	20.5	45.7	..	..	..	40 34.06	14.96	0.00	V.	6	7.1	27 29.64	6.27	3.23	40 49.02	42 39.14
55	8	12.	26.	40.	..	..	..	..	45 53.87	14.92	0.00	IV.	7	5.16	31 36.74	5.89	3.58	46 8.79	46 46.21
56	5	0.8	15.	29.	43.2	..	..	..	51 42.96	14.89	0.00	V.	9	7.20	42 36.92	5.49	4.52	51 57.85	57 46.93
57	10	..	..	..	..	..	..	3.5	52 21.89	14.89	0.00	VII.	7	6.46	32 21.77	5.45	3.65	52 36.78	47 30.87
58	10	..	..	..	..	..	..	33.2	22 53 51.66	14.88	0.00	VII.	8	9.31	38 43.66	5.36	4.19	22 54 6.54	53 53.21
59	9	27.	41.2	55.5	..	..	..	..	23 13 9.05	14.77	0.00	IV.	6	5.5	26 31.19	4.27	3.13	23 13 23.82	41 38.59
60	8	..	..	47.5	1.4	15.	..	..	15 0.77	14.76	0.00	VI.	3	0.18	10 7.01	4.18	1.73	15 15.53	25 12.92
61	11	..	..	31.	..	59.	..	..	23 44.76	14.71	0.00	VI.	2	6.36	8 18.53	3.80	1.67	23 59.47	23 24.00
62	8	..	23.3	37.5	51.4	..	..	..	25 51.20	14.70	0.00	V.	6	4.22	26 9.47	3.72	3.10	26 5.90	41 16.29
63	8	3.	17.	31.1	45.2	..	..	..	29 14.95	14.68	0.00	V.	6	6.33	27 18.53	3.56	3.20	29 59.63	42 22.29
64	12	..	..	..	58.5	12.5	26.3	..	32 58.55	14.67	0.00	VII.	9	7.18	42 35.59	3.45	4.53	33 13.22	57 43.57
65	10	..	..	40.	54.2	8.	21.5	..	35 53.92	14.65	0.00	VII.	4	11.21	20 40.12	3.36	2.62	36 8.57	35 46.10
66	10	..	..	9.	23.2	37.	..	..	39 22.83	14.63	−0.01	VI.	1	7.14	3 38.68	3.25	1.19	39 37.45	18 43.12
67	6	..	12.4	26.5	41.	..	..	..	44 40.49	14.61	0.00	V.	7	10.8	34 3.94	3.12	3.79	44 55.10	49 10.85
68	4	49.	3.2	17.3	31.4	..	..	..	48 30.98	14.59	0.00	V.	4	8.27	19 12.71	3.03	2.48	48 45.57	34 18.22
69	9	4.5	..	33.	47.	0.5	14.	28.2	51 46.50	14.57	0.00	VII.	3	7.31	13 45.15	2.96	2.01	52 1.07	28 50.12
70	8	..	..	0.5	14.	28.	..	..	55 13.97	14.55	0.00	VI.	2	10.42	10 22.57	2.91	1.75	55 28.52	25 27.23
71	9	..	..	31.5	45.5	59.3	..	..	23 58 45.25	+14.54	0.00	VI.	2	14.3	−12 3.90	− 2.85	−1.89	23 58 59.79	−25 27 8.64

CORRECTIONS.

Date.		Corr. of Clock.	Hourly rate.	m	n	c
1847.	h.	s.	s.	s.	s.	s.

INSTRUMENT READINGS.

Date.		Barom.	THERMOM. At.	Ex.
1847.	h. m.	in.	°	°

REMARKS.

(129) 24. Micrometer reading assumed as $7^r.40$ instead of $5^r.40$.
(129) 32. Hor. thread assumed at 9 instead of 8.
(129) 39. Micrometer reading assumed as $3^r.00$ instead of $2^r.00$.
(129) 46. Micrometer reading assumed as $13^r.15$ instead of $11^r.15$.
(129) 56. Transit over T. I assumed as at $0^s.8$ instead of $10^s.8$.

ZONE 129. SEPTEMBER 6. B. BELT, −25° 39′. D_0 = −25° 15′ 0″—Continued.

No.	Mag.	SECONDS OF TRANSIT. I.	II.	III.	IV.	V.	VI.	VII.	T.	a_1	a_2	MICROMETER.			i	d_1	d_2	Mean Right Ascension, 1850.0.	Mean Declination, 1850.0.
									h. m. s.	s.	s.			r.	′ ″	″	″	h. m. s.	° ′ ″
72	8	26.8	..	54.5	9.	..	..	..	0 12 8.73	+14.48	0.00	V.	9	6.1	−41 57.09	− 2.76	−4.48	0 12 23.21	−25 57 4.33
73	9	..	..	31.5	45.5	..	..	..	13 45.40	14.46	0.00	VI.	8	10.48	39 22.69	2.76	4.26	13 59.86	54 29.71
74	10	..	..	31.3	45.4	59.2	13.2	..	17 45.33	14.45	0.00	VI.	7	8.6	33 2.31	2.77	3.70	17 59.78	48 8.78
75	10	..	..	54.2	7.8	21.6	35.	49.2	20 7.63	14.44	0.00	VII.	5	11.31	25 44.37	2.77	3.07	20 22.07	40 50.21
76	10	..	..	..	..	..	..	12.	21 30.36	14.44	0.00	VII.	6	12.28	30 14.20	2.78	3.46	21 44.80	45 20.44
77	9	0.6	14.5	..	..	..	12.	26.4	23 42.40	14.43	0.00	VII.	7	11.24	34 41.94	2.80	3.85	23 56.83	49 48.59
78	8	..	..	..	..	..	..	..	26 44.53	14.41	0.00	III.	5	12.45	26 22.00	2.83	3.12	26 58.94	41 27.95
79	8	..	..	..	..	21.5	35.5	..	28 7.75	14.41	0.00	VII.	6	6.21	27 9.15	2.84	3.19	28 22.16	42 15.18
80	5	..	..	..	..	20.	34.	48.4	31 6.46	14.40	0.00	VII.	9	3.46	40 48.72	2.88	4.39	31 20.86	55 55.99
81	10	..	..	..	..	..	..	52.	32 10.43	14.39	0.00	VII.	8	4.16	36 4.83	2.90	3.98	32 24.82	51 11.71
82	9	..	..	..	..	16.5	30.6	44.2	34 2.76	14.38	0.00	VII.	4	9.52	19 55.25	2.93	2.56	34 17.14	35 0.74
83	9	..	52.4	..	20.	34.	..	..	36 20.04	14.37	0.00	VI.	5	12.58	26 28.43	2.97	3.13	36 34.41	41 34.53
84	10	..	..	10.8	24.5	38.3	52.	..	40 24.52	14.36	0.00	VII.	8	9.30	38 43.16	3.05	4.20	40 38.88	54 20.41
85	7	54.1	8.3	22.4	36.	..	..	..	45 35.97	14.34	0.00	V.	4	11.9	20 34.40	3.18	2.61	45 50.31	35 40.19
86	9	..	..	..	..	..	51.	5.4	46 23.59	14.33	0.00	VII.	8	4.40	36 16.93	3.21	3.99	46 37.92	51 24.13
87	11	..	..	..	..	49.5	3.	17.3	51 35.56	14.32	0.00	VII.	7	4.39	31 17.73	3.35	3.55	51 49.88	46 24.63
88	11	..	..	..	34.	48.3	1.5	15.4	53 33.93	14.31	0.00	VII.	4	6.35	18 15.92	3.42	2.41	53 48.24	33 21.75
89	11	..	..	55.	9.5	23.4	37.	51.2	0 59 9.38	14.29	0.00	VII.	9	4.29	41 10.38	3.60	4.45	0 59 23.67	56 18.43
90	12	..	..	..	..	11.5	25.	39.5	1 2 57.61	14.29	0.00	VII.	6	6.45	27 21.26	3.75	3.21	1 3 11.89	42 28.22
91	9	22.7	37.	51.	5.	..	..	..	13 4.74	14.25	0.00	V.	6	9.59	28 59.40	4.19	3.35	12 18.99	44 6.94
92	8	..	..	..	23.	36.5	50.	4.	13 22.58	14.25	0.00	VII.	8	8.40	38 17.95	4.20	4.19	13 36.83	53 26.34
93	8	..	39.4	54.	8.	21.6	..	..	18 7.54	14.23	0.00	VI.	3	9.30	15 45.87	4.43	2.18	18 21.77	30 52.48
94	12	..	..	..	..	..	24.	38.	19 56.32	14.22	0.00	VII.	4	13.59	21 59.79	4.52	2.74	20 10.54	37 7.05
95	7	..	..	..	..	..	54.	8.5	21 26.56	14.22	0.00	VII.	4	8.55	19 26.51	4.60	2.51	21 40.78	31 33.62
96	9	10.	24.3	38.4	52.5	..	..	..	29 52.17	14.20	0.00	V.	7	6.45	31 51.32	5.07	3.60	30 6.37	47 0.99
97	8	..	..	19.4	33.	47.	..	..	31 33.07	14.20	0.00	VI.	7	6.13	32 5.32	5.18	3.62	31 47.27	47 14.12
98	5	..	..	9.	23.	37.	50.	4.2	38 22.76	14.17	0.00	VII.	7	8.18	33 8.15	5.62	3.72	38 36.93	48 17.49
99	10	..	41.	55.	9.	..	..	..	45 8.84	14.16	0.00	V.	7	6.26	32 12.00	6.09	3.62	45 23.00	47 21.71
100	8	11.	25.6	40.	..	..	..	..	47 53.30	14.15	0.00	IV.	5	5.16	22 35.64	6.29	2.79	48 7.45	25 37 44.72
101	9	..	..	52.5	6.5	20.2	..	..	49 6.43	14.15	−0.01	VI.	10	3.40	45 46.77	6.38	4.87	49 20.57	26 0 58.02
102	12	..	..	..	56.	9.5	..	..	53 55.75	14.14	0.00	VII.	7	9.32	33 45.47	6.75	3.79	54 9.89	25 48 56.01
103	11	37.5	51.5	5.4	..	..	..	..	1 59 19.07	14.13	+0.01	IV.	2	7.9	8 35.32	7.18	1.55	1 59 33.21	23 44.05
104	10	..	..	..	..	20.3	34.	48.4	2 1 6.42	14.13	+0.01	VII.	2	8.43	9 22.36	7.32	1.61	2 1 20.56	24 31.29
115	12	..	..	..	..	..	15.	29.3	2 47.51	14.12	0.00	VII.	6	8.33	28 15.72	7.47	3.28	3 1.63	43 26.47
106	9	50.	4.2	18.	..	..	..	..	6 31.74	14.12	+0.01	IV.	3	9.11	14 35.93	7.79	2.06	6 45.87	29 45.78
107	9	..	15.	29.4	43.2	..	..	..	8 42.89	14.11	0.01	V.	3	6.40	13 19.76	7.98	1.98	8 57.01	28 29.72
108	10	..	..	..	..	..	..	30.	9 48.03	14.11	+0.01	VII.	1	9.17	4 40.50	8.07	1.21	10 2.15	19 49.78
109	11	..	..	32.	46.	59.	..	..	12 45.69	14.11	−0.01	VII.	9	9.3	43 28.54	8.35	4.67	12 59.79	58 41.56
110	10	..	43.	57.	11.	..	..	..	? 10.89	14.10	−0.01	V.	8	7.41	37 48.51	8.57	4.15	15 24.98	53 1.23
111	11	..	..	..	..	..	17.	31.5	21 49.58	14.10	0.00	VII.	5	8.20	24 8.06	9.20	2.92	22 3.68	39 20.18
112	7	..	..	..	..	..	..	56.	23 14.43	14.10	−0.01	VII.	8	4.8	36 0.79	9.34	3.99	23 28.52	51 14.12
113	10	..	..	38.	52.	5.8	19.	..	26 51.74	14.09	0.00	VII.	6	10.17	29 8.16	9.71	3.36	27 5.83	44 21.23
114	9	10.3	24.5	38.7	..	..	..	..	29 52.29	14.09	0.00	IV.	5	10.37	25 27.59	10.01	3.04	30 6.38	25 40 40.64
115	10	..	26.	40.2	54.	..	..	..	31 54.02	14.09	−0.01	V.	10	4.34	45 43.86	10.22	4.82	32 8.10	26 0 58.90
116	10	..	..	1.3	..	..	..	..	35 15.29	14.08	0.01	IV.	9	5.40	41 46.54	10.58	4.51	35 29.36	25 57 1.63
117	10	..	..	..	..	..	..	32.	35 50.51	14.08	−0.01	VII.	9	7.3	42 28.03	10.64	4.54	36 4.56	57 43.21
118	9	..	..	..	..	41.	54.5	9.4	37 27.18	14.08	+0.01	VII.	3	8.20	14 9.86	10.81	2.02	37 41.27	29 22.69
119	7	..	..	..	..	..	..	57.	39 15.51	14.08	−0.01	VII.	9	7.26	42 39.63	11.01	4.60	39 29.58	57 55.24
120	8	..	..	..	..	17.	30.5	45.	2 41 3.03	+14.08	+0.01	VII.	2	10.30	−10 16.32	−11.20	−1.67	2 41 17.12	−25 25 29.19

CORRECTIONS.

Date.		Corr. of Clock.	Hourly rate.	m	n	c
1847.	h.	s.	s.	s.	s.	s.

INSTRUMENT READINGS.

Date.		Barom.	THERMOM. At.	Ex.
1847.	h. m.	in.	°	°

REMARKS.

(129) 77. Transit observations upon T.'s VI and VII assumed to belong to following star.
(129) 91. Minutes assumed as 12 instead of 13.
(129) 93. Micrometer reading assumed as 11ʳ.30 instead of 9ʳ.30.
(129) 96. Micrometer reading assumed as 5ʳ.45 instead of 6ʳ.45.
(129) 110. Minutes assumed as 15.
(129) 115. Micrometer reading assumed as 3ʳ.34 instead of 4ʳ.34.

ZONE 129. SEPTEMBER 6. B. BELT, −25° 39′. D₀ = −25° 15′ 0″—Continued.

No.	Mag.	Seconds of Transit. I.	II.	III.	IV.	V.	VI.	VII.	T.	a_1	a_2	Micrometer.			i	d_1	d_2	Mean Right Ascension, 1850.0.	Mean Declination, 1850.0.
									h. m. s.	s.	s.			r.	′ ″	″	″	h. m. s.	° ′ ″
121	12	26.2	..	55.	..	..	..	..	2 45 8.38	+14.07	0.00	IV.	5	7.25	−23 40.68	−11.66	−2.88	2 45 22.45	−25 38 55.22
122	7	57.6	11.5	25.5	..	..	..	..	48 39.48	14.07	−0.01	IV.	8	10.46	39 21.85	12.06	4.30	48 53.54	54 38.21
123	7.8	41.	55.2	9.3	23.	..	..	..	50 22.88	14.07	0.00	V.	4	9.3	19 30.86	12.25	2.52	50 36.95	34 45.63
124	6	17.5	31.5	45.7	..	..	..	..	52 45.44	14.07	−0.01	V.	8	6.42	37 18.77	12.56	4.13	52 59.50	52 35.46
125	12	..	..	..	..	..	15.5	30.	53 48.09	14.07	0.00	VII.	5	10.1	24 58.99	12.66	3.00	54 2.16	40 14.65
126	11	..	..	51.	5.	19.	32.5	..	? 4.86	14.07	+0.01	VII.	4	5.20	17 38.09	12.93	2.34	56 18.94	32 53.36
127	11	..	42.	56.	10.2	..	..	..	2 59 9.88	+14.07	0.00	V.	6	9.5	−28 32.17	−13.29	−3.31	2 59 23.95	−25 43 48.77

ZONE 130. SEPTEMBER 13. B. BELT, −31° 16′. D₀ = −30° 51′ 40″.

No.	Mag.	Seconds of Transit. I.	II.	III.	IV.	V.	VI.	VII.	T.	a_1	a_2	Micrometer.			i	d_1	d_2	Mean Right Ascension, 1850.0.	Mean Declination, 1850.0.
1	9	..	49.	4.	..	..	..	..	20 6 18.28	+12.52	−0.01	IV.	4	3.00	−16 27.86	−16.53	−1.87	20 6 30.79	−31 8 26.26
2	9	..	..	..	..	38.	52.3	7.	7 23.26	12.51	+0.02	VII.	10	7.32	47 43.45	16.39	6.15	7 35.79	39 45.99
3	9	10.4	25.4	..	..	..	..	..	10 54.76	12.48	+0.01	III.	8	3.57	35 55.55	15.94	4.49	11 7.25	27 55.98
4	11	..	..	13.	28.	..	..	..	12 27.64	12.47	0.00	V.	5	3.56	21 55.25	15.75	2.58	12 40.11	31 13 53.58
5	8	..	9.5	24.5	..	53.3	..	..	14 38.61	12.45	−0.02	VI.*	2	2.16	6 7.37	15.49	0.47	14 51.04	30 58 3.33
6	8	..	39.	53.5	8.2	..	..	..	17 8.04	12.43	−0.01	V.	4	5.16	17 36.39	15.19	2.02	17 20.46	31. 9 33.60
7	7	..	..	..	..	27.	41.3	56.	18 12.25	12.42	0.00	VII.	6	8.8	28 3.00	15.06	3.42	18 24.67	20 1.48
8	10	52.	7.	22.	..	..	..	..	21 36.25	12.40	0.00	IV.	4	10.40	20 19.82	14.66	2.35	21 48.65	12 16.83
9	11	27.8	42.5	57.5	..	..	..	..	23 11.78	12.38	−0.01	IV.	3	9.53	14 57.11	14.48	1.66	23 24.15	6 53.25
10	10	32.	47.	..	..	..	..	..	25 16.21	12.37	0.00	III.	5	7.7	23 31.56	14.24	2.79	25 28.58	15 28.59
11	9	..	..	54.	..	23.	..	..	26 8.34	12.36	−0.01	VI.	4	3.3	16 42.78	14.15	1.87	26 20.69	8 38.80
12	8	..	..	..	..	..	..	12.	26 27.84	12.36	−0.01	VII.	3	3.34	11 45.54	14.11	1.22	26 40.19	3 40.82
13	9	..	12.	27.	..	..	..	..	28 41.29	12.34	−0.01	IV.	3	5.24	12 41.47	13.86	1.36	28 53.62	4 36.69
14	10	..	..	55.	9.8	24.	..	..	30 9.56	12.33	+0.01	VI.	7	11.46	34 53.19	13.69	4.39	30 21.90	26 51.27
15	9	..	21.2	36.	50.6	..	..	..	32 50.37	12.31	−0.01	V.	3	6.52	13 25.80	13.41	1.45	33 2.67	5 20.66
16	9	1.5	17.3	..	..	..	..	..	35 46.27	12.29	+0.01	IV.	8	6.56	37 25.87	13.10	4.75	35 58.57	29 23.72
17	7	..	..	..	..	..	28.	43.3	35 59.03	12.28	−0.01	VII.	2	8.17	9 9.14	13.08	0.87	36 11.30	1 3.09
18	10	..	..	..	..	..	11.4	26.5	37 42.38	12.27	0.00	VII.	5	6.51	23 23.07	12.89	2.78	37 54.65	15 18.74
19	11	..	..	..	..	..	13.	28.	39 43.92	12.25	0.00	VII.	5	3.15	21 34.15	12.68	2.52	39 56.17	31 13 29.35
20	11	..	..	..	..	..	23.	38.5	40 54.12	12.24	−0.02	VII.	2	3.32	6 45.43	12.56	0.57	41 6.34	30 58 38.56
21	8	..	..	..	51.	5.4	19.5	34.5	42 50.59	12.23	−0.01	VII.	4	6.47	23 21.07	12.37	2.77	43 2.81	31 15 16.21
22	7	..	..	..	..	..	30.5	46.	44 1.68	12.21	0.00	VII.	5	9.49	24 52.83	12.25	2.98	44 13.89	16 48.06
23	9	..	..	0.5	15.3	29.3	..	..	46 15.00	12.19	+0.01	VI.	8	3.35	35 44.30	12.02	4.50	46 27.20	27 40.82
24	12	..	..	40.	54.5	..	..	..	48 54.37	12.17	−0.01	V.	4	8.54	19 26.32	11.76	2.24	49 6.53	11 20.32
25	8	..	..	..	5.	20.	34.5	49.2	51 5.15	12.15	−0.01	VII.	3	6.52	13 25.38	11.55	1.43	51 17.29	5 18.36
26	10	34.	49.	4.	..	..	..	..	53 18.26	12.13	0.00	IV.	6	12.13	30 7.00	11.34	3.72	53 30.39	22 2.06
27	12	..	..	..	..	36.5	51.	6.	54 21.91	12.13	0.00	VII.	5	13.9	26 33.67	11.24	3.22	54 34.04	18 28.13
28	10	..	..	38.3	53.2	..	..	..	56 53.08	12.10	−0.02	V.	10	3.59	45 56.46	11.00	5.92	57 5.16	37 53.38
29	8	..	8.5	23.5	..	..	..	..	58 38.19	12.09	−0.02	IV.	10	7.27	47 41.39	10.83	6.15	20 58 50.26	39 38.37
30	11	..	..	..	48.5	2.5	17.	..	20 59 48.02	12.08	+0.01	VII.	8	4.1	35 57.15	10.72	4.53	21 0 0.11	27 52.40
31	8	..	22.5	37.4	52.5	..	..	..	21 2 51.96	12.05	0.00	VI.	4	9.51	19 54.89	10.44	2.32	3 4.01	11 47.65
32	9	31.4	47.	1.5	..	..	..	..	5 16.03	12.03	0.00	IV.	6	13.1	30 31.20	10.23	3.77	5 28.06	22 25.20
33	8	..	..	..	..	..	..	30.	5 46.23	12.03	−0.01	VII.	9	10.7	44 0.71	10.19	5.66	5 58.25	35 56.56
34	7	..	..	57.	12.	26.5	..	..	8 11.77	12.01	0.00	VI.	6	12.26	30 13.35	9.97	3.75	8 23.78	22 7.07
35	9	19.	34.	49.	..	..	..	..	11 3.52	11.98	−0.01	IV.	9	4.50	41 21.33	9.72	5.26	11 15.49	33 16.31
36	8	..	..	..	..	58.	..	27.	11 43.29	11.98	−0.01	VII.	9	1.52	39 51.10	9.67	5.07	11 55.26	31 45.84
37	9	..	..	..	50.6	5.4	19.5	34.	13 50.42	11.96	+0.01	VII.	8	1.37	35 14.79	9.49	4.37	14 2.39	27 8.65
38	8	48.	2.3	17.3	..	..	..	..	21 17 31.73	+11.93	−0.01	IV.	4	1.20	−16 7.68	−9.22	−1.73	21 17 43.65	−31 7 58.63

CORRECTIONS.

Date.	Corr. of Clock.	Hourly rate.	m	n	c
1847. h.	s.	s.	s.	s.	s.
Sept. 13. 18	f 5.14	g 0.022	− 0.006	+ 0.251	0.000

INSTRUMENT READINGS.

	Date.	Barom.	Thermom. At.	Thermom. Ex.
	1847. h. m.	in.	°	°
Zone 130	Sept. 13. 9 5	29.960	69.	61.

REMARKS.

(129) 124. Transits over T.'s I, II, and III assumed as recorded over T.'s II, III, and IV.
(129) 126. Minutes assumed as 56.
(130) 11. Micrometer divisions assumed as 30 instead of 3.
(130) 21. Hor. thread assumed as 5 instead of 4.
(130) 37. Micrometer reading assumed as 2ʳ.37 instead of 1ʳ.37.
(130) 38. Micrometer reading assumed as 2ʳ.20 instead of 1ʳ.20.

ZONE 130. SEPTEMBER 13. B. BELT, −31° 16'. D₀=−30° 51' 40"—Continued.

No.	Mag.	Seconds of Transit. I.	II.	III.	IV.	V.	VI.	VII.	T.	a_1	a^2	Micrometer.			i	d_1	d_2	Mean Right Ascension, 1850.0.	Mean Declination, 1850.0.
									h. m. s.	s.	s.			r.	' "	"	"	h. m. s.	° ' "
39	8	32.	47.	..	..	..	..	..	21 20 16.07	+11.91	−0.01	III.	2	13.33	−11 48.89	− 8.96	−1.23	21 20 27.97	−31 3 39.06
40	11	26.	41.	..	..	..	..	..	27 10.27	11.85	0.00	III.	6	9.13	28 36.19	8.42	3.50	27 22.12	20 28.11
41	9	..	..	..	40.5	55.	..	..	27 40.36	11.84	0.00	VI.	7	5.19	31 38.04	8.38	3.92	27 52.20	31 23 30.34
42	8	32.4	47.2	2.4	..	..	..	..	30 16.41	11.82	−0.01	IV.	2	3.51	6 55.47	8.18	0.57	30 28.22	30 58 44.22
43	5	58.	13.3	28.	43.	..	..	..	38 42.70	11.75	+0.01	V.	9	9.3	43 28.85	7.57	5.58	38 54.46	31 35 22.00
44	7	..	31.5	46.3	1.2	..	..	..	43 0.73	11.71	−0.01	VI.	2	8.35	9 18.48	7.28	0.62	43 12.43	31 1 6.38
45	11	..	58.4	13.2	..	..	..	..	45 27.43	11.69	−0.01	IV.	1	10.49	5 27.26	7.12	0.37	45 39.11	30 57 14.75
46	6	..	..	43.4	58.	12.5	27.	..	21 46 57.87	+11.68	0.00	VII.	6	5.48	−26 52.40	− 7.02	−3.26	21 47 9.55	−31 18 42.68

ZONE 131. SEPTEMBER 13. B. BELT, −31° 16'. D₀=−30° 51' 40".

No.	Mag.	I.	II.	III.	IV.	V.	VI.	VII.	T.	a_1	a^2	Micrometer.			i	d_1	d_2	Mean Right Ascension, 1850.0.	Mean Declination, 1850.0.
1	6	36.	51.	..	..	..	..	..	22 55 20.20	+11.10	0.00	III.	5	5.19	−22 37.10	−49.24	−2.68	22 55 31.30	−31 15 9.02
2	10	..	..	..	..	59.5	13.5	..	55 44.61	11.09	0.00	VII.	5	3.49	21 51.30	49.24	2.57	55 55.70	14 23.11
3	11	..	7.5	22.5	..	..	..	..	58 36.84	11.07	0.00	IV.	5	3.53	21 53.78	49.23	2.57	22 58 47.91	31 14 25.58
4	9	..	..	30.5	45.	59.	13.5	..	23 0 44.53	11.05	−0.01	VII.	1	8.21	4 12.15	49.23	0.15	23 0 55.57	30 56 41.53
5	12	..	21.	36.	51.	..	..	..	4 50.52	11.02	0.00	V.	5	6.42	23 18.96	49.22	2.76	5 1.54	31 15 50.94
6	8	54.3	9.3	24.	..	..	..	..	10 38.59	10.97	0.00	IV.	6	11.25	29 42.80	49.21	3.66	10 49.56	22 15.67
7	9	19.8	35.	49.5	4.5	..	..	..	13 4.21	10.96	0.00	V.	7	4.56	31 26.61	49.23	3.90	13 15.17	23 59.74
8	9	..	20.	..	..	..	..	..	14 49.38	10.94	0.00	III.	7	7.12	32 35.18	49.24	4.06	15 0.32	25 8.48
9	12	..	..	..	..	..	..	36.	14 51.90	10.94	0.00	VII.	4	3.57	16 56.13	49.24	1.90	15 2.84	9 27.27
10	10	..	..	..	55.4	9.4	..	..	19 55.04	10.90	+0.01	VI.	10	6.25	47 9.92	49.28	6.12	20 5.95	31 39 45.32
11	11	..	..	49.	3.3	..	..	..	22 3.15	10.88	+0.01	V.	1	6.42	3 22.65	49.31	0.05	22 14.02	30 55 52.01
12	8	9.	24.	38.5	..	..	..	..	27 53.29	10.84	0.00	IV.	8	3.49	35 51.57	49.40	4.52	28 4.13	31 28 25.49
13	10	..	..	..	55.	9.5	24.	..	29 54.85	10.82	0.00	VII.	7	3.43	30 49.37	49.43	3.82	30 5.67	23 22 62
14	12	..	..	15.	..	44.	..	..	37 29.46	10.76	0.00	VII.	6	8.48	28 23.17	49.60	3.47	37 40.22	20 56.24
15	8	..	..	..	..	15.	29.	44.	46 0.05	10.70	0.00	VII.	4	7.41	18 49.08	49.85	2.15	46 10.75	11 21.08
16	9	11.5	26.5	41.2	..	..	..	..	49 55.69	10.67	0.00	IV.	5	7.52	23 54.30	49.97	2.85	50 6.36	16 27.12
17	10	26.	41.	..	..	..	..	..	54 10.38	10.64	0.00	IV.	8	9.15	38 35.95	50.13	4.91	54 21.02	31 31 10.99
18	10	..	..	..	..	..	..	46.5	55 2.28	10.63	−0.01	VII.	2	3.15	6 36.85	50.17	0.47	55 12.90	30 59 7.49
19	9	..	..	..	..	16.	30.5	45.4	57 1.34	10.62	0.00	VII.	4	6.6	18 1.19	50.24	2.05	57 11.96	31 10 33.48
20	8	..	..	..	..	..	..	47.5	23 58 3.42	10.61	0.00	VII.	4	8.37	19 17.33	50.28	2.21	23 58 14.03	11 49.82
21	11	..	8.	..	37.	..	..	..	0 2 36.84	10.58	0.00	VII.	6	4.31	26 13.58	50.48	3.17	0 2 47.42	18 47.23
22	10	..	..	..	45.5	0.0	..	..	4 45.39	10.57	0.00	VII.	10	9.9	48 32.30	50.58	6.33	4 55.96	41 9.27
23	10	..	..	..	..	54.5	8.4	23.5	7 39.48	10.55	0.00	VII.	1	16.20	7 43.41	50.72	0.72	7 50.03	0 14.85
24	11	..	28.	42.5	..	..	..	..	11 57.30	10.52	0.00	VI.	8	6.8	37 1.45	50.94	4.68	12 7.82	29 37.07
25	10	..	..	..	..	..	13.5	29.	12 44.71	10.51	0.00	VII.	7	3.35	30 45.34	50.98	3.81	12 55.22	23 20.13
26	8	..	..	..	22.	37.	..	..	17 22.12	10.48	0.00	VI.	8	8.53	38 24.65	51.21	4.89	17 32.60	31 31 31.07
27	10	..	..	33.	48.	..	..	..	20 47.50	10.45	0.00	V.	1	8.10	4 7.03	51.43	0.10	20 57.95	30 56 38.56
28	10	40.	55.	9.5	..	..	..	..	27 24.03	10.41	0.00	IV.	4	2.45	16 20.30	51.78	1.80	26 34.44	31 8 53.88
29	9	..	..	..	54.	..	23.	..	28 53.88	10.40	0.00	VII.	4	4.15	17 5.21	51.87	1.92	29 4.28	9 39.00
30	10	..	40.	55.	..	..	..	..	36 9.30	10.35	0.00	IV.	4	6.11	23 3.37	52.44	2.78	36 19.65	15 38.59
31	11	..	..	..	..	53.	7.	22.3	36 38.14	10.35	0.00	VII.	4	3.15	21 34.16	52.47	2.64	36 48.49	14 9.27
32	5	..	..	..	..	..	37.	52.	39 7.88	10.34	0.00	VII.	2	6.16	8 8.13	52.65	0.66	39 18.22	0 41.44
33	6	43.2	58.	13.	..	..	..	..	45 27.29	10.30	0.00	IV.	4	5.57	17 57.11	53.14	2.02	45 37.59	31 10 32.27
34	12	..	..	..	..	..	..	6.	46 21.78	10.29	0.00	VII.	2	4.2	7 0.56	53.21	0.52	46 32.07	30 59 34.29
35	9	..	47.5	2.3	17.	..	..	..	50 16.82	10.27	0.00	V.	6	2.53	25 24.58	53.52	3.06	50 27.09	31 18 1.16
36	12	18.5	33.5	..	..	..	..	..	55 2.66	10.24	0.00	IV.	4	8.47	19 22.84	53.94	2.20	55 12.90	11 58.98
37	11	..	..	8.	23.	..	..	..	56 22.62	10.23	0.00	V.	4	9.16	19 37.41	54.05	2.23	56 32.85	12 13.69
38	9	..	..	..	..	..	45.	0.2	0 57 16.03	+10.23	0.00	VII.	5	10.23	−25 9.97	−54.13	−3.02	0 57 26.26	−31 17 47.12

CORRECTIONS.

Date.		Corr. of Clock.	Hourly rate.	m	n	c
1847.	h.	s.	s.	s.	s.	s.

INSTRUMENT READINGS.

Date.		Barom.	Thermom. At.	Thermom. Ex.
1847.	h. m.	in.	°	°

REMARKS.

(130) 44. Micrometer reading assumed as 12r.35 instead of 8r.35.
(131) 23. Micrometer reading assumed as 15r.20 instead of 16r.20.
(131) 30. Hor. thread assumed as 5 instead of 4.

ZONE 132. SEPTEMBER 14. K. BELT, −28° 8'. D_0 = −27° 43' 40".

No.	Mag.	Seconds of Transit I.	II.	III.	IV.	V.	VI.	VII.	T.	a_1	a_2	Micrometer.			i	d_1	d_2	Mean Right Ascension, 1850.0.	Mean Declination, 1850.0.
									h. m. s.	s.	s.			r.	' "	"	"	h. m. s	° ' "
1	9	..	..	..	..	..	..	22.6	19 59 40.12	+11.87	+0.01	VII.	8	9.44	−38 50.17	−21.33	−4.47	19 59 52.00	−28 22 55.97
2	9	..	..	..	..	..	..	24.0	20 0 41.33	11.86	0.00	VII.	5	7.47	23 51.37	21.21	2.88	20 0 53.19	7 55.46
3	10	..	..	..	20.1	..	..	..	3 19.94	11.84	+0.01	IV.	8	7.19	37 37.46	20.89	4.35	3 31.79	21 42.70
4	9	..	..	..	..	..	44.1	..	6 15.76	11.82	+0.02	VI.	10	7.44	47 49.79	20.53	5.43	6 27.60	31 55.75
5	8	..	..	..	..	..	40.5	..	7 12.29	11.81	−0.01	VI.	4	8.58	19 28.19	20.42	2.43	7 24.09	28 3 31.04
6	9	..	..	..	28.8	..	..	..	10 28.64	11.79	−0.02	IV.	1	9.11	4 37.83	20.02	0.90	10 40.41	27 48 38.75
7	8	..	..	..	..	..	18.0	..	10 49.72	11.78	+0.01	VI.	7	10.26	34 12.88	19.98	3.97	11 1.51	28 18 16.83
8	9	..	..	..	..	..	..	4.5	11 21.77	11.78	−0.01	VII.	4	7.7	18 32.00	19.92	2.33	11 33.54	2 34.25
9	9	..	..	..	..	18.2	..	..	13 4.01	11.77	0.00	V.	6	7.45	27 51.82	19.72	3.30	13 15.78	11 54.84
10	9	..	..	..	46.4	..	14.4	..	14 46.22	11.76	−0.01	IV.	4	7.49	18 53.59	19.53	2.37	14 57.97	28 2 55.49
11	9	..	..	..	..	..	..	43.9	18 0.91	11.73	−0.01	VII.	3	3.58	11 57.69	19.14	1.66	18 12.63	27 55 58.49
12	9	..	..	..	17.9	..	45.7	..	20 17.56	11.71	+0.02	IV.	9	8.2	42 58.15	18.88	4.86	20 29.29	28 27 1.92
13	6.7	..	38.1	..	6.7	..	..	..	22 6.55	11.70	0.00	IV.	7	3.52	30 54.38	18.67	3.61	22 18.25	28 14 56.66
14	9	..	..	..	..	26.8	..	..	26 12.53	11.67	−0.02	V.	2	5.40	7 50.39	18.21	1.22	26 24.18	27 51 49.82
15	9	..	..	..	48.0	..	..	..	28 47.84	11.65	+0.01	IV.	8	2.34	35 13.76	17.92	4.06	28 59.50	28 19 15.74
16	5.6	57.1	11.3	25.7	..	..	..	..	30 39.85	11.64	0.00	IV.	7	2.54	30 25.14	17.72	3.55	30 51.49	14 26.41
17	5.6	..	..	..	20.4	..	48.5	..	31 20.26	11.63	0.00	VI.	5	5.36	22 45.55	17.65	2.75	31 31.89	6 45.96
18	5.6	..	..	..	..	..	..	59.3	32 16.93	11.62	+0.02	VII.	10	6.57	47 25.86	17.55	5.41	32 28.57	31 28.82
19	7	..	..	25.4	39.8	..	..	..	34 39.68	11.60	+0.02	IV.	9	8.7	43 0.67	17.29	4.91	34 51.30	28 27 2.87
20	8	..	..	..	..	38.1	..	..	35 23.86	11.60	−0.01	V.	3	8.48	14 24.29	17.21	1.90	35 35.45	27 58 23.40
21	5	..	..	..	..	..	28.3	..	36 0.17	11.59	0.02	VI.	1	6.30	3 16.48	17.15	0.75	36 11.74	47 14.38
22	6.7	..	..	..	..	42.3	..	..	37 28.01	11.58	−0.02	V.	1	5.10	2 36.26	16.99	0.68	37 39.57	27 46 33.93
23	5	..	..	..	..	..	31.5	..	38 3.24	11.57	0.00	VJ.	7	3.14	30 35.04	16.93	3.57	38 14.81	28 14 35.54
24	5	..	..	..	..	..	7.1	21.6	38 38.94	11.57	+0.01	VII.	7	9.48	33 53.48	16.86	3.93	38 50.52	28 17 54.27
25	5	..	25.3	..	53.7	..	..	..	40 53.54	11.55	−0.01	IV.	2	12.15	11 9.61	16.63	1.57	41 5.08	27 55 7.81
26	6.7	..	..	8.0	..	36.8	..	..	42 22.56	11.54	+0.02	IV.	10	10.24	49 10.65	16.48	5.58	42 34.12	28 33 12.71
27	4.5	..	..	..	..	8.6	..	..	44 54.68	11.52	0.02	V.	9	12.33	45 14.75	16.22	5.14	45 6.22	29 16.11
28	7	..	..	..	..	..	..	6.1	45 23.71	11.52	+0.02	VII.	9	13.29	45 42.61	16.16	5.20	45 35.25	29 43.97
29	8.9	..	1.5	..	30.1	..	..	..	48 29.97	11.49	+0.01	IV.	7	12.10	35 5.49	15.84	4.06	48 41.47	28 19 5.39
30	6	..	42.3	56.8	..	..	..	..	51 10.59	11.47	−0.01	III.	2	12.33	11 18.64	15.58	1.57	51 22.05	27 55 15.79
31	7	..	..	..	11.8	..	..	..	52 11.64	11.46	+0.01	IV.	7	8.56	33 27.68	15.48	3.87	52 23.11	28 17 27.03
32	5	..	..	..	..	..	..	6.0	52 23.53	11.46	0.01	VII.	8	10.33	39 14.88	15.45	4.51	52 35.00	23 14.84
33	4	..	..	..	1.3	15.6	..	..	54 1.29	11.45	+0.01	IV.	7	12.11	35 5.99	15.30	4.06	54 12.75	19 5.35
34	7	..	..	58.9	13.3	..	..	..	55 13.03	11.44	0.00	V.	5	6.52	23 24.00	15.18	2.82	55 24.47	28 7 22.00
35	9	..	..	..	8.3	..	..	..	58 8.14	11.41	0.00	IV.	3	9.46	14 53.59	14.90	1.95	58 19.55	27 58 50.44
36	6	..	..	..	..	..	..	59.3	20 58 16.46	11.41	−0.01	VII.	2	8.43	9 22.31	14.89	1.38	20 58 27.86	53 18.58
37	9	..	..	..	13.5	..	..	..	21 14 13.34	11.29	−0.01	IV.	2	10.42	10 22.73	13.43	1.48	21 14 24.62	27 54 17.61
38	8	..	..	5.1	..	33.5	..	..	16 19.32	11.27	+0.01	IV.	7	12.38	35 19.61	13.25	4.09	16 30.60	28 19 16.95
39	7	..	..	18.6	..	47.0	..	..	18 32.85	11.25	0.01	IV.	8	8.48	38 22.34	13.06	4.43	18 44.11	22 19.83
40	9	..	..	..	48.8	2.5	..	..	19 48.50	11.24	+0.01	IV.	8	10.27	39 12.26	12.95	4.52	19 59.75	28 23 9.73
41	8	..	..	..	..	..	..	52.2	20 9.39	11.24	−0.01	VII.	2	14.10	12 7.18	12.92	1.65	20 20.62	27 56 1.75
42	8	..	..	..	..	18.2	..	..	22 3.91	11.22	−0.02	V.	1	7.26	3 44.84	12.76	0.77	22 15.11	27 47 38.37
43	6	..	..	16.1	30.0	..	..	..	24 30.18	11.20	+0.02	IV.	10	10.2	48 59.56	12.56	5.58	24 41.40	28 32 57.70
44	8	..	..	..	..	32.8	..	..	25 18.56	11.20	−0.01	V.	3	12.20	16 11.18	12.50	2.08	25 29.75	0 5.76
45	9	..	..	..	..	40.5	..	..	28 26.34	11.17	+0.01	V.	7	9.45	33 52.33	12.24	3.94	28 37.52	17 48.51
46	7	..	..	..	..	48.8	..	..	29 34.70	11.16	+0.02	V.	10	12.6	50 2.03	12.15	5.68	29 45.88	28 33 59.86
47	4	..	..	..	..	..	39.5	..	30 11.31	11.16	−0.01	VI.	3	9.26	14 43.32	12.10	1.93	30 22 46	27 58 37.35
48	9	..	42.1	56.2	..	..	..	..	33 10.53	11.14	+0.01	IV.	8	7.59	37 57.63	11.87	4.38	33 21.68	28 21 53.88
49	8	..	..	..	..	..	..	0.9	21 33 18.25	+11.14	0.00	VII.	5	10.18	−25 7.51	−11.86	−3.01	21 33 29.39	−28 9 2.38

CORRECTIONS.

Date.		Corr. of Clock.	Hourly rate.	m	n	c
1847.	h.	s.	s.	s.	s.	s.
Sept. 14,	18	1 4.47	g 0.020	+ 0.057	+ 0.173	0.000

INSTRUMENT READINGS.

	Date.		Barom.	Thermom. At.	Thermom. Ex.
	1847.	h. m.	in.	°	°
Zone 132	Sept. 14,	9 5	30.150	65.	54.2

REMARKS.

ZONE 132. SEPTEMBER 14. K. BELT, −28° 8′. D_0=−27° 43′ 40″—Continued.

No.	Mag.	SECONDS OF TRANSIT. I.	II.	III.	IV.	V.	VI.	VII.	T.	a_1	a_2	MICROMETER.			i	d_1	d_2	Mean Declination, 1850.0.	Mean Right Ascension, 1850.0.
									h. m. s.	s.	s.			r.	′ ″	″	″	° ′ ″	h. m. s.
50	7	..	20.6	..	49.3	..	..	..	21 38 49.15	+11.10	+0.01	IV.	9	7.14	−42 33.94	−11.45	−4.88	21 39 0.26	−28 26 30.27
51	9	..	..	..	..	..	..	51.8	39 8.96	11.09	0.00	VII.	2	9.32	9 47.02	11.42	1.42	39 20.05	27 53 39.86
52	5	..	..	49.0	3.0	..	..	..	41 2.92	11.08	0.00	IV.	5	4.18	22 6.39	11.29	2.69	41 14.00	28 6 0.37
53	8	..	..	..	45.0	59.0	..	..	41 44.81	11.08	−0.01	V.	4	5.58	17 57.56	11.24	2.26	41 55.88	1 51.06
54	8	..	..	..	41.8	56.0	..	..	42 41.71	11.07	0.00	V.	5	4.6	22 0.30	11.17	2.68	42 52.78	5 54.15
55	7	..	..	..	40.0	..	..	..	43 39.84	11.06	−0.01	IV.	4	4.16	17 6.19	11.11	2.17	43 50.89	0 59.47
56	9	..	23.2	37.3	..	..	..	..	46 51.39	11.04	0.00	IV.	4	9.26	19 42.50	10.89	2.45	47 2.43	28 3 35.84
57	9	..	..	..	..	45.5	..	..	21 47 31.26	+11.03	−0.01	V.	3	11.57	−15 59.59	−10.84	−2.06	21 47 42.28	−27 59 52.49

ZONE 133. SEPTEMBER 15. B. BELT, −26° 53′. D_0=−26° 28′ 30″.

No.	Mag.	SECONDS OF TRANSIT. I.	II.	III.	IV.	V.	VI.	VII.	T.	a_1	a_2	MICROMETER.			i	d_1	d_2	Mean Declination, 1850.0.	Mean Right Ascension, 1850.0.
1	11	..	..	..	..	30.5	44.	..	20 10 16.28	+11.63	−0.01	VII.	4	8.41	−19 19.42	−30.08	−2.49	20 10 27.90	−26 48 21.99
2	6	..	49.	3.2	17.3	..	..	..	14 17.25	11.60	+0.02	V.	9	9.14	43 34.41	29.55	4.78	14 28.87	27 12 38.74
3	7	51.	5.1	19.3	..	..	..	..	16 33.32	11.58	0.01	IV.	7	8.33	33 16.08	29.26	3.79	16 44.91	2 19.13
4	10	..	..	..	16.	30.	..	57.	17 15.92	11.58	+0.01	VII.	8	7.15	37 35.05	29.16	4.22	17 27.51	27 6 38.43
5	10	..	..	28.	..	56.	..	..	19 41.78	11.56	−0.01	VI.	2	10.41	10 22.05	28.84	1.64	19 53.33	26 39 22.53
6	10	..	..	..	49.	..	17.	..	21 49.01	11.55	−0.02	VII.	1	11.59	6 2.16	28.57	1.24	22 0.54	26 35 1.97
7	6	..	..	26.	40.3	54.	..	..	26 40.02	11.51	0.00	VII.	6	8.29	48 13.27	27.95	5.79	26 51.53	27 17 17.01
8	10	..	..	..	36.	50.2	..	..	28 36.01	11.49	0.00	VI.	6	7.24	27 41.11	27.71	3.27	28 47.50	26 56 42.09
9	12	0.2	14.2	28.3	..	..	..	..	31 42.46	11.48	+0.01	IV.	8	4.9	36 1.66	27.44	4.09	30 53.95	27 5 3.19
10	5	..	..	..	..	38.2	52.	6.3	31 24.20	11.47	+0.01	VII.	9	4.28	41 9.84	27.36	4.54	31 35.68	27 10 11.74
11	9	..	3.5	18.	32.	..	..	..	34 31.71	11.45	−0.01	V.	4	6.12	18 4.64	26.98	2.35	34 43.15	26 47 3.97
12	9	..	..	34.	48.	2.3	..	..	35 47.95	11.44	0.01	VI.	4	5.54	17 55.43	26.83	2.34	35 59.38	46 50.60
13	9	..	..	..	12.	26.2	40.	..	37 12.03	11.43	−0.01	VII.	3	8.12	14 5.80	26.65	2.00	37 23.45	43 4.45
14	5	..	..	..	..	..	39.	53.3	38 11.14	11.42	0.00	VII.	6	9.4	28 31.32	26.53	3.35	38 22.56	57 31.20
15	9	..	..	..	14.	28.	41.5	..	40 13.80	11.41	0.00	VII.	5	9.3	24 29.73	26.28	2.95	40 25.21	53 28.96
16	8	7.	21.3	35.2	..	..	..	..	44 49.24	11.37	0.00	IV.	5	7.35	23 45.73	25.75	2.88	45 0.61	52 44.36
17	9	..	..	2.	16.	30.	34.	..	46 15.81	11.36	−0.01	VII.	3	3.36	11 46.63	25.59	1.76	46 27.16	40 43.98
18	5	..	..	..	..	55.	8.4	23.	47 40.75	11.35	0.00	VII.	5	5.55	22 51.91	25.41	2.83	47 52.10	51 50.15
19	9	..	..	..	..	..	..	24.	48 41.64	11.34	−0.01	VII.	2	9.36	9 49.06	25.29	1.56	48 52.97	26 38 45.91
20	9	..	58.5	12.5	..	..	..	..	20 59 26.71	11.26	+0.01	IV.	9	9.1	43 27.89	24.09	4.79	20 59 37.98	27 12 26.77
21	9	..	..	42.	56.	..	..	..	21 0 55.81	11.25	−0.01	V.	3	13.42	16 52.54	23.93	2.25	21 1 7.05	26 45 48.72
22	9	..	..	..	..	23.3	37.5	51.3	2 9.38	11.24	+0.01	VII.	8	6.34	36 13.87	23.81	4.11	2 20.63	27 5 11.78
23	8	..	..	33.	47.	0.5	14.	..	5 46.49	11.21	−0.02	VII.	1	5.28	2 45.24	23.43	0.86	5 57.68	26 31 39.53
24	10	..	10.5	25.	39.	..	..	..	8 38.87	11.19	+0.01	V.	8	7.9	37 32.38	23.13	4.24	8 50.07	27 6 29.75
25	11	48.	2.2	16.3	..	..	..	..	10 30.13	11.18	−0.01	IV.	3	3.32	11 45.00	22.94	1.74	10 41.27	26 40 39.68
26	7	..	..	..	..	..	21.3	36.	10 53.65	11.17	+0.01	VII.	7	10.10	29 44.59	22.90	3.40	11 4.84	26 58 9.89
27	9	..	..	..	..	..	..	53.	12 10.94	11.16	+0.01	VII.	7	7.25	33 11.64	22.77	3.80	12 22.11	27 2 8.21
28	10	33.2	47.3	1.2	16.	..	..	..	16 15.37	11.13	−0.01	V.	3	11.37	15 49.52	22.37	2.12	16 26.49	26 44 44.01
29	8	..	55.	9.	23.2	..	..	..	19 23.14	11.10	+0.01	V.	9	6.35	42 14.24	22.05	4.68	19 34.25	27 11 10.97
30	12	..	19.	33.	48.	..	..	..	22 47.29	11.08	0.00	V.	6	8.41	28 20.07	21.74	3.33	22 58.37	26 57 15.14
31	12	..	..	..	..	21.	..	48.	24 6.55	11.07	−0.01	VII.	9	7.9	42 31.03	21.62	4.71	24 17.61	27 11 27.36
32	5	35.	49.3	3.4	18.	..	..	..	27 17.42	11.04	0.00	V.	5	2.35	21 14.42	21.32	2.55	27 28.46	26 50 8.29
33	8	..	43.	57.5	11.5	..	..	..	32 11.08	11.00	−0.02	V.	1	5.06	2 34.26	20.88	0.86	32 22.06	31 26.00
34	5	..	..	..	..	29.	43.	..	33 15.03	11.00	0.02	VII.	1	7.2	3 32.40	20.79	0.95	33 26.01	32 24.14
35	11	..	..	..	..	..	0.2	14.	34 32.02	10.99	−0.01	VII.	3	6.42	12 50.17	20.68	1.81	34 43.00	41 42.66
36	9	..	..	..	..	14.5	28.	42.5	36 0.32	10.98	0.00	VII.	6	12.17	30 8.63	20.55	3.48	36 11.30	59 2.66
37	10	..	20.	34.2	..	..	..	..	21 38 47.86	+10.96	0.00	IV.	2	4.00	−7 0.02	−20.30	−1.30	21 38 58.82	−26 35 51.62

CORRECTIONS.

Date.		Corr. of Clock.	Hourly rate.	m	n	c
1847.	h.	s.	s.	s.	s.	s.
Sept. 15,	18	s 4.27	s 0.017	+ 0.109	+ 0.108	0.000

INSTRUMENT READINGS.

	Date.	Barom.	THERMOM. At.	THERMOM. Ex.
	1847. h. m.	In.	°	°
Zone 133	Sept. 15, 9 5	30.150	64.5	53.

REMARKS.

(133) 7. Hor. thread assumed as 10 instead of 6.
(133) 9. Minutes of transit assumed as 30 instead of 31.
(133) 14. Transits discordant; observation of T. VI assumed as 39s.0 instead of 37s.0.
(133) 22. Micrometer reading assumed as 4r.34 instead of 6r.34.
(133) 26. Hor. thread assumed as 6 instead of 7.
(133) 27. Micrometer reading assumed as 8r.25 instead of 7r.25.
(133) 35. Micrometer reading assumed as 5r.42 instead of 6r.42.

ZONE 133. SEPTEMBER 15. B. BELT, −26° 53'. D_0=−26° 28' 30"—Continued.

No.	Mag.	Seconds of Transit. I.	II.	III.	IV.	V.	VI.	VII.	T.	a_1	a_2	Micrometer.			i	d_1	d_2	Mean Right Ascension, 1850.0.	Mean Declination, 1850.0.
									h. m. s.	s.	s.			r.	' "	"	"	h. m. s.	° ' "
38	10	. .	. .	. .	. .	55.	8.5	23.	21 39 40.81	+10.95	0.00	VII.	6	5.15	−26 35.84	−20.23	−3.16	21 39 51.76	−26 55 29.23
39	10	. .	. .	. .	. .	. .	48.	2.5	41 20.23	10.94	0.00	VII.	6	9.12	24 34.25	20.09	2.96	41 31.17	53 27.30
40	10	6.4	21.	35.2	. .	. .	. .	. .	43 48.83	10.92	−0.01	IV.	3	8.52	14 26.36	19.90	1.98	43 59.74	26 43 18.24
41	8	35.	49.	3.	. .	. .	. .	. .	48 17.29	10.88	+0.01	IV.	9	6.43	42 18.31	19.54	4.69	48 28.18	27 11 12.54
42	10	. .	. .	54.2	9.	23.	. .	. .	52 8.77	10.85	+0.02	V.	10	3.7	45 30.25	19.24	5.01	52 19.64	27 14 24.50
43	6	. .	20.	34.5	48.	. .	. .	. .	21 55 47.96	10.82	−0.01	V.	2	5.14	7 47.38	18.98	1.38	21 55 58.77	26 36 37.74
44	7	. .	. .	52.	6.	20.	. .	. .	22 0 5.74	10.79	−0.02	VI.	1	2.30	1 15.47	18.66	0.75	22 0 16.51	26 30 4.88
45	6	36.	51.	5.	19.	. .	. .	. .	4 18.91	10.76	+0.01	V.	8	12.10	40 4.15	18.39	4.47	4 29.68	27 8 57.01
46	8	. .	6.	20.	34.	. .	. .	. .	6 33.92	10.74	−0.01	V.	3	7.37	13 48.50	18.24	1.94	6 44.65	26 42 38.63
47	6	. .	46.5	0.5	14.5	. .	. .	. .	8 0.17	10.73	−0.01	V.	2	9.28	9 45.36	18.14	6.53	8 10.89	26 38 40.03
48	9	49.	3.2	17.5	. .	. .	. .	. .	15 31.48	10.68	+0.01	IV.	8	9.40	38 48.56	17.69	4.36	15.42.17	27 7 40.61
49	7	. .	38.	52.	6.	. .	. .	. .	19 6.07	10.65	+0.01	IV.	9	6.4	41 58.65	17.47	4.66	19 16.73	27 10 50.78
50	7	40.	54.	8.2	. .	. .	. .	. .	22 22.06	10.62	0.00	IV.	4	3.6	16 30.89	17.29	2.19	22 32.68	26 45 20.37
51	10	. .	. .	. .	. .	. .	43.	58.	24 15.40	10.61	−0.01	VII.	2	6.36	8 18.30	17.19	1.43	24 26.00	37 6.92
52	10	. .	. .	15.	. .	43.5	57.	. .	28 29.12	10.58	0.00	VII.	4	11.50	20 54.71	16.96	2.61	28 39.70	26 49 44.23
53	12	. .	2.	16.	. .	. .	. .	. .	38 30.12	10.50	0.00	IV.	8	5.2	36 28.38	16.45	4.13	38 40.62	27 5 18.96
54	9	8.	22.	36.4	. .	. .	. .	. .	48 50.31	10.43	0.00	IV.	7	5.20	31 38.75	15.94	3.64	49 0.74	27 0 28.33
55	8	. .	. .	. .	. .	40.3	54.	8.	49 26.04	10.42	0.00	VII.	4	9.58	20 58.24	15.91	2.62	49 36.46	26 49 46.77
56	8	16.5	31.	45.	. .	. .	. .	. .	51 58.96	10.41	0.00	IV.	6	8.26	28 12.55	15.79	3.32	52 9.37	57 1.66
57	10	. .	52.	6.2	20.5	. .	. .	. .	22 56 20.21	10.37	0.00	V.	7	3.44	30 50.31	15.59	3.56	22 56 30.58	59 39.46
58	6	. .	11.	25.	38.7	52.5	. .	. .	23 0 38.63	10.34	0.00	VI.	2	9.13	9 37.67	15.39	1.52	23 0 48.97	26 38 24.58
59	11	. .	. .	. .	1.	15.	29.	. .	5 0.96	10.31	0.00	VI.	9	10.15	44 5.04	15.20	4.86	5 11.27	27 12 55.10
60	9	. .	12.5	25.7	40.5	55.	. .	. .	10 40.31	10.27	0.00	VI.	4	3.54	16 54.92	14.95	2.22	10 50.58	26 45 42.09
61	6	. .	. .	. .	51.	5.	. .	. .	23 23 50.87	+10.19	−0.01	VI.	1	11.1	− 5 33.13	−14.42	−1.15	23 24 1.05	−26 34 18.70

ZONE 134. SEPTEMBER 16. K. BELT, −30° 38'. D_0=−30° 14' 20".

No.	Mag.	I.	II.	III.	IV.	V.	VI.	VII.	T.	a_1	a_2	Micrometer.			i	d_1	d_2	Mean Right Ascension, 1850.0.	Mean Declination, 1850.0.
1	9	43.8	. .	13.3	. .	. .	. .	. .	19 35 27.62	+11.44	−0.01	III.	4	10.42	−20 20.78	−20.32	−2.40	19 35 39.05	−30 35 3.50
2	9	. .	. .	. .	2.9	. .	31.5	. .	37 2.60	11.43	+0.02	IV.	10	6.17	47 6.10	20.15	5.83	37 14.05	31 1 52.08
3	8	41.5	56.2	. .	. .	. .	. .	. .	39 25.42	11.41	+0.01	II.	7	5.26	31 41.58	19.88	3.84	39 36.84	30 46 25.30
4	8	. .	. .	49.8	4.6	. .	. .	. .	40 4.31	11.41	0.00	V.	5	7.34	23 45.18	19.81	2.85	40 15.72	38 27.84
5	10	. .	. .	9.8	. .	. .	. .	. .	42 24.34	11.39	+0.01	III.	7	8.51	33 25.10	19.56	4.08	42 35.74	48 8.74
6	8.9	. .	. .	17.0	31.8	. .	. .	. .	43 31.53	11.38	0.00	IV.	5	11.4	25 31.12	19.44	3.07	43 42.91	40 13.63
7	8	. .	. .	. .	. .	. .	. .	. .	44	11.38	0.00	V.	5	8.25	24 10.89	19.40	2.90	44	38 53.19
8	10	51.0	5.8	. .	. .	. .	. .	. .	46 35.05	11.36	+0.01	III.	8	10.38	39 17.76	19.12	4.85	46 46.42	54 1.73
9	6	. .	. .	56.7	11.4	25.8	. .	. .	47 11.32	11.35	+0.02	IV.	9	8.7	43 0.67	19.05	5.33	47 22.69	57 45.05
10	7	. .	. .	. .	. .	. .	7.2	22.1	47 38.30	11.35	−0.01	VII.	4	11.56	20 57.69	19.00	2.56	47 49.64	35 39.25
11	8	. .	. .	. .	. .	. .	22.3	37.2	48 53.39	11.34	−0.01	VII.	3	6.56	13 27.42	18.87	1.56	49 4.72	28 7.85
12	10	. .	. .	. .	. .	. .	25.8	. .	50 56.80	11.32	+0.01	VI.	8	7.43	37 49.36	18.65	4.64	51 8.13	52 32.65
13	10	. .	. .	44.3	. .	. .	. .	. .	52 58.76	11.31	0.00	IV.	6	8.5	28 1.96	18.45	3.40	53 10.07	42 43.81
14	6	. .	21.6	36.2	50.4	. .	. .	. .	54 50.46	11.29	−0.01	IV.	4	10.29	20 14.27	18.26	2.39	55 1.74	34 54.92
15	8	. .	. .	. .	39.0	. .	. .	. .	55 38.84	11.28	0.02	IV.	1	13.39	6 52.97	18.17	0.72	55 50.10	21 31.86
16	9	. .	: .	2.0	16.5	. .	. .	. .	57 16.22	11.27	−0.02	V.	1	10.59	5 32.25	18.01	0.55	57 27.47	30 20 10.81
17	7	. .	22.8	37.0	. .	. .	. .	. .	58 51.93	11.26	+0.02	IV.	10	2.56	45 24.75	17.85	5.64	59 3.21	31 0 8.24
18	8	. .	. .	. .	. .	. .	51.8	6.3	19 59 22.77	11.26	+0.01	V.	8	6.16	37 5.65	17.79	4.55	19 59 34.04	30 51 47.99
19	5	. .	. .	5.9	20.4	. .	. .	. .	20 6 20.17	17.21	−0.01	V.	3	5.40	12 49.49	17.11	1.47	20 6 31.37	27 28.07
20	9	. .	13.0	27.5	. .	. .	. .	. .	8 42.26	11.19	+0.02	III.	9	10.2	43 58.61	16.88	5.46	8 53.47	58 40.95
21	8	. .	20.0	34.7	. .	. .	. .	. .	20 9 49.10	+11.18	0.00	III.	5	11.15	−25 36.61	−16.77	−3.08	20 10 0.28	−30 40 16.46

CORRECTIONS.

Date.	Corr. of Clock.	Hourly rate.	m	n	c
1847. h.	s.	s.	s.	s.	s.
Sept. 16, 18	s 3.21	g 0.010	+ 0.324	− 0.159	0.000

INSTRUMENT READINGS.

	Date.	Barom.	Thermom. At.	Thermom. Ex.
	1847. h. m.	In.	°	°
Zone 134	Sept. 16, 9 5	30.050	66.5	59.3

REMARKS.

(133) 39. Hor. thread assumed as 5 instead of 6.
(133) 47. Transits over T.'s III–V assumed as recorded over T.'s II–IV.
(134) 7. Double.
(134) 11. A star of 6th magnitude passed the upper part of field.

ZONE 134. SEPTEMBER 16. K. BELT, −30° 38′. D_0 = −30° 14′ 20″—Continued.

No.	Mag.	SECONDS OF TRANSIT. I.	II.	III.	IV.	V.	VI.	VII.	T.	a_1	a_2	MICROMETER.			i	d_1	d_2	Mean Right Ascension, 1850.0.	Mean Declination, 1850.0.
									h. m. s.	s.	s.			r.	′ ″	″	″	h. m. s.	° ′ ″
22	5	. .	55.8	10.4	25.1	. .	. .	. .	20 12 24.98	+11.16	+0.01	IV.	8	4.38	−36 16.28	−16.52	−4.45	20 12 36.15	−30 50 57.25
23	8	. .	11.9	26.3	. .	. .	. .	. .	14 41.10	11.14	+0.02	III.	9	8.43	43 18.77	16.31	5.37	14 52.26	58 0.45
24	11	. .	. .	. .	13.4	. .	. .	. .	15 13.24	11.14	0.02	IV.	9	12.8	45 2.18	16.26	5.60	15 24.40	59 44.04
25	10	. .	. .	. .	. .	. .	. .	57.9	15 14.42	11.14	+0.01	VII.	6	10.47	29 23.20	16.26	3.57	15 25.57	44 3.03
26	10	. .	15.5	30.2	. .	. .	. .	. .	17 44.64	11.12	0.00	III.	6	8.19	28 8.97	16.02	3.40	17 55.76	42 48.39
27	9	. .	33.0	. .	2.7	. .	. .	. .	20 2.40	11.09	+0.01	IV.	8	9.14	38 35.45	15.81	4.73	20 13.50	53 15.99
28	10	. .	59.9	14.6	. .	. .	. .	. .	22 29.00	11.08	0.00	IV.	5	11.10	25 34.14	15.59	3.07	22 40.08	40 12.80
29	9	. .	. .	. .	17.1	31.5	. .	. .	23 16.92	11.07	−0.01	IV.	3	5.45	12 52.06	15.52	1.47	23 27.98	27 29.05
30	6	. .	. .	. .	12.1	26.5	. .	. .	24 11.95	11.06	0.00	IV.	6	5.22	26 39.76	15.44	3.20	24 23.01	41 18.40
31	5.6	. .	. .	. .	. .	35.5	49.4	. .	25 20.70	11.05	+0.01	IV.	9	10.31	34 13.28	15.33	4.18	25 31.76	30 48 52.79
32	10	. .	. .	. .	. .	. .	58.7	. .	26 29.64	11.04	0.02	VI.	10	10.15	49 5.91	15.23	6.13	26 40.70	31 3 47.27
33	8	. .	. .	. .	43.6	. .	12.0	. .	28 43.29	11.02	0.02	IV.	10	12.7	50 2.58	15.03	6.26	28 54.33	31 4 43.87
34	9	. .	37.0	44.6	. .	. .	. .	. .	31 59.24	10.99	+0.01	IV.	8	10.40	39 18.82	14.76	4.84	32 10.24	30 53 58.42
35	10	. .	49.6	. .	. .	. .	. .	. .	33 18.62	10.98	−0.01	IV.	4	7.26	18 41.99	14.65	2.23	33 29.59	33 18.87
36	7.8	. .	45.8	0.4	15.0	. .	. .	. .	34 14.90	10.98	+0.01	IV.	7	4.17	31 6.99	14.57	3.78	34 25.89	30 45 45.34
37	7	. .	31.8	46.7	1.2	. .	. .	. .	36 1.21	10.96	0.02	IV.	10	4.54	46 24.25	14.42	5.78	36 12.19	31 1 4.45
38	10	. .	. .	. .	. .	. .	4.9	. .	38 35.92	10.94	+0.01	VI.	7	6.40	32 18.90	14.21	3.94	38 46.87	30 46 57.05
39	6.7	. .	. .	. .	. .	. .	25.3	40.2	39 56.44	10.93	0.00	VII.	6	11.39	29 49.42	14.09	3.62	40 7.37	44 27.13
40	7	. .	. .	8.0	22.3	. .	. .	. .	42 22.12	10.91	−0.02	IV.	1	11.10	5 37.84	13.90	0.56	42 33.01	20 12.30
41	9	. .	28.6	. .	57.9	. .	. .	. .	43 57.74	10.90	0.00	IV.	6	9.42	28 50.87	13.78	3.48	44 8.64	43 28.13
42	9.8	. .	. .	41.1	55.6	. .	. .	. .	45 55.50	10.88	0.00	IV.	6	7.22	27 40.27	13.62	3.35	46 6.38	42 17.24
43	8.9	. .	. .	. .	. .	. .	55.0	9.9	46 26.13	10.88	0.00	VI.	6	5.52	26 54.69	13.58	3.25	46 37.01	41 31.52
44	9.10	. .	. .	. .	. .	. .	16.7	. .	47 47.73	10.87	0.00	VI.	7	2.59	30 27.46	13.47	3.70	47 58.60	45 4.63
45	7	. .	19.4	34.3	. .	. .	. .	. .	51 48.32	10.84	−0.02	IV.	1	7.54	3 59.01	13.17	0.35	51 59.14	18 32.53
46	7.8	. .	. .	. .	. .	. .	20.2	35.3	51 51.36	10.84	−0.02	IV.	1	10.29	5 17.17	13.17	0.51	52 2.18	19 50.85
47	9	. .	. .	. .	. .	. .	. .	22.9	52 39.34	10.83	+0.01	VII.	8	7.8	37 31.46	13.11	4.60	52 50.18	52 9.17
48	9	. .	. .	58.2	12.9	. .	. .	. .	56 12.78	10.80	+0.01	IV.	8	10.38	39 17.81	12.85	4.85	56 23.59	53 55.51
49	5.6	. .	. .	. .	46.7	1.1	15.2	. .	56 46.45	10.80	0.00	IV.	6	8.45	28 22.13	12.81	3.44	56 57.25	42 58.38
50	9	. .	12.4	27.0	. .	. .	. .	. .	20 59 41.43	10.77	0.00	IV.	5	8.19	24 7.92	12.60	2.89	20 59 52.20	38 43.41
51	6	10.6	25.6	40.2	. .	. .	. .	. .	21 0 54.33	10.76	+0.02	IV.	1	9.55	5 0.02	12.52	0.48	21 1 5.11	19 33.02
52	9	. .	. .	. .	. .	. .	6.0	. .	4 37.01	10.73	0.01	VI.	7	10.0	33 59.75	12.28	4.15	4 47.75	48 36.18
53	9	. .	. .	15.5	. .	44.3	. .	. .	7 29.97	10.70	0.01	IV.	8	11.57	39 57.64	12.09	4.94	7 40.68	53 34.67
54	9	. .	44.1	58.4	. .	. .	. .	. .	9 13.20	10.69	0.01	IV.	8	12.15	40 6.71	11.98	4.95	9 23.90	54 43.64
55	8	. .	. .	41.8	56.8	. .	. .	. .	9 56.55	10.68	0.01	IV.	8	14.24	41 11.75	11.93	5.01	10 7.24	30 55 48.69
56	7	. .	. .	53.9	9.0	. .	. .	. .	11 8.76	10.67	+0.02	IV.	10	11.31	49 44.43	11.86	6.23	11 19.45	31 4 22.52
57	9	. .	. .	. .	. .	. .	. .	0.6	11 16.81	10.67	−0.01	VII.	4	8.32	19 14.83	11.86	2.26	11 27.48	30 33 48.95
58	9	. .	. .	. .	. .	50.1	. .	19.0	12 35.48	10.66	+0.01	VII.	7	6.55	32 26.21	11.75	3.95	12 46.15	47 1.94
59	9	. .	. .	. .	. .	4.0	. .	33.0	13 49.36	10.65	0.00	VII.	5	9.33	24 44.78	11.71	2.97	14 0.01	39 19.46
60	8	. .	. .	23.8	38.2	. .	. .	. .	16 38.05	10.63	−0.01	IV.	3	12.5	16 3.67	11.55	1.87	16 48.67	30 37.09
61	8	. .	. .	. .	. .	. .	26.4	41.6	16 57.65	10.62	0.00	VI.	4	9.34	19 46.09	11.53	2.33	17 8.27	34 19.95
62	9	. .	. .	39.2	53.3	. .	. .	. .	18 53.21	10.61	−0.02	IV.	1	8.36	4 20.20	11.41	0.41	19 3.80	30 18 52.02
63	7.8	. .	. .	3.4	18.2	. .	. .	. .	20 18.11	10.60	+0.02	IV.	10	10.34	49 15.70	11.33	6.16	20 28.73	31 3 53.19
64	9	. .	. .	. .	59.3	. .	28.0	. .	20 59.08	10.59	0.00	IV.	7	5.36	31 46.83	11.30	3.87	21 9.67	30 46 22.00
65	9	12.5	. .	41.8	56.4	. .	. .	. .	23 56.35	10.56	+0.01	IV.	7	8.52	33 25.66	11.15	4.08	24 6.92	48 0.89
66	8	. .	33.1	47.4	. .	. .	. .	. .	25 2.08	10.55	0.00	IV.	7	3.24	30 40.26	11.10	3.73	25 12.63	45 15.09
67	9	. .	. .	. .	37.6	51.8	. .	. .	25 37.35	10.55	0.00	IV.	6	7.54	27 56.41	11.07	3.38	25 47.90	42 30.86
68	8	. .	54.9	9.6	24.0	. .	. .	. .	29 23.76	10.51	−0.02	IV.	1	8.28	4 16.15	10.88	0.40	29 34.25	18 47.43
69	7.8	. .	. .	4.0	18.8	33.0	. .	. .	30 18.62	10.51	+0.01	IV.	9	10.22	44 8.74	10.84	5.50	30 29.14	58 45.08
70	9	. .	20.6	35.4	. .	. .	. .	. .	20 31 49.56	+10.49	−0.01	IV.	2	13.57	−11 34.82	−10.77	−1.31	21 32 0.04	−30 26 6.90

CORRECTIONS.

Date.		Corr. of Clock.	Hourly rate.	m	n	c
1847.	h.	s.	s.	s.	s.	s.

INSTRUMENT READINGS.

Date.		Barom.	THERMOM. At.	THERMOM. Ex.
1847.	h. m.	in.	°	°

REMARKS.

ZONE 134. SEPTEMBER 16. K. BELT, −30° 38′. D₀=−30° 14′ 20″—Continued.

No.	Mag.	SECONDS OF TRANSIT. I.	II.	III.	IV.	V.	VI.	VII.	T.	a_1	a_2	MICROMETER.			i	d_1	d_2	Mean Right Ascension, 1850.0.	Mean Declination, 1850.0.
									h. m. s.	s.	s.			r.	′ ″	″	″	h. m. s.	° ′ ″
71	9.10	. .	. .	. .	. .	18.3	. .	. .	20 33 3.67	+10.48	−0.01	V.	2	3.13	− 6 36.26	−10.73	−0.69	21 33 14.14	−30 21 7.68
72	9	. .	. .	. .	. .	. .	. .	27.3	33 43.50	10.48	0.01	VII.	10	5.34	17 45.07	10.70	2.10	33 53.97	32 17.87
73	10	. .	. .	. .	. .	. .	59.3	. .	37 30.46	10.44	−0.01	VI.	1	11.13	5 39.15	10.54	0.55	37 40.89	20 10.24
74	9	. .	3.6	18.2	. .	. .	. .	. .	40 32.59	10.42	0.00	IV.	4	12.7	21 3.68	10.42	2.50	40 43.01	35 36.60
75	10	. .	. .	. .	. .	2.9	. .	. .	40 48.39	10.41	+0.01	IV.	7	10.28	34 14.06	10.41	4.18	40 58.81	48 48.65
76	9	. .	. .	. .	36.4	51.2	. .	. .	41 36.45	10.41	0.00	IV.	5	12.20	26 9.43	10.39	3.15	41 46.86	30 40 42.97
77	8	. .	33.6	48.2	2.7	. .	. .	. .	43 2.81	10.40	+0.02	IV.	10	5.10	46 32.31	10.34	5.80	43 13.23	31 1 8.45
78	8	45.9	. .	15.2	. .	. .	. .	. .	45 29.93	10.38	+0.01	IV.	9	7.23	42 38.47	10.26	5.30	45 40.32	30 57 14.03
79	7.8	. .	59.2	13.9	28.6	. .	. .	. .	46 28.39	10.36	0.00	IV.	7	1.57	29 56.39	10.23	3.64	46 38.75	44 30.26
80	8	. .	. .	. .	. .	. .	17.2	32.6	46 48.50	10.36	−0.01	VII.	2	−0.10	10 2.70	10.22	1.12	46 58.85	30 24 34.04
81	9	. .	. .	. .	. .	56.8	. .	. .	21 48 42.20	10.35	0.01	IV.	4	2.46	16 20.81	10.15	1.90	48 52.54	30 52.86
82	8	. .	. .	. .	. .	51.6	5.8	. .	49 36.96	10.34	−0.01	IV.	3	5.12	12 35.42	10.12	1.46	49 47.29	27 7.00
83	8	. .	. .	. .	52.7	7.0	. .	. .	50 52.50	10.33	0.00	IV.	6	6.32	27 15.07	10.09	3.29	51 2.83	41 48.45
84	7	42.1	57.0	11.6	. .	. .	. .	. .	54 25.96	10.30	0.00	IV.	5	6.4	22 59.85	10.00	2.74	54 36.29	37 32.59
85	6	50.2	5.3	19.8	. .	. .	. .	. .	21 55 34.20	+10.29	0.00	IV.	5	7.54	−23 55.31	− 9.98	−2.87	21 55 44.49	−30 38 28.16

ZONE 135. SEPTEMBER 16. K. BELT, −26° 53′. D₀=−26° 28′ 20″.

No.	Mag.	I.	II.	III.	IV.	V.	VI.	VII.	T.	a_1	a_2	MICROMETER.			i	d_1	d_2	Mean Right Ascension, 1850.0.	Mean Declination, 1850.0.
1	9	. .	. .	. .	51.3	. .	. .	. .	23 23 51.14	+ 9.51	0.00	IV.	2	11.32	− 5 48.94	−13.11	−1.12	23 24 0.65	−26 34 23.17
2	9	7.7	28.1	. .	. .	. .	. .	. .	25 49.92	9.50	0.00	IV.	2	4.58	7 29.26	12.87	1.29	25 59.42	36 3.42
3	9	. .	. .	54.3	8.1	. .	. .	. .	29 7.98	9.48	0.00	IV.	3	6.13	13 6.18	12.47	1.84	29 17.46	26 41 40.49
4	7.8	. .	56.7	10.2	24.8	. .	. .	. .	32 24.56	9.45	0.00	IV.	7	8.9	33 3.98	12.08	3.80	32 34.01	27 1 39.86
5	6.7	48.7	2.8	16.9	. .	. .	. .	. .	36 31.02	9.42	0.00	IV.	8	4.9	36 1.66	11.61	4.10	36 40.44	4 37.37
6	9	. .	19.2	33.3	. .	. .	. .	. .	39 47.33	9.40	0.00	IV.	7	7.13	32 35.73	11.24	3.76	39 56.73	1 10.73
7	9	. .	. .	. .	23.2	. .	50.8	. .	40 22.92	9.40	0.00	IV.	10	3.10	45 31.80	11.17	5.06	40 32.32	27 14 8.03
8	9	4.4	18.8	32.7	. .	. .	. .	. .	42 46.62	9.38	0.00	IV.	4	1.52	15 53.57	10.91	2.10	42 56.00	26 44 26.58
9	9	. .	. .	2.5	. .	30.5	. .	. .	45 16.41	9.37	0.00	IV.	5	10.11	25 4.39	10.65	3.01	45 25.78	53 38.05
10	9	31.1	45.5	. .	. .	. .	. .	. .	48 13.41	9.35	0.00	IV.	3	12.3	16 2.66	10.33	2.12	48 22.76	44 35.11
11	9	42.1	56.2	10.2	. .	. .	. .	. .	51 24.25	9.33	0.00	IV.	5	9.3	24 30.10	10.00	2.95	51 33.58	26 53 3.05
12	9	. .	46.8	. .	14.8	. .	. .	. .	52 14.79	9.33	0.00	IV.	7	6.44	32 21.12	9.92	3.75	52 24.12	27 0 54.79
13	8	. .	. .	. .	. .	48.9	2.4	. .	55 34.68	9.32	+0.01	IV.	10	5.9	46 31.81	9.59	5.16	55 44.01	27 15 6.56
14	8	. .	. .	. .	. .	. .	44.4	. .	23 56 16.53	9.31	0.00	VII.	3	10.6	15 3.29	9.52	2.02	23 56 25.84	26 43 34.83
15	8	. .	13.1	27.2	41.4	. .	. .	. .	0 0 41.09	9.28	0.00	IV.	4	10.29	20 14.27	9.10	2.54	0 0 50.37	26 48 45.91
16	9	. .	. .	. .	. .	21.4	35.2	. .	1 7.33	9.27	0.00	IV.	7	11.54	34 57.43	9.06	4.00	1 16.60	27 3 30.49
17	7	. .	. .	. .	11.2	25.4	. .	. .	2 11.18	9.27	0.00	IV.	3	8.7	14 5.67	8.97	1.92	2 20.45	26 42 36.56
18	9	. .	14.1	. .	42.2	. .	. .	. .	3 42.16	9.26	0.00	IV.	8	5.0	36 27.37	8.83	4.15	3 51.42	27 5 0.35
19	7	. .	. .	15.5	29.8	. .	. .	. .	5 29.64	9.25	0.00	IV.	9	4.8	41 0.15	8.67	4.60	5 38.89	9 33.42
20	6	. .	. .	. .	59.1	. .	26.4	40.8	5 58.73	9.24	0.00	IV.	8	9.18	38 37.46	8.63	4.36	6 7.97	27 7 10.45
21	10	. .	. .	. .	. .	. .	22.0	. .	6 54.13	9.24	0.00	VI.	3	9.9	14 34.76	8.55	1.98	7 3.37	26 43 5.29
22	8	. .	44.4	58.9	. .	. .	. .	. .	10 12.52	9.22	0.00	IV.	3	12.48	16 25.35	8.25	2.16	10 21.74	26 44 55.76
23	9	15.8	30.2	44.4	. .	. .	. .	. .	12 58.39	9.20	0.00	IV.	8	10.33	39 15.29	8.03	4.43	13 7.59	27 7 47.75
24	9	. .	. .	. .	24.4	. .	. .	5.3	13 23.77	9.20	0.00	IV.	8	6.18	37 6.70	8.00	4.21	13 32.97	27 5 38.91
25	9	14.0	. .	42.8	. .	. .	. .	. .	23 56.47	9.14	0.00	IV.	5	4.54	22 24.55	7.18	2.75	24 5.61	26 50 54.48
26	9	. .	59.7	. .	27.8	. .	. .	. .	25 27.63	9.13	0.00	IV.	2	5.41	7 59.94	7.07	1.32	25 36.76	36 19.33
27	7	. .	. .	. .	. .	13.3	27.3	41.8	25 59.44	6.13	0.00	IV.	6	5.17	26 37.24	7.03	3.16	26 8.57	55 7.43
28	7.8	. .	. .	. .	. .	39.9	. .	8.0	27 25.89	9.12	0.00	V.	6	9.16	28 37.72	6.93	3.36	27 35.01	57 8.01
29	8	12.7	26.7	41.2	. .	. .	. .	. .	29 54.71	9.11	0.00	IV.	1	11.55	6 0.53	6.73	1.14	30 3.82	34 28.40
30	8.9	. .	. .	36.3	50.3	. .	. .	. .	0 31 50.13	+ 9.10	0.00	IV.	4	10.17	−20 8.22	− 6.62	−2.51	0 31 59.23	−26 48 37.35

CORRECTIONS. INSTRUMENT READINGS.

Date.		Corr. of Clock.	Hourly rate.	m	n	c		Date.		Barom.	THERMOM. At.	THERMOM. Ex.
1847.	h.	s.	s.	s.	s.	s.		1847.	h. m.	in.	°	°

REMARKS.

(134) 80. Hor. thread assumed as 3 instead of 2.
(135) 1. Micrometer thread assumed as 1 instead of 2.

ZONE 135. SEPTEMBER 16. K. BELT, −26° 53′. D_o = −26° 28′ 20″—Continued.

No.	Mag.	SECONDS OF TRANSIT. I.	II.	III.	IV.	V.	VI.	VII.	T.	a_1	a_2	MICROMETER.			i	d_1	d_2	Mean Right Ascension, 1850.0.	Mean Declination, 1850.0.
									h. m. s.	s.	s.			r.	′ ″	″	″	h. m. s.	° ′ ″
31	9	..	..	..	34.1	..	..	..	0 32 33.94	+ 9.10	0.00	IV.	4	10.46	−20 22.85	− 6.57	−2.52	0 32 43.04	−26 48 51.94
32	8	..	..	..	..	..	..	16.9	32 34.72	9.10	0.00	VII.	5	7.56	23 55.93	6.57	2.90	32 43.82	52 25.40
33	8	10.3	..	38.6	..	..	..	..	35 52.44	9.08	0.00	IV.	3	12.25	16 13.75	6.36	2.18	36 1.52	26 44 42.29
34	9	..	14.2	28.2	..	..	..	..	40 42.40	9.06	0.00	IV.	9	7.31	42 42.51	6.95	4.78	40 51.46	27 11 13.34
35	7.8	..	..	..	1.6	15.2	28.8	..	41 1.16	9.05	0.00	IV.	9	7.45	42 49.57	6.03	4.79	41 10.21	11 20.39
36	7.8	..	21.6	35.6	49.6	..	..	..	44 49.69	9.04	0.00	IV.	9	10.40	34 17.82	5.82	3.93	44 58.73	2 47.57
37	9	..	..	..	11.8	..	39.0	..	46 11.32	9.03	+0.01	IV.	10	7.20	47 37.86	5.74	5.27	46 20.36	27 16 8.87
38	9	32.9	47.2	1.5	15.3	..	..	..	48 15.20	9.02	0.00	IV.	4	8.20	19 9.22	5.62	2.42	48 24.22	26 47 37.26
39	8.9	..	..	..	..	53.0	6.2	..	48 38.63	9.02	0.00	VII.	9	4.27	41 9.34	5.60	4.60	48 47.65	27 9 39.54
40	7.8	..	17.9	31.9	..	..	..	..	50 45.68	9.01	0.00	IV.	3	5.34	12 46.52	5.48	1.81	50 54.79	26 41 13.81
41	8	..	56.3	10.7	24.6	..	..	..	52 24.36	9.00	0.00	IV.	3	3.48	11 53.06	5.40	1.72	52 33 36	40 20.18
42	7	..	..	..	7.3	..	35.0	..	53 7.16	8.99	−0.01	IV.	1	9.41	4 52.96	5.37	1.03	53 16.14	33 19.36
43	9	..	..	..	..	..	52.0	..	54 24.14	8.99	0.00	IV.	3	6.23	13 11.22	5.31	1.85	54 33.13	41 38.38
44	7	..	..	36.0	..	5.0	..	..	55 50.48	8.98	0.00	IV.	7	3.10	30 33.20	5.24	3.55	55 59.46	26 59 1.99
45	9	..	..	..	..	39.1	52.9	..	0 58 25.03	8.97	0.00	VII.	8	8.2	37 58.76	5.12	4.30	0 58 34.00	27 6 28.18
46	9	..	55.0	9.2	..	..	..	..	1 1 22.90	8.95	0.00	IV.	2	12.24	11 14.15	4.98	1.65	1 1 31.85	26 39 40.78
47	7	0.2	14.6	28.8	42.8	..	..	..	2 42.68	8.95	0.00	IV.	6	14.1	31 1.45	4.93	3.60	2 51.63	26 59 29.98
48	8	..	..	37.8	52.0	..	..	..	4 51.86	8.94	0.00	IV.	8	7.28	37 41.99	4.85	4.27	5 0.80	27 6 11.11
49	8	..	..	37.0	51.2	..	..	..	5 51.12	8.94	−0.01	IV.	10	3.45	45 49.45	4.80	5.10	6 0.06	27 14 19.35
50	7.8	..	..	5.2	19.2	33.2	..	..	11 19.02	8.91	0.00	IV.	3	5.44	12 51.56	4.60	1.81	11 27.93	26 41 17.97
51	9	..	58.7	..	26.8	..	..	..	16 26.73	8.89	0.00	IV.	7	1.54	29 54.88	4.43	3.49	16 35.62	26 58 22.80
52	9	..	..	..	34.2	..	..	..	17 34.04	8.88	0.00	IV.	7	9.57	33 58.43	4.39	3.90	17 42.92	27 2 26.72
53	9	..	..	..	..	..	..	58.6	18 16.55	8.88	0.00	V.	7	9.57	33 58.39	4.37	3.90	18 25.43	2 26.66
54	9	..	..	59.5	13.5	..	..	..	20 13.54	8.87	−0.01	VII.	10	3.43	45 48.05	4.30	5.08	20 22.40	27 14 17.43
55	9	..	1.8	16.4	29.9	..	..	..	21 29.88	8.86	0.00	IV.	3	8.3	14 1.65	4.27	1.92	21 38.74	26 42 27.84
56	9	27.8	41.8	55.9	..	..	..	..	23 9.68	8.86	+0.01	IV.	1	7.6	3 34.81	4.23	0.90	23 18.55	31 59.94
57	9	..	..	..	..	..	15.8	..	29 47.87	8.83	0.00	IV.	6	9.48	28 53.89	4.07	3.39	29 56.70	57 21.35
58	9	..	..	..	..	24.7	..	52.8	34 10.63	8.82	0.00	IV.	5	6.49	23 22.54	3.99	2.85	34 19.45	26 51 49.38
59	9	..	..	53.8	..	..	..	..	38 7.89	8.81	0.00	IV.	8	8.22	38 9.23	3.92	4.31	38 16.70	27 6 37.46
60	8	..	..	..	..	41.1	..	..	38 27.12	8.80	0.00	IV.	8	12.44	40 21.34	3.92	4.53	38 35.92	8 49.79
61	8	..	..	..	..	..	..	20.5	38 38.47	8.80	0.00	VII.	7	11.56	34 58.05	3.91	4.00	38 47.27	27 3 25.96
62	9	30.9	45.0	..	..	..	..	..	41 13.11	8.80	0.00	IV.	5	4.50	22 22.53	3.88	2.75	41 21.91	26 50 49.16
63	8	..	..	..	31.2	..	..	..	41 31.05	8.80	0.00	IV.	7	5.5	31 31.19	3.88	3.65	41 39.85	59 58.72
64	5.6	..	..	..	..	..	14.7	..	41 46.79	8.79	0.00	IV.	5	10.42	25 20.03	3.88	3.03	41 55.58	53 46.94
65	8	..	..	..	..	..	..	38.2	41 55.95	8.79	0.00	VII.	4	6.34	18 15.39	3.87	2.34	42 4.74	46 41.60
66	8	..	18.3	..	46.9	..	..	..	45 46.46	8.78	+0.01	IV.	1	10.41	5 23.22	3.84	1.08	45 55.25	26 33 48.14
67	9	42.1	56.4	10.6	..	..	..	..	47 24.59	8.78	0.00	IV.	7	14.35	36 18.60	3.83	4.13	47 33.37	27 4 46.56
68	9	..	10.5	24.4	..	..	..	..	48 38.67	8.77	−0.01	IV.	9	11.52	44 54.12	3.82	5.00	48 47.43	27 13 22.94
69	9	..	..	4.2	18.3	..	..	..	50 18.15	8.77	0.00	IV.	6	8.38	28 18.60	3.81	3.33	50 26.92	26 56 45.74
70	9	..	25.6	39.8	..	..	..	..	51 53.93	8.76	−0.01	IV.	10	3.33	45 43.40	3.81	5.08	52 2.68	27 14 12.29
71	5	..	..	..	..	32.0	45.8	0.1	52 18.00	8.76	0.00	IV.	9	4.19	41 5.70	3.81	4.61	52 26.76	27 9 34.12
72	9	..	..	..	..	..	54.5	..	53 26.57	8.76	0.00	VI.	6	9.8	28 33.55	3.81	3.35	53 35.33	26 57 0.71
73	9	..	..	..	..	55.8	..	..	1 54 41.84	8.76	−0.01	VI.	9	11.7	44 46.39	3.81	4.98	1 54 50.59	27 13 15.15
74	9	22.2	..	50.8	..	..	..	..	2 0 4.52	8.74	0.00	IV.	4	6.37	18 17.29	3.80	2.34	2 0 13.26	26 46 43.43
75	9	57.0	11.2	..	..	..	..	..	2 1 39.37	+ 8.74	0.00	IV.	7	6.6	−32 1.96	− 3.81	−3.70	2 1 48.11	−27 0 29.47

CORRECTIONS.

Date.		Corr. of Clock.	Hourly rate.	m	n	c
1847.	h.	s.	s.	s.	s.	s.

INSTRUMENT READINGS.

Date.		Barom.	THERMOM. At.	Ex.
1847.	h. m.	in.	°	°

REMARKS.

ZONE 136. SEPTEMBER 21. II. BELT, −26° 16′. D_0 = −25° 51′ 50″.

No.	Mag.	Seconds of Transit. I.	II.	III.	IV.	V.	VI.	VII.	T.	a_1	a_2	Micrometer.			i	d_1	d_2	Mean Right Ascension, 1850.0.	Mean Declination, 1850.0.
									h. m. s.	s.	s.			r.	′ ″	″	″	h. m. s.	° ′ ″
1	9			13.2	27.0				22 19 26.85	+ 8.59	−0.01	II.	3	6.10	−13 4.51	+29.19	−1.90	22 19 35.43	−26 4 27.22
2	9.10				14.0				20 13.85	8.59	0.00	IV.	5	8.05	24 0.86	29.26	2.91	20 22.44	15 24.51
3	9			11.8	25.2		53.		21 25.28	8.56	0.00	VI.	4	5.40	17 48.38	29.35	2.33	21 33.86	9 11.36
4	9				39.0	53.	6.2		22 38.77	8.57	0.00	V.	4	4.08	17 2.11	29.44	2.27	22 47.34	8 24.94
5	8			4.2	18.2				25 18.11	8.55	0.00	IV.	7	11.43	34 51.88	29.64	3.92	25 26.66	26 16.16
6	9.10		12.5	27.0					26 40.73	8.54	0.00	III.	7	10.55	34 27.64	29.74	3.89	26 49.27	25 51.79
7	9				19.0				27 18.85	8.53	0.00	V.	5	4.13	22 4.05	29.79	2.73	27 27.38	13 26.99
8	8.9					32.8			28 18.88	8.52	0.00	VII.	7	12.01	35 0.57	29.86	3.93	28 27.40	26 24.64
9	7.8							43.8	29 43.65	8.52	0.00	VII.	7	11.07	34 33.35	29.92	3.89	29 52.17	25 57.32
10	7.8		18.2	32.2	46.2				30 51.05	8.50	0.00	IV.	5	8.31	24 13.97	30.04	2.92	30 59.55	15 36.85
11	7		23.	37.	51.				32 55.79	8.49	0.00	IV.	3	13.02	16 32.41	30.18	2.22	33 4.26	7 54.45
12	9		0.2	14.2	28.2				34 28.15	8.47	+0.01	IV.	7	14.40	36 21.12	30.29	4.05	34 36.63	27 44.88
13	8				53.0	7.0	20.2		34 52.78	8.47	+0.01	V.	7	13.13	35 37.21	30.32	3.98	35 1.26	27 0.87
14	6.7	42.0	56.	10.0					37 9.73	8.45	−0.01	III.	2	10.10	10 6.55	30.46	1.63	37 18.17	1 27.72
15	9				54.2	7.5	21.2		37 53.68	8.45	+0.01	VI.	9	5.03	41 27.72	30.52	4.55	38 2.14	32 51.75
16	9.10				26.0				39 25.85	8.43	−0.01	IV.	3	8.02	14 1.15	30.62	1.99	39 34.27	5 22.52
17	9						26.2		40 58.42	8.43	0.00	VI.	6	10.09	29 4.33	30.72	3.37	41 6.85	20 26.98
18	9					59.2	13.0		42 45.24	8.41	+0.01	VII.	9	5.07	41 29.52	30.82	4.55	42 53.66	32 53.25
19	9				13.2				44 13.05	8.40	0.00	IV.	7	4.21	31 9.00	30.91	3.56	44 21.45	22 31.65
20	7.8				29.2	43.1			45 29.10	8.39	0.00	V.	6	6.12	27 4.94	30.93	3.18	44 37.49	26 18 27.19
21	8							31.0	45 48.82	8.39	0.00	VII.	1	10.25	5 14.76	30.94	1.20	44 57.21	25 56 35.02
22	8.9				11.0	25.0			46 10.92	8.39	0.00	V.	3	11.46	15 54.06	31.02	2.15	46 19.31	26 7 15.19
23	8.9	38.0	52.0	6.1	20.1				49 19.94	8.36	0.00	III.	5	5.57	23 26.57	31.21	2.04	49 28.30	26 14 47.40
24	9		58.2	12.8					50 26.13	8.36	−0.01	III.	1	11.06	5 35.79	31.27	1.24	50 34.48	25 56 55.76
25	9				55.5				50 55.35	8.35	+0.01	IV.	9	14.37	46 17.31	31.30	5.01	51 3.71	26 37 41.02
26	7			53.5	8.				52 7.65	8.34	0.00	IV.	7	10.42	34 21.13	31.36	3.87	52 15.99	25 43.64
27	9	46.2	0.	14.2					55 28.05	8.32	0.00	III.	5	6.26	23 10.89	31.52	2.83	55 36.37	14 32.20
28	8				40.			22.	54 40.07	8.32	0.00	IV.	9	6.50	42 21.84	31.48	4.64	54 48.39	33 45.00
29	8				19.5		47.2		56 19.37	8.31	0.00	IV.	8	8.14	38 5.20	31.57	4.22	56 27.68	26 29 27.85
30	9			10.5	24.2				57 24.05	8.30	−0.01	IV.	2	3.24	6 41.86	31.62	1.34	57 32.34	25 58 1.58
31	9	23.0		51.0					59 4.88	8.29	0.00	III.	4	7.29	18 43.46	31.70	2.43	59 13.17	26 10 4.19
32	8.9					26.0	39.2		22 59 11.74	8.29	+0.01	V.	7	7.58	32 58.38	31.71	4.68	22 59 20.04	24 21.35
33	6			27.	41.	54.8			23 0 40.97	8.27	+0.01	IV.	10	6.15	47 5.09	31.78	5.07	23 0 49.25	38 28.38
34	9			46.2					2 0.05	8.26	0.00	III.	6	5.27	26 42.24	31.83	3.15	2 8.31	26 18 3.56
35	9		45.						3 12.73	8.25	−0.01	II.	2	2.53	6 26.06	31.88	1.32	3 20.97	25 57 45.50
36	8.9			45.	58.5	12.2			3 58.44	8.25	0.00	IV.	5	3.29	21 41.68	31.92	2.69	4 6.69	26 13 2.45
37	9			47.8	1.8	15.2			5 1.54	8.24	0.00	IV.	7	3.07	30 31.69	31.96	3.51	5 9.78	21 53.24
38	8	52.5	7.1	21.1	35.0				6 34.90	8.23	0.00	III.	7	4.45	31 21.07	32.03	3.53	6 43.13	22 42.62
39	9					59.8			6 45.77	8.23	0.00	V.	2	11.02	10 32.77	32.03	1.67	6 54.00	1 52.41
40	9		46.5						8 14.35	8.21	0.00	II.	4	8.42	19 20.16	32.10	2.48	8 22.56	10 40.54
41	9				15.2	29.5			8 15.30	8.21	0.00	IV.	5	11.43	25 50.78	32.10	3.07	8 23.51	17 11.75
42	8		10.	23.	37.				9 37.19	8.20	0.00	III.	5	10.06	25 1.84	32.15	3.00	9 45.39	16 22.69
43	8.9	14.		42.1	56.0				10 55.97	8.19	0.00	III.	6	4.50	26 23.59	32.20	3.12	11 4.16	17 44.51
44	9.10						23.0		10 55.23	8.19	0.00	VI.	6	5.59	26 58.26	32.20	3.18	11 3.42	26 18 19.24
45	8.9						59.2	7.	11 31.51	8.19	0.00	VI.	2	6.43	8 22.05	32.22	1.49	11 39.70	25 59 41.32
46	9					53.2			12 39.16	8.18	0.00	V.	2	4.16	7 8.05	32.26	1.38	12 47.34	26 58 27.17
47	8						26.0	40.4	12 58.30	8.18	0.00	VI.	3	8.13	13 6.02	32.27	1.90	13 6.48	4 25.65
48	9	32.0	46.2						16 13.97	8.16	0.00	II.	3	3.06	11 31.73	32.39	1.76	16 22.13	2 51.10
49	8.9				26.4	41.0			23 16 26.63	+ 8.15	0.00	IV.	4	10.13	−20 6.20	+32.39	−2.55	23 16 34.78	−26 11 26.36

CORRECTIONS.

Date.		Corr. of Clock.	Hourly rate.	m	n	c
1847.	h.	s.	s.	s.	s.	s.
Sept. 21,	19	[illegible] 1.89	[illegible] 0.028			

INSTRUMENT READINGS.

	Date.		Barom.	Thermom. At.	Thermom. Ex.
	1847.	h. m.	In.	°	°
Zone 136	Sept. 21,	9 5	30.00	67.	57.5

REMARKS.

(136) 9. Transit over T. IV assumed as recorded over T. VII.
(136) 10. Transits all assumed as 5s too small.
(136) 11. Transits all assumed as 5s too small.
(136) 20. Minutes of transit assumed as 44 instead of 45.
(136) 21. Minutes of transit assumed as 44 instead of 45.
(136) 23. Micrometer reading assumed as $6^{r}.57$ instead of $5^{r}.57$.
(136) 47. Micrometer reading assumed as $6^{r}.13$ instead of $8^{r}.13$.

ZONE 136. SEPTEMBER 21. II. BELT, −26° 16'. D_0 = −25° 51' 50"—Continued.

No.	Mag.	I.	II.	III.	IV.	V.	VI.	VII.	T.	a_1	a_2	MICROMETER.			i	d_1	d_4	Mean Right Ascension, 1850.0.	Mean Declination, 1850.0.
									h. m. s.	s.	s.			r.	' "	"	"	h. m. s.	° ' "
50	9					12.0	26.0		23 16 58.14	+ 8.15	0.00	V.	5	7.32	−23 44.18	+32.41	−2.89	23 17 6.29	−26 15 4.66
51	9		21.8	35.2			17.0		17 49.42	8.15	0.00	III.	8	11.22	39 39.95	32.44	4.37	17 57.57	31 1.88
52	7	39.2	54.	8.					20 21.65	8.13	0.00	III.	5	6.47	23 21.49	32.53	2.85	20 29.78	14 41.81
53	9		24.0	38.2					20 52.00	8.12	0.00	III.	6	7.08	27 33.18	32.55	3.23	21 0.12	18 53.86
54	10			24.5	38.2				21 38.25	8.12	0.00	IV.	7	8.47	33 23.14	32.56	3.76	21 46.37	24 44.34
55	6.7		25.2						23 53.28	8.10	0.00	II.	9	8.00	42 56.98	32.63	4.67	24 1.38	34 19.02
56	8				52.5				23 52.35	8.10	+0.01	IV.	10	8.11	48 3.58	32.63	5.18	24 0.46	39 26.13
57	9						23.2	37.5	23 55.55	8.10	0.00	VI.	7	12.06	35 3.32	32.63	3.93	24 3.65	26 24.62
58	8.9		23.2	37.2					25 51.30	8.08	0.00	III.	9	11.27	44 41.47	32.68	4.85	25 59.38	26 36 3.64
59	9					19.0			26 4.94	8.08	−0.01	V.	1	8.11	4 7.54	32.69	1.08	26 13.01	25 55 25.93
60	9			5.	18.6				27 18.68	8.07	0.00	IV.	7	4.17	31 6.99	32.72	3.56	27 26.75	26 22 27.83
61	9			50.8					28 4.45	8.07	0.00	III.	3	8.00	14 0.10	32.75	1.99	28 12.52	5 19.34
62	7.8			26.6	40.7				28 40.39	8.06	0.00	IV.	3	5.39	12 49.04	32.76	1.89	28 48.45	4 8.17
63	9					16.			29 2.04	8.06	0.00	V.	5	9.13	24 35.10	32.77	2.96	29 10.10	15 55.29
64	9					43.			29 29.10	8.05	0.00	V.	8	10.53	39 25.33	32.79	4.36	29 37.15	26 30 46.90
65	9						24.2	38.8	29 56.71	8.05	0.00	VI.	8	8.03	37 59.49	32.80	4.22	30 4.76	29 20.91
66	9			23.5	37.				31 36.98	8.04	0.00	IV.	2	11.37	11 0.54	32.83	1.71	31 45.02	2 19.42
67	10				18.				32 17.85	8.03	0.00	IV.	4	5.50	17 53.58	32.84	2.34	32 25.88	26 9 13.08
68	8.9			52.2	6.1				33 5.84	8.03	0.00	IV.	2	2.32	6 16.65	32.86	1.28	33 13.87	25 57 35.07
69	8					45.2	59.		33 31.24	8.03	0.00	V.	8	9.33	38 44.99	32.86	4.22	33 39.27	26 30 6.35
70	9							23.2	33 41.18	8.02	0.00	VII.	4	6.06	18 1.28	32.87	2.35	33 49.20	9 20.76
71	9					14.0			35 23.85	8.02	0.00	V.	3	12.28	16 15.22	32.89	2.19	36 21.87	7 34.52
72	8						51.6	5.5	35 23.68	8.01	0.00	VI.	4	10.28	20 13.60	32.90	2.56	35 31.69	11 33.26
73	8			39.2	53.5				36 53.28	8.00	0.00	IV.	8	9.56	37 56.12	32.92	2.54	37 1.28	26 29 15.74
74	9			54.2					41 7.72	7.97	0.00	III.	1	10.55	5 30.24	32.99	1.21	41 15.69	25 56 48.46
75	9						30.0	44.5	41 2.33	7.97	0.00	VII.	1	14.29	7 17.79	32.99	1.38	41 10.30	25 58 36.18
76	6.7						23.	37.2	41 55.22	7.97	0.00	VI.	4	5.07	19 32.24	33.00	2.39	42 3.19	26 9 51.63
77	9					33.	46.2		43 18.74	7.96	0.00	V.	2	7.10	6 34.76	33.01	1.31	43 26.70	25 57 53.06
78	9		33.	47.0					45 0.64	7.91	0.00	II.	2	3.43	6 51.28	33.03	1.34	45 8.58	25 58 9.59
79	9			14.	28.				45 27.94	7.94	0.00	IV.	9	0.56	39 23.34	33.03	4.34	45 35.88	26 30 44.65
80	9						18.2		45 50.43	7.94	0.00	VI.	5	7.42	27 50.20	33.04	3.27	45 58.37	19 10.43
81	9				28.	41.2			47 27.55	7.93	0.00	IV.	6	3.39	25 47.83	33.05	3.06	47 35.48	17 7.84
82	9				15.6				48 15.45	7.93	0.00	IV.	4	2.13	16 34.41	33.06	2.22	48 23.37	7 53.57
83	9					39.2	53.0		48 25.24	7.92	0.00	V.	5	10.22	25 9.90	33.06	3.01	48 33.16	16 29.85
84	9				41.5				50 41.35	7.91	0.00	IV.	4	7.36	18 47.04	33.08	2.43	50 49.26	10 6.39
85	8.9			21.5	36.0				51 35.54	7.90	0.00	IV.	4	7.50	18 54.09	33.08	2.44	51 43.44	10 13.45
86	9				20.	33.5			52 19.68	7.89	0.00	IV.	4	6.28	18 12.74	33.09	2.37	52 27.57	9 32.02
87	8.9						9.6	24.0	52 42.00	7.89	0.00	VI.	7	13.59	36 0.29	33.09	4.02	52 49.89	27 21.22
88	9				1.8	15.2			54 1.49	7.88	0.00	IV.	9	13.37	45 47.06	33.09	4.95	54 9.37	37 8.92
89	8.9		36.0	51.					55 4.33	7.87	0.00	III.	5	5.03	22 29.04	33.10	2.76	55 12.20	13 48.70
90	8.9				18.5	31.4			55 17.87	7.87	0.00	IV.	3	8.39	14 19.81	33.10	2.02	55 25.74	5 38.73
91	9	15.	29.2						56 57.10	7.86	0.00	II.	5	10.39	25 18.36	33.10	3.03	57 4.96	16 38.29
92	9			35.0					58 48.69	7.85	0.00	III.	3	13.00	16 31.36	33.11	2.23	58 56.54	7 50.48
93	8		5.2	19.2	33.2				23 59 33.01	7.84	0.00	IV.	4	9.40	19 49.56	33.11	2.53	23 59 40.85	26 11 8.98
94	9		29.0	43.0					0 0 56.62	7.84	0.00	III.	1	10.21	5 13.09	33.11	1.19	0 1 4.46	25 56 31.17
95	9			38.2	52.2				41 52.03	7.59	0.00	IV.	5	8.17	24 6.91	32.32	2.92	41 59.62	26 15 27.51
96	8				48.0	1.3			47 47.61	7.55	0.00	IV.	7	9.46	33 52.89	32.09	3.83	47 55.16	25 14.63
97	8			38.3	52.8				48 52.43	7.55	0.00	III.	6	13.08	30 34.69	32.05	3.51	48 59.98	21 56.15
98	8				22.2	36.1			0 49 22.08	+ 7.54	0.00	IV.	4	8.30	−19 14.27	+32.03	−2.47	0 49 29.62	−26 10 34.71

CORRECTIONS.

Date.		Corr. of Clock.	Hourly rate.	m	n	c
1847.	h.	s.	s.	s.	s.	s.

INSTRUMENT READINGS.

Date.		Barom.	THERMOM. At.	THERMOM. Ex.
1847.	h. m.	in.	°	°

REMARKS.

(136) 71. Minutes assumed as 36 instead of 35, and transit over T. IV as recorded over T. V.
(136) 73. Micrometer reading assumed as 7r.56 instead of 9r.56.
(136) 76. Micrometer reading assumed as 7r.07 instead of 5r.07.
(136) 77. Micrometer reading assumed as 3r.10 instead of 7r.10.
(136) 82. Micrometer reading assumed as 3r.13 instead of 2r.13.
(136) 94. Interval of 40 minutes.

ZONE 136. SEPTEMBER 21. II. BELT, −26° 16′. D₀ = −25° 51′ 50″—Continued.

No.	Mag.	SECONDS OF TRANSIT. I. II. III. IV. V. VI. VII.	T.	a_1	a_2	MICROMETER.			i	d_1	d_2	Mean Right Ascension, 1850.0.	Mean Declination, 1850.0.
			h. m. s.	s.	s.			r.	′ ″	″	″	h. m. s.	° ′ ″
99	9	. . 59.2 13.	0 51 27.06	+ 7.53	0.00	III.	7	7.05	−32 31.66	+31.94	−3.69	0 51 34.59	−26 23 53.41
100	9	. . . . 3.5	51 49.58	7.53	0.00	V.	7	10.12	34 5.96	31.92	3.84	51 57.11	25 27.88
101	8	. . . . 36.2 50.2	53 8.43	7.52	0.00	VI.	8	16.02	42 1.00	31.86	4.60	53 15.95	33 23.74
102	9	. . . . 59.2 . . 27.2	55 13.04	7.51	0.00	IV.	3	11.05	15 33.42	31.75	2.13	55 20.55	6 53.80
103	9	. . 33.8 48.2	57 2.96	7.50	0.00	III.	7	7.03	32 30.65	31.66	3.69	57 10.46	23 53.68
104	9	47.5 1.5 16.0	57 29.61	7.50	0.00	III.	5	11.47	25 52.76	31.64	3.07	57 37.11	26 17 14.19
105	9	. . 49.2 3.2	0 59 16.82	7.49	+0.01	III.	1	9.58	5 1.49	31.55	1.15	0 59 24.32	25 56 21.09
106	9	59.3 16.6	1 0 41.40	7.48	0.00	II.	6	9.37	28 48.19	31.48	3.35	1 1 48.88	26 20 10.06
107	9	. . 11.2 . . 40.2	2 39.57	7.48	0.00	IV.	5	5.23	22 39.16	31.37	2.78	2 47.05	14 0.57
108	9	. . . . 34.0 47.2 1.8	3 47.39	7.47	0.00	IV.	4	9.29	19 44.01	31.31	2.52	3 54.86	11 5.22
109	9	43. . . . . . 25.2	5 25.00	7.46	0.00	IV.	2	11.18	10 40.87	31.22	1.66	5 32.46	2 1.31
110	8.9	39.2 53.2 8.0	9 21.54	7.44	0.00	III.	8	5.14	36 34.39	31.00	4.06	9 28.98	27 57.47
111	9	. . . . 58.2	12 12.17	7.43	0.00	III.	8	3.21	35 37.40	30.82	3.98	12 19.60	26 27 0.56
112	8	. . . . 18.5 30.0	13 30.91	7.42	+0.01	IV.	1	4.07	2 4.55	30.74	0.86	13 38.34	25 53 24.69
113	9	. . 0. 14.2	15 27.90	7.41	0.00	III.	5	0.04	19 58.28	30.62	2.54	15 35.31	26 11 20.20
114	9	35.2 49.6 3.2 . . . . 59.1	17 17.16	7.40	0.00	III.	3	9.40	14 50.52	30.51	2.07	17 24.56	26 6 12.08
115	9	. . . . 12.5 26.	19 25.93	7.39	+0.01	IV.	1	8.58	4 31.25	30.38	1.10	19 33.33	25 55 52.00
116	8.9	. . . . 8.2 22.1 36.5	20 22.09	7.39	0.00	IV.	3	10.39	15 20.32	30.32	2.11	20 29.48	26 6 42.11
117	9	. . . . 17.2 31.2 45.1	21 31.17	7.38	−0.01	IV.	9	3.30	40 40.99	30.23	4.47	21 38.54	32 5.23
118	7	. . . . 16.2 30.2 44.	22 30.08	7.38	0.00	IV.	7	6.40	32 19.10	30.16	3.66	22 37.46	23 42.60
119	9	. . . . . . 58.2 12.	22 58.02	7.38	0.00	IV.	3	9.05	14 32.91	30.13	2.03	23 5.40	5 54.81
120	9	. . . . 46.2 0.4	23 18.43	7.37	0.00	VI.	4	6.57	18 27.20	30.11	2.37	23 25.80	9 49.46
121	9	. . . . 57.0	24 15.05	7.37	0.00	VII.	5	7.16	23 35.77	30.04	2.87	24 22.42	14 58.60
122	9	. . . . 16.0	24 34.10	7.37	0.00	VII.	6	5.01	26 28.79	30.02	3.12	24 41.47	17 51.89
123	9	. . . . 52.2	26 10.12	7.36	0.00	VII.	3	6.21	13 9.83	29.90	1.90	25 17.48	4 31.83
124	9	. . . . 18.5 32.1 46.2	27 4.43	7.35	0.00	VII.	7	11.39	31 47.98	29.84	2.92	26 11.78	26 11.06
125	9	. . 51.0 5.2	30 18.91	7.34	0.00	III.	4	11.24	20 41.96	29.61	2.60	30 26.25	12 4.95
126	9	8.2 22.8 36.4	29 50.50	7.34	0.00	III.	7	7.27	32 42.75	29.64	3.70	29 57.84	24 6.81
127	9	. . 4.3 18.0	30 32.06	7.34	0.00	III.	6	9.02	28 30.66	29.59	3.53	30 39.40	19 54.40
128	9	. . . . 8.0 . .	30 40.26	7.34	0.00	VI.	4	9.33	19 45.87	29.56	2.52	30 47.60	11 8.81
129	9	. . . . 49.2	33 2.81	7.33	+0.01	III.	2	13.37	11 50.92	29.39	1.77	33 10.15	3 12.30
130	9	. . 34.2 48.1 3.	38 2.45	7.31	−0.01	IV.	9	11.46	44 52.10	28.99	4.94	38 9.75	36 18.01
131	9	. . 37.2 51.5	40 5.34	7.30	0.00	III.	7	12.39	35 20.08	28.82	3.98	40 12.64	26 45.24
132	9	. . 43.2 57.2 11.2 25.8	41 11.29	7.30	0.00	IV.	6	11.11	29 35.74	28.72	3.42	41 18.59	21 0.44
133	9	9. 23.0	42 50.95	7.29	0.00	II.	4	9.58	19 58.47	28.56	2.52	42 58.24	11 22.41
134	8	. . . . 34.0 48.0 1.8	45 47.94	7.28	−0.01	IV.	9	7.25	42 39.48	28.32	4.70	45 55.21	34 5.86
135	9	. . . . 4.2 18.	47 17.97	7.27	0.00	IV.	7	0.03	28 58.91	28.19	3.37	47 25.24	20 24.09
136	8	. . 45. 59.0 12.8 26.2	48 12.58	7.27	+0.01	IV.	3	9.38	14 49.55	28.11	2.07	48 19.86	6 13.51
137	8	. . . . 0.2 14. 28.0	49 13.86	7.26	+0.01	IV.	2	7.01	9 31.79	28.02	1.58	49 21.13	0 55.35
138	7.8	. . . . . . 58.5 12.1	49 58.26	7.26	0.00	IV.	6	11.41	29 50.87	27.95	3.45	50 5.52	21 16.37
139	9	. . . . . . 31. 45.2 58.3 . . .	50 30.88	7.26	0.00	IV.	5	6.02	22 58.84	27.90	2.82	50 38.14	14 23.76
140	9	. . . . 45. 59.2	51 59.08	7.25	−0.01	IV.	9	11.04	44 29.92	27.76	4.83	52 6.32	35 56.99
141	9	. . 9. 23.2 37.0	53 36.81	7.25	+0.01	IV.	2	12.25	11 14.65	27.61	1.70	53 44.07	2 38.74
142	8	. . . . 21.6 35.4	54 35.37	7.25	0.00	IV.	6	12.14	30 7.51	27.51	3.47	54 42.62	21 33.47
143	9	. . . . . . 2.	1 55 1.85	7.24	0.00	IV.	4	5.35	18 46.53	27.47	2.42	1 55 9.09	10 11.48
144	8.9	0.8 14.5 29. 43.0 56.5 10.2 24.2	2 1 42.68	+ 7.22	−0.01	IV.	8	3.43	−35 48.54	+26.82	−4.00	2 1 49.89	−26 27 15.72
11*	7	. . 29.5 43.2 58.0	22 32	. .	. . .	IV.	3	13.02					

CORRECTIONS.

Date.		Corr. of Clock.	Hourly rate.	m	n	c
1847.	h.	s.	s.	s.	s.	s.

INSTRUMENT READINGS.

Date.		Barom.	THERMOM. At.	THERMOM. Ex.
1847.	h. m.	in.	°	°

REMARKS.

(136) 106. Transit observations discordant by 3ˢ; observation of T. II used as 13ˢ.6 instead of 16ˢ.6; minutes assumed as 1 instead of 0.
(136) 123. Minutes assumed as 25 instead of 26.
(136) 124. Minutes assumed as 26 instead of 27.
(136) 137. Micrometer reading assumed as 9ʳ.01 instead of 7ʳ.01.
(136) 143. Micrometer reading assumed as 7ʳ.35 instead of 5ʳ.35.

* Omitted by copyist.

ZONE 137. OCTOBER 4. K. BELT, −24° 23'. D_0 = −23° 58' 0".

No.	Mag.	SECONDS OF TRANSIT. I.	II.	III.	IV.	V.	VI.	VII.	T.	σ_1	σ_2	MICROMETER.		i	d_1	d_2	Mean Right Ascension, 1850.0.	Mean Declination, 1850.0.
									h. m. s.	s.	s.		r.	' "	"	"	h. m. s.	° ' "
1	10	..	..	..	..	..	45.5	..	21 57 18.20	− 3.25	0.00	VI.	4	4.18	−17 7.03	− 6.91	−2.40	21 57 14.95	−24 15 16.34
2	8	..	26.4	40.3	..	..	..	..	21 59 53.95	3.27	0.00	IV.	6	11.3	29 31.71	6.81	3.35	21 59 50.68	27 41.87
3	6.7	..	40.8	54.3	8.4	..	..	..	22 1 8.14	3.29	0.00	IV.	5	10.35	25 16.50	6.78	3.02	22 1 4.85	23 26.30
4	9.10	..	..	57.9	..	..	..	..	1 11.41	3.29	0.00	IV.	4	8.1	18 59.64	6.78	2.53	2 8.12	17 8.95
5	9	..	7.0	21.0	34.7	..	..	..	5 34.56	3.33	0.00	IV.	6	5.28	26 42.78	6.65	3.14	5 31.23	24 52.57
6	6	..	..	..	26.8	40.3	..	..	6 26.65	3.34	+0.01	IV.	10	4.11	46 32.81	6.63	4.70	6 23.32	44 44.14
7	9	30.7	44.9	58.6	12.3	..	..	..	9 12.25	3.36	0.00	IV.	7	0.38	29 16.56	6.55	3.33	9 8.89	27 26.44
8	8.9	..	42.8	..	10.5	..	..	..	11 10.36	3.38	+0.01	IV.	8	1.48	34 50.55	6.51	3.78	11 6.99	33 0.84
9	7	..	..	6.0	19.2	..	46.6	..	16 19.25	3.43	−0.01	IV.	2	8.31	9 16.67	6.40	1.76	16 15.81	7 24.83
10	6	..	..	42.3	56.3	..	..	..	17 56.05	3.44	0.00	IV.	6	8.42	28 20.62	6.36	3.26	17 52.61	26 30.24
11	8	57.8	11.8	25.7	39.3	53.0	6.4	20.5	21 39.31	3.47	+0.01	IV.	8	9.46	38 51.59	6.30	4.09	21 35.85	37 1.98
12	9	..	..	45.2	..	..	..	..	25 58.53	3.51	−0.01	IV.	1	11.14	5 39.86	6.24	1.48	25 55.01	3 47.58
13	6	..	56.8	..	24.7	..	51.3	..	27 24.30	3.53	+0.01	IV.	10	7.31	47 43.41	6.22	4.83	27 20.76	45 54.46
14	9	..	..	..	28.4	..	..	..	29 28.25	3.55	−0.01	IV.	1	8.16	4 10.11	6.19	1.37	29 24.69	2 17.66
15	9.10	..	..	16.8	..	..	..	..	31 30.47	3.56	0.00	IV.	6	11.3	29 31.71	6.17	3.35	31 26.91	27 41.23
16	7	..	27.5	..	55.3	..	..	..	32 55.04	3.58	0.00	IV.	4	9.34	19 46.54	6.17	2.58	32 51.46	17 55.29
17	9	..	49.8	..	17.5	..	..	..	36 17.36	3.61	0.00	IV.	7	11.40	34 50.37	6.15	3.75	36 13.75	33 0.30
18	7.8	30.9	44.8	58.9	12.8	26.4	39.5	53.5	43 12.47	3.67	0.00	IV.	7	12.43	35 22.13	6.13	3.82	43 8.80	33 32.08
19	8.9	..	51.6	5.0	18.8	32.5	..	..	45 18.90	3.69	+0.01	IV.	9	12.32	45 14.29	6.14	4.62	45 14.22	43 25.05
20	9	..	..	..	6.1	..	..	..	47 5.95	3.70	0.01	IV.	10	7.56	47 56.02	6.14	4.84	47 2.26	46 7.00
21	9	..	..	..	15.9	..	..	..	49 15.75	3.72	+0.01	IV.	9	2.24	40 7.71	6.14	4.19	49 12.04	38 18.04
22	9	..	..	20.2	..	..	..	..	50 33.90	3.73	0.00	IV.	7	5.26	31 41.78	6.14	3.53	50 30.17	29 51.45
23	9.8	..	..	..	..	..	..	25.9	22 50 44.50	3.74	0.00	VII.	4	3.53	16 54.23	6.15	2.36	22 50 40.76	15 2.74
24	8	..	..	..	..	..	59.3	..	23 4 31.92	3.86	+0.01	VII.	9	1.56	39 53.24	6.31	4.19	23 4 28.07	38 3.74
25	7	..	7.4	..	35.3	..	..	..	8 34.93	3.90	−0.01	IV.	2	−1.14	4 19.19	6.38	1.43	8 31.02	2 27.00
26	7.8	..	20.1	34.5	47.8	1.5	..	..	11 47.63	3.92	+0.01	IV.	1	8.58	4 31.28	6.45	1.38	11 43.72	2 39.11
27	9	15.5	..	43.5	57.0	..	..	..	14 56.90	3.94	0.00	IV.	4	6.18	18 7.70	6.53	2.48	14 52.96	16 16.71
28	9	..	..	..	..	..	..	53.6	15 12.17	3.95	0.00	IV.	3	9.17	14 38.96	6.54	2.18	15 8.22	12 47.68
29	9	..	..	49.7	..	17.6	..	..	31 3.65	4.09	0.00	IV.	7	6.49	32 23.64	7.04	3.58	30 59.56	30 34.26
30	8	..	..	..	58.8	..	..	..	33 58.65	4.11	0.00	IV.	6	8.33	28 16.08	7.16	3.26	33 54.54	26 26.50
31	9	..	57.5	..	25.7	..	..	..	35 25.30	4.12	+0.01	IV.	7	6.31	32 14.56	7.22	4.38	35 21.19	30 26.16
32	7.8	..	41.0	55.1	..	..	..	..	37 8.55	4.14	0.00	IV.	5	4.5	21 59.84	7.29	2.77	37 4.41	20 9.90
33	9	..	59.9	..	..	..	..	..	41 27.20	4.17	0.00	IV.	2	0.31	5 14.63	7.48	1.43	41 23.03	3 23.54
34	9	..	..	..	39.0	..	..	..	52 38.85	4.26	0.00	IV.	4	8.14	19 6.20	8.07	2.53	52 34.59	17 16.80
35	10	..	..	..	..	39.5	..	..	53 25.72	4.27	0.00	IV.	4	5.52	17 54.59	8.12	2.45	53 21.45	16 5.16
36	9	..	..	..	..	53.9	..	..	55 40.18	4.29	0.00	IV.	6	9.30	28 44.82	8.25	3.29	55 35.89	26 56.36
37	9	..	26.4	..	54.0	..	..	..	22 59 53.94	− 4.32	0.00	IV.	9	4.37	−41 14.78	− 8.50	−4.32	23 59 49.62	−24 39 27.60

ZONE 138. OCTOBER 15. H. BELT, −28° 46'. D_0 = −27° 53' 50".

No.	Mag.	I.	II.	III.	IV.	V.	VI.	VII.	T.	σ_1	σ_2	MICROMETER.		i	d_1	d_2	Mean Right Ascension, 1850.0.	Mean Declination, 1850.0.
1	8.9	20.2	35.	49.2	..	..	..	..	22 30 3.24	−16.85	0.00	.	4	6.39	−19 18.81	−10.23	−2.38	22 29 46.39	−28 13 21.42
2	9.10	..	..	..	..	18.2	..	..	30 3.91	16.85	−0.01	.	3	5.32	12 45.51	10.23	1.58	30 47.05	6 47.32
3	9	..	..	..	..	50.2	4.4	..	30 35.96	16.86	−0.01	.	2	8.22	9 12.13	10.20	1.21	30 19.09	3 13.54
4	9	..	..	..	..	..	23.2	38.1	30 54.99	16.86	0.00	.	6	5.313	26 44.45	10.18	3.21	30 38.13	20 47.84
5	9	..	..	..	8.2	23.	..	..	32 8.37	16.87	−0.01	.	2	10.53	10 28.27	10.11	1.35	31 51.49	4 29.73
6	9	..	..	..	..	..	..	50.2	32 7.27	16.87	0.00	.	7	7.205	32 39.51	10.11	3.91	31 50.40	26 43.53
7	8.9	..	..	11.2	25.2	39.2	..	..	33 25.05	16.89	−0.01	.	2	6.458	8 23.62	10.04	1.13	33 8.15	2 24.79
8	8.9	..	..	3.7	17.3	31.8	..	..	22 34 17.46	−16.90	−0.01	.	3	11.44	−15 53.09	− 9.99	−1.95	22 34 0.55	−28 9 55.03

CORRECTIONS.

Date.		Corr. of Clock.	Hourly rate.	m	n	c
1847.	h.	s.	s.	s.	s.	s.
Oct. 4,	19	ƒ 9.91	ζ 0.044			
Oct. 15,	19	ƒ 23.20	ζ 0.040			

INSTRUMENT READINGS.

	Date.		Barom.	THERMOM. At.	THERMOM. Ex.
	1847.	h. m.	in.	°	°
Zone 137	Oct. 4,	9 5	30.040	63.5	52.5
Zone 138	Oct. 15,	9 5	30.276	62.0	37.5

REMARKS.

(137) 4. Minutes assumed as 2 instead of 1.
(137) 6. Micrometer reading assumed as $5^r.11$ instead of $4^r.11$.
(137) 25. Micrometer reading assumed as $1^t\ 8^r.34$, to agree with W. Transit 1847, Z. 33, October 18, and Argel. 257, 91.
(138) 1. Micrometer reading assumed as $8^r.39$ instead of $6^r.39$.

ZONE 138. OCTOBER 15. II. BELT, −28° 46′. D_s = −27° 53′ 50″—Continued.

No.	Mag.	I.	II.	III.	IV.	V.	VI.	VII.	T.	a_1	a_4	MICROMETER.			l	d_1	d_2	Mean Right Ascension, 1850.0.	Mean Declination, 1850.0.
									h. m. s.	s.	s.			r	′ ″	″	″	h. m. s.	° ′ ″
9	9	. .	. .	. .	. .	4.2	19.2	. .	22 34 50.36	−16.90	+0.01	.	10	4.205	−46 7.35	− 9.96	−5.50	22 34 33.47	−28 40 12.81
10	9	. .	. .	44.	58.	12.1	. .	. .	35 57.89	16.92	−0.01	.	1	10.194	5 12.33	9.90	0.77	35 40.96	27 59 13.00
11	9	. .	. .	51.2	5.2	. .	. .	. .	37 5.15	16.93	−0.01	.	2	11.185	10 41.12	9.83	1.39	36 48.21	28 4 42.34
12	9	. .	. .	. .	3.2	. .	. .	. .	38 3.04	16.94	0.00	IV.	4	3.350	16 45.51	9.78	2.04	37 46.10	10 47.33
13	9	42.8	57.	10.8	. .	. .	. .	. .	39 25.27	16.95	0.00	IV.	4	1.438	16 19.69	9.70	2.00	39 8.32	10 21.39
14	9	. .	. .	25.2	38.5	53.	. .	. .	39 38.76	16.96	0.00	.	3	10.378	15 19.71	9.69	1.88	39 21.80	9 21.28
15	7.8	. .	. .	16.0	30.2	44.2	. .	. .	40 29.98	16.96	0.00	.	6	4.03	27 0.44	9.65	3.24	40 13.02	21 3.33
16	9	14.2	29.	43.2	. .	11.2	. .	. .	42 57.17	16.99	−0.01	.	2	11.186	10 41.17	9.52	1.39	42 40.17	4 42.08
17	9	. .	3.0	17.8	. .	46.	. .	. .	44 31.65	17.01	0.00	.	3	13.390	16 51.07	9.44	2.07	44 14.64	10 52.58
18	9	. .	30.8	. .	59.2	. .	. .	41.2	46 59.09	17.04	+0.01	IV.	9	11.354	44 45.75	9.32	5.33	46 42.06	38 50.40
19	10	. .	29.0	. .	. .	. .	. .	. .	48 57.61	17.06	0.00	II.	6	12.27	30 13.87	9.22	3.62	48 40.55	24 16.71
20	9	. .	. .	. .	. .	30.2	. .	. .	49 15.98	17.06	+0.01	V.	8	11.073	39 32.53	9.21	4.70	48 58.93	33 36.44
21	10	. .	. .	. .	. .	. .	. .	22.8	49 39.90	17.07	0.00	VII.	5	11.42	25 49.86	9.19	3.10	49 22.83	19 52.15
22	10	. .	. .	. .	. .	. .	. .	47.2	50 4.18	17.07	0.00	VII.	3	12.20	16 10.81	9.17	1.99	49 47.11	10 11.97
23	9	. .	. .	58.4	. .	. .	. .	. .	52 12.26	17.09	−0.01	III.	2	4.34	7 17.12	9.08	0.99	51 55.16	1 17.19
24	9	. .	. .	. .	. .	32.5	. .	. .	52 18.32	17.10	+0.01	V.	10	11.082	49 32.88	9.07	5.69	52 1.23	43 37.84
25	9	. .	. .	. .	. .	. .	. .	24.0	52 41.35	17.10	+0.01	VII.	10	4.580	46 25.84	9.05	5.55	52 24.26	40 30.44
26	9.10	. .	. .	3.5	18.0	. .	. .	0.2	54 17.50	17.12	0.00	IV.	3	10.454	15 23.54	8.99	1.89	54 0.38	28 9 25.42
27	9	. .	. .	. .	3.3	. .	. .	. .	56 3.14	17.14	−0.01	III.	.	6.133	3 8.18	8.91	0.53	55 45.99	27 57 7.62
28	6.7	. .	. .	46.1	0.1	. .	. .	. .	57 59.92	17.16	−0.01	IV.	1	5.158	2 39.23	8.83	0.40	57 42.75	27 56 38.54
29	9	. .	. .	. .	38.0	. .	. .	20.	22 58 37.45	17.17	0.00	IV.	5	4.123	22 3.52	8.86	2.66	22 58 20.28	28 16 4.98
30	9	. .	41.5	56.2	. .	24.2	. .	. .	23 [1] 9.89	17.19	−0.01	III.	1	7.452	3 54.52	8.70	0.63	23 0 52.69	27 57 53.85
31	9	. .	58.3	12.4	26.2	41.2	55.2	. .	2 26.56	17.21	0.00	IV.	3	4.000	11 59.12	8.65	1.52	2 9.35	28 5 59.29
32	9	. .	. .	. .	. .	. .	21.	. .	2 52.49	17.21	+0.01	VI.	10	3.058	45 29.49	8.63	5.41	2 35.29	39 33.53
33	10	. .	. .	. .	20.2	. .	. .	1.8	4 19.54	17.23	0.00	IV.	8	3.29	35 11.48	8.58	4.26	4 2.31	29 44.32
34	9	. .	29.0	13.	58.	. .	. .	40.2	5 57.33	17.25	0.00	III.	3	3.398	11 48.88	8.52	1.50	5 40.08	5 48.90
35	9	. .	12.5	27.	. .	. .	. .	. .	7 41.09	17.27	0.00	III.	5	10.125	25 5.10	8.46	3.01	7 23.82	19 6.57
36	9	. .	. .	46.2	1.0	. .	. .	. .	8 0.73	17.27	0.00	IV.	9	9.145	43 34.70	8.44	5.21	8 43.46	28 37 38.35
37	9	. .	. .	. .	. .	. .	25.2	. .	7 56.89	17.27	−0.01	VI.	1	11.264	5 45.92	8.45	0.82	7 39.61	27 59 45.19
38	9	27.2	42.	56.2	10.5	. .	. .	. .	13 10.52	17.33	0.00	IV.	8	13.400	35 50.87	8.27	4.29	12 53.19	28 29 53.43
39	9	. .	. .	. .	32.5	46.	. .	. .	13 32.07	17.33	0.00	IV.	9	10.158	44 5.61	8.26	5.23	13 14.74	37 9.10
40	10	. .	. .	. .	57.5	. .	. .	. .	14 57.34	17.34	0.00	IV.	5	6.20	23 7.91	8.22	2.76	14 40.00	17 8.69
41	8	. .	51.2	6.	20.	. .	. .	. .	16 19.80	17.36	0.00	IV.	3	4.362	12 17.38	8.17	1.55	16 2.44	6 17.10
42	8	. .	33.0	47.4	1.2	. .	. .	. .	17 1.30	17.37	0.00	IV.	4	2.475	16 21.56	8.15	1.99	16 43.93	10 21.70
43	9	. .	. .	. .	. .	. .	28.0	. .	16 59.53	17.37	0.00	VI.	8	9.398	39 48.27	8.15	4.62	16 42.16	32 51.04
44	9	. .	. .	. .	. .	. .	58.2	. .	17 29.73	17.37	0.00	VI.	8	9.402	38 48.47	8.14	4.62	17 12.36	32 51.23
45	6	. .	. .	. .	30.3	44.2	. .	. .	18 30.02	17.38	0.00	IV.	4	11.592	20 59.75	8.11	2.54	18 12.64	15 0.40
46	8	. .	23.8	38.2	. .	6.2	20.2	. .	18 52.02	17.39	0.00	III.	3	4.300	12 14.20	8.10	1.54	18 34.63	6 13.84
47	9	. .	9.0	23.1	37.2	. .	5.3	. .	21 37.21	17.42	0.00	VII.	7	4.445	31 20.44	8.02	3.74	21 19.79	25 22.20
48	9	. .	. .	. .	42.5	. .	. .	. .	22 42.34	17.43	0.00	IV.	2	8.210	9 11.12	7.99	1.21	22 24.91	3 10.32
49	9	. .	. .	. .	10.2	24.2	. .	. .	23 9.95	17.43	0.00	IV.	3	6.41	13 20.30	7.98	1.69	22 52.52	7 19.97
50	9	. .	. .	. .	. .	48.6	. .	. .	23 34.28	17.44	0.00	V.	4	6.162	18 6.75	7.97	2.19	23 16.84	28 12 6.91
51	9	. .	. .	. .	. .	36.8	. .	. .	24 22.42	17.45	−0.01	V.	1	8.185	4 11.31	7.95	0.64	24 4.96	27 58 9.90
52	9	24.8	. .	. .	8.0	. .	. .	. .	27 7.82	17.48	0.00	IV.	4	11.238	20 41.89	7.88	2.50	26 50.34	28 14 42.27
53	6.7	. .	44.	58.2	3.8	. .	. .	. .	28 2.42	17.49	0.00	IV.	5	9.343	24 45.89	7.86	2.97	27 44.93	18 46.72
54	9	. .	. .	. .	. .	21.8	. .	. .	28 7.46	17.49	0.00	V.	3	5.183	12 38.54	7.86	1.61	27 49.97	6 38.01
55	10	. .	. .	. .	22.2	. .	. .	. .	30 22.04	17.51	0.00	IV.	2	12.130	11 8.60	7.80	1.42	30 4.53	28 5 7.82
56	9	. .	19.2	33.2	. .	2.3	. .	0.3	31 47.50	17.53	0.00	III.	1	10.375	5 21.41	7.78	0.77	31 29.97	27 59 19.96
57	9	. .	14.2	28.2	42.0	. .	. .	. .	23 33 42.26	−17.55	0.00	IV.	4	9.552	−19 57.22	− 7.74	−2.42	23 33 24.71	−28 13 57.38

CORRECTIONS.

Date.		Corr. of Clock.	Hourly rate.	m	n	c
1847.	h.	s.	s.	s.	s.	s.

INSTRUMENT READINGS.

Date.		Barom.	THERMOM. At.	THERMOM. Ex.
1847.	h. m.	in.	°	°

REMARKS.

(138) 13. Micrometer reading assumed as 2r.438 instead of 1r.438.
(138) 15. Micrometer reading assumed as 6r.03 instead of 4r.03.
(138) 38. Hor. thread assumed as 7 instead of 8.
(138) 53. Times of transit over T.'s II and III assumed to have been recorded 10s too late.

ZONE 138. OCTOBER 15. II. BELT, −28° 46′. D_o = −27° 53′ 50″—Continued.

No.	Mag.	SECONDS OF TRANSIT. I.	II.	III.	IV.	V.	VI.	VII.	T.	a_1	a_2	MICROMETER.			i	d_1	d_2	Mean Right Ascension, 1850.0.	Mean Declination, 1850.0.
									h. m. s.	s.	s.			r.	′ ″	″	″	h. m. s.	° ′ ″
58	9	0.2	14.	28.8	. . .	57.2	11.	. . .	23 34 42.92	−17.56	0.00	III.	7	10.150	−34 7.46	−7.73	−4.07	23 34 25.36	−28 28 9.26
59	9	. . .	. . .	. . .	50.	. . .	. . .	. . .	35 49.84	17.57	0.00	IV.	6	10.073	29 3.63	7.71	3.47	35 31.27	23 4.81
60	9	. . .	. . .	. . .	35.	49.2	. . .	. . .	36 34.84	17.58	0.00	V.	2	5.342	7 47.47	7.69	1.05	36 17.26	1 46.21
61	9	24.2	38.3	53.	7.3	. . .	. . .	. . .	38 7.12	17.59	0.00	IV.	6	6.298	27 13.95	7.66	3.27	37 49.53	21 14.88
62	9	42.8	57.0	11.5	. . .	. . .	. . .	. . .	39 25.56	17.60	0.00	III.	3	10.322	15 16.84	7.64	1.89	39 7.96	9 16.37
63	9	. . .	. . .	34.5	49.3	. . .	. . .	. . .	40 49.07	17.62	+0.01	IV.	10	11.302	49 44.03	7.62	5.93	40 31.46	43 47.58
64	9	. . .	. . .	. . .	. . .	. . .	44.6	. . .	41 16.38	17.62	0.00	VI.	6	10.070	29 3.29	7.61	3.48	40 58.76	23 4.38
65	7	. . .	. . .	. . .	. . .	. . .	18.2	32.2	41 49.63	17.63	+0.01	VI.	10	6.30	47 12.47	7.61	5.62	41 32.01	28 41 15.70
66	9	. . .	. . .	. . .	. . .	. . .	8.2	. . .	42 39.89	17.64	−0.01	VI.	1	8.05	4 4.38	7.60	0.62	42 22.24	27 58 2.60
67	9	. . .	. . .	. . .	. . .	. . .	18.0	. . .	46 49.52	17.69	0.00	VI.	8	12.390	40 18.63	7.54	4.84	46 31.83	28 34 21.01
68	9	. . .	. . .	. . .	4.0	18.2	32.2	. . .	48 3.85	17.70	0.00	V.	5	4.532	22 24.09	7.53	2.71	47 46.15	26 24.23
69	9	. . .	56.3	10.8	24.5	39.2	. . .	. . .	51 24.89	17.73	0.00	IV.	9	7.07	42 30.41	7.49	5.06	51 7.16	36 32.96
70	9	49.2	. . .	18.5	32.3	. . .	. . .	. . .	52 32.46	17.75	0.00	IV.	8	8.536	38 25.16	7.49	4.59	52 14.71	32 27.24
71	9	12.2	. . .	40.	55.5	. . .	. . .	. . .	53 54.95	17.76	0.00	IV.	6	9.153	28 37.40	7.48	3.41	53 37.19	28 22 38.29
72	7.8	. . .	23.	37.8	. . .	. . .	. . .	. . .	54 51.48	17.77	0.00	III.	1	8.430	4 23.67	7.48	0.66	54 33.71	27 58 21.81
73	9	. . .	51.	. . .	6.2	. . .	50.	. . .	57 20.40	17.79	0.00	IV.	3	3.313	11 44.64	7.47	1.45	57 2.61	28 5 43.56
74	9	. . .	. . .	. . .	51.3	5.3	. . .	34.2	57 51.08	17.80	0.00	IV.	3	4.46	12 22.32	7.47	1.54	57 33.28	6 21.33
75	9	. . .	10.5	. . .	. . .	. . .	. . .	. . .	59 39.12	17.82	0.00	II.	7	4.382	31 17.49	7.46	3.74	59 21.30	25 18.69
76	8.9	. . .	. . .	41.2	55.3	9.6	. . .	. . .	23 59 55.29	17.82	0.00	IV.	4	9.54	19 56.62	7.46	2.42	23 59 37.47	13 56.50
77	9	. . .	. . .	. . .	. . .	. . .	58.0	. . .	0 0 29.61	17.82	0.00	VI.	5	7.10	23 32.93	7.46	2.82	0 0 11.79	17 33.21
78	6	. . .	46.2	0.4	15.2	. . .	. . .	. . .	4 14.94	17.86	0.00	IV.	9	10.262	44 10.86	7.49	5.24	3 57.08	38 13.59
79	8.9	. . .	. . .	38.2	53.1	7.5	. . .	. . .	4 52.95	17.87	0.00	IV.	9	8.285	43 11.50	7.50	5.14	4 35.08	37 14.14
80	9	. . .	21.8	36.2	51.	. . .	. . .	. . .	5 50.68	17.88	0.00	IV.	9	11.475	44 51.85	7.50	5.32	5 32.80	38 54.67
81	8	19.6	33.8	17.8	2.0	. . .	. . .	. . .	7 2.23	17.89	0.00	IV.	6	7.166	27 37.55	7.51	3.31	6 44.34	21 38.37
82	9	. . .	. . .	. . .	. . .	. . .	22.	. . .	6 53.69	17.89	0.00	VI.	2	2.270	6 12.93	7.51	0.85	6 35.80	0 11.29
83	9	. . .	. . .	. . .	45.	59.2	. . .	. . .	7 44.88	17.90	0.00	V.	6	4.334	26 15.21	7.51	3.15	7 26.98	20 15.87
84	9	. . .	. . .	41.	. . .	. . .	. . .	. . .	8 55.41	17.91	0.00	III.	9	7.152	42 34.49	7.52	5.07	8 37.50	28 36 37.08
85	9	. . .	. . .	. . .	. . .	18.2	. . .	. . .	9 3.83	17.91	0.00	V.	1	9.310	4 47.87	7.52	0.69	8 45.92	27 58 46.08
86	9	. . .	. . .	. . .	. . .	. . .	. . .	49.0	9 5.90	17.91	0.00	VII.	2	8.349	9 18.22	7.52	1.20	8 47.99	28 3 16.94
87	10	. . .	22.5	. . .	. . .	. . .	. . .	. . .	11 51.18	17.94	0.00	III.	8	6.46	37 20.78	7.56	4.46	11 33.21	31 22.80
88	9	. . .	9.2	23.2	. . .	. . .	. . .	. . .	12 37.68	17.95	0.00	III.	7	12.58	35 29.64	7.57	4.24	12 19.73	29 31.45
89	9	. . .	15.2	. . .	. . .	. . .	12.2	. . .	13 43.74	17.96	0.00	II.	2	9.180	9 40.17	7.59	1.25	13 25.78	3 39.01
90	9	. . .	. . .	40.2	54.4	. . .	. . .	. . .	14 54.24	17.97	0.00	IV.	4	9.556	19 57.42	7.61	2.42	14 36.27	13 57.45
91	9	2.5	17.2	31.1	. . .	. . .	. . .	. . .	16 45.46	17.99	0.00	III.	4	11.310	20 45.48	7.64	2.51	16 27.47	14 45.63
92	9	. . .	. . .	23.3	. . .	. . .	. . .	. . .	17 51.99	18.00	0.00	III.	8	8.445	38 20.53	7.66	4.57	17 33.99	32 22.76
93	6.7	. . .	4.0	18.	32.8	. . .	0.3	. . .	18 32.47	18.01	0.00	IV.	8	8.504	38 23.55	7.67	4.57	18 14.46	32 25.79
94	9	. . .	22.0	36.1	50.6	. . .	. . .	. . .	21 50.27	18.04	0.00	IV.	4	8.492	19 23.95	7.74	2.34	21 32.23	13 24.03
95	9	. . .	17.2	32.1	46.	. . .	. . .	. . .	22 46.11	18.05	0.00	IV.	2	11.065	10 35.08	7.76	1.34	22 28.06	4 34.18
96	9	. . .	. . .	25.	. . .	. . .	. . .	. . .	24 39.45	18.07	0.00	III.	9	13.005	45 28.60	7.82	5.43	24 21.38	39 31.85
97	9	. . .	. . .	28.5	. . .	. . .	. . .	. . .	26 42.76	18.09	0.00	III.	7	7.110	32 34.67	7.87	3.89	26 24.67	26 36.43
98	8.9	. . .	5.	19.8	. . .	. . .	. . .	. . .	28 33.58	18.11	0.00	III.	3	4.186	12 8.44	7.93	1.50	29 15.47	6 7.87
99	8	. . .	. . .	. . .	35.8	. . .	3.8	. . .	28 35.53	18.11	0.00	VI.	5	−0.372	45.13	7.93	2.51	?28 17.42	14 45.57
100	8.9	. . .	. . .	. . .	0.4	14.2	28.1	. . .	31 59.95	18.14	0.00	IV.	7	7.402	32 49.45	8.03	3.92	31 41.81	28 26 51.40
101	9	. . .	. . .	. . .	. . .	. . .	5.5	. . .	31 37.19	18.13	0.00	VI.	1	11.40	5 52.78	8.02	0.80	31 19.06	27 59 51.60
102	9	. . .	. . .	. . .	. . .	. . .	35.2	. . .	32 6.88	18.14	0.00	VI.	2	4.436	7 21.81	8.04	0.97	31 48.74	28 1 20.82
103	8.9	. . .	. . .	12.	26.2	. . .	. . .	. . .	33 26.02	18.15	0.00	IV.	4	3.162	11 37.03	8.08	1.47	33 7.87	5 36.58
104	8.7	. . .	. . .	. . .	. . .	56.6	9.8	. . .	33 41.86	18.15	0.00	V.	5	5.538	17 55.45	8.09	2.18	33 23.71	11 55.72
105	10	. . .	. . .	. . .	. . .	48.8	. . .	. . .	34 34.45	18.16	0.00	V.	2	10.031	10 3.11	8.12	1.27	34 16.29	4 2.50
106	9	. . .	. . .	. . .	47.5	1.5	. . .	. . .	0 35 47.31	−18.17	0.00	IV.	8	8.360	−38 16.30	−8.17	−4.57	0 35 29.14	−28 32 19.04

CORRECTIONS.

Date.		Corr. of Clock.	Hourly rate.	m	n	c
1847.	h.	s.	s.	s.	s.	s.

INSTRUMENT READINGS.

Date.		Barom.	THERMOM. At.	Ex.
1847.	h. m.	in.	°	°

REMARKS.

(138) 73. Transit over T. III assumed as recorded over T. IV.
(138) 98. Minutes assumed as 29 instead of 28.
(138) 99. Micrometer reading assumed as 1 instead of 0.
(138) 103. Micrometer thread assumed as 3 instead of 4.
(138) 104. Micrometer thread assumed as 4 instead of 5.

ZONE 138. OCTOBER 15. II. BELT, −28° 46′. D_o = −27° 53′ 50″—Continued.

No.	Mag.	Seconds of Transit. I.	II.	III.	IV.	V.	VI.	VII.	T.	a_1	a_2	Micrometer.			i	d_1	d_2	Mean Right Ascension, 1850.0.	Mean Declination, 1850.0.
									h. m. s.	s.	s.			r.	′ ″	″	″	h. m. s.	° ′ ″
107	9	. .	29.2	43.2	. .	. .	. .	. .	0 37 57.82	−18.19	0.00	III.	10	5.112	−46 32.86	− 8.25	−5.56	0 37 39.63	−28 40 36.67
108	9	. .	. .	. .	7.3	. .	. .	. .	38 7.14	18.19	0.00	IV.	9	4.485	41 20.57	8.25	4.93	37 48.95	35 23.75
109	9	. .	. .	. .	. .	45.8	. .	. .	38 31.62	18.20	0.00	V.	10	8.156	48 5.85	8.27	5.75	38 13.42	42 9.87
110	9.10	. .	. .	. .	. .	. .	20.	. .	38 51.53	18.20	0.00	VI.	8	10.30	39 13.59	8.28	4.69	38 33.33	33 16.56
111	9	. .	52.9	7.	20.9	. .	. .	. .	42 21.16	18.23	0.00	IV.	7	4.500	31 23.63	8.43	3.74	42 2.93	25 25.80
112	9	. .	49.5	4.2	. .	. .	. .	. .	43 18.14	18.24	0.00	III.	5	2.552	21 24.59	8.47	2.59	42 59.90	15 25.65
113	9	. .	. .	. .	35.0	49.	. .	. .	43 34.77	18.24	0.00	IV.	5	2.212	21 7.49	8.48	2.55	43 16.53	15 8.52
114	9	. .	29.	43.6	58.	. .	. .	. .	44 57.77	18.25	0.00	IV.	7	6.392	32 18.70	8.55	3.87	44 39.52	26 21.12
115	9	. .	. .	36.5	. .	. .	. .	. .	45 50.56	18.26	0.00	III.	4	11.195	20 39.68	8.59	2.50	45 32.30	14 40.77
116	6.7	. .	28.2	43.	57.1	. .	. .	39.5	48 57.01	18.29	0.00	IV.	8	14.331	41 16.34	8.73	4.93	48 38.72	35 20.00
117	9	8.0	. .	37.	. .	. .	. .	. .	53 50.86	18.33	0.00	III.	2	4.372	7 18.73	8.98	0.96	53 32.53	28 1 18.67
118	9	. .	. .	. .	. .	57.0	. .	. .	53 42.61	18.33	0.00	V.	1	3.559	1 58.75	8.98	0.37	53 24.28	27 55 58.10
119	9	. .	. .	. .	34.0	48.2	. .	. .	54 33.87	18.34	0.00	IV.	5	3.070	21 30.59	9.02	2.60	54 15.53	28 15 32.21
120	9	. .	32.	46.4	. .	. .	. .	. .	56 0.63	18.35	0.00	III.	7	4.410	31 19.04	9.10	3.74	55 42.28	25 21.88
121	9	. .	. .	. .	. .	25.5	. .	. .	56 11.28	18.36	0.00	V.	8	8.53	38 24.81	9.11	4.59	55 52.92	32 28.51
122	8	. .	. .	. .	. .	. .	51.5	6.	56 23.18	18.36	0.00	VI.	10	6.265	47 10.70	9.12	5.64	56 4.82	41 15.46
123	9	. .	. .	. .	. .	. .	51.3	. .	57 23.00	18.37	0.00	VI.	1	7.306	3 47.02	9.17	0.55	57 4.63	27 57 46.74
124	9	. .	35.0	. .	3.5	. .	31.3	. .	0 59 3.28	18.38	0.00	IV.	6	11.393	29 50.02	9.26	3.57	0 58 44.90	28 23 52.85
125	6.7	2.1	16.2	30.	44.8	59.	13.	27.1	1 0 44.69	18.39	0.00	IV.	8	6.56	37 25.87	9.35	4.48	1 0 26.30	31 29.70
126	9	. .	16.2	30.1	45.1	. .	. .	. .	2 0 44.75	−18.39	0.00	IV.	8	7.440	−37 50.07	− 9.35	−4.53	1 0 26.36	−28 31 53.95

ZONE 139. OCTOBER 15. II. BELT, −28° 46′. D_o = −27° 53′ 50″.

No.	Mag.	I.	II.	III.	IV.	V.	VI.	VII.	T.	a_1	a_2	Micrometer.			i	d_1	d_2	Mean Right Ascension, 1850.0.	Mean Declination, 1850.0.
1	7.8	. .	12.	26.8	41.	. .	. .	. .	2 1 40.76	−18.87	0.00	IV.	5	6.226	−23 9.22	. .	2.7	2 1 21.89	−28 17 9.
2	9	. .	. .	. .	30.2	. .	. .	. .	2 30.04	18.88	+0.01	IV.	3	6.578	13 28.77	. .	1.68	2 11.17	7 21.
3	9	. .	. .	. .	. .	. .	14.2	29.1	2 45.99	18.88	0.00	VI.	5	5.515	22 53.34	. .	2.75	2 27.11	
4	9	. .	. .	. .	2.1	16.	. .	. .	4 1.86	18.89	−0.01	IV.	8	6.525	37 24.10	. .	4.47	3 42.96	31 28.
5	7.8	. .	. .	31.0	45.	. .	13.1	28.	5 44.90	18.90	0.00	IV.	4	8.070	19 17.80	. .	2.33	5 26.00	13 5.
6	9	. .	. .	. .	. .	. .	. .	27.	6 44.30	18.91	−0.01	VII.	9	4.460	41 18.90	. .	4.94	6 25.38	
7	7	. .	. .	. .	55.2	9.2	22.8	. .	7 54.78	18.92	−0.01	IV.	9	6.408	42 17.20	. .	5.05	7 35.85	36 20.
8	10	. .	. .	. .	48.2	. .	16.8	. .	8 48.24	18.92	+0.01	IV.	3	9.17	14 38.96	. .	1.80	8 29.33	
9	10	. .	. .	7.5	. .	. .	. .	. .	13 21.76	18.95	0.00	III.	7	7.450	32 51.82	. .	3.92	13 2.81	
10	8.9	. .	. .	45.0	59.1	. .	. .	. .	13 58.87	18.95	+0.01	IV.	1	7.340	3 48.93	. .	0.55	13 39.93	
11	9	. .	. .	. .	. .	. .	33.1	48.	14 4.92	18.95	0.00	VII.	6	10.474	29 23.43	. .	3.51	13 45.97	
12	7.8	. .	. .	27.3	41.8	56.1	. .	24.1	15 41.63	18.96	−0.01	IV.	8	9.280	38 42.50	. .	4.63	15 22.66	
13	9	. .	. .	40.3	. .	. .	. .	. .	19 54.32	18.99	+0.01	III.	4	5.590	17 58.07	. .	2.18	19 35.34	
14	9	. .	41.3	55.2	9.2	. .	38.0	. .	21 9.45	19.00	0.00	IV.	5	7.480	23 52.28	. .	2.87	20 50.45	
15	9	4.4	. .	33.1	. .	. .	. .	. .	22 47.16	19.01	+0.01	III.	2	11.13	10 38.30	. .	1.34	22 28.16	
16	9	. .	. .	. .	. .	9.	. .	. .	22 54.61	19.01	0.02	V.	1	3.14	1 37.77	. .	0.30	22 35.62	
17	9	29.5	. .	. .	. .	. .	. .	. .	24 12.39	19.02	0.01	I.	2	8.188	9 10.09	. .	1.18	23 53.38	
18	8.9	. .	. .	1.0	15.	. .	43.2	. .	24 14.89	19.02	+0.01	IV.	3	12.195	16 10.98	. .	1.97	24 55.88	
19	8.9	. .	. .	. .	56.2	10.	. .	. .	25 55.90	19.03	−0.01	IV.	7	8.200	33 9.52	. .	3.95	25 36.86	
20	9	. .	. .	. .	. .	. .	48.	. .	26 19.61	19.03	0.00	VI.	5	3.410	21 47.54	. .	2.63	26 0.58	
21	9	. .	. .	10.2	. .	. .	. .	. .	28 24.04	19.04	+0.01	III.	2	2.202	6 9.64	. .	0.80	28 5.01	
22	9	. .	. .	52.0	6.2	. .	. .	. .	28 6.26	19.04	−0.01	IV.	10	7.230	47 39.37	. .	5.72	27 47.21	
23	9	. .	. .	. .	. .	41.0	. .	. .	29 26.68	19.05	+0.01	V.	4	3.422	16 49.09	. .	2.06	29 7.64	
24	9	. .	2.6	17.	. .	. .	. .	. .	32 31.12	19.06	0.01	III.	3	11.168	15 39.32	. .	1.91	32 12.07	
25	9	. .	. .	. .	. .	3.0	. .	. .	2 ?32 48.63	−19.06	+0.02	V.	2	1.045	− 5 31.47	. .	−0.74	2 32 29.59	

CORRECTIONS.

Date.		Corr. of Clock.	Hourly rate.	m	n	c
1847.	h.	s.	s.	s.	s.	s.

INSTRUMENT READINGS.

Date.		Barom.	Thermom. At.	Thermom. Ex.
1847.	h. m.	in.	°	°

REMARKS.

(139) Telescope not well clamped.
(139) 18. Minutes assumed as 25 instead of 24.
[(138) 126. This star belongs to the following Zone.]

ZONE 139. OCTOBER 15. II. BELT, −28° 46′. D_o = −27° 53′ 50″—Continued.

No.	Mag.	SECONDS OF TRANSIT. I.	II.	III.	IV.	V.	VI.	VII.	T.	a_1	a_2	MICROMETER.			i	d_1	d_2	Mean Right Ascension, 1850.0.	Mean Declination, 1850.0.
									h. m. s.	s.	s.			r.	′ ″	″	″	h. m. s.	° ′ ″
26	9	..	..	..	45.	59.3	..	..	2 33 44.98	−19.07	−0.02	IV.	10	9.305	−48 43.67	. .	−5.87	2 33 25.89	
27	9	..	49.	3.8	..	..	..	..	35 17.62	19.08	0.01	III.	3	11.050	15 33.37	. .	1.90	34 58.53	
28	9	..	..	30.5	..	..	..	..	35 44.89	19.08	−0.01	III.	9	4.580	41 25.31	. .	4.95	35 25.79	
29	9	..	..	..	..	9.3	..	..	35 54.93	19.08	+0.02	V.	2	2.226	6 10.85	. .	0.80	35 35.87	
30	8	49.6	3.5	18.0	32.2	..	0.3	..	37 32.24	19.09	−0.01	IV.	8	8.145	38 5.45	. .	4.55	37 13.14	
31	9	..	..	..	..	31.0	..	..	40 16.76	19.10	−0.01	VI.	7	11.142	34 37.17	. .	4.14	39 57.65	
32	9	..	..	1.3	..	..	44.2	..	42 15.57	19.11	+0.01	IV.	3	12.026	16 2.46	. .	1.96	41 56.47	
33	6	..	14.6	29.1	43.3	..	11.1	..	43 43.13	19.12	−0.01	IV.	8	11.346	39 46.35	. .	4.75	43 24.00	
34	9	..	50.3	4.4	19.	..	..	..	45 18.71	19.13	0.00	IV.	4	12.033	21 1.81	. .	2.54	44 59.58	
35	8.9	..	54.1	8.4	22.8	36.8	..	..	46 22.70	19.13	−0.01	IV.	8	12.350	40 16.80	. .	4.83	46 3.56	
36	9	..	56.4	10.	24.4	..	..	..	47 24.52	19.14	0.01	IV.	7	10.00	33 59.95	. .	4.06	47 5.37	
37	9	..	12.8	..	41.	..	9.2	..	48 41.02	19.14	−0.01	IV.	8	4.145	36 4.43	. .	4.30	48 21.87	
38	9	..	..	..	15.4	..	..	..	51 15.24	19.16	+0.01	IV.	3	8.350	14 17.79	. .	1.76	50 56.09	
39	9	..	..	..	..	46.2	..	..	51 31.90	19.16	0.00	V.	5	4.228	22 8.76	. .	2.67	51 12.74	
40	9	..	..	..	..	..	12.0	..	51 43.60	19.16	+0.01	VI.	5	8.400	24 18.32	. .	2.91	51 24.45	
41	9	..	..	..	..	37.	51.	..	52 22.65	19.16	−0.01	V.	7	13.023	35 31.81	. .	4.24	52 3.48	
42	8	..	34.8	48.3	3.1	..	..	..	54 3.09	19.17	−0.02	IV.	10	5.098	46 32.21	. .	5.60	53 43.90	
43	8	..	..	..	36.8	40.2	..	..	54 36.21	19.17	0.00	IV.	4	9.572	19 58.23	. .	2.42	54 17.09	
44	9	..	..	..	10.2	24.1	..	..	55 9.90	19.18	+0.01	IV.	2	12.045	11 4.32	. .	1.34	54 50.73	
45	7	..	..	..	..	..	56.9	..	55 28.39	19.18	−0.02	VI.	9	13.430	45 53.89	. .	5.53	55 9.19	
46	9	..	..	..	..	..	26.8	..	55 58.32	19.18	0.01	VI.	9	3.54	40 52.90	. .	4.91	55 39.13	
47	9	..	..	..	..	..	..	9.5	57 26.70	19.19	0.01	VII.	7	6.49	33 23.71	. .	3.98	57 7.50	
48	9	..	..	55.	9.6	..	..	..	59 9.39	19.20	−0.01	IV.	8	8.010	37 58.64	. .	4.56	58 50.18	
49	9	..	..	..	..	40.2	..	..	2 59 25.89	19.20	0.00	V.	4	10.522	20 25.92	. .	2.47	59 6.69	
50	9	..	..	..	..	24.0	37.8	..	3 0 9.55	19.20	0.00	V.	4	10.055	20 2.37	. .	2.42	2 59 50.35	
51	7	2.5	17.1	31.4	..	0.0	13.8	28.	0 45.63	19.21	0.00	III.	6	12.282	30 14.61	. .	3.62	3 0 26.42	
52	9	..	..	..	56.5	..	..	..	4 56.34	19.22	+0.01	IV.	3	8.162	14 8.31	. .	1.72	4 37.13	
53	9	..	33.	..	..	15.2	..	..	5 1.16	19.22	0.01	II.	3	9.205	14 40.53	. .	1.78	5 41.95	
54	7	35.2	49.6	4.1	..	..	46.	..	7 17.97	19.23	0.01	IV.	3	8.228	14 11.63	. .	1.72	6 58.75	
55	9	16.	30.	..	..	..	..	..	8 58.64	19.24	+0.01	II.	2	9.370	9 49.76	. .	1.22	8 39.41	
56	9	..	4.5	..	..	..	..	..	9 33.21	19.24	−0.01	II.	8	11.123	39 34.91	. .	4.75	9 13.96	
57	9	..	34.3	..	3.4	..	..	..	16 3.09	19.26	0.01	IV.	7	7.55	32 56.91	. .	3.96	15 43.82	
58	7	57.5	11.8	26.4	40.2	..	..	..	17 40.46	19.27	−0.01	IV.	7	9.188	33 39.16	. .	4.05	17 21.18	
59	9	8.4	22.	..	..	..	..	..	18 51.00	19.28	0.00	II.	5	7.38	23 47.05	. .	2.85	18 31.72	
60	10	..	..	..	..	..	17.0	..	18 48.56	19.28	−0.01	VI.	7	7.06	32 32.02	. .	3.91	18 29.27	
61	10	..	..	..	..	..	..	4.0	19 21.22	19.28	−0.01	VII.	7	12.46	35 23.23	. .	4.25	19 1.93	
62	9	..	0.6	..	29.	..	..	..	21 28.98	19.28	0.00	IV.	5	4.006	21 57.72	. .	2.66	21 9.70	
63	9	..	..	..	..	56.5	..	..	21 42.11	19.29	+0.02	V.	1	4.30	2 16.10	. .	0.37	21 22.84	
64	9	..	..	33.	46.6	..	..	..	22 47.31	19.29	0.01	IV.	3	9.50	14 55.60	. .	1.82	22 28.03	
65	9	..	..	..	..	19.	32.5	..	23 4.41	19.29	+0.01	V.	2	9.03	9 32.75	. .	1.19	22 45.13	
66	8.9	0.	14.	28.2	42.2	..	..	..	24 42.52	19.29	0.00	IV.	6	6.439	27 21.06	. .	3.28	24 23.23	
67	9	..	..	18.5	..	..	..	..	25 32.50	19.30	+0.01	III.	4	2.52	?16 23.78	. .	1.97	25 13.21	
68	9	52.2	7.0	..	..	..	..	..	27 35.46	19.30	0.00	II.	7	0.552	29 25.04	. .	3.53	27 16.16	
69	9	..	..	..	..	34.5	..	..	28 20.19	19.31	0.00	V.	4	8.92	19 13.86	. .	2.31	28 0.88	
70	9	..	..	..	10.	..	..	..	29 9.84	19.31	+0.01	IV.	4	7.556	18 56.91	. .	2.29	28 50.54	
71	10	..	1.0	..	..	..	..	..	30 29.60	19.31	0.00	II.	6	9.392	28 49.27	. .	3.45	30 9.29	
72	9	..	..	..	4.9	..	..	..	31 4.74	19.31	−0.01	IV.	9	5.410	41 47.04	. .	5.01	30 45.42	
73	9	..	56.	..	..	..	..	..	32 24.35	19.32	+0.02	II.	1	9.170	4 40.67	. .	0.62	32 5.05	
74	6	..	..	38.5	52.2	..	..	..	3 32 52.39	−19.32	−0.01	IV.	7	5.49	−31 53.38	. .	−3.84	3 32 33.06	

CORRECTIONS.

Date.		Corr. of Clock.	Hourly rate.	m	n	c
1847.	h.	s.	s.	s.	s.	s.

INSTRUMENT READINGS.

Date.		Barom.	THERMOM. At.	Ex.
1847.	h. m.	in.	°	°

REMARKS.

(139) 30. Transit over T. IV assumed 32ˢ.2 instead of 22ˢ.2.
(139) 43. Transit observation of T.V assumed as 50ˢ.2 instead of 40ˢ.2.
(139. 53. Minutes assumed as 6 instead of 5.

ZONE 139. OCTOBER 15. H. BELT, −28° 46′. $D_0 = -27° 53′ 50″$—Continued.

No.	Mag.	Seconds of Transit. I.	II.	III.	IV.	V.	VI.	VII.	T.	a_1	a_2	Micrometer.			i	d_1	d_2	Mean Right Ascension, 1850.0.	Mean Declination, 1850.0.
									h. m. s.	s.	s.			r.	′ ″	″	″	h. m. s.	° ′ ″
75	8.9	. .	. .	. .	. .	23.4	36.8	. .	3 33 8.76	−19.32	−0.01	V.	8	4.42	−36 18.25	. .	−4.36	3 32 49.43	
76	7	. .	27.	. .	56.	. .	. .	. .	34 55.74	19.32	−0.01	IV.	7	7.580	32 58.42	. .	3.94	34 36.41	
77	9	. .	. .	. .	. .	39.	. .	. .	35 24.72	19.33	0.00	V.	5	11.470	25 52.75	. .	3.10	35 5.39	
78	9	. .	50.0	. .	. .	. .	. .	. .	37 18.74	19.33	+0.02	II.	9	7.516	42 52.80	. .	5.16	36 59.43	
79	9	. .	. .	21.5	. .	. .	. .	. .	38 35.38	19.34	0.02	III.	2	7.278	8 44.74	. .	1.09	38 16.06	
80	9	. .	. .	56.2	. .	. .	. .	. .	39 10.10	19.34	+0.01	III.	2	10.58	10 30.74	. .	1.29	38 50.77	
81	9	. .	. .	. .	. .	23.0	. .	. .	39 8.71	19.34	0.02	V.	5	6.00	22 57.78	. .	2.76	38 49.37	
82	9	. .	. .	. .	. .	. .	56.2	. .	39 27.79	19.34	0.00	VI.	6	4.170	26 6.80	. .	3.13	39 8.45	
83	9	. .	. .	36.0	50.0	. .	. .	. .	40 50.14	19.34	−0.02	IV.	9	11.575	44 56.89	. .	5.38	40 30.75	
84	9	. .	53.0	. .	. .	. .	. .	. .	43 21.36	19.35	+0.02	II.	1	12.462	6 26.16	. .	0.81	43 2.03	
85	9	. .	. .	2.0	15.5	. .	. .	. .	45 15.83	19.35	−0.01	IV.	8	5.395	36 47.29	. .	4.39	44 56.47	
86	6.7	. .	56.6	11.2	25.2	39.3	. .	. .	47 25.04	19.35	+0.01	IV.	3	5.123	12 35.57	. .	1.51	48 5.70	
87	9	. .	. .	4.0	18.	. .	. .	. .	48 17.87	19.36	0.01	IV.	2	9.174	9 40.06	. .	1.20	47 58.52	
88	9	. .	. .	. .	. .	. .	53.1	. .	49 24.74	19.36	+0.01	VI.	3	11.440	15 52.90	. .	1.89	49 5.39	
89	9	. .	. .	. .	. .	. .	3.0	. .	50 34.56	19.36	−0.01	VI.	7	6.102	32 3.88	. .	3.83	50 15.19	
90	9	. .	33.0	48.2	2.0	. .	. .	. .	52 1.90	19.36	0.00	IV.	5	8.060	24 1.37	. .	2.88	51 42.54	
91	10	. .	. .	. .	57.0	. .	. .	. .	52 56.84	19.37	+0.01	IV.	3	12.00	16 1.15	. .	1.91	52 37.48	
92	9	50.8	5.2	19.4	. .	. .	. .	. .	54 33.68	19.37	+0.01	II.	3	5.075	12 32.96	. .	1.51	54 14.32	
93	9	. .	. .	2.3	16.6	. .	. .	. .	55 16.61	19.37	−0.02	IV.	10	6.460	47 20.73	. .	5.71	54 57.22	
94	9	. .	. .	. .	47.8	. .	. .	. .	55 47.64	19.37	+0.01	IV.	3	7.468	13 53.48	. .	1.67	55 28.28	
95	9	. .	4.8	19.2	. .	. .	. .	. .	58 33.16	19.38	0.01	III.	2	11.125	10 38.05	. .	1.29	58 13.79	
96	9	. .	. .	42.1	. .	. .	. .	. .	58 56.09	19.38	+0.01	III.	3	11.118	15 36.79	. .	1.87	58 36.72	
97	5.6	. .	17.3	32.1	46.0	0.2	14.	29.0	3 59 45.82	−19.38	−0.02	IV.	2	9.095	− 9 36.08	. .	−1.17	3 59 26.42	

ZONE 140. OCTOBER 16. K. BELT, −30° 2′. $D_0 = -29° 39′ 0″$.

No.	Mag.	I.	II.	III.	IV.	V.	VI.	VII.	T.	a_1	a_2	Micrometer.			i	d_1	d_2	Mean Right Ascension, 1850.0.	Mean Declination, 1850.0.
1	7.8	33.2	. .	2.7	. .	. .	. .	. .	23 46 17.04	−19.38	0.00	IV.	7	11.30	−34 45.33	− 2.74	−4.25	23 45 57.66	−30 13 52.32
2	10	21.9	. .	. .	. .	. .	. .	. .	49 5.42	19.41	0.00	IV.	4	5.31	17 44.00	2.73	2.10	48 46.01	29 56 48.83
3	8	. .	. .	. .	. .	. .	. .	49.2	49 5.97	19.41	0.00	VII.	9	5.5	41 28.45	2.73	5.11	48 46.56	30 20 36.29
4	10	. .	47.9	. .	. .	. .	. .	. .	51 16.97	19.43	0.00	IV.	9	4.21	41 6.70	2.74	5.07	50 57.54	20 14.51
5	9	. .	. .	. .	. .	. .	. .	52.5	51 9.04	19.43	0.00	VII.	5	7.33	23 44.28	2.74	2.85	50 49.61	2 49.87
6	7.8	. .	. .	. .	. .	. .	. .	47.4	52 3.91	19.44	0.00	VII.	9	2.21	40 5.75	2.74	4.94	51 44.47	30 19 13.43
7	10	. .	35.2	. .	. .	. .	. .	. .	56 3.96	19.48	0.00	IV.	2	11.57	11 0.54	2.76	1.24	55 44.48	29 50 4.54
8	7	. .	32.5	47.2	. .	. .	. .	. .	57 1.57	19.49	0.00	IV.	7	9.50	33 54.90	2.77	4.15	56 42.08	30 13 1.82
9	7	. .	. .	. .	. .	. .	. .	42.0	57 58.60	19.49	0.00	IV.	6	6.15	27 6.40	2.77	3.25	56 39.11	30 6 12.51
10	7.8	. .	. .	34.2	48.9	3.0	. .	. .	23 58 48.56	19.51	0.00	IV.	4	10.6	20 2.68	2.77	2.37	23 58 29.05	29 59 7.82
11	10	2.3	. .	. .	. .	. .	. .	. .	0 0 46.03	19.53	0.00	IV.	6	3.55	35 54.59	2.79	4.39	0 0 26.50	30 15 1.77
12	10	38.9	. .	. .	. .	. .	. .	. .	5 22.58	19.58	0.00	IV.	7	5.3	31 30.18	2.85	3.80	5 3.00	10 36.83
13	9	. .	49.0	3.4	. .	. .	. .	. .	6 18.08	19.59	0.00	IV.	10	3.7	45 30.29	2.86	5.64	5 58.49	30 24 38.79
14	9	. .	. .	53.5	. .	. .	. .	. .	7 7.53	19.60	0.00	IV.	2	3.37	6 49.42	2.87	0.69	6 47.93	29 45 51.98
15	9	. .	. .	. .	52.1	. .	. .	. .	7 51.94	19.61	0.00	IV.	3	2.46	11 21.81	2.88	1.27	7 32.33	50 25.96
16	6	. .	. .	. .	18.5	32.8	46.8	. .	14 18.25	19.67	0.00	IV.	2	9.8	9 35.32	3.01	1.06	13 58.58	29 48 39.39
17	9	42.8	57.2	. .	. .	. .	. .	. .	16 26.34	19.69	0.00	IV.	7	8.40	33 19.61	3.06	4.08	16 6.65	30 12 26.75
18	9	. .	. .	. .	. .	. .	. .	8.6	16 25.35	19.69	0.00	IV.	8	12.9	40 3.69	3.06	4.93	16 5.66	30 19 11.68
19	9	. .	. .	9.0	23.2	. .	. .	. .	18 23.13	19.71	0.00	IV.	4	8.34	19 16.29	3.10	2.28	18 3.42	29 58 21.67
20	9	. .	34.9	. .	4.2	. .	. .	. .	20 4.03	19.73	0.00	IV.	10	3.27	45 40.37	3.14	5.66	19 44.30	30 24 49.17
21	9	. .	. .	. .	44.8	. .	. .	. .	20 44.64	19.74	0.00	IV.	8	7.34	37 45.03	3.16	4.66	20 24.90	16 52.85
22	7	. .	. .	. .	. .	31.6	45.5	. .	0 21 16.94	−19.74	0.00	IV.	6	7.22	−27 40.27	− 3.18	−3.32	0 20 57.20	−30 6 46.77

CORRECTIONS.

Date.		Corr. of Clock.	Hourly rate.	m	n	c
1847.	h.	s.	s.	s.	s.	s.
Oct. 16.	19	f 24.95	g 0.030			

INSTRUMENT READINGS.

	Date.		Barom.	Thermom. At.	Thermom. Ex.
	1847.	h. m.	in.	°	°
Zone 140	Oct. 16	9 8	30.230	64.5	44.

REMARKS.

(139) 86. Minutes assumed as 48 instead of 47.

ZONE 140. OCTOBER 16. K. BELT, −30° 2'. D_{o} = −29° 39' 0"—Continued.

No.	Mag.	I.	II.	III.	IV.	V.	VI.	VII.	T.	a_1	a_2	MICROMETER.			i	d_1	d_2	Mean Right Ascension, 1850.0.	Mean Declination, 1850.0.
									h. m. s.	s.	s.			r.	' "	"	"	h. m. s.	° ' "
23	9	. .	28.6	43.0	. .	. .	. .	. .	0 25 57.28	−19.79	0.00	IV.	3	9.45	−14 53.08	− 3.33	−1.73	0 25 37.49	−29 53 58.14
24	6	. .	. .	. .	35.5	49.6	3.7	. .	26 35.14	19.80	0.00	IV.	9	10.1	43 58.15	3.35	5.49	26 15.34	30 23 6.99
25	9	. .	. .	56.0	10.2	24.8	. .	. .	28 10.22	19.81	0.00	IV.	5	6.0	22 57.83	3.40	2.76	27 50.41	30 2 3.99
26	7	. .	. .	14.1	. .	. .	. .	. .	29 28.05	19.82	0.00	IV.	1	3.19	1 40.34	3.44	0.07	29 8.23	29 40 43.85
27	9	14.4	29.0	43.6	. .	. .	. .	. .	30 58.14	10.84	0.00	IV.	9	4.24	41 8.22	3.50	5.08	30 38.30	30 20 16.80
28	9	. .	58.5	12.8	27.0	. .	. .	. .	32 27.03	19.85	0.00	IV.	3	9.57	14 59.13	3.50	1.74	32 7.18	29 54 4.43
29	6.7	. .	. .	15.9	30.6	44.8	. .	. .	33 30.40	19.87	0.00	IV.	7	13.12	35 36.75	3.60	4.35	33 10.53	30 14 44.70
30	9	. .	. .	21.8	. .	. .	. .	. .	39 36.19	19.93	0.00	IV.	6	11.48	29 54.40	3.85	3.61	39 16.26	30 9 1.86
31	9	. .	. .	. .	. .	22.5	. .	. .	40 8.01	19.93	0.00	IV.	4	6.16	18 6.70	3.88	2.10	39 48.08	29 57 12.68
32	8	. .	. .	. .	. .	11.3	25.5	. .	40 56.80	19.94	0.00	IV.	7	3.39	30 47.83	3.92	3.71	40 36.86	30 9 55.46
33	9	. .	. .	. .	10.2	. .	. .	. .	43 10.04	19.96	0.00	IV.	2	10.22	10 12.64	4.03	1.12	42 50.08	29 49 17.79
34	10	. .	. .	. .	41.8	. .	. .	. .	46 41.64	19.99	0.00	IV.	2	4.48	7 24.22	4.20	0.76	46 21.65	29 46 29.18
35	6	. .	58.8	. .	27.9	. .	. .	. .	0 56 27.80	20.08	0.00	IV.	9	3.31	40 41.4	4.73	5.03	56 7.72	30 19 51.25
36	9	. .	. .	56.8	. .	. .	. .	. .	1 0 11.44	20.11	+0.01	IV.	10	4.00	45 57.02	4.96	5.72	0 59 51.34	25 7.70
37	9	. .	. .	. .	. .	59.8	. .	. .	0 45.35	20.12	0.00	IV.	6	4.27	27 12.03	5.00	3.26	1 0 25.23	6 20.29
38	8	. .	. .	48.5	3.2	. .	. .	. .	2 3.09	20.13	−0.01	IV.	10	4.28	46 11.13	5.09	5.71	1 42.95	25 21.93
39	9	. .	. .	24.5	. .	. .	. .	. .	4 39.02	20.15	0.00	IV.	8	8.24	38 10.24	5.27	4.71	4 18.87	30 17 20.22
40	8.9	. .	. .	37.8	6.5	. .	. .	. .	6 52.20	20.16	0.00	IV.	4	8.50	19 24.35	5.36	2.29	5 32.04	29 58 32.00
41	9	17.4	. .	46.8	. .	. .	. .	. .	7 1.16	20.17	0.00	IV.	7	6.4	32 0.95	5.43	3.86	6 40.99	30 11 10.24
42	9	. .	. .	. .	31.0	. .	. .	. .	8 30.84	20.19	0.00	IV.	9	7.41	42 47.55	5.53	5.26	8 10.65	30 21 58.34
43	9	. .	44.1	58.5	. .	. .	. .	. .	12 12.84	20.22	0.00	IV.	4	9.18	19 38.46	5.80	2.31	11 52.62	29 58 46.57
44	9	58.2	12.9	27.4	42.1	. .	. .	. .	13 41.99	20.23	+0.01	IV.	9	10.15	44 5.21	5.91	5.49	13 21.75	30 23 16.61
45	9	. .	. .	. .	. .	23.8	. .	. .	20 9.39	20.29	0.00	VI.	7	13.12	35 36.55	6.41	4.37	19 49.19	14 47.33
46	9	. .	17.8	32.8	47.5	. .	. .	. .	1 23 47.03	−20.32	0.00	IV.	5	6.55	−23 25.56	− 6.72	−2.82	1 23 26.71	−30 2 35.10

ZONE 141. OCTOBER 18. H. BELT, −23° 46'. D_{o} = −23° 21' 0".

No.	Mag.	I.	II.	III.	IV.	V.	VI.	VII.	T.	a_1	a_2	MICROMETER.			i	d_1	d_2	Mean Right Ascension, 1850.0.	Mean Declination, 1850.0.
1	9	23.	30.5	. .	4.	. .	. .	. .	22 59 3.98	−20.18	−0.01	II.	3	7.33	−13 46.37	−13.45	−2.18	22 58 43.79	−23 35 2.00
2	9	. .	. .	. .	23.2	36.5	. .	. .	22 59 22.96	20.18	+0.01	IV.	8	4.135	36 3.92	13.42	3.84	59 2.79	57 21.18
3	9	. .	. .	. .	. .	. .	42.0	. .	23 0 14.78	20.19	0.00	VI.	7	6.060	32 1.81	13.35	3.53	22 59 54.59	53 18.69
4	9	. .	. .	. .	. .	14.4	27.8	. .	1 0.66	20.20	0.00	V.	7	3.158	30 36.08	13.29	3.44	23 0 40.46	51 52.81
5	9	. .	. .	45.0	59.	. .	. .	. .	1 58.75	20.21	0.00	IV.	7	8.520	33 25.66	13.21	3.65	1 38.54	54 42.52
6	9	. .	. .	. .	17.5	30.8	. .	. .	2 17.25	20.21	0.00	IV.	7	9.318	33 45.72	13.19	3.67	1 57.04	55 2.58
7	8.9	. .	27.5	41.6	55.	. .	. .	. .	4 54.97	20.24	0.00	IV.	6	8.050	28 1.96	12.98	3.23	4 34.73	49 18.17
8	9	. .	. .	. .	. .	. .	25.2	39.2	4 57.97	20.24	−0.01	VI.	1	9.510	4 57.85	12.98	1.51	4 37.72	26 12.34
9	9	. .	. .	. .	. .	. .	. .	44.2	5 2.86	20.24	−0.01	VII.	1	10.318	5 18.24	12.97	1.54	4 42.62	26 32.75
10	9	. .	. .	. .	23.3	36.6	. .	. .	6 23.06	20.25	0.00	IV.	8	7.500	37 53.09	12.87	3.99	6 2.81	23 59 9.95
11	7	. .	24.0	37.5	51.5	4.8	. .	. .	8 51.32	20.28	+0.01	IV.	9	4.326	41 12.56	12.67	4.25	8 31.05	24 2 29.48
12	8	. .	23.3	37.1	. .	. .	. .	. .	9 50.54	20.29	0.00	III.	4	3.290	16 42.4	12.59	2.40	9 30.25	23 37 57.43
13	9	. .	. .	. .	1.2	. .	. .	. .	10 1.05	20.29	−0.01	IV.	1	9.030	4 33.80	12.58	1.47	9 40.75	25 47.85
14	9.10	. .	. .	. .	27.8	41.8	. .	. .	11 27.91	20.30	0.00	IV.	7	14.070	36 4.48	12.47	3.84	11 7.61	23 57 20.79
15	8	. .	. .	. .	4.1	17.6	. .	. .	12 3.97	20.31	+0.01	IV.	9	4.530	41 22.84	12.43	4.25	11 43.67	24 2 39.52
16	8	. .	. .	. .	. .	38.8	52.1	. .	12 25.01	20.31	0.00	V.	4	4.385	17 17.49	12.40	2.44	12 4.70	23 38 32.33
17	8.9	. .	53.0	6.6	20.2	. .	. .	. .	15 20.20	20.34	0.00	IV.	6	8.558	29 28.07	12.19	3.35	14 59.86	50 43.61
18	8.9	. .	41.2	55.2	9.	. .	. .	. .	16 8.56	20.35	−0.01	IV.	1	9.180	4 41.36	12.13	1.48	15 48.20	25 54.97
19	9	. .	. .	. .	55.5	. .	. .	. .	17 55.35	20.36	0.00	IV.	8	6.222	37 8.82	12.00	3.92	17 34.99	23 58 24.74
20	9	. .	. .	. .	12.0	. .	. .	. .	18 11.85	20.36	0.00	IV.	9	3.050	40 28.38	11.98	4.20	17 51.49	24 1 44.56
21	8.9	. .	. .	32.2	46.1	. .	. .	. .	23 18 45.89	−20.37	0.00	IV.	7	6.533	−32 25.80	−11.94	−3.57	23 18 25.52	−23 53 41.31

CORRECTIONS.

Date.		Corr. of Clock.	Hourly rate.	m	n	c
1847.	h.	s.	s.	s.	s.	s.
Oct. 18,	20	ƒ 26.14	g 0.037	− 0.086	+ 0.153	0.000

INSTRUMENT READINGS.

	Date.		Barom.	THERMOM. At.	THERMOM. Ex.
	1847.	h. m.	in.	°	°
Zone 141	Oct. 18,	9 5	30.110	65.5	54.

REMARKS.

(140) 40. Observation of T. V assumed to have been recorded as of T. IV, and minutes assumed as 5 instead of 6.
(141) 17. Micrometer reading assumed as 10ʳ.558 instead of 8ʳ.558.

ZONE 141. OCTOBER 18. II. BELT, −23° 46′. D_o = −23° 21′ 0″—Continued.

No.	Mag.	Seconds of Transit. I.	II.	III.	IV.	V.	VI.	VII.	T.	a_1	a_2	Micrometer.			i	d_1	d_2	Mean Right Ascension, 1850.0.	Mean Declination, 1850.0.
									h. m. s.	s.	s.			r.	′ ″	″	″	h. m. s.	° ′ ″
22	9						14.9		23 18 47.75	−20.37	0.00	VI.	3	5.240	−12 41.33	−11.94	−2.09	23 18 27.38	−23 33 55.36
23	8					40.4	54.0		19 26.76	20.37	0.00	V.	2	13.11	11 37.80	11.89	1.99	19 6.39	32 51.66
24	9				19.3				20 19.15	20.38	0.00	IV.	4	6.338	18 15.67	11.83	2.48	19 58.77	23 39 29.98
25	9				23.				21 22.85	20.39	+0.01	IV.	10	10.238	49 10.54	11.75	4.87	21 2.47	24 10 27.16
26	9					51.1	4.6		21 37.42	20.40	0.00	V.	9	7.480	42 51.04	11.74	4.37	21 17.02	24 4 7.15
27	9						25.3	39.	21 58.04	20.39	0.00	V.	7	5.27	31 42.24	11.77	3.52	20 37.65	23 52 57.53
28	9						58.1	12.1	22 30.88	20.41	0.00	V.	2	5.450	7 52.94	11.66	1.73	22 10.47	29 6.33
29	9						39.		23 11.80	20.41	0.00	VI.	6	3.482	25 52.31	11.61	3.07	22 51.39	47 6.99
30	9					54.	7.		23 40.05	20.42	0.00	V.	4	1.461	15 50.56	11.59	2.32	22 19.63	37 4.47
31	9			5.	18.2				26 18.19	20.44	0.00	IV.	2	11.148	10 39.26	11.41	1.95	25 57.75	31 52.62
32	9					40.	53.1		26 26.11	20.44	0.00	V.	5	9.485	24 53.00	11.40	2.99	26 5.67	46 7.39
33	9						19.3		26 52.09	20.45	0.00	VI.	6	5.100	26 33.71	11.37	3.12	26 31.64	47 48.20
34	9				3.7	17.2			28 3.49	20.46	0.00	IV.	1	10.130	5 9.10	11.29	1.51	27 43.03	26 21.90
35	8.9		36.8	50.2	4.2				33 3.89	20.51	0.00	IV.	3	7.348	13 47.43	10.96	2.18	32 43.38	35 0.57
36	9						23.8	37.8	32 56.68	20.51	0.00	VI.	6	11.028	29 1.21	10.97	3.31	32 36.17	23 50 15.49
37	9						5.5	19.2	33 28.04	20.51	0.00	VI.	9	7.330	42 43.37	10.93	4.37	33 7.53	24 3 58.67
38	9				15.1				35 14.95	20.53	0.00	IV.	8	6.452	37 20.42	10.82	3.95	34 54.42	23 58 35.19
39	9				33.1	47.			35 33.16	20.53	0.00	IV.	7	10.500	34 55.41	10.80	3.73	35 12.63	23 56 9.94
40	9					14.2			36 0.59	20.53	0.00	V.	9	7.110	42 32.38	10.77	4.35	35 40.06	24 3 47.50
41	8						32.8	46.4	36 5.44	20.53	0.00	VII.	5	5.220	22 38.32	10.77	2.86	35 44.91	23 43 51.95
42	9				19.				37 18.85	20.55	0.00	IV.	9	4.480	41 20.32	10.69	4.26	36 58.30	24 2 35.27
43	9						47.5		37 20.25	20.55	0.00	VI.	8	10.450	39 21.19	10.69	4.13	36 59.70	24 0 36.01
44	9							21.8	37 40.47	20.55	0.00	VII.	1	12.470	6 26.41	10.67	1.61	37 19.92	23 27 38.69
45	9							40.	38 58.90	20.56	0.00	VII.	5	9.465	24 51.70	10.58	2.99	38 38.34	46 5.27
46	9				1.5		28.2		39 1.17	20.56	0.00	IV.	5	12.250	. . .	10.58	. .	38 40.61	
47	8.9		56.8	10.6	24.1				40 24.08	20.57	0.00	IV.	5	12.312	26 15.08	10.50	3.10	40 3.51	47 28.68
48	8.9					40.1			40 26.33	20.57	0.00	V.	1	10.330	5 19.15	10.50	1.52	40 5.76	26 31.17
49	9					19.8			41 6.05	20.58	0.00	V.	2	8.352	9 18.75	10.46	1.82	40 45.47	23 30 31.03
50	9							54.2	41 13.42	20.58	0.00	VII.	10	8.050	48 0.22	10.45	4.80	40 52.84	24 9 15.47
51	8						10.2	24.2	41 43.07	20.59	0.00	VI.	9	6.230	42 8.07	10.42	4.34	41 22.48	24 3 22.83
52	9						44.8	59.2	42 17.91	20.59	0.00	VI.	8	1.300	34 41.33	10.39	3.77	41 56.32	23 55 55.49
53	9							40.8	42 59.68	20.60	0.00	VII.	5	5.300	22 42.36	10.35	2.83	42 39.08	23 43 55.54
54	9							6.8	43 25.99	20.60	0.00	VII.	10	3.060	45 29.45	10.33	4.60	43 5.39	24 6 44.38
55	9		41.0						46 8.36	20.63	0.00	II.	6	3.578	25 57.15	10.18	3.08	45 47.73	23 47 10.41
56	9			32.0					46 45.55	20.64	0.00	III.	6	3.433	25 49.95	10.14	3.06	46 24.91	47 3.15
57	8.9			1.8	15.5				47 15.38	20.64	0.00	IV.	7	3.405	30 48.58	10.11	3.45	46 54.74	52 2.14
58	9			55.					48 8.60	20.65	0.00	III.	6	11.198	29 40.14	10.06	3.37	47 47.95	50 53.57
59	9				25.0	38.1			48 24.64	20.65	0.00	IV.	6	6.308	27 14.46	10.05	3.18	48 3.99	48 27.67
60	9				16.5				49 16.35	20.66	0.00	IV.	5	5.261	22 40.73	10.00	2.81	48 55.69	43 53.54
61	9					24.0			50 10.34	20.67	0.00	V.	6	8.550	28 27.12	9.95	3.27	49 49.67	49 40.34
62	9			21.1	34.6	48.1			51 34.53	20.68	0.00	IV.	6	11.383	29 49.51	9.88	3.38	51 13.85	51 2.77
63	8.9			35.	48.				52 48.36	20.70	0.00	IV.	1	10.03	4 33.81	9.83	1.49	52 27.66	25 45.13
64	8.9			6.8	20.				53 19.98	20.70	0.00	IV.	2	7.425	8 52.21	9.80	1.79	52 59.28	30 3.80
65	8.9				40.	53.5			53 39.81	20.70	0.00	IV.	3	10.339	15 17.69	9.78	2.26	53 19.11	36 29.73
66	9				43.2	57.			54 43.15	20.71	0.00	IV.	2	8.354	9 18.89	9.73	1.81	54 22.44	23 30 30.43
67	9			56.2	10.2	23.5			56 9.99	20.73	0.00	IV.	9	11.458	44 50.99	9.66	4.55	55 49.26	24 6 5.20
68	9				58.7	12.2			56 58.58	20.73	0.00	IV.	9	10.058	44 0.57	9.63	4.48	56 37.85	24 5 14.68
69	5.6		2.8	16.2	30.				59 29.99	20.76	0.00	IV.	7	12.148	35 7.91	9.50	3.77	59 9.23	23 56 21.18
70	9				48.2				23 59 48.05	−20.76	0.00	IV.	7	3.555	−30 56.14	−9.49	−3.31	23 59 27.29	−23 52 8.94

CORRECTIONS.

Date.		Corr. of Clock.	Hourly rate.	m	n	c
1847.	h.	s.	s.	s.	s.	s.

INSTRUMENT READINGS.

Date.		Barom.	Thermom. At.	Thermom. Ex.
1847.	h. m.	In.	°	°

REMARKS.

(141) 37. One of the transit observations (supposed T. VII) is 10ˢ in error.
(141) 39. Micrometer reading assumed as 11ʳ.500 instead of 10ʳ.500.
(141) 46. Declination 4′ discordant from Arg. Z. 270, 8, and 6′ from Mural, 1847, October 28.
(141) 63. Micrometer reading assumed as 9ʳ.03 instead of 10ʳ.03.

ZONE 141. OCTOBER 18. II. BELT, −23° 46′. D_0 = −23° 21′ 0″—Continued.

No.	Mag.	Seconds of Transit. I.	II.	III.	IV.	V.	VI.	VII.	T.	a_1	a_2	Micrometer.			i	d_1	d_2	Mean Right Ascension, 1850.0.	Mean Declination, 1850.0.
									h. m. s.	s.	s.			r.	′ ″	″	″	h. m. s.	° ′ ″
71	9	. .	. .	. .	. .	. .	19.2	33.7	23 59 52.36	−20.76	0.00	VI.	8	3.268	−35 40.22	− 9.49	−3.83	23 59 31.60	−23 56 43.54
72	9	. .	36.0	49.4	3.1	. .	. .	. .	0 3 3.00	20.79	0.00	IV.	3	10.078	15 4.58	9.35	2.25	0 2 42.21	36 16.18
73	8	. .	. .	23.9	37.0	51.	. .	. .	4 37.15	20.80	0.00	IV.	4	5.420	17 49.55	9.29	2.45	4 16.35	23 39 1.29
74	9	. .	. .	. .	. .	14.4	. .	. .	5 0.81	20.81	0.00	V.	10	5.520	46 53.45	9.27	4.71	4 40.00	24 8 7.43
75	9	. .	. .	38.2	52.2	6.1	. .	. .	5 52.66	20.81	0.00	IV.	5	9.485	24 53.04	9.24	2.99	5 31.25	23 46 5.27
76	8	. .	. .	. .	. .	41.2	54.1	. .	6 27.21	20.82	0.00	V.	9	5.050	41 28.84	9.22	4.28	6 6.39	24 2 42.34
77	9	. .	. .	. .	. .	. .	16.8	. .	6 49.54	20.82	0.00	VI.	9	4.268	41 9.48	9.20	4.26	6 28.72	24 2 22.94
78	8	. .	. .	46.8	59.	. .	. .	. .	7 59.57	20.83	0.00	IV.	5	4.275	22 11.18	9.15	2.78	7 38.74	23 43 23.11
79	9	. .	33.8	47.8	. .	. .	28.0	. .	9 0.99	20.84	0.00	III.	1	6.500	3 26.70	9.11	1.37	8 40.15	24 37.18
80	8	. .	. .	. .	. .	21.8	. .	19.0	9 7.84	20.84	0.00	VII.	1	8.010	4 2.19	9.11	1.41	8 47.00	25 12.71
81	9	. .	24.8	38.5	52.0	. .	. .	. .	11 52.00	20.86	0.00	IV.	5	6.528	23 24.45	9.01	2.88	11 31.14	44 36.34
82	9	. .	. .	. .	. .	. .	. .	21.2	11 40.29	20.86	0.00	VII.	8	9.122	38 34.20	9.02	4.04	11 19.43	59 47.26
83	6.7	. .	. .	24.0	38.1	. .	. .	. .	14 37.77	20.89	0.00	IV.	6	9.440	28 51.88	8.92	3.30	14 16.88	50 4.10
84	9	. .	. .	. .	. .	52.0	. .	. .	14 38.24	20.89	0.00	V.	2	4.230	7 11.57	8.92	1.65	14 17.35	28 22.14
85	9	. .	. .	. .	. .	. .	20.2	. .	14 53.00	20.89	0.00	VI.	5	10.08	25 2.73	8.91	3.00	14 32.11	46 14.64
86	9	. .	. .	. .	. .	. .	. .	52.1	15 11.08	20.89	0.00	VII.	6	11.589	29 59.55	8.90	3.39	14 50.19	51 11.84
87	9	. .	. .	. .	. .	. .	19.8	. .	15 52.58	20.90	0.00	VI.	7	7.160	32 37.10	8.88	3.59	15 31.68	53 49.57
88	9	. .	. .	. .	. .	. .	46.0	. .	16 18.79	20.90	0.00	VI.	6	11.534	29 56.97	8.87	3.39	15 57.89	23 51 9.23
89	9	. .	. .	26.5	40.	. .	. .	. .	18 40.09	20.92	0.00	IV.	10	3.223	45 38.00	8.79	4.61	18 19.17	24 6 51.40
90	9	. .	. .	. .	24.2	. .	. .	. .	19 24.05	20.93	0.00	IV.	10	2.513	45 22.37	8.77	4.59	19 3.12	6 35.73
91	8.9	. .	. .	. .	47.	. .	. .	. .	19 46.85	20.93	0.00	IV.	10	4.330	46 13.66	8.76	4.66	19 25.92	24 7 27.08
92	9	. .	. .	. .	. .	18.5	. .	. .	20 4.76	20.93	0.00	V.	3	5.188	12 38.80	8.75	2.07	19 43.83	23 33 49.62
93	9	. .	. .	. .	. .	54.2	. .	. .	20 40.58	20.94	0.00	V.	8	9.590	38 58.10	8.74	4.08	20 19.64	24 0 10.92
94	8	. .	. .	. .	. .	25.2	38.6	. .	21 11.46	20.94	0.00	V.	4	9.432	19 51.13	8.73	2.61	20 50.52	23 41 2.47
95	9	. .	. .	. .	. .	. .	53.	. .	22 25.75	20.95	0.00	VI.	8	10.410	39 19.17	8.70	4.10	22 4.80	24 0 31.97
96	9	. .	. .	26.5	. .	. .	. .	. .	23 39.82	20.95	0.00	III.	2	10.512	10 27.32	8.68	1.90	23 18.87	23 31 37.90
97	9	. .	. .	. .	11.2	. .	. .	. .	24 11.05	20.97	0.00	IV.	8	12.128	40 5.59	8.67	4.16	23 50.08	24 1 18.42
98	9	. .	. .	. .	. .	56.2	. .	. .	24 42.52	20.97	0.00	V.	5	10.478	25 22.91	8.66	3.03	24 21.55	23 46 34.60
99	6.7	. .	32.5	46.1	0.1	. .	. .	. .	28 59.76	21.01	0.00	IV.	4	7.350	18 46.53	8.57	2.52	28 38.75	39 57.62
100	10	. .	. .	. .	. .	25.	38.3	. .	29 11.51	21.01	0.00	V.	4	9.570	19 58.09	8.57	2.61	28 50.50	23 41 9.27
101	9	. .	. .	. .	. .	19.2	. .	. .	30 5.60	21.02	0.00	V.	9	11.230	44 39.44	8.55	4.54	29 44.58	24 5 52.53
102	9	. .	. .	. .	. .	51.5	4.5	. .	30 37.56	21.03	0.00	V.	6	5.322	26 44.87	8.54	3.14	30 16.53	23 47 56.55
103	9	. .	. .	. .	. .	. .	. .	36.5	30 55.60	21.03	0.00	VII.	8	10.08	39 2.34	8.54	4.09	30 34.57	24 0 14.97
104	9	. .	. .	. .	. .	. .	. .	49.5	31 8.63	21.03	0.10	VII.	9	4.565	41 24.27	8.54	4.28	30 47.60	24 2 37.09
105	9	. .	31.8	. .	. .	. .	. .	. .	32 59.18	21.05	0.00	II.	6	8.480	28 23.49	8.51	3.26	32 38.13	23 49 35.26
106	9	. .	. .	58.2	12.0	. .	. .	. .	34 58.21	21.05	0.00	IV.	8	6.161	37 5.75	8.50	3.96	33 37.16	58 18.21
107	9	. .	. .	53.1	7.1	. .	. .	. .	35 6.81	21.06	0.00	IV.	6	7.165	27 37.50	8.48	3.21	34 45.75	23 48 49.19
108	9	. .	. .	. .	. .	31.0	44.5	. .	35 17.22	21.06	0.00	V.	9	12.195	45 7.94	8.48	4.57	34 56.26	24 6 20.99
109	9	. .	. .	. .	. .	. .	17.0	. .	35 49.86	21.07	0.00	VI.	2	10.550	10 29.13	8.47	1.91	35 28.79	23 31 39.51
110	8.9	. .	. .	. .	. .	37.3	51.	. .	36 23.55	21.07	0.00	V.	2	12.246	6 15.41	8.47	1.55	36 2.48	27 25.43
111	9	. .	. .	. .	. .	25.2	38.6	. .	37 11.46	21.08	0.00	V.	7	6.020	31 59.90	8.46	3.53	36 50.38	53 11.89
112	9	. .	. .	. .	53.3	7.1	. .	. .	37 53.27	21.08	0.00	IV.	4	9.060	19 33.42	8.45	2.59	37 32.19	40 44.46
113	9	. .	29.2	43.	56.4	. .	. .	. .	39 56.35	21.10	0.00	IV.	3	6.420	13 20.81	8.42	2.12	39 35.25	34 31.35
114	9	. .	. .	. .	23.2	. .	. .	. .	40 23.05	21.10	0.00	IV.	7	11.280	34 44.31	8.42	3.75	40 1.95	55 56.48
115	9	. .	18.6	32.6	. .	. .	. .	. .	41 46.12	21.11	0.00	III.	7	5.490	31 53.34	8.41	3.54	41 25.01	23 53 5.29
116	8	. .	51.	5.	. .	44.	. .	. .	42 18.62	21.12	0.00	III.	9	3.528	40 52.44	8.41	4.23	41 57.50	24 2 5.08
117	9	. .	. .	. .	. .	33.	46.2	. .	42 19.17	21.12	0.00	V.	10	7.180	47 36.81	8.41	4.76	41 58.05	8 49.98
118	9	. .	. .	. .	. .	. .	. .	12.0	42 31.23	21.12	0.00	VII.	10	11.118	49 34.40	8.41	4.93	42 10.11	24 10 47.74
119	9	. .	. .	. .	6.3	. .	. .	. .	0 44 6.15	−21.13	0.00	IV.	3	10.292	−15 15.37	− 8.40	−2.27	0 43 45.02	−23 36 26.04

CORRECTIONS.

Date.		Corr. of Clock.	Hourly rate.	m	n	c
1847.	h.	s.	s.	s.	s.	s.

INSTRUMENT READINGS.

Date.		Barom.	Thermom. At.	Thermom. Ex.
1847.	h. m.	in.	°	°

REMARKS.

(141) 106. Transits over T.'s III and IV as recorded over T.'s IV and V, and minutes assumed as 33 instead of 34.
(141) 110. Hor. thread assumed as 1 instead of 2.

ZONE 141. OCTOBER 18. H. BELT, −23° 46′. D₀ = −23° 21′ 0″—Continued.

No.	Mag.	Seconds of Transit I.	II.	III.	IV.	V.	VI.	VII.	T.	a_1	a_2	Micrometer.			i	d_1	d_2	Mean Right Ascension, 1850.0.	Mean Declination, 1850.0.
									h. m. s.	s.	s.			r.	′ ″	″	″	h. m. s.	° ′ ″
120	8.9					39.1	52.6		0 44 25.41	−21.14	0.00	V.	5	3.28	−21 41.13	−8.40	−2.75	0 44 4.27	−23 42 52.28
121	6.7					26.0			45 12.22	21.14	0.00	V.	1	8.315	4 17.87	8.40	1.42	44 51.06	25 27.69
122	9			4.0	18.0				46 17.68	21.15	0.00	IV.	5	8.35	24 15.99	8.39	2.94	45 56.53	45 27.32
123	9			59.	12.6				47 12.46	21.16	0.00	IV.	4	11.264	20 43.21	8.39	2.67	46 51.30	41 54.27
124	9				58.0				47 57.85	21.16	0.00	IV.	5	12.038	26 1.26	8.39	3.08	47 36.69	47 12.73
125	9					17.2			48 3.53	21.16	0.00	V.	6	5.308	26 44.16	8.39	3.14	47 42.37	47 55.69
126	8.9			47.6	1.0	14.6			51 0.83	21.19	0.00	IV.	1	6.188	3 11.00	8.39	1.34	50 39.64	24 20.73
127	9			17.3	31.0				54 30.78	21.21	0.00	IV.	3	11.095	15 35.69	8.40	2.29	54 9.57	23 36 46.38
128	8.9					5.2			54 51.62	21.21	0.00	V.	10	12.493	50 23.87	8.40	5.01	54 30.41	24 11 37.28
129	9					54.			55 40.33	21.22	0.00	V.	5	13.252	26 42.26	8.41	3.14	55 19.11	23 47 53.81
130	8.9				16.	29.2	42.2		56 15.46	21.22	0.00	IV.	8	7.522	37 54.20	8.41	4.03	55 54.24	59 6.64
131	8		25.8	39.6	53.2	7.			0 59 53.13	21.25	0.00	IV.	4	6.130	18 5.18	8.43	2.48	59 31.68	23 39 16.09
132	9					18.	31.1		1 0 4.12	21.25	0.00	V.	9	5.282	41 40.54	8.43	4.31	0 59 42.87	24 2 53.28
133	10		49.						2 16.30	21.27	0.00	II.	4	9.395	19 49.16	8.46	2.60	1 1 55.03	23 41 0.22
134	9			15.3	29.	42.6			3 28.79	21.28	0.00	IV.	3	6.333	13 16.42	8.47	2.11	3 7.51	34 27.00
135	8.9			21.2	35.				4 34.69	21.29	0.00	IV.	2	10.410	10 22.22	8.49	1.89	4 13.40	31 32.60
136	9			5.2	19.				5 18.67	21.29	0.00	IV.	1	13.07	6 36.83	8.50	1.59	4 57.38	27 46.92
137	9			3.2					7 16.49	21.31	0.00	III.	2	5.406	7 50.70	8.52	1.70	6 55.18	29 0.92
138	9				29.2				7 29.05	21.31	0.00	IV.	6	4.43	26 20.10	8.53	3.10	7 7.74	47 31.73
139	8.9			1.0	14.6				8 14.46	21.32	0.00	IV.	4	10.448	20 22.24	8.54	2.64	7 53.14	41 33.42
140	9				28.2	42.			8 28.17	21.32	0.00	IV.	5	3.273	21 40.82	8.54	2.75	8 6.85	42 52.11
141	9					5.			8 51.29	21.32	0.00	V.	4	8.410	19 19.77	8.55	2.55	8 29.97	23 40 30.87
142	9			0.	14.				10 13.80	21.33	0.00	IV.	8	13.035	40 31.16	8.57	4.21	9 52.47	24 1 43.94
143	8		6.	20.1	34.				11 33.63	21.34	0.00	IV.	6	5.510	26 54.38	8.59	3.15	11 12.29	23 48 6.12
144	9						57.8		11 30.56	21.34	0.00	VI.	8	3.152	35 34.37	8.59	3.83	11 9.22	56 46.79
145	9						15.	29.1	11 47.86	21.34	0.00	VI.	4	1.060	15 30.23	8.60	2.29	11 26.52	36 41.12
146	9						58.0	12.1	12 30.95	21.35	0.00	VI.	7	7.152	32 36.69	8.62	4.38	12 9.60	53 49.69
147	9							57.0	13 15.91	21.35	0.00	VII.	6	10.395	29 39.53	8.63	3.36	12 54.56	23 50 51.52
148	9				56.1				15 55.95	21.37	0.00	IV.	9	4.250	41 8.72	8.69	4.25	15 34.58	24 2 21.66
149	9				23.2	37.0			16 23.23	21.37	−0.01	IV.	10	1.582	44 55.59	8.70	4.57	16 1.85	24 6 8.86
150	9				17.0	30.3			17 16.70	21.38	0.00	IV.	3	3.552	11 56.69	8.72	2.02	16 55.32	23 33 7.43
151	7.8	34.1	48.1	1.8	15.1				18 15.17	21.39	0.00	IV.	3	7.030	13 31.39	8.74	2.13	18 53.78	34 42.26
152	9			41.3					18 54.93	21.39	0.00	III.	7	6.270	32 12.50	8.76	3.57	18 33.54	23 53 24.83
153	9				14.4				19 14.25	21.39	0.00	IV.	9	9.470	43 51.09	8.77	4.48	18 52.86	24 5 4.34
154	9		13.1	27.4	41.				21 40.63	21.41	0.00	IV.	2	12.422	11 23.33	8.83	1.96	21 19.22	23 32 34.12
155	9				8.0				22 7.85	21.41	0.00	IV.	10	9.542	48 55.62	8.84	4.89	21 46.44	24 10 9.35
156	9					35.3	48.2		22 21.31	21.42	0.00	V.	6	4.38	26 17.54	8.85	3.10	21 59.89	23 47 29.49
157	8.9		17.2	31.	45.				23 44.75	21.43	0.00	IV.	8	11.340	39 46.05	8.89	4.16	23 23.32	24 0 59.10
158	9		41.	54.2					25 54.36	21.44	0.00	III.	8	10.130	39 5.16	8.93	4.10	24 32.92	24 0 18.19
159	9						24.0		24 56.76	21.43	0.00	VI.	8	7.560	37 55.97	8.93	4.02	24 35.33	23 59 8.92
160	9	39.0	53.	6.8					1 26 20.15	−21.44	0.00	III.	2	12.410	−11 32.76	−8.97	−1.96	1 25 58.71	−23 32 43.69

CORRECTIONS.

Date.		Corr. of Clock.	Hourly rate.	m	n	c
1847.	h.	s.	s.	s.	s.	s.

INSTRUMENT READINGS.

Date.		Barom.	Thermom. At.	Thermom. Ex.
1847.	h. m.	In.	°	°

REMARKS.

(141) 151. Minutes assumed as 19 instead of 18.
(141) 153. Minutes of transit assumed as 20 instead of 19.
(141) 158. Transits over T.'s III and IV assumed as recorded over T.'s II and III, and minutes assumed as 24 instead of 25.
(141) 160. Minutes assumed as 27 instead of 26, and Micrometer reading assumed as 12r.61 instead of 12r.41.

ZONE 142. OCTOBER 18. H. BELT, −23° 46′. D_0 = −23° 21′ 0″.

No.	Mag.	Seconds of Transit. I.	II.	III.	IV.	V.	VI.	VII.	T.	a_1	a_2	Micrometer.			i	d_1	d_2	Mean Right Ascension, 1850.0.	Mean Declination, 1850.0.
									h. m. s.	s.	s.			r.	′ ″	″	″	h. m. s.	° ′ ″
1	9					26.8	40.2		2 28 13.06	−21.80	0.00	V.	6	5.37	−26 47.29	−18.30	−3.14	2 27 51.26	−23 48 8.73
2	9				35.2				29 35.05	21.81	−0.01	IV.	10	10.528	49 25.17	18.55	4.96	29 13.23	24 10 48.68
3	9		36.2	50.3					31 3.55	21.81	+0.01	III.	3	6.466	13 23.09	18.79	2.11	30 41.75	23 34 43.99
4	8			5.9	19.0	42.9			31 19.09	21.81	0.00	IV.	4	4.572	17 26.96	18.83	2.43	30 57.28	38 48.22
5	9			4.2	18.				34 17.67	21.83	+0.01	IV.	2	3.42	6 50.94	19.31	1.60	33 55.85	28 11.85
6	9					37.2	50.8		34 23.56	21.83	0.01	V.	3	9.24	14 42.45	19.33	2.21	34 1.74	36 3.99
7	9				48.2				38 4.05	21.84	+0.01	IV.	1	11.378	5 51.86	19.93	1.50	37 42.22	27 13.29
8	10				17.2		43.5		40 17.05	21.85	0.00	IV.	5	4.123	22 3.52	20.29	2.77	39 55.20	43 26.58
9	9			53.1	6.7	20.2			42 6.54	21.86	0.00	IV.	5	3.442	21 49.35	20.59	2.70	41 44.68	23 43 12.70
10	9		33.	46.8					44 0.63	21.87	−0.01	III.	10	13.598	50 59.41	20.92	5.08	43 38.75	24 12 25.41
11	9					41.8	55.0		44 27.96	21.87	0.00	V.	4	5.082	17 32.46	21.00	2.43	44 6.09	23 38 55.89
12	9			15.2	29.2		55.8		45 28.80	21.87	0.00	IV.	6	4.340	26 15.57	21.17	3.10	45 6.93	23 47 39.84
13	9				16.2				46 16.05	21.87	−0.01	IV.	9	8.215	43 7.97	21.30	4.44	45 54.17	24 4 33.71
14	9					40.2			46 26.61	21.88	−0.01	V.	10	4.060	46 0.01	21.33	4.67	46 4.72	24 7 26.01
15	8			10.8	24.1	38.2			47 24.26	21.88	0.00	IV.	5	8.202	24 8.52	21.49	2.93	47 2.38	23 45 32.94
16	8					7.5	21.4		47 54.01	21.88	0.00	V.	4	4.489	17 22.73	21.58	2.42	47 32.13	38 46.73
17	9							42.5	48 1.51	21.88	0.00	VII.	7	6.190	32 8.17	21.59	3.56	47 39.63	53 33.32
18	9						8.	22.	48 40.91	21.88	−0.01	VI.	7	13.29	35 45.17	21.70	3.85	48 19.02	57 10.72
19	8						3.3	17.2	49 36.12	21.89	0.00	VI.	6	8.362	28 17.55	21.86	3.25	49 14.23	49 42.66
20	8.9					45.2	58.6		50 31.45	21.89	+0.01	V.	3	8.306	14 15.53	22.02	2.15	50 9.57	35 39.70
21	9							25.0	50 44.08	21.89	−0.01	VII.	8	7.02	37 28.55	22.05	3.99	50 22.18	58 54.59
22	9						50.8	4.5	51 23.55	21.90	0.00	VI.	7	5.318	31 44.55	22.17	3.53	51 1.65	53 10.25
23	9		57.5	11.5	25.2				52 25.07	21.90	−0.01	IV.	8	6.488	. . .	22.35	4.	53 3.16	
24	9		0.2	14.	27.5				54 27.57	21.91	−0.01	IV.	8	5.488	36 51.97	22.70	3.94	54 5.65	58 18.61
25	9					54.2	7.8		54 40.56	21.91	+0.01	V.	2	11.382	10 51.02	22.73	1.89	54 18.66	32 15.64
26	9				48.6				55 48.45	21.91	+0.01	IV.	2	8.482	9 25.34	22.93	1.78	55 26.55	23 30 50.05
27	9				37.2	50.8			56 37.13	21.92	−0.01	IV.	10	5.390	46 46.94	23.08	4.75	56 15.20	24 8 14.77
28	9			42.3	56.1	9.4			2 57 55.82	21.92	0.00	IV.	5	5.025	22 28.83	23.31	2.81	57 33.90	23 43 54.95
29	9			3.3	17.				3 0 16.90	21.93	0.00	IV.	7	9.11	33 35.23	23.73	3.67	2 59 54.97	23 55 2.67
30	9		55.	8.3					1 22.30	21.94	−0.01	III.	9	9.163	41 4.30	23.92	4.28	3 1 0.35	24 3 32.50
31	9					52.2			1 38.51	21.94	0.00	V.	5	8.370	24 16.96	23.97	2.94	1 16.57	23 45 43.87
32	9							36.5	1 55.49	21.94	0.00	VII.	6	13.508	30 55.97	24.03	3.47	1 33.55	52 23.47
33	8.9	32.6	46.6	0.3	13.2				6 13.50	21.95	+0.01	IV.	3	3.50	12 54.58	24.80	2.07	5 51.56	23 34 21.45
34	9				55.5				6 55.35	21.96	−0.01	IV.	10	6.404	47 17.90	24.93	4.77	6 33.38	24 8 47.60
35	9					16.1	29.5		7 2.37	21.96	−0.01	V.	7	7.380	32 48.30	24.95	3.61	6 40.40	23 54 16.86
36	9							56.	7 14.71	21.96	+0.01	VII.	2	8.045	14 2.07	24.98	2.15	6 52.76	35 29.20
37	9						9.		8 41.82	21.96	0.00	VI.	4	8.060	19 2.02	25.25	2.53	8 19.86	40 29.80
38	9						49.2		9 21.99	21.97	0.00	VI.	6	10.045	29 2.07	25.37	3.31	9 0.02	50 30.75
39	9		28.		56.1				10 55.60	21.97	+0.01	IV.	3	5.271	12 43.08	25.67	2.04	10 33.64	23 34 10.79
40	7.8					18.2	31.6		11 4.47	21.97	−0.01	V.	9	7.582	42 56.18	25.70	4.44	10 42.49	24 4 26.32
41	9							3.	11 21.71	21.97	+0.01	VII.	2	8.503	9 26.06	25.75	1.78	10 59.75	23 30 53.59
42	9		46.1		13.1				13 13.13	21.98	+0.01	IV.	2	11.261	10 44.96	26.11	1.88	12 51.16	32 12.95
43	9		1.8	15.5	29.3				14 29.15	21.98	0.00	IV.	6	11.372	29 48.96	26.35	3.37	14 7.17	51 18.68
44	9				51.				14 50.85	21.98	+0.01	IV.	3	11.40	15 51.07	26.43	2.29	14 28.88	23 37 19.79
45	6				12.5				15 11.85	21.98	−0.01	IV.	10	10.150	49 6.11	26.49	4.94	14 49.86	24 10 37.54
46	9		3.0	16.6					16 30.16	21.98	+0.01	III.	2	6.365	8 18.90	26.73	1.68	16 8.19	23 29 47.31
47	9					41.2			16 27.60	21.98	−0.01	V.	9	8.050	42 59.62	26.72	4.45	16 5.61	24 4 30.79
48	9					10.5			16 56.88	21.98	0.01	V.	8	8.162	38 6.27	26.83	4.05	16 34.89	23 59 37.15
49	9					28.2			3 17 14.58	−21.98	−0.01	V.	8	9.532	−38 55.17	−26.87	−4.12	3 16 52.59	−24 0 26.16

CORRECTIONS.

Date.	Corr. of Clock.	Hourly rate.	m	n	c
1847. h.	s.	s.	s.	s.	s.

INSTRUMENT READINGS.

Date.	Barom.	Thermom. At.	Thermom. Ex.
1847. h. m.	in.	°	°

REMARKS.

(142) 4. Time of transit over T. V assumed as 32ˢ.9 instead of 42ˢ.9.
(142) 23. Minutes assumed as 54 instead of 52.
(142) 30. Micrometer reading assumed as 4ʳ.163 instead of 9ʳ.163.
(142) 33. Micrometer reading assumed as 5ʳ.50 instead of 3ʳ.50.
(142) 36. Hor. thread assumed as 3 instead of 2.

ZONE 142. OCTOBER 18. H. BELT, −23° 46′. D_o=−23° 21′ 0″—Continued.

No.	Mag.	SECONDS OF TRANSIT. I.	II.	III.	IV.	V.	VI.	VII.	T.	a_1	a_2	MICROMETER.			i	d_1	d_2	Mean Right Ascension, 1850.0.	Mean Declination, 1850.0.
									h. m. s.	s.	s.			r.	′ ″	″	″	h. m. s.	° ′ ″
50	9						52.1		3 17 24.84	−21.98	−0.01	VI.	8	4.220	−36 8.06	−26.90	−3.89	3 17 2.85	−23 57 38.83
51	9			57.0					19 10.72	21.99	−0.01	III.	8	9.310	38 43.98	27.23	4.10	18 50.72	24 0 15.31
52	9		44.2				25.2		20 11.39	21.99	+0.02	IV.	1	8.322	4 18.28	27.43	1.37	19 49.42	23 25 47.08
53	6.7		56.4		24.4		51.2		23 24.02	22.00	−0.01	IV.	8	8.310	38 13.77	28.04	4.07	23 2.01	59 45.88
54	9		33.	46.6					25 0.18	22.01	0.00	VI.	4	10.360	20 17.66	28.39	2.63	24 38.17	41 48.68
55	9				10.5				25 10.35	22.01	+0.01	IV.	2	7.025	8 32.04	28.42	1.70	24 48.35	23 30 2.16
56	9				27.0				26 26.85	22.01	−0.02	IV.	10	5.518	46 53.38	28.65	4.76	26 4.82	24 8 26.79
57	9		0.0	13.8					27 27.12	22.01	+0.01	III.	1	11.460	5 55.96	28.85	1.51	27 5.12	23 27 26.32
58	9		14.0	27.5					28 41.14	22.02	0.00	III.	4	11.038	20 31.77	29.07	2.65	28 19.12	42 3.49
59	9			39.5	53.2	6.3			29 52.85	22.02	+0.01	IV.	4	5.27	17 41.98	29.32	2.43	29 30.84	23 39 13.73
60	9			38.2	52.1				29 51.94	22.02	−0.01	IV.	8	10.530	39 25.37	29.32	4.13	29 29.91	24 0 58.82
61	9				19.5				30 19.35	22.02	−0.01	IV.	7	12.128	35 6.90	29.41	3.80	29 57.32	23 56 40.11
62	9					37.1	50.8		30 23.52	22.02	0.00	V.	6	13.033	30 32.32	29.42	3.43	30 1.50	52 5.17
63	9		52.1	5.8					32 19.25	22.02	+0.01	III.	3	6.313	13 15.37	29.81	2.07	31 57.24	23 34 47.25
64	9			33.0					32 46.77	22.03	−0.01	III.	9	5.445	41 48.77	29.89	4.36	32 24.73	24 3 23.02
65	9					38.0			33 24.38	22.03	−0.01	V.	8	9.100	38 33.39	30.02	4.09	33 2.34	24 0 7.50
66	9						9.6		33 42.38	22.03	0.00	VI.	7	6.53	32 25.50	30.08	3.59	33 20.35	23 53 59.17
67	9						46.		34 18.88	22.03	+0.02	VI.	1	5.030	2 32.63	30.20	1.23	33 56.87	24 4.06
68	9			34.1					35 47.54	22.03	+0.01	III.	4	7.155	13 37.65	30.50	2.13	35 25.52	35 10.28
69	9				58.1				35 57.95	22.03	0.00	IV.	5	4.262	22 10.53	30.54	2.76	35 35.92	23 43 43.83
70	9						20.8		35 53.52	22.03	−0.02	VI.	10	4.216	46 7.75	30.53	4.71	35 31.47	24 7 42.99
71	9							57.5	36 16.18	22.03	+0.01	VII.	2	4.206	7 10.06	30.59	1.60	35 54.16	23 28 42.25
72	6.7		18.2	32.1	45.8				38 45.65	22.04	0.00	IV.	6	11.132	29 36.85	31.09	3.35	38 23.61	51 11.29
73	9				32.				39 3.05	22.04	+0.02	IV.	1	6.40	3 21.70	31.15	1.29	38 41.03	24 54.14
74	9		35.8	49.2					40 2.79	22.04	+0.01	III.	3	3.13	11 35.37	31.35	1.94	39 40.76	33 8.66
75	3.4		18.5	32.	46.				40 45.70	22.04	0.00	IV.	4	10.148	20 7.11	31.50	2.62	40 23.66	41 41.23
76	9			21.8					41 35.33	22.04	0.00	III.	5	10.404	24 48.92	31.68	2.99	41 13.29	46 23.59
77	9			42.2	56.				41 55.74	22.04	+0.01	IV.	4	6.230	18 10.22	31.76	2.46	41 33.71	23 39 44.44
78	9				22.				42 21.85	22.04	−0.02	IV.	10	8.06	48 1.07	31.83	4.87	41 59.79	24 9 37.77
79	9					36.	50.		42 22.56	22.04	−0.01	V.	9	10.355	44 15.51	31.83	4.55	42 0.51	24 5 51.89
80	9					21.2			43 7.46	22.05	+0.01	V.	2	13.10	11 37.30	32.00	1.94	42 45.42	23 33 11.24
81	9					8.3			43 54.57	22.05	+0.01	V.	3	8.192	14 9.78	32.17	2.13	43 32.53	35 44.08
82	9					8.0			44 54.37	22.05	−0.01	V.	7	13.250	35 43.26	32.38	3.84	44 32.31	57 19.48
83	7.8		22.4	36.2					47 49.60	22.06	+0.01	III.	3	5.24	12 41.43	32.97	2.03	47 27.55	34 16.43
84	9							34.5	47 53.51	22.06	0.00	VII.	7	5.152	31 35.99	32.98	3.52	47 31.45	53 12.49
85	9	10.5							49 51.86	22.06	−0.01	I.	8	7.292	37 42.26	33.40	4.03	49 29.79	59 19.69
86	9			38.4					49 51.68	22.06	+0.01	III.	2	4.44	7 22.17	33.40	1.61	49 29.63	28 57.19
87	9				7.	20.6			50 6.90	22.06	0.00	IV.	7	2.33	30 14.55	33.45	3.41	49 44.84	51 51.41
88	9			33.2	47.				50 46.73	22.06	+0.01	IV.	3	11.230	15 42.49	33.59	2.29	50 24.68	37 18.37
89	9				12.8				51 12.65	22.06	−0.01	IV.	7	13.286	35 45.11	33.68	3.85	50 50.58	57 22.64
90	9		18.5						52 45.69	22.06	+0.02	II.	2	5.010	7 30.62	34.01	1.62	52 23.65	29 6.25
91	9					6.2			52 52.46	22.06	0.01	V.	3	5.462	12 52.63	34.03	2.05	52 30.41	34 28.71
92	9		36.2						54 49.59	22.06	0.01	II.	3	9.185	14 39.56	34.23	2.19	53 27.54	36 15.99
93	9		0.						54 13.38	22.06	0.01	II.	3	8.315	14 15.87	34.33	2.15	53 51.33	35 52.35
94	9				19.6				54 19.45	22.06	+0.01	IV.	3	7.398	13 49.95	34.35	2.12	53 57.40	35 26.42
95	9			45.3	59.2				54 58.91	22.06	0.00	IV.	4	11.402	20 50.17	34.48	2.68	54 36.85	23 42 27.33
96	9				3.3				57 3.15	22.07	−0.02	IV.	9	13.243	45 40.65	34.91	4.68	56 41.06	24 7 20.24
97	9			59.0					58 12.42	22.07	+0.01	III.	4	4.355	17 15.98	35.17	2.39	57 50.36	23 38 53.54
98	9	12.							3 58 52.48	−22.07	0.00	I.	7	1.522	−29 53.63	−35.50	−3.38	3 59 30.41	−23 51 32.51

CORRECTIONS.

Date.	Corr. of Clock.	Hourly rate.	m	n	c
1847. h.	s.	s.	s.	s.	s.

INSTRUMENT READINGS.

Date.	Barom.	THERMOM. At.	Ex.
1847. h. m.	in.	°	°

REMARKS.

(142) 53. Transit observation on T. VI assumed as 51ˢ.2 instead of 57ˢ.2.
(142) 68. Hor. thread assumed as 3 instead of 4.
(142) 92. Transit over T. III assumed as recorded over T. II, and minutes assumed as 53 instead of 54.
(142) 93. Transit over T. III assumed as recorded over T. II.
(142) 98. Transit over T. I assumed as 11ˢ.2 instead of 1ˢ.2, and minutes assumed as 59 instead of 58.

ZONE 142. OCTOBER 18. H. BELT, −23° 46′. D₀=−23° 21′ 0″—Continued.

No.	Mag.	I.	II.	III.	IV.	V.	VI.	VII.	T.	a_1	a_2	MICROMETER.			i	d_1	d_2	Mean Right Ascension, 1850.0.	Mean Declination, 1850.0.
									h. m. s.	s.	s.			r.	′ ″	″	″	h. m. s.	° ′ ″
99	9	. .	41.0	. .	. .	. .	. .	. .	3 59 8.60	−22.07	0.00	II.	4	9.218	−19 40.22	−35.56	−2.59	3 59 46.53	−23 41 18.37
100	8.9	. .	. .	. .	10.	23.5	. .	. .	3 59 9.85	22.07	0.00	IV.	7	5.261	31 41.83	35.56	3.53	59 47.78	53 20.92
101	9	. .	. .	. .	. .	. .	. .	50.6	4 0 9.60	22.07	+0.01	VII.	4	4.265	17 11.14	35.58	2.39	3 59 47.54	38 49.11
102	9	. .	. .	. .	. .	32.1	. .	. .	1 23.36	22.07	−0.01	V.	7	8.110	33 4.94	35.86	3.63	4 1 1.28	54 44.43
103	9	. .	. .	. .	44.1	. .	. .	. .	1 43.95	22.07	+0.02	IV.	2	7.450	8 53.47	35.93	1.72	1 21.90	30 31.12
104	9	. .	. .	. .	. .	. .	30.2	. .	2 2.97	22.07	−0.01	VI.	7	12.360	32 18.46	35.99	3.57	1 40.89	53 58.02
105	9	. .	55.2	. .	. .	. .	. .	. .	3 22.63	22.07	−0.01	II.	7	9.585	33 59.03	36.27	3.71	3 0.55	55 39.01
106	9	. .	. .	54.5	. .	. .	. .	. .	4 8.08	22.07	0.00	III.	6	8.340	28 16.54	36.44	3.25	3 46.01	49 56.23
107	9	. .	29.	43.	. .	. .	. .	. .	4 56.44	22.07	0.00	III.	5	10.130	25 5.36	36.64	3.01	4 34.37	46 45.01
108	9	. .	. .	. .	. .	59.	. .	. .	5 45.27	22.08	+0.01	V.	3	9.152	14 38.01	36.80	2.18	5 23.20	36 16.99
109	8	. .	34.2	48.	2.	. .	. .	. .	6 1.52	22.08	+0.02	IV.	2	8.205	9 11.37	36.85	1.73	5 39.46	30 49.95
110	9	. .	. .	. .	. .	. .	. .	23.6	6 42.47	22.08	0.00	VII.	5	4.072	22 0.61	36.99	2.77	6 20.39	43 40.37
111	9	. .	. .	. .	. .	. .	39.2	. .	7 12.08	22.08	+0.02	VI.	1	9.262	4 45.35	37.11	1.39	6 50.02	23 26 23.85
112	9	. .	. .	. .	. .	. .	. .	34.	7 53.16	22.08	−0.02	VII.	9	8.425	43 18.23	37.26	4.48	7 31.06	24 4 59.97
113	9	. .	. .	. .	. .	. .	. .	55.2	8 14.36	22.08	−0.02	VII.	9	8.082	43 0.93	37.32	4.46	7 52.26	24 4 42.71
114	9	. .	. .	23.	. .	. .	. .	. .	9 22.85	22.08	0.00	III.	5	6.082	23 1.92	37.57	2.84	9 10.77	23 44 42.33
115	9	. .	. .	. .	. .	55.5	. .	. .	9 41.83	22.08	0.00	V.	6	4.161	26 6.50	37.65	3.09	9 19.75	47 47.24
116	7.8	. .	. .	. .	. .	. .	11.4	25.2	9 44.11	22.08	+0.01	VI.	3	10.430	15 22.18	37.66	2.26	9 22.04	37 2.10
117	9	. .	. .	. .	. .	. .	56.	. .	10 28.78	22.08	−0.01	VI.	7	7.510	32 54.74	37.83	3.63	10 6.69	23 54 36.20
118	9	. .	. .	. .	42.	. .	. .	. .	11 41.85	22.08	−0.02	IV.	10	9.088	48 32.72	38.08	4.92	11 19.75	24 10 15.72
119	9	. .	. .	. .	. .	3.5	. .	. .	11 49.81	22.08	0.00	V.	5	8.128	24 4.75	38.11	2.92	11 27.73	23 45 45.78
120	9	. .	. .	. .	14.5	28.	. .	55.2	13 14.25	22.08	0.00	IV.	5	10.288	25 13.36	38.42	3.02	12 52.17	23 46 54.80
121	9	. .	. .	52.2	. .	. .	. .	. .	15 6.00	22.09	−0.02	III.	9	10.288	44 12.12	38.82	4.57	14 43.89	24 25 55.51
122	9	. .	. .	19.2	. .	. .	. .	. .	16 33.01	22.09	−0.02	III.	9	11.570	44 56.60	39.15	4.63	16 10.90	24 6 40.38
123	9	. .	. .	52.6	6.0	. .	. .	. .	17 5.88	22.09	+0.02	IV.	2	9.480	9 55.49	39.28	1.79	16 43.81	23 31 36.56
124	9	. .	28.8	42.2	56.4	. .	. .	. .	17 56.18	22.09	−0.02	IV.	9	9.122	43 33.54	39.47	4.51	17 34.07	24 5 17.52
125	9	. .	. .	. .	30.	. .	. .	. .	18 29.85	22.09	+0.02	IV.	1	10.390	5 22.22	39.60	1.44	18 7.78	23 27 3.26
126	9	. .	. .	53.0	. .	. .	. .	. .	20 6.27	22.09	0.02	III.	1	11.40	5 52.93	39.95	1.48	19 44.20	27 34.36
127	8.9	. .	. .	. .	31.8	. .	. .	. .	20 31.65	22.09	+0.02	IV.*	2	3.555	6 57.74	40.04	1.57	20 9.58	28 39.35
128	9	. .	. .	4.2	18.1	. .	. .	. .	21 17.89	22.09	−0.01	IV.	7	5.255	31 41.52	40.22	3.53	20 55.79	53 25.27
129	9	. .	. .	. .	. .	46.4	. .	. .	21 32.65	22.09	+0.01	V.	3	2.280	11 12.68	40.27	1.90	21 10.57	32 54.85
130	9	. .	. .	. .	15.2	. .	. .	. .	22 15.05	22.09	+0.02	IV.	1	9.112	4 37.93	40.43	1.44	21 52.98	26 19.80
131	9	. .	. .	. .	. .	. .	50.3	4.3	22 23.16	22.09	0.00	VI.	5	10.02	24 59.71	40.47	3.00	22 1.07	46 43.18
132	9	. .	. .	. .	42.5	. .	. .	. .	23 42.35	22.09	0.00	IV.	6	8.565	28 27.93	40.74	3.27	23 20.26	50 11.94
133	9	. .	. .	. .	8.2	. .	. .	. .	24 8.05	22.09	0.00	IV.	6	11.248	29 42.70	40.84	3.36	23 45.96	51 26.90
134	9	. .	. .	. .	25.0	. .	. .	. .	24 25.35	22.09	0.00	IV.	6	7.172	27 37.85	40.91	3.21	24 3.26	49 21.97
135	9	. .	. .	. .	. .	. .	55.2	. .	24 27.97	22.09	−0.01	VI.	8	2.55	35 24.19	40.94	3.83	24 5.87	57 8.96
136	9	. .	. .	. .	. .	. .	. .	31.	24 49.87	22.09	0.00	VII.	5	4.575	22 25.97	41.01	2.73	24 27.78	44 9.76
137	9	. .	. .	. .	. .	. .	. .	57.5	25 16.35	22.09	+0.01	VII.	5	1.362	20 44.47	41.11	2.67	24 54.27	42 28.25
138	9	. .	36.6	50.2	. .	. .	. .	. .	27 3.76	22.09	+0.01	III.	4	6.090	18 3.13	41.51	2.43	26 41.68	49 47.07
139	9	. .	. .	8.	21.2	35.	. .	. .	27 21.33	22.09	−0.01	IV.	7	6.320	32 15.07	41.57	3.57	26 59.23	54 0.21
140	9	. .	57.1	11.	. .	. .	. .	. .	28 24.46	22.09	0.00	III.	5	5.196	22 37.41	41.80	2.82	28 2.37	44 22.03
141	9	. .	52.6	6.2	. .	. .	. .	. .	29 19.71	22.09	+0.01	III.	3	8.480	14 24.30	42.02	2.17	28 57.63	36 8.49
142	9	. .	. .	. .	. .	. .	. .	43.2	28 2.21	22.09	−0.01	VII.	7	6.518	32 24.70	41.72	3.59	27 40.11	54 10.01
143	9	. .	. .	. .	. .	. .	17.	. .	29 49.78	22.09	−0.01	VI.	7	5.43	31 50.20	42.13	3.54	29 27.68	53 35.87
144	9	. .	. .	. .	. .	. .	44.2	58.5	31 17.16	22.09	+0.01	VI.	3	10.535	15 27.47	42.47	2.25	30 53.08	23 37 12.19
145	9	. .	. .	. .	48.2	. .	. .	. .	32 48.05	22.09	−0.02	IV.	9	4.280	41 10.23	42.81	4.32	32 25.94	24 2 57.36
146	9	. .	. .	. .	5.3	. .	. .	. .	33 5.15	22.09	0.02	IV.	9	6.078	42 20.73	42.88	4.38	32 43.04	24 4 7.99
147	9	. .	49.2	3.0	. .	. .	. .	. .	4 36 16.61	−22.08	−0.01	III.	7	4.000	−30 58.38	−43.62	−3.47	4 35 54.52	−23 52 45.47

CORRECTIONS.

Date.		Corr. of Clock.	Hourly rate.	m	n	c
1847.	h.	s.	s.	s.	s.	s.

INSTRUMENT READINGS.

Date.		Barom.	THERMOM. At.	THERMOM. Ex.
1847.	h. m.	in.	°	°

REMARKS.

(142) 99. Minutes assumed as 0 instead of 59.
(142) 100. Minutes assumed as 0 instead of 59.
(142) 114. Transit over T. IV assumed as recorded over T. III.
(142) 144. Declination 16′ greater than that of Mural Z., 1849, January 27.
(142) 146. Micrometer reading assumed as 6r.478 instead of 6r.078.

ZONE 142. OCTOBER 18. H. BELT, −23° 46′. D₀ = −23° 21′ 0″—Continued.

No.	Mag.	Seconds of Transit. I.	II.	III.	IV.	V.	VI.	VII.	T.	a_1	a_2	Micrometer.			i	d_1	d_2	Mean Right Ascension, 1850.0.	Mean Declination, 1850.0.
									h. m. s.	s.	s.			r.	′ ″	″	″	h. m. s.	° ′ ″
148	8	..	..	11.	24.1	..	..	..	4 36 24.11	−22.08	+0.02	IV.	1	11.143?	− 5 40.01	−43.65	−1.47	4 36 2.05	−23 27 25.13
149	9	..	53.2	7.	21.	..	..	..	37 20.70	22.08	−0.01	IV.	7	8.06	33 2.47	43.86	3.63	36 58.61	23 54 49.96
150	9	..	..	..	..	..	40.6	..	37 13.32	22.08	0.02	VI.	10	4.260	46 9.98	43.84	4.72	36 51.22	24 7 58.54
151	9	..	17.2	31.	..	..	..	..	38 44.67	22.08	0.01	III.	8	5.192	36 37.01	44.17	3.94	38 22.58	24 58 25.12
152	9	..	..	..	48.2	2.1	..	..	38 48.26	22.08	0.01	IV.	8	4.470	36 20.82	44.18	3.91	38 26.17	23 58 8.91
153	9	..	..	..	..	32.	45.1	..	39 17.62	22.08	−0.02	V.	9	11.566	44 56.40	44.30	4.63	38 55.52	24 6 45.33
154	9	..	..	..	..	17.	..	..	40 3.26	22.08	+0.01	V.	3	4.158	12 8.55	44.47	1.99	39 41.19	23 33 55.01
155	9	..	..	..	..	..	53.3	..	40 26.10	22.08	0.00	VI.	6	4.140	26 5.33	44.55	3.09	40 4.02	47 52.97
156	9	..	..	..	13.2	..	..	..	41 13.05	22.08	−0.01	IV.	8	2.212	35 7.29	44.74	3.80	40 50.96	56 55.83
157	9	..	..	..	..	55.3	..	..	41 41.63	22.08	0.00	V.	6	3.410	25 48.79	44.85	3.06	41 19.55	47 36.70
158	9	..	21.8	..	..	..	..	..	42 49.25	22.08	−0.01	II.	8	4.205	36 7.30	45.13	3.88	42 27.16	57 56.31
159	9	..	..	39.2	53.	..	..	..	42 52.69	22.08	+0.02	IV.	2	11.033	10 33.46	45.14	1.85	42 30.63	32 20.45
160	9	..	..	..	..	..	5.5	..	44 38.36	22.08	+0.01	VI.	2	12.288	11 16.41	45.53	1.91	44 16.29	33 3.85
161	9	..	29.0	..	..	..	..	..	47 56.32	22.08	0.00	II.	4	13.040	21 32.27	46.30	2.73	47 34.24	43 21.30
162	9	..	..	55.0	9.0	..	..	..	48 8.68	22.08	0.00	IV.	5	6.215	23 8.66	46.34	2.84	47 46.60	44 57.84
163	9	..	..	..	34.8	..	..	..	48 34.65	22.08	0.00	IV.	5	6.322	23 14.07	46.46	2.84	48 12.57	45 3.37
164	9	..	..	..	..	7.	..	..	48 53.32	22.08	0.00	V.	5	9.065	24 31.83	46.52	2.96	48 31.24	46 21.31
165	9	..	..	..	..	30.	43.6	..	49 16.36	22.08	+0.01	V.	4	3.49	16 52.53	46.62	2.38	48 54.29	38 41.53
166	9	..	..	..	..	..	10.2	..	49 43.07	22.08	0.02	VI.	2	5.19	7 39.70	46.72	1.60	49 21.01	29 28.02
167	9	..	..	..	31.0	44.5	..	..	50 30.80	22.08	+0.02	IV.	2	9.58	10 0.53	46.91	1.79	50 8.74	31 49.23
168	9	..	..	..	..	..	..	8.1	50 27.09	22.08	−0.02	VII.	7	3.38	30 46.98	46.90	3.45	50 4.99	52 37.33
169	9	..	..	..	49.	..	..	..	54 48.85	22.07	−0.01	IV.	8	3.118	35 32.80	47.93	3.86	54 26.77	57 24.59
170	8.9	..	..	..	..	38.1	..	..	55 24.46	22.07	0.01	V.	7	10.	33 59.90	48.07	3.73	55 2.38	23 55[51.70]
171	9	..	..	19.8	..	..	..	..	56 33.52	22.07	−0.01	III.	8	8.350	38 15.74	48.34	4.08	56 11.44	24 0 8.16
172	9	..	..	..	20.6	..	..	..	57 20.45	22.07	+0.02	IV.	1	13.021	6 34.36	48.54	1.51	56 58.40	23 28 24.41
173	9	..	..	..	10.8	..	..	..	58 10.65	22.06	−0.01	IV.	7	8.092	33 4.08	48.74	3.65	57 48.58	23 54 56.47
174	9	..	..	..	..	34.1	..	..	58 20.51	22.06	−0.02	V.	10	4.542	46 24.30	48.78	4.75	57 58.43	24 8 17.83
175	9	..	..	57.	..	..	..	..	59 10.43	22.06	+0.01	III.	4	5.065	37 31.60	48.98	2.40	58 48.38	23 39 22.98
176	9	..	..	..	34.5	48.2	..	..	59 34.44	22.06	0.00	IV.	6	8.550	28 27.17	49.08	3.27	59 12.38	50 19.52
177	9	..	..	..	..	..	20.	..	4 59 38.68	−22.06	+0.02	VI.	2	6.527	− 8 26.95	−49.09	−1.67	4 59 16.64	−23 30 17.71

ZONE 143. OCTOBER 28. K. BELT, −23° 8′. D₀ = −22° 43′ 30″.

No.	Mag.	I.	II.	III.	IV.	V.	VI.	VII.	T.	a_1	a_2	Micrometer.			i	d_1	d_2	Mean Right Ascension, 1850.0.	Mean Declination, 1850.0.
1	9	..	..	..	..	..	27.2	..	22 8 0.06	−28.92	+0.01	I.	9	9.7	−43 30.59	− 5.96	−4.38	22 7 31.15	−23 27 10.93
2	6	41.0	54.6	8.5	..	..	..	..	10 21.87	28.94	0.00	IV.	4	8.14	19 6.20	5.79	2.56	9 52.93	2 44.55
3	9	..	..	55.1	..	22.3	..	..	11 8.85	28.95	+0.01	IV.	10	10.4	49 0.57	5.74	4.82	10 39.91	23 32 41.13
4	6	..	..	..	12.2	..	39.2	..	12 12.05	28.96	−0.01	IV.	1	4.47	2 24.72	5.67	1.33	11 43.08	22 46 1.72
5	9	..	..	21.1	..	..	..	..	16 34.45	29.01	0.00	IV.	4	4.36	17 16.28	5.37	2.44	16 5.44	23 0 54.09
6	9	..	11.8	..	..	..	..	..	19 39.09	29.05	+0.01	IV.	7	10.10	34 4.99	5.18	3.70	19 10.05	23 17 43.87
7	5	21.7	..	49.2	..	..	..	..	21 2.49	29.06	−0.01	IV.	1	12.48	6 27.25	5.10	1.65	20 33.42	22 50 4.00
8	9	..	..	..	49.6	..	..	..	23 49.45	29.09	+0.01	IV.	9	7.37	42 45.54	4.94	4.33	23 20.37	23 26 24.81
9	9	..	..	..	41.3	..	..	..	30 41.15	29.16	0.01	IV.	10	6.50	47 22.74	4.56	4.69	30 12.00	31 1.99
10	9	..	..	..	23.2	..	..	..	31 23.05	29.16	0.01	IV.	9	4.34	41 13.27	4.52	4.21	30 53.90	24 52.00
11	9	..	..	4.8	..	..	..	..	41 18.53	29.26	0.01	IV.	9	8.17	43 5.71	4.03	4.36	40 49.28	26 44.10
12	9	..	..	48.2	..	..	..	..	43 1.98	29.28	+0.01	IV.	10	5.19	46 36.85	3.95	4.63	42 32.71	30 15.43
13	5.6	..	50.2	4.0	17.3	..	..	..	47 17.35	29.33	0.00	IV.	6	3.27	25 51.85	3.75	3.06	46 48.02	9 28.66
14	9	..	..	..	10.5	..	..	..	50 10.35	29.36	+0.01	IV.	9	4.10	41 1.16	3.62	4.20	49 41.00	24 38.98
15	8	..	24.0	37.8	..	..	..	..	22 52 51.37	−29.39	0.00	IV.	8	4.18	−36 6.19	− 3.53	−3.86	22 52 21.98	−23 19 43.58

CORRECTIONS.

Date.	Corr. of Clock.	Hourly rate.	m	n	c
1847. h.	s.	s.	s.	s.	s.
Oct. 28, 20	f 35.70	g 0.038	+ 0.090	+ 0.131	0.000

INSTRUMENT READINGS.

	Date.	Barom.	Thermom. At.	Thermom. Ex.
	1847. h. m.	in.	°	°
Zone 143	Oct. 28, 9 5	30.754	78.8	33.2

REMARKS.

(142) 177. Transit over T. VII assumed to have been recorded as over T. VI.
(143) 13. Micrometer reading assumed as 3r.47 instead of 3r.27.

ZONE 143. OCTOBER 28. K. BELT, −23° 8′. D₀ = −22° 43′ 30″—Continued.

No.	Mag.	I.	II.	III.	IV.	V.	VI.	VII.	T.	a_1	a_2	MICROMETER.			i	d_1	d_2	Mean Right Ascension, 1850.0.	Mean Declination, 1850.0.
									h. m. s.	s.	s.			r.	′ ″	″	″	h. m. s.	° ′ ″
16	8	. .	31.5	45.4	59.3	. .	. .	. .	22 53 58.95	−29.40	0.00	IV.	7	1.30	−29 42.78	− 3.48	−3.36	22 53 29.55	−23 13 19.62
17	7	. .	11.2	24.9	. .	. .	. .	. .	56 38.31	29.43	0.00	IV.	4	7.33	18 45.52	3.38	2.54	56 8.88	2 21.44
18	6	11.1	24.8	38.8	52.5	5.1	19.2	32.4	22 57 52.05	29.44	0.00	IV.	7	10.12	34 6.00	3.34	3.70	22 57 22.61	17 43.04
19	4.5	42.3	56.3	9.8	23.5	36.8	50.0	4.1	23 2 23.31	29.48	0.00	IV.	7	7.12	32 35.23	3.19	3.58	23 1 53.83	16 12.00
20	9	36.7	50.4	4.6	18.0	31.9	44.8	58.3	16 17.93	29.63	+0.01	IV.	9	6.42	42 17.81	2.85	4.29	15 48.31	25 54.95
21	8	. .	. .	. .	32.7	45.8	59.3	13.4	17 32.41	29.64	0.00	IV.	8	3.47	35 50.56	2.83	3.84	17 2.77	23 19 27.23
22	9	8.4	21.9	35.6	49.1	2.8	. .	30.2	21 49.09	29.69	0.00	IV.	3	9.34	14 47.54	2.76	2.25	21 19.40	22 58 22.55
23	8.9	. .	. .	. .	7.1	20.8	33.7	48.2	23 6.95	29.70	−0.01	IV.	1	12.55	6 30.78	2.74	1.65	22 37.24	22 50 5.17
24	9	. .	22.7	. .	. .	3.2	16.2	30.3	37 49.62	29.85	+0.01	IV.	9	10.12	44 3.70	2.64	4.42	37 19.78	23 27 40.76
25	9	28.0	42.0	55.5	9.2	. .	36.0	50.3	40 9.08	29.87	0.00	IV.	5	5.24	22 39.67	2.64	2.82	39 39.21	6 15.13
26	9	29.1	42.8	56.8	10.2	23.7	36.7	51.1	43 10.03	29.90	0.00	IV.	5	2.03	20 58.32	2.66	2.70	42 40.13	23 4 33.68
27	6.7	. .	26.0	40.0	53.2	7.2	20.4	34.4	48 53.24	29.96	0.00	IV.	1	12.6	6 6.08	2.70	1.61	48 23.28	22 49 40.39
28	9	. .	12.9	26.4	40.0	53.4	. .	21.0	23 52 40.00	30.00	0.00	IV.	7	6.35	32 16.58	2.75	3.56	23 52 10.00	23 15 52.89
29	5.6	55.8	9.4	23.2	36.8	50.2	3.2	17.6	0 0 36.68	−30.07	0.00	IV.	8	5.57	−36 56.11	− 2.90	−3.93	0 0 6.61	−23 20 32.94

ZONE 144. OCTOBER 28. K. BELT, −23° 8′. D₀ = −22° 43′ 30″.

No.	Mag.	I.	II.	III.	IV.	V.	VI.	VII.	T.	a_1	a_2	MICROMETER.			i	d_1	d_2	Mean Right Ascension, 1850.0.	Mean Declination, 1850.0.
1	8	. .	. .	. .	. .	. .	35.3	. .	1 30 8.22	−30.86	0.00	VI.	6	9.58	−28 58.78	. .	−3.30	1 29 37.36	
2	9	. .	21.3	. .	. .	. .	. .	. .	34 48.38	30.69	0.00	IV.	2	10.18	10 10.62	. .	1.85	34 17.49	
3	9	. .	31.0	. .	58.9	. .	. .	. .	36 58.57	30.91	0.00	IV.	10	3.50	45 51.97	. .	4.63	36 27.66	
4	9	11.8	25.2	39.7	. .	. .	. .	. .	41 52.69	30.94	0.00	IV.	3	9.3	14 31.90	. .	2.18	41 21.75	
5	9	. .	. .	24.1	. .	. .	. .	. .	42 37.51	30.95	0.00	IV.	4	12.52	21 26.37	. .	2.74	42 6.56	
6	9	. .	. .	36.3	. .	. .	. .	. .	43 50.08	30.96	−0.01	IV.	10	6.23	47 9.12	. .	4.75	43 19.11	
7	9	25.5	. .	53.0	. .	. .	. .	. .	45 6.34	30.97	+0.01	IV.	2	12.11	11 7.59	. .	1.92	44 35.39	
8	9	. .	. .	55.4	. .	. .	. .	. .	46 8.88	30.97	0.00	IV.	5	13.48	26 53.80	. .	3.15	45 37.91	
9	9	. .	55.9	. .	. .	. .	. .	. .	49 23.20	30.99	0.00	IV.	7	12.28	35 14.56	. .	3.82	48 52.21	
10	4.5	. .	42.7	56.0	10.2	. .	. .	. .	50 9.86	30.99	0.00	VII.	7	6.5	32 1.12	. .	3.54	49 38.87	
11	9	. .	. .	. .	. .	. .	. .	14.1	52 33.02	31.01	+0.01	VII.	3	2.47	11 21.98	. .	1.94	52 2.02	
12	8.7	50.7	4.6	18.6	. .	. .	. .	. .	1 59 31.70	31.07	+0.01	IV.	2	7.13	8 37.33	. .	1.70	1 59 0.64	
13	9	. .	. .	. .	16.0	. .	. .	. .	2 9 15.85	31.13	0.00	IV.	7	10.26	34 13.06	. .	3.71	2 8 44.72	
14	9	. .	. .	. .	13.8	. .	. .	. .	11 13.65	31.14	0.00	IV.	2	4.37	7 18.68	. .	1.64	10 42.51	
15	9	. .	. .	. .	. .	3.9	. .	. .	11 50.33	31.15	0.00	IV.	7	8.31	33 15.07	. .	3.63	11 19.18	
16	9	. .	. .	48.3	. .	. .	. .	. .	14 1.48	31.16	+0.01	IV.	1	9.41	4 52.96	. .	1.44	13 30.33	
17	9	. .	. .	. .	. .	. .	47.9	. .	14 20.80	31.16	0.00	IV.	7	9.46	33 52.89	. .	3.68	13 49.64	
18	6	. .	8.4	. .	. .	. .	. .	. .	23 35.71	31.21	0.00	IV.	8	6.58	37 26.87	. .	4.00	23 4.50	
19	5.4	. .	. .	0.0	13.9	. .	. .	. .	24 13.64	31.22	0.00	IV.	6	10.11	29 5.49	. .	3.30	23 42.42	
20	9	22.9	. .	. .	. .	. .	. .	. .	31 4.07	31.26	0.00	IV.	8	9.44	38 50.58	. .	4.09	30 32.81	
21	9	. .	. .	30.8	. .	. .	. .	. .	2 31 43.96	−31.26	0.00	IV.	1	6.24	− 3 13.63	. .	−1.30	2 31 12.70	

ZONE 145. NOVEMBER 2. K. BELT, −22° 31′. D₀ = −22° 6′ 30″.

No.	Mag.	I.	II.	III.	IV.	V.	VI.	VII.	T.	a_1	a_2	MICROMETER.			i	d_1	d_2	Mean Right Ascension, 1850.0.	Mean Declination, 1850.0.
1	8	. .	. .	. .	. .	. .	. .	14.4	22 16 33.56	−32.43	−0.01	VII.	3	9.12	−14 36.12	− 6.90	−2.31	22 16 1.12	−22 21 15.33
2	9	. .	. .	. .	. .	9.8	. .	. .	17 56.20	32.44	−0.01	VI.	3	9.34	14 47.39	6.83	2.32	17 23.75	21 26.54
3	9	28.2	. .	. .	. .	. .	. .	. .	20 9.11	32.46	0.00	IV.	7	4.29	31 13.04	6.70	3.44	19 36.65	37 53.18
4	8	25.3	39.4	. .	. .	. .	. .	. .	21 6.48	32.47	+0.01	IV.	9	8.53	43 23.86	6.66	4.26	20 34.02	50 4.78
5	10	. .	. .	. .	. .	28.4	. .	. .	21 14.80	32.47	−0.01	IV.	3	10.57	15 29.34	6.65	2.37	20 42.32	22 8.41
6	10	22.4	. .	. .	. .	. .	. .	. .	22 24 3.33	−32.50	0.00	IV.	7	7.26	−32 42.29	− 6.52	−3.54	22 23 30.83	−22 39 22.35

CORRECTIONS.

Date.	Corr. of Clock.	Hourly rate.	m	n	c
1847. h.	s.	s.	s.	s.	s.
Nov. 2, 20	f 39.31	g 0.020	+ 0.209	+ 0.209	0.000

INSTRUMENT READINGS.

	Date.	Barom.	THERMOM. At.	THERMOM. Ex.
	1847. h. m.	in.	°	°
Zone 145	Nov. 2, 9 5	30.026	76.5	51.

REMARKS.

ZONE 145. NOVEMBER 2. K. BELT, −22° 31′. D_o = −22° 6′ 30″—Continued.

No.	Mag.	SECONDS OF TRANSIT. I.	II.	III.	IV.	V.	VI.	VII.	T.	a_1	a_2	MICROMETER.			i	d_1	d_2	Mean Right Ascension, 1850.0.	Mean Declination, 1850.0.
									h. m. s.	s.	s.			r.	′ ″	″	″	h. m. s.	° ′ ″
7	10			56.8					22 24 10.38	−32.51	−0.01	IV.	8	7.45	−37 50.57	−6.51	−3.88	22 23 37.86	−22 44 0.96
8	10						50.4		24 23.48	32.51	−0.01	IV.	3	10.46	15 23.85	6.50	2.36	23 50.96	22 2.71
9	9		47.3						26 14.57	32.53	+0.01	IV.	10	4.49	45 51.47	6.42	4.47	25 42.05	52 32.36
10	9	46.8							27 27.83	32.54	+0.01	IV.	10	1.13	44 32.80	6.36	4.35	26 55.30	51 13.51
11	10	55.3							28 36.11	32.55	0.00	IV.	5	0.59	20 26.05	6.31	2.68	28 3.56	27 5.04
12	10			39.2					28 52.46	32.56	−0.01	IV.	3	9.50	14 55.60	6.29	2.33	28 19.89	21 34.22
13	9						38.2		29 11.29	32.56	−0.01	VI.	2	13.33	11 48.79	6.28	2.12	28 38.72	18 27.19
14	8						23.7		29 56.75	32.57	0.00	VI.	6	7.29	27 43.65	6.24	3.20	29 23.18	34 23.09
15	9				5.3				31 5.15	32.58	0.00	IV.	4	8.46	19 22.34	6.20	2.60	30 32.57	26 1.14
16	9	50.0							34 30.87	32.62	0.00	IV.	6	5.2	26 29.68	6.06	3.10	33 58.25	33 8.84
17	7	44.8	58.4						35 25.52	32.62	0.00	IV.	4	9.34	19 46.54	6.03	2.63	34 52.90	26 25.20
18	10		56.8						37 23.93	32.64	0.00	IV.	6	11.10	29 35.24	5.95	3.32	36 51.29	36 14.51
19	9	8.7		36.0					39 49.56	32.67	0.00	IV.	7	5.24	31 40.77	5.86	3.47	39 16.89	38 20.10
20	8				31.9				40 31.75	32.68	0.00	IV.	4	6.58	18 27.87	5.83	2.54	39 59.07	25 6.24
21	9				47.9				42 47.75	32.70	0.00	IV.	3	11.30	15 46.03	5.76	2.38	42 15.05	22 24.17
22	7	9.9	23.5						44 50.65	32.72	0.00	IV.	5	6.57	23 26.57	5.70	2.89	44 17.93	30 5.16
23	11		58.8						46 26.00	32.73	+0.01	IV.	8	7.3	37 29.39	5.65	3.88	45 53.28	44 8.92
24	9				58.9				46 58.75	32.74	−0.01	IV.	1	8.4	4 4.06	5.63	1.60	46 26.00	10 41.29
25	9	47.9	1.8						50 28.99	32.78	+0.01	IV.	9	9.8	43 31.42	5.53	4.28	49 56.22	50 11.23
26	10						31.3		56 4.36	32.84	0.00	IV.	5	6.43	23 19.51	5.41	2.89	55 31.52	29 57.81
27	8	5.5	19.4						58 46.54	32.87	+0.01	IV.	8	8.5	38 0.66	5.35	3.91	58 13.68	44 39.92
28	9			14.0					22 59 27.19	32.88	−0.01	IV.	2	8.4	9 3.06	5.33	1.93	22 58 54.30	15 40.32
29	10	9.1							23 0 49.89	32.88	0.00	IV.	4	6.16	18 6.70	5.31	2.51	23 0 17.01	24 44.52
30	9			17.3					3 30.61	32.92	0.00	IV.	4	6.29	18 13.25	5.27	2.55	2 57.69	24 51.07
31	9			3.7					4 17.12	32.92	0.00	IV.	6	4.33	26 15.06	5.26	3.09	3 44.20	32 53.41
32	9		53.6	7.1					5 20.75	32.93	0.00	IV.	8	9.1	38 28.89	5.25	3.22	4 47.82	45 8.06
33	10	32.8							7 13.59	32.95	0.00	IV.	4	6.5	18 1.15	5.22	2.51	6 40.64	24 38.88
34	9	33.3							8 14.23	32.96	0.00	IV.	7	8.31	33 15.07	5.20	3.58	7 41.27	39 53.85
35	8				15.1				8 14.95	32.96	−0.01	IV.	1	3.44	1 52.95	5.20	1.44	7 41.98	8 29.59
36	9	55.5							15 36.27	33.04	0.00	IV.	3	9.3	14 31.90	5.15	2.28	15 3.23	21 9.33
37	8				44.5				15 44.35	33.04	−0.01	IV.	1	3.4	1 32.78	5.15	1.41	15 11.30	3 9.34
38	7			29.9	43.2				16 43.20	33.05	0.00	IV.	6	9.58	28 58.93	5.15	3.28	16 10.15	35 37.36
39	9					30.5			17 17.02	33.05	+0.01	IV.	9	5.20	41 36.45	5.14	4.12	16 43.98	48 15.71
40	9				11.9				18 11.75	33.06	0.00	IV.	5	4.24	22 9.42	5.14	2.80	17 38.69	28 47.36
41	6		47.0						19 14.11	33.07	0.00	IV.	6	6.29	27 13.55	5.13	3.15	18 41.04	33 51.83
42	9				42.8				19 42.65	33.08	+0.01	IV.	10	6.25	47 10.13	5.13	4.52	19 9.59	53 49.78
43	9	40.1	53.4						21 20.60	33.09	0.00	IV.	2	11.58	11 1.04	5.14	2.06	20 47.51	17 38.24
44	8							50.9	23 10.45	33.11	0.00	IV.	9	9.2	43 28.40	5.14	4.27	22 37.34	50 7.81
45	7				36.9				24 36.75	33.13	0.00	IV.	7	4.28	31 12.53	5.15	3.43	24 3.62	37 51.11
46	8			18.8					25 32.19	33.14	0.00	IV.	5	8.33	24 14.98	5.15	2.95	24 59.05	30 53.08
47	10			29.0					27 42.47	33.16	0.00	IV.	6	11.5	29 32.72	5.16	3.32	27 9.31	36 11.20
48	10						11.2		27 44.23	33.16	0.00	IV.	7	11.6	34 33.23	5.16	3.67	27 11.07	41 12.06
49	10					52.8			28 39.25	33.17	0.00	IV.	4	13.18	21 39.47	5.16	2.77	28 6.08	28 17.40
50	9					59.2			29 45.63	33.18	0.00	IV.	5	7.29	23 42.70	5.17	2.91	29 12.45	30 20.78
51	9		44.2	58.1					31 11.38	33.19	0.00	IV	5	7.0	23 28.08	5.18	2.89	30 38.19	30 6.15
52	8		32.3		59.6				31 59.48	33.20	0.00	IV.	8	9.20	38 38.47	5.20	3.93	31 26.28	45 17.60
53	9	27.8		55.5					36 8.78	33.24	0.00	IV.	5	11.21	25 39.68	5.25	3.05	35 35.54	32 17.98
54	9	20.4	33.8						40 1.12	33.29	0.00	IV.	6	11.26	29 43.31	5.30	3.33	39 27.84	36 21.94
55	7.6	32.8	46.7						23 42 13.68	−33.30	0.00	IV.	4	10.27	−20 13.26	−5.35	−2.66	23 41 40.38	−22 26 51.27

CORRECTIONS.

Date.		Corr. of Clock.	Hourly rate.	m	n	c
1847.	h.	s.	s.	s.	s.	s.

INSTRUMENT READINGS.

Date.		Barom.	THERMOM. At.	Ex.
1847.	h. m.	In.	°	°

REMARKS.

(145) 9. Micrometer reading assumed as $3^r.49$ instead of $4^r.49$

ZONE 145. NOVEMBER 2. K. BELT, −22° 31′. D_o = −22° 6′ 30″—Continued.

No.	Mag.	SECONDS OF TRANSIT. I.	II.	III.	IV.	V.	VI.	VII.	T.	a_1	n_1	MICROMETER.			i	d_1	d_2	Mean Right Ascension, 1850.0.	Mean Declination, 1850.0.
									h. m. s.	s.	s.			r.	′ ″	″	″	h. m. s.	° ′ ″
56	9	..	..	31.8	45.0	..	..	..	23 45 44.94	−33.34	0.00	IV.	3	4.20	−12 9.20	−5.43	−2.13	23 45 11.60	−22 13 46.76
57	6	..	..	57.2	..	..	23.4	..	48 56.72	33.37	0.00	IV.	9	8.4	42 59.16	5.51	4.25	48 23.35	49 38.92
58	9	..	31.8	45.3	..	..	..	..	50 58.94	33.39	0.00	IV.	8	7.0	37 27.88	5.56	3.85	49 25.55	44 7.29
59	10	..	..	..	..	28.6	..	..	51 15.10	33.39	0.00	IV.	8	6.50	37 22.84	5.57	3.84	50 41.71	44 2.25
60	9	..	..	..	..	19.1	..	..	54 5.47	33.42	0.00	IV.	2	5.30	7 45.40	5.66	1.82	53 32.05	14 22.88
61	9	..	9.7	..	..	..	..	..	55 36.89	33.43	0.00	IV.	8	5.56	36 55.61	5.70	3.82	55 3.46	43 35.13
62	9	..	..	..	14.0	..	..	..	58 3.85	33.45	0.00	IV.	1	4.39	2 20.69	5.79	1.45	57 30.40	8 57.93
63	9	..	58.0	..	..	..	..	..	59 24.99	33.46	0.00	IV.	3	7.6	13 32.91	5.82	2.23	58 51.53	20 10.96
64	9	..	..	41.7	..	..	..	..	23 59 55.22	33.47	0.00	IV.	7	8.31	33 15.07	5.84	3.57	23 59 21.75	39 54.48
65	9	..	45.9	59.1	..	..	..	..	0 1 12.84	33.48	0.00	IV.	7	9.37	33 48.35	5.89	3.60	0 0 39.36	40 27.84
66	9	..	..	..	..	..	32.2	..	1 5.30	33.48	0.00	IV.	2	6.10	8 5.57	5.88	1.86	0 31.82	14 43.31
67	9	..	..	..	30.5	..	..	..	2 30.35	33.50	0.00	IV.	2	2.46	6 22.71	5.94	1.75	1 56.85	13 0.40
68	9	..	..	28.9	42.3	..	..	..	3 42.33	35.51	0.00	IV.	8	10.45	39 21.34	5.99	3.98	3 8.82	46 1.31
69	9	..	..	10.2	..	..	..	..	5 23.91	33.52	0.00	IV.	10	6.38	47 16.69	6.06	4.55	4 50.39	53 57.30
70	9	..	..	..	16.8	..	..	..	7 16.65	33.54	0.00	IV.	4	6.40	17 48.55	6.13	2.53	6 43.11	24 27.21
71	10.9	54.0	..	..	..	..	..	..	9 34.78	33.57	0.00	IV.	4	2.37	16 16.27	6.22	2.40	9 1.21	22 54.89
72	9	..	..	..	..	48.2	..	..	9 34.62	33.57	0.00	IV.	4	8.46	19 22.34	6.22	2.60	9 1.05	26 1.16
73	9	..	..	46.9	..	..	..	..	11 0.42	33.58	0.00	IV.	7	8.35	33 17.09	6.29	3.58	10 26.84	39 56.96
74	8.9	52.0	..	..	..	..	..	..	12 32.92	33.60	0.00	IV.	7	5.34	31 45.82	6.37	3.47	11 59.32	38 25.66
75	8	..	16.0	29.9	..	..	..	..	18 43.15	33.66	0.00	IV.	5	2.22	21 7.90	6.68	2.73	18 9.49	27 47.31
76	8	..	..	..	17.1	30.3	..	..	0 19 16.84	−33.66	0.00	IV.	5	7.15	−23 35.64	−6.70	−2.90	0 18 43.18	−22 30 15.24

ZONE 146. NOVEMBER 2. K. BELT, −22° 31′. D_o = −22° 7′ 0″.

No.	Mag.	I.	II.	III.	IV.	V.	VI.	VII.	T.	a_1	n_1	MICROMETER.			i	d_1	d_2	Mean Right Ascension, 1850.0.	Mean Declination, 1850.0.
1	9	..	..	..	..	13.1	..	..	3 8 59.60	−34.86	+0.01	IV.	8	4.16	−36 5.19	−25.82	−3.80	3 8 24.73	−22 43 34.81
2	9	..	..	..	..	..	8.7	..	8 41.80	34.86	+0.01	IV.	3	2.29	11 13.23	25.83	2.00	8 6.95	18 41.06
3	9	3.2	17.0	..	..	..	..	..	10 44.12	34.87	0.00	IV.	7	2.30	30 13.04	25.75	3.36	10 9.25	37 42.15
4	10	..	45.0	..	..	..	..	..	12 12.06	34.88	0.00	IV.	5	3.28	21 41.17	25.70	2.77	11 37.18	29 9.64
5	5.4	45.1	59.0	12.7	26.2	..	..	..	13 25.94	34.88	+0.01	IV.	3	2.7	11 2.14	25.65	2.00	12 51.07	18 29.79
6	9	..	..	44.9	58.2	..	..	..	15 58.22	34.89	0.00	IV.	7	4.8	31 2.45	25.55	3.42	15 23.33	38 31.42
7	9	..	52.9	..	20.0	..	..	..	18 19.83	34.90	+0.01	IV.	1	8.55	4 29.77	25.45	1.55	17 44.91	11 56.77
8	9	13.2	27.1	..	..	..	..	..	19 54.15	34.91	0.00	IV.	6	8.59	28 29.19	25.40	3.25	19 19.24	35 57.84
9	10	..	..	..	5.9	..	..	..	20 5.75	34.91	−0.01	IV.	7	10.13	34 6.50	25.40	3.63	19 30.83	41 35.53
10	10	..	..	24.5	..	..	..	..	28 37.99	34.94	0.00	IV.	7	4.59	31 28.17	25.12	3.45	28 3.05	38 56.74
11	10	..	..	54.4	..	..	..	..	31 7.59	34.94	+0.01	IV.	2	9.31	9 46.92	25.05	1.90	30 32.66	17 13.87
12	9	..	..	41.8	..	..	..	..	31 55.24	34.95	0.00	IV.	6	7.52	27 55.40	25.03	3.21	31 20.29	35 23.64
13	9	..	..	..	17.5	..	..	..	32 17.35	34.95	−0.01	IV.	8	8.8	38 2.17	25.03	3.91	31 42.39	45 31.11
14	9	..	..	3.8	17.4	..	..	..	33 17.34	34.95	−0.01	IV.	9	4.38	41 15.28	25.00	4.17	32 42.38	48 44.45
15	9	..	..	19.4	2.9	..	..	..	34 2.83	34.96	0.00	IV.	7	6.1	31 59.43	24.96	3.49	33 27.87	39 27.88
16	9	58.3	12.1	25.9	..	..	..	..	36 39.16	34.97	0.00	IV.	4	10.19	20 9.23	24.91	2.66	36 4.19	27 36.80
17	9	..	..	6.1	..	..	..	..	37 19.69	34.97	−0.01	IV.	8	8.12	38 4.19	24.88	3.93	36 44.71	45 33.00
18	9	..	..	42.9	..	..	..	..	37 56.19	34.97	0.00	IV.	4	4.45	17 20.81	24.87	2.45	37 21.22	24 48.13
19	10	..	..	..	..	21.7	..	..	38 8.14	34.97	0.00	IV.	5	6.0	22 57.83	24.86	2.86	37 33.17	30 25.55
20	9	..	3.8	..	..	..	..	..	39 30.92	34.98	0.00	IV.	6	7.12	27 35.23	24.83	3.18	38 55.94	35 3.24
21	9	..	..	..	..	43.2	..	..	39 29.72	34.98	−0.01	IV.	9	4.58	41 25.36	24.83	4.18	38 54.73	48 54.37
22	9	..	57.0	..	..	..	..	..	42 24.15	34.98	0.00	IV.	7	5.54	31 55.90	24.78	3.49	41 49.17	39 21.17
23	9	..	..	..	..	..	51.8	..	42 24.82	34.98	−0.01	IV.	8	3.6	35 29.89	24.78	3.75	41 49.81	42 58.42
24	10	..	..	48.9	..	..	..	..	3 44 2.19	−34.99	+0.01	IV.	4	4.24	−17 10.22	−24.74	−2.44	3 43 27.21	−22 24 37.40

CORRECTIONS.

Date.		Corr. of Clock.	Hourly rate.	m	n	c
1847.	h.	s.	s.	s.	s.	s.

INSTRUMENT READINGS.

Date.		Barom.	THERMOM. At.	THERMOM. Ex.
1847.	h. m.	in.	°	°

REMARKS.

(145) 57. The transit observations assumed to have been on T.'s IV and VI.
(145) 58. Minutes assumed as 49 instead of 48.
(145) 62. Transit over T. IV assumed as 4ˢ.0 instead of 14ˢ.0.
(145) 70. Micrometer reading assumed as 5ʳ.40 instead of 6ʳ.40.
(146) 1. Minutes of transit assumed as 7 instead of 8.
(146) 15. Transit over T. III assumed as 59ˢ.4 instead of 19ˢ.4.

ZONE 146. NOVEMBER 2. K. BELT, −22° 31′. D₀=−22° 7′ 0″.—Continued.

No.	Mag.	I.	II.	III.	IV.	V.	VI.	VII.	T.	a_1	a_2	MICROMETER.			i	d_1	d_2	Mean Right Ascension, 1850.0.	Mean Declination, 1850.0.
									h. m. s.	s.	s.			r.	′ ″	″	″	h. m. s.	° ′ ″
25	9			37.7					3 45 51.41	−34.99	−0.01	IV.	10	5.35	−46 44.92	−24.72	−4.59	3 45 16.41	−22 54 14.23
26	9			40.3	54.6				46 54.16	34.99	0.01	IV.	8	4.13	36 3.67	24.69	3.79	46 19.16	43 32.15
27	9	32.3	46.2						48 13.37	35.00	−0.01	IV.	9	6.13	42 3.18	24.66	4.24	47 38.36	49 32.08
28	10				30.3				48 20.15	35.00	+0.01	IV.	2	8.44	9 23.23	24.65	1.88	47 45.16	16 49.76
29	9				34.7				49 47.97	35.00	+0.01	IV.	2	12.46	11 25.25	24.64	2.02	49 12.98	18 51.91
30	7	37.1	50.9						55 18.06	35.01	−0.01	IV.	7	10.56	34 28.19	24.56	3.67	54 43.04	41 56.42
31	9.10		53.1						3 59 20.17	35.02	0.00	IV.	5	6.5	23 0.35	24.50	2.86	58 45.15	30 27.71
32	9			55.7					4 0 9.09	35.02	0.00	IV.	5	8.5	24 0.86	24.49	2.93	3 59 34.07	31 28.28
33	9		36.9	50.8	4.0				2 3.94	35.02	+0.01	IV.	3	11.15	16 38.96	24.48	2.42	4 1 28.93	24 5.86
34	9			46.8					2 59.95	35.03	+0.01	IV.	1	11.29	6 47.92	24.47	1.74	2 24.93	14 14.13
35	6.5	18.9	32.9						9 59.85	35.04	0.00	IV.	5	6.27	23 11.44	24.41	2.87	9 24.81	30 38.72
36	9							30.8	18 49.83	35.06	+0.01	IV.	1	9.39	5 52.47	24.37	1.68	18 14.78	13 18.52
37	9		2.8						4 23 16.44	−35.06	−0.01	IV.	9	6.22	−43 8.23	−24.37	−4.27	4 23	−22 50 36.87

ZONE 147. NOVEMBER 5. B. BELT, −21° 53′. D₀=−21° 29′ 0″.

No.	Mag.	I.	II.	III.	IV.	V.	VI.	VII.	T.	a_1	a_2	MICROMETER.			i	d_1	d_2	Mean Right Ascension, 1850.0.	Mean Declination, 1850.0.
1	8	37.	51.						22 28 17.71	−34.06	+0.01	III.	3	5.30	−12 44.46	−30.45	−2.22	22 27 43.66	−21 42 17.13
2	8					24.	37.1	51.	29 10.40	34.07	0.00	VII.	5	4.385	22 16.42	30.46	2.84	28 36.33	51 49.72
3	8						30.	44.	35 3.27	34.13	+0.01	VII.	3	8.10	14 4.87	30.59	2.29	34 29.15	43 37.75
4	8		51.	4.5	18.				39 17.74	34.17	+0.01	IV.	1	9.22	5 13.64	30.68	1.72	38 43.58	34 46.14
5	8.9						20.5	31.3	40 53.70	34.19	0.00	VII.	4	11.40	20 49.76	30.73	2.74	40 19.51	50 23.23
6	9			54.	7.				44 7.02	34.22	0.00	IV.	3	8.57	14 28.88	30.82	2.32	43 32.80	44 2.02
7	9						24.	38.	44 57.35	34.22	0.00	VII.	6	10.9	29 4.18	30.85	3.25	43 23.13	58 38.28
8	9		59.	12.5	26.				51 25.79	34.30	+0.01	IV.	3	3.51	11 54.57	31.06	2.16	50 51.50	40 27.79
9	9						32.	46.	53 5.24	34.31	0.01	VII.	2	4.46	7 22.91	31.13	1.88	52 30.94	36 55.92
10	6			5.	18.	31.3	44.4	58.3	55 17.73	34.34	+0.01	VII.	2	11.26	10 44.60	31.22	2.08	54 43.40	21 40 17.90
11	9			14.	28.				22 59 27.75	34.38	−0.01	V.	10	4.21	46 7.56	31.38	4.40	22 58 53.36	22 15 43.34
12	4	20.	34.	48.	1.3	14.			23 2 0.96	34.41	0.00	VII.	6	10.58	29 28.88	31.49	3.28	23 1 26.55	21 59 3.65
13	9					40.	53.5	7.	3 26.53	34.42	0.00	VII.	4	10.23	20 10.93	31.56	2.69	2 52.11	21 49 45.18
14	9		49.	2.5	16.	29.			8 15.88	34.47	−0.01	VI.	8	9.58	38 57.49	31.78	3.92	7 41.40	22 8 33.19
15	9						40.	54.	9 13.36	34.48	0.00	VII.	6	12.53	30 26.86	31.83	3.35	8 38.88	0 2.01
16	8							7.	9 26.57	34.48	0.00	VII.	6	14.10	31 35.90	31.84	3.43	8 52.09	1 12.17
17	9					40.		7.	17 26.68	34.56	−0.01	VII.	10	4.48	46 20.91	32.28	4.41	16 52.11	15 57.60
18	8				24.5		51.		24 24.24	34.63	−0.01	VII.	9	5.32	42 12.43	32.74	4.13	23 49.60	22 11 49.30
19	6					13.5	27.	40.5	23 26 0.00	−34.65	+0.01	VII.	3	10.4	−15 2.36	−32.84	−2.36	23 25 25.36	−21 44 37.56

ZONE 148. DECEMBER 4.* K. BELT, −25° 39′. D₀=−24° 0′ 6″.

No.	Mag.	I.	II.	III.	IV.	V.	VI.	VII.	T.	a_1	a_2	MICROMETER.			i	d_1	d_2	Mean Right Ascension, 1850.0.	Mean Declination, 1850.0.
1	10		4.3	18.2					22 52 31.90	+ 2.97	0.00	IV.	4	4.26	−17 11.23	. .	−2.38	23 52 34.87	
2	10				18.3				53 18.15	2.96	0.00	IV.	4	2.5	16 0.13	. .	2.29	53 21.11	
3	9						0.3		22 59 46.44	2.87	0.00	IV.	8	10.45	39 21.34	. .	4.15	23 59 49.31	
4	9		55.2	9.2					23 3 22.95	2.84	0.00	IV.	5	11.35	25 46.75	. .	3.06	0 3 25.79	
5	8				55.0	8.8			3 54.88	2.83	0.00	IV.	6	7.49	27 53.89	. .	3.23	3 57.71	
6	8.9	56.0	9.7						5 37.76	2.81	0.00	IV.	7	10.50	33 54.90	. .	3.77	5 40 57	
7	6.7				4.2	17.8			6 3.93	2.81	0.00	IV.	1	5.17	2 39.84	. .	1.22	6 6.74	
8	6	30.1	44.0	58.0					13 11.86	2.72	0.00	IV.	6	7.21	27 39.76	. .	3.21	13 14.58	
9	10		46.9						23 14 14.55	+ 2.71	0.00	IV.	3	6.1	−13 0.13	. .	−2.05	0 14 17.26	

CORRECTIONS.

Date.		Corr. of Clock.	Hourly rate.	m	n	c
1847.	h.	s.	s.	s.	s.	s.
Nov. 5.	21	f 40.52	g 0.025	− 0.134	+ 0.198	0.000
Dec. 4.	23	f 3.88	g 0.020			

INSTRUMENT READINGS.

	Date.	Barom.	THERMOM. At.	THERMOM. Ex.
	1847. h. m.	in.	°	°
Zone 148	Dec. 4. 9 5	30.114	71.	33.5

REMARKS.

(146) 28. Transit over T. IV assumed as 20s.3 instead of 30s.3.
(146) 33. Micrometer reading assumed as 13r.15 instead of 11r.15.
(146) 34. Micrometer reading assumed as 13r.29 instead of 11r.29.
(146) 36. Micrometer reading assumed as 11r.39 instead of 9r.39.
(146) 37. Micrometer reading assumed as 8r.22 instead of 6r.22.
(147) 4. Micrometer reading assumed as 10r.[illegible] instead of 9r.[illegible].
(147) 16. Micrometer reading assumed as 15r.10 instead of 14r.10.
(147) 18. Micrometer reading assumed as 6r.32 instead of 5r.32
(148) 1. Hour assumed as 23 instead of 22.
(148) 3. Transit over T. V assumed to have been recorded as over T. VI.
(148) 4. Hour assumed as 0 instead of 23.
(148) 6. Micrometer reading assumed as 9r.50 instead of 10r.50

* The hour of this Zone was not recorded, and the declination was erroneously given. There can, however, be little doubt that the Zone has now been identified.

ZONE 148. DECEMBER 4. K. BELT, −25° 39'. D_o = −24° 0' 6"—Continued.

No.	Mag.	SECONDS OF TRANSIT. I.	II.	III.	IV.	V.	VI.	VII.	T.	a_1	a_2	MICROMETER.			i	d_1	d_2	Mean Right Ascension, 1850.0.	Mean Declination, 1850.0.
									h. m. s.	s.	s.			r.	' "	"	"	h. m. s.	° ' "
10	6	. .	. .	29.6	. .	57.2	. .	. .	23 14 43.47	+ 2.70	0.00	IV.	9	2.50	−40 20.82	. .	−4.23	0 14 46.17	
11	7.8	55.9	9.9	24.0	. .	. .	. .	. .	16 37.60	2.68	0.00	IV.	3	7.55	13 57.61	. .	2.13	16 40.28	
12	9	. .	. .	59.8	. .	27.2	. .	. .	17 13.34	2.67	0.00	IV.	4	3.49	16 52.57	. .	2.36	17 16.01	
13	10	42.0	. .	10.2	. .	. .	. .	. .	23 19 23.68	+ 2.65	0.00	IV.	2	4.41	− 7 20.69	. .	−1.59	0 19 26.33	

ZONE 149. DECEMBER 6. B. BELT, −25° 1'. D_o = −24° 36" 30'.

No.	Mag.	I.	II.	III.	IV.	V.	VI.	VII.	T.	a_1	a_2	MICROMETER.			i	d_1	d_2	Mean Right Ascension, 1850.0.	Mean Declination, 1850.0.
1	8	34.3	48.3	. .	. .	. .	. .	. .	0 30 16.21	+ 3.15	0.00	II.	9	7.42	−42 47.90	− 6.77	−4.53	0 29 19.36	−25 19 29.2
2	10	. .	. .	. .	57.	. .	23.5	. .	31 56.41	3.13	0.00	VII.	9	10.42	44 18.47	6.81	4.66	31 59.54	20 59.9
3	10	. .	. .	. .	31.	45.	59.	. .	38 31.19	3.05	0.00	V.	7	3.50	30 53.33	6.94	3.50	38 34.24	7 33.8
4	6	. .	. .	. .	. .	. .	0.0	14.2	39 32.62	3.04	0.00	VII.	6	2.54	25 24.78	6.96	3.03	39 35.66	2 4.8
5	8	. .	. .	. .	19.	. .	. .	. .	40 18.85	3.03	0.00	V.	5	2.4	25 58.79	6.98	3.08	40 21.88	25 2 38.8
6	5	. .	. .	. .	. .	1.	14.5	28.5	41 47.05	3.01	0.00	VII.	5	1.13	20 32.74	7.02	2.62	41 50.06	24 57 12.4
7	5	. .	48.	2.	. .	. .	. .	. .	45 15.60	2.97	0.00	IV.	3	5.35	12 47.02	7.13	1.97	45 18.57	49 26.1
8	9	. .	. .	56.	. .	23.5	. .	. .	46 9.56	2.96	0.00	VI.	3	7.20	13 39.80	7.15	2.06	46 12.52	50 19.0
9	10	. .	. .	. .	. .	33.	47.	. .	47 19.35	2.95	0.00	VII.	2	4.58	7 28.90	7.19	1.52	47 22.30	44 7.6
10	9	. .	. .	15.	29.	. .	. .	. .	52 28.58	2.89	0.00	VI.	1	1.58	0 59.33	7.37	1.00	52 31.47	24 37 37.7
11	9	. .	. .	. .	. .	. .	. .	5.	53 23.62	2.88	0.00	VI.	7	8.26	28 12.39	7.41	3.25	53 26.50	25 4 53.0
12	9	. .	. .	45.5	59.	. .	. .	. .	56 59.04	2.84	0.00	IV.	6	8.10	28 4.48	7.55	3.24	57 1.88	25 4 45.3
13	.	. .	. .	. .	53.0	7.	. .	. .	0 59 52.98	2.80	0.00	VI.	2	12.26	11 15.00	7.67	1.84	0 59 55.76	24 47 54.5
14	8	. .	. .	. .	. .	. .	. .	50.	1 1 8.51	2.79	0.00	VII.	5	11.55	25 56.47	7.72	3.08	1 1 11.30	25 2 37.3
15	8	. .	. .	. .	4.	18.	. .	. .	7 4.03	2.72	0.00	VI.	6	10.18	29 9.02	8.02	3.34	7 6.75	25 5 50.4
16	8	. .	5.	19.	33.	. .	. .	. .	15 32.65	2.63	0.00	V.	4	6.41	18 19.26	8.48	2.43	14 35.28	24 55 0.2
17	6	. .	. .	10.	23.5	. .	. .	. .	16 23.56	2.62	0.00	V.	7	5.15	31 36.19	8.53	3.55	16 26.18	25 8 18.3
18	8	. .	. .	. .	28.	. .	55.	. .	20 27.72	2.57	0.00	VII.	2	12.2	11 2.70	8.77	1.80	20 30.29	24 47 43.3
19	8	. .	. .	. .	. .	50.	13.	17.	21 45.55	2.56	0.00	VII.	6	12.31	30 15.72	8.85	3.44	21 48.11	25 6 58.0
20	8	. .	36.	50.	. .	17.5	. .	. .	24 3.81	2.54	0.00	VI.	9	4.2	40 56.97	9.00	4.39	24 6.35	25 17 40.4
21	7	23.	37.	51.	5.	. .	. .	. .	26 4.65	2.51	0.00	V.	4	9.50	19 54.56	9.14	2.57	26 7.16	24 56 36.3
22	9	. .	. .	14.	28.	. .	. .	. .	28 27.87	2.49	0.00	V.	8	11.19	39 38.44	9.29	4.24	28 30.36	25 16 22.0
23	9	. .	. .	. .	43.	57.	. .	. .	30 42.97	2.46	0.00	V.	1	11.55	10 59.49	9.45	1.82	30 45.43	24 47 41.76
24	9	. .	. .	16.	29.5	43.	. .	. .	31 29.36	2.45	0.00	VI.	4	7.27	18 42.33	9.51	2.48	36 31.81	55 24.3
25	8	55.	9.	23.	. .	. .	. .	. .	39 36.47	2.36	0.00	IV.	2	9.4	9 33.31	10.10	1.68	39 38.83	24 46 15.1
26	9	. .	43.	56.5	12.	. .	. .	. .	43 10.83	2.33	0.00	V.	8	12.50	40 24.32	10.40	4.34	43 13.16	25 17 9.1
27	8	. .	. .	. .	. .	. .	. .	51.	44 9.68	2.32	0.00	VII.	8	9.3	33 30.86	10.48	3.76	44 12.00	25 10 15.1
28	8	. .	53.	7.	. .	. .	. .	. .	46 20.58	2.30	0.00	IV.	4	10.27	20 13.26	10.66	2.60	46 22.88	24 56 56.5
29	8	. .	49.	3.	. .	. .	. .	. .	49 16.53	2.27	0.00	IV.	3	10.51	15 26.36	10.90	2.20	49 18.80	52 9.5
30	7	. .	. .	. .	. .	. .	. .	0.0	49 18.45	2.27	0.00	VII.	4	12.52	21 26.01	10.90	2.70	49 20.72	24 58 9.6
31	8	3.	17.	31.	. .	. .	. .	. .	53 44.73	2.21	0.00	VII.	7	6.06	32 1.60	11.30	3.59	53 46.94	25 8 46.5
32	8	. .	45.	59.	12.5	. .	. .	. .	55 12.42	2.20	0.00	V.	2	8.32	9 17.14	11.44	2.08	55 14.62	24 46 0.7
33	8.9	. .	13.	27.	41.	. .	. .	. .	1 58 40.74	2.16	0.00	V.	6	9.39	28 49.32	11.76	3.33	1 58 42.90	25 5 34.4
34	9	. .	. .	5.	19.	. .	. .	. .	2 1 18.92	2.14	−0.01	V.	10	7.36	47 45.90	12.01	5.00	2 1 21.05	24 32.9
35	6.7	. .	. .	. .	. .	54.5	8.	. .	2 40.61	2.12	0.00	VII.	6	5.12	26 34.36	12.15	3.12	2 42.73	3 19.7
36	9	13.	27.2	41.3	. .	. .	. .	. .	4 54.83	2.10	0.00	IV.	6	4.46	26 21.62	12.37	3.10	4 56.93	3 7.1
37	8	. .	45.	59.	. .	. .	. .	. .	10 12.66	2.05	0.00	IV.	6	4.2	25 59.43	12.91	3.08	10 14.71	2 45.4
38	9	27.	41.2	55.	. .	. .	. .	. .	13 8.81	2.02	0.00	IV.	7	8.57	33 28.18	13.23	3.73	13 10.83	25 10 15.1
39	7	. .	. .	. .	. .	57.	10.	. .	18 42.84	1.97	+0.01	VII.	1	4.36	2 18.82	13.85	1.08	18 44.82	24 39 3.7
40	9	. .	. .	8.2	22.	. .	. .	. .	32 21.84	1.84	0.00	V.	5	4.52	22 23.50	15.46	2.77	32 23.68	59 11.7
41	7	. .	. .	. .	. .	0.3	13.5	25.	2 33 46.25	+ 1.83	+0.01	VII.	1	10.6	−10 4.22	−15.65	−1.70	2 33 48.09	−24 46 51.6

CORRECTIONS.

Date.		Corr. of Clock.	Hourly rate.	m	n	c
1847.	h.	s.	s.	s.	s.	s.
Dec. 6,	23	f 3.43	g 0.000	+ 0.468	− 0.549	0.000

INSTRUMENT READINGS.

	Date.	Barom.	THERMOM. At.	Ex.
	1847. h. m.	in.	°	°
Zone 149	Dec. 6, 6 5	30.294	77.	34.

REMARKS.

(149) 1. Minutes assumed as 29 instead of 30.
(149) 11. Hor. thread assumed as 6 instead of 7.
(149) 16. Minutes assumed as 14 instead of 15, and micrometer reading as 6r.41 instead of 5r.41.
(149) 23. Hor. thread assumed as 2 instead of 1.
(149) 24. Minutes assumed as 36 instead of 31, to agree with Arg. Z. 327, 27.
(149) 26. Transit observations discordant.
(149) 27. Hor. thread assumed as 7 instead of 8.
(149) 41. Micrometer thread assumed as 2 instead of 1.

ZONE 149. DECEMBER 6. B. BELT, $-25° 1'$. $D_0 = -24° 36' 30''$—Continued.

No.	Mag.	SECONDS OF TRANSIT. I.	II.	III.	IV.	V.	VI.	VII.	T.	a_1	a_2	MICROMETER.			i	d_1	d_2	Mean Right Ascension, 1850.0.	Mean Declination, 1850.0.
									h. m. s.	s.	s.			r.	′ ″	″	″	h. m. s.	° ′ ″
42	9	5.	19.	33.					2 37 46.64	+ 1.79	0.00	IV.	5	9.22	−24 39.68	−16.15	−2.97	2 37 48.43	−25 1 28.8
43	9				40.	54.			38 40.05	1.78	−0.01	VI.	8	12.56	40 27.23	16.26	4.35	38 41.82	17 17.8
44	9		47.	1.	15.				42 14.81	1.75	−0.01	V.	8	6.43	37 19.27	16.73	4.08	42 16.55	14 10.1
45	7						38.	52.	43 10.56	1.74	0.00	VII.	7	9.48	33 53.53	16.85	3.75	43 12.30	25 10 44.1
46	9	2.	16.	29.5					49 43.40	1.68	0.00	IV.	4	5.42	17 49.55	17.73	2.39	49 45.08	24 54 39.7
47	9						1.	15.	50 33.58	1.68	−0.01	VII.	8	5.50	36 52.22	17.84	4.02	50 35.25	25 13 44.1
48	9				5.2	19.	32.	47.	2 59 5.02	1.61	+0.01	VII.	2	11.21	10 42.02	19.07	1.75	2 59 6.64	24 47 32.8
49	8		9.2	23.3	37.				3 2 36.73	1.58	+0.01	V.	1	11.48	5 56.96	19.60	1.34	3 2 38.32	24 42 47.9
50	9					47.	0.0	14.5	4 33.01	1.56	−0.02	VII.	10	7.24	47 39.52	19.89	5.01	4 34.55	25 24 34.4
51	7			38.	52.				11 51.60	1.50	+0.02	V.	1	4.11	2 6.52	21.01	1.03	11 53.12	24 38 58.6
52	6					15.	28.	42.	13 0.64	1.50	0.02	VII.	1	6.34	3 18.32	21.20	1.12	13 2.16	40 10.6
53	7	2.	16.2	30.2	44.				17 43.69	1.46	+0.01	IV.	3	6.26	13 12.74	21.95	1.99	17 45.16	24 50 6.7
54	8							14.	22 32.81	1.42	−0.02	VII.	10	8.12	48 3.73	22.74	5.05	22 34.21	25 25 1.5
55	7	16.	30.	44.	58.				25 57.75	1.40	0.00	V.	6	13.21	30 41.24	23.31	3.48	25 59.15	25 7 38.0
56	6	53.3	7.	21.2	35.				28 34.71	1.38	+0.01	V.	3	3.30	11 43.94	23.75	2.09	28 36.10	24 48 39.8
57	9							12.	29 30.49	1.37	0.00	VII.	5	8.2	23 58.99	23.91	2.92	29 31.86	25 0 55.8
58	8		24.	38.	52.				33 51.70	1.34	0.00	V.	5	7.15	23 35.60	24.66	2.89	33 53.04	0 33.1
59	7		3.5	17.5	32.				35 31.43	1.33	0.00	V.	7	4.14	31 5.44	24.95	3.52	35 32.76	25 8 3.9
60	9						45.	59.	36 17.49	1.33	+0.01	VII.	4	7.52	18 54.74	25.10	2.46	36 18.83	24 55 52.3
61	8	5.	19.2	33.1					39 46.63	1.30	+0.01	IV.	3	9.28	14 44.50	25.70	2.10	39 47.94	24 51 42.35
62	9		13.2		41.				42 40.87	1.28	0.00	V.	7	4.58	31 27.62	26.22	3.55	42 42.15	25 8 27.4
63	9							20.	43 38.67	1.28	−0.01	VII.	8	7.10	37 32.56	26.39	4.08	43 40.94	14 33.0
64	5		51.	5.	19.	32.5	46.		47 18.71	1.26	0.00	VII.	6	5.12	26 34.36	27.06	3.12	47 19.97	25 3 34.5
65	9						12.	26.	48 44.61	1.25	−0.01	VII.	9	7.5	42 29.04	27.32	4.58	48 45.85	19 30.9
66	7		22.	35.5	49.5	3.			50 49.31	1.24	0.00	VI.	6	6.08	27 2.80	27.70	3.16	50 50.55	4 3.7
67	9				22.	36.			53 22.02	1.22	0.00	VI.	5	6.57	23 26.41	28.18	2.88	53 23.24	0 27.5
68	9	37.5		5.1	19.				3 57 19.04	+ 1.20	−0.01	V.	8	12.48	−40 23.31	−28.91	−4.39	3 57 20.23	−25 17 26.6

ZONE 150. DECEMBER 6. H. BELT, $-25° 1'$. $D_0 = -24° 37' 0''$.

No.	Mag.	I.	II.	III.	IV.	V.	VI.	VII.	T.	a_1	a_2	MICROMETER.			i	d_1	d_2	Mean Right Ascension, 1850.0.	Mean Declination, 1850.0.
1	8	41.	55.	9.2	22.				7 8 22.49	+ 0.99	0.00	IV.	5	10.35	−25 16.50	−25.45	−3.02	7 8 23.48	−25 2 44.97
2	8	56.	9.5	24.	37.4				10 37.34	0.99	+0.01	IV.	3	12.19	16 10.73	26.83	2.18	10 38.34	24 53 39.74
3	5		1.	15.					12 28.39	1.00	0.02	IV.	1	7.2	3 32.79	28.03	1.06	12 29.41	41 1.88
4	6					12.5	26.	40.5	12 58.64	1.00	0.02	VII.	1	7.2	3 32.43	28.35	1.06	12 59.66	41 1.84
5	10			18.	32.	45.5			16 31.67	1.02	+0.01	VI.	3	11.25	15 43.34	30.65	2.15	16 32.70	24 53 16.14
6	10					18.	32.	46.	18 4.47	1.02	−0.01	VII.	8	8.31	38 13.41	31.66	4.19	18 5.48	25 15 49.26
7	6						38.5	53.	19 11.23	1.03	+0.01	VII.	4	5.45	17 50.70	32.39	2.31	19 12.27	24 55 25.40
8	8		49.	3.	16.3				22 16.34	1.04	0.02	V.	2	4.34	7 17.13	34.39	1.37	22 17.40	44 52.88
9	8					23.	36.5		22 9.10	1.04	0.02	VII.	3	3.52	11 54.72	34.30	1.75	22 10.16	49 30.77
10	7				47.	1.	14.5		23 47.02	1.05	+0.01	VII.	3	5.21	12 39.59	35.36	1.84	23 48.08	24 50 16.79
11	8		16.5		44.				25 44.15	1.06	0.00	V.	6	8.12	28 5.45	36.62	3.25	25 45.21	25 5 45.32
12	9		31.						29 58.58	1.08	0.00	III.	4	12.0	21 0.11	39.38	2.59	29 59.66	24 58 42.08
13	8					29.	42.5		29 15.11	1.08	0.00	VII.	5	12.48	26 23.19	38.91	3.11	29 16.19	25 4 5.21
14	7						44.	58.	30 16.50	1.08	0.00	VII.	5	5.2	22 28.22	39.56	2.75	30 17.58	0 10.53
15	6				3.	16.5			32 2.76	1.09	0.00	V.	5	8.12	24 4.34	40.70	2.92	32 3.65	1 47.96
16	6					11.	24.3		32 57.01	1.09	0.00	VII.	5	9.26	25 41.45	41.29	3.22	32 58.10	6 25.96
17	9						20.	34.	7 34 52.58	+ 1.10	−0.01	VII.	8	9.57	−38 56.77	−42.53	−4.26	7 34 53.67	−25 16 43.56

CORRECTIONS.

Date.	Corr. of Clock.	Hourly rate.	m	n	c
1847. h.	s.	s.	s.	s.	s.

INSTRUMENT READINGS.

Date.	Barom.	THERMOM. At.	Ex.
1847. h. m.	in.	°	°

REMARKS.

(149) 64. Observed transit over T. IV assumed to have been at 19s instead of 17s.
(150) 16. Hor. thread assumed as 6 instead of 5.

ZONE 151. DECEMBER 15. B. DELT, −24° 23′. D_0 = −23° 58′ 40″.

No.	Mag.	Seconds of Transit. I.	II.	III.	IV.	V.	VI.	VII.	T.	a_1	a_2	Micrometer.		r.	i	d_1	d_2	Mean Right Ascension, 1850.0.	Mean Declination, 1850.0.
									h. m. s.	s.	s.			r.	′ ″	″	″	h. m. s.	° ′ ″
1	8	. .	57.	11.	. .	. .	. .	. .	1 52 24.69	− 3.51	0.00	IV.	8	9.12	−43 33.44	−15.43	−4.50	1 52 21.18	−24 42 33.37
2	9	. .	. .	. .	. .	7.5	21.	. .	54 53.72	3.54	0.00	IX.	8	6.29	37 10.74	15.66	3.97	54 50.18	36 10.37
3	6	. .	. .	. .	. .	. .	27.	40.5	55 59.50	3.55	0.00	VII.	6	7.10	37 32.57	15.76	4.00	55 55.95	36 32.33
4	8	57.	11.2	25.	. .	. .	. .	. .	1 59 38.47	3.60	0.00	IV.	4	1.10	18 5.67	16.11	2.44	1 59 34.87	17 2.22
5	9	. .	. .	. .	15.3	29.	43.	. .	2 2 15.36	3.62	0.00	VII.	6	11.21	29 40.43	16.38	3.38	2 2 11.74	28 40.19
6	7	. .	45.2	59.3	13.	. .	. .	. .	8 12.73	3.69	0.00	V.	3	10.42	15 21.79	17.00	2.23	8 9.04	14 21.02
7	9	. .	. .	. .	21.	35.	48.	. .	10 20.93	3.71	−0.01	VII.	10	5.27	47 40.53	17.23	4.85	10 17.21	46 42.61
8	6	. .	. .	. .	. .	. .	57.	. .	11 29.69	3.73	0.00	VII.	4	10.42	20 20.48	17.36	2.63	11 25.96	19 20.47
9	9	. .	. .	53.5	7.	21.	. .	. .	13 7.03	3.74	0.00	VI.	4	10.44	20 21.68	17.54	2.63	13 3.29	19 21.85
10	5	3.	17.	31.	45.	. .	. .	. .	15 44.64	3.77	0.00	V.	7	4.2	30 59.39	17.87	3.48	15 40.87	30 0.74
11	8	. .	. .	53.	7.5	. .	. .	. .	19 7.02	3.82	0.00	V.	7	6.29	32 13.51	18.21	3.58	19 3.20	31 15.30
12	9	. .	. .	. .	. .	. .	59.	13.	19 31.74	3.82	0.00	VII.	7	10.52	34 25.82	18.26	3.76	19 27.92	33 27.84
13	8	. .	19.5	33.2	. .	. .	. .	. .	22 47.15	3.85	−0.01	V.	10	7.15	47 35.30	18.64	4.83	22 43.29	46 38.77
14	9	. .	. .	. .	. .	30.	43.	. .	2 25 15.97	3.88	0.00	VII.	7	5.01	31 28.82	18.93	3.52	2 25 12.09	30 31.27
15	6	22.	36.	50.	. .	. .	. .	. .	3 4 3.39	4.28	+0.01	IV.	3	9.3	14 31.90	24.23	2.16	3 3 59.12	13 38.29
16	9	. .	. .	. .	. .	52.	6.	. .	4 38.47	4.28	−0.01	VII.	6	5.38	36 56.26	24.31	3.97	4 34.18	36 4.54
17	7	. .	19.	34.	47.5	. .	. .	. .	10 46.99	4.34	+0.01	V.	1	10.32	5 18.65	25.27	1.43	10 42.66	4 25.35
18	9	. .	. .	44.	58.	. .	. .	. .	11 57.83	4.35	−0.01	V.	8	11.51	39 54.57	25.46	4.20	11 53.47	39 4.23
19	6	. .	. .	53.	7.	21.	. .	. .	13 7.00	4.36	0.01	VI.	8	13.51	40 54.95	25.65	4.27	13 2.63	40 4.87
20	9	. .	. .	. .	. .	13.5	27.	41.	13 59.78	4.37	−0.01	VII.	8	12.36	40 16.96	25.79	4.23	13 55.40	39 26.98
21	6	. .	. .	. .	. .	. .	. .	37.	14 55.53	4.37	+0.01	VII.	2	12.53	11 28.42	25.94	1.93	14 51.17	10 36.29
22	9	. .	1.	15.	28.	. .	. .	. .	23 28.24	4.45	+0.01	IV.	4	3.23	16 39.45	27.33	2.32	23 23.80	15 49.10
23	9	. .	. .	. .	. .	. .	. .	29.	23 47.79	4.45	0.00	IV.	7	3.70	16 32.90	27.39	2.34	23 43.34	15 42.59
24	9	. .	. .	. .	. .	. .	. .	46.	25 4.68	4.47	0.00	VII.	5	5.16	22 35.29	27.59	2.80	25 0.21	21 45.68
25	8	. .	. .	. .	. .	54.	. .	22.	26 40.41	4.48	+0.01	VII.	4	4.58	17 27.01	27.87	2.38	26 35.94	16 37.26
26	9	. .	. .	40.	54.	8.	. .	. .	28 53.92	4.50	0.00	VI.	6	8.31	28 14.91	28.23	3.26	28 49.42	27 26.40
27	9	. .	. .	. .	. .	50.	3.	17.3	32 36.03	4.53	−0.01	VII.	8	3.10	35 31.55	28.86	3.87	32 31.49	34 44.28
28	9	. .	9.	23.	37.	. .	. .	. .	35 36.51	4.56	+0.01	V.	2	6.52	8 56.97	29.38	1.67	35 31.96	8 8.02
29	5	36.	50.	4.	17.5	. .	. .	. .	41 17.43	4.61	0.00	V.	5	2.37	21 15.43	30.38	2.70	41 12.82	20 28.51
30	7	. .	. .	. .	. .	. .	. .	45.5	42 4.14	4.62	0.00	VII.	5	0.43	20 17.63	30.52	2.62	41 59.52	19 30.77
31	9	. .	. .	. .	43.5	57.	10.	. .	44 43.10	4.64	−0.01	VII.	7	11.26	34 42.96	31.00	3.81	3 44 38.47	33 57.77
32	8	33.	47.2	1.	. .	. .	. .	. .	52	4.70	+0.01	IV.	3	8.47	14 23.84	32.34	2.11	. . .	13 38.29
33	5	. .	9.	23.	37.	. .	. .	. .	3 53	− 4.72	0.00	V.	6	6.50	−27 24.10	−32.53	−3.19	. . .	−24 26 39.82

CORRECTIONS.

Date.	Corr. of Clock.	Hourly rate.	m	n	c
1847. h.	s.	s.	s.	s.	s.
Dec. 15, 0	/ 9.77	ɛ 0.063	+ 0.760	− 1.280	0.000

INSTRUMENT READINGS.

	Date.	Barom.	Thermom. At.	Thermom. Ex.
	1847. h. m.	in.	°	°
Zone 151	Dec. 15, 9 5	30.042	76.	35.

REMARKS.

(151) 1. Hor. thread assumed as 9 instead of 8.
(151) 2. Transit observations incongruous. Observed transit over T. VI assumed as $21^s.0$ instead of $27^s.0$.
(151) 4. Micrometer reading assumed as 4^t $6^r.10$ instead of 4^t $1^r.10$.
(151) 23. Micrometer reading assumed to have been 4^t $2^r.7$ instead of 7^t $3^r.70$.
(151) 28. Micrometer reading assumed as $7^r.52$ instead of $6^r.52$.

ZONE 152. JANUARY 18. K. BELT, −25° 38′. D₀=−25° 13′ 50″.

No.	Mag.	I.	II.	III.	IV.	V.	VI.	VII.	T.	a_1	a_2	MICROMETER.			i	d_1	d_2	Mean Right Ascension, 1850.0.	Mean Declination, 1850.0.
									h. m. s.	s.	s.			r.	′ ″	″	″	h. m. s.	° ′ ″
1	9		27.8						4 17 55.60	−49.20	0.00	.	7	2.26	−30 11.02	− 6.07	−2.48	4 17 6.40	−25 44 9.57
2	9.10		7.1	21.3					25 34.69	49.26	+0.02	.	1	8.45	4 24.73	7.50	0.16	24 45.45	18 22.39
3	9.10		46.3		14.3				27 14.06	49.28	0.00	.	4	5.12	17 34.42	7.82	1.34	26 24.78	31 33.58
4	9					1.7	15.0		27 47.60	49.28	−0.01	.	8	9.29	38 43.01	7.93	3.28	26 58.31	52 44.22
5	9.10		3.2	17.4					29 30.83	49.30	+0.01	.	2	4.9	7 4.56	8.25	0.39	28 41.54	25 21 3.20
6	9		7.3	21.3					33 35.29	49.34	−0.02	..	10	4.42	46 18.20	9.03	3.96	32 45.93	26 0 21.19
7	9	0.5	14.7	28.5					40 42.20	49.45	+0.01	.	3	4.21	12 9.70	10.41	0.83	39 52.76	25 26 10.94
8	8.9	29.9	44.0	57.9					50 11.92	49.49	−0.02	.	9	10.30	44 12.78	12.30	3.78	49 22.41	58 18.86
9	8.9			23.8	37.3				4 54 37.18	−49.51	+0.03	.	1	5.49	− 2 55.98	−13.22	−0.02	4 53 47.70	−25 16 59.22

ZONE 153. JANUARY 18. K. BELT, −25° 1′. D₀=−24° 36′ 20″.

No.	Mag.	I.	II.	III.	IV.	V.	VI.	VII.	T.	a_1	a_2	MICROMETER.			i	d_1	d_2	Mean Right Ascension, 1850.0.	Mean Declination, 1850.0.
1	9						33.8		7 18 6.41	−50.14	+0.02	F.	1	7.2	− 3 32.63	−10.38	−0.04	7 17 16.29	−24 40 3.11
2	9.10	25.9	39.9						22 7.37	50.14	0.02	.	2	6.27	3 14.14	11.99	0.02	21 17.25	39 46.15
3	10			46.8					22 0.29	50.14	0.01	.	3	5.49	12 54.08	11.94	0.89	21 10.16	49 26.91
4	9	56.9	10.9						24 38.38	50.14	+0.02	.	2	7.15	8 38.34	13.06	0.52	23 48.16	24 45 11.92
5	10	53.3							26 34.99	50.14	−0.01	.	6	10.6	29 2.93	13.89	2.37	25 44.84	25 5 39.19
6	8	26.0	40.2						31 7.72	50.15	0.00	.	5	7.1	23 28.58	15.84	1.86	30 17.57	0 6.28
7	6.7	12.1	26.1	40.0					32 53.70	50.15	0.00	.	5	9.55	24 56.32	16.59	1.99	32 3.55	1 34.90
8	8		46.8	0.6					7 38 14.43	−50.15	−0.01	.	7	6.40	−32 19.10	−18.88	−2.68	7 37 24.27	−25 9 0.66

ZONE 154. JANUARY 19. B. BELT, −23° 46′. D₀=−23° 21′ 50″.

No.	Mag.	I.	II.	III.	IV.	V.	VI.	VII.	T.	a_1	a_2	MICROMETER.			i	d_1	d_2	Mean Right Ascension, 1850.0.	Mean Declination, 1850.0.
1	7							5.	6 8 23.95	−50.61	0.00	VII.	6	7.27	−27 42.45	− 4.55	−2.22	6 7 33.34	−23 49 39.2
2	7			38.	52.	5.			11 51.63	50.63	−0.01	V.	8	10.42	39 19.79	5.15	3.21	11 1.01	24 1 18.2
3	9		12.	26.					13 39.23	50.63	+0.02	IV.	1	13.10	6 38.34	5.46	0.48	12 48.62	23 28 34.3
4	7		15.	59.					15 12.29	50.64	0.01	IV.	2	14.26	12 15.66	5.71	0.95	14 21.66	34 12.3
5	8		19.	33.					16 0.14	50.64	+0.02	VI.	2	8.29	9 15.51	5.99	0.70	16 9.54	23 31 12.2
6	9				26.5	40.5			19 26.62	50.66	−0.02	V.	9	8.03	42 58.61	6.44	3.53	18 35.94	24 4 58.6
7	10			33.5	47.				21 47.02	50.66	0.01	V.	8	7.45	37 50.53	6.85	3.10	20 56.35	23 59 50.5
8	10	22.	6.	50.					24 3.44	50.67	0.01	IV.	7	4.16	31 6.49	7.25	2.52	23 12.76	24 53 6.3
9	9			49.	3.				25 2.80	50.68	−0.02	IV.	9	3.23	40 37.45	7.41	3.33	24 12.10	24 2 38.2
10	10				16.	29.8			26 15.96	50.68	+0.01	V.	3	7.30	13 55.05	7.64	1.08	25 25.29	23 35 53.8
11	10				54.	7.3			27 53.71	50.69	0.01	VI.	3	13.9	16 35.79	7.91	1.31	27 3.03	38 34.9
12	11				4.		27.	41.	29 59.85	50.69	+0.01	VII.	5	7.29	23 42.36	8.28	1.89	29 9.17	45 42.5
13	8				43.2	57.			30 43.21	50.70	−0.01	VI.	8	7.49	37 52.44	8.40	3.10	29 52.50	59 53.9
14	6		32.	45.5	59.				32 58.92	50.70	+0.02	V.	1	10.8	5 6.54	8.81	0.34	32 8.24	27 5.7
15	6					26.	39.3		34 12.21	50.71	+0.02	VII.	3	3.25	11 41.12	9.02	0.90	33 21.52	23 33 41.0
16	10		18.	32.					36 45.65	50.72	−0.02	IV.	9	9.56	43 55.63	9.47	3.61	35 54.91	24 5 58.7
17	10						28.5	42.9	37 1.65	50.72	−0.02	VII.	9	8.55	43 24.53	9.51	3.57	36 10.91	24 5 27.6
18	10						47.	1.2	38 19.92	50.72	+0.01	VII.	3	11.43	15 52.24	9.73	1.25	37 29.21	23 37 53.2
19	9							10.3	39 29.14	50.73	+0.01	VII.	4	11.04	20 31.58	9.94	1.63	38 38.42	42 33.2
20	9					29	42.		41 15.07	50.73	−0.01	VII.	8	6.13	37 3.84	10.26	3.03	40 24.33	59 7.1
21	7		48.						44 15.45	50.74	0.01	III.	8	4.45	36 19.77	10.79	2.96	43 24.70	58 23.5
22	6							2.	44 21.01	50.74	−0.01	VII.	7	6.46	32 21.79	10.81	2.63	43 30.26	54 25.2
23	8					31.	44.		46 17.06	50.75	+0.01	VII.	3	11.4	15 32.58	11.15	1.21	45 26.32	37 34.9
24	8				52.5	6.			6 47 52.35	−50.75	−0.01	VI.	7	10.6	−34 2.83	−11.43	−2.76	6 47 1.59	−23 56 7.0

CORRECTIONS.

Date.		Corr. of Clock.	Hourly rate.	m	n	c
1848.	h.	s.	s.	s.	s.	s.
Jan. 18.	6	− 55.49	g 0.039	+ 1.47	− 1.61	+ 0.01
19.	6	− 56.19	g 0.049			

INSTRUMENT READINGS.

Date.		Barom.	THERMOM. At.	THERMOM. Ex.
1848.	h. m.	in.	°	°

REMARKS.

(154) 5. Transits over T.'s I and II assumed as recorded over T.'s II and III, and minutes assumed as 17 instead of 16.
(154) 8. Transit over T. II assumed as 36ˢ instead of 34ˢ.

ZONE 154. JANUARY 19. B. BELT, −23° 46′. D_o = −23° 21′ 50″—Continued.

No.	Mag.	Seconds of Transit I.	II.	III.	IV.	V.	VI.	VII.	T.	a_1	a_2	Micrometer.			i	d_1	d_2	Mean Right Ascension, 1850.0.	Mean Declination, 1850.0.
									h. m. s.	s.	s.			r.	′ ″	″	″	h. m. s.	° ′ ″
25	5						12.	26.	6 48 44.92	−50.75	−0.01	VII.	8	7.57	−37 56.28	−11.37	−3.11	6 47 54.16	−24 0 1.0
26	9	58.	12.	25.5					51 39.05	50.76	+0.02	IV.	3	2.41	11 19.28	12.09	0.85	50 48.31	23 33 22.2
27	9	30.	44.	58.					53 11.30	50.77	0.01	IV.	4	7.31	18 44.51	12.37	1.48	52 20.54	40 48.4
28	10			26.4	40.2				54 39.95	50.77	0.01	V.	4	7.31	18 44.47	12.63	1.48	53 49.19	40 48.6
29	4		9.	22.8	36.5				57 36.26	50.78	+0.01	V.	3	9.50	14 55.56	13.15	1.16	56 45.39	23 36 59.9
30	10						48.5	2.	6 58 21.19	50.78	−0.02	VII.	9	8.46	43 20.0	13.28	3.57	6 57 30.39	24 5 26.9
31	8	16.							7 1 57.08	50.78	+0.01	.	3	8.51	14 25.85	13.92	1.12	7 1 6.31	23 36 30.9
32	10		35.	49.					3 2.48	50.79	0.00	IV.	6	8.49	28 24.15	14.10	2.29	2 11.69	50 30.5
33	7					38.	51.5	5.5	3 24.35	50.79	0.00	VII.	6	4.42	26 19.26	14.17	2.11	2 33.56	48 25.5
34	10							59.	4 17.96	50.79	0.00	VII.	6	9.58	28 58.59	14.33	2.34	3 27.17	51 5.3
35	11				8.	22.			6 8.07	50.79	+0.01	VI.	4	11.39	21 19.71	14.65	1.68	5 17.29	43 26.0
36	8				39.	53.			7 39.11	50.80	−0.01	VI.	8	5.12	36 33.27	14.98	3.00	6 48.30	58 11.3
37	7							17.	8 35.63	50.80	+0.02	VII.	1	7.14	3 38.50	15.09	0.18	7 44.85	25 43.8
38	7					34.	47.5		10 20.30	50.80	0.02	VII.	1	13.17	6 41.53	15.39	0.46	9 29.52	28 47.4
39	8		3.5	17.3	31.			51.	11 10.74	50.80	0.02	VII.	2	13.30	11 47.09	15.61	0.89	10 19.96	33 53.6
40	9					39.	52.5		13 30.31	50.80	+0.02	IV.	2	6.12	8 6.58	15.04	0.57	12 39.53	30 13.1
41	9			2.	15.6				16 15.56	50.81	−0.01	IV.	7	11.48	34 54.40	16.46	2.84	15 24 74	57 3.7
42	10				48.5	1.5			17 48.10	50.81	0.01	VI.	6	12.49	30 25.01	16.72	2.47	16 57.28	23 52 34.2
43	10				34.	47.5			19 33.88	50.81	−0.02	VI.	10	3.50	45 51.82	17.04	3.81	18 43.05	24 8 2.7
44	7			59.		26.			21 12.21	50.82	+0.02	VI.	1	5.40	2 51.29	17.32	0.10	20 21.41	23 24 58.7
45	9		10.	24.					36 37.42	50.82	0.00	IV.	5	7.22	23 39.17	20.04	1.89	35 46.60	23 45 51.1
46	11					24.	37.2	51.	37 10.18	50.83	−0.02	VII.	10	5.40	46 47.10	20.14	3.90	36 19.33	24 9 1.1
47	10			0.0	14.	27.			39 13.60	50.83	0.01	VI.	7	3.10	30 33.05	20.49	2.48	38 22.76	23 52 46.0
48	9				23.				40 22.85	50.83	−0.01	V.	7	4.55	31 26.11	20.70	2.56	39 32.01	53 39.4
49	10					17.	30.2		41 3.16	50.83	0.00	VI.	5	10.02	24 59.71	20.82	2.00	40 12.33	47 12.5
50	9						11.		42 43.88	50.83	+0.02	VI.	1	12.28	6 17.01	21.10	0.40	41 53.07	28 28.5
51	8			14.2	28.				44 27.64	50.83	+0.02	V.	1	6.17	3 10.06	21.41	0.14	43 36.83	23 25 21.6
52	8							11.	45 30.20	50.83	−0.02	VII.	10	5.20	46 37.01	21.59	3.89	44 39.35	24 8 52.5
53	9		50.3	4.2					49 17.78	50.83	0.01	IV.	7	7.13	32 35.73	22.25	2.66	48 26.91	23 54 50.64
54	10							35.5	51 54.67	50.83	−0.02	VII.	9	11.54	44 54.79	22.83	3.74	51 3.82	24 7 11.36
55	10				46.	0.0			53 46.05	50.82	+0.02	VI.	2	10.42	10 22.58	23.03	3.76	52 55.25	23 32 39.4
56	9						11.		55 43.88	50.82	+0.02	VII.	1	5.51	2 56.64	23.36	0.11	54 53.08	23 25 10.1
57	8				38.	51.5	5.		57 37.82	50.82	−0.02	VII.	10	9.35	48 45.60	23.68	4.04	56 46.98	24 11 3.3
58	9				6.	19.7			7 59 5.96	50.82	0.01	VI.	7	10.30	34 14.93	23.94	2.79	7 58 15.13	23 56 31.7
59	4.5	19.	33.	46.9	0.5	14.			8 2 0.37	50.82	0.01	VI.	6	12.22	30 11.39	24.44	2.44	8 1 9.54	52 28.3
60	10				7.5	21.			3 7.35	50.82	0.01	VI.	7	12.45	34 52.74	24.63	2.90	2 16.52	23 57 10.3
61	11					33.	46.2	0.0	8 4 19.16	−50.82	−0.02	VII.	9	7.07	−42 30.07	−24.83	−3.54	8 3 28.32	−24 4 48.4

ZONE 155. JANUARY 20. K. BELT, −26° 16′. D_o = −25° 52′ 0″.

No.	Mag.	I.	II.	III.	IV.	V.	VI.	VII.	T.	a_1	a_2	Micrometer.			i	d_1	d_2	Mean Right Ascension, 1850.0.	Mean Declination, 1850.0.
1	9				25.1	39.0			2 38 24.97	−50.37	0.00	.	3	11.39	−15 50.57	−10.29	−2.13	2 37 34.60	−26 8 2.99
2	9		52.2	6.5					40 19.98	50.40	+0.01	.	1	11.17	5 41.37	10.52	1.20	39 29.59	25 57 53.09
3	9				27.3				44 27.15	50.45	−0.01	.	10	7.15	47 35.34	11.03	5.17	43 36.69	26 39 51.54
4	8				50.3				53 50.15	50.56	+0.01	.	1	0.43	0 21.68	12.24	1.69	52 59.60	25 32 35.61
5	9	13.9	28.2	42.3					55 55.89	50.58	0.01	.	2	8.3	9 2.55	12.50	1.51	55 5.32	26 1 16.56
6	9					23.7	37.6		56 23.46	50.58	+0.01	.	2	11.39	10 51.47	12.53	1.69	55 32.89	3 5.69
7	7.8	20.8	35.0	49.2					2 59 3.00	50.63	0.00	.	6	11.8	29 34.23	12.91	3.44	2 58 12.37	21 50.58
8	9		41.0	55.1					3 1 9.09	−50.65	−0.01	.	8	11.31	−39 44.53	−13.19	−4.39	3 0 18.43	−26 32 2.11

CORRECTIONS.

Date.		Corr. of Clock.	Hourly rate.	m	n	c
1848.	h.	s.	s.	s.	s.	s.
Jan. 20,	6	− 57.81	0.048			

INSTRUMENT READINGS.

°	Date.	Barom.	Thermom. At.	Thermom. Ex.
	1848. h. m.	in.	°	°

REMARKS.

(154) 31. One of the threads 10ˢ wrong. The first assumed correct.
(154) 35. Micrometer reading assumed as 12ʳ.39 instead of 11ʳ.39.
(154) 39. Transits assumed as 43ˢ.5, 57ˢ.3, and 11ˢ instead of 3ˢ.5, 17ˢ.3, and 11ˢ.3.
(154) 40. Threads V and VI assumed as 44ˢ and 57ˢ.5 instead of 39ˢ and 52ˢ.5.
(154) 58. Time of transit over T. II assumed as 19ˢ.7 instead of 17ˢ.7.
(154) 60. Micrometer reading assumed as 11ʳ.45 instead of 12ʳ.45.
(155) 6. Transits over T.'s IV and V assumed as recorded over T.'s V and VI.

ZONE 155. JANUARY 20. K. BELT, −26° 16'. D₀ = −25° 52' 0"—Continued.

No.	Mag.	Seconds of Transit. I.	II.	III.	IV.	V.	VI.	VII.	T.	a_1	a_2	Micrometer.			i	d_1	d_2	Mean Right Ascension, 1850.0.	Mean Declination, 1850.0.
									h. m. s.	s.	s.			r.	' "	"	"	h. m. s.	° ' "
9	9	. .	. .	. .	40.9	54.8	. .	. .	3 1 50.77	−50.65	0.00	.	3	11.32	−15 47.04	−13.22	−2.14	3 1 0.12	−26 8 2.40
10	8.9	. .	. .	. .	. .	. .	2.8	. .	4 35.11	50.69	+0.01	F.	2	7.28	8 44.72	13.70	1.46	3 44.43	0 59.88
11	9	. .	. .	. .	49.3	3.0	. .	. .	5 49.07	50.70	0.00	.	3	7.51	13 55.59	13.85	1.96	4 58.37	6 11.40
12	9	. .	2.0	15.5	30.2	. .	. .	. .	7 29.91	50.73	−0.01	.	9	9.37	43 46.05	14.09	4.80	6 39.17	36 4.94
13	5.6	. .	. .	9.9	. .	38.0	. .	. .	9 24.08	50.75	−0.01	.	10	6.46	47 20.73	14.33	5.16	8 33.32	26 39 40.22
14	8	24.5	38.8	. .	. .	. .	. .	. .	13 6.47	50.79	+0.01	.	2	2.41	6 20.18	14.87	1.25	12 15.69	25 58 36.30
15	9	. .	33.0	47.3	1.9	. .	. .	. .	14 1.17	50.80	0.00	.	4	0.20	15 7.18	15.02	2.07	13 10.37	26 7 24.27
16	5.6	. .	12.7	26.7	40.8	. .	. .	. .	16 40.51	50.83	0.00	.	3	10.32	15 16.79	15.40	2.09	15 49.68	7 34.28
17	7	. .	. .	. .	36.4	. .	. .	. .	17 36.25	50.84	−0.01	.	10	6.26	47 10.64	15.53	5.15	16 45.40	39 31.32
18	7	. .	. .	15.9	30.2	. .	. .	. .	18 30.01	50.85	−0.01	.	9	5.41	41 47.04	15.67	4.61	17 39.15	34 7.32
19	8	. .	. .	3.2	17.3	. .	. .	. .	19 17.08	50.86	0.00	.	5	9.17	24 37.16	15.79	2.96	18 26.22	16 55.91
20	9	. .	. .	. .	. .	45.2	59.4	. .	19 45.32	50.86	−0.01	.	8	6.30	37 12.76	15.82	4.17	18 54.45	29 32.75
21	9	15.1	29.0	. .	. .	. .	. .	. .	21 57.00	50.91	0.00	.	4	11.37	20 48.56	16.34	2.62	22 6.09	13 7.52
22	9	. .	. .	. .	. .	. .	54.8	. .	22 27.07	50.94	0.00	.	3	13.28	16 45.51	16.76	2.23	24 36.13	9 4.50
23	9	. .	. .	. .	. .	. .	. .	38.5	26 56.55	50.95	0.00	F.	5	8.2	23 59.18	16.98	2.90	26 5.60	16 19.06
24	6	. .	. .	. .	. .	. .	. .	1.9	26 19.84	50.95	+0.01	*g	3	10.25	15 12.87	16.89	2.08	25 28.90	7 31.84
25	9	. .	. .	. .	. .	. .	. .	35.9	26 53.87	50.95	0.00	*g	4	5.38	17 47.15	16.97	2.33	26 2.92	10 6.45
26	5	. .	. .	5.9	19.8	. .	. .	. .	29 19.59	50.98	+0.01	.	3	5.51	12 55.08	17.34	1.86	28 28.62	5 14.28
27	9.8	. .	48.8	2.8	. .	. .	. .	. .	33 16.52	51.03	+0.01	.	3	7.15	13 37.44	17.94	1.92	32 25.50	5 57.30
28	9	. .	8.9	23.2	. .	. .	. .	. .	37 37.03	51.07	−0.01	.	7	11.45	34 52.89	18.66	3.95	36 45.95	27 15.50
29	6.7	. .	. .	51.0	5.0	. .	. .	. .	40 4.88	51.10	0.00	.	7	3.47	30 51.86	19.08	3.56	39 13.78	23 14.50
30	6.7	51.2	5.6	19.5	. .	. .	. .	. .	43 33.51	51.14	−0.01	.	8	6.22	37 8.72	19.66	4.16	42 42.36	26 29 32.54
31	9	. .	. .	12.2	26.2	. .	. .	. .	44 25.89	51.14	+0.01	.	1	10.51	5 28.26	19.82	1.16	43 34.76	25 57 49.24
32	10	. .	2.6	. .	. .	. .	. .	. .	46 30.37	51.17	+0.01	.	2	9.56	9 59.53	20.17	1.59	45 39.21	26 2 21.29
33	9	. .	. .	. .	. .	13.0	26.8	. .	46 59.04	51.17	0.00	.	6	13.7	30 34.23	20.25	3.53	46 7.87	22 58.01
34	9	. .	. .	. .	14.8	28.7	. .	. .	48 14.69	51.18	0.00	.	5	7.57	23 56.82	20.48	2.89	47 23.51	16 20.17
35	7.8	27.3	41.8	. .	. .	. .	. .	. .	50 9.60	51.20	0.00	.	6	11.31	29 45.83	20.80	3.45	49 18.40	22 10.08
36	9	. .	. .	. .	21.4	. .	. .	. .	50 21.25	51.20	0.00	.	6	13.19	30 40.28	20.84	3.54	49 30.05	23 4.66
37	6.7	. .	. .	. .	. .	. .	. .	53.5	50 11.84	51.20	−0.01	.	9	15.47	46 52.61	20.81	5.13	49 20.63	39 18.55
38	9	. .	. .	. .	. .	19.1	. .	. .	52 5.11	51.22	0.00	.	4	6.10	18 3.67	21.14	2.34	51 13.89	10 27.15
39	9	. .	. .	. .	. .	. .	24.3	. .	52 56.48	51.23	−0.01	F.	9	2.20	40 5.52	21.30	4.45	52 5.24	32 31.27
40	10	. .	. .	. .	. .	. .	31.6	. .	54 3.83	51.24	0.00	F.	6'	7.59	27 58.76	21.48	3.28	53 12.59	20 23.52
41	10	. .	. .	7.9	. .	. .	. .	. .	57 21.91	51.28	−0.01	.	8	8.27	38 11.75	22.07	4.26	56 30.62	30 38.08
42	10	. .	. .	. .	57.3	. .	. .	. .	57 57.15	51.29	−0.01	.	7	9.4	33 31.71	22.18	3.81	57 5.85	25 57.70
43	5.6	. .	. .	. .	. .	. .	37.1	51.4	58 9.39	51.29	0.00	F.	5	6.54	23 24.89	22.21	2.85	57 18.10	26 15 49.95
44	9	. .	. .	. .	. .	42.3	. .	. .	3 59 28.23	−51.30	+0.01	.	1	5.49	− 2 55.98	−22.46	−0.91	3 58 36.94	−25 55 19.35

ZONE 156. JANUARY 20. K. BELT, −26° 16'. D₀ = −25° 51' 20".

No.	Mag.	I.	II.	III.	IV.	V.	VI.	VII.	T.	a_1	a_2	Micrometer.			i	d_1	d_2	Mean Right Ascension, 1850.0.	Mean Declination, 1850.0.
1	9	19.0	33.1	. .	. .	. .	. .	. .	6 23 0.90	−52.36	+0.02	.	2	8.37	− 9 19.70	−43.13	−1.46	6 22 8.56	−26 1 24.3
2	9	. .	. .	. .	9.5	23.3	. .	. .	24 9.36	52.37	−0.01	.	7	6.28	32 13.04	43.44	3.71	23 16.98	24 20.2
3	9	. .	10.8	24.9	. .	. .	. .	. .	25 37.65	52.37	+0.01	.	4	12.21	21 10.73	43.78	2.63	24 45.29	13 17.1
4	10.9	. .	. .	. .	6.9	. .	. .	. .	26 6.75	52.37	−0.02	.	8	9.33	33 46.33	43.91	3.87	25 14.36	25 54.1
5	10	. .	1.8	. .	29.8	. .	. .	. .	27 29.73	52.38	−0.01	.	7	11.11	34 35.74	44.26	3.94	26 37.34	26 43.9
6	8	. .	. .	15.9	29.8	. .	. .	. .	28 29.65	52.39	+0.01	.	4	11.55	20 57.63	44.51	2.60	27 37.27	26 13 4.7
7	9	. .	. .	. .	. .	7.5	. .	. .	28 53.43	52.39	0.03	.	1	5.41	2 51.94	44.61	0.91	28 1.07	25 54 57.5
8	9	. .	58.1	12.8	. .	. .	. .	. .	30 26.05	52.39	+0.02	.	1	6.28	3 15.64	44.98	0.95	29 33.68	26 55 21.6
9	8	. .	42.6	. .	. .	. .	. .	. .	6 31 10.58	−52.39	−0.01	.	7	7.19	−32 38.76	−45.18	−4.74	6 30 18.18	−26 24 48.7

CORRECTIONS.

Date.		Corr. of Clock.	Hourly rate.	m	n	c
1848.	h.	s.	s.	s.	s.	s.

INSTRUMENT READINGS.

Date.		Barom.	Thermom. At.	Thermom. Ex.
1848.	h. m.	In.	°	°

REMARKS.

(155) 9. Transits over T.'s IV and V assumed as 50s.9 and 4s.8 instead of 40s.9 and 54s 8.
(155) 20. Transits over T.'s IV and V assumed as recorded over T.'s V and VI.
(156) 4. Hor. thread assumed as 7 instead of 8.

In Mr. Keith's observations, when the transit wire on which the declination was observed is not recorded, the observation was made on the IV. thread. I know of no signification for the letters *g g*. It is assumed that F denotes the VI. and *g* the VII. thread.

ZONE 156. JANUARY 20. K. BELT, −26° 16'. D_0=−25° 51' 20"—Continued.

No.	Mag.	SECONDS OF TRANSIT. I.	II.	III.	IV.	V.	VI.	VII.	T.	a_1	a_2	MICROMETER.			i	d_1	d_2	Mean Right Ascension, 1850.0.	Mean Declination, 1850.0.
									h. m. s.	s.	s.			r.	' "	"	"	h. m. s.	° ' "
10	9	. .	. .	. .	. .	19.1	. .	. .	6 31 5.06	−52.39	+0.02	.	2	6.12	− 8 6.58	−45.15	−1.35	6 30 12.69	−26 0 13.1
11	9	. .	. .	. .	. .	59.8	. .	. .	31 45.93	52.40	−0.02	.	10	4.1	45 57.52	45.32	5.09	30 53.51	26 38 7.9
12	10	. .	. .	52.5	. .	. .	. .	. .	33 6.05	52.40	+0.02	.	2	4.48	7 24.22	45.64	1.29	32 13.07	25 59 31.2
13	9	. .	. .	48.8	. .	. .	. .	. .	34 2.75	52.41	−0.01	.	7	10.47	34 23.65	45.84	3.93	33 10.33	26 26 33.4
14	9	. .	. .	. .	17.9	. .	. .	. .	34 17.75	52.41	0.00	.	5	9.31	24 44.22	45.94	2.97	33 25.34	16 53.1
15	10	. .	. .	4.4	. .	. .	. .	. .	35 18.11	52.41	+0.01	.	4	6.29	18 13.25	46.18	2.33	34 25.71	26 10 21.8
16	9	17.3	32.0	. .	. .	. .	. .	. .	37 59.47	52.42	0.02	.	2	3.27	6 43.37	46.85	1.22	37 7.07	25 58 51.4
17	9	. .	. .	6.8	. .	. .	. .	. .	38 20.36	52.42	+0.02	.	2	6.53	8 27.25	46.93	1.39	37 27.96	26 0 35.6
18	8.9	. .	. .	. .	. .	. .	36.8	51.2	38 9.16	52.42	0.00	F.	6	6.35	27 16.41	46.90	3.22	37 16.74	19 26.5
19	10	. .	. .	. .	13.1	. .	. .	. .	40 12.95	52.43	+0.02	.	2	5.58	7 59.51	47.39	1.34	39 20.54	0 8.2
20	10	. .	. .	. .	51.8	. .	. .	. .	40 51.65	52.43	−0.02	.	9	5.36	41 44.53	47.55	4.66	39 59.20	33 56.7
21	10	6.0	20.1	. .	. .	. .	. .	. .	42 48.11	52.44	−0.01	.	7	2.27	30 11.52	48.02	3.52	41 55.66	26 22 23.6
22	7.8	. .	. .	. .	54.1	. .	. .	. .	43 53.95	52.44	+0.03	.	1	3.5	1 33.28	48.28	0.79	43 1.54	25 53 42.3
23	9.10	. .	. .	52.4	. .	. .	. .	. .	45 6.48	52.45	−0.02	.	9	9.44	43 49.58	48.57	4.87	44 14.01	26 36 3.0
24	8	44.6	59.0	13.2	. .	. .	. .	. .	46 26.92	52.45	0.01	.	6	10.42	29 21.13	48.89	3.44	45 34.46	21 33.5
25	8	. .	. .	. .	39.1	. .	. .	. .	46 38.95	52.45	−0.01	.	7	6.19	32 8.51	48.94	4.71	45 46.49	26 24 22.2
26	7.8	. .	. .	. .	. .	42.6	56.4	. .	47 28.63	52.45	+0.02	.	1	8.5	4 4.56	49.14	0.97	46 36.20	25 56 14.7
27	10	. .	19.0	33.7	. .	. .	. .	. .	51 47.38	52.47	−0.02	.	8	10.14	39 5.71	50.20	4.36	50 54.89	26 31 20.3
28	9	. .	12.7	26.9	. .	. .	. .	. .	52 40.63	52.47	0.00	.	5	5.11	22 33.11	50.41	2.75	51 48.16	14 46.3
29	9	. .	. .	11.4	. .	. .	. .	. .	53 25.41	52.47	−0.02	.	8	9.14	38 35.45	50.59	4.34	52 32.92	30 50.4
30	8.9	. .	. .	. .	49.0	. .	. .	. .	53 48.85	52.47	0.00	.	5	8.13	24 4.89	50.69	2.91	52 56.38	16 18.5
31	10	. .	. .	35.5	. .	. .	. .	. .	54 49.28	52.47	0.00	.	5	6.5	23 0.35	50.93	2.81	53 56.81	15 14.1
32	9	. .	. .	. .	10.7	. .	. .	. .	55 10.55	52.48	−0.01	.	6	8.25	28 12.04	51.02	3.31	54 18.06	20 26.4
33	9	3.8	17.9	. .	. .	. .	. .	. .	56 45.97	52.48	−0.01	.	8	5.20	36 37.45	51.41	4.15	55 53.48	28 53.0
34	9	. .	. .	. .	47.0	. .	. .	. .	56 46.85	52.48	0.00	.	5	10.32	25 14.99	51.41	3.02	55 54.37	17 29.4
35	10	. .	. .	. .	. .	. .	19.2	. .	56 51.42	52.48	−0.01	F.	6	10.7	29 3.31	51.43	3.37	55 58.93	21 18.1
36	9	. .	18.7	. .	. .	. .	. .	. .	58 46.74	52.49	0.02	.	8	11.30	39 44.03	51.90	4.47	57 54.23	32 0.4
37	10	. .	. .	. .	. .	52.8	. .	. .	6 58 38.88	52.49	−0.01	.	7	8.23	33 11.03	51.88	3.81	57 46.38	25 26.7
38	9	. .	53.2	. .	. .	. .	. .	. .	7 0 20.97	52.49	+0.02	.	2	9.13	9 37.84	52.28	1.48	59 28.50	1 51.6
39	10	. .	. .	. .	. .	45.0	. .	. .	0 30.96	52.49	+0.02	.	2	7.59	9 0.53	52.32	1.43	6 59 38.49	1 14.3
40	10	. .	. .	. .	25.9	. .	. .	. .	1 25.75	52.49	−0.01	.	7	4.55	31 26.15	52.54	3.64	7 0 33.25	23 42.3
41	6.7	. .	. .	. .	. .	2.0	. .	. .	1 48.08	52.49	−0.01	F.	7	8.49	33 23.98	52.64	3.83	0 55.58	25 40.5
42	9	. .	. .	. .	. .	. .	50.2	. .	2 22.48	52.49	+0.01	F.	3	10.48	15 24.68	52.79	2.06	1 30.00	7 39.5
43	2.1	. .	. .	. .	. .	. .	37.9	52.3	3 10.22	52.50	+0.01	F.	3	14.32	17 17.62	52.97	2.24	2 17.73	9 32.8
44	9	23.8	. .	. .	. .	. .	. .	. .	7 5.86	52.50	0.00	.	5	4.21	22 7.90	53.93	2.70	6 13.36	14 24.5
45	10	. .	. .	29.8	. .	. .	. .	. .	8 43.52	52.51	+0.01	.	4	8.20	19 9.22	54.32	2.42	7 51.02	11 26.0
46	6	. .	. .	. .	1.0	15.1	. .	. .	9 0.97	52.51	0.01	.	3	7.13	13 36.43	54.42	1.89	8 8.47	5 52.7
47	4.5	. .	. .	. .	. .	50.1	3.8	. .	9 36.09	52.51	0.02	F.	8	9.20	38 38.30	54.54	4.37	8 43.60	30 57.2
48	9	. .	. .	. .	53.7	. .	. .	. .	10 53.55	52.51	+0.02	.	9	4.58	41 25.36	54.84	4.65	10 1.06	33 44.9
49	10	45.8	. .	. .	. .	. .	. .	. .	12 27.97	52.51	−0.01	.	7	6.24	32 11.03	55.24	3.71	11 35.45	24 30.0
50	6	. .	. .	21.0	35.2	. .	. .	. .	12 35.04	52.51	−0.02	.	8	11.2	39 29.91	55.27	4.46	11 42.51	31 49.6
51	5	. .	. .	23.8	37.9	. .	. .	. .	13 37.70	52.51	0.00	.	6	5.9	26 33.21	55.51	3.15	12 45.19	18 51.9
52	9	. .	. .	. .	26.5	. .	. .	. .	14 26.35	52.51	0.00	.	5	11.9	25 33.64	55.72	3.06	13 33.84	17 52.4
53	7.8	. .	. .	5.4	. .	. .	. .	. .	15 19.34	52.51	−0.01	.	7	8.23	33 11.03	55.93	3.81	14 26.82	25 30.8
54	7	. .	. .	. .	. .	55.9	. .	. .	15 42.04	52.51	−0.03	.	10	9.51	48 54.00	56.02	5.39	14 49.50	41 15.4
55	10	. .	5.2	. .	. .	. .	. .	. .	17 32.99	52.52	+0.02	.	3	4.57	12 27.86	56.47	1.76	16 40.49	4 46.1
56	10	. .	. .	. .	14.4	. .	. .	. .	18 14.25	52.52	0.00	.	5	8.17	24 6.91	56.63	2.91	17 21.73	16 26.4
57	9	. .	54.2	. .	. .	. .	. .	. .	19 22.17	52.52	−0.01	.	7	4.4	31 0.44	56.93	3.60	18 29.64	23 21.0
58	8	36.8	51.1	. .	. .	. .	. .	. .	7 20 18.99	−52.52	0.00	.	6	7.3	−27 30.69	−57.14	−3.25	7 19 26.47	−26 19 51.1

CORRECTIONS.

Date.		Corr. of Clock.	Hourly rate.	m	n	c
1848.	h.	s.	s.	s.	s.	s.

INSTRUMENT READINGS.

	Date.		Barom.	THERMOM. At.	Ex.
	1848.	h. m.	in.	°	°

REMARKS.

ZONE 156. JANUARY 20. K. BELT, −26° 16′. D_0 = −25° 51′ 20″—Continued.

No.	Mag.	Seconds of Transit. I.	II.	III.	IV.	V.	VI.	VII.	T.	a_1	a_2	Micrometer.			i	d_1	d_2	Mean Right Ascension, 1850.0.	Mean Declination, 1850.0.
									h. m. s.	s.	s.			r.	′ ″	″	″	h. m. s.	° ′ ″
59	8	. .	23.4	37.0	. .	. .	. .	. .	7 20 51.20	−52.52	−0.01	.	8	6.50	−37 22.84	−57.28	−4.23	7 19 58.67	−26 29 44.4
60	9	. .	. .	11.3	. .	. .	. .	. .	21 25.01	52.52	+0.01	.	4	11.6	20 32.93	57.41	2.75	20 32.50	12 53.1
61	7	5.3	19.4	. .	. .	. .	. .	. .	22 47.50	52.52	−0.02	.	8	11.43	39 50.58	57.75	4.49	21 54.96	32 12.8
62	10	. .	. .	. .	57.2	. .	. .	. .	22 57.05	52.52	+0.01	.	3	11.34	15 48.05	57.79	2.10	22 4.54	8 7.9
63	9	39.8	53.8	. .	. .	. .	. .	. .	26 21.75	52.53	+0.01	.	4	9.22	19 40.48	58.61	2.47	25 29.23	12 1.6
64	9	. .	. .	. .	26.3	. .	. .	. .	26 26.15	52.53	−0.02	.	9	7.46	42 50.08	58.63	4.79	25 33.60	35 13.5
65	6	. .	52.3	. .	20.8	. .	. .	. .	29 20.54	52.53	−0.03	.	10	10.22	49 9.64	59.33	5.43	28 27.98	41 34.4
66	6.7	. .	. .	3.8	17.4	. .	. .	. .	30 17.44	52.53	0.00	.	5	11.27	25 42.71	59.56	3.07	29 24.91	18 5.3
67	7	5.1	19.7	33.8	. .	. .	. .	. .	31 47.61	52.53	−0.01	.	8	5.2	36 28.38	59.92	4.15	30 54.07	28 52.4
68	8	18.8	32.8	. .	. .	. .	. .	. .	32 0.96	52.53	0.02	.	9	4.45	41 18.81	59.97	4.64	31 8.41	33 43.4
69	6	. .	. .	19.5	. .	47.1	. .	. .	33 33.32	52.53	0.01	.	8	2.58	35 25.85	60.34	4.04	32 40.78	27 50.2
70	7	. .	. .	. .	. .	. .	38.2	. .	34 10.38	52.53	−0.02	F.	8	9.58	38 57.46	60.49	4.41	33 17.83	31 22.4
71	9	. .	. .	. .	. .	. .	28.9	. .	35 1.15	52.53	0.00	F.	5	5.2	22 28.41	60.69	2.76	34 8.62	26 14 51.9
72	5.6	47.3	1.3	. .	. .	. .	. .	. .	37 29.14	52.53	+0.02	.	2	5.0	7 30.27	61.29	1.28	36 36.63	25 59 52.8
73	10	. .	. .	42.8	. .	. .	. .	. .	38 56.65	52.53	0.00	.	6	5.28	27 43.29	61.64	3.27	38 4.12	26 20 8.2
74	10	. .	. .	. .	40.0	. .	. .	. .	39 39.85	52.53	−0.02	.	8	11.32	39 45.04	61.81	3.48	38 47.30	32 11.3
75	10	. .	. .	25.7	. .	. .	. .	. .	40 39.80	52.53	−0.02	.	9	11.36	44 46.06	62.05	4.99	39 47.25	37 13.1
76	9	. .	. .	. .	. .	. .	33.8	. .	41 6.10	52.53	+0.02	F.	2	8.21	9 11.45	62.15	1.43	40 13.59	1 35.0
77	10	. .	. .	2.6	. .	. .	. .	. .	43 16.75	52.53	−0.03	.	10	8.21	48 8.62	62.72	5.32	42 24.19	40 36.7
78	10	. .	. .	. .	. .	. .	45.4	. .	43 17.56	52.53	−0.02	F.	9	12.7	45 1.51	62.72	5.01	42 25.01	37 29.2
79	10	. .	30.6	. .	. .	. .	. .	. .	44 58.65	52.53	+0.01	.	4	7.22	18 39.97	63.08	2.37	44 6.13	11 5.4
80	10	. .	2.2	. .	. .	. .	. .	. .	46 30.24	52.53	−0.02	.	8	9.3	38 29.90	63.44	4.34	45 37.69	30 57.7
81	9	2.7	16.5	. .	. .	. .	. .	. .	47 44.45	52.53	+0.02	.	2	8.37	9 19.70	63.73	1.43	46 51.94	26 1 44.9
82	8	55.5	9.9	. .	. .	. .	. .	. .	48 37.51	52.53	0.02	.	1	10.21	5 13.13	63.95	1.05	47 45.00	25 57 38.1
83	9	. .	12.0	. .	. .	. .	. .	. .	49 39.79	52.53	0.02	.	2	14.29	12 17.17	64.20	1.73	48 47.28	26 4 43.1
84	8	. .	. .	. .	55.1	9.0	. .	. .	49 54.96	52.53	+0.02	.	2	9.1	9 31.79	64.26	1.45	49 2.45	1 57.5
85	9	. .	. .	. .	45.5	59.6	. .	. .	50 45.50	52.53	0.00	.	6	6.38	27 18.09	64.46	3.22	49 52.97	26 19 45.8
86	9	. .	. .	. .	. .	. .	47.1	. .	51 19.43	52.53	+0.03	.	1	5.36	2 49.43	64.59	0.82	50 26.93	25 55 14.8
87	9	. .	. .	. .	52.2	. .	. .	. .	53 52.05	52.53	−0.03	.	10	10.25	49 11.15	65.20	5.44	52 59.49	26 41 41.8
88	9	. .	29.2	43.5	. .	. .	. .	. .	54 57.14	52.53	+0.01	.	4	7.4	18 30.90	65.46	1.35	54 4.62	10 57.7
89	10	. .	29.6	. .	. .	. .	. .	. .	55 57.71	52.52	−0.03	.	10	5.21	46 37.85	65.70	5.18	55 5.16	39 8.7
90	8.9	58.8	13.1	. .	. .	. .	. .	. .	58 41.11	52.52	−0.02	.	9	3.44	40 48.05	66.35	4.60	57 48.57	33 19.0
91	10	. .	. .	19.1	. .	. .	. .	. .	7 59 32.74	52.52	+0.02	.	3	7.12	13 5.68	66.55	1.79	58 40.24	5 34.0
92	9	. .	. .	. .	0.9	13.8	. .	. .	8 0 0.30	52.52	0.00	.	6	7.31	27 44.81	66.66	3.29	7 59 7.78	20 14.8
93	9	. .	2.6	16.5	. .	. .	. .	. .	1 30.69	52.52	−0.03	.	10	9.30	48 43.42	67.02	0.39	8 0 38.14	26 41 15.8
94	10	. .	. .	4.3	. .	. .	. .	. .	2 17.81	52.52	+0.03	.	1	8.27	4 15.65	67.22	0.96	1 25.32	25 56 43.8
95	10	. .	29.6	. .	. .	. .	. .	. .	4 57.59	52.52	+0.02	.	3	6.12	13 5.68	67.85	1.82	4 5.09	26 5 35.4
96	10	. .	0.5	. .	. .	. .	. .	. .	6 28.61	52.51	−0.03	.	10	6.41	47 18.20	68.21	5.25	5 36.07	39 51.7
97	10	45.8	. .	. .	. .	. .	. .	. .	11 27.87	52.51	0.00	.	5	5.42	22 48.75	69.40	2.76	10 35.36	15 20.9
98	10	. .	. .	7.8	21.6	. .	. .	. .	12 21.70	52.51	−0.03	.	10	8.24	48 10.14	69.61	5.34	11 29.16	40 45.1
99	9	. .	. .	. .	. .	58.0	. .	. .	12 44.10	52.50	0.02	F.	8	11.12	39 34.78	69.70	4.48	11 51.58	32 9.0
100	8.9	. .	50.2	4.6	. .	. .	. .	. .	8 14 18.49	−52.50	−0.02	.	9	10.40	−44 17.82	−70.08	−4.95	8 13 25.97	−26 36 52.9

CORRECTIONS.

Date.	Corr. of Clock.	Hourly rate.	m	n	c
1848. h.	s.	s.	s.	s.	s.

INSTRUMENT READINGS.

Date.	Barom.	Thermom. At.	Thermom. Ex.
1848. h. m.	in.	°	°

REMARKS.

(156) 73. Micrometer reading assumed as 7r.28 instead of 5r.28.
(156) 91. Micrometer reading assumed as 6r.12 instead of 7r.12.

ZONE 157. JANUARY 22. K. BELT, −26° 53′. D$_\circ$=−26° 58′ 40″.

No.	Mag.	SECONDS OF TRANSIT. I.	II.	III.	IV.	V.	VI.	VII.	T.	a_1	n_1	MICROMETER.			i	d_1	d_2	Mean Right Ascension, 1850.0.	Mean Declination, 1850.0.
									h. m. s.	s.	s.			r.	′ ″	″	″	h. m. s.	° ′ ″
1	10				26.1				2 57 25.94	. .	. .	.	9	12.28	−35 14.56	−32.73	−3.00		
2	9				3.1				3 3 2.95	. .	. .	.	1	4.22	2 12.11	31.49	+0.20		
3	8.9				40.1	54.0			3 39.94	. .	. .	.	4	6.33	18 15.27	31.35	−1.35		
4	10					23.9			9 9.76	. .	. .	.	1	4.19	2 10.60	30.12	+0.20		
5	9				20.5	37.3			10 23.32	. .	. .	.	9	5.56	41 54.61	29.85	−3.67		
6	9					11.8	25.3		10 57.57	. .	. .	.	1	8.32	4 18.18	29.74	0.00		
7	7.6	31.9	46.3	0.6	14.5	28.3	42.2	57.0	15 14.24	. .	. .	.	2	8.05	9 3.56	28.82	0.45		
8	9		48.9	3.0					17 17.19	. .	. .	.	10	5.23	46 38.86	28.35	4.14		
9	9		15.2	29.5					19 43.56	. .	. .	.	9	12.28	45 12.26	27.86	4.00		
10	8.9			54.8	9.3				20 9.08	. .	. .	.	10	5.46	46 50.47	27.77	4.16		
11	6.7			41.9	56.4				20 56.20	. .	. .	.	10	11.11	49 34.34	27.60	4.44		
12	10				46.2				21 46.04	. .	. .	.	5	9.34	24 45.74	27.45	1.98		
13	9		6.9	20.9					25 34.67	. .	. .	.	2	6.20	8 10.61	26.66	0.37		
14	10	42.5	56.6						30 24.67	. .	. .	.	4	5.19	17 37.95	25.68	1.29		
15	10							6.3	41 25.28	−53.58	−0.01	.	8	4.33	36 13.76	23.50	3.10	3 40 31.69	−27 5 20.36
16	6					49.3	2.9		42 35.13	53.59	0.00	.	4	7.4	18 30.90	23.27	1.37	41 41.54	26 47 35.54
17	9			58.9					45 13.00	53.62	−0.01	.	8	10.58	34 29.19	22.75	2.93	44 19.37	27 3 34.96
18	9		32.2						49 0.31	53.66	0.00	.	6	12.06	30 3.48	21.98	2.49	49 6.65	26 59 7.95
19	7				14.3	28.2			50 14.13	53.67	+0.01	.	2	10.26	10 14.66	21.77	−0.56	49 20.47	39 16.99
20	10			45.3					52 58.86	53.70	+0.01	.	1	6.51	3 27.24	21.28	+0.09	52 5.17	26 32 28.43
21	9		38.9						54 7.16	53.72	−0.01	.	10	5.30	46 42.40	21.10	−4.16	53 13.43	27 15 47.66
22	8	17.6	31.9	46.0					55 59.95	53.74	0.00	.	6	6.39	27 18.60	20.74	2.22	55 6.21	26 56 21.56
23	10			58.6					57 12.61	53.75	−0.01	.	7	9.51	33 55.40	20.52	2.87	56 18.85	27 2 58.79
24	9				11.8	25.6			3 58 11.60	−53.76	0.00	.	5	9.26	−14 43.50	−20.35	−0.98	3 57 17.84	−26 43 44.83

ZONE 158. JANUARY 22. K. BELT, −23° 46′. D$_\circ$=−23° 20′ 40″.

No.	Mag.	I.	II.	III.	IV.	V.	VI.	VII.	T.	a_1	n_1	MICROMETER.			i	d_1	d_2	Mean Right Ascension, 1850.0.	Mean Declination, 1850.0.
1	9					4.2			5 55 50.56	−54.65	−0.01	.	7	6.57	−33 28.18	−3.61	−2.68	5 54 55.90	−23 54 14.50
2	9					20.5			5 57 20.35	54.65	0.00	.	6	11.31	29 45.83	3.91	2.38	56 25.90	23 50 32.12
3	7			15.4	29.5				6 0 29.47	54.67	−0.03	.	10	13.1	50 29.80	4.62	4.07	5 59 34.77	24 11 18.49
4	8			6.8	20.1				1 20.04	54.68	+0.02	.	2	11.47	10 55.50	4.79	0.87	6 0 25.38	23 31 41.16
5	9					3.3			1 49.54	54.68	0.02	.	2	8.11	9 6.58	4.89	0.73	0 54.88	29 52.20
6	9					52.2		19.4	2 38.33	54.68	+0.01	.	3	10.43	15 22.33	5.06	1.23	1 43.66	36 8.62
7	9						56.9		3 29.71	54.69	0.00	.	5	9.7	24 32.12	5.25	1.96	2 35.02	45 19.33
8	9							48.8	4 7.54	54.69	+0.01	.	3	3.36	11 47.02	5.38	0.95	3 12.86	32 33.35
9	8		46.3	0.2					6 13.67	54.70	0.00	.	5	8.43	24 20.02	5.82	1.94	5 18.97	45 7.78
10	9		57.2	10.8					7 24.29	54.71	+0.01	.	3	4.2	12 0.13	6.07	0.96	6 29.59	32 47.16
11	7	46.9	0.9	14.8					8 28.28	54.71	0.00	.	6	9.43	28 51.37	6.30	2.31	7 33.57	49 39.98
12	9		49.3						9 16.60	54.72	+0.01	.	4	11.5	20 32.42	6.47	1.64	8 21.89	23 41 20.53
13	10				36.9				9 36.75	54.72	−0.02	.	9	6.58	42 25.87	6.54	3.41	8 42.01	24 3 15.82
14	7.8	14.6	28.3						11 55.88	54.73	−0.02	.	9	3.9	40 30.40	7.03	3.25	11 1.13	24 1 20.68
15	9						23.2		11 55.96	54.73	+0.01	.	7	8.29	33 14.06	7.03	2.66	11 1.24	23 54 3.75
16	8		16.8	30.7					13 43.99	54.73	+0.02	.	2	5.29	7 44.89	7.41	0.61	12 49.28	28 32.91
17	9	3.8	17.6						14 44.96	54.74	0.00	.	5	4.20	22 7.40	7.62	1.77	13 50.22	42 56.79
18	7		49.6	3.2					15 16.70	54.74	+0.01	.	3	6.54	13 26.66	7.75	1.07	14 21.97	34 15.68
19	7.8		37.9	51.8					17 5.12	54.75	+0.02	.	2	10.47	10 25.25	8.14	0.83	16 10.39	23 31 14.22
20	9		3.2	16.9					19 30.71	54.76	−0.02	.	9	10.15	44 5.21	8.65	3.55	18 35.93	24 4 57.41
21	9					18.4			6 20 4.79	−54.76	−0.02	.	9	3.52	−40 52.08	−8.77	−3.29	6 19 10.01	−24 1 44.14

CORRECTIONS.

Date.		Corr. of Clock.	Hourly rate.	m	n	c
1848.	h.	s.	s.	s.	s.	s.
Jan. 22,	6	−60.25	+0.051			

INSTRUMENT READINGS.

Date.		Barom.	THERMOM. At.	Ex.
1848.	h. m.	in.	°	°

REMARKS.

(157) 1–14. Found instrument insecurely clamped; declination of stars consequently doubtful. Clamped firmly.
(157) 5. Transit observations discordant; 23^s.5 used instead of 20^s.5.
(157) 17. Hor. thread assumed as 7 instead of 8.
(157) 18. Minutes assumed as 50 instead of 49.
(157) 24. Hor. thread assumed as 3 instead of 5.
(158) 2. Transit over T. IV assumed as recorded over T. V.

ZONE 158. JANUARY 22. K. BELT, −23° 46′. D₀ = −23° 20′ 40″—Continued.

No.	Mag.	SECONDS OF TRANSIT. I.	II.	III.	IV.	V.	VI.	VII.	T.	a_1	a_2	MICROMETER.			i	d_1	d_2	Mean Right Ascension, 1850.0.	Mean Declination, 1850.0.
									h. m. s.	s.	s.			r.	′ ″	″	″	h. m. s.	° ′ ″
22	9	...	23.9	37.5	...	...	...	...	6 21 51.30	−54.77	−0.01	.	8	10.4	−39 0.67	− 9.16	−3.14	6 20 56.52	−23 59 52.97
23	9	...	...	...	...	6.8	...	...	21 53.22	54.77	−0.02	.	10	8.40	48 18.21	9.16	3.90	20 58.43	24 9 11.27
24	9	...	...	...	...	39.3	...	...	22 25.59	54.77	0.00	.	4	12.42	21 21.33	9.28	1.71	21 30.82	23 42 12.32
25	9	26.6	40.4	...	...	...	...	...	24 7.86	54.78	−0.01	.	7	6.32	32 15.07	9.65	2.59	23 13.07	53 7.31
26	10	...	...	...	...	...	17.1	...	23 49.88	54.78	0.01	.	7	7.45	32 51.87	9.58	2.64	22 55.09	23 53 44.09
27	9	...	...	53.3	...	...	...	...	25 7.07	54.79	−0.02	.	9	5.34	41 43.52	9.86	3.36	24 12.26	24 2 36.74
28	10	...	...	39.9	...	...	...	...	25 53.43	54.79	0.00	.	5	11.12	25 35.15	10.03	2.05	24 58.64	23 46 27.23
29	9	...	...	...	20.1	...	...	...	26 19.95	54.79	+0.01	.	3	9.56	14 58.63	10.12	1.20	26 25.17	35 49.95
30	10	...	...	3.8	...	...	...	...	27 17.13	54.80	0.02	.	2	11.13	10 38.35	10.34	0.84	26 22.35	31 29.53
31	8.9	...	...	44.6	58.0	...	...	...	27 57.94	54.80	+0.01	.	4	5.26	17 41.49	10.48	1.41	27 3.15	23 38 33.37
32	9	...	...	...	...	...	38.8	...	28 11.52	54.80	−0.02	.	10	3.49	45 51.47	10.53	3.70	27 16.70	24 6 45.70
33	9	...	...	...	...	17.9	...	...	29 4.21	54.80	0.00	.	5	9.38	24 47.75	10.72	1.98	28 9.40	23 45 40.45
34	8	6.2	20.5	...	...	...	...	...	30 47.77	54.80	−0.02	.	8	10.10	39 3.69	11.09	3.14	29 52.95	59 57.92
35	9	...	...	35.7	...	...	...	...	32 49.03	54.81	+0.02	.	2	11.3	10 33.31	11.53	1.84	31 54.24	31 25.68
36	7	...	...	...	...	16.8	30.2	...	33 3.06	54.82	0.02	.	1	12.27	6 16.66	11.58	0.49	32 8.26	27 8.73
37	9.10	...	...	...	1.2	...	...	...	34 1.05	54.82	0.01	.	4	9.59	19 59.14	11.79	1.60	33 6.24	40 52.53
38	6.7	...	...	...	...	30.3	...	...	34 16.56	54.82	0.01	.	3	5.51	12 55.08	11.85	1.02	33 21.75	33 47.95
39	9	...	...	...	...	...	...	17.5	34 36.26	54.82	+0.01	.	3	6.42	13 20.81	11.92	1.05	33 41.45	23 34 13.78
40	9	8.3	...	...	...	...	...	...	36 49.73	54.83	−0.02	.	9	12.13	45 4.70	12.40	3.64	35 54.88	24 6 0.74
41	9	...	38.9	...	...	...	...	...	37 6.41	54.83	−0.02	.	9	11.11	44 33.44	12.46	3.60	36 11.56	24 5 29.50
42	10	...	...	...	...	19.4	...	...	37 5.74	54.83	0.00	.	6	9.59	28 59.44	12.46	2.32	36 10.91	23 49 54.22
43	9	...	...	...	...	...	2.4	...	37 35.22	54.83	0.00	.	4	12.31	21 15.78	12.57	1.70	36 44.39	42 10.05
44	9	...	...	...	...	38.6	...	...	38 24.88	54.84	+0.01	.	3	14.1	17 2.15	12.74	1.34	37 30.05	37 56.23
45	8	...	...	20.2	...	...	...	...	39 33.68	54.84	0.00	.	4	13.21	21 40.98	12.99	1.74	38 38.84	23 42 35.71
46	9	...	...	6.2	...	...	...	...	40 19.95	54.84	−0.02	.	8	13.47	40 53.10	13.16	3.29	39 25.09	24 1 49.55
47	9	...	52.4	...	...	...	...	...	41 19.86	54.84	0.01	.	8	8.33	38 14.78	13.38	3.07	40 24.01	23 59 11.23
48	9	36.2	...	...	...	...	...	...	42 17.52	54.85	−0.01	.	7	9.38	33 31.60	13.59	2.69	41 22.68	54 27.88
49	10	...	...	...	15.8	...	...	...	42 15.65	54.85	+0.01	.	4	11.53	20 56.62	13.59	1.67	41 20.81	41 51.88
50	8	38.2	...	...	...	...	...	...	44 19.56	54.86	−0.02	.	8	7.10	37 32.92	14.03	3.02	43 24.68	58 29.97
51	6.7	...	...	...	25.5	...	...	...	44 25.35	54.85	−0.01	.	7	9.2	33 30.70	14.05	2.69	43 30.49	54 27.44
52	9	...	...	...	...	...	39.8	...	45 12.67	54.86	+0.02	.	2	6.46	8 23.73	14.22	0.66	44 17.83	29 18.61
53	9	...	...	...	...	18.4	...	...	46 4.63	54.86	0.02	.	2	4.45	7 22.71	14.42	0.57	45 9.79	28 17.70
54	8.7	...	...	...	...	...	48.4	...	46 21.23	54.86	+0.01	.	4	3.30	16 42.99	14.47	1.33	45 26.38	37 38.79
55	8	...	...	40.8	...	...	...	...	47 54.52	54.86	−0.01	.	8	8.26	38 11.25	14.81	3.08	46 59.65	59 9.14
56	9	...	...	...	...	10.3	...	...	47 56.66	54.86	0.01	.	7	12.24	35 12.55	14.81	2.84	47 1.79	23 56 10.20
57	5.6	...	...	...	49.4	...	...	...	48 49.25	54.87	−0.02	.	8	10.11	39 4.19	15.00	3.16	47 54.36	24 0 2.35
58	8	...	...	...	...	...	20.1	...	48 52.91	54.87	0.00	.	5	7.25	23 40.68	15.02	1.90	47 58.04	23 44 37.60
59	9	...	...	...	...	...	53.2	...	49 25.92	54.87	−0.02	.	10	5.58	46 56.51	15.14	3.77	48 31.03	24 7 55.42
60	9	2.0	15.9	...	...	...	...	...	51 43.12	54.88	+0.01	.	3	4.57	12 27.86	15.64	0.99	50 48.25	23 33 24.49
61	9	...	...	...	...	...	53.2	...	51 26.01	54.88	0.00	.	5	10.43	25 20.53	15.59	2.02	50 31.13	23 46 18.14
62	9	...	...	43.4	...	...	...	...	52 57.21	54.88	−0.02	.	9	12.40	45 48.58	15.92	3.70	53 2.31	24 6 48.21
63	9	...	...	...	...	...	...	56.6	53 14.43	54.88	+0.01	.	4	9.50	19 54.60	15.97	1.59	52 19.56	23 40 52.16
64	9	...	...	...	...	57.6	...	...	54 43.89	54.88	+0.01	.	4	9.49	19 54.10	16.30	1.59	53 49.02	23 40 51.99
65	9	...	...	...	...	...	...	57.3	55 16.43	54.89	−0.02	.	9	9 57	43 56.13	16.43	3.55	54 21.54	24 4 56.11
66	9	...	15.2	...	...	...	...	...	57 42.68	54.89	−0.02	.	8	12.10	40 4.19	16.96	3.24	56 47.77	24 1 4.39
67	4.5	...	...	...	...	54.3	...	...	57 40.57	54.89	+0.01	.	3	12.10	16 6.19	16.96	1.28	56 45.69	23 37 4.43
68	9	...	...	...	...	39.7	...	...	6 58 26.10	54.90	−0.02	.	9	11.16	44 35.97	17.11	3.61	6 57 31.18	24 5 36.69
69	7.8	20.2	34.2	...	...	...	...	...	7 1 1.40	54.90	+0.01	.	3	11.15	15 38.46	17.68	1.24	7 0 6.51	23 36 37.38
70	9	17.2	...	...	...	...	...	...	7 2 58.58	−54.90	−0.02	.	8	11.50	−39 54.11	−18.11	−3.22	7 2 3.66	−24 0 55.44

CORRECTIONS.

Date.		Corr. of Clock.	Hourly rate.	m	n	c
1848.	h.	s.	s.	s.	s.	s.

INSTRUMENT READINGS.

Date.		Barom.	THERMOM. At.	Ex.
1848.	h. m.	in.	°	°

REMARKS.

(158) 62. Minutes assumed as 53 instead of 52, and Micrometer reading assumed as 13ʳ.40 instead of 12ʳ.40.

ZONE 158. JANUARY 22. K. BELT, −23° 46′. D₀ = −23° 20′ 40″—Continued.

No.	Mag.	SECONDS OF TRANSIT. I.	II.	III.	IV.	V.	VI.	VII.	T.	a_1	a_2	MICROMETER.			i	d_1	d_2	Mean Right Ascension, 1850.0.	Mean Declination, 1850.0.
									h. m. s.	s.	s.			r.	′ ″	″	″	h. m. s.	° ′ ″
71	8.9	...	...	53.1	...	...	...	...	7 3 6.70	−54.91	0.00	.	6	11.4	−29 32.22	−18.14	−2.36	7 2 11.79	−23 50 32.72
72	7	...	...	...	28.9	...	...	...	3 28.75	54.91	0.00	.	6	6.54	27 26.16	18.22	2.19	2 33.84	48 26.57
73	9	...	...	8.8	...	...	...	...	4 22.27	54.91	0.00	.	5	2.23	21 8.40	18.42	1.69	3 27.36	23 42 8.51
74	8	...	...	...	...	...	...	55.5	4 14.65	54.91	−0.02	.	9	8.28	43 11.25	18.39	3.49	3 19.72	24 4 13.13
75	9	...	...	...	...	...	33.7	...	5 6.43	54.91	−0.02	.	9	12.38	45 17.31	18.59	3.67	4 11.50	24 6 19.57
76	9	...	...	...	12.3	...	...	...	6 12.15	54.91	0.00	.	5	5.7	22 31.10	18.83	1.80	5 17.24	23 43 31.73
77	8.9	2.2	15.8	...	...	...	...	...	7 43.41	54.92	−0.01	.	8	7.34	37 45.03	19.16	3.04	6 48.48	58 47.23
78	8.9	...	...	...	57.6	...	...	...	7 57.45	54.92	0.01	.	7	12.45	35 23.14	19.22	2.85	7 2.52	23 56 25.21
79	9	53.7	...	...	...	...	...	...	9 35.13	54.92	−0.02	.	9	12.13	45 4.70	19.58	3.65	8 40.19	24 6 7.93
80	9	...	43.2	...	...	...	...	...	10 10.37	54.92	+0.02	.	1	11.37	5 51.46	19.71	0.45	9 15.47	23 26 51.62
81	8.7	...	...	11.4	...	...	...	...	10 24.76	54.92	+0.01	.	2	15.27	7 47.41	19.76	0.60	9 29.85	28 47.77
82	9	...	...	...	27.8	...	...	...	11 27.65	54.92	0.00	.	6	9.32	28 45.83	19.99	2.31	10 32.73	23 49 48.13
83	9	...	...	...	...	...	16.8	...	11 49.53	54.92	−0.02	.	9	10.24	41 9.75	20.07	3.57	10 54.59	24 5 13.39
84	8.9	54.2	...	...	...	...	...	...	13 35.28	54.93	+0.02	.	2	8.21	9 11.62	20.47	0.72	12 40.37	23 30 12.81
85	9	48.4	...	...	...	...	...	...	14 29.65	54.93	0.00	.	6	5.3	26 30.18	20.67	2.12	13 34.72	47 32.97
86	9	...	16.7	...	...	...	...	...	14 44.08	54.93	0.00	.	6	7.9	27 33.72	20.72	2.21	13 49.15	23 48 36.65
87	9	...	...	...	...	55.8	...	...	14 42.21	54.93	−0.02	.	10	7.3	47 29.29	20.71	3.84	13 47.26	24 8 33.84
88	7.8	38.7	...	...	...	...	...	...	16 20.04	54.93	−0.01	.	8	4.10	36 2.16	21.08	2.90	15 25.10	23 57 6.14
89	9	...	20.6	...	...	...	...	...	16 47.93	54.93	0.00	.	5	8.55	24 26.07	21.18	1.95	15 53.00	45 29.20
90	9	...	...	...	...	...	56.4	...	16 29.26	54.93	+0.02	.	2	9.31	9 46.92	21.11	0.75	15 34.35	30 48.78
91	9	...	...	38.9	...	...	...	...	17 52.53	54.94	−0.01	.	7	5.5	31 31.19	21.42	2.52	16 57.58	52 35.13
92	9	...	...	...	3.8	...	...	...	18 3.65	54.94	+0.01	.	4	5.5	17 30.89	21.46	1.39	17 8.72	38 33.74
93	9	...	...	...	31.6	...	...	...	18 31.45	54.94	+0.01	.	4	5.20	17 38.45	21.56	1.40	17 36.52	23 38 41.41
94	8	...	10.7	...	...	...	...	...	19 38.23	54.94	−0.02	.	10	6.13	47 4.08	21.60	3.82	18 43.27	24 8 9.70
95	9	...	...	...	...	50.9	...	...	19 37.16	54.94	+0.02	.	3	3.14	11 35.92	21.80	0.91	18 42.24	23 32 38.63
96	8	...	49.4	...	...	...	...	...	21 16.55	54.94	+0.02	.	1	7.44	3 53.97	22.7	0.28	20 21.63	23 24 56.42
97	9	...	...	...	...	32.6	...	...	21 19.02	54.94	−0.03	.	10	9.56	48 56.53	22.18	3.97	20 24.05	24 10 2.68
98	9	...	...	...	...	13.7	...	...	21 59.98	54.94	+0.01	.	4	4.59	17 27.87	22.33	1.39	21 5.05	23 38 31.59
99	9	...	...	...	...	...	...	37.2	22 56.19	54.94	−0.01	.	7	3.58	30 57.40	22.53	2.48	22 1.24	52 2.41
100	9	...	...	...	...	12.9	...	...	23 59.17	54.94	+0.01	.	3	11.52	15 57.12	22.76	1.27	23 4.24	37 1.15
101	9	...	...	...	...	...	...	4.2	24 23.05	54.94	0.01	.	4	11.21	20 40.48	22.85	1.66	23 28.12	41 44.99
102	9	...	...	...	50.0	...	...	...	25 49.85	54.95	+0.02	.	2	10.27	10 15.16	23.15	0.79	24 54.92	31 19.10
103	9	...	33.3	...	...	...	...	...	26 0.62	54.95	0.00	.	5	4.50	22 22.53	23.20	1.79	25 5.67	43 27.52
104	9	...	10.5	...	...	...	...	...	27 37.76	54.95	+0.01	.	3	11.41	15 51.57	23.55	1.26	26 42.82	36 56.38
105	9	...	...	57.6	...	...	...	...	28 11.32	54.95	−0.01	.	8	9.29	38 43.01	23.68	3.13	27 16.36	59 49.82
106	9	...	...	...	...	37.8	...	...	28 24.06	54.95	+0.01	.	3	6.34	13 16.78	23.72	1.04	27 29.12	34 21.54
107	9	...	22.6	...	...	...	...	...	30 49.80	54.95	0.02	.	2	8.54	9 28.27	24.24	0.73	29 54.87	30 33.24
108	9	...	7.6	...	...	...	...	...	31 34.90	54.95	+0.01	.	4	9.29	19 44.01	24.40	1.58	30 39.96	40 49.99
109	9	...	...	...	...	...	46.8	...	31 19.61	54.95	0.00	.	5	11.9	25 33.64	24.34	2.05	30 24.66	46 40.03
110	8.9	...	...	...	...	...	...	14.3	31 33.37	54.95	−0.01	.	8	5.51	36 53.08	24.40	2.98	30 38.41	58 0.46
111	8	25.8	...	...	...	...	...	...	34 7.10	54.95	−0.01	.	7	6.29	32 13.55	24.93	2.58	33 12.14	53 21.06
112	9	...	58.2	...	...	...	...	...	34 25.49	54.95	+0.01	.	4	7.36	18 47.04	24.99	1.50	33 30.55	39 53.53
113	8.9	...	...	...	23.2	...	...	...	34 23.05	54.95	+0.02	.	2	10.29	10 16.17	24.99	0.79	33 28.12	31 21.95
114	8	...	...	12.7	...	...	...	...	35 26.37	54.96	−0.01	.	7	11.18	34 39.27	25.20	2.79	34 31.40	55 47.26
115	9	...	...	...	...	...	50.2	...	35 23.01	54.96	0.00	.	5	7.3	23 29.59	25.19	1.85	34 28.05	44 36.66
116	9	...	...	...	...	...	...	31.2	35 50.16	54.96	0.00	.	6	8.20	28 9.52	25.30	2.26	34 55.20	49 17.08
117	9	...	...	...	...	55.3	...	...	36 41.61	54.96	0.00	.	5	9.42	24 49.77	25.48	1.98	35 46.65	23 45 57.23
118	9	...	...	...	...	28.3	...	...	37 14.72	54.96	−0.03	.	10	8.7	48 1.57	25.58	3.89	36 19.73	24 9 11.04
119	9	...	...	20.4	...	...	...	...	7 38 33.96	−54.96	0.00	.	6	6.15	−27 6.49	−25.86	−2.17	7 37 39.00	−23 48 14.52

CORRECTIONS.

Date.		Corr. of Clock.	Hourly rate.	m	n	c
1848.	h.	s.	s.	s.	s.	s.

INSTRUMENT READINGS.

Date.		Barom.	THERMOM. At.	THERMOM. Ex.
1848.	h. m.	in.	°	°

REMARKS.

(158) 81. Micrometer thread assumed as 1 instead of 2.

ZONE 158. JANUARY 22. K. BELT, −23° 46′. D₀=−23° 20′ 40″—Continued.

No.	Mag.	I.	II.	III.	IV.	V.	VI.	VII.	T.	a_1	a_4	Micrometer			i	d_1	d_2	Mean Right Ascension, 1850.0.	Mean Declination, 1850.0.
									h. m. s.	s.	s.			r.	′ ″	″	″	h. m. s.	° ′ ″
120	9					5.9			7 38 52.15	−54.96	+0.02	.	3	1.52	−10 54.57	−25.94	−0.84	7 37 57.21	−23 32 1.35
121	9							40.3	38 59.02	54.96	+0.02	.	3	0.58	10 27.34	25.96	0.81	38 4.08	31 34.11
122	7.8				26.9				40 26.75	54.96	−0.01	.	7	7.11	32 34.72	26.26	2.61	39 31.79	53 43.59
123	9				7.7				41 7.55	54.96	0.00	.	6	4.15	26 5.98	26.40	2.09	40 12.59	23 47 14.47
124	9						41.0		41 13.72	54.96	−0.02	.	10	4.47	46 20.72	26.42	3.75	40 18.74	24 7 30.89
125	9			20.2					7 42 33.83	−54.96	−0.01	.	7	6.43	−32 20.61	−26.72	−2.58	7 41 38.86	−23 53 29.91

ZONE 159. JANUARY 24. B. BELT, −25° 1′. D₀=−24° 38′ 0″.

No.	Mag.	I.	II.	III.	IV.	V.	VI.	VII.	T.	a_1	a_4	Micrometer			i	d_1	d_2	Mean Right Ascension, 1850.0.	Mean Declination, 1850.0.
1	8			24.	38.2				7 42 37.98	−57.02	+0.02	V.	9	5.01	−41 26.83	−0.89	−4.50	7 41 40.98	−25 19 32.22
2	9						35.	49.	48 7.59	57.02	+0.02	VII.	8	12.02	39 59.80	2.70	4.36	48 10.59	18 6.86
3	8				17.	30.5			51 16.76	57.02	0.00	VII.	5	5.57	22 55.95	3.74	2.80	50 19.74	1 2.49
4	9					53.5	7.	21.	52 39.62	57.02	+0.01	VII.	7	12.35	34 47.49	4.21	3.87	51 42.61	25 12 55.57
5	9		32.	45.2					54 59.19	57.02	−0.01	IV.	4	11.8	20 23.93	4.98	2.59	54 2.16	24 58 41.50
6	7				52.5	6.			56 52.26	57.02	0.00	VI.	5	4.10	22 2.20	5.57	2.72	54 55.24	25 0 10.49
7	10			9.5	23.	37.			58 23.08	57.02	0.00	VI.	6	4.30	26 13.39	6.11	3.11	57 26.06	4 22.61
8	6					25.	38.2	52.	7 59 10.87	57.02	+0.01	VII.	8	9.4	38 30.05	6.37	4.24	58 13.86	25 16 40.66
9	8							51.	8 0 9.29	57.02	−0.02	VII.	2	9.26	9 44.04	6.69	1.61	7 59 12.25	24 47 52.34
10	10						31.	45.	2 3.60	57.02	+0.02	VII.	9	7.16	42 34.59	7.32	4.61	8 1 6.60	25 20 46.52
11	8							28.	2 46.76	57.02	0.02	VII.	9	11.26	44 40.65	7.57	4.82	1 49.76	22 53.04
12	9				31.	44.			4 30.57	57.02	0.02	VI.	10	5.2	46 28.12	8.12	4.99	3 33.57	24 41.23
13	10			27.	41.				8 40.85	57.01	+0.01	V.	8	6.52	37 23.81	9.48	4.14	7 43.85	25 15 37.43
14	8	56.	10.						10 37.47	57.01	−0.02	III.	2	6.41	8 21.16	10.11	1.50	9 40.44	24 46 32.77
15	9						5.	19.5	10 37.68	57.01	0.03	VII.	1	12.11	1 5.69	10.11	0.84	9 40.64	39 16.64
16	9						56.	10.5	12 28.72	57.01	0.01	VII.	3	11.32	15 46.68	10.71	2.16	11 31.70	53 59.55
17	8						42.	56.3	14 14.64	57.01	0.01	VII.	4	8.12	19 4.83	11.35	2.46	13 17.62	57 18.64
18	9				55.5	9.5			15 55.51	57.00	−0.01	VI.	4	9.49	19 53.94	11.52	2.53	14 58.50	24 58 8.29
19	10						12.5	27.	16 45.30	57.00	+0.01	VII.	7	4.21	31 8.64	12.12	3.56	15 48.31	25 9 24.32
20	10					22.	36.		18 8.35	57.00	−0.01	VII.	4	7.58	18 57.76	12.51	2.45	17 11.34	24 57 12.72
21	10		8.2	22.	36.				20 35.87	57.00	+0.01	IV.	8	5.35	36 45.02	13.36	4.08	19 38.88	25 15 2.46
22	10		13.	27.5					22 40.64	56.99	−0.02	IV.	1	8.59	4 31.79	13.99	1.14	21 43.63	24 42 46.92
23	10		28.5	42.					24 55.72	56.99	−0.02	IV.	2	11.34	10 48.95	14.78	1.70	23 58.71	24 49 5.43
24	9				0.	14.			26 0.04	56.99	+0.01	VI.	7	7.5	32 31.54	15.12	3.69	25 3.06	25 10 50.35
25	10			56.	10.				8 28 9.84	−56.98	+0.01	IV.	8	3.47	−35 50.56	−15.82	−3.99	8 27 12.87	−25 14 10.37

ZONE 160. JANUARY 29. B. BELT, −25° 38′. D₀=−25° 14′ 0″.

No.	Mag.	I.	II.	III.	IV.	V.	VI.	VII.	T.	a_1	a_4	Micrometer			i	d_1	d_2	Mean Right Ascension, 1850.0.	Mean Declination, 1850.0.
1	6				59.	11.			5 44 11.61	−61.11	−0.01	X.	7	12.50	−35 29.26	−14.31	−2.99	5 44 10.49	−25 49 47.1
2	9							47.	46 5.35	61.12	0.00	IX.	6	11.42	29 50.38	17.08	2.45	45 4.23	44 9.9
3	7							44.5	47 3.03	61.13	−0.02	VII.	9	11.31	44 43.17	18.21	3.84	46 1.88	59 5.2
4	9			1.5	15.				49 15.00	61.14	0.00	IV.	4	8.12	19 5.19	20.84	1.47	48 13.86	33 27.5
5	9						13.	27.3	49 45.43	61.14	+0.01	VII.	3	8.00	13 59.78	21.44	0.99	48 44.30	28 22.2
6	10			33.2	47.				53 46.78	61.16	0.01	V.	2	8.3	9 2.51	26.27	0.55	52 45.63	23 29.3
7	9			56.	10.				55 9.70	61.18	0.01	IV.	3	5.33	12 46.01	27.93	0.89	54 8.53	27 14.8
8	10						15.4		55 47.81	61.18	+0.01	IV.	4	5.33	17 45.01	28.69	1.34	54 46.64	32 15.0
9	9							47.5	5 57 5.75	61.19	0.00	VII.	5	7.22	23 38.81	30.25	1.87	5 56 4.56	38 10.9
10	9		5.	19.					6 1 32.90	−61.22	−0.01	IV.	8	9.28	−38 42.50	−35.58	−3.28	6 0 31.67	−25 53 21.4

CORRECTIONS.

Date.		Corr. of Clock.	Hourly rate.	m	n	c
1848.	h.	s.	s.	s.	s.	s.
Jan. 24,	6	− 62.14	g 0.047	+ 1.35	− 1.47	+ 0.01
Jan. 29,	6	− 66.67	g 0.038			

INSTRUMENT READINGS.

Date.	Barom.	Thermom. At.	Thermom. Ex.
1848. h. m.	in.	°	°

REMARKS.

(159) 2. Minutes assumed as 49 instead of 48.
(159) 4. Micrometer reading assumed as 11ʳ.35 instead of 12ʳ.35, to agree with Arg. Z. 360, 222, and Arg. Z. 362, 159.
(159) 6. Minutes assumed as 55 instead of 56.
(160) 1. Transits over T.'s III and IV assumed to have been recorded over T.'s IV and V, and minutes as 45 instead of 44.

ZONE 160. JANUARY 29. B. BELT, −25° 38′. D_0 = −25° 14′ 0″—Continued.

No.	Mag.	SECONDS OF TRANSIT. I.	II.	III.	IV.	V.	VI.	VII.	T.	a_1	a_2	MICROMETER.			i	d_1	d_2	Mean Right Ascension, 1850.0.	Mean Declination, 1850.0.
									h. m. s.	s.	s.			r.	′ ″	″	″	h. m. s.	° ′ ″
11	10	. .	. .	. .	. .	. .	. .	14.	6 1 32.39	−61.22	−0.01	VII.	7	8.18	−33 8.15	−35.56	−2.76	6 0 31.16	−25 47 46.5
12	8	. .	. .	. .	. .	18.	31.5	. .	3 3.99	61.23	+0.02	VII.	2	8.49	9 25.39	37.40	0.59	2 2.78	24 3.4
13	9	. .	. .	. .	53.	. .	. .	. .	4 52.85	61.23	0.00	IV.	4	13.9	21 34.94	39.58	1.69	3 51.62	36 16.2
14	9	. .	. .	. .	. .	. .	. .	0.	5 18.23	61.24	0.00	VII.	4	13.35	21 47.69	40.09	1.71	4 16.99	36 29.5
15	10	. .	. .	. .	. .	. .	27.	41.3	7 59.50	61.25	0.00	VII.	6	5.29	26 42.93	43.33	2.16	5 58.25	41 28.4
16	9	8.	22.	. .	. .	. .	. .	. .	9 49.91	61.26	−0.01	III.	8	3.2	35 27.83	45.53	2.99	8 48.64	50 16.4
17	10	. .	22.	36.	. .	. .	. .	. .	6 11 49.59	−61.27	+0.01	IV.	3	6.32	−13 15.77	−47.90	−0.93	6 10 48.33	−25 28 4.6

ZONE 161. JANUARY 29. B. BELT, −25° 38′. D_0 = −25° 14′ 20″.

No.	Mag.	I.	II.	III.	IV.	V.	VI.	VII.	T.	a_1	a_2	MICROMETER.			i	d_1	d_2	Mean Right Ascension, 1850.0.	Mean Declination, 1850.0.
1	7.8	. .	. .	43.	. .	10.	. .	. .	7 15 56.34	−61.52	+0.01	VI.	4	4.9	−22 1.70	−26.67	−1.54	7 14 54.83	−25 36 50.01
2	7	. .	. .	. .	35.	. .	. .	. .	17 34.85	61.52	0.00	IV.	6	9.37	28 48.35	26.94	2.36	16 33.33	43 37.65
3	7	. .	. .	. .	. .	. .	. .	31.	17 49.14	61.52	+0.01	VII.	3	8.21	14 10.36	27.11	1.00	17 47.63	28 58.47
4	9	. .	49.	. .	17.	. .	. .	. .	20 30.85	61.52	−0.01	IV.	7	4.41	31 19.09	27.33	2.61	19 29.32	46 9.03
5	10	. .	6.	20.	. .	. .	. .	. .	22 33.94	61.53	0.02	IV.	9	4.38	41 15.28	27.68	3.55	21 32.39	56 6.51
6	7	37.5	52.	. .	. .	. .	. .	. .	29 19.63	61.54	−0.01	III.	7	7.28	32 43.25	28.71	2.73	28 18.08	47 34.69
7	10	. .	. .	. .	. .	48.5	. .	16.	29 34.47	61.54	0.00	VII.	6	7.11	27 34.36	28.76	2.25	28 32.93	42 25.37
8	7	. .	. .	16.	30.3	44.	. .	. .	33 30.11	61.54	−0.02	VI.	9	9.42	43 48.41	29.35	3.79	32 28.55	58 41.55
9	7	. .	. .	24.	38.2	. .	. .	. .	37 38.04	61.55	0.02	IV.	10	1.14	44 33.31	29.98	3.86	36 36.47	25 59 27.15
10	9	. .	47.	1.	15.	. .	. .	. .	41 14.94	61.55	−0.02	IV.	10	5.25	46 39.87	30.53	4.05	40 13.37	26 1 34.45
11	6	. .	25.	39.	53.	. .	. .	. .	42 52.73	61.55	+0.01	IV.	4	8.275	19 12.75	30.78	1.45	41 51.19	25 34 4.98
12	10	. .	. .	. .	. .	. .	7.	21.2	43 39.42	61.55	0.00	VII.	5	6.33	23 14.11	30.90	1.84	42 37.87	38 6.85
13	7	. .	. .	23.4	36.5	. .	. .	. .	48 36.84	61.56	−0.02	IV.	8	7.10	37 32.92	31.66	3.20	47 35.26	52 27.78
14	8	. .	. .	. .	. .	. .	. .	39.	49 57.42	61.56	−0.01	VII.	8	3.47	35 50.20	31.86	3.04	48 55.85	50 45.10
15	9	. .	. .	. .	. .	. .	. .	46.	50 4.18	61.56	+0.01	VII.	4	5.10	17 33.05	31.88	1.31	49 2.63	32 26.24
16	10	. .	. .	14.	28.	. .	. .	. .	52 27.84	61.56	−0.01	IV.	7	6.19	32 8.51	32.24	2.69	51 26.27	47 3.44
17	10	. .	. .	31.	45.	. .	. .	. .	54 44.78	61.56	0.00	IV.	5	9.23	24 40.18	32.58	1.97	53 43.22	39 34.73
18	8	. .	35.3	49.3	. .	. .	. .	. .	57 3.02	61.56	0.00	IV.	5	8.14	24 5.40	32.93	1.92	56 1.46	39 0.25
19	9	. .	0.	14.2	. .	. .	. .	. .	7 58 27.79	61.56	+0.01	IV.	4	12.36	21 18.31	33.14	1.65	57 26.24	36 33.10
20	9	. .	59.	13.5	. .	. .	. .	. .	8 0 26.75	61.56	+0.02	IV.	1	10.54	5 29.78	33.44	0.19	7 59 25.21	20 23.41
21	9	. .	37.	51.	. .	. .	. .	. .	2 4.94	61.56	−0.02	IV.	9	6.48	42 20.83	33.69	3.66	8 1 3.36	57 18.28
22	10	. .	. .	. .	. .	. .	56.5	. .	2 28.86	61.56	0.01	VII.	7	5.15	31 35.87	33.75	2.64	1 27.29	46 32.26
23	9	. .	. .	. .	. .	5.	19.	. .	3 51.25	61.56	−0.01	VII.	7	7.32	32 44.96	33.95	2.75	2 49.68	47 41.66
24	9	30.	. .	58.	. .	. .	. .	. .	6 11.75	61.56	0.00	IV.	5	3.1	21 27.56	34.30	1.67	5 10.19	36 23.53
25	8	. .	49.	3.2	. .	. .	. .	. .	8 16.87	61.56	0.00	IV.	6	7.11	27 34.72	34.61	2.26	7 15.31	42 31.59
26	8	. .	. .	. .	22.	. .	. .	. .	9 21.85	61.56	+0.01	IV.	4	5.28	17 42.48	34.77	1.32	8 20.30	32 38.57
27	9	. .	. .	. .	. .	21.	. .	. .	10 7.12	61.56	0.00	VI.	6	5.8	26 32.54	34.89	2.14	9 5.56	41 29.57
28	7	. .	43.5	57.5	. .	. .	. .	. .	12 11.36	61.56	−0.01	IV.	7	12.37	35 19.11	35.19	2.99	11 9.79	50 17.29
29	6	. .	. .	. .	3.	. .	. .	. .	13 2.85	61.56	−0.01	.	8	7.24	37 39.98	35.31	3.22	12 1.28	52 38.51
30	10	27.	41.5	. .	. .	. .	. .	. .	15 9.03	61.55	0.00	III.	5	6.33	23 14.43	35.62	1.84	14 7.48	38 11.89
31	9	. .	. .	. .	55.	. .	. .	. .	15 54.85	61.55	−0.01	V.	8	5.7	36 30.86	35.73	3.10	14 53.29	51 29.69
32	9	. .	. .	. .	. .	. .	56.	. .	16 28.40	61.55	+0.01	VII.	4	8.6	19 1.81	35.82	1.44	15 26.86	33 59.07
33	6	. .	. .	. .	. .	45.	58.5	. .	17 31.00	61.55	−0.01	VII.	8	6.32	37 13.41	35.98	3.17	16 29.44	52 12.56
34	10	. .	. .	. .	26.	. .	. .	. .	19 12.09	61.55	0.00	IV.	5	6.42	23 19.01	36.24	1.85	18 10.54	38 17.10
35	10	. .	48.	2.	. .	. .	. .	. .	21 15.59	61.55	+0.01	IV.	3	5.23	12 40.96	36.51	0.85	20 14.05	27 38.32
36	6	. .	. .	. .	. .	47.	1.	15.	22 33.24	61.55	0.00	VII.	5	6.34	23 14.62	36.70	1.84	21 31.69	38 13.16
37	7	. .	30.	44.	. .	. .	. .	. .	24 57.64	61.55	+0.01	IV.	4	4.58	17 27.36	37.05	1.30	23 56.10	32 25.71
38	8	. .	. .	4.	18.	. .	. .	. .	8 26 17.72	−61.54	+0.01	IV.	3	11.22	−15 41.99	−37.23	−1.13	8 25 16.19	−25 30 40.35

CORRECTIONS.

Date.		Corr. of Clock.	Hourly rate.	m	n	c
1848.	h.	s.	s.	s.	s.	s.

INSTRUMENT READINGS.

Date.		Barom.	THERMOM. At.	Ex.
1848.	h. m.	in.	°	°

REMARKS.

(160) 15. Minutes assumed as 6 instead of 7.
(161) 1. Hor. thread assumed as 5 instead of 4.
(161) 3. Minutes of transit assumed as 18 instead of 17.
(161) 4. Transits over T.'s I and III assumed to be recorded as over T.'s II and IV.

Zone 161. January 29. B. Belt, −25° 38′. $D_0 = -25° 14′ 20″$—Continued.

No.	Mag.	Seconds of Transit. I.	II.	III.	IV.	V.	VI.	VII.	T.	a_1	a_2	Micrometer.			i	d_1	d_2	Mean Right Ascension, 1850.0.	Mean Declination, 1850.0.
									h. m. s.	s.	s.			r.	′ ″	″	″	h. m. s.	° ′ ″
39	9		27.	41.2					8 27 54.84	−61.54	0.00	IV.	5	11.10	−25 34.14	−37.46	−2.06	8 26 53.30	−25 40 33.66
40	9		16.	30.					29 29.93	61.54	−0.02	IV.	9	9.59	43 57.14	37.72	3.82	28 28.37	58 58.68
41	7		14.	27.9					31 41.86	61.54	−0.01	IV.	8	9.34	38 45.54	38.00	3.34	30 40.31	53 46.88
42	6				29.7	43.	57.		32 29.35	61.53	0.00	VII.	6	9.52	28 55.55	38.11	2.38	31 27.82	43 56.04
43	8		40.	54.2					37 7.92	61.53	−0.01	IV.	7	5.9	31 33.21	38.75	2.64	36 6.38	46 34.60
44	9			53.	7.				39 6.93	61.52	0.02	IV.	9	8.48	43 21.34	39.03	3.77	38 5.39	58 24.14
45	9							13.5	39 31.91	61.52	0.01	VII.	7	12.15	35 7.65	39.09	2.98	38 30.38	25 50 9.72
46	8						30.		41 2.29	61.52	−0.03	VII.	10	11.49	49 53.15	39.30	4.38	40 0.74	26 4 56.83
47	9			51.	5.				43 4.76	61.51	0.00	IV.	4	13.23	21 41.99	39.58	1.67	42 3.25	25 36 43.24
48	8	38.5	52.3	6.5					45 20.14	61.51	+0.01	IV.	4	7.15	18 36.44	39.89	1.39	44 18.64	33 37.72
49	8				11.	25.			46 10.97	61.51	0.00	VI.	5	7.33	23 44.56	40.00	1.88	45 9.46	38 46.44
50	10			59.	13.2				48 12.79	61.50	+0.02	IV.	3	3.44	11 51.05	40.28	0.76	47 11.31	26 52.09
51	10			15.	29.				50 28.79	61.50	0.00	IV.	5	11.25	25 41.70	40.59	2.07	49 27.29	40 44.36
52	9	4.	18.3						52 46.02	61.49	−0.01	III.	7	4.29	31 13.00	40.90	2.61	51 44.52	25 46 16.51
53	7			29.	43.	56.			55 42.72	61.48	0.03	VI.	10	11.00	49 28.64	41.29	4.35	54 41.21	26 4 34.28
54	8							4.4	56 22.92	61.48	−0.02	VII.	9	10.22	44 8.38	41.38	3.85	55 21.42	25 59 13.61
55	8						37.	51.2	8 58 9.40	61.48	+0.01	VII.	4	6.58	18 27.51	41.62	1.38	57 7.93	33 30.51
56	9				30.	43.7			9 0 29.82	61.47	0.00	VI.	5	6.40	23 17.84	41.93	1.85	8 59 28.35	38 21.62
57	9	35.	49.4	3.					3 17.01	61.46	−0.01	IV.	7	10.11	34 5.49	42.30	2.89	9 2 15.54	49 10.68
58	8	14.	28.						9 4 55.80	−61.46	0.00	III.	5	10.54	−25 26.04	−42.51	−2.04	9 3 54.34	−25 40 30.59

Zone 162. March 29. B. Belt, −25° 38′. $D_0 = -25° 15′ 0″$.

No.	Mag.	I.	II.	III.	IV.	V.	VI.	VII.	T.	a_1	a_2	Micrometer.			i	d_1	d_2	Mean Right Ascension, 1850.0.	Mean Declination, 1850.0.
1	8	4.	19.	32.					8 36 46.19	−40.86	−0.01	IV.	7	5.04	−31 30.69	−4.98	−3.57	8 36 5.32	−25 46 39.2
2	8	29.	44.	57.					37 11.17	40.86	0.00	IV.	7	2.18	30 6.98	5.05	3.44	36 30.31	25 45 15.5
3	9			11.	25.				40 24.97	40.87	−0.02	IV.	10	9.57	48 57.03	5.62	5.14	39 44.08	26 4 7.8
4	9			30.	43.5				42 43.51	40.88	0.00	IV.	4	12.58	21 29.39	6.03	2.70	42 2.63	25 36 38.1
5	8					45.	59.	13.	43 31.31	40.88	−0.01	VII.	8	6.17	37 5.84	6.17	4.08	42 50.42	52 16.1
6	9					13.2	27.		44 59.34	40.88	0.00	VII.	4	7.4	18 30.54	6.43	2.44	44 18.46	33 39.4
7	8						17.4	31.5	45 49.76	40.89	0.00	VII.	5	7.25	23 40.32	6.57	2.89	45 8.87	38 49.8
8	9	28.3	43.5	56.5					48 10.73	40.89	−0.01	V.	9	5.36	41 44.49	6.99	4.48	47 29.83	25 56 56.0
9	9			55.3	9.				50 9.12	40.90	−0.02	IV.	10	10.49	49 23.26	7.34	5.18	49 28.20	26 4 35.8
10	9	43.5	58.	11.2					52 25.53	40.90	0.00	IV.	7	4.19	31 8.00	7.73	3.54	51 44.63	25 46 19.3
11	7		55.	7.5	21.				55 21.80	40.91	−0.02	IV.	10	10.53	49 25.27	8.25	5.18	54 40.87	26 4 38.7
12	7					16.	29.5	43.5	56 2.01	40.91	−0.02	VII.	9	10.20	44 7.37	8.35	4.70	55 21.08	25 59 20.4
13	9							31.	8 57 49.19	40.92	0.00	VII.	4	6.42	18 19.45	8.67	2.42	57 8.27	33 30.5
14	10			55.	9.				9 0 8.74	40.92	0.00	V.	4	6.23	18 10.18	9.07	2.41	8 59 27.82	33 21.7
15	10			42.5	56.2				2 56.21	40.93	−0.01	IV.	7	10.1	34 0.45	9.56	3.80	9 2 15.27	49 13.8
16	8				35.	49.	2.5		4 34.95	40.93	0.01	V.	7	10.58	25 28.05	9.84	3.04	3 54.01	40 40.9
17	9			9.5	23.				6 23.13	40.93	−0.01	IV.	8	5.12	36 33.42	10.14	4.03	5 42.19	51 47.6
18	9			24.5	38.				8 37.90	40.94	+0.02	IV.	1	11.41	5 53.47	10.52	1.35	7 56.98	21 5.3
19	9			56.	9.6	23.5			13 9.58	40.95	0.00	VI.	5	6.00	22 57.67	11.30	2.82	12 28.63	38 11.8
20	6		4.5	18.2	31.5				15 31.68	40.95	+0.02	IV.	1	9.7	4 35.82	11.68	1.24	14 50.75	19 48.7
21	9		57.4	11.	25.				17 24.83	40.95	+0.01	IV.	3	12.35	16 18.80	12.00	2.25	16 43.89	31 33.1
22	10		14.5	28.	41.5				20 41.90	40.96	−0.01	IV.	8	12.55	40 26.88	12.56	4.38	20 0.93	55 43.8
23	9						12.	26.	21 44.34	40.96	0.00	VII.	6	6.38	27 17.73	12.72	3.21	21 3.38	42 33.7
24	8					56.	10.	24.	23 42.33	40.96	−0.01	VII.	9	4.13	41 2.31	13.05	4.43	23 1.36	56 19.8
25	9		4.	17.					9 26 31.31	−40.97	0.00	IV.	7	3.12	−30 34.21	−13.51	−3.49	9 25 50.34	−25 45 51.2

CORRECTIONS.

Date.		Corr. of Clock.	Hourly rate.	m	n	c
1848.	h.	s.	s.	s.	s.	s.
Mar. 29.	7	− 45.96	+0.015			

INSTRUMENT READINGS.

	Date.		Barom.	Thermom. At.	Thermom. Ex.
	1848.	h. m.	in.	°	°

REMARKS.

(162) 16. Hor. thread assumed as 5 instead of 7.
(162) 24. Double; observed the first.

ZONE 162. MARCH 29. B. BELT, −25° 38'. D_o = −25° 15' 0"—Continued.

No.	Mag.	SECONDS OF TRANSIT. I.	II.	III.	IV.	V.	VI.	VII.	T.	a_1	a_2	MICROMETER.			i	d_1	d_2	Mean Right Ascension, 1850.0.	Mean Declination, 1850.0.
									h. m. s.	s.	s.			r.	' "	"	"	h. m. s.	° ' "
26	7	...	...	54.	7.5	...	...	...	9 30 7.37	−40.97	+0.02	IV.	1	3.37	− 2 19.68	−14.11	−0.02	9 29 26.42	−25 17 33.8
27	9	...	...	...	...	16.	30.	...	31 2.24	40.97	+0.01	VII.	2	10.5	10 3.71	11.25	1.69	30 21.28	25 19.7
28	8	...	...	...	...	...	36.	...	32 8.38	40.97	0.00	VII.	6	5.17	26 36.88	14.43	3.14	31 27.41	25 41 54.5
29	9	...	...	...	...	2.5	16.	30.	46 48.51	40.98	−0.02	VII.	10	7.47	47 51.12	16.78	5.05	46 7.51	26 3 13.0
30	9	...	...	46.	0.	14.	...	...	48 59.95	40.98	0.01	V.	7	10.52	34 26.13	17.13	3.85	48 18.96	25 49 47.1
31	7	...	12.	25.3	39.5	...	...	...	50 39.46	40.98	0.01	IV.	7	12.00	35 0.45	17.40	3.90	49 58.47	50 21.8
32	10	...	...	...	...	...	45.	59.1	51 17.46	40.98	−0.01	VII.	9	9.5	38 30.54	17.49	4.11	50 36.47	53 52.2
33	9	...	15.2	28.5	42.3	...	...	...	53 42.44	40.98	0.00	IV.	5	11.3	25 30.61	17.86	3.05	53 1.46	40 51.5
34	8	...	...	21.	35.	...	...	...	55 34.82	40.99	0.00	IV.	6	10.40	29 20.12	18.15	3.39	54 53.83	25 44 41.7
35	10	...	...	...	...	...	41.	55.	57 13.42	40.99	−0.02	VII.	9	11.58	44 56.78	18.39	4.79	56 32.41	26 0 20.0
36	10	...	...	32.	46.	...	...	...	9 59 45.71	40.99	+0.01	IV.	3	9.10	14 35.43	18.79	2.08	59 4.73	25 29 56.3
37	9	...	...	...	...	...	...	55.	10 0 13.43	40.99	−0.01	VII.	8	5.35	36 44.66	18.85	4.05	9 59 32.43	52 7.6
38	10	...	19.	32.2	46.	...	...	...	9 45.97	40.98	−0.01	IV.	8	3.15	35 34.42	20.28	3.95	10 9 4.98	50 58.7
39	9	...	...	54.2	7.7	...	...	...	12 7.78	40.98	0.00	IV.	6	11.16	29 38.27	20.61	3.41	11 26.80	45 2.3
40	9	...	39.3	52.4	6.0	20.2	...	...	15 6.26	40.98	+0.01	IV.	3	10.9	15 5.19	21.03	2.13	14 25.29	30 28.4
41	9	4.	18.3	32.	...	...	...	...	18 45.83	40.98	0.00	IV.	4	10.55	20 27.38	21.56	2.60	18 4.85	35 51.5
42	7	49.	3.2	16.5	...	...	...	...	24 30.72	40.98	0.00	IV.	6	7.10	27 34.22	22.33	3.23	23 49.74	42 59.8
43	8	47.	2.	5.2	...	...	...	...	26 29.29	40.97	−0.01	IV.	7	11.42	34 51.38	22.60	3.89	25 48.31	25 50 17.9
44	9	...	...	...	...	...	29.	43.	27 1.43	40.97	−0.02	VII.	10	5.53	46 53.63	22.67	4.98	26 20.44	26 2 21.3
45	10	...	...	...	...	...	...	46.	28 4.19	40.97	+0.01	VII.	4	6.12	18 4.32	22.82	2.39	27 23.23	25 33 29.5
46	7	...	...	...	41.	...	...	...	29 40.85	40.97	−0.01	IV.	8	8.48	38 22.34	23.03	4.22	28 59.87	53 49.6
47	10	17.	31.	...	...	...	...	...	32 58.72	40.97	+0.01	III.	3	11.36	15 49.02	23.45	2.20	32 17.76	25 31 14.7
48	9	...	55.5	8.4	22.3	...	...	...	34 22.67	40.97	−0.02	IV.	9	11.28	44 42.01	23.63	4.78	33 41.68	26 0 10.4
49	9	...	32.	45.	...	...	...	...	36 59.48	40.96	0.02	IV.	9	13.2	45 29.41	23.97	4.85	36 18.50	26 0 58.2
50	8	8.	23.	36.4	...	...	...	...	39 50.42	40.96	−0.01	IV.	8	13.39	40 49.07	24.34	4.43	39 9.45	25 56 17.9
51	9	...	25.	38.5	...	...	...	...	41 38.50	40.96	0.00	IV.	4	11.58	20 59.14	24.58	2.64	40 57.94	36 26.4
52	9	...	45.3	58.5	12.3	...	...	...	46 12.53	40.95	0.00	IV.	7	5.13	31 35.22	25.09	3.60	45 31.58	47 3.9
53	8	...	...	...	...	42.	56.	...	47 28.25	40.95	−0.01	VII.	9	4.59	41 25.51	25.25	4.49	46 47.29	25 56 55.3
54	10	...	...	...	...	...	...	50.	48 8.53	40.94	0.02	VII.	10	2.29	45 10.77	25.32	4.84	47 27.57	26 0 40.9
55	9	9.5	24.4	37.4	...	...	...	...	52 51.74	40.93	0.01	IV.	9	7.00	42 26.88	25.89	4.59	52 10.80	25 57 57.4
56	8	...	...	35.5	49.5	3.4	...	...	10 55 49.49	40.93	0.02	V.	9	12.30	45 43.49	26.23	4.88	55 8.54	26 1 14.6
57	7	...	53.3	6.2	20.1	...	...	...	11 5 20.46	40.91	−0.01	IV.	9	10.9	44 2.19	27.32	4.73	4 39.54	25 59 34.2
58	9	...	...	20.	33.5	48.	...	...	6 33.72	40.91	0.00	V.	5	7.11	23 33.58	27.47	2.88	5 52.81	39 3.9
59	9	...	...	8.	22.	...	...	...	10 21.63	40.90	+0.01	IV.	1	6.30	3 16.66	27.90	1.09	9 40.74	18 45.6
60	9	...	23.5	36.5	51.	...	...	...	11 50.83	40.90	0.00	IV.	7	4.17	31 6.99	28.06	3.56	11 9.93	46 38.6
61	9	...	...	20.	34.	...	...	...	15 33.86	40.89	−0.01	IV.	7	11.35	34 47.85	28.48	3.89	14 52.96	50 20.2
62	8	...	32.	45.5	...	...	...	...	21 59.38	40.87	+0.01	IV.	4	2.52	16 23.83	29.17	2.24	21 18.52	31 55.2
63	9	...	...	...	...	...	...	49.	22 7.20	40.87	0.00	VII.	4	8.16	14 7.84	29.18	2.03	21 26.33	29 39.0
64	6	...	24.5	37.5	51.	...	...	...	41 51.56	40.80	−0.01	IV.	8	10.43	39 20.33	31.25	4.32	41 10.75	25 54 55.9
65	9	...	...	24.5	38.	52.5	...	...	46 38.35	40.79	−0.01	V.	9	11.46	44 51.06	31.71	4.83	45 57.55	26 0 27.6
66	9	...	8.2	21.3	35.1	...	...	...	49 35.31	40.78	0.00	IV.	5	12.51	26 25.06	32.02	3.13	48 54.53	25 42 0.2
67	10	...	...	...	...	49.	3.	...	50 35.25	40.78	0.00	VII.	5	12.20	26 9.07	32.11	3.10	49 54.47	41 44.3
68	9	...	58.	11.5	...	...	...	...	53 25.38	40.77	+0.01	IV.	3	12.50	16 26.36	32.37	2.23	52 44.62	32 1.0
69	8	...	...	...	...	15.	...	43.	54 1.17	40.76	0.00	VII.	5	6.16	23 5.54	32.43	2.83	53 20.41	38 40.8
70	9	...	53.	6.	19.5	...	...	...	11 59 19.88	−40.74	0.00	IV.	4	6.24	−18 10.73	−32.93	−2.38	10 58 39.14	−25 33 46.0

CORRECTIONS.

Date.		Corr. of Clock.	Hourly rate.	m	n	c
1848.	h.	s.	s.	s.	s.	s.

INSTRUMENT READINGS.

Date.		Barom.	THERMOM. At.	Ex.
1848.	h. m.	In.	°	°

REMARKS.

(162) 26. Micrometer reading assumed as $4^r.37$ instead of $3^r.37$.
(162) 32. Hor. thread assumed as 8 instead of 9.
(162) 43. Transit over T. III assumed as $15^s.2$ instead of $5^s.2$.
(162) 51. Transits over T.'s III and IV assumed as recorded over T.'s II and III.
(162) 56. Micrometer reading assumed as $13^r.30$ instead of $12^r.30$.
(162) 63. Hor. thread assumed as 3 instead of 4.

ZONE 163. MAY 3. B. BELT, −25° 38'. D₀ = −25° 15' 10".

No.	Mag.	SECONDS OF TRANSIT. I.	II.	III.	IV.	V.	VI.	VII.	T.	a_1	a_2	MICROMETER.			i	d_1	d_2	Mean Right Ascension, 1850.0.	Mean Declination, 1850.0.
									h. m. s.	s.	s.			r.	' "	"	"	h. m. s.	° ' "
1	8	..	..	..	52.	5.	..	..	14 3 51.51	+47.15	0.00	VI.	8	9.48	−38 52.43	−8.52	−3.24	14 4 38.66	−25 54 4.2
2	9	..	..	..	..	..	45.	59.5	5 17.60	47.16	0.00	VII.	6	5.1	26 28.81	8.60	2.14	6 4.76	41 49.6
3	9	..	..	..	..	32.	45.	0.	14 17.93	47.22	0.00	VII.	6	3.1	25 28.30	9.06	2.05	15 5.15	25 40 49.4
4	9	..	..	..	20.5	34.4	..	..	16 20.48	47.23	0.00	V.	10	7.33	47 44.38	9.16	4.04	17 7.71	26 3 7.6
5	7	..	38.4	52.	5.5	..	..	..	20 5.58	47.25	0.00	IV.	1	10.11	5 8.09	9.33	0.24	20 52.83	25 20 27.7
6	9	..	..	..	..	25.5	39.	..	21 11.50	47.26	0.00	VII.	7	8.44	33 21.27	9.38	2.75	21 58.76	48 43.4
7	9	..	..	55.	9.	..	..	..	25 8.67	47.28	0.00	IV.	2	6.59	8 30.28	9.52	0.54	25 55.95	25 23 50.3
8	7	..	..	55	9.4	23.3	..	..	30 9.28	47.31	0.00	V.	10	9.53	48 54.97	9.72	4.17	30 56.59	26 4 18.9
9	7	..	..	..	..	..	23.	37.	30 55.31	47.32	0.00	VII.	4	11.53	20 56.26	9.74	1.62	31 42.63	25 36 17.6
10	8	38.	51.5	5.5	..	..	..	..	34 19.24	47.34	0.00	IV.	2	3.48	6 53.96	9.84	0.40	35 6.58	22 14.2
11	5	31.	45.1	58.4	12.5	26.5	..	..	38 12.45	47.37	0.00	V.	3	3.53	11 55.54	9.96	0.84	39 0.22	25 27 16.3
12	9	..	..	..	38.6	52.4	..	..	42 38.52	47.40	0.00	V.	9	11.28	44 41.97	10.09	3.78	43 25.92	26 0 5.8
13	7	40.5	55.	8.4	22.	..	..	..	47 22.27	47.42	0.00	V.	5	10.8	25 2.84	10.21	2.00	14 48 9.69	25 40 25.1
14	8	..	51.	3.2	17.3	..	..	..	14 59 17.72	47.51	0.00	V.	8	11.24	39 40.96	10.46	3.33	15 0 5.23	55 4.8
15	7	..	..	..	38.	52.	15.5	19.5	15 0 37.92	47.51	0.00	VII.	6	12.2	30 1.10	10.48	2.46	1 25.43	45 24.0
16	7	31.	46.	59.	13.	..	..	..	4 13.02	47.54	0.00	IV.	5	4.32	22 13.46	10.52	1.74	5 0.56	37 35.7
17	8	18.	32.3	46.	59.3	..	..	..	7 59.72	47.57	0.00	V.	6	3.54	25 55.35	10.57	2.08	8 47.29	41 18.0
18	7	..	32.5	46.	59.5	..	..	..	10 59.66	47.59	0.00	IV.	2	11.31	10 47.43	10.61	0.73	11 47.25	26 8.8
19	7	..	52.	5.	..	..	..	..	27 19.47	47.70	0.00	IV.	9	10.6	44 0.68	10.66	3.74	28 7.17	59 25.1
20	5.6	..	..	..	..	..	8.	22.5	27 40.61	47.70	0.00	VII.	7	4.56	31 26.30	10.66	2.59	28 28.31	46 49.6
21	9	..	32.3	46.	..	..	..	..	36 59.76	47.77	0.00	IV.	3	9.16	14 38.46	10.61	1.05	37 47.53	30 0.1
22	7	..	..	..	..	..	3.	16.9	37 35.23	47.77	0.00	VII.	3	6.50	8 25.38	10.61	0.52	38 23.00	23 46.4
23	8	..	18.	31.	45.	..	..	..	40 45.18	47.80	0.00	IV.	7	10.29	34 14.57	10.58	2.85	41 32.98	25 49 38.0
24	8	..	..	59.	12.5	..	..	..	42 12.72	47.81	+0.01	IV.	10	9.21	48 38.87	10.57	4.16	43 0.54	26 4 3.6
25	5	56.5	11.5	24.3	..	52.	6.	20.	15 45 38.47	+47.83	0.00	VII.	7	9.40	−33 49.50	−10.53	−2.81	15 46 26.30	−25 49 12.8

ZONE 164. JUNE 6. B. BELT, −25° 38'. D₀ = −25° 13' 30".

No.	Mag.	I.	II.	III.	IV.	V.	VI.	VII.	T.	a_1	a_2	MICROMETER.			i	d_1	d_2	Mean Right Ascension, 1850.0.	Mean Declination, 1850.0.
1	8	..	..	..	..	54.	7.3	21.5	15 4 39.84	+20.87	0.00	VII.	5	7.29	−23 42.34	−25.89	−2.89	15 5 0.71	−25 37 41.1
2	9	45.	59.5	13.	..	..	..	..	8 26.97	20.88	0.00	V.	6	6.43	27 20.57	25.88	3.21	8 47.85	41 19.7
3	10	..	..	4.	17.5	..	..	..	10 17.52	20.88	0.00	IV.	5	4.8	22 1.35	25.88	2.74	10 38.40	36 0.0
4	8	..	..	..	..	41.	54.	..	11 26.74	20.88	0.00	VI.	3	4.41	12 19.63	25.87	1.90	11 47.62	26 17.4
5	9	..	..	..	25.	39.1	52.	..	13 24.81	20.89	−0.01	VII.	1	7.6	3 34.45	25.86	1.15	13 45.69	17 31.5
6	9	..	..	2.5	16.2	30.3	..	..	15 16.24	20.90	0.00	VI.	6	4.44	26 20.45	25.85	3.12	15 37.14	40 19.4
7	7	..	..	36.2	50.	4.	..	..	23 49.84	20.92	−0.01	V.	1	6.36	3 19.65	25.80	1.12	24 10.75	17 16.6
8	9	..	..	6.3	20.5	34.	..	..	26 20.22	20.93	0.00	V.	7	9.32	33 45.79	25.77	3.78	26 41.15	47 45.3
9	8	..	..	54.4	8.2	22.	..	..	28 8.14	20.93	0.00	V.	7	7.39	32 48.81	25.76	3.69	28 29.07	46 48.3
10	10	..	1.	14.5	28.	..	..	..	32 28.28	20.95	0.00	IV.	6	3.30	25 43.29	25.71	3.06	32 49.23	39 42.1
11	10	..	..	40.	54.	..	..	..	33 53.94	20.95	+0.01	IV.	10	3.22	45 37.85	25.69	4.82	34 14.90	59 38.4
12	10	..	..	50.	3.5	17.	..	..	36 3.46	20.96	0.00	V.	8	2.45	35 19.26	25.66	3.92	36 24.42	49 18.8
13	10	..	..	25.	..	53.2	..	..	37 39.13	20.96	0.00	V.	8	5.44	36 49.52	25.64	4.05	38 0.09	50 49.2
14	7	..	..	..	..	..	1.	14.5	38 33.04	20.97	0.00	VII.	4	4.22	17 8.85	25.63	2.32	38 54.01	31 6.8
15	8	30.	45.	58.2	..	..	..	..	41 12.30	20.97	0.00	IV.	8	3.17	35 35.43	25.59	3.94	41 33.27	49 35.0
16	9	..	22.	35.5	49.	..	..	..	45 49.12	20.99	−0.01	IV.	1	12.3	6 4.56	25.50	1.36	46 10.10	20 1.4
17	8	..	..	..	..	55.3	9.	23.	46 41.31	20.99	0.00	VII.	3	10.59	15 30.04	25.48	2.18	47 2.30	29 27.7
18	4	44.5	59.	12.5	26.	40.	..	..	49 26.28	21.00	0.00	V.	6	5.23	26 40.22	25.43	3.15	49 47.28	40 38.8
19	10	..	..	..	..	0.	13.5	27.2	50 45.85	21.00	0.00	VII.	7	3.39	30 47.47	25.40	3.51	51 6.85	44 46.4
20	9	..	36.	49.	..	16.5	..	..	15 53 3.07	+21.01	0.00	V.	6	10.58	−29 29.15	−25.35	−3.39	15 53 24.08	−25 43 27.9

CORRECTIONS.

Date.		Corr. of Clock.	Hourly rate.	m	n	c
1848.	h.	s.	s.	s.	s.	s.
May 3,	11	+ 42.83	/0.157	− 0.08	+ 0.41	0.00
June 6,	15	+ 17.09	/0.116	− 0.07	+ 0.06	0.00

INSTRUMENT READINGS.

	Date.	Barom.	THERMOM. At.	THERMOM. Ex.
	1848. h. m.	in.	°	°

REMARKS.

(163) 15. Transit over T. VI assumed as 5ˢ.5 instead of 15ˢ.5.
(163) 22. Hor. thread assumed as 2 instead of 3.
(163) 23. Minutes of transit assumed as 40 instead of 45.
(163) 24. Minutes of transit assumed as 4[illegible] instead of 47.

ZONE 164. JUNE 6. B. BELT, −25° 38′. D_0 = −25° 13′ 30″—Continued.

No.	Mag.	Seconds of Transit. I.	II.	III.	IV.	V.	VI.	VII.	T.	a_1	a_2	Micrometer.			i	d_1	d_2	Mean Right Ascension, 1850.0.	Mean Declination, 1850.0.
									h. m. s.	s.	s.			r.	′ ″	″	″	h. m. s.	° ′ ″
21	5	..	..	..	..	..	23.5	37.5	15 53 55.78	+21.01	0.00	VII.	3	5.12	−12 35.06	−25.33	−1.91	15 54 16.79	−25 26 32.3
22	10	..	..	..	..	..	5.3	19.2	55 37.64	21.02	0.00	VII.	8	5.32	36 43.15	25.29	4.04	55 58.66	50 42.5
23	7	56.5	11.5	24.4	38.2	..	..	..	15 58 38.64	21.03	+0.01	IV.	9	4.22	41 7.21	25.22	4.43	15 58 59.68	25 55 6.9
24	9	..	..	..	17.	31.	..	..	16 0 17.03	21.03	+0.01	VII.	10	8.43	48 20.[illegible]	25.18	5.08	16 0 38.07	26 2 20.6
25	9	..	58.5	12.2	25.5	40.	..	..	2 25.83	21.04	0.00	VI.	3	10.17	15 9.06	25.13	2.14	2 46.87	25 29 6.3
26	10	..	..	11.	25.	39.	..	..	4 24.99	21.05	+0.01	V.	8	11.51	39 54.57	25.07	4.33	4 46.05	53 54.0
27	9	41.	..	9.	..	..	..	..	7 22.66	21.06	0.00	IV.	3	9.58	14 59.63	24.99	2.12	7 43.72	28 56.7
28	8	14.	28.3	41.7	55.5	..	..	..	9 55.54	21.07	−0.01	V.	2	8.32	9 17.14	24.92	1.62	10 16.60	23 13.7
29	9	54.5	9.	22.	36.	..	..	..	17 36.30	21.10	+0.01	IV.	9	3.35	40 43.51	24.69	4.40	17 57.41	54 42.6
30	9	..	..	9.2	23.	37.	..	..	22 22.98	21.12	0.00	VII.	6	5.55	26 56.04	24.54	3.17	22 44.10	40 53.8
31	8	..	6.	19.	33.	..	..	..	30 33.16	21.14	0.00	V.	7	5.3	31 30.14	24.25	3.59	30 54.30	25 45 28.0
32	9	..	..	..	..	..	14.3	28.	34 46.58	21.16	+0.01	VII.	10	8.32	48 13.82	24.10	5.08	35 7.75	26 2 13.0
33	6	..	..	..	19.	33.	..	..	37 18.92	21.17	−0.01	V.	1	2.4	1 2.49	24.01	0.0	37 40.08	25 14 56.6
34	9	..	..	33.5	47.	..	..	..	39 46.88	21.18	0.01	IV.	1	6.40	3 21.70	23.94	1.11	40 8.05	17 16.8
35	8	..	..	4.	17.5	..	..	..	42 17.37	21.19	0.01	IV.	1	4.55	2 28.75	23.83	1.03	42 38.55	16 23.6
36	7	..	..	..	..	..	10.	24.2	42 42.35	21.19	−0.01	VII.	2	5.5	6 31.92	23.81	1.47	43 3.53	20 27.2
37	7	..	..	..	..	..	..	29.	44 47.22	21.20	0.00	VII.	5	1.10	20 31.23	23.73	2.61	45 8.42	34 27.6
38	9	..	..	..	..	..	1.5	16.	46 34.12	21.21	0.00	VII.	7	5.3	31 29.82	23.67	3.59	46 55.33	45 27.1
39	9	..	..	27.	41.1	..	..	..	48 40.92	21.21	0.00	V.	7	12.25	35 13.01	23.59	3.92	49 2.13	49 10.5
40	8	..	58.9	12.5	26.	..	..	..	54 26.15	21.24	0.00	IV.	3	7.33	14 47.03	23.37	2.00	54 47.39	28 42.4
41	7	..	..	..	..	..	36.	50.2	16 55 8.37	21.24	−0.01	VII.	2	13.11	11 37.48	23.34	1.80	16 55 29.60	25 32.6
42	9	..	..	..	..	..	..	44.	17 0 1.98	21.26	0.01	VII.	1	3.26	6 42.59	23.15	1.37	17 0 23.23	20 37.1
43	9	..	..	11.5	25.2	..	..	..	2 25.07	21.27	0.01	V.	3	10.32	15 16.75	23.05	2.14	2 46.33	29 11.9
44	9	..	..	..	..	..	..	26.	2 44.13	21.27	−0.01	VII.	3	7.28	13 43.63	23.04	1.99	3 5.39	27 38.7
45	10	..	..	8.	22.	..	..	..	6 21.93	21.29	+0.01	VI.	9	9.26	43 40.34	22.88	4.68	6 43.23	57 37.9
46	10	..	..	..	..	26.3	40.	54.	10 12.42	21.30	0.01	VII.	8	8.10	38 2.82	22.72	4.18	10 33.73	25 51 59.7
47	9	..	..	..	43.5	57.2	..	..	12 43.38	21.32	+0.01	V.	10	11.21	49 39.34	22.61	5.23	13 4.71	26 3 37.2
48	9	..	50.2	3.1	17.	..	..	..	17 17.28	21.34	0.00	IV.	7	10.55	34 27.68	22.40	3.86	17 38.62	25 48 23.9
49	8	21.	35.	48.5	2.	..	..	..	19 2.28	21.35	−0.01	VII.	2	7.41	8 51.09	22.32	1.56	19 23.62	22 45.0
50	10	..	30.	43.	57.	..	..	..	30 57.05	21.39	0.00	IV.	4	6.40	18 18.80	21.74	2.41	31 18.44	32 13.0
51	9	34.	48.5	..	..	..	..	..	49 16.13	21.48	0.00	IV.	7	8.42	33 20.62	20.78	3.76	49 37.61	47 15.2
52	8	..	..	..	..	..	4.2	18.1	49 36.52	21.48	0.00	VII.	7	10.9	34 4.13	20.76	3.83	49 58.00	47 58.7
53	9	..	15.	28.	42.2	..	..	..	53 42.30	21.50	+0.01	IV.	8	13.24	40 41.50	20.53	4.43	54 3.81	54 36.5
54	8	..	..	..	10.	24.	37.2	..	55 9.84	21.51	0.00	VI.	5	5.2	22 28.42	20.45	2.78	55 31.35	36 21.7
55	8	..	..	..	54.5	9.	..	..	56 54.70	21.51	−0.01	V.	3	10.5	15 3.13	20.35	2.11	57 16.20	28 55.6
56	8	..	..	..	..	10.	24.	38.	57 56.23	21.52	0.00	VII.	4	11.55	20 57.27	20.30	2.65	58 17.75	34 50.2
57	6	..	..	..	..	..	43.	57.5	17 59 15.54	21.53	−0.01	VII.	3	10.53	15 27.01	20.22	2.15	17 59 37.06	29 19.4
58	8	..	..	12.	26.2	..	..	..	18 1 25.95	21.54	0.00	IV.	7	8.40	33 19.61	20.10	3.76	18 1 47.44	47 13.5
59	9	..	4.	17.	30.5	..	..	..	4 30.80	21.55	−0.01	V.	2	3.45	6 52.41	19.92	1.39	4 52.34	20 43.7
60	8	..	..	..	..	46.	0.	14.	5 32.33	21.56	+0.01	VII.	9	3.54	35 53.73	19.87	3.99	5 53.90	49 47.6
61	9	..	..	45.	58.6	..	..	..	7 58.48	21.57	−0.01	V.	2	9.00	9 31.25	19.72	1.61	8 20.04	23 22.6
62	8	..	..	..	3.	17.1	..	..	9 3.02	21.58	0.00	V.	5	10.46	25 22.01	19.66	3.04	9 24.60	39 14.7
63	10	..	..	..	..	..	..	12.	9 30.22	21.58	0.00	VII.	4	11.6	20 32.57	19.64	2.60	9 51.80	34 24.8
64	9	..	..	..	5.	19.	..	..	11 4.96	21.59	0.00	V.	4	5.50	17 53.54	19.55	2.36	11 26.55	31 45.5
65	9	..	..	..	..	..	20.	34.	11 52.29	21.59	−0.01	VII.	3	11.00	15 30.54	19.50	2.15	12 13.87	29 22.2
66	10	..	..	..	..	..	44.	58.	13 16.28	21.60	−0.01	VII.	3	9.5	14 32.55	19.42	2.05	13 37.87	25 28 24.0
67	10	..	..	..	..	35.5	49.5	..	15 21.75	21.61	+0.01	VII.	10	6.2	46 58.18	19.30	5.00	15 43.37	26 0 52.5
68	3.4	..	54.	7.	21.2	..	..	..	18 21.10	21.62	−0.01	IV.	4	2.10	16 2.65	19.12	2.19	18 42.71	25 29 54.0
69	7	..	..	..	45.	59.2	..	..	18 19 45.04	+21.63	−0.01	VI.	2	3.48	−6 53.80	−19.04	−1.39	18 20 6.66	−25 20 44.2

CORRECTIONS.

Date.	Corr. of Clock.	Hourly rate.	m	n	c
1848. h.	s.	s.	s.	s.	s.

INSTRUMENT READINGS.

Date.	Barom.	Thermom. At.	Thermom. Ex.
1848. h. m.	in.	°	°

REMARKS.

(164) 36. Micrometer reading assumed as 3r.5 instead of 5r.5.
(164) 40. Micrometer reading assumed as 9r.33 instead of 7r.33.
(164) 42. Hor. thread assumed as 2 instead of 1.
(164) 56. Double.
(164) 59. Minutes of transit assumed as 4.
(164) 60. Hor. thread assumed as 8 instead of 9.

ZONE 164. JUNE 6. B. BELT, −25° 38'. D₀ = −25° 13' 30"—Continued.

No.	Mag.	SECONDS OF TRANSIT. I.	II.	III.	IV.	V.	VI.	VII.	T.	a_1	a_2	MICROMETER.			i	d_1	d_2	Mean Right Ascension. 1850.0.	Mean Declination. 1850.0.
									h. m. s.	s.	s.			r.	' "	"	"	h. m. s.	° ' "
70	9	. .	. .	. .	. .	. .	9.	23.2	18 20 41.44	+21.63	0.00	VII.	6	1.33	−24 43.93	−18.99	−2.97	18 21 3.07	−25 38 35.9
71	9	. .	. .	14.3	. .	42.	. .	. .	23 27.86	21.65	−0.01	V.	1	6.33	3 18.13	18.62	1.06	23 49.50	17 8.0
72	9	. .	. .	. .	. .	. .	35.	49.	24 7.30	21.65	0.00	VII.	4	8.48	19 22.98	18.78	2.50	24 28.95	33 14.3
73	9	. .	32.	45.5	59.	. .	. .	. .	26 59.17	21.66	−0.01	.	2	13.48	11 56.50	18.61	1.81	27 20.82	25 46.9
74	8	. .	. .	. .	. .	. .	5.5	19.5	28 37.80	21.67	0.00	VII.	4	7.33	18 45.16	18.51	2.43	28 59.47	32 36.1
75	8	. .	. .	. .	. .	29.	42.5	57.	30 15.08	21.68	0.00	VII.	5	8.19	24 7.56	18.41	2.92	30 36.76	37 58.9
76	10	. .	. .	. .	15.5	29.7	. .	. .	33 15.57	21.70	0.00	V.	5	8.49	24 23.01	18.23	2.95	33 37.27	38 14.2
77	10	. .	37.	50.	4.5	. .	. .	. .	35 4.20	21.71	−0.01	IV.	4	3.59	16 57.61	18.11	2.28	35 25.90	30 48.0
78	10	. .	. .	. .	. .	56.	9.5	25.	37 42.35	21.72	−0.01	VII.	2	6.58	8 29.41	17.95	1.51	38 4.06	22 18.9
79	10	. .	35.5	48.5	3.	. .	. .	. .	40 2.84	21.72	0.00	IV.	7	7.39	32 48.84	17.81	3.72	40 24.56	46 40.4
80	10	. .	. .	. .	. .	. .	13.	27.3	41 45.40	21.74	−0.01	VII.	2	5.55	7 57.64	17.70	1.47	42 7.13	21 46.8
81	10	. .	6.2	19.5	33.3	. .	. .	. .	47 33.45	21.77	0.00	V.	6	3.00	25 28.12	17.33	3.05	47 55.22	39 18.5
82	10	. .	54.	8.	21.5	. .	. .	. .	49 21.46	21.77	−0.01	V.	2	2.2	6 0.48	17.22	1.29	49 43.22	19 49.0
83	10	. .	. .	. .	. .	36.	50.	. .	51 22.25	21.79	+0.01	VI.	9	5.38	41 45.37	17.09	4.54	51 44.05	55 37.0
84	10	. .	. .	. .	. .	. .	15.	29.3	52 47.55	21.79	+0.01	VII.	8	12.3	40 0.30	17.00	4.39	53 9.35	53 51.7
85	7.8	. .	44.3	54.7	. .	. .	39.	. .	55 11.68	21.81	−0.01	VII.	3	6.4	13 1.29	16.84	0.92	55 33.48	26 49.1
86	10	. .	. .	. .	. .	2.3	16.	30.	18 57 48.31	21.82	−0.01	VII.	3	9.35	14 47.68	16.67	2.08	18 58 10.12	28 36.4
87	10	. .	. .	. .	. .	. .	35.	49.	19 3 7.40	21.85	+0.01	VII.	9	4.20	41 5.84	16.30	4.49	19 3 29.26	54 56.6
88	7	17.	31.3	45.	58.8	. .	. .	. .	5 58.75	21.86	0.00	V.	4	3.47	16 51.52	16.09	2.27	6 20.61	30 39.9
89	10	. .	. .	. .	. .	4.	18.	32.	6 50.24	21.86	0.00	VII.	5	4.32	22 13.10	16.04	2.74	7 12.10	36 1.9
90	10	. .	. .	. .	. .	. .	. .	47.	8 5.72	21.87	+0.01	VII.	8	4.31	36 12.39	15.94	4.04	8 27.60	50 2.4
91	10	. .	. .	. .	. .	. .	. .	5.	9 43.19	21.88	0.00	VII.	4	6.32	18 14.41	15.85	2.38	10 5.07	32 2.6
92	10	. .	. .	. .	24.	38.	. .	. .	11 23.99	21.89	+0.01	V.	7	11.7	34 33.69	15.71	3.89	11 45.89	48 23.3
93	10	. .	. .	. .	. .	22.5	36.6	50.5	13 8.82	21.90	0.00	VII.	7	4.54	31 25.29	15.59	3.61	13 30.72	45 14.5
94	10	. .	. .	. .	. .	. .	. .	15.4	14 33.66	21.90	0.00	VII.	5	8.19	24 7.56	15.48	2.92	14 55.56	37 56.0
95	10	30.	44.4	57.6	11.3	. .	. .	. .	17 11.76	21.92	+0.01	IV.	8	9.22	38 39.48	15.30	4.27	17 33.69	52 29.1
96	10	1.	16.2	29.	43.	. .	. .	. .	19 42.87	21.93	0.00	IV.	7	6.51	32 24.64	15.12	3.70	20 4.80	46 13.5
97	10	. .	. .	. .	40.5	54.3	8.3	. .	20 40.49	21.93	0.00	VI.	7	3.42	30 49.18	15.05	3.54	21 2.42	44 37.8
98	10	. .	. .	3.	16.3	30.5	. .	. .	23 16.48	21.95	0.00	VI.	5	6.12	23 3.72	14.86	2.83	23 38.43	36 51.4
99	10	. .	. .	27.	40.3	. .	. .	. .	25 40.35	21.96	−0.01	IV.	3	5.53	12 56.09	14.7	1.90	26 2.30	26 42.7
100	10	. .	. .	. .	32.	46.3	. .	. .	28 32.10	21.97	−0.01	VII.	2	10.58	10 30.43	14.5	1.66	28 54.06	24 16.6
101	10	. .	. .	. .	. .	42.	55.	9.	33 27.65	22.00	+0.01	VII.	8	10.26	39 11.40	14.1	4.33	33 49.66	52 59.8
102	10	. .	. .	. .	41.5	55.5	. .	. .	35 41.49	22.01	0.00	VII.	6	12.16	30 8.16	14.0	3.49	36 3.50	43 55.7
103	9	. .	. .	42.5	. .	9.5	23.5	. .	37 55.99	22.02	+0.01	VII.	9	4.30	41 10.89	13.8	4.51	38 18.02	54 59.2
104	9	. .	. .	. .	. .	25.	39.	. .	41 11.25	22.05	0.00	VII.	6	6.3	27 0.08	13.6	3.18	41 33.30	40 46.9
105	10	. .	49.3	3.	16.3	30.5	. .	. .	43 16.52	22.06	−0.01	VI.	2	7.31	8 46.25	13.4	1.51	43 38.57	22 31.2
106	10	. .	. .	. .	. .	53.3	. .	. .	19 47 39.36	+22.08	−0.01	VI.	3	9.21	−15 11.06	−13.1	−2.06	19 48 1.43	−25 28 56.2

ZONE 165. JUNE 13. B. BELT, −26° 16'. D₀ = −25° 51' 50".

No.	Mag.	I.	II.	III.	IV.	V.	VI.	VII.	T.	a_1	a_2	MICROMETER.			i	d_1	d_2	Mean Right Ascension. 1850.0.	Mean Declination. 1850.0.
1	9	. .	. .	. .	18.	32.	. .	. .	16 4 17.89	+27.37	−0.01	VI.	1	3.42	− 1 51.77	−23.4	−0.97	16 4 45.25	−25 54 6.1
2	9	. .	. .	56.	10.	. .	37.	. .	14 9.56	27.40	−0.01	VI.	1	5.15	2 38.66	22.6	1.02	14 36.95	25 54 52.3
3	8	. .	. .	. .	. .	. .	. .	43.1	15 1.30	27.41	0.00	VII.	8	2.8	35 0.26	22.6	3.92	15 28.71	26 27 16.8
4	8	. .	. .	. .	. .	. .	. .	53.5	16 11.51	27.41	0.00	VII.	5	1.54	20 53.40	22.5	2.63	16 38.92	13 8.5
5	.	. .	. .	. .	45.4	. .	. .	. .	45.25										
6	4	. .	18.	32.1	45.	59.3	13.	. .	19 45.33	27.42	−0.01	VI.	.	.	. . .	22.2	. .	16 30 12.74	
7	7	. .	16.	29.3	43.	. .	. .	. .	21 43.25	27.43	0.00	V.	4	10.15	20 7.17	22.0	2.56	22 10.68	12 21.7
8	10	40.	54.	. .	. .	. .	. .	. .	16 26 21.90	+27.44	0.00	III.	3	9.9	−14 34.89	−21.7	−2.06	16 25 49.34	−26 6 48.7

CORRECTIONS.

Date.		Corr. of Clock.	Hourly rate.	m	n	c
1848.	h.	s.	s.	s.	s.	s.
June 13.	15	+ 23.60	/ 0.130			

INSTRUMENT READINGS.

	Date.		Barom.	THERMOM. At.	THERMOM. Ex.
	1848.	h. m.	in.	°	°

REMARKS.

(164) 75. Triplet.
(164) 78. Minutes not certain.
(164) 91. Transit over T. VII assumed as 15 instead of 5.
(164) 106. Micrometer reading assumed as 10r.21 instead of 9r.21.
(165) 6. Minutes assumed as 23 instead of 26.

ZONE 165. JUNE 13. B. BELT, −26° 16′. D_o = −25° 51′ 50″—Continued.

No.	Mag.	SECONDS OF TRANSIT. I.	II.	III.	IV.	V.	VI.	VII.	T.	a_1	a_2	MICROMETER.			i	d_1	d_2	Mean Right Ascension, 1850.0.	Mean Declination, 1850.0.
									h. m. s.	s.	s.			r.	° ′ ″	″	″	h. m. s.	° ′ ″
9	8	. .	. .	. .	. .	49.	. .	17.	16 25 34.98	+27.44	0.00	VII.	4	4.52	−17 23.96	. .	−2.32	16 26 2.42	
10	9	. .	3.5	17.	. .	. .	. .	. .	30 31.01	27.45	0.00	IV.	4	3.46	16 51.06	−21.3	2.27	30 58.46	−26 9 4.6
11	8	. .	. .	. .	. .	. .	. .	14.	30 31.87	27.45	−0.01	VII.	2	8.4	9 2.68	21.3	1.57	30 59 31	1 15.6
12	9	42.5	57.	. .	. .	. .	. .	. .	33 24.70	27.46	0.00	III.	4	5.24	17 40.43	21.1	2.34	33 52.16	9 54.9
13	8	. .	. .	. .	8.5	22.	36.	. .	34 8.21	27.47	+0.01	VII.	8	9.36	38 46.17	21.0	4.27	34 35.69	31 1.4
14	7	. .	38.8	52.0	6.	. .	. .	. .	36 6.16	27.47	0.00	IV.	6	11.29	29 44.82	20.8	3.42	36 33.63	21 59.0
15	10	6.5	22.	. .	49.	. .	. .	. .	37 49.25	27.48	+0.01	IV.	9	12.1	44 58.65	20.7	4.83	38 16.74	37 14.2
16	9	45.	0.0	13.2	. .	. .	. .	. .	40 27.47	27.49	0.00	IV.	8	4.30	36 12.25	20.5	4.03	40 54.96	28 26.8
17	9	. .	. .	. .	. .	. .	26.	40.	40 58.18	27.49	0.00	VII.	6	10.33	29 16.21	20.5	3.39	41 25.67	21 30.1
18	8	. .	. .	58.	11.5	26.	. .	. .	43 11.81	27.50	0.00	V.	8	7.5	37 30.36	20.3	4.16	43 39.31	29 44.7
19	8	25.	39.8	53.	7.	. .	. .	. .	45 7.04	27.50	0.00	IV.	4	4.57	17 26.86	20.1	2.31	45 34.54	9 39.3
20	9	. .	15.7	29.	. .	. .	. .	. .	47 43.32	27.51	0.00	IV.	7	10.36	34 18.11	19.0	3.80	48 10.83	26 31.9
21	10	. .	. .	31.5	45.	. .	. .	. .	48 45.00	27.52	0.00	IV.	3	9.45	14 53.08	19.9	2.08	49 12.52	7 5.0
22	9	. .	. .	. .	. .	58.	12.	26.	49 44.07	27.52	0.00	VII.	4	2.13	16 3.78	19.7	2.19	50 11.59	8 15.7
23	9	. .	. .	. .	. .	. .	7.5	21.	51 39.44	27.53	0.00	VII.	7	7.32	32 44.94	19.5	3.71	52 6.97	24 58.2
24	9	. .	. .	38.	51.5	. .	. .	. .	54 51.48	27.54	−0.01	V.	2	11.30	10 46.89	19.2	1.72	55 19.01	2 57.8
25	8	. .	. .	6.5	20.	. .	. .	. .	56 20.12	27.55	0.00	V.	6	11.52	29 56.38	19.1	3.44	56 47.67	22 8.9
26	7.	. .	. .	. .	. .	22.	35.5	50.	57 7.95	27.55	0.00	VII.	5	12.7	26 2.50	19.0	3.09	57 35.50	18 14.6
27	9	. .	. .	. .	50.	. .	17.5	. .	16 58 49.75	27.56	+0.01	VII.	10	3.47	45 50.08	18.9	4.91	16 59 17.32	38 3.9
28	9	. .	39.5	53.	6.5	. .	. .	. .	17 1 6.84	27.56	0.00	IV.	5	6.32	23 13.97	18.7	2.84	17 1 34.40	15 25.5
29	9	. .	. .	. .	. .	. .	18.	32.5	1 50.46	27.57	+0.01	VII.	8	8.55	38 25.49	18.6	4.25	2 18.04	30 38.3
30	8	45.	0.0	. .	. .	. .	. .	. .	5 27.54	27.58	0.00	IV.	6	9.48	28 53.89	18.2	3.36	5 55.12	21 5.5
31	7	. .	. .	. .	. .	. .	8.	22.	5 40.18	27.58	0.00	VII.	6	12.42	30 21.25	18.2	3.50	6 7.76	22 33.0
32	8	. .	. .	. .	. .	7.	20.5	34.5	7 52.83	27.59	0.00	VII.	7	11.31	35 15.70	18.0	3.91	8 20.42	27 27.6
33	9	. .	47.	0.5	. .	. .	. .	. .	10 14.73	27.60	0.00	IV.	7	11.4	34 32.22	17.8	3.88	10 42.33	26 43.9
34	8	. .	. .	. .	. .	58.	12.	26.	10 44.29	27.60	0.00	VII.	7	4.26	31 11.15	17.7	3.57	11 11.89	23 22.4
35	8	. .	. .	. .	. .	. .	. .	21.3	11 39.42	27.61	0.00	VII.	6	8.45	28 21.75	17.6	3.32	12 7.03	20 32.7
36	10	. .	. .	. .	17.	. .	45.	. .	14 17.07	27.62	−0.01	VI.	3	4.50	12 24.16	17.4	1.86	14 44.68	4 33.4
37	8	. .	37.	50.5	4.4	. .	. .	. .	17 4.44	27.63	0.00	IV.	4	8.49	19 23.85	17.1	2.48	17 32.07	11 33.4
38	9	. .	34.5	47.4	1.5	. .	. .	. .	19 1.80	27.63	+0.01	IV.	9	9.27	43 41.00	16.9	4.73	19 29.44	35 52.6
39	8	16.	30.9	44.	58.	. .	. .	. .	21 58.07	27.64	0.00	IV.	4	3.30	16 42.99	16.6	2.23	22 25.71	8 51.8
40	9	15.	30.5	. .	57.9	. .	. .	. .	24 57.89	27.65	+0.01	IV.	10	3.32	45 42.90	16.3	4.92	25 25.55	37 54.1
41	8	. .	. .	55.5	9.4	23.5	. .	. .	34 9.34	27.69	0.00	V.	5	3.27	21 40.63	15.4	2.69	34 37.03	13 48.7
42	9	. .	11.	24.5	38.5	. .	. .	. .	36 38.45	27.70	0.00	V.	3	11.46	15 54.06	15.1	2.16	37 6.15	8 1.3
43	10	. .	. .	. .	. .	0.5	13.5	28.	37 46.11	27.71	0.00	VII.	5	9.59	24 57.96	15.0	3.00	38 13.82	17 6.0
44	10	. .	49.	. .	16.	. .	. .	. .	40 16.45	27.72	+0.01	IV.	9	6.45	42 19.32	14.7	4.61	40 44.18	26 34 28.6
45	10	. .	. .	20.5	34.	. .	. .	. .	43 33.94	27.73	−0.01	IV.	1	11.9	5 37.34	14.4	1.24	44 1.66	25 57 43.0
46	10	. .	. .	. .	. .	. .	5.	19.	17 44 37.12	27.74	0.00	VII.	4	9.32	19 45.15	14.3	2.52	17 45 4.86	26 11 52.0
47	9	. .	. .	9.	22.5	. .	. .	. .	18 5 22.61	27.83	0.00	IV.	6	9.41	28 50.36	12.0	3.36	18 5 50.44	20 55.7
48	8	. .	. .	. .	. .	. .	19.	33.	5 51.18	27.83	0.00	VII.	6	11.27	29 43.43	11.9	3.44	6 19.01	26 21 48.8
49	8	. .	. .	. .	55.	. .	. .	. .	9 54.85	27.84	−0.01	V.	1	15.9	7 38.30	11.4	1.42	10 22.68	25 59 41.1
50	9	. .	. .	. .	. .	. .	0.0	14.2	12 32.31	27.86	+0.01	VII.	8	5.31	36 42.62	11.1	4.10	13 0.18	26 28 47.8
51	8	. .	. .	. .	. .	. .	. .	28.	13 46.02	27.86	0.00	VII.	5	4.11	22 2.48	11.0	2.71	14 13.88	26 14 6.2
52	9	. .	. .	39.5	53.5	. .	. .	. .	16 53.21	27.88	−0.01	IV.	2	5.22	7 41.36	10.6	1.42	17 21.08	25 59 43.4
53	9	. .	. .	. .	21.	35.	. .	. .	18 20.94	27.88	0.00	V.	5	7.53	23 54.76	10.4	2.90	18 48.82	26 15 58.1
54	8	. .	. .	. .	. .	. .	. .	50.5	19 8.86	27.89	+0.01	VII.	10	8.22	48 8.75	10.3	5.18	19 36.76	40 14.2
55	9	. .	59.	12.	25.5	. .	. .	. .	22 26.17	27.90	+0.01	VI.	9	8.53	43 24.20	9.9	4.64	22 54.08	35 28.7
56	8	. .	. .	. .	. .	32.	46.	0.0	23 18.11	27.91	0.00	VII.	5	10.42	25 19.65	9.8	3.03	23 46.02	17 22.5
57	10	. .	. .	. .	. .	16.5	30.	. .	25 2.39	27.92	+0.01	VII.	9	8.8	43 0.79	9.6	4.70	25 30.32	35 5.1
58	9	. .	. .	47.5	1.5	. .	. .	. .	18 27 1.39	+27.92	0.00	V.	7	6.31	−32 14.52	− 9.3	−3.68	18 27 29.31	−26 24 17.5

CORRECTIONS.

Date.		Corr. of Clock.	Hourly rate.	m	n	c
1848.	h.	s.	s.	s.	s.	s.

INSTRUMENT READINGS.

Date.		Barom.	THERMOM. At.	Ex.
1848.	h. m.	in.	°	°

REMARKS.

(165) 31. Double.
(165) 32. Micrometer reading assumed as 12r.31 instead of 11r.31.
(165) 54. Double.
(165) 55. Micrometer reading assumed as 10r.53 instead of 8r.53.

ZONE 166. JUNE 15. B. BELT, −25° 1′. D_o = −24° 37′ 50″.

No.	Mag.	SECONDS OF TRANSIT. I.	II.	III.	IV.	V.	VI.	VII.	T.	a_1	a_2	MICROMETER.			i	d_1	d_2	Mean Right Ascension, 1850.0.	Mean Declination, 1850.0.
									h. m. s.	s.	s.			r.	′ ″	″	″	h. m. s.	° ′ ″
1	8	..	..	38.2	51.4	6.	..	..	15 4 51.78	+29.94	0.00	VI.	6	5.15	−26 36.19	− 6.9	−2.13	15 5 21.62	−25 4 35.2
2	8	..	..	..	..	..	..	38.	5 56.32	29.94	−0.01	VII.	3	2.42	11 19.43	7.2	0.94	6 26.25	24 49 17.6
3	8	..	..	5.	19.	32.	..	..	9 18.42	29.94	−0.01	VI.	1	3.53	1 57.32	8.3	0.22	9 48.35	39 55.8
4	9	..	0.0	13.	26.5	..	..	..	11 26.87	29.94	0.00	V.	5	8.50	24 23.51	8.9	1.95	11 56.81	25 2 24.4
5	8	..	..	45.5	59.	..	..	..	12 58.96	29.94	−0.01	IV.	3	13.14	16 38.46	9.4	1.36	13 28.89	24 54 39.2
6	8	34.	48.5	2.	15.3	..	..	..	15 15.68	29.94	0.00	IV.	6	10.46	29 23.15	10.1	2.35	15 45.62	25 7 25.6
7	9	..	..	27.5	41.2	..	..	..	16 41.13	29.94	0.00	V.	6	5.21	26 39.21	10.6	2.13	17 11.07	25 4 41.9
8	9	..	36.	49.2	2.5	..	..	..	21 2.81	29.93	−0.01	V.	2	6.38	8 19.65	11.9	0.70	21 32.73	24 46 22.3
9	8	..	14.5	27.3	41.	..	..	..	23 41.32	29.93	+0.01	IV.	8	9.44	38 50.58	12.7	3.09	24 11.26	25 16 56.4
10	8	7.5	22.	35.4	49.	..	..	..	25 49.25	29.93	0.01	IV.	8	2.40	35 16.78	13.4	2.82	26 19.19	13 23.0
11	8	..	..	20.8	34.5	..	..	..	27 34.54	29.93	+0.01	IV.	9	8.42	43 18.32	14.0	3.46	28 4.48	25 21 25.8
12	9	27.6	41.5	55.	9.	..	..	..	30 8.87	29.93	−0.01	IV.	3	7.33	13 46.52	14.7	1.12	30 38.79	24 51 52.3
13	8	..	..	..	10.	24.	37.4	51.5	31 10.01	29.93	0.00	VII.	6	7.32	27 44.96	15.1	2.22	31 39.94	25 5 52.3
14	7	1.5	16.	29.5	43.2	..	..	..	33 43.19	29.93	0.00	IV.	4	5.96	17 31.40	15.8	1.42	34 13.12	24 55 38.6
15	8	36.	50.8	4.	17.3	..	..	..	36 17.70	29.94	0.00	V.	5	5.48	22 51.73	16.6	1.83	36 47.64	25 1 0.2
16	8	..	..	39.	53.	..	..	..	37 52.90	29.94	+0.01	VII.	9	13.1	45 28.54	17.1	3.65	38 22.85	23 39.3
17	5.6	40.5	1.	14.	27.3	41.5	..	..	41 27.76	29.94	+0.01	VI.	8	10.9	39 3.03	18.1	3.12	41 57.71	25 17 14.3
18	6	25.	40.	53.	6.5	20.5	34.	..	44 6.68	29.94	−0.01	VII.	3	8.18	14 8.85	18.9	1.15	44 36.61	24 52 18.9
19	6	..	..	..	..	..	..	50.5	45 8.80	29.94	0.01	VII.	2	8.35	9 18.33	19.2	0.77	45 38.73	47 28.3
20	8	..	..	..	55.	9.	..	..	51 54.97	29.94	−0.01	V.	2	5.42	7 51.41	21.3	0.66	52 24.90	24 46 3.4
21	6	..	..	33.	46.5	1.	14.	..	53 46.77	29.94	+0.01	VI.	10	8.31	48 13.51	21.8	3.87	54 16.72	25 26 29.2
22	10	..	13.5	27.	40.3	..	..	..	57 40.66	29.94	0.00	IV.	6	5.10	26 33.71	23.0	2.13	58 10.60	25 4 48.8
23	10	..	..	..	10.	24.	..	..	15 59 10.01	29.94	0.00	V.	4	9.52	19 55.57	23.5	1.59	15 59 39.95	24 58 10.7
24	10	..	..	25.4	37.	53.	..	..	16 0 39.00	29.94	0.00	V.	4	9.19	19 38.93	23.9	1.57	16 1 8.94	24 57 54.4
25	10	..	..	..	..	..	..	2.	1 20.56	29.94	0.00	VII.	6	10.12	29 5.64	24.1	2.33	1 50.50	25 7 22.1
26	8	36.5	51.2	4.2	18.	..	..	..	5 18.19	29.94	0.00	V.	6	6.32	27 15.03	25.2	2.18	5 48.13	25 5 32.4
27	9	..	..	..	..	27.	40.5	54.	6 12.81	29.94	−0.01	VII.	2	1.41	5 49.56	25.5	0.50	6 42.74	24 44 5.6
28	9	..	19.5	32.4	46.	..	..	..	9 46.50	29.94	+0.01	V.	10	2.8	45 0.50	26.5	3.60	10 16.45	25 23 20.6
29	6	53.	7.	21.	34.	48.	..	..	11 34.47	29.94	+0.01	VI.	8	2.53	35 23.17	27.1	2.83	12 4.42	13 43.1
30	10	..	..	13.	26.	..	..	..	14 26.26	29.94	0.00	V.	6	1.31	24 43.24	27.9	1.98	14 56.20	3 3.1
31	10	..	..	..	..	..	50.	4.	15 22.50	29.95	0.00	VII.	5	3.50	21 51.91	28.2	1.75	15 52.45	0 11.9
32	8	..	..	..	37.	51.2	..	..	18 37.13	29.95	0.00	VII.	6	8.59	28 28.83	29.1	2.28	19 7.08	25 6 50.2
33	9	..	..	..	23.	..	..	..	20 22.85	29.95	−0.01	IV.	2	10.39	10 21.22	29.6	0.84	20 52.79	24 48 41.7
34	6	..	..	..	..	..	..	17.	20 35.28	29.95	0.01	VII.	2	6.54	8 27.40	29.7	0.69	21 5.22	46 47.8
35	9	..	..	22.	35.5	..	..	..	24 35.35	29.95	−0.01	V.	1	4.49	2 25.69	30.8	0.22	25 5.29	24 40 46.7
36	10	..	..	0.5	14.	..	..	..	16 26 14.02	+29.95	0.00	V.	6	3.13	−25 34.67	−31.3	−2.05	16 26 43.97	−25 3 58.0

ZONE 167. JUNE 16. P. BELT, −9° 20′.

No.	Mag.	I.	II.	III.	IV.	V.	VI.	VII.	T.	a_1	a_2	MICROMETER.			i	d_1	d_2	Mean Right Ascension, 1850.0.	Mean Declination, 1850.0.
1	9	35.5	..	1.4	13.6	..	..	..	18 57 13.66	+29.23	0.00	III.	4	4.4	−17 0.13	. .	−1.71	18 57 42.89	
2	10	43.	56.1	..	..	..	..	..	19 0 21.26	29.24	−0.01	II.	3	3.56	11 57.05	. .	1.51	19 0 50.49	
3	10	..	..	..	23.9	..	49.	..	1 23.76	29.24	0.00	IV.	6	8.4	28 1.46	. .	2.11	1 52.00	
4	10	..	..	..	..	..	51.9	..	4 26.69	29.25	0.00	VI.	4	10.12	20 5.65	. .	1.81	4 55.94	
5	8	..	..	..	35.5	..	0.9	..	5 35.53	29.25	−0.01	VI.	2	5.20	7 40.30	. .	1.35	6 4.77	
6	9	..	..	23.9	..	49.9	..	..	8 36.63	29.26	−0.01	V.	1	3.2	1 31.76	. .	1.12	9 5.88	
7	10	..	12.9	..	..	..	..	..	10 38.23	29.27	0.00	I.	5	3.2[illegible]	21 37.02	. .	1.87	11 7.50	
8	10	..	..	..	19.9	..	34.6	..	11 19.57	29.27	0.00	IV.	4	6.2	17 59.64	. .	1.73	11 48.84	
9	10	..	..	..	27.4	..	..	..	19 14 27.26	+29.28	0.00	IV.	8	6.00	−36 57.63	. .	−2.46	19 14 56.54	

CORRECTIONS.

Date.		Corr. of Clock.	Hourly rate.	m	n	c
1848.	h.	s.	s.	s.	s.	s.
June 15,	15	+ 26.30	0.008	− 0.19	+ 0.71	0.00
16,	15	+ 25.76	0.032	− 0.37	+ 0.94	0.00

INSTRUMENT READINGS.

	Date.	Barom.	THERMOM. At.	THERMOM. Ex.
	1848. h. m.	in.	°	°

REMARKS.

(166) 17. Transit over T. I assumed as 45ˢ.5 instead of 40ˢ.5.
(166) 24. Transit over T. IV assumed as 39ˢ instead of 37ˢ.
(166) 29. Double; time of transit over T. I assumed as 53ˢ instead of 55ˢ.
(167) 8. One of these transits is erroneous by 10ˢ. T. IV is assumed to be correctly observed.

ZONE 167. JUNE 16. P. BELT, −9° 20′—Continued.

No.	Mag.	SECONDS OF TRANSIT. I.	II.	III.	IV.	V.	VI.	VII.	T.	a_1	a_2	MICROMETER.			i	d_1	d_2	Mean Right Ascension, 1850.0.	Mean Declination, 1850.0.
									h. m. s.	s.	s.			r.	′ ″	″	″	h. m. s.	° ′ ″
10	10	38.1	51.4	. .	. .	. .	. .	. .	19 18 16.56	+29.29	0.00	II.	6	7.44	−27 51.32	. .	−2.11	19 18 45.85	
11	9	35.2	48.1	. .	. .	. .	. .	. .	19 13.50	29.29	0.00	II.	7	9.51	33 55.35	. .	2.34	19 42.79	
12	[illegible]	. .	. .	. .	. .	6.4	. .	. .	19 53.74	29.29	0.00	VII.	6	12.12	30 6.38	. .	2.19	20 23.03	
13	11	. .	. .	. .	33.9	. .	. .	. .	25 33.76	29.31	−0.01	V.	1	7.44	3 53.96	. .	1.20	26 3.06	
14	10	. .	. .	. .	. .	. .	. .	2.9	27 24.95	29.32	0.00	VII.	7	11.30	34 55.29	. .	2.38	27 54.27	
15	10	. .	. .	. .	8.9	. .	. .	46.3	32 8.58	29.33	+0.01	IV.	8	8.25	38 10.74	. .	2.50	32 37.92	
16	10	. .	. .	. .	. .	17.2	. .	. .	34 4.61	29.34	+0.01	V.	9	8.30	43 12.26	. .	2.70	34 33.96	
17	10	46.9	. .	. .	. .	. .	. .	. .	36 25.03	29.34	−0.01	I.	2	8.14	9 7.98	. .	1.40	36 54.36	
18	11	. .	. .	. .	. .	. .	33.	. .	40 7.77	29.36	0.00	VI.	6	9.15	28 37.20	. .	2.14	40 37.13	
19	10	. .	. .	. .	. .	. .	. .	13.9	40 35.84	29.36	0.00	VII.	6	5.50	26 53.76	. .	2.07	41 5.20	
20	10	. .	. .	. .	. .	. .	48.1	. .	42 22.89	29.36	0.00	VI.	4	7.52	18 55.05	. .	1.77	42 52.25	
21	10	. .	. .	38.5	. .	. .	. .	33.5	44 50.62	29.37	0.00	III.	4	5.23	17 39.95	. .	1.72	45 19.99	
22	10	. .	36.5	. .	. .	. .	. .	. .	47 1.81	29.38	0.00	III.	4	5.54	17 55.59	. .	1.73	47 31.19	
23	10	. .	. .	. .	. .	. .	. .	40.4	47 2.20	29.38	0.00	VII.	4	6.2	17 59.52	. .	1.73	47 31.58	
24	10	36.5	. .	. .	. .	. .	. .	. .	49 14.81	29.39	0.00	I.	8	6.2	36 58.52	. .	2.46	49 44.20	
25	10	. .	. .	. .	. .	. .	46.5	. .	49 21.26	29.39	+0.01	VI.	8	8.8	38 2.12	. .	2.50	49 50.66	
26	10	. .	. .	50.4	. .	. .	15.8	. .	50 1.77?	29.39	0.00	IV.	6	6.21	27 9.51	. .	2.09	50 31.16	
27	10	. .	. .	. .	. .	. .	7.8	. .	52 42.58	29.40	0.00	VI.	5	9.42	24 49.72	. .	1.99	53 11.98	
28	10	. .	. .	. .	. .	. .	43.5	. .	55 18.28	29.41	0.00	VI.	5	7.37	23 46.69	. .	1.95	55 47.69	
29	10	. .	14.6	. .	. .	. .	. .	. .	57 39.83	29.41	−0.01	VII.	1	9.12	4 38.22	. .	1.22	58 9.23	
30	10	. .	. .	. .	. .	. .	. .	55.5	19 58 17.34	29.42	0.00	VII.	4	12.9	21 4.57	. .	1.85	19 58 46.76	
31	10	. .	. .	. .	. .	. .	24.9	. .	20 1 59.66	29.43	+0.01	VI.	8	13.35	40 47.00	. .	2.61	20 2 29.10	
32	5	. .	. .	. .	. .	. .	. .	11.9	2 33.82	29.43	0.00	VI.	5	13.15	26 37.11	. .	2.06	3 3.25	
33	9	. .	. .	. .	. .	. .	. .	26.9	4 48.48	29.44	−0.01	VII.	1	7.27	3 45.27	. .	1.17	5 17.91	
34	10	. .	. .	. .	. .	. .	. .	16.9	6 39.00	29.45	0.00	VII.	8	7.3	37 29.27	. .	2.48	7 8.45	
35	10	. .	. .	. .	. .	. .	30.1	. .	9 4.87	29.46	0.00	VI.	6	9.47	28 53.34	. .	2.16	9 34.33	
36	10	. .	45.6	. .	10.8	. .	. .	. .	11 10.74	29.46	−0.01	IV.	1	8.29	4 16.66	. .	1.19	11 40.19	
37	9	19.5	. .	. .	. .	10.5	. .	. .	13 57.79	29.47	0.00	I.	6	7.3	27 30.57	. .	2.10	14 27.26	
38	9	30.1	. .	55.6	. .	. .	. .	. .	16 8.23	29.48	0.00	III.	5	9.10	24 33.62	. .	1.98	16 37.71	
39	9	. .	. .	. .	. .	. .	. .	58.9	24 20.81	29.51	0.00	VII.	5	12.25	26 11.88	. .	2.05	24 50.32	
40	10	44.5	. .	10.4	. .	. .	. .	. .	30 22.73	29.54	0.00	III.	3	6.2	13 0.63	. .	1.52	30 52.27	
41	10	. .	. .	. .	22.5	. .	. .	. .	31 22.36	29.54	−0.01	IV.	1	9.40	4 52.46	. .	1.21	31 51.89	
42	9	. .	. .	7.8	. .	32.3	. .	. .	32 19.93	29.54	0.00	V.	4	9.10	19 34.42	. .	1.78	32 49.47	
43	8	. .	. .	. .	. .	32.5	. .	. .	33 19.83	29.55	0.00	V.	6	7.58	27 58.41	. .	2.12	33 49.38	
44	10	. .	. .	2.3	. .	. .	. .	. .	35 14.60	29.56	−0.01	III.	1	10.27	5 16.15	. .	1.22	35 44.15	
45	10	. .	. .	. .	. .	. .	. .	30.5	35 52.22	29.56	0.00	VII.	2	15.45	12 55.37	. .	1.53	36 21.78	
46	10	. .	. .	. .	. .	2.9	. .	. .	38 50.13	29.57	−0.01	V.	2	9.12	9 37.33	. .	1.39	39 19.69	
47	10	20.2	. .	. .	. .	. .	. .	. .	40 58.47	29.58	0.00	II.	6	12.16	30 8.47	. .	2.20	41 28.05	
48	9	. .	. .	. .	. .	12.1	. .	. .	44 59.39	29.58	0.00	V.	4	12.58	21 39.38	. .	1.87	41 28.97	
49	6	25.4	. .	. .	4.9	. .	. .	. .	44 4.25	29.59	+0.01	I.	9	6.15	42 4.07	. .	2.67	44 33.85	
50	10	. .	33.9	. .	. .	. .	. .	. .	45 59.33	29.60	0.00	II.	8	5.29	36 41.94	. .	2.46	46 28.93	
51	10	. .	34.9	. .	. .	. .	. .	. .	49 0.31	29.61	0.00	II.	8	0.30	34 11.18	. .	2.36	49 29.92	
52	10	. .	53.9	. .	14.8	. .	. .	. .	52 16.90	29.63	−0.01	II.	2	1.20	5 39.28	. .	1.23	52 46.52	
53	9	. .	. .	59.1	. .	. .	. .	. .	20 56 11.70	+29.64	0.00	III.	7	0.16	−29 5.46	. .	−2.17	20 56 41.34	

CORRECTIONS.

Date.		Corr. of Clock.	Hourly rate.	m	n	c
1848.	h.	s.	s.	s.	s.	s.

INSTRUMENT READINGS.

Date.		Barom.	THERMOM. At.	Ex.
1848.	h. m.	in.	°	°

REMARKS.

(167) 21. The two transits are discordant by about 5ˢ; transit over T. VII assumed as 28ˢ.5 instead of 33ˢ.5.
(167) 52. The transits are discordant by about 5ˢ; time of transit over T. IV assumed as 19ˢ.8 instead of 14ˢ.8.

ZONE 168. JUNE 20. B. BELT, −24° 23′. D_o = −24° 0′ 0″.

No.	Mag.	I.	II.	III.	IV.	V.	VI.	VII.	T.	a_1	a_2	MICROMETER.			i	d_1	d_2	Mean Right Ascension, 1850.0.	Mean Declination, 1850.0.
									h. m. s.	s.	s.			r.	′ ″	″	″	h. m. s.	° ′ ″
1	10					11.	24.5		15 10 57.22	+26.14	0.00	VII.	5	10.35	−25 16.15	−24.5	−2.02	15 11 23.36	−24 25 42.7
2	9					13.	26.5		16 59.22	26.13	0.00	VII.	6	6.1	26 59.08	24.1	2.15	17 25.35	27 25.3
3	8			15.	28.4	42.2			18 28.52	26.13	0.00	V.	8	8.16	38 6.17	24.0	3.01	18 54.65	38 33.2
4	10				51.	5.			19 51.03	26.18	0.00	V.	3	10.2	15 1.62	23.9	1.24	20 17.16	15 26.8
5	9				6.4	20.2			21 6.40	26.12	0.00	VI.	10	4.41	46 17.53	23.8	3.64	21 32.52	46 45.0
6	10		32.	44.5	58.2				23 58.59	26.12	0.00	V.	7	3.4	30 30.14	23.6	2.41	24 24.71	30 56.2
7	8					48.2	1.6	15.5	24 34.36	26.12	0.00	VII.	8	3.18	35 35.58	23.5	2.82	25 0.48	36 1.9
8	9						13.	26.3	25 45.39	26.12	0.00	VII.	8	3.3	35 28.62	23.4	2.82	26 11.51	35 54.2
9	8			48.2	1.2	15.5			29 1.43	26.11	0.00	VI.	2	8.12	9 6.93	23.2	0.80	29 27.53	9 30.9
10	10					29.3	43.3		33 15.77	26.11	0.00	VI.	9	11.11	34 33.28	22.8	2.74	33 41.88	34 58.8
11	8	47.	1.2	14.4	28.2				36 28.20	26.10	0.00	IV.	3	8.3	14 1.65	22.6	1.16	36 54.30	14 25.4
12	9					38.5	52.		37 24.70	26.10	0.00	VII.	1	3.36	1 48.57	22.5	0.27	37 50.80	2 11.3
13	7					52.2	6.		38 38.57	26.10	0.00	VI.	9	10.45	34 20.18	22.4	2.72	39 4.67	34 45.3
14	9				35.	48.5	2.		40 34.76	26.10	0.00	X.	9	3.47	40 48.16	22.3	3.22	41 0.86	41 13.7
15	10			39.5	53.5				42 53.27	26.09	0.00	V.	7	4.46	31 21.58	22.1	2.48	43 19.36	31 46.2
16	6.7				13.	27.	40.5		45 13.10	26.09	0.00	VI.	10	6.45	47 20.06	21.9	3.71	45 39.19	47 45.7
17	8					2.	15.5		47 48.22	26.09	0.00	IV.	10	8.465	48 21.49	21.7	3.79	48 14.31	48 47.0
18	6.7					23.	36.5	50.5	49 9.20	26.08	0.00	VII.	5	6.385	23 16.89	21.6	1.86	49 35.28	23 40.4
19	9			15.	28.4				51 28.39	26.08	0.00	V.	5	2.53	21 23.49	21.4	1.73	51 54.47	21 46.6
20	8							40.	52 58.98	26.08	0.00	VII.	10	3.49	45 51.12	21.2	3.60	52 25.06	46 15.9
21	8						52.	6.	53 24.74	26.08	0.00	VII.	8	2.14	35 3.32	21.2	2.77	53 50.82	35 27.3
22	6.7							9.3	54 27.92	26.08	0.00	VII.	4	6.7	18 1.81	21.1	1.47	54 54.00	18 24.4
23	9	17.	31.5						56 58.77	26.08	0.00	II.	7	4.12	31 4.31	20.9	2.46	57 24.85	31 27.7
24	7		59.7	12.2	25.5				15 58 25.60	26.07	−0.01	IV.	1	5.50	2 56.48	20.7	0.36	15 58 51.66	3 17.5
25	7.8			29.3	43.	56.3			16 0 42.67	26.07	0.01	V.	2	10.54	10 28.74	20.5	0.90	16 1 8.73	10 50.1
26	6.7			4.5	18.	32.			4 17.92	26.07	−0.01	V.	1	3.23	1 42.31	20.2	0.27	4 43.98	2 2.8
27	9				55.				6 54.85	26.07	0.00	IV.	8	4.26	36 10.23	20.0	2.86	7 20.92	36 33.1
28	10						46.	59.5	7 18.47	26.06	0.00	VII.	7	5.585	31 57.82	19.9	2.54	7 44.53	32 20.3
29	9		0.	13.	26.4				12 26.87	26.06	0.00	IV.	8	6.13	37 4.18	19.5	2.93	12 52.93	37 26.6
30	10				50.	15.5			14 49.84	26.06	+0.01	V.	8	11.41	39 49.53	19.3	3.14	15 15.91	40 12.0
31	9					8.3	22.	35.	15 54.25	26.06	−0.01	VII.	2	13.2	6 34.04	19.2	0.61	16 20.30	6 53.8
32	9	13.	27.5	41.	55.	8.4			18 54.69	26.05	0.00	V.	5	8.15	24 5.86	18.9	1.93	19 20.74	24 26.7
33	9				22.3				20 22.15	26.05	+0.01	IV.	8	8.30	38 13.27	18.8	3.02	20 48.21	38 35.1
34	6							20.5	20 39.49	26.05	0.01	VII.	10	4.59	46 26.42	18.7	3.65	21 5.55	46 48.8
35	9	58.	13.						24 40.10	26.05	+0.01	II.	9	2.42	40 16.63	18.3	3.18	25 6.16	40 38.1
36	7		37.2	51.	4.5	18.			32 4.36	26.04	−0.01	V.	2	9.52	9 57.47	17.6	0.86	32 30.39	10 15.9
37	10					25.	39.		37 11.47	26.04	0.00	VII.	8	1.20	35 6.59	17.1	2.78	37 37.51	35 26.5
38	9						8.	21.5	38 40.39	26.04	0.00	VII.	3	9.40	14 50.21	16.9	1.22	39 6.43	15 8.3
39	7						38.	52.	40 10.65	26.04	0.00	VII.	4	3.585	21 56.20	16.8	1.76	40 36.69	22 14.8
40	8					0.	13.5	27.5	41 46.25	26.04	0.00	VII.	7	9.54	33 56.57	16.6	2.69	42 12.29	34 15.9
41	9			7.5	21.	35.			44 21.21	26.04	+0.01	V.	10	4.11	48 33.77	16.4	3.71	44 47.26	48 53.9
42	9				23.4	37.	50.5		45 23.22	26.04	0.00	VI.	3	10.22	15 11.58	16.3	1.26	45 49.26	15 29.1
43	9							42.	46 0.60	26.03	0.00	VII.	4	4.43	17 19.45	16.2	1.41	48 26.63	17 37.1
44	8		5.5	18.2	32.				50 32.36	26.03	+0.01	IV.	9	11.52	44 54.12	15.7	3.53	50 58.40	45 13.4
45	10	0.	14.5	27.					52 41.23	26.03	0.00	IV.	3	12.55	16 28.88	15.5	1.36	54 7.26	16 45.7
46	10		50.5	3.3	14.				54 18.00	26.03	0.00	IV.	8	6.24	37 9.73	15.3	2.94	54 44.03	37 28.0
47	9				55.8	10.5			55 56.19	26.03	0.00	VI.	4	11.2	20 30.75	15.1	1.66	56 22.22	20 47.5
48	10					31.	44.	58.2	16 57 16.93	26.03	0.00	VII.	4	11.33	20 46.19	15.0	1.68	16 57 42.96	21 2.9
49	8		46.	58.5	12.5	26.			17 1 12.35	+26.03	+0.01	VI.	9	11.42	−44 48.92	−14.6	−3.52	17 1 38.39	−24 45 7.0

CORRECTIONS.

Date.		Corr. of Clock.	Hourly rate.	m	n	c
1848.	h.	s.	s.	s.	s.	s.
June 20,	17	+ 22.47	+ 0.062	− 0.26	+ 0.77	0.00

INSTRUMENT READINGS.

Date.		Barom.	THERMOM. At.	THERMOM. Ex.
1848.	h. m.	in.	°	°

REMARKS.

(168) 20. Minutes assumed as 51 instead of 52.
(168) 30. Transit over T. V is assumed to have been at 3ˢ.5 instead of 15ˢ.5.
(168) 31. Hor. thread assumed as 1 instead of 2.
(168) 37. Micrometer reading assumed as 2ʳ.20 instead of 1ʳ.20.
(168) 39. Micrometer thread assumed as 5 instead of 4.
(168) 41. Micrometer reading assumed as 9ʳ.11.
(168) 43. Minutes assumed as 48 instead of 46.
(168) 45. Minutes assumed as 53 instead of 52.
(168) 46. Transit over T. IV assumed as 19ˢ instead of 14ˢ.
(168) 49. Double; observed second; other, first magnitude.

ZONE 168. JUNE 20. B. BELT, −24° 23'. D₀ = −24° 0' 0"—Continued.

No.	Mag.	SECONDS OF TRANSIT. I.	II.	III.	IV.	V.	VI.	VII.	T.	a_1	n_1	MICROMETER.			i	d_1	d_2	Mean Right Ascension, 1850.0.	Mean Declination, 1850.0.
									h. m. s.	s.	s.			r.	' "	"	"	h. m. s.	° ' "
50	10	..	..	6.	19.5	33.5	..	..	17 3 19.50	+26.03	0.00	VI.	3	12.34	−16 18.14	−14.3	−1.34	17 3 45.53	−24 16 33.8
51	10	..	..	9.5	23.	..	..	..	5 23.03	26.03	0.00	IV.	7	7.14	32 36.24	14.1	2.58	5 49.06	32 52.9
52	10	..	..	..	45.	..	..	..	6 44.85	26.03	+0.01	IV.	10	8.19	48 7.62	14.0	3.78	7 10.89	48 25.4
53	7.9	..	..	..	..	..	59.5	13.5	7 32.04	26.03	+0.01	VII.	8	8.39	38 17.46	13.9	3.02	7 58.05	38 34.4
54	10	..	..	..	..	51.3	..	19.	9 37.58	26.02	0.00	VII.	4	9.52	19 55.26	13.6	1.61	10 3.60	20 10.5
55	9	..	..	..	29.	45.	..	..	11 30.07	26.02	0.00	V.	7	4.32	31 14.52	13.4	2.47	11 56.09	31 30.4
56	9	..	..	..	..	..	6.4	19.5	12 38.75	26.02	+0.01	VII.	10	7.38	47 46.59	13.3	3.75	13 4.78	48 3.6
57	7	49.	3.	16.	30.	..	..	..	15 29.93	26.02	−0.01	IV.	1	11.29	5 47.42	13.0	0.56	15 55.94	6 1.0
58	6	..	..	..	..	0.	13.	..	15 45.95	26.02	0.01	V.	1	3.25	1 43.32	13.0	0.25	17 11.96	1 56.6
59	10	..	56.5	..	23.5	..	..	..	19 23.61	26.02	0.01	V.	3	4.125	12 5.38	12.6	1.03	19 49.62	12 19.0
60	10	..	..	10.	23.2	37.2	..	..	23 23.28	26.02	−0.01	VI.	3	3.585	11 58.19	12.1	1.01	23 49.29	12 11.3
61	10	..	..	..	..	49.	2.5	..	24 35.22	26.02	+0.01	VII.	9	3.34	40 42.66	12.0	3.21	25 1.25	40 57.9
62	9	..	..	..	..	10.	23.5	37.5	17 25 56.24	+26.02	0.00	VII.	7	4.005	−30 58.32	−11.9	−2.45	17 26 22.26	−24 31 12.7

ZONE 169. JUNE 26. P. BELT, −20°. D₀ = −19° 35' 0".

No.	Mag.	I.	II.	III.	IV.	V.	VI.	VII.	T.	a_1	n_1	MICROMETER.			i	d_1	d_2	Mean Right Ascension, 1850.0.	Mean Declination, 1850.0.
1	9	37.4	51.9	..	17.4	..	..	..	15 42 17.82	+29.09	0.00	IV.	7	5.3	−31 30.16	−64.0	−2.37	15 42 46.91	−20 7 36.6
2	4	..	..	56.2	..	22.9	..	..	44 9.31	29.09	−0.01	V.	2	4.29	6 44.33	63.6	0.98	44 38.39	19 42 49.1
3	10	..	..	..	..	48.9	..	..	45 35.67	29.08	+0.01	V.	10	4.14	46 4.05	63.6	3.21	46 4.76	20 22 10.9
4	10	..	..	54.9	..	21.9	..	..	47 8.50	29.08	+0.01	V.	10	0.43	44 47.90	63.4	3.11	47 37.59	20 54.4
5	9	43.4	..	9.9	..	..	..	..	56 23.32	29.07	0.00	III.	5	11.24	25 41.17	62.3	2.04	56 52.39	1 45.5
6	3	..	..	19.9	..	..	..	13.	15 57 33.24	29.07	+0.01	III.	8	11.33	39 15.26	62.1	2.81	15 58 2.32	15 20.2
7	11	..	53.5	..	19.9	..	..	..	16 0 19.99	29.07	0.00	IV.	8	5.32	36 43.51	61.8	2.66	16 0 49.06	12 48.0
8	10	..	..	..	42.9	..	9.9	..	1 43.08	29.07	0.00	VI.	5	7.59	23 57.71	61.6	1.94	2 12.15	20 0 1.2
9	11	..	..	..	..	..	..	7.	5 26.83	29.06	0.00	VII.	3	9.30	14 45.24	61.1	1.43	5 55.89	19 50 47.8
10	8	..	18.4	31.9	44.2	..	..	..	7 44.60	29.06	−0.01	IV.	2	5.15	7 37.83	60.9	1.01	8 13.65	43 39.7
11	8	12.4	25.4	..	..	..	..	..	9 52.21	29.06	0.00	II.	3	9.31	14 45.88	60.6	1.43	10 21.27	50 47.9
12	8	..	23.	..	49.1	..	..	..	10 49.21	29.06	−0.01	II.	1	9.17	4 40.72	60.5	0.85	11 18.26	40 42.1
13	10	..	..	..	..	..	..	55.	11 14.77	29.06	0.01	VII.	2	8.9	9 5.30	60.4	1.09	11 43.82	45 6.8
14	3	11.9	24.2	37.9	..	..	..	..	14 51.13	29.06	−0.01	VI.	1	10.42	4 53.36	60.0	0.88	15 20.18	19 40 54.2
15	10	..	..	53.9	..	20.	..	..	23 7.10	29.05	+0.01	V.	10	11.1	49 29.27	58.9	3.44	23 36.16	20 25 31.6
16	11	39.	..	..	..	..	..	..	26 19.12	29.05	0.00	I.	4	11.0	20 29.62	58.5	1.75	26 48.17	19 56 29.9
17	10	..	57.9	..	24.9	..	..	..	27 24.69	29.05	0.00	IV.	8	8.21	38 8.72	58.3	2.77	27 53.74	20 14 9.8
18	6	35.1	49.2	3.1	15.2	..	..	..	31 15.65	29.05	0.00	II.	6	13.10	30 35.62	57.8	2.32	31 41.70	20 6 35.7
19	10	..	..	45.5	..	12.2	..	..	32 58.71	29.04	0.00	III.	4	5.20	17 38.42	57.6	1.57	33 27.75	19 53 37.6
20	10	..	..	..	..	..	..	57.	34 17.31	29.04	+0.01	VII.	10	8.39	48 17.43	57.4	3.37	34 46.36	20 24 18.2
21	9	58.1	12.2	25.3	..	..	..	..	37 38.41	29.04	0.00	III.	3	7.44	13 21.79	56.9	1.33	38 7.45	19 49 20.0
22	9	..	2.9	15.	..	..	..	..	39 28.72	29.04	0.00	III.	3	4.33	11 45.46	56.7	1.23	39 57.76	47 43.4
23	10	..	..	..	..	..	..	8.9	39 28.70	29.04	0.00	VII.	3	4.33	11 45.24	56.7	1.23	39 57.74	19 47 43.2
24	10	..	..	33.9	..	0.1	..	..	41 47.11	29.04	+0.01	V.	10	3.46	45 49.93	56.4	3.22	42 16.16	20 21 49.6
25	8	24.9	39.8	51.8	..	..	..	..	44 5.53	29.04	0.00	II.	6	9.2	33 30.58	56.0	2.50	44 34.57	20 9 29.1
26	11	..	54.5	..	20.1	..	..	..	46 20.50	29.04	0.00	IV.	3	10.48	15 24.85	55.7	1.44	46 49.54	19 51 22.0
27	11	..	..	..	22.9	..	..	..	48 9.46	29.04	−0.01	IV.	1	6.19	2 40.85	55.4	0.71	48 38.49	38 37.0
28	11	..	..	..	..	39.9	..	..	50 26.54	29.04	0.00	V.	4	6.30	18 13.73	55.1	1.60	50 55.58	19 54 10.4
29	10	13.5	27.6	..	..	..	..	..	51 54.04	29.04	0.00	II.	8	5.3	36 28.76	54.9	2.67	52 23.08	20 12 26.3
30	9	..	..	..	..	..	..	34.9	52 54.98	29.04	0.00	VII.	7	5.3	31 29.90	54.8	2.38	53 24.02	7 27.1
31	9	55.2	..	22.9	..	..	..	..	54 35.88	29.04	0.00	III.	8	11.2	39 29.87	54.6	2.85	55 4.92	15 27.3
32	10	..	..	..	..	..	..	8.9	16 54 29.02	+29.04	0.00	VII.	7	11.7	−34 33.45	−54.6	−2.56	16 54 58.06	−20 10 30.6

CORRECTIONS.

Date.		Corr. of Clock.	Hourly rate.	m	n	c
1848.	h.	s.	s.	s.	s.	s.
June 26,	17	+ 25.51	/ 0.010	− 0.22	+ 0.75	

INSTRUMENT READINGS.

Date.		Barom.	THERMOM. At.	Ex.
1848.	h. m.	in.	°	°

REMARKS.

(168) 58. Minutes assumed as 16 instead of 17.
(169) 2. Micrometer reading assumed as 3r.29 instead of 4r.29.
(169) 4. Micrometer reading assumed as 1r.43 instead of 0r.43.
(169) 6. Micrometer reading assumed as 10r.33 instead of 11r.33.
(169) 12. Double.
(169) 14. Micrometer reading assumed as 11r.42 instead of 12r.42.
(169) 19. Double; other two near.
(169) 21. Micrometer reading assumed as 6r.44 instead of 7r.44.
(169) 22. Micrometer reading assumed as 3r.33 instead of 4r.33. } Evidently the same star.
(169) 23. Micrometer reading assumed as 3r.33 instead of 4r.33. }
(169) 25. Hor. thread assumed as 7 instead of 6.
(169) 27. Transit over T. V assumed as recorded over T. IV, and micrometer reading as 5r.19 instead of 6r.19.
(169) 32. Right ascension differs 3s.9 from Arg. Z. 211, 60; 305, 71, and declination 20'.

ZONE 169. JUNE 26. P. BELT, −20°. D_o = −19° 35′ 0″—Continued.

No.	Mag.	SECONDS OF TRANSIT. I.	II.	III.	IV.	V.	VI.	VII.	T.	a_1	a_2	MICROMETER.			i	d_1	d_2	Mean Right Ascension, 1850.0.	Mean Declination, 1850.0.
									h. m. s.	s.	s.			r.	′ ″	″	″	h. m. s.	° ′ ″
33	10						49.9	3.5	16 55 23.54	+29.04	+0.01	VI.	8	13.37	−40 47.94	−54.4	−2.93	16 55 52.59	−20 16 45.3
34	10					49.9		17.9	57 37.04	29.04	−0.01	V.	1	8.4	4 4.03	54.1	0.79	58 6.07	19 39 58.9
35	9					24.9		51.2	16 59 11.37	29.04	0.00	V.	5	10.1	24 59.32	53.9	2.00	16 59 40.41	20 0 55.2
36	9			46.4		13.1			17 7 59.80	29.04	0.00	V.	8	8.11	38 3.65	53.5	2.77	17 2 28.84	20 13 59.9
37	10			43.5		8.8			3 56.29	29.04	+0.01	V.	10	8.28	48 12.12	53.2	3.38	4 25.34	24 8.7
38	10		29.5		55.5				5 55.76	29.04	0.00	II.	6	11.45	29 52.77	52.9	2.28	6 24.80	5 48.0
39	10		58.5			38.1			9 25.01	29.04	0.00	V.	7	12.5	36 3.44	52.4	2.58	9 54.05	20 11 58.4
40	9					53.4			10 40.01	29.04	0.00	V.	3	7.0	13 29.85	52.2	1.33	11 9.05	19 49 23.4
41	10			3.					12 16.24	29.04	0.00	III.	6	10.15	29 7.48	52.0	2.24	12 45.28	20 5 1.7
42	10				20.				13 19.85	29.04	0.00	IV.	6	9.0	28 29.69	51.8	2.20	13 48.89	4 23.7
43	10							14.5	13 34.52	29.04	0.00	VII.	6	7.52	27 55.12	51.8	2.16	14 3.56	20 3 49.1
44	11				48.1				15 47.95	29.04	0.00	IV.	5	4.1	21 57.82	51.4	1.83	16 16.99	19 57 51.1
45	11					3.			17 49.72	29.04	0.00	V.	7	12.4	35 2.44	51.1	2.59	18 18.76	20 10 56.1
46	11					14.9			19 1.64	29.04	0.00	V.	8	10.28	39 12.73	50.9	2.84	19 30.68	15 6.5
47	10					35.9			22 22.59	29.04	0.00	V.	6	10.27	29 13.56	50.4	2.24	22 51.63	5 6.2
48	10			2.9					25 16.22	29.04	0.00	III.	7	12.10	34 50.34	50.0	2.60	25 45.26	10 42.9
49	9					58.1			26 44.84	29.04	+0.01	V.	8	12.16	40 7.19	49.8	2.90	27 13.89	20 15 59.9
50	10		20.2						28 46.76	29.04	0.00	II.	3	13.47	16 54.98	49.5	1.52	29 15.80	19 52 46.0
51	10					34.5			32 34.35	29.04	0.00	VII.	4	4.50	17 23.05	48.9	1.55	33 3.39	19 53 13.5
52	10						35.8		33 9.31	29.04	0.00	VI.	6	3.13	25 34.59	48.8	2.04	33 38.35	20 1 25.4
53	10			2.9		30.			36 16.45	29.04	0.00	V.	7	7.145	32 21.59	48.4	2.46	36 45.49	20 8 12.5
54	10				19.4		45.4		39 19.10	29.04	0.00	VI.	3	5.8	12 33.28	47.9	1.27	39 48.14	19 48 42.5
55	10							52.9	40 12.82	29.04	0.00	VII.	5	2.54	21 23.76	47.8	1.78	40 41.86	19 57 13.3
56	10					59.9		27.1	42 46.84	29.04	0.00	VII.	6	3.34	25 45.03	47.4	2.05	43 15.88	20 1 34.5
57	10					38.1		4.9	44 24.73	29.04	0.00	VII.	3	10.13	15 6.92	47.1	1.42	44 53.77	19 50 55.4
58	11		24.1				17.		46 50.64	29.04	0.00	VI.	7	11.30	31 45.21	46.8	2.58	47 19.68	20 10 34.6
59	11			27.4		53.5			49 40.38	29.05	0.00	III.	5	13.27	26 43.18	46.3	2.10	50 9.43	20 2 31.6
60	10			30.					51 43.12	29.05	0.00	III.	4	11.27	15 44.19	46.0	1.44	52 12.17	19 51 30.9
61	10							25.	51 44.92	29.05	0.00	VII.	4	12.31	21 15.50	46.0	1.78	52 13.97	57 3.3
62	10	2.5		29.5					54 42.54	29.05	0.00	III.	3	5.22	12 40.43	45.5	1.26	55 11.59	48 27.2
63	9		45.5						56 12.01	29.05	−0.01	II.	3	0.42	10 19.16	45.3	1.12	56 41.05	46 5.6
64	10							14.	56 33.74	29.05	−0.01	VII.	2	4.23	7 11.33	45.3	0.94	57 2.78	42 57.6
65	10	24.9							59 5.00	29.05	0.00	I.	4	4.10	17 2.88	44.8	1.52	17 59 34.05	52 49.2
66	6				42.9				17 59 42.75	29.05	0.00	IV.	4	5.28	17 42.48	44.7	1.56	18 0 11.80	53 28.7
67	7	12.9	26.2						18 1 52.88	29.05	0.00	I.	4	2.22	16 8.42	44.4	1.47	2 21.93	51 54.3
68	10	55.	9.						3 35.25	29.05	−0.01	II.	2	2.55	10 59.41	44.1	1.16	4 4.29	46 44.7
69	10	19.5		16.4					5 59.59	29.06	0.00	III.	5	5.16	22 35.61	43.7	1.86	6 28.65	58 21.2
70	10							1.	6 20.87	29.06	0.00	VII.	4	4.24	17 9.94	43.7	1.53	6 49.93	52 55.2
71	9					31.9		58.	8 18.11	29.06	−0.01	V.	2	4.59	7 29.74	43.4	0.96	8 47.16	43 14.1
72	9				12.		38.1		10 11.76	29.06	0.01	VI.	2	5.11	11 37.72	43.1	1.20	10 40.81	47 22.0
73	10							19.	11 38.74	29.06	0.01	VII.	2	4.57	7 28.48	42.8	0.95	12 7.79	43 12.2
74	11			19.4		45.5			13 32.27	29.06	0.01	.	3	8.30	14 15.27	42.5	1.36	14 1.32	49 59.1
75	10					1.5		28.1	14 47.93	29.06	−0.01	.	1	7.48	1 54.96	42.3	0.62	15 16.98	37 37.9
76	10	33.5		59.9					19 13.27	29.07	0.00	I.	3	12.7	16 4.40	41.6	1.46	19 42.34	51 47.5
77	10					34.9		1.	20 21.06	29.07	−0.01	VII.	1	2.58	5 58.21	41.4	0.89	20 50.12	19 41 40.5
78	10				37.9				22 37.75	29.07	0.00	IV.	6	6.52	27 25.15	41.1	2.14	23 6.82	20 3 8.4
79	10			2.1		26.9			24 15.39	29.07	0.00	V.	4	12.32	21 16.26	40.8	1.78	24 44.46	19 56 58.8
80	8						29.9	45.8	25 3.56	29.07	0.00	VII.	4	5.42	17 49.27	40.6	1.57	25 32.63	53 31.4
81	10	4.5	18.2						18 27 44.69	+29.08	0.00	II.	4	5.35	−17 45.90	−40.2	−1.57	18 28 13.77	−19 53 27.7

CORRECTIONS.

Date.		Corr. of Clock.	Hourly rate.	m	n	c
1848.	h.	s.	s.	s.	s.	s.

INSTRUMENT READINGS.

Date.		Barom.	THERMOM. At.	THERMOM. Ex.
1848.	h. m.	in.	°	°

REMARKS.

(169) 39. Micrometer reading assumed as 14ʳ.5 instead of 12ʳ.5.
(169) 48. Micrometer reading assumed as 11ʳ.40 instead of 12ʳ.10.
(169) 51. Transit over T. III assumed as recorded over T. IV, to agree with Mural Z., 1849, June 22.
(169) 53. Micrometer reading assumed as 6ʳ.45 instead of 7ʳ.145.
(169) 60. Hor. thread assumed as 3 instead of 4.
(169) 68. Micrometer reading assumed as 11ʳ.55 instead of 2ʳ.55.
(169) 72. Micrometer reading assumed as 13ʳ.11 instead of 5ʳ.11.
(169) 75. Micrometer reading assumed as 3ʳ.48 instead of 7ʳ.48.
(169) 77. Hor. thread assumed as 2 instead of 1, and micrometer reading as 1ʳ.58 instead of 2ʳ.58.
(169) 80. T. VII assumed as 43ˢ.8 instead of 45ˢ.8.

ZONE 169. JUNE 26. P. BELT, −20°. D₀=−19° 35′ 0″—Continued.

No.	Mag.	Seconds of Transit I.	II.	III.	IV.	V.	VI.	VII.	T.	a_1	a_2	Micrometer.			i	d_1	d_2	Mean Right Ascension, 1850.0.	Mean Declination, 1850.0.
									h. m. s.	s.	s.			r.	′ ″	″	″	h. m. s.	° ′ ″
82	9	..	..	..	31.9	..	57.9	..	18 28 31.61	+29.08	−0.01	IV.	2	5.55	− 7 58.00	−40.1	−0.97	18 29 0.66	−19 43 39.1
83	9	..	..	..	..	44.5	..	10.1	30 30.73	29.08	+0.01	V.	8	4.396	36 17.06	39.7	2.69	30 59.82	20 11 59.5
84	9	..	..	..	..	15.5	..	..	33 2.30	29.08	+0.01	V.	10	13.40	51 19.69	39.3	3.55	33 31.39	20 27 2.5
85	6	..	14.9	..	40.2	..	..	..	36 40.73	29.08	−0.01	IV.	2	9.239	9 43.33	38.7	1.07	37 9.80	19 45 23.1
86	9	..	..	..	..	30.1	..	..	38 16.77	29.09	0.00	V.	5	11.10	25 34.11	38.4	2.04	38 45.86	20 1 14.6
87	10	..	..	12.	..	38.	..	..	40 25.02	29.09	0.00	V.	7	11.30	34 45.30	38.0	2.60	40 54.11	20 10 25.9
88	10	..	..	..	..	0.9	..	27.2	41 47.29	29.09	0.00	V.	3	12.50	11 27.22	37.8	1.18	42 16.38	19 47 6.2
89	10	..	1.9	..	..	40.	..	..	44 27.58	29.10	0.00	II.	5	5.31	22 43.08	37.3	1.86	44 56.68	19 58 22.2
90	11	..	..	..	..	..	..	6.	45 26.02	29.10	0.00	VII.	6	7.4	27 0.67	37.2	2.12	45 55.12	20 2 40.0
91	10	..	..	..	..	33.5	..	..	47 20.11	29.10	−0.01	V.	3	7.13	13 36.40	36.9	1.30	47 49.20	19 49 14.6
92	10	..	..	..	..	..	54.5	..	48 28.03	29.10	0.00	VI.	4	6.49	18 23.22	36.7	1.59	48 57.13	54 1.5
93	9	..	50.1	3.1	..	..	..	..	51 16.41	29.10	0.00	III.	4	2.9	16 2.12	36.2	1.44	51 45.51	51 39.8
94	9	..	19.5	..	..	59.9	..	..	53 46.19	29.11	−0.01	V.	1	1.54	0 57.45	35.8	0.55	54 15.29	36 33.8
95	9	..	..	..	51.	..	..	..	55 50.85	29.11	0.00	IV.	4	7.42	18 50.06	35.4	1.62	56 19.96	54 27.1
96	10	..	..	..	..	..	41.	..	57 14.53	29.11	0.00	VI.	4	4.45	17 20.69	35.2	1.52	57 43.64	52 57.4
97	9	..	..	..	53.5	..	20.	..	18 58 53.45	29.11	−.010	IV.	3	9.46	14 53.59	34.9	1.37	18 59 22.55	19 50 29.9
98	9	..	..	..	28.4	..	..	..	19 0 28.35	29.12	0.00	IV.	5	12.475	26 23.30	34.6	2.08	19 0 57.47	20 2 0.0
99	10	..	..	3.9	59.9	..	..	..	2 17.18	29.12	0.00	IV.	7	10.19	34 9.53	34.3	2.57	2 46.30	20 9 46.4
100	9	..	..	..	..	..	..	..	3 46.48	29.12	−0.01	III.	2	7.5	8 15.14	34.1	1.00	4 15.59	19 43 53.2
101	9	27.2	..	54.1	..	..	..	..	6 7.34	29.12	0.00	III.	6	5.35	26 46.29	33.7	2.11	6 36.46	20 2 22.1
102	10	..	..	15.1	..	41.9	..	..	8 28.26	29.13	0.00	V.	4	5.40	17 48.51	33.3	1.56	8 57.49	19 53 23.4
103	10	..	..	..	..	..	56.6	..	9 30.11	29.13	0.00	VI.	5	7.30	23 43.09	33.1	1.92	9 59.24	19 59 18.1
104	10	..	..	..	5.	..	31.	..	12 4.67	29.13	0.00	VI.	7	9.54	33 56.80	32.6	2.55	12 33.80	20 9 32.0
105	10	..	..	..	..	..	18.9	..	13 52.37	29.14	+0.01	VI.	8	9.11	38 33.81	32.3	2.84	14 21.52	14 9.0
106	10	..	..	..	..	..	..	46.5	15 6.46	29.14	0.00	VII.	5	9.2	24 29.32	32.1	1.97	15 35.60	20 0 3.4
107	9	..	..	49.	..	15.	..	..	26 1.86	29.16	0.00	V.	4	6.46	18 21.80	30.3	1.59	26 31.02	19 53 53.7
108	8	..	..	30.5	..	57.	..	..	31 43.87	29.17	+0.01	V.	10	4 34	45 43.88	29.3	3.29	31 13.05	20 21 16.5
109	10	..	..	..	..	..	..	53.5	32 13.51	29.17	0.00	VII.	6	5.15	27 6.21	29.2	2.12	32 42.68	2 37.5
110	10	..	..	..	..	11.5	..	..	34 58.17	29.17	0.00	V.	5	9.36	24 46.72	28.8	1.99	35 27.34	0 17.5
111	4	..	41.5	54.	..	20.1	..	..	37 7.42	29.18	0.00	V.	6	15.54	31 28.15	28.4	2.44	37 36.60	6 59.0
112	8	57.1	..	..	..	..	50.	..	42 37.90	29.19	0.00	V.	6	9.36	28 47.82	27.5	2.25	43 7.09	20 4 17.6
113	10	..	35.2	..	1.2	..	..	..	45 1.39	29.19	−0.01	IV.	2	12.56	11 30.29	27.0	1.15	45 30.57	19 46 58.4
114	8	..	..	..	..	23.4	..	..	46 9.97	29.19	−0.01	V.	1	11.16	5 40.84	26.9	0.81	46 39.15	41 8.5
115	9	..	..	..	26.6	..	..	..	48 26.45	29.20	0.00	IV.	4	9.27	19 43.00	26.5	1.66	48 55.65	19 55 11.2
116	10	..	..	..	..	16.9	..	..	50 3.57	29.20	0.00	V.	5	9.15	24 36.12	26.2	1.98	50 32.77	20 0 4.3
117	9	..	..	..	..	15.5	..	..	57 2.14	29.20	0.00	V.	4	8.40	19 19.28	25.0	1.64	57 31.34	19 54 45.9
118	9	..	..	..	..	..	..	18.	19 58 37.68	29.20	−0.01	VII.	1	5.47	1 54.19	24.7	0.57	59 6.87	37 19.5
119	10	..	..	..	..	..	..	47.4	20 0 7.25	29.22	0.01	VII.	3	11.37	15 49.28	24.4	1.41	19 0 36.46	51 15.1
120	9	..	..	..	..	..	..	54.9	1 14.72	29.22	−0.01	VII.	3	7.2	13 30.61	24.2	1.27	20 1 43.93	19 48 56.1
121	11	..	..	..	..	..	..	16.9	3 37.05	29.23	+0.01	VII.	8	6.12	37 3.40	23.8	2.75	4 6.29	20 12 30.0
122	8	..	..	4.9	..	31.9	..	..	11 18.52	29.25	+0.01	V.	10	5.55	46 54.97	22.5	3.39	11 47.78	20 22 20.9
123	9	..	..	..	..	..	30.	..	25 3.55	29.28	−0.01	VI.	3	4.58	12 28.24	20.1	1.20	25 32.82	19 47 49.5
124	10	..	9.9	..	..	48.5	..	..	20 27 35.73	+29.28	+0.01	.	9	3.41	−45 18.82	−19.7	−3.30	20 28 5.02	−20 20 41.8

CORRECTIONS.

Date.	Corr. of Clock.	Hourly rate.	m	n	c
1848. h.	s.	s.	s.	s.	s.

INSTRUMENT READINGS.

Date.	Barom.	Thermom. At.	Thermom. Ex.
1848. h. m.	in.	°	°

REMARKS.

(169) 84. Micrometer reading assumed as 14ʳ.40 instead of 13ʳ.40.
(169) 88. Hor. thread assumed as 2 instead of 3.
(169) 90. Micrometer reading assumed as 6ʳ.4 instead of 7ʳ.4.
(169) 99. Transits entirely discordant and incongruous; transit at 59ˢ.9 assumed to belong to following star and to T. V.
(169) 100. Micrometer reading assumed as 6ʳ.35 instead of 7ʳ.5.
(169) 108. Minutes assumed as 30 instead of 31, and micrometer reading as 3ʳ.34 instead of 4ʳ.34.
(169) 109. Micrometer reading assumed as 6ʳ.15 instead of 5ʳ.15.
(169) 111. Micrometer reading assumed as 14ʳ.54 instead of 15ʳ.54.
(169) 112. Transit over T. VI is assumed to have been 5ˢ instead of 50ˢ.
(169) 118. Micrometer reading assumed as 3ʳ.47 instead of 5ʳ.47.
(169) 124. Micrometer reading assumed as 12ʳ.41 instead of 3ʳ.41.

ZONE 170. JULY 3. P. BELT, −20° 30′. D₀ = −20° 8′ 0″.

No.	Mag.	SECONDS OF TRANSIT. I.	II.	III.	IV.	V.	VI.	VII.	T.	a_1	a_2	MICROMETER.			i	d_1	d_2	Mean Right Ascension, 1850.0.	Mean Declination, 1850.0.
									h. m. s.	s.	s.			r.	′ ″	″	″	h. m. s.	° ′ ″
1	9	54.5	. .	21.3	34.9	. .	. .	. .	16 34 34.62	+11.04	0.00	III.	3	12.39	−16 20.79	− 2.82	−1.50	16 34 45.66	−20 24 25.11
2	10	. .	. .	. .	. .	21.5	. .	. .	39 8.17	11.05	0.00	VII.	7	6.57	32 27.38	2.48	2.45	39 19.22	40 32.31
3	10	. .	. .	. .	. .	17.9	. .	. .	42 4.47	11.05	0.00	V.	3	7.506	13 55.36	2.25	1.35	42 15.52	21 58.96
4	7	. .	. .	. .	22.9	. .	49.5	. .	44 22.88	11.06	−0.01	VI.	1	3.7	1 34.17	2.07	0.63	44 33.93	9 36.87
5	10	. .	57.4	. .	23.4	. .	. .	. .	49 23.68	11.06	0.00	IV.	5	9.255	24 41.44	1.68	1.98	49 34.74	32 45.10
6	10	. .	4.5	17.1	. .	. .	. .	57.4	51 30.71	11.06	−0.01	VI.	3	5.51	12 54.96	1.51	1.29	51 41.76	20 57.76
7	10	. .	. .	. .	. .	11.9	. .	38.8	52 58.49	11.07	0.00	V.	3	8.22	14 11.20	1.39	1.36	53 9.56	22 13.95
8	8	. .	. .	. .	41.5	. .	7.9	. .	16 55 41.37	11.07	−0.01	V.	2	7.19	8 40.35	1.16	1.04	55 52.43	16 42.53
9	10	. .	. .	. .	14.5	. .	. .	. .	17 5 14.35	+11.08	0.00	IV.	4	2.26	−16 10.72	− 0.37	−1.48	17 5 25.43	−20 24 12.57

ZONE 171. JULY 10. P. BELT, −20° 30′. D₀ = −20° 7′ 30″.

No.	Mag.	I.	II.	III.	IV.	V.	VI.	VII.	T.	a_1	a_2	MICROMETER.			i	d_1	d_2	Mean Right Ascension, 1850.0.	Mean Declination, 1850.0.
1	10	. .	. .	26.3	. .	53.5	. .	. .	17 16 39.88	+15.67	0.00	V.	7	3.13	−30 34.71	−13.16	−2.34	17 16 55.55	−20 38 20.2
2	10	. .	. .	. .	. .	. .	38.1	. .	18 11.56	15.67	0.00	V.	3	10.5	15 3.14	13.12	1.40	18 27.23	22 47.7
3	11	15.5	. .	42.3	. .	. .	. .	. .	22 55.60	15.66	0.00	III.	4	10.55	20 2.39	13.01	1.70	23 11.26	27 47.1
4	10	. .	47.4	0.5	13.4	27.4	. .	. .	24 13.83	15.66	0.00	V.	6	16.46	32 24.62	12.99	2.45	24 29.49	40 10.1
5	11	. .	. .	. .	. .	. .	23.1	. .	26 56.56	15.66	0.00	V.	3	6.48	8 24.70	12.93	0.99	27 12.22	16 8.6
6	8	. .	. .	. .	. .	. .	. .	59.2	28 19.10	15.66	0.00	VII.	6	7.21	27 39.48	12.91	2.16	28 34.76	35 24.6
7	10	. .	. .	. .	. .	19.1	. .	. .	30 5.76	15.65	0.00	V.	6	11.5	29 32.69	12.86	2.29	30 21.41	37 17.8
8	10	. .	. .	. .	. .	. .	. .	18.5	31 38.47	15.65	0.00	VII.	7	9.49	38 53.22	12.82	2.75	31 54.12	46 38.8
9	9	38.9	52.5	5.4	18.2	. .	. .	. .	38 18.87	15.64	+0.01	IV.	8	10.12	39 4.70	12.66	2.86	38 34.52	46 50.2
10	9	. .	48.5	0.9	14.4	27.4	. .	. .	40 14.50	15.64	+0.01	V.	6	10.38	39 17.78	12.61	2.87	40 30.15	47 3.3
11	10	. .	. .	30.9	44.4	. .	. .	. .	42 44.12	15.64	0.00	IV.	3	11.35	15 48.55	12.53	1.44	42 59.76	23 32.5
12	10	. .	. .	. .	. .	. .	16.4	. .	43 49.78	15.64	+0.01	VII.	8	14.34	41 16.52	12.50	3.00	44 5.43	49 2.0
13	10	. .	. .	. .	50.4	. .	17.4	. .	45 50.56	15.63	0.00	VII.	3	9.7	14 33.64	12.44	1.37	46 6.19	22 17.5
14	9	45.4	59.1	. .	. .	38.5	. .	. .	48 25.57	15.63	0.00	V.	6	8.52	28 25.63	12.39	2.20	48 41.20	36 10.2
15	8	9.2	22.4	. .	48.5	. .	. .	. .	50 48.90	15.63	−0.01	IV.	3	3.5	11 31.38	12.31	1.17	51 4.52	19 14.9
16	10	. .	. .	. .	. .	. .	21.4	. .	51 54.83	15.63	0.00	V.	5	5.26	22 40.65	12.28	1.86	52 10.46	30 24.8
17	8	. .	. .	. .	. .	. .	51.4	3.9	53 24.35	15.62	0.00	VI.	8	4.37	36 15.66	12.24	2.70	53 39.97	44 0.6
18	10	. .	. .	23.4	. .	50.1	. .	. .	55 36.79	15.62	0.00	V.	8	5.24	41 38.44	12.18	3.02	55 52.41	49 23.6
19	10	. .	. .	. .	21.4	. .	. .	. .	57 21.25	15.62	0.00	IV.	4	10 30	20 14.78	12.14	1.71	57 36.87	27 58.6
20	9	. .	. .	. .	38.5	. .	. .	. .	17 59 38.35	15.62	+0.01	IV.	10	8.43	48 19.72	12.05	3.45	17 59 53.98	56 5.2
21	10	31.4	. .	58.5	. .	. .	. .	. .	18 2 11.84	15.61	0.00	III.	8	5.8	36 31.37	11.99	2.71	18 2 27.45	44 16.1
22	9	. .	. .	. .	. .	. .	11.9	. .	2 45.34	15.61	0.00	VI.	4	8.39	19 18.69	11.98	1.66	3 0.95	27 2.3
23	10	. .	. .	. .	52.5	. .	. .	. .	4 52.35	15.61	0.00	IV.	7	9.17	33 38.26	11.90	2.53	5 7.96	41 22.7
24	5	. .	. .	. .	. .	13.5	. .	40.1	6 0.17	15.61	+0.01	VII.	8	8.56	38 26.10	11.87	2.82	6 15.79	46 10.8
25	7	. .	. .	. .	. .	. .	50.	3.5	7 23.41	15.61	0.00	VII.	6	7.14	27 35.96	11.83	2.16	7 39.02	35 20.0
26	11	42.5	36.1	9.2	. .	. .	. .	. .	18 11 22.51	+15.60	−0.01	III.	2	7.31	− 8 46.38	−11.71	−1.01	18 11 38.10	−20 16 29.1

ZONE 172. JULY 15. K. BELT, −25° 1′. D₀ = −24° 37′ 20″.

No.	Mag.	I.	II.	III.	IV.	V.	VI.	VII.	T.	a_1	a_2	MICROMETER.			i	d_1	d_2	Mean Right Ascension, 1850.0.	Mean Declination, 1850.0.
1	9	53.6	. .	20.8	. .	. .	. .	. .	16 53 34.71	+16.47	0.00	.	3	5.2	−12 30.38	− 5.4	−1.02	16 52 51.18	−24 49 56.9
2	9	. .	. .	37.4	. .	5.2	. .	. .	53 51.06	16.47	−0.01	.	2	7.57	8 59.52	5.6	0.74	54 7.52	24 46 25.9
3	8.9	. .	. .	0.4	. .	27.1	. .	. .	55 13.90	16.47	+0.01	.	10	8.19	48 7.62	6.3	3.87	55 38.38	25 25 37.8
4	10	. .	. .	. .	. .	6.9	20.4	. .	55 53.02	16.47	0.01	.	8	8.36	38 16.30	6.6	3.06	56 9.50	15 46.0
5	10	. .	. .	. .	. .	35.2	. .	. .	16 58 21.44	+16.47	+0.01	.	8	7.22	−37 38.97	− 7.9	−3.01	16 58 37.92	−25 15 9.9

CORRECTIONS.

Date.		Corr. of Clock.	Hourly rate.	m	n	c
1848.	h.	s.	s.	s.	s.	s.
July 3,	17	+ 7.60	l 0.141			
July 10,	17	+ 12.22	g 0.016			
July 15,	16	+ 12.91	l 0.060	− 0.28	+ 0.82	0.00

INSTRUMENT READINGS.

Date.		Barom.	THERMOM. At.	THERMOM. Ex.
1848.	h. m.	in.	°	°

REMARKS.*

(170) 6. Transit over T. VI is assumed to have been erroneously recorded as over T. VII.
(171) 1. Nos. 4, 7, 9, 10, and 12 seem to be in error by quantities from 20″ to 40″ in declination.
(171) 5. Hor. thread assumed as 2 instead of 3.
(171) 8. Hor. thread assumed as 8 instead of 7.
(171) 18. Hor. thread assumed as 9 instead of 8.
(172) 1. Minutes of transit assumed as 52 instead of 53.

ZONE 173. JULY 17. K. BELT, −25° 1′. $D_o = -24° 37′ 20″$.

No.	Mag.	SECONDS OF TRANSIT. I.	II.	III.	IV.	V.	VI.	VII.	T.	a_1	a_2	MICROMETER.		i	d_1	d_2	Mean Right Ascension, 1850.0.	Mean Declination, 1850.0.
									h. m. s.	s.	s.		r.	′ ″	″	″	h. m. s.	° ′ ″
1	10	. .	. .	. .	. .	. .	50.3	. .	18 14 22.88	+23.71	−0.01	VI.	3	6.52	−12 55.44	−13.3	−1.04	18 14 46.58	−24 50 29.8
2	8	. .	. .	. .	. .	. .	. .	24.2	19 42.94	23.71	+0.01	VII.	9	8.37	43 15.44	12.9	3.55	20 6.66	25 20 51.0
3	9	. .	. .	. .	. .	38.7	. .	6.5	21 24.91	23.71	0.00	V.	5	3.30	21 42.15	12.7	1.72	21 48.62	24 59 16.6
4	9	. .	52.3	5.7	. .	. .	. .	. .	29 19.50	23.70	−0.01	.	3	5.00	12 29.37	12.0	0.96	29 43.19	24 50 2.3
5	9	. .	32.3	45.6	. .	. .	. .	. .	31 59.65	23.70	0.00	.	6	11.35	29 47.85	11.8	2.41	32 23.35	25 7 22.1
6	10	. .	49.2	2.1	. .	. .	. .	. .	33 16.39	23.69	0.00	.	7	7.34	32 46.33	11.7	2.67	33 40.06	10 20.7
7	6.7	30.9	45.5	58.6	. .	. .	. .	. .	35 12.73	23.69	0.00	.	7	7.39	31 48.34	11.5	2.67	35 36.42	25 9 22.5
8	9	52.7	7.0	. .	. .	. .	. .	. .	38 34.43	23.69	0.00	.	4	8.29	18 13.25	11.2	1.51	38 58.12	24 55 46.0
9	10	. .	28.8	. .	55.9	. .	. .	. .	39 55.99	23.69	−0.01	.	1	9.28	4 46.40	11.1	0.32	40 19.67	42 17.8
10	9	. .	. .	. .	. .	50.1	. .	17.9	40 36.22	23.69	0.01	.	2	11.28	10 45.91	11.1	0.82	40 59.90	48 17.8
11	9	. .	. .	. .	16.3	. .	44.2	. .	42 16.47	23.68	−0.01	.	2	13.51	11 58.01	10.9	0.92	42 40.14	49 29.8
12	9	5.9	20.2	. .	. .	. .	. .	. .	44 47.63	23.68	0.00	.	4	8.38	19 18.30	10.7	1.51	45 11.31	24 56 50.5
13	9	. .	56.5	9.4	. .	. .	. .	. .	46 23.73	23.68	+0.01	.	8	5.5	36 29.89	10.5	2.98	46 47.42	25 14 3.4
14	9	. .	. .	. .	8.4	22.2	. .	. .	47 8.28	23.68	−0.01	.	2	11.59	11 1.55	10.5	0.82	47 31.95	24 48 32.9
15	7.8	. .	17.3	30.8	44.5	. .	. .	. .	48 44.59	23.68	0.00	.	5	13.30	26 44.73	10.3	2.14	49 8.27	25 4 17.2
16	7.8	7.0	21.3	. .	. .	. .	. .	. .	50 48.65	23.68	0.00	.	6	14.14	31 8.01	10.2	2.52	51 12.53	25 8 40.7
17	9	. .	. .	. .	. .	. .	. .	7.2	55 25.58	23.68	−0.01	.	3	11.11	15 36.44	9.8	1.21	55 49.25	24 53 7.5
18	8.9	. .	. .	. .	. .	. .	. .	44.8	56 3.18	23.68	0.00	.	3	12.15	16 8.71	9.7	1.25	56 26.86	24 53 39.7
19	8	. .	. .	. .	. .	. .	59.9	. .	57 32.42	23.67	0.00	.	6	10.13	29 6.50	9.6	2.35	57 56.09	25 6 38.5
20	6.7	. .	. .	. .	. .	54.5	8.2	. .	18 58 40.70	23.67	−0.01	VII.	3	11.21	15 41.12	9.5	1.22	18 59 4.36	24 53 11.6
21	9	31.5	. .	59.0	. .	. .	. .	. .	19 3 12.66	23.67	−0.01	.	1	12.43	6 24.73	9.1	0.44	19 3 36.32	43 54.3
22	8	48.5	. .	15.7	. .	. .	. .	. .	6 29.70	23.66	0.00	.	4	11.40	20 50.07	8.8	1.64	6 53.36	58 20.5
23	9	. .	. .	31.4	. .	. .	. .	. .	10 44.93	23.66	−0.01	.	3	11.20	16 41.49	8.4	1.22	11 8.58	54 11.1
24	5.6	. .	. .	. .	. .	58.5	. .	. .	15 44.58	23.66	0.01	VII.	1	8.19	10 10.77	8.0	0.75	16 8.23	47 31.5
25	6.7	. .	. .	. .	. .	. .	. .	32.8	15 51.10	23.66	−0.01	VII.	2	9.5	4 34.45	8.0	0.29	16 14.75	24 42 2.7
26	5	. .	. .	. .	. .	. .	. .	52.2	19 27 10.83	+23.65	+0.01	VII.	7	12.16	−35 8.16	− 7.0	−2.88	19 27 34.49	−25 12 38.0

ZONE 174. JULY 18. P. BELT, −25° 1′. $D_o = -24° 40′ 20″$.

No.	Mag.	I.	II.	III.	IV.	V.	VI.	VII.	T.	a_1	a_2	MICROMETER.		i	d_1	d_2	Mean Right Ascension, 1850.0.	Mean Declination, 1850.0.
1	6	34.5	49.6	. .	. .	. .	. .	. .	16 37 16.77	+24.22	0.00	II.	7	10.44	−34 21.98	−19.5	−2.74	16 37 40.99	−25 15 4.2
2	10	. .	51.4	4.3	. .	. .	. .	. .	39 18.60	24.21	0.00	III.	7	9.6	33 32.68	19.4	2.68	39 42.81	25 14 14.8
3	9	. .	. .	. .	1.2	. .	28.4	. .	40 1.03	24.21	−0.01	VI.	2	5.11	7 35.65	19.4	0.63	40 25.23	24 48 15.7
4	10	. .	. .	1.1	. .	28.4	. .	. .	42 14.78	24.21	0.00	V.	8	4.37	35 45.48	19.3	3.28	42 38.99	25 16 28.1
5	10	. .	. .	. .	. .	47.4	. .	15.1	43 33.55	24.21	0.00	VII.	4	10.3	20 0.80	19.2	1.61	43 57.76	0 41.6
6	9	26.1	41.2	54.2	8.1	. .	. .	. .	46 8.20	24.20	0.00	IV.	7	15.2	36 32.21	19.1	2.93	46 32.40	25 17 14.2
7	7	41.4	55.6	. .	. .	. .	. .	. .	50 23.00	24.19	−0.01	II.	2	11.38	10 50.80	18.9	0.68	50 47.18	24 51 30.6
8	6	. .	. .	. .	. .	. .	. .	15.5	50 33.73	24.19	0.01	II.	1	9.34	4 49.28	18.9	0.39	50 57.91	45 28.6
9	10	. .	. .	. .	. .	40.9	. .	9.9	52 27.60	24.19	−0.01	V.	2	8.44	9 23.19	18.8	0.76	52 51.78	24 50 2.8
10	8	. .	. .	. .	22.5	. .	50.5	. .	54 22.65	24.19	+0.01	IV.	10	8.15	48 5.60	18.7	3.88	54 46.85	25 28 48.2
11	10	. .	. .	. .	. .	. .	. .	27.1	55 45.73	24.18	0.00	VII.	7	12.9	35 4.63	18.6	2.81	56 9.91	15 46.0
12	9	. .	46.4	59.9	. .	. .	. .	. .	58 13.91	24.18	0.00	III.	7	11.3	34 31.67	18.5	2.76	16 58 38.09	15 12.9
13	10	17.4	32.5	. .	. .	. .	. .	. .	16 59 59.73	24.18	+0.01	II.	8	11.50	39 53.95	18.4	3.21	17 0 23.92	25 20 35.6
14	10	. .	46.9	. .	14.9	. .	. .	. .	17 1 14.54	24.18	−0.01	IV.	1	8.58	4 31.28	18.4	0.37	1 38.71	24 45 10.1
15	7	. .	. .	. .	6.5	. .	. .	. .	3 37.04	24.17	0.00	IV.	5	6.18	23 6.90	18.3	1.85	3 1.21	25 3 47.1
16	9	. .	. .	. .	. .	. .	50.5	. .	3 23.06	24.17	0.00	V.	4	7.8	18 32.87	18.3	1.49	3 47.23	24 59 12.7
17	10	20.	. .	53.4	. .	. .	. .	. .	6 4.17	24.17	−0.01	III.	2	6.6	8 3.52	18.1	0.66	6 28.33	24 48 42.3
18	8	. .	. .	. .	. .	3.9	17.1	. .	6 49.86	24.17	0.00	V.	6	6.235	27 10.73	18.1	2.18	7 14.03	25 7 51.0
19	11	. .	. .	59.9	. .	. .	. .	. .	17 9 13.49	+24.16	0.00	IV.	4	10.4	−20 1.67	−17.9	−1.61	17 9 37.65	−25 0 41.2

CORRECTIONS.

Date.		Corr. of Clock.	Hourly rate.	m	n	c
1848.	h.	s.	s.	s.	s.	s.
July 17,	19	+ 20.31	/ 0.001			
July 18,	19	+ 20.73	/ 0.009	− 0.36	+ 0.75	0.00

INSTRUMENT READINGS.

	Date.		Barom.	THERMOM. At.	Ex.
	1848.	h. m.	in.	°	°

REMARKS.

(173) 1. Micrometer reading assumed as 5r.52 instead of 6r.52.
(173) 7. Micrometer reading assumed as 5r.39 instead of 7r.39.
(173) 8. Micrometer reading assumed as 6r.29 instead of 8r.29.
(173) 23. Micrometer reading assumed as 13r.20 instead of 11r.20.
(173) 24. Hor. thread assumed as 2 instead of 1, and micrometer reading as 10r.19 instead of 8r.19.
(173) 25. Hor. thread assumed as 1 instead of 2.
(174) 4. Micrometer reading assumed as 3r.37 instead of 4r.37.
(174) 15. Transit over T. VI assumed as recorded over T. IV, and minutes as 4.5 instead of 6.5.
(174) 17. The two transits are discordant by 5s, and the right ascension therefore probably erroneous by 2s.5.

ZONE 174. JULY 18. P. BELT, −25° 1'. D_0=−24° 40' 20"—Continued.

No.	Mag.	SECONDS OF TRANSIT. I.	II.	III.	IV.	V.	VI.	VII.	T.	a_1	a_2	MICROMETER.			i	d_1	d_2	Mean Right Ascension, 1850.0.	Mean Declination, 1850.0.
									h. m. s.	s.	s.			r.	' "	"	"	h. m. s.	° ' "
20	11	. .	51.4	. .	18.4	. .	. .	. .	17 11 18.56	+24.16	−0.01	IV.	2	6.26	− 8 13.64	−17.8	−0.66	17 11 42.71	−24 48 52.1
21	2	. .	. .	. .	23.5	37.7	. .	. .	12 23.58	24.16	−0.01	IV.	2	10.3	10 3.06	17.8	0.80	12 47.73	50 41.7
22	9	. .	. .	. .	. .	46.5	. .	13.9	13 32.46	24.16	0.00	V.	3	12.12	16 7.16	17.7	1.30	13 56.62	24 56 46.2
23	11	. .	. .	. .	. .	. .	. .	3.5	17 22.03	24.15	0.00	VII.	6	6.49	27 23.28	17.5	2.20	17 46.18	25 8 3.0
24	10	. .	. .	. .	. .	. .	26.8	40.9	19 59.45	24.14	+0.01	VI.	9	5.58	41 55.45	17.4	3.38	29 23.60	25 22 36.2
25	11	. .	59.8	. .	25.9	. .	. .	. .	23 25.58	24.14	−0.01	IV.	3	10.39	15 20.32	17.2	1.23	23 49.71	24 55 58.8
26	10	. .	. .	. .	. .	51.5	. .	19.1	24 37.46	24.14	0.01	V.	1	10.44	5 24.70	17.1	0.43	25 1.59	46 2.2
27	10	36.1	49.9	. .	. .	. .	. .	. .	28 17.51	24.13	0.01	II.	3	3.7	11 32.23	16.9	0.92	28 41.63	52 10.1
28	11	. .	. .	. .	50.5	. .	18.9	. .	32 50.91	24.13	−0.01	VI.	3	8.51	14 25.69	16.6	1.16	33 15.03	24 55 3.5
29	10	57.1	12.2	26.1	. .	. .	. .	. .	35 39.58	24.12	+0.01	III.	7	15.9	36 35.70	16.4	2.94	36 3.71	25 17 15.0
30	9	23.5	37.8	. .	. .	. .	. .	. .	38 5.31	24.12	0.00	I.	6	5.50	26 53.52	16.3	2.15	38 29.43	25 7 32.0
31	11	. .	. .	18.4	. .	. .	. .	. .	39 31.80	24.12	−0.01	III.	2	3.26	6 42.83	16.2	0.54	39 55.91	24 47 19.6
32	10	. .	. .	. .	. .	. .	19.1	33.9	40 51.94	24.12	0.01	VI.	2	5.00	7 30.11	16.1	0.60	41 16.05	48 6.8
33	10	. .	23.5	. .	50.5	. .	. .	. .	43 50.63	24.11	0.01	IV.	1	6.37	3 20.19	15.9	0.26	44 14.73	13 56.4
34	8	. .	48.9	2.3	15.9	29.9	. .	. .	45 15.97	24.11	−0.01	IV.	2	10.55	10 29.28	15.9	0.83	45 40.07	24 51 6.0
35	10	. .	. .	. .	. .	. .	55.9	9.9	46 28.44	24.11	0.00	VI.	6	10.44	29 21.98	15.8	2.36	46 52.55	25 10 0.1
36	10	. .	. .	. .	. .	9.9	23.1	. .	47 55.86	24.11	0.00	VI.	5	5.28	22 41.52	15.7	1.82	48 19.97	3 19.0
37	9	39.9	. .	. .	. .	. .	. .	. .	50 21.53	24.10	0.00	I.	5	7.5	23 30.24	15.5	1.88	50 45.63	25 4 7.6
38	10	. .	. .	. .	. .	. .	42.9	56.6	50 15.29	24.10	0.00	VI.	6	8.11	28 4.82	15.5	2.26	50 39.39	8 42.6
39	10	. .	. .	. .	. .	. .	40.5	55.8	51 13.60	24.10	−0.01	VII.	2	11.47	5 56.14	15.5	0.86	51 37.69	24 46 32.5
40	9	. .	. .	. .	. .	. .	9.9	23.4	52 42.13	24.10	0.00	VII.	3	12.6	11 4.72	15.4	1.29	53 6.23	24 51 41.4
41	10	. .	. .	. .	1.9	. .	28.5	. .	55 1.38	24.10	0.00	VI.	7	4.43	31 19.94	15.2	2.52	55 25.48	25 11 57.7
42	10	. .	. .	38.3	. .	6.1	. .	. .	56 52.34	24.10	+0.01	V.	10	9.12	48 34.30	15.1	3.94	57 16.45	29 13.3
43	10	. .	. .	. .	. .	. .	. .	30.1	17 57 48.62	24.09	0.00	VII.	6	4.30	26 13.19	15.0	2.10	17 58 12.71	25 6 50.3
44	9	. .	37.4	50.9	. .	. .	. .	. .	18 0 4.53	24.09	−0.01	II.	1	6.53	3 28.09	14.9	0.27	18 0 28.61	24 44 3.3
45	11	. .	. .	. .	. .	27.4	. .	54.9	2 13.48	24.09	0.00	VII.	5	8.29	24 12.60	14.8	1.94	2 37.57	25 4 49.3
46	9	. .	. .	. .	. .	. .	10.2	24.8	3 43.05	24.09	0.00	VI.	6	12.26	30 13.40	14.7	2.43	4 7.14	25 10 50.5
47	10	. .	. .	. .	41.5	. .	. .	. .	5 41.35	24.08	0.00	IV.	5	9.52	24 54.81	14.5	1.99	6 5.43	5 31.3
48	11	. .	. .	. .	. .	. .	39.	. .	7 11.55	24.08	0.00	VI.	5	4.54	22 24.39	14.4	1.79	7 35.63	3 0.6
49	10	. .	. .	. .	. .	. .	49.5	. .	8 22.03	24.08	0.00	VI.	6	6.18	27 7.84	14.4	2.17	8 46.11	25 7 44.4
50	10	50.4	4.5	18.1	. .	. .	. .	. .	10 31.90	24.08	0.00	III.	4	8.35	19 16.75	14.2	1.54	10 55.98	24 59 52.5
51	5	. .	26.2	39.2	53.4	. .	. .	. .	10 53.25	24.08	0.00	IV.	4	5.54	17 55.60	14.2	1.43	12 17.33	58 31.2
52	10	. .	. .	. .	. .	49.5	. .	17.4	12 35.72	24.08	0.00	VII.	4	5.6	17 31.04	14.1	1.40	12 59.80	58 6.5
53	10	. .	. .	. .	. .	. .	50.5	4.5	14 22.96	24.08	−0.01	VI.	3	1.2	9 58.95	13.9	0.83	14 47.03	50 33.7
54	10	. .	. .	23.3	36.5	. .	. .	. .	17 36.58	24.07	−0.01	IV.	3	7.40	13 50.05	13.7	1.10	18 0.64	24 54 24.9
55	8	. .	. .	4.5	18.1	. .	. .	. .	19 18.08	24.07	0.00	IV.	6	6.255	27 11.78	13.6	2.18	19 42.15	25 7 47.6
56	9	. .	. .	. .	. .	. .	. .	24.5	19 43.20	24.07	+0.01	VII.	9	2.27	40 8.86	13.6	3.26	20 7.28	25 20 45.7
57	9	. .	. .	. .	38.5	. .	. .	6.5	21 38.76	24.07	0.00	VII.	4	7.23	18 40.11	13.5	1.50	22 2.84	24 59 15.1
58	11	. .	. .	2.5	. .	30.5	. .	. .	23 16.36	24.07	0.00	V.	4	6.47	18 22.29	13.3	1.47	23 40.43	24 58 57.1
59	10	. .	. .	. .	. .	12.9	. .	. .	23 59.15	24.06	+0.01	V.	8	8.37	38 16.76	13.3	3.11	25 23.22	25 18 53.2
60	10	. .	. .	43.5	57.4	10.9	. .	. .	26 57.29	24.06	+0.01	V.	9	12.23	45 9.70	13.1	3.68	27 31.36	25 25 46.5
61	9	38.1	52.4	5.5	. .	. .	. .	. .	29 19.47	24.06	−0,01	III.	2	8.41	9 21.67	12.9	0.73	29 43.52	24 49 55.3
62	10	. .	42.9	55.9	. .	. .	. .	. .	31 10.08	24.06	0.00	III.	6	7.54	27 56.37	12.8	2.24	31 34.14	25 8 31.4
63	10	. .	. .	. .	59.9	. .	26.6	. .	31 59.44	24.06	0.00	VI.	6	5.39	26 48.18	12.7	2.15	32 23.50	25 7 23.0
64	10	. .	. .	. .	27.4	. .	. .	. .	34 27.25	24.06	−0.01	IV.	2	10.29	10 16.17	12.6	0.80	34 51.30	24 50 49.6
65	7	. .	. .	. .	. .	. .	39.9	53.6	35 12.29	24.06	0.00	VII.	6	10.33	28 45.97	12.5	2.36	35 36.35	25 9 0.8
66	10	. .	. .	. .	16.1	. .	43.5	. .	37 16.00	24.05	0.00	VI.	5	7.18	23 36.99	12.4	1.89	37 40.05	25 4 11.3
67	10	. .	. .	. .	. .	47.5	. .	15.6	38 33.80	24.05	−0.01	VII.	3	10.39	15 19.96	12.3	1.21	38 57.84	24 55 53.5
68	9	. .	. .	22.5	. .	49.9	. .	. .	18 40 35.95	+24.05	−0.01	V.	2	4.25	− 7 42.83	−12.1	−0.61	18 40 59.99	−24 48 15.5

CORRECTIONS.

Date.		Corr. of Clock.	Hourly rate.	m	n	c
1848.	h.	s.	s.	s.	s.	s.

INSTRUMENT READINGS.

Date.		Barom.	THERMOM. At.	Ex.
1848.	h. m.	in.	°	°

REMARKS.

(174) 24. Minutes assumed as 18 instead of 19.
(174) 25. Transit over T. II assumed to have been at 57s.8 instead of 59s.8.
(174) 39. Micrometer thread assumed as 1 instead of 2.
(174) 40. Micrometer thread assumed as 2 instead of 3.
(174) 51. Minutes assumed as 11 instead of 12.
(174) 53. Micrometer reading assumed as 0r.2 instead of 1r.2.
(174) 57. Transit over T. VI assumed to have been recorded as over T. VII.
(174) 59. Minutes assumed as 24 instead of 23.
(174) 65. Micrometer reading assumed as 9r.33 instead of 10r.33.
(174) 68. Micrometer reading assumed as 5r.25 instead of 4r.25.

ZONE 174. JULY 18. P. BELT, −25° 1'. D_o=−24° 40' 20"—Continued.

No.	Mag.	SECONDS OF TRANSIT. I.	II.	III.	IV.	V.	VI.	VII.	T.	a_1	a_2	MICROMETER.			i	d_1	d_2	Mean Right Ascension, 1850.0.	Mean Declination, 1850.0.
									h. m. s.	s.	s.			r.	' "	"	"	h. m. s.	° ' "
69	8	. .	. .	3.8	16.4	. .	. .	. .	18 42 16.74	+24.05	−0.01	IV.	2	7.53	− 8 57.50	−12.0	−0.70	18 42 40.78	−24 49 30.2
70	10	. .	20.5	33.6	47.5	. .	. .	. .	43 47.42	24.05	−0.01	IV.	1	10.2	5 3.56	11.9	0.39	44 11.46	45 35.9
71	8	. .	. .	. .	46.5	. .	14.5	. .	44 46.71	24.05	0.00	VI.	3	12.51	16 26.70	11.8	1.30	45 10.76	24 56 59.8
72	8	. .	. .	. .	. .	36.8	50.9	. .	46 23.22	24.04	0.00	VI.	7	9.1	33 30.03	11.7	2.71	46 47.26	25 14 4.4
73	7	3.9	17.5	. .	. .	. .	. .	. .	48 45.32	24.04	0.00	II.	5	7.26	23 41.03	11.5	1.90	49 9.36	4 14.4
74	10	. .	. .	. .	. .	33.5	. .	1.1	49 19.77	24.04	+0.01	VII.	8	10.26	39 11.40	11.5	3.19	49 43.82	19 46.1
75	7	. .	. .	. .	. .	. .	15.9	30.	50 48.48	24.04	0.00	VII.	6	8.25	28 11.68	11.4	2.27	51 12.52	8 45.4
76	4	. .	. .	. .	52.9	. .	20.5	. .	52 52.90	24.04	0.00	VI.	5	4.51	22 22.87	11.3	1.78	53 16.94	2 56.0
77	10	32.5	47.1	. .	. .	. .	43.9	. .	54 14.40	24.04	0.00	II.	5	3.57	21 25.39	11.2	1.72	55 38.44	1 58.3
78	10	. .	5.5	18.9	31.4	. .	. .	. .	55 32.34	24.04	0.00	III.	6	7.105	23 34.47	11.1	1.89	55 56.38	4 7.5
79	8	. .	. .	. .	30.5	. .	. .	. .	57 30.35	24.04	0.00	IV.	6	3.52	25 54.38	10.9	2.07	18 57 54.39	25 6 27.4
80	9	. .	. .	. .	. .	54.5	. .	21.9	18 59 40.43	24.04	−0.01	VI.	3	5.30	12 44.34	10.8	1.00	19 59 4.46	24 53 16.1
81	10	29.9	44.2	. .	. .	. .	. .	. .	19 3 11.72	24.03	0.00	II.	6	9.4	28 31.55	10.5	2.30	3 35.75	25 9 4.4
82	11	. .	. .	. .	12.4	. .	. .	39.5	4 12.15	24.03	0.00	IV.	4	8.10	19 4.18	10.5	1.52	4 36.18	24 59 36.2
83	9	48.2	2.9	15.9	. .	. .	. .	. .	6 29.90	24.03	0.00	III.	4	5.42	17 49.51	10.3	1.43	6 53.93	24 58 21.2
84	10	. .	1.5	14.2	27.4	41.4	. .	. .	8 28.05	24.03	+0.01	V.	8	10.59	39 28.36	10.2	3.22	8 52.09	25 20 1.8
85	10	4.1	17.9	31.4	. .	. .	. .	. .	10 45.32	24.03	−0.01	I.	3	7.16	13 37.59	10.0	1.07	11 9.34	24 54 8.7
86	11	. .	. .	. .	. .	51.5	. .	. .	11 37.79	24.03	+0.01	V.	10	8.17	48 6.57	9.9	3.92	12 0.83	25 28 40.4
87	11	. .	34.5	. .	. .	. .	. .	. .	15 2.10	24.02	0.00	II.	5	8.22	24 9.27	9.7	1.94	15 26.12	25 4 40.9
88	5	. .	. .	30.5	44.5	58.5	. .	. .	15 44.28	24.02	−0.01	V.	2	4.18	7 9.05	9.7	0.55	16 8.29	24 47 39.3
89	11	. .	44.5	. .	. .	. .	39.9	. .	19 12.25	24.02	−0.01	VI.	4	4.5	17 0.48	9.4	1.34	19 36.26	57 31.2
90	10	. .	. .	. .	. .	48.5	. .	16.1	20 34.59	24.02	0.00	VII.	4	8.47	19 22.48	9.3	1.54	20 58.61	24 59 53.3
91	11	. .	. .	. .	. .	. .	46.7	. .	22 19.20	24.02	+0.01	VI.	7	10.27	34 13.40	9.2	2.78	22 43.23	25 14 45.4
92	10	9.1	23.5	36.9	. .	. .	. .	. .	24 50.72	24.02	−0.01	III.	4	3.45	16 50.51	9.0	1.32	25 14.73	24 57 20.8
93	6	. .	5.	. .	30.9	. .	. .	. .	26 30.67	24.02	0.00	IV.	5	4.9	22 1.86	8.9	1.75	26 54.69	25 2 32.5
94	5	. .	. .	. .	. .	. .	. .	51.9	27 10.50	24.02	0.00	VII.	7	6.9	32 3.11	8.8	2.59	27 34.52	25 12 34.5
95	11	. .	. .	. .	. .	. .	6.6	19.9	28 38.71	24.02	−0.01	VI.	3	4.23	12 10.55	8.7	0.94	29 2.72	24 52 40.2
96	10	. .	. .	. .	. .	. .	26.6	. .	29 59.18	24.02	−0.01	VI.	3	5.19	12 38.79	8.6	0.97	30 23.19	24 53 8.4
97	10	. .	. .	51.4	5.5	. .	. .	. .	32 5.25	24.02	0.00	IV.	6	12.40	30 20.62	8.5	2.44	32 29.27	25 10 51.6
98	8	. .	. .	. .	. .	. .	19.1	. .	32 51.61	24.02	0.00	VII.	6	15.25	31 43.44	8.4	2.57	33 15.63	25 12 14.4
99	10	. .	. .	. .	. .	. .	. .	50.1	35 8.34	24.02	−0.01	VII.	1	10.36	5 20.35	8.3	0.38	35 32.35	24 45 49.0
100	9	. .	. .	. .	. .	9.9	. .	36.9	19 36 55.83	+24.02	0.00	VII.	7	10.36	−33 47.50	− 8.1	−2.79	19 37 19.85	−25 14 18.4

ZONE 175. JULY 19. K. BELT, −25° 1'. D_o=−24° 38' 20".

No.	Mag.	I.	II.	III.	IV.	V.	VI.	VII.	T.	a_1	a_2	MICROMETER.			i	d_1	d_2	Mean Right Ascension, 1850.0.	Mean Declination, 1850.0.
1	9	. .	. .	38.5	51.3	. .	. .	. .	17 33 51.63	+23.77	0.00	V.	5	3.15	−21 34.58	− 6.5	−1.73	17 34 15.40	−25 0 2.8
2	9	. .	. .	. .	38.9	52.5	. .	. .	35 38.73	23.77	0.00	VI.	6	8.8	28 3.31	6.3	2.25	36 2.50	25 6 31.9
3	9	. .	24.0	. .	51.2	. .	. .	. .	43 51.24	23.76	−0.01	.	1	11.10	5 37.84	5.5	0.45	44 14.99	24 44 3.8
4	7	. .	49.3	. .	16.4	. .	. .	. .	45 16.53	23.75	−0.01	.	3	5.22	12 40.46	5.4	1.02	45 40.27	24 51 6.9
5	9	. .	1.6	14.4	. .	. .	. .	. .	46 28.73	23.75	0.00	.	7	5.14	31 35.73	5.3	2.54	46 52.48	25 10 3.6
6	9	. .	. .	. .	. .	56.6	. .	24.3	46 42.89	23.75	+0.01	.	8	3.02	35 27.87	5.3	2.86	47 6.65	13 56.0
7	9	. .	. .	. .	. .	. .	23.2	37.3	47 55.77	23.75	0.00	VII.	5	10.0	24 58.49	5.2	2.00	48 19.52	3 25.7
8	9	. .	. .	. .	. .	. .	. .	22.8	48 41.35	23.75	0.00	VII.	6	9.30	28 44.46	5.1	2.30	49 5.10	7 11.9
9	9	. .	. .	. .	15.2	. .	. .	. .	50 15.05	23.74	0.00	.	6	12.27	30 14.66	4.9	2.43	50 38.79	8 41.4
10	9	. .	. .	. .	34.2	. .	. .	. .	50 20.38	23.74	0.00	.	5	11.33	25 45.74	4.9	2.06	50 44.12	25 4 12.7
11	9	. .	. .	. .	. .	. .	40.9	. .	51 13.50	23.74	−0.01	.	2	6.5	8 3.05	4.9	0.64	51 37.23	24 46 28.6
12	9	. .	. .	. .	. .	37.7	. .	5.4	52 23.86	23.74	0.00	.	5	3.56	21 55.30	4.8	1.75	52 47.60	25 0 21.9
13	9	. .	. .	. .	46.3	0.6	. .	. .	17 53 46.50	+23.74	+0.01	.	8	6.40	−37 17.80	− 4.6	−3.01	17 54 10.25	−25 15 45.4

CORRECTIONS.

Date.	Corr. of Clock.	Hourly rate.	m	n	c
1848. h.	s.	s.	s.	s.	s.
July 19. 19	+ 20.36	/ 0.001			

INSTRUMENT READINGS.

Date.	Barom.	THERMOM. At.	THERMOM. Ex.
1848. h. m.	In.	°	°

REMARKS.

(174) 77. Minutes assumed as 55 instead of 54, and micrometer reading as 2r.57 instead of 3r.57.
(174) 78. Micrometer reading uncertain; differs 4' from Arg. 308, 110.
(174) 80. Minutes assumed as 58 instead of 59.
(174) 82. Transit over T. VI assumed to have been recorded as over T. VII.
(174) 93. T. II assumed to have been observed at 3s instead of 5s.
(174) 100. Micrometer reading assumed as 9r.36 instead of 10r.36.

ZONE 175. JULY 19. K. BELT, −25° 1'. D₀ = −24° 38' 20"—Continued.

No.	Mag.	SECONDS OF TRANSIT. I.	II.	III.	IV.	V.	VI.	VII.	T.	a_1	a_2	MICROMETER.			i	d_1	d_2	Mean Right Ascension, 1850.0.	Mean Declination, 1850.0.
									h. m. s.	s.	s.			r.	' "	"	"	h. m. s.	° ' "
14	8						48.4		17 55 21.01	+23.74	−0.01	VII.	1	5.8	− 2 34.94	− 4.5	−0.19	17 55 44.74	−24 40 59.6
15	9	6.6	21.4						57 48.67	23.73	0.00	.	6	8.57	28 28.18	4.3	2.29	58 12.40	25 6 54.8
16	9			13.4	27.0				58 27.09	23.73	+0.01	.	9	8.45	43 19.83	4.2	3.51	58 50.83	21 47.5
17	9							27.1	17 58 13.40	23.73	0.01	.	10	13.58	50 58.54	4.2	4.14	17 59 37.14	29 26.9
18	9					30.9	44.3		18 0 16.97	23.73	+0.01	.	8	11.50	39 54.11	4.0	3.23	18 0 40.71	18 21.3
19	10				13.7				2 13.55	23.73	0.00	.	5	12.58	26 28.59	3.8	2.12	2 37.28	4 54.6
20	8		16.2	29.5					3 43.59	23.72	0.00	.	7	7.2	32 30.19	3.7	2.61	4 7.31	10 56.5
21	9			14.5	28.4				4 28.33	23.72	−0.01	.	9	6.47	42 20.33	3.6	3.43	4 52.06	20 47.4
22	9						49.7		8 22.22	23.72	0.00	VI.	6	10.48	29 23.99	3.3	2.36	8 45.94	7 49.7
23	9						51.0		9 23.46	23.72	+0.01	VI.	9	10.30	44 12.62	3.2	3.58	9 47.19	22 39.4
24	10					45.8			10 31.97	23.72	0.00	.	4	13.16	21 38.47	3.1	1.73	10 55.69	25 0 3.3
25	7			39.8	53.3				11 53.27	23.71	0.00	.	4	10.24	20 11.75	2.9	1.62	12 16.98	24 58 36.3
26	8	36.4	51.2						19 18.48	23.70	0.00	.	6	10.50	29 25.16	2.2	2.36	19 42.18	25 7 49.7
27	7.8				42.7		10.5		19 42.76	23.70	+0.01	.	9	6.55	42 24.36	2.2	3.45	20 6.47	25 20 50.0
28	8		57.8		24.8				21 25.01	23.70	0.00	.	4	11.42	20 51.08	2.0	1.67	21 48.71	24 59 14.8
29	9			17.4	31.3				22 31.01	23.70	−0.01	.	2	12.40	11 22.22	1.9	0.90	22 54.70	24 49 45.0
30	9				14.6				23 14.45	23.70	0.00	.	7	8.15	33 7.00	1.9	2.66	23 38.15	25 11 31.6
31	9							7.7	18 23 26.38	+23.70	+0.01	.	8	9.29	−38 43.01	− 1.8	−3.14	18 23 50.09	−25 17 8.0

ZONE 176. JULY 19. K. BELT, −22° 31'. D₀ = −22° 7' 10".

No.	Mag.	I.	II.	III.	IV.	V.	VI.	VII.	T.	a_1	a_2	MICROMETER.			i	d_1	d_2	Mean Right Ascension, 1850.0.	Mean Declination, 1850.0.
1	9					6.4			20 58 52.69	+23.59	+0.01	VII.	10	9.18	−48 37.03	−14.63	−3.76	20 59 16.29	−22 56 5.4
2	9			39.4		6.6			21 4 53.01	23.59	0.00	.	7	8.31	33 15.07	14.29	2.61	21 5 16.60	40 42.0
3	9							41.0	5 59.72	23.59	+0.01	VII.	9	5.26	42 9.42	14.23	3.26	5 23.32	49 36.9
4	9							44.1	6 2.51	23.59	0.00	VII.	4	7.0	18 28.55	14.23	1.53	6 26.10	25 54.3
5	10			15.0	28.6				7 28.55	23.59	0.00	.	5	9.1	24 29.09	14.13	1.95	7 52.14	31 55.2
6	9				17.8				8 17.65	23.59	−0.01	.	3	8.25	14 12.74	14.10	1.22	8 41.23	21 38.1
7	9					50.6			8 36.68	23.59	−0.01	.	1	11.58	6 2.04	14.07	0.62	9 0.26	13 26.7
8	8			39.7	53.0				9 53.16	23.59	0.00	.	7	6.48	32 23.13	14.00	2.54	10 16.75	39 49.7
9	10				47.0				10 46.85	23.60	0.00	.	6	4.24	26 10.52	13.95	2.08	11 10.45	33 36.6
10	11				3.8				14 3.65	23.60	+0.01	.	8	7.36	37 46.01	13.78	2.94	14 27.26	45 12.8
11	9			10.7	24.2				15 24.18	23.60	0.00	.	5	4.16	22 5.39	13.70	1.79	15 47.78	29 30.9
12	9		18.7	32.3					16 45.97	23.60	−0.01	.	3	1.53	10 55.07	13.64	0.97	17 9.56	18 19.7
13	8				17.2	31.4			17 17.30	23.60	0.00	.	4	12.00	21 0.15	13.60	1.71	17 40.90	28 25.5
14	9	35.4	49.6						19 17.02	23.60	−0.01	.	3	8.47	14 23.84	13.49	1.23	19 40.61	21 48.6
15	5			32.7	46.3				19 46.22	23.60	0.00	.	4	9.54	19 56.62	13.46	1.63	20 9.82	27 21.7
16	9	55.9	10.3						22 37.81	23.60	0.00	.	7	7.12	32 35.23	13.32	2.56	23 1.41	40 1.1
17	10	3.6	17.5						23 45.12	23.60	0.00	.	4	7.22	18 39.97	13.27	1.55	24 8.72	26 4.8
18	9			29.3					24 42.77	23.60	−0.01	.	3	3.57	11 57.60	13.20	1.05	25 6.36	19 21.9
19	9							26.0	24 44.57	23.60	0.00	VII.	7	2.33	30 14.22	13.20	2.38	25 8.17	37 39.8
20	9			37.3		4.9			27 50.81	23.61	−0.01	.	1	5.31	2 46.90	13.03	0.38	28 14.41	10 10.3
21	8	56.3		23.6					32 37.65	23.61	0.00	.	6	9.54	28 56.92	12.80	2.28	33 1.26	36 22.0
22	9						7.6	21.6	32 40.07	23.61	−0.01	VII.	3	6.11	13 4.84	12.80	1.13	33 3.67	20 28.8
23	10				16.0				38 15.85	23.62	−0.01	.	3	7.6	13 32.91	12.53	1.15	38 39.46	20 56.6
24	10					51.			39 37.26	23.62	+0.01	.	9	5.39	41 46.04	12.47	3.25	40 0.89	49 11.8
25	10						58.6		40 31.10	23.62	0.00	VI.	7	9.34	33 46.69	12.42	2.66	40 54.72	41 11.8
26	10				47.6				44 47.45	23.62	0.00	.	4	5.48	17 52.57	12.23	1.48	45 11.07	25 16.3
27	10						8.8		21 47 41.27	+23.62	+0.01	.	9	4.33	−41 12.76	−12.09	−3.21	21 48 4.90	−22 48 39.1

CORRECTIONS.

Date.	Corr. of Clock.	Hourly rate.	m	n	c
1848. h.	s.	s.	s.	s.	s.

INSTRUMENT READINGS.

Date.	Barom.	THERMOM. At.	Ex.
1848. h. m.	in.	°	°

REMARKS.

(175) 17. Transit over T. V assumed as recorded over T. VII, and minutes as 59 instead of 58.
(176) 3. Minutes assumed as 4 instead of 5, and micrometer reading as 6ʳ.26 instead of 5ʳ.26.

ZONE 176. JULY 19. K. BELT, −22° 31′. D₀ = −22° 7′ 10″.

No.	Mag.	SECONDS OF TRANSIT. I.	II.	III.	IV.	V.	VI.	VII.	T.	a_1	a_4	MICROMETER.			i	d_1	d_2	Mean Right Ascension, 1850.0.	Mean Declination, 1850.0.
									h. m. s.	s.	s.			r.	′ ″	″	″	h. m. s.	° ′ ″
28	10	..	..	..	..	..	39.8	..	21 53 12.27	+23.63	+0.01	.	9	5.50	−41 51.58	−11.83	−3.26	22 53 35.91	−22 49 16.7
29	9	44.2	58.2	..	..	..	..	..	55 25.81	23.63	0.00	.	5	5.44	22 49.76	11.74	1.83	55 40.44	30 13.3
30	9	25.8	39.8	..	..	..	..	..	59 7.31	23.64	−0.01	.	3	3.55	11 56.59	11.57	1.03	59 30.94	19 19.2
31	9	..	31.3	..	57.9	..	..	..	21 59 58.30	+23.64	0.00	.	4	5.43	−17 50.05	−11.53	−1.46	22 0 21.94	−22 25 13.0

ZONE 177. JULY 20. P. BELT, −26° 53′. D₀ = −26° 31′ 30″.

No.	Mag.	I.	II.	III.	IV.	V.	VI.	VII.	T.	a_1	a_4	MICROMETER.			i	d_1	d_2	Mean Right Ascension, 1850.0.	Mean Declination, 1850.0.
1	9	..	..	48.1	..	15.8	..	..	17 25 1.69	+23.70	−0.01	V.	1	12.55	− 6 30.74	− 6.93	−0.33	17 25 25.38	−26 38 8.00
2	9	..	..	..	17.8	..	45.5	..	27 17.61	23.69	0.00	VI.	6	8.275	28 13.13	7.33	2.30	27 41.30	59 52.76
3	8	..	..	6.2	20.1	34.1	..	..	29 19.99	23.69	0.00	V.	4	7.37	18 47.50	7.67	1.44	29 43.68	26 50 26.61
4	10	..	..	..	4.5	18.5	..	47.4	31 44.45	23.69	+0.01	V.	10	5.54	46 54.46	7.97	4.05	31 28.15	27 18 36.48
5	10	..	..	..	..	..	..	2.5	32 21.20	23.68	−0.01	VII.	3	8.18	14 8.82	8.19	1.00	32 44.87	26 45 48.01
6	10	..	31.2	44.3	58.1	..	..	..	34 58.57	23.68	+0.01	IV.	8	7.375	37 46.79	8.63	3.18	35 22.26	27 9 28.60
7	9	..	..	..	15.2	29.2	43.2	..	36 15.17	23.68	0.00	V.	5	5.125	22 33.83	8.86	1.78	36 38.85	26 54 14.47
8	11	..	..	..	..	..	..	58.1	37 16.89	23.68	0.00	VII.	5	4.11	22 2.47	9.03	1.73	37 40.57	53 43.23
9	8	..	..	..	..	55.4	..	23.4	38 41.78	23.67	0.00	VII.	5	6.39	23 17.11	9.28	1.84	39 5.45	26 54 58.23
10	8	..	..	..	34.5	48.2	2.9	..	40 34.50	23.67	0.00	VI.	6	9.30	25 44.65	9.59	2.34	40 58.17	27 0 26.58
11	9	..	..	18.5	32.5	..	..	..	42 18.40	23.67	0.00	IV.	5	8.31	19 14.77	9.91	1.46	42 42.07	26 50 56.14
12	9	..	..	..	..	59.9	..	28.6	43 46.80	23.66	+0.01	VII.	9	7.58	42 55.73	10.15	3.68	44 10.47	27 14 39.56
13	10	..	..	37.	50.9	5.5	..	..	45 51.01	23.66	0.00	V.	5	4.29	22 11.90	19.52	1.74	46 14.67	26 53 54.16
14	10	..	..	..	..	23.9	37.4	51.9	47 10.07	23.66	0.00	VII.	8	. .	34	10.74	2.84	47 33.73	27
15	11	..	..	50.1	..	18.5	..	46.1	17 49 4.35	+23.66	0.00	VII.	4	3.51	−16 53.18	−11.08	−1.26	17 49 28.01	−26 48 35.52

ZONE 178. JULY 20. P. BELT, −26° 53′. D₀ = −26° 15′ 30″.

No.	Mag.	I.	II.	III.	IV.	V.	VI.	VII.	T.	a_1	a_4	MICROMETER.			i	d_1	d_2	Mean Right Ascension, 1850.0.	Mean Declination, 1850.0.
1	10	..	..	..	58.4	..	25.9	..	17 52 58.07	+23.65	+0.01	VI.	10	8.39	−48 17.54	−19.9	−4.20	17 53 21.73	−27 4 11.6
2	10	..	..	..	44.5	58.1	11.9	..	54 44.13	23.65	0.00	VI.	7	11.49	34 54.74	19.7	2.93	55 7.78	26 50 47.4
3	10	..	..	..	..	..	..	20.5	55 39.29	23.65	0.00	VII.	4	14.1	22 0.76	19.6	1.72	56 2.94	37 52.1
4	11	..	..	..	..	5.5	..	33.5	57 51.90	23.64	0.00	VII.	5	12.41	26 19.63	19.3	2.12	58 15.54	42 11.1
5	10	..	..	..	..	..	32.9	46.5	17 59 5.14	23.64	0.00	VII.	4	6.31	18 13.87	19.1	1.37	59 28.78	34 4.3
6	11	..	..	18.5	..	46.1	59.9	..	18 1 32.14	23.64	0.00	VI.	5	8.16	24 6.24	18.8	1.92	59 55.78	39 57.0
7	10	..	..	..	..	..	..	29.	2 47.64	23.64	−0.01	VII.	2	8.48	9 24.85	18.7	0.56	3 11.27	25 14.1
8	9	45.5	59.8	13.6	..	..	..	..	5 27.55	23.63	0.01	II.	2	10.20	5 12.46	18.3	0.15	5 51.17	21 0.9
9	10	..	..	..	..	9.9	23.4	37.8	5 55.91	23.63	−0.01	III.	1	12.14	6 10.07	18.3	0.26	17 6 19.53	26 21 58.6
10	5	..	..	..	..	29.9	..	58.1	8 16.60	23.63	+0.01	VII.	10	11.3	49 29.92	18.0	4.32	18 8 40.24	27 5 22.2
11	9	..	..	..	..	..	9.5	23.5	9 41.88	23.63	−0.01	VI.	1	13.34	6 50.28	17.8	0.32	10 5.50	26 22 38.4
12	8	55.1	9.4	23.1	..	..	..	..	12 37.15	23.62	0.01	III.	3	5.45	12 52.02	17.4	0.88	13 0.76	28 40.3
13	11	..	..	..	..	..	..	35.9	12 54.60	23.62	0.01	VII.	3	9.21	14 40.58	17.4	1.04	13 18.21	30 29.0
14	10	..	..	..	..	39.5	53.5	7.9	14 25.89	23.62	−0.01	VII.	3	10.23	15 11.85	17.2	1.09	14 49.50	31 0.1
15	8	17.1	31.5	44.9	..	..	..	..	17 59.29	23.61	0.00	VII.	6	7.11	27 4.08	16.7	2.18	18 22.90	42 53.0
16	7	..	..	..	..	..	48.5	3.1	18 21.31	23.61	0.00	VII.	7	11.19	34 39.39	16.7	2.90	18 44.92	26 50 29.0
17	11	..	..	..	..	55.5	..	25.2	20 41.54	23.61	+0.01	VII.	9	12.57	45 26.50	16.4	3.93	21 5.16	27 1 16.8
18	9	..	..	16.1	29.8	43.7	..	..	22 29.73	23.61	0.00	V.	4	9.20	19 39.43	16.1	1.50	22 53.34	26 35 27.0
19	10	..	..	0.5	14.5	28.4	..	..	24 14.45	23.61	+0.01	V.	8	8.43	38 19.78	15.9	3.26	24 38.07	54 8.9
20	11	..	..	11.1	..	..	..	..	26 24.98	23.60	0.00	III.	5	10.30	25 13.94	15.6	2.02	26 48.58	41 1.6
21	9	..	..	..	..	20.	33.4	..	27 5.73	23.60	−0.01	VI.	2	7.22	8 41.70	15.5	0.49	27 29.32	24 27.7
22	9	39.9	54.5	7.9	21.3	..	..	..	18 31 23.03	+23.60	+0.01	IV.	8	5.29	−36 41.99	−15.0	−3.08	18 31 45.64	−26 52 30.1

CORRECTIONS.

Date.		Corr. of Clock.	Hourly rate.	m	n	c
1848.	h.	s.	s.	s.	s.	s.
July 20,	19	+ 20.24	g 0.001			

INSTRUMENT READINGS.

Date.		Barom.	THERMOM. At.	THERMOM. Ex.
1848.	h. m.	in.	°	°

REMARKS.

(177) 11. Transits over T.'s IV and V assumed as recorded over T.'s III and IV, to agree with Arg. Z. 223, 16, and Mural, 1848, June 26. Hor. thread assumed as 4 instead of 5.
(178) 8. Hor. thread assumed as 1 instead of 2.
(178) 15. Micrometer reading assumed as 6r.11 instead of 7r.11.
(178) 17. Transits discordant by nearly 3s. T. V only employed.

ZONE 178. JULY 20. P. BELT, −26° 53′. D₀=−26° 15′ 30″—Continued.

No.	Mag.	Seconds of Transit. I.	II.	III.	IV.	V.	VI.	VII.	T.	a_1	a_2	Micrometer.			i	d_1	d_2	Mean Right Ascension, 1850.0.	Mean Declination, 1850.0.
									h. m. s.	s.	s.			r.	′ ″	″	″	h. m. s.	° ′ ″
23	10	...	...	36.8	50.9	...	...	...	18 32 50.75	+23.60	0.00	IV.	6	9.27	−28 43.30	−14.8	−2.36	18 33 14.35	−26 44 30.5
24	10	...	...	...	...	14.9	28.5	42.9	34 0.98	23.59	−0.01	VII.	1	11.59	6 2.16	14.7	0.25	34 24.56	21 47.1
25	10	...	...	...	48.9	2.8	16.9	...	35 48.83	23.59	+0.01	VI.	9	9.56	43 25.21	14.4	3.75	36 12.43	59 13.4
26	10	...	...	...	28.5	...	57.4	...	37 28.96	23.59	−0.01	VI.	2	4.54	6 26.57	14.2	0.26	37 52.54	26 22 11.0
27	10	...	...	...	...	12.4	26.1	40.1	38 58.58	23.59	+0.01	VII.	10	2.54	44 53.10	14.0	3.94	39 22.18	27 0 41.0
28	8	...	...	...	...	7.5	21.8	35.4	39 53.93	23.59	+0.01	VII.	9	2.53	40 21.94	13.9	3.46	41 17.53	26 56 9.3
29	6	...	...	...	...	59.1	12.2	26.9	42 45.07	23.58	0.00	VII.	7	9.2	33 30.31	13.5	2.80	43 8.65	49 16.6
30	7	52.1	6.8	20.5	33.5	48.5	2.5	16.5	45 34.40	23.58	−0.01	VII.	3	6.1	12 59.74	13.2	0.88	45 57.97	28 43.8
31	7	41.8	55.4	9.4	...	...	...	...	49 23.42	23.58	−0.01	III.	2	5.24	7 42.33	12.7	0.40	49 46.99	23 25.4
32	10	...	...	...	...	...	...	26.5	49 45.35	23.57	0.00	VII.	6	3.59	25 57.52	12.6	2.09	50 8.92	41 42.2
33	11	...	...	...	11.5	...	...	...	53 11.35	23.57	0.00	IV.	4	5.24	17 40.47	12.2	1.32	53 34.92	33 24.0
34	8	...	4.1	17.4	31.5	...	...	...	55 31.44	23.57	−0.01	IV.	1	11.40	5 52.97	11.9	0.23	55 55.00	21 35.1
35	10	...	...	...	...	...	52.5	7.1	56 25.21	23.57	−0.01	VII.	2	15.6	12 35.44	11.8	0.85	56 48.77	28 18.1
36	10	...	...	...	15.5	...	43.4	...	18 58 15.44	23.56	0.00	V.	4	7.8	18 32.87	11.5	1.40	18 58 39.00	34 15.8
37	10	...	...	57.4	...	...	...	...	19 0 11.47	23.56	+0.01	VI.	8	6.20	37 7.54	11.3	3.14	19 0 35.04	52 52.0
38	10	...	...	...	...	...	...	49.9	0 8.82	23.56	0.00	VII.	7	4.41	30 48.44	11.3	2.55	0 32.38	26 46 32.3
39	10	...	...	0.9	...	27.4	...	...	4 14.32	23.56	+0.01	V.	10	14.59	51 29.26	10.8	4.53	4 37.89	27 7 14.6
40	9	...	43.9	...	21.4	...	...	...	6 21.62	23.56	+0.01	IV.	7	16.7	36 34.73	10.5	3.08	6 45.19	26 52 18.3
41	10	...	36.4	...	4.8	...	...	...	9 4.46	23.55	−0.01	IV.	1	11.34	5 49.95	10.2	0.21	9 28.00	21 30.4
42	11	...	...	...	...	...	55.4	9.9	10 28.13	23.55	0.00	VII.	6	7.2	27 29.80	10.0	2.24	10 51.68	43 12.0
43	9	...	...	...	...	20.5	33.	48.5	12 6.23	23.55	+0.01	VII.	2	11.43	10 53.09	9.8	0.68	12 29.77	26 33.6
44	9	22.5	36.8	50.1	3.9	...	...	...	17 4.31	23.55	0.00	IV.	4	11.285	21 4.43	9.1	1.61	17 27.86	36 45.1
45	11	...	...	...	...	...	...	29.5	17 48.19	23.55	−0.01	VII.	3	7.575	13 58.48	9.0	0.98	18 11.73	29 38.5
46	10	...	...	...	...	...	...	37.4	26 56.21	23.54	0.00	VII.	5	6.29	23 12.06	7.9	1.83	27 19.75	38 51.8
47	11	...	...	...	...	18.9	...	47.9	29 5.96	23.54	+0.01	VII.	9	9.125	43 33.30	7.6	3.78	29 29.51	59 14.7
48	11	...	38.5	...	6.1	...	...	...	32 6.27	23.53	0.00	IV.	6	7.53	27 55.90	7.2	2.25	32 29.80	43 35.4
49	10	...	...	...	...	...	15.5	29.9	32 48.20	23.53	0.00	VII.	7	5.26	31 41.39	7.1	2.63	33 11.73	47 21.1
50	10	...	...	...	50.1	...	18.5	...	34 50.26	23.53	0.00	VI.	6	7.47	32 52.71	6.8	2.76	35 13.79	48 32.3
51	11	...	...	...	...	52.5	...	...	36 38.50	23.53	0.00	V.	7	9.7	33 33.18	6.6	2.81	37 2.03	49 12.6
52	9	...	...	...	...	...	...	7.1	37 26.07	23.53	+0.01	VII.	7	12.38	35 19.22	6.5	2.98	37 49.61	26 50 58.7
53	10	...	...	...	...	...	11.9	26.1	38 44.55	23.53	+0.01	VII.	9	14.38	45 47.17	6.3	4.02	39 8.09	27 1 27.5
54	10	...	...	...	...	...	50.5	4.8	40 23.13	23.53	0.00	VII.	6	8.395	28 18.97	6.1	2.31	40 46.66	26 43 57.4
55	11	...	45.5	59.5	13.8	...	...	...	43 13.47	23.53	0.00	IV.	4	5.45	17 51.06	5.7	1.31	43 37.00	26 33 28.1
56	8	...	...	21.5	...	...	...	...	44 35.75	23.52	+0.01	III.	10	10.14	49 5.57	5.5	4.32	44 59.28	27 4 45.4
57	9	...	...	...	...	...	18.5	33.0	44 51.29	23.52	+0.01	VII.	9	5.51	41 51.69	5.5	3.61	45 14.82	26 57 30.8
58	5	...	...	...	...	...	...	56.9	46 15.76	23.52	0.00	VII.	6	4.1	25 58.53	5.3	2.08	46 39.28	41 35.9
59	10	...	...	...	57.4	...	...	25.4	47 57.38	23.52	0.00	VII.	5	3.41	21 47.34	5.1	1.69	48 20.90	37 24.1
60	6	...	...	24.9	39.4	...	...	...	19 50 38.98	+23.52	0.00	V.	5	1.31	−20 42.14	−4.7	−1.59	19 51 2.50	−26 36 18.4

ZONE 179. JULY 20. P. BELT, −22° 31′. D₀=−22° 8′ 10″.

No.	Mag.	I.	II.	III.	IV.	V.	VI.	VII.	T.	a_1	a_2	Micrometer.			i	d_1	d_2	Mean Right Ascension, 1850.0.	Mean Declination, 1850.0.
1	9	...	...	55.4	...	21.9	...	...	20 13 55.12	+23.45	0.00	V.	4	5.14	−17 35.40	−0.27	−1.47	20 13 18.57	−22 25 47.14
2	10	...	43.4	56.1	9.8	...	...	...	15 9.86	23.45	0.00	IV.	5	7.11	23 33.62	+0.01	1.96	15 33.31	31 45.57
3	10	...	...	10.5	23.8	...	...	...	16 23.66	23.45	−0.01	IV.	2	5.51	0 25.71	0.17	0.78	16 47.10	8 36.32
4	10	...	...	4.9	18.5	...	...	...	18 18.28	23.45	0.00	IV.	4	9.35	18 46.53	0.43	1.55	18 41.73	26 57.65
5	10	...	...	...	...	8.2	21.5	35.5	18 54.65	23.45	0.00	VII.	4	15.2	22 31.58	0.51	1.81	19 18.10	30 42.88
6	11	8.1	22.4	...	...	...	...	...	21 49.21	23.45	0.00	II.	5	9.315	24 44.32	0.90	1.98	22 12.66	32 55.40
7	11	...	...	...	...	...	...	24.5	20 21 43.85	+23.45	0.00	VII.	6	9.15	−28 36.92	+0.89	−2.26	20 22 7.30	−22 36 48.29

CORRECTIONS.

Date.		Corr. of Clock.	Hourly rate.	m	n	c
1848.	h.	s.	s.	s.	s.	s.

INSTRUMENT READINGS.

Date.		Barom.	Thermom. At.	Thermom. Ex.
1848.	h. m.	in.	°	°

REMARKS.

(178) 25. Micrometer reading assumed as 8r.56 instead of 9r.56.
(178) 26. Micrometer reading assumed as 2r.54 instead of 4r.54.
(178) 27. Micrometer reading assumed as 1r.54 instead of 2r.54.
(178) 28. Minutes assumed as 40 instead of 39.
(178) 38. Micrometer reading assumed as 3r.41 instead of 4r.41.
(178) 40. Transit observations discordant by 10s. T. IV assumed as correct, and micrometer reading assumed 15r.7 instead of 16r.7.
(178) 43. Transit observations discordant.
(178) 44. Micrometer reading assumed as 12r.8¼ instead of 11r.28¼.
(178) 50. Hor. thread assumed as 7 instead of 6.
(178) 53. Micrometer reading assumed as 13r.38 instead of 14r.38.
(178) 59. Observations in right ascension incongruous. Transit over T. VI assumed as at 25s.4 instead of over T. VII at 5s.4.
(179) 1. Minutes assumed as 12 instead of 13, and transits over T.'s IV and VI as recorded over T.'s III and V.
(179) 3. Micrometer reading assumed as 0 0r.51 instead of 2 5r.51.

ZONE 179. JULY 20. P. BELT, −22° 31′. D₀ = −22° 8′ 10″—Continued.

No.	Mag.	I.	II.	III.	IV.	V.	VI.	VII.	T.	a_1	a_2	MICROMETER.			i	d_1	d_2	Mean Right Ascension, 1850.0.	Mean Declination, 1850.0.
									h. m. s.	s.	s.			r.	′ ″	″	″	h. m. s.	° ′ ″
8	10	. .	. .	. .	. .	. .	29.9	43.5	20 23 2.90	+23.45	0.00	VII.	7	4.315	−31 13.97	+ 1.06	−2.45	20 23 26.35	−22 39 25.36
9	9	. .	. .	. .	. .	. .	48.4	2.5	24 21.69	23.45	0.00	VII.	8	4.1	35 57.29	1.23	2.80	24 45.14	44 8.86
10	11	. .	30.1	42.9	. .	. .	. .	. .	26 56.76	23.45	0.00	III.	6	4.7	26 1.92	1.57	2.07	27 20.21	31 12.42
11	10	. .	21.9	35.9	48.4	. .	. .	. .	28 48.72	23.45	−0.01	IV.	2	5.45	7 52.96	1.82	0.78	29 12.16	16 1.92
12	10	. .	. .	10.5	24.6	37.4	. .	. .	32 13.92	23.45	−0.01	V.	1	1.2	0 31.23	2.29	0.24	31 47.36	8 39.18
13	10	. .	. .	46.8	0.1	14.5	. .	. .	34 0.34	23.45	0.00	V.	5	3.14	21 34.09	2.50	1.75	34 23.79	29 43.34
14	10	. .	. .	. .	. .	. .	15.5	29.5	34 48.75	23.45	+0.01	VI.	8	10.23	39 10.09	2.60	3.04	35 12.21	47 20.53
15	10	. .	. .	. .	. .	. .	40.5	. .	36 23.51	23.44	+0.01	VI.	8	11.12	39 34.80	2.78	3.07	36 36.96	47 45.09
16	8	. .	. .	. .	. .	. .	49.8	3.5	37 22.87	23.44	0.00	VII.	7	10.115	34 5.41	2.93	2.67	37 46.31	42 15.15
17	10	. .	. .	. .	. .	. .	48.9	2.6	39 22.05	23.44	0.00	VII.	6	10.29	29 14.24	3.19	2.31	39 45.49	37 23.36
18	9	. .	. .	. .	. .	. .	. .	25.5	40 44.71	23.44	0.00	VII.	4	7.185	18 37.87	3.36	1.54	41 8.15	26 46.05
19	11	. .	. .	. .	. .	. .	. .	49.8	42 22.87	23.44	0.00	VII.	4	10.55	20 27.05	3.54	1.66	42 46.31	28 35.17
20	11	. .	. .	4.9	17.9	31.8	. .	. .	44 18.06	23.44	0.00	V.	5	6.12	23 3.85	3.82	1.86	44 41.52	31 11.89
21	10	. .	. .	32.4	. .	59.9	. .	. .	45 46.09	23.44	0.00	V.	6	5.48	31 52.83	4.00	2.50	46 9.53	40 1.33
22	9	. .	28.4	. .	54.8	8.2	. .	. .	50 54.83	23.45	−0.01	V.	1	6.3	3 3.01	4.65	0.42	51 18.27	11 8.79
23	10	. .	. .	18.5	. .	45.5	. .	. .	53 32.12	23.45	+0.01	V.	10	4.0	45 56.99	4.98	3.54	53 55.58	54 5.55
24	11	. .	. .	46.4	. .	13.5	. .	. .	55 59.76	23.45	0.00	V.	3	5.39	12 49.01	5.29	1.12	56 23.21	20 54.84
25	9	. .	39.9	. .	. .	6.7	. .	. .	20 58 6.86	23.45	+0.01	V.	10	7.56	47 55.99	5.55	3.68	20 58 30.32	56 4.12
26	12	. .	31.9	. .	. .	11.5	. .	38.5	21 1 58.14	23.45	0.00	VII.	3	7.35	13 47.20	6.04	1.18	21 2 21.59	21 52.34
27	9	20.9	34.5	47.4	. .	. .	. .	. .	4 1.55	23.45	+0.01	III.	9	5.00	41 26.34	6.29	3.20	5 25.01	49 33.25
28	9	. .	. .	49.9	3.4	16.9	. .	. .	6 3.25	23.45	0.00	V.	4	5.37	17 47.00	6.55	1.48	6 26.70	25 51.93
29	11	. .	. .	. .	. .	42.9	. .	9.9	7 29.25	23.45	0.00	VII.	5	7.40	23 47.92	6.73	2.91	7 52.70	31 54.10
30	9	. .	. .	. .	. .	. .	. .	58.9	8 18.04	23.45	0.00	VII.	3	7.21	13 40.13	6.84	1.18	8 41.49	21 44.47
31	10	. .	. .	. .	. .	7.2	. .	34.6	9 53.83	23.45	0.00	VII.	7	5.21	31 38.92	7.04	2.49	10 17.28	39 44.37
32	11	. .	. .	. .	. .	54.5	. .	20.9	11 40.35	23.45	−0.01	VII.	1	0.32	0 15.81	7.26	0.22	12 3.79	8 18.77
33	11	. .	. .	. .	4.9	17.6	31.2	. .	14 4.35	23.46	0.00	VI.	8	6.11	37 3.02	7.56	2.89	14 27.81	45 8.35
34	10	. .	. .	. .	24.5	38.5	51.5	5.9	15 24.74	23.46	0.00	VII.	4	10.53	20 26.04	7.72	1.66	15 48.20	28 29.98
35	8	. .	. .	. .	18.4	. .	44.9	58.9	17 18.12	23.46	0.00	VII.	4	10.37	20 17.98	7.96	1.65	17 41.58	28 21.67
36	5	19.5	32.5	46.4	59.9	. .	. .	. .	19 46.23	23.46	0.00	IV.	4	8.29	19 13.76	8.30	1.57	20 9.69	27 17.03
37	11	. .	. .	. .	. .	3.9	16.4	30.5	20 49.90	23.46	0.00	VII.	7	6.23	32 10.19	8.40	2.53	21 13.36	40 14.32
38	8	. .	. .	23.6	36.8	50.5	. .	. .	22 36.88	23.46	0.00	V.	6	6.8	32 2.93	8.62	2.52	23 0.34	40 6.83
39	11	. .	. .	. .	. .	58.4	11.5	25.	23 44 53	23.46	0.00	VII.	4	6.21	18 8.88	8.75	1.50	24 7.99	26 11.63
40	10	. .	. .	. .	. .	. .	. .	26.5	24 45.87	23.46	0.00	VII.	6	11.4	29 31.89	8.88	2.33	25 9.33	37 35.34
41	11	. .	. .	. .	18.5	32.6	. .	. .	26 18.72	23.46	0.00	V.	7	12.57	35 29.16	9.07	2.78	26 42.18	43 32.87
42	9	. .	. .	37.9	. .	4.9	17.9	. .	27 51.09	23.46	−0.01	VI.	1	4.00	2 0.87	9.26	0.34	28 14.54	10 1.95
43	9	. .	49.5	2.9	15.9	. .	. .	. .	30 16.14	23.46	0.00	IV.	3	10.30	15 15.78	9.56	1.29	30 39.60	23 17.51
44	10	. .	. .	. .	. .	. .	29.1	43.5	31 2.56	23.46	+0.01	VII.	8	15.35	41 47.22	9.64	3.24	31 26.03	49 50.82
45	8	. .	. .	. .	. .	. .	4.5	18.4	32 37.65	23.46	0.00	VII.	6	8.32	28 15.25	9.83	2.23	33 1.11	36 17.65
46	11	48.9	3.1	16.1	29.4	42.1	. .	. .	36 29.44	23.47	0.00	V.	5	11.9	25 33.61	10.29	2.04	36 52.91	33 35.36
47	10	. .	3.4	16.4	30.1	. .	. .	. .	38 29.99	23.47	0.00	IV.	3	5.41	12 50.04	10.52	1.11	38 53.46	20 50.63
48	10	. .	. .	. .	38.1	51.9	. .	. .	39 38.12	23.47	0.00	V.	3	7.31	13 45.48	10.65	1.18	40 1.59	21 46.01
49	10	. .	. .	. .	. .	. .	58.9	12.4	40 31.86	23.47	0.00	VII.	7	8.7	33 2.64	10.76	2.59	40 55.33	41 4.47
50	11	. .	. .	. .	33.5	. .	0.1	14.2	42 33.31	23.47	0.00	VII.	4	7.435	18 50.48	10.99	1.54	42 56.78	26 51.03
51	11	. .	33.5	47.5	. .	. .	. .	. .	44 0.80	23.47	0.00	III.	6	11.54	29 57.40	11.15	2.36	44 24.27	37 58.61
52	10	. .	15.9	. .	41.9	. .	. .	. .	47 42.44	23.48	+0.01	IV.	8	13.5	40 31.92	11.58	3.15	48 5.93	48 33.49
53	11	. .	0.1	. .	26.4	40.	53.9	. .	53 26.77	23.48	+0.01	VI.	8	14.25	41 12.10	12.22	3.20	53 50.26	49 13.08
54	8	. .	. .	11.4	24.9	. .	. .	. .	55 24.75	23.48	0.00	IV.	4	14.11	22 6.19	12.43	1.78	55 48.23	30 5.54
55	9	. .	. .	14.5	28.9	43.4	. .	. .	58 28.98	23.49	+0.01	V.	10	12.25	50 11.62	12.77	4.24	21 58 52.48	58 13.09
56	11	. .	. .	. .	. .	12.5	24.6	39.9	21 59 58.55	+23.49	0.00	VII.	3	12.325	−17 17.72	+12.94	−1.43	22 0 22.04	−22 25 16.21

CORRECTIONS.

Date.		Corr. of Clock.	Hourly rate.	m	n	c
1848.	h.	s.	s.	s.	s.	s.

INSTRUMENT READINGS.

Date.		Barom.	THERMOM. At.	THERMOM. Ex.
1848.	h. m.	In.	°	°

REMARKS.

(179) 8. Double.
(179) 12. Minutes assumed as 31 instead of 32.
(179) 19. Transits over T. VI assumed as recorded over T. VII.
(179) 21. Hor. thread assumed as 7 instead of 6.
(179) 25. Record of one of the transit threads misplaced. T. IV assumed to have been recorded as T. V. If III was recorded as II the seconds of T. are 53ˢ.43.
(179) 27. Minutes assumed as 5 instead of 4.
(179) 36. Transits over T.'s II–V assumed as recorded over T.'s I–IV.
(179) 38. Micrometer thread assumed as 7 instead of 6.
(179) 56. Micrometer reading assumed as 14ʳ.325 instead of 12ʳ.325.

ZONE 179. JULY 20. P. BELT, −22° 31′. D_0 = −22° 8′ 10″—Continued.

No.	Mag.	SECONDS OF TRANSIT. I.	II.	III.	IV.	V.	VI.	VII.	T.	a_1	a_2	MICROMETER.			i	d_1	d_2	Mean Right Ascension, 1850.0.	Mean Declination, 1850.0.
									h. m. s.	s.	s.			r.	′ ″	″	″	h. m. s.	° ′ ″
57	9	. .	12.4	24.6	38.4	51.9	. .	. .	22 2 38.68	+23.49	+0.01	VI.	10	9.27	−48 41.75	+13.22	−3.75	22 3 2.18	−22 56 42.28
58	10	. .	. .	. .	. .	25.9	. .	53.9	4 12.59	23.49	−0.01	VI.	1	10.52	5 28.62	13.39	0.59	4 36.07	13 25.82
59	9	. .	7.4	20.5	34.5	47.8	. .	. .	10 34.09	23.50	−0.01	V.	1	5.2	2 32.25	14.07	0.37	9 57.58	10 28.55
60	9	. .	53.4	6.2	19.5	. .	46.9	. .	11 19.91	23.50	+0.01	VI.	8	8.4	38 0.01	14.15	2.96	11 43.42	45 58.82
61	6	. .	. .	. .	56.5	9.9	23.5	. .	12 56.41	23.50	0.00	V.	3	5.52	12 55.56	14.32	1.12	13 19.91	20 52.36
62	7	. .	11.2	24.5	37.4	51.2	. .	. .	15 37.92	23.51	0.00	V.	3	6.38	13 18.76	14.60	1.15	16 1.43	21 15.31
63	10	. .	. .	. .	. .	13.5	26.4	40.1	17 59.57	23.51	0.00	VII.	4	6.51	13 25.02	14.85	1.12	17 23.08	21 21.29
64	9	. .	. .	. .	13.1	26.2	39.5	53.5	19 12.75	23.51	0.00	VII.	6	11.54	29 57.10	14.98	2.36	19 36.26	37 54.48
65	9	. .	. .	. .	. .	. .	. .	50.5	20 10.16	23.51	+0.01	VII.	10	16.16	42 7.90	15.08	3.29	20 33.68	50 6.11
66	10	. .	. .	53.5	. .	20.5	34.9	47.9	23 7.30	23.52	0.00	VII.	6	15.44	31 53.06	15.37	2.51	23 30.82	39 50.20
67	10	. .	. .	4.5	17.9	31.6	. .	. .	25 18.02	23.52	+0.01	V.	9	10.48	44 21.82	15.60	3.44	25 41.55	52 19.66
68	10	. .	. .	26.1	40.5	53.5	. .	. .	27 39.89	23.52	0.00	V.	4	7.52	18 55.07	15.83	1.55	28 3.41	26 50.79
69	9	. .	. .	. .	0.9	. .	27.4	. .	29 0.61	23.52	0.00	VI.	5	12.43	26 20.88	15.97	2.09	29 24.13	34 17.00
70	10	. .	. .	. .	. .	. .	35.6	49.9	22 30 6.89	+23.52	0.00	.	4	6.7	−18 2.16	+16.08	−1.49	22 30 32.41	−22 25 57.57

ZONE 180. JULY 24. P. BELT, −21° 16′. D_0 = −20° 52′ 20″.

No.	Mag.	SECONDS OF TRANSIT. I.	II.	III.	IV.	V.	VI.	VII.	T.	a_1	a_2	MICROMETER.			i	d_1	d_2	Mean Right Ascension, 1850.0.	Mean Declination, 1850.0.
1	12	. .	. .	. .	. .	18.5	. .	. .	19 18 4.96	+23.28	−0.01	V.	1	9.205	− 4 42.59	− 8.1	−0.69	19 18 28.23	−20 57 11.4
2	11	45.5	. .	. .	25.4	39.1	52.5	6.1	20 25.79	23.28	+0.01	VI.	9	13.41	45 48.93	8.0	3.37	20 49.08	21 38 20.3
3	10	. .	. .	. .	. .	21.6	. .	48.5	22 8.19	23.28	0.00	VII.	6	8.31	28 14.77	8.0	2.21	22 31.47	20 45.0
4	12	. .	18.5	31.5	44.9	. .	. .	. .	24 44.94	23.28	0.00	IV.	5	4.55	22 25.05	7.9	1.83	25 8.22	14 54.8
5	8	. .	53.5	7.4	. .	. .	. .	. .	26 20.39	23.27	−0.01	III.	3	6.30	13 14.73	7.9	1.24	26 43.65	5 43.9
6	11	. .	. .	. .	. .	. .	. .	16.5	26 36.10	23.27	0.00	VII.	5	4.18	22 6.09	7.9	1.81	26 59.37	14 35.8
7	11	. .	. .	. .	. .	. .	56.9	10.9	28 30.32	23.27	−0.01	VII.	3	12.11	16 6.39	7.8	1.43	28 53.58	8 35.6
8	9	. .	34.5	46.4	. .	. .	. .	. .	30 0.74	23.27	+0.01	III.	9	13.19	45 37.95	7.8	3.36	30 24.02	38 9.1
9	10	. .	. .	. .	. .	. .	1.5	15.5	31 34.93	23.27	0.00	VII.	4	6.6	18 1.36	7.8	1.54	31 58.20	10 30.7
10	11	. .	. .	. .	. .	. .	. .	15.9	32 35.45	23.27	0.00	VII.	4	6.5	18 0.85	7.7	1.54	32 58.72	10 30.1
11	12	. .	. .	. .	30.1	43.4	. .	. .	34 29.99	23.26	+0.01	V.	8	7.56	37 56.09	7.7	2.85	34 53.26	30 36.6
12	10	43.5	57.4	10.5	. .	. .	. .	. .	37 23.99	23.26	0.00	III.	5	6.18	23 6.87	7.6	1.87	37 47.25	15 36.3
13	10	. .	. .	. .	. .	. .	. .	45.	37 24.12	23.26	0.00	.	5	6.18	23 6.90	7.6	1.87	37 47.38	15 36.4
14	8	. .	. .	. .	. .	59.8	12.6	26.6	38 46.18	23.26	0.00	.	6	5.2	26 29.68	7.6	2.10	39 9.44	18 59.4
15	11	. .	. .	. .	31.5	45.5	58.6	. .	40 31.76	23.26	−0.01	VI.	3	8.28	14 14.11	7.5	1.30	40 55.01	6 42.9
16	11	. .	. .	. .	6.8	20.5	33.6	. .	42 6.88	23.26	+0.01	VI.	8	9.23	38 39.84	7.5	2.90	42 30.15	31 10.2
17	10	. .	. .	. .	38.9	. .	. .	. .	43 38.75	23.26	0.00	IV.	7	10.6	34 2.98	7.5	2.59	44 2.01	21 26 33.1
18	12	. .	. .	. .	. .	. .	. .	29.9	47 49.29	23.25	−0.01	VII.	1	12.2	6 3.76	7.4	0.77	48 12.53	20 58 31.9
19	10	. .	. .	57.5	. .	24.6	. .	. .	51 10.95	23.25	0.00	V.	5	6.25	23 10.40	7.3	1.88	51 34.20	21 15 39.6
20	10	. .	35.4	. .	. .	. .	. .	. .	53 2.35	23.25	+0.01	II.	8	6.58	37 26.73	7.2	2.83	53 25.61	29 56.8
21	10	. .	. .	. .	46.1	59.8	. .	. .	53 46.14	23.25	0.00	V.	4	4.49	17 22.80	7.2	1.50	54 9.39	21 9 51.5
22	11	. .	. .	22.9	. .	49.5	. .	. .	55 35.94	23.25	−0.01	V.	1	9.15	4 39.82	7.2	0.68	55 59.18	20 57 7.7
23	10	. .	. .	. .	56.4	. .	. .	. .	56 56.25	23.24	0.00	IV.	6	7.13	32 35.73	7.1	2.50	57 19.49	21 25 5.3
24	11	48.1	2.6	15.4	. .	. .	. .	. .	19 58 28.88	23.24	0.00	V.	5	5.56	22 55.78	7.1	1.86	19 58 52.12	15 24.7
25	10	43.9	57.2	. .	. .	. .	. .	. .	20 0 24.08	23.24	−0.01	II.	2	7.34	8 47.79	7.1	0.93	20 0 47.31	21 1 15.8
26	10	. .	. .	. .	. .	. .	41.9	. .	1 15.26	23.24	0.01	VI.	1	4.24	2 12.98	7.0	0.51	1 38.49	20 54 40.6
27	11	. .	. .	. .	. .	. .	41.5	. .	6 14.85	23.24	−0.01	VI.	1	8.5	4 4.42	7.0	0.63	6 38.08	20 56 32.1
28	12	. .	. .	. .	. .	. .	4.9	18.5	7 38.21	23.24	0.00	VII.	6	13.175	30 39.22	6.9	2.39	8 1.45	21 23 8.5
29	9	. .	50.9	3.5	16.4	. .	. .	. .	10 16.98	23.24	0.00	IV.	7	6.27	32 12.54	6.9	2.49	10 40.22	24 41.9
30	11	. .	. .	. .	37.4	50.9	. .	. .	11 37.35	23.23	0.00	V.	5	7.32	23 44.19	6.9	1.91	12 0.58	16 13.0
31	9	. .	. .	10.3	23.9	. .	. .	. .	20 14 23.55	+23.23	−0.01	IV.	2	5.29	− 7 44.89	− 6.8	−1.86	20 14 46.77	−21 0 13.6

CORRECTIONS.

Date.		Corr. of Clock.	Hourly rate.	m	n	c
1848.	h.	s.	s.	s.	s.	s.
July 24,	19	+ 20.16	g 0.001	− 0.41	+ 0.81	0.00

INSTRUMENT READINGS.

Date.		Barom.	THERMOM. At.	Ex.
1848.	h. m.	in.	°	°

REMARKS.

(179) 58. Double.
(179) 59. Minutes assumed as 9 instead of 10.
(179) 63. Minutes assumed as 16 instead of 17, and hor. thread as 3 instead of 4.
(179) 65. Hor. thread assumed as 8 instead of 10.
(180) 23. Hor. thread assumed as 7 instead of 6.

ZONE 180. JULY 24. P. BELT, −21° 16′. D_0 = −20° 52′ 20″—Continued.

No.	Mag.	Seconds of Transit I.	II.	III.	IV.	V.	VI.	VII.	T.	a_1	a_2	MICROMETER.			i	d_1	d_2	Mean Right Ascension, 1850.0.	Mean Declination, 1850.0.
									h. m. s.	s.	s.			r.	′ ″	″	″	h. m. s.	° ′ ″
32	9	..	..	..	..	26.4	39.5	..	20 15 12.69	+23.23	0.00	VI.	7	4.14	−31 5.34	−6.8	−2.42	20 15 36.12	−21 23 34.6
33	11	..	..	..	..	..	45.9	..	16 19.14	23.23	+0.01	.	10	3.51	45 52.47	6.8	3.40	16 42.38	38 22.7
34	11	..	..	..	..	57.5	..	..	17 44.07	23.23	0.00	V.	6	2.0	24 57.88	6.7	2.00	18 7.30	17 26.6
35	12	..	..	..	..	36.6	..	..	19 23.10	23.23	−0.01	V.	2	13.30	11 47.40	6.7	1.12	19 46.32	4 15.2
36	8	..	8.9	21.8	..	..	..	..	21 35.36	23.23	0.00	III.	4	9.32	19 45.50	6.7	1.65	21 58.59	12 13.9
37	11	..	52.5	..	19.9	..	47.4	..	24 19.98	23.23	0.00	VI.	7	5.25	31 41.13	6.6	2.46	24 43.21	24 10.2
38	10	34.6	48.4	..	..	..	..	..	27 15.28	23.23	+0.01	II.	8	6.4	36 59.51	6.6	2.81	27 38.52	29 28.9
39	9	..	..	..	..	48.5	..	..	27 21.83	23.23	−0.01	V.	3	6.50	13 24.81	6.6	1.23	27 45.05	5 52.6
40	8	..	53.9	6.1	19.6	..	..	..	29 19.96	23.23	+0.01	IV.	8	8.2	37 59.15	6.6	2.88	29 43.20	30 28.6
41	10	..	..	..	32.6	..	..	..	30 32.45	23.23	+0.01	IV.	7	12.5	35 2.97	6.6	2.68	30 55.69	27 32.3
42	12	..	41.8	..	..	..	..	..	34 8.68	23.22	0.00	VI.	6	8.16	28 7.37	6.5	2.21	34 31.90	20 36.1
43	11	..	..	32.6	45.5	..	..	..	36 45.68	23.22	0.00	IV.	7	8.39	33 19.11	6.5	2.56	37 8.90	25 48.2
44	12	..	..	..	32.1	..	..	12.9	39 32.27	23.22	0.00	VII.	6	6.34	27 15.78	6.5	2.16	39 55.49	19 44.4
45	11	..	..	..	..	..	52.5	..	42 25.74	23.22	+0.01	VII.	9	7.12	42 32.63	6.4	3.18	42 48.97	35 2.2
46	12	1.8	16.1	28.5	..	..	..	..	45 42.48	23.22	+0.01	III.	8	8.49	38 22.82	6.4	2.91	46 5.71	30 52.1
47	9	45.6	57.6	..	..	..	..	..	50 25.13	23.22	−0.01	II.	2	7.36	8 48.80	6.4	1.91	50 48.34	21 1 17.1
48	10	..	..	6.9	..	34.2	..	..	53 20.27	23.22	−0.01	V.	1	5.28	2 45.35	6.3	0.52	53 43.48	20 55 12.2
49	9	..	..	..	54.1	..	..	..	54 53.95	23.22	0.00	IV.	6	7.34	27 46.33	6.3	2.20	55 17.17	21 20 14.8
50	10	..	18.5	..	44.6	..	..	..	56 44.97	23.22	+0.01	IV.	8	12.29	40 13.77	6.3	3.03	57 8.20	32 43.1
51	9	..	55.9	..	22.4	..	48.5	..	20 58 22.24	23.22	−0.01	V.	3	6.49	13 24.31	6.3	1.22	20 58 45.45	5 51.8
52	8	54.5	8.2	21.4	..	..	..	..	21 0 34.82	23.22	0.00	III.	4	3.20	16 37.91	6.3	1.43	21 0 58.04	21 9 5.6
53	9	..	40.9	53.5	7.5	..	..	..	3 7.14	23.22	−0.01	IV.	1	8.5	4 4.56	6.3	0.60	3 30.35	20 56 31.5
54	10	21.5	35.8	..	..	..	..	..	5 2.38	23.22	0.00	II.	7	5.22	31 39.62	6.2	2.47	5 25.60	21 24 8.3
55	6	..	..	..	41.5	..	8.9	..	6 41.77	23.22	0.00	VI.	5	7.25	23 40.54	6.2	1.91	7 4.99	21 16 8.7
56	7	51.9	..	18.8	32.6	45.9	..	..	9 32.21	23.22	−0.01	V.	1	10.14	5 9.58	6.2	0.67	9 55.42	20 57 36.5
57	8	27.4	41.9	54.9	8.	21.5	..	..	12 8.21	23.22	0.00	V.	7	10.55	34 27.65	6.1	2.66	12 31.43	21 26 56.4
58	5	..	..	3.5	15.2	28.5	42.6	55.5	15 15.66	23.22	0.00	.	7	15.14	36 38.26	6.1	2.80	15 38.88	29 7.2
59	9	..	1.9	14.3	..	..	..	..	17 28.40	23.22	+0.01	III.	9	14.3	46 0.13	6.0	3.44	17 51.63	21 38 29.6
60	9	..	..	..	38.9	..	5.5	..	18 38.80	23.22	−0.01	VI.	1	4.47	2 24.58	6.0	0.51	19 2.01	20 54 51.1
61	10	..	8.9	..	34.5	47.4	..	..	24 34.14	23.23	+0.01	V.	9	9.32	44 13.76	6.0	3.31	23 57.38	21 36 43.1
62	9	25.5	..	..	..	..	..	..	41 5.93	23.23	0.00	I.	4	6.14	18 5.39	6.0	1.53	41 29.16	10 32.9
63	10	..	..	..	..	..	..	33.5	40 53.07	23.23	0.00	VII.	4	9.28	19 43.21	6.0	1.64	41 16.30	12 10.9
64	9	..	..	..	..	..	31.2	44.5	42 4.30	23.23	0.00	.	4	13.46	21 53.60	6.0	1.79	42 27.53	14 21.4
65	10	11.4	25.4	38.1	..	..	17.9	..	21 46 51.61	+23.24	0.00	.	3	10.31	−15 16.28	−6.1	−1.36	21 47 14.85	−21 57 43.7

ZONE 181. AUGUST 4. P. BELT, −19° 23′. D_0 = −18° 52′ 30″.

No.	Mag.	I.	II.	III.	IV.	V.	VI.	VII.	T.	a_1	a_2	MICROMETER.			i	d_1	d_2	Mean Right Ascension, 1850.0.	Mean Declination, 1850.0.
1	10	..	..	..	..	..	..	8.5	20 29 28.53	+25.25	−0.01	.	2	4.2	− 7 1.03	−20.9	−0.96	20 29 53.77	−18 59 52.9
2	10	..	..	35.8	..	2.8	..	28.9	31 49.17	25.24	0.00	.	6	3.65	26 0.93	20.5	2.06	32 14.41	19 18 53.5
3	10	..	9.8	22.5	35.4	..	..	..	35 35.86	25.24	+0.01	IV.	9	6.32	42 12.77	19.9	3.02	36 1.11	35 5.7
4	9	..	..	15.9	29.4	..	..	..	37 29.17	25.24	0.00	IV.	6	8.215	28 10.27	19.6	2.19	37 54.41	21 2.1
5	9	..	..	..	..	38.5	51.5	4.9	38 25.08	25.24	0.00	VI.	4	4.8	17 2.03	19.4	1.52	38 50.32	9 53.0
6	9	42.8	56.5	9.4	..	..	..	..	43 22.79	25.23	0.00	III.	6	2.51	25 23.59	18.6	2.02	43 48.02	18 14.2
7	8	..	..	20.9	33.5	47.4	..	..	44 33.97	25.23	+0.01	IV.	10	7.39	47 47.45	18.4	3.36	44 59.21	40 39.2
8	8	..	..	..	48.8	2.9	15.4	..	45 49.10	25.23	0.00	.	6	9.15	28 37.25	18.2	2.22	46 14.33	21 27.7
9	11	..	..	3.9	..	29.9	..	..	48 16.86	25.22	0.00	.	6	9.3	28 31.20	17.8	2.21	48 42.08	21 21.2
10	9	..	40.9	..	6.8	..	33.1	..	50 6.98	25.22	+0.01	.	9	5.22	41 37.46	17.5	2.98	50 32.21	34 27.9
11	10	..	4.9	..	30.5	..	56.7	..	20 52 30.72	+25.22	+0.01	.	8	6.37	−37 16.29	−17.1	−2.72	20 52 55.95	−19 30 6.1

CORRECTIONS.

Date.		Corr. of Clock.	Hourly rate.	m	n	c
1848.	h.	s.	s.	s.	s.	s.
Aug. 4.	19	+ 22.04	/ 0.000			

INSTRUMENT READINGS.

Date.		Barom.	THERMOM. At.	THERMOM. Ex.
1848.	h. m.	in.	°	°

REMARKS.

(180) 39. Transit over T. VI assumed as recorded over T. V.
(180) 61. Minutes assumed as 23 instead of 24, and micrometer readings as 10r.32 instead of 9r.32.
(181) 1. Micrometer reading uncertain 10′ discordant from Argelander 252, 140; wrong.
(181) 2. Micrometer reading uncertain 1′ discordant from Argelander 243, 93.

ZONE 181. AUGUST 4. P. BELT, −19° 23′. D_0 = −18° 52′ 30″—Continued.

No.	Mag.	SECONDS OF TRANSIT. I.	II.	III.	IV.	V.	VI.	VII.	T.	a_1	a_2	MICROMETER.			i	d_1	d_2	Mean Right Ascension, 1850.0.	Mean Declination, 1850.0.
									h. m. s.	s.	s.			r.	′ ″	″	″	h. m. s.	° ′ ″
12	10	29.4	52.9	5.9	18.4	. .	. .	. .	20 55 18.92	+25.22	−0.01	IV.	3	8.17	−14 8.71	−16.7	−1.35	20 55 44.13	−19 6 56.8
13	10	. .	46.4	. .	12.5	. .	38.5	. .	58 12.46	25.21	−0.01	VI.	5	10.32	15 14.87	16.2	1.41	58 37.66	8 2.5
14	9	. .	. .	55.9	. .	. .	22.4	. .	20 59 9.07	25.21	0.00	VI.	5	13.57	26 58.22	16.1	2.12	20 59 34.28	19 46.4
15	11	. .	. .	. .	6.4	. .	32.5	. .	21 2 6.17	25.21	+0.01	VI.	9	9.36	43 45.43	15.6	3.13	21 2 31.39	36 34.2
16	10	40.1	54.4	7.4	20.5	. .	. .	0.0	9 20.41	25.20	0.00	VII.	5	11.57	25 57.57	14.5	2.06	9 45.61	18 44.1
17	10	. .	0.9	13.	25.8	. .	. .	. .	11 26.52	25.20	−0.01	IV.	8	14.10	41 4.69	14.1	2.96	11 51.73	33 51.8
18	10	. .	. .	. .	53.4	. .	19.5	. .	12 53.18	25.20	+0.01	.	6	10.38	29 19.11	13.9	2.26	13 18.38	19 22 5.3
19	10	. .	. .	. .	36.4	49.5	2.8	. .	14 36.28	25.20	0.00	VI.	1	5.2)	2 45.77	13.6	0.70	15 1.47	18 55 30.1
20	10	. .	. .	. .	. .	19.4	32.6	46.4	16 6.40	25.20	+0.01	VII.	9	8.8	43 0.90	13.4	3.06	16 31.61	19 35 47.4
21	9	. .	46.4	. .	. .	. .	. .	. .	18 13.12	25.20	+0.01	II.	10	10.42	49 19.61	13.1	3.44	18 38.53	12 6.2
22	10	. .	15.9	28.3	. .	. .	. .	. .	19 41.98	25.20	0.00	III.	6	12.29	26 13.94	12.8	2.07	20 7.18	18 58.8
23	11	. .	35.9	. .	. .	. .	. .	. .	21 2.28	25.20	−0.01	II.	2	4.54	7 27.13	12.6	0.98	21 27.47	0 10.7
24	10	20.5	35.4	. .	0.3	. .	. .	. .	24 0.87	25.20	0.00	.	5	14.10	27 4.89	12.2	2.13	24 26.07	19 49.2
25	9	. .	. .	. .	. .	. .	. .	5.8	24 25.88	25.20	0.00	.	4	11.55	20 57.63	12.1	1.76	24 51.08	13 41.5
26	10	4.5	18.4	30.5	. .	. .	. .	. .	26 44.43	25.20	0.00	III.	7	8.59	33 29.16	11.8	2.51	27 9.63	26 13.5
27	10	. .	. .	22.9	. .	49.8	. .	. .	27 36.14	25.19	−0.01	III.	2	11.46	10 54.97	11.6	1.16	28 1.32	3 37.7
28	10	. .	28.5	. .	54.5	. .	. .	. .	29 54.70	25.19	0.00	V.	6	8.39	28 19.08	11.3	2.20	30 19.89	21 2.6
29	9	. .	59.5	. .	25.6	. .	. .	. .	31 25.69	25.19	−0.01	IV.	3	7.25	13 42.48	11.0	1.32	31 50.87	6 24.8
30	4	11.5	25.6	38.1	. .	. .	. .	. .	33 51.75	25.19	+0.01	III.	8	12.205	40 9.45	10.7	2.90	34 16.95	32 53.1
31	11	. .	. .	53.9	6.5	20.4	. .	. .	35 6.82	25.19	0.00	IV.	5	7.195	23 37.91	10.5	1.92	35 32.01	16 20.3
32	11	. .	. .	54.9	. .	. .	. .	. .	38 7.92	25.19	0.00	.	4	4.54	17 25.35	10.0	1.55	38 33.11	19 10 6.9
33	5	. .	. .	. .	. .	. .	26.5	. .	39 0.18	25.19	−0.01	VI.	1	3.21	1 41.22	9.9	0.65	39 25.36	18 54 21.8
34	9	. .	. .	. .	. .	49.5	. .	15.5	40 35.81	25.19	−0.01	VII.	3	4.30	12 13.98	9.7	1.25	41 0.99	19 4 54.9
35	6.7	. .	30.9	. .	. .	. .	. .	. .	42 57.44	25.19	0.00	III.	6	4.55	26 26.12	9.3	2.08	43 22.63	19 7.5
36	10	. .	. .	. .	. .	. .	. .	43.4	43 3.75	25.19	0.00	VII.	8	11.30	29 43.76	9.3	2.28	43 28.94	19 22 25.3
37	9	0.5	14.6	26.8	40.5	. .	. .	. .	46 40.43	25.19	−0.01	IV.	1	8.15	4 9.60	8.7	0.78	47 5.61	18 56 49.1
38	10	54.9	9.8	21.9	. .	. .	. .	. .	48 35.60	25.19	+0.01	III.	9	9.39	43 47.03	8.5	3.11	49 0.80	19 36 28.6
39	10	. .	. .	. .	44.5	. .	. .	. .	49 44.35	25.19	0.00	IV.	5	7.21	23 38.66	8.3	1.92	50 9.54	19 16 18.9
40	10	. .	. .	. .	54.9	. .	. .	. .	50 54.75	25.19	−0.01	IV.	1	4.31	2 16.65	8.2	0.69	51 19.93	18 54 55.5
41	9	28.1	41.9	54.5	. .	. .	. .	. .	21 57 8.12	25.19	0.00	III.	7	4.15	31 5.95	7.3	2.37	21 57 33.31	−19 23 45.6
42	7	40.5	54.4	7.1	. .	. .	. .	. .	22 0 20.50	+25.19	0.00	.	5	. .	−22	−6.8	−1.82	22 0 45.69	

ZONE 182. AUGUST 7. P. BELT, −21° 53′. D_0 = −21° 30′ 0″.

No.	Mag.	I.	II.	III.	IV.	V.	VI.	VII.	T.	a_1	a_2	MICROMETER.			i	d_1	d_2	Mean Right Ascension, 1850.0.	Mean Declination, 1850.0.
1	9	. .	. .	24.5	. .	. .	. .	. .	18 25 37.98	+27.81	+0.01	III.	7	12.35	−35 18.07	−23.3	−2.68	18 26 5.80	−22 5 44.1
2	11	. .	. .	49.5	. .	. .	. .	. .	28 3.23	27.80	−0.01	III.	4	7.45	18 51.54	23.1	1.61	28 31.02	21 49 16.3
3	10	20.5	35.6	. .	. .	. .	. .	. .	30 1.97	27.79	+0.01	II.	7	12.28	35 14.42	22.9	2.67	30 29.77	22 5 40.0
4	10	. .	. .	. .	. .	. .	. .	41.5	30 1.11	27.79	+0.01	VII.	7	12.29	35 14.75	22.9	2.67	30 28.91	22 5 40.3
5	11	. .	. .	. .	. .	. .	. .	52.5	31 11.74	27.79	−0.01	VII.	1	14.36	7 21.39	22.8	0.87	31 39.52	21 37 45.1
6	11	. .	8.9	. .	. .	. .	. .	. .	35 35.74	27.78	−0.01	II.	2	10.21	10 12.00	22.5	1.03	36 3.51	21 40 35.5
7	11	. .	. .	49.9	. .	. .	. .	. .	37 3.38	27.78	+0.01	II.	7	12.14	35 7.37	22.4	2.68	37 31.17	22 5 32.5
8	9	. .	55.5	. .	. .	. .	. .	. .	41 22.66	27.76	0.01	II.	10	10.26	49 11.52	22.0	3.61	41 50.43	19 37.1
9	10	. .	. .	26.4	. .	. .	. .	. .	42 39.88	27.76	+0.01	III.	7	12.8	35 4.45	21.9	2.67	43 7.65	22 5 29.0
10	12	. .	. .	38.3	. .	6.6	. .	. .	43 52.35	27.76	0.00	V.	5	7.19	23 37.63	21.8	1.90	44 20.11	21 54 1.3
11	8	. .	. .	. .	. .	. .	0.4	. .	44 33.65	27.76	−0.01	VI.	1	3.26	1 43.73	21.8	0.49	45 1.40	32 6.0
12	9	39.9	54.5	. .	. .	. .	. .	. .	48 21.03	27.75	0.00	II.	6	3.14	25 35.08	21.5	2.04	48 48.78	21 55 58.6
13	10	27.9	. .	. .	. .	. .	. .	. .	50 8.63	27.74	0.00	I.	7	4.23	31 9.69	21.3	2.41	50 36.37	22 1 33.4
14	10	. .	. .	17.4	. .	44.6	. .	. .	18 52 31.13	+27.74	+0.01	V.	10	7.37	−47 46.41	−21.1	−3.52	18 52 58.88	−22 18 11.0

CORRECTIONS.

Date.		Corr. of Clock.	Hourly rate.	m	n	c
1848.	h.	s.	s.	s.	s.	s.
Aug. 7.	19	+ 24.45	/ 0.021			

INSTRUMENT READINGS.

Date.		Barom.	THERMOM. At.	THERMOM. Ex.
1848.	h. m.	in.	°	°

REMARKS.

(181) 12. The first thread assumed to have been observed at 39s.4 instead of 29s.4.
(181) 14. T. V assumed to have been recorded as T. VI. If T. IV was recorded as T. III the seconds of T. are 55s.89.
(181. 22. Hor. thread assumed as 5 instead of 6.

ZONE 182. AUGUST 7. P. BELT, −21° 53′. D_0 = −21° 30′ 0″.

No.	Mag.	Seconds of Transit. I.	II.	III.	IV.	V.	VI.	VII.	T.	a_1	a_2	Micrometer.			i	d_1	d_2	Mean Right Ascension, 1850.0.	Mean Declination, 1850.0.
									h. m. s.	s.	s.			r.	′ ″	″	″	h. m. s.	° ′ ″
15	9	13.5	27.4	..	..	..	..	..	18 54 54.17	+27.73	−0.01	II.	3	7.34	−13 46.89	−21.0	−1.27	18 55 21.89	−21 44 9.2
16	4	..	..	13.5	..	..	39.9	..	55 13.21	27.73	0.00	VI.	6	5.52	26 54.75	20.9	2.13	55 40.94	57 17.8
17	11	..	..	45.5	..	..	..	..	57 58.77	27.73	0.00	III.	4	10.1	20 0.12	20.7	1.67	58 26.50	50 22.5
18	11	..	..	..	..	..	1.9	..	18 58 35.08	27.72	0.00	VI.	5	10.2	24 59.72	20.7	2.00	18 59 2.80	55 22.4
19	10	..	..	..	17.1	..	..	..	19 0 16.95	27.72	−0.01	IV.	2	11.315	10 47.68	20.5	1.07	19 0 44.66	41 9.3
20	10	..	..	..	..	..	..	48.9	1 8.36	27.72	0.00	IV.	5	8.42	24 19.52	20.5	1.95	1 36.08	54 42.0
21	10	..	..	48.5	..	15.8	..	..	3 2.05	27.71	0.00	V.	5	7.26	23 41.16	20.3	1.91	3 29.76	21 54 3.4
22	10	..	..	27.9	..	55.1	..	..	4 41.64	27.71	+0.01	V.	10	8.25	48 10.61	20.2	3.55	5 9.36	22 18 34.4
23	10	..	..	17.4	30.5	..	..	..	6 30.62	27.70	+0.01	IV.	7	12.8	35 4.48	20.0	2.68	6 58.33	22 5 27.2
24	10	..	..	..	33.5	..	..	..	7 33.35	27.70	−0.01	IV.	3	10.2	15 1.66	20.0	1.35	8 1.04	21 45 23.0
25	10	..	..	..	..	45.5	58.5	12.9	8 31.97	27.70	−0.01	.	3	11.40	15 51.07	19.9	1.36	8 59.66	21 46 12.3
26	11	..	34.6	..	..	..	..	..	11 1.67	27.69	+0.01	II.	8	7.58	37 56.98	19.7	2.86	11 29.37	22 8 19.5
27	9	..	..	..	..	42.8	..	..	11 29.27	27.69	0.00	V.	4	6.15	18 6.16	19.7	1.55	11 56.96	21 48 27.4
28	11	..	..	..	..	..	55.6	..	12 28.77	27.69	0.00	VI.	6	2.575	24 56.51	19.6	2.03	12 56.46	21 55 18.1
29	10	..	..	..	34.6	47.9	..	..	14 34.45	27.68	0.00	V.	7	5.50	31 53.85	19.4	2.47	15 2.13	22 2 15.7
30	11	2.9	17.4	..	..	..	..	..	16 43.92	27.68	0.00	II.	4	9.17	19 37.82	19.2	1.63	17 11.60	21 49 58.7
31	6	..	..	..	..	..	20.9	34.5	16 54.07	27.68	0.00	VI.	7	9.11	34 5.35	19.2	2.62	17 21.75	22 4 27.2
32	11	40.5	54.6	8.1	..	..	..	..	20 21.27	27.67	−0.01	III.	2	6.6	8 3.53	19.0	0.90	20 48.93	21 38 23.4
33	10	..	..	24.6	..	..	..	..	21 37.78	27.17	0.01	III.	3	6.31	13 45.48	18.9	1.27	22 5.44	44 5.6
34	6	..	..	..	..	..	..	12.5	21 31.67	27.67	−0.01	VII.	1	3.52	6 55.69	18.9	0.91	21 59.33	37 15.5
35	9	..	..	..	..	..	19.1	33.5	22 52.59	27.67	0.00	VII.	4	8.52	19 25.04	18.8	1.63	23 20.26	49 45.5
36	12	..	..	..	..	..	8.5	..	24 41.69	27.66	0.00	VI.	4	11.52	20 55.98	18.6	1.73	25 9.35	51 16.3
37	10	..	15.5	..	41.6	..	..	..	27 41.97	27.66	0.00	IV.	6	6.185	27 8.25	18.4	2.14	28 9.63	21 57 28.8
38	9	..	..	..	..	..	..	42.9	28 2.67	27.66	+0.01	VII.	10	5.36	46 45.11	18.3	3.46	28 30.34	22 17 6.9
39	11	..	..	8.5	..	..	..	..	30 21.86	27.65	0.00	III.	6	4.325	26 14.78	18.1	2.08	30 49.51	21 56 35.0
40	9	..	..	..	..	..	22.9	..	31 56.13	27.65	+0.01	VI.	2	6.21	8 10.97	18.0	1.59	32 23.79	38 29.9
41	10	..	..	..	..	..	49.6	..	32 22.78	27.65	0.00	VI.	5	6.52	23 23.91	18.0	1.89	32 50.43	53 43.8
42	10	28.5	42.6	..	..	..	..	..	37 9.35	27.64	0.00	II.	5	5.10	22 32.47	17.6	1.83	37 36.99	21 52 51.9
43	11	..	..	..	..	21.6	..	..	38 8.18	27.64	+0.01	V.	9	4.23	41 7.68	17.5	3.09	38 35.83	22 11 28.3
44	11	..	32.9	..	..	..	..	..	41 59.92	27.63	0.00	II.	7	4.14	31 5.34	17.2	2.42	41 27.55	22 1 25.0
45	11	..	45.9	..	..	..	..	..	46 12.85	27.62	0.00	II.	5	7.6	23 30.97	16.9	1.90	46 40.47	21 53 49.8
46	11	..	..	..	..	..	..	52.9	46 12.35	27.62	0.00	VII.	5	6.385	23 16.92	16.9	1.88	46 39.97	21 53 35.7
47	10	..	..	39.5	..	6.4	..	..	49 53.10	27.61	+0.01	V.	10	10.15	49 6.08	16.6	3.64	50 20.72	22 19 26.3
48	11	..	23.5	36.8	..	..	..	..	51 50.27	27.61	0.00	III.	5	4.1	21 57.79	16.4	1.79	52 17.88	21 52 16.0
49	10	0.9	13.9	..	40.5	..	..	..	55 40.86	27.60	−0.01	II.	3	6.56	13 27.73	16.1	1.23	56 8.45	43 45.1
50	10	..	49.8	2.9	16.4	31.5	..	..	19 57 16.89	27.60	+0.01	V.	7	13.1	35 31.17	16.0	2.72	19 57 44.50	22 5 49.9
51	11	..	..	21.5	..	49.5	..	..	20 1 35.33	27.59	−0.01	V.	3	8.13	14 6.66	15.6	1.27	20 2 2.91	21 44 23.5
52	12	..	..	44.6	..	..	..	..	3 57.90	27.59	0.00	III.	5	5.495	22 52.50	15.5	1.86	4 25.49	53 9.9
53	11	..	..	1.9	..	..	..	..	6 15.22	27.58	0.00	III.	5	9.18	24 37.63	15.3	1.97	6 42.80	54 54.9
54	8	..	25.4	37.6	..	..	..	..	7 51.65	27.58	−0.01	III.	3	12.11	16 6.66	15.2	1.40	8 19.22	21 46 23.2
55	8	..	..	..	44.5	..	11.6	24.5	8 44.44	27.58	+0.01	VII.	9	13.45	45 50.77	15.1	3.41	9 12.03	22 16 9.3
56	11	39.5	33.4	..	..	..	..	..	12 20.24	27.58	0.00	II.	5	5.37	22 46.09	14.8	1.85	12 47.82	21 53 2.7
57	11	35.5	49.6	2.9	..	..	..	..	14 16.25	27.57	0.00	III.	4	5.37	17 47.00	14.7	1.51	14 43.82	48 3.2
58	10	..	..	..	..	..	19.4	33.6	14 52.74	27.57	−0.01	VI.	2	7.24	8 42.74	14.6	1.91	15 20.30	21 38 59.3
59	10	..	..	..	..	31.6	..	59.8	16 18.82	27.57	+0.01	VII.	8	8.37	38 16.48	14.5	3.59	16 46.40	22 8 34.6
60	12	39.6	53.6	7.4	..	..	..	..	19 20.34	27.56	−0.01	III.	2	4.00	6 59.99	14.3	0.81	19 47.89	21 37 15.1
61	11	..	..	..	24.6	..	..	..	20 24.45	27.56	−0.01	IV.	3	4.1	7 0.52	14.2	0.89	20 52.00	37 15.6
62	10	..	..	..	24.9	..	51.6	..	29 24.77	27.55	0.00	VI.	5	2.3	20 58.18	13.6	1.72	29 52.32	21 51 13.5
63	11	..	..	6.4	..	..	16.5	..	20 32 19.73	+27.54	+0.01	VI.	8	9.405	−38 48.67	−13.4	−2.95	20 31 47.33	−22 9 5.0

CORRECTIONS.

Date.		Corr. of Clock.	Hourly rate.	m	n	c
1848.	h.	s.	s.	s.	s.	s.

INSTRUMENT READINGS.

Date.		Barom.	Thermom. At.	Ex.
1848.	h. m.	in.	°	°

REMARKS.

(182) 16. One of the transit threads wrongly recorded. T. IV assumed as observed instead of T. III.
(182) 28. Micrometer reading assumed as 1 57.5.
(182) 31. Micrometer reading assumed as 10r.11 instead of 9r.11.
(182) 33. Micrometer reading assumed as 7r.31 instead of 6r.31.
(182) 34. Hor. thread assumed as 2 instead of 1.
(182) 44. Minutes assumed as 40 instead of 41.
(182) 59. Hor. thread assumed as 6 instead of 8.
(182) 61. Hor. thread assumed as 2 instead of 3.
(182) 63. Minutes assumed as 31 instead of 32.

ZONE 182. AUGUST 7. P. BELT, −21° 53′. D_0 = −21° 30′ 0″—Continued.

No.	Mag.	Seconds of Transit I.	II.	III.	IV.	V.	VI.	VII.	T.	a_1	a_2	Micrometer		i	d_1	d_2	Mean Right Ascension, 1850.0.	Mean Declination, 1850.0.
									h. m. s.	s.	s.		r.	′ ″	″	″	h. m. s.	° ′ ″
64	11	58.5	..	..	..	..	..	..	20 34 39.09	+27.54	−0.01	I.	3 11.47	+15 54.28	−13.2	−1.38	20 35 6.62	−21 46 8.9
65	10	..	..	..	..	15.5	..	42.9	35 2.09	27.54	−0.01	V.	3 7.2	13 30.86	13.2	1.22	35 29.62	21 43 45.3
66	7	..	..	46.4	59.5	13.6	..	..	36 59.78	27.54	0.00	V.	7 8.16	33 8.48	13.0	2.57	37 27.32	22 3 24.1
67	10	5.6	19.4	..	45.6	..	..	..	41 46.01	27.53	0.00	IV.	4 13.4	21 32.42	12.7	1.75	42 13.54	21 51 46.9
68	9	..	47.5	..	14.6	..	..	..	43 14.43	27.53	0.00	IV.	3 14.11	17 7.19	12.6	1.46	43 41.96	21 47 21.3
69	10	..	..	..	..	30.5	..	..	45 17.04	27.53	0.00	V.	7 4.55	31 26.12	12.5	2.45	45 44.57	22 1 41.1
70	11	..	52.1	..	17.9	..	..	..	54 18.46	27.52	+0.01	IV.	8 6.345	37 15.03	11.8	2.86	54 45.99	22 7 29.7
71	10	..	..	54.6	..	21.8	..	..	56 8.05	27.52	0.00	V.	4 4.24	16 39.93	11.7	1.42	56 35.57	21 46 53.0
72	5	..	3.5	16.4	29.4	42.6	56.6	..	20 59 29.63	27.51	0.00	VI.	4 4.30	17 13.11	11.5	1.45	20 59 57.14	47 26.1
73	12	..	..	..	..	..	..	12.1	21 0 31.36	27.51	−0.01	VII.	2 7.27	8 44.07	11.4	1.89	21 0 58.86	38 57.4
74	10	54.6	8.5	21.4	34.6	..	..	..	3 34.84	27.51	−0.01	IV.	2 4.21	7 10.60	11.2	0.78	4 2.34	37 22.6
75	10	..	45.9	..	11.9	..	..	..	6 12.30	27.50	0.00	IV.	5 9.7	24 32.12	11.0	1.97	6 39.80	21 54 45.1
76	10	..	..	..	..	..	..	30.5	6 50.09	27.50	0.00	VII.	7 8.53	33 25.84	11.0	2.60	7 17.59	22 3 39.4
77	11	..	..	..	..	..	31.6	..	10 4.79	27.50	0.00	VI.	5 8.35	24 15.85	10.8	1.95	10 32.29	21 54 28.6
78	10	..	..	27.4	..	54.9	..	..	16 41.29	27.50	+0.01	V.	10 8.35	48 15.66	10.3	3.61	17 8.80	22 18 29.6
79	7	..	..	3.6	16.5	29.6	..	..	18 16.43	27.50	0.00	V.	4 10.32	20 15.76	10.2	1.67	18 43.93	21 50 27.6
80	10	..	3.6	16.4	..	..	..	..	23 30.01	27.49	−0.01	III.	3 3.4	11 30.85	9.9	1.07	23 57.49	36 42.4
81	11	..	..	..	..	..	..	8.5	24 28.00	27.49	0.00	VII.	6 4.18	26 4.15	9.8	2.08	24 55.49	56 16.0
82	11	..	..	..	54.6	..	..	..	27 54.45	27.49	−0.01	IV.	3 8.56	14 28.38	9.6	1.27	28 21.93	21 44 39.3
83	10	..	..	..	36.4	..	3.9	..	32 36.62	27.49	+0.01	VI.	10 11.53	49 55.38	9.3	3.73	33 4.12	22 20 8.4
84	8	7.4	21.4	34.6	48.3	..	..	..	47 48.09	27.48	0.00	IV.	4 11.3	20 31.41	8.4	1.67	48 15.57	21 50 41.5
85	8	12.5	27.4	40.6	..	..	..	..	49 53.81	27.48	0.00	III.	5 7.16	23 36.12	8.3	1.90	50 21.29	53 46.3
86	9	..	..	27.4	..	56.6	..	..	51 40.72	27.48	−0.01	V.	1 4.4	2 3.01	8.2	0.42	52 8.19	21 32 11.6
87	10	..	..	..	2.9	..	29.9	..	21 59 2.87	27.48	+0.01	VI.	10 10.23	49 10.00	7.8	3.71	21 59 30.36	22 19 21.5
88	10	..	..	..	32.6	..	59.9	..	22 1 32.78	27.48	0.00	VI.	3 7.49	13 24.20	7.7	1.23	22 2 0.26	21 43 33.1
89	8	..	..	..	..	..	..	55.6	3 15.12	27.48	0.00	VII.	6 7.30	27 43.99	7.6	2.18	2 42.60	21 57 53.8
90	10	..	..	..	22.6	..	..	50.	4 22.79	27.48	+0.01	VII.	8 9.3	48 29.58	7.5	3.66	4 50.28	22 18 40.7
91	9	..	..	..	..	..	..	1.5	5 21.16	27.48	0.00	VII.	8 9.49	38 52.78	7.5	2.99	5 48.64	9 3.3
92	10	..	..	..	..	8.8	..	..	6 55.34	27.48	0.00	VII.	6 14.56	30 58.64	7.4	2.42	7 22.82	22 1 8.5
93	9	..	..	..	..	..	9.5	22.6	9 42.32	27.48	0.00	V.	3 9.33	14 47.00	7.3	1.28	10 9.80	21 44 55.6
94	6	..	..	38.5	..	..	4.9	..	12 52.20	27.48	+0.01	VI.	10 13.1	50 29.66	7.1	3.80	13 19.69	22 20 40.6
95	9	..	..	..	..	33.6	..	..	22 15 33.45	+27.48	+0.01	V.	10 14.12	−51 5.57	−7.0	−3.84	22 16 0.94	−22 21 16.4

ZONE 183. AUGUST 14. P. BELT, −19° 23′. D_0 = −19° 1′ 0″.

No.	Mag.	I.	II.	III.	IV.	V.	VI.	VII.	T.	a_1	a_2	Micrometer		i	d_1	d_2	Mean Right Ascension, 1850.0.	Mean Declination, 1850.0.
1	8	..	..	..	..	56.4	..	22.9	18 53 43.04	+31.38	0.00	VII.	5 1.22	−20 37.37	−13.5	−1.75	18 54 14.42	−19 21 52.6
2	9	..	10.5	23.4	..	..	..	..	56 36.95	31.38	+0.01	III.	8 13.48	40 53.57	13.3	2.87	57 8.34	42 9.7
3	7	16.4	30.4	42.9	..	..	..	..	58 56.50	31.37	0.00	III.	6 11.59	29 59.92	13.2	2.25	59 27.87	31 15.4
4	9	..	..	..	..	..	..	5.2	18 59 25.12	31.37	−0.01	VII.	2 9.47	9 54.72	13.2	1.17	18 59 56.48	11 9.1
5	10	..	59.9	..	25.4	..	..	..	19 2 25.84	31.36	0.00	IV.	5 9.24	24 40.69	13.0	1.98	19 2 57.20	25 55.7
6	10	..	..	..	25.4	..	51.8	..	3 25.33	31.36	0.00	VI.	6 6.29	27 13.43	12.9	2.12	3 56.69	28 28.5
7	10	..	..	13.2	..	39.5	..	..	5 26.39	31.35	+0.01	V.	8 5.10	36 32.38	12.8	2.63	5 57.75	37 47.8
8	11	12.2	..	..	..	..	..	..	7 52.11	31.34	−0.01	I.	3 7.1	13 30.11	12.6	1.37	8 23.44	14 44.1
9	4	..	..	..	19.5	33.5	46.4	..	8 19.86	31.34	−0.01	VI.	3 3.29	11 43.36	12.6	1.28	8 51.19	12 57.2
10	9	59.9	..	..	..	..	..	..	10 40.00	31.34	+0.01	I.	8 5.505	36 52.56	12.4	2.66	11 11.35	38 7.6
11	10	..	..	..	18.5	21.9	..	..	11 8.44	31.33	−0.01	V.	2 3.385	6 49.14	12.4	1.00	11 39.76	8 2.5
12	8	..	..	..	18.4	31.5	44.6	..	12 18.24	31.33	0.00	VI.	7 1.57	29 26.02	12.3	2.26	12 49.57	30 40.6
13	9	..	..	..	..	..	..	57.8	19 13 17.82	+31.33	0.00	VI.	4 3.49	−16 52.45	−12.2	−1.56	19 13 49.15	−19 18 6.2

CORRECTIONS.

Date.	Corr. of Clock.	Hourly rate.	m	n	c
1848.	h. s.	s.	s.	s.	s.
Aug. 14.	21 + 28.12	/ 0.021	− 0.27	+ 0.87	0.00

INSTRUMENT READINGS.

Date.		Barom.	Thermom. At.	Thermom. Ex.
1848.	h. m.	in.	°	°

REMARKS.

(182) 71. Micrometer reading assumed as $3^r.24$ instead of $4^r.24$.
(182) 77. Double.
(182) 80. Hor. thread assumed as 2 instead of 3.
(182) 86. Time of transit over T. V assumed as $54^s.6$ instead of $56^s.6$.
(182) 88. Micrometer reading assumed as $6^r.49$ instead of $7^r.49$.
(182) 89. T. VI assumed to have been recorded as T. VII, and minutes as 2 instead of 3.
(182) 92. Micrometer reading assumed as $13^r.56$ instead of $14^r.56$.
(182) 95. Transit over T. IV assumed as recorded over T. V.
(183) 1. Declination 5′ 30″ less than Mural Z. 1848, August 15, and 1849, June 21.
(183) 11. An error of 10^s in transit over one of the threads; T. IV assumed to be $8^s.5$ instead of $18^s.5$.
(183) 12. Micrometer reading assumed as $0^r.57$ instead of $1^r.57$.

ZONE 183. AUGUST 14. P. BELT, −19° 23′. D_o = −19° 1′ 0″—Continued.

No.	Mag.	Seconds of Transit. I.	II.	III.	IV.	V.	VI.	VII.	T.	a_1	a_2	Micrometer.			i	d_1	d_2	Mean Right Ascension, 1850.0.	Mean Declination, 1850.0.
									h. m. s.	s.	s.			r.	′ ″	″	″	h. m. s	° ′ ″
14	10	..	..	59.8	..	..	..	..	19 15 12.75	+31.32	−0.01	VII.	3	3.11	+11 34.13	−12.1	−1.27	19 15 44.06	−19 12 47.5
15	11	..	..	..	..	36.9	..	..	16 23.64	31.32	0.00	V.	6	5.9	26 33.18	12.0	2.09	16 54.96	27 47.3
16	10	4.9	..	..	..	..	..	..	18 45.00	31.31	+0.01	I.	8	6.36	37 15.52	11.9	2.68	19 16.32	38 30.1
17	10	..	..	..	..	..	56.9	..	18 30.57	31.31	−0.01	VI.	1	5.17	2 39.72	11.9	0.75	19 1.87	3 52.4
18	10	..	..	37.5	..	..	..	..	20 50.83	31.31	+0.01	III.	9	2.43	40 17.26	11.7	2.85	21 22.15	41 31.8
19	9	..	..	10.5	24.1	36.9	..	..	22 23.83	31.30	+0.01	IV.	8	12.57	40 27.89	11.6	2.86	22 55.14	41 42.4
20	11	..	..	..	..	42.4	..	..	24 29.04	31.30	−0.01	V.	2	8.555	9 28.99	11.5	1.14	25 0.33	10 41.6
21	10	..	..	..	..	53.9	..	20.4	25 40.38	31.29	0.01	V.	1	8.12	4 8.06	11.4	0.83	26 11.66	5 20.3
22	8	..	..	..	..	23.6	..	..	27 10.24	31.29	−0.01	V.	2	8.59	9 30.76	11.3	1.14	27 41.52	10 43.2
23	11	..	..	31.9	45.1	8.5	..	..	28 45.11	31.28	0.00	V.	6	14.19	31 10.50	11.2	2.35	29 16.39	32 24.1
24	10	..	..	16.4	..	..	..	..	30 29.63	31.28	0.00	III.	7	7.29	32 43.77	11.1	2.43	31 0.91	33 57.3
25	10	..	..	..	55.9	..	21.9	..	33 55.63	31.27	0.00	VI.	6	5.01	26 29.05	10.9	2.08	34 26.90	27 42.0
26	9	25.4	38.2	51.4	4.9	..	..	..	38 4.70	31.26	−0.01	V.	1	9.185	5 1.50	10.6	0.90	38 35.95	6 13.0
27	11	..	6.8	..	..	..	..	..	40 33.47	31.25	+0.01	II.	9	7.31	42 42.39	10.4	2.99	41 4.73	43 55.8
28	11	..	..	..	..	..	..	32.9	40 52.89	31.25	−0.01	VII.	3	9.1	14 30.62	10.4	1.42	41 24.13	15 42.4
29	8	..	..	..	..	10.6	..	37.4	42 57.42	31.25	0.00	VII.	5	8.20	24 8.15	10.3	1.95	43 28.67	25 20.4
30	9	..	50.4	3.6	16.6	..	..	..	45 16.65	31.24	0.00	IV.	4	5.44	17 50.56	10.1	1.60	45 47.89	19 2.3
31	9	..	..	..	..	21.6	34.6	..	46 8.29	31.24	0.00	VI.	8	1.30	34 41.36	10.1	2.51	46 39.53	35 54.0
32	9	7.4	21.6	33.8	47.4	..	..	..	52 47.46	31.22	0.00	IV.	6	10.38	29 19.11	9.6	2.24	53 18.68	30 31.0
33	10	10.4	24.2	37.4	..	..	..	..	55 50.40	31.22	−0.01	III.	2	10.365	10 19.93	9.4	1.18	56 21.61	11 30.5
34	11	..	..	38.4	51.5	4.9	..	..	56 51.38	31.22	0.01	VII.	1	12.51	6 28.49	9.4	0.96	57 22.59	7 38.9
35	9	..	..	48.4	1.5	14.9	28.	..	19 59 1.48	31.21	−0.01	VI.	3	5.345	12 46.65	9.2	1.32	19 59 32.68	13 57.2
36	9	32.6	46.2	58.9	..	..	..	..	20 1 12.68	31.20	+0.01	III.	10	7.39	47 47.42	9.1	3.29	20 1 43.89	48 59.8
37	11	..	9.9	22.8	..	..	..	..	3 35.99	31.20	−0.01	III.	2	6.52	8 26.72	8.9	1.07	4 7.18	9 36.7
38	10	50.9	5.4	17.5	30.9	..	..	..	5 31.15	31.20	+0.01	IV.	8	8.31	38 13.77	8.8	2.74	6 2.36	39 25.3
39	10	..	..	..	31.6	44.5	57.8	11.4	6 31.39	31.19	0.00	VII.	4	11.20	20 39.71	8.7	1.76	7 2.58	21 50.2
40	10	..	..	..	8.9	..	34.6	..	8 8.46	31.19	+0.01	VI.	9	8.20	43 7.10	8.6	3.02	8 39.66	44 18.7
41	4	..	46.5	59.5	12.8	25.4	38.5	52.5	10 12.59	31.18	0.00	VI.	7	9.406	33 50.04	8.5	2.49	10 43.77	35 1.0
42	11	33.1	46.4	59.9	..	..	..	..	13 12.83	31.18	−0.01	III.	2	7.15	8 38.31	8.3	1.08	13 44.00	9 47.7
43	12	..	..	..	..	36.4	..	..	15 23.06	31.17	−0.01	V.	3	7.36	13 48.01	8.2	1.38	15 54.22	14 57.6
44	9	..	..	58.5	12.5	..	38.5	..	17 12.08	31.17	+0.01	VI.	8	6.11	37 3.05	8.1	2.67	17 43.26	38 13.8
45	10	54.6	8.9	21.6	..	..	..	..	20 35.01	31.16	0.01	III.	7	10.3	34 1.43	7.9	2.50	21 6.18	35 11.8
46	11	..	..	38.5	52.1	5.6	18.4	..	21 52.02	31.16	0.01	VI.	7	14.385	36 20.24	7.8	2.63	22 23.1	37 30.7
47	10	..	..	40.9	53.5	7.8	..	..	23 54.11	31.16	+0.01	V.	10	7.165	47 36.07	7.7	3.29	24 25.28	48 47.1
48	9	55.4	9.4	21.4	34.5	48.6	..	..	26 35.17	31.15	0.00	V.	7	11.35	34 47.82	7.5	2.55	27 6.32	35 57.9
49	11	..	16.4	..	42.6	..	..	..	28 42.65	31.15	0.00	IV.	4	3.6	16 30.89	7.4	1.53	29 13.80	17 39.8
50	10	..	..	..	..	..	..	45.4	29 5.44	31.14	0.00	VII.	4	5.13	17 34.65	7.4	1.59	29 36.58	18 43.6
51	10	3.9	17.4	29.9	..	..	..	..	31 43.53	31.14	0.00	III.	4	3.52	16 54.05	7.2	1.55	32 14.67	18 2.8
52	10	..	44.6	..	..	..	..	..	33 10.97	31.14	−0.01	II.	2	2.49	6 24.10	7.1	0.94	33 42.10	7 32.1
53	11	20.4	35.4	..	..	..	..	..	35 1.28	31.13	+0.01	II.	8	11.36	39 46.94	7.0	2.84	35 32.42	40 56.8
54	11	..	..	..	..	..	49.5	2.9	20 37 23.05	+31.13	0.00	VII.	4	9.58	−19 58.36	− 6.9	−1.72	20 37 54.18	−19 21 7.0

ZONE 184. AUGUST 15. II. BELT, −20° 38′. D_o = −20° 13′ 40″.

No.	Mag.	I.	II.	III.	IV.	V.	VI.	VII.	T.	a_1	a_2	Micrometer.			i	d_1	d_2	Mean Right Ascension, 1850.0.	Mean Declination, 1850.0.
1	8	..	..	..	42.5	56.3	9.	..	18 0 42.55	+33.02	−0.01	VI.	3	10.9	−15 5.05	−18.1	−1.41	18 1 15.56	−20 29 4.6
2	7	..	..	..	28.	41.5	..	..	2 27.95	33.01	0.01	V.	3	6.14	13 6.66	17.9	1.32	3 0.95	27 5.9
3	8	..	..	..	..	55.	8.	..	3 41.50	33.01	0.01	VI.	3	5.15	12 36.79	17.8	1.29	4 14.50	26 35.9
4	7	..	18.	31.	..	..	..	..	18 5 41.34	+33.00	−0.01	III.	3	3.28	−11 42.94	−17.6	−1.23	18 6 17.33	−20 25 41.8

CORRECTIONS.

Date.		Corr. of Clock.	Hourly rate.	m	n	c
1848.	h.	s.	s.	s.	s.	s.
Aug. 15.	21	+ 29.69	/0.051			

INSTRUMENT READINGS.

Date.		Barom.	Thermom. At.	Thermom. Ex.
1848.	h. m.	in.	°	°

REMARKS.

(183) 23. T. V assumed as 58ˢ.5 instead of 8ˢ.5.
(183) 26. Micrometer reading assumed as 9ʳ.58 instead of 9ʳ.185.

ZONE 184. AUGUST 15. B. BELT, −20° 38′. D_v = −20° 13′ 40″—Continued.

No.	Mag.	SECONDS OF TRANSIT. I.	II.	III.	IV.	V.	VI.	VII.	T.	a_1	a_2	MICROMETER.			δ	d_1	d_2	Mean Right Ascension, 1850.0.	Mean Declination, 1850.0.
									h. m. s.	s.	s.			r.	′ ″	″	″	h. m. s.	° ′ ″
5	8	. .	39.5	52.5	. .	. .	. .	. .	18 7 6.06	+33.00	0.00	III.	5	2.47	−21 20.48	−17.4	−1.79	18 7 39.06	−20 35 19.7
6	9	. .	. .	. .	. .	19.4	32.5	. .	8 5.96	32.99	0.00	VI.	5	8.59	24 27.95	17.3	1.97	8 38.95	38 27.2
7	9	. .	. .	. .	. .	. .	42.4	55.	9 15.39	32.99	+0.01	VII.	8	11.32	39 44.78	17.2	2.85	9 48.39	53 44.8
8	8	. .	. .	. .	. .	53.	6.	. .	11 39.50	32.99	−0.01	VI.	2	4.2	7 0.89	17.0	0.96	12 12.48	20 58.9
9	8	. .	24.	. .	50.	. .	. .	. .	13 50.31	32.98	0.00	IV.	6	11.49	30 25.16	16.8	2.33	14 23.29	44 24.3
10	7	. .	1.5	14.	. .	. .	. .	. .	15 27.85	32.98	+0.01	III.	8	7.4	37 29.87	16.6	2.75	16 0.87	51 29.2
11	5.6	. .	. .	. .	. .	. .	18.2	32.	15 51.70	32.98	0.00	VII.	5	6.225	23 8.89	16.5	1.89	16 24.68	20 37 7.3
12	8	. .	. .	. .	20.	33.2	. .	. .	20 19.89	32.97	+0.01	V.	10	9.25	48 40.86	16.1	3.44	20 52.87	21 2 40.4
13	7	. .	. .	. .	. .	. .	. .	45.5	21 5.11	32.96	−0.01	VII.	2	8.4	9 2.78	16.0	1.06	21 38.06	20 22 59.8
14	9	. .	. .	. .	. .	. .	. .	14.	22 33.92	32.96	0.00	VII.	7	6.31	32 14.28	15.8	2.42	23 6.88	46 12.5
15	9	. .	. .	. .	. .	. .	29.5	43.	24 2.89	32.96	0.00	VII.	6	11.39	29 49.59	15.7	2.28	24 35.85	43 47.6
16	8	. .	. .	. .	50.5	4.	. .	. .	25 50.53	32.95	+0.01	V.	9	8.25	43 9.71	15.5	3.08	26 23.49	57 8.3
17	8	. .	. .	. .	. .	25.	38.	51.3	27 11.31	32.95	−0.01	VII.	3	4.1	11 59.34	15.3	1.24	27 44.25	25 55.9
18	7	. .	. .	. .	. .	11.	24.	. .	32 57.50	32.94	0.01	VI.	3	6.13	13 6.04	14.7	1.31	33 30.43	27 2.1
19	9	. .	. .	. .	. .	4.2	17.	. .	34 50.59	32.93	−0.01	VI.	2	5.5	7 32.65	14.5	0.98	35 23.51	21 28.1
20	9	. .	. .	. .	. .	. .	. .	37.5	35 57.56	32.93	+0.01	VII.	9	6.3	42 7.86	14.4	3.02	36 30.50	56 5.3
21	8	. .	. .	. .	30.3	14.	. .	. .	37 30.39	32.93	0.00	V.	6	6.43	27 20.58	14.2	2.14	38 3.32	41 16.9
22	6.7	. .	. .	59.5	13.2	26.	. .	. .	40 12.73	32.92	−0.01	V.	3	10.005	15 31.12	13.9	1.42	40 45.64	29 26.4
23	7	. .	. .	. .	. .	46.	. .	. .	41 32.55	32.92	−0.01	V.	3	7.49	13 54.56	13.8	1.35	42 5.46	27 49.7
24	7	. .	. .	. .	. .	. .	0.	14.	42 33.58	32.91	0.00	VII.	4	7.49	18 53.31	13.6	1.64	43 6.49	32 48.6
25	9	. .	. .	. .	20.	34.	. .	. .	44 20.22	32.91	0.00	IV.	5	3.1	21 27.56	13.4	1.80	44 53.13	35 22.8
26	8	. .	. .	. .	. .	44.5	. .	11.	45 30.87	32.90	−0.01	VII.	3	10.025	15 1.66	13.3	1.42	46 3.76	28 56.3
27	6	11.5	26.	39.	52.	. .	. .	. .	47 52.12	32.90	+0.01	IV.	8	5.445	36 49.81	13.1	2.70	48 25.03	50 45.6
28	8	. .	. .	. .	. .	57.	10.	. .	48 43.51	32.90	0.00	VI.	5	6.29	23 12.33	13.0	1.89	49 16.41	37 7.2
29	9	. .	. .	58.	11.	. .	. .	. .	51 11.12	32.89	+0.01	VII.	8	3.18	35 35.65	12.7	2.63	51 44.02	49 31.0
30	9	. .	. .	. .	37.	50.5	. .	. .	52 36.97	32.89	0.00	V.	4	10.10	20 4.66	12.5	1.72	53 9.86	33 58.9
31	8	. .	. .	. .	. .	. .	. .	55.	53 14.79	32.89	0.00	VII.	5	7.47	23 21.25	12.5	1.90	53 47.68	37 15.6
32	8	. .	. .	. .	. .	. .	19.	32.5	54 52.27	32.88	−0.01	VII.	2	5.6	7 33.02	12.3	0.98	55 25.14	21 26.3
33	9	. .	. .	. .	50.5	4.2	. .	. .	57 50.60	32.88	0.00	VI.	7	3.10	30 33.06	12.0	2.34	58 23.48	44 27.4
34	9	. .	. .	. .	. .	8.	. .	35.	18 58 54.63	32.87	−0.01	VII.	4	2.56	16 25.57	11.9	1.49	18 59 27.49	30 19.0
35	8	. .	. .	. .	. .	. .	51.2	4.2	19 0 24.32	32.87	0.00	VII.	6	6.41	27 19.32	11.7	2.14	19 0 57.19	41 13.2
36	9	. .	. .	. .	. .	7.	20.	33.2	1 53.32	32.87	0.00	VII.	4	12.35	21 17.52	11.5	1.78	2 26.19	35 10.8
37	9	. .	. .	. .	. .	56.3	9.5	. .	3 42.92	32.86	0.00	VI.	5	12.24	26 11.31	11.3	2.07	4 15.78	40 4.7
38	9	. .	. .	29.5	43.	56.5	. .	. .	6 42.92	32.86	0.00	V.	6	9.40	28 49.83	11.0	2.23	7 15.78	42 43.1
39	8	. .	. .	. .	. .	59.5	12.5	26.	7 46.00	32.86	+0.01	VII.	7	11.8	34 33.94	10.9	2.57	8 18.87	48 27.4
40	8	. .	. .	. .	. .	. .	22.2	35.5	8 55.48	32.85	0.00	VII.	6	9.1	28 29.91	10.8	2.21	9 28.33	42 22.9
41	9	. .	. .	. .	. .	. .	. .	43.5	10 3.35	32.85	0.00	VII.	6	6.50	27 23.86	10.6	2.14	10 36.20	41 16.6
42	9	. .	. .	41.5	54.2	. .	. .	. .	12 54.37	32.85	0.00	IV.	5	3.54	21 54.29	10.3	1.81	13 27.22	35 46.4
43	9	. .	. .	. .	30.	43.5	. .	. .	14 29.96	32.84	−0.01	V.	3	10.33	15 17.26	10.1	1.42	15 2.79	29 8.8
44	8	. .	. .	. .	. .	21.5	34.3	. .	17 7.93	32.84	+0.01	VI.	8	9.51	38 53.96	9.8	2.83	17 40.78	52 46.6
45	7	. .	. .	. .	. .	. .	37.	50.3	18 10.29	32.83	0.00	VII.	6	11.37	25 47.48	9.7	2.04	18 43.12	39 39.2
46	9	. .	. .	. .	. .	58.	11.	. .	19 44.52	32.83	0.00	VI.	5	8.18	24 7.27	9.5	1.95	20 17.35	37 58.7
47	9	. .	49.5	. .	6.	. .	. .	. .	22 16.06	32.82	0.00	IV.	6	11.39	29 49.87	9.2	2.29	22 48.88	43 41.4
48	9	. .	. .	. .	. .	. .	9.5	22.3	22 42.53	32.82	0.00	VII.	6	9.33	28 46.05	9.2	2.23	23 15.35	42 37.5
49	9	8.	22.3	. .	. .	. .	. .	. .	27 48.71	32.81	0.00	II.	6	7.30	27 44.17	8.6	2.16	28 21.52	41 34.9
50	9	. .	. .	50.	3.	. .	. .	. .	29 2.80	32.81	0.00	IV.	5	9.1	24 29.09	8.5	1.97	29 42.61	38 19.6
51	9	20.5	33.	. .	. .	. .	. .	. .	30 46.86	32.81	+0.01	II.	8	10.35	39 16.16	8.4	2.86	31 19.68	53 7.4
52	9	. .	. .	36.	. .	3.	. .	. .	31 49.45	32.81	0.00	V.	6	7.50	27 54.36	8.2	2.17	32 22.26	41 44.7
53	9	. .	. .	12.7	26.	. .	. .	. .	19 33 25.85	+32.80	0.00	IV.	4	7.25	−18 41.48	− 8.0	−1.62	19 33 58.65	−20 32 31.1

CORRECTIONS.

Date.		Corr. of Clock.	Hourly rate.	m	n	c
1848.	h.	s.	s.	s.	s.	s.

INSTRUMENT READINGS.

Date.		Barom.	THERMOM. At.	Ex.
1848.	h. m.	in.	°	°

REMARKS.

(184) 5. Micrometer reading assumed as 2r.47 instead of 24r.7.
(184) 10. Micrometer reading assumed as 12r.49 instead of 11r.49.
(184) 22. Micrometer reading assumed as 11r.005 instead of 10r.005.
(184) 23. Double star; observed the first.
(184) 31. Micrometer reading assumed as 6r.47 instead of 7r.47.
(184) 45. Hor. thread assumed as 5 instead of 6.
(184) 47. One of the transits erroneous by 10s; T. IV assumed as observed at 16s.0.
(184) 50. Transit over T. VII assumed as recorded over T. III.
(184) 51. Transits over T.'s II and III assumed as recorded over T.'s I and II; other transits rejected.

ZONE 184. AUGUST 15. B. BELT, −20° 38′. D₀ = −20° 13′ 40″—Continued.

No.	Mag.	SECONDS OF TRANSIT. I.	II.	III.	IV.	V.	VI.	VII.	T.	a_1	a_4	MICROMETER.			i	d_1	d_2	Mean Right Ascension, 1850.0.	Mean Declination, 1850.0.
									h. m. s.	s.	s.			r.	′ ″	″	″	h. m. s.	° ′ ″
54	9	..	..	..	..	..	28.5	42.	19 34 1.82	+32.80	0.00	VII.	4	3.59	−16 57.33	−7.9	−1.52	19 34 34.62	−20 30 46.8
55	9	..	..	..	..	..	..	32.8	34 52.57	32.80	0.00	VII.	5	3.35	21 44.43	7.8	1.80	35 25.37	35 34.0
56	9	..	38.5	51.5	..	..	..	..	37 4.90	32.79	0.00	III.	4	5.37	17 47.00	7.6	1.57	37 37.69	31 36.2
57	10	..	..	..	52.	5.6	..	..	38 52.08	32.79	+0.01	V.	9	4.30	41 11.22	7.4	2.98	39 24.86	55 1.6
58	10	..	7.	20.	33.	..	..	..	40 33.23	32.79	0.00	IV.	4	10.2	20 0.66	7.2	1.70	41 6.02	33 49.6
59	10	11.4	25.	38.	..	..	..	..	42 51.50	32.78	0.00	III.	4	7.35	18 46.50	6.9	1.63	43 24.28	32 35.0
60	9	..	24.	36.8	49.5	..	..	..	45 50.07	32.78	0.00	IV.	6	11.52	29 56.42	6.6	2.29	46 22.85	43 45.3
61	10	..	..	..	38.4	51.6	..	..	47 38.28	32.78	+0.01	V.	9	11.42	44 49.05	6.4	3.21	48 11.07	58 38.7
62	9	..	..	28.2	41.1	..	..	..	49 41.12	32.78	−0.01	IV.	3	10.34	15 17.80	6.2	1.41	50 13.89	29 5.4
63	9	..	..	12.5	26.	..	..	..	55 25.02	32.76	+0.01	IV.	9	9.1	43 27.89	5.6	3.13	55 58.69	57 16.6
64	9	..	44.	57.	10.	..	..	..	19 57 10.24	32.75	0.00	IV.	4	10.25	20 12.25	5.4	1.71	19 57 42.99	20 33 59.4
65	8	..	48.	0.5	13.5	..	..	..	20 0 14.11	32.75	+0.01	IV.	10	7.29	47 42.40	5.1	3.40	20 0 46.87	21 1 30.9
66	8	..	29.	40.	..	..	..	..	1 53.52	32.75	0.00	III.	6	8.57	28 28.15	4.9	2.21	2 26.27	20 42 15.3
67	8	..	..	..	..	31.	44.4	..	2 17.71	32.74	0.00	VI.	6	3.15	25 35.58	4.8	2.04	2 50.45	39 22.4
68	8	..	..	..	7.	20.4	..	..	4 6.89	32.74	−0.01	V.	2	2.38	6 18.64	4.6	0.87	4 39.62	20 4.1
69	9	..	..	51.5	4.5	..	..	..	6 4.67	32.74	+0.01	IV.	9	7.57	42 55.62	4.4	3.10	6 37.42	56 43.1
70	10	..	..	9.	22.3	..	..	..	7 22.18	32.74	0.00	IV.	5	8.48	24 22.54	4.3	1.96	7 54.92	38 8.8
71	8	..	..	..	..	..	..	40.	8 59.63	32.73	−0.01	VII.	3	2.34	11 15.48	4.1	1.15	8 32.35	25 0.7
72	8	..	..	5.	18.2	..	..	..	10 18.06	32.73	0.01	IV.	3	7.37	13 48.54	4.0	1.32	10 50.78	27 33.9
73	9	..	..	..	..	..	..	19.4	10 39.06	32.73	−0.01	VII.	3	7.20	13 39.68	4.0	1.31	11 11.78	27 25.0
74	9	26.	40.5	53.3	..	..	..	..	13 6.73	32.73	0.00	III.	6	8.43	28 21.09	3.7	2.20	13 39.46	20 42 7.0
75	9	..	..	0.	13.2	27.	..	..	14 13.44	32.73	+0.01	V.	10	5.8	46 31.27	3.6	3.33	14 46.18	21 0 18.2
76	9	..	..	30.	43.	..	..	..	15 42.98	32.73	−0.01	IV.	3	11.32	15 47.04	3.4	1.44	16 15.70	20 29 31.9
77	9	..	..	38.8	51.6	5.	..	..	17 51.59	32.72	−0.01	V.	2	3.1	6 30.23	3.2	0.87	18 24.30	20 14.3
78	9	..	..	..	..	..	15.3	29.	18 48.81	32.72	+0.01	VII.	7	9.54	33 56.64	3.1	2.55	19 21.54	47 42.3
79	9	27.5	41.8	54.2	..	..	..	..	21 8.01	32.72	0.00	III.	7	7.40	32 49.32	2.9	2.48	21 40.73	46 34.7
80	9	..	..	55.2	8.2	22.	..	..	22 8.24	32.72	−0.01	V.	1	8.28	4 16.12	2.8	0.76	22 40.95	17 59.7
81	9	..	..	..	33.7	47.1	..	..	23 33.62	32.71	0.00	V.	4	11.5	20 32.39	2.6	1.72	24 6.33	20 34 16.7
82	8	..	..	..	40.5	53.	6.2	..	24 39.87	32.71	+0.01	VI.	10	9.57	48 56.89	2.5	3.49	25 12.59	21 2 42.9
83	9	..	..	..	..	..	18.	31.3	25 51.26	32.71	0.00	VII.	5	11.46	25 52.02	2.4	2.05	26 23.97	20 39 36.5
84	9	..	..	..	31.	44.5	..	..	27 30.94	32.71	−0.01	V.	2	3.54	6 56.96	2.2	0.91	28 3.64	20 40.1
85	9	..	..	..	..	..	..	51.2	28 10.79	32.71	−0.01	VII.	2	6.34	8 17.40	2.2	0.98	28 43.49	22 0.6
86	10	..	58.	11.2	24.3	..	..	..	20 30 24.42	+32.70	0.00	IV.	5	6.18	−23 6.90	−1.9	−1.89	20 30 57.12	−20 36 50.7

ZONE 185. AUGUST 16. P. BELT, −18° 46′. D₀ = −18° 21′ 20″.

No.	Mag.	I.	II.	III.	IV.	V.	VI.	VII.	T.	a_1	a_4	MICROMETER.			i	d_1	d_2	Mean Right Ascension, 1850.0.	Mean Declination, 1850.0.
1	10	27.4	..	..	..	..	..	..	17 57 7.40	+33.96	+0.01	I.	9	7.23	−42 38.20	−29.6	−2.91	17 57 41.37	−19 4 30.7
2	9	..	..	..	..	32.4	..	58.5	17 57 19.11	33.96	+0.01	VII.	8	7.15	37 35.17	29.6	2.66	17 57 53.08	18 59 27.4
3	10	55.6	8.4	21.5	..	..	..	..	18 0 21.64	33.95	0.00	III.	6	11.47	29 53.87	28.2	2.25	18 0 55.59	51 44.3
4	10	..	..	..	..	..	..	11.9	1 32.29	33.95	0.00	VII.	7	5.33	31 45.04	28.1	2.35	2 6.24	53 35.5
5	10	..	..	..	..	..	32.9	..	3 6.59	33.94	0.00	VI.	7	5.20	31 38.64	27.9	2.34	3 40.53	53 28.9
6	9	..	..	..	..	43.6	56.4	9.5	4 30.07	33.94	0.00	VII.	6	4.12	26 4.20	27.7	2.06	5 4.01	47 54.0
7	9	..	..	..	..	..	..	1.5	5 21.82	33.94	0.00	VII.	6	6.49	27 23.37	27.6	2.13	5 55.76	49 13.1
8	8	..	..	..	..	27.4	..	54.5	7 14.53	33.93	0.00	VII.	6	10.24	29 11.78	27.4	2.21	7 48.46	51 1.4
9	9	..	..	..	..	..	..	15.5	8 35.76	33.93	0.00	VII.	5	6.42	23 18.74	27.3	1.91	9 9.69	45 8.0
10	9	..	..	..	..	..	..	16.9	9 37.00	33.92	−0.01	VII.	3	3.58	11 57.83	27.2	1.34	10 10.91	33 46.4
11	3	20.4	34.5	46.5	..	..	..	..	11 0.32	33.92	0.00	III.	7	6.57	32 24.61	27.0	2.38	11 34.24	54 14.0
12	11	..	..	..	..	..	..	58.5	15 12 18.62	+33.92	−0.01	VII.	3	6.26	−13 12.47	−26.9	−1.41	18 12 52.53	−18 35 0.8

CORRECTIONS.

Date.		Corr. of Clock.	Hourly rate.	m	n	c
1848.	h.	s.	s.	s.	s.	s.
Aug. 16.	21	+ 30.60	/ 0.052			

INSTRUMENT READINGS.

Date.		Barom.	THERMOM. At.	Ex.
1848.	h. m.	in.	°	°

REMARKS.

(184) 66. Transit over T. II assumed as 27ˢ instead of 29ˢ.
(184) 71. Minutes assumed as 7 instead of 8.
(185) 3. Transits over T.'s II, III, and IV assumed to have been recorded as over T.'s I, II, and III.

ZONE 185. AUGUST 16. P. BELT, −18° 46′. D_0 = −18° 21′ 20″—Continued.

No.	Mag.	SECONDS OF TRANSIT. I.	II.	III.	IV.	V.	VI.	VII.	T.	a_1	a_2	MICROMETER.			i	d_1	d_2	Mean Right Ascension, 1850.0.	Mean Declination, 1850.0.
									h. m. s.	s.	s.			r.	′ ″	″	″	h. m. s.	° ′ ″
13	11	41.5							18 15 21.37	+33.91	0.00	I.	6	5.42	−26 49.58	−26.5	−2.09	18 15 55.28	−18 48 38.2
14	9					51.4		17.9	15 38.09	33.91	−0.01	VII.	3	10.00	15 0.38	26.5	1.48	16 11.99	36 48.4
15	10						16.0	29.8	16 49.85	33.90	0.01	VII.	3	10.32	15 16.52	26.4	1.50	17 23.74	37 4.4
16	10							19.7	17 39.82	33.90	0.01	VII.	3	5.495	12 54.06	26.3	1.38	18 13.71	34 41.7
17	11				25.6		52.4		19 25.80	33.90	−0.01	VI.	3	7.58	13 59.01	26.1	1.43	19 59.69	35 46.5
18	8					2.9	15.4	28.4	20 49.17	33.89	0.00	VII.	6	6.44	27 20.85	25.9	2.12	21 23.06	49 8.9
19	7					18.5	31.5	45.6	22 5.37	33.89	−0.01	VII.	2	6.22	8 11.35	25.8	1.14	22 39.25	29 58.3
20	8				49.4	2.9	15.5	28.5	23 49.15	33.88	−0.01	VII.	2	3.115	6 35.28	25.6	1.07	24 23.02	28 22.0
21	8		49.5	2.1	15.4				26 15.40	33.88	0.00	IV.	4	6.00	17 58.63	25.3	1.64	26 49.28	39 45.6
22	9						36.5	49.9	27 10.24	33.87	0.00	VII.	7	6.595	32 28.66	25.2	2.39	27 44.11	54 16.3
23	10					12.9	25.6		28 59.47	33.87	−0.00	VI.	2	7.22	8 41.76	25.0	1.17	29 33.33	30 27.9
24	11					26.9		53.9	32 13.98	33.86	0.00	VII.	6	11.24	29 42.03	24.6	2.24	32 47.84	51 28.9
25	9		17.4	29.5	42.8				34 43.07	33.85	0.00	IV.	7	7.18	32 38.25	24.3	2.40	35 16.92	55 25.0
26	8	53.5	7.4	19.9					36 33.23	33.85	−0.01	III.	2	8.12	9 7.06	24.1	1.18	37 7.07	30 52.3
27	9		45.5	58.5					38 11.84	33.84	0.00	III.	7	7.40	32 49.32	23.9	2.44	38 45.68	51 35.7
28	8						50.4		39 24.12	33.84	0.00	VI.	5	7.38	23 47.13	23.8	1.93	39 57.96	45 32.9
29	11						18.9	32.5	41 52.77	33.83	+0.01	VII.	8	7.00	37 27.61	23.5	2.67	42 26.61	59 13.8
30	9				45.1	58.5	11.5		43 45.15	33.83	0.00	VI.	6	6.4	27 0.84	23.2	2.11	44 18.98	18 48 46.2
31	10		31.4	43.6					45 57.50	33.82	+0.01	III.	10	8.6	48 1.04	23.0	3.22	46 31.33	19 9 47.3
32	8			39.9	52.9	6.1			46 52.91	33.82	0.00	V.	7	6.3	32 0.41	22.9	2.37	47 26.73	18 53 45.7
33	9	27.8							49 7.62	33.82	0.00	I.	5	3.49	21 51.50	22.6	1.84	49 41.44	43 35.9
34	10					20.4		46.9	49 7.15	33.82	0.00	VII.	5	3.59	21 56.54	22.6	1.84	49 40.97	43 41.0
35	8					19.4	27.5	40.9	50 1.19	33.81	0.00	VII.	6	1.29	24 42.00	22.5	1.98	50 35.00	46 26.5
36	9			36.4	49.4				53 49.29	33.80	−0.01	IV.	3	7.425	13 51.31	22.0	1.41	54 23.08	18 35 34.7
37	8		46.4	58.5	11.8				54 12.12	33.80	+0.01	IV.	8	10.455	39 21.59	22.0	2.76	54 45.93	19 1 6.4
38	10			14.5		40.5			55 27.51	33.80	0.00	V.	7	9.425	33 51.09	21.8	2.47	56 1.31	18 55 35.4
39	10				43.5		9.9		56 43.52	33.79	−0.01	VI.	1	3.2	1 31.66	21.7	0.81	57 17.30	23 14.2
40	7						13.5	27.5	57 47.57	33.79	+0.01	VII.	8	4.20	36 6.93	21.5	2.59	18 58 21.37	57 51.0
41	8		23.8	36.4					59 36.61	33.79	0.00	III.	7	8.13	33 5.96	21.3	2.48	19 0 10.40	54 49.7
42	11							29.5	18 59 49.74	33.79	0.00	VII.	5	4.14	21 33.85	21.3	1.80	0 23.53	43 17.0
43	12					30.5			19 2 17.17	33.78	−0.01	V.	1	10.453	5 25.36	21.0	1.00	2 50.94	18 27 7.4
44	10				4.5		30.5		4 4.26	33.77	+0.01	VI.	9	11.3	44 29.30	20.8	3.03	4 38.04	19 6 13.1
45	10					43.9			5 30.67	33.77	0.00	V.	5	8.6	24 1.34	20.6	1.95	6 4.44	18 45 43.9
46	9			37.5	50.5				6 50.52	33.77	0.00	IV.	7	8.6	33 2.47	20.5	2.42	7 24.29	54 45.4
47	10		41.5		7.4		33.6		8 7.50	33.77	0.00	VI.	6	9.42	28 50.76	20.3	2.20	8 41.27	18 50 33.3
48	10						50.5	3.5	9 24.02	33.76	0.00	.	7	.	38 28.39	20.2	2.72	9 57.78	19 0 11.3
49	10					18.9	31.4	44.6	11 5.38	33.76	+0.01	VII.	10	5.005	46 27.25	20.0	3.14	11 39.15	19 8 10.4
50	10						49.5		13 23.21	33.75	0.00	VI.	6	7.37	27 47.73	19.7	2.15	13 56.96	18 49 29.6
51	9	52.5	6.4	18.5					15 32.33	33.75	0.00	III.	7	7.4	32 31.17	19.4	2.40	16 6.08	54 13.0
52	9						15.2	28.5	15 48.86	33.75	0.00	VII.	6	10.38	29 18.84	19.4	2.22	16 22.61	51 0.5
53	9			9.5				27.5	18 48.48	33.74	0.00	III.	4	4.12	17 34.40	19.0	1.61	19 22.22	39 15.0
54	9								19	33.74	0.00	VII.	4	5.36	17 46.26	. .	1.62		
55	10		25.4	38.5					20 51.55	33.74	−0.01	III.	3	2.45	11 21.27	18.8	1.28	21 25.28	33 1.4
56	10						5.4	18.5	22 39.02	33.73	+0.01	VII.	8	4.51	36 24.08	18.6	2.61	23 12.76	58 5.3
57	10					29.9	42.6	57.5	24 16.98	33.73	0.00	VII.	7	10.65	34 2.96	18.4	2.48	24 50.71	55 43.8
58	10							8.5	25 28.49	33.72	−0.01	VII.	1	7.185	3 40.83	18.3	0.91	26 2.20	18 25 20.0
59	8					20.5	33.6	46.6	27 7.30	33.72	+0.01	VII.	10	10.24	49 10.38	18.1	3.30	27 41.03	19 12 51.8
60	9	22.5	36.4						31 2.53	33.71	0.00	II.	4	5.6	17 31.29	17.6	1.60	31 30.24	18 39 10.5
61	10	56.5	9.9	22.9					19 33 36.08	+33.70	−0.01	III.	2	12.5	−11 4.54	−17.3	−1.27	19 34 9.77	−18 32 43.1

CORRECTIONS.

Date.		Corr. of Clock.	Hourly rate.	m	n	c
1848.	h.	s.	s.	s.	s.	s.

INSTRUMENT READINGS.

Date.		Barom.	THERMOM. At.	THERMOM. Ex.
1848.	h. m.	in.	°	°

REMARKS.

(185) 28. Double.
(185) 35. Transit over T. V assumed as 14^s.4 instead of 19^s.4.
(185) 41. Transits over T.'s III and IV assumed to have been recorded as over T.'s II and III.
(185) 42. Micrometer reading assumed as 3^r.14 instead of 4^r.14.
(185) 48. Hor. thread assumed as 8, and rev. as 9.
(185) 53. Transit observations incongruous, and can be approximately reconciled only by assuming the transit over T. I to have been recorded as over T. III.
(185) 54. Micrometer reading assumed as 5^r.12 instead of 4^r.12.

Zone 185. August 16. P. Belt, −18° 46'. D_o = −18° 21' 20"—Continued.

No.	Mag.	Seconds of Transit. I.	II.	III.	IV.	V.	VI.	VII.	T.	a_1	a_2	Micrometer.			i	d_1	d_2	Mean Right Ascension, 1850.0.	Mean Declination, 1850.0.
									h. m. s.	s.	s.			r.	' "	"	"	h. m. s.	° ' "
62	10		20.4						19 35 46.69	+33.70	−0.01	II.	2	7.9	− 8 35.21	−17.1	−1.14	19 36 20.38	−18 30 13.5
63	10							25.9	35 45.96	33.70	0.01	VII.	2	7.31	8 46.14	17.1	1.15	36 19.65	30 24.4
64	10						37.4	51.5	37 11.36	33.70	−0.01	VII.	2	8.115	9 6.56	16.9	1.17	37 45.05	18 30 44.6
65	10					2.9	15.2	28.4	38 49.17	33.70	+0.01	VII.	8	9.28	38 42.23	16.7	2.73	39 22.88	19 0 21.7
66	9	40.5	54.9	6.6					43 20.41	33.69	0.00	III.	4	13.20	21 40.46	16.2	1.82	43 54.10	18 43 18.5
67	11					35.			45 21.69	33.68	−0.01	V.	2	6.126	8 5.85	15.9	1.12	45 55.36	29 42.9
68	10	55.5	8.4	21.5					47 34.83	33.68	0.00	.	4	3.28	16 41.97	15.7	1.55	48 8.51	38 19.2
69	10							24.6	47 44.86	33.68	0.00	.	5	7.32	23 44.22	15.7	1.93	48 18.54	45 21.9
70	9							29.9	48 50.11	33.68	0.00	VII.	4	8.58	19 28.11	15.6	1.70	49 23.79	41 5.4
71	9	49.5	3.4						51 29.55	33.67	0.00	II.	4	10.00	19 59.54	15.2	1.73	52 3.22	41 36.5
72	7	0.9	14.9	27.4	40.4				53 40.78	33.67	+0.01	IV.	8	3.39	35 46.53	15.0	2.58	54 14.46	18 57 24.1
73	9						13.6		55 47.26	33.66	+0.01	VI.	10	11.50	49 53.90	14.7	3.35	56 20.93	19 11 32.0
74	10				16.9	30.5	44.5	56.4	58 17.21	33.66	0.00	.	5	4.45	22 20.01	14.5	1.86	19 58 50.87	18 43 [56.4]
75	11				36.4		3.8		19 59 36.92	33.65	−0.01	VI.	1	1.40	0 50.31	14.3	0.76	20 0 10.56	22 25.4
76	10							9.9	20 0 30.19	33.65	0.00	VII.	6	2.37	25 16.30	14.2	2.02	1 3.84	46 52.5
77	9					17.4	30.5	43.5	3 3.99	33.65	−0.01	VII.	3	7.1	13 30.11	13.9	1.38	3 37.63	35 5.4
78	10					43.6	56.8	10.2	4 30.49	33.64	0.00	VII.	6	11.12	29 35.95	13.8	2.25	5 4.13	51 12.0
79	11						39.4	53.5	6 13.29	33.64	0.00	VII.	5	5.25	22 28.56	13.6	1.87	6 46.93	44 4.0
80	8				48.4	2.	15.5		7 48.75	33.64	0.00	VI.	7	5.32	31 44.67	13.4	2.36	8 22.39	53 20.4
81	9						16.4	29.8	8 50.08	33.63	0.00	VII.	5	3.54	21 54.02	13.3	1.83	9 23.71	43 29.2
82	8	3.9	17.4	30.1					11 43.49	33.63	−0.01	III.	3	11.446	15 53.36	13.0	1.51	12 17.11	37 27.9
83	9							30.5	11 50.79	33.63	0.00	VII.	5	11.59	25 58.58	13.0	2.05	12 24.42	47 33.6
84	8					10.4	24.3	37.4	13 57.64	33.63	0.00	VII.	6	6.495	27 23.62	12.7	2.13	14 31.27	48 58.5
85	10	3.9	18.5						16 44.44	33.62	+0.01	II.	8	5.00	36 27.26	12.4	2.62	17 18.07	58 2.3
86	9			0.3	13.8	27.4	40.5		17 13.87	33.62	0.00	VI.	6	10.585	29 29.33	12.4	2.24	17 47.49	51 4.0
87	2.3						36.4	49.9	18 10.13	33.62	0.00	VII.	4	10.46	20 22.58	12.3	1.75	18 43.75	41 56.6
88	8					6.6			19 53.25	33.62	−0.01	V.	1	0.9	0 4.51	12.1	0.71	20 26.86	21 37.3
89	10				16.4		42.6		22 16.30	33.61	0.01	VI.	3	7.00	13 29.77	11.8	1.38	22 49.90	35 3.0
90	11	8.8	22.6						26 48.70	33.60	0.01	II.	2	4.17	7 8.48	11.3	1.05	27 22.29	28 40.8
91	11					59.4			27 46.11	33.60	0.01	V.	3	4.56	12 27.33	11.2	1.32	28 19.70	33 59.9
92	10						1.2		28 45.78	33.60	−0.01	VI.	1	3.005	1 30.90	11.1	0.78	29 19.37	23 2.8
93	4					9.4	22.9	36.5	30 56.49	33.60	0.00	VII.	4	6.34	18 15.51	10.9	1.63	31 30.09	39 48.0
94	7	54.6	8.4	20.7					33 34.26	33.59	−0.01	III.	4	3.585	16 57.32	10.6	1.57	34 7.84	38 29.5
95	9					37.4	50.4		34 24.15	33.59	0.00	VI.	5	8.24	24 10.33	10.5	1.96	34 57.74	45 42.8
96	11				53.6		19.5		36 53.32	33.59	0.00	VI.	6	10.115	29 5.63	10.2	2.23	37 26.91	18 50 38.1
97	9					29.8			38 16.70	33.59	+0.01	V.	10	9.15	48 35.82	10.1	3.28	38 50.30	19 10 9.2
98	9						45.5	59.5	40 19.49	33.58	0.00	VII.	5	6.42	23 18.74	9.9	1.91	39 53.07	18 44 50.6
99	10				43.9		10.5		41 43.99	33.58	0.00	VI.	5	11.23	25 40.58	9.7	2.04	42 17.57	47 12.3
100	8	5.5	19.6	32.6					45 45.52	33.58	−0.01	III.	2	5.40	7 50.41	9.3	1.08	46 19.09	18 29 20.8
101	10				39.5	52.6	5.7		46 39.40	33.58	+0.01	VI.	10	6.54	47 24.65	9.2	3.23	47 12.99	19 8 57.1
102	10				20.0				48 19.85	33.57	−0.01	IV.	1	7.15	3 32.53	9.0	0.86	48 53.41	18 25 2.4
103	10		36.6						50? 2.86	33.57	0.01	II.	1	10.35	5 20.09	8.9	0.95	50 36.42	26 49.9
104	10							12.3	50 2.30	33.57	−0.01	VII.	1	9.44	4 54.21	8.9	0.94	50 35.86	26 24.1
105	10	11.5							52 51.33	33.57	0.00	I.	5	5.395	22 47.22	8.6	1.87	53 24.90	44 17.7
106	9				26.5				53 26.35	33.57	−0.01	IV.	3	3.255	11 41.71	8.5	1.27	53 59.91	33 11.5
107	8		33.8	46.8	59.9				54 59.91	33.56	0.00	IV.	4	10.525	20 26.12	8.3	1.74	55 33.47	41 56.2
108	8	54.9	8.9	21.5	34.6				20 58 34.69	33.56	−0.01	III.	3	6.7	13 3.13	8.0	1.35	20 59 8.24	18 34 32.5
109	10	49.9	3.9		29.9				21 0 30.02	33.56	+0.01	IV.	8	11.56	39 57.14	7.8	2.83	21 1 3.59	19 1 27.8
110	10						57.7		21 1 31.39	+33.56	0.00	VI.	7	11.42	−34 51.27	− 7.7	−2.54	21 2 4.95	−18 56 21.5

CORRECTIONS.

Date.		Corr. of Clock.	Hourly rate.	m	n	c
1848.	h.	s.	s.	s.	s.	s.

INSTRUMENT READINGS.

Date.		Barom.	Thermom. At.	Thermom. Ex.
1848.	h. m.	in.	°	°

REMARKS.

(185) 83. Double.
(158) 98. Minutes assumed as 39 instead of 40; beautiful double star.
(185) 107. Double.

ZONE 185. AUGUST 16. P. BELT, −18° 46′. D_0 = −18° 21′ 20″—Continued.

No.	Mag.	Seconds of Transit I.	II.	III.	IV.	V.	VI.	VII.	T.	a_1	a_2	Micrometer.			i	d_1	d_2	Mean Right Ascension, 1850.0.	Mean Declination, 1850.0.
									h. m. s.	s.	s.			r.	′ ″	″	″	h. m. s.	° ′ ″
111	12					55.5		22.2	21 3 42.48	+33.55	0.00	VII.	7	11.54	−39 55.86	− 7.5	−2.83	21 4 16.03	−19 1 26.2
112	10						58.5	11.9	5 32.11	33.55	−0.01	VII.	2	9.39	9 50.69	7.3	1.17	6 5.65	18 31 19.2
113	10	5.5							8 45.36	33.55	0.00	I.	5	12.31	26 14.71	7.0	2.06	9 18.91	47 43.8
114	4.3				58.9		24.6	38.5	8 58.58	33.55	−0.01	VII.	3	10.16	15 8.45	6.9	1.46	9 32.12	36 36.8
115	11				17.4				11 17.25	33.55	0.01	IV.	1	7.535	3 58.75	6.7	0.89	11 50.79	25 26.3
116	10			25.5	38.5	51.6	4.6		12 38.35	33.54	−0.01	VI.	2	12.55	11 4.71	6.6	1.24	13 11.85	32 32.6
117	11			14.2					14 27.34	33.54	0.00	III.	6	10.1	29 0.42	6.4	2.23	15 0.68	50 29.1
118	9	39.9	53.7	6.6					17 19.87	33.54	0.00	III.	6	4.22	26 9.48	6.1	2.07	17 53.41	47 37.7
119	10				2.9				18 2.75	33.54	0.00	IV.	6	3.42	25 49.34	6.1	2.04	18 36.29	18 47 17.5
120	10		27.4	39.8					20 53.50	33.54	+0.01	III.	8	9.28	38 42.47	5.8	2.76	21 27.05	19 0 11.0
121	11			53.6					22 6.75	33.54	0.00	III.	6	11.18	29 39.24	5.7	2.26	22 40.29	18 51 7.2
122	10	18.4	31.6						24 58.02	33.54	−0.01	II.	2	10.31	10 17.07	5.4	1.19	25 31.55	31 43.7
123	12							18.5	25 38.54	33.54	−0.01	VII.	2	4.305	7 15.13	5.3	1.04	26 12.07	28 41.5
124	10			20.8	33.8	47.5			27 33.96	33.53	0.00	V.	6	8.54	28 26.64	5.2	2.17	28 7.49	18 49 54.0
125	11						31.6	44.5	29 5.14	33.53	+0.01	VII.	8	10.8	39 2.41	5.0	2.79	29 38.68	19 0 30.2
126	10			3.6		29.5			31 16.66	33.53	0.01	V.	9	11.57	44 56.61	4.8	3.10	31 50.20	19 6 24.5
127	11	19.9	34.0						33 0.20	33.53	+0.01	II.	8	7.22	37 38.86	4.7	2.69	34 33.74	18 59 6.3
128	10				7.4		33.6		35 7.28	33.53	0.00	VI.	7	3.6	30 31.08	4.5	2.32	35 40.81	51 57.9
129	9			45.6					36 58.59	33.53	0.00	III.	4	7.14	18 35.91	4.3	1.63	37 32.12	40 1.8
130	8.9					48.9		15.6	37 35.68	33.53	−0.01	VII.	3	9.45	15 2.64	4.2	1.41	38 9.20	18 36 28.3
131	10	47.4							41 27.41	33.52	0.00	I.	9	9.4	43 29.14	3.9	3.03	41 0.93	19 4 56.1
132	11				21.9				41 21.75	33.52	0.00	IV.	7	5.45	31 51.36	3.9	2.39	41 55.27	18 53 17.7
133	11							27.5	41 47.73	33.52	0.00	VII.	5	2.21	21 7.12	3.9	1.77	42 21.25	18 42 32.8
134	11					8.9			21 43 55.76	+33.52	+0.01	V.	9	7.21	−42 37.43	− 3.7	−2.98	21 44 29.29	−19 4 4.1

ZONE 186. AUGUST 18. B. BELT, −23° 46′. D_0 = −23° 19′ 40″.

No.	Mag.	I.	II.	III.	IV.	V.	VI.	VII.	T.	a_1	a_2	Micrometer.			i	d_1	d_2	Mean Right Ascension, 1850.0.	Mean Declination, 1850.0.
1	9			10.2	24.2				19 14 23.83	+37.22	−0.01	IV.	3	11.43	−15 52.58	−20.1	−1.34	19 15 1.04	−23 35 54.0
2	9					41.8	55.		15 27.96	37.22	0.00	VI.	4	7.22	16 39.82	20.0	1.54	16 5.18	38 41.4
3	9	57.5	12.	25.					17 38.79	37.22	0.00	III.	4	6.53	18 25.31	19.8	1.52	18 16.01	38 26.6
4	10							26.	18 45.10	37.22	+0.01	VII.	9	0.20	39 4.84	19.7	3.03	19 22.33	59 7.6
5	10			7.	20.5				21 20.30	37.22	−0.01	VII.	1	11.42	5 53.64	19.5	0.64	21 57.51	25 53.8
6	7					16.5	30.		22 2.80	37.22	−0.01	VI.	1	9.1	4 32.64	19.4	0.53	22 40.01	24 32.6
7	8	57.2	11.5						24 38.58	37.22	0.00	II.	4	7.16	18 36.80	19.2	1.54	25 15.80	38 37.5
8	8	19.	33.						25 0.22	37.22	0.00	II.	4	5.52	17 54.44	19.2	1.48	26 37.44	37 55.1
9	10				9.	23.			27 9.05	37.22	−0.01	V.	2	10.53	10 28.23	19.0	0.96	27 46.26	30 28.2
10	10						46.3	0.	28 18.95	37.22	−0.01	VII.	3	5.45	12 51.72	18.9	1.12	28 56.16	32 51.7
11	6			57.	11.				30 10.70	37.22	0.00	IV.	6	3.45	25 50.85	18.7	2.06	30 47.92	45 51.6
12	5.6						55.3		30 28.10	37.22	0.00	VII.	6	4.14	26 5.14	18.7	2.08	31 5.32	46 5.9
13	8						29.5	43.5	32 2.42	37.22	+0.01	VII.	8	9.27	38 41.66	18.6	3.00	32 39.65	58 43.3
14	10				44.2	58.			33 44.21	37.22	+0.01	VI.	8	7.51	37 53.44	18.4	2.94	34 21.44	57 54.8
15	7.8				3.	17.			35 3.06	37.21	−0.01	V.	4	4.3	16 59.59	18.3	1.42	35 40.26	36 59.3
16	9.10		10.	23.					36 37.08	37.21	+0.01	III.	8	2.1	34 57.07	18.2	2.74	37 14.30	54 58.0
17	9			20.	33.	47.			37 33.23	37.21	0.00	V.	6	1.11	24 33.14	18.1	1.97	38 10.44	44 33.2
18	9							28.5	38 47.19	37.21	−0.01	VII.	2	6.53	8 26.91	18.0	0.80	39 24.39	23 28 25.7
19	9						45.	59.	40 17.95	37.21	+0.01	VII.	9	12.34	45 14.96	17.9	3.47	40 55.17	24 5 16.3
20	9							10.5	41 29.39	37.21	0.00	VII.	5	8.38	24 17.16	17.8	1.95	42 6.60	23 44 16.9
21	8					29.	42.		19 43 15.06	+37.21	−0.01	VI.	3	4.36	−12 17.13	−17.7	−1.06	19 43 52.26	−23 32 15.9

CORRECTIONS.

Date.		Corr. of Clock.	Hourly rate.	m	n	c
1848.	h.	s.	s.	s.	s.	s.
Aug. 18,	21	+ 34.21	/ 0.181	− 0.27	+ 0.86	0.00

INSTRUMENT READINGS.

	Date.		Barom.	Thermom. At.	Thermom. Ex.
	1848.	h. m.	in.	°	°

REMARKS.

(185) 111. Hor. thread assumed as 8 instead of 7.
(185) 130. Micrometer reading assumed as 10ʳ.45 instead of 9ʳ.45.
(185) 131. Minutes of transit assumed as 40 instead of 41.
(186) 8. Minutes assumed as 26 instead of 25.

ZONE 186. AUGUST 18. B. BELT, −23° 46'. D_0 = −23° 19' 40"—Continued.

No.	Mag.	SECONDS OF TRANSIT. I.	II.	III.	IV.	V.	VI.	VII.	T.	a_1	a_4	MICROMETER.			i	d_1	d_2	Mean Right Ascension, 1850.0.	Mean Declination, 1850.0.
									h. m. s.	s.	s.			r.	' "	"	"	h. m. s.	° ' "
22	9							36.2	19 44 55.13	+37.21	0.00	VII.	6	4.40	−26 18.25	−17.5	−2.10	19 44 32.34	−23 46 17.9
23	9				56.4	10.			45 56.33	37.21	+0.01	V.	10	3.56	45 54.96	17.5	3.54	46 33.55	24 5 56.0
24	7						32.5	46.3	47 5.38	37.21	−0.01	V.	2	5.11	7 35.77	17.4	0.75	47 42.58	23 27 33.9
25	7					0.1	14.2		48 46.71	37.21	0.00	VI.	5	7.19	23 37.51	17.2	1.90	49 23.92	43 36.6
26	7							7.2	49 26.09	37.21	0.00	VII.	5	8.35	24 15.65	17.2	1.95	50 3.30	23 44 14.8
27	9			24.3	38.2				53 38.07	37.21	+0.01	IV.	9	9.48	43 51.59	16.8	3.39	54 15.29	24 3 51.8
28	6						56.	10.2	53 28.93	37.21	0.00	VII.	4	7.10	18 33.58	16.8	1.54	54 6.14	23 38 31.9
29	9				5.	18.8			58 5.01	37.21	+0.01	VI.	7	11.43	34 51.73	16.5	2.73	58 42.23	23 54 51.0
30	7						29.	43.6	19 59 2.23	37.21	+0.01	VII.	8	14.6	41 2.34	16.4	3.18	19 59 39.45	24 1 1.9
31	9		20.6	33.5	47.				20 2 47.31	37.21	0.00	IV.	6	9.30	28 44.82	16.1	2.26	20 3 24.52	23 48 43.2
32	7						34.8	49.	3 7.80	37.20	0.00	VII.	7	8.16	33 7.17	16.1	2.61	3 45.00	53 5.9
33	9		11.6	25.					6 38.82	37.20	0.00	III.	7	4.42	31 19.56	15.8	2.47	7 16.02	51 17.8
34	7							20.	6 39.08	37.20	+0.01	VII.	8	7.51	37 53.25	15.8	2.95	7 16.29	57 52.0
35	9			14.	28.2				9 27.70	37.20	−0.01	IV.	3	5.2	12 30.39	15.6	1.08	10 4.89	32 27.1
36	8		32.	45.1	58.5				11 58.87	37.20	+0.01	IV.	8	5.33	36 44.01	15.4	2.88	12 36.08	56 42.3
37	8	8.	22.3	35.3					14 49.37	37.20	0.01	III.	8	6.45	37 20.28	15.2	2.92	15 26.58	57 18.4
38	8			37.5	51.	5.			15 51.14	37.20	+0.01	V.	8	7.44	37 50.03	15.1	2.96	16 28.35	57 48.1
39	9.10			8.5	22.	36.			17 22.10	37.20	0.00	V.	6	11.26	29 43.27	15.0	2.35	17 59.30	49 40.6
40	8			40.3	54.5				18 40.45	37.20	−0.01	IV.	2	10.46	10 24.75	14.8	0.93	19 17.64	30 20.5
41	8.9				28.3	42.			20 28.23	37.20	0.00	V.	6	4.2	25 59.39	14.7	2.07	21 5.43	23 45 56.2
42	8				38.	51.5			21 37.88	37.20	+0.01	V.	10	7.39	47 47.41	14.6	3.70	22 15.09	24 7 45.7
43	9						8.	22.	22 40.83	37.20	0.00	VII.	4	9.32	19 45.19	14.6	1.62	23 18.03	23 39 41.4
44	8		13.	26.3	40.2				24 40.08	37.20	0.00	IV.	5	11.8	25 33.13	14.4	2.04	25 17.28	23 45 29.6
45	8			46.5	0.0				26 0.07	37.20	+0.01	IV.	9	7.45	42 49.57	14.3	3.33	26 37.28	24 2 47.2
46	8						49.	3.	26 21.94	37.20	+0.01	VII.	9	7.45	42 49.23	14.3	3.33	26 59.15	24 2 46.9
47	9			14.5	28.3				30 27.98	37.20	−0.01	IV.	2	7.42	8 51.96	14.0	0.82	31 5.17	23 28 46.8
48	8			13.3	26.5	40.5			32 26.63	37.20	0.00	V.	4	10.10	20 4.65	13.9	1.64	33 3.83	23 40 0.2
49	8		47.6	59.5	13.				34 13.73	37.20	+0.01	IV.	9	2.52	40 21.83	13.7	3.15	34 50.94	24 0 18.7
50	9					27.5	41.		35 13.81	37.20	0.00	VI.	5	7.27	23 41.54	13.7	1.90	35 51.01	23 43 37.1
51	9			39.	52.5	6.3			36 52.62	37.20	+0.01	V.	9	10.14	44 4.67	13.5	3.42	37 29.83	24 4 1.6
52	9					38.	51.3		38 24.22	37.20	+0.01	VI.	9	3.33	41 12.60	13.4	3.17	39 1.43	24 1 9.2
53	9					55.	8.		42 41.05	37.20	. .	VI.							
54	9		16.	29.3	42.8				45 42.79	37.20	−0.01	IV.	1	10.22	5 13.64	12.9	0.54	46 19.98	23 25 7.1
55	9.10							13.6	46 46.38	37.20	0.00	VII.	7	4.18	31 7.15	12.8	2.45	47 23.58	51 2.4
56	9					58.3	11.5	25.5	48 44.38	37.20	−0.01	VII.	3	4.8	12 2.81	12.7	1.03	49 21.57	31 56.5
57	9			45.7	59.2	13.			50 59.23	37.21	0.00	V.	6	11.10	29 35.20	12.5	2.34	51 36.44	49 30.0
58	7				4.4	18.3	32.		52 4.55	37.21	0.00	VI.	4	9.34	19 46.39	12.4	1.61	52 41.76	39 40.4
59	9				19.3	33.5	47.		56 19.59	37.21	−0.01	VI.	2	6.36	8 18.54	12.1	0.77	56 56.79	28 11.4
60	6						58.2	12.	57 30.98	37.21	0.00	VII.	6	9.50	28 54.56	12.1	2.29	20 58 8.19	48 49.0
61	8							12.	20 59 30.80	37.21	0.00	VII.	4	4.11	12 4.31	11.9	1.05	21 0 8.01	31 57.3
62	8					34.5	48.		21 1 20.82	37.21	0.00	VI.	7	11.43	34 51.73	11.8	2.74	1 58.03	54 46.3
63	7				8.5	22.5	35.5		4 8.48	37.21	−0.01	VI.	1	5.23	2 42.71	11.6	0.35	4 45.68	22 34.7
64	8						42.	56.	6 14.78	37.21	−0.01	VII.	2	9.33	9 47.59	11.5	0.88	5 51.98	29 40.0
65	9						12.5	26.3	6 45.28	37.21	0.00	VII.	6	7.32	27 44.98	11.4	2.20	7 22.49	23 47 38.6
66	9.10			49.	2.5				12 2.56	37.21	+0.01	IV.	9	6.7	42 0.16	11.0	3.28	12 39.78	24 1 54.4
67	9.10				40.	53.8			13 40.03	37.22	+0.01	V.	10	5.5	46 29.75	11.0	3.62	14 17.26	24 6 24.4
68	7			54.	8.	21.3			15 7.73	37.22	0.00	V.	8	3.58	35 56.06	10.9	2.83	15 44.95	23 55 49.8
69	8						30.	44.	16 2.91	37.22	0.00	VII.	7	11.18	34 38.93	10.8	2.73	16 40.13	54 32.5
70	9			39.	53.5				19 52.87	37.22	0.00	IV.	3	9.52	14 56.61	10.6	1.24	20 30.09	34 48.5
71	9		14.	27.					21 20 41.04	−37.22	0.00	III.	7	9.51	−33 55.36	−10.5	−2.68	21 22 18.26	−23 53 48.5

CORRECTIONS.

Date.		Corr. of Clock.	Hourly rate.	m	n	c
1848.	h.	s.	s.	s.	s.	s.

INSTRUMENT READINGS.

Date.		Barom.	THERMOM. At.	THERMOM. Ex.
1848.	h. m.	in.	°	°

REMARKS.

(186) 22. Minute assumed as 43 instead of 44.
(186) 40. Transits over T's IV and V assumed as recorded over T's III and IV.
(186) 52. Micrometer reading assumed as 4r.33 instead of 3r.33.
(186) 53. Omitted micrometer reading.
(186) 61. Hor. thread assumed as 3 instead of 4.
(186) 64. Minute assumed as 5 instead of 6.
(186) 71. Minute assumed as 21 instead of 22.

ZONE 186. AUGUST 18. B. BELT, 23° 46'. $D_o = -23° 19' 40''$—Continued.

No.	Mag.	I.	II.	III.	IV.	V.	VI.	VII.	T.	a_1	a_2	MICROMETER.			i	d_1	d_2	Mean Right Ascension, 1850.0.	Mean Declination, 1850.0.
									h. m. s.	s.	s.			r.	' "	"	"	h. m. s.	° ' "
72	9.10						19.	33.	21 20 51.89	+37.22	0.00	VII.	7	6.2	−31 59.60	−10.5	−2.53	21 22 29.11	−23 51 52.6
73	8		36.2	49.	2.5				23 2.79	37.22	0.00	IV.	5	6.30	23 12.96	10.4	1.87	23 40.01	43 5.2
74	8					25.3	39.		26 11.71	37.22	0.00	VI.	6	6.21	27 9.36	10.2	2.16	26 48.93	47 1.7
75	8		21.	34.2	47.8				28 47.90	37.23	0.00	IV.	5	8.48	24 22.54	10.0	1.96	29 25.13	44 14.5
76	8		20.5	33.3	46.5				30 47.06	37.23	0.00	IV.	7	7.18	32 38.25	9.9	2.58	31 24.29	52 30.7
77	9.10					58.			31 44.	37.23	. .	.	. .	.	. . .	9.8			
78	4.5							31.	32 50.06	37.23	0.00	VII.	8	4.51	36 22.49	9.8	2.86	33 27.29	56 15.2
79	8		34.	47.	0.5	14.			21 40 0.53	+37.24	−0.01	V.	2	9.53	−10 58.45	− 9.4	−0.86	21 40 37.76	−23 30 48.6

ZONE 187. AUGUST 24. B. BELT, −21° 16'. $D_o = -20° 49' 10''$.

No.	Mag.	I.	II.	III.	IV.	V.	VI.	VII.	T.	a_1	a_2	MICROMETER.			i	d_1	d_2	Mean Right Ascension, 1850.0.	Mean Declination, 1850.0.
1	9					12.5	26.		22 37 59.20	+43.99	+0.01	VI.	10	3.1	−45 27.12	− 9.3	−3.39	22 38 43.20	−21 34 49.8
2	9			53.5	7.	20.3			41 6.77	43.98	0.00	V.	3	10.26	15 13.73	9.4	1.34	41 50.75	21 4 34.5
3	7			29.5	42.3	56.2		23.	46 42.44	43.98	−0.01	V.	2	3.44	6 51.92	9.7	0.80	47 26.41	20 56 12.4
4	8			36.7	50.	3.4			48 49.87	43.97	0.00	IV.	3	10.14	15 7.71	9.8	1.33	49 33.84	21 4 28.8
5	8			45.5		12.			22 53 58.90	43.97	+0.01	V.	10	14.44	50 51.46	10.1	3.78	22 54 42.88	40 15.3
6	10				58.	11.5			23 3 57.98	43.96	0.00	V.	7	8.41	33 20.08	10.7	2.56	23 4 41.94	21 22 43.3
7	10		4.	17.3	30.5				6 30.48	43.96	0.00	IV.	2	7.57	8 39.52	10.8	0.92	7 14.44	20 58 21.2
8	9					31.	44.3	58.	7 17.56	43.96	0.00	VII.	4	4.14	17 4.89	10.8	1.46	8 1.52	21 6 27.2
9	10			51.2	4.5				9 4.37	43.96	0.00	IV.	4	6.11	18 4.17	10.9	1.53	9 48.33	7 26.6
10	9		27.	39.5	53.				10 53.28	43.96	0.00	IV.	9	5.50	41 51.58	11.1	3.15	11 37.24	21 31 15.8
11	7					1.			11 47.44	43.95	0.00	V.	1	2.51	1 26.19	11.1	0.41	12 31.39	20 50 47.7
12	8					8.5	21.5	35.	12 54.80	43.95	0.00	VII.	4	5.10	17 33.12	11.2	1.50	13 38.75	21 6 55.8
13	4.5						47.	01.	14 20.37	43.95	0.00	VII.	1	11.32	5 48.65	11.3	0.70	15 4.32	20 55 10.7
14	7						25.5	39.	15 58.70	43.95	0.00	VII.	5	6.48	23 21.74	11.4	1.89	15 42.65	21 12 45.0
15	5.6					39.	52.	5.2	16 25.31	43.95	0.00	VII.	8	8.47	38 21.55	11.4	2.92	18 9.26	27 45.9
16	8			38.3	52.	5.5			19 51.86	43.95	0.00	V.	6	8.4	28 1.43	11.6	2.20	20 35.81	17 25.2
17	8			4.	17.5				21 17.38	43.95	0.00	IV.	7	11.33	34 46.84	11.7	2.68	22 1.33	21 24 11.2
18	6				30.	44.	57.		24 30.22	43.95	0.00	V.	2	11.1	5 33.27	11.9	0.68	25 14.17	20 54 56.0
19	9					16.2	29.5		23 26 2.76	+43.95	0.00	VI.	2	11.5	−10 34.19	−12.1	−1.02	23 26 46.71	−20 59 47.3

ZONE 188. AUGUST 29. B. BELT, −21° 16'. $D_o = -20° 48' 40''$.

No.	Mag.	I.	II.	III.	IV.	V.	VI.	VII.	T.	a_1	a_2	MICROMETER.			i	d_1	d_2	Mean Right Ascension, 1850.0.	Mean Declination, 1850.0.
1	9				10.	23.3	36.5		22 35 9.84	+44.55	+0.01	VI.	9	7.28	−42 40.85	−9.92	−3.17	22 35 54.40	−21 31 33.9
2	8		20.	33.	46.5				37 46.42	44.54	0.00	IV.	3	8.19	14 9.72	9.76	1.31	38 30.96	3 0.8
3	9			10.5	23.5	38.			40 23.83	44.54	0.00	V.	3	8.32	14 16.25	9.65	1.31	41 8.37	3 7.2
4	7.8								41 5.79	44.54	0.00	VII.	3	11.20	15 40.69	9.61	1.41	41 50.33	21 4 31.7
5	9		21.5	34.5	48.	1.3		46.3	42 47.72	44.53	−0.01	VI.	1	5.49	2 55.84	9.53	0.56	43 32.24	20 51 46.9
6	8					27.	40.		45 13.43	44.53	0.00	VI.	5	3.7	21 30.45	9.42	1.78	45 57.96	21 10 21.7
7	6.7					56.	9.	22.	46 42.07	44.52	−0.01	VII.	2	4.52	7 25.95	9.33	0.86	47 26.58	20 56 16.1
8	7					3.	16.	29.5	48 49.31	44.52	0.00	VII.	4	11.19	20 39.19	9.24	1.73	49 33.83	21 9 30.2
9	9					48.		15.	50 34.72	44.51	0.00	VII.	8	4.12	36 2.88	9.16	2.73	51 19.23	24 54.8
10	9		40.	53.3	6.5				53 6.50	44.51	0.00	IV.	3	4.19	12 8.70	9.06	1.17	53 51.01	0 58.9
11	9						7.2	21.	53 40.61	44.51	0.00	VII.	7	3.45	30 50.56	9.04	2.37	54 25.11	19 42.0
12	8							19.5	54 39.30	44.51	0.00	VII.	8	3.30	35 41.70	9.00	2.71	55 23.81	24 33.4
13	9			24.	37.5	51.			56 37.48	44.50	+0.01	V.	8	7.46	37 51.04	8.90	2.85	57 21.99	26 42.8
14	9				19.	32.6			22 58 19.01	+44.50	0.00	V.	6	7.36	−27 47.31	−8.83	−2.18	22 59 3.51	−21 16 38.3

CORRECTIONS.

Date.		Corr. of Clock.	Hourly rate.	m	n	c
1848.	h.	s.	s.	s.	s.	s.
Aug. 24.	21	+ 41.40	*l* 0.020			
29.	21	+ 41.90	*g* 0.021	− 0.31	+ 0.84	0.00

INSTRUMENT READINGS.

Date.		Barom.	THERMOM. At.	THERMOM. Ex.
1848.	h. m.	In.	°	°

REMARKS.

(186) 72. Minutes assumed as 21 instead of 22.
(186) 77. Star disappeared before micrometer reading was obtained.
(186) 79. Micrometer reading assumed as 11r.53 instead of 9r.53.
(187) 5. Micrometer reading assumed as 13r.44 instead of 14r.44.
(187) 15. Minutes assumed as 17, not 6.
(187) 18. Micrometer thread assumed as 1 instead of 2.
(188) 4. Transit over VII assumed to have been recorded against following star.

ZONE 188. AUGUST 29. B. BELT, −21° 16′. D_0 = −20° 48′ 40″—Continued.

No.	Mag.	Seconds of Transit I.	II.	III.	IV.	V.	VI.	VII.	T.	a_1	a_2	Micrometer.			i	d_1	d_2	Mean Right Ascension, 1850.0.	Mean Declination, 1850.0.
									h. m. s.	s.	s.			r.	′ ″	″	″	h. m. s.	° ′ ″
15	9	. .	. .	. .	. .	32.7	46.	59.	22 59 19.12	+44.50	0.00	VII.	7	12.36	−35 18.32	−8.79	−2.68	23 0 3.62	−21 24 9.8
16	9	. .	31.5	44.	. .	. .	. .	. .	23 3 57.92	44.49	0.00	III.	7	9.52	33 55.88	8.60	2.59	4 42.41	22 47.1
17	9	. .	. .	23.5	37.	. .	. .	. .	6 36.86	44.48	0.00	IV.	6	13.13	30 37.25	8.50	2.36	7 21.34	19 28.1
18	8	. .	1.7	14.	27.	. .	. .	. .	8 27.68	44.48	0.00	IV.	8	14.23	41 11.24	8.40	3.07	9 12.16	30 2.7
19	9	12.	26.5	39.	52.4	. .	. .	. .	10 52.72	44.47	0.00	IV.	9	6.56	42 24.87	8.32	3.15	11 37.19	31 16.3
20	9	. .	. .	41.	54.2	8.	. .	. .	12 54.25	44.47	0.00	V.	4	6.8	18 2.63	8.24	1.55	13 38.72	21 6 52.4
21	4.5	. .	. .	7.	20.3	34.	. .	. .	14 20.22	44.47	0.00	V.	1	12.27	6 16.63	8.20	0.79	15 4.69	20 55 5.6
22	8	. .	. .	45.2	58.	. .	. .	. .	15 58.16	44.46	0.00	IV.	5	7.53	23 54.80	8.13	1.93	16 42.62	21 12 44.9
23	8	. .	. .	. .	. .	. .	. .	53.3	16 12.82	44.46	0.00	VII.	4	2.56	16 25.56	8.13	1.45	16 57.28	5 15.1?
24	5	. .	. .	. .	. .	. .	. .	5.	17 24.84	44.46	0.00	VII.	8	9.51	38 53.81	8.06	2.92	18 9.30	27 44.8
25	8	. .	. .	. .	51.3	5.	. .	. .	19 51.37	44.46	0.00	V.	6	8.2	28 30.66	7.99	2.22	20 35.83	17 20.9
26	8	. .	. .	. .	10.	23.2	. .	. .	21 9.79	44.45	0.00	V.	4	7.42	18 50.03	7.96	1.60	21 54.24	21 7 39.6
27	8	50.	. .	16.3	30.	. .	. .	. .	24 29.80	44.45	−0.01	III.	1	1.53	5 55.94	7.86	0.50	25 14.24	20 54 44.3
28	9	. .	. .	49.	2.5	. .	. .	. .	26 2.23	44.45	0.00	IV.	3	2.16	11 6.68	7.79	1.09	26 46.68	59 55.6
29	9	. .	. .	5.	18.5	32.	. .	. .	27 18.27	44.44	0.00	V.	1	7.34	3 48.90	7.76	0.61	28 2.71	20 52 37.3
30	9	. .	. .	51.3	4.5	. .	. .	. .	29 4.51	44.44	0.00	IV.	7	3.26	31 11.53	7.72	2.39	29 48.95	21 20 1.6
31	9	. .	. .	. .	54.3	8.	. .	. .	30 54.38	44.44	0.00	V.	7	11.7	34 33.70	7.67	2.63	31 38.82	23 24.0
32	9	32.	46.	. .	. .	. .	. .	. .	33 12.56	44.44	0.00	II.	2	12.28	11 16.02	7.60	1.11	33 57.00	0 4.7
33	8	. .	29.3	43.1	55.5	. .	. .	. .	40 55.99	44.43	0.00	IV.	6	7.15	27 36.74	7.41	2.17	41 40.42	16 26.3
34	7	. .	. .	. .	. .	17.1	30.	44.	42 3.48	44.42	0.00	VII.	3	10.14	15 7.42	7.39	1.36	42 47.90	3 56.2
35	8	29.	. .	56.	9.	. .	. .	. .	45 9.26	44.42	0.00	IV.	7	6.17	32 7.50	7.33	2.46	45 53.68	20 57.3
36	8	. .	. .	. .	. .	. .	. .	45.6	48 5.07	44.42	0.00	VII.	3	4.7	12 2.36	7.24	1.14	48 49.49	0 50.7
37	8	. .	. .	. .	54.	7.5	. .	. .	49 53.95	44.42	0.00	V.	4	10.32	20 15.76	7.21	1.70	50 38.37	9 4.7
38	8	. .	. .	. .	. .	. .	6.	19.	50 39.04	44.41	0.00	VII.	8	9.32	38 44.24	7.20	2.92	51 23.45	27 34.4
39	9	. .	. .	. .	29.	42.	. .	. .	52 28.72	44.41	0.00	V.	6	5.15	26 36.20	7.17	2.10	53 13.13	21 15 25.5
40	7	. .	. .	. .	30.	43.8	57.	. .	23 54 30.15	+44.41	0.00	VI.	1	8.27	− 4 15.51	−7.14	−0.63	23 55 14.56	−20 53 3.3

ZONE 189. AUGUST 31. K. BELT, −21° 53′. D_0 = −21° 26′ 20″.

No.	Mag.	I.	II.	III.	IV.	V.	VI.	VII.	T.	a_1	a_2	Micrometer.			i	d_1	d_2	Mean Right Ascension, 1850.0.	Mean Declination, 1850.0.
1	10	. .	. .	. .	. .	. .	. .	21.5	22 30 40.95	+44.40	0.00	VII.	5	7.30	−23 42.89	− 3.4	−1.91	22 31 25.35	−21 50 8.2
2	9	. .	. .	53.	6.3	20.	. .	. .	33 6.23	44.40	0.00	V.	2	5.1	7 30.74	3.3	0.84	33 50.63	33 54.9
3	9	. .	. .	. .	. .	. .	12.	25.4	33 44.99	44.40	0.00	VII.	4	4.35	17 15.45	3.2	1.48	34 29.39	43 40.1
4	9	. .	. .	29.5	43.	. .	. .	. .	35 42.79	44.39	0.00	IV.	4	4.57	17 26.86	3.1	1.48	36 27.18	43 51.4
5	9	. .	. .	. .	26.3	40.	. .	. .	36 26.30	44.39	0.00	V.	3	6.3	13 1.11	3.1	1.21	37 10.69	39 25.4
6	8	. .	32.4	45.5	59.	. .	. .	. .	37 58.89	44.38	0.00	V.	2	6.53	8 27.22	3.0	0.90	38 43.27	34 51.1
7	8	. .	. .	. .	. .	. .	2.	15.5	38 35.07	44.38	0.00	VII.	5	7.59	23 57.51	3.0	1.93	39 19.45	50 22.4
8	9	. .	. .	11.5	25.	38.	. .	. .	41 24.70	44.38	0.00	V.	4	9.33	19 46.00	2.9	1.65	42 9.08	46 10.6
9	9	. .	22.	35.	48.2	2.	. .	. .	42 48.41	44.37	0.00	V.	4	5.21	17 38.92	2.8	1.56	43 32.78	44 3.2
10	9	. .	. .	. .	. .	52.	5.	. .	43 38.34	44.37	0.00	VI.	7	6.39	32 17.45	2.8	2.50	44 22.71	21 58 42.8
11	9	. .	. .	. .	15.	28.5	. .	. .	45 14.97	44.37	0.00	V.	9	4.42	41 17.27	2.7	3.11	45 59.34	22 7 43.1
12	9	. .	. .	. .	. .	. .	58.	11.5	46 31.11	44.36	0.00	VII.	7	7.32	32 45.00	2.7	2.53	47 15.47	21 59 10.2
13	9	. .	. .	8.	21.7	. .	. .	. .	48 21.50	44.36	0.00	IV.	7	6.15	32 6.49	2.6	2.48	49 5.86	58 31.6
14	9	. .	. .	. .	32.5	46.	. .	. .	49 32.45	44.36	0.00	V.	7	8.58	33 28.65	2.6	2.58	50 16.81	59 53.8
15	8	. .	. .	. .	. .	. .	. .	47.	50 6.34	44.36	0.00	VII.	3	10.12	15 6.38	2.5	1.33	50 50.70	41 30.2
16	9	. .	. .	. .	. .	0.0	13.	. .	51 46.33	44.35	0.00	VI.	2	11.19	10 41.24	2.5	1.05	52 30.68	37 4.8
17	7	18.	32.2	45.	. .	. .	. .	. .	22 53 58.61	44.35	0.00	III.	3	7.41	13 50.52	2.4	1.24	22 54 42.96	40 14.2
18	8	. .	. .	. .	. .	. .	. .	29.	23 15 48.47	44.30	0.00	VII.	5	10.15	25 6.09	1.6	2.01	23 16 32.77	21 51 29.7
19	8	. .	. .	39.	52.5	6.	. .	. .	23 17 52.46	+44.30	0.00	V.	7	10.57	−34 28.66	− 1.6	−2.65	23 18 36.76	−22 0 52.9

CORRECTIONS.

Date.		Corr. of Clock.	Hourly rate.	m	n	c
1848.	h.	s.	s.	s.	s.	s.
Aug. 31,	21	+ 41.78	0.000	− 0.36	+ 0.86	0.00

INSTRUMENT READINGS.

Date.			Barom.	Thermom. At.	Ex.
1848.	h.	m.	In.	°	°

REMARKS.

(188) 25. Micrometer reading assumed as $9^r.2$ instead of $8^r.2$.
(188) 27. Micrometer thread assumed as 2 instead of 1.
(188) 30. Micrometer reading assumed as $4^r.26$ instead of $3^r.26$.
(188) 33. Transit over T. II assumed as $29^s.3$.

ZONE 189. AUGUST 31. K. BELT, −21° 53′. D_o = −21° 26′ 20″—Continued.

No.	Mag.	Seconds of Transit. I.	II.	III.	IV.	V.	VI.	VII.	T.	a_1	a_2	Micrometer.			i	d_1	d_2	Mean Right Ascension, 1850.0.	Mean Declination, 1850.0.
									h. m. s.	s.	s.			r.	′ ″	″	″	h. m. s.	° ′ ″
20	7	24.2	39.	51.5					23 23 5.43	+44.30	0.00	III.	9	12.51	−45 23.83	−1.4	−3.39	23 23 49.73	−22 11 48.6
21	7							52.5	23 12.16	44.30	0.00	VII.	8	8.15	38 5.38	1.4	2.90	23 56.46	22 4 29.7
22	6						7.3	21.	24 40.44	44.29	0.00	VII.	4	6.30	18 13.44	1.4	1.54	25 24.73	21 44 36.4
23	8	27.2	41.	54.2	7.3				30 7.55	44.28	0.00	IV.	3	10.52	15 26.87	1.2	1.35	30 51.83	41 49.5
24	9	5.3	19.5	32.2	45.4				34 45.84	44.28	0.00	IV.	6	9.50	28 54.90	1.1	2.27	35 30.12	55 18.3
25	10			23.2	37.	50.5			40 36.79	44.27	0.00	V.	5	9.54	21 55.79	1.0	1.99	41 21.06	51 18.8
26	8					59.	12.		41 45.33	44.27	0.00	VI.	4	3.4	16 29.74	1.0	1.41	42 29.60	42 52.2
27	8					16.5	29.5		43 2.83	44.27	0.00	VI.	2	11.3	10 33.17	1.0	1.03	43 47.10	21 36 55.2
28	7		15.7		41.2				46 40.95	44.26	0.00	IV.	10	4.48	46 21.22	0.9	3.48	47 25.21	22 12 45.6
29	7		16.2	29.4	42.5				48 42.67	44.26	0.00	IV.	3	7.25	13 42.48	0.9	1.23	49 26.93	21 40 4.6
30	8				39.	53.			50 39.12	44.26	0.00	V.	1	2.31	1 16.11	0.9	0.39	51 23.38	27 37.4
31	9			30.	43.	57.			43.12	. .	0.00	V.	2	2.16	6 7.55	. .	0.72		
32	9				43.5	59.			54 43.41	44.25	0.00	V.	4	1.50	15 52.53	0.8	1.37	54 27.66	21 42 14.7
33	8			42.	55.3				23 56 55.37	44.25	0.00	V.	9	7.18	42 35.92	0.8	3.22	23 57 39.62	22 3 59.9
34	8			46.2	0.0				0 0 59.86	44.25	0.00	IV.	10	8.44	48 20.23	0.7	3.60	0 1 44.11	14 44.5
35	7						2.	15.5	0 15.42	44.25	0.00	VII.	8	4.6	35 59.83	0.7	2.76	1 59.67	2 23.3
36	8		29.	42.	55.				2 55.49	44.25	0.00	IV.	8	9.2	38 29.40	0.7	2.94	3 39.74	22 4 53.0
37	8				55.	9.			4 55.12	44.25	0.00	V.	1	1.52	0 56.44	0.7	0.36	5 39.37	21 27 17.5
38	6						26.4	40.	6 59.58	44.24	0.00	VII.	8	1.54	34 53.26	0.7	2.68	6 43.82	22 1 16.6
39	7				56.	10.			9 55.20	44.24	0.00	V.	7	5.45	31 51.33	0.7	2.48	10 39.44	21 58 14.5
40	7			22.	35.3	49.			18 25.20	44.24	0.00	V.	1	6.44	3 23.69	0.6	0.52	19 9.44	21 29 44.8
41	9						14.		19 47.36	44.24	0.00	VI.	9	4.14	41 3.04	0.6	3.12	20 31.60	22 7 26.8
42	9			42.2	55.3	9.			22 55.52	44.24	0.00	V.	9	10.52	44 23.84	0.6	3.35	23 39.76	22 10 47.8
43	7			39.	52.	6.			27 52.13	44.24	0.00	V.	2	7.29	8 45.37	0.6	0.90	28 36.37	21 35 6.9
44	9				31.4	45.			0 30 31.38	+44.23	0.00	V.	5	9.55	−25 26.55	−0.6	−2.03	0 31 15.61	−21 51 49.2

ZONE 190. SEPTEMBER 1. B. BELT, −20° 38′. D_o = −20° 10′ 0″.

No.	Mag.	I.	II.	III.	IV.	V.	VI.	VII.	T.	a_1	a_2	Micrometer.			i	d_1	d_2	Mean Right Ascension, 1850.0.	Mean Declination, 1850.0.
1	10					5.5	19.		20 3 52.25	+47.78	−0.01	VI.	2	9.2	−9 32.16	−34.5	−1.10	20 4 40.02	−20 20 7.8
2	10				33.	46.4			5 32.91	47.77	−0.01	V.	3	11.53	15 57.59	34.2	1.46	6 20.68	26 33.3
3	8					58.	11.		6 44.51	47.77	0.00	V.	4	5.46	17 51.54	34.0	1.57	7 32.25	28 27.1
4	9					59.	12.		7 45.50	47.77	−0.01	VI.	3	9.12	14 36.30	33.8	1.38	8 33.26	25 11.5
5	8		37.						10 3.63	47.76	0.01	II.	3	4.10	12 4.02	33.4	1.26	10 51.38	22 38.7
6	9					38.			10 24.55	47.76	−0.01	V.	3	4.1	11 59.59	33.3	1.25	11 12.30	22 34.1
7	8							38.5	10 58.18	47.76	−0.01	VII.	3	10.58	15 29.61	33.2	1.43	11 45.93	26 4.2
8	9						18.	32.	12 51.65	47.76	0.00	VII.	7	5.18	31 37.46	32.8	2.40	13 39.41	42 12.7
9	8						4.5		13 37.94	47.76	−0.01	VI.	3	5.40	12 49.49	32.7	1.27	14 25.69	23 23.4
10	8					42.	54.5		15 28.25	47.75	0.00	VI.	4	8.7	19 2.53	32.3	1.64	16 16.00	29 36.5
11	10	57.	10.5						17 37.15	47.75	−0.01	II.	2	9.41	9 51.82	31.9	1.11	18 24.89	20 24.8
12	9				34.				18 33.85	47.75	+0.01	IV.	8	6.3	36 59.14	31.8	2.73	19 21.61	47 33.7
13	8							31.	18 51.04	47.75	0.01	VII.	9	3.17	40 34.15	31.7	2.94	19 38.80	51 8.8
14	10				52.5	6.			20 52.51	47.74	+0.01	V.	8	4.20	36 7.17	31.3	2.67	21 40.26	46 41.1
15	10					6.5			21 53.03	47.74	−0.01	V.	2	4.56	7 28.23	31.2	0.96	22 40.76	18 0.4
16	9					19.3			23 5.88	47.74	0.00	V.	4	5.27	17 41.95	30.9	1.53	23 53.62	28 14.4
17	11		37.						25 3.76	47.73	0.00	II.	6	7.12	27 35.09	30.6	2.17	25 51.49	38 7.9
18	10					50.	3.		25 36.52	47.73	0.00	VI.	6	10.16	29 7.88	30.5	2.24	26 24.25	39 40.6
19	10								27	47.73	−0.01	IV.	2	10.23	10 13.14	30.3	1.13	27	20 44.6
20	10			3.	16.3				20 28 16.10	+47.73	−0.01	VII.	3	3.10	−11 33.62	−30.0	−1.22	20 29 3.82	−20 22 4.8

CORRECTIONS.

Date.		Corr. of Clock.	Hourly rate.	m	n	c
1848.	h.	s.	s.	s.	s.	s.
Sept. 1,	21	+ 44.80	*l* 0.123	− 0.37	+ 0.92	0.00

INSTRUMENT READINGS.

Date.		Barom.	Thermom. At.	Thermom. Ex.
1848.	h. m.	in.	°	°

REMARKS.

(189) 28. Transit over II assumed as 13ˢ.7 instead of 15ˢ.7.
(189) 32. Transit observations discordant; thread V assumed to have been observed at 57ˢ.0 instead of 59ˢ IV; minutes assumed as 53 instead of 54.
(189) 35. Minutes assumed as 1 instead of 0; and transits over T.'s III and IV assumed as recorded over T.'s VI and VII.
(189) 38. Minutes assumed as 5 instead of 6.
(189) 40. Transits III–V assumed as 12ˢ, 25ˢ.3, and 39ˢ, instead of 22ˢ, 35ˢ.3, and 49ˢ.
(189) 44. Micrometer reading assumed as 10ʳ.55 instead of 9ʳ.55.

ZONE 190. SEPTEMBER 1. B. BELT, −20° 38'. D_o = −20° 10' 0"—Continued.

No.	Mag.	SECONDS OF TRANSIT. I.	II.	III.	IV.	V.	VI.	VII.	T.	a_1	a_2	MICROMETER.			i	d_1	d_2	Mean Right Ascension, 1850.0.	Mean Declination, 1850.0.
									h. m. s.	s.	s.			r.	' "	"	"	h. m. s.	° ' "
21	11	...	...	...	...	9.	...	39.	20 29 56.23	+47.72	0.00	V.	6	4.50	−26 23.60	−29.7	−2.09	20 30 43.95	−20 36 55.4
22	11	...	56.5	9.5	...	...	...	...	33 22.94	47.72	0.00	III.	5	4.21	22 7.87	29.1	1.81	34 10.66	32 38.8
23	9	...	25.	37.	...	...	...	...	35 51.22	47.71	+0.01	III.	10	4.125	46 3.29	28.6	3.28	36 38.94	56 35.2
24	10	...	47.3	...	...	...	...	...	37 14.02	47.71	0.00	II.	5	7.58	23 57.18	28.4	1.93	38 1.73	34 27.5
25	10	...	...	...	...	...	...	8.2	37 28.09	47.71	0.00	VII.	6	12.13	30 6.72	28.3	2.31	38 15.80	40 37.3
26	9	...	9.8	21.2	34.3	...	...	...	42 34.29	47.70	−0.01	IV.	3	7.15	13 37.44	27.4	1.32	43 21.98	24 6.2
27	10	...	47.	...	...	...	...	...	44 13.60	47.70	0.01	IV.	2	8.55	9 28.77	27.1	1.09	45 1.29	19 57.0
28	9	...	...	...	16.5	30.	...	...	46 16.45	47.69	−0.01	V.	3	5.34	12 46.49	26.7	1.28	47 4.13	23 14.5
29	10	...	...	...	12.7	...	...	...	48 12.55	47.69	0.00	IV.	6	3.32	25 44.30	26.4	2.04	49 0.24	36 12.7
30	8	...	...	...	...	...	...	14.	48 33.87	47.69	0.00	VII.	6	8.17	28 7.73	26.3	2.20	49 21.56	38 36.2
31	9	...	14.	27.	...	...	...	...	50 40.46	47.68	0.00	III.	5	6.10	23 2.84	26.0	1.89	51 28.14	33 30.7
32	8	...	...	19.3	32.7	...	...	...	51 32.50	47.68	0.00	III.	4	9.4	19 31.38	25.8	1.67	52 20.18	29 58.9
33	10	...	29.	41.3	...	...	...	...	52 55.35	47.68	+0.01	III.	9	11.30	44 43.00	25.6	3.20	53 43.04	55 11.8
34	8	...	...	...	59.2	13.	...	...	53 59.29	47.68	−0.01	V.	2	6.36	8 18.66	25.4	1.00	54 46.96	18 45.1
35	4.5	...	...	...	...	17.	30.3	44.	55 3.67	47.68	−0.01	VII.	3	12.30	16 16.00	25.2	1.48	55 51.34	26 42.7
36	7	...	54.5	7.3	...	...	...	...	57 21.01	47.67	+0.01	III.	8	4.23	36 8.68	24.8	2.67	58 8.69	46 36.2
37	9	...	...	...	...	...	55.2	9.	58 28.73	47.67	0.00	VII.	6	7.21	27 39.48	24.6	2.17	20 59 16.40	38 6.3
38	7	...	...	...	54.	8.	...	...	20 59 54.27	47.67	+0.01	V.	8	6.54	37 24.83	24.3	2.75	21 0 41.95	47 51.9
39	8	...	16.2	29.	...	...	...	...	21 2 42.82	47.66	+0.01	III.	10	4.13	46 3.54	23.9	3.29	3 30.49	56 30.7
40	7	58.	12.	...	...	...	...	...	4 38.58	47.66	0.00	II.	7	5.45	31 51.22	23.5	2.42	5 26.24	42 17.1
41	8	...	11.6	24.5	...	...	...	...	5 38.10	47.66	0.00	III.	7	3.59	30 57.88	23.4	2.37	6 25.76	41 23.7
42	8	41.	55.1	...	...	...	...	...	7 21.68	47.66	+0.01	II.	8	6.22	37 8.58	23.1	2.74	8 9.35	47 34.4
43	10	...	...	...	...	14.	27.2	...	8 0.60	47.65	0.00	VI.	3	9.19	14 39.83	23.0	1.37	8 48.25	25 4.2
44	10	...	...	...	36.	...	...	...	9 35.85	47.65	−0.01	IV.	3	2.10	11 3.65	22.7	1.17	10 23.49	21 27.5
45	9	...	...	...	...	...	...	28.	9 47.94	47.65	0.00	VII.	7	9.3	33 30.92	22.6	2.52	10 35.59	43 56.0
46	9	...	...	...	...	...	25.	29.	10 58.57	47.65	0.00	VII.	4	3.37	16 46.24	22.4	1.50	11 46.22	27.10.1
47	9	...	...	...	...	1.6	15.	...	12 48.31	47.65	0.00	VI.	5	7.19	23 37.52	22.1	1.92	13 35.96	34 1.5
48	10	17.	30.5	...	...	...	...	...	14 57.19	47.64	0.00	II.	3	11.10	15 35.80	21.8	1.43	15 44.83	25 59.0
49	9	...	...	...	13.	...	...	...	18 12.85	47.64	+0.01	IV.	9	4.8	41 0.15	21.2	2.98	19 0.50	51 24.3
50	9	...	52.	5.	18.2	...	...	...	20 18.26	47.64	0.00	IV.	3	9.9	14 34.93	20.9	1.37	21 5.90	24 57.2
51	10	...	...	...	43.	45.3	57.	...	22 30.27	47.63	+0.01	VI.	8	9.20	38 38.44	20.5	2.83	23 17.91	49 1.8
52	11	...	...	...	...	...	17.	30.5	23 50.40	47.63	0.00	VII.	7	8.5	33 1.68	20.3	2.49	24 38.03	43 24.5
53	7	...	...	24.	37.	...	...	...	25 37.11	47.63	0.00	IV.	7	11.15	34 37.76	20.0	2.59	26 24.74	45 0.4
54	7	...	...	...	...	...	...	20.	25 40.09	47.63	+0.01	VII.	9	10.52	44 23.59	20.0	3.19	26 27.73	54 46.8
55	11	...	...	...	...	...	25.	39.	26 58.60	47.63	0.00	VII.	5	7.2	23 28.81	19.7	1.90	27 46.23	33 50.4
56	11	...	...	...	...	...	56.3	...	28 29.76	47.62	−0.01	VI.	2	3.52	6 55.84	19.5	0.91	29 17.37	17 16.3
57	9	...	...	...	...	...	4.	17.5	29 37.27	47.62	−0.01	VII.	2	5.4	7 32.01	19.3	0.94	30 24.88	17 52.3
58	8	54.	8.	21.	...	...	...	...	32 34.36	47.62	0.00	IV.	4	7.28	18 42.99	18.8	1.61	33 21.98	29 3.4
59	11	...	...	37.7	50.	...	...	...	33 50.32	47.62	0.00	IV.	3	8.52	14 26.36	18.6	1.35	34 37.94	24 46.3
60	7	...	...	...	...	...	...	42.	34 1.59	47.62	−0.01	VII.	2	5.43	7 51.67	18.6	0.97	34 49.20	18 11.2
61	11	...	...	...	52.6	...	...	...	35 52.45	47.62	0.00	IV.	8	6.55	37 25.36	18.2	2.76	36 40.07	47 46.3
62	8	...	5.	18.	...	...	...	...	38 31.27	47.61	−0.01	III.	1	11.28	5 46.88	17.8	0.84	39 18.87	16 5.5
63	11	...	26.	38.8	...	...	...	...	39 52.55	47.61	+0.01	III.	8	12.15	40 6.68	17.6	2.93	40 40.17	50 27.2
64	11	...	47.	0.0	...	...	...	...	41 23.36	47.61	0.00	III.	3	8.11	14 5.65	17.4	1.34	42 10.97	24 24.4
65	9	...	...	...	...	...	33.	46.5	42 6.35	47.61	0.00	VII.	5	5.26	22 40.40	17.2	1.86	42 53.96	32 59.5
66	.	...	...	...	...	39.	52.2	...	43 25.6	47.61	. .	.	.	. . .	. . .	17.0			
67	9	...	...	...	...	...	3.5	17.	44 36.91	47.61	0.00	VII.	7	7.17	32 37.47	16.8	3.47	45 24.52	42 56.7
68	8.	...	...	...	...	...	37.	...	46 10.45	47.60	−0.01	VI.	2	10.45	10 24.12	16.5	1.11	46 58.04	20 41.7
69	10	...	...	...	...	...	...	5.	21 47 24.94	+47.60	0.00	VII.	7	9.26	−33 42.52	−16.3	−2.53	21 48 12.54	−20 44 1.4

CORRECTIONS.

Date.		Corr. of Clock.	Hourly rate.	m	n	c
1848.	h.	s.	s.	s.	s.	s.

INSTRUMENT READINGS.

Date.		Barom.	THERMOM. At.	THERMOM. Ex.
1848.	h. m.	in.	°	°

REMARKS.

(190) 21. Transit over thread VII assumed as at 37s instead of 39s.
(190) 26. Transit over thread II assumed as at 7s.8 instead of 9s.8.
(190) 46. Transit over VII assumed as observed at 39s instead of 29s.
(190) 51. Transits incongruous; thread VI assumed as correctly observed; IV and V used as 30s.4 and 43s.5 instead of 43s and 45s.3.
(190) 64. Transits over T.'s II, III, assumed as 57s and 10s instead of 47s and 0s.
(190) 66. Omitted micrometer reading.

ZONE 191. SEPTEMBER 1. B. BELT, −20° 38′. D_o = −20° 10′ 0″.

No.	Mag.	SECONDS OF TRANSIT. I.	II.	III.	IV.	V.	VI.	VII.	T.	a_1	a_2	MICROMETER.			i	d_1	d_2	Mean Right Ascension, 1850.0.	Mean Declination, 1850.0.
									h. m. s.	s.	s.			r.	′ ″	″	″	h. m. s.	° ′ ″
1	11	..	..	..	..	..	..	57.	23 7 16.77	+47.57	0.00	VII.	5	4.19	−22 6.62	−57.6	−1.80	23 8 4.34	−20 33 6.0
2	10	..	17.5	30.3	43.5	..	..	..	11 43.81	47.57	+0.01	IV.	8	11.40	39 49.07	57.6	2.94	12 31.39	50 49.6
3	10	24.	37.6	50.5	..	..	..	..	14 4.07	47.57	0.00	III.	4	8.44	19 21.30	57.7	1.64	14 51.64	30 20.6
4	5.6	..	..	..	..	..	..	57.3	14 17.39	47.57	0.00	VII.	9	10.14	44 4.43	57.7	3.20	15 4.96	55 5.3
5	11	..	..	50.	3.	..	..	..	19 3.09	47.57	0.00	IV.	7	5.7	31 32.20	57.7	2.42	19 50.66	42 32.3
6	10	..	..	..	..	..	..	55.0	19 14.76	47.57	0.00	VII.	5	2.50	21 24.74	57.7	1.76	20 2.33	
7	10	..	..	..	..	38.7	52.	..	21 25.38	47.58	0.00	VI.	10	9.50	48 53.38	57.8	3.51	22 12.96	59 54.7
8	10	..	..	..	..	..	48.5	1.7	22 21.62	47.58	0.00	VII.	2	6.21	8 10.83	57.8	0.95	23 9.20	19 9.6
9	10	..	..	..	..	40.	..	..	24 26.71	47.58	0.00	V.	9	9.52	43 53.58	57.8	3.20	25 14.29	54 54.6
10	10	..	..	..	..	..	..	39.	24 59.11	47.58	0.00	VII.	10	3.43	45 48.16	57.9	3.32	25 46.69	56 49.4
11	10	..	..	..	..	25.2	38.2	..	30 11.70	47.58	0.00	VI.	4	2.5	16 0.0	58.0	1.43	30 59.28	26 59.4
12	11	..	..	..	..	..	..	15.5	31 35.27	47.58	0.00	VII.	5	4.29	22 11.66	58.0	1.81	32 22.85	33 11.5
13	9	..	..	..	9.8	23.2	36.3	..	33 9.72	47.58	0.00	VI.	5	8.59	21 27.97	58.0	1.97	33 57.30	35 27.9
14	10	..	44.	56.5	10.	..	..	..	36 10.17	47.59	0.00	IV.	7	9.57	33 58.43	58.1	2.57	36 57.76	44 59.1
15	10	42.	56.	8.5	22.	..	..	..	38 22.15	47.59	0.00	IV.	5	9.32	24 44.73	58.2	1.98	39 9.74	35 44.9
16	10	..	..	13.3	27.	..	..	..	40 26.68	47.59	0.00	IV.	5	6.56	23 26.07	58.2	1.90	41 14.27	34 26.2
17	10	..	..	13.	26.	..	53.	6.5	46 26.31	47.59	0.00	IV.	9	5.43	41 48.05	58.4	3.07	47 13.90	52 49.5
18	10	32.8	46.2	..	12.3	..	..	..	50 12.82	47.60	0.00	IV.	9	3.27	40 39.47	58.5	3.00	51 0.42	51 41.0
19	9	..	..	..	..	6.	19.4	..	50 52.73	47.60	0.00	VI.	9	3.7	40 29.27	58.6	2.99	51 40.33	51 30.9
20	11	..	..	..	1.3	15.	..	..	53 1.42	47.60	0.00	V.	8	7.7	37 31.38	58.6	2.79	53 49.02	48 32.8
21	8	..	..	13.3	27.3	41.5	..	..	54 27.38	47.60	0.00	V.	9	6.15	42 4.16	58.7	3.09	55 14.98	53 6.0
22	10	..	..	..	7.5	21.	..	..	56 7.43	47.60	−0.01	V.	1	1.1	5 29.72	58.7	0.77	56 55.02	16 29.2
23	10	..	..	..	..	31.3	45.	..	57 18.17	47.61	0.00	VI.	4	7.50	18 53.95	58.8	1.60	58 5.78	29 54.4
24	11	..	..	..	..	..	50.	3.5	23 58 23.41	47.61	0.00	VII.	7	9.53	33 56.13	58.8	2.56	23 59 11.02	44 57.5
25	11	..	17.5	30.	43.5	..	..	..	0 1 43.70	47.61	0.00	IV.	8	8.14	38 5.20	59.0	2.84	0 2 31.31	49 7.0
26	10	..	..	9.5	23.	36.5	..	..	3 22.87	47.61	0.00	V.	5	2.46	21 19.98	59.1	1.74	4 10.48	32 20.8
27	8	..	..	..	57.5	11.	..	..	7 57.43	47.62	0.00	V.	1	5.26	2 44.35	59.3	0.60	8 45.05	13 44.3
28	11	..	..	..	59.	12.2	..	..	10 58.89	47.62	0.00	V.	10	5.26	46 40.35	59.4	3.39	11 46.51	57 43.1
29	10	..	..	..	29.	42.5	55.5	..	12 28.96	47.63	0.00	VI.	7	11.29	34 44.70	59.5	2.62	13 16.59	45 46.8
30	6.7	..	..	..	..	..	..	3.	13 23.06	47.63	0.00	VII.	9	6.54	42 23.58	59.5	3.12	14 10.69	53 26.2
31	10	..	58.	11.	..	..	..	..	16 24.54	47.63	0.00	IV.	6	11.45	29 52.89	59.7	2.33	17 12.17	40 54.9
32	9	..	..	..	..	..	..	14.3	16 34.27	47.63	0.00	VII.	7	13.4	35 32.44	59.7	2.67	17 21.90	46 34.8
33	7	..	..	..	..	29.	42.5	..	19 15.79	47.64	0.00	VI.	10	6.00	46 57.39	59.8	3.40	20 3.43	58 0.6
34	10	..	35.4	48.4	1.5	..	..	..	22 1.74	47.64	0.00	IV.	6	11.41	29 50.87	60.0	2.33	22 49.38	40 53.2
35	11	..	57.5	..	23.4	37.	..	..	24 23.71	47.65	0.00	V.	6	6.41	27 19.57	60.1	2.15	25 11.36	38 21.8
36	10	..	..	42.	55.3	..	..	..	32 55.17	47.66	0.00	IV.	5	3.13	21 33.61	60.6	1.77	33 42.83	32 36.0
37	9	..	..	6.5	19.5	..	..	..	0 30 19.53	+47.66	0.00	IV.	5	6.18	−23 6.90	−60.4	−1.86	0 31 7.19	−20 34 9.2

ZONE 192. SEPTEMBER 2. K. BELT, −22° 31′. D_o = −22° 5′ 30″.

No.	Mag.	I.	II.	III.	IV.	V.	VI.	VII.	T.	a_1	a_2	MICROMETER.			i	d_1	d_2	Mean Right Ascension, 1850.0.	Mean Declination, 1850.0.
1	8.9	..	..	..	..	..	48.3	..	18 54 21.30	+48.82	+0.01	F.	9	3.30	−40 40.84	−17.2	−3.03	18 55 10.13	−22 46 31.1
2	9	..	..	..	..	44.1	..	..	56 30.63	48.81	+0.01	E.	10	0.47	43 20.80	17.0	3.26	18 57 19.45	49 11.1
3	9	..	14.8	..	41.4	..	..	..	18 59 41.60	48.79	0.00	.	6	13.41	30 51.37	16.7	2.38	19 0 29.39	36 40.5
4	10	..	..	..	9.9	..	..	..	19 1 9.75	48.78	+0.01	.	9	6.20	42 6.71	16.6	3.12	1 58.54	47 56.4
5	8	..	..	..	17.9	..	..	..	2 17.75	48.78	−0.01	.	1	8.53	4 28.76	16.5	0.69	3 6.52	10 16.0
6	9	..	..	..	..	1.9	..	..	2 48.32	48.78	0.00	E.	4	10.21	20 10.19	16.5	1.69	3 37.10	25 58.4
7	6	..	..	..	48.8	2.4	..	..	3 48.79	48.77	+0.01	.	9	8.7	43 0.67	16.4	3.18	4 37.57	48 50.3
8	8.9	..	..	..	..	24.6	..	..	19 4 10.99	+48.77	−0.01	E.	3	5.36	− 2 47.49	−16.3	−1.21	19 4 59.75	−22 18 35.0

CORRECTIONS.

Date.		Corr. of Clock.	Hourly rate.	m	n	c
1848.	h.	s.	s.	s.	s.	s.
Sept. 2,	21	+ 45.36	/ 0.011			

INSTRUMENT READINGS.

Date.		Barom.	THERMOM. At.	THERMOM. Ex.
1848.	h. m.	in.	°	°

REMARKS.

(191) 22. Hor. thread assumed as 2 instead of 1.
(192) 2. Micrometer reading assumed as 9t 8r.47 instead of 10t 0r.47, to agree with Arg. Z. 220, 81, W. Mer. Cir., July 17, and Mural July 17.

ZONE 192. SEPTEMBER 2. K. BELT, −22° 31′. D₀=−22° 5′ 30″—Continued.

No.	Mag.	Seconds of Transit I.	II.	III.	IV.	V.	VI.	VII.	T.	a_1	a_2	Micrometer.		r.	i	d_1	d_2	Mean Right Ascension, 1850.0.	Mean Declination, 1850.0.
									h. m. s.	s.	s.			r.	′ ″	″	″	h. m. s.	° ′ ″
9	10	34.5	. .	1.4	. .	. .	. .	. .	19 7 14.87	+48.76	−0.01	.	2	3.15	− 6 37.32	−16.0	−0.82	19 8 3.62	−22 12 24.1
10	9	. .	. .	53.4	6.8	. .	. .	. .	9 6.66	48.75	−0.01	.	3	12.47	16 24.85	15.9	1.45	9 55.40	22 12.3
11	9	. .	. .	. .	56.4	9.9	. .	. .	9 56.29	48.74	0.00	.	4	12.50	21 25.36	15.8	1.77	10 45.03	27 12.9
12	6.7	. .	. .	. .	49.7	3.2	. .	. .	10 49.62	48.74	+0.01	.	7	11.33	34 46.84	15.7	2.64	11 38.37	40 35.2
13	10	. .	1.9	15.2	. .	. .	. .	. .	12 28.68	48.73	−0.01	.	3	10.32	15 16.79	15.6	1.38	13 17.40	21 3.8
14	9	. .	23.0	35.8	. .	. .	. .	. .	14 49	48.73	+0.01	.	10	3.34	45 43.91	15.3	3.36	15 38.61	51 32.6
15	8.9	. .	28.8	41.4	. .	. .	. .	. .	15 55.50	48.72	+0.01	.	8	9.9	38 32.93	15.2	3.54	16 44.23	44 21.7
16	10	. .	. .	19.4	. .	. .	. .	. .	16 32.57	48.72	−0.01	.	2	6.13	8 7.08	15.2	0.91	17 21.28	13 53.2
17	11	. .	. .	. .	23.9	. .	. .	. .	20 23.75	48.70	0.01	.	2	12.20	11 12.13	14.8	1.11	21 12.44	16 58.0
18	10	. .	34.0	57.3	. .	. .	. .	. .	23 0.76	48.69	−0.01	.	3	6.15	13 7.19	14.6	1.23	23 49.44	18 53.0
19	9	. .	. .	. .	. .	. .	48.6	. .	24 21.67	48.68	0.00	VI.	4	9.10	19 34.28	14.5	1.65	25 10.35	25 20.4
20	10	. .	. .	41.8	55.3	. .	. .	. .	25 55.16	48.67	0.00	.	5	7.9	23 32.62	14.3	1.90	26 43.83	29 18.8
21	9	. .	43.8	57.0	. .	. .	. .	. .	27 10.46	48.67	−0.01	.	2	6.3	8 3.04	14.2	0.90	27 59.12	13 47.1
22	9	. .	. .	28.2	41.7	. .	. .	. .	27 41.48	48.66	−0.01	.	2	12.22	11 13.14	14.2	1.11	28 30.13	16 58.5
23	8.9	30.2	57.5	. .	. .	. .	. .	. .	30 10.90	48.66	0.00	.	4	6.25	18 11.23	13.9	1.56	30 59.56	23 56.7
24	10	. .	. .	. .	. .	. .	42.3	. .	32 15.31	48.65	+0.01	.	8	7.41	37 48.55	13.7	2.64	33 3.97	43 35.1
25	10	. .	. .	39.3	52.7	. .	. .	. .	33 52.53	48.64	−0.01	.	2	12.9	11 6.59	13.6	1.10	34 41.16	16 51.8
26	10	. .	. .	42.5	55.8	. .	. .	. .	34 55.94	48.64	+0.01	.	10	8.28	48 12.15	13.5	3.53	35 44.59	53 59.2
27	9	. .	20.6	. .	47.4	. .	. .	. .	37 47.38	48.62	−0.01	.	1	11.20	5 42.88	13.2	0.75	38 35.99	11 26.8
28	9	. .	24.9	37.7	. .	. .	. .	. .	48 51.64	48.57	0.00	.	7	8.25	33 12.04	12.2	2.54	49 40.21	38 56.8
29	8	. .	25.4	38.3	. .	. .	. .	. .	49 52.17	48.57	0.00	.	6	14.1	31 1.45	12.1	2.39	50 40.71	36 45.9
30	9	. .	. .	. .	. .	10.6	23.8	. .	51 56.95	48.56	0.00	V.	5	10.42	25 19.99	11.9	2.02	52 45.51	31 4.0
31	9	. .	. .	31.9	45.2	. .	. .	. .	19 54 45.22	48.55	0.00	.	6	13.9	30 35.24	11.7	2.36	19 55 33.77	36 19.3
32	10	4.9	18.8	. .	. .	. .	. .	. .	20 1 45.76	48.52	0.00	. .	4	5.27	17 41.98	11.1	1.53	20 2 34.28	23 24.6
33	9.10	. .	12.6	. .	39.4	. .	. .	. .	4 39.45	48.51	0.00	.	5	7.5	23 30.60	10.8	1.90	5 27.96	29 13.3
34	7	43.1	56.9	. .	. .	. .	. .	. .	8 23.84	48.49	−0.01	.	2	10.48	10 25.75	10.5	1.05	9 12.32	16 7.3
35	9	. .	. .	. .	. .	41.3	54.2	. .	8 27.52	48.49	0.00	V.	4	13.29	21 44.98	10.5	1.79	9 16.01	27 27.3
36	10	. .	. .	. .	. .	47.3	. .	. .	9 33.77	48.48	0.00	V.	6	11.31	29 45.79	10.4	2.31	10 22.25	35 28.5
37	9	49.1	3.3	. .	. .	. .	. .	. .	12 30.12	48.47	0.00	.	4	9.57	19 58.13	10.1	1.68	13 18.59	25 39.9
38	9	. .	. .	45.7	59.4	. .	. .	. .	12 59.31	48.47	+0.01	.	9	8.56	43 25.38	10.1	3.22	13 47.79	49 8.7
39	9	. .	. .	30.9	. .	58.2	. .	. .	14 44.49	48.46	0.00	.	6	4.5	26 0.94	9.9	2.06	15 32.95	31 42.9
40	10	. .	. .	44.4	. .	. .	. .	. .	15 57.50	48.46	−0.01	.	1	5.57	3 0.01	9.8	0.57	16 45.95	8 40.4
41	11	. .	. .	. .	. .	. .	25.5	. .	15 58.60	48.46	−0.01	.	2	10.35	10 19.20	9.8	1.04	16 47.05	16 0.0
42	9	. .	2.8	15.8	. .	. .	. .	. .	18 29.53	48.45	0.00	.	5	10.1	24 59.35	9.6	2.00	19 17.98	30 41.0
43	6.7	. .	27.9	. .	53.8	. .	. .	. .	19 54.41	48.44	+0.01	.	10	6.56	47 25.77	9.5	3.50	20 42.86	53 8.8
44	9	56.9	11.2	. .	. .	. .	. .	. .	22 38.10	48.43	0.00	.	7	9.29	33 44.37	9.3	2.57	23 26.53	39 26.2
45	9	. .	. .	. .	39.8	53.3	. .	. .	22 39.72	48.43	+0.01	.	7	10.21	34 10.53	9.3	2.61	23 28.16	39 52.4
46	9	. .	29.5	. .	56.2	. .	. .	. .	23 56.38	48.42	0.01	.	8	8.57	38 26.88	9.1	2.90	24 44.81	44 8.9
47	10.11	. .	46.9	. .	. .	. .	. .	. .	29 14.15	48.40	+0.01	.	9	11.38	44 47.06	8.7	3.33	30 2.56	50 29.1
48	10	. .	. .	. .	58.8	. .	. .	. .	30 58.65	48.39	−0.01	.	1	6.37	3 20.19	8.6	0.58	31 47.03	8 59.4
49	10	54.3	8.8	. .	. .	. .	. .	. .	33 35.51	48.38	0.00	.	5	8.17	24 6.91	8.3	1.94	34 23.89	29 47.2
50	10	. .	. .	. .	. .	. .	15.4	. .	20 33 48.39	+48.38	+0.01	.	9	11.25	−44 40.50	− 8.3	−3.32	20 34 36.78	−22 50 22.1

CORRECTIONS.

Date.		Corr. of Clock.	Hourly rate.	m	k	c
1848.	h.	s.	s.	s.	s.	s.

INSTRUMENT READINGS.

Date.		Barom.	Thermom. At.	Thermom. Ex.
1848.	h. m.	in.	°	°

REMARKS.

(192) 18. One of the transits wrong by 10ˢ; thread III assumed to have been observed at 47ˢ.3.
(192) 23. Transit across thread III assumed as recorded against thread II.

ZONE 193. SEPTEMBER 7. B. BELT, −18° 46'. D₀=−18° 21' 0".

No.	Mag.	SECONDS OF TRANSIT. I.	II.	III.	IV.	V.	VI.	VII.	T.	a_1	a_2	MICROMETER.			i	d_1	d_2	Mean Right Ascension, 1850.0.	Mean Declination, 1850.0.
									h. m. s.	s.	s.			r.	' "	'	"	h. m. s.	° ' "
1	9		27.	39.	25.3				20 13 51.89	− 8.48	+0.01	V.	10	8.30	−48 13.14	− 18.1	−3.27	20 13 43.42	−19 9 34.5
2	9					53.3	6.		14 39.90	8.48	0.00	VI.	6	7.32	27 45.21	18.1	2.16	14 31.42	18 49 5.5
3	9							46.	15 6.22	8.48	0.00	VII.	4	10.31	20 15.01	18.1	1.74	14 57.74	41 34.9
4	9		0.0	12.9	26.				17 26.17	8.49	0.00	IV.	8	5.20	36 37.45	17.9	2.64	17 17.68	57 58.0
5	9						22.		17 55.70	8.50	0.00	VI.	6	11.40	29 50.26	17.9	2.27	17 47.20	51 10.4
6	5							32.	18 52.22	8.50	0.00	VII.	4	11.25	20 42.23	17.9	1.76	18 43.72	18 42 1.8
7	8			11.	24.				21 24.06	8.51	+0.01	IV.	8	8.48	43 21.34	17.7	3.03	21 15.56	19 54 42.1
8	9			45.6	59.				23 58.69	8.52	0.00	IV.	3	7.40	13 50.05	17.5	1.39	22 50.17	18 35 8.9
9	11			58.	11.				25 11.02	8.52	0.00	IV.	7	7.41	32 49.85	17.4	2.44	25 2.50	54 9.7
10	11						15.	28.5	26 48.75	8.53	0.00	VII.	5	11.8	25 32.86	17.3	2.03	26 40.22	18 46 52.2
11	10			48.3	1.5				30 1.52	8.54	+0.01	IV.	10	8.50	48 23.25	17.1	3.28	29 52.99	19 9 43.6
12	5.6	59.	13.	26.	38.5	52.			31 38.85	8.54	0.00	V.	4	6.00	18 28.85	17.0	1.61	31 30.31	18 39 47.5
13	11							48.	32 8.55	8.55	+0.01	VII.	9	8.13	43 3.42	17.0	3.00	32 0.01	19 4 23.4
14	11				49.5	3.			33 49.61	8.56	+0.01	VI.	10	4.15	46 4.47	16.9	3.16	33 41.06	19 7 24.5
15	10					19.	32.		35 5.75	8.56	0.00	VI.	5	8.54	24 25.46	16.8	1.97	34 57.19	18 45 44.2
16	11		9.5	22.					37 35.55	8.57	0.00	III.	6	10.21	29 10.50	16.7	2.23	37 26.98	18 50 29.4
17	9		33.	45.	58.2				38 58.69	8.57	+0.01	IV.	10	9.7	48 31.82	16.6	3.29	38 50.13	19 9 51.7
18	9				1.5				40 1.35	8.58	0.00	IV.	5	7.17	23 36.65	16.5	1.92	39 52.77	18 44 55.1
19	7						24.5	38.	40 58.19	8.58	0.00	VII.	3	7.53	13 56.33	16.4	1.39	40 49.61	35 14.1
20	10							6.	42 26.29	8.59	0.00	VII.	5	11.48	25 53.03	16.4	2.05	42 17.70	47 11.5
21	11			6.	19.6				45 19.33	8.60	0.00	IV.	7	12.52	35 26.67	16.2	2.58	45 10.73	56 45.5
22	11					33.			46 19.76	8.60	0.00	VI.	4	13.12	21 36.34	16.1	1.81	46 11.16	18 42 54.3
23	9						47.5	1.	47 21.38	8.60	+0.01	VII.	10	7.10	47 32.55	16.1	3.25	47 12.79	19 8 51.9
24	11	5.	19.						50 44.98	8.61	−0.01	II.	1	9.55	4 59.91	15.9	0.91	50 36.36	18 26 16.7
25	11	53.5	7.						53 33.37	8.62	0.00	II.	5	6.12	23 3.77	15.7	1.90	53 24.75	44 21.4
26	11					57.	10.		54 43.75	8.63	0.00	VI.	6	6.15	27 6.38	15.7	2.12	54 35.12	48 24.2
27	7						8.	21.3	55 41.63	8.63	0.00	VII.	4	11.33	20 46.27	15.6	1.77	55 33.00	42 3.6
28	9	37.		3.5	16.4				20 59 16.48	8.64	0.00	IV.	3	6.30	13 14.76	15.4	1.35	20 59 7.84	18 34 31.5
29	10			58.	11.2				21 1 11.17	8.65	+0.01	IV.	8	12.15	40 6.71	15.3	2.85	21 1 2.53	19 1 24.9
30	8						40.	53.	2 13.56	8.65	0.00	VII.	7	12.1	35 0.68	15.3	2.56	2 4.91	18 56 18.5
31	9		48.	1.2	14.2				6 14.14	8.67	−0.01	IV.	2	10.4	10 3.57	15.1	1.19	6 5.48	31 19.9
32	10						18.0	31.	6 51.49	8.67	0.00	VII.	5	4.3	21 58.56	15.0	1.83	6 42.82	43 15.4
33	5.6	1.	14.5	27.5	40.5				9 40.60	8.68	0.00	IV.	3	10.36	15 18.81	14.9	1.47	9 31.92	36 35.2
34	10				20.3	34.			13 20.43	8.69	−0.01	VI.	2	12.25	11 14.54	14.7	1.24	13 11.73	32 30.5
35	10				9.	23.	36.		15 9.45	8.70	0.00	VI.	7	10.15	34 7.40	14.6	2.51	16 0.75	55 24.5
36	9				1.5	15.			18 1.57	8.71	0.00	VI.	6	4.44	26 20.50	14.5	2.07	17 52.86	18 47 37.1
37	10							25.	18 45.52	8.71	+0.01	VII.	9	4.00	40 55.85	14.4	2.89	18 36.82	19 2 13.1
38	10		9.5	22.					21 35.65	8.72	+0.01	III.	8	9.48	38 52.56	14.3	2.79	21 26.94	19 0 9.7
39	10			45.5	58.5	12.			22 58.56	8.72	0.00	V.	5	11.41	25 49.74	14.2	2.04	22 49.84	18 47 6.0
40	10							33.	23 53.40	8.72	0.00	VII.	7	7.15	32 36.47	14.2	2.43	23 44.68	53 53.1
41	10					53.5	6.		25 39.98	8.73	−0.01	VI.	2	10.50	10 31.19	14.1	1.20	25 31.24	18 31 46.5
42	9	29.	43.6	56.	9.				28 9.33	8.74	+0.01	IV.	9	6.43	42 18.31	14.0	2.97	28 0.60	19 3 35.3
43	8					20.5	33.	47.	29 7.20	8.74	+0.01	VII.	10	2.16	45 4.31	13.9	3.13	28 58.47	19 6 21.3
44	10			7.2	20.	33.3			31 20.07	8.75	0.00	V.	6	3.41	25 48.80	13.8	2.05	31 11.32	18 47 4.7
45	10							38.	31 58.57	8.75	+0.01	VII.	9	12.1	44 58.38	13.8	3.12	31 49.83	19 6 15.3
46	10		16.	28.3	41.5				34 41.81	8.76	+0.01	IV.	8	7.46	37 51.08	13.7	2.73	34 33.06	18 59 7.5
47	10					2.5	15.2		35 49.11	8.76	0.00	VI.	7	3.33	30 44.69	13.6	2.32	35 40.35	52 0.6
48	9			27.6	40.5	54.			37 40.56	8.77	0.00	VI.	4	7.41	18 49.44	13.5	1.65	37 31.79	18 40 4.6
49	8						44.	57.5	21 38 17.70	− 8.77	0.00	VII.	3	10.36	−15 18.54	− 13.5	−1.46	21 38 8.93	−18 36 33.5

CORRECTIONS.							INSTRUMENT READINGS.				
Date.		Corr. of Clock.	Hourly rate.	m	n	c	Date.		Barom.	THERMOM. At.	Ex.
1848.	h.	s.	s.	s.	s.	s.	1848.	h. m.	in.	°	°
Sept. 7.	21	− 11.38	/ 0.041								

REMARKS.

(193) 1. Transit over T. II assumed as recorded over T. IV. Other transits rejected.
(193) 7. Hor. thread assumed as 9 instead of 8.
(193) 8. Minutes assumed as 22 instead of 23.
(193) 12. Micrometer reading assumed as 7r.00 instead of 6r.00.
(193) 35. Minutes assumed as 16 instead of 15.

Zone 193. September 7. B. Belt, −18° 46′. D₀ = −18° 21′ 0″—Continued.

No.	Mag.	SECONDS OF TRANSIT. I.	II.	III.	IV.	V.	VI.	VII.	T.	a_1	a_2	MICROMETER.		i	d_1	d_2	Mean Right Ascension, 1850.0.	Mean Declination, 1850.0.
									h. m. s.	s.	s.		r.	′ ″	″	″	h. m. s.	° ′ ″
50	8	..	..	..	..	..	0.4	14.2	21 39 34.35	− 8.77	0.00	VII.	7 8.15	−33 6.73	−13.5	−2.45	21 39 25.58	−18 54 22.7
51	8	..	..	..	9.	22.5	36.	..	41 9.30	8.78	+0.01	VI.	9 9.10	43 32.32	13.4	3.04	41 0.53	19 4 48.8
52	8	..	..	..	..	..	56.	9.5	42 29.81	8.78	0.00	VII.	8 2.53	35 23.06	13.3	2.59	42 21.03	18 56 39.0
53	10	..	48.	..	13.5	..	..	..	47 13.93	8.79	0.00	IV.	8 3.5	35 29.38	13.1	2.60	47 5.14	56 45.1
54	8	..	13.	25.4	39.	52.	..	..	48 38.81	8.80	0.00	V.	3 10.19	15 10.20	13.1	1.45	48 30.01	36 24.8
55	9	41.4	55.2	8.	21.	..	..	..	51 21.17	8.81	0.00	IV.	4 12.37	21 18.81	13.0	1.79	51 12.36	42 33.6
56	7	25.	39.	51.4	4.5	..	..	..	54 4.73	8.81	0.00	IV.	3 11.58	16 0.14	12.9	1.50	53 55.92	37 14.5
57	8	..	40.	52.5	5.5	19.	..	..	56 5.69	8.82	−0.01	V.	2 11.56	11 0.01	12.8	1.21	55 56.86	32 14.0
58	8	8.	22.	35.	..	..	..	..	21 59 48.04	8.83	0.00	III.	4 6.48	18 22.80	12.6	1.62	21 59 39.21	39 37.0
59	8	..	..	..	..	..	35.5	49.	22 0 9.18	− 8.83	−0.01	VII.	3 5.5	−12 31.62	−12.6	−1.30	22 0 0.34	−18 33 45.5

Zone 194. September 19. P. Belt, −17° 30′. D₀ = −17° 6′ 10″.

1	11	6.4	19.4	..	45.5	..	..	..	20 32 45.56	+ 3.12	−0.01	IV.	2 3.28	− 6 43.87	−11.71	−1.10	20 32 48.67	−17 13 6.7
2	10	..	..	..	..	14.5	..	39.9	38 0.76	3.10	0.01	VII.	2 6.6	8 3.33	12.34	1.17	38 3.75	14 26.8
3	8	..	..	..	..	55.5	8.9	22.6	39 42.70	3.09	−0.01	VII.	2 11.8	10 35.60	12.55	1.29	39 45.88	16 59.4
4	7	46.5	0.1	12.6	..	..	..	..	42 26.17	3.08	+0.01	III.	9 10.37	44 16.29	12.87	2.96	42 29.26	56 42.1
5	11	..	..	58.5	..	..	..	..	45 11.42	3.07	0.00	III.	4 10.32	20 15.77	13.20	1.76	45 14.49	26 40.7
6	9	5.1	18.4	31.4	43.9	..	..	..	49 44.25	3.05	−0.01	IV.	2 9.12	9 37.34	13.75	1.24	49 47.29	16 2.3
7	9	..	..	45.4	57.9	11.5	..	..	50 58.14	3.04	0.00	V.	4 11.575	20 58.87	13.90	1.80	51 1.18	27 24.6
8	10	23.5	36.5	49.5	..	..	..	..	53 2.71	3.03	0.00	III.	4 9.34	19 46.52	14.15	1.74	53 5.74	26 12.4
9	6	41.9	56.4	8.5	21.6	..	..	..	56 21.86	3.02	+0.01	V.	8 9.53	38 55.09	14.56	2.70	56 24.89	45 22.4
10	4	..	..	..	..	..	53.5	6.5	57 27.35	3.02	+0.01	VII.	9 8.15	43 4.47	14.70	2.91	57 30.38	49 32.1
11	8	..	..	..	45.9	..	12.5	25.9	20 59 46.16	3.01	−0.01	VII.	3 6.28	13 13.51	14.98	1.41	20 59 49.16	19 39.9
12	9	..	..	..	..	8.9	..	34.8	21 1 55.59	3.00	0.00	VII.	6 6.48	27 22.90	15.25	2.12	21 1 58.59	33 50.3
13	10	..	..	58.5	..	25.6	..	..	11 11.91	2.96	0.00	V.	4 5.43	17 50.03	16.42	1.64	11 14.87	24 18.1
14	6	..	24.5	37.4	49.8	3.9	..	..	21 13 50.36	+ 2.95	0.00	V.	5 3.43	−21 48.72	−16.78	−1.84	21 13 53.31	−17 28 17.3

Zone 195. September 23. P. Belt, −16° 53½′.

1	10	15.5	29.7	41.8	..	..	..	..	19 47 55.24	+19.06	0.00	III.	7 6.4	−32 0.93	..	−2.33	19 48 14.30	
2	11	..	..	..	..	41.5	..	7.8	49 28.28	19.05	−0.01	VII.	2 3.35	6 47.19	..	1.13	49 47.32	
3	11	..	6.4	..	..	..	..	..	52 32.58	19.04	0.00	III.	6 11.32	29 46.32	..	2.22	52 51.62	
4	9	..	..	..	..	..	..	16.8	52 37.84	19.04	+0.01	VII.	10 8.12	48 3.87	..	3.15	52 56.89	
5	10	..	21.8	34.4	47.5	..	..	..	54 47.66	19.03	+0.01	IV.	9 4.45	41 18.81	..	2.80	55 6.70	
6	9	..	4.8	17.3	30.9	..	..	..	19 56 30.61	19.02	0.00	IV.	4 8.4	19 1.16	..	1.72	19 56 49.63	
7	10	28.4	41.8	54.5	..	..	..	..	20 5 7.86	18.99	+0.01	III.	8 8.41	38 18.79	..	2.65	20 5 26.86	
8	9	..	..	..	21.5	34.5	47.5	..	6 21.45	18.98	+0.01	VI.	10 11.13	49 35.25	..	3.23	6 40.44	
9	9	..	..	..	..	..	..	51.8	7 12.38	18.98	0.00	VII.	3 12.40	16 21.10	..	1.59	7 31.36	
10	10	..	..	..	2.9	..	29.9	..	12 3.30	18.96	+0.01	VI.	10 10.58	49 27.67	..	3.23	12 22.27	
11	10	..	..	..	..	56.5	9.8	22.9	13 43.64	18.95	0.00	VII.	6 14.26	31 13.84	..	2.30	14 2.59	
12	10	..	..	37.5	50.4	3.7	16.4	..	21 50.46	18.92	0.00	VI.	7 10.41	34 20.52	..	2.46	22 9.38	
13	8	..	..	29.5	42.4	55.4	..	..	23 42.42	18.91	+0.01	V.	8 8.27	38 11.73	..	2.64	24 1.34	
14	7	4.6	17.6	30.6	43.	..	..	..	26 43.86	18.90	0.00	IV.	7 9.34	33 46.84	..	2.44	27 2.76	
15	11	..	..	22.6	..	48.5	..	..	28 35.29	18.89	−0.01	V.	1 8.38	4 21.18	..	1.00	28 54.17	
16	8	..	..	2.4	15.5	28.8	41.8	54.9	30 15.63	18.89	+0.01	VII.	8 4.19	36 21.61	..	2.57	30 34.53	
17	7	..	..	..	..	0.5	13.4	26.5	20 31 47.26	+18.88	−0.01	VII.	3 1.33	−10 44.77	..	−1.30	20 32 6.13	

CORRECTIONS.

Date.		Corr. of Clock.	Hourly rate.	m	n	c
1848.	h.	s.	s.	s.	s.	s.
Sept. 19,	22	− 0.30	/ 0.053	− 0.52	+ 1.17	0.00
Sept. 23,	22	− 16.07	/ 0.087	− 0.53	+ 1.18	0.00

INSTRUMENT READINGS.

Date.		Barom.	THERMOM. At.	Ex.
1848.	h. m.	in.	°	°

REMARKS.

ZONE 195. SEPTEMBER 23. P. BELT, −16° 53½′—Continued.

No.	Mag.	Seconds of Transit. I.	II.	III.	IV.	V.	VI.	VII.	T.	a_1	a_2	Micrometer.			i	d_1	d_2	Mean Right Ascension, 1850.0.	Mean Declination, 1850.0.
									h. m. s.	s.	s.			r.	′ ″	″	″	h. m. s.	° ′ ″
18	10				44.6		10.5		20 37 44.46	+18.86	+0.01	VI.	10	4.10	−46 1.96	..	−3.05	20 38 3.33	
19	10				27.4		53.5		39 27.37	18.85	+0.01	VI.	8	3.3	35 28.27	..	2.52	39 46.23	
20	10		54.6	7.4	20.5	33.8			42 20.57	18.84	0.00	V.	6	11.22	29 41.27	..	2.23	42 39.41	
21	10								43	18.84	−0.01	II.	2	9.47	9 54.89	..	1.27		
22	8		39.2						50 5.21	18.81	−0.01	VII.	2	5.41	7 50.72	..	1.15	50 24.01	
23	10					10.4	22.2	36.5	20 52 56.95	+18.80	0.00	VII.	7	7.55	−32 56.69	..	−2.40	20 53 15.75	

ZONE 196. SEPTEMBER 25. B. BELT, −16° 53½′. D_0 = −16° 27′ 0″.

No.	Mag.	I.	II.	III.	IV.	V.	VI.	VII.	T.	a_1	a_2	Micrometer.		r.	i	d_1	d_2	Mean Right Ascension, 1850.0.	Mean Declination, 1850.0.
1	9	23.3	37.2	49.2					20 5 2.88	+23.19	+0.01	III.	8	11.15	−39 35.43	−14.4	−2.70	20 5 26.08	−17 6 52.5
2	10							3.5	5 24.37	23.19	+0.01	VII.	8	4.35	36 14.55	14.3	2.54	5 47.57	17 3 31.4
3	7			55.	8.2	21.			7 7.93	23.18	0.00	V.	4	5.26	17 41.45	14.1	1.66	7 31.11	16 44 57.2
4	10							14.	7 34.65	23.18	0.00	VII.	4	11.32	20 45.82	14.0	1.80	7 57.83	48 1.6
5	10			18.5	31.	44.5			9 31.12	23.17	−0.01	V.	1	9.57	5 1.01	13.8	1.06	9 54.28	32 15.9
6	10						49.	2.	10 22.88	23.16	0.00	VII.	6	12.50	30 25.44	13.7	2.26	10 46.04	16 57 41.4
7	9				59.1	12.4			11 59.20	23.16	+0.01	V.	10	13.11	50 34.82	13.4	3.23	12 22.37	17 17 51.5
8	9				16.5				13 16.36	23.15	+0.01	IV.	10	7.39	47 47.45	13.3	3.09	13 39.52	17 15 3.8
9	9							18.3	13 39.12	23.15	0.00	VII.	7	7.7	32 32.49	13.2	2.36	14 2.27	16 59 48.1
10	10				25.5	39.			16 25.63	23.14	0.00	V.	5	7.20	23 38.14	12.8	1.93	16 48.77	50 52.9
11	10		51.	3.	17.5				21 17.22	23.11	0.00	IV.	4	3.2	16 28.87	12.2	1.60	21 40.33	16 43 42.7
12	8							25.	21 45.86	23.11	+0.01	VII.	8	3.1	35 27.14	12.1	2.51	22 8.98	17 2 41.8
13	9			25.5	38.4	51.			23 38.29	23.10	0.01	V.	8	10.58	39 27.87	11.9	2.70	24 1.40	6 42.5
14	7			26.	39.	52.5			26 39.13	23.09	+0.01	V.	7	12.00	35 0.43	11.5	2.48	27 2.23	17 2 14.4
15	10		5.2	18.	31.				28 30.91	23.08	−0.01	IV.	1	11.12	5 38.85	11.2	1.08	28 53.98	16 32 51.1
16	8	32.	46.2						30 11.99	23.07	+0.01	II.	8	7.14	37 34.84	11.0	2.62	30 35.07	17 4 48.5
17	10							55.	30 15.80	23.07	0.00	VII.	7	4.48	31 22.40	11.0	2.31	30 38.87	16 58 35.7
18	6.7					55.2	9.	23.	31 42.87	23.07	−0.01	VII.	3	4.15	12 6.46	10.8	1.39	32 5.93	39 18.7
19	10		27.	39.5		5.2			33 52.57	23.06	0.00	V.	6	3.30	25 43.27	10.5	2.03	34 15.63	52 55.8
20	10			50.5	3.4				36 3.36	23.04	0.00	IV.	6	4.14	26 5.48	10.3	2.05	36 26.40	53 17.6
21	10							4.	36 24.46	23.04	−0.01	VII.	2	4.53	7 26.52	10.2	1.16	36 47.49	34 37.9
22	10		37.	50.					39 3.00	23.03	0.00	III.	5	2.32	21 12.93	9.9	1.82	39 26.03	48 24.7
23	7						44.	57.4	39 18.06	23.03	0.00	VII.	6	4.45	26 20.89	9.8	2.06	39 41.09	53 32.8
24	10	37.5	51.5	4.					42 17.23	23.02	0.00	III.	7	3.56	30 56.38	9.4	2.29	42 40.25	16 58 8.1
25	10				7.	20.5			43 7.17	23.01	+0.01	VI.	8	6.12	37 3.58	9.3	2.59	43 30.19	17 4 15.5
26	9						27.	40.5	44 1.05	23.01	0.00	VII.	3	12.34	16 18.08	9.2	1.58	44 24.06	16 43 28.9
27	10			9.	21.5				46 21.58	23.00	−0.01	IV.	3	9.11	14 35.93	8.9	1.50	46 44.57	41 46.3
28	10							16.	46 36.49	23.00	0.01	VII.	2	9.38	9 50.23	8.9	1.27	46 59.48	37 0.4
29	10				24.	37.5			48 24.09	22.99	0.01	V.	3	4.53	17 24.84	8.7	1.64	48 47.07	44 35.2
30	7							32.5	48 52.99	22.98	−0.01	VII.	2	8.19	9 10.40	8.6	1.24	49 15.96	38 20.2
31	10					53.			51 39.94	22.97	0.00	V.	6	11.33	29 46.82	8.3	2.23	52 2.91	16 56 57.4
32	9					2.	14.5	28.	52 48.76	22.96	0.00	VII.	7	10.5	34 2.25	8.1	2.44	53 11.72	17 1 12.8
33	10	55.	9.						55 34.86	22.96	0.00	II.	7	9.57	33 58.33	7.8	2.44	55 57.82	17 1 8.6
34	10			5.4	18.5				56 18.32	22.95	0.00	IV.	5	2.43	21 18.49	7.7	1.82	56 41.27	16 48 28.0
35	9		4.	16.5					58 29.89	22.94	0.00	III.	7	10.5	34 2.45	7.4	2.44	58 52.83	17 1 12.3
36	10						7.	20.5	20 58 41.18	22.94	+0.01	VII.	8	7.33	37 44.30	7.4	2.63	20 59 4.13	17 4 54.3
37	10			12.	25.				21 0 24.86	22.93	0.00	IV.	4	7.7	18 32.41	7.2	1.69	21 0 47.79	16 45 41.3
38	10			40.	53.				1 52.86	22.92	0.00	IV.	4	9.11	19 34.93	7.0	1.74	2 15.78	46 43.7
39	9					51.5	4.5		21 2 38.43	+22.92	−0.01	VI.	2	11.45	−10 54.39	− 6.9	−1.32	21 3 1.34	−16 38 2.6

CORRECTIONS.

Date.		Corr. of Clock.	Hourly rate.	m	n	c
1848.	h.	s.	s.	s.	s.	s.
Sept. 25,	22	+ 20.21	/ 0.047	− 0.56	+ 1.22	0.00

INSTRUMENT READINGS.

Date.		Barom.	Thermom. At.	Thermom. Ex.
1848.	h. m.	in.	°	°

REMARKS.

(195) 19. Fine double star.
(196) 29. Hor. thread assumed as 4 instead of 3.

ZONE 196. SEPTEMBER 25. B. BELT, −16° 53½'. D_0 = −16° 27' 0"—Continued.

No.	Mag.	SECONDS OF TRANSIT. I.	II.	III.	IV.	V.	VI.	VII.	T.	a_1	a_2	MICROMETER.		i	d_1	d_2	Mean Right Ascension, 1850.0.	Mean Declination, 1850.0.
									h. m. s.	s.	s.		r.	' "	"	"	h. m. s.	° ' "
40	9	12.3	25.3	38.3	. .	. .	. .	. .	21 6 51.38	+22.90	0.00	III.	3 10.59	−15 30.38	− 6.4	−1.54	21 7 14.28	−16 42 38.3
41	9	. .	43.	56.	. .	. .	. .	. .	9 8.82	22.89	−0.01	III.	1 7.28	3 45.67	6.1	0.98	9 31.70	30 53.0
42	8	52.2	6.	18.5	. .	. .	. .	. .	10 31.70	22.89	0.00	III.	4 12.14	21 7.19	5.9	1.81	10 54.59	16 48 14.9
43	10	. .	. .	. .	26.5	39.5	. .	. .	11 26.43	22.88	+0.01	V.	8 9.41	38 49.04	5.8	2.68	11 49.32	17 5 57.5
44	8	. .	41.3	53.8	6.5	. .	. .	. .	13 6.66	22.88	0.00	IV.	6 5.54	26 55.90	5.6	2.09	13 29.74	16 54 3.6
45	10	. .	. .	. .	. .	. .	. .	38.5	13 59.47	22.88	+0.01	VII.	9 8.42	43 18.10	5.5	2.90	14 22.36	17 10 26.5
46	10	. .	. .	. .	. .	. .	. .	50.	15 10.89	22.87	+0.01	VII.	8 7.6	37 30.69	5.3	2.62	15 33.77	4 38.6
47	10	. .	. .	. .	. .	. .	54.	7.	16 27.91	22.86	0.00	VII.	8 2.55	35 24.12	5.2	2.51	16 50.77	2 31.8
48	10	. .	. .	. .	23.	. .	. .	. .	18 22.86	22.86	+0.01	IV.	8 7.25	37 40.48	5.0	2.63	18 45.73	17 4 48.1
49	10	. .	. .	. .	. .	34.5	. .	. .	19 21.42	22.85	0.00	V.	6 4.55	26 26.13	4.8	2.07	19 44.27	16 53 33.0
50	10	. .	10.5	23.	36.	. .	. .	. .	21 36.19	22.84	0.00	IV.	6 12.46	30 23.65	4.6	2.26	21 59.03	57 30.5
51	9	20.5	34.2	46.5	59.5	. .	. .	. .	24 59.76	22.83	0.00	IV.	5 8.45	24 21.03	4.2	1.97	25 22.59	51 27.2
52	10	. .	7.2	. .	. .	. .	. .	. .	26 33.38	22.82	0.00	II.	6 9.16	28 37.66	4.0	2.17	26 56.20	55 43.8
53	10	10.3	25.	37.5	. .	. .	. .	. .	28 50.50	22.80	0.00	III.	7 4.41	31 19.07	3.7	2.31	29 13.30	58 25.1
54	9	. .	. .	. .	. .	. .	. .	48.	29 8.76	22.80	0.00	VII.	6 8.36	28 17.38	3.7	2.16	29 31.56	55 23.2
55	10	. .	. .	. .	10.	. .	. .	. .	31 9.86	22.80	0.00	.	3 7.32	13 46.02	3.5	1.46	31 32.66	16 40 51.0
56	10	. .	. .	42.	. .	. .	. .	. .	33 55.23	22.78	+0.01	III.	10 6.24	47 9.61	3.2	3.10	34 18.02	17 14 15.9
57	9	2.	15.5	28.3	41.5	54.	. .	. .	35 41.03	22.78	−0.01	IV.	3 4.24	12 11.22	3.0	1.38	36 3.80	16 39 15.6
58	10	. .	. .	. .	40.	. .	. .	. .	37 39.84	22.77	0.00	IV.	3 7.25	13 42.45	2.7	1.45	38 2.61	40 46.6
59	4	. .	. .	. .	. .	. .	48.	1.5	38 21.10	22.76	0.00	VII.	6 2.16	21 4.61	2.7	1.81	38 43.86	48 9.1
60	9	8.	21.5	34.2	47.	. .	. .	. .	41 47.28	22.75	0.00	IV.	6 4.22	26 9.51	2.3	2.06	42 10.03	53 13.9
61	9	. .	32.2	45.	58.	. .	. .	. .	43 58.09	22.74	0.00	IV.	7 2.59	30 27.66	2.0	2.26	44 20.83	57 31.9
62	10	. .	31.5	44.2	. .	. .	. .	. .	47 57.20	22.72	−0.01	III.	2 3.34	6 46.89	1.6	1.11	48 19.91	33 49.6
63	10	. .	. .	. .	. .	59.	12.	25.4	48 46.03	22.72	0.00	VII.	6 11.17	29 38.55	1.5	2.22	49 8.75	16 56 42.3
64	10	. .	. .	. .	4.	. .	. .	. .	51 3.86	22.71	0.00	IV.	7 12.40	35 20.62	1.3	2.51	51 26.57	17 2 24.4
65	10	. .	. .	. .	. .	. .	. .	17.	51 37.91	22.71	+0.01	VII.	8 10.55	39 26.16	1.2	2.73	52 0.63	17 6 30.1
66	10	. .	28.	40.5	. .	. .	. .	. .	53 53.82	22.70	0.00	III.	6 6.39	27 18.58	1.0	2.11	54 16.52	16 54 21.7
67	10	. .	. .	. .	47.	. .	. .	. .	54 46.86	22.70	0.00	IV.	6 5.15	26 36.23	0.9	2.08	55 9.56	53 39.2
68	9	. .	. .	. .	38.5	51.5	. .	. .	55 38.39	22.69	0.00	V.	6 4.12	26 4.45	0.8	2.05	56 1.08	53 7.3
69	10	. .	. .	13.5	26.2	. .	. .	. .	57 26.19	22.69	0.00	IV.	4 2.37	16 16.27	0.6	1.57	57 48.88	43 18.4
70	9	40.6	54.3	7.	. .	. .	. .	. .	21 59 20.06	+22.68	0.00	III.	4 3.42	−16 49.02	− 0.4	−1.60	21 59 42.74	−16 43 51.0

ZONE 197. SEPTEMBER 25. B. BELT, −16° 53½'. D_0 = −16° 26' 50".

No.	Mag.	I.	II.	III.	IV.	V.	VI.	VII.	T.	a_1	a_2	MICROMETER.		i	d_1	d_2	Mean Right Ascension, 1850.0.	Mean Declination, 1850.0.
1	9	. .	. .	. .	36.2	49.5	. .	. .	23 6 36.27	+22.44	0.00	V.	8 4.27	−36 10.71	−10.8	−2.58	23 6 58.71	−17 3 14.1
2	10	. .	. .	. .	. .	. .	. .	56.	7 16.54	22.44	0.00	VII.	3 6.45	13 22.10	10.7	1.39	7 38.98	16 40 24.2
3	9	. .	47.	59.2	12.2	. .	. .	. .	9 12.50	22.43	0.00	IV.	7 6.435	32 20.86	10.6	2.38	9 34.93	16 59 23.8
4	11	. .	. .	. .	. .	. .	. .	25.	9 45.97	22.43	0.00	VII.	9 8.2	42 57.93	10.5	2.94	10 8.40	17 10 1.4
5	9	. .	. .	. .	25.5	38.3	. .	. .	11 25.35	22.43	0.00	V.	2 8.12	9 7.07	10.4	1.18	11 47.78	16 36 8.7
6	10	. .	12.	24.5	. .	. .	. .	. .	13 38.00	22.43	0.00	III.	9 13.34	40 46.53	10.3	2.82	14 0.43	17 7 49.6
7	11	. .	. .	. .	33.5	. .	. .	. .	15 33.36	22.42	0.00	IV.	3 8.58	14 29.38	10.2	1.45	15 55.78	16 41 31.0
8	10	. .	. .	. .	. .	. .	34.	47.5	16 8.14	22.42	0.00	VII.	7 7.48	32 53.16	10.1	2.41	16 30.56	59 55.7
9	10	. .	. .	. .	47.5	2.	. .	. .	17 48.18	22.41	0.00	V.	7 4.2	30 59.41	10.0	2.31	18 10.57	16 58 1.7
10	10	. .	. .	. .	. .	. .	. .	1.	18 21.91	22.41	0.00	VII.	8 10.21	39 9.01	10.0	2.74	18 44.32	17 6 11.8
11	9	2.1	16.	28.5	41.5	. .	. .	. .	22 41.59	22.40	0.00	IV.	4 12.46	21 23.35	9.8	1.82	23 3.99	16 48 25.0
12	10	. .	. .	. .	. .	. .	. .	35.	33 55.90	22.37	0.00	VII.	8 9.40	38 48.34	9.1	2.72	34 18.27	17 5 50.2
13	9	. .	. .	. .	53.5	7.	. .	. .	35 53.58	22.36	0.00	V.	1 5.25	2 43.85	9.0	0.87	36 15.94	16 29 43.7
14	9	48.5	2.2	14.8	28.	. .	. .	. .	39 27.94	22.36	0.00	IV.	5 4.9	22 1.86	8.8	1.85	39 50.30	49 2.5
15	8.9	45.	58.4	11.	24.	. .	. .	. .	23 41 24.12	+22.35	0.00	IV.	3 9.165	−14 38.71	− 8.7	−1.46	23 41 46.47	−16 41 38.9

CORRECTIONS.

Date.		Corr. of Clock.	Hourly rate.	m	n	c
1848.	h.	s.	s.	s.	s.	s.

INSTRUMENT READINGS.

Date.		Barom.	THERMOM. At.	Ex.
1848.	h. m.	in.	°	°

REMARKS.

(196) 59. Hor. thread assumed as 5 instead of 6.
(197) 6. Hor. thread assumed as 8 instead of 9.

Zone 197. September 25. B. Belt, −16° 53½′. D₀ = −16° 26′ 50″—Continued.

No.	Mag.	Seconds of Transit. I.	II.	III.	IV.	V.	VI.	VII.	T.	a_1	a_2	Micrometer.			i	d_1	d_2	Mean Right Ascension, 1850.0.	Mean Declination, 1850.0.
									h. m. s.	s.	s.			r.	′ ″	″	″	h. m. s.	° ′ ″
16	10	. .	. .	. .	43.	56.	. .	. .	23 42 42.90	+22.35	0.00	V.	6	12.53	−30 27.15	− 8.7	−2.29	23 43 5.25	−16 57 28.1
17	10	. .	. .	54.	7.	. .	. .	. .	44 7.05	22.34	0.00	IV.	10	6.95	47 2.32	8.6	3.17	44 29.39	17 14 4.1
18	9	. .	. .	. .	. .	. .	. .	14.	44 35.00	22.34	0.00	VII.	10	3.4	45 28.56	8.6	3.09	44 57.34	12 30.3
19	10	. .	. .	. .	40.2	53.3	. .	. .	48 40.19	22.33	0.00	V.	10	5.7	46 30.78	8.4	3.14	49 2.52	17 13 32.3
20	10	. .	. .	. .	. .	. .	. .	57.	49 17.76	22.33	0.00	VII.	6	9.40	28 49.64	8.4	2.20	49 40.09	16 55 50.2
21	10	. .	4.3	17.	30.2	. .	. .	. .	54 30.06	22.32	0.00	IV.	3	6.00	12 59.63	8.2	1.37	54 52.38	39 59.2
22	10	. .	. .	3.5	16.2	. .	. .	. .	56 16.17	22.32	0.00	IV.	3	5.44	12 51.56	8.1	1.37	56 38.49	39 51.0
23	10	. .	. .	. .	. .	. .	. .	12.5	56 33.08	22.32	0.00	VII.	4	2.57	16 26.13	8.1	1.55	56 55.40	43 25.8
24	10	. .	38.8	51.5	4.5	. .	. .	. .	23 59 4.50	22.31	0.00	IV.	3	8.9	14 4.68	8.0	1.42	59 26.81	41 4.1
25	10	. .	. .	. .	8.	. .	33.5	. .	0 0 7.69	22.31	0.00	.	1	9.33	4 48.93	7.9	0.96	23 59 30.00	31 47.8
26	10	. .	. .	. .	. .	. .	. .	45.	4 5.55	22.30	0.00	VII.	3	7.14	13 36.72	7.8	1.39	0 4 27.85	40 35.9
27	10	. .	34.5	47.	0.2	. .	. .	. .	7 0.23	22.30	0.00	IV.	6	4.45	26 21.11	7.7	2.06	7 22.53	53 20.9
28	10	. .	. .	. .	59.2	12.5	. .	. .	10 59.23	22.29	0.00	V.	5	5.435	22 49.48	7.5	1.88	11 21.52	49 48.9
29	9	. .	. .	34.	47.	0.2	. .	. .	12 46.83	22.29	0.00	V.	1	5.35	2 48.90	7.5	0.86	13 9.12	16 29 47.3
30	9	. .	. .	. .	. .	. .	. .	16.	13 36.86	22.29	0.00	VII.	7	13.495	35 25.14	7.5	2.58	13 59.15	17 2 25.2
31	8.9	. .	. .	. .	. .	18.3	31.	44.3	15 5.01	22.29	0.00	VII.	4	9.10	19 34.21	7.4	1.71	15 27.33	16 46 33.3
32	9	. .	. .	42.3	55.	8.5	. .	. .	16 55.16	22.28	0.00	V.	5	9.14	24 35.63	7.4	1.98	17 17.44	51 35.0
33	9	. .	. .	. .	. .	. .	. .	12.	17 32.80	22.28	0.00	VII.	7	4.00	30 58.20	7.3	2.32	17 55.08	16 57 57.8
34	8.9	. .	15.5	28.	41.	. .	. .	. .	19 41.30	22.28	0.00	IV.	10	7.8	47 31.81	7.3	3.22	20 3.58	17 14 32.3
35	10	. .	. .	. .	. .	. .	. .	20.2	19? 41.06	22.28	0.00	VII.	8	3.48	35 50.84	7.3	2.57	21 3.34	17 2 50.7
36	9	. .	. .	15.5	28.3	41.3	. .	. .	22 28.23	22.28	0.00	V.	4	5.34	17 45.50	7.2	1.62	22 50.51	16 44 44.3
37	10	. .	3.	15.	28.2	. .	. .	. .	27 28.41	22.27	0.00	IV.	6	8.55	28 27.17	7.1	2.19	27 50.68	55 26.5
38	9	. .	. .	. .	. .	. .	23.	37.	27 57.32	22.27	0.00	VII.	4	11.505	20 55.14	7.1	1.76	28 19.59	16 47 54.0
39	9	. .	52.	4.3	17.2	30.2	. .	. .	30 17.46	22.26	0.00	V.	8	6.29	37 12.23	7.0	2.65	30 39.72	17 4 11.9
40	10	. .	. .	40.	53.	. .	. .	. .	31 53.02	22.26	0.00	IV.	9	6.15	42 4.19	7.0	2.92	32 15.28	17 9 4.1
41	10	. .	29.	41.5	. .	. .	. .	. .	34 54.84	22.26	0.00	III.	6	11.58	29 59.42	6.9	2.27	35 17.10	16 56 58.6
42	10	. .	. .	. .	. .	. .	. .	43.2	35 3.90	22.26	0.00	VII.	5	10.52	25 24.85	6.9	2.02	35 26.16	16 52 23.8
43	5.6	. .	. .	. .	49.	2.2	15.	. .	37 49.01	22.26	0.00	VI.	10	7.37	47 46.34	6.8	3.24	38 11.27	17 14 46.4
44	10	. .	. .	. .	10.	. .	. .	. .	41 9.86	22.25	0.00	IV.	10	3.18	45 35.83	6.8	3.11	41 32.11	17 12 35.7
45	10	. .	. .	. .	. .	. .	32.2	46.2	42 6.59	22.25	0.00	VII.	7	6.12	32 4.76	6.8	2.38	42 28.84	16 59 3.9
46	10	. .	. .	. .	. .	. .	. .	51.	44? 11.78	22.25	0.00	VII.	7	2.52	30 23.91	6.7	2.29	44 34.03	57 22.9
47	9	. .	. .	3.	16.3	29.3	. .	. .	46 16.15	22.25	0.00	V.	7	6.48	32 23.11	6.7	2.39	46 38.40	16 59 22.2
48	10	. .	35.	47.	0.	. .	. .	. .	48 0.44	22.25	0.00	IV.	9	9.35	43 45.04	6.7	3.02	48 22.69	17 10 44.8
49	9	. .	. .	14.2	27.3	40.5	. .	. .	49 27.22	22.25	0.00	V.	5	4.21	22 7.88	6.6	1.84	49 49.47	16 49 6.3
50	9	. .	. .	. .	. .	. .	. .	58.	50 18.40	22.24	0.00	.	1	5.32	2 47.41	6.6	0.83	50 40.64	16 29 44.8
51	9	. .	. .	. .	32.	. .	. .	. .	53 31.86	22.24	0.00	IV.	8	9.29	38 43.01	6.6	2.74	53 54.10	17 5 42.4
52	7.8	. .	. .	. .	. .	. .	15.3	29.	53 49.57	22.24	0.00	VII.	8	6.38	37 16.57	6.6	2.65	0 54 11.81	17 4 15.8
53	10	. .	29.2	42.	55.	. .	. .	. .	0 59 55.00	22.24	0.00	IV.	4	6.6	18 1.66	6.5	1.63	1 0 17.24	16 44 59.8
54	9	. .	12.	24.2	37.3	. .	. .	. .	1 2 37.58	22.24	0.00	IV.	8	10.50	39 23.86	6.5	2.78	2 59.82	17 6 23.1
55	10	. .	41.	54.	6.5	. .	. .	. .	4 6.80	22.24	0.00	IV.	5	8.47	24 22.04	6.5	1.97	4 29.04	16 51 20.5
56	11	. .	. .	13.5	26.	. .	. .	. .	6 26.18	22.24	0.00	IV.	6	9.11	28 35.23	6.5	2.19	6 48.42	55 33.9
57	8	. .	. .	. .	. .	. .	33.2	47.	7 7.36	22.24	0.00	VII.	2	9.39	9 50.74	6.5	1.18	7 29.60	36 48.4
58	8	. .	47.	0.	13.	. .	. .	. .	12 12.87	22.23	0.00	IV.	2	8.2	9 2.05	6.5	1.14	12 35.10	16 35 59.7
59	10	. .	24.2	36.6	49.8	. .	. .	. .	15 49.90	22.23	0.00	IV.	7	8.58	33 28.68	6.5	2.46	16 12.13	17 0 27.6
60	9	. .	. .	. .	. .	. .	. .	34.5	16 55.52	22.23	0.00	VII.	10	5.10	46 32.09	6.5	3.18	17 17.75	17 13 31.8
61	10	. .	. .	29.4	42.3	55.	. .	. .	20 42.07	22.23	0.00	V.	3	8.15	14 7.68	6.5	1.41	21 4.30	16 41 5.6
62	10	. .	. .	. .	48.	1.2	. .	. .	21 48.00	22.23	0.00	V.	6	7.27	27 42.79	6.5	2.15	22 10.23	54 41.4
63	9	. .	31.5	44.	57.	. .	. .	. .	24 57.17	22.23	0.00	VI.	6	8.43	28 21.00	6.5	2.19	25 19.40	55 19.7
64	9	. .	. .	27.5	40.4	53.2	. .	. .	1 28 40.24	+22.23	0.00	VII.	4	10.3	−20 0.87	− 6.5	−1.72	1 29 2.47	−16 46 59.1

CORRECTIONS.

Date.	Corr. of Clock.	Hourly rate.	m	n	c
1848. h.	s.	s.	s.	s.	s.

INSTRUMENT READINGS.

Date.	Barom.	Thermom. Att.	Thermom. Ex.
1848. h. m.	In.	°	°

REMARKS.

(197) 25. Minutes assumed as 59 instead of 0.
(197) 30. Micrometer reading assumed as 12ʳ.495 instead of 13ʳ.495.
(197) 35. Minutes assumed as 20 instead of 19.
(197) 46. Minutes doubtful.
(197) 37. Double

ZONE 197. SEPTEMBER 25. D. BELT, −16° 53½′. D_0 = −16° 26′ 50″—Continued.

No.	Mag.	SECONDS OF TRANSIT. I.	II.	III.	IV.	V.	VI.	VII.	T.	a_1	a_2	MICROMETER.			i	d_1	d_2	Mean Right Ascension, 1850.0.	Mean Declination, 1850.0.
									h. m. s.	s.	s.			r.	′ ″	″	″	h. m. s.	° ′ ″
65	9	..	..	..	..	59.	11.5	25.	1 29 45.67	+22.22	0.00	VII.	4	7.48	−18 52.86	−6.5	−1.66	1 30 7.89	−16 45 51.0
66	9	..	..	..	..	..	11.	25.	30 45.27	22.23	0.00	VII.	2	12.15	11 9.39	6.5	1.24	31 7.50	16 38 7.1
67	10	..	..	..	56.	9.2	..	..	32 56.02	22.24	0.00	V.	8	3.57	35 55.58	6.5	2.59	33 18.26	17 2 54.7
68	4.5	5.	18.	31.	44.	57.	..	..	36 44.00	22.24	0.00	VII.	4	3.32	16 43.78	6.6	1.55	37 6.24	16 43 41.9
69	10	..	..	..	..	..	1.	14.5	37 35.02	22.24	0.00	VII.	3	3.16	11 36.71	6.6	1.27	37 57.26	38 34.6
70	10	..	..	..	24.	..	..	..	42 23.86	22.24	0.00	IV.	1	8.65	4 4.55	6.6	0.85	42 46.10	16 31 1.7
71	10	24.	37.5	50.	3.	..	..	..	44 3.29	22.24	0.00	IV.	7	13.33	35 47.34	6.6	2.58	44 25.53	17 2 46.5
72	10	..	..	12.3	25.	38.5	..	..	45 25.31	22.24	0.00	V.	10	5.45	46 49.94	6.7	3.22	45 47.55	17 13 49.9
73	10	..	..	..	..	..	51.2	..	46 25.21	22.24	0.00	VII.	4	8.12	16 33.69	6.7	1.54	46 47.45	16 43 31.9
74	10	..	..	..	35.2	..	..	..	48 35.06	22.24	0.00	IV.	1	6.17	3 10.10	6.7	0.83	48 57.30	30 7.6
75	9	..	..	58.	11.	..	..	..	50 10.89	22.24	0.00	IV.	5	9.43	24 50.27	6.7	1.99	50 33.13	16 51 49.0
76	11	..	..	..	..	55.4	8.3	21.5	52 42.3	22.24	0.00	VII.	8	6.55	37 25.14	6.8	2.67	53 4.58	17 4 24.6
77	9	18.3	32.	44.5	..	..	..	..	56 57.82	22.25	0.00	III.	6	10.13	29 6.48	6.9	2.23	57 20.07	16 56 5.6
78	9	..	..	..	..	..	..	46.	1 57 6.83	22.25	0.00	VII.	7	8.55	33 26.95	6.9	2.46	1 57 29.08	17 0 26.3
79	10	..	37.5	50.	..	..	..	..	2 0 3.39	+22.25	0.01	III.	7	11.6	−34 33.21	−6.9	−2.52	2 0 25.64	−17 1 32.6
ZONE 198. OCTOBER 5. P. BELT, −20° 1′. D_0 = −19° 35′ 0″.																			
1	10	..	..	14.6	27.4	40.9	..	..	22 35 27.59	+39.45	0.00	V.	7	8.36	−33 17.57	−8.2	−2.51	22 36 7.04	−20 8 28.3
2	10	9.4	22.8	35.7	48.7	..	..	..	38 49.08	39.44	0.00	IV.	4	13.425	21 51.83	8.0	1.82	39 28.52	19 57 1.7
3	10	24.4	..	51.6	4.6	17.8	..	..	43 4.67	39.42	+0.01	.	8	13.13	40 35.95	7.7	2.95	43 44.10	20 15 46.6
4	8	..	..	..	..	..	..	34.6	43 54.43	39.42	0.00	VII.	3	9.24	14 42.21	7.7	1.38	44 33.85	19 49 51.3
5	7	..	..	..	..	..	49.8	3.4	45 23.33	39.42	0.00	VII.	5	6.275	23 11.41	7.6	1.89	46 2.75	58 20.9
6	11	..	..	..	23.1	..	49.9	3.5	49 23.23	39.40	−0.01	VII.	3	2.345	11 15.73	7.4	1.18	50 2.62	19 46 24.3
7	10	..	..	2.3	15.4	28.8	41.9	..	53 15.44	39.39	0.00	VI.	8	0.585	34 25.47	7.1	2.57	53 54.83	20 9 35.1
8	11	..	..	..	..	..	..	25.8	54 45.88	39.38	0.00	VII.	7	5.49	31 53.10	7.1	2.42	55 25.26	20 7 2.6
9	11	..	..	..	..	30.5	43.6	57.4	56 17.19	39.38	0.00	VII.	4	7.41	18 49.26	7.0	1.63	56 56.57	19 53 57.9
10	11	..	..	..	..	13.8	25.9	39.6	57 59.91	39.37	+0.01	VII.	8	13.305	40 44.78	6.9	2.96	22 58 39.29	20 15 54.6
11	11	..	..	..	..	..	51.4	5.6	22 59 25.24	39.37	0.00	VII.	5	11.53	25 55.53	6.8	2.05	23 0 4.61	19 1 4.4
12	11	..	50.4	2.9	..	..	..	..	23 2 16.38	39.36	0.00	III.	2	8.15	9 8.57	6.7	1.05	2 55.74	19 44 16.3
13	10	17.4	30.6	43.5	57.4	..	..	..	5 57.14	39.35	0.00	IV.	2	2.24	6 11.61	6.5	0.87	6 36.49	41 19.0
14	7	..	58.5	11.4	24.6	37.8	50.5	..	8 24.44	39.34	0.00	VI.	1	12.33	6 19.57	6.4	0.88	9 3.78	41 26.9
15	9	..	..	..	..	9.4	21.7	35.6	9 55.51	39.33	0.00	VII.	1	9.35	4 33.77	6.3	0.78	10 34.84	39 40.9
16	8	..	55.8	9.4	21.6	35.6	48.8	..	12 22.15	39.33	0.00	VI.	3	6.54	13 26.74	6.2	1.29	13 1.48	48 34.2
17	8	32.6	46.	58.8	12.3	25.6	..	..	15 12.32	39.32	0.00	V.	4	11.28	20 43.98	6.1	1.73	15 51.64	19 55 51.8
18	11	56.4	10.5	23.1	..	..	..	..	17 36.77	39.31	0.00	III.	8	5.205	36 37.67	6.0	2.72	18 16.08	20 11 46.4
19	11	12.5	26.4	39.5	52.8	..	..	..	23 52.61	39.29	0.00	IV.	4	5.5	17 30.89	5.7	1.55	24 31.90	19 52 38.1
20	11	..	..	..	..	14.6	27.5	..	25 1.14	39.29	0.00	VI.	5	6.42	23 18.89	5.6	1.90	25 40.43	58 26.4
21	11	..	..	..	1.9	15.5	28.3	41.6	31 1.80	39.27	0.00	VII.	4	6.45	19 21.55	5.4	1.59	31 41.07	54 28.5
22	9	..	..	..	..	..	..	46.5	32 6.33	39.27	0.00	VII.	3	8.55	14 2.63	5.4	1.33	32 45.60	49 9.4
23	10	..	..	..	8.4	21.5	..	..	36 8.15	39.26	0.00	V.	1	3.455	1 53.67	5.2	0.60	36 47.41	19 36 59.5
24	11	1.9	15.9	28.3	..	..	..	..	38 42.14	39.25	0.00	III.	8	4.27	36 10.70	5.1	2.69	39 21.39	20 11 18.5
25	11	..	..	..	..	..	..	21.4	38 41.38	39.25	0.00	VII.	6	2.125	25 3.93	5.1	2.01	39 20.63	20 0 11.0
26	10	5.5	18.8	..	..	..	..	..	42 45.49	39.24	0.00	II.	4	5.495	17 53.21	5.0	1.56	43 24.73	19 52 59.8
27	7	..	..	..	..	..	23.6	37.5	42 57.20	39.24	0.00	VII.	2	8.48	9 24.96	5.0	1.05	43 36.44	44 31.0
28	11	..	5.6	18.5	31.7	..	..	..	46 31.80	39.23	0.00	IV.	5	5.44	22 49.76	4.9	1.86	47 11.03	19 57 56.5
29	10	8.8	23.1	35.5	..	2.9	15.5	..	51 49.31	39.22	0.00	VI.	9	11.57	39 57.52	4.8	2.93	52 28.53	20 15 5.2
30	11	..	..	..	..	45.5	58.8	12.6	23 56 32.43	+39.21	0.00	VII.	9	1.118	−39 31.01	−4.7	−2.91	23 57 11.64	−20 14 38.6

CORRECTIONS.

Date.			Corr. of Clock.	Hourly rate.	m	n	c
1848.		h.	s.	s.	s.	s.	s.
Oct.	5.	21	+ 36.68	/ 0.119	− 0.41	+ 0.99	0.00

INSTRUMENT READINGS.

Date.		Barom.	THERMOM. At.	THERMOM. Ex.
1848.	h. m.	In.	°	°

REMARKS.

(197) 70. Micrometer reading assumed as $7^{r}.65$ instead of $8^{r}.65$.
(198) 19. Transit over T. II assumed as at $26^{s}.4$ instead of $24^{s}.4$.
(198) 21. Micrometer reading assumed as $8^{r}.45$ instead of $6^{r}.45$.
(198) 29. Hor. thread assumed as 8 instead of 9.

ZONE 198. OCTOBER 5. P. BELT, −20° 1'. D_0 = −19° 35' 0"—Continued.

No.	Mag.	I.	II.	III.	IV.	V.	VI.	VII.	T.	a_1	a_2	MICROMETER.			i	d_1	d_2	Mean Right Ascension, 1850.0.	Mean Declination, 1850.0.
									h. m. s.	s.	s.			r.	' "	"	"	h. m. s.	° ' "
31	12	..	..	..	4.8	17.9	31.6	..	0 0 4.78	+39.20	0.00	VI.	6	8.47	−28 23.02	− 4.6	−2.21	0 0 43.98	−20 3 29.8
32	3	..	55.6	8.4	21.8	35.6	48.4	2.	6 21.85	39.18	0.00	VII.	2	11.16	10 39.59	4.5	1.12	7 1.03	19 45 45.2
33	12	..	..	..	6.4	..	32.8	45.9	8 6.20	39.18	0.00	VII.	8	8.56	38 26.10	4.4	2.84	8 45.38	20 13 33.3
34	7	..	..	..	..	..	43.2	56.4	9 16.51	39.18	0.00	VII.	4	5.44	17 50.28	4.4	1.54	9 55.69	19 52 56.2
35	10	..	44.6	57.6	10.9	23.9	36.9	..	15 10.64	39.17	0.00	VI.	2	6.20	8 10.49	4.3	0.97	15 49.81	43 15.8
36	10	..	55.4	8.5	21.7	35.4	..	..	0 17 21.69	+39.16	0.00	V.	1	7.26	− 3 4.86	− 4.3	−0.66	0 18 0.85	−19 38 9.8

ZONE 199. OCTOBER 6. H. BELT, −17° 31'. D_0 = −17° 5' 0".

No.	Mag.	I.	II.	III.	IV.	V.	VI.	VII.	T.	a_1	a_2	MICROMETER.			i	d_1	d_2	Mean Right Ascension, 1850.0.	Mean Declination, 1850.0.
1	8	..	50.	3.	16.	..	..	..	22 38 15.90	+42.66	−0.01	IV.	2	3.41	− 6 50.43	− 3.01	−1.06	22 38 58.55	−17 11 54.5
2	9	..	..	..	..	..	42.	55.	39 15.68	42.65	0.00	VII.	3	10.59	15 30.17	2.97	1.51	39 58.33	20 34.6
3	8	..	..	..	..	..	..	49.	40 9.82	42.65	+0.01	VII.	9	6.1	41 56.90	2.93	2.89	40 52.48	47 2.7
4	8	5.2	18.5	31.5	..	..	..	..	43 44.63	42.64	0.00	III.	5	3.56	21 55.28	2.81	1.84	44 27.27	26 59.9
5	9	..	..	21.	34.	..	..	..	45 34.05	42.63	+0.01	IV.	9	11.16	44 35.97	2.74	3.04	46 16.69	49 41.8
6	9	..	..	32.5	..	..	..	..	48 45.68	42.62	+0.01	III.	8	12.52	40 25.35	2.63	2.81	49 28.31	45 30.8
7	8	..	..	57.	9.5	..	..	..	50 9.61	42.62	0.00	VI.	3	11.31	15 46.43	2.59	1.52	50 52.23	20 50.5
8	8	..	..	58.	11.	..	..	..	52 10.99	42.61	0.00	V.	8	4.5	35 59.62	2.53	2.58	52 53.60	41 4.7
9	6.7	..	16.2	28.7	42.	..	..	..	56 42.07	42.60	0.00	IV.	8	6.37	37 16.29	2.39	2.65	57 24.67	42 21.3
10	9	..	..	..	..	16.	..	..	58 2.85	42.59	0.00	VI.	5	3.59	21 56.71	2.35	1.84	22 58 45.44	27 0.9
11	8	..	..	27.3	40.	53.	..	..	22 59 40.02	42.59	0.00	V.	6	8.56	28 27.66	2.30	2.18	23 0 22.61	33 32.1
12	9	..	..	22.5	35.5	..	..	..	23 2 35.33	42.58	0.00	V.	3	6.48	13 23.81	2.23	1.40	3 17.91	18 27.4
13	9	..	38.	51.5	..	..	..	..	5 4.51	42.57	0.00	III.	8	10.16	39 6.70	2.16	2.74	5 47.08	44 11.6
14	7	..	..	..	..	..	6.	19.	5 39.82	42.57	0.00	VII.	8	8.55	38 25.64	2.15	2.71	6 22.39	43 30.5
15	9	..	..	..	..	..	6.	19.	6 39.73	42.56	0.00	VII.	5	9.21	24 38.94	2.12	1.98	7 22.29	29 43.0
16	9	..	0.0	13.	..	..	..	..	9 25.90	42.56	−0.01	III.	1	9.34	4 49.42	2.05	0.95	10 8.45	9 52.4
17	9	..	..	7.5	20.5	..	..	..	11 20.40	42.55	0.00	IV.	5	4.41	22 17.99	2.01	1.86	12 2.95	27 21.9
18	8	..	27.5	40.2	..	..	..	..	12 53.47	42.55	0.00	III.	6	3.9	25 32.68	1.98	2.03	13 36.02	30 36.7
19	9	..	..	..	..	..	44.	..	13 17.89	42.54	0.00	VI.	6	5.2	26 29.58	1.97	2.08	14 0.43	31 33.6
20	9	..	..	..	..	52.	5.	..	14 38.87	42.54	0.00	VI.	4	2.59	16 27.26	1.95	1.56	15 21.41	21 30.8
21	9	..	..	..	..	17.	..	..	21 3.89	42.52	0.00	V.	6	10.37	29 18.59	1.83	2.22	21 46.41	34 22.6
22	8	..	48.	0.2	13.	..	..	..	24 13.08	42.52	0.00	IV.	7	11.31	34 45.83	1.78	2.52	24 55.60	39 50.1
23	8	..	..	..	20.3	33.	..	..	25 19.96	42.51	0.00	IV.	1	6.55	3 29.26	1.77	0.87	26 2.47	8 31.9
24	8	..	..	..	..	..	..	42.	26 2.56	42.51	0.00	VII.	9	11.12	44 33.72	1.76	3.05	26 45.37	49 38.5
25	9	..	..	40.3	54.	..	..	..	28 53.06	42.50	0.00	IV.	8	8.41	38 18.81	1.71	2.70	29 36.16	43 23.2
26	9	..	..	51.	4.	..	..	..	33 3.90	42.49	0.00	IV.	5	10.51	25 24.56	1.67	2.02	33 46.39	30 28.3
27	9	..	..	..	23.	36.2	..	..	35 22.92	42.49	0.00	V.	1	9.32	4 48.41	1.65	0.94	36 5.41	9 51.0
28	7.8	3.	17.	29.5	42.	..	..	..	39 42.60	42.48	0.00	IV.	6	5.33	26 45.31	1.61	2.09	40 25.28	31 49.0
29	9.10	38.	52.	5.	17.5	..	..	..	43 17.88	42.47	0.00	IV.	8	7.26	37 40.99	1.60	2.68	44 0.35	42 45.3
30	9.10	..	..	..	..	36.3	49.	3.	44 23.23	42.47	0.00	VII.	6	8.00	27 59.21	1.59	2.17	45 5.70	33 3.0
31	9	..	..	..	20.3	33.	..	..	48 19.97	42.46	0.00	VII.	2	6.58	8 29.50	1.58	1.13	49 2.43	13 32.2
32	9	11.3	25.	37.5	51.	..	..	..	50 50.98	42.45	0.00	IV.	9	5.6	41 29.40	1.57	2.89	51 33.43	46 33.9
33	8	..	..	..	0.5	13.9	26.	..	52 0.34	42.45	0.00	VI.	6	5.31	26 44.20	1.57	2.09	52 42.79	31 47.9
34	8	..	..	..	19.	32.	..	59.	53 19.05	42.45	0.00	VII.	4	4.1	16 58.39	1.57	1.57	54 1.50	22 1.5
35	5.6	..	..	43.	55.2	9.	..	..	55 55.55	42.44	0.00	VI.	4	3.245	16 40.11	1.57	1.56	56 37.99	21 43.2
36	7	..	..	..	..	..	..	9.	56 29.74	42.44	0.00	VII.	8	4.17	36 5.46	1.57	2.60	57 12.18	41 9.6
37	9	..	..	..	..	17.	30.	43.5	23 58 3.97	+42.44	0.00	VII.	6	10.38	−29 18.88	− 1.57	−2.23	23 58 46.41	−17 34 22.7

CORRECTIONS.

Date.	Corr. of Clock.	Hourly rate.	m	n	c
1848. h.	s.	s.	s.	s.	s.
Oct. 6, 21	+ 40.01	+ 0.116	− 0.50	+ 1.10	0.00

INSTRUMENT READINGS.

Date.	Barom.	THERMOM. At.	THERMOM. Ex.
1848. h. m.	in.	°	°

REMARKS.

(199) 10. Micrometer reading assumed as 3r.59 instead of 5r.59.

ZONE 200. OCTOBER 7. P. BELT, −19° 23.

No.	Mag.	I.	II.	III.	IV.	V.	VI.	VII.	T.	a_1	a_2	Micrometer			i	d_1	d_2	Mean Right Ascension, 1850.0.	Mean Declination, 1850.0.
									h. m. s.	s.	s.			r.	′ ″	″	″	h. m. s.	° ′ ″
1	11	27.6	41.8	54.6	..	..	..	..	22 42 7.97	+45.07	0.00	III.	7	8.2	−33 0.42	. .	−2.46	22 42 53.04	
2	10	..	..	10.4	23.2	36.5	49.5	..	43 22.27	45.07	0.00	VI.	7	13.26	35 43.69	. .	2.62	44 8.34	
3	11	39.4	..	5.6	18.4	32.4	46.5	..	49 19.16	45.05	0.00	VI.	7	5.5	31 31.07	. .	2.37	50 4.21	
4	11	43.2	..	9.4	23.9	..	..	..	53 23.10	45.03	0.00	IV.	4	4.47	17 21.82	. .	1.57	54 8.13	
5	7	..	..	..	..	..	..	33.2	53 53.29	45.03	0.00	VII.	5	3.55	21 54.51	. .	1.82	54 38.32	
6	11	..	..	..	..	..	23.6	..	22 56 57.26	45.02	−0.01	VI.	3	2.43	11 20.17	. .	1.21	22 57 42.27	
7	9	21.4	35.6	48.5	..	..	..	..	23 0 1.81	45.01	0.00	III.	7	11.40	34 50.34	. .	2.58	23 0 46.82	
8	9	..	..	..	22.6	..	48.8	2.9	1 22.66	45.00	0.00	VII.	7	6.19	32 8.24	. .	2.42	2 7.66	
9	8	..	9.4	22.2	35.4	..	..	..	4 35.44	44.99	0.00	IV.	4	3.27	16 41.47	. .	1.52	5 20.43	
10	9	..	..	..	..	4.6	17.8	30.9	5 51.35	44.99	0.00	VII.	9	6.56	42 24.60	. .	2.92	6 36.34	
11	8	..	52.6	5.6	18.8	31.6	44.7	..	8 18.71	44.98	0.00	VI.	9	7.24	42 38.86	. .	2.94	9 3.69	
12	9	..	..	..	49.8	2.9	16.8	29.9	9 50.00	44.97	0.00	VII.	8	13.50	40 54.34	. .	2.83	10 34.97	
13	9	..	..	4.5	17.4	30.8	43.8	57.5	12 17.48	44.97	0.00	VII.	5	6.11	23 3.10	. .	1.89	13 2.45	
14	6	48.4	2.8	14.9	..	..	..	..	15 28.66	44.96	0.00	III.	7	6.45	32 1.17	. .	2.41	16 13.62	
15	11	..	..	..	10.5	23.6	37.6	..	21 10.63	44.94	0.00	VI.	5	3.34	21 44.09	. .	1.81	21 55.57	
16	10	57.5	11.5	24.1	37.4	..	..	..	26 37.50	44.92	0.00	IV.	5	10.26	25 11.96	. .	2.01	27 22.42	
17	10	..	..	..	..	..	27.8	..	32 1.35	44.91	+0.01	VI.	10	12.15	50 6.49	. .	3.49	32 46.27	
18	5	0.2	14.5	26.5	40.5	53.6	6.4	..	35 40.16	44.90	0.00	VI.	2	5.399	7 50.27	. .	1.00	36 25.06	
19	9	..	..	15.4	28.	41.8	54.6	7.7	37 28.23	44.89	0.00	VII.	7	6.15	42 6.22	. .	2.91	38 13.12	
20	10	..	..	..	59.9	13.5	26.2	39.9	39 0.04	44.89	+0.01	VII.	10	1.50	44 51.19	. .	3.17	40 44.94	
21	6.7	..	..	37.3	51.1	4.8	17.8	..	42 51.16	44.88	+0.01	VI.	10	4.3	45 58.41	. .	3.24	43 36.05	
22	8	..	..	..	..	..	36.6	49.9	44 10.13	44.88	0.00	VII.	6	2.15	25 5.20	. .	2.01	44 55.01	
23	9	..	..	..	..	..	..	56.9	45 16.94	44.87	0.00	VII.	4	6.29	18 12.98	. .	1.60	46 1.81	
24	12	..	..	..	..	..	..	6.6	46 26.89	44.87	0.00	VII.	8	3.49	35 51.30	. .	2.64	47 11.76	
25	11	..	22.9	..	..	..	..	..	48 49.34	44.86	0.00	II.	3	10.13	15 7.08	. .	1.43	49 34.20	
26	9	..	17.8	30.5	43.5	57.6	..	..	50 43.94	44.86	0.00	V.	5	7.37	23 46.71	. .	1.93	51 28.80	
27	10	..	..	..	..	..	..	3.9	51 23.99	44.86	0.00	VII.	5	3.50	21 5.00	. .	1.77	52 8.85	
28	10	..	..	..	18.4	..	44.6	..	56 18.27	44.84	−0.01	VI.	1	4.39	2 20.57	. .	0.67	57 3.10	
29	10	..	..	..	37.4	50.4	3.9	..	58 37.31	44.84	0.00	V.	8	13.14	40 36.43	. .	2.83	23 59 22.15	
30	10	..	..	..	..	..	..	14.6	23 59 34.84	44.84	0.00	VII.	7	6.50	32 23.87	. .	2.43	0 0 19.68	
31	10	..	..	..	34.6	47.5	1.	..	0 2 34.43	44.83	0.00	VI.	5	11.12	25 35.03	. .	2.04	3 19.26	
32	9	..	..	..	..	..	..	50.9	3 11.07	44.83	0.00	VII.	6	6.38	27 17.82	. .	2.13	3 55.90	
33	11	..	39.9	52.2	..	..	..	..	7 5.79	44.82	0.00	III.	4	4.22	17 9.18	. .	1.53	6 50.61	
34	5	..	..	..	..	..	..	55.5	6 15.95	44.82	+0.01	VII.	10	6.35	47 14.91	. .	3.32	7 0.78	
35	11	..	..	..	..	20.5	..	46.6	0 8 6.81	+44.82	0.00	VII.	2	3.9	− 6 34.03	. .	−0.93	0 8 51.63	

ZONE 201. OCTOBER 10. P. BELT, −16° 16′.

No.	Mag.	I.	II.	III.	IV.	V.	VI.	VII.	T.	a_1	a_2	Micrometer			i	d_1	d_2	Mean Right Ascension, 1850.0.	Mean Declination, 1850.0.
1	10	..	..	37.4	49.9	3.8	15.5	..	22 59 50.00	+54.07	−0.01	VI.	1	2.31	− 1 16.05	. .	−0.88	23 0 44.06	
2	10	47.7	1.5	14.6	..	..	..	..	23 2 27.22	54.06	0.00	III.	2	10.16	10 9.60	. .	1.29	3 21.28	
3	9	..	..	37.5	..	3.8	16.4	29.4	7 50.48	54.04	0.00	VII.	6	7.51	27 54.67	. .	2.14	8 44.52	
4	10	..	..	..	31.6	..	58.4	10.8	9 31.77	54.04	0.00	VII.	1	6.38	3 20.47	. .	0.97	10 25.81	
5	9	..	..	..	..	5.6	18.4	31.6	10 52.61	54.03	0.00	VII.	10	4.41	46 17.47	. .	3.02	11 46.64	
6	3.4	..	..	40.4	52.4	6.6	18.5	..	13 52.83	54.03	0.00	VI.	1	3.54	1 57.90	. .	0.91	14 46.86	
7	11	..	..	..	..	4.6	..	..	15 51.51	54.02	0.00	V.	4	4.36	17 16.26	. .	1.63	16 45.53	
8	8.9	..	..	59.9	11.9	25.6	38.2	..	18 12.30	54.01	0.00	VI.	3	9.16	14 38.37	. .	1.50	19 6.31	
9	9	..	..	..	..	..	..	49.9	23 19 10.63	+54.01	0.00	VII.	3	15.11	−17 37.22	. .	−1.64	23 20 4.64	

CORRECTIONS.

Date.		Corr. of Clock.	Hourly rate.	m	n	c
1848.	h.	s.	s.	s.	s.	s.
Oct. 7,	22	+ 42.48	/ 0.119	− 0.45	+ 1.03	0.00
10,	22	+ 52.53	/ 0.133	− 0.50	+ 1.02	0.00

INSTRUMENT READINGS.

Date.		Barom.	Thermom. At.	Thermom. Ex.
1848.	h. m.	in.	°	°

REMARKS.

(200) 13. Double star.
(200) 19. Fine double star; same magnitude.
(200) 20. Minutes assumed as 40 instead of 39.
(200) 28. Transits over T.'s IV and VI assumed as 28s.4 and 54s.6 instead of 18s.4 and 44s.6.
(200) 33. Minutes assumed as 6 instead of 7.

ZONE 202. OCTOBER 10. P. BELT, −16° 16′. D_0=−15° 50′ 0″.

No.	Mag.	Seconds of Transit. I.	II.	III.	IV.	V.	VI.	VII.	T.	a_1	a_2	Micrometer.			i	d_1	d_2	Mean Right Ascension, 1850.0.	Mean Declination, 1850.0.
									h. m. s.	s.	s.			r.	′ ″	″	″	h. m. s.	° ′ ″
1	10	..	..	..	..	48.8	1.8	14.9	1 43 35.84	+54.85	0.00	VII.	7	8.40	−33 19.40	−4.67	−2.43	1 44 30.69	−16 23 26.5
2	10	..	..	..	5.5	18.4	..	..	51 5.30	54.85	0.00	V.	1	2.23	1 12.08	5.78	0.78	52 0.15	15 51 18.6
3	10	12.4	25.5	38.3	51.4	4.1	16.4	30.6	54 51.16	54.86	0.00	VI.	3	3.275	11 42.64	6.36	1.31	55 46.02	16 1 50.3
4	12	..	..	..	..	..	17.4	29.9	56 51.16	54.86	0.00	VII.	7	9.46	33 52.68	6.67	2.46	57 46.02	24 1.8
5	10	..	..	..	..	..	39.2	52.8	1 58 13.42	54.86	0.00	VII.	4	9.13	19 35.73	6.86	1.72	1 59 8.28	9 44.3
6	10	..	..	..	..	35.5	48.8	..	2 7 22.70	54.87	−0.01	VI.	10	12.57	50 27.70	8.30	3.35	2 8 17.56	40 39.4
7	11	..	30.6	41.9	55.6	..	..	..	9 55.79	54.87	−0.01	IV.	10	4.56	46 25.26	8.72	3.13	10 50.65	16 36 37.1
8	11	..	18.2	..	43.8	56.9	..	..	12 43.85	54.87	0.00	V.	2	4.35	7 17.65	9.18	1.08	13 38.72	15 57 27.9
9	10	..	..	..	23.6	36.6	49.9	..	14 23.64	54.88	+0.01	VI.	2	0.29	5 13.53	9.45	0.97	15 18.53	15 55 24.0
10	8.9	19.5	32.7	45.6	58.8	..	..	..	17 58.58	54.88	0.00	IV.	2	11.47	10 55.50	10.03	1.26	18 53.46	16 1 6.8
11	11	..	..	32.6	..	58.8	..	24.6	20 45.38	54.89	0.00	.	2	6.17	8 9.10	10.48	1.12	21 40.27	15 58 20.7
12	11	..	16.4	28.4	42.	..	..	..	23 41.77	54.89	0.00	IV.	2	2.54	5 56.48	10.95	1.02	24 36.66	56 8.4
13	6	..	..	..	..	..	30.6	44.1	24 4.66	54.89	+0.01	VII.	1	0.19	4 11.36	11.01	1.00	24 59.56	15 54 23.4
14	10	..	..	..	..	31.6	44.6	..	34 18.65	54.91	−0.01	VI.	10	11.23	49 40.30	12.72	3.32	35 13.55	16 39 56.3
15	11	..	..	..	58.8	..	25.6	37.9	35 59.06	54.91	0.00	VII.	7	12.30	35 15.37	13.01	2.55	36 53.97	16 25 30.9
16	10	..	28.6	40.4	53.8	..	..	..	38 53.77	54.92	+0.01	IV.	1	9.49	4 57.00	13.49	0.95	39 48.70	15 55 11.4
17	10	..	..	..	..	..	..	49.9	39 10.52	54.92	0.00	VII.	2	10.55	10 29.07	13.52	1.24	40 5.44	16 0 43.8
18	11	..	..	43.4	56.6	9.8	22.6	..	42 56.54	54.93	0.00	VI.	2	10.27	10 15.07	14.18	1.23	43 51.47	0 30.5
19	10	..	..	50.8	3.8	16.2	..	..	49 3.53	54.94	0.00	V.	6	11.39	29 49.84	15.24	2.26	49 58.47	20 7.3
20	9	..	..	..	..	17.8	29.9	43.8	50 4.52	54.94	0.00	VII.	8	5.12	36 33.21	15.42	2.62	50 59.46	26 51.2
21	10	..	..	55.4	8.5	21.5	34.4	..	52 8.49	54.95	−0.01	VI.	10	11.46	49 51.91	15.76	3.34	53 3.43	16 40 11.0
22	10	..	..	23.5	..	48.5	1.6	..	2 54 35.72	+54.95	+0.01	VI.	1	3.28	− 1 44.65	−16.22	−0.78	2 55 30.68	−15 52 1.7

ZONE 203. OCTOBER 10. P. BELT, −19° 23′. D_0=−18° 57′ 40″.

No.	Mag.	I.	II.	III.	IV.	V.	VI.	VII.	T.	a_1	a_2	Micrometer.			i	d_1	d_2	Mean Right Ascension, 1850.0.	Mean Declination, 1850.0.
1	8	16.4	29.8	43.4	55.6	9.8	..	..	3 9 56.13	+54.86	−0.01	V.	2	7.195	− 8 40.58	− 7.48	−0.98	3 10 50.98	−19 6 29.0
2	9	..	..	..	..	15.6	29.8	42.6	12 2.83	54.87	0.00	VII.	6	3.27	25 41.50	7.94	2.04	12 57.70	23 31.5
3	10	22.6	36.4	48.6	..	..	..	..	15 2.57	54.88	+0.01	III.	9	3.00	40 25.83	8.61	2.99	15 57.46	38 17.4
4	9	..	13.5	25.5	39.	..	..	..	16 39.01	54.88	−0.01	IV.	1	5.1	3 31.77	8.98	0.66	17 33.88	1 21.4
5	10	52.8	7.4	19.4	..	..	..	..	18 33.11	54.89	0.00	III.	6	6.37	27 17.56	9.40	2.15	19 28.00	25 9.1
6	10	..	..	16.5	29.5	43.6	..	..	19 29.70	54.89	0.00	V.	3	10.47	15 24.32	9.63	1.38	20 24.59	13 15.3
7	10	..	..	..	..	..	23.6	37.5	20 57.36	54.89	0.00	VII.	2	15.4	12 34.56	9.96	1.22	21 52.25	19 10 25.7
8	10	..	..	..	8.4	20.6	33.4	..	23 7.51	54.90	−0.01	VI.	1	2.18	1 9.46	10.45	0.52	24 2.40	18 59 0.4
9	12	..	..	41.6	..	..	..	..	25 54.89	54.91	0.00	III.	6	5.1	36 27.84	11.08	2.75	26 49.80	19 34 21.7
10	8	52.6	5.8	19.	32.6	..	..	..	28 32.22	54.91	−0.01	IV.	1	9.50	4 57.50	11.68	0.75	29 27.12	2 49.9
11	12	..	23.4	36.4	..	2.9	..	..	30 49.61	54.92	0.00	V.	3	13.12	16 37.42	12.23	1.46	31 44.53	14 31.1
12	10	..	..	..	..	..	16.8	30.6	31 50.62	54.92	0.00	VII.	6	14.37	31 19.31	12.46	2.41	32 45.54	29 14.2
13	10	..	6.2	18.9	32.4	45.6	..	..	34 32.37	54.93	0.00	V.	7	4.3	30 59.90	13.05	2.39	35 27.30	28 55.3
14	10	..	..	..	..	..	..	45.6	35 5.58	54.93	0.00	VII.	3	7.6	13 32.64	13.22	1.26	36 0.51	11 27.1
15	10	..	..	..	..	..	49.9	3.8	37 23.80	54.94	0.00	VII.	8	9.1	38 28.62	13.76	2.87	38 18.74	36 25.3
16	10	..	..	15.8	..	42.8	..	..	39 29.35	54.95	0.00	V.	8	9.32	39 14.19	14.24	2.90	40 24.30	37 11.9
17	9	56.6	10.4	23.6	36.2	..	..	..	42 36.49	54.96	0.00	IV.	3	8.36	14 18.30	14.97	1.31	43 31.45	12 14.6
18	12	..	..	..	..	47.4	0.	14.6	43 34.15	54.96	0.00	VII.	5	7.54	23 55.04	15.21	1.93	44 29.11	21 52.2
19	8	..	..	..	25.4	..	51.6	..	49 25.26	54.98	−0.01	VI.	1	2.57	1 29.13	16.59	0.53	50 20.23	59 26.2
20	9	44.6	58.8	10.8	23.8	..	..	..	3 52 24.51	54.99	+0.01	.	9	7.3	42 28.39	17.31	3.14	3 53 19.51	40 28.8
21	11	..	..	7.7	20.2	34.	46.6	..	4 1 20.49	55.03	0.00	VI.	7	10.3	34 1.34	19.45	2.58	4 2 15.52	32 3.4
22	9	..	..	..	..	..	50.4	4.6	4 2 24.38	+55.03	0.00	VII.	5	12.58	−26 28.32	−19.72	−2.10	4 3 19.41	−19 24 30.1

CORRECTIONS.

Date.		Corr. of Clock.	Hourly rate.	m	n	c
1848.	h.	s.	s.	s.	s.	s.
Oct. 10,	22	+ 52.53	/ 0.133	− 0.50	+ 1.02	0.00

INSTRUMENT READINGS.

Date.		Barom.	Thermom. At.	Thermom. Ex.
1848.	h. m.	in.	°	°

REMARKS.

(202) 12. Micrometer reading assumed as 1r.54 instead of 2r.54.
(202) 13. Micrometer reading assumed as 1t 8r.19 instead of 1t 0r.19.
(203) 1. Double.
(203) 16. Micrometer reading assumed as 10r.32 instead of 9r.32.

ZONE 204. OCTOBER 12. P. BELT, −15° 38½′. D₀=−15° 13′ 10″.

No.	Mag.	SECONDS OF TRANSIT. I.	II.	III.	IV.	V.	VI.	VII.	T.	a_1	a_2	MICROMETER.		i	d_1	d_2	Mean Right Ascension, 1850.0.	Mean Declination, 1850.0.
									h. m. s.	s.	s.		r.	′ ″	″	″	h. m. s	° ′ ″
1	9	. .	45.4	58.5	. .	. .	. .	. .	21 33 11.29	+ 4.80	0.00	III.	4 6.5	−18 1.13	− 3.40	−1.67	21 33 16.09	−15 31 16.2
2	8	6.4	19.4	32.6	45.4	58.2	. .	. .	35 45.29	4.79	−0.01	V.	3 5.275	12 43.21	3.36	1.43	35 50.07	25 58.0
3	11	. .	. .	. .	. .	. .	21.6	34.9	38 55.71	4.78	−0.01	VII.	2 6.22	8 11.41	3.31	1.24	39 0.48	21 26.0
4	11	. .	. .	. .	55.4	8.4	21.6	34.6	40 55.52	4.77	0.00	VI.	6 13.16	30 38.66	3.28	2.27	41 0.29	43 54.2
5	8	. .	. .	31.6	43.7	56.8	. .	. .	46 44.06	4.74	+0.01	V.	9 11.14	44 34.94	3.21	2.92	46 48.81	57 51.1
6	9	53.1	6.4	19.2	. .	. .	. .	. .	49 32.35	4.74	+0.01	III.	8 5.48	36 51.55	3.18	2.56	49 37.10	50 7.3
7	12	. .	. .	. .	. .	. .	. .	35.5	49 56.60	4.73	0.00	VII.	8 1.59	34 55.89	3.17	2.47	50 1.33	48 11.5
8	10	30.9	44.6	56.8	9.6	. .	. .	. .	53 9.96	4.71	0.00	IV.	6 6.36	27 17.09	3.13	2.11	55 14.67	40 32.3
9	10	19.5	33.1	45.6	58.4	. .	. .	. .	57 58.62	4.70	0.00	IV.	5 8.24	24 10.44	3.11	1.96	58 3.32	37 25.5
10	8	. .	. .	11.4	24.5	37.8	. .	. .	21 59 24.32	4.70	−0.01	V.	1 0.51	0 20.49	3.10	0.85	21 59 29.01	13 14.4
11	10	. .	0.2	12.8	25.4	. .	. .	. .	22 6 25.78	4.67	0.00	IV.	8 6.27	37 11.24	3.08	2.57	22 6 30.45	50 26.9
12	11	. .	9.4	21.8	. .	. .	. .	. .	8 35.06	4.66	0.00	III.	6 10.44	29 22.12	3.07	2.21	8 39.72	42 37.4
13	11	. .	. .	. .	. .	. .	. .	58.9	11 19.96	4.66	0.00	VII.	7 6.29	32 13.34	3.06	2.34	11 24.62	45 28.7
14	12	. .	. .	. .	51.6	. .	. .	. .	12 51.46	4.65	0.00	IV.	6 6.12	27 4.98	3.06	2.10	12 56.11	40 20.1
15	11	. .	. .	. .	. .	. .	. .	46.9	13 7.93	4.65	0.00	VII.	7 2.33	30 14.34	3.06	2.25	13 12.58	43 29.6
16	11	. .	. .	58.3	11.5	24.7	37.4	50.8	15 11.53	4.95	0.00	VII.	6 9.12	28 35.53	3.06	2.17	15 16.18	41 50.8
17	11	12.4	25.5	38.5	. .	. .	. .	. .	17 51.51	4.64	0.00	III.	6 10.48	29 24.13	3.06	2.21	17 56.15	42 39.4
18	8	. .	. .	. .	. .	. .	. .	. .	. . . .	. .	−0.01	VII.	2 .	4 58.79	. .	1.07		
19	8	. .	. .	. .	. .	. .	. .	49.9	22 10.61	4.61	−0.01	VII.	2 5.28	7 44.17	3.07	1.19	22 15.21	20 58.4
20	11	. .	. .	. .	. .	. .	. .	4.9	23 25.91	4.61	0.00	VII.	6 8 1	27 59.73	3.08	2.14	23 30.52	41 14.9
21	10	. .	. .	. .	19.4	32.6	45.6	. .	22 26 19.55	+ 4.60	+0.01	.	9 2.30	−40 10.74	− 3.09	−2.72	22 26 24.16	−15 53 26.6

ZONE 205. OCTOBER 14. P. BELT, −15° 38½′. D₀=−15° 14′ 20″.

No.	Mag.	I.	II.	III.	IV.	V.	VI.	VII.	T.	a_1	a_2	MICROMETER.		i	d_1	d_2	Mean Right Ascension, 1850.0.	Mean Declination, 1850.0.
1	9	45.9	59.8	12.	24.9	37.9	. .	. .	20 46 25.15	+ 8.71	0.00	V.	8 5.35	−36 45.00	+ 9.57	−1.53	20 46 33.86	−15 50 56.96
2	9	54.9	8.4	20.2	33.6	. .	. .	. .	49 33.78	8.69	0.00	IV.	6 11.53	29 56.92	9.90	2.22	49 42.47	44 9.24
3	10	5.6	19.6	. .	. .	. .	. .	. .	51 45.20	8.68	0.00	II.	6 7.9	27 33.63	10.12	2.13	51 53.88	41 45.64
4	9	32.6	45.7	58.9	11.8	. .	. .	. .	53 11.57	8.68	−0.01	IV.	1 7.16	3 39.85	10.27	1.05	53 20.24	15 17 50.63
5	10	. .	. .	. .	. .	16.4	29.9	42.9	54 3.91	8.67	+0.01	VII.	10 10.39	49 18.01	10.36	3.12	54 12.59	16 3 30.77
6	10	32.5	45.9	57.8	. .	. .	. .	. .	57 11.31	8.66	0.00	III.	3 11.36	15 49.04	10.68	1.59	20 57 19.97	15 29 59.95
7	10	20.2	33.6	46.8	59.2	10.9	. .	. .	20 59 59.35	8.64	0.00	V.	3 12.12	16 7.18	10.96	1.60	21 0 7.99	30 17.82
8	11	49.4	3.6	15.4	. .	. .	. .	. .	21 3 28.75	8.63	0.00	III.	4 9.46	19 52.57	11.31	1.77	3 37.38	34 3.03
9	9	. .	. .	. .	. .	7.9	20.5	33.7	3 54.71	8.63	0.00	VII.	4 14.375	22 19.29	11.35	1.86	4 3.34	36 52.48
10	10	4.9	17.6	. .	. .	. .	. .	. .	6 43.88	8.62	0.00	II.	6 12.44	30 22.55	11.63	2.25	6 52.50	44 33.17
11	6.5	. .	. .	. .	. .	30.2	43.6	56.7	7 17.58	8.61	0.00	VII.	7 8.32	33 15.37	11.69	2.38	7 26.19	47 26.06
12	9	24.5	38.6	50.8	3.9	16.5	. .	. .	13 3.73	8.59	0.00	V.	7 8.9	33 3.96	12.24	2.37	13 12.32	47 14.09
13	10	. .	. .	. .	. .	. .	18.5	31.5	13 52.52	8.58	0.00	VII.	4 8.10	19 3.97	12.33	1.73	14 1.10	33 13.37
14	11	. .	. .	. .	. .	. .	. .	47.9	15 8.88	8.58	0.00	VII.	6 4.2	25 59.22	12.44	2.05	15 17.46	40 8.83
15	9	22.6	36.5	48.8	1.9	. .	. .	. .	18 2.02	8.56	+0.01	II.	8 10.46	39 21.76	12.72	2.66	18 10.59	53 31.70
16	12	. .	. .	. .	. .	. .	. .	19.9	19 40.70	8.56	0.00	VII.	3 9.28	14 44.29	12.88	1.53	19 49.26	28 52.94
17	10	. .	. .	. .	33.6	46.2	58.8	. .	21 33.21	8.55	0.00	VI.	7 10.45	34 22.55	13.05	2.43	21 41.76	48 31.93
18	10	. .	. .	4.9	. .	25.6	. .	. .	26 12.47	8.53	−0.01	V.	1 6.43	3 23.19	13.49	1.03	26 20.99	17 30.73
19	10	13.5	26.5	39.2	. .	. .	. .	. .	29 52.36	8.51	0.00	III.	5 1.37	20 45.19	13.82	1.81	30 0.87	34 53.18
20	9	28.5	41.6	53.8	. .	. .	. .	. .	33 7.23	8.50	0.00	III.	4 4.21	17 8.68	14.10	1.64	33 15.73	31 16.22
21	10	. .	. .	. .	. .	. .	. .	25.9	33 46.56	8.49	−0.01	VII.	1 7.43	3 53.25	14.16	1.05	33 55.04	18 0.14
22	8	. .	. .	28.3	40.8	54.6	7.5	20.6	35 41.24	8.49	−0.01	VII.	2 13.39	11 52.04	14.33	1.41	35 49.72	25 59.12
23	10	. .	. .	. .	. .	39.4	. .	. .	37 26.45	8.48	0.00	V.	7 12.3	35 1.94	14.48	2.46	37 34.93	49 9.92
24	10	. .	. .	. .	. .	5.8	18.2	31.6	21 38 52.47	+ 8.47	−0.01	VII.	2 6.43	− 8 22.00	+14.60	−1.25	21 39 0.93	−15 22 28.65

CORRECTIONS.

Date.		Corr. of Clock.	Hourly rate.	m	n	c
1848.	h.	s.	s.	s.	s.	s.
Oct. 12,	22	+ 1.93	/0.126			
Oct. 14,	22	+ 5.59	/0.107			

INSTRUMENT READINGS.

Date.		Barom.	THERMOM. At.	Ex.
1848.	h. m.	In.	°	°

REMARKS.

(204) 10. Micrometer reading assumed as 0ʳ.01 instead of 0ʳ.51.
(205) 9. Micrometer reading assumed as 14ʳ.375 instead of 10ʳ.375.
(205) 18. One of the transit observations (assumed T. V) in error by 5ˢ.

ZONE 205. OCTOBER 14. P. BELT, −15° 38¼′. D_c = −15° 14′ 20″—Continued.

No.	Mag.	SECONDS OF TRANSIT. I.	II.	III.	IV.	V.	VI.	VII.	T.	a_1	a_2	MICROMETER.			i	d_1	d_2	Mean Right Ascension, 1850.0.	Mean Declination, 1850.0.
									h. m. s.	s.	s.			r.	′ ″	″	″	h. m. s.	° ′ ″
25	10	. .	26.5	38.6	51.8	5.2	17.5	. .	21 40 51.93	+ 8.46	0.00	VI.	7	11.43	−34 51.79	+14.78	−2.46	21 41 0.39	−15 48 59.47
26	12	. .	57.4	10.4	. .	. .	. .	. .	45 23.40	8.44	0.00	III.	7	9.12	33 35.72	15.16	2.40	45 31.84	47 42.96
27	9	. .	. .	27.5	40.7	53.6	. .	. .	46 40.62	8.41	+0.01	V.	9	9.20	43 37.45	15.27	2.86	46 49.07	57 45.04
28	9	49.8	3.7	15.5	. .	. .	. .	. .	49 29.12	8.43	0.00	III.	8	4.6	36 0.13	15.51	2.51	49 37.55	50 7.13
29	10	. .	. .	. .	. .	. .	18.9	31.9	49 53.01	8.42	0.00	VII.	8	0.14	34 2.95	15.54	2.42	50 1.43	48 9.83
30	12	. .	26.8	39.	. .	. .	. .	. .	53 52.35	8.41	0.00	III.	6	7.41	27 49.83	15.85	2.13	54 0.76	41 56.11
31	9	. .	40.8	52.9	6.4	19.4	. .	. .	55 6.31	8.40	0.00	V.	6	4.43	26 20.08	15.95	2.06	55 14.71	40 26.19
32	8	15.6	29.9	41.9	54.8	. .	. .	. .	21 57 54.72	8.39	0.00	IV.	5	6.43	23 19.51	16.18	1.92	21 58 3.11	37 25.25
33	11	. .	. .	. .	. .	. .	53.1	. .	22 0 27.29	8.38	−0.01	VI.	2	6.46	8 23.64	16.37	1.24	22 0 35.66	22 28.51
34	12	. .	. .	. .	. .	. .	. .	36.4	7 57.15	8.34	−0.01	VII.	2	11.52	10 57.81	16.94	1.35	8 5.48	25 2.22
35	8	43.4	56.4	9.8	22.4	31.9	. .	. .	6 22.43	8.35	0.00	V.	8	4.46	36 20.30	16.83	2.53	6 30.78	50 26.00
36	9	51.4	5.8	18.5	31.4	. .	. .	. .	8 31.04	8.34	0.00	IV.	6	8.43	28 21.12	16.98	2.16	8 39.38	42 26.30
37	10	. .	50.8	3.9	16.4	29.5	41.9	. .	10 16.50	8.34	0.00	VI.	7	5.11	31 34.12	17.11	2.31	10 24.84	45 39.32
38	11	8.8	. .	35.8	. .	. .	. .	. .	12 48.34	8.33	0.00	III.	6	4.27	26 12.01	17.29	2.06	12 56.67	40 16.78
39	10	. .	. .	. .	. .	. .	29.9	42.9	13 3.98	8.32	0.00	VII.	7	1.55	29 24.87	17.30	2.22	13 12.30	43 29.79
40	9	. .	42.9	55.5	8.4	21.4	. .	. .	15 8.49	8.32	0.00	V.	6	7.39	27 48.83	17.45	2.13	15 16.81	41 53.51
41	10	8.4	22.6	34.6	47.5	0.9	. .	. .	17 47.80	8.30	0.00	V.	6	9.14	28 36.73	17.65	2.17	17 56.10	42 41.25
42	9	28.5	41.6	54.5	7.6	. .	. .	. .	22 7.39	8.29	−0.01	IV.	2	3.49	6 54.47	17.94	1.16	22 15.67	20 57.69
43	10	. .	. .	. .	. .	. .	13.7	26.8	22 47.69	8.28	−0.01	VII.	1	11.35	5 50.20	17.98	1.08	22 55.96	19 53.33
44	11	. .	50.5	2.9	13.9	. .	41.4	. .	26 15.97	8.27	+0.01	VI.	8	10.29	39 13.18	18.21	2.67	26 24.25	53 17.64
45	12	. .	. .	52.9	. .	. .	. .	. .	29 5.82	8.26	0.00	III.	6	13.50	30 55.89	18.40	2.28	29 14.08	44 59.77
46	11	8.9	21.9	34.8	. .	. .	. .	. .	31 47.71	8.25	−0.01	III.	2	10.49	5 27.25	18.57	1.09	31 55.98	19 29.77
47	11	. .	. .	. .	. .	26.4	39.2	52.9	32 13.47	8.25	−0.01	VI.	2	7.44	8 52.88	18.59	1.24	32 21.71	22 55.53
48	9	35.4	48.8	. .	. .	. .	. .	. .	36 14.64	8.23	0.00	II.	4	7.19	13 39.36	18.83	1.47	36 22.87	27 42.10
49	10	. .	. .	. .	. .	. .	. .	49.9	36 10.63	8.23	−0.01	VII.	2	9.10	9 36.12	18.83	1.28	36 18.85	23 38.57
50	10	. .	58.9	11.	24.4	37.5	50.6	. .	39 24.52	8.22	+0.01	VI.	9	4.35	41 13.69	19.02	2.77	39 32.75	55 17.43
51	10	. .	. .	. .	27.4	40.9	53.8	. .	42 27.70	8.21	0.00	VI.	5	4.54	22 24.46	19.19	1.88	42 35.91	36 27.15
52	10	39.9	52.6	5.6	18.5	. .	. .	. .	46 18.68	8.19	0.00	IV.	7	10.44	34 22.14	19.42	2.44	46 26.87	48 25.16
53	8	45.5	59.9	12.4	. .	. .	. .	. .	48 25.34	8.18	0.00	III.	7	8.43	33 21.10	19.53	2.40	48 33.52	47 23.97
54	9	. .	. .	. .	39.9	52.9	. .	. .	49 39.77	8.18	−0.01	V.	1	1.34	0 47.38	19.60	0.87	49 47.94	14 48.65
55	9	. .	. .	. .	. .	49.9	. .	. .	52 36.77	8.17	−0.01	V.	1	1.19	0 39.81	19.75	0.86	52 44.93	14 40.92
56	11	18.6	32.6	44.8	. .	. .	. .	. .	54 58.15	8.16	+0.01	III.	9	4.45	41 18.79	19.87	2.78	55 6.32	55 21.70
57	11	5.8	18.6	31.5	43.6	. .	. .	. .	56 44.24	8.15	−0.01	IV.	2	10.0	10 1.55	19.96	1.29	56 52.38	24 2.88
58	11	. .	. .	. .	. .	41.6	54.5	7.9	57 28.78	8.15	0.00	VII.	8	8.10	38 2.97	20.01	2.62	57 36.93	52 5.58
59	11	. .	. .	. .	. .	. .	. .	14.6	22 58 35.58	8.14	0.00	VII.	6	4.56	26 26.45	20.06	2.07	22 58 43.72	40 28.46
60	9	. .	10.3	22.6	35.7	. .	. .	. .	23 0 35.81	8.14	0.00	IV.	8	6.26	37 10.74	20.17	2.58	23 0 43.95	51 13.15
61	9	. .	. .	. .	. .	1.9	14.6	27.7	1 48.63	8.13	−0.01	VII.	1	0.33	0 16.43	20.23	0.83	1 56.75	14 17.03
62	10	. .	. .	. .	. .	. .	38.8	51.5	3 12.81	8.13	0.00	VII.	9	3.39	40 45.32	20.30	2.76	3 20.94	54 47.78
63	10	. .	. .	. .	. .	55.6	9.4	. .	5 43.04	8.12	−0.01	VI.	1	3.8	1 34.70	20.42	0.89	5 51.15	15 35.17
64	11	. .	. .	. .	. .	. .	. .	9.9	7 30.60	8.11	0.00	VII.	2	4.32	7 15.95	20.51	1.16	7 38.71	21 16.60
65	12	. .	56.2	. .	. .	. .	. .	. .	10 17.28	8.10	0.00	II.	8	10.31	39 14.19	20.64	2.68	10 25.38	53 16.23
66	9	. .	13.4	25.5	38.5	51.4	4.6	17.4	14 38.68	8.09	0.00	VII.	8	7.6	37 30.70	20.83	2.60	14 46.77	51 32.47
67	12	. .	. .	. .	20.4	. .	. .	. .	18 20.26	8.07	0.00	IV.	4	5.41	17 49.04	21.01	1.65	18 28.33	31 49.68
68	12	. .	5.6	18.6	31.5	. .	. .	. .	23 31.41	8.06	0.00	IV.	4	4.34	17 15.27	21.22	1.63	23 39.47	31 15.68
69	11	. .	. .	. .	. .	. .	. .	27.4	23 48.33	8.06	0.00	VII.	5	7.40	23 48.04	21.23	1.94	23 56.39	37 48.75
70	12	. .	. .	. .	. .	. .	39.9	51.5	25 12.31	8.05	0.00	VII.	7	11.18	34 39.06	21.29	2.46	25 20.36	15 48 40.23
71	10	. .	. .	. .	. .	4.9	17.6	. .	26 1.86	8.05	+0.01	VI.	10	12.18	50 8.03	21.35	3.22	27 9.92	16 4 9.90
72	9	27.1	41.5	. .	. .	. .	. .	. .	30 6.99	8.04	0.00	II.	8	14.24	41 11.66	21.48	2.79	30 15.03	15 55 12.97
73	5.4	10.5	23.7	36.5	49.8	2.9	. .	. .	23 34 49.54	+ 8.02	0.00	V.	2	6.48	− 8 24.71	+21.65	−1.21	23 34 57.56	−15 22 24.27

CORRECTIONS.

Date.	Corr. of Clock.	Hourly rate.	m	n	c
1848. h.	s.	s.	s.	s.	s.

INSTRUMENT READINGS.

Date.	Barom.	THERMOM. At.	Ex.
1848. h. m.	In.	°	°

REMARKS.

(205) 39. Micrometer reading assumed as 0r.55 instead of 1r.55.
(205) 44. Transit observations over T. IV assumed as 15s.9 instead of 13s.9.
(205) 46. Hor. thread assumed as 1 instead of 2.
(205) 48. Hor. thread assumed as 3 instead of 4.
(205) 65. Transit over T. II assumed as 51s.2 instead of 56s.2.
(205) 70. Transit over T. VI assumed as 37s.9 instead of 39s.9.
(205) 71. Transits over T.'s V and VI assumed as 14s.9 and 27s.6 instead of 4s.9 and 17s.6, and minutes as 27 instead of 26.

ZONE 205. OCTOBER 14. P. BELT, −15° 38½′. D₀ = −15° 14′ 20″—Continued.

No.	Mag.	Seconds of Transit. I.	II.	III.	IV.	V.	VI.	VII.	T.	a_1	a_2	Micrometer.			i	d_1	d_2	Mean Right Ascension, 1850.0.	Mean Declination, 1850.0.
									h. m. s.	s.	s.			r.	′ ″	″	″	h. m. s.	° ′ ″
74	9	. .	. .	23.6	36.4	49.5	. .	. .	23 36 36.26	+ 8.02	−0.01	V.	1	2.1	− 1 0.99	+21.70	−0.86	23 36 44.27	−15 15 0.15
75	10	. .	. .	. .	59.9	12.6	75.8	. .	37 59.76	8.01	0.00	VI.	5	2.25	21 9.32	21.75	1.81	38 7.77	35 9.38
76	12	. .	. .	. .	. .	45.8	58.4	. .	39 32.65	8.01	0.00	VI.	2	8.56	9 29.19	21.80	1.26	39 40.66	23 28.65
77	12	. .	. .	. .	. .	32.6	45.5	58.4	41 19.57	8.00	0.00	VII.	6	13.53	30 57.21	21.86	2.29	41 27.57	44 57.64
78	9	. .	. .	. .	. .	. .	0.9	13.9	42 35.01	7.99	0.00	VII.	7	11.49	34 54.70	21.90	2.48	42 43.00	48 55.28
79	12	. .	14.6	. .	39.5	. .	. .	. .	47 39.98	7.98	0.00	IV.	6	5.42	26 49.85	22.05	2.09	47 47.96	40 49.89
80	11	55.5	. .	21.6	34.6	47.9	. .	. .	52 34.45	7.97	0.00	V.	2	1.2	5 30.24	22.18	1.05	52 42.42	19 20.11
81	10	. .	32.6	45.5	. .	. .	. .	. .	54 58.61	7.96	0.00	III.	8	9.00	38 28.37	22.23	2.67	55 6.57	52 28.81
82	7	. .	. .	. .	38.8	. .	. .	. .	23 57 38.66	7.96	−0.01	IV.	1	2.42	1 21.69	22.29	0.86	23 57 46.61	15 20.26
83	9	49.9	3.6	15.9	28.8	. .	. .	. .	0 0 29.02	7.95	0.00	IV.	6	3.5	25 30.68	22.35	2.02	0 0 36.97	39 30.35
84	10	13.6	26.5	39.5	51.8	. .	. .	. .	2 39.13	7.94	0.00	IV.	1	6.23	3 13.12	22.39	0.95	2 47.07	17 11.68
85	5	. .	. .	13.8	26.4	39.5	. .	. .	4 26.60	7.94	0.00	.	10	4.4	45 59.04	22.41	3.05	4 34.54	59 59.68
86	9	14.6	27.4	40.	52.6	. .	. .	. .	6 53.11	7.93	0.00	IV.	6	0.34	24 14.55	22.46	1.96	7 1.04	38 14.05
87	10	. .	. .	20.5	33.5	. .	. .	. .	8 33.23	7.93	0.00	IV.	1	8.44	4 24.23	22.48	0.99	8 41.16	18 22.74
88	11	. .	. .	. .	. .	. .	33.5	46.8	9 7.67	7.93	0.00	VII.	4	8.4	19 0.95	22.49	1.71	9 15.60	33 0.17
89	10	58.5	. .	21.6	37.4	. .	. .	. .	0 11 37.46	+ 7.92	0.00	.	5	11.55	−25 56.83	+22.53	−2.04	0 11 45.38	−15 39 56.34

ZONE 206. OCTOBER 18. P. BELT, −18° 46′. D₀ = −18° 18′ 10″.

No.	Mag.	I.	II.	III.	IV.	V.	VI.	VII.	T.	a_1	a_2	Micrometer.			i	d_1	d_2	Mean Right Ascension, 1850.0.	Mean Declination, 1850.0.
1	8	48.8	3.5	15.5	28.9	. .	. .	. .	23 33 28.72	+17.97	0.00	IV.	7	7.58	−32 58.42	− 8.78	−2.45	23 33 46.69	−18 51 19.7
2	7	. .	40.4	52.5	5.7	18.8	32.3	. .	36 5.80	17.96	0.00	VI.	10	8.26	48 11.04	9.07	3.31	36 23.76	19 6 33.4
3	10	. .	. .	29.9	42.9	56.4	9.8	. .	42 43.20	17.94	+0.01	VI.	10	10.22	49 9.53	9.81	3.36	43 1.15	19 7 32.7
4	11	. .	. .	. .	16.4	29.8	. .	. .	44 16.44	17.94	0.00	V.	7	8.26	33 12.52	9.98	2.46	44 34.38	18 51 35.0
5	10	45.6	59.6	11.9	24.9	38.7	. .	3.9	47 25.11	17.93	0.00	VII.	4	12.47	21 23.58	10.35	1.79	47 43.04	39 45.7
6	11	. .	. .	. .	. .	18.9	32.7	45.8	23 52 6.10	+17.92	0.00	VII.	7	6.52	−32 24.88	−10.91	−2.42	23 52 24.02	−18 50 48.2

ZONE 207. OCTOBER 18. P. BELT, −18° 46′. D₀ = −18° 19′ 40″.

No.	Mag.	I.	II.	III.	IV.	V.	VI.	VII.	T.	a_1	a_2	Micrometer.			i	d_1	d_2	Mean Right Ascension, 1850.0.	Mean Declination, 1850.0.
1	10	. .	. .	7.4	20.5	33.6	46.6	59.9	1 49 20.33	+17.73	0.00	VI.	5	9.185	−24 37.80	− 7.06	−1.98	1 49 38.06	−18 44 26.8
2	10	. .	. .	23.1	36.4	49.5	2.9	. .	51 36.34	17.74	0.00	VI.	6	7.12	27 35.12	7.95	2.15	51 54.08	47 25.2
3	12	. .	. .	23.6	35.8	49.5	. .	. .	1 53 36.25	17.74	0.00	V.	7	6.17	32 7.47	8.75	2.41	1 53 53.99	51 58.6
4	5	. .	. .	8.8	21.6	35.4	. .	. .	2 1 21.74	17.75	0.00	V.	2	9.10	9 36.30	11.82	1.11	2 1 39.49	29 29.2
5	8	. .	43.6	56.5	9.4	23.5	. .	. .	4 9.66	17.75	+0.01	V.	2	4.19	7 9.57	12.95	0.97	4 27.42	27 3.5
6	9	. .	. .	. .	. .	. .	. .	20.5	4 40.52	17.75	+0.01	VII.	2	2.28	6 13.35	13.15	0.92	4 58.28	26 7.4
7	10	. .	. .	. .	28.5	41.9	54.9	8.8	6 28.66	17.75	0.00	VII.	5	0.58	20 25.27	13.89	1.73	6 46.41	40 20.9
8	11	. .	. .	. .	23.8	37.4	49.9	. .	8 23.82	17.75	0.00	VI.	8	4.9	36 1.55	14.64	2.64	8 41.57	55 58.8
9	11	. .	11.9	24.5	37.8	. .	. .	. .	12 37.83	17.75	0.00	IV.	5	5.15	22 35.13	16.34	1.86	12 55.58	42 33.3
10	8	. .	. .	. .	. .	56.9	. .	. .	14 43.55	17.74	+0.01	V.	1	1.39	0 49.89	17.19	0.61	15 1.30	20 47.7
11	10	. .	. .	. .	. .	. .	. .	42.9	2 17 3.27	+17.74	0.00	VII.	7	3.25	−30 40.49	−18.15	−2.33	2 17 21.01	−18 50 41.0

CORRECTIONS.

Date.		Corr. of Clock.	Hourly rate.	m	n	c
1848.	h.	s.	s.	s.	s.	s.
Oct. 18,	22	+ 15.52	/0.155			

INSTRUMENT READINGS.

Date.		Barom.	Thermom. At.	Thermom. Ex.
1848.	h. m.	In.	°	°

REMARKS.

(205) 84. Transits over T.'s II–V assumed as recorded over T.'s I–IV.

ZONE 208. OCTOBER 20. P. BELT, −18° 46′. D_c = −18° 20′ 0″.

No.	Mag.	SECONDS OF TRANSIT. I.	II.	III.	IV.	V.	VI.	VII.	T.	a_1	a_2	MICROMETER.			i	d_1	d_2	Mean Right Ascension, 1850.0.	Mean Declination, 1850.0.
									h. m. s.	s.	s.			r.	′ ″	″	″	h. m. s.	° ′ ″
1	11	..	..	..	14.6	..	..	..	21 35 14.45	+25.88	0.00	VII.	8	5.15	−31 35.97	−17.50	−2.34	21 35 40.33	−18 51 55.8
2	11	..	..	..	..	56.6	9.9	23.7	37 43.61	25.87	0.00	VII.	4	2.22	16 8.43	16.92	1.52	38 9.48	36 26.9
3	10	..	..	..	..	..	..	38.8	38 59.22	25.86	0.00	VII.	8	0.10	34 0.87	16.85	2.49	39 25.08	54 20.2
4	10	..	..	..	..	17.4	30.8	14.6	48 4.48	25.82	0.00	VII.	3	12.4	16 2.90	14.55	1.51	48 30.30	36 19.0
5	10	50.5	4.6	17.4	29.9	44.6	..	..	53 30.55	25.80	0.00	V.	4	3.49	16 52.54	13.34	1.56	53 56.35	37 7.4
6	12	..	..	..	..	..	57.9	11.4	55 31.58	25.80	−0.01	VII.	3	3.54	11 55.82	12.89	1.30	21 55 57.37	32 10.0
7	10	..	..	..	35.5	48.8	1.9	14.6	21 59 35.50	25.78	0.00	VII.	3	7.0	13 29.61	11.96	1.38	22 0 1.28	33 43.0
8	11	..	..	..	27.5	40.5	53.6	..	22 1 27.38	25.77	0.00	VI.	5	6.46	23 20.92	11.55	1.91	1 53.15	43 34.4
9	10	..	29.9	41.4	55.6	8.9	..	..	8 55.57	25.75	0.00	V.	7	10.38	34 19.08	9.89	2.50	9 21.32	18 54 31.5
10	10	..	..	..	29.9	13.4	..	..	22 17 30.02	+25.71	+0.01	V.	10	5.36	−46 45.40	− 8.01	−3.18	22 17 55.74	−19 6 56.6

ZONE 209. OCTOBER 24. D. BELT, −18° 46′. D_c = −18° 19′ 0″.

No.	Mag.	I.	II.	III.	IV.	V.	VI.	VII.	T.	a_1	a_2	MICROMETER.			i	d_1	d_2	Mean Right Ascension, 1850.0.	Mean Declination, 1850.0.
1	7	40.	58.	6.5	19.	..	..	..	0 0 19.28	+35.22	0.00	IV.	1	11.6	− 5 35.83	− 8.53	−0.95	0 0 54.50	−18 24 45.31
2	9	43.	57.	9.2	..	..	..	..	3 22.90	35.21	0.00	III.	6	10.8	29 3.95	8.56	2.22	3 58.11	48 14.73
3	6	..	..	..	55.	8.3	21.3	..	3 54.98	35.20	0.00	VI.	6	6.20	27 8.90	8.57	2.12	4 30.18	46 19.59
4	10	..	..	..	..	..	10.	23.5	4 43 70	35.20	0.00	VII.	3	12.15	16 8.44	8.58	1.51	5 18.90	35 18.53
5	10	39.	53.	5.5	..	..	..	..	7 18.95	35.19	0.00	III.	5	11.28	25 43.18	8.59	2.04	7 54.14	44 53.81
6	9	..	..	8.	21.2	..	..	..	8 21.06	35.18	0.00	IV.	5	9.29	24 43.21	8.60	1.98	8 56.24	43 53.79
7	10	..	..	..	..	..	21.	34.	8 54.46	35.18	0.00	VII.	4	3.27	16 41.20	8.61	1.55	9 29.64	35 51.36
8	9	38.	51.5	4.5	..	..	..	..	11 17.65	35.17	0.00	III.	3	3.31	11 44.46	8.64	1.28	11 52.82	30 54.38
9	7	..	..	..	51.3	5.	17.5	..	11 51.37	35.17	0.00	VI.	3	5.39	12 48.93	8.65	1.34	12 26.54	31 58.92
10	8	..	22.5	35.3	48.5	..	..	..	15 48.51	35.15	0.00	IV.	4	9.00	19 29.39	8.71	1.70	16 23.66	38 39.80
11	10	..	..	18.	31.3	..	..	..	17 30.98	35.15	0.00	IV.	1	9.39	4 51.96	8.74	0.90	18 6.13	24 1.60
12	10	..	48.	1.	14.	..	..	..	19 14.09	35.14	0.00	IV.	5	3.3	21 28.57	8.77	1.81	19 49.23	18 40 39.15
13	9	16.	30.	42.3	..	..	..	..	23 56.07	35.12	0.00	III.	9	9.44	43 49.55	8.87	3.04	24 31.19	19 3 1.46
14	9	..	..	..	..	39.5	53.	6.	24 26.49	35.12	0.00	VII.	8	6.16	37 5.43	8.88	2.68	25 1.61	18 56 16.99
15	9	..	..	..	..	..	59.	12.	25 32.57	35.11	0.00	VII.	8	6.49	37 22.07	8.91	2.69	26 7.68	18 56 33.67
16	10	..	..	..	..	24.5	..	..	27 11.36	35.11	0.00	V.	9	7.16	42 34.93	8.94	2.98	27 46.47	19 1 46.85
17	8	..	..	24.	37.	..	..	..	28 36.92	35.10	0.00	IV.	4	6.5	18 1.15	8.98	1.62	29 12.02	18 37 11.75
18	10	..	..	..	33.	46.2	59.	..	32 32.85	35.08	0.00	VI.	5	6.59	23 27.47	9.09	1.92	33 7.93	42 38.48
19	10	..	..	49.	2.5	..	..	..	34 2.14	35.08	0.00	IV.	3	9.21	14 40.97	9.13	1.43	34 37.22	33 51.53
20	3.4	..	..	15.	28.	41.3	54.	..	35 27.95	35.07	0.00	VI.	6	10.52	29 26.06	9.18	2.24	36 3.02	48 37.48
21	10	..	..	22.	35.	..	..	..	37 34.95	35.06	0.00	IV.	5	7.18	23 37.15	9.24	1.92	38 10.01	18 42 48.31
22	10	..	..	..	35.	45.	..	..	38 34.85	35.06	0.00	V.	9	4.46	41 19.29	9.28	2.91	39 9.91	19 0 31.48
23	6.7	..	..	..	39.	52.2	5.2	19.	39 39.02	35.06	0.00	VII.	7	9.29	33 44.04	9.30	2.50	40 14.08	18 52 55.84
24	8	27.	40.5	53.5	..	..	..	..	45 6.63	35.04	0.00	III.	2	7.52	8 56.97	9.50	0.84	45 41.67	28 7.31
25	10	..	..	53.	6.	19.	..	..	46 5.94	35.03	0.00	V.	7	4.49	31 23.10	9.54	2.35	46 40.97	50 34.99
26	10	..	..	28.3	41.	..	..	..	50 41.15	35.02	0.00	IV.	6	11.24	29 42.30	9.71	2.25	51 16.17	48 54.26
27	9	..	..	..	57.4	..	..	..	51 57.25	35.01	0.00	IV.	7	8.22	33 10.53	9.76	2.46	52 32.26	52 22.75
28	8	..	..	..	..	51.	4.	..	52 37.76	35.01	0.00	VI.	8	7.8	37 31.80	9.78	2.70	53 12.77	56 44.28
29	10	..	14.	27.	40.	..	..	..	0 56 40.13	35.00	0.00	IV.	6	4.20	26 8.50	9.96	2.06	0 57 15.13	45 20.52
30	10	..	37.	50.	3.	..	..	..	1 4 3.06	34.98	0.00	IV.	4	5.3	17 29.88	10.29	1.58	1 4 38.04	36 41.75
31	8	..	39.5	51.9	5.	..	..	..	7 5.35	34.96	0.00	IV.	8	10.36	39 16.81	10.43	2.80	7 40.31	58 30.04
32	10	9.	23.	35.5	49.	..	..	..	8 49.02	34.96	0.00	IV.	8	7.4	37 29.90	10.52	2.71	9 23.98	56 43.13
33	9	48.3	2.5	15.	..	..	..	..	15 28.37	34.94	0.00	III.	5	8.015	23 59.06	10.86	1.94	16 3.31	18 43 11.86
34	10	..	17.	30.	43.	..	..	..	16 43.23	34.94	0.00	IV.	9	4.00	40 56.12	10.92	2.90	17 18.17	19 0 9.91
35	10	..	..	..	..	..	36.3	49.5	1 19 10.04	+34.93	0.00	VII.	10	9.2	−48 29.03	−11.04	−3.34	1 19 44.97	−19 7 43.41

CORRECTIONS.

Date.	Corr. of Clock.	Hourly rate.	m	n	c
1848. h.	s.	s.	s.	s.	s.
Oct. 20, 22	+ 22.95	/ 0.180	− 0.51	+ 1.13	0.00
Oct. 24, 22	+ 32.96	/ 0.082			

INSTRUMENT READINGS.

Date.	Barom.	THERMOM. At.	Ex.
1848. h. m.	in.	°	°

REMARKS.

(208) 1. Hor. thread assumed as 7 instead of 8.
(208) 3. Minutes assumed as 38 instead of 37.
(209) 1. Transit over T. II is assumed to have been at 53ˢ instead of 58ˢ.
(209) 17. Declination differs 1′ from Arg. Z. 321, 1, and 6′ from Mer. Cir., 1848, October 17.
(209) 22. Transit observations discordant by 3ˢ; that over T. V used as 48ˢ instead of 45ˢ.

Zone 209. October 24. B. Belt, −18° 46′. D_o = −18° 19′ 0″—Continued.

No.	Mag.	Seconds of Transit. I.	II.	III.	IV.	V.	VI.	VII.	T.	a_1	a_2	Micrometer.			i	d_1	d_2	Mean Right Ascension, 1850.0.	Mean Declination, 1850.0.
									h. m. s.	s.	s.			r.	′ ″	″	″	h. m. s.	° ′ ″
36	9	. .	. .	. .	. .	33.	46.	. .	1 20 19.75	+34.93	0.00	VI.	7	4.27	−31 11.92	−11.11	−2.34	1 20 54.68	−18 50 25.37
37	7	. .	. .	. .	. .	. .	31.	45.	21 4.99	34.93	0.00	VII.	5	7.25	23 40.41	11.15	1.93	21 39.92	42 53.49
38	10	. .	0.4	13.	26.	. .	. .	. .	23 26.13	34.92	0.00	IV.	2	7.12	8 36.83	11.29	1.08	24 1.05	27 49.20
39	8	. .	. .	. .	. .	18.7	31.	45.	24 5.07	34.92	0.00	VII.	2	8.35	9 18.42	11.32	1.12	24 39.99	28 30.86
40	8	. .	. .	. .	. .	. .	. .	43.	25 2.99	34.92	0.00	VII.	1	7.41	3 52.18	11.37	0.82	25 37.91	23 4.37
41	10	. .	. .	55.3	8.3	21.5	. .	. .	28 8.31	34.91	0.00	V.	7	5.27	31 42.25	11.55	2.38	28 43.22	50 56.18
42	10	. .	. .	. .	36.4	50.	. .	. .	29 36.50	34.90	0.00	V.	4	10.41	20 20.30	11.64	1.74	30 11.40	39 33.68
43	9	42.3	56.4	9.	22.	. .	. .	. .	41 22.22	34.88	0.00	IV.	5	10.43	25 20.53	12.37	2.02	41 57.10	44 34.92
44	9	. .	37.5	50.2	3.	. .	. .	. .	49 3.31	34.86	0.00	IV.	4	10.26	20 12.76	12.89	1.73	49 38.17	39 27.38
45	7	. .	. .	. .	. .	. .	58.	. .	49 31.77	34.86	0.00	VII.	1	10.55	5 30.01	12.93	0.90	50 6.63	24 43.84
46	10	. .	. .	. .	20.	33.2	. .	. .	51 19.92	34.86	0.00	VI.	6	8.32	28 15.47	13.06	2.19	51 54.78	47 30.72
47	10	39.5	54.	6.3	19.	. .	. .	. .	53 19.25	34.85	0.00	IV.	7	7.33	32 45.82	13.21	2.44	53 54.10	52 1.47
48	10	. .	10.	23.2	36.5	. .	. .	. .	55 36.24	34.85	0.00	IV.	2	10.24	10 13.65	13.38	1.16	56 11.09	29 28.19
49	9	. .	. .	32.	45.	. .	. .	. .	56 44.96	34.85	0.00	IV.	5	10.17	25 7.42	13.46	2.01	1 57 19.81	44 22.89
50	8	. .	. .	. .	38.	52.	5.	. .	1 59 38.43	34.84	0.00	VI.	1	3.27	1 44.26	13.68	0.69	2 0 13.27	20 58.63
51	6.7	. .	. .	. .	5.	18.	. .	. .	2 1 4.78	34.84	0.00	V.	2	10.32	10 17.66	13.79	1.17	1 39.62	29 32.62
52	7	. .	. .	. .	. .	. .	. .	3.3	4 23.33	34.83	0.00	VII.	2	3.39	6 49.16	14.06	0.96	4 58.16	26 4.18
53	8	. .	. .	. .	. .	25.3	38.0	. .	6 11.89	34.83	0.00	VI.	5	2.20	21 6.78	14.20	1.77	6 46.72	40 22.75
54	9	. .	41.5	54.2	7.2	. .	. .	. .	8 7.50	34.83	0.00	IV.	8	5.34	36 44.52	14.35	2.67	8 42.33	56 1.54
55	10	. .	. .	. .	. .	1.	14.5	27.5	10 47.98	34.82	0.00	VII.	7	11.27	34 43.54	14.57	2.56	11 22.80	54 0.67
56	9	. .	. .	. .	. .	34.	47.	. .	12 20.74	34.82	0.00	VI.	5	6.21	23 8.30	14.70	1.89	12 55.56	18 42 24.89
57	7	. .	1.	14.	27.	. .	. .	. .	14 26.95	34.82	+0.01	IV.	1	3.24	1 4.	14.87	0.67	15 1.78	
58	6	. .	. .	. .	. .	. .	. .	57.5	15 18.04	34.82	0.00	VII.	9	7.39	42 46.28	14.95	3.02	15 52.86	19 2 4.25
59	8	. .	. .	. .	. .	59.5	12.	. .	16 46.01	34.82	0.00	VI.	7	4.52	31 24.53	15.07	2.37	17 20.83	18 50 41.97
60	10	. .	. .	. .	51.	4.5	18.	. .	25 51.29	34.81	0.00	VI.	6	4.18	36 6.08	15.86	2.64	26 26.10	55 24.58
61	10	. .	. .	. .	. .	36.	49.	. .	31 22.76	34.80	0.00	VI.	7	4.40	31 18.48	16.36	2.36	31 57.56	50 37.20
62	10	58.	12.	24.5	. .	. .	. .	. .	33 37.92	34.80	0.00	IV.	5	5.25	22 40.17	16.57	1.87	34 12.72	41 58.61
63	7	. .	. .	. .	. .	. .	. .	18.	33 38.06	34.80	0.00	VII.	2	7.4	8 32.53	16.57	1.06	34 12.86	18 27 50.16
64	10	. .	. .	27.	40.4	. .	. .	. .	35 40.27	34.80	0.00	IV.	9	4.3	40 57.63	16.76	2.93	36 15.07	19 0 17.32
65	8	. .	. .	. .	25.5	39.	. .	. .	37 25.51	34.80	0.00	V.	2	1.40	5 49.40	16.93	0.90	38 0.31	18 25 7.33
66	10	. .	. .	47.5	0.6	. .	. .	. .	39 0.49	34.80	0.00	IV.	5	2.37	21 15.47	17.08	1.78	39 35.29	40 34.33
67	10	. .	. .	. .	. .	32.	45.	. .	40 18.76	34.80	0.00	VI.	8	2.35	36 14.66	17.20	2.64	40 53.56	18 55 34.50
68	10	. .	. .	15.	28.	41.	. .	. .	42 28.03	34.80	0.00	V.	10	4.20	46 7.07	17.41	3.23	43 2.83	19 5 27.71
69	10	. .	3.	15.5	28.3	. .	. .	. .	44 28.82	34.80	0.00	IV.	8	11.7	39 32.43	17.61	2.84	45 3.62	18 58 52.88
70	10	. .	. .	. .	. .	. .	. .	53.	45 13.56	34.80	0.00	VII.	9	10.58	44 26.62	17.69	3.15	45 48.36	19 3 47.46
71	10	. .	. .	. .	56.	9.5	. .	. .	51 56.06	34.79	0.00	V.	5	9.48	24 52.76	18.34	1.99	52 30.85	18 44 13.09
72	9	. .	. .	. .	37.5	51.	. .	. .	53 37.52	34.79	0.00	V.	2	7.00	8 30.75	18.52	1.05	54 12.31	27 50.32
73	8	. .	. .	. .	. .	47.5	0.3	14.	54 34.21	34.79	0.00	VII.	6	9.24	28 41.52	18.62	2.22	55 9.00	48 2.36
74	10	. .	. .	. .	. .	. .	. .	13.5	2 55 33.86	−34.79	0.00	VII.	6	12.00	−30 0.18	−18.72	−2.29	2 56 8.65	−18 49 21.19

CORRECTIONS.

Date.		Corr. of Clock.	Hourly rate.	m	n	c
1848.	h.	s.	s.	s.	s.	s.

INSTRUMENT READINGS.

Date.		Barom.	Thermom. At.	Thermom. Ex.
1848.	h. m.	in.	°	°

REMARKS.

(209) 67. Micrometer reading assumed as 8 4r.35 instead of 8 2r.35, to agree with Arg. Z. 318, 50.

ZONE 210. JANUARY 23. K. BELT, −26° 53′. D_o = −26° 27′ 30″.

No.	Mag.	Seconds of Transit I.	II.	III.	IV.	V.	VI.	VII.	T.	a_1	a_2	Micrometer			i	d_1	d_2	Mean Right Ascension, 1850.0.	Mean Declination, 1850.0.
									h. m. s.	s.	s.			r.	′ ″	″	″	h. m. s.	° ′ ″
1	10	..	..	2.9	..	..	..	..	6 10 16.90	− 0.96	−0.98	.	1	8.27	− 4 15.65	− 3.24	−3.91	6 10 14.96	−26 31 52.80
2	8	..	48.8	2.8	..	..	..	..	11 16.83	0.96	1.00	.	5	10.8	25 2.88	3.56	6.10	11 14.87	52 42.54
3	8	..	..	..	42.8	..	..	..	13 42.81	0.97	1.00	.	6	6.58	27 28.17	4.34	6.25	13 40.84	26 55 8.76
4	8	12.8	26.8	..	..	..	..	..	38 54.91	1.06	1.01	.	10	0.46	44 19.20	12.41	8.02	38 52.84	27 12 9.63
5	9	..	9.3	..	..	..	..	..	42 37.35	1.07	1.01	.	9	12.13	45 4.70	13.61	8.10	42 35.27	12 56.41
6	8.9	..	..	54.4	..	..	..	..	44 8.43	1.08	1.01	.	9	6.1	42 1.67	14.10	7.78	44 6.34	9 53.55
7	8.9	..	..	58.6	..	..	..	..	45 12.63	1.08	1.01	.	8	6.38	37 16.79	14.44	7.29	45 10.54	27 5 8.52
8	10	..	..	..	49.0	..	..	..	45 49.06	1.08	0.99	.	3	10.37	15 19.31	14.64	5.00	45 46.99	26 43 8.95
9	6.7	..	32.8	46.8	..	..	..	..	47 0.82	1.09	0.99	.	4	7.8	18 32.91	15.02	5.33	46 58.74	26 46 23.26
10	10	..	..	..	..	28.8	..	..	48 14.76	1.09	1.01	V.	9	7.52	42 53.06	15.42	7.88	48 12.66	27 10 46.36
11	9	..	..	17.4	31.8	..	..	..	49 31.60	1.09	1.00	IV.	6	9.58	28 58.93	15.84	6.42	49 29.51	26 56 51.19
12	9.10	..	..	..	..	4.3	..	..	49 50.29	1.09	1.00	.	6	9.5	28 32.21	15.94	6.37	49 48.20	56 24.52
13	9.10	..	58.1	..	..	..	..	..	51 26.14	1.09	1.00	.	7	5.15	31 36.23	16.44	6.69	51 24.05	59 29.36
14	7.8	..	..	54.3	8.1	..	..	..	52 8.20	1.10	1.00	.	6	12.3	30 1.96	16.68	6.52	52 6.10	57 55.16
15	10	..	..	32.5	..	..	..	..	54 46.52	1.10	1.00	.	6	6.51	27 24.64	17.53	6.24	54 44.42	26 55 18.41
16	8.9	..	..	20.3	34.3	..	..	..	55 34.28	1.10	1.00	.	7	7.44	32 51.37	17.79	6.82	55 32.18	27 0 45.98
17	10	..	..	..	..	11.8	..	..	57 57.81	1.11	1.00	V.	4	10.9	20 4.15	18.56	5.49	57 55.70	26 47 58.20
18	8.9	..	..	..	59.9	..	..	..	6 58 59.94	1.11	1.00	.	5	11.14	25 36.16	18.90	6.06	6 58 57.83	26 53 31.12
19	10	..	47.8	..	..	..	..	..	7 3 15.85	1.12	1.01	.	8	11.13	39 35.45	20.28	7.53	7 3 13.72	27 7 33.26
20	5.6	..	51.9	..	20.0	..	..	..	4 19.90	1.12	1.02	.	10	6.19	47 7.11	20.62	8.33	4 17.76	15 6.06
21	7	23.3	37.3	..	..	..	..	..	6 5.38	1.12	1.01	.	8	6.41	37 18.30	21.19	7.29	6 3.25	5 16.78
22	7	..	..	..	24.8	..	..	..	6 24.63	1.12	1.02	.	10	3.27	45 40.37	21.29	8.11	6 22.49	13 39.77
23	6	..	..	22.1	36.2	..	..	..	7 36.12	1.13	1.01	.	8	8.43	38 19.82	21.68	7.40	7 33.98	27 6 18.90
24	4	..	17.4	31.8	..	..	..	..	8 45.61	1.13	0.98	.	1	6.14	3 8.59	22.06	3.77	8 43.50	26 31 4.42
25	5.6	..	..	..	..	..	17.3	..	8 49.19	1.13	0.99	VI.	4	7.46	18 51.91	22.08	5.36	8 47.07	46 49.35
26	7	..	16.8	30.8	..	..	..	..	11 44.81	1.13	0.98	.	1	7.51	3 57.49	23.03	3.85	11 42.70	31 54.37
27	6.7	9.9	23.9	..	..	..	..	..	14 51.86	1.14	0.99	.	3	6.33	13 16.27	24.04	4.79	14 49.73	41 15.10
28	10	..	5.3	..	..	..	..	..	16 33.34	1.14	1.00	.	5	6.39	23 17.50	24.59	5.81	16 31.20	51 17.90
29	10	..	7.5	..	..	..	..	..	25 35.52	1.15	0.99	.	2	4.34	7 17.17	27.53	4.19	25 33.38	35 18.89
30	5	48.1	2.4	..	..	..	..	..	28 30.22	1.15	0.99	.	3	7.5	13 32.40	28.48	4.80	28 28.08	41 35.68
31	9.10	..	3.3	..	..	..	..	..	29 31.34	1.15	1.00	.	7	4.52	31 24.64	28.81	6.67	29 29.19	26 59 30.12
32	8	..	..	50.9	..	..	..	..	30 4.93	1.15	1.01	.	9	7.11	42 32.42	29.02	7.84	30 2.77	27 10 39.28
33	8	..	33.0	46.3	..	..	..	..	31 0.69	1.15	1.01	.	8	6.49	37 22.34	29.32	7.80	30 58.53	27 5 28.96
34	9	..	..	..	9.9	..	..	..	32 9.92	1.15	0.98	.	1	11.28	5 46.91	29.70	4.04	32 7.79	26 33 50.65
35	7	..	..	..	20.2	..	..	..	33 20.22	1.15	0.98	.	1	6.46	3 24.73	30.08	3.78	33 18.09	31 28.59
36	7	..	..	..	..	..	..	5.3	33 23.50	1.15	0.98	VII.	2	3.34	6 46.52	30.10	4.12	33 21.37	34 50.74
37	10	31.4	..	..	..	..	..	..	7 36 13.38	− 1.16	−1.00	.	5	4.9	−22 1.86	−31.02	−5.68	7 36 11.22	−26 50 8.56

ZONE 211. JANUARY 27. K. BELT, −26° 53′. D_o = −26° 28′ 30″.

No.	Mag.	I.	II.	III.	IV.	V.	VI.	VII.	T.	a_1	a_2	Micrometer			i	d_1	d_2	Mean Right Ascension, 1850.0.	Mean Declination, 1850.0.
1	9.10	..	..	..	27.2	..	..	..	4 4 27.17	− 4.31	−1.00	.	7	4.43	−31 20.10	− 7.47	−6.68	4 4 21.86	−27 0 4.3
2	9	35.5	..	58.8	..	..	..	..	20 17.58	4.44	0.99	.	2	7.22	8 41.87	8.53	4.25	20 12.15	26 37 24.7
3	8.9	..	..	55.2	..	23.8	..	..	22 9.50	4.45	1.00	.	7	4.39	31 18.09	8.65	6.68	22 4.05	27 0 3.4
4	8.9	..	..	..	..	..	1.0	..	22 32.83	4.45	0.99	VI.	2	7.58	8 59.85	8.67	4.27	22 27.39	26 37 42.8
5	7	..	..	..	..	..	56.3	10.2	23 28.26	4.46	0.99	VI.	2	5.34	7 47.25	8.73	4.16	23 22.61	36 30.1
6	10	..	..	..	59.2	..	..	..	26 59.22	4.49	0.99	.	1	8.39	4 21.71	8.93	3.80	26 53.74	33 4.4
7	10	..	..	..	..	57.8	..	..	27 43.88	4.49	0.99	V.	2	3.57	6 58.46	8.99	4.08	27 38.40	35 41.5
8	9.10	38.0	52.0	..	..	..	..	..	4 30 20.02	− 4.51	−1.00	.	5	7.3	−23 29.59	− 9.16	−5.84	4 30 14.51	−26 52 14.6

CORRECTIONS.

Date.		Corr. of Clock.	Hourly rate.	m	n	c
1849.	h.	s.	s.	s.	s.	s.

INSTRUMENT READINGS.

Date.		Barom.	Thermom. At.	Thermom. Ex.
1849.	h. m.	In.	°	°

REMARKS.

(211) 2. Transit over T. III assumed to have been at 3ˢ.8 instead of 58ˢ.8.

ZONE 211. JANUARY 27. K. BELT, −26° 53'. D₀ = −26° 28' 30"—Continued.

No.	Mag.	I.	II.	III.	IV.	V.	VI.	VII.	T.	a_1	a_2	MICROMETER.			i	d_1	d_2	Mean Right Ascension, 1850.0.	Mean Declination, 1850.0.
									h. m. s.	s.	s.			r.	° ' "	"	"	h. m. s.	° ' "
9	9.10	..	..	..	37.8	..	..	..	4 30 37.86	−4.52	−1.00	.	3	11.10	−15 35.94	−9.18	−4.98	4 30 32.34	−26 44 20.1
10	9	..	..	28.8	42.6	..	..	..	31 42.67	4.52	1.00	.	8	4.27	36 10.73	9.26	7.21	31 37.15	27 4 57.2
11	9	2.3	16.3	..	..	..	..	..	30 44.35	4.57	1.00	III.	6	12.39	30 20.08	9.73	6.57	38 38.7?	26 59 6.4
12	9.8	..	..	..	..	1.6	..	..	38 47.59	4.57	1.00	V.	7	10.19	34 9.49	9.74	7.00	38 42.02	27 2 56.2
13	8	..	20.2	34.2	..	..	..	..	40 48.21	4.59	0.99	.	1	11.28	5 46.91	9.88	3.94	40 42.63	26 34 30.7
14	10	..	..	..	26.2	..	..	..	41 26.14	4.59	1.00	.	7	9.49	33 54.40	9.92	6.42	41 20.55	27 2 40.7
15	10	..	22.0	..	..	..	..	..	47 50.02	4.64	0.99	II.	3	3.43	11 50.37	10.40	4.58	47 44.39	26 40 35.4
16	9	..	..	20.4	..	..	..	..	4 48 34.42	−4.64	−1.00	.	6	11.30	−29 45.33	−10.44	−6.52	4 48 28.78	−26 58 32.3

ZONE 212. FEBRUARY 10. B. BELT, −26° 16'. D₀ = −25° 50' 10".

No.	Mag.	I.	II.	III.	IV.	V.	VI.	VII.	T.	a_1	a_2	MICROMETER.			i	d_1	d_2	Mean Right Ascension, 1850.0.	Mean Declination, 1850.0.
1	8	..	..	..	..	56.	10.	24.	4 15 42.16	−23.38	−1.00	IV.	3	10.295	−14 45.26	−7.55	−4.07	4 15 17.78	−26 5 6.89
2	8	..	..	22.3	36.5	50.3	..	..	20 36.31	23.42	1.01	IV.	10	6.13	47 4.08	8.00	7.15	20 11.88	37 29.23
3	9	..	..	3.	17.	31.	..	..	22 17.00	23.43	1.00	V.	6	10.39	29 19.58	8.16	6.41	21 52.57	19 44.15
4	7	..	..	..	47.5	1.	..	..	23 47.18	23.45	1.01	IV.	10	4.15	46 4.58	8.31	8.05	23 22.72	36 30.94
5	9	..	..	55.	9.	..	..	..	25 9.00	23.46	0.99	IV.	3	8.46	14 23.34	8.44	4.00	24 44.55	4 45.78
6	9	36.2	50.5	4.3	..	..	..	..	27 18.29	23.47	1.01	III.	10	7.16	47 35.81	8.65	8.20	26 53.81	38 2.66
7	9	..	..	3.	17.	..	..	..	29 16.98	23.49	1.00	IV.	6	5.25½	26 41.52	8.84	6.16	28 52.49	26 17 6.52
8	9	..	..	9.	23.	..	..	..	31 22.96	23.50	0.99	IV.	1	2.45	1 23.20	9.05	3.79	30 58.47	25 51 46.04
9	10	..	..	..	40.	53.8	..	..	32 39.87	23.51	1.01	V.	9	2.10	40 0.61	9.18	7.45	32 15.35	26 30 27.24
10	10	..	..	..	52.2	6.4	..	..	34 51.39	23.53	0.99	V.	3	7.19	13 39.42	9.40	4.91	34 26.87	4 3.73
11	11	..	..	1.5	16.	..	..	..	37 15.66	23.55	1.01	V.	9	5.52	41 52.55	9.64	7.64	36 51.10	32 19.83
12	9	..	..	..	..	..	45.6	59.3	38 17.62	23.56	1.00	VI.	5	3.23	21 38.48	9.75	5.68	37 53.06	12 3.91
13	9	..	..	..	..	..	3.	17.	39 35.15	23.57	1.00	VII.	6	9.11	28 34.84	9.88	6.36	39 10.58	19 1.08
14	9	..	..	..	7.	21.	..	..	41 6.94	23.58	1.01	V.	9	10.8	44 1.64	10.04	7.85	40 42.35	34 29.53
15	11	..	59.8	13.5	27.5	..	..	..	44 27.57	23.60	1.00	IV.	4	6.29	18 13.25	10.38	4.34	44 2.97	8 37.97
16	9	..	..	..	..	25.6	39.2	..	45 11.40	23.61	1.00	VI.	7	7.34	32 46.16	10.46	6.76	44 46.79	23 13.38
17	11	..	..	..	..	..	20.	34.	46 52.16	23.62	1.00	VII.	6	8.34	28 16.20	10.63	6.33	46 27.54	18 43.16
18	8	..	..	..	..	..	23.	37.2	47 55.25	23.63	1.00	VII.	6	11.2	29 30.82	10.74	6.45	47 30.62	26 19 58.01
19	7	..	..	..	47.	0.5	14.5	..	49 46.72	23.65	0.99	V.	2	5.45	7 52.92	10.94	4.37	49 22.08	25 58 18.23
20	11	..	..	17.4	31.4	..	..	..	52 31.30	23.66	1.01	IV.	10	4.21	46 7.60	11.23	8.05	52 6.63	26 36 36.88
21	10	..	..	36.5	50.5	..	..	..	55 50.51	23.69	1.00	V.	4	9.48	19 53.55	11.58	5.51	55 25.82	10 20.64
22	6	..	..	..	..	..	..	10.	56 28.08	23.70	1.01	VII.	8	12.6	39 1.28	11.65	7.45	56 3.37	29 30.38
23	9	..	..	..	27.	40.5	..	..	58 12.81	23.72	1.00	IV.	3	11.50	15 56.11	11.86	4.13	57 48.09	6 22.10
24	8	..	..	..	36.	49.5	3.4	17.	4 59 35.56	23.72	1.00	VII.	7	4.13	31 4.58	11.99	6.59	4 59 10.84	21 33.16
25	7	22.	36.	50.	3.5	..	..	..	5 5 3.77	23.76	1.00	IV.	3	11.6	15 33.93	12.59	4.10	5 4 39.01	6 0.62
26	11	..	59.5	13.3	27.4	..	..	..	7 27.37	23.77	1.00	IV.	4	12.4	21 2.07	12.85	5.62	7 2.60	11 30.64
27	9	..	23.	37.	50.8	..	..	..	8 50.87	23.79	1.00	IV.	7	6.55½	32 26.91	13.00	6.72	8 26.08	22 56.63
28	10	..	..	..	..	15.	29.	..	10 1.12	23.79	1.01	VI.	10	4.39	46 16.52	13.13	8.07	9 36.32	36 47.72
29	10	..	..	..	..	..	..	45.	11 3.09	23.88	1.01	VII.	9	1.48	39 49.16	13.25	7.43	10 38.28	30 19.04
30	8	24.2	39.	52.	6.	..	..	..	5 14 5.94	−23.82	−1.00	IV.	3	9.31½	−15 16.55	−13.60	−4.06	5 13 41.12	−26 5 44.19

CORRECTIONS.

Date.	Corr. of Clock.	Hourly rate.	m	n	c
1849. h.	s.	s.	s.	s.	s.

INSTRUMENT READINGS.

Date.	Barom.	THERMOM. At.	THERMOM. Ex.
1849. h. m.	in.	°	

REMARKS.

(212) 1. Micrometer reading assumed as 9ʳ.295 instead of 10ʳ.295.
(212) 22. Micrometer reading assumed as 10ʳ.6 instead of 12ʳ.6.
(212) 23. Transit over T.'s V and VI assumed as recorded over T.'s IV and V.
(212) 30. Micrometer reading assumed as 10ʳ.31½ instead of 9ʳ.31½.

ZONE 213. FEBRUARY 13. B. BELT, −33° 8′. D₀ = −32° 43′ 50″.

No.	Mag.	SECONDS OF TRANSIT. I.	II.	III.	IV.	V.	VI.	VII.	T.	a_1	a_2	MICROMETER.		i	d_1	d_2	Mean Right Ascension, 1850.0.	Mean Declination, 1850.0.
									h. m. s.	s.	s.		r.	′ ″	″	″	h. m. s.	° ′ ″
1	9				27.5		57.		7 55 27.25	−26.38	−1.01	VI. 9	8.20	−43 7.05	−11.22	−9.32	7 54 59.86	−33 27 17.59
2	8		14.	29.					7 59 43.90	26.39	1.00	III. 5	11.41	25 49.73	11.77	6.14	59 16.51	9 57.64
3	7				25.	40.			8 1 25.02	26.39	1.00	V. 5	8.53	24 25.02	11.98	5.89	7 59 57.63	33 8 32.89
4	9	8.5	23.						2 52.96	26.40	1.00	II. 2	10.25	10 13.90	12.16	3.38	8 2 25.56	32 54 19.52
5	9				33.5	48.3			3 33.51	26.40	0.98	V. 1	9.00	4 32.23	12.25	2.38	3 6.13	32 48 36.86
6	8		48.2	3.					8 18.00	26.41	1.00	III. 5	5.58	22 56.75	12.85	5.63	7 50.59	33 7 5.23
7	10				57.	12.			11 57.03	26.41	1.00	V. 6	7.15	27 36.68	13.30	6.47	11 29.62	11 46.45
8	10				5.3	20.4			13 5.44	26.41	0.99	V. 4	6.26	18 11.68	13.45	4.78	12 38.04	2 19.91
9	9					16.5	32.		14 1.85	26.42	1.00	VII. 7	5.56	31 56.41	13.56	7.26	13 34.43	16 7.23
10	9						29.3	44.5	15 59.54	26.42	1.01	VII. 7	10.9	34 3.99	13.81	7.65	15 32.11	33 18 15.45
11	10					53.	8.		16 38.09	26.42	0.99	VI. 2	7.27	8 44.17	13.69	3.12	16 10.68	32 52 51.18
12	10					40.	54.5		19 24.84	26.42	1.00	VI. 4	12.13	21 6.48	14.24	5.30	18 57.42	33 5 16.02
13	10							11.	20 26.41	26.42	0.99	VII. 3	10.45	15 22.84	14.37	4.28	19 59.00	32 59 31.49
14	10							59.5	22 14.98	26.43	0.99	VII. 2	8.33	9 17.18	14.59	3.22	21 47.56	32 53 26.99
15	7.8		53.5	8.4	23.3				25 23.35	26.43	1.00	IV. 5	8.50	24 23.55	14.98	5.89	24 55.92	33 8 33.42
16	8		48.5	3.3					27 18.29	26.43	0.99	III. 4	5.485	17 52.76	15.22	4.72	26 50.87	33 2 2.70
17	10			9.	23.5				28 23.74	26.43	0.99	IV. 3	9.02½	14 31.65	15.36	4.12	27 56.32	32 58 41.13
18	8						13.	28.3	29 43.25	26.43	1.02	VII. 10	2.53	45 22.73	15.53	9.74	29 15.80	33 29 38.00
19	8				0.0	15.	30.		32 0.06	26.44	1.00	VI. 6	10.14	29 6.79	15.81	6.75	31 32.62	13 19.35
20	8			26.3	41.3				33 41.30	26.44	1.00	IV. 4	11.48	20 54.10	16.02	5.26	33 13.86	5 5.38
21	8							18.4	34 33.42	26.44	1.02	VII. 9	12.25	45 10.25	16.12	9.71	34 5.96	33 29 26.08
22	10		45.	59.5					38 14.62	26.44	0.99	III. 2	11.23	10 43.33	16.58	3.46	37 47.19	32 54 53.37
23	10						57.5	12.5	38 28.74	26.44	0.99	VII. 2	11.15½	10 39.11	16.61	3.45	38 1.31	32 54 49.17
24	8			35.	49.8	4.5			40 49.74	26.44	1.01	V. 8	6.17	37 6.14	16.90	8.20	40 22.29	33 21 21.24
25	7.8					31.	46.		42 16.09	26.44	0.98	VI. 2	3.29	6 44.16	17.08	2.76	41 48.67	32 50 54.00
26	10		57.	12.3					49 27.01	26.44	0.98	III. 2	2.3	6 0.96	17.98	2.63	48 59.59	50 11.57
27	7.8						17.	32.	49 47.27	26.44	0.98	VII. 2	2.31	6 14.64	18.02	2.66	49 19.85	32 50 25.32
28	10				10.	25.			54 9.92	26.44	1.02	V. 10	3.32	45 42.84	18.57	9.81	53 42.46	33 30 1.22
29	8		6.	21.2	36.				57 36.01	26.44	0.99	IV. 4	2.34	16 14.76	18.99	4.43	57 8.58	0 28.18
30	9						28.	43.	8 58 58 20	26.44	1.00	VII. 5	3.55	21 54.29	19.16	5.44	8 58 30.76	33 6 8.89
31	9		31.5	46.4	1.2				9 1 1.30	26.44	0.99	IV. 3	7.32	13 46.02	19.42	4.00	9 0 33.87	32 57 59.44
32	9						8.8	24.	1 39.10	26.44	1.00	VII. 4	10.5	20 1.67	19.49	5.10	1 11.66	33 4 16.26
33	8					57.	13.	27.5	8 42.74	26.43	0.98	VII. 1	8.9½	4 6.33	20.37	2.30	8 15.33	32 48 19.00
34	9		23.	38.	53.				12 52.94	26.43	0.99	IV. 3	9.45	14 53.08	20.89	4.18	12 25.52	32 59 8.15
35	7.8					17.	32.	47.	14 2.07	16.43	1.01	VII. 9	10.1	43 57.65	21.03	9.19	13 34.63	33 28 18.17
36	8		4.	19.2	34.				16 33.99	26.42	1.01	IV. 8	7.18	37 36.95	21.34	8.30	16 6.56	21 56.59
37	10			43.	58.				17 57.97	26.42	1.00	IV. 6	6.41	27 19.60	21.52	6.41	17 30.55	11 37.53
38	9						7.2	22.3	18 37.46	26.42	0.99	VII. 4	5.46	17 51.07	21.60	4.71	18 10.05	2 7.38
39	8				46.2	1.3			20 46.27	26.42	1.00	V. 7	3.16	30 36.17	21.87	7.03	20 18.85	14 55.07
40	8						54.	9.	21 24.19	26.42	1.00	VII. 5	8.15	24 5.40	21.95	5.84	20 56.77	8 23.19
41	9				27.	42.2			23 27.19	26.41	0.99	V. 4	3.34	16 44.95	22.21	4.51	22 59.79	33 1 1.67
42	10		11.3	26.4					25 41.22	26.41	0.99	III. 2	6.49	8 25.18	22.49	3.05	25 13.82	32 52 40.72
43	9						12.		26 11.97	26.41	0.98	VI. 1	4.23	2 12.39	22.56	1.96	25 44.58	32 46 26.91
44	10				29.	44.			28 29.09	26.40	0.99	V. 4	9.38	19 48.49	22.05	5.05	28 1.70	33 4 6.39
45	10					45.	59.4	4.4	30 29.61	26.40	0.99	VII. 3	3.20	11 38.44	23.10	3.59	30 2.42	32 55 55.13
46	10		33.		0.5	15.4			33 0.54	26.40	1.00	V. 4	10.30	20 14.72	23.43	5.13	32 33.14	33 4 33.28
47	10				34.4	49.5			34 34.51	26.39	1.00	V. 5	10.27	25 12.40	23.63	6.04	34 7.12	9 32.07
48	10		50.5	5.3	20.2				38 20.28	26.39	1.00	IV. 5	10.55½	25 26.83	24.12	6.08	37 52.89	33 9 47.03
49	.							58.7	9 39 14.12	−26.38	−0.99	VII. 3	8.35	−14 17.29	−24.23	−4.08	9 38 46.75	−32 58 35.60

CORRECTIONS.

Date.	Corr. of Clock.	Hourly rate.	m	n	c
1849. h.	s.	s.	s.	s.	s.

INSTRUMENT READINGS.

Date.	Barom.	THERMOM. At.	Ex.
1849. h. m.	in.	°	°

REMARKS.

(213) 3. Minutes of transit assumed as 0 instead of 1.
(213) 45. Transit over T. VII assumed to have been at 14ˢ.4 instead of 4ˢ.4.

ZONE 213. FEBRUARY 13. B. BELT, −33° 8'. D_0 = −32° 43' 50"—Continued.

No.	Mag.	Seconds of Transit. I. II. III. IV. V. VI. VII.	T.	a_1	a_2	Micrometer.			i	d_1	d_2	Mean Right Ascension, 1850.0.	Mean Declination, 1850.0.
			h. m. s.	s.	s.			r.	' "	"	"	h. m. s.	° ' "
50	10	. . . . 27. . .	9 43 56.98	−26.37	−0.98	VI.	1	6.47	− 3 25.01	−24.85	− 2.16	9 43 29.63	−32 47 42.02
51	10	. . . . 7. 22.	47 21.92	26.36	1.01	IV.	8	12.57	40 27.89	25.30	8.84	46 54.55	33 24 52.03
52	8	. . . . 52. 7.5 22.5	49 7.37	26.36	0.99	V.	3	9.21½	14 41.16	25.55	4.16	48 40.02	32 59 0.85
53	8.9	. . . . . . 34. 49.2	50 34.19	26.36	0.99	V.	4	5.53	17 55.03	25.73	4.72	50 6.84	33 2 15.48
54	9.10	. . . . 23. 38.	52 37.98	26.35	0.99	IV.	3	5.43	12 51.05	20.01	3.80	52 10.64	32 57 10.86
55	8.9	. . 25. 40. 55.2	9 55 54.98	−26.34	−1.01	.	9	7.34	−42 44.03	−26.45	− 9.28	9 55 27.63	−33 27 9.76

ZONE 214. FEBRUARY 13. H. BELT, −36° 16'. D_0 = −35° 53' 40".

No.	Mag.	Seconds of Transit. I. II. III. IV. V. VI. VII.	T.	a_1	a_2	Micrometer.			i	d_1	d_2	Mean Right Ascension, 1850.0.	Mean Declination, 1850.0.
1	9	. . 35. 50. 5.5	11 44 5.68	−25.94	−0.99	IV.	3	3.44	−12 21.31	− 0.87	− 3.73	11 43 38.75	−36 6 5.91
2	11	. . 48. 3.5 19.	48 19.01	25.91	1.00	IV.	7	8.37	33 18.10	1.80	9.17	47 52.10	27 9.07
3	11	. . 20.5 . . 50.5	50 50.96	25.89	1.01	IV.	9	9.44	43 49.58	2.35	12.00	50 24.06	37 43.93
4	10	50.3 6. 21.	53 36.89	25.87	1.01	III.	9	11.31	44 43.47	2.94	12.24	53 10.01	38 38.65
5	10	46. 1.5	55 31.94	25.85	0.99	II.	3	11.2	15 31.67	3.35	4.55	55 5.10	9 19.57
6	11	. . . . 31.5 . . 1.5	58 15.37	25.83	1.01	VII.	9	5.27	41 39.42	3.94	11.41	57 48.53	35 34.77
7	11	. . 16. 32. 47.5	11 58 47.55	25.83	1.00	IV.	5	2.33	21 13.45	4.05	6.01	11 58 20.72	15 3.51
8	11	. . 27.3 42.3 58.2	12 2 58.12	25.80	1.00	IV.	6	7.20	27 39.26	4.95	7.68	12 2 31.32	21 31.89
9	11	. . . . 26.2	6 39.96	25.76	0.99	VII.	2	11.44	10 53.43	5.72	3.38	6 13.21	4 42.53
10	8	. . . . . . 48. 3.5 19. . .	15 47.92	25.60	1.01	VI.	9	13.3	45 29.66	7.61	12.45	15 21.23	39 29.72
11	11	. . . . 50. 5.5 . .	17 34.48	25.66	1.01	VI.	8	6.49	37 22.09	7.98	10.27	17 7.81	36 31 20.34
12	11	. . . . 18.	23 33.47	25.60	0.99	V.	1	6.37	3 20.13	9.12	1.47	23 6.88	35 57 10.72
13	11	. . . . 3.5	24 16.89	25.60	1.01	VII.	8	2.23½	35 7.89	9.25	9.66	23 50.28	36 29 6.80
14	11	. . . . . . 19.3	28 19.21	25.56	1.01	IV.	8	5.8	36 31.40	9.99	10.04	27 52.64	30 31.43
15	11	. . 33.2 49.	32 4.39	25.53	1.00	III.	7	7.56	32 57.36	10.62	9.09	31 37.86	26 57.07
16	10	. . 4. 19.5 35.	36 35.02	25.49	1.00	IV.	7	3.38	30 47.32	11.36	8.52	36 8.53	24 47.20
17	9	. . . . . . 47.3 2.5	37 47.04	25.48	1.01	V.	9	9.14	43 34.39	11.55	11.94	37 20.55	37 37.88
18	9	. . 39. 54.2 9.5	41 9.42	25.44	0.99	IV.	3	8.35	14 17.79	12.12	4.25	40 42.99	8 14.16
19	9	. . . . 42.0 57.4 . .	42 26.43	25.43	1.01	V.	10	3.30	45 41.83	12.35	12.52	41 59.99	36 39 46.70
20	9	. . . . . . 45.5	46 55.51	25.39	0.99	I.	1	1.37	0 48.35	13.14	0.85	46 29.13	35 54 42.34
21	9	. . . . 52. 7.5	50 7.50	25.35	0.99	IV.	1	10.12	5 8.60	13.71	1.92	49 41.16	35 59 4.23
22	10	. . 6.5 22. 37.	54 37.33	25.32	1.01	IV.	9	10.30	44 12.78	14.54	12.12	54 11.00	36 38 19.44
23	10	. . . . . . 45. 0.5	12 59 45.06	25.26	0.99	V.	3	10.35	15 18.24	15.46	4.49	12 59 18.81	9 18.21
24	8	. . 14. 29.5 45.	13 3 45.00	−25.23	−1.01	IV.	9	5.37	−41 45.03	−16.28	−11.46	13 3 18.76	−36 35 52.77

ZONE 215. FEBRUARY 19. K. BELT, −28° 46'. D_0 = −28° 21' 50".

No.	Mag.	Seconds of Transit. I. II. III. IV. V. VI. VII.	T.	a_1	a_2	Micrometer.			i	d_1	d_2	Mean Right Ascension, 1850.0.	Mean Declination, 1850.0.
1	10	. . . . . . 41.8	6 58 41.77	−24.15	−1.00	IV.	7	4.8	−31 2.45	− 8.12	− 6.74	6 58 16.62	−28 53 7.31
2	10	. . . . . . 12.8	7 2 12.84	24.17	1.00	.	5	7.9	23 32.62	8.77	5.82	7 1 47.67	45 37.21
3	10	. . . . . . 52.4	2 52.43	24.17	1.00	.	5	9.57	24 57.33	8.87	5.99	2 27.26	47 2.19
4	9	. . . . 15.8 29.9	4 30.00	24.18	0.99	.	2	6.59	8 30.28	9.13	4.00	4 4.83	28 30 33.41
5	10	. . . . 16.5	5 2.21	24.18	1.01	.	9	10.39	44 17.32	9.22	8.40	4 37.02	29 6 24.94
6	9	15.8 30.0	7 58.59	24.20	1.00	.	7	9.32	33 45.83	9.68	7.10	7 33.39	28 55 52.61
7	9	. . . . 10.3 24.8	9 24.72	24.20	1.00	.	4	10.47	20 23.35	9.91	5.43	8 59.52	42 28.69
8	9.10	. . . . 14.5 . .	11 45.94	24.21	1.00	VI.	5	9.20	24 38.49	10.29	5.95	11 20.73	46 44.72
9	10	. . . . 54.8	13 40.56	24.22	1.00	V.	5	4.46	22 20.47	10.58	5.66	13 15.34	28 44 26.71
10	.	38.8 53.7	18 21.96	24.24	1.01	.	8	8.4	38 0.16	11.31	7.61	17 56.71	29 0 9.08
11	5	. . . . 49.5 4.2 18.6	7 18 35.55	−24.24	−1.00	.	8	9.43	−38 50.07	−11.35	− 7.71	7 18 10.31	−29 0 59.13

CORRECTIONS.

Date.		Corr. of Clock.	Hourly rate.	m	n	c
1849.	h.	s.	s.	s.	s.	s.

INSTRUMENT READINGS.

Date.		Barom.	Thermom. At.	Thermom. Ex.
1849.	h. m.	In.	°	°

REMARKS.

(214) 1. Micrometer reading assumed as 4r.44 instead of 3r.44.
(214) 16. Time of transit over T. III assumed as 19s.5 instead of 17s.5.

ZONE 215. FEBRUARY 19. K. BELT, −28° 46′. D₀ = −28° 21′ 50″—Continued.

No.	Mag.	Seconds of Transit. I.	II.	III.	IV.	V.	VI.	VII.	T.	a_1	a_2	Micrometer.			i	d_1	d_2	Mean Right Ascension, 1850.0.	Mean Declination, 1850.0.
									h. m. s.	s.	s.			r.	′ ″	″	″	h. m. s.	° ′ ″
12	7	58.9	13.1	...	...	...	...	...	7 22 27.40	−24.26	−1.00	.	6	10.13	−29 6.50	−11.99	−6.51	7 22 2.14	−28 51 15.00
13	7.8	...	...	...	8.3	...	...	...	23 8.31	24.26	0.99	.	6	5.27	26 42.28	12.06	6.20	22 43.06	48 50.54
14	9	...	...	...	...	37.8	...	...	24 23.60	24.27	0.99	V.	3	8.52	14 26.31	12.25	4.71	23 59.34	36 33.27
15	9	...	...	...	...	...	...	34.9	24 52.30	24.27	0.99	VII.	2	10.47	10 24.83	12.32	4.22	24 27.04	32 31.37
16	8.9	...	36.5	...	5.3	...	...	...	34 5.16	24.31	1.00	.	6	9.40	28 49.86	13.77	6.49	33 39.85	51 0.11
17	10	...	...	3.2	...	...	...	...	37 17.44	24.32	0.98	.	1	5.0	2 31.27	14.27	3.27	36 52.14	24 38.81
18	5.4	...	44.1	...	12.5	...	...	...	38 12.58	24.32	0.99	.	3	7.45	13 52.57	14.41	4.64	37 47.27	36 1.62
19	9	0.7	14.6	...	...	...	...	...	40 43.21	24.33	0.99	.	2	11.23	10 43.39	14.80	4.25	40 17.89	32 52.44
20	10	...	...	32.6	...	...	...	...	44 46.89	24.34	0.99	.	4	8.27	19 12.75	15.45	5.28	44 21.56	41 23.48
21	10	2.8	16.9	...	...	...	...	...	53 45.49	24.37	1.01	.	7	10.55	34 27.68	16.86	7.17	53 20.11	28 56 41.71
22	9	...	...	14.9	29.8	...	...	...	54 29.41	24.38	1.01	.	9	10.21	44 8.23	16.98	8.40	54 4.02	29 6 23.61
23	9	...	...	17.8	...	17.4	...	...	55 32.61	24.38	1.00	.	5	5.30	22 42.70	17.15	5.70	55 7.23	28 44 55.55
24	9	...	...	25.9	40.3	...	...	...	56 40.25	24.38	1.00	.	5	6.55	23 25.56	17.32	5.81	56 14.87	45 39.69
25	9	20.9	35.1	...	...	...	...	...	58 3.71	24.39	1.01	.	8	7.28	37 41.99	17.54	7.59	57 38.31	59 57.12
26	9	...	...	...	25.6	...	...	...	7 58 25.65	24.39	1.00	.	5	5.54	22 54.80	17.60	5.73	7 58 0.26	45 8.13
27	9	34.0	48.8	3.3	...	...	...	...	8 1 17.25	24.40	1.01	.	8	2.28	35 10.72	18.05	7.27	8 0 51.84	57 26.04
28	9	37.6	...	...	...	...	...	...	6 20.22	24.41	0.99	.	3	2.23	11 10.20	18.83	4.30	5 54.82	33 23.33
29	9	...	...	...	29.2	...	...	...	8 28.25	24.42	0.99	.	3	5.56	12 57.61	19.16	4.52	8 2.84	35 11.29
30	9	...	...	...	...	...	...	26.1	11 43.46	24.42	0.99	VII.	4	5.40	17 48.13	19.67	5.12	11 18.05	28 40 2.92
31	9	...	...	...	10.8	...	...	...	13 10.70	24.43	1.01	.	9	1.58	39 54.59	19.89	7.87	12 45.26	29 2 12.35
32	9	...	...	...	13.5	...	...	...	14 13.48	24.43	1.00	.	7	2.37	30 16.57	20.05	6.66	13 48.05	28 52 33.28
33	9	...	...	...	...	...	5.2	...	14 36.73	24.43	1.01	VI.	9	6.0	41 56.45	20.11	8.13	14 11.29	29 4 14.69
34	9	...	...	34.3	...	2.8	...	...	16 48.59	24.43	0.99	.	2	4.49	7 24.73	20.46	3.86	16 23.17	28 29 39.05
35	8.9	...	...	49.7	4.0	...	...	...	18 3.99	24.41	0.99	.	2	4.7	7 3.55	20.65	3.80	17 38.55	29 18.00
36	9	...	...	4.3	18.3	...	...	...	19 18.43	24.44	0.99	.	2	2.10	6 4.55	20.84	3.68	18 53.00	28 19.07
37	9	22.5	...	51.2	...	...	...	...	21 5.33	24.41	1.00	.	4	12.23	21 11.74	21.12	5.51	20 39.89	43 28.47
38	9	...	...	...	15.3	29.8	...	...	26 15.41	24.45	1.00	.	6	10.36	29 18.11	21.94	6.54	25 49.96	51 36.59
39	9	...	...	...	25.3	...	...	...	27 25.27	24.45	1.00	.	7	3.52	30 54.38	22.10	6.74	26 59.82	53 13.22
40	9	...	...	25.6	...	54.2	...	...	28 39.93	24.46	0.99	.	3	4.19	12 8.70	22.29	4.41	28 14.48	28 34 25.40
41	9	...	...	...	43.2	...	...	...	29 43.11	24.46	1.01	.	8	8.33	38 14.78	22.46	7.67	29 17.64	29 0 34.91
42	9	32.1	...	...	...	...	...	...	37 14.95	24.47	1.01	.	8	3.46	35 50.06	23.63	7.35	36 49.47	28 58 11.04
43	9	...	36.2	...	...	...	...	...	38 4.70	24.47	0.98	.	2	3.21	6 40.34	23.76	3.74	37 39.25	28 57.84
44	9	...	20.3	...	...	...	...	...	38 48.81	24.47	0.99	.	3	4.55	12 26.85	23.87	4.44	38 23.35	28 34 45.16
45	9	...	...	32.5	...	...	...	...	40 46.79	24.47	1.02	.	10	9.10	48 33.33	24.18	8.99	40 21.30	29 10 56.52
46	7.8	27.9	42.1	56.4	...	...	...	...	44 10.68	24.48	1.00	.	7	6.12	32 4.98	24.69	6.88	43 45.20	28 54 26.55
47	10	...	27.2	...	...	...	...	...	45 55.74	24.48	1.00	.	7	5.27	31 42.28	24.97	6.83	45 30.26	51 4.08
48	9	...	56.5	...	...	...	...	...	50 25.03	24.48	1.00	.	7	2.48	30 22.11	25.66	6.68	49 59.55	52 44.45
49	9	...	...	...	...	48.8	...	...	50 34.55	24.48	0.99	.	5	7.89?	1	25.69	. .	50 9.08	
50	10	...	11.9	...	...	...	...	...	53 10.42	24.49	1.00	.	4	8.45	19 21.83	26.08	5.29	52 44.93	41 43.20
51	7	...	...	54.4	...	22.8	...	...	8 59 8.65	24.49	0.98	.	1	5.52	2 57.49	27.01	3.29	8 58 43.18	25 17.79
52	10	9.1	...	...	...	...	...	...	9 0 51.80	−24.19	−0.99	.	4	8.15	−19 6.70	−27.27	−5.25	9 0 26.32	−28 41 29.22

CORRECTIONS.

Date.		Corr. of Clock.	Hourly rate.	m	n	c
1849.	h.	s.	s.	s.	s.	s.

INSTRUMENT READINGS.

Date.		Barom.	Thermom. At.	Thermom. Ex.
1849.	h. m.	In.	°	°

REMARKS.

ZONE 216. FEBRUARY 19. K. BELT, −38° 46′ for Mural.

No.	Mag.	I.	II.	III.	IV.	V.	VI.	VII.	T.	a_1	a_2	MICROMETER.	r.	i	d_1	d_2	Mean Right Ascension, 1850.0.	Mean Declination, 1850.0.
									h. m. s.	s.	s.		r.	′ ″	″	″	h. m. s.	° ′ ″
1	9	. .	. .	. .	. .	51.0	. .	22.9	9 38 34.82	−25.17	−1.00	VI.	6 8.26	−28 12.27	. .	− 7.20	9 38 8.65	
2	9.10	. .	33.3	49.4	5.3	. .	. .	. .	41 5.37	25.16	0.99	.	3 6.5	13 2.15	. .	1.50	48 39.22	
3	10	. .	48.5	. .	. .	. .	. .	. .	44 20.59	25.16	1.00	.	6 7.32	27 45.32	. .	7.04	43 54.43	
4	10	. .	. .	. .	35.2	. .	. .	. .	47 35.06	25.15	1.02	.	9 4.31	41 11.75	. .	12.31	47 8.89	
5	8.9	. .	38.8	54.8	. .	. .	. .	. .	50 10.86	25.14	1.00	.	5 11.2	25 30.11	. .	6.19	49 44.72	
6	10	. .	35.8	. .	. .	. .	24.0	. .	51 7.92	25.14	1.00	.	6 5.23	26 40.26	. .	6.64	50 41.78	
7	10	. .	51.2	. .	. .	. .	. .	. .	55 23.25	25.13	0.97	.	4 7.52	18 55.10	. .	3.68	54 57.13	
8	10	. .	. .	. .	. .	. .	. .	0.9	57 12.98	25.12	0.99	.	4 4.16	17 6.19	. .	− 3.02	56 46.87	
9	9.10	. .	. .	. .	. .	. .	. .	39.3	9 57 51.54	−25.12	−0.98	.	2 5.56	− 7 58.51	. .	+ 0.39	9 57 25.44	
10	9	14.0	30.2	46.3	2.3	18.3	34.4	50.6	10 11	. .	. .	.	Mid.	. .	. .	. .	10 14	

ZONE 217. FEBRUARY 23. B. BELT, −31° 53′. D_c = −31° 28′ 30″.

No.	Mag.	I.	II.	III.	IV.	V.	VI.	VII.	T.	a_1	a_2	MICROMETER.	r.	i	d_1	d_2	Mean Right Ascension, 1850.0.	Mean Declination, 1850.0.
1	8	3.	18.5	. .	. .	. .	. .	. .	5 59 47.60	+33.23	+0.99	II.	7 8.46	−33 22.43	− 0.23	− 7.34	6 0 21.82	−32 2 0.00
2	8	. .	28.	43.5	. .	. .	. .	. .	6 0 57.84	33.24	1.00	III.	5 8.46	24 21.49	0.42	5.90	1 32.08	31 52 57.81
3	8	. .	40.2	55.	. .	. .	. .	. .	3 9.68	33.21	1.00	III.	5 4.8	22 1.30	0.78	5.54	3 43.89	31 50 37.62
4	9	. .	. .	. .	. .	. .	54.	9.2	3 24.76	33.21	0.99	VII.	7 9.50	33 54.42	0.82	7.42	3 58.96	32 2 32.66
5	8	. .	. .	. .	33.	48.	. .	. .	5 33.14	33.20	1.00	V.	6 7.50	27 54.34	1.17	6.46	6 7.34	31 56 31.97
6	9	. .	. .	. .	. .	. .	58.	13.2	6 28.72	33.19	0.99	VII.	9 7.8	42 30.43	1.32	8.79	7 2.90	32 11 10.54
7	9	. .	. .	. .	37.6	52.8	. .	. .	10 37.91	33.16	1.01	V.	1 5.12	2 37.27	2.00	2.52	11 12.08	31 31 11.79
8	10	. .	. .	19.	34.2	. .	. .	. .	12 33.92	33.15	0.99	IV.	8 11.49	39 53.61	2.31	8.37	13 8.06	32 8 34.29
9	9	. .	. .	. .	58.5	13.5	. .	. .	13 58.70	33.14	1.01	V.	2 12.44	11 24.19	2.53	3.89	14 32.85	31 40 0.61
10	10	. .	. .	55.	10.	. .	. .	. .	16 9.86	33.12	1.01	IV.	2 .	4 59.0	2.88	2.90	16 43.99	33 34.78
11	7	. .	. .	. .	. .	. .	33.	48.	17 3.74	33.12	1.01	VII.	3 8.44	14 21.85	3.02	4.34	17 37.87	42 59.21
12	9	. .	. .	. .	. .	. .	19.	. .	18 49.47	33.11	1.00	VII.	4 3.20	16 37.46	3.30	4.70	19 23.58	45 15.46
13	8	. .	. .	23.8	38.	53.	. .	. .	20 38.30	33.09	1.00	V.	4 6.40	18 18.75	3.59	4.95	21 12.39	31 46 57.29
14	7	. .	. .	. .	30.	45.	. .	. .	22 30.02	33.08	0.99	V.	10 7.45	47 50.42	3.88	9.67	23 4.09	32 16 33.97
15	10	. .	. .	. .	0.	14.	. .	. .	25 59.71	33.06	1.01	V.	2 8.44	9 23.18	4.43	3.57	26 33.78	31 38 1.18
16	7	. .	. .	13.	28.	43.	. .	. .	26 28.01	33.05	1.00	V.	6 5.18	26 37.69	4.50	6.26	27 2.06	55 18.45
17	10	. .	. .	58.	12.3	. .	. .	. .	28 12.54	33.04	1.01	IV.	3 7.20	13 39.96	4.77	4.25	28 46.59	42 18.98
18	9	. .	. .	. .	. .	. .	15.	. .	28 45.46	33.04	1.01	VI.	3 9.56	14 58.42	4.86	4.43	29 19.51	43 37.71
19	7	. .	. .	. .	16.	30.5	. .	. .	30 15.95	33.03	1.00	V.	4 4.22	17 9.16	5.09	4.78	30 49.98	31 45 49.03
20	7	. .	. .	. .	36.4	51.3	. .	. .	31 36.39	33.02	0.99	V.	9 10.23	44 9.19	5.29	9.07	32 10.40	32 12 53.55
21	9	. .	. .	. .	2.5	17.3	. .	. .	33 2.60	33.01	1.01	V.	3 7.4	13 31.85	5.51	4.2	33 36.62	31 42 11.56
22	10	. .	. .	. .	46.	1.	. .	. .	34 46.18	33.01	1.00	V.	5 3.3	21 28.52	5.78	5.46	35 20.19	31 50 9.76
23	9	. .	. .	. .	. .	. .	. .	12.	35 27.70	33.01	0.99	VII.	9 2.26	40 8.24	5.89	8.41	36 1.70	32 8 52.53
24	9	. .	23.3	38.2	53.	. .	. .	. .	37 52.90	33.98	1.00	IV.	6 8.9½	28 4.23	6.26	6.49	38 27.88	31 56 46.98
25	9	. .	. .	. .	30.	44.5	. .	. .	39 29.82	32.97	0.99	V.	8 8.27	38 11.70	6.50	8.11	40 3.78	32 6 56.31
26	9	. .	. .	. .	11.	25.8	. .	. .	41 11.08	32.96	1.00	V.	5 6.20	23 7.86	6.76	5.72	41 44.04	31 51 50.34
27	9	. .	. .	. .	28.7	43.5	. .	. .	42 28.60	32.95	0.98	V.	10 12.4½	50 1.27	6.96	10.03	43 2.53	32 18 48.26
28	7	. .	. .	54.	9.	. .	. .	. .	44 8.86	32.94	1.01	IV.	1 6.51	3 27.24	7.22	2.65	44 42.81	31 32 7.11
29	9	. .	. .	. .	. .	. .	. .	56.	44 12.16	32.91	1.01	VII.	1 6.12	3 7.10	7.23	2.60	44 46.11	31 46.93
30	9	. .	. .	. .	. .	21.4	36.7	. .	46 6.95	32.93	1.00	VI.	5 2.49½	21 21.56	7.53	5.44	46 40.88	31 50 4.53
31	9	. .	. .	. .	. .	. .	55.	10.	47 25.60	32.92	0.99	VII.	10 6.59½	47 27.05	7.73	9.62	47 59.51	32 16 14.40
32	10	. .	. .	. .	. .	. .	26.	40.5	48 56.37	32.91	0.99	VII.	9 7.38	42 45.56	7.97	8.85	49 30.27	11 32.38
33	10	. .	. .	. .	54.	9.2	. .	. .	50 54.13	32.90	0.99	V.	10 3.58	45 55.95	8.27	9.35	51 28.02	32 14 43.57
34	10	. .	. .	. .	59.	. .	28.	. .	53 58.72	32.88	1.01	VI.	1 10.12	5 8.39	8.75	2.92	54 32.61	31 33 50.16
35	10	. .	. .	1.5	16.	30.5	. .	. .	6 57 16.04	+32.87	+1.01	V.	1 12.47	− 6 26.70	− 9.26	− 3.11	6 57 49.92	−31 35 9.07

CORRECTIONS.

Date.	Corr. of Clock.	Hourly rate.	m	n	c
1849. h.	s.	s.	s.	s.	s.

INSTRUMENT READINGS.

Date.	Barom.	THERMOM. At.	THERMOM. Ex.
1849. h. m.	in.	°	°

REMARKS.

(216) 6. Transit over T. V assumed to have been recorded as over T. VI.

ZONE 217. FEBRUARY 23. B. BELT, −31° 53′. D_o=−31° 28′ 30″—Continued.

No.	Mag.	Seconds of Transit. I.	II.	III.	IV.	V.	VI.	VII.	T.	a_1	a_2	Micrometer.			i	d_1	d_2	Mean Right Ascension, 1850.0.	Mean Declination, 1850.0.
									h. m. s.	s.	s.			r.	′ ″	″	″	h. m. s.	° ′ ″
36	9						50.3		6 58 20.78	+32.86	+1.00	VI.	4	6.38	−18 17.58	−9.42	−4.95	6 58 54.64	−31 47 1.95
37	9				19.	33.5			7 0 18.93	32.85	1.00	V.	5	5.54½	22 55.00	9.73	5.68	7 0 52.78	51 40.41
38	9						37.	51.5	1 7.49	32.84	1.01	VII.	3	7.26	13 42.51	9.86	4.25	1 41.34	42 26.62
39	8				53.	8.			2 53.20	32.83	1.01	V.	3	5.47	12 53.02	10.13	4.12	3 27.04	41 37.27
40	9						29.	44.	4 59.76	32.82	1.01	VII.	2	5.4	7 31.81	10.46	3.29	5 33.59	36 15.56
41	7.8			8.	23.	37.5			6 22.87	32.82	1.01	V.	3	10.20½	15 10.93	10.68	4.46	6 56.70	43 56.07
42	9					47.	2.		7 32.40	32.82	1.00	VI.	6	11.50	29 55.20	10.86	6.78	8 6.22	58 42.84
43	7.8			48.5	3.	18.			10 3.20	32.80	1.01	V.	4	5.43	17 50.0	11.25	4.89	10 37.01	46 36.14
44	10					34.	48.5		11 19.14	32.79	1.00	VI.	5	4.18	22 6.16	11.45	5.56	11 52.93	31 50 53.19
45	7							1.	12 16.75	32.79	0.99	VII.	8	3.38½	35 45.79	11.60	7.72	12 50.53	32 4 35.11
46	9				33.				14 33.07	32.78	1.01	IV.	4	3.51	16 53.57	11.93	4.73	15 6.86	31 45 40.23
47	9				53.				15 53.06	32.77	1.01	IV.	4	2.14	16 4.67	12.16	4.60	16 26.84	44 51.43
48	8						48.		16 18.47	32.77	1.01	VI.	4	3.36	16 45.81	12.23	4.71	16 52.25	45 32.75
49	6.7						44.	59.	18 13.70	32.76	1.00	VII.	6	3.50	25 52.89	12.53	6.13	18 47.46	31 54 41.55
50	10					36.	50.5		19 21.17	32.75	0.99	VI.	9	9.33	43 43.82	12.71	9.01	19 54.91	32 12 35.54
51	9						3.	18.	20 33.70	32.75	1.00	VII.	5	11.29	25 43.24	12.90	6.11	21 7.45	31 54 32.25
52	7				31.	46.	1.		22 31.27	32.74	1.01	VI.	1	7.36	3 49.73	13.21	2.70	23 5.02	31 32 35.64
53	10							34.2	23? 49.98	32.73	0.99	VII.	7	8.21½	33 9.79	13.41	7.30	24 23.70	32 2 0.50
54	10		39.	54.					28 8.58	32.71	1.00	III.	4	10.5	20 2.12	14.11	5.22	28 42.29	31 48 51.45
55	9			14.	29.	43.5			29 28.86	32.70	1.00	V.	5	7.36	23 43.16	14.33	5.79	30 2.56	31 52 33.28
56	10					4.	19.		30 49.41	32.70	0.99	VI.	8	6.16	37 5.49	14.54	7.94	31 23.10	32 5 57.97
57	7							21.5	31 37.48	32.70	1.00	VII.	4	7.21½	18 39.23	14.67	5.00	32 11.16	31 47 28.90
58	9						0.	15.	33 30.61	32.69	0.99	VII.	10	2.36	45 14.19	14.98	9.25	34 4.29	32 14 8.42
59	9				13.	27.5			40 12.89	32.66	1.00	VI.	6	6.55	27 26.45	16.06	6.37	40 46.55	31 56 18.90
60	9				37.	51.5			41 36.83	32.65	0.99	VI.	8	5.5	36 29.68	16.30	7.84	42 10.47	32 5 23.82
61	9		49.5	4.	19.				43 18.88	32.64	0.99	IV.	7	9.48	33 53.69	16.58	7.42	43 52.51	32 2 47.89
62	9						28.	43.	44 58.73	32.64	1.00	VII.	4	6.49½	18 23.11	16.86	4.96	45 32.37	31 47 14.93
63	9		45.	59.5	14.3				46 14.34	32.64	1.01	IV.	3	10.10	15 5.69	17.07	4.45	46 47.99	31 43 57.21
64	9					21.	33.5		48 6.15	32.63	0.99	VI.	8	6.17	37 5.99	17.37	7.94	48 39.77	32 6 1.30
65	9				39.				51 39.05	32.62	1.01	IV.	3	6.50	13 24.84	17.96	4.17	52 12.68	31 42 16.97
66	9						37.	52.	52 7.70	32.62	1.00	VII.	5	9.23	24 39.70	18.04	5.95	52 41.32	53 33.69
67	9	3.	18.						54 47.31	32.61	1.00	II.	6	8.11	28 4.77	18.48	6.50	55 20.92	56 59.75
68	8	16.	30.5						55 0.04	32.60	1.00	II.	5	8.54	24 25.36	18.52	5.90	55 33.64	31 53 19.78
69	7.8				39.	54.	8.5		56 39.08	32.60	0.99	VI.	7	9.22	33 40.57	18.79	7.39	57 12.67	32 2 36.75
70	7.8						21.	36.	7 57 51.61	32.60	0.98	VII.	10	4.34	46 13.69	18.99	9.41	7 58 25.19	32 15 12.09
71	9		29.	43.7	58.3				8 0 58.38	32.59	1.01	IV.	2	1.42	5 50.43	19.51	3.02	8 1 31.98	31 34 42.96
72	9				48.5	3.			3 48.26	32.58	0.98	V.	10	10.10	49 3.54	19.98	9.89	4 21.82	32 18 3.41
73	7.8						53.5	8.	4 23.93	32.57	1.00	VII.	6	6.59	27 28.20	20.07	6.40	4 57.50	31 56 24.67
74	10						7.	22.	5 37.70	32.57	1.00	VII.	6	.	23 56.92	20.28	5.87	6 11 27	52 53.07
75	9						50.	4.5	7 20.44	32.57	1.00	VII.	5	6.34	23 14.50	20.36	5.73	7 54.01	52 10.79
76	10						23.	37.5	10 53.52	32.56	1.02	VII.	1	9.20½	4 46.17	21.15	2.85	11 27.10	33 40.17
77	9							49.	12 4.83	32.55	1.00	VII.	6	9.3	28 30.72	21.34	6.56	12 38.38	57 28.62
78	9				45.				14 45.01	32.54	1.00	IV.	6	6.12	27 4.98	21.79	6.33	15 18.55	56 3.10
79	10						35.5	50.	15 5.96	32.54	1.00	VII.	5	6.41	23 18.02	21.85	5.73	15 39.50	52 15.60
80	8.9				49.	4.			16 49.18	32.54	1.00	V.	5	3.47	21 50.71	22.13	5.52	17 22.72	31 50 43.36
81	9		5.	19.5					18 34.36	32.54	0.99	III.	8	7.57	37 56.57	22.12	8.07	19 7.89	32 6 57.06
82	10						15.	30.	18 45.68	32.54	1.00	VII.	6	9.44½	28 51.65	22.45	6.61	19 19.22	31 57 50.71
83	7.8				31.	46.			20 31.20	32.53	1.01	V.	3	5.17	12 37.84	22.75	4.07	21 4.74	41 34.71
84	9		35.	49.5					8 24 4.33	+32.53	+1.00	III.	5	3.17	−21 35.56	−23.33	−5.45	8 24 37.86	−31 50 34.39

CORRECTIONS.

Date.	Corr. of Clock.	Hourly rate.	m	n	c
1849. h.	s.	s.	s.	s.	s.

INSTRUMENT READINGS.

Date.		Barom.	Thermom. At.	Ex.
1849.	h. m.	in.	°	°

REMARKS.

(217) 49. Minutes of transit assumed as 17 instead of 18.

ZONE 217. FEBRUARY 23. B. BELT, −31° 53′. D₀ = −31° 28′ 30″—Continued.

No.	Mag.	SECONDS OF TRANSIT. I.	II.	III.	IV.	V.	VI.	VII.	T.	a_1	a_2	MICROMETER.			i	d_1	d_2	Mean Right Ascension, 1850.0.	Mean Declination, 1850.0.
									h. m. s.	s.	s.			r.	′ ″	″	″	h. m. s.	° ′ ″
85	7.8	..	..	..	..	..	..	38.	8 23 54.08	+32.53	+1.01	VII.	3	1.19	−10 37.45	−23.30	−3.75	8 24 27.62	−31 39 34.50
86	9	..	..	..	..	..	24.	39.	25 54.74	32.52	1.01	VII.	3	9.8	14 33.94	23.63	4.35	26 28.27	31 43 31.92
87	7.8	..	..	..	..	12.5	27.3	..	27 57.81	32.52	0.99	VII.	8	3.32½	35 42.77	23.97	7.71	28 31.32	32 4 44.45
88	8	..	..	..	46.	1.	..	..	8 29 46.20	+32.51	+1.01	V.	3	11.21	−15 41.43	−24.27	−4.53	8 30 19.72	−31 44 40.23

ZONE 218. MARCH 7. B. BELT, −34° 23½′. D₀ = −34° 0′ 0″.

No.	Mag.	I.	II.	III.	IV.	V.	VI.	VII.	T.	a_1	a_2	MICROMETER.			i	d_1	d_2	Mean Right Ascension, 1850.0.	Mean Declination, 1850.0.
1	9	..	..	..	31.	46.	..	..	7 25 30.96	+27.20	+1.00	V.	5	8.21	−24 8.86	−11.96	−5.83	7 25 59.16	−34 24 26.65
2	10	..	..	..	..	7.	22.	..	26 51.76	27.19	1.01	VI.	3	11.15	15 38.23	12.18	4.17	27 19.96	13 54.58
3	10	..	..	..	..	..	..	27.	27 41.75	27.18	1.01	VII.	3	11.34	15 47.52	12.32	4.20	28 9.04	16 4.04
4	10	..	..	..	55.	10.5	..	..	29 55.85	27.17	1.01	V.	1	10.56	5 30.74	12.69	2.26	30 23.43	5 45.69
5	6	..	..	..	20.4	36.	51.	6.	31 20.56	27.16	0.99	VII.	8	7.24	37 33.45	12.93	8.52	31 48.71	38 0.90
6	8	..	..	..	15.	30.	..	..	33 14.92	27.15	1.00	V.	6	9.3	28 0.89	13.26	6.60	33 43.07	28 20.75
7	9	..	..	..	..	..	27.5	43.	34 57.36	27.13	1.00	VII.	6	6.5½	27 1.17	13.55	6.40	35 25.49	27 21.12
8	9	..	..	37.5	52.5	..	..	..	36 52.57	27.12	1.00	V.	7	3.54½	30 55.58	13.86	7.17	37 20.69	31 16.61
9	10	..	..	..	12.5	27.	..	..	38 12.23	27.12	1.00	V.	4	7.1½	18 29.58	14.10	4.74	38 40.35	18 48.42
10	10	..	..	..	..	..	31.	46.	41 0.72	27.10	1.01	VII.	1	11.24	5 44.37	14.59	2.29	41 28.83	6 1.25
11	9	10.	25.3	..	..	..	..	..	43 55.64	27.08	0.99	II.	9	8.45½	43 19.85	15.11	9.67	44 23.71	43 44.63
12	10	..	..	..	..	28.5	..	..	44 13.35	27.08	1.00	VI.	6	11.18	29 39.04	15.16	6.92	44 41.43	30 1.12
13	9	..	..	..	..	..	19.	34.5	44 48.91	27.08	1.00	VII.	4	9.58	19 58.10	15.27	5.02	45 16.99	20 18.39
14	6.7	..	..	..	..	26.	41.5	57.	46 11.23	27.07	1.00	VII.	4	9.5	19 31.38	15.52	4.93	46 39.30	19 51.83
15	7	..	2.5	17.5	33.3	..	..	..	48 32.93	27.05	1.00	V.	6	5.47½	26 52.56	15.93	6.30	49 0.98	27 14.07
16	10	..	..	47.	2.	..	..	..	50 2.11	27.04	1.00	IV.	4	7.3	18 30.39	16.20	4.74	50 30.15	18 51.33
17	9	..	11.	26.	..	..	..	..	51 41.25	27.04	1.00	III.	7	4.24	31 10.47	16.52	7.23	52 9.29	31 34.22
18	9	..	..	..	34.	49.	..	..	52 33.89	27.03	0.99	V.	7	7.5	32 31.64	16.67	7.49	53 1.91	32 55.80
19	.	..	..	..	19.	..	..	..	54 19.08	27.02	1.00	.	.	.	. . .	. . .	. .	54 47.10	
20	9	..	..	..	..	16.	31.	..	55 0.76	27.02	1.01	VI.	3	9.2	14 31.17	17.13	3.96	55 28.79	14 52.26
21	9	..	..	..	..	..	..	15.	55 29.74	27.01	1.01	VII.	3	12.24½	16 12.97	17.22	4.28	56 57.76	16 34.47
22	10	..	..	..	..	28.	43.	..	12.75	27.01	1.00	VI.	4	6.9	18 2.94	17.54	4.63	57 40.76	18 25.11
23	10	..	..	..	..	45.5	0.05	..	58 30.26	27.00	1.00	V.	6	8.12	28 5.43	17.78	6.62	7 58 58.26	28 29.83
24	9	..	..	..	..	8.	23.	..	7 59 52.76	26.99	1.00	VI.	4	8.26½	19 12.27	18.04	4.87	8 0 20.75	19 35.18
25	8	..	..	35.5	51.	..	..	..	8 1 50.75	26.98	0.98	IV.	10	4.18	46 6.09	18.40	10.23	2 18.71	46 34.72
26	10	..	..	..	18.	..	..	..	3 17.90	26.97	0.99	IV.	8	7.59	37 57.63	18.68	8.59	3 45.86	38 24.90
27	9	..	..	19.	..	..	..	..	4 34.17	26.97	1.00	III.	6	8.29	28 14.00	18.93	6.64	5 2.14	28 39.57
28	10	..	..	..	..	10.	25.	..	4 54.76	26.96	1.00	VI.	6	4.37	26 16.85	18.98	6.25	5 22.72	26 42.08
29	10	..	..	..	..	26.	..	..	6 10.82	26.96	0.99	V.	8	5.22	36 38.40	19.22	8.33	6 38.77	37 5.95
30	8.9	..	..	18.	33.	..	..	..	7 33.10	26.95	1.01	IV.	4	3.37	16 46.52	19.48	4.38	8 1.06	17 10.39
31	10	..	..	..	..	..	28.	43.5	8 57.90	26.94	1.00	VII.	4	10.1	19 59.62	19.74	5.02	9 25.84	20 24.38
32	9	..	..	..	..	..	..	45.5	9 0.29	26.94	1.01	VII.	2	14.11	12 7.56	19.75	3.49	9 28.24	12 30.80
33	10	..	..	..	..	..	..	27.	10 41.63	26.93	1.00	VII.	5	7.00	23 27.55	20.07	5.69	11 9.56	23 53.31
34	11	..	..	..	..	..	22.	..	11 51.62	26.93	1.00	VI.	4	7.28	18 42.76	20.28	4.78	12 19.55	19 7.82
35	6.7	..	..	..	..	29.	44.	59.	13.79	26.92	1.01	VII.	2	4.00	6 59.49	20.52	2.52	13 41.62	7 22.53
36	9	..	..	..	52.5	8.	23.	..	14 52.68	26.91	1.00	VII.	5	9.38	24 47.22	20.83	5.96	15 20.59	25 14.01
37	9	..	..	..	..	4.	19.	..	16 47.76	26.91	1.00	VI.	6	9.13	28 36.01	21.18	6.72	17 15.67	29 3.91
38	9	..	..	..	..	5.	20.5	..	17 50.01	26.90	1.00	VI.	6	9.13	28 36.01	21.36	6.72	18 17.91	29 4.09
39	10	..	..	..	..	41.	56.	..	19 25.76	26.89	1.01	VI.	3	11.47	15 54.37	21.65	4.21	19 53.66	16 20.23
40	9	..	..	..	..	..	1.	16.	20 30.72	26.89	1.01	VII.	1	12.10	6 7.56	21.84	2.35	20 58.62	6 31.75
41	5.6	..	..	..	..	..	13.	28.	8 21 43.57	+26.88	+0.99	VII.	8	5.23	−36 38.43	−22.05	−8.33	8 22 11.44	−34 37 8.81

CORRECTIONS.

Date.		Corr. of Clock.	Hourly rate.	m	n	c
1849.	h.	s.	s.	s.	s.	s.

INSTRUMENT READINGS.

Date.		Barom.	THERMOM. At.	THERMOM. Ex.
1849.	h. m.	in.	°	°

REMARKS.

(218) 6. Micrometer reading assumed as 8r.3 instead of 9r.3.
(218) 35. Transit over T. V assumed as at 29s instead of 27s.

ZONE 218. MARCH 7. B. BELT, −34° 23½'. D₀=−34° 0' 0"—Continued.

No.	Mag.	SECONDS OF TRANSIT. I.	II.	III.	IV.	V.	VI.	VII.	T.	a_1	a_2	MICROMETER.			i	d_1	d_2	Mean Right Ascension, 1850.0.	Mean Declination, 1850.0.
									h. m. s.	s.	s.			r.	' "	"	"	h. m. s.	° ' "
42	9	...	...	...	...	...	...	53.	8 23 7.87	+26.88	+1.01	VII.	2	3.26	− 6 42.34	−22.29	− 2.46	8 23 35.76	−34 7 7.09
43	7	...	...	23.	38.	...	...	...	26 38.07	26.86	1.01	IV.	2	4.12	7 6.07	22.88	2.54	27 5.94	7 31.49
44	9	...	...	...	...	53.0	8.	...	27 38.01	26.86	0.99	VI.	7	7.46	32 52.15	23.05	7.57	28 5.86	33 22.77
45	8	...	...	9.3	24.3	...	...	...	36 24.38	26.82	1.01	IV.	3	2.58	11 27.85	24.45	3.37	36 52.21	11 55.67
46	10	...	...	...	...	43.	58.	...	37 27.76	26.82	1.01	VI.	1	10.45	5 25.01	24.50	2.22	37 55.59	5 51.73
47	10	...	...	31.	46.	...	...	...	46.04	26.81	0.99	IV.	8	8.35	38 15.79	24.81	8.65	39 13.84	38 49.25
48	10	...	...	...	...	...	...	51.	40 5.61	26.81	1.00	VII.	5	10.28	25 12.43	25.01	6.04	40 33.42	25 43.48
49	10	...	...	...	...	59.	...	...	41 43.80	26.80	0.98	V.	9	11.38½	44 47.25	25.30	9.98	42 11.58	45 22.53
50	10	...	...	...	14.5	30.	45.	...	43 14.70	26.80	1.01	VI.	4	4.20	17 7.97	25.56	4.44	43 42.51	17 37.97
51	10	...	...	28.	43.5	...	...	...	45 43.28	26.79	0.99	IV.	8	9.53	38 55.11	25.99	8.78	46 11.06	39 29.88
52	10	...	...	...	...	...	57.	12.	46 26.61	26.79	1.00	VII.	6	8.57½	28 27.90	26.12	6.69	46 54.40	29 0.71
53	10	...	...	...	...	...	...	20.	47 34.75	26.78	1.01	VII.	3	10.6	15 3.15	26.33	4.05	48 2.54	15 33.53
54	10	...	40.	55.5	10.5	...	...	...	50 10.50	26.77	1.00	IV.	5	9.3½	24 30.35	26.78	5.90	50 38.27	25 3.03
55	10	...	...	4.5	19.5	...	...	...	51 19.61	26.77	1.00	IV.	4	6.39	19 18.81	26.99	4.88	51 47.38	19 50.68
56	10	...	...	32.5	48.	...	...	...	52 47.76	26.76	0.99	IV.	9	8.19	43 6.72	27.27	9.62	53 14.51	43 43.61
57	9	...	...	...	57.	12.	...	...	54 56.79	26.76	0.98	V.	10	11.55	49 56.47	27.66	11.01	55 24.53	50 35.14
58	9	...	...	...	...	25.	40.2	...	56 9.86	26.75	1.01	VI.	3	6.3½	13 1.16	27.89	3.66	56 37.62	13 32.71
59	9	...	...	...	...	57.	12.3	...	57 41.90	26.75	1.00	VI.	6	5.1	26 28.94	28.18	6.30	58 9.65	27 3.42
60	10	...	...	...	...	16.4	31.5	...	8 59 1.21	26.74	1.01	VI.	3	3.29	11 43.25	28.42	3.41	8 59 28.96	12 15.08
61	10	...	...	...	...	46.	...	...	9 0 24.93	26.74	1.01	V.	3	2.13	11 5.10	28.68	3.30	9 0 52.68	11 37.08
62	8	...	...	...	...	33.	...	...	2 17.80	26.73	0.98	V.	9	10.5	44 0.11	29.05	9.83	2 45.51	44 38.99
63	9	...	...	...	...	...	59.5	14.5	4 29.22	26.73	1.01	VII.	1	12.47	6 26.22	29.41	2.39	4 56.96	6 58.02
64	10	...	...	...	0.5	15.5	...	...	8 0.48	26.72	1.01	V.	3	11.40	15 51.01	30.03	4.19	8 28.21	16 25.23
65	9	...	...	...	17.6	33.	...	...	11 17.63	26.71	0.99	V.	9	8.52	43 23.30	30.60	9.69	11 45.33	44 3.59
66	11	...	...	...	...	15.5	31.	...	13 0.51	26.71	1.01	VI	3	6.50½	13 24.86	30.89	3.72	13 28.23	13 59.47
67	8	...	...	...	33.3	48.2	3.	...	14 32.97	26.70	0.99	V.	8	7.24½	37 40.17	31.14	8.55	15 0.66	38 19.86
68	8	...	...	...	...	...	13.	...	15 42.63	26.70	1.00	VI.	4	9.36	19 47.32	31.34	4.97	16 10.33	20 23.62
69	8	...	...	...	...	...	...	0.	16 14.77	26.70	1.01	V.	3	7.9	13 34.36	31.42	3.76	16 42.48	14 9.54
70	8	...	...	...	...	...	...	56.	17 10.45	26.70	0.99	VII.	8	3.29	35 40.95	31.58	8.14	17 38.14	36 20.67
71	9	...	...	37.3	...	8.	...	...	19 52.66	26.69	1.00	V.	7	2.45	30 20.55	32.03	7.08	20 20.35	30 59.66
72	6.7	...	...	...	...	12.	27.	...	20 56.75	26.69	1.00	VI.	5	1.43	20 48.00	32.20	5.17	21 24.44	21 25.37
73	9	...	...	...	4.	19.	...	...	23 3.84	26.69	0.99	V.	9	3.17	40 34.37	32.56	9.13	23 31.52	41 16.06
74	10	37.	52.3	7.2	...	...	...	...	26 22.53	26.68	0.99	III.	8	7.17	37 36.39	33.12	8.53	26 50.20	38 18.04
75	9	...	14.5	30.	45.	...	...	...	30 44.99	26.67	1.01	IV.	3	11.37	15 49.56	33.84	4.18	31 12.67	16 27.58
76	9	...	...	...	...	...	46.5	2.	31 16.39	26.67	1.00	VII.	5	6.58	23 26.54	33.94	5.70	31 44.06	24 6.18
77	9	...	...	12.	27.	...	...	...	33 27.08	26.67	1.00	IV.	6	11.14½	29 37.51	34.29	6.93	33 54.75	30 18.73
78	8.9	...	...	...	...	40.5	56.	...	34 25.51	26.67	0.98	VI.	10	7.7	47 31.08	34.46	10.54	34 53.16	48 16.08
79	6.7	...	...	...	54.	9.2	...	...	35 53.90	26.66	0.98	V.	10	8.52½	48 24.45	34.71	10.72	36 21.54	49 9.88
80	10	...	...	...	...	26.5	41.7	...	37 11.35	26.66	1.00	VI.	5	6.59	23 27.35	34.93	5.70	37 39.01	24 7.98
81	11	...	...	5.	20.	...	...	...	40 20.07	26.66	1.00	IV.	7	3.34	30 45.31	35.45	7.16	40 47.73	31 27.92
82	6.7	...	23.2	38.4	...	...	...	...	41 53.52	26.66	1.00	IV.	4	7.40	18 49.05	35.72	4.80	42 21.18	19 29.57
83	10	...	...	...	...	...	...	7.5	42 22.28	26.66	1.01	VII.	3	6.44	13 21.29	35.79	3.71	42 49.95	14 0.79
84	9	...	...	35.5	51.	...	...	...	44 50.80	26.65	0.99	IV.	7	10.9	34 4.49	36.22	7.63	45 18.44	34 48.54
85	9	...	...	...	...	...	0.5	15.5	45 30.14	26.65	1.00	VII.	5	7.34	23 44.70	36.33	5.77	45 57.79	24 26.80
86	6.7	...	...	22.	37.2	52.4	...	...	47 37.24	26.65	1.01	V.	2	12.51	11 27.70	36.68	3.36	48 4.90	12 7.74
87	7.8	...	32.	47.	2.2	...	...	...	50 2.20	26.65	1.01	IV.	1	12.33	6 19.69	37.09	2.36	50 29.86	6 59.14
88	7.6	...	...	...	...	...	...	3.4	50 18.06	26.65	1.00	VII.	4	13.14	21 36.93	37.14	5.33	50 45.71	22 19.40
89	11	...	...	...	52.	...	...	...	52 52.07	26.65	1.01	IV.	4	5.00	17 28.37	37.57	4.50	53 19.73	18 10.44
90	7.8	...	...	...	...	...	...	2.	9 53 16.46	+26.64	+0.99	VII.	7	12.7½	−35 3.70	−37.65	− 8.02	9 53 44.09	−34 35 49.37

CORRECTIONS.

Date.	Corr. of Clock.	Hourly rate.	m	n	c
1849.	h. s.	s.	s.	s.	s.

INSTRUMENT READINGS.

Date.	Barom.	THERMOM. At.	Ex.
1849. h. m.	In.	°	°

REMARKS.

ZONE 218. MARCH 7. B. BELT, −34° 23½′. D₀=−34° 0′ 0″—Continued.

No.	Mag.	Seconds of Transit I.	II.	III.	IV.	V.	VI.	VII.	T.	a_1	a_2	Micrometer.			i	d_1	d_2	Mean Right Ascension, 1850.0.	Mean Declination, 1850.0.
									h. m. s.	s.	s.			r.	′ ″	″	″	h. m. s.	° ′ ″
91	9	..	..	..	9.	24.	..	..	9 55 8.98	+26.64	+1.01	V.	3	6.50	−13 25.03	−37.96	−3.73	9 55 36.63	−34 14 6.72
92	9	..	..	39.5	55.	..	..	..	56 54.83	26.64	1.01	IV.	2	6.28	8 14.64	38.26	2.73	57 22.48	8 55.63
93	9	..	..	..	..	..	26.	41.	57 55.63	26.64	1.00	VII.	5	9.38	24 47.22	38.43	5.96	9 58 22.27	25 31.61
94	9	..	..	..	32.5	48.	3.	..	9 59 32.68	26.64	1.00	VI.	6	4.48	26 22.39	38.70	6.28	10 0 0.32	27 7.37
95	7	..	..	6.	21.	..	..	..	10 1 21.10	+26.64	+1.01	IV.	4	2.23	−16 9.20	−39.01	−4.24	10 1 48.75	−34 16 52.45

ZONE 219. MARCH 12. K. BELT, −35° 38′. D₀=−35° 14′ 40″.

No.	Mag.	I.	II.	III.	IV.	V.	VI.	VII.	T.	a_1	a_2	Micrometer.			i	d_1	d_2	Mean Right Ascension, 1850.0.	Mean Declination, 1850.0.
1	8	..	..	54.8	..	..	..	..	8 33 10.24	+26.46	+0.98	.	10	11.45	−49 51.40	−4.95	−11.84	8 33 37.68	−36 4 48.28
2	9.10	..	..	7.8	..	..	..	..	34 23.17	26.46	1.01	.	4	4.36	17 16.28	5.19	4.25	34 50.64	35 32 5.72
3	8.9	..	..	..	4.5	19.6	..	..	35 4.43	26.45	1.01	.	3	1.3	10 29.86	5.32	2.74	35 31.89	25 17.92
4	8	..	..	23.8	39.0	54.5	..	..	36 39.14	26.45	1.01	.	2	9.5	9 33.81	5.63	2.53	37 6.60	24 21.97
5	9	..	51.9	7.1	..	..	..	..	38 22.58	26.44	1.00	.	5	5.36	24 45.73	5.97	5.94	38 50.02	39 37.64
6	11	..	18.9	..	..	..	..	..	41 49.71	26.43	0.99	II.	8	5.12	36 33.18	6.65	8.67	42 16.13	35 51 28.50
7	9	..	..	..	..	42.8	..	..	42 27.36	26.43	0.98	V.	10	7.55	47 55.45	6.77	11.38	42 54.77	36 2 53.60
8	9.8	..	..	..	38.8	..	..	..	43 38.59	26.42	0.98	.	10	9.21	48 38.87	7.01	11.55	44 5.99	36 3 37.43
9	8	..	..	..	..	..	..	47.9	44 2.06	26.42	1.01	VII.	2	4.59	7 29.22	7.08	2.08	44 29.49	35 22 18.38
10	7.8	..	47.8	3.3	18.6	..	..	..	46 18.60	26.42	0.99	.	10	0.7	43 59.53	7.53	10.44	46 45.01	58 57.50
11	9.10	..	..	..	..	..	51.8	..	47 20.92	26.41	1.01	VI.	3	4.36	12 17.04	7.73	3.13	47 48.34	27 7.90
12	9	..	..	..	29.6	..	0.7	..	48 29.75	26.40	1.01	.	3	8.30	14 15.27	7.96	3.55	48 57.16	29 6.78
13	9.10	..	..	..	5.9	..	..	..	53 5.85	26.39	0.99	.	7	8.8	33 3.47	8.87	7.85	53 33.23	48 0.19
14	8.9	..	58.9	14.2	..	..	..	..	8 59 29.61	26.37	1.01	.	3	9.46	14 53.59	10.12	3.72	8 59 56.99	29 47.43
15	9.10	..	..	..	24.5	..	..	..	9 7 24.49	26.35	1.00	.	6	11.44	29 52.39	11.66	7.12	9 7 51.84	44 51.17
16	7.8	..	..	..	..	..	..	10.	8 15.19	26.35	1.02	VII.	1	11.6	5 35.28	11.82	1.65	8 42.56	20 28.75
17	9	..	..	..	1.2	..	..	..	11 1.19	26.34	1.00	.	6	11.30	29 45.33	12.35	7.09	11 28.53	44 44.77
18	9.10	..	..	..	58.6	..	..	..	11 58.63	26.34	1.02	.	1	12.49	6 27.76	12.54	1.81	12 25.99	21 22.14
19	9	..	..	18.1	33.7	..	..	..	9 13 33.55	+26.33	+0.99	.	9	3.48	−40 50.06	−12.84	−9.63	9 14 0.87	−35 55 52.58

ZONE 220. MARCH 16. K. BELT, −36° 53′.

No.	Mag.	I.	II.	III.	IV.	V.	VI.	VII.	T.	a_1	a_2	Micrometer.			i	d_1	d_2	Mean Right Ascension, 1850.0.	Mean Declination, 1850.0.
1	8	..	..	..	..	..	43.3	..	7 34 12.06	+23.38	+0.99	VI.	9	11.27	−44 41.26	..	−11.37	7 34 36.43	
2	10	..	15.8	31.2	..	..	..	..	37 46.93	23.36	1.01	III.	3	9.23	14 41.92	..	3.31	38 11.30	
3	10	..	..	25.3	..	..	..	..	38 40.99	23.36	0.98	.	10	9.50	48 53.50	..	12.54	39 5.33	
4	8.9	..	..	24.7	40.3	..	..	..	39 40.26	23.35	0.99	.	9	11.20	44 37.98	..	11.34	40 4.60	
5	11	..	5.1	..	..	..	..	..	41 36.38	23.34	1.00	II.	6	7.50	27 54.14	..	6.78	42 0.72	
6	10.9	..	..	10.4	26.3	..	..	..	43 26.15	23.33	0.99	.	7	9.4	33 31.71	..	8.27	43 50.47	
7	10	..	10.7	26.3	..	..	..	..	45 42.01	23.32	0.99	.	10	3.34	45 43.91	..	11.65	46 6.32	
8	8.9	..	..	40.6	56.1	..	..	..	46 56.16	23.31	1.01	.	1	4.36	2 19.18	..	0.14	47 20.49	
9	8.9	..	..	..	41.3	56.5	..	..	47 41.16	23.30	1.01	.	1	7.19	3 41.36	..	0.49	48 5.47	
10	8	..	1.3	17.1	32.9	..	..	..	51 32.71	23.29	1.00	V.	6	7.38	27 48.28	..	6.76	51 57.00	
11	8	..	..	..	..	29.3	45.0	..	52 13.68	23.28	0.99	VI.	9	4.31	41 11.50	..	10.38	52 37.95	
12	7.8	..	..	..	37.8	53.3	..	..	53 37.77	23.27	1.00	IV.	5	8.42	24 19.52	..	5.81	54 2.04	
13	7	56.0	12.6	28.2	43.6	..	..	..	55 43.78	23.26	1.00	.	5	5.39	22 47.24	..	5.41	56 8.03	
14	7	..	..	..	..	26.8	42.3	57.9	56 11.17	23.26	1.01	VII.	2	7.32	8 46.36	..	1.76	56 35.44	
15	9.10	..	..	..	39.6	55.3	..	..	57 39.67	23.25	1.00	VI.	5	6.44	23 19.77	..	5.56	58 3.92	
16	9	..	..	18.7	34.3	..	..	..	7 59 34.32	23.24	1.01	.	2	2.36	6 17.67	..	1.14	7 59 58.57	
17	8.9	..	..	..	..	..	..	30.3	8 1 43.37	+23.23	+1.00	VII.	7	3.36	−30 45.76	..	−7.54	8 2 7.60	

CORRECTIONS.

Date.		Corr. of Clock.	Hourly rate.	m	n	c
1849.	h.	s.	s.	s.	s.	s.

INSTRUMENT READINGS.

Date.		Barom.	Thermom. At.	Thermom. Ex.
1849.	h. m.	in.	°	°

REMARKS.

(219) 16. Transit over T. VII assumed as 1ˢ.0 instead of 10ˢ.

Zone 220. March 16. K. Belt, $-36°$ 53′—Continued.

No.	Mag.	Seconds of Transit. I.	II.	III.	IV.	V.	VI.	VII.	T.	a_1	a_2	Micrometer.			i	d_1	d_2	Mean Right Ascension, 1850.0.	Mean Declination, 1850.0.
									h. m. s.	s.	s.			r.	′ ″	″	″	h. m. s.	° ′ ″
18	7						14.7	30.3	8 2 43.31	+23.23	+0.98	VI.	10	2.38	−45 15.42	. .	−11.53	8 3 7.52	
19	9.10				23.8				4 23.87	23.22	1.01	.	3	9.55	14 58.12	. .	3.37	4 48.10	
20	6.7					20.3	35.9		5 4.63	23.21	1.00	VI.	5	3.16	21 34.88	. .	5.10	5 28.84	
21	6.7				28.4		59.8		6 28.40	23.21	1.01	VI.	1	6.22	3 12.37	. .	0.36	6 52.62	
22	10					6.3			7 50.60	23.20	0.98	.	10	5.58	46 56.51	. .	12.00	8 14.78	
23	9		59.2	14.2					9 30.15	23.19	1.00	.	6	2.56	25 26.15	. .	6.11	9 54.34	
24	7.8	26.9	42.7	58.2					12 13.70	23.18	1.00	.	5	10.19	25 8.43	. .	6.03	12 37.88	
25	11.10							59.3	12 12.36	23.18	0.99	VII.	7	4.34	31 15.92	. .	7.67	12 36.53	
26	7.8				46.1	1.6			13 46.04	23.17	1.00	.	6	6.55	27 26.66	. .	6.66	14 10.21	
27	10				5.4				17 5.45	23.16	1.01	.	2	8.44	9 23.23	. .	1.93	17 29.62	
28	10	19.2	34.3						25 5.90	23.12	0.99	.	7	10.35	34 17.60	. .	8.50	25 29.01	
29	9.10		7.8	23.7					26 39.22	23.12	1.00	.	6	8.26	28 12.55	. .	6.86	27 3.34	
30	6.7		53.7	9.2	25.0				28 24.93	23.11	0.99	.	8	4.45	36 19.81	. .	9.04	28 49.03	
31	9						24.5		28 53.17	23.11	1.00	VI.	4	10.23	20 10.99	. .	4.73	29 17.28	
32	10	41.3	56.9						31 28.29	23.10	0.97	III.	9	4.18	41 5.13	. .	10.37	31 52.38	
33	7.8		36.2	51.9	7.4				33 7.48	23.09	1.00	.	6	6.21	27 9.51	. .	6.58	33 31.57	
34	9				49.9				35 49.94	23.08	1.01	.	2	7.34	8 47.93	. .	1.78	36 14.03	
35	8					44.4	0.1	15.8	36 28.86	23.08	1.00	VI.	4	12.16	21 7.97	. .	4.96	36 52.94	
36	7.8			54.2	10.2	25.6			38 10.00	23.07	1.00	V.	6	10.53	29 26.61	. .	7.18	38 34.07	
37	6.7						12.6	28.1	38 41.37	23.07	1.01	VII.	1	13.39	6 52.41	. .	1.28	39 5.45	
38	9		2.2	17.5					42 33.35	23.06	0.99	III.	9	6.13	42 3.12	. .	10.63	42 57.40	
39	9	48.4	3.9						44 35.14	23.05	1.01	III.	3	8.46	14 23.28	. .	3.20	44 57.20	
40	9		49.3	4.9					45 20.55	23.05	1.00	.	5	11.31	25 44.73	. .	6.19	45 44.60	
41	10		11.5						46 42.73	23.04	1.01	III.	3	7.3	13 31.33	. .	2.99	47 6.78	
42	8			50.6	6.2				50 6.20	23.03	1.02	.	1	6.55	3 29.26	. .	0.43	50 30.25	
43	9				44.2	59.9			51 44.16	23.03	0.99	V.	8	9.28	38 42.44	. .	9.69	52 8.18	
44	9.10		3.7	19.2					53 34.87	23.02	1.01	.	3	5.34	12 46.52	. .	2.79	53 58.90	
45	9.10		20.8	36.2					54 52.00	23.02	0.99	.	9	5.36	41 44.53	. .	10.55	55 16.01	
46	9.10	14.2	29.8						56 1.00	23.01	1.01	.	3	12.12	16 7.20	. .	3.67	8 56 25.02	
47	8.9		18.8	34.3					8 59 49.95	23.00	1.02	.	1	7.15	3 39.34	. .	0.47	9 0 13.97	
48	9.10					23.8			9 0 8.22	23.00	1.01	VI.	4	4.01	16 58.37	. .	3.89	0 32.23	
49	8.9		15.8	31.2					1 46.93	23.00	1.01	III.	4	1.26	15 40.40	. .	3.56	2 10.94	
50	9			43.2	58.8				2 58.79	23.00	0.99	.	8	6.8	37 1.66	. .	9.23	3 22.78	
51	9.10	27.9		59.2					8 14.88	22.98	0.99	.	8	3.51	35 52.57	. .	8.94	8 38.85	
52	6.7					50.0			8 34.36	22.98	1.00	V.	6	10.32	29 16.03	. .	7.14	8 58.34	
53	6						52.2		9 20.86	22.98	1.00	VII.	4	5.34	17 44.97	. .	4.09	9 44.84	
54	8.9		1.9	17.6					11 33.18	22.98	1.01	.	4	3.37	16 46.52	. .	3.84	11 57.17	
55	10.9			4.8	20.4				12 20.45	22.97	1.00	.	4	4.22	17 9.21	. .	3.93	12 44.42	
56	7	17.4	32.9						13 4.23	22.97	1.00	III.	6	6.32	27 15.01	. .	6.60	14 28.20	
57	9						29.6		14 58.24	22.97	1.01	VI.	3	5.44	12 51.31	. .	2.81	15 22.22	
58	7	32.2	47.8	3.3	19.2				17 19.10	22.96	0.99	.	8	6.42	37 18.81	. .	9.32	17 43.05	
59	9.10		56.5						22 27.83	22.95	0.99	III.	9	4.32	41 12.20	. .	10.40	22 51.77	
60	9			55.9		26.9			26 11.45	22.95	1.01	.	1	8.28	4 16.15	. .	0.61	26 35.41	
61	10		24.1	39.5					27 55.23	22.94	1.01	III.	3	11.25	15 43.44	. .	3.55	28 19.18	
62	10				17.2				28 17.23	22.94	1.01	.	1	12.1	6 3.55	. .	1.06	28 41.18	
63	10		46.8	2.8					33 18.28	22.93	0.99	III.	7	6.51	32 24.58	. .	7.99	33 42.20	
64	9			15.0	30.5				35 30.59	22.93	1.01	.	3	6.29	13 14.25	. .	2.91	35 54.53	
65	11		9.4	24.7					36 40.47	22.93	1.01	III.	3	6.52	13 25.79	. .	2.95	37 4.41	
66	9	27.3	42.8						9 38 14.09	+22.93	+1.00	.	5	5.31	−22 43.20	. .	−5.38	9 38 38.02	

CORRECTIONS.

Date.		Corr. of Clock.	Hourly rate.	m	n	c
1849.	h.	s.	s.	s.	s.	s.

INSTRUMENT READINGS.

Date.		Barom.	Thermom. At.	Thermom. Ex.
1849.	h. m.	in.	°	°

REMARKS.

(220) 56. Minutes of transit assumed as 14 instead of 13.

ZONE 220. MARCH 16. K. BELT, −36° 53′—Continued.

No.	Mag.	I.	II.	III.	IV.	V.	VI.	VII.	T.	a_1	a_2	MICROMETER.			i	d_1	d_2	Mean Right Ascension, 1850.0.	Mean Declination, 1850.0.
									h. m. s.	s.	s.			r.	′ ″	″	″	h. m. s.	° ′ ″
67	8.9	. .	. .	35.8	51.3	. .	. .	. .	9 38 51.35	+22.92	+0.99	.	7	8.10	−33 4.48	. .	− 8.17	9 39 15.26	
68	8.9	. .	. .	. .	. .	. .	31.3	46.8	39 59.90	22.92	0.99	.	8	7.29	37 42.50	. .	9.43	40 23.81	
69	9.8	. .	. .	33.5	49.1	. .	. .	. .	40 49.11	22.92	0.99	.	7	6.9	32 3.47	. .	7.89	41 12.02	
70	7.8	. .	. .	32.3	47.5	. .	. .	. .	9 41 47.73	+22.92	+1.00	.	6	5.28	−26 42.78	. .	− 6.46	9 42 11.65	

ZONE 221. MARCH 16. K. BELT, −34° 23′. $D_0 = -33° 57′ 40″$.

No.	Mag.	I.	II.	III.	IV.	V.	VI.	VII.	T.	a_1	a_2	MICROMETER.			i	d_1	d_2	Mean Right Ascension, 1850.0.	Mean Declination, 1850.0.
1	9	. .	18.2	33.8	. .	. .	. .	. .	11 31 48.74	+23.22	+1.00	.	6	4.44	−26 20.61	− 1.61	− 6.27	11 32 12.96	−34 24 8.49
2	9	. .	. .	. .	. .	. .	25.0	40.0	31 54.70	23.22	1.01	VII.	3	2.35	11 15.73	1.60	3.26	32 18.93	9 0.59
3	8	. .	. .	. .	32.0	47.2	. .	. .	33 31.89	23.23	0.98	V.	10	9.0	48 28.23	1.84	10.86	33 56.10	46 20.93
4	9	. .	. .	. .	. .	58.9	. .	. .	37 43.69	23.24	0.98	V.	10	9.7	48 31.77	2.43	10.86	38 7.91	46 25.06
5	8	. .	. .	0.7	15.9	31.0	. .	. .	39 15.84	23.25	0.99	.	8	3.53	35 53.58	2.63	8.24	39 40.08	33 44.45
6	7.8	. .	. .	. .	9.9	25.7	40.2	55.6	40 10.13	23.25	1.00	.	6	2.58	25 27.15	2.77	6.09	40 34.38	23 16.01
7	10	44.5	. .	14.8	29.8	. .	. .	. .	44 29.85	23.26	1.01	.	3	3.11	11 34.40	3.35	3.31	44 54.12	9 21.06
8	7	. .	. .	13.8	28.7	44.0	59.1	14.1	45 28.83	23.27	1.00	.	3	12.11	16 6.69	3.49	4.21	45 53.10	13 54.39
9	9	28.9	44.3	59.5	14.5	29.6	44.4	59.8	49 14.43	23.28	1.00	.	5	7.25	23 40.68	4.01	5.74	49 38.71	21 30.43
10	9	44.1	59.3	14.3	29.7	44.6	0.0	14.9	11 51 29.56	23.29	1.00	.	7	3.14	30 35.22	4.31	7.14	11 51 53.85	28 26.67
11	9	. .	. .	. .	28.5	. .	. .	. .	12 4 28.57	23.35	1.00	.	4	5.50	17 53.58	6.04	4.57	12 4 52.92	15 44.19
12	8	. .	22.2	37.2	52.5	7.3	22.8	38.0	13 52.38	23.39	0.99	.	9	7.48	42 51.08	7.27	9.69	14 16.76	40 48.04
13	6	. .	. .	. .	18.7	33.8	49.0	3.9	15 18.57	23.40	0.99	.	8	5.46	36 50.57	7.47	8.44	15 42.96	34 46.48
14	6.7	. .	. .	48.9	3.5	19.3	34.1	49.2	17 3.87	23.41	1.00	.	5	6.54	23 25.06	7.67	5.68	17 28.28	21 18.41
15	8	. .	. .	. .	47.8	. .	. .	. .	20 47.82	23.42	1.01	.	1	4.35	2 18.67	8.15	1.47	21 12.25	0 8.31
16	9	. .	. .	. .	12.1	. .	. .	. .	27 12.13	23.46	1.01	.	1	10.15	5 10.11	8.95	2.06	27 36.60	3 1.12
17	8.7	. .	. .	. .	. .	. .	. .	59.8	30 14.44	23.47	1.00	VII.	5	5.4	22 29.06	9.34	5.49	30 38.91	20 23.89
18	9	25.4	10.4	55.7	11.0	26.0	41.2	56.5	12 35 10.88	+23.50	+0.99	.	8	7.8	−37 31.91	− 9.93	− 8.58	12 35 35.37	−34 35 30.42

ZONE 222. MARCH 19. B. BELT, −41° 15′. $D_0 = -40° 51′ 50″$.

No.	Mag.	I.	II.	III.	IV.	V.	VI.	VII.	T.	a_1	a_2	MICROMETER.			i	d_1	d_2	Mean Right Ascension, 1850.0.	Mean Declination, 1850.0.
1	8	. .	39.	55.3	12.5	. .	. .	. .	8 2 12.25	+21.86	+1.00	IV.	7	3.28	−30 42.27	− 6.55	− 9.07	8 2 35.11	−41 22 47.89
2	10	. .	20.	36.	. .	. .	. .	. .	5 53.00	21.84	0.99	III.	8	3.42	35 47.97	6.87	11.85	6 15.83	27 56.69
3	7	. .	. .	. .	. .	. .	54.	11.	6 20.91	21.84	1.00	VII.	6	8.53	28 25.79	6.91	7.84	6 43.75	20 30.54
4	9	. .	. .	. .	. .	. .	. .	17.3	7 27.38	21.84	1.00	VII.	6	10.7	29 2.81	7.01	8.16	7 50.22	21 7.98
5	9	. .	. .	. .	. .	. .	. .	15.5	8 25.81	21.83	1.01	VII.	4	4.34	17 14.60	7.10	1.90	8 48.65	9 13.60
6	9	. .	15.	32.	49.2	. .	. .	. .	10 48.71	21.82	0.99	IV.	7	8.12	33 5.49	7.31	9.38	11 11.52	25 12.18
7	9	. .	. .	. .	3.4	20.	. .	. .	12 3.42	21.81	1.00	V.	5	5.17	22 36.07	7.43	4.74	12 26.23	14 38.24
8	9	. .	. .	. .	. .	26.	43.	. .	13 9.53	21.80	1.00	VI.	6	10.57	29 28.39	7.53	8.41	13 32.33	21 34.33
9	9	. .	. .	. .	. .	. .	. .	44.	14 53.89	21.79	0.99	VII.	8	12.36	40 16.64	7.69	14.31	15 16.67	32 28.64
10	9	. .	. .	12.5	29.	45.3	. .	. .	16 28.97	21.79	1.00	V.	4	11.48	20 54.03	7.83	3.84	16 51.76	12 55.70
11	8	. .	37.	53.5	10.3	. .	. .	. .	18 10.24	21.78	0.99	IV.	9	3.56	40 54.10	7.98	14.61	18 33.01	33 6.69
12	9	. .	. .	. .	. .	. .	17.	33.5	18 43.74	21.78	1.01	VII.	3	10.55	15 27.71	8.03	0.99	19 6.53	7 26.73
13	6	. .	. .	. .	. .	31.	47.5	4.	19 14.10	21.77	0.98	VII.	10	7.32	47 43.25	8.08	17.94	19 36.85	39 59.27
14	10	. .	. .	. .	. .	. .	12.	29.	21 38.95	21.76	1.00	VII.	5	4.13	22 3.20	8.30	4.45	22 1.71	14 5.95
15	10	. .	30.	47.	. .	. .	. .	. .	23 3.46	21.75	1.00	III.?	6	2.50	25 23.05	8.43	− 6.22	23 26.21	17 27.70
16	6.7	. .	. .	47.	3.5	20.5	. .	. .	25 3.71	21.74	1.01	V.	2	7.38	8 49.87	8.61	+ 2.42	25 26.46	0 46.06
17	9	. .	. .	. .	. .	51.3	8.	. .	26 34.68	21.74	1.00	VI.	6	7.3	27 30.39	8.76	− 7.35	26 57.42	19 36.50
18	6.7	. .	. .	3.	20.	37.	. .	. .	28 20.04	21.73	1.00	VI.	4	5.56	17 56.31	8.92	− 2.27	28 42.77	9 57.50
19	10	. .	. .	59.	15.5	32.	. .	. .	8 30 15.54	+21.72	+1.01	V.	2	9.53	− 9 57.94	− 9.09	+ 1.83	8 30 38.27	−41 1 55.20

CORRECTIONS.

Date.	Corr. of Clock.	Hourly rate.	m	n	c
1849. h.	s.	s.	s.	s.	s.

INSTRUMENT READINGS.

Date.	Barom.	Thermom. At.	Thermom. Ex.
1849. h. m.	in.	°	°

REMARKS.

(222) 13. Minutes of transit assumed as 20 instead of 19.

ZONE 222. MARCH 19. B. BELT, −41° 15′. D_o = −40° 51′ 50″—Continued.

No.	Mag.	SECONDS OF TRANSIT. I.	II.	III.	IV.	V.	VI.	VII.	T.	a_1	a_2	MICROMETER.			i	d_1	d_2	Mean Right Ascension, 1850.0.	Mean Declination, 1850.0.
									h. m. s.	s.	s.			r.	′ ″	″	″	h. m. s.	° ′ ″
20	9.10				23.	40.			8 32 23.09	+21.71	+0.99	V.	9	4.7	−40 59.58	− 9.29	−14.66	8 32 45.79	−41 33 13.53
21	9				30.	47.			34 30.25	21.70	1.00	VI.	4	10.25	20 11.95	9.49	3.47	34 52.95	12 14.91
22	9.10			57.5	14.				36 14.09	21.70	1.00	IV.	5	11.10	25 34.14	9.66	6.31	36 36.79	17 40.11
23	9					37.	54.		37 20.53	21.69	1.00	VI.	6	7.4	27 30.90	9.76	7.35	37 43.22	19 38.01
24	9						17.	34.	38 43.97	21.69	1.00	VII.	4	7.26	18 41.32	9.89	2.68	39 6.66	10 43.89
25	9			43.	59.5				40 59.56	21.68	0.99	IV.	7	6.27	32 12.54	10.11	9.89	41 22.23	24 22.54
26	6						23.	40.	43 49.97	21.67	1.01	VII.	4	7.14	18 35.27	10.38	2.63	44 12.65	10 38.28
27	8				56.				44 56.09	21.66	1.00	IV.	4	8.20	19 9.22	10.48	2.92	45 18.75	11 12.62
28	9				18.5	35.			45 18.51	21.66	1.01	V.	3	8.19	14 9.65	10.52	0.33	45 41.18	6 10.50
29	8		30.	47.					48 3.45	21.65	1.00	III.	5	6.56	23 26.00	10.78	5.18	48 26.10	15 31.96
30	9						46.	2.5	48 12.61	21.65	0.99	VII.	7	7.32	32 44.65	10.80	10.19	48 34.25	24 55.64
31	10					40.			50 23.31	21.65	0.98	V.	9	7.30	42 41.94	11.01	15.50	50 45.94	34 58.45
32	10			24.	41.				53 40.75	21.64	0.98	IV.	10	5.47	46 50.97	11.33	17.55	54 3.37	39 9.85
33	7				24.	40.5	58.		56 24.20	21.63	1.00	VI.	5	9.8	24 32.32	11.59	− 5.75	55 46.83	41 16 39.66
34	8		18.	35.	52.				57 51.60	21.62	1.02	V.	1	11.28	5 46.84	11.73	+ 3.93	58 14.24	40 57 44.66
35	9					42.	59.		8 59 25.52	21.61	0.98	VI.	9	5.54	41 53.30	11.88	−15.11	8 59 48.11	41 34 10.29
36	9							22.	9 0 32.03	21.61	0.99	VII.	7	6.21	32 8.84	11.99	9.86	9 0 54.63	24 20.69
37	9		37.	53.	10.				7 9.98	21.60	0.99	IV.	7	5.58	31 57.91	12.63	9.77	7 31.57	24 10.31
38	8			13.	29.5				8 29.59	21.59	1.00	IV.	5	8.26	24 11.45	12.76	5.58	8 52.18	16 19.79
39	9				49.	5.5			10 48.93	21.59	1.00	V.	6	7.16	27 37.18	12.99	7.40	11 11.52	19 47.57
40	9			10.5	28.	44.5			13 27.70	21.58	1.00	V.	5	4.58	22 26.49	13.24	4.65	13 50.28	14 34.38
41	6.7		9.	26.	43.5				15 42.81	21.58	0.98	IV.	9	4.6	40 59.15	13.46	−14.67	16 5.37	41 33 17.28
42	9					44.	0.5		17 27.31	21.57	1.02	VI.	1	8.3	4 3.25	13.63	+ 4.82	17 49.90	40 56 2.06
43	9					15.	31.5		19 58.26	21.57	0.98	VI.	9	10.19	44 6.93	13.87	−16.20	20 20.81	41 36 27.00
44	8						48.	4.5	20 14.52	21.57	0.98	VII.	9	9.43½	43 48.65	13.90	16.05	20 37.07	36 8.60
45	9		48.	4.5					23 21.20	21.56	1.00	III.	5	6.18	23 6.83	14.20	4.99	23 43.76	15 16.02
46	9				21.				24 20.90	21.56	0.99	IV.	8	7.38	37 47.04	14.30	12.96	24 43.45	30 4.30
47	9					31.	47.5		25 14.28	21.56	1.00	VI.	5	5.12½	22 33.57	14.39	4.72	25 36.84	14 42.68
48	9							6.5	26 16.34	21.55	0.98	VII.	9	8.7	43 0.00	14.49	15.65	26 38.87	35 20.14
49	9.10					16.5	33.3		28 59.95	21.55	1.01	VI.	3	11.31	15 46.23	14.75	1.14	29 22.51	7 52.12
50	10		23.	39.5					30 56.11	21.55	0.99	III.	6	9.22	28 40.71	14.94	7.97	31 18.65	20 53.62
51	9		9.5	26.	43.				33 42.81	21.54	0.99	IV.	6	10.12	29 6.00	15.21	8.19	34 5.34	21 19.40
52	8			46.	2.5				35 2.59	21.54	1.00	IV.	5	11.1	25 29.60	15.34	6.27	35 25.13	17 41.21
53	10							56.5	35 6.51	21.54	0.99	VII.	7	8.53	33 25.49	15.35	−10.55	35 29.04	25 41.39
54	9						13.5	29.5	36 40.03	21.54	1.01	VII.	2	9.57	9 59.36	15.50	+ 1.82	37 2.58	41 2 3.04
55	8			3.2	20.				39 19.91	21.53	1.02	IV.	1	12.59	6 32.80	15.76	+ 3.57	39 42.46	40 58 34.99
56	10			52.	9.				42 8.66	21.53	1.00	IV.	4	11.40	20 50.07	16.04	− 3.80	42 31.39	41 12 59.91
57	10					30.5	47.5		43 14.02	21.53	0.99	VI.	7	8.40	33 19.31	16.14	10.50	43 36.54	25 35.95
58	9	26.	43.	59.5					47 16.02	21.52	1.01	III.	4	2.41½	16 18.46	16.53	1.42	47 38.55	8 26.41
59	9	54.2	11.	27.5	44.				51 44.17	21.52	0.99	IV.	8	5.3	36 28.88	16.96	12.24	52 6.68	28 48.08
60	10				47.	4.			53 17.16	21.52	0.99	V.	7	4.25	31 10.95	17.15	9.34	54 9.67	23 27.44
61	9			9.	25.5				55 25.60	21.52	1.00	IV.	5	3.44	21 49.25	17.31	4.31	55 47.12	14 0.87
62	9				6.5				58 6.40	21.52	0.99	IV.	8	7.8	37 31.91	17.57	12.81	58 28.91	29 52.29
63	8						10.	26.8	9 58 36.75	21.52	0.99	VII.	7	10.45	34 21.97	17.62	11.07	9 58 59.26	26 40.66
64	10					51.5			10 0 34.80	21.52	0.98	V.	9	12.36	45 16.24	17.81	−16.78	10 0 56.30	37 40.83
65	10				18.3	35.			2 18.41	21.52	1.01	V.	2	10.58	10 30.72	17.99	+ 1.56	2 40.94	41 2 37.15
66	7				57.	13.3	30.		3 56.82	21.52	1.02	VI.	1	12.32½	6 19.14	18.15	+ 3.69	4 19.36	40 58 23.60
67	9			54.	10.5	27.3			10.63	21.52	1.00	V.	5	5.13½	22 34.30	. .	− 4.72		
68	5.6					20.	37.	53.5	10 8 3.53	+21.52	+0.99	VII.	7	3.7	−30 31.02	−18.58	− 8.96	10 8 26.04	−41 22 48.56

CORRECTIONS.

Date.		Corr. of Clock.	Hourly rate.	m	n	c
1849.	h.	s.	s.	s.	s.	s.

INSTRUMENT READINGS.

Date.		Barom.	THERMOM. At.	THERMOM. Ex.
1849.	h. m.	in.	°	°

REMARKS.

(222) 33. Minutes of transit assumed as 55 instead of 56.

ZONE 222. MARCH 19. B. BELT, −41° 15′. D₀ = −40° 51′ 50″—Continued.

No.	Mag.	I.	II.	III.	IV.	V.	VI.	VII.	T.	a_1	a_2	Micrometer.		i	d_1	d_2	Mean Right Ascension, 1850.0.	Mean Declination, 1850.0.
									h. m. s.	s.	s.		r.	′ ″	″	″	h. m. s.	° ′ ″
69	6	..	..	26.	42.5	58.7	..	..	10 11 42.45	+21.52	+1.02	V.	1 6.5	− 3 3.98	−18.97	+ 5.32	10 12 4.99	−40 55 7.63
70	5.6	..	..	..	..	47.5	3.5	..	15 30.57	21.52	1.02	VI.	1 3.31	1 46.09	19.39	+ 5.96	15 53.11	40 53 49.52
71	6	..	..	..	..	10.	26.5	43.	16 53.29	21.52	1.00	VII.	4 8.57	19 27.21	19.54	− 3.07	17 15.81	41 11 39.82
72	9	29.5	46.	2.5	..	..	..	..	20 19.26	21.52	1.00	III.	5 10.8	25 2.81	19.91	6.03	20 41.76	17 18.75
73	9	..	..	..	..	24.	40.5	..	21 7.28	21.52	1.00	VI.	5 5.3	22 28.78	20.00	4.68	21 29.50	14 43.46
74	10	..	..	..	59.	16.	..	..	22 59.08	21.53	0.98	VI.	9 6.50	42 21.54	20.19	15.34	23 21.59	34 47.07
75	8	..	..	..	..	36.5	53.	..	24 19.77	21.53	0.99	VI.	7 12.26	35 13.26	20.33	−11.54	24 42.29	41 27 35.13
76	8	..	10.	26.4	43.	..	..	..	27 43.07	21.53	1.02	IV.	1 12.45	6 25.74	20.66	+ 3.64	28 5.62	40 58 32.78
77	9	..	..	14.5	32.	48.5	..	..	31 31.65	21.53	0.99	V.	7 10.12½	34 6.18	21.07	−10.93	31 53.17	41 26 28.18
78	9	..	..	..	..	3.	20.	..	32 46.51	21.53	0.98	VI.	9 11.38½	44 47.01	21.21	16.54	33 8.02	37 14.76
79	9	..	0.5	16.5	33.3	..	..	..	35 33.41	21.54	1.01	IV.	3 9.44	14 52.58	21.49	0.68	35 55.96	7 4.75
80	10	..	..	..	3.	20.	..	..	37 3.22	21.54	1.00	V.	5 7.35	23 45.66	21.64	5.35	37 25.76	16 2.65
81	10	..	46.5	3.3	..	..	..	..	39 19.83	21.54	1.01	III.	4 4.35	17 15.71	21.87	1.91	39 42.38	9 29.48
82	10	..	..	..	..	..	2.5	19.	39 29.16	21.54	1.00	VII.	6 5.35	26 45.65	21.89	6.97	39 51.70	19 4.51
83	10	..	..	..	16.	32.5	..	..	41 15.98	21.55	1.00	V.	5 3.56	21 25.23	22.07	4.10	41 38.53	13 41.40
84	10	..	..	..	..	37.	53.	..	42 20.02	21.55	0.98	VI.	9 6.12	42 2.38	22.17	−15.16	42 42.55	41 34 29.71
85	9	..	..	..	..	41.	57.5	..	44 24.30	21.55	1.02	VI.	2 4.41	7 20.39	22.38	+ 3.18	44 46.87	40 59 29.59
86	8	43.	59.3	16.	..	..	..	..	47 32.66	21.56	1.00	III.	4 11.25	20 42.43	22.69	− 3.73	47 55.22	41 12 58.85
87	9	..	..	..	..	..	21.5	38.	47 48.26	21.56	1.01	VII.	3 5.26	12 41.81	22.72	+ 0.45	48 10.83	4 54.08
88	9	..	..	..	..	45.	1.4	..	49 28.23	21.56	1.00	VI.	5 4.19	22 6.60	22.89	− 4.47	49 50.79	14 23.96
89	8	..	..	..	..	2.	18.3	34.5	50 45.03	21.56	1.00	VII.	5 3.55½	21 54.37	23.01	4.37	51 7.59	14 11.75
90	5	..	..	37.	54.	10.3	..	..	52 53.75	21.57	0.99	V.	7 7.54	32 56.34	23.22	10.28	53 16.31	25 19.84
91	9	..	..	39.	56.	12.4	..	..	54 55.72	21.57	0.98	V.	10 9.58	48 57.46	23.42	−18.59	55 18.27	41 29.47
92	9	43.	0.	16.3	33.	..	..	..	58 32.97	21.58	1.01	IV.	3 7.7	13 33.41	23.78	+ 0.02	58 55.56	5 47.17
93	10	..	..	..	33.	49.5	..	..	10 59 32.91	+21.58	+0.99	V.	7 4.24	−31 10.45	−23.88	− 9.34	10 59 55.48	−41 23 33.67

ZONE 223. MARCH 22. K. BELT, −38° 8′. D₀ = −37° 45′ 10″.

No.	Mag.	I.	II.	III.	IV.	V.	VI.	VII.	T.	a_1	a_2	Micrometer.		i	d_1	d_2	Mean Right Ascension, 1850.0.	Mean Declination, 1850.0.
1	10	..	..	40.2	..	..	..	..	8 14 56.10	+22.06	+1.00	III.	5 5.54	−22 54.73	− 4.33	− 5.32	8 15 19.16	−38 8 14.38
2	10	30.9	46.2	..	..	..	..	..	16 18.21	22.06	1.00	III.	4 4.14	17 5.11	4.57	− 3.47	16 41.27	38 2 23.17
3	7.8	..	..	5.2	21.0	..	..	..	17 21.04	22.06	1.02	.	1 5.58	3 0.51	4.76	+ 0.86	17 44.12	37 48 14.41
4	9.10	..	..	..	..	58.0	13.9	..	17 42.09	22.05	1.01	.	3 5.32	12 45.51	4.83	− 2.12	18 5.15	37 58 2.46
5	8.9	..	52.6	8.5	..	..	..	..	20 24.48	22.04	0.98	.	10 9.26	48 41.40	5.31	13.93	20 47.50	38 34 10.64 *
6	9.10	..	..	..	23.8	..	..	..	22 23.63	22.03	0.99	.	9 8.21	43 7.72	5.67	12.01	22 46.65	28 35.40
7	9.10	..	..	22.2	37.9	..	..	..	23 38.04	22.03	1.00	.	4 12.4	21 2.17	5.89	4.72	24 1.07	6 22.78
8	9	..	38.5	54.5	..	..	..	..	25 10.43	22.02	0.98	.	10 8.2	47 59.05	6.17	13.69	25 33.43	33 28.91
9	7.8	..	..	..	29.2	45.0	..	..	25 29.00	22.02	0.98	.	10 8.26	48 11.15	6.23	12.76	25 52.00	33 40.14
10	9.10	..	..	..	..	..	33.0	49.1	26 1.18	22.02	0.98	.	10 8.29	48 12.66	6.32	12.76	26 24.18	33 41.74
11	6.7	..	..	..	41.0	57.2	12.9	..	27 41.10	22.01	0.99	.	7 11.43	34 51.88	6.63	− 9.24	28 4.10	38 20 17.75
12	9.10	..	..	..	..	..	..	1.2	28 13.89	22.01	1.01	VII.	1 10.11	5 7.47	6.72	+ 0.22	28 36.91	37 50 23.97
13	9	..	..	..	52.9	..	..	..	29 52.91	22.00	1.02	.	1 2.55	1 28.24	7.02	+ 1.33	30 15.93	46 43.93
14	9	..	..	..	..	38.2	53.9	..	30 22.19	22.00	1.01	.	3 2.1	10 59.11	7.11	− 1.57	30 45.20	37 56 17.79
15	8	32.4	48.4	4.4	20.4	..	..	..	37 20.27	21.97	0.99	.	8 4.3	35 58.63	8.36	9.62	37 43.23	38 21 26.61
16	8.9	..	..	..	58.2	14.3	..	..	37 58.29	21.97	1.00	.	6 10.3	29 1.46	8.47	7.31	38 21.26	14 27.24
17	9	..	17.5	3.3	..	..	..	..	40 19.28	21.96	0.99	.	7 7.23	32 40.77	8.89	8.53	40 42.23	18 8.19
18	8	53.1	8.9	..	..	..	..	..	41 40.83	21.96	0.99	I.	8 7.57	37 56.00	9.14	10.29	42 3.78	23 25.43
19	8	..	..	..	..	53.2	9.1	..	41 37.28	21.96	0.99	V.	9 4.7	40 59.59	9.13	11.31	42 0.23	26 30.03
20	9	..	..	34.8	50.2	..	..	..	8 42 50.41	+21.95	+0.99	.	8 10.59	−39 28.40	− 9.36	−10.79	8 43 13.35	−38 24 58.55

CORRECTIONS.

Date.		Corr. of Clock.	Hourly rate.	m	n	c
1849.	h.	s.	s.	s.	s.	s.

INSTRUMENT READINGS.

Date.		Barom.	Thermom. At.	Thermom. Ex.
1849.	h. m.	in.	°	°

REMARKS.

ZONE 223. MARCH 22. K. BELT, −38° 8′. D_o = −37° 45′ 10″—Continued.

No.	Mag.	SECONDS OF TRANSIT. I.	II.	III.	IV.	V.	VI.	VII.	T.	a_1	a_2*	MICROMETER.			i	d_1	d_2*	Mean Right Ascension, 1850.0.	Mean Declination, 1850.0.
									h. m. s.	s.	s.			r.	′ ″	″	″	h. m. s.	° ′ ″
21	9	. . .	. . .	. . .	23.2	. . .	. . .	. . .	8 43 23.15	+21.95	+0.99	.	7	8.5	−33 1.96	− 9.46	− 8.64	8 43 46.09	−38 18 30.06
22	8	. . .	. . .	. . .	. . .	5.8	. . .	. . .	43 49.82	21.94	0.98	V.	10	11.32	49 44.87	9.53	14.29	44 12.74	35 18.69
23	8	. . .	. . .	51.7	7.6	. . .	. . .	. . .	45 7.63	21.94	1.00	.	4	7.51	18 54.59	9.77	4.05	45 30.57	4 18.41
24	7	53.8	9.8	. . .	. . .	. . .	. . .	. . .	46 41.52	21.94	1.00	II.	5	8.28	24 12.18	10.06	5.74	47 4.46	9 37.98
25	10	. . .	53.2	. . .	. . .	. . .	. . .	. . .	48 25.02	21.93	1.00	.	6	8.22	28 10.53	10.37	7.05	48 47.95	13 37.95
26	10	. . .	. . .	. . .	. . .	. . .	. . .	28.0	48 40.69	21.93	0.98	VII.	1	9.24	44 43.77	10.42	12.58	49 3.60	30 16.77
27	9	. . .	3.1	18.9	35.0	. . .	. . .	. . .	51 34.92	21.92	1.00	.	5	8.55	24 26.07	10.95	5.82	51 57.84	9 52.84
28	9	. . .	. . .	. . .	0.2	15.6	. . .	. . .	55 59.91	21.91	0.99	V.	7	11.30	34 45.26	11.76	9.22	56 22.81	20 16.24
29	10	. . .	. . .	. . .	. . .	3.9	. . .	. . .	55 48.00	21.91	1.00	.	6	10.53	29 26.67	11.72	7.46	56 10.91	14 55.65
30	9	. . .	. . .	41.7	57.3	. . .	. . .	. . .	56 57.48	21.91	1.01	.	3	11.23	15 42.49	11.94	3.03	57 20.40	1 7.46
31	9	. . .	. . .	. . .	. . .	. . .	. . .	48.3	57 0.78	21.91	1.01	VII.	4	6.1	17 58.51	11.95	3.77	57 23.70	3 24.23
32	10	. . .	. . .	. . .	. . .	. . .	. . .	26.4	57 38.73	21.91	1.00	VII.	6	5.28	26 42.16	12.06	6.56	58 1.64	12 10.73
33	10	. . .	. . .	. . .	. . .	50.9	. . .	. . .	8 59 35.05	21.90	1.01	.	4	5.6	17 31.40	12.43	3.61	8 59 57.96	38 2 57.44
34	10	. . .	. . .	. . .	. . .	44.5	. . .	. . .	9 0 28.69	21.90	1.01	.	3	3.55	11 56.59	12.59	1.87	9 0 51.60	37 57 21.05
35	9	. . .	. . .	. . .	. . .	20.8	36.4	. . .	1 4.74	21.89	1.00	V.	6	7.8	27 33.14	12.70	6.84	1 27.63	38 13 2.68
36	9	. . .	. . .	8.2	23.4	. . .	. . .	. . .	2 23.77	21.89	1.00	.	5	10.33	25 15.49	12.94	6.08	2 46.66	10 44.51
37	9	7.8	. . .	39.6	. . .	. . .	. . .	. . .	3 55.46	21.89	1.00	.	5	5.39	22 47.24	13.22	5.29	4 18.35	8 15.75
38	9	. . .	. . .	58.9	14.8	. . .	. . .	. . .	6 14.78	21.88	0.99	.	7	10.10	34 4.99	13.65	8.99	6 37.65	38 19 37.63
39	8	. . .	. . .	52.4	8.4	. . .	. . .	. . .	7 8.37	21.88	1.01	.	3	8.43	14 21.82	13.81	2.60	7 31.26	37 59 48.23
40	10	. . .	. . .	47.2	2.8	. . .	. . .	. . .	8 2.96	21.88	1.01	.	2	8.9	9 5.58	13.98	0.98	8 25.85	54 30.54
41	5	. . .	47.5	3.2	19.2	. . .	. . .	. . .	9 19.19	21.87	1.01	.	3	2.47	11 22.31	14.21	1.68	9 42.07	37 56 48.20
42	10	. . .	. . .	. . .	9.0	. . .	. . .	. . .	10 9.08	21.87	1.01	.	4	3.28	16 41.97	14.36	3.35	10 31.96	38 2 9.68
43	8	. . .	42.5	58.1	. . .	. . .	. . .	. . .	12 14.20	21.87	0.99	.	8	10.40	39 18.82	14.75	10.73	12 37.06	24 54.30
44	10	. . .	. . .	. . .	58.5	. . .	. . .	. . .	12 58.50	21.86	1.00	.	6	8.59	28 29.19	14.88	− 7.14	13 21.36	38 14 1.21
45	10	. . .	39.8	55.3	. . .	. . .	. . .	. . .	16 11.35	21.86	1.02	.	1	8.5	4 4.56	15.46	+ 0.55	16 34.23	37 49 29.47
46	10	. . .	. . .	56.6	11.8	. . .	. . .	. . .	17 12.16	21.86	1.01	.	2	9.10	9 36.33	15.65	− 1.14	17 35.03	37 55 3.12
47	9	. . .	13.5	29.1	. . .	. . .	. . .	. . .	22 45.19	21.84	0.99	.	8	3.24	35 38.96	16.65	9.51	23 8.02	38 21 15.12
48	10	. . .	. . .	. . .	48.9	. . .	. . .	. . .	24 48.87	21.84	0.99	.	7	3.58	30 57.40	17.01	7.96	25 11.70	16 32.37
49	9	. . .	. . .	19.1	34.8	. . .	. . .	. . .	25 34.90	21.84	0.99	.	7	4.12	31 4.47	17.16	8.00	25 57.73	16 39.63
50	8	. . .	10.5	26.6	. . .	. . .	. . .	. . .	26 42.46	21.84	0.99	.	9	6.57	42 25.37	17.35	11.79	27 5.29	38 28 4.51
51	9	. . .	. . .	. . .	. . .	. . .	23.4	39.1	26 51.58	21.84	1.01	VII.	3	6.10	13 4.05	17.37	2.21	27 14.43	37 58 33.63
52	10	. . .	24.9	40.8	. . .	. . .	. . .	. . .	28 56.72	21.83	1.00	.	6	8.54	28 26.67	17.75	− 7.14	29 19.55	38 14 1.56
53	10	. . .	. . .	48.6	4.3	. . .	. . .	. . .	35 4.48	21.83	1.02	.	1	7.8	3 35.81	18.82	+ 0.72	35 27.33	37 49 3.91
54	10	. . .	. . .	. . .	. . .	. . .	54.3	. . .	35 22.50	21.83	0.99	.	8	7.26	37 40.99	18.87	−10.22	35 45.32	38 23 20.08
55	10	. . .	. . .	32.8	. . .	. . .	. . .	. . .	37 48.70	21.82	1.00	.	5	3.30	21 42.19	19.30	4.93	38 11.52	38 7 16.42
56	10	. . .	. . .	. . .	30.0	46.0	. . .	. . .	38 30.13	21.82	1.01	.	2	6.49	8 25.24	19.42	0.76	38 52.96	37 53 55.42
57	9	37.9	54.1	. . .	. . .	. . .	. . .	. . .	43 25.77	21.81	1.00	.	6	11.14	29 37.26	20.28	7.51	43 48.58	38 15 15.05
58	9	. . .	. . .	58.0	13.8	. . .	. . .	. . .	44 13.88	21.81	1.00	.	5	5.43	22 49.25	20.42	− 5.29	44 36.69	38 8 24.96
59	9	. . .	. . .	19.5	35.3	. . .	. . .	. . .	49 35.33	21.81	1.02	.	1	4.33	2 17.66	21.37	+ 1.11	49 58.16	37 47 47.92
60	9	20.8	36.1	52.3	. . .	. . .	. . .	. . .	55 8.20	21.80	1.00	.	6	4.36	26 16.58	22.33	− 6.43	55 31.00	38 11 55.34
61	10	. . .	. . .	. . .	. . .	53.0	. . .	. . .	55 37.06	21.80	0.99	.	9	2.57	40 24.35	22.42	11.11	55 59.85	26 7.88
62	7	. . .	. . .	46.1	2.3	18.3	. . .	. . .	57 2.16	21.80	0.98	.	10	7.12	47 33.83	22.67	13.57	57 24.94	33 20.07
63	10	. . .	. . .	. . .	12.5	. . .	. . .	. . .	9 58 12.40	21.80	0.99	.	8	7.35	37 45.53	22.88	10.24	9 58 35.10	23 28.65
64	7	6.0	22.3	37.9	. . .	. . .	. . .	. . .	10 3 53.85	21.80	1.00	.	5	9.38	24 47.75	23.86	5.94	10 4 16.65	10 27.55
65	9	. . .	. . .	21.0	. . .	. . .	. . .	. . .	4 36.93	21.80	0.99	.	7	11.31	34 45.83	23.99	9.22	4 59.72	20 29.04
66	9	. . .	. . .	. . .	49.2	. . .	. . .	. . .	8 48.99	21.80	0.98	.	10	5.57	46 56.01	24.71	13.36	9 11.77	32 44.08
67	10	. . .	4.8	20.7	. . .	. . .	. . .	. . .	15 36.62	21.80	1.00	.	6	10.22	29 11.04	25.88	− 7.38	15 59.42	38 14 54.30
68	9	. . .	. . .	11.5	27.6	. . .	. . .	. . .	17 27.48	21.80	1.02	.	1	7.20	3 41.86	26.20	+ 0.68	17 50.30	37 49 17.38
69	10	. . .	. . .	8.9	. . .	. . .	. . .	. . .	10 18 24.77	+21.80	+1.01	.	2	6.16	− 9 9.11	−26.36	− 0.99	10 18 47.58	−37 54 46.48

CORRECTIONS.

Date.		Corr. of Clock.	Hourly rate.	m	n	c
1849.	h.	s.	s.	s.	s.	s.

INSTRUMENT READINGS.

Date.		Barom.	THERMOM. At.	Ex.
1849.	h. m.	in.	°	°

REMARKS.

ZONE 223. MARCH 22. K. BELT, −38° 8′. D₀=−37° 45′ 10″—Continued.

No.	Mag.	I.	II.	III.	IV.	V.	VI.	VII.	T.	a_1	a_2	MICROMETER.			i	d_1	d_2	Mean Right Ascension, 1850.0.	Mean Declination, 1850.0.
									h. m. s.	s.	s.			r.	′ ″	″	″	h. m. s.	° ′ ″
70	10	..	..	..	..	..	3.0	..	10 18 58.09	+21.80	+1.01	VI.	2	8.15	− 9 8.33	−26.46	− 0.99	10 19 20.90	−37 54 45.78
71	7	..	..	..	..	44.2	..	..	19 28.22	21.80	0.98	V.	10	12.19	50 8.56	26.55	14.46	19 51.00	38 35 59.57
72	9	..	..	33.2	49.0	..	..	..	20 49.06	21.80	1.01	.	2	11.2	10 32.81	26.78	1.42	21 11.87	37 56 11.01
73	9	..	..	25.3	41.3	..	..	..	21 41.28	21.80	1.00	.	4	13.27	21 44.01	26.93	4.94	22 4.08	38 7 25.88
74	9	..	..	..	38.9	..	..	..	22 38.87	21.80	1.00	.	6	13.9	30 35.24	27.09	7.84	23 1.67	16 20.17
75	9	..	..	..	33.2	..	..	..	23 33.16	21.80	0.99	.	7	6.2	31 59.94	27.25	8.30	23 55.95	38 17 45.49
76	9	..	..	..	..	39.8	..	..	25 23.98	21.81	1.01	V.	3	5.19	12 38.88	27.56	2.07	25 46.80	37 58 18.51
77	10	..	..	..	..	..	37.7	..	26 5.82	21.81	1.01	.	3	10.36	15 18.81	27.68	2.90	26 28.64	38 0 59.39
78	11	..	..	..	..	..	42.8	..	32 11.03	21.81	0.99	VII.	9	8.1	42 57.02	28.72	11.99	32 33.83	28 47.73
79	8	39.3	55.3	11.2	..	..	..	..	36 27.09	21.81	1.00	.	6	13.3	30 32.21	29.44	− 7.82	36 49.90	38 16 19.47
80	9	..	..	..	56.9	..	..	..	40 56.93	21.82	1.02	.	1	9.55	5 1.02	30.20	+ 0.29	41 19.77	37 50 40.93
81	7.8	..	..	52.8	9.0	..	..	..	42 8.82	21.82	0.99	.	8	5.44	36 49.56	30.40	− 9.91	42 31.63	38 22 39.87
82	9	..	..	35.7	51.8	..	..	..	44 51.66	21.83	0.99	.	8	12.23	40 10.74	30.85	11.05	45 14.48	26 2.64
83	8	..	48.2	3.8	20.2	..	..	..	46 19.97	21.83	0.99	.	7	9.16	33 37.76	31.09	8.83	46 42.79	38 19 27.68
84	8	..	..	..	19.1	35.3	..	..	47 19.32	21.83	1.01	.	2	13.14	11 39.36	31.26	1.76	47 42.16	37 57 22.38
85	10	..	6.6	22.2	..	..	..	..	49 38.21	21.83	1.01	.	3	3.0	11 28.86	31.65	1.71	50 1.05	57 12.22
86	9	..	16.9	32.6	..	..	..	..	50 48.56	21.83	1.01	.	2	7.56	8 59.02	31.84	0.94	51 11.40	37 54 41.80
87	10	..	28.9	..	..	..	..	..	54 0.75	21.84	0.99	III.	8	7.50	37 53.02	32.37	10.29	54 23.58	38 23 45.68
88	10	..	..	..	..	30.9	..	..	55 14.96	21.84	0.99	V.	8	10.00	38 58.58	32.57	10.66	55 37.79	24 51.81
89	10	..	..	..	..	..	40.7	..	10 56 8.87	21.85	1.00	VI.	6	6.12	27 4.71	32.73	6.69	10 56 31.72	12 54.13
90	10	..	..	10.2	26.8	..	..	..	11 2 26.42	21.86	0.99	.	8	5.55	36 55.10	33.76	9.96	11 2 49.27	22 48.82
91	10	..	..	..	57.0	..	..	..	3 56.69	21.86	0.99	.	8	8.33	38 14.78	34.02	−10.42	4 19.74	38 24 9.22
92	9	..	..	..	..	..	51.2	..	4 19.27	21.86	1.01	VI.	1	10.23	5 13.87	34.08	+ 0.23	4 42.14	37 50 57.72
93	10	..	..	..	6.1	..	..	..	6 6.15	21.87	1.01	.	2	11.53	10 58.52	34.37	− 1.54	6 29.03	37 56 44.43
94	10	..	..	16.7	..	..	..	..	7 32.65	21.87	0.98	.	9	11.49	44 52.61	34.61	12.68	7 55.50	38 30 49.90
95	8	..	..	5.0	21.0	..	..	..	8 20.88	21.87	0.98	.	9	11.44	44 50.09	34.74	12.66	8 43.73	30 47.49
96	8	..	..	30.9	46.9	..	..	..	11 14 46.84	+21.89	+0.99	.	7	4.18	−31 7.49	−35.78	− 8.01	11 15 9.72	−38 17 1.28

ZONE 224. MARCH 22. K. BELT, −35° 1′. D₀=−34° 37′ 0″.

No.	Mag.	I.	II.	III.	IV.	V.	VI.	VII.	T.	a_1	a_2	MICROMETER.			i	d_1	d_2	Mean Right Ascension, 1850.0.	Mean Declination, 1850.0.
1	9	..	..	29.4	44.7	..	..	..	12 34 44.67	+22.41	+0.99	.	7	9.56	−33 57.93	−14.52	− 8.03	12 35 8.07	−35 11 20.48
2	9	..	29.9	45.5	..	..	..	..	35 0.66	22.41	0.99	.	10	5.8	46 31.30	14.56	10.94	35 24.06	35 23 56.80
3	9	..	4.8	19.9	..	..	..	..	39 35.23	22.43	1.01	.	3	11.52	15 57.12	15.23	4.00	39 58.67	34 53 16.35
4	9	..	..	..	..	29.7	..	..	42 15.40	22.45	0.99	V.	9	2.0	39 55.55	15.61	9.40	42 38.84	35 17 20.56
5	10	..	..	..	..	..	..	24.9	42 39.12	22.45	1.00	VI.	6	7.22	27 40.03	15.64	6.62	43 2.57	5 2.29
6	10	..	..	..	..	..	12.7	..	43 42.14	22.46	1.00	VI.	6	11.32	29 46.10	15.81	7.09	44 5.60	7 9.00
7	9	..	..	..	..	..	..	5.9	44 20.26	22.46	1.00	VII.	4	7.10	18 33.38	15.89	4.56	44 43.72	5 53.83
8	10	..	..	..	19.9	..	..	..	47 19.74	22.48	0.99	.	9	8.55	43 24.87	16.33	10.21	47 43.21	35 20 51.41
9	9	..	..	..	..	13.4	..	..	47 58.22	22.49	1.01	V.	3	3.58	11 58.04	16.39	3.12	48 21.72	34 49 17.55
10	10	..	..	51.7	..	..	..	..	50 6.99	22.50	0.99	.	7	10.27	34 13.56	16.69	8.10	50 30.48	35 11 38.35
11	9	..	..	40.8	56.2	..	..	..	50 56.16	22.50	1.01	.	3	6.7	13 3.16	16.80	3.35	51 19.67	34 50 23.31
12	8	..	..	..	..	38.8	54.4	..	51 24.19	22.51	0.99	.	9	11.44	44 50.09	16.84	10.56	51 47.69	35 22 17.49
13	10	..	..	..	..	..	..	39.7	51 53.96	22.51	1.00	.	5	12.28	26 13.46	16.93	6.28	52 17.47	3 36.67
14	9	..	..	..	..	26.9	42.2	..	53 12.14	22.52	0.99	.	9	4.12	41 2.17	17.12	9.66	53 35.65	18 28.95
15	10	..	..	..	..	32.0	..	..	54 17.67	22.52	0.99	V.	10	8.2	47 58.99	17.28	11.29	54 41.18	25 27.56
16	5	..	..	55.2	10.8	25.3	..	..	58 10.46	22.55	1.00	.	5	11.56	25 57.34	17.82	6.22	58 34.01	35 3 21.38
17	10	..	..	11.8	27.3	..	..	..	59 27.18	22.56	1.01	.	1	9.12	4 38.34	18.00	1.50	12 59 50.75	34 41 57.84
18	7	..	..	..	..	..	9.5	24.6	12 59 38.77	+22.56	+0.99	VI.	10	7.57	−47 56.28	−18.01	−11.28	13 0 2.32	−35 25 25.57

CORRECTIONS.

Date.	Corr. of Clock.	Hourly rate.	m	n	c
1849. h.	s.	s.	s.	s.	s.

INSTRUMENT READINGS.

Date.	Barom.	THERMOM. At.	THERMOM. Ex.
1849. h. m.	in.	°	°

REMARKS.

ZONE 224. MARCH 22. K. BELT, −35° 1′. D₀ = −34° 37′ 0″—Continued.

No.	Mag.	SECONDS OF TRANSIT. I.	II.	III.	IV.	V.	VI.	VII.	T.	a_1	a_4	MICROMETER.			i	d_1	d_2	Mean Right Ascension, 1850.0.	Mean Declination, 1850.0.
									h. m. s.	s.	s.			r.	′ ″	″	″	h. m. s.	° ′ ″
19	8				58.8	14.3			13 0 59.30	+22.56	+0.99	.	9	10.20	−44 7.73	−18.20	−10.39	13 1 22.85	−35 21 36.32
20	10				58.2				1 58.26	22.57	1.01	.	3	6.7	13 3.16	18.36	3.35	2 21.84	34 50 34.87
21	9			44.5					2 59.75	22.58	1.01	.	3	3.28	11 42.97	18.52	3.05	3 23.34	34 49 4.54
22	9	10.3	25.6						4 56.16	22.59	1.00	II.	7	5.3	31 29.94	18.81	7.49	5 19.75	35 8 56.24
23	9					53.9			5 38.79	22.59	1.01	V.	1	2.33	1 17.09	18.90	0.77	6 2.39	34 38 36.76
24	8			42.8	58.2				12 58.07	22.64	0.99	.	9	6.10	42 1.67	19.99	9.89	13 21.60	35 19 31.54
25	10					37.2			13 21.98	22.64	1.00	V.	4	4.34	17 15.21	20.04	4.27	13 45.62	34 54 39.52
26	10				10.8				15 10.84	22.65	1.01	.	2	7.49	8 55.49	20.32	2.44	15 34.50	34 46 18.25
27	10				59.3				15 59.37	22.66	1.00	.	5	5.53	22 54.29	20.35	5.52	16 23.03	35 0 20.16
28	10		36.2	52.2					18 7.08	22.67	1.00	.	3	5.47	12 51.05	20.75	3.32	18 30.75	34 50 15.12
29	9			46.8	1.8				19 1.96	22.68	1.00	.	3	11.58	16 0.14	20.88	4.00	19 25.64	53 25.02
30	10				36.8				19 36.68	22.68	1.00	.	4	7.19	18 38.46	20.97	4.57	20 0.56	56 4.00
31	9				11.1	26.1			20 11.01	22.69	1.00	.	5	4.50	22 22.53	21.05	5.41	20 34.70	59 48.49
32	9						58.2	13.5	20 27.76	22.69	1.01	.	2	10.9	10 6.09	21.09	2.70	20 51.46	47 29.88
33	9				14.8				26 14.87	22.73	1.00	.	3	10.34	15 17.80	21.89	3.54	26 38.60	52 43.53
34	10					36.4			28 21.17	22.74	1.00	.	4	7.46	18 52.08	22.20	4.64	28 44.91	56 18.92
35	9					41.2			29 26.05	22.75	1.01	.	2	4.14	7 7.08	22.37	2.03	29 49.81	34 44 31.48
36	10		1.2						33 31.78	22.78	0.99	.	8	10.12	39 4.70	22.91	9.23	33 55.55	35 16 36.84
37	10		10.6						35 41.11	22.79	1.00	.	3	7.39	13 49.55	23.21	3.51	36 4.90	34 51 16.27
38	8.7		52.8	8.3	23.6				36 23.51	22.80	1.00	.	7	7.19	32 38.76	23.30	7.75	36 47.31	35 10 9.81
39	6	8.2	23.8	39.0					39 54.16	22.83	1.00	.	4	8.48	19 23.34	23.79	4.74	40 17.99	34 56 51.85
40	10				26.8				40 26.86	22.83	1.00	.	4	3.47	16 51.56	23.87	4.18	40 50.69	34 54 19.61
41	10			2.4					42 17.70	22.84	0.99	.	8	6.21	37 8.21	24.11	8.78	42 41.53	35 14 41.10
42	6.7	38.2	53.5						44 23.95	22.86	1.00	.	4	5.42	17 49.55	24.36	4.39	44 47.81	34 55 18.30
43	9				23.6				44 23.66	22.86	1.00	.	4	3.34	16 45.01	24.36	4.15	44 47.52	54 13.52
44	9				11.0				45 11.02	22.86	1.01	.	1	3.56	1 59.00	24.49	0.92	45 34.89	34 39 24.41
45	9			9.6					13 46 24.89	+22.87	+1.00	.	6	8.11	−28 4.98	−24.65	−6.71	13 46 48.76	−35 5 36.34

ZONE 225. MARCH 23. B. BELT, −31° 53′. D₀ = −31° 28′ 20″.

No.	Mag.	I.	II.	III.	IV.	V.	VI.	VII.	T.	a_1	a_4	MICROMETER.			i	d_1	d_2	Mean Right Ascension, 1850.0.	Mean Declination, 1850.0.
1	9						6.6	15.	8 21 31.00	+23.54	+1.00	VII.	5	10.3	−24 59.88	−1.24	−6.00	8 21 55.54	−31 53 27.12
2	9						28.	42.4	23 58.43	23.52	1.00	VII.	5	3.48	21 50.78	1.55	5.50	24 22.95	31 50 17.83
3	9	46.	1.	15.5					25 30.32	23.52	0.99	IV.	8	4.29	36 7.20	1.75	7.73	25 54.83	32 4 36.68
4	9						24.	38.5	25 54.40	23.52	0.99	VII.	8	3.6	35 29.41	1.80	7.63	26 18.91	3 58.84
5	7		37.	52.					28 6.60	23.51	0.99	IV.	8	4.48	36 21.32	2.09	7.75	28 30.10	4 51.16
6	9						54.3	9.	28 24.82	23.51	1.00	VII.	7	6.52½	32 24.92	2.13	7.16	28 49.33	32 0 54.21
7	9							59.5	29 15.54	23.50	1.01	VII.	3	6.51	13 24.86	2.23	4.25	29 40.05	31 41 51.34
8	10						40.	55.	31 10.69	23.50	1.00	VII.	6	5.35	26 45.84	2.48	6.29	31 35.19	55 14.61
9	9		10.	24.5					33 39.32	23.49	1.01	IV.	4	2.26	16 10.72	2.81	4.66	34 3.82	44 38.19
10	8						18.	33.	33 48.75	23.49	1.01	VII.	2	9.27½	9 44.67	2.82	3.70	34 13.25	38 11.19
11	8				43.	58.	12.5		35 43.11	23.48	1.01	VI.	3	6.50	13 24.63	3.07	4.25	36 7.60	31 41 51.95
12	10			1.	16.				38 15.83	23.47	0.99	IV.	8	5.59½	36 57.37	3.40	7.86	38 40.29	32 5 28.63
13	8		35.	49.5	4.				40 4.20	23.46	0.99	IV.	9	4.2	40 57.13	3.64	8.47	40 28.65	9 29.24
14	9				37.	51.5			41 37.03	23.46	0.99	V.	8	6.29	37 12.20	3.84	7.90	42 1.48	5 43.94
15	8						40.	55.	42 10.65	23.46	0.99	VII.	8	4.42	36 17.82	3.91	7.76	42 35.10	32 4 49.49
16	9				33.				44 33.02	23.45	1.02	IV.	1	4.49	2 25.73	4.23	2.59	44 57.49	31 30 52.55
17	7			37.	54.				45 53.87	23.44	1.00	IV.	6	8.26	28 12.55	4.40	6.49	46 18.31	31 56 43.44
18	9							1.8	8 46 17.60	+23.44	+1.00	VII.	7	5.28	−31 42.30	−4.45	−7.05	8 46 42.04	−32 0 13.80

CORRECTIONS.

Date.		Corr. of Clock.	Hourly rate.	m	n	c
1849.	h.	s.	s.	s.	s.	s.

INSTRUMENT READINGS.

	Date.		Barom.	THERMOM. At.	Ex.
	1849.	h. m.	in.	°	°

REMARKS.

(225) 1. Time of transit over T. VI assumed as 0ˢ.6 instead of 6ˢ.6.
(225) 17. Time of transit over T. III assumed as 39ˢ instead of 35ˢ.

ZONE 225. MARCH 23. B. BELT, −31° 53'. D_0=−31° 28' 20"—Continued.

No.	Mag.	SECONDS OF TRANSIT. I.	II.	III.	IV.	V.	VI.	VII.	T.	a_1	a_2	MICROMETER.			i	d_1	d_2	Mean Right Ascension, 1850.0.	Mean Declination, 1850.0.
									h. m. s.	s.	s.			r.	' "	"	"	h. m. s	° ' "
19	7.8	. .	. .	. .	. .	. .	. .	21.	8 47 37.12	+23.43	+1.01	VII.	2	3.47	− 6 52.98	− 4.63	−3.26	8 48 1.56	−31 35 20.87
20	9	. .	. .	. .	. .	. .	37.5	52.	49 8.01	23.43	1.01	VII.	2	6.14	8 7.11	4.83	3.45	49 32.45	36 35.39
21	9	. .	. .	. .	43.	57.7	. .	. .	43.05	23.43	1.01	V.	3	7.19	13 39.41	. .	4.27		
22	10	. .	. .	. .	. .	. .	. .	9.5	52 25.40	23.42	1.00	VII.	5	8.11	24 3.40	5.26	5.84	52 49.82	52 34.50
23	9	. .	. .	. .	. .	. .	22.	. .	53 52.49	23.42	1.00	VI.	4	11.55	20 57.42	5.45	5.37	54 16.91	49 28.24
24	9	. .	. .	37.	53.3	. .	. .	. .	55 52.54	23.41	1.01	IV.	3	7.18	13 38.95	5.71	4.28	56 16.96	42 8.94
25	7.8	. .	. .	. .	. .	. .	35.	49.8	56 5.61	23.41	1.00	VII.	5	5.4	22 29.11	5.74	5.60	56 30.02	51 0.45
26	8	. .	. .	. .	. .	7.4	22.3	37.	57 52.80	23.40	1.00	VII.	5	5.33	22 43.73	5.97	5.64	58 17.20	31 51 15.34
27	10	. .	. .	. .	. .	. .	52.4	7.	8 59 22.81	23.40	0.99	VII.	10	3.8	45 30.31	6.17	9.20	8 59 47.20	32 14 5.66
28	8	. .	. .	. .	11.5	26.	. .	. .	9 1 11.39	23.40	1.00	V.	6	6.56	27 27.12	6.40	6.39	9 1 35.79	31 55 59.91
29	9	. .	. .	. .	13.5	28.2	. .	. .	2 13.52	23.39	1.00	V.	5	4.59	22 27.01	6.54	5.60	2 37.91	31 50 59.15
30	10	. .	. .	. .	. .	31.	46.	. .	3 16.40	23.39	0.99	VI.	7	7.59	32 58.72	6.68	7.24	3 40.78	32 1 32.64
31	10	. .	. .	16.5	31.0	. .	. .	. .	6 31.13	23.38	1.01	IV.	3	6.41	13 20.30	7.10	4.23	6 55.52	31 41 51.63
32	7.8	. .	. .	. .	40.8	55.4	. .	. .	7 40.67	23.38	1.00	V.	6	7.17	27 37.69	7.26	6.42	8 5.05	56 11.37
33	9	. .	. .	. .	8.2	22.8	. .	. .	9 8.21	23.37	1.01	VI.	2	5.54	7 57.29	7.45	3.42	9 32.59	31 36 28.16
34	9	. .	. .	27.	42.	. .	. .	. .	10 41.85	23.37	1.00	IV.	7	5.56	31 56.91	7.65	7.09	11 6.22	32 0 31.65
35	9	. .	. .	. .	. .	. .	4.5	19.	11 34.93	23.37	1.00	VII.	6	7.38	27 47.86	7.77	6.45	11 59.30	31 56 22.08
36	9	. .	. .	5.5	20.3	. .	. .	. .	13 20.24	23.36	0.99	IV.	7	10.34½	34 17.35	8.00	7.44	13 44.59	32 2 52.79
37	9	. .	. .	. .	. .	. .	. .	14.3	13 30.06	23.36	0.99	VII.	7	12.18	35 9.04	8.02	7.57	13 54.41	32 3 44.63
38	9	. .	. .	35.	49.8	. .	. .	. .	15 49.76	23.36	1.00	IV.	6	11.36½	29 48.61	8.32	6.75	16 14.12	31 58 23.68
39	9	. .	52.5	7.2	22.	. .	. .	. .	17 21.93	23.36	0.99	IV.	9	8.53½	43 24.11	8.52	8.87	17 46.28	32 12 1.50
40	9	. .	. .	31.	46.	. .	. .	. .	18 45.87	23.35	1.01	IV.	2	6.34	8 17.68	8.71	3.48	19 10.23	31 36 49.87
41	8	. .	. .	. .	. .	. .	50.	. .	19 20.48	23.35	1.01	VI.	4	7.57	18 57.41	8.75	5.07	19 44.84	47 31.26
42	10	. .	. .	. .	36.4	51.	. .	. .	20 36.33	23.35	1.00	VI.	6	11.15	29 37.55	8.95	6.72	21 0.68	58 13.22
43	10	. .	. .	. .	59.5	. .	. .	. .	22 59.57	23.34	1.00	IV.	4	6.11	18 4.17	9.26	4.93	23 23.91	46 38.36
44	10	. .	. .	49.4	4.	. .	. .	. .	24 4.07	23.34	1.00	IV.	6	4.30	26 13.55	9.40	6.19	24 28.41	54 49.14
45	9	. .	. .	. .	53.3	8.	. .	. .	24 53.27	23.34	1.00	IV.	7	3.42	30 49.34	9.50	6.91	25 17.61	59 25.75
46	10	. .	. .	11.	25.5	. .	. .	. .	26 25.61	23.34	1.01	IV.	2	2.58	6 28.75	9.70	3.20	26 49.96	35 1.65
47	10	. .	. .	17.	31.7	. .	. .	. .	27 31.72	23.34	1.00	IV.	6	3.27	25 41.77	9.85	6.11	27 56.06	54 17.73
48	9	. .	. .	. .	24.3	39.	. .	. .	28 24.35	23.33	1.00	V.	4	3.52½	16 54.28	9.96	4.76	28 48.68	45 29.00
49	9	. .	. .	. .	47.	2.	. .	. .	29 47.21	23.33	1.01	V.	1	11.48	5 56.95	10.14	3.11	30 11.55	34 30.20
50	9	. .	. .	43.3	58.	. .	. .	. .	31 58.02	23.33	1.01	IV.	2	10.59	10 31.30	10.43	3.80	32 22.36	39 5.53
51	9	. .	. .	. .	5.	19.5	. .	. .	33 4.95	23.32	1.01	V.	2	9.34	9 48.39	10.57	3.69	33 29.26	38 22.65
52	9	. .	. .	. .	. .	. .	23.	. .	33 53.45	23.32	1.01	VI.	3	5.55	12 56.89	10.68	4.15	34 17.78	41 31.72
53	7.8	. .	. .	. .	. .	11.	26.	. .	33 56.39	23.32	1.01	VI.	1	4.11	2 6.35	10.68	2.54	34 21.72	31 30 39.57
54	9	. .	. .	. .	. .	. .	34.3	49.	36 4.77	23.32	0.99	VII.	10	2.7	44 59.56	10.96	9.13	36 29.05	32 13 39.65
55	8	. .	3.	17.5	32.	. .	. .	. .	38 32.22	23.31	1.00	IV.	7	4.1	30 58.92	11.28	6.94	38 56.53	31 59 37.14
56	10	. .	. .	. .	. .	. .	33.	48.	[illegible] 3.67	23.31	1.00	VII.	7	3.3½	30 29.41	11.35	6.86	39 27.98	59 7.65
57	9	. .	. .	. .	. .	43.	57.5	12.	40 28.14	23.31	1.01	VII.	1	10.26	5 15.18	11.53	3.01	40 52.46	33 49.72
58	10	. .	. .	. .	. .	. .	10.	25.	41 40.68	23.31	1.00	VII.	6	11.48½	29 54.17	11.68	6.76	42 4.99	38 32.61
59	9	. .	42.3	57.	12.	. .	. .	. .	44 11.85	23.31	1.00	IV.	4	11.44½	20 52.35	12.01	5.36	44 36.16	49 29.71
60	9	. .	. .	. .	4.	19.	. .	. .	45 4.19	23.31	1.01	V.	3	10.15	15 8.16	12.12	4.49	45 28.51	43 44.77
61	9	. .	. .	. .	44.	. .	. .	. .	44.03	23.30	1.01	IV.	1	10.44	5 24.74	. .	3.02		
82	9	. .	. .	. .	48.3	3.	. .	. .	48 48.32	23.30	1.00	V.	5	7.59	23 57.78	12.60	5.82	49 12.62	52 36.20
63	9	. .	54.	9.	23.5	. .	. .	. .	50 23.56	23.30	1.00	VI.	6	11.7	29 33.52	12.80	6.71	50 47.86	31 58 13.03
64	10	. .	. .	. .	. .	. .	4.	. .	51 34.54	23.30	0.99	VI.	7	6.25	32 11.32	12.95	7.13	51 58.83	32 0 51.40
65	8	. .	. .	. .	. .	10.5	25.	. .	52 55.64	23.30	1.00	VI.	5	4.23	22 8.70	13.12	5.55	53 19.94	31 50 47.37
66	10	. .	. .	58.5	13.2	. .	. .	. .	56 13.21	23.29	1.00	IV.	7	2.5	30 0.43	13.54	6.78	56 37.50	58 40.75
67	9	. .	. .	. .	. .	. .	40.	54.7	9 57 10.53	+23.29	+1.00	VII.	6	9.2	−28 30.22	−13.66	−6.54	9 57 34.82	−31 57 10.42

CORRECTIONS.

Date.		Corr. of Clock.	Hourly rate.	m	n	c
1849.	h.	s.	s.	s.	s.	s.

INSTRUMENT READINGS.

Date.		Barom.	THERMOM. At.	Ex.
1849.	h. m.	in.	°	°

REMARKS.

ZONE 225. MARCH 23. B. BELT, −31° 53'. D_o = −31° 28' 20"—Continued.

No.	Mag.	SECONDS OF TRANSIT. I.	II.	III.	IV.	V.	VI.	VII.	T.	a_1	a_2	MICROMETER.			i	d_1	d_2	Mean Right Ascension, 1850.0.	Mean Declination, 1850.0.
									h. m. s.	s.	s.			r.	' "	"	"	h. m. s.	° ' "
68	9	. .	. .	. .	. .	. .	43.5	58.5	9 58 14.16	+23.29	+0.99	VII.	7	9.26	−33 42.32	−13.79	−7.35	9 58 38.44	−32 2 23.46
69	9	. .	20.5	35.3	. .	. .	. .	. .	10 0 49.99	23.29	1.00	III.	5	6.54	23 24.50	14.12	5.76	10 1 14.28	31 52 4.38
70	9	. .	. .	. .	. .	3.	. .	. .	1 48.25	23.29	0.99	V.	8	11.16	39 36.92	14.25	8.29	2 12.53	32 8 19.46
71	8	. .	. .	. .	. .	. .	. .	13.5	2 29.22	23.29	0.99	VII.	8	8.24½	38 10.01	14.33	8.07	2 53.50	6 52.41
72	8	. .	. .	. .	. .	20.5	35.	. .	4 5.65	23.29	1.00	VI.	7	6.51	32 24.43	14.53	7.16	4 29.94	32 1 6.12
73	7.8	. .	. .	. .	. .	35.	. .	. .	5 20.27	23.29	1.00	VI.	6	8.46	28 22.43	14.69	6.54	5 44.56	31 57 3.66
74	7.8	. .	. .	. .	. .	. .	. .	27.6	5 43.59	23.29	1.00	VII.	4	5.51	17 53.60	14.74	4.90	6 7.85	46 33.24
75	10	. .	. .	. .	. .	44.	. .	. .	7 29.34	23.29	1.01	VI.	3	5.26	12 42.27	14.96	4.12	7 53.64	41 21.35
76	10	. .	. .	. .	. .	58.3	13.	. .	8 43.55	23.29	1.00	VI.	6	4.41	26 18.88	15.12	6.21	9 7.84	55 0.21
77	9	. .	. .	. .	5.	19.5	. .	. .	10 4.89	23.29	1.00	V.	6	7.35	27 46.78	15.29	6.45	10 29.18	56 28.52
78	8	. .	. .	13.7	28.3	. .	. .	. .	11 28.40	23.29	1.00	IV.	4	7.21	18 39.46	15.46	5.02	11 52.69	47 19.94
79	9	. .	. .	. .	35.	50.	. .	. .	12 35.12	23.28	1.00	V.	6	12.31	30 16.03	15.60	6.82	12 59.40	58 58.45
80	9	7.	21.5	. .	. .	. .	. .	. .	14 51.03	23.28	1.00	II.	5	6.58	23 26.86	15.88	5.75	15 15.31	52 8.49
81	9	. .	. .	. .	. .	. .	24.	. .	14 54.48	23.28	1.00	VI.	4	8.40	19 19.10	15.89	5.12	15 18.76	48 0.11
82	8	. .	. .	. .	. .	10.	24.8	. .	15 55.29	23.28	1.00	VI.	4	8.35	19 19.58	16.02	5.12	16 19.57	47 57.72
83	9	. .	. .	16.7	31.3	. .	. .	. .	18 31.39	23.28	1.01	IV.	3	9.58	14 59.63	16.35	4.47	18 55.68	43 40.45
84	9	. .	. .	. .	35.	49.5	. .	. .	19 34.94	23.28	1.01	V.	3	10.52	15 26.82	16.48	4.54	19 59.23	44 7.84
85	8	. .	. .	31.6	46.2	1.	. .	. .	21 46.31	23.28	1.01	V.	1	5.35	2 48.87	16.73	2.63	22 10.60	31 28.23
86	8	. .	. .	. .	58.8	13.4	. .	. .	22 58.74	23.28	1.00	V.	6	8.4	28 1.41	16.90	6.48	23 23.02	31 56 44.79
87	8	. .	. .	10.	24.5	39.	. .	. .	24 24.43	23.28	0.98	V.	10	10.22	49 9.59	17.08	9.81	24 48.69	32 17 56.48
88	9	. .	. .	. .	. .	. .	51.	. .	25 21.49	23.28	1.00	VI.	4	10.7	20 2.97	17.19	5.24	25 45.77	31 48 45.40
89	10	. .	. .	. .	. .	. .	. .	35.	26 51.12	23.28	1.01	VII.	2	4.42	7 10.72	17.38	3.28	27 15.41	35 51.38
90	9	. .	. .	. .	. .	48.	2.4	. .	28 33.10	23.28	1.01	VI.	3	10.1½	15 1.19	17.59	4.47	28 57.39	43 43.25
91	9	. .	. .	. .	. .	. .	11.5	26.5	29 42.17	23.28	1.00	VII.	7	4.14	31 5.00	17.73	6.96	30 6.45	31 59 49.69
92	9	. .	38.	52.5	. .	. .	. .	. .	32 7.35	23.28	0.99	III.	7	8.37	33 18.05	18.03	7.29	32 31.62	32 2 3.37
93	9.10	. .	. .	. .	. .	46.	1.	. .	32 31.40	23.28	1.00	VI.	6	9.12	28 35.53	28.08	6.56	32 55.68	31 57 20.17
94	9.10	. .	. .	. .	. .	52.	7.	. .	33 37.39	23.28	1.00	VI.	4	5.49	17 52.87	18.22	4.90	34 1.67	46 35.99
95	7	. .	. .	7.	22.	36.5	. .	. .	35 21.84	23.29	1.01	VI.	6	6.31	27 14.35	18.43	6.35	35 46.14	31 55 59.13
96	9.10	. .	. .	. .	. .	50.	4.5	. .	36 35.15	23.29	0.99	VI.	9	7.18	42 35.74	18.58	8.78	36 59.43	32 11 23.10
97	9	. .	. .	. .	. .	31.5	46.	. .	38 16.67	23.29	0.98	VI.	10	10.11	49 3.68	18.79	9.80	38 40.94	17 52.47
98	9	. .	. .	. .	. .	. .	7.5	22.	39 37.89	23.29	0.99	VII.	8	8.39½	38 17.58	18.95	8.09	40 2.17	7 4.62
99	9.10	. .	. .	. .	30.5	45.3	. .	. .	41 30.41	23.29	0.98	VI.	10	9.44	48 50.27	19.18	9.76	41 54.68	32 17 39.21
100	9	. .	55.	10.	. .	. .	. .	. .	43 24.60	23.29	1.00	III.	6	8.8	28 3.42	19.41	6.49	43 48.89	31 56 49.32
101	9	. .	. .	. .	. .	15.3	30.	. .	44 0.54	23.29	1.00	VI.	5	4.15	22 4.67	19.48	5.54	44 24.83	31 50 49.69
102	9	. .	. .	33.5	48.3	. .	. .	. .	45 48.20	23.29	0.99	IV.	9	7.40	42 47.05	19.69	8.81	46 12.48	32 11 35.55
103	9	. .	. .	. .	. .	. .	9.	23.4	46 39.37	23.29	1.00	VII.	7	3.59	30 57.44	19.79	6.93	47 3.66	31 59 44.16
104	9	. .	. .	. .	27.4	42.3	. .	. .	48 27.37	23.30	0.99	V.	10	6.00	46 57.48	20.01	9.47	48 51.66	32 15 46.96
105	8.9	. .	. .	. .	. .	38.5	53.	. .	49 23.64	23.30	1.00	VI.	4	3.13	16 34.20	20.12	4.70	49 47.94	31 45 19.02
106	9	. .	. .	. .	. .	1.3	16.	. .	50 46.54	23.30	1.00	VI.	4	8.16	19 7.00	20.28	5.10	51 10.84	31 37 52.38
107	8.9	. .	. .	. .	. .	8.	23.	. .	51 53.42	23.30	0.98	VI.	10	7.13	47 34.12	20.41	9.58	52 17.70	32 16 24.11
108	9	. .	. .	. .	54.3	9.	. .	. .	53 54.36	23.30	1.01	V.	2	3.42	6 50.89	20.65	3.22	54 18.67	31 35 34.76
109	9	49.	4.	18.5	. .	. .	. .	. .	56 33.24	23.30	1.00	III.	4	6.2	17 59.59	20.96	4.92	56 57.54	31 46 45.47
110	9	. .	15.2	30.	45.	. .	. .	. .	57 44.75	23.31	0.98	IV.	10	7.24½	47 40.13	21.10	9.59	58 9.04	32 16 30.82
111	7	50.4	5.	20.	34.7	. .	. .	. .	10 59 34.60	23.31	1.00	IV.	4	5.31	17 44.00	21.32	4.87	10 59 58.91	31 46 30.19
112	9	. .	. .	. .	39.5	54.2	. .	. .	11 0 39.53	23.31	1.00	V.	5	3.49	21 51.72	21.45	5.51	11 1 3.84	50 38.68
113	9.10	. .	. .	. .	57.5	12.3	. .	. .	1 57.59	23.31	1.01	V.	3	6.23	13 11.17	21.60	4.18	2 21.91	41 56.95
114	7.8	. .	. .	. .	. .	13.	28.	. .	3 58.39	23.31	1.00	VI.	3	12.29	16 15.56	21.84	4.65	4 22.70	45 2.05
115	7.8	. .	. .	. .	37.	51.5	6.2	. .	4 36.84	23.32	1.01	VI.	2	6.48	8 24.52	21.91	3.45	5 1.17	37 9.88
116	9	. .	. .	. .	. .	36.5	51.	. .	11 6 21.65	+23.32	+1.00	VI.	6	10.37	−29 18.40	−22.12	−6.67	11 6 45.97	−31 58 7.19

CORRECTIONS.

Date.		Corr. of Clock.	Hourly rate.	m	n	c
1849.	h.	s.	s.	s.	s.	s.

INSTRUMENT READINGS.

Date.		Barom.	THERMOM. At.	Ex.
1849.	h. m.	In.	°	°

REMARKS.

(225) 95. Transit over T. IV assumed to have been at 22s instead of 42s.

ZONE 225. MARCH 23. B. BELT, −31° 53'. D_0=−31° 28' 20"—Continued.

No.	Mag.	Seconds of Transit. I.	II.	III.	IV.	V.	VI.	VII.	T.	a_1	a_2	Micrometer.			i	d_1	d_2	Mean Right Ascension, 1850.0.	Mean Declination, 1850.0.
									h. m. s.	s.	s.			r.	' "	"	"	h. m. s.	° ' "
117	9	..	..	1.3	16.	..	..	..	11 8 16.04	+23.32	+1.00	IV.	5	8.20	−24 8.42	−22.35	− 5.86	11 8 40.36	−31 52 56.63
118	9.10	..	37.3	52.	..	..	..	..	10 6.74	23.33	1.00	III.	4	5.56½	17 56.81	22.56	4.90	10 31.07	46 44.27
119	9	..	15.	29.6	44.4	..	..	..	11 44.40	23.33	1.01	IV.	3	4.45	12 21.81	22.75	4.05	12 8.74	41 8.61
120	9	29.2	44.	58.5	..	..	..	..	15 13.36	23.34	1.00	III.	6	8.51	28 25.10	23.16	6.52	15 37.70	57 14.78
121	9.10	..	..	17.4	32.3	..	..	..	16 32.23	23.34	1.00	IV.	3	8.35½	14 18.04	23.31	4.35	16 56.57	43 5.70
122	9.10	46.	0.5	15.4	..	..	..	..	18 30.05	23.34	1.00	III.	4	7.20	18 38.91	23.54	5.01	18 54.39	47 27.46
123	9.10	19.	34.	..	..	..	..	..	19 3.26	23.34	1.00	II.	4	10.28	20 13.55	23.61	5.26	19 27.60	49 2.42
124	10	..	..	49.	3.2	..	..	..	21 3.49	23.35	1.00	IV.	4	6.15	18 6.19	23.84	4.93	21 27.84	46 54.96
125	10	..	..	9.	23.5	38.	..	..	22 23.54	23.35	1.01	V.	2	3.17½	6 38.53	23.99	3.18	22 47.90	35 26.70
126	10	..	..	..	..	..	6.5	21.	23 36.98	23.35	1.00	VII.	4	8.25	19 11.26	24.12	5.11	24 1.33	31 48 0.49
127	9	..	..	..	..	..	18.	32.5	24 48.39	23.36	0.99	VII.	8	8.41	38 18.33	24.26	8.10	25 12.74	32 7 10.69
128	7.8	..	..	33.3	48.	3.	..	..	26 48.08	23.36	1.00	V.	7	8.54	33 26.62	24.48	7.33	27 12.44	2 18.43
129	8	5.	20.	..	..	..	..	..	28 49.37	23.37	0.99	II.	8	5.52	36 53.38	24.71	7.88	29 13.73	5 45.97
130	7.8	..	..	..	..	25.	40.	..	29 10.41	23.37	0.99	VI.	9	3.3½	40 27.41	24.75	8.45	29 34.77	32 9 20.61
131	9.10	..	16.5	31.2	46.	..	..	..	31 45.97	23.37	1.00	IV.	4	4.35	17 15.77	25.03	4.80	32 10.34	31 46 5.60
132	7	..	21.	35.5	..	..	..	..	33 50.32	23.38	1.01	III.	3	2.32	11 14.70	25.26	3.87	34 14.71	40 3.83
133	9	..	..	..	..	..	..	38.	33 54.08	23.38	1.01	VII.	3	0.50	10 22.83	25.27	3.74	34 18.47	39 11.84
134	9	..	48.	3.	17.5	..	..	..	39 17.56	23.39	1.00	IV.	9	2.43	30 19.59	25.86	6.83	39 41.95	59 12.28
135	10	46.5	1.	..	..	..	..	..	41 30.56	23.40	1.00	II.	6	8.38	28 18.39	26.10	6.51	41 54.96	57 11.00
136	10	..	..	..	..	57.4	12.	..	41 42.60	23.40	1.00	VI.	5	7.20	23 39.95	26.12	5.78	42 7.00	52 31.85
137	10	..	..	28.5	43.2	..	..	..	43 43.21	23.41	1.00	IV.	6	12.50	30 25.66	26.34	6.84	44 7.62	59 18.84
138	10	..	55.	..	..	..	..	..	45 24.44	23.41	1.00	II.	4	13.2	21 31.20	26.52	5.46	45 48.85	31 50 23.18
139	9	..	27.5	42.2	57.	..	..	..	46 56.93	23.42	0.99	IV.	9	11.17	44 36.46	26.68	9.12	47 21.34	32 13 32.26
140	10	..	..	..	..	..	15.	29.5	47 45.48	23.42	1.00	VII.	4	6.38	18 17.31	26.77	4.95	48 9.90	31 47 9.03
141	10	..	..	..	..	..	52.2	7.	49 22.84	23.42	1.00	VII.	3	7.55	13 57.13	26.94	4.28	49 47.26	31 42 48.35
142	10	..	..	..	31.2	..	..	..	51 30.90	23.43	0.99	IV.	8	7.51	37 53.59	27.17	8.04	51 55.32	32 6 48.80
143	9	..	..	49.3	4.3	..	..	..	53 4.16	23.44	1.01	IV.	1	5.59	3 1.02	27.33	2.62	53 28.61	31 31 50.97
144	9	..	..	..	..	..	8.2	23.	53 38.83	23.44	1.00	VII.	4	7.13	18 34.95	27.39	5.00	54 3.27	47 27.34
145	9	..	..	..	..	..	..	25.5	54 41.35	23.44	1.00	VII.	6	6.30½	27 13.83	27.50	6.35	55 5.79	56 7.68
146	9	..	15.	29.5	44.2	..	..	..	56 44.30	23.45	1.00	IV.	6	10.20	29 10.03	27.72	6.65	57 8.75	58 4.40
147	9	..	..	..	15.	..	..	..	58 15.06	23.45	1.00	IV.	3	12.8	16 5.18	27.88	4.61	58 39.51	44 57.67
148	8.9	..	..	..	15.2	30.	..	..	11 59 15.29	+23.46	+1.00	V.	4	11.19	−20 39.43	−27.98	− 5.33	11 59 39.75	−31 49 32.74

ZONE 226. MARCH 23. B. BELT, −35° 38'. D_0=−35° 15' 20".

No.	Mag.	I.	II.	III.	IV.	V.	VI.	VII.	T.	a_1	a_2	Micrometer.			i	d_1	d_2	Mean Right Ascension, 1850.0.	Mean Declination, 1850.0.
1	9	..	..	..	9.	24.	..	..	13 27 8.73	+23.72	+0.99	V.	8	10.11	−39 4.13	− 2.57	− 9.31	13 27 33.44	−35 54 36.01
2	8	..	..	..	28.	43.3	..	..	28 27.93	23.72	1.00	V.	7	7.37	32 47.78	2.82	7.82	28 52.65	48 18.42
3	9	..	51.	6.2	21.2	..	..	..	30 21.52	23.74	1.00	IV.	3	6.11	13 5.17	3.17	3.30	30 46.26	28 31.64
4	8.9	..	36.6	52.	7.6	..	..	..	35 7.45	23.78	1.00	IV.	7	7.15	32 36.74	4.04	7.78	35 32.23	48 8.56
5	10	..	..	1.	16.5	..	..	..	36 16.47	23.78	1.00	IV.	5	4.24	22 9.42	4.26	5.33	36 41.25	37 39.01
6	9	..	..	22.	37.3	..	..	..	37 37.37	23.79	1.00	IV.	5	7.3	23 29.59	4.50	5.65	38 2.16	38 59.74
7	8.9	..	..	..	34.	49.	..	..	38 33.76	23.80	1.00	V.	8	3.21½	35 37.63	4.67	8.49	38 58.56	51 10.79
8	9	..	0.4	16.	31.2	..	..	..	40 31.25	23.81	1.00	IV.	8	3.49	35 51.57	5.03	8.55	40 56.06	51 25.15
9	8.9	..	55.3	11.	26.	..	..	..	42 26.17	23.83	1.00	IV.	6	2.55	25 25.64	5.37	6.09	42 51.00	40 57.10
10	9.10	..	..	..	..	..	..	31.	42 44.73	23.83	0.99	VII.	8	6.00	36 57.08	5.42	8.82	43 9.55	52 31.32
11	9	..	..	..	..	32.3	47.5	..	44 16.80	23.84	1.00	VI.	4	12.19	21 9.49	5.71	5.10	44 41.64	36 40.30
12	10	..	..	26.5	42.	..	..	..	48 41.95	23.87	1.00	IV.	6	10.11	29 5.49	6.49	6.97	49 6.82	44 38.95
13	10	28.2	43.2	59.	..	..	..	..	13 51 14.30	+23.89	+0.99	III.	8	7.11	−37 3.36	− 6.94	− 8.84	13 51 39.18	−35 52 39.14

CORRECTIONS.

Date.		Corr. of Clock.	Hourly rate.	m	n	c
1849.	h.	s.	s.	s.	s.	s.

INSTRUMENT READINGS.

Date.			Barom.	Thermom. At.	Thermom. Ex.
1849.	h.	m.	in.	°	°

REMARKS.

ZONE 226. MARCH 23. B. BELT, −35° 38′. D₀=−35° 15′ 20″—Continued.

No.	Mag.	SECONDS OF TRANSIT. I.	II.	III.	IV.	V.	VI.	VII.	T.	a_1	a_2	MICROMETER.			i	d_1	d_2	Mean Right Ascension, 1850.0.	Mean Declination, 1850.0.
									h. m. s.	s.	s.			r.	′ ″	″	″	h. m. s.	° ′ ″
14	10	. . .	. . .	. . .	. . .	49.	5.	. . .	13 51 33.91	+23.90	+1.00	VI.	7	11.18	−34 39.03	− 6.99	− 8.27	13 51 58.81	−35 50 14.29
15	9	. . .	. . .	. . .	5.2	20.8	36.	. . .	53 5.29	23.91	1.01	VI.	2	13.12	11 38.11	7.26	2.95	53 30.21	27 8.32
16	9	. . .	. . .	. . .	. . .	32.	47.	. . .	54 16.40	23.92	1.00	VI.	6	9.32½	28 45.84	7.47	6.86	54 41.32	44 20.17
17	9	. . .	. . .	55.	11.	. . .	. . .	. . .	56 10.67	23.93	1.00	IV.	7	10.31	34 15.58	7.80	6.18	56 35.60	49 51.56
18	9.10	. . .	. . .	. . .	. . .	. . .	8.	23.4	13 56 37.18	23.93	1.00	VII.	8	5.36	36 44.98	7.87	8.76	13 57 2.11	35 52 20.71
19	10	. . .	36.4	52.	7.4	. . .	. . .	. . .	14 0 7.30	23.97	0.99	IV.	10	6.23	47 9.12	8.49	11.26	14 0 32.26	36 2 48.87
20	10	. . .	. . .	22.2	37.3	. . .	. . .	. . .	1 37.48	23.97	1.00	IV.	4	5.27½	17 42.23	8.74	4.32	2 2.45	35 33 15.29
21	9	. . .	. . .	7.	22.3	37.2	. . .	. . .	4 22.13	23.99	1.00	V.	8	10.25	39 11.20	9.21	9.35	4 47.12	54 49.76
22	10	. . .	. . .	. . .	. . .	38.	53.	. . .	5 22.40	24.01	1.00	VI.	4	7.18	18 37.71	9.38	4.53	5 47.41	34 11.62
23	10	. . .	. . .	. . .	9.	24.	. . .	. . .	7 8.73	24.02	1.00	V.	8	12.6	40 2.12	9.66	9.54	7 33.75	55 41.32
24	10.11	. . .	. . .	31.5	47.	. . .	. . .	. . .	8 46.93	24.03	1.00	IV.	7	4.33	31 15.06	9.95	7.47	9 11.96	46 52.48
25	10	. . .	. . .	. . .	. . .	. . .	. . .	10.	9 23.74	24.04	1.00	VII.	8	4.33	36 13.21	10.05	8.64	9 48.78	51 51.90
26	10	. . .	. . .	. . .	. . .	. . .	13.5	29.	10 42.83	24.05	1.00	VII.	4	6.17	18 6.65	10.26	4.41	11 7.88	33 41.32
27	9	. . .	. . .	20.	35.2	. . .	. . .	. . .	13 35.28	24.07	1.00	IV.	1	3.53	1 57.48	10.73	0.75	14 0.35	17 28.96
28	10.11	. . .	. . .	. . .	0.	15.5	. . .	. . .	17 0.03	24.10	1.00	VI.	7	8.37	33 17.86	11.28	8.91	17 25.13	48 58.08
29	10	. . .	. . .	. . .	21.	36.4	52.	. . .	18 21.06	24.11	1.00	VI.	6	13.48	30 54.66	11.48	7.39	18 46.17	46 53.53
30	10	. . .	50.	6.	22.	37.	. . .	. . .	21 21.44	24.13	1.00	V.	7	3.44	30 50.29	11.98	7.37	21 46.57	46 29.64
31	10.11	. . .	. . .	53.	8.3	24.	. . .	. . .	23 8.47	24.15	1.00	V.	4	11.20	20 39.92	12.26	4.99	23 33.62	36 17.17
32	11	. . .	. . .	25.	40.	55.	. . .	. . .	25 40.04	24.17	1.00	V.	3	4.11	12 4.60	12.67	3.05	26 5.21	35 27 40.32
33	10	. . .	. . .	54.	9.5	25.	. . .	. . .	27 9.44	24.18	0.99	VI.	10	1.42	44 47.19	12.90	10.71	27 34.61	36 0 30.80
34	10	. . .	8.4	24.	39.	. . .	. . .	. . .	29 39.21	24.20	1.00	IV.	4	8.14	19 6.20	13.29	4.63	30 4.41	35 34 44.12
35	8.9	. . .	53.5	9.	24.	. . .	. . .	. . .	31 24.22	24.22	1.00	IV.	3	7.20	13 39.96	13.57	3.41	31 49.44	29 16.91
36	11	. . .	. . .	. . .	33.5	49.	. . .	. . .	35 33.62	24.25	1.00	V.	2	12.34	11 19.14	14.21	2.87	35 58.87	26 56.22
37	8.9	. . .	. . .	. . .	40.	56.2	12.	. . .	35 40.69	24.25	1.00	VI.	3	10.3	15 1.92	14.23	3.69	36 5.94	35 30 39.84
38	8	9.	24.3	39.5	55.	. . .	. . .	. . .	14 38 55.06	+24.28	+0.99	IV.	9	10.55	−44 25.38	−14.73	−10.62	14 39 20.33	−36 0 10.73

ZONE 227. MARCH 29. B. BELT, −30° 38′. D₀=−30° 14′ 0″.

No.	Mag.	I.	II.	III.	IV.	V.	VI.	VII.	T.	a_1	a_2	MICROMETER.			i	d_1	d_2	Mean Right Ascension, 1850.0.	Mean Declination, 1850.0.
1	9	. . .	45.	59.	13.	. . .	. . .	. . .	11 55 13.54	+23.22	+1.01	IV.	3	3.56	−11 57.10	− 5.68	− 4.24	11 55 37.77	−30 26 7.02
2	6	. . .	. . .	. . .	27.5	42.2	. . .	. . .	56 27.55	23.22	1.01	V.	2	4.10	7 5.01	5.71	3.57	56 51.78	21 14.29
3	9	. . .	. . .	1.5	16.4	. . .	. . .	. . .	58 16.21	23.22	1.00	IV.	7	1.10	29 32.69	5.75	6.63	58 40.43	43 45.07
4	7.8	. . .	. . .	. . .	29.	43.5	. . .	. . .	11 59 29.03	23.22	1.00	VI.	4	10.8	20 3.48	5.77	5.33	11 59 53.25	34 14.58
5	10	. . .	. . .	. . .	9.2	24.	. . .	. . .	12 1 9.34	23.22	1.00	V.	6	7.21	27 39.71	5.80	6.37	12 1 33.56	41 51.88
6	10	. . .	. . .	. . .	20.5	35.	. . .	. . .	3 20.48	23.23	1.00	V.	6	10.22	29 10.99	5.83	6.59	3 44.71	43 23.41
7	8	. . .	25.	40.	54.2	. . .	. . .	. . .	4 54.27	23.23	1.00	IV.	6	5.52	26 54.89	5.86	6.27	5 18.50	30 41 7.02
8	9	. . .	. . .	32.	46.	. . .	. . .	. . .	6 46.19	23.23	0.99	IV.	10	4.24	46 9.12	5.89	8.86	7 10.31	31 0 23.87
9	7.8	. . .	. . .	. . .	. . .	37.	51.5	. . .	7 22.46	23.23	0.99	V.	8	6.23	37 9.17	5.90	7.70	7 46.68	30 51 22.77
10	7	. . .	. . .	. . .	33.3	48.	. . .	. . .	9 33.45	23.24	1.01	V.	2	6.6	8 3.51	5.93	3.72	9 57.70	22 13.16
11	7.8	. . .	. . .	. . .	. . .	. . .	42.5	57.	10 13.43	23.24	1.00	VII.	6	2.48	25 21.66	5.94	6.05	10 37.67	39 33.65
12	8	. . .	. . .	. . .	15.3	0.	. . .	. . .	12 45.45	23.24	1.01	VI.	1	10.29	5 16.97	5.97	3.34	13 9.70	19 26.28
13	8	. . .	. . .	. . .	58.	12.5	. . .	. . .	13 58.05	23.24	1.01	V.	1	13.5	6 35.77	5.98	3.51	14 22.30	20 45.26
14	9	. . .	. . .	. . .	. . .	. . .	24.	. . .	14 54.92	23.25	1.00	VI.	6	8.38	28 18.40	5.99	6.46	15 19.17	42 30.85
15	9	. . .	. . .	17.4	32.0	. . .	. . .	. . .	16 31.97	23.25	1.00	IV.	6	5.5	26 31.19	6.01	6.21	16 56.22	40 43.41
16	8	. . .	. . .	. . .	44.4	59.	. . .	. . .	17 44.35	23.25	0.99	V.	9	6.26	42 9.69	6.02	8.35	18 8.59	56 24.06
17	8	. . .	. . .	. . .	56.5	11.	. . .	. . .	18 56.54	23.25	1.00	V.	3	10.15	15 8.16	6.03	4.66	19 20.79	29 18.85
18	9	. . .	23.	37.5	52.	. . .	. . .	. . .	20 52.04	23.26	1.00	IV.	6	8.38	28 18.60	6.04	6.46	21 16.30	42 31.10
19	9	. . .	. . .	. . .	. . .	. . .	55.	9.3	21 25.90	23.26	1.01	VII.	1	10.5	5 4.62	6.05	3.30	21 50.17	19 13.97
20	8	. . .	. . .	. . .	. . .	3.1	18.	. . .	12 22 48.75	+23.26	+1.00	VI.	7	10.36	−34 17.91	− 6.05	− 7.30	12 23 13.01	−30 48 31.26

CORRECTIONS.

Date.		Corr. of Clock.	Hourly rate.	m	n	c
1849.	h.	s.	s.	s.	s.	s.

INSTRUMENT READINGS.

Date.		Barom.	THERMOM. At.	Ex.
1849.	h. m.	in.	°	°

REMARKS.

ZONE 227. MARCH 29. B. BELT, −30° 38′. D_o = −30° 14′ 0″—Continued.

No.	Mag.	Seconds of Transit. I.	II.	III.	IV.	V.	VI.	VII.	T.	a_1	a_2	Micrometer.			i	d_1	d_2	Mean Right Ascension, 1850.0.	Mean Declination, 1850.0.
									h. m. s.	s.	s.			r.	′ ″	″	″	h. m. s.	° ′ ″
21	9	. .	. .	. .	43.	58.	. .	. .	12 24 43.20	+23.26	+1.00	V.	7	11.33	−34 46.79	−6.06	−7.37	12 25 7.46	−30 49 0.22
22	9	. .	2.5	17.	. .	. .	. .	. .	26 31.54	23.27	1.00	III.	4	12.11	21 5.64	6.07	5.47	26 55.61	35 17.18
23	8	. .	33.	47.4	. .	. .	. .	. .	29 2.01	23.27	1.00	III.	6	9.4	28 31.66	6.08	6.49	29 26.28	42 41.23
24	7.8	. .	. .	21.3	36.	50.5	. .	. .	32 35.89	23.28	0.99	V.	8	13.4	40 31.37	6.08	8.15	33 0.16	54 45.60
25	8	. .	. .	. .	6.	20.5	. .	. .	34 5.92	23.29	0.99	V.	8	10.5	39 1.12	6.08	7.94	34 30.20	53 15.14
26	9	. .	44.	58.5	13.	. .	. .	. .	36 13.02	23.29	1.00	IV.	7	8.12	33 5.49	6.09	7.13	36 37.31	47 18.71
27	5	. .	7.5	22.	. .	. .	. .	. .	37 36.57	23.29	0.99	III.	9	6.45	42 19.27	6.09	8.38	38 0.85	56 33.74
28	9	. .	. .	. .	. .	. .	27.	41.5	37 57.87	23.30	0.99	VII.	9	11.1	44 27.95	6.09	8.65	38 22.16	58 42.69
29	8	. .	41.5	56.	. .	. .	. .	. .	40 10.54	23.30	1.00	III.	4	11.55	20 57.58	6.09	5.45	40 34.84	35 9.12
30	8	. .	. .	. .	. .	. .	. .	8.	40 24.58	23.30	1.00	VII.	3	12.7	16 4.23	6.09	4.79	40 48.88	30 15.11
31	6.7	37.	51.5	5.9	. .	. .	. .	. .	43 20.45	23.31	1.01	III.	2	7.48	8 54.93	6.08	3.82	43 44.77	23 4.83
32	9	. .	19.5	34.2	. .	. .	. .	. .	44 48.64	23.31	1.00	III.	4	5.30	17 43.45	6.08	5.02	45 12.95	31 54.55
33	9	23.	38.	. .	. .	. .	. .	. .	47 6.81	23.32	1.00	II.	5	9.25	24 40.99	6.07	5.96	47 31.13	38 53.02
34	7.8	. .	. .	. .	29.	43.5	. .	. .	47 29.01	23.32	1.00	.	5	11.46	25 52.30	6.07	6.12	47 53.33	40 4.49
35	9	. .	. .	55.3	10.	. .	. .	. .	49 9.91	23.33	1.01	IV.	1	6.13	3 8.08	6.06	3.04	49 34.25	17 17.18
36	8	. .	. .	. .	. .	. .	12.	26.5	49 42.97	23.33	1.00	VII.	3	7.53	13 56.15	6.06	4.51	50 7.30	28 6.72
37	8	40.5	55.	. .	. .	. .	. .	. .	52 24.02	23.34	1.00	II.	4	11.30	20 44.83	6.04	5.42	52 48.36	34 56.29
38	9	. .	31.	45.5	0.5	. .	. .	. .	54 0.16	23.34	0.99	IV.	9	10.9	44 2.19	6.03	8.60	54 24.49	58 16.82
39	9	. .	4.	18.3	. .	. .	. .	. .	55 32.96	23.35	1.00	III.	6	10.24	29 12.00	6.02	6.59	55 57.31	43 24.61
40	9	37.	51.2	6.	. .	. .	. .	. .	57 20.47	23.35	1.00	III.	7	4.16	31 6.44	6.00	6.85	57 44.82	45 19.29
41	6.7	. .	. .	. .	0.	14.3	29.	. .	58 59.88	23.36	1.00	VI.	7	10.17	34 8.32	5.99	7.27	12 58 24.24	18 21.58
42	9	. .	. .	44.3	59.	. .	. .	. .	12 59 58.93	23.36	1.00	IV.	3	6.20	13 9.71	5.98	4.39	13 0 23.29	27 20.08
43	10	. .	. .	. .	56.7	11.2	. .	. .	13 1 56.69	23.37	1.00	V.	6	3.58	25 57.35	5.96	6.13	2 21.06	40 9.44
44	10	. .	. .	. .	8.2	23.	. .	. .	3 8.38	23.37	1.00	V.	4	5.53	17 55.04	5.95	5.05	3 32.75	32 6.04
45	11	. .	. .	. .	14.	28.7	. .	. .	4 14.07	23.38	1.00	V.	7	3.16	30 36.18	5.93	6.78	4 38.45	44 48.89
46	10	. .	36.5	51.	. .	. .	. .	. .	6 5.56	23.38	1.00	III.	7	3.51	30 53.82	5.91	6.83	6 29.94	45 6.56
47	10	. .	27.	41.5	55.2	. .	. .	. .	7 55.78	23.39	1.00	IV.	4	4.53	17 24.84	5.89	4.98	8 20.17	31 35.71
48	5	. .	. .	. .	. .	. .	. .	53.	8 9.43	23.39	1.00	VII.	6	8.50	28 24.20	5.89	6.48	8 33.82	42 36.57
49	9	. .	. .	37.	51.3	. .	. .	. .	12 51.43	23.41	1.00	IV.	2	10.23	10 13.14	5.83	3.98	13 15.84	24 22.95
50	10	. .	. .	. .	. .	. .	53.	8.	13 24.13	23.41	1.00	VII.	8	9.27	38 41.55	5.82	7.93	13 48.54	52 55.30
51	9	. .	. .	8.4	23.	. .	. .	. .	16 22.08	23.42	1.00	IV.	2	11.6	10 34.83	5.78	4.02	16 46.50	24 44.63
52	8	. .	. .	33.	48.	2.	. .	. .	17 47.68	23.42	1.00	V.	5	11.45	25 51.74	5.76	6.12	18 12.10	40 3.62
53	8	. .	. .	10.	25.	39.	. .	. .	19 24.65	23.43	1.00	V.	7	6.21	32 9.46	5.74	7.00	19 49.08	46 22.20
54	9	. .	. .	38.	52.4	. .	. .	. .	20 52.50	23.44	1.00	IV.	5	4.37	22 15.98	5.72	5.63	21 16.94	36 27.33
55	10	. .	. .	17.	31.4	. .	. .	. .	22 31.50	23.44	1.00	IV.	4	7.44	18 51.07	5.69	5.17	22 55.94	33 1.93
56	10	. .	. .	. .	2.	17.	. .	. .	24 2.20	23.45	1.00	V.	7	11.50	34 55.36	5.67	7.40	24 26.65	49 8.43
57	10	. .	. .	. .	. .	. .	. .	28.	24 44.50	23.45	1.00	VII.	5	5.51	22 52.83	5.66	5.73	25 8.95	37 4.22
58	10	. .	. .	. .	24.	38.4	. .	. .	26 23.96	23.46	1.00	V.	5	8.18	24 7.36	5.63	5.87	26 48.42	38 18.86
59	11	. .	. .	37.5	52.	. .	. .	. .	52.04	23.47	1.00	IV.	5	7.12	23 34.13	. .	5.82		
60	11	. .	. .	. .	55.2	9.8	. .	. .	30 55.29	23.48	1.00	V.	4	3.26	16 40.92	5.56	4.87	31 19.77	30 51.35
61	11	. .	3.	18.	32.5	. .	. .	. .	33 32.36	23.49	1.00	IV.	7	5.40	31 48.84	5.51	6.95	33 56.85	46 1.30
62	11	. .	. .	. .	31.	45.5	. .	. .	34 30.99	23.49	1.00	V.	6	5.49	26 53.33	5.49	6.27	34 55.48	41 5.09
63	10	17.5	32.2	47.	1.5	. .	. .	. .	40 1.34	23.52	1.00	IV.	5	6.35	23 15.48	5.38	5.75	40 25.86	37 26.64
64	10	. .	. .	. .	. .	. .	4.4	19.	40 35.36	23.52	1.00	VII.	7	3.7	30 31.24	5.37	6.77	40 59.88	44 43.38
65	7	. .	. .	. .	. .	. .	10.3	25.	41 41.37	23.52	1.00	VII.	3	6.34	13 16.33	5.34	4.41	42 5.89	27 26.08
66	8	. .	. .	. .	. .	. .	11.	26.	42 42.20	23.53	1.00	VII.	4	10.5	20 1.72	5.32	5.32	43 6.73	34 12.36
67	9	. .	. .	. .	. .	22.	36.8	. .	44 7.62	23.53	0.99	VI.	9	10.57	44 26.19	5.29	8.66	44 32.14	58 40.14
68	6.7	. .	14.	28.8	43.	. .	. .	. .	46 43.15	23.54	1.00	IV.	4	7.9	18 33.42	5.23	5.13	47 7.69	32 43.78
69	7.8	. .	11.5	26.	40.5	. .	. .	. .	13 48 40.52	+23.55	+1.01	IV.	1	11.50	−5 58.01	−5.18	−3.39	13 49 5.08	−30 20 6.58

CORRECTIONS.

Date.		Corr. of Clock.	Hourly rate.	m	n	c
1849.	h.	s.	s.	s.	s.	s.

INSTRUMENT READINGS.

Date.		Barom.	Thermom. At.	Thermom. Ex.
1849.	h. m.	In.	°	°

REMARKS.

(227) 41. Minutes assumed as 57 instead of 58.
(227) 51. Time for T. IV assumed at 22ˢ.8 instead of 28ˢ.

ZONE 227. MARCH 29. B. BELT, −30° 38′. D₀ = −30° 14′ 0″—Continued.

No.	Mag.	I.	II.	III.	IV.	V.	VI.	VII.	T.	a_1	a_2	MICROMETER.			i	d_1	d_2	Mean Right Ascension, 1850.0.	Mean Declination, 1850.0.
									h. m. s.	s.	s.			r.	′ ″	″	″	h. m. s.	° ′ ″
70	9	..	49.5	4.	18.5	..	..	..	13 50 18.49	+23.56	+0.99	IV.	10	4.47	−46 20.72	− 5.14	− 8.92	13 50 42.94	−31 0 34.78
71	10	35.	49.5	..	..	..	..	..	52 18.54	23.57	1.00	II.	5	5.17	22 35.94	5.09	5.68	52 43.11	30 36 46.71
72	9	..	..	..	..	..	50.	5.	52 21.14	23.57	1.00	VII.	8	10.10	39 3.24	5.09	7.98	52 45.71	53 16.31
73	7	..	..	..	55.	9.3	24.1	..	53 54.86	23.58	1.00	VI.	9	8.51	43 22.65	5.04	8.54	54 19.44	57 36.23
74	8	10.	24.5	39.	..	..	..	..	57 53.48	23.60	1.00	III.	2	7.23	8 42.32	4.94	3.76	58 18.08	22 51.02
75	8	..	..	..	..	..	..	44.	13 58 0.44	+23.60	+1.00	.	6	5.19½	−26 38.25	− 4.93	− 6.23	13 58 25.04	−30 40 49.41

ZONE 228. MARCH 30. K. BELT, −39° 23′. D₀ = −39° 0′ 50″.

No.	Mag.	I.	II.	III.	IV.	V.	VI.	VII.	T.	a_1	a_2	MICROMETER.			i	d_1	d_2	Mean Right Ascension, 1850.0.	Mean Declination, 1850.0.
1	10	..	..	..	..	..	53.0	..	11 57 20.63	+23.84	+0.99	VI.	7	9.10	−33 34.44	− 2.74	− 9.34	11 57 45.46	−39 34 36.52
2	10	..	..	..	0.2	..	..	..	11 59 0.26	23.85	1.00	.	5	6.59	23 27.58	2.86	5.41	11 59 25.11	24 25.85
3	9	..	5.0	21.2	..	..	..	..	12 2 37.44	23.85	0.99	III.	9	7.56	42 55.05	3.11	13.03	12 3 2.28	44 1.19
4	10	..	41.3	57.4	..	..	..	..	5 13.59	23.86	1.01	.	3	7.31	13 45.51	3.28	1.77	5 38.46	14 40.56
5	10	..	..	..	44.6	..	..	..	5 44.43	23.86	0.99	.	9	9.36	43 45.55	3.32	13.37	6 9.28	44 52.24
6	9	..	46.0	2.3	..	..	..	..	7 18.39	23.87	1.01	III.	3	11.12	15 36.88	3.43	− 2.44	7 43.27	16 32.75
7	9	..	..	..	50.3	..	..	..	8 50.32	23.87	1.01	.	1	6.55	3 29.26	3.53	+ 2.07	9 15.20	4 20.72
8	8	..	..	..	..	36.8	53.2	..	9 20.74	23.87	1.01	.	2	8.41	9 21.71	3.56	− 0.11	9 45.62	10 15.38
9	8	..	..	..	37.6	53.9	..	..	10 37.72	23.87	1.01	.	3	13.11	16 36.94	3.65	2.82	11 2.60	17 33.41
10	9	..	..	..	..	..	..	38.5	10 50.13	23.88	1.00	VII.	4	7.46	18 51.43	3.66	3.65	11 15.01	19 48.74
11	10	..	50.6	6.5	..	..	..	..	15 22.85	23.89	0.99	.	7	3.20	30 38.24	3.96	8.17	15 47.73	31 40.37
12	9	..	46.5	2.8	..	..	..	..	16 18.92	23.89	1.00	.	5	5.56	22 55.81	4.02	5.19	16 43.81	23 55.02
13	9	..	..	..	..	49.9	..	..	16 33.64	23.89	0.98	.	10	10.42	49 19.73	4.04	−15.62	16 58.51	50 29.39
14	9	..	..	..	..	59.5	..	..	17 43.46	23.89	1.01	.	1	12.42	6 24.23	4.11	+ 0.98	18 8.36	7 17.36
15	10	..	43.0	..	15.2	..	..	..	20 15.27	23.90	1.01	.	2	4.48	7 24.22	4.28	0.61	20 40.18	8 17.89
16	9	..	..	..	..	56.8	..	..	20 40.78	23.90	1.01	VI.	1	7.50	3 56.70	4.30	+ 1.89	21 5.69	4 49.11
17	9	..	..	48.8	5.0	..	..	..	22 5.01	23.91	1.00	.	5	10.45	25 21.54	4.39	− 6.14	22 29.92	26 22.07
18	9	..	..	49.4	5.5	..	..	..	23 5.55	23.91	1.00	.	5	13.30	26 44.73	4.45	− 6.66	23 30.46	27 45.84
19	7	..	..	..	34.5	50.6	..	..	27 34.56	23.92	1.02	.	1	3.18	1 39.83	4.72	+ 2.74	27 59.50	2 31.81
20	9	..	51.8	7.8	..	..	..	..	30 24.07	23.93	1.00	.	5	9.37	24 47.25	4.89	− 5.92	30 49.00	25 48.06
21	9	..	..	..	41.4	57.3	..	..	30 41.23	23.94	1.00	.	7	5.15	31 36.23	4.91	8.55	31 6.17	32 39.69
22	9	..	..	..	37.4	53.6	..	..	31 37.47	23.94	1.00	.	4	11.50	20 55.11	4.96	− 9.44	32 2.41	21 59.51
23	9	..	..	32.2	48.2	..	..	..	33 48.28	23.94	1.01	.	1	11.21	5 43.38	5.09	+ 1.24	34 13.23	6 37.23
24	7	..	21.0	37.1	53.2	..	..	..	34 53.30	23.95	1.00	.	4	10.28	20 13.76	5.17	− 4.18	35 18.25	21 13.11
25	9	..	..	..	..	35.2	..	..	35 19.04	23.95	1.00	V.	5	7.50	23 53.22	5.18	5.58	35 43.99	24 53.98
26	10	..	..	52.3	8.8	..	..	..	37 8.65	23.96	1.00	.	6	4.39	26 18.09	5.28	6.50	37 33.61	27 19.87
27	8	..	41.3	0.8	17.0	..	..	..	38 16.89	23.96	1.00	.	6	7.13	27 35.73	5.35	7.01	38 41.65	28 38.09
28	9	..	..	..	..	0.0	16.4	..	38 43.92	23.96	1.00	.	4	4.52	17 24.34	5.37	− 3.12	39 8.88	18 22.83
29	9	..	36.2	52.2	..	..	..	..	42 8.40	23.98	1.01	.	1	4.35	2 18.67	5.57	+ 2.51	42 33.39	3 11.73
30	5	56.8	13.2	29.3	45.3	1.6	17.8	33.8	44 45.41	23.99	1.00	.	4	11.19	20 39.48	5.71	− 4.35	45 10.40	21 39.54
31	9	..	..	..	..	11.0	27.2	..	45 54.81	23.99	1.00	V.	4	10.24	20 11.68	5.77	4.18	46 19.80	21 11.63
32	9	12.4	28.5	..	..	..	..	..	47 0.89	24.00	1.00	III.	5	11.19	25 38.61	5.84	6.23	47 25.89	26 40.68
33	9	..	41.5	57.5	..	..	..	..	50 13.75	24.01	1.01	.	3	11.47	15 54.60	6.01	2.55	50 38.77	16 53.16
34	9	..	16.0	..	..	..	..	..	51 48.45	24.02	0.99	II.	9	10.59	44 27.11	6.10	13.69	52 13.46	45 36.90
35	9	..	..	..	..	..	13.0	..	51 40.61	24.02	1.00	VI.	6	9.32	28 45.54	6.09	7.46	52 5.63	29 49.09
36	9.8	..	28.0	44.1	..	..	..	..	58 0.29	24.04	1.01	.	3	6.53	13 26.35	6.42	1.63	12 58 25.34	14 24.40
37	9.8	..	..	27.7	43.4	..	..	..	59 43.68	24.05	1.00	.	4	11.22	20 40.99	6.50	− 4.36	13 0 8.73	21 41.85
38	9	..	..	..	..	..	..	27.9	12 59 39.76	24.05	1.01	VII.	1	11.37	5 50.81	6.50	+ 1.21	0 4.82	6 46.10
39	9	..	..	..	..	..	51.0	..	13 0 18.51	+24.05	+1.01	VI.	1	6.17	− 3 9.80	− 6.53	+ 2.20	13 0 43.57	−39 4 4.13

CORRECTIONS.

Date.		Corr. of Clock.	Hourly rate.	m	n	c
1849.	h.	s.	s.	s.	s.	s.

INSTRUMENT READINGS.

Date.		Barom.	THERMOM. At.	THERMOM. Ex.
1849.	h. m.	in.	°	°

REMARKS.

ZONE 228. MARCH 30. K. BELT, −39° 23′. D_0 = −39° 0′ 50″—Continued.

No.	Mag.	SECONDS OF TRANSIT. I.	II.	III.	IV.	V.	VI.	VII.	T.	a_1	a_2	MICROMETER.			i	d_1	d_2	Mean Right Ascension, 1850.0.	Mean Declination, 1850.0.
									h. m. s.	s.	s.			r.	′ ″	″	″	h. m. s.	° ′ ″
40	9	59.3	15.7	...	...	...	...	...	13 3 47.93	+24.07	+1.00	II.	5	9.6	−24 31.33	− 6.65	− 5.81	13 4 13.00	−39 25 33.79
41	9.10	...	...	...	31.9	...	...	...	4 31.75	24.07	0.99	.	9	6.34	42 13.78	6.68	12.77	4 56.81	43 23.23
42	8	31.4	47.7	...	...	...	...	...	6 20.07	24.08	0.99	III.	7	10.55	34 27.61	6.74	9.68	6 45.14	35 34.03
43	10	...	...	28.9	...	...	...	...	6 45.12	24.09	0.99	III.	8	8.12	38 4.12	6.75	11.13	7 10.20	39 12.00
44	9	...	...	...	...	...	26.7	...	6 54.26	24.09	1.01	VII.	3	9.10	14 34.78	6.76	2.05	7 19.36	15 33.59
45	9	...	...	54.1	...	...	...	...	10 10.27	24.10	1.00	.	4	12.19	21 9.73	6.87	4.53	10 35.37	22 11.13
46	9	20.8	36.9	...	...	...	...	...	12 9.35	24.11	1.00	II.	7	5.25	31 40.98	6.92	8.60	12 34.46	32 46.50
47	9	...	...	...	...	39.8	56.0	...	12 23.63	24.11	1.01	V.	2	12.04	11 4.00	6.93	0.74	12 48.75	12 1.67
48	10	...	...	...	...	...	53.4	...	13 21.02	24.12	1.00	VI.	7	4.20	31 8.21	6.95	8.38	13 46.14	32 13.54
49	7	...	39.1	55.2	...	...	...	...	18 11.41	24.14	1.00	III.	5	3.45	21 49.68	7.07	4.77	18 36.55	22 51.52
50	9	...	...	46.0	...	...	...	...	19 2.16	24.15	1.01	.	4	1.48	15 51.55	7.09	2.52	19 27.32	16 51.16
51	8	6.3	22.5	38.7	...	...	...	...	23 54.75	24.17	1.01	.	2	11.42	10 52.98	7.19	0.65	24 19.93	11 50.82
52	7.8	...	...	24.9	40.9	...	...	...	24 41.00	24.17	1.01	.	2	8.51	9 26.75	7.21	0.12	25 6.18	10 24.08
53	9	...	...	...	35.1	51.3	...	...	25 35.11	24.18	1.00	.	6	6.26	27 12.04	7.23	6.86	26 0.29	28 16.13
54	6	...	5.3	21.3	...	...	...	...	30 37.54	24.20	1.01	.	3	12.08	16 5.18	7.33	2.62	31 2.75	17 5.13
55	9	14.7	30.7	...	...	...	...	...	32 3.27	24.21	0.99	.	8	11.20	39 38.98	7.36	11.76	32 28.47	40 48.10
56	8	47.2	3.3	19.3	...	...	...	...	34 35.61	24.22	1.00	.	5	8.14	24 5.40	7.41	5.66	35 0.83	25 8.47
57	10	4.8	21.0	...	...	...	...	...	38 53.47	24.25	0.99	II.	8	12.15	40 6.42	7.49	11.94	39 18.71	41 15.85
58	9	...	...	0.8	17.2	...	...	...	39 17.13	24.25	1.00	.	4	7.26	18 41.99	7.50	3.61	39 42.38	19 43.10
59	8	...	...	...	...	14.6	...	...	40 58.36	24.26	0.99	V.	9	11.42	44 49.01	7.52	13.84	41 23.61	46 6.37
60	8	...	...	...	...	...	35.4	...	40 3.06	24.26	0.99	VII.	9	11.38	44 46.41	7.51	−13.83	40 28.31	45 57.75
61	7	...	...	...	...	...	...	49.3	41 1.11	24.26	1.01	VII.	1	16.22	8 14.49	7.52	+ 0.33	41 26.38	9 11.68
62	10	...	7.8	23.4	...	...	...	...	48 39.84	24.31	1.01	.	3	5.24	12 41.47	7.63	− 1.35	49 5.16	13 40.45
63	9	2.6	19.0	35.2	...	...	...	...	49 51.26	24.32	1.00	.	5	4.44	22 19.51	7.65	4.96	50 16.58	23 22.12
64	9	...	...	4.9	21.0	...	...	...	50 21.05	24.32	1.01	.	3	1.4	10 30.37	7.65	0.51	50 46.38	11 28.53
65	7	...	20.9	37.0	...	...	...	...	51 53.24	24.33	1.01	.	6	8.57	28 28.18	7.66	7.38	52 18.58	29 33.22
66	10	24.9	40.9	...	...	...	...	...	54 13.27	24.34	1.00	II.	4	6.49	18 23.05	7.68	3.49	54 38.61	19 24.22
67	9	11.9	28.1	...	...	...	...	...	56 0.34	24.36	1.01	II.	3	9.48	14 54.30	7.69	2.16	56 25.71	15 54.15
68	10	...	...	...	6.1	...	...	...	57 6.15	24.36	1.01	.	2	11.16	10 39.87	7.70	0.57	57 31.52	11 38.14
69	10	...	...	47.7	...	...	...	...	13 58 3.90	24.37	1.00	.	6	13.4	30 32.72	7.71	8.15	13 58 29.27	31 38.58
70	10	...	19.5	35.2	...	...	...	...	14 2 51.58	24.40	1.01	III.	2	11.49	10 56.44	7.73	− 0.66	14 3 16.99	11 54.83
71	9	...	...	...	12.7	...	...	...	3 12.72	24.40	1.01	.	1	7.55	3 59.51	7.73	+ 1.90	3 38.13	4 55.34
72	9	...	...	...	...	5.3	21.6	...	3 49.17	24.40	1.00	.	3	10.30	15 15.78	7.73	− 2.29	4 14.57	16 15.80
73	8	...	9.7	25.8	...	...	...	...	5 42.02	24.41	1.00	.	5	5.27	22 41.18	7.74	5.11	6 7.43	23 44.03
74	10	...	...	...	23.9	...	...	...	9 23.98	24.44	1.00	.	4	5.56	17 56.61	7.75	3.31	9 49.42	18 57.67
75	10	...	15.5	...	...	...	...	...	14 15 47.85	+24.48	+1.00	II.	4	10.19	−20 8.94	− 7.75	− 4.14	14 16 13.33	−39 21 10.83

ZONE 229. APRIL 2. B. BELT, −24° 23′. D_0 = −23° 57′ 40″.

No.	Mag.	I.	II.	III.	IV.	V.	VI.	VII.	T.	a_1	a_2	MICROMETER.			i	d_1	d_2	Mean Right Ascension, 1850.0.	Mean Declination, 1850.0.
1	8	...	...	...	54.	7.6	21.	...	10 9 53.82	+24.55	+1.00	VI.	5	4.07	−22 0.69	− 1.52	− 5.76	10 10 19.37	−24 19 47.97
2	9	18.7	32.	46.	...	...	...	...	11 59.70	24.55	0.99	III.	7	10.33	34 16.55	1.71	6.78	12 25.24	32 5.07
3	9	...	...	...	...	32.	45.5	...	12 18.16	24.55	1.00	VI.	6	13.45	30 53.23	1.78	6.47	12 43.71	28 41.48
4	7.8	...	...	...	56.6	10.	...	...	13 56.38	24.55	0.99	V.	8	10.32	39 14.75	1.95	7.18	14 21.92	37 3.88
5	8	41.6	55.2	9.	...	...	...	...	17 22.67	24.54	1.00	III.	3	9.45	14 53.04	2.32	5.17	17 48.21	12 40.53
6	10	...	15.3	29.	...	...	...	...	18 42.75	24.54	0.99	III.	9	10.37	44 16.27	2.46	7.62	19 8.28	42 6.35
7	10	...	...	...	...	...	27.	41.	18 59.68	24.54	0.99	VII.	8	6.53	37 24.25	2.60	7.03	19 25.21	35 13.88
8	11	...	...	...	59.5	...	...	...	20 59.56	24.54	1.00	IV.	3	9.33	14 47.03	2.70	5.16	21 25.10	12 34.89
9	9	...	...	57.5	11.5	25.	...	...	10 22 11.37	+24.54	+1.00	V.	4	11.22	−20 40.95	− 2.83	− 5.65	10 22 36.91	−24 18 29.43

CORRECTIONS.

Date.	Corr. of Clock.	Hourly rate.	m	n	c
1849. h.	s.	s.	s.	s.	s.

INSTRUMENT READINGS.

Date.	Barom.	THERMOM. At.	Ex.
1849. h. m.	in.	°	°

REMARKS.

ZONE 229. APRIL 2. B. BELT, −24° 23′. D_o = −23° 57′ 40″—Continued.

No.	Mag.	Seconds of Transit. I.	II.	III.	IV.	V.	VI.	VII.	T.	a_1	a_2	Micrometer.		i	d_1	d_2	Mean Right Ascension, 1850.0.	Mean Declination, 1850.0.
									h. m. s.	s.	s.		r.	′ ″	″	″	h. m. s.	° ′ ″
10	9	..	..	..	19.	32.4	..	..	10 23 18.84	+24.53	+1.00	V.	6 10.495	−29 24.87	− 2.94	− 6.35	10 23 44.37	−24 27 14.16
11	9	37.	51.	4.5	..	..	..	..	25 18.28	24.53	1.00	III.	6 9.29	28 44.31	3.15	6.30	25 43.81	26 33.76
12	8	..	..	..	..	..	52.5	6.	25 24.95	24.53	1.00	VII.	5 9.17	24 36.81	3.16	5.97	25 50.48	22 25.94
13	9	..	..	5.	19.	32.7	..	..	27 16.93	24.53	1.00	V.	5 3.00½	21 27.27	3.36	5.71	27 44.46	19 16.34
14	9	..	25.	38.4	52.2	..	..	..	29 52.28	24.52	1.00	IV.	4 8.27	19 12.75	3.63	5.53	30 17.80	17 1.91
15	10	..	..	..	..	..	44.	57.5	30 16.48	24.52	1.00	VII.	3 12.3	16 2.31	3.67	5.26	30 42.00	13 51.24
16	9	..	11.	24.3	38.	..	..	..	32 38.17	24.52	1.00	VI.	5 7.50	23 53.29	3.90	5.91	33 3.69	21 43.10
17	8	3.	17.	30.6	..	..	..	..	34 44.26	24.52	1.01	III.	3 4.51	12 24.79	4.11	4.96	35 9.79	10 13.86
18	8	..	35.5	49.2	3.	..	..	..	36 2.98	24.51	1.00	IV.	4 10.16½	20 7.97	4.24	5.60	36 28.49	17 57.81
19	10	..	..	10.	23.3	..	..	..	37 23.54	24.51	1.00	III.	4 12.41	21 20.78	4.37	5.70	37 49.05	19 10.85
20	7.8	..	..	..	..	..	25.3	39.2	37 57.97	24.51	1.00	VII.	3 12.52	16 27.02	4.43	5.29	38 23.48	14 16.74
21	6.7	..	..	..	..	..	40.	53.2	39 12.33	24.51	0.99	.	. .	. . .	. .	. .	39 37.83	
22	10	..	..	46.3	60.2	..	..	..	41 0.13	24.51	1.00	IV.	5 6.30	24 13.47	4.72	5.94	41 25.64	22 4.13
23	8	..	..	51.4	5.3	..	..	..	43 5.14	24.51	0.99	IV.	10 3.46	45 49.96	4.92	7.75	43 30.64	43 42.63
24	8	..	36.4	50.5	4.	..	..	..	46 4.02	24.51	1.01	IV.	1 6.47	3 25.23	5.20	4.22	46 29.54	1 14.65
25	8	..	..	..	..	16.	30.	..	47 2.38	24.50	1.01	VI.	1 5.44	2 53.30	5.29	4.18	47 27.89	0 42.77
26	9	..	16.5	30.	44.	..	..	..	49 43.89	24.50	1.00	IV.	7 5.22	31 39.76	5.51	6.54	50 9.39	29 31.84
27	8	..	46.5	0.	13.8	..	..	..	52 13.78	24.50	0.99	IV.	10 3.23	45 38.35	5.77	7.74	52 39.27	43 31.86
28	10	..	..	..	27.	40.3	..	..	55 26.82	24.50	1.00	V.	5 8.27	24 11.91	6.06	5.94	55 52.32	22 3.91
29	11	10.3	24.2	..	..	..	..	..	10 57 51.50	24.50	1.00	IV.	3 9.17	14 38.96	6.28	5.14	10 58 17.00	12 30.38
30	8	..	..	..	..	35.	48.4	..	11 2 21.14	24.50	0.99	VI.	10 11.20½	49 38.97	6.69	8.11	11 2 46.63	47 33.77
31	10	..	..	..	9.	22.5	..	..	4 8.83	24.50	1.00	V.	6 11.16	29 38.23	6.86	6.38	4 34.38	27 31.46
32	10	..	..	21.2	35.	..	..	..	6 34.90	24.49	0.99	IV.	9 5.28	41 40.48	7.07	7.41	7 0.38	39 34.96
33	10	..	49.7	3.	17.	..	..	..	8 16.95	24.49	1.00	IV.	7 5.50	31 53.88	7.22	6.56	8 42.44	29 47.66
34	9	33.	46.5	59.5	14.	..	..	..	11 13.84	24.49	1.00	IV.	6 4.41	26 19.09	7.49	6.11	11 39.33	24 12.69
35	9	17.	31.	44.3	..	..	..	..	12 58.22	24.49	1.00	III.	6 8.29	28 14.02	7.64	6.26	13 23.71	26 7.92
36	10	..	..	..	..	46.	0.	13.5	13 32.42	24.49	1.00	VII.	3 12.52	16 27.02	7.69	5.29	13 57.91	14 20.00
37	8	..	..	..	..	..	54.5	8.	14 26.95	24.49	1.00	VII.	5 8.51	24 23.70	7.77	5.95	14 52.44	22 17.42
38	8	..	..	..	..	..	57.3	11.	15 29.80	24.49	0.99	VII.	10 6.57	46 55.67	7.86	7.90	15 55.28	44 51.43
39	10	..	10.5	24.	37.8	..	..	..	18 37.83	24.49	1.00	.	6 6.27	27 12.54	8.13	6.18	19 3.32	25 6.85
40	9	..	..	18.	32.	..	..	..	20 31.86	24.50	1.01	IV.	1 10.19	5 12.13	8.29	4.37	20 57.37	3 4.79
41	6.7	..	26.	39.3	53.	7.	..	..	21 53.21	24.50	1.00	V.	3 5.4	12 31.35	8.41	4.95	22 18.71	10 24.71
42	10	..	..	22.	36.	..	..	..	23 35.87	24.50	1.00	IV.	6 3.13	25 34.71	8.55	6.05	24 1.37	23 29.31
43	10	..	6.5	20.5	34.2	..	..	..	26 34.12	24.50	1.00	IV.	7 4.20	31 6.50	8.60	6.50	26 59.62	29 3.80
44	10	..	..	..	22.5	36.	50.	..	30 22.40	24.50	0.99	VI.	10 7.14	47 34.68	9.11	7.93	30 47.89	45 31.72
45	7.8	..	..	..	..	..	54.	..	32 26.38	24.50	1.01	VII.	1 11.53	5 59.17	9.28	4.42	32 51.89	3 52.87
46	9	..	57.3	11.	24.5	..	..	..	37 24.67	24.50	1.00	IV.	3 5.11	12 34.91	9.69	4.96	37 50.17	10 29.56
47	7.8	..	..	..	..	40.	53.5	7.	38 26.10	24.50	1.01	VII.	2 9.00	10 31.45	9.77	4.71	38 51.61	8 25.93
48	8	..	..	17.	31.	..	..	..	40 30.90	24.50	1.00	IV.	3 14.6	17 4.68	9.94	5.33	40 56.40	14 59.95
49	10	..	..	..	10.	23.5	..	..	44 9.94	24.51	1.00	V.	4 11.52	20 56.08	10.25	5.66	44 35.45	18 51.99
50	8	..	17.5	31.2	45.	..	..	..	47 41.98	24.51	1.00	IV.	4 11.7	20 33.43	10.54	5.62	48 10.49	18 29.59
51	9	..	32.	46.	59.5	..	..	..	48 59.53	24.51	0.99	IV.	8 13.14	40 36.46	10.68	7.33	49 25.03	38 34.44
52	11	..	14.	27.5	..	..	..	..	50 41.33	24.51	1.00	III.	3 12.5	16 3.63	10.78	5.25	51 6.84	13 59.66
53	10	..	41.	55.	8.5	..	..	..	52 8.56	24.52	1.00	V.	6 8.48	28 23.60	10.89	6.29	52 34.08	26 20.77
54	10	..	13.5	27.3	41.	..	..	..	53 41.01	24.52	1.00	IV.	3 11.26½	15 44.26	11.01	5.22	54 6.53	13 40.49
55	9	..	..	21.3	35.	49.	..	..	56 35.14	24.52	1.01	V.	1 11.48½	5 59.21	11.24	4.41	57 0.67	3 52.86
56	9	..	..	5.	19.	..	46.	..	58 18.74	24.52	0.99	VI.	9 8.39	43 16.65	11.37	7.58	11 58 44.25	41 15.60
57	9	..	..	..	39.5	53.	..	..	11 59 39.36	24.52	1.00	V.	7 6.32	32 15.03	11.47	6.60	12 0 4.88	30 13.10
58	8	..	53.	6.6	20.3	34.	..	..	12 2 20.36	+24.53	+1.01	V.	2 8.46	− 9 24.20	−11.67	− 4.69	12 2 45.90	−24 7 20.56

CORRECTIONS.

Date.		Corr. of Clock.	Hourly rate.	m	n	c
1849.	h.	s.	s.	s.	s.	s.

INSTRUMENT READINGS.

Date.		Barom.	Thermom. At.	Thermom. Ex.
1849.	h. m.	in.	°	°

REMARKS.

(229) 38. Micrometer reading assumed as 5r.57 instead of 6r.57.
(229) 47. Micrometer reading assumed as 11r.00 instead of 9r.00.

ZONE 229. APRIL 2. B. BELT, −24° 23'. D_0 = −23° 57' 40"—Continued.

No.	Mag.	Seconds of Transit. I.	II.	III.	IV.	V.	VI.	VII.	T.	a_1	a_2	Micrometer.			i	d_1	d_2	Mean Right Ascension, 1850.0.	Mean Declination, 1850.0.
									h. m. s.	s.	s.			r.	' "	"	"	h. m. s.	° ' "
59	8						31.4		12 3 3.77	+24.53	+1.01	VI.	1	5.17	− 2 39.68	−11.73	− 4.15	12 3 29.31	−24 0 35.56
60	8				43.3	57.			4 43.36	24.53	1.01	V.	1	3.3	1 32.23	11.85	4.05	5 8.90	23 59 28.13
61	10					50.	4.		5 36.42	24.53	1.00	VI.	7	7.38	32 48.18	11.91	6.64	6 1.95	24 30 46.73
62	10		5.3	19.	33.				8 32.83	24.54	1.00	IV.	6	8.9	28 3.98	12.13	6.25	8 58.37	26 2.36
63	11					35.	49.		11 21.42	24.54	1.00	VI.	7	7.17	32 37.59	12.33	6.63	11 46.96	30 36.55
64	11		41.5	55.	9.				14 8.91	24.55	1.00	IV.	5	7.31	23 43.71	12.53	5.88	14 34.46	21 42.12
65	9				0.6	14.	27.7		15 0.37	24.55	1.00	VI.	6	9.00	28 29.53	12.59	6.28	15 25.92	26 28.40
66	9				8.3	22.	36.		16 8.38	24.55	1.00	VI.	5	9.154	24 36.24	12.66	5.97	16 33.93	22 34.87
67	9					41.	55.	8.5	17 27.44	24.55	1.01	VII.	2	6.14	8 7.24	12.75	4.59	17 53.00	6 0.56
68	9		9.4	23.	36.5				19 36.69	24.56	1.00	IV.	7	5.19	31 38.25	12.90	6.55	20 2.25	29 37.70
69	9			11.3	25.				21 25.00	24.56	1.00	IV.	7	3.26	30 41.27	13.02	6.47	21 50.56	28 40.76
70	10							26.4	21 45.25	24.56	1.00	VII.	6	6.58	27 27.82	13.04	6.20	22 10.81	25 27.06
71	9	19.4	33.5	47.2	1.2				25 0.91	24.57	1.00	IV.	6	8.20	28 9.52	13.25	6.25	25 26.48	26 9.02
72	9	38.	51.4	5.					28 18.92	24.58	1.00	III.	7	4.32	31 4.56	13.49	6.51	28 44.50	29 4.56
73	9						13.4	27.	28 45.90	24.58	1.00	VII.	6	2.58	25 26.80	13.52	6.03	29 11.48	23 26.35
74	8					29.	43.	56.5	30 15.36	24.58	0.99	VII.	9	4.45	41 18.46	13.63	7.39	30 40.93	39 19.48
75	10					6.	19.2		31 52.01	24.58	1.00	VI.	6	4.41	26 18.93	13.73	6.10	32 17.59	24 18.76
76	9						34.	47.7	33 6.58	24.59	1.00	VII.	3	4.12	12 4.82	13.82	4.90	33 32.17	10 13.54
77	7			37.5	51.2	5.			34 51.27	24.59	1.00	VI.	3	3.39	11 48.37	13.93	4.87	35 16.86	9 47.17
78	7		56.	9.3	23.				37 23.16	24.60	1.00	IV.	3	2.30	11 13.74	14.09	4.83	37 48.76	9 12.66
79	9		35.	49.	2.5				41 2.55	24.61	1.00	IV.	7	2.45	30 20.60	14.32	6.45	41 28.16	28 21.37
80	9				40.	54.			42 40.16	24.61	1.00	V.	6	4.3	25 59.89	14.43	6.08	43 5.77	24 0.40
81	9		58.5	12.3	26.				46 26.03	24.62	0.99	IV.	8	9.18	35 37.46	14.66	7.16	46 51.64	36 39.28
82	11				20.5	33.8			48 20.24	24.63	0.99	V.	8	4.40	36 17.25	14.78	6.97	48 45.86	34 19.00
83	8			38.	51.4	5.			49 51.49	24.63	1.00	V.	5	6.14	23 4.83	14.87	5.83	50 17.12	21 5.55
84	9			42.	56.				51 55.89	24.64	1.00	IV.	3	10.38	15 19.81	14.99	5.17	52 21.53	13 19.97
85	10	58.5	12.2	26.					55 39.70	24.65	1.00	III.	7	8.47	33 23.10	15.21	6.68	12 56 5.35	31 24.99
86	10		15.	29.3	43.				12 59 42.83	24.66	1.00	IV.	6	8.33	28 16.08	15.41	6.26	13 0 8.49	26 17.78
87	10				45.	59.	12.5		13 0 45.11	24.66	1.00	V.	6	5.6	26 31.66	15.50	6.12	1 10.77	24 33.28
88	9					3.7	17.	31.	1 49.82	24.67	1.00	VI.	3	11.12	15 30.79	15.56	5.20	2 15.49	13 37.55
89	9		18.	31.8	45.3				5 45.43	24.68	1.00	IV.	6	5.42	26 49.85	15.77	6.14	6 11.11	24 51.76
90	9	56.	9.8	24.	37.				8 37.27	24.69	1.00	IV.	3	11.48	15 55.10	15.93	5.22	9 2.96	13 56.25
91	9		56.	9.4					10 23.29	24.70	1.00	III.	6	10.31	29 15.54	16.02	6.36	10 48.99	27 17.92
92	9					44.3	58.	12.	10 30.67	24.70	1.00	VII.	4	8.46	19 21.99	16.02	5.52	10 56.37	17 23.53
93	9	14.4	28.4	42.					12 55.66	24.71	1.00	III.	2	10.37	10 20.17	16.14	4.74	13 21.37	8 21.05
94	7.8	6.5	20.4	34.					15 47.73	24.71	1.00	III.	5	5.8	22 31.56	16.27	5.77	16 13.44	20 33.66
95	10			21.	35.				16 34.90	24.72	1.00	IV.	4	6.2	17 59.64	16.31	5.39	17 0.62	16 1.34
96	10		31.	44.5	58.				17 58.23	24.72	1.09	IV.	6	2.23	25 9.50	16.37	6.01	18 23.95	23 11.88
97	8	7.	20.3	34.					19 47.88	24.73	1.00	III.	6	7.39	27 48.81	16.48	6.22	20 13.61	25 51.51
98	11			47.	0.4				0.56	24.74	1.01	IV.	1	7.42	3 52.96	. .	4.23	26.31	
99	10		9.	22.5	36.4				36.33	24.75	1.00	IV.	8	8.44	38 20.33	16.69	7.14	26 2.08	36 24.16
100	11					37.	50.3		26 23.06	24.76	1.00	VI.	7	9.51	33 55.24	16.74	6.75	26 48.82	31 58.73
101	8	38.	51.3	5.	19.				30 18.87	24.77	1.00	IV.	2	5.14	7 37.33	16.91	4.51	30 44.64	5 38.75
102	9						37.		31 9.60	24.77	1.00	VI.	9	4.45	41 18.65	16.95	7.43	31 35.37	39 23.03
103	10				54.				32 53.99	24.78	1.00	IV.	6	9.45	28 52.38	17.01	6.35	33 19.77	26 55.74
104	10		19.5	33.2					46.93	24.78	1.00	III.	3	9.8	14 34.38	. .	5.11		
105	7			46.3	0.2	14.			36 0.11	24.79	0.99	V.	10	4.58	46 26.22	17.13	7.87	36 25.89	44 31.22
106	10							56.3	37 15.02	24.80	1.00	VII.	9	7.13	42 33.08	17.18	7.51	37 40.82	40 37.77
107	10		33.5	47.	0.5				13 41 0.70	+24.81	+1.00	IV.	8	8.5	−38 0.66	−17.33	− 7.12	13 41 26.51	−24 36 5.11

CORRECTIONS.

Date.		Corr. of Clock.	Hourly rate.	m	n	c
1849.	h.	s.	s.	s.	s.	s.

INSTRUMENT READINGS.

Date.		Barom.	Thermom. At.	Thermom. Ex.
1849.	h. m.	in.	°	°

REMARKS.

(229) 99. Minutes assumed as 25.

ZONE 229. APRIL 2. B. BELT, −24° 23'. D_o = −23° 57' 40"—Continued.

No.	Mag.	SECONDS OF TRANSIT. I.	II.	III.	IV.	V.	VI.	VII.	T.	a_1	a_2	MICROMETER.		i	d_1	d_2	Mean Right Ascension, 1850.0.	Mean Declination, 1850.0.
									h. m. s.	s.	s.		r.	' "	"	"	h. m. s.	° ' "
108	9	..	...	...	0.	13.6	27.3	...	13 42 59.91	+24.82	+1.00	VI.	3 11.35	−15 48.39	−17.39	− 5.20	13 43 25.73	−24 13 50.98
109	9	..	0.5	14.	28.	...	...	...	46 27.89	24.83	1.00	IV.	6 10.48	29 24.15	17.52	6.39	46 53.72	27 28.06
110	11	..	..	19.5	33.	...	...	...	33.12	24.84	1.00	IV.	1 13.11	6 38.84	..	4.44		
111	6	..	..	28.	42.	55.5	9.	...	50 41.76	24.85	1.00	VI.	3 12.8	16 5.02	17.66	5.29	50 7.61	14 7.97
112	6	..	..	58.	11.4	25.	39.	...	52 11.49	24.86	1.00	VI.	4 6.40	18 18.64	17.71	5.40	51 37.35	16 21.75
113	9	..	3.4	17.	30.5	44.	...	...	54 30.61	24.87	1.01	V.	1 11.2	5 33.77	17.78	4.34	54 56.40	3 35.89
114	8	..	52.	5.3	19.	...	...	...	57 19.16	24.88	1.01	IV.	1 5.58½	3 0.76	17.86	4.15	57 45.05	1 2.77
115	9	22.	36.	49.4	3.	...	...	...	13 59 3.19	+24.89	+1.00	IV.	5 2.41	−21 17.48	−17.91	− 5.66	13 59 29.08	−24 19 21.05

ZONE 230. APRIL 5. K. BELT, −25° 1'. D_o = −24° 34' 30".

No.	Mag.	I.	II.	III.	IV.	V.	VI.	VII.	T.	a_1	a_2	MICROMETER.		i	d_1	d_2	Mean Right Ascension, 1850.0.	Mean Declination, 1850.0.
1	5.6	..	56.3	10.3	24.4	...	...	...	9 14 24.08	+26.13	+0.99	.	10 2.14	−45 3.57	− 8.13	− 8.70	9 14 51.20	−25 19 50.40
2	9	39.9	53.4	...	...	...	...	...	17 21.06	26.12	1.01	III.	3 3.5	11 31.34	8.48	4.88	17 48.19	24 46 14.70
3	10	..	49.6	3.2	...	...	...	...	18 17.10	26.11	1.00	III.	5 6.10	23 2.83	8.60	5.85	18 44.21	57 47.28
4	6	..	...	...	...	...	56.4	9.9	18 28.71	26.11	1.01	VI.	2 3.35	6 47.25	8.62	4.48	18 55.83	41 30.35
5	10	..	...	5.9	19.8	...	...	...	20 19.76	26.11	1.01	.	3 2.30	11 13.74	8.84	4.85	20 46.88	45 57.43
6	7	..	2.0	15.8	29.6	...	...	...	21 29.60	26.11	1.01	.	3 4.8	12 3.15	8.96	4.93	21 56.72	46 47.04
7	9	..	...	...	3.3	...	...	...	22 3.34	26.10	1.00	.	5 10.6	25 1.88	9.03	6.00	22 30.44	59 46.91
8	7	..	...	...	...	50.2	4.3	...	22 36.54	26.10	1.00	.	4 5.9	17 32.91	9.09	5.36	23 3.64	52 17.36
9	8	..	...	...	...	45.2	58.9	...	23 31.34	26.10	1.00	VI.	5 5.32	22 43.55	9.19	5.82	23 58.44	57 28.56
10	10	..	...	...	0.0	...	...	...	25 0.08	26.09	1.00	.	4 11.57	20 58.64	9.35	5.67	25 27.17	24 55 43.66
11	9	50.1	4.3	...	...	...	...	...	26 31.69	26.09	1.00	III.	6 12.0	30 0.41	9.52	6.42	26 58.78	25 4 46.35
12	9	..	...	17.1	...	...	...	...	27 30.91	26.09	0.99	.	8 11.51	39 54.61	9.63	7.27	27 57.99	14 41.51
13	10	..	8.1	...	...	...	...	...	28 35.70	26.08	0.99	II.	8 8.4	38 0.00	9.74	7.13	29 2.77	12 46.87
14	7	..	...	...	52.7	...	...	...	28 52.59	26.08	0.99	.	8 12.0	39 59.15	9.78	7.28	29 19.66	14 46.21
15	7	..	...	...	...	...	...	41.7	29 0.19	26.08	0.99	.	9 7.44	42 59.07	9.79	7.55	29 27.26	25 17 46.41
16	6	..	...	...	...	...	...	29.8	29 48.70	26.08	1.02	VII.	1 5.48	2 55.12	9.88	4.17	30 15.80	24 37 39.17
17	9	39.1	...	6.8	...	...	...	...	34 20.58	26.07	0.99	.	8 8.15	38 5.70	10.38	7.14	34 47.64	25 12 53.22
18	9	..	...	...	...	...	58.2	...	34 30.67	26.07	0.99	VI.	9 6.36	42 14.63	10.40	7.48	34 57.73	17 2.51
19	10	..	...	23.9	...	...	...	...	36 37.71	26.06	0.99	.	9 5.40	41 46.54	10.63	7.44	37 4.76	16 34.61
20	9	..	43.8	57.2	...	...	...	...	38 11.20	26.06	1.00	.	7 5.18	31 37.74	10.80	6.56	38 38.26	6 25.10
21	9	..	...	...	...	...	40.8	...	38 13.19	26.06	1.00	VI.	6 5.27	26 42.12	10.80	6.15	38 40.25	25 1 29.07
22	9	..	...	...	...	28.5	42.2	...	39 14.65	26.05	1.00	VI.	5 8.19	24 7.76	10.92	5.92	39 41.70	24 58 54.60
23	9	..	...	...	...	...	45.6	...	40 17.82	26.05	1.01	VI.	1 3.38	1 49.76	11.01	4.89	40 44.88	24 36 35.69
24	10	..	33.4	...	...	...	...	...	43 1.00	26.05	0.99	II.	8 10.31	39 14.12	11.36	7.22	43 28.04	25 14 2.70
25	9	..	...	...	14.8	...	...	...	44 14.72	26.04	0.99	.	8 4.52	36 23.34	11.51	6.99	44 41.75	11 11.84
26	5	15.9	30.1	43.8	...	...	...	...	46 57.56	26.04	0.99	.	8 9.50	38 53.60	11.84	7.21	47 24.59	25 13 42.65
27	9	..	...	...	...	...	...	30.2	46 49.02	26.04	1.01	VII.	2 12.6	11 4.73	11.82	4.83	47 16.07	24 45 51.38
28	9	..	...	...	...	...	42.6	...	48 14.92	26.03	1.01	VI.	3 8.57	14 28.72	11.99	5.11	48 41.96	49 15.82
29	9	..	...	...	...	35.5	...	...	49 21.81	26.03	1.01	V.	1 5.11	2 36.77	12.12	4.12	49 48.85	37 23.01
30	9	30.9	44.9	...	...	...	...	...	55 12.32	26.02	1.01	II.	2 9.31	9 46.76	12.77	4.71	55 39.35	44 31.24
31	7	..	...	...	20.4	...	...	...	56 20.41	26.02	1.01	.	1 1.36	0 48.41	12.90	3.98	56 47.44	35 35.29
32	9	..	...	...	...	...	...	15.0	56 33.88	26.02	1.01	.	2 0.5	5 1.52	12.92	4.33	57 0.91	39 48.77
33	9	..	...	...	...	...	34.9	...	9 58 7.27	26.01	1.00	VI.	5 5.15	22 34.97	13.09	5.81	9 58 34.28	24 57 23.87
34	9	..	16.6	...	...	...	...	...	10 1 44.20	26.01	1.00	II.	6 9.14	28 36.59	13.47	6.32	10 2 11.21	25 3 26.38
35	9	..	0.6	...	28.0	...	...	...	3 28.10	26.00	1.01	.	2 4.27	7 13.63	13.65	4.50	3 55.11	24 42 1.78
36	9	48.8	2.4	...	...	...	...	...	7 30.02	26.00	1.01	II.	3 7.29	13 44.34	14.05	5.05	7 57.03	48 33.44
37	9	..	...	...	...	49.2	...	...	10 7 35.52	+26.00	+1.01	V.	1 3.12	− 1 36.77	−14.06	− 4.05	10 8 2.53	−24 36 24.88

CORRECTIONS.

Date.		Corr. of Clock.	Hourly rate.	m	n	c
1849.	h.	s.	s.	s.	s.	s.

INSTRUMENT READINGS.

Date.		Barom.	THERMOM. At.	Ex.
1849.	h. m.	in.	°	°

REMARKS.

(229) 111. Minutes assumed as 49 instead of 50.
(229) 112. Minutes assumed as 51 instead of 52.

Zone 230. April 5. K. Belt, −25° 1′. $D_0 = -24° 34′ 30″$—Continued.

No.	Mag.	Seconds of Transit. I.	II.	III.	IV.	V.	VI.	VII.	T.	a_1	a_2	Micrometer.			i	d_1	d_2	Mean Right Ascension, 1850.0.	Mean Declination, 1850.0.
									h. m. s.	s.	s.			r.	′ ″	″	″	h. m. s.	° ′ ″
38	9	..	32.5	46.4	..	..	..	..	10 10 0.15	+25.99	+1.00	.	5	6.17	−23 6.40	−14.30	− 5.85	10 10 27.14	−24 57 56.55
39	9	..	..	..	27.8	..	..	..	10 27.84	25.99	1.01	.	2	9.55	9 59.02	14.35	4.73	10 54.84	44 48.10
40	10	..	..	22.8	..	..	..	..	11 36.58	25.99	1.01	.	3	11.56	15 59.14	14.46	5.23	12 3.58	50 48.83
41	7.6	..	..	..	53.8	8.5	..	..	13 54.24	25.99	1.01	.	1	4.34	2 18.17	14.69	4.09	14 21.24	24 37 6.95
42	7	21.8	36.4	50.3	..	..	..	..	16 3.80	25.98	0.99	.	9	8.20	43 7.22	14.91	7.58	16 30.77	25 17 59.71
43	6	..	..	44.9	58.7	..	..	..	16 58.72	25.98	1.01	.	3	7.21	13 40.46	15.00	5.05	17 25.71	24 48 30.51
44	9	..	..	..	..	..	49.5	..	17 21.90	25.98	1.00	VI.	6	6.34	27 15.92	15.04	6.21	17 48.88	25 2 7.17
45	9	..	..	..	41.7	55.2	..	..	18 41.60	25.98	1.00	.	5	2.50	21 22.02	15.17	5.69	19 8.58	24 56 12.88
46	9	0.3	14.7	..	..	..	..	..	20 41.98	25.98	1.00	III.	6	7.36	27 47.30	15.37	6.25	21 8.96	25 2 38.92
47	9	..	..	..	..	45.2	..	..	21 31.47	25.98	1.01	V.	3	6.21	13 10.17	15.45	5.01	21 58.46	24 48 0.63
48	9	..	..	..	..	38.2	52.4	..	23 24.62	25.97	0.99	VI.	9	11.35	44 45.39	15.64	7.71	23 51.58	25 19 38.74
49	10	..	..	..	..	47.7	..	..	25 33.96	25.97	1.01	.	3	8.27	14 13.75	15.86	5.08	26 0.94	24 49 4.69
50	9	..	..	26.4	..	..	..	..	27 40.18	25.97	1.01	.	1	13.8	6 37.33	16.07	4.44	28 7.16	24 41 27.84
51	10	..	..	..	56.8	..	..	..	28 56.79	25.96	1.00	.	6	11.20	29 40.28	16.19	6.40	29 23.75	25 4 32.87
52	10	..	..	..	58.8	..	..	..	30 58.65	25.96	0.99	.	9	9.50	43 52.60	16.39	7.64	31 25.60	18 46.63
53	7	..	..	..	36.0	49.9	..	..	31 36.05	25.96	1.00	.	6	10.35	29 17.60	16.44	6.37	32 3.01	4 10.41
54	6	..	51.2	4.9	..	..	..	..	34 18.76	25.96	0.99	.	9	3.52	40 52.08	16.69	7.38	34 45.71	15 46.15
55	9	..	..	..	..	..	..	54.9	34 13.51	25.96	1.00	VII.	7	2.13	30 4.10	16.68	6.44	34 40.47	4 57.22
56	7	..	..	..	35.2	48.9	..	..	35 35.14	25.96	1.00	.	7	2.1	29 58.41	16.80	6.42	36 2.10	4 51.63
57	6.7	..	40.8	54.6	..	..	..	..	39 8.40	25.95	0.99	.	9	3.37	40 44.52	17.12	7.37	39 35.34	25 15 39.01
58	7	..	..	..	..	..	..	51.7	39 10.53	25.95	1.01	VII.	2	9.40	9 51.10	17.13	4.69	39 37.49	24 44 42.92
59	9	..	35.5	49.6	..	..	..	..	43 3.23	25.95	1.01	.	2	7.42	8 51.96	17.47	4.63	43 30.19	24 43 44.06
60	9	..	..	..	..	15.1	..	..	45 1.31	25.95	1.00	V.	6	12.30	30 15.54	17.65	6.45	45 28.26	25 5 9.64
61	9	..	..	..	2.0	..	..	..	47 2.07	25.95	1.00	.	4	6.27	18 12.24	17.83	5.41	47 29.02	24 53 5.48
62	9	..	8.2	21.9	..	..	..	..	48 35.73	25.95	1.01	.	3	4.30	12 14.25	17.97	4.91	49 2.69	24 47 7.13
63	9	..	32.0	..	..	..	..	..	50 59.60	25.95	0.99	.	9	9.20	43 37.47	18.19	7.63	51 26.54	25 18 33.29
64	9	..	44.2	..	..	..	..	..	52 11.78	24.95	1.01	.	2	7.32	8 46.92	18.30	4.61	52 37.74	24 43 39.83
65	8	..	..	38.7	52.1	..	..	..	52 52.30	24.95	1.00	.	6	8.18	28 8.51	18.36	6.28	53 18.25	25 3 3.15
66	7	..	..	..	..	..	29.3	43.2	53 1.79	25.95	1.00	.	4	9.39	19 49.06	18.37	5.55	53 28.74	24 54 42.98
67	7	..	..	45.5	59.4	..	..	..	54 59.34	25.95	1.00	.	6	11.58	29 59.44	18.55	6.43	55 26.29	25 4 54.42
68	5	..	..	..	..	36.3	..	..	55 22.58	25.95	1.01	.	2	12.28	11 16.16	18.58	4.82	55 49.54	24 46 9.56
69	8	..	..	..	..	..	26.0	..	55 58.32	25.95	1.00	.	3	12.33	16 17.79	18.64	5.25	56 25.27	24 51 11.68
70	9	43.3	..	..	..	..	..	..	10 59 24.78	+25.94	+0.99	.	9	5.28	−41 40.48	−18.95	− 7.48	10 59 51.71	−25 16 36.91

Zone 231. April 5. K. Belt, −25° 1′. $D_0 = -24° 35′ 50″$.

No.	Mag.	I.	II.	III.	IV.	V.	VI.	VII.	T.	a_1	a_2	Micrometer.			i	d_1	d_2	Mean Right Ascension, 1850.0.	Mean Declination, 1850.0.
1	9	..	..	..	53.8	..	..	..	13 6 53.84	+26.15	+1.00	.	5	8.50	−24 23.55	−25.64	− 5.95	13 7 20.99	−25 0 45.14
2	9	..	13.5	27.2	..	..	..	..	9 41.03	26.16	1.00	.	2	4.51	7 25.73	25.70	4.44	10 8.19	24 43 45.87
3	9	..	..	7.2	20.8	..	..	..	10 20.91	26.17	1.00	.	3	3.42	11 50.04	25.72	4.82	10 48.08	48 10.58
4	9	..	..	..	2.2	15.7	..	..	11 2.11	26.17	1.00	.	4	2.00	15 57.61	25.73	5.20	11 29.28	24 52 18.54
5	8	..	21.3	34.9	..	..	..	..	13 48.80	26.18	1.00	.	6	5.50	26 53.88	25.80	6.17	14 15.98	25 3 15.85
6	9	..	..	..	29.9	..	..	..	17 29.73	26.19	0.99	.	10	7.49	47 52.49	25.88	8.05	17 56.91	25 24 16.42
7	9	..	..	..	..	58.2	..	..	18 44.50	26.20	1.00	V.	1	12.2	6 4.02	25.91	4.33	19 11.70	24 42 24.26
8	9	..	..	58.4	..	25.8	..	..	21 12.10	26.21	1.00	.	9	3.5	40 28.38	25.97	7.38	21 39.31	25 16 51.73
9	7	..	..	10.6	24.2	..	..	..	22 24.32	26.21	1.00	.	4	2.54	16 24.84	26.00	5.23	22 51.53	24 52 46.07
10	8	..	..	..	..	22.5	..	..	23 8.68	26.22	0.99	.	9	10.13	44 4.20	26.01	7.74	23 35.89	25 20 27.95
11	9	..	..	..	..	..	..	7.5	23 25.84	26.22	1.00	.	8	7.26	37 40.99	26.02	7.14	23 53.06	14 4.15
12	10	..	6.9	20.8	..	..	..	..	13 29 34.55	+26.24	+1.00	.	6	8.38	−28 18.60	−26.17	− 6.30	13 30 1.79	−25 4 41.07

CORRECTIONS.

Date.		Corr. of Clock.	Hourly rate.	m	n	c
1849.	h.	s.	s.	s.	s.	s.

INSTRUMENT READINGS.

Date.		Barom.	Thermom. At.	Thermom. Ex.
1849.	h. m.	in.	°	°

REMARKS.

ZONE 231. APRIL 5. K. BELT, −25° 1′. D₀ = −24° 35′ 50″—Continued.

No.	Mag.	I.	II.	III.	IV.	V.	VI.	VII.	T.	a_1	a_2	Micrometer.			i	d_1	d_2	Mean Right Ascension, 1850.0.	Mean Declination, 1850.0.
									h. m. s.	s.	s.			r.	′ ″	″	″	h. m. s.	° ′ ″
13	9	..	..	38.8	..	6.8	..	..	13 30 52.80	+26.25	+1.00	.	8	6.27	−37 11.24	−26.20	−7.09	13 31 20.05	−25 13 34.53
14	9	46.6	..	14.3	..	..	..	..	34 28.04	26.26	1.00	III.	6	12.25	30 13.01	26.28	6.48	34 55.30	25 6 35.77
15	5	..	..	..	58.2	12.3	..	..	35 58.41	26.27	1.00	.	2	6.35	8 18.18	26.32	4.48	36 25.68	24 44 38.98
16	5	..	..	..	..	..	15.8	29.6	36 48.27	26.27	1.00	VII.	10	2.41	45 16.82	26.34	7.82	37 15.54	25 21 40.98
17	9	..	35.4	49.4	..	..	..	..	43 3.10	26.30	1.00	.	8	10.16	39 6.72	26.46	7.27	43 30.40	25 15 30.45
18	9	..	..	11.3	..	..	..	..	46 25.08	26.32	1.01	.	1	5.37	2 49.93	26.53	4.04	46 52.41	24 39 10.50
19	9	..	..	..	..	..	..	38.7	46 57.46	26.32	1.00	.	3	13.31	16 16.78	26.54	5.24	47 24.78	24 52 38.52
20	5	..	37.3	50.9	4.7	..	..	..	49 4.74	26.33	1.00	.	8	10.51	39 24.36	26.57	7.31	49 32.07	25 15 48.24
21	9	..	..	..	14.8	..	..	..	50 14.85	26.33	1.00	.	2	14.40	12 22.72	26.59	4.85	50 42.15	24 48 44.16
22	9	..	30.2	..	58.2	..	..	..	13 57 57.96	26.37	1.00	.	8	7.56	37 56.12	26.68	7.18	13 58 25.33	25 14 19.98
23	7	..	48.8	2.4	16.2	..	..	..	14 1 16.26	26.39	1.00	.	1	9.46	4 55.49	26.71	4.21	14 1 43.65	24 41 16.41
24	9	..	52.8	6.3	..	..	..	..	14 20.25	26.45	1.00	.	7	6.13	32 5.48	26.79	6.64	14 47.70	25 8 28.91
25	9	..	..	..	53.4	..	..	..	19 53.44	26.48	1.00	.	5	9.50	24 53.80	26.81	5.98	20 20.92	25 1 16.59
26	7	..	..	..	45.9	59.4	..	..	20 45.81	26.49	1.00	.	3	11.27	15 44.51	26.81	5.18	21 13.30	24 52 6.50
27	7	..	..	..	57.2	..	..	..	21 57.21	26.50	1.00	.	1	4.30	2 16.15	26.81	4.00	22 24.71	24 38 36.96
28	9	..	..	..	..	..	57.2	..	26 29.67	26.52	0.99	.	9	7.21	42 37.46	26.81	7.62	26 57 18	25 19 1.89
29	8	..	..	44.1	58.0	..	..	..	27 57.92	26.53	1.00	.	7	6.0	31 58.93	26.81	6.64	28 25.45	8 22.38
30	9	..	..	..	..	51.9	..	..	32 38.10	26.55	1.00	.	8	3.29	35 41.48	26.80	6.97	33 5.65	12 5.25
31	9	..	11.9	25.9	..	..	..	..	34 39.60	26.57	1.00	.	9	13.41	45 49.07	26.79	7.92	35 7.17	22 13.78
32	9	..	26.1	39.3	..	..	..	..	35 53.40	26.58	1.00	.	7	12.3	35 1.96	26.78	6.91	36 20.98	25 11 25.65
33	4	..	..	38.7	52.6	6.1	..	..	36 52.50	26.58	1.00	.	2	13.25	11 41.90	26.78	4.78	37 20.08	24 48 6.46
34	5	..	..	58.7	12.7	..	..	..	38 12.62	26.59	1.00	.	5	6.7	23 1.36	26.77	5.81	38 40.21	59 23.94
35	6	..	..	..	20.4	34.4	..	..	39 20.56	26.60	1.00	.	3	10.51	15 26.36	26.76	5.27	39 48.16	51 48.39
36	8	46.8	0.6	14.4	..	..	..	..	43 28.14	26.62	1.00	.	3	6.44	13 21.82	26.73	4.91	43 55.76	24 49 43.46
37	9	32.9	46.9	..	..	..	..	..	46 14.36	26.64	1.00	III.	5	7.24	23 40.14	26.70	5.88	46 42.00	25 0 2.72
38	8.7	37.8	51.4	5.1	..	..	..	..	48 18.97	26.65	1.00	.	3	7.7	13 33.41	26.68	4.92	48 46.62	24 49 55.01
39	9	..	..	..	..	13.6	..	..	53 59.79	26.68	1.00	.	8	10.25	39 11.25	26.60	7.31	54 27.47	25 15 35.16
40	2.3	..	..	..	..	4.4	18.0	32.1	54 50.64	26.69	1.00	VII.	1	9.54	4 59.16	26.59	4.18	55 18.33	24 41 19.93
41	8	..	..	..	..	45.3	59.0	..	56 31.46	26.70	1.00	VI.	7	13.22	35 41.63	26.57	6.97	56 59.16	25 12 5.17
42	9	..	..	..	..	..	4.0	..	14 58 36.23	26.71	1.00	.	1	5.54	2 58.50	26.54	4.03	14 59 3.94	24 39 19.07
43	9	..	..	..	3.9	..	..	..	15 0 3.89	26.72	1.00	.	6	11.33	29 46.84	26.52	6.44	15 0 31.61	25 6 9.80
44	9	..	..	10.8	..	..	..	..	1 24.61	26.73	1.00	.	10	6.12	47 3.58	26.49	8.03	1 52.34	23 28.10
45	8	44.7	58.8	..	..	..	..	..	3 26.24	26.74	1.00	II.	7	3.11	30 33.54	26.45	6.53	3 53.98	25 6 56.52
46	7	..	48.8	..	16.0	..	..	..	15 4 16.21	+26.75	+1.00	.	2	6.0	−8 0.53	−26.43	−4.42	15 4 43.96	−24 44 21.38

*ZONE 232. APRIL 10. B. BELT, −30° 30′. D₀ = −30° 5′ 20″.

No.	Mag.	I.	II.	III.	IV.	V.	VI.	VII.	T.	a_1	a_2	Micrometer.			i	d_1	d_2	Mean Right Ascension, 1850.0.	Mean Declination, 1850.0.
1	7	2.	17.	31.2	46.	..	..	..	11 29 45.80	+33.18	+1.00	IV.	4	5.24	−17 40.47	−1.66	−2.05	11 30 19.98	−30 23 4.18
2	8	..	52.5	7.	21.7	..	..	..	31 21.59	33.18	1.00	IV.	4	4.24	17 10.22	1.80	2.00	31 55.77	22 34.02
3	7	..	..	10.5	25.	39.6	..	..	35 25.07	33.18	1.01	V.	2	2.15	6 7.00	2.17	0.60	35 59.26	11 29.77
4	8	..	..	..	..	..	44.	58.	36 14.67	33.18	0.99	VII.	9	6.25	42 8.61	2.24	5.25	36 48.84	47 36.13
5	8	..	..	..	..	..	..	3.7	37 19.96	33.18	0.98	VII.	10	6.41	47 17.61	2.34	5.95	37 54.12	52 45.90
6	7	..	..	5.3	19.8	..	..	..	39 19.85	33.18	1.00	IV.	4	9.38	19 48.55	2.52	2.32	39 54.03	25 13.39
7	9	..	..	..	..	..	..	5.3	39 21.95	33.18	1.01	VII.	3	11.25	15 42.91	2.52	1.80	39 56.14	21 7.23
8	9	..	16.2	30.5	45.	..	..	..	41 45.01	33.18	0.99	IV.	7	8.33	33 16.08	2.73	4.07	42 19.18	38 42.88
9	8	..	50.4	5.	..	..	..	..	43 19.50	33.18	0.99	III.	9	6.20	42 6.89	2.87	5.25	43 53.67	47 35.01
10	10	..	..	..	..	27.5	..	..	44 12.99	33.18	1.01	V.	7	3.28	30 42.20	2.95	3.74	44 47.18	36 8.89
11	8	..	..	..	30.5	45.2	..	..	11 45 30.52	+33.18	+0.99	V.	9	6.56	−42 24.80	−3.06	−5.29	11 46 4.69	−30 47 53.15

CORRECTIONS.

Date.		Corr. of Clock.	Hourly rate.	m	n	c
1849.	h.	s.	s.	s.	s.	s.

INSTRUMENT READINGS.

Date.		Barom.	Thermom. At.	Thermom. Ex.
1849.	h. m.	in.	°	°

REMARKS.

(231) 19. Micrometer reading assumed as 12r.31 instead of 13r.31.

* In the hand-book this Zone is distinctly written as −37° 31′, three several times. May it not be −39° 31′? If so, the stars of the 5 and 6.7 magnitudes are found in Lacaille.—J. FERGUSON.

ZONE 232. APRIL 10. B. BELT, −37° 31′. D_o = −30° 5′ 20″—Continued.

No.	Mag.	Seconds of Transit. I.	II.	III.	IV.	V.	VI.	VII.	T.	a_1	a_2	Micrometer.			i	d_1	d_2	Mean Right Ascension, 1850.0.	Mean Declination, 1850.0.
									h. m. s.	s.	s.			r.	′ ″	″	″	h. m. s.	° ′ ″
12	9				37.	51.4			11 46 36.86	+33.18	+0.98	V.	9	11.4	−44 29.85	−3.16	−5.58	11 47 11.02	−30 49 58.59
13	8			57.	11.5				48 11.53	33.18	1.01	IV.	3	11.40	15 51.07	3.30	1.83	48 45.72	21 16.20
14	10					30.5			49 14.98	33.18	0.99	V.	7	12.19	35 9.96	3.39	4.32	49 49.15	40 37.68
15	10							27.5	49 44.01	33.18	1.00	VII.	5	13.30	26 44.13	3.44	3.22	50 18.19	32 10.79
16	8		34.4	49.2	4.				55 3.74	33.19	1.00	IV.	4	10.59	20 29.40	3.89	2.42	55 37.93	25 55.71
17	8							54.	55 10.40	33.19	0.99	VII.	7	12.8	35 3.89	3.90	4.32	55 44.58	40 32.11
18	9			0.3	15.				57 14.83	33.19	0.99	IV.	9	13.7	45 31.93	4.08	5.72	57 49.01	51 1.73
19	7						36.	50.4	58 6.88	33.19	0.99	VII.	8	6.29	38 12.17	4.16	4.73	58 41.06	43 41.06
20	6.7				33.7	48.5	3.		11 59 19.34	33.19	1.00	VI.	6	9.41	28 50.10	4.28	3.50	11 59 53.53	34 17.88
21	9						28.5		12 0 59.49	33.19	0.99	VI.	8	4.41	36 17.53	4.42	4.48	12 1 33.67	41 46.43
22	9						28.5	43.	1 59.49	33.19	1.00	VII.	5	4.21	22 7.31	4.51	2.62	2 33.68	27 34.44
23	8	50.5	5.	19.3					4 33.94	33.19	1.00	III.	5	5.37	22 46.16	4.76	2.71	5 8.13	28 13.63
24	6.7							28.	4 44.39	33.20	0.97	VII.	8	3.14	35 33.93	4.79	4.38	5 18.58	41 3.10
25	7			58.3	12.6				7 12.63	33.20	0.99	IV.	10	3.49	45 51.47	5.03	5.75	7 46.82	51 22.25
26	10							30.3	7 47.07	33.20	1.01	VII.	2	0.50	5 23.62	5.09	0.49	8 21.28	10 49.20
27	8					38.4	53.		9 23.91	33.20	1.01	VI.	4	3.16	16 35.67	5.24	1.93	9 58.12	22 2.84
28	10						7.5	22.	10 38.45	33.20	0.99	VII.	8	0.59	34 25.26	5.37	4.22	11 12.64	39 54.85
29	10					50.			12 35.55	33.20	1.01	V.	3	7.45	13 52.50	5.57	1.58	13 9.76	19 19.65
30	9				38.	53.			13 38.29	33.21	1.00	V.	4	9.21	19 39.90	5.69	2.31	14 12.50	25 7.90
31	8							31.	13 47.65	33.21	1.01	VII.	3	10.23	15 11.65	5.70	1.74	14 21.87	20 39.09
32	8		41.	55.3	10.				16 9.97	33.21	1.00	IV.	4	7.28	18 42.99	5.95	2.19	16 44.18	24 11.13
33	8					12.	26.5		16 57.46	33.21	1.01	VI.	3	5.7	12 32.64	6.04	1.40	17 31.68	18 0.08
34	9				29.				18 29.07	33.21	1.00	IV.	4	5.9	17 32.91	6.20	2.04	19 3.28	23 1.15
35	9						16.	30.5	18 46.99	33.21	1.00	VII.	5	7.35	23 45.14	6.23	2.84	19 21.20	29 14.21
36	9			28.5	42.2				20 42.58	33.21	0.99	IV.	8	5.49	36 52.08	6.42	4.55	21 16.78	42 23.05
37	8						45.	59.5	21 16.02	33.22	1.01	VII.	3	7.29	13 43.91	6.46	1.55	21 50.25	19 11.92
38	9					53.5	8.		22 38.99	33.22	0.99	VI.	9	7.50	42 51.83	6.58	5.34	23 13.20	48 23.75
39	9				3.4	18.			24 3.48	33.22	1.00	V.	5	4.25	22 11.38	6.70	2.64	24 37.70	27 40.72
40	9						37.		25 8.03	33.22	0.99	VI.	10	4.4	45 58.78	6.79	5.77	25 42.24	51 31.34
41	9				22.	36.4			26 21.94	33.22	1.00	V.	7	1.43	29 49.26	6.89	3.63	26 56.16	35 19.78
42	9	9.	23.5						28 52.59	33.23	0.99	II.	8	6.16	37 5.44	7.11	4.58	29 26.81	42 37.13
43	8			11.5	26.3	40.5			32 26.03	33.23	0.99	V.	10	10.20	49 8.56	7.39	6.17	33 0.25	54 42.12
44	9			41.5	56.	10.5			33 55.94	33.23	0.99	V.	10	7.21	47 38.29	7.51	5.98	34 30.16	53 11.78
45	9					13.3	28.		34 58.87	33.24	1.00	VI.	6	8.29	28 13.80	7.60	3.42	35 33.11	33 44.82
46	9						32.5	47.	36 3.42	33.24	0.99	VII.	9	5.28	41 39.89	7.68	5.19	36 37.65	47 12.76
47	9			42.5	57.				38 57.00	33.25	1.01	IV.	1	4.54	2 28.25	7.92	0.11	39 31.26	7 56.28
48	8			46.	0.5	15.			40 0.50	33.25	1.00	V.	6	11.14	29 37.19	8.00	3.60	40 34.75	35 8.79
49	9							0.	40 16.71	33.25	1.01	VII.	3	.	9 57.51	8.02	1.06	40 50.97	15 26.59
50	8						5.	19.5	41 36.01	33.25	1.00	VII.	4	7.41	18 48.96	8.12	2.20	42 10.26	24 19.28
51	7					25.3	40.		43 10.87	33.26	1.00	VI.	4	5.14	17 35.17	8.25	2.05	43 45.13	23 5.47
52	8				39.	53.3			44 38.89	33.26	1.00	V.	6	5.1	26 29.10	8.36	3.20	45 13.15	32 0.66
53	8			42.5	57.3				46 57.13	33.26	0.99	IV.	7	8.43	33 21.12	8.54	4.08	47 31.38	38 53.74
54	7						48.3	3.	47 19.34	33.27	0.99	VII.	7	11.5	34 32.13	8.57	4.24	47 53.60	40 4.94
55	8		4.5	19.2	33.4				49 33.57	33.27	1.00	IV.	5	5.13	22 34.12	8.75	2.69	50 7.84	28 5.56
56	10						22.5	37.	49 53.47	33.27	1.00	VII.	6	6.20	27 8.42	8.77	3.27	50 27.74	32 40.46
57	9	31.	45.5	59.5					52 14.37	33.28	1.00	III.	6	10.43	29 21.56	8.95	3.57	52 48.65	34 54.08
58	11					52.5			53 38.08	33.28	1.01	V.	2	8.40	9 21.14	9.03	0.99	54 12.37	14 51.16
59	10		54.5	9.					55 23.54	33.29	0.99	III.	8	8.34	38 15.22	9.18	4.73	55 57.82	43 49.13
60	10			56.	10.5				12 57 10.46	+33.29	+0.99	IV.	8	11.28	−39 43.01	−9.31	−4.93	12 57 44.74	−30 45 17.25

CORRECTIONS.

Date.		Corr. of Clock.	Hourly rate.	m	n	c
1849.	h.	s.	s.	s.	s.	s.

INSTRUMENT READINGS.

Date.		Barom.	Thermom. At.	Thermom. Ex.
1849.	h. m.	in.	°	°

REMARKS.

(232) 20. Transits over T.'s V–VII assumed as recorded over T.'s IV–VI.

Zone 232. April 10. B. Belt, −37° 31′. D₀ = −30° 5′ 20″—Continued.

No.	Mag.	Seconds of Transit. I.	N.	III.	IV.	V.	VI.	VII.	T.	a_1	a_2	Micrometer.			i	d_1	d_2	Mean Right Ascension, 1850.0.	Mean Declination, 1850.0.
									h. m. s.	s.	s.			r.	′ ″	″	″	h. m. s.	° ′ ″
61	7					4.5	19.		12 57 49.99	+33.29	+0.99	VI.	9	7.33	−42 43.26	−9.37	−5.31	12 58 24.27	−30 48 17.94
62	9			34.5	49.				12 59 49.04	33.30	1.00	IV.	5	3.37	20 45.72	9.52	2.45	13 0 23.34	26 17.69
63	10		18.	33.	47.2				13 1 47.23	33.30	0.99	IV.	8	1.8	34 30.38	9.66	4.24	2 21.52	40 4.28
64	10			44.	58.5				2 58.51	33.31	1.00	IV.	6	5.1	26 29.17	9.74	3.20	3 32.82	32 2.11
65	9						28.		3 58.81	33.31	1.01	VI.	1	5.18	2 40.08	9.81	0.13	4 33.13	8 10.07
66	9				20.	34.4			5 19.86	33.31	0.99	V.	9	8.24	43 9.17	9.90	5.37	5 54.16	48 44.44
67	9						25.	39.8	5 56.07	33.32	0.99	VII.	8	11.1	39 28.81	9.95	4.90	6 30.38	45 3.66
68	5			45.	59.4	14.			7 59.44	33.32	0.99	V.	8	6.12	37 3.61	10.10	−4.57	8 33.75	42 38.28
69	8				23.	27.4			10 23.02	33.33	1.01	V.	1	0.25	0 12.53	10.27	+0.18	10 57.36	5 42.62
70	10				42.	56.3			12 41.93	33.33	1.00	V.	5	7.30	23 43.14	10.42	−2.83	13 16.26	29 16.39
71	8		44.3	58.8	13.				16 13.23	33.35	1.00	IV.	4	8.17	19 7.71	10.66	2.24	16 47.58	24 40.61
72	10			15.5	30.				17 30.03	33.35	1.00	IV.	5	8.11	19 4.68	10.75	2.23	18 4.38	24 37.66
73	9							21.6	17 38.01	33.35	1.00	VII.	7	10.51	34 25.07	10.76	4.22	18 12.36	40 0.05
74	8				15.	29.5			19 14.93	33.36	0.99	V.	8	13.21	40 39.91	10.87	5.06	19 49.28	46 15.84
75	9		13.7	28.5	42.5				20 42.74	33.36	1.00	IV.	6	13.54	30 57.93	10.96	3.77	21 17.10	36 32.66
76	10			7.	22.0				22 21.76	33.37	1.00	IV.	6	7.5	27 31.70	11.07	3.32	22 56.13	33 6.09
77	9							31.3	22 48.00	33.37	1.01	VII.	2	12.22	11 12.55	11.10	1.21	23 22.38	16 44.86
78	10							36.	23 52.29	33.37	0.99	VII.	9	9.10	43 31.84	11.17	5.42	24 26.65	49 8.43
79	11	31.	45.2						26 14.41	33.38	1.00	II.	7	7.26	32 42.63	11.32	3.99	26 48.79	38 17.94
80	11				0.5				27 0.54	33.38	1.01	IV.	2	9.52	9 57.51	11.37	1.05	27 34.93	15 29.93
81	10			28.	42.5				29 42.53	33.39	1.00	IV.	7	6.19	32 8.51	11.54	3.69	30 16.92	37 43.74
82	10				45.6	0.			30 45.57	33.40	1.00	V.	5	10.37	25 17.44	11.61	3.04	31 19.97	30 52.09
83	9				19.				32 18.02	33.40	1.01	IV.	1	7.27	3 45.39	11.70	0.26	32 52.43	9 17.35
84	9		52.	6.7					34 21.14	33.41	1.00	III.	7	13.0	35 30.63	11.82	4.37	34 55.55	41 6.82
85	9						52.5	7.	34 23.43	33.41	0.99	VII.	8	7.56	37 55.53	11.82	4.68	34 57.83	43 32.03
86	9		22.5	37.	51.5				39 51.51	33.43	1.00	IV.	7	5.52	31 54.89	12.16	3.89	40 25.94	37 30.94
87	7				13.				41 13.05	33.43	1.00	IV.	3	7.17	13 38.45	12.24	1.53	41 47.48	19 12.22
88	7						0.5		41 31.42	33.44	1.00	VI.	5	3.54	21 54.03	12.26	2.60	42 5.86	27 28.89
89	7						1.5	16.	42 32.47	33.44	1.00	VII.	6	9.22	28 40.19	12.31	3.48	43 6.91	34 15.98
90	7	50.	4.3	19.	33.3				46 33.42	33.46	1.00	IV.	6	6.25	27 11.53	12.54	3.29	47 7.88	32 47.36
91	7		2.	16.3					47 30.90	33.46	1.00	III.	3	8.51	14 25.78	12.60	1.63	48 5.36	20 0.01
92	7						12.4	27.	48 43.49	33.46	1.01	VII.	2	6.27	8 13.55	12.67	0.85	49 17.96	13 47.07
93	9		40.	54.3					52 8.93	33.48	1.00	III.	7	4.34	31 15.50	12.85	3.81	52 43.41	36 52.16
94	9				5.	19.6	34.		53 4.98	33.48	0.99	VI.	9	7.8	42 30.65	12.90	5.29	53 39.45	48 8.64
95	9				53.5	8.			55 53.56	33.49	1.01	V.	2	2.29	6 14.06	13.04	0.59	56 28.06	11 47.69
96	9	0.	15.	29.3					57 43.74	33.50	1.00	III.	4	4.30	17 13.18	13.14	1.97	58 18.24	22 48.29
97	8						19.5	34.2	57 50.54	33.50	1.00	VII.	8	2.41	35 16.69	13.15	4.34	13 58 25.04	40 54.18
98	9			35.	49.5				13 59 49.73	+33.51	+1.00	IV.	4	2.4	−15 59.63	−13.25	−1.83	14 0 24.24	−30 21 34.71

CORRECTIONS.

Date.		Corr. of Clock.	Hourly rate.	m	n	c
1849.	h.	s.	s.	s.	s.	s.

INSTRUMENT READINGS.

Date.		Barom.	Thermom. At.	Ex.
1849.	h. m.	In.	°	°

REMARKS.

(232) 69. Transit over T. V assumed as 73ˢ.4 instead of 27ˢ.4.
(232) 72. Hor. thread assumed as 4 instead of 5.

ZONE 233. APRIL 11. K. BELT, −26° 16′. $D_0 = -25°\ 49'\ 40''$.

No.	Mag.	Seconds of Transit. I.	II.	III.	IV.	V.	VI.	VII.	T.	a_1	a_2	Micrometer.			i	d_1	d_2	Mean Right Ascension, 1850.0.	Mean Declination, 1850.0.
									h. m. s.	s.	s.			r.	′ ″	″	″	h. m. s.	° ′ ″
1	9					22.5			10 12 8.86	+34.66	+1.00	V.	8	4.14	−36 4.14	− 8.58	−7.02	10 12 44.52	−26 25 59.74
2	9			2.1	16.0				15 16.02	34.66	1.01	.	1	10.10	5 7.59	9.02	4.21	15 51.69	25 55 0.82
3	9	29.2	42.8	57.0					17 10.93	34.65	0.99	.	9	4.16	41 4.70	9.29	7.48	17 46.57	26 31 1.46
4	9							48.8	17 7.10	34.65	1.00	VII.	4	9.2	19 30.01	9.28	5.50	17 42.75	9 24.79
5	9				29.4				18 29.45	34.65	1.00	.	5	7.20	23 38.16	9.47	5.87	19 5.10	26 13 33.50
6	10				49.4				19 49.42	34.65	1.01	.	1	8.19	4 11.62	9.66	4.13	20 25.08	25 54 5.41
7	9		35.6	49.8					21 3.64	34.65	0.99	.	10	3.0	45 26.76	9.83	7.89	21 39.28	26 35 24.48
8	9				36.4	50.4			21 36.48	34.65	1.00	.	4	2.20	16 7.69	9.90	5.19	22 12.13	6 2.78
9	9				53.0				22 53.02	34.65	1.00	.	6	4.30	26 13.55	10.07	6.11	23 28.67	16 9.73
10	9					37.2			23 23.31	34.65	1.00	.	4	1.51	15 53.06	10.14	5.17	23 58.96	5 48.37
11	9			25.6	39.7				24 39.64	34.64	1.00	.	6	1.59	24 57.40	10.32	6.00	25 15.28	14 53.72
12	9		17.3	31.1					25 45.10	34.64	1.00	.	3	4.41	12 19.79	10.47	4.85	26 20.74	2 15.11
13	8							55.4	26 13.44	34.64	0.99	.	10	1.28	44 40.37	10.53	7.82	26 49.07	26 34 38.72
14	5			11.0	24.8				28 24.87	34.64	1.01	V.	1	7.44	3 53.93	10.83	4.10	29 0.52	25 53 48.86
15	9							27.7	28 46.12	34.64	1.01	.	2	5.8	7 34.30	10.86	4.43	29 21.77	25 57 29.61
16	5							17.8	29 35.80	34.64	0.99	.	10	8.39	48 17.71	10.99	8.18	30 11.13	26 38 16.88
17	9		2.2		30.0				31 30.04	34.64	1.01	.	1	7.12	3 37.83	11.25	4.07	32 5.69	25 53 33.15
18	10						20.8		31 52.96	34.63	0.99	.	8	8.58	38 27.38	11.30	7.23	32 28.58	26 28 25.91
19	9				5.4	19.7			33 5.64	34.63	1.01	.	2	10.33	10 18.19	11.47	4.67	33 41.28	0 14.33
20	9					5.8			33 51.87	34.63	1.00	.	6	7.53	27 55.90	11.57	6.26	34 27.50	17 53.73
21	9	57.6							35 39.51	34.63	0.99	.	9	3.44	40 48.05	11.81	7.46	36 15.13	30 47.32
22	9				42.3				35 42.34	34.63	1.01	V.	2	12.5	11 4.53	11.82	4.74	36 17.98	1 1.09
23	9				42.1				37 42.16	34.63	1.00	.	3	10.15	15 8.21	12.09	5.10	38 17.79	26 5 5.40
24	7					47.8			38 33.96	34.63	1.01	.	1	12.41	6 23.72	12.21	4.32	39 9.60	25 56 20.25
25	8		2.7	16.5					41 30.50	34.62	1.01	.	3	3.1	11 29.36	12.60	4.77	42 6.13	26 1 26.73
26	9		19.5		47.3				43 47.37	34.62	1.00	.	3	12.24	16 13.25	12.90	5.20	44 22.99	6 11.35
27	9		40.8						45 8.71	34.62	0.99	.	9	13.1	45 28.90	13.08	7.91	45 44.32	26 35 29.89
28	6.7		44.1	57.3					46 11.59	34.62	1.01	.	1	13.54	7 0.53	13.22	4.37	46 47.22	25 56 58.12
29	7			37.8	51.4				46 51.58	34.62	1.01	.	2	11.13	10 38.35	13.31	4.69	47 27.21	26 0 36.35
30	7			27.2	41.0				47 41.11	34.62	1.00	.	4	11.31	20 45.53	13.42	5.62	48 16.73	26 10 44.57
31	8	53.3	7.3						56 35.06	34.61	1.01	.	2	5.47	7 53.97	14.59	4.45	52 10.68	25 57 53.01
32	9		46.8	0.8					57 14.73	34.61	0.99	.	8	9.19	38 37.97	14.68	7.24	52 50.33	26 28 39.89
33	6		5.5	19.5	33.5				54 33.45	34.61	1.01	.	2	12.18	11 11.12	14.33	4.74	55 9.07	1 10.19
34	9			11.9					56 25.86	34.61	1.00	.	8	9.28	38 42.50	14.57	7.25	57 1.47	28 44.32
35	5		3.2	16.9					57 30.98	34.61	1.00	.	8	10.7	39 2.18	14.71	7.30	58 6.59	29 4.19
36	5.6			51.8	5.8				10 58 5.73	34.61	1.00	.	8	9.18	38 37.46	14.79	7.24	10 58 41.34	26 28 29.49
37	8				3.7	17.8			11 4 3.84	34.61	1.01	.	2	9.4	9 33.31	15.56	4.59	11 4 39.46	25 59 33.46
38	9		55.3	9.0					6 23.06	34.61	1.00	.	5	9.26	24 41.70	15.86	5.97	6 58.67	26 14 43.53
39	9				50.3	4.0			6 50.12	34.61	0.99	.	8	11.0	39 28.90	15.91	7.34	7 25.72	29 32.15
40	9				55.2				7 55.26	34.61	1.00	.	3	11.17	15 39.47	16.05	5.14	8 30.87	5 40.66
41	9			42.2	55.8				8 55.94	34.61	1.00	.	7	12.28	35 14.56	16.18	6.95	9 31.55	25 17.69
42	10		40.4						11 8.28	34.61	1.00	.	3	9.16	14 38.46	16.46	5.05	11 43.69	4 39.97
43	10	43.5							14 25.37	34.61	1.00	.	7	12.17	35 9.02	16.88	6.94	15 0.98	25 12.84
44	9			40.3	54.2				15 54.26	34.61	1.00	.	4	6.15	18 6.19	17.06	5.36	16 29.87	8 8.61
45	9	34.0	48.9						17 16.68	34.61	1.00	.	3	5.34	12 46.52	17.21	4.87	17 52.29	26 2 48.63
46	8		43.4		21.1				24 21.19	34.61	1.01	.	1	10.14	5 9.61	18.12	4.19	24 56.81	25 55 11.92
47	9		19.8	33.8					25 47.70	34.61	1.01	.	2	8.17	9 9.61	18.30	4.55	26 23.32	59 12.46
48	9			4.4	18.3				28 18.32	34.61	1.01	.	1	11.17	5 41.37	18.61	4.23	28 53.94	25 55 44.21
49	9					13.2			11 29 59.25	+34.61	+0.99	.	8	12.9	−40 3.69	−18.82	−7.41	11 30 34.85	−26 30 9.92

CORRECTIONS.

Date.		Corr. of Clock.	Hourly rate.	m	n	c
1849.	h.	s.	s.	s.	s.	s.

INSTRUMENT READINGS.

Date.		Barom.	Thermom. At.	Thermom. Ex.
1849.	h. m.	in.	°	°

REMARKS.

(233) 3. Micrometer reading assumed as 6r.16 instead of 4r.16.
(233) 31. Minutes assumed as 51 instead of 56.
(233) 32. Minutes assumed as 52 instead of 57.

ZONE 233. APRIL 11. K. BELT, −26° 16′. D_0=−25° 49′ 40″—Continued.

No.	Mag.	I.	II.	III.	IV.	V.	VI.	VII.	T.	a_1	a_2	MICROMETER.			i	d_1	d_2	Mean Right Ascension, 1850.0.	Mean Declination, 1850.0.
									h. m. s.	s.	s.			r.	′ ″	″	″	h. m. s.	° ′ ″
50	8	..	..	58.8	12.7	..	..	..	11 31 12.74	+34.61	+1.01	.	2	11.28	−10 45.91	−18.97	−4.69	11 31 48.36	−26 0 49.57
51	10	..	28.0	..	..	..	..	..	32 56.07	34.61	1.01	.	2	8.55	9 28.77	19.18	4.57	33 31.69	25 59 32.52
52	10	..	..	..	..	3.9	..	..	32 50.04	34.61	1.01	.	2	7.20	8 40.86	19.17	4.50	33 25.66	25 58 44.53
53	9	..	28.0	41.8	..	..	..	..	35 55.81	34.62	1.00	.	3	6.1	13 0.13	19.54	4.89	36 31.43	26 3 4.56
54	9	..	..	..	11.3	..	..	..	38 11.35	34.62	1.00	.	5	7.36	23 46.24	19.82	5.89	38 46.97	26 13 51.95
55	5	..	7.6	..	35.6	..	..	..	40 35.54	34.62	1.01	.	1	9.42	4 53.47	20.11	4.16	41 11.17	25 54 57.74
56	6	45.4	59.4	..	..	..	..	..	42 27.29	34.62	1.00	.	8	5.1	36 27.87	20.34	7.06	43 2.01	26 26 35.27
57	6	..	..	40.7	54.7	..	..	..	42 54.69	34.62	1.00	.	5	9.18	24 37.66	20.39	5.96	43 30.31	14 44.01
58	8	..	28.2	41.8	..	..	..	..	44 55.93	34.62	0.99	.	8	8.36	38 16.30	20.64	7.22	45 31.54	28 24.16
59	8	..	..	..	22.3	36.3	..	..	45 22.24	34.62	1.01	.	2	10.49	10 26.26	20.70	4.66	45 57.87	0 31.62
60	9	..	..	..	40.3	..	..	..	47 40.12	34.62	0.99	.	10	7.8	47 31.81	20.98	8.14	48 15.73	37 40.93
61	8	36.0	..	3.8	..	..	..	..	49 17.75	34.62	1.00	.	5	5.57	22 56.31	21.17	5.80	49 53.37	13 3.28
62	9	12.5	26.4	..	..	..	..	..	51 54.28	34.63	1.00	.	5	11.45	25 51.79	21.49	6.08	52 29.91	15 59.36
63	9	..	..	16.4	..	..	..	..	52 30.37	34.63	0.99	.	9	7.54	42 54.11	21.56	7.70	53 5.99	26 33 3.37
64	8	..	..	..	..	..	14.0	28.1	53 46.26	34.63	1.01	.	1	11.49	5 57.51	21.71	4.25	54 21.90	25 56 3.47
65	9	..	..	23.6	..	..	..	..	57 37.56	34.63	0.99	.	8	6.7	37 1.16	22.17	7.11	58 13.18	26 27 10.44
66	9	..	..	..	2.0	15.9	..	..	11 59 1.90	34.61	0.99	.	9	8.23	43 8.73	22.34	7.72	11 59 37.53	33 18.79
67	10	..	..	48.8	..	..	..	..	12 2 2.74	34.64	1.00	.	3	11.24	15 43.00	22.70	5.14	12 2 38.38	5 50.84
68	8	..	..	59.6	13.6	..	..	..	3 13.59	34.64	1.00	.	5	8.7	24 1.87	22.84	5.91	3 49.23	14 10.62
69	8.7	37.9	51.9	..	..	..	..	..	5 19.80	34.64	0.99	.	8	10.5	39 1.17	23.09	7.32	5 55.43	29 11.58
70	9	14.8	28.7	..	..	..	..	..	7 56.58	34.65	1.00	.	5	11.51	25 54.81	23.39	6.08	8 32.23	16 4.28
71	8	..	..	..	..	5.9	..	..	7 51.94	34.65	0.99	.	9	13.45	45 51.09	23.38	7.99	8 27.58	36 2.46
72	9	..	..	..	44.2	..	..	..	12 8 44.19	+34.65	+1.00	.	6	10.7	−29 3.48	−23.50	−6.37	12 9 19.84	−26 19 13.35

ZONE 234. APRIL 11. K. BELT, −27° 31′. D_0=−27° 7′ 0″.

No.	Mag.	I.	II.	III.	IV.	V.	VI.	VII.	T.	a_1	a_2	MICROMETER.			i	d_1	d_2	Mean Right Ascension, 1850.0.	Mean Declination, 1850.0.
1	9	..	..	..	..	..	28.3	..	12 29 0.09	+34.63	+1.00	.	6	8.12	−28 5.49	−15.62	−6.31	12 29 35.72	−27 35 27.42
2	9	..	..	..	57.2	..	..	..	34 57.26	34.64	1.00	.	3	8.51	14 25.85	17.38	4.90	35 32.90	21 28.13
3	5.4	..	..	..	..	10.1	54.0	..	35 25.89	34.64	1.00	.	5	5.22	22 38.66	17.53	5.74	36 1.53	30 1.93
4	8	..	..	..	41.0	..	..	..	36 40.81	34.65	0.99	.	10	10.40	49 18.72	17.86	8.61	37 16.45	56 45.19
5	10	..	..	..	39.3	..	..	..	38 39.22	34.65	0.99	.	8	4.24	36 9.22	18.50	7.19	39 14.86	43 34.91
6	9	..	..	..	..	22.9	..	..	42 8.82	34.67	1.00	.	5	5.35	22 45.22	19.49	5.76	42 44.49	30 10.47
7	9	..	..	10.5	24.8	..	..	..	43 24.62	34.67	0.99	.	10	6.45	47 20.22	19.88	8.40	44 0.28	54 48.50
8	9	..	..	..	1.5	..	..	..	44 1.33	34.67	0.99	.	10	4.17	46 5.59	20.05	8.27	44 36.99	53 33.91
9	8	..	..	22.4	..	..	..	..	45 36.48	34.67	1.01	.	1	2.32	1 16.65	20.49	3.54	46 12.16	8 40.68
10	9	..	..	..	..	49.2	..	..	46 35.22	34.68	1.01	.	1	11.14	5 39.86	20.79	3.98	47 10.91	13 4.63
11	9	..	1.6	..	..	..	..	..	49 29.79	34.69	1.00	.	4	11.15	20 37.46	21.66	5.54	50 5.48	28 4.66
12	9	..	59.3	..	..	..	..	..	50 27.50	34.69	1.00	.	6	7.4	27 31.20	21.94	6.25	51 3.19	34 59.39
13	8.7	57.4	..	25.8	..	..	..	..	53 39.75	34.70	1.00	.	4	12.36	21 18.31	22.82	5.61	54 15.45	28 46.74
14	9	..	..	..	2.4	..	..	..	54 2.37	34.70	1.00	.	6	13.1	30 31.20	22.97	6.57	54 38.07	38 0.74
15	8	57.5	11.7	..	..	..	..	..	55 39.84	34.70	1.00	.	6	8.55	28 27.17	23.50	6.35	56 15.54	35 57.02
16	8	..	59.4	13.6	..	..	..	..	56 27.65	34.71	1.00	.	7	12.36	35 18.61	23.66	7.09	57 3.36	42 49.36
17	9	..	..	59.4	..	..	..	..	57 13.48	34.71	1.01	.	1	8.42	4 23.22	23.89	3.85	12 57 49.20	11 50.96
18	7	..	57.3	..	..	..	..	..	59 25.48	34.72	1.00	.	4	3.28	16 41.97	24.54	5.13	13 0 1.20	24 11.64
19	9	..	..	..	48.1	..	..	..	12 59 48.12	+34.72	+1.01	.	1	5.45	− 2 53.96	−24.63	−3.70	13 0 23.85	−27 10 22.29

CORRECTIONS.

Date.		Corr. of Clock.	Hourly rate.	m	n	c
1849.	h.	s.	s.	s.	s.	s.

INSTRUMENT READINGS.

Date.		Barom.	THERMOM. At.	THERMOM. Ex.
1849.	h. m.	in.	°	°

REMARKS.

ZONE 235. APRIL 12. B. BELT, −25° 38½′. D₀=−25° 13′ 10″.

No.	Mag.	SECONDS OF TRANSIT. I.	II.	III.	IV.	V.	VI.	VII.	T.	a_1	a_2	MICROMETER.			i	d_1	d_2	Mean Right Ascension, 1850.0.	Mean Declination, 1850.0.
									h. m. s.	s.	s.			r.	′ ″	″	″	h. m. s.	° ′ ″
1	9	35.2	49.3	..	..	..	..	..	11 48 16.92	+37.06	+1.00	II.	6	9.35	−25 47.18	−4.04	−6.34	11 48 54.98	−25 42 7.56
2	10	..	..	..	..	..	..	55.5	48 13.99	37.06	1.00	VII.	5	6.49	23 22.18	4.03	5.85	48 52.05	35 42.06
3	9	..	..	..	..	..	..	58.5	49 16.93	37.06	1.00	VII.	6	8.46	28 22.28	4.15	6.30	49 54.99	41 42.73
4	9	..	39.	53.	..	..	..	..	52 6.80	37.07	1.00	III.	4	7.12	18 34.89	4.48	5.43	52 44.87	31 54.80
5	8	..	..	..	43.	57.	..	..	52 43.10	37.07	1.00	V.	5	10.32	25 14.95	4.55	6.02	53 21.17	38 35.52
6	10	..	..	..	..	51.	4.	..	53 36.68	37.07	1.00	VI.	5	9.45	24 51.12	4.66	5.99	54 14.75	38 11.77
7	9	23.	36.7	..	..	..	..	..	57 4.56	37.07	0.99	II.	8	6.45	37 20.16	5.06	7.13	57 42.62	50 42.35
8	9	..	..	..	..	..	59.	..	57 31.28	37.07	0.99	VI.	7	9.9	33 34.07	5.11	6.78	11 58 9.34	46 55.96
9	8	..	12.5	26.2	..	..	..	..	11 59 40.14	37.07	1.01	III.	2	7.24	8 42.81	5.36	4.54	12 0 18.22	22 2.74
10	10	..	..	..	52.	..	..	..	12 3 51.96	37.08	1.00	IV.	7	5.18	31 37.74	5.84	6.59	4 30.04	45 0.17
11	9	..	..	52.	..	..	..	..	6 5.89	37.08	0.99	III.	9	7.30	42 41.97	6.09	7.61	6 43.96	56 5.67
12	7	..	8.3	22.	..	..	..	..	7 35.96	37.08	1.00	III.	7	2.45	30 20.56	6.26	6.45	8 14.04	43 43.30
13	7	..	..	..	27.	41.	55.	..	8 27.12	37.08	1.00	VI.	7	6.50	32 23.98	6.35	6.67	9 5.20	45 47.00
14	9	..	..	56.	10.	24.	..	..	10 10.00	37.09	1.00	V.	7	3.9	30 32.66	6.55	6.50	10 48.09	43 55.71
15	7	..	..	..	..	14.5	..	..	11 0.75	37.09	1.01	V.	1	5.34	2 48.38	6.64	4.02	11 38.85	16 9.04
16	9	..	..	..	..	..	..	7.	11 25.63	37.09	1.00	VII.	2	7.23	8 42.01	6.69	4.54	12 3.72	22 3.24
17	7	..	..	..	..	3.7	18.	..	12 50.06	37.09	1.00	VI.	8	3.10	35 31.74	6.84	6.96	13 28.15	48 55.54
18	8	..	37.3	41.	..	..	..	..	14 54.94	37.10	1.00	III.	6	1.53	24 54.33	7.07	5.99	15 33.04	38 17.39
19	9	..	..	..	49.4	3.	..	..	15 49.26	37.10	1.00	V.	7	5.29	31 43.25	7.18	6.60	16 27.36	45 7.03
20	9	..	..	..	..	..	0.4	14.4	16 32.74	37.10	1.00	VII.	6	10.9	29 4.13	7.26	6.37	17 10.84	42 27.76
21	10	..	..	..	..	..	2.3	..	18 34.64	37.10	0.99	VI.	9	9.4	43 29.25	7.47	7.68	19 12.73	56 54.40
22	9	..	29.3	43.5	..	..	..	..	20 57.19	37.10	1.00	III.	3	10.5	15 3.13	7.74	5.11	21 35.29	28 25.98
23	9	..	..	..	..	..	53.	6.3	25.00	37.11	1.00	VII.	4	13.15	21 37.60	. .	5.69		
24	9	26.	40.	..	..	..	..	..	24 7.65	37.11	1.00	II.	5	8.50	24 23.39	8.08	5.95	24 45.76	37 47.42
25	9	..	8.	22.	..	..	..	..	25 35.80	37.11	1.00	III.	5	5.30	22 42.66	8.24	5.79	26 13.91	36 6.69
26	9	..	..	..	..	..	46.3	0.3	26 18.62	37.11	0.99	VII.	8	11.19	39 38.12	8.32	7.34	26 56.72	53 3.78
27	9	..	..	..	..	..	26.4	40.	28 58.51	37.12	0.99	VII.	9	7.18	42 35.59	8.61	7.61	29 36.62	56 1.81
28	8	..	47.5	1.5	..	..	..	..	30 15.30	37.12	1.00	III.	6	8.5	28 1.92	8.75	6.27	30 53.42	25 41 26.94
29	8	..	..	55.	9.	..	..	..	32 8.85	37.12	0.99	IV.	10	6.3	46 59.04	8.94	8.03	32 46.96	26 0 26.01
30	9	10.2	24.	..	..	..	..	..	34 51.76	37.13	1.00	II.	6	4.19	26 7.84	9.23	6.10	35 29.89	25 39 33.17
31	6.7	..	..	..	..	10.	24.	..	34 56.18	37.13	1.00	VI.	3	10.33	15 17.13	9.24	5.13	35 34.31	28 41.50
32	10	..	..	..	..	..	23.	36.5	36 55.10	37.14	1.00	VII.	5	4.37	22 15.62	9.45	5.75	37 33.24	35 40.82
33	9	..	10.	23.5	37.5	..	..	..	40 37.53	37.14	1.00	IV.	6	10.6	29 2.98	9.63	6.37	41 15.67	42 29.18
34	8	..	..	46.3	0.5	14.4	..	..	42 50.37	37.15	0.99	V.	8	6.9	37 2.13	9.97	7.10	42 28.51	50 29.20
35	8	..	..	..	..	..	54.	8.	42 26.34	37.15	1.00	VII.	5	9.20	24 38.31	10.01	5.96	43 4.49	38 4.28
36	8	..	..	..	..	..	56.	10.3	43 28.50	37.15	1.00	VII.	5	9.49	24 52.91	10.11	5.99	44 6.65	38 19.04
37	8	..	..	..	..	..	..	8.5	44 26.92	37.15	1.00	VII.	6	9.39	28 49.00	10.21	6.34	45 5.07	42 15.55
38	7	7.	21.2	34.5	48.5	..	..	..	47 48.60	37.16	1.00	IV.	5	10.39	25 18.52	10.54	6.03	48 26.76	38 45.09
39	10	..	..	48.3	2.2	..	..	..	50 2.22	37.17	1.00	IV.	4	12.57	21 28.89	10.76	5.67	51 40.39	34 45.32
40	10	20.	31.	..	..	..	..	..	54 1.67	37.18	1.00	II.	6	10.39	29 19.46	11.12	6.40	54 39.85	42 46.98
41	8	..	..	..	52.	..	..	..	54 52.08	27.18	1.00	IV.	4	11.0	20 29.90	11.20	5.59	55 20.26	33 56.69
42	8	..	..	..	..	..	9.	23.	55 41.34	27.18	1.00	VII.	6	5.59	26 58.06	11.27	6.18	56 9.52	40 25.51
43	9	..	..	..	..	..	..	45.	12 57 3.60	27.19	1.00	VII.	3	3.45	11 51.19	11.39	4.81	12 57 31.79	25 17.39
44	9	..	42.	56.	..	..	..	..	13 2 9.80	27.20	1.00	III.	7	3.19	30 37.70	11.85	6.51	13 2 38.00	44 6.06
45	7	..	..	..	52.	6.	19.5	33.8	2 52.02	27.20	1.00	VII.	7	5.28	31 42.42	11.91	6.61	3 20.22	45 10.94
46	8	..	..	42.	56.	10.	..	..	4 56.00	27.21	1.00	V.	6	10.53	29 26.63	12.09	6.41	5 24.21	42 55.13
47	10	..	..	..	49.	..	..	..	6 49.07	27.21	1.00	IV.	4	4.46	17 21.32	12.26	5.31	7 17.28	30 48.89
48	9	..	..	38.	52.	..	..	..	7 51.90	27.22	1.00	IV.	8	6.20	37 7.71	12.35	7.11	8 20.12	25 50 37.17
49	8	..	..	..	25.	39.	..	..	9 24.95	27.22	0.99	V.	10	11.1	49 29.26	12.49	8.26	9 53.16	26 3 0.01
50	8	..	..	..	36.	50.	..	..	13 10 46.12	+27.23	+1.00	V.	3	4.45	−12 21.77	−12.59	−4.85	13 11 14.35	−25 25 49.21

CORRECTIONS.

Date.	Corr. of Clock.	Hourly rate.	m	n	c
1849. h.	s.	s.	s.	s.	s.

INSTRUMENT READINGS.

Date.	Barom.	THERMOM. At.	THERMOM. Ex.
1849. h. m.	in.	°	°

REMARKS.

(235) 18. One of the transit in error by 10ˢ. T. III assumed as correct, otherwise T=4ˢ.94

(235) 34. Transits over T.'s III–V assumed as 36ˢ.3, 50ˢ.5, and 4ˢ.4 instead of 46ˢ.3, 0ˢ.5, and 14ˢ.4, and minutes as 41 instead of 42.

(235) 39. Minutes assumed as 51 instead of 50.

(235) 50. Transits over T.'s IV and V assumed as 46ˢ and 0ˢ instead of 36ˢ and 50ˢ.

ZONE 235. APRIL 12. B. BELT, −25° 38½'. D$_0$ −25° 13' 10"—Continued.

No.	Mag.	SECONDS OF TRANSIT. I.	II.	III.	IV.	V.	VI.	VII.	T.	a_1	a_6	MICROMETER.			i	d_1	d_2	Mean Right Ascension, 1850.0.	Mean Declination, 1850.0.
									h. m. s.	s.	s.			r.	' "	"	"	h. m. s.	° ' "
51	10	..	19.	33.	..	..	..	..	13 13 46.78	+27.23	+1.00	III.	2	4.13	− 7 6.53	−12.87	−4.38	13 14 15.01	−25 20 33.78
52	9	..	52.	5.5	19.5	..	..	..	15 19.54	27.24	1.00	IV.	6	4.51	26 24.13	13.00	6.13	15 47.78	39 53.26
53	8	31.4	45.5	59.3	..	..	..	..	17 23.16	27.25	1.00	III.	8	3.40	35 46.99	13.17	6.99	17 51.41	49 17.15
54	7.8	..	26.	40.	53.5	..	..	..	18 53.72	27.26	1.00	IV.	5	7.44	23 50.27	13.40	5.88	19 21.98	37 19.55
55	8	..	26.5	40.3	54.2	..	..	..	20 54.19	27.26	1.00	IV.	7	10.0	33 59.95	13.49	6.83	21 22.45	47 30.27
56	7.8	16.	29.5	43.	57.	..	..	..	22 57.12	37.27	1.01	IV.	2	3.51	6 55.47	13.66	4.36	23 35.40	20 23.49
57	7.8	..	..	..	57.5	11.5	..	..	23 57.53	37.27	1.00	V.	8	2.35	35 14.22	13.75	6.94	24 35.80	48 44.91
58	7.8	..	..	37.5	51.	5.	..	..	27 51.16	37.28	1.00	V.	7	2.22	30 8.96	14.08	6.47	28 29.44	43 39.51
59	10	..	..	..	..	..	..	20.	49 38.28	37.37	0.99	VII.	9	7.31	42 42.15	15.85	7.64	50 16.64	56 15.64
60	6.7	..	..	..	..	43.3	57.2	..	51 29.42	37.38	1.00	VI.	4	6.27	18 12.08	15.99	5.39	52 7.80	31 43.46
61	9	..	..	48.5	2.5	..	..	..	53 2.38	37.39	0.99	IV.	9	7.46	42 50.08	16.10	7.66	53 40.76	56 23.84
62	9	..	..	..	16.	30.	..	..	55 16.07	37.40	1.00	V.	6	7.9	27 33.68	16.27	6.23	55 54.47	41 6.18
63	8	..	..	16.	30.	..	..	..	56 29.89	37.40	1.00	IV.	8	11.45	39 51.59	16.36	7.39	57 8.29	53 25.34
64	5	..	..	..	12.	26.	39.7	..	57 12.00	37.41	0.99	VI.	9	9.44	43 49.42	16.41	7.98	57 50.40	57 23.81
65	9	..	..	47.	1.	15.	..	..	58 14.88	37.41	1.00	V.	5	8.49	24 23.01	16.47	5.95	13 59 53.29	37 55.43
66	9	..	..	..	47.	1.3	..	..	13 59 47.24	37.42	1.00	IV.	6	4.49	26 23.13	16.60	6.13	14 0 25.66	39 55.86
67	9	..	..	..	..	55.	..	..	14 0 41.16	37.42	1.00	V.	4	6.41	18 19.26	16.67	5.39	1 19.58	31 51.32
68	8	..	..	56.	10.	..	..	..	2 9.95	37.43	1.00	IV.	5	8.59	24 28.09	16.77	5.95	2 48.38	38 0.81
69	9	..	..	..	..	55.	9.	..	2 41.18	37.43	1.00	VI.	4	5.37	17 46.87	16.81	5.34	3 19.61	31 19.02
70	7	..	..	..	1.	14.5	28.5	..	4 0.77	37.44	1.00	VI.	9	3.40	40 45.87	16.90	7.48	4 39.21	54 20.25
71	9	..	..	..	27.	49.4	..	..	5 26.77	37.44	1.00	V.	6	8.54	28 26.63	16.99	6.31	6 5.21	25 41 59.93
72	9	..	..	11.3	25.2	..	..	..	7 25.10	37.45	0.99	IV.	10	7.10	47 32.82	17.13	8.12	8 3.54	26 1 8.07
73	9	..	41.5	55.	9.	..	..	..	9 9.05	37.46	1.00	IV.	4	4.22	17 9.21	17.25	5.28	9 47.51	25 30 41.74
74	9	..	..	..	43.	57.	..	..	9 43.04	37.46	1.00	V.	7	6.31	32 14.52	17.29	6.67	10 21.50	45 48.48
75	9	45.4	59.2	13.	..	..	..	..	14 26.93	37.48	1.00	III.	6	6.38	27 18.05	17.58	6.21	15 5.41	40 51.84
76	9	..	..	..	8.5	22.3	..	..	15 8.52	37.49	1.00	V.	3	8.55	14 27.83	17.62	5.04	15 47.01	25 28 0.49
77	8	..	..	15.	29.2	43.	..	..	16 29.00	37.49	0.99	V.	10	11.15	49 36.32	17.71	8.31	17 7.48	26 3 12.34
78	9	..	28.	42.	..	..	..	..	17 55.82	37.50	1.00	III.	8	13.56	40 57.60	17.79	7.50	18 34.32	25 54 32.89
79	7	..	..	0.5	14.5	28.6	..	..	20 14.57	37.51	1.00	V.	1	13.46	6 56.46	17.93	4.33	20 53.08	20 28.72
80	9	..	..	..	..	34.5	48.5	..	21 20.70	37.52	1.00	VI.	7	12.20	35 10.37	18.00	6.95	21 59.22	48 45.32
81	9	..	..	..	..	..	..	52.	22 10.30	37.52	1.00	VII.	8	14.7	41 2.82	18.05	7.50	22 48.82	54 38.37
82	8	..	50.4	4.3	..	..	..	..	25 18.14	37.53	1.00	III.	3	10.30	15 15.74	18.23	5.12	25 56.67	28 49.09
83	9	..	..	..	..	..	4.	..	25 36.14	37.53	1.00	VI.	2	10.6	10 4.42	18.25	4.62	26 14.67	23 37.29
84	9	..	..	..	19.	..	46.5	..	27 18.85	37.55	1.00	VI.	9	5.31	41 41.84	18.35	7.58	27 57.40	55 17.77
85	7	..	36.5	..	..	..	..	..	31 4.24	37.57	1.00	II.	6	5.37	22 46.07	18.57	5.77	31 42.81	36 20.41
86	9	..	..	..	..	..	37.	..	31 9.23	37.57	1.00	VI.	6	4.52	26 24.48	18.58	6.13	31 47.80	39 59.19
87	8	..	..	..	..	..	..	16.3	32 34.68	37.58	1.00	VII.	7	8.44	33 21.27	18.66	6.77	32 13.26	46 56.70
88	8	..	60.2	14.	..	..	..	..	35 27.88	37.59	1.00	III.	2	7.23	8 42.33	18.82	4.49	35 6.47	22 15.64
89	7	40.	53.5	7.3	..	..	..	..	38 21.28	37.61	1.00	III.	3	7.25	13 42.44	18.98	4.95	38 59.89	25 27 16.37
90	7	..	..	..	..	..	..	14.5	38 32.74	37.61	0.99	VII.	10	6.41	47 17.84	18.99	8.12	39 11.34	26 0 54.95
91	9	..	..	50.	..	..	..	..	41 3.88	37.63	1.00	III.	8	12.8	40 3.14	19.14	7.42	41 42.51	25 53 39.70
92	8	..	19.5	33.4	47.5	..	..	..	42 47.29	37.63	0.99	IV.	10	5.0	46 27.27	19.23	8.04	43 25.91	26 0 4.54
93	9	..	10.	23.5	..	..	..	..	45 37.55	37.65	1.00	III.	6	3.37	25 46.78	19.38	6.07	46 16.20	25 39 22.23
94	8	49.5	3.3	17.	..	..	..	..	47 30.99	37.66	1.00	III.	6	5.45	26 51.32	19.48	6.17	48 9.65	40 26.97
95	8	37.5	51.3	5.	..	..	..	..	52 18.98	37.69	1.00	III.	5	4.50	22 22.49	19.72	5.74	52 57.67	35 57.95
96	8	..	..	..	..	..	16.	..	52 48.06	37.69	1.00	VI.	1	4.9	2 5.40	19.74	3.64	53 26.77	15 38.78
97	8	..	49.	3.	..	..	..	..	56 16.80	37.71	1.00	III.	6	3.37	25 46.78	19.91	6.07	56 55.51	39 22.76
98	7.8	..	..	..	52.5	6.3	..	..	56 52.52	37.71	1.00	V.	4	5.18	17 37.40	19.93	5.31	57 31.23	31 12.64
99	9	..	..	..	..	24.	..	..	58 10.19	37.72	1.00	V.	3	6.39	13 19.26	19.99	4.92	14 58 48.91	26 54.17
100	7	..	..	..	27.	40.	54.	..	14 59 26.43	+37.73	+0.99	VI.	10	5.7	−41 29.74	−20.05	−7.57	15 0 5.15	−25 55 7.36

CORRECTIONS.

Date.	Corr. of Clock.	Hourly rate.	m	n	c
1849. h.	s.	s.	s.	s.	s.

INSTRUMENT READINGS.

Date.	Barom.	THERMOM. At.	Ex.
1849. h. m.	In.	°	°

REMARKS.

(235) 53. Transits over T.'s I–III assumed as 41ˢ.4, 55ˢ.5, and 9ˢ.3 instead of 31ˢ.4, 45ˢ.5, and 59ˢ.3.
(235) 54. Right ascension small 24ˢ by Mural, 1848, April 20, Mer. Circle, 1848, May 27, and Arg. Z., 380, 60.
(235) 65. Minutes assumed as 59 instead of 58, and transits over T.'s II–IV as recorded over T.'s III–V.
(235) 85. Hor. thread assumed as 5 instead of 6.
(235) 87. Minutes assumed as 31 instead of 32.
(235) 88. Minutes assumed as 34 instead of 35.
(235) 100. Hor. thread assumed as 9 instead of 10.

ZONE 236. APRIL 14. B. BELT, −26° 53½′. D₀ = −26° 28′ 40″.

No.	Mag.	SECONDS OF TRANSIT. I.	II.	III.	IV.	V.	VI.	VII.	T.	a_1	a_2	MICROMETER.			i	d_1	d_2	Mean Right Ascension, 1850.0.	Mean Declination, 1850.0.
									h. m. s.	s.	s.			r.	′ ″	″	″	h. m. s.	° ′ ″
1	8	. .	. .	18.	31.8	. .	. .	. .	10 22 31.94	+40.87	+1.01	IV.	4	2.25	−16 10.21	− 9.76	− 5.13	10 23 13.82	−26 45 5.10
2	9	. .	. .	37.	51.	. .	. .	. .	25 51.04	40.87	0.99	IV.	7	11.2	34 31.21	11.29	6.96	26 32.90	27 3 29.46
3	6	. .	. .	. .	. .	. .	. .	49.5	26 7.72	40.67	1.01	VII.	1	11.14	5 39.47	11.42	4.09	26 49.60	26 34 34.98
4	7	. .	30.5	44.3	54.5	. .	. .	. .	28 58.43	40.86	1.00	IV.	5	7.41	23 48.75	12.73	5.88	29 40.29	26 52 47.36
5	7	. .	. .	. .	. .	6.5	20.2	. .	29 52.34	40.86	0.99	VI.	10	9.2	48 29.13	13.14	8.39	30 34.19	27 17 30.66
6	8	. .	. .	24.5	38.4	. .	. .	. .	31 38.40	40.86	0.99	IV.	9	4.30	41 11.25	13.95	7.65	32 20.25	27 10 12.85
7	9	. .	29.	43.	. .	. .	. .	. .	33 57.03	40.85	1.00	III.	6	4.40	26 18.55	15.02	6.13	34 38.88	26 55 19.70
8	8	. .	. .	. .	. .	. .	. .	14.	34 32.15	40.85	1.01	VII.	3	3.43	11 50.15	15.28	4.69	35 14.01	40 50.12
9	8	. .	. .	. .	. .	. .	24.2	38.	37 56.10	40.85	1.01	VII.	3	9.7	14 33.53	16.85	4.96	38 37.96	43 35.34
10	9	. .	. .	. .	. .	. .	54.	8.	39 25.96	40.85	1.00	VII.	6	11.5	29 32.33	17.54	6.45	40 7.81	26 58 36.32
11	7	. .	. .	. .	. .	23.	37.	. .	41 8.98	40.85	1.00	VI.	8	9.2	38 29.23	18.32	7.37	41 50.83	27 7 34.92
12	9	. .	58.5	12.	26.2	. .	. .	. .	43 26.23	40.85	1.00	IV.	8	4.53	36 23.84	19.38	7.16	44 8.08	27 5 30.38
13	9	. .	. .	. .	. .	24.	38.	. .	44 9.95	40.84	1.00	VI.	4	5.17	17 35.76	19.72	5.26	44 41.79	26 46 40.74
14	9	. .	. .	. .	. .	. .	. .	44.	45 2.20	40.84	1.01	VII.	2	3.3	6 30.88	20.12	4.16	45 44.05	26 35 35.16
15	8	. .	31.3	45.	59.2	. .	. .	. .	47 59.17	40.84	0.99	IV.	7	6.24	32 11.03	21.47	6.72	48 41.00	27 1 19.22
16	9	. .	. .	35.	48.5	2.3	. .	. .	50 48.63	40.84	1.00	V.	3	12.39	16 20.78	22.77	5.13	51 30.47	26 45 28.68
17	7	. .	. .	. .	. .	. .	5.	18.5	51 36.75	40.84	1.01	VII.	2	13.28	11 46.02	23.14	4.65	52 18.60	40 53.84
18	8	39.	53.	7.3	. .	. .	. .	. .	55 21.08	40.84	1.01	III.	3	7.2	13 30.85	24.86	4.85	56 2.93	42 40.56
19	9	. .	. .	. .	58.	. .	. .	. .	56 58.05	40.83	1.01	IV.	3	4.38	12 18.28	25.61	4.73	57 39.89	26 41 28.62
20	8	. .	. .	. .	7.7	21.3	. .	. .	10 58 7.48	40.83	1.00	V.	7	5.23	31 40.22	26.14	6.67	10 58 49.31	27 0 53.03
21	5.6	. .	. .	. .	47.	1.	15.	. .	11 0 46.94	40.83	0.99	VI.	10	6.6	47 0.39	27.36	8.25	11 1 28.76	27 16 16.00
22	9	. .	. .	. .	. .	7.	21.	. .	1 52.96	40.83	1.00	V.	5	7.9	23 32.58	27.86	5.85	2 34.79	26 52 46.29
23	8	. .	. .	. .	. .	. .	. .	51.	11 2 9.13	+40.83	+1.01	VII.	3	8.20	−14 9.83	−27.99	− 4.92	11 2 50.97	−26 43 22.74

ZONE 237. APRIL 16. B. BELT, −22° 31′. D₀ = −22° 6′ 0″.

No.	Mag.	I.	II.	III.	IV.	V.	VI.	VII.	T.	a_1	a_2	MICROMETER.			i	d_1	d_2	Mean Right Ascension, 1850.0.	Mean Declination, 1850.0.
1	7	. .	. .	. .	. .	. .	24.5	38.	10 11 57.46	+46.18	+1.01	VII.	2	4.12	− 7 5.76	− 1.09	− 4.74	10 12 44.65	−22 13 11.59
2	10	. .	. .	58.2	11.5	. .	. .	. .	14 11.61	46.18	1.00	IV.	6	9.38	28 48.85	1.31	6.27	14 58.79	34 56.43
3	9	5.6	19.	. .	. .	. .	. .	. .	15 46.15	46.17	0.99	II.	7	14.48	36 25.01	1.46	6.82	16 33.31	42 33.29
4	9	. .	58.	11.4	. .	. .	. .	. .	17 25.00	46.17	0.99	III.	8	13.0	40 29.37	1.63	7.10	18 12.16	46 38.10
5	9	. .	. .	14.5	28.	. .	. .	. .	18 28.00	46.17	1.00	IV.	7	8.50	33 24.65	1.74	6.58	19 15.17	39 32.97
6	9	. .	. .	24.	37.5	. .	. .	. .	19 37.56	46.16	1.00	IV.	4	10.20	20 9.73	1.85	5.67	20 24.72	26 17.25
7	10	. .	42.3	56.	. .	. .	. .	. .	21 9.44	46.16	1.00	III.	6	6.41	27 19.57	2.00	6.16	21 56.60	33 27.73
8	10	15.	28.4	. .	. .	. .	. .	. .	55.52	46.16	1.00	II.	6	5.13	26 35.08	2.18	6.11	23 42.68	32 43.37
9	7	. .	. .	17.	30.4	. .	. .	. .	24 30.40	46.16	0.99	IV.	9	9.32	43 43.53	2.33	7.33	25 17.55	49 53.19
10	7	. .	35.7	49.2	. .	. .	. .	. .	27 2.74	46.15	1.00	III.	4	6.2	17 59.61	2.58	5.50	27 49.89	24 7.69
11	10	. .	. .	. .	. .	. .	. .	56.	27 15.47	46.15	1.00	VII.	5	7.27	23 41.38	2.59	5.92	28 2.62	29 49.89
12	8	58.7	12.2	26.	. .	. .	. .	. .	29 39.36	46.15	1.00	III.	6	12.5	30 2.94	2.84	6.34	30 26.51	36 12.12
13	10	. .	19.8	33.2	. .	. .	. .	. .	31 46.80	46.14	1.00	III.	7	7.49	32 53.86	3.02	6.55	32 33.94	39 3.43
14	10	. .	. .	. .	. .	. .	24.	37.5	31 56.92	46.14	1.00	VII.	7	10.3	34 1.15	3.06	6.64	32 44.06	40 10.85
15	9	. .	. .	23.3	37.	50.4	. .	. .	33 36.89	46.14	1.00	V.	7	4.14	31 5.45	3.21	6.42	34 24.03	37 15.08
16	8	. .	. .	40.	54.	. .	. .	. .	34 53.72	46.14	0.99	IV.	9	1.25	39 37.95	3.34	7.04	35 40.85	45 48.33
17	9	. .	. .	. .	. .	56.2	10.	23.	36 42.73	46.14	1.01	VII.	1	8.1	4 2.23	3.51	4.51	37 29.88	10 10.25
18	7.8	. .	. .	. .	. .	22.5	36.	. .	38 8.96	46.13	1.00	VI.	7	6.56	32 27.13	3.61	6.52	38 56.09	38 37.26
19	9	. .	52.	5.7	19.2	. .	. .	. .	41 19.18	46.13	1.00	IV.	3	6.23	13 11.22	3.95	5.16	42 6.31	19 20.33
20	10	. .	. .	6.5	20.3	. .	. .	. .	43 20.08	46.13	0.99	IV.	10	6.51	47 23.24	4.13	7.61	44 7.20	53 34.98
21	10	. .	. .	. .	20.5	34.	. .	. .	44 20.54	46.13	1.00	V.	4	6.31	18 14.23	4.23	5.51	45 7.67	24 23.97
22	9	. .	. .	. .	40.5	. .	. .	. .	10 45 40.51	+46.12	+1.00	IV.	6	5.41	−26 49.34	− 4.35	− 6.13	10 46 27.63	−22 32 59.82

CORRECTIONS.

Date.		Corr. of Clock.	Hourly rate.	m	n	c
1849.	h.	s.	s.	s.	s.	s.

INSTRUMENT READINGS.

Date.	Barom.	THERMOM. At.	THERMOM. Ex.
1849. h. m.	in.	°	°

REMARKS.

(237) 21. Transit over middle thread assumed as at 20ˢ.5 instead of 28ˢ.5.

ZONE 237. APRIL 16. B. BELT, −22° 31′. D_0 = −22° 6′ 0″—Continued.

No.	Mag.	SECONDS OF TRANSIT. I.	II.	III.	IV.	V.	VI.	VII.	T.	a_1	a_2	MICROMETER.			i	d_1	d_2	Mean Right Ascension, 1850.0.	Mean Declination, 1850.0.
									h. m. s.	s.	s.			r.	′ ″	″	″	h. m. s.	° ′ ″
23	8	..	..	..	..	..	30.	43.2	10 46 2.80	+46.12	+1.00	VII.	4	3.23	−16 39.14	− 4.38	− 5.41	10 46 49.92	−22 22 48.93
24	10	..	..	..	..	..	52.	5.6	47 25.00	46.12	1.00	VII.	4	4.51	17 23.51	4.51	5.46	48 12.12	23 33.48
25	10	..	..	..	..	4.5	18.	..	49 50.95	46.12	1.00	VI.	6	11.45	29 52.75	4.64	6.34	48 38.07	36 3.73
26	8	..	34.4	48.	..	..	..	..	51 1.50	46.12	1.00	III.	4	11.26	20 42.98	4.84	5.70	51 47.62	26 53.52
27	8	..	55.5	9.	..	..	..	..	52 22.54	46.12	1.00	III.	6	13.11	30 36.21	4.96	6.39	53 9.66	36 47.56
28	9	..	54.2	6.	..	..	..	..	53 21.40	46.12	1.00	III.	6	12.51	30 26.13	5.05	6.38	54 8.52	36 37.56
29	10	..	..	55.	8.8	..	..	..	54 8.66	46.12	1.00	IV.	6	11.33	29 46.84	5.12	6.34	54 55.78	35 58.30
30	9	..	..	..	..	..	3.	16.5	54 35.90	46.11	0.99	VII.	9	6.38	42 15.48	5.16	7.25	55 23.00	48 27.89
31	10	..	33.4	46.5	..	..	..	..	57 0.24	46.11	1.00	III.	4	6.44	18 20.79	5.38	5.51	10 57 47.35	24 31.68
32	9	..	53.	6.5	..	..	..	..	10 59 20.04	46.11	1.00	III.	6	11.21	29 40.75	5.58	6.33	11 0 7.15	35 52.66
33	9	..	2.	15.8	..	..	..	..	11 1 29.20	46.11	0.99	III.	8	8.6	38 1.14	5.77	6.94	2 16.30	44 13.85
34	10	..	..	..	..	10.8	24.2	..	1 57.19	46.11	1.00	VI.	5	11.37	25 47.62	5.82	6.06	2 44.30	31 59.50
35	8	..	4.5	18.1	41.5	..	..	..	5? 31.57	46.11	1.01	IV.	1	9.36	4 50.45	6.13	4.55	6 18.69	11 1.13
36	10	..	17.	30.3	43.6	..	..	..	10 43.84	46.10	1.00	IV.	2	12.58	11 31.29	6.59	5.03	11 30.94	17 42.91
37	8	..	..	53.2	7.	20.5	..	..	12 6.83	46.10	0.99	V.	10	8.3	47 59.52	6.72	7.67	12 53.92	54 13.91
38	9	..	..	..	..	..	..	22.	12 21.74	46.10	1.00	VII.	4	4.22	17 8.90	6.74	5.43	13 8.84	23 21.07
39	10	..	..	..	..	..	4.5	16.1	13 37.48	46.10	1.00	VII.	7	2.32	30 13.75	6.85	6.36	14 24.58	36 26.95
40	7.8	..	..	..	6.	19.5	33.	..	15 5.93	46.10	0.99	VI.	10	5.50	46 52.34	6.98	7.59	15 53.02	53 6.91
41	10	41.5	55.1	..	..	..	..	..	17 22.16	46.10	0.99	II.	8	9.27	38 41.86	7.19	6.99	18 9.25	44 56.04
42	9	..	..	..	46.	59.5	..	..	19 46.04	46.10	1.00	V.	4	11.4	20 31.89	7.40	5.69	20 33.14	26 44.98
43	10	..	..	..	..	..	32.4	..	21 5.16	46.10	1.01	VI.	1	11.35	5 50.31	7.51	4.62	21 52.27	12 2.43
44	10	..	..	..	..	..	27.	40.3	22 59.82	46.10	1.00	VII.	7	8.7	33 2.66	7.68	6.56	23 46.92	39 16.90
45	9	..	46.5	0.	..	..	..	..	24 13.54	46.10	1.00	III.	6	6.20	27 8.98	7.78	6.15	25 0.64	33 22.91
46	7	..	..	..	..	55.2	8.5	..	24 41.55	46.10	1.00	VI.	7	3.23	30 39.61	7.82	6.40	25 28.65	36 53.83
47	8	..	..	..	..	..	..	12.	25 31.40	46.10	1.00	VII.	7	4.26	31 11.22	7.89	6.43	26 18.50	37 25.54
48	9	..	..	..	..	..	2.5	16.	26 35.41	46.10	0.99	VII.	8	8.22	38 8.92	7.99	6.94	27 22.50	44 23.85
49	10	..	..	..	7.	20.5	..	..	28 6.90	46.10	0.99	V.	9	6.26	42 9.71	8.11	7.24	28 53.99	48 25.06
50	7.8	..	..	..	..	21.2	35.	..	29 7.78	46.10	1.00	VI.	4	6.54	18 25.72	8.20	5.52	29 54.88	24 39.44
51	10	..	..	..	29.	..	..	..	32 29.02	46.10	1.01	IV.	2	3.56	6 58.00	8.49	4.70	33 16.13	13 11.19
52	9	..	..	..	..	..	..	11.	32 30.62	46.10	1.01	VII.	2	7.36	8 48.63	8.49	4.84	33 17.73	15 1.96
53	9	..	23.3	37.	..	..	..	..	35 50.44	46.10	1.01	III.	1	9.59	5 2.01	8.77	4.56	36 37.55	11 15.34
54	10	..	36.3	50.	..	..	..	..	37 3.44	46.10	1.01	III.	1	9.16	4 40.33	8.87	4.54	37 50.55	10 53.74
55	9	..	..	54.6	7.3	..	..	..	38 7.72	46.10	1.01	IV.	2	6.20	8 10.61	8.96	4.78	38 54.83	14 24.35
56	9	..	..	..	..	14.	27.3	..	39 0.32	46.10	1.01	VI.	2	2.38	6 18.53	9.04	4.65	39 47.43	12 32.22
57	7.8	..	..	..	..	..	34.	17.4	40 6.90	46.10	1.01	VII.	2	8.24	9 12.83	9.13	4.85	40 54.01	15 26.81
58	9	..	..	..	..	48.	1.2	..	41 34.26	46.10	1.01	VI.	1	12.17	6 11.48	9.25	4.64	42 21.37	12 25.37
59	10	..	18.	31.2	45.	..	..	..	45 44.94	46.10	1.01	IV.	2	9.8	9 35.32	9.59	4.89	46 32.05	15 49.80
60	8	..	..	..	7.2	21.	..	..	47 7.32	46.10	1.00	V.	6	13.17	30 39.24	9.70	6.40	47 54.42	36 55.34
61	10	..	..	40.5	54.	..	..	..	48 53.99	46.10	1.00	IV.	7	9.17	33 38.26	9.85	6.60	49 41.09	39 54.71
62	8	55.2	9.	22.3	..	..	..	..	50 35.89	46.10	1.00	III.	6	8.43	28 21.09	9.99	6.24	51 22.99	34 37.32
63	10	0.	13.6	..	..	..	..	..	52 40.60	46.10	1.00	II.	5	8.20	24 8.28	10.15	5.94	53 27.70	30 24.37
64	10	..	..	..	..	..	20.8	34.2	11 52 53.70	46.10	1.01	VII.	2	11.32	10 47.63	10.17	4.97	11 53 40.81	17 2.77
65	9	..	..	..	..	..	..	31.	12 0 50.45	46.11	1.00	VII.	5	12.29	26 13.66	10.81	6.08	12 1 37.56	32 30.55
66	6.7	..	..	20.2	34.	47.4	..	..	2 33.83	46.11	0.99	V.	8	11.16	39 36.94	10.94	7.07	3 20.93	45 54.95
67	8	..	47.	0.3	14.2	28.	..	..	7 14.15	46.11	1.00	V.	5	9.32	24 44.70	11.32	5.98	8 1.26	31 2.00
68	10	..	38.	51.7	..	..	..	..	10 5.14	46.12	1.01	III.	1	9.5	4 34.78	11.56	4.52	10 52.27	10 50.86
69	9	..	..	53.6	7.2	..	..	..	11 7.16	46.12	1.00	IV.	6	11.15	29 37.76	11.64	6.33	11 54.28	35 55.73
70	9	..	10.2	23.2	37.	..	..	..	12 36.95	46.12	0.99	IV.	9	9.49	43 52.10	11.76	7.40	13 24.06	50 11.26
71	9	..	..	55.	8.4	..	..	..	12 14 8.50	+46.12	+1.00	IV.	4	8.00	−18 59.14	−11.88	− 5.57	12 14 55.62	−22 25 16.59

CORRECTIONS.

Date.	Corr. of Clock.	Hourly rate.	m	n	c
1849. h.	s.	s.	s.	s.	s.

INSTRUMENT READINGS.

Date.	Barom.	THERMOM. At.	Ex.
1849. h. m.	In.	°	°

REMARKS.

(237) 25. Minutes assumed as 47 instead of 48.
(237) 35. Time of transit over middle thread assumed as 31ˢ.5 instead of 41ˢ.5.

ZONE 237. APRIL 16. B. BELT, −22° 31′. $D_0 = -22° 6' 0''$—Continued.

No.	Mag.	Seconds of Transit. I.	II.	III.	IV.	V.	VI.	VII.	T.	a_1	a_2	Micrometer.			i	d_1	d_2	Mean Right Ascension, 1850.0.	Mean Declination, 1850.0.
									h. m. s.	s.	s.			r.	′ ″	″	″	h. m. s.	° ′ ″
72	7	..	48.8	2.2	15.5	..	..	..	12 59 15.66	+46.20	+0.99	IV.	9	3.34	−40 43.01	−15.22	− 7.16	13 0 2.85	−22 47 5.39
73	5	..	..	..	..	25.2	39.	52.2	13 1 11.78	46.20	1.00	VII.	3	5.11	12 34.60	15.34	5.08	0 58.98	18 55.02
74	8	..	..	..	..	..	44.	58.	2 17.17	46.21	1.00	VII.	7	2.56	30 25.84	15.41	6.39	3 4.38	36 47.64
75	8	..	..	5.5	19.3	..	..	..	4 19.16	46.21	1.01	IV.	1	3.36	1 48.92	15.54	4.31	5 6.38	8 8.77
76	10	39.3	53.	6.2	..	..	..	..	7 19.92	46.22	1.00	V.	8	9.40	38 48.53	15.72	7.02	8 7.14	45 11.27
77	10	..	..	..	33.2	46.3	0.0	..	8 32.97	46.22	1.00	.	3	4.53	12 25.84	15.80	5.06	9 20.19	18 46.70
78	5	..	..	46.	59.	13.	26.	40.	9 59.29	46.22	1.00	VII.	4	2.53	16 24.02	15.89	5.37	10 46.51	22 45.28
79	9	..	9.5	23.	..	..	..	..	13 36.54	46.23	1.01	III.	1	7.15	3 39.31	16.09	4.44	14 23.78	9 59.84
80	9	..	..	..	..	..	5.	19.	13 38.21	46.23	1.01	VII.	1	7.22	3 42.56	16.09	4.44	14 25.45	10 3.09
81	10	..	..	20.	33.4	..	..	..	15 33.40	46.24	1.00	IV.	9	5.5	41 28.89	16.19	7.22	16 20.64	47 52.30
82	10	..	..	..	..	..	..	37.	15 56.43	46.24	1.00	VII.	6	7.12	27 34.92	16.22	6.18	16 43.67	33 57.32
83	9	..	..	..	..	32.	45.4	..	17 18.38	46.25	1.00	VI.	4	12.21	21 10.59	16.29	5.72	18 5.63	27 32.60
84	8	3.5	17.	30.5	..	..	..	..	20 44.05	46.26	1.00	III.	5	7.47	23 51.75	16.48	5.92	21 31.31	30 14.15
85	8	..	..	..	..	43.	56.5	..	29.48	46.27	0.99	VI.	10	4.7	46 0.41	16.	7.57		
86	8	28.	41.5	55.3	8.8	..	..	..	28 8.69	46.28	0.99	IV.	9	10.48	44 21.85	16.86	7.46	28 55.96	50 46.17
87	8	..	..	..	..	..	22.5	36.	28 55.44	46.28	1.00	VII.	4	11.44	20 51.78	16.90	5.69	29 42.72	27 14.37
88	8	..	0.5	13.8	27.5	..	..	..	32 27.45	46.29	1.00	IV.	7	11.48	34 54.40	17.09	6.74	33 14.74	41 18.23
89	10	..	..	..	..	..	25.	38.7	32 58.05	46.29	1.00	VII.	3	10.3	15 1.84	17.11	5.26	33 45.34	21 24.21
90	9	26.4	40.	..	..	..	..	..	38 7.06	46.31	1.00	II.	8	8.1	37 58.50	17.36	6.96	38 54.37	44 22.82
91	9	..	..	..	38.5	52.	..	..	38 38.45	46.31	1.00	V.	7	9.13	33 36.21	17.39	6.65	39 25.76	40 0.25
92	8	..	31.6	45.	58.7	..	..	..	43 58.60	46.33	1.00	IV.	8	11.32	39 45.04	17.64	7.11	44 45.93	46 9.79
93	9	..	..	..	..	8.1	21.6	..	44 54.57	46.33	1.00	VI.	9	5.33	41 42.87	17.66	7.25	45 41.90	48 7.78
94	8	..	..	37.6	51.	..	..	..	47 51.09	46.35	1.00	IV.	3	11.1	15 31.40	17.81	5.30	48 38.44	21 54.51
95	7	..	..	..	..	33.5	47.	0.5	48 19.98	46.35	1.00	IV.	2	11.46	10 55.00	17.83	4.94	49 7.33	17 17.77
96	8	13.	26.4	40.	..	..	..	..	50 53.54	46.36	1.00	III.	7	12.37	35 19.08	17.94	6.77	51 40.90	41 43.79
97	9	..	..	53.	..	20.5	..	..	52 6.75	46.36	1.00	V.	6	13.48	30 54.87	17.99	6.43	52 54.11	37 19.29
98	10	..	..	0.2	..	..	..	..	56 13.73	46.37	1.00	III.	4	10.25	20 12.22	18.14	5.64	57 1.10	26 36.00
99	9	..	..	..	..	..	53.2	7.	56 26.29	46.37	1.00	VII.	4	7.27	18 42.18	18.15	5.52	57 13.66	25 5.85
100	9	56.	9.5	23.2	..	..	..	..	13 58 36.62	46.39	1.00	III.	6	8.55	28 27.14	18.24	6.26	13 59 24.01	34 51.64
101	9	46.	59.7	13.3	..	..	..	..	14 1 26.67	46.40	1.00	III.	2	5.8	7 34.27	18.34	4.69	14 2 14.07	13 57.30
102	9	..	..	..	..	..	29.3	..	2 2.06	46.40	1.00	VI.	1	10.57	0 28.60	18.36	3.50	2 49.46	6 50.46
103	8	25.4	39.	52.3	6.	..	..	..	5 5.97	46.42	1.00	IV.	4	8.27	19 12.75	18.47	5.56	5 53.39	25 36.78
104	8	..	24.2	37.5	51.	..	..	..	9 51.11	46.43	1.00	IV.	2	8.12	9 7.09	18.64	4.81	10 38.54	15 30.54
105	8	..	..	3.	16.5	..	..	..	12 16.52	46.44	1.00	IV.	1	4.16	2 9.09	18.71	4.31	13 3.96	8 32.11
106	8	..	..	..	6.5	20.	..	..	19 6.54	46.48	1.00	V.	3	7.19	13 39.43	18.93	5.16	19 54.02	20 3.52
107	8	7.	21.	34.	..	..	..	..	22 47.76	46.49	1.00	III.	9	2.30	40 10.71	19.03	7.15	23 35.25	46 36.89
108	8	58.	11.5	25.	..	..	..	..	28 38.55	46.52	1.00	III.	5	8.13	24 4.86	19.18	5.94	29 26.07	30 29.98
109	8	..	..	21.	35.	..	..	..	14 29 34.75	+46.53	+1.00	IV.	7	8.18	−33 8.51	−19.20	− 6.60	14 30 22.28	−22 39 34.31

ZONE 238. APRIL 20. B. BELT, −23° 46′. $D_0 = -23° 20' 50''$.

No.	Mag.	I.	II.	III.	IV.	V.	VI.	VII.	T.	a_1	a_2	Micrometer.			i	d_1	d_2	Mean Right Ascension, 1850.0.	Mean Declination, 1850.0.
1	8	..	..	..	..	56.4	10.2	..	10 24 56.49	− 3.12	−1.00	VI.	6	4.8	−26 2.30	− 3.37	− 6.08	10 24 52.37	−23 47 1.75
2	8	..	..	..	..	..	..	0.	25 19.08	3.12	1.00	VII.	5	10.11	25 4.06	3.40	6.01	25 14.96	46 3.47
3	8	..	..	..	..	..	38.	51.3	27 10.54	3.13	0.99	VII.	2	11.35	10 49.12	3.46	4.91	27 6.42	31 47.49
4	9	..	53.	6.3	20.	..	..	..	29 20.11	3.13	1.00	IV.	5	12.52	26 25.57	3.55	6.11	29 15.98	23 47 25.23
5	10	..	..	18.5	32.2	..	..	..	30 32.11	3.14	1.01	IV.	9	9.41	43 48.06	3.59	7.45	30 27.96	24 4 49.10
6	8	..	..	..	29.	42.3	56.	..	31 28.78	3.14	1.00	VI.	5	9.8	24 32.47	3.62	5.96	31 24.64	23 45 32.05
7	8	..	..	..	..	..	35.3	49.1	10 32 8.08	− 3.14	−1.00	VII.	4	8.18	−19 7.88	− 3.65	− 5.55	10 32 3.94	−23 40 7.08

CORRECTIONS.

Date.	Corr. of Clock.	Hourly rate.	m	n	c
1849. h.	s.	s.	s.	s.	s.

INSTRUMENT READINGS.

Date.	Barom.	Thermom. At.	Thermom. Ex.
1849. h. m.	in.	°	°

REMARKS.

(237) 73. Minutes assumed as 0 instead of 1.
(237) 102. Micrometer reading assumed as 0r.57 instead of 10r.57.
(238) 1. Transits over T.'s IV and V assumed as recorded over T.'s V and VI.

ZONE 238. APRIL 20. B. BELT, −23° 46′. D₀=−23° 20′ 50″—Continued.

No.	Mag.	Seconds of Transit. I.	II.	III.	IV.	V.	VI.	VII.	T.	a_1	a_2	Micrometer.			i	d_1	d_2	Mean Right Ascension, 1850.0.	Mean Declination, 1850.0.
									h. m. s.	s.	s.			r.	′ ″	″	″	h. m. s.	° ′ ″
8	9	. .	. .	. .	. .	. .	46.	. .	10 33 18.69	− 3.14	−1.00	VI.	7	7.32	−32 45.17	− 3.69	− 6.61	10 33 14.55	−23 53 45.47
9	9	. .	. .	. .	. .	. .	. .	29.	33 48.09	3.14	1.00	VII.	5	8.57	24 26.75	3.71	5.95	33 43.95	23 45 26.41
10	8	. .	. .	. .	. .	26.	40.	. .	35 12.54	3.14	1.01	VI.	10	10.36	49 16.56	3.76	7.89	35 8.39	24 10 18.21
11	10	. .	. .	. .	. .	. .	48.3	. .	36 21.01	3.15	1.00	VI.	8	4.28	36 11.08	3.80	6.87	36 16.86	23 57 11.75
12	10	. .	. .	. .	53.4	. .	. .	. .	37 53.41	3.15	1.00	IV.	6	5.42	26 49.85	3.85	6.14	37 49.26	47 49.84
13	10	. .	. .	17.4	. .	. .	. .	. .	39 31.06	3.15	1.00	III.	6	4.55	26 26.12	3.91	6.11	39 26.91	47 26.14
14	8	. .	. .	. .	. .	15.	29.	. .	40 1.50	3.15	1.00	VI.	5	9.6	24 31.47	3.93	5.96	39 57.35	45 31.36
15	11	. .	17.	31.	. .	. .	. .	. .	42 44.50	3.16	1.01	III.	8	6.45	37 20.29	4.03	6.95	42 40.33	58 21.27
16	8	. .	23.4	37.	. .	. .	. .	. .	44 50.69	3.16	0.99	III.	3	7.1	13 30.35	4.10	5.13	44 46.54	34 29.58
17	8	. .	. .	26.4	. .	. .	. .	. .	45 40.06	3.16	1.00	III.	7	10.41	34 20.59	4.13	6.72	45 35.90	55 21.44
18	7.8	. .	. .	. .	. .	. .	14.3	27.7	45 46.86	3.16	1.00	VII.	8	4.40	36 16.96	4.13	6.87	45 42.69	23 57 17.96
19	9	. .	. .	. .	31.5	45.	59.	. .	47 31.49	3.17	1.01	VI.	8	11.27	39 42.36	4.19	7.15	47 27.31	24 0 43.70
20	8	40.3	54.	7.4	. .	. .	. .	. .	51 21.20	3.17	1.00	III.	5	6.55	23 25.53	4.33	5.87	51 17.03	23 44 25.73
21	9	. .	34.	48.	1.5	. .	. .	. .	10 55 1.47	3.18	1.00	IV.	7	10.4	34 1.97	4.45	6.70	10 54 57.29	54 3.12
22	9	. .	. .	57.2	11.	. .	. .	. .	11 1 10.96	3.19	1.00	IV.	5	4.41	22 17.99	4.66	5.80	11 1 6.77	43 18.45
23	10	. .	. .	58.2	12.	. .	. .	. .	6 11.96	3.19	1.00	IV.	3	9.25	14 42.99	4.82	5.20	6 7.77	35 43.01
24	9	. .	10.	23.5	37.2	. .	. .	. .	7 37.25	3.19	1.00	IV.	4	9.18	19 38.46	4.86	5.58	7 33.06	40 38.90
25	9	. .	. .	. .	. .	. .	30.	15.	8 3.37	3.19	1.00	VII.	5	5.25	22 39.84	4.88	5.83	7 59.18	43 40.55
26	9	. .	. .	. .	3.3	. .	30.5	. .	10 3.21	3.20	0.99	IV.	3	5.13	12 35.92	4.94	5.05	9 59.02	33 35.91
27	6.7	. .	. .	. .	17.	30.7	44.	. .	11 16.90	3.20	0.99	VI.	2	10.43	10 23.08	4.98	4.87	11 12.71	31 22.93
28	8	4.5	18.3	32.	. .	. .	. .	. .	14 45.60	3.20	1.00	III.	8	4.24	36 9.19	5.08	6.87	14 41.40	57 11.14
29	10	. .	. .	. .	47.5	60.5	. .	. .	16 47.21	3.20	1.00	VI.	5	5.30	22 42.55	5.14	5.83	16 43.01	43 43.52
30	8	. .	. .	6.3	20.	. .	. .	. .	17 20.02	3.20	1.00	IV.	4	7.10	18 33.92	5.15	5.50	17 15.82	23 39 34.57
31	9	. .	. .	47.	0.7	. .	. .	. .	21 0.62	3.21	1.01	IV.	9	6.2	41 57.64	5.26	7.34	20 56.40	24 3 0.24
32	8	. .	. .	. .	. .	49.	2.7	. .	21 35.37	3.21	1.00	VI.	8	2.48	35 20.66	5.27	6.80	21 31.16	23 56 22.73
33	5.6	. .	. .	. .	. .	. .	. .	56.5	22 15.66	3.21	1.00	VII.	4	4.26½	17 11.15	5.29	5.40	22 11.45	38 11.84
34	8	. .	. .	10.3	24.	. .	. .	. .	24 24.01	3.21	1.00	IV.	4	3.17½	21 35.88	5.35	5.73	24 19.80	42 36.96
35	11	. .	. .	55.5	9.3	. .	. .	. .	26 9.20	3.21	1.00	IV.	7	7.34	32 46.33	5.40	6.61	26 4.99	53 48.34
36	10	. .	31.2	45.	. .	. .	. .	. .	27 58.59	3.21	1.00	III.	5	6.47	23 21.49	5.45	5.88	27 54.38	44 22.82
37	8	. .	52.5	6.2	. .	. .	. .	. .	29 19.84	3.21	1.00	III.	3	10.38	15 19.77	5.49	5.24	29 15.63	36 20.50
38	10	. .	. .	. .	37.2	51.	. .	. .	30 37.23	3.21	1.00	VI.	8	5.52	36 53.44	5.53	6.93	30 33.02	57 55.90
39	7.8	. .	. .	. .	. .	46.3	0.	13.7	31 32.68	3.21	1.00	VII.	7	5.31	31 43.97	5.55	6.53	31 28.47	52 46.05
40	10	. .	. .	31.	45.	. .	. .	. .	33 44.84	3.21	0.99	IV.	2	10.22	10 12.64	5.60	4.86	33 40.64	31 13.10
41	7.8	. .	. .	. .	. .	43.3	57.2	. .	34 29.73	3.21	0.99	VI.	3	4.17½	12 7.79	5.62	5.01	34 25.53	33 8.42
42	9	. .	. .	. .	52.3	6.	19.3	. .	35 52.19	3.22	1.00	VI.	7	10.47	34 23.50	5.66	6.73	35 47.97	23 55 25.89
43	9	. .	. .	. .	53.5	. .	. .	. .	37 53.31	3.22	1.01	IV.	10	11.6	49 31.83	5.71	7.93	37 49.08	24 10 35.47
44	7.8	. .	. .	. .	. .	54.	7.6	. .	38 40.33	3.22	1.01	VI.	9	4.28	41 10.08	5.73	7.27	38 36.10	24 2 13.08
45	9	36.3	51.1	4.	. .	. .	. .	. .	45 17.75	3.22	1.00	IV.	4	5.14	17 35.43	5.88	5.43	45 13.53	23 38 36.74
46	8	. .	. .	. .	. .	1.7	15.	. .	45 47.85	3.22	1.00	VI.	5	8.20	24 8.27	5.89	5.94	45 43.63	45 10.10
47	8	. .	. .	. .	. .	. .	. .	0.5	46 19.67	3.22	1.00	VII.	4	2.38	16 16.44	5.90	5.32	46 15.45	23 37 17.66
48	7	. .	. .	. .	. .	. .	25.	. .	47 57.73	3.22	1.01	VI.	9	2.41	40 16.13	5.94	7.20	47 53.50	24 1 19.27
49	8	. .	18.	31.5	45.	. .	. .	. .	49 58.81	3.22	1.00	IV.	6	9.59	28 59.44	5.97	6.31	49 54.59	23 50 1.72
50	8	. .	. .	. .	. .	. .	35.	. .	50 21.37	3.22	1.00	VI.	4	7.57	18 59.47	5.98	5.53	50 17.15	39 58.98
51	8	. .	26.	39.3	53.	. .	. .	. .	52 53.11	3.22	1.00	IV.	3	12.26	16 14.26	6.04	5.32	52 48.89	23 37 15.62
52	10	. .	. .	. .	. .	. .	40.	. .	12.57	3.22	0.99	VI.	3	6.55	13 27.21	. .	5.12		
53	10	. .	. .	22.	. .	. .	. .	. .	11 58 35.67	3.22	1.01	III.	9	3.85	40 31.10	6.16	7.27	11 58 31.44	24 1 34.53
54	8	. .	59.2	13.	. .	. .	. .	. .	12 0 26.59	3.22	1.00	III.	4	8.31	19 14.74	6.20	5.55	12 0 22.37	23 40 16.49
55	5.6	. .	. .	. .	. .	59.	12.	26.	0 45.02	3.22	1.00	VII.	7	6.48½	32 23.05	1.21	6.58	0 40.80	23 53 25.84
56	8	. .	. .	35.2	49.	. .	. .	. .	12 2 48.85	− 3.22	−1.01	IV.	10	4.35	−46 14.67	− 6.25	− 7.68	12 2 44.62	−24 7 18.60

CORRECTIONS.

Date.		Corr. of Clock.	Hourly rate.	m	n	c
1849.	h.	s.	s.	s.	s.	s.

INSTRUMENT READINGS.

Date.		Barom.	Thermom. At.	Thermom. Ex.
1849.	h. m.	In.	°	°

REMARKS.

(238) 34. Hor. thread assumed as 5 instead of 4.
(238) 49. Transits over T.'s I–III assumed as recorded over T.'s II–IV.
(238) 50. Transits over T. V assumed as recorded over T. VI.
(238) 53. Micrometer reading assumed as 9ᵗ 3ʳ 8ᵈ instead of 9ᵗ 3ʳ 85ᵈ.

ZONE 238. APRIL 20. B. BELT, −23° 46′. D_0 = −23° 20′ 50″—Continued.

No.	Mag.	SECONDS OF TRANSIT. I.	II.	III.	IV.	V.	VI.	VII.	T.	a_1	a_2	MICROMETER.			i	d_1	d_2	Mean Right Ascension, 1850.0.	Mean Declination, 1850.0.
									h. m. s.	s.	s.			r.	′ ″	″	″	h. m. s.	° ′ ″
57	10	. .	47.2	0.5	. .	. .	. .	. .	12 5 14.34	− 3.22	−1.00	III.	4	7.34	−18 45.99	− 6.30	− 5.51	12 5 10.12	−23 39 47.80
58	8	. .	27.	40.3	54.2	. .	. .	. .	6 4.07	3.22	1.00	IV.	8	2.31	35 12.24	6.33	6.81	6 59.85	56 15.38
59	10	. .	. .	. .	47.4	. .	. .	. .	12 47.40	3.21	1.00	IV.	6	6.49	27 23.64	6.45	6.19	12 43.19	23 48 26.28
60	10	. .	. .	7.2	20.5	34.	. .	. .	20.56	3.21	1.00	V.	7	3.22	30 39.21	6.46	6.45	14 16.35	23 51 42.14
61	8	. .	. .	. .	. .	30.	. .	. .	15 16.33	3.21	1.01	V.	9	4.31	41 11.71	6.49	7.29	15 12.11	24 2 15.49
62	6.7	. .	. .	. .	. .	. .	. .	18.	15 36.95	3.21	1.01	VII.	9	0.43	39 16.45	6.50	7.13	15 32.73	0 20.08
63	10	. .	. .	. .	. .	9.7	. .	. .	17 56.02	3.21	1.01	V.	10	1.50	44 51.42	6.54	7.57	17 51.80	24 5 55.53
64	8	. .	. .	. .	6.	19.7	. .	. .	19 5.98	3.21	1.00	V.	8	3.4	35 28.84	6.56	6.83	19 1.77	23 56 32.23
65	7.8	. .	. .	. .	. .	8.	21.5	. .	19 54.27	3.21	1.00	VI.	7	7.10	32 34.07	6.58	6.61	19 50.06	53 37.26
66	10	. .	. .	. .	. .	38.7	. .	. .	24 25.11	3.20	0.99	V.	2	9.5	9 33.77	6.65	4.80	24 20.92	30 35.22
67	8	. .	. .	13.	26.6	40.2	. .	. .	26 26.63	3.20	0.99	V.	1	12.41	6 23.68	6.69	4.55	26 22.44	27 24.91
68	11	. .	. .	35.	49.	. .	. .	. .	28 48.80	3.20	1.00	IV.	7	9.48	33 53.89	6.72	6.71	28 44.60	23 54 57.32
69	8	. .	. .	. .	. .	. .	. .	3.	29 21.92	3.20	1.01	VII.	9	7.38	42 45.71	6.73	7.41	29 17.71	24 3 49.85
70	7.8	. .	. .	. .	36.	49.2	. .	. .	33 35.66	3.19	1.01	V.	10	10.2	48 59.52	6.79	7.91	33 31.46	24 10 4.22
71	9	. .	. .	. .	. .	56.4	10.	. .	34 42.71	3.19	1.00	VI.	6	4.25	26 10.87	6.81	6.09	34 38.52	23 47 13.77
72	9	. .	. .	. .	. .	11.1	. .	. .	35 57.53	3.19	0.99	V.	2	3.12	6 35.77	6.82	4.57	35 53.35	27 37.16
73	8	. .	22.	35.5	49.	. .	. .	. .	37 49.18	3.19	1.00	IV.	5	3.20	21 37.14	6.85	5.73	37 44.99	23 42 39.72
74	8	. .	32.5	46.	. .	. .	. .	. .	39 59.75	3.19	1.01	II.	9	3.51	40 51.42	6.88	7.25	39 55.55	24 1 55.55
75	9	. .	. .	. .	. .	. .	54.	7.3	26.49	3.19	1.00	VII.	8	6.57	37 26.04	6.88	6.98	40 22.30	23 58 29.90
76	8	. .	. .	. .	. .	38.5	52.	. .	42 24.79	3.18	1.01	VI.	10	5.13	46 33.67	6.91	7.72	42 20.60	24 7 38.30
77	8	. .	. .	. .	3.	. .	. .	. .	44 2.96	3.18	1.00	IV.	7	6.14	32 5.99	6.92	6.57	43 58.78	23 53 9.48
78	9	. .	3.5	17.2	. .	. .	. .	. .	46 30.84	3.18	1.00	III.	5	9.31	24 44.18	6.95	5.98	46 26.66	23 45 47.11
79	8	. .	. .	. .	13.5	27.2	. .	. .	47 13.41	3.18	1.01	V.	10	6.25	47 10.09	6.96	7.76	47 9.22	24 8 14.81
80	7.8	. .	. .	32.	45.4	. .	. .	. .	48 45.56	3.17	1.00	IV.	4	3.57	16 56.60	6.98	5.36	48 41.39	23 37 58.94
81	8	. .	. .	. .	. .	. .	33.5	47.	49 6.12	3.17	1.00	VII.	5	6.31	23 13.13	6.98	5.86	49 1.95	44 15.97
82	8	56.	9.3	23.1	. .	. .	. .	. .	53 36.75	3.17	1.00	III.	7	2.17	30 6.44	7.02	6.41	53 32.58	51 9.87
83	9	. .	29.	42.3	56.	. .	. .	. .	12 55 56.12	− 3.17	−1.00	IV.	5	3.00	−21 27.06	− 7.03	− 5.72	12 55 51.95	−23 42 29.81

ZONE 239. MAY 2. B. BELT, −23° 46′. D_0 = −23° 20′ 20″.

No.	Mag.	I.	II.	III.	IV.	V.	VI.	VII.	T.	a_1	a_2	MICROMETER.			i	d_1	d_2	Mean Right Ascension, 1850.0.	Mean Declination, 1850.0.
1	8	. .	. .	. .	29.	42.	56.	9.5	14 55 28.68	+ 9.25	+1.00	V.	3	5.21	−12 39.91	− 6.69	− 5.01	14 55 38.93	−23 33 11.61
2	7	. .	. .	. .	16.3	30.	43.5	. .	57 16.27	9.26	1.00	VI.	3	2.3	12 0.48	6.75	4.97	57 26.53	32 32.20
3	7	. .	. .	. .	. .	35.	49.	2.5	14 58 21.54	9.26	1.00	VII.	3	12.15	16 8.38	6.79	5.29	14 58 31.80	23 36 40.46
4	9.10	. .	. .	. .	31.4	45.	. .	. .	15 0 31.28	9.27	1.00	V.	9	10.48	44 21.81	6.87	7.59	15 0 41.55	24 4 56.27
5	10	. .	. .	. .	59.6	13.	. .	. .	1 59.53	9.27	1.00	VI.	1	12.34	6 20.05	6.92	4.50	2 9.80	23 26 51.47
6	10	. .	. .	. .	. .	. .	19.	33.	2 51.83	9.27	1.00	VII.	9	8.50	43 22.02	6.95	7.49	3 2.10	24 3 56.46
7	9	. .	29.3	43.	56.5	. .	. .	. .	4 56.59	9.28	1.00	IV.	6	8.24	28 11.54	7.01	6.26	5 6.87	23 48 44.81
8	9	. .	. .	. .	47.	0.5	. .	. .	5 46.89	9.28	1.00	V.	7	9.37	33 48.31	7.04	6.71	5 57.17	54 22.06
9	8	10.3	24.	37.5	. .	. .	. .	. .	7 51.18	9.29	1.00	III.	1	13.8	6 37.29	7.11	4.53	8 1.47	27 8.93
10	8	. .	. .	24.5	38.2	. .	. .	. .	9 38.22	9.30	1.00	IV.	4	12.9	21 4.69	7.17	5.68	9 48.52	41 37.54
11	7	. .	. .	. .	. .	25.	39.	. .	10 11.49	9.30	1.00	VI.	4	14.22	22 11.59	7.19	5.77	10 21.79	42 44.55
12	6.7	. .	. .	. .	. .	. .	. .	23.	10 42.19	9.30	1.00	VII.	4	15.3	22 32.08	7.20	5.79	10 52.40	43 5.07
13	8	49.8	3.3	17.	. .	. .	. .	. .	14 30.64	9.31	1.00	III.	3	5.11	12 34.87	7.31	5.00	14 40.95	33 7.18
14	8	. .	. .	20.	33.7	47.3	. .	. .	15 33.67	9.32	1.00	V.	6	10.38	29 19.07	7.35	6.35	15 43.99	49 52.77
15	9	8.7	22.5	36.	. .	. .	. .	. .	17 49.68	9.33	1.00	III.	3	12.32	16 17.25	7.41	5.29	18 0.01	36 49.95
16	9	. .	. .	. .	42.4	56.	. .	. .	18 42.36	9.33	1.00	V.	6	14.31	31 16.54	7.44	6.51	18 52.69	51 50.49
17	8	. .	. .	55.	. .	22.4	. .	. .	20 8.70	9.34	1.00	V.	7	12.17	35 8.98	7.48	6.83	20 19.04	23 55 43.29
18	9.10	. .	. .	. .	. .	19.5	33.	. .	15 21 5.78	+ 9.34	+1.00	VI.	8	11.48	−39 52.95	− 7.51	− 7.22	15 21 16.12	−24 0 27.68

CORRECTIONS.

Date.		Corr. of Clock.	Hourly rate.	m	n	c
1849.	h.	s.	s.	s.	s.	s.

INSTRUMENT READINGS.

Date.		Barom.	THERMOM. At.	Ex.
1849.	h. m.	in.	°	°

REMARKS.

(238) 58. Transits over T.'s II–IV assumed as 37ˢ, 50ˢ.3, and 4ˢ.2 instead of 27ˢ, 40ˢ.3, and 54ˢ.2, and minutes as 7 instead of 6.
(238) 60. Minutes assumed as 14.
(239) 2. Micrometer reading assumed as 4ʳ.30 instead of 2ʳ.30.

Zone 239. May 2. B. Belt, −23° 46′. D_o = −23° 20′ 20″—Continued.

No.	Mag.	SECONDS OF TRANSIT.							T.	a_1	a_2	MICROMETER.			i	d_1	d_2	Mean Right Ascension, 1850.0.	Mean Declination, 1850.0.
		I.	II.	III.	IV.	V.	VI.	VII.											
									h. m. s.	s.	s.			r.	′ ″	″	″	h. m. s.	° ′ ″
19	9	. .	. .	. .	. .	44.	. .	. .	15 22 30.36	+ 9.35	+1.00	V.	4	12.48	−21 24.31	− 7.54	−5.70	15 22 40.71	−23 41 57.55
20	8	. .	. .	. .	. .	47.3	0.5	. .	23 33.38	9.35	1.00	VI.	1	7.52	3 57.85	7.57	4.31	23 43.73	24 29.73
21	9	. .	. .	. .	. .	28.3	42.	. .	25 14.65	9.36	1.00	VI.	5	5.49	22 52.13	7.62	5.82	25 25.01	43 25.57
22	7.8	. .	. .	. .	. .	. .	. .	40.8	26 59.95	9.37	1.00	VII.	4	6.35	18 15.95	7.66	5.45	27 10.32	38 49.06
23	8	. .	. .	. .	36.5	. .	. .	. .	27 36.55	9.37	1.00	IV.	3	5.50	12 54.58	7.67	5.02	27 46.92	37 27.27
24	9	. .	. .	. .	. .	. .	24.3	38.	28 57.01	9.38	1.00	VII.	6	8.30	28 14.24	7.71	6.26	29 7.39	23 48 48.21
25	8	. .	. .	. .	. .	31.3	15.	. .	29 17.69	9.38	1.00	VI.	10	9.50	48 53.35	7.72	7.96	29 28.07	24 9 29.03
26	8	. .	44.	57.5	. .	. .	. .	. .	32 11.24	9.39	1.00	III.	7	10.6	34 2.94	7.73	6.73	32 21.63	23 54 37.45
27	8	. .	. .	. .	54.2	8.	. .	. .	32 54.27	9.39	1.00	V.	6	8.19	28 8.98	7.80	6.25	33 4.66	48 43.03
28	8	. .	. .	. .	55.4	9.	22.5	. .	33 55.31	9.39	1.00	VI.	6	4.40	26 18.44	7.82	6.10	34 5.70	46 52.36
29	8	. .	. .	. .	13.	26.6	40.2	. .	35 12.93	9.40	1.00	VI.	7	10.44	34 21.99	7.84	6.77	35 23.33	23 54 56.60
30	8	. .	. .	34.5	48.2	. .	. .	. .	36 48.11	9.41	1.00	IV.	9	10.57	44 26.39	7.88	7.60	36 58.52	24 5 1.87
31	9	. .	. .	. .	. .	. .	. .	59.3	37 18.49	9.41	1.00	VII.	3	7.12	13 35.60	7.80	5.08	37 28.90	23 34 8.57
32	7	. .	. .	9.3	23.	. .	. .	. .	39 22.98	9.42	1.00	IV.	1	2.51	1 26.22	7.93	4.10	39 33.40	21 58.25
33	8	59.	12.5	. .	. .	. .	. .	. .	41 39.82	9.43	1.00	II.	3	4.5	12 1.49	7.97	4.95	41 50.25	32 34.41
34	7	9.4	23.	37.	50.	. .	. .	. .	44 50.31	9.44	1.00	IV.	3	2.4	11 0.63	8.02	4.87	45 0.75	31 33.52
35	9	. .	. .	51.6	5.2	19.	. .	. .	46 5.25	9.45	1.00	V.	7	8.1	32 59.90	8.04	6.65	46 15.70	53 34.59
36	9.10	. .	. .	16.2	30.	. .	. .	. .	47 29.90	9.46	1.00	IV.	7	8.9	33 3.98	8.07	6.66	47 40.36	53 38.71
37	9.10	. .	. .	. .	. .	44.5	. .	. .	48 30.83	9.47	1.00	V.	8	7.6	37 30.87	8.09	7.02	48 41.30	23 58 5.98
38	9.10	. .	. .	. .	49.5	. .	. .	. .	49 49.38	9.48	1.00	IV.	9	4.58	41 25.36	8.11	7.34	49 59.86	24 2 0.81
39	8.9	. .	. .	. .	. .	. .	52.2	6.	50 24.96	9.48	1.00	VII.	6	2.22	25 8.67	8.12	6.01	50 35.44	23 45 42.80
40	8	. .	. .	. .	59.8	13.2	. .	. .	51 59.67	9.49	1.00	V.	7	1.37	29 46.27	8.14	6.39	52 10.16	50 20.80
41	9	. .	. .	. .	33.4	. .	. .	. .	53 33.39	9.49	1.00	IV.	6	9.21	28 40.27	8.16	6.30	53 43.88	49 14.73
42	8	. .	. .	. .	. .	. .	. .	30.	49.08	9.50	1.00	VII.	5	9.1	24 28.75	. .	5.96		
43	9	. .	. .	. .	2.	. .	. .	. .	56 2.06	9.51	1.00	IV.	4	4.0	16 58.12	8.20	5.35	56 12.57	37 31.67
44	7.8	. .	. .	. .	. .	. .	54.	7.8	56 26.76	9.51	1.00	VII.	6	7.41	27 49.52	8.21	6.22	56 37.27	48 23.95
45	9	. .	. .	. .	. .	. .	13.7	27.2	57 46.41	9.51	1.00	VII.	5	7.58	23 56.99	8.23	5.90	57 56.92	23 44 31.12
46	6.7	. .	. .	. .	. .	. .	. .	22.3	15 58 41.22	9.52	1.00	VII.	9	7.18	42 35.62	8.24	7.45	15 58 51.74	24 3 11.31
47	7	. .	. .	45.	58.8	12.4	. .	. .	16 0 58.66	9.53	1.00	V.	10	12.39	50 18.68	8.27	8.10	16 1 9.19	24 10 55.05
48	8	. .	33.5	47.2	1.	. .	. .	. .	3 0.90	9.54	1.00	IV.	2	12.5	11 4.57	8.30	4.87	3 11.44	23 31 37.74
49	9	. .	10.	23.7	37.5	. .	. .	. .	4 37.39	9.55	1.00	IV.	6	14.46	41 22.85	8.32	7.34	4 47.94	24 1 58.51
50	6	. .	. .	. .	. .	. .	. .	15.	33.93	9.55	1.00	VII.	8	14.41	41 19.99	. .	7.34		
51	8	. .	. .	2.	15.4	. .	. .	. .	7 15.50	9.56	1.00	IV.	7	9.10	33 34.73	8.35	6.70	7 26.06	23 54 9.78
52	9	. .	. .	23.	37.	. .	. .	. .	8 36.83	9.57	1.00	IV.	5	11.20	25 39.18	8.36	6.05	8 47.40	46 13.59
53	9	. .	25.5	39.2	. .	. .	. .	. .	10 52.85	9.58	1.00	III.	8	10.29	39 13.23	8.39	7.18	11 3.43	59 48.80
54	7	. .	. .	. .	. .	40.	54.	. .	11 26.51	9.58	1.00	VI.	6	7.14	27 36.09	8.39	6.21	11 37.09	23 48 10.69
55	8	. .	43.2	57.	. .	. .	. .	. .	16 16 10.60	+ 9.61	+1.00	IV.	10	4.42	−46 18.20	− 8.43	−7.77	16 16 21.21	−24 6 54.40

Zone 240. May 11. K. Belt, −23° 46′. D_o = −23° 21′ 0″.

No.	Mag.	I.	II.	III.	IV.	V.	VI.	VII.	T.	a_1	a_2	MICROMETER.			i	d_1	d_2	Mean Right Ascension, 1850.0.	Mean Declination, 1850.0.
1	10	. .	34.3	. .	. .	. .	. .	. .	13 42 1.62	+18.94	+1.00	II.	3	8.10	−14 5.03	− 3.83	−5.22	13 42 21.56	−23 35 14.08
2	6.7	. .	. .	. .	45.5	. .	. .	. .	42 45.56	18.94	1.00	V.	4	3.46	16 51.02	3.85	5.41	43 5.50	38 0.31
3	9	. .	. .	. .	14.1	. .	. .	. .	44 14.02	18.94	1.00	.	8	4.20	36 7.20	3.96	6.82	44 33.96	57 17.98
4	9	. .	. .	41.8	. .	. .	. .	. .	46 55.46	18.94	1.00	.	8	2.47	35 20.31	4.14	6.76	47 15.40	56 31.21
5	8.7	41.4	55.0	8.8	. .	. .	. .	. .	48 22.36	18.94	1.00	.	4	8.53	19 25.86	4.23	5.60	48 42.30	40 35.69
6	9	. .	. .	. .	33.2	. .	. .	. .	50 33.24	18.95	1.00	.	2	11.18	10 40.87	4.36	4.97	50 53.19	31 50.20
7	10	. .	4.9	. .	. .	. .	. .	. .	53 32.22	18.95	1.00	II.	3	7.49	13 54.44	4.54	5.19	53 52.17	35 4.17
8	8	. .	. .	. .	42.7	56.2	. .	. .	13 53 42.59	+18.95	+1.00	.	7	11.36	−34 48.36	− 4.55	−6.73	13 54 2.54	−23 55 59.64

CORRECTIONS.

Date.		Corr. of Clock.	Hourly rate.	m	n	c
1849.	h.	s.	s.	s.	s.	s.

INSTRUMENT READINGS.

Date.		Barom.	THERMOM.	
			At.	Ex.
1849.	h. m.	in.	°	°

REMARKS.

(239) 44. Minutes assumed as 27 instead of 28.
(239) 49. Micrometer thread assumed as 8 instead of 6.

ZONE 240. MAY 11. K. BELT, −23° 46′. D₀ = −23° 21′ 0″—Continued.

No.	Mag.	Seconds of Transit. I.	II.	III.	IV.	V.	VI.	VII.	T.	a_1	a_2	Micrometer.			i	d_1	d_4	Mean Right Ascension, 1850.0.	Mean Declination, 1850.0.
									h. m. s.	s.	s.			r.	′ ″	″	″	h. m. s.	° ′ ″
9	8	..	..	..	35.9	..	3.4	..	13 54 35.95	+18.95	+1.00	.	9	7.3	−42 28.39	− 4.61	− 7.28	13 54 55.90	−24 3 40.28
10	8	..	..	..	..	37.7	51.2	..	13 57 23.98	18.95	1.00	VI.	9	2.15	40 3.02	4.78	7.11	13 57 43.93	24 1 14.91
11	9	54.4	..	..	..	..	..	..	14 0 35.39	18.95	1.00	III.	7	5.18	31 37.70	4.97	6.48	14 0 55.34	23 52 49.15
12	6	54.9	8.9	22.3	..	..	..	..	3 35.98	18.95	1.00	.	4	6.8	18 2.66	5.14	5.50	3 55.93	39 13.30
13	9	9.8	23.6	..	..	..	..	..	8 50.85	18.96	1.00	II.	6	12.22	30 11.39	5.43	6.38	9 10.81	51 23.20
14	10	..	..	..	38.2	..	..	..	10 38.25	18.96	1.00	.	3	5.41	12 50.04	5.53	5.12	10 58.21	34 0.69
15	8	..	17.5	31.2	..	..	..	..	11 44.84	18.96	1.00	.	6	7.22	27 40.27	5.59	6.19	12 4.80	48 52.05
16	9	..	..	..	..	17.0	..	..	12 3.34	18.96	1.00	.	7	11.23	34 41.79	5.61	6.70	12 23.30	23 55 54.10
17	5	..	..	41.9	..	9.2	..	..	15 55.54	18.97	1.00	.	10	4.22	46 8.11	5.82	7.56	16 15.51	24 7 21.49
18	9	..	..	..	..	48.2	..	..	16 34.54	18.97	1.00	.	7	7.35	32 46.83	5.86	6.57	16 54.51	23 53 59.26
19	9	..	..	..	..	40.1	53.2	..	17 26.13	18.97	1.00	.	2	11.25	10 44.40	5.90	4.97	17 46.10	31 55.27
20	9	..	30.2	44.2	..	..	..	..	21 57.69	18.97	1.00	.	7	9.51	33 55.40	6.15	6.64	22 17.66	55 8.19
21	8	..	24.8	38.2	..	..	..	..	22 51.99	18.97	1.00	.	7	6.27	32 12.54	6.19	6.52	23 11.96	53 25.25
22	9	..	..	27.7	..	..	..	..	23 41.35	18.97	1.00	.	2	10.7	10 5.08	6.24	4.93	24 1.32	31 16.25
23	8	..	..	..	..	11.2	24.9	..	23 57.55	18.97	1.00	.	5	8.50	24 23.55	6.25	5.96	24 17.52	45 35.76
24	9	..	..	..	..	6.1	..	..	24 52.46	18.98	1.00	.	4	9.16	19 37.46	6.30	5.61	25 12.44	40 49.37
25	9	..	7.9	21.0	..	..	..	..	26 34.94	18.98	1.00	.	4	9.27	19 43.00	6.40	5.62	26 54.92	40 55.02
26	9	..	..	..	7.8	..	..	..	27 7.74	18.98	1.00	.	7	10.41	34 20.62	6.42	6.69	27 27.72	55 33.73
27	10	..	..	..	..	0.2	..	..	27 46.55	18.98	1.00	.	6	7.45	27 51.87	6.46	6.20	28 6.53	49 4.53
28	9	..	..	52.8	..	..	..	..	31 6.45	18.98	1.00	.	1	11.51	5 58.51	6.64	4.63	31 26.43	27 9.78
29	7.8	..	..	40.9	..	8.2	..	..	31 54.60	18.98	1.00	.	1	6.44	3 23.72	6.68	4.45	32 14.58	24 34.85
30	8	..	..	42.6	56.4	..	..	..	33 56.32	18.99	1.00	.	6	11.8	29 34.23	6.79	6.34	34 16.31	50 47.36
31	9	..	..	..	32.5	..	..	..	34 32.50	18.99	1.00	.	6	7.39	27 48.85	6.82	6.20	34 52.49	49 1.87
32	9	..	..	..	..	9.2	..	..	34 55.64	18.99	1.00	.	1	10.13	5 9.10	6.84	4.57	35 15.63	26 20.51
33	7	..	..	..	..	..	..	1.0	35 29.25	18.99	1.00	.	2	6.25	8 13.13	6.88	4.78	35 49.24	29 24.79
34	9	..	..	..	26.3	..	..	..	38 26.32	19.00	1.00	.	1	8.55	4 29.77	7.03	4.52	38 46.32	25 41.32
35	4	..	51.9	5.2	..	..	..	..	40 19.04	19.00	1.00	.	3	12.18	16 10.22	7.14	5.36	40 39.04	37 22.72
36	8	..	50.9	4.9	..	..	..	..	41 18.39	19.00	1.00	.	6	5.51	26 54.38	7.19	6.13	41 38.39	48 7.70
37	8	..	23.4	36.8	..	..	..	..	42 50.59	19.00	1.00	.	4	7.8	18 32.91	7.26	5.53	43 10.59	39 45.70
38	7.8	..	..	24.3	37.8	..	..	..	43 37.89	19.00	1.00	.	2	7.9	8 35.32	7.30	4.82	43 57.89	23 29 47.44
39	9	22.4	..	..	..	..	..	..	45 3.46	19.01	1.00	.	9	4.5	40 58.64	7.37	7.18	45 23.47	24 2 13.19
40	4	..	..	4.2	17.8	31.2	..	..	45 17.70	19.01	1.00	.	9	2.41	40 16.28	7.38	7.13	45 37.71	24 1 30.79
41	9	..	19.9	33.6	..	..	..	..	46 47.24	19.01	1.00	.	4	7.39	18 48.55	7.46	5.55	47 7.25	23 40 1.56
42	8	..	..	16.9	30.7	..	..	..	47 31.66	19.01	1.00	.	4	5.11	17 33.91	7.50	4.46	47 51.67	38 46.87
43	9	30.4	..	..	..	..	..	..	51 11.33	19.02	1.00	.	5	7.28	23 42.19	7.67	5.90	51 31.35	44 55.76
44	10	..	..	..	13.9	..	..	..	52 13.90	19.02	1.00	V.	6	6.12	27 4.94	7.72	6.15	52 33.92	48 18.81
45	9	..	..	35.0	..	..	..	..	53 48.65	19.02	1.00	.	1	5.25	3 14.13	7.79	4.43	54 8.67	24 26.35
46	9	..	..	5.1	..	..	..	..	55 18.75	19.03	1.00	.	3	3.59	11 58.61	7.86	5.06	55 38.78	33 11.53
47	7	25.3	39.0	52.8	..	..	..	..	57 6.30	19.03	1.00	.	2	12.30	11 17.18	7.95	5.01	57 26.33	32 30.14
48	7.8	30.8	44.4	..	..	..	..	..	14 58 11.69	+19.03	+1.00	.	3	10.53	−15 27.37	− 8.00	− 5.31	14 58 31.72	−23 36 40.68

ZONE 241. MAY 19. K. BELT, −29° 23′. D₀ = −28° 57′ 50″.

No.	Mag.	I.	II.	III.	IV.	V.	VI.	VII.	T.	a_1	a_2	Micrometer.			i	d_1	d_4	Mean Right Ascension, 1850.0.	Mean Declination, 1850.0.
1	10	..	..	38.1	..	..	..	..	12 5 52.45	+31.82	+1.00	.	5	6.23	−23 9.42	− 0.29	− 5.80	12 6 25.27	−29 21 5.51
2	10	..	..	..	..	..	50.2	..	6 21.49	31.82	1.00	.	6	10.45	29 22.64	0.35	6.50	6 54.31	27 19.49
3	9.10	44.4	58.8	12.9	..	..	..	..	10 27.32	31.81	1.01	III.	2	9.9	9 35.78	0.88	4.26	11 0.14	7 30.92
4	10	28.9	43.4	..	..	..	..	..	12 12.07	31.80	1.00	II.	8	8.53	38 24.67	1.10	7.55	12 44.87	36 23.32
5	9	..	..	..	26.3	..	..	..	12 12 26.13	+31.80	+0.99	.	10	5.30	−46 42.40	− 1.13	− 8.53	12 12 58.92	−29 44 42.06

CORRECTIONS.

Date.		Corr. of Clock.	Hourly rate.	m	n	c
1849.	h.	s.	s.	s.	s.	s.

INSTRUMENT READINGS.

Date.		Barom.	Thermom. At.	Thermom. Ex.
1849.	h. m.	in.	°	°

REMARKS.

(240) 45. Micrometer reading assumed as 6ʳ.25 instead of 5ʳ.25.

ZONE 241. MAY 19. K. BELT, −29° 23'. D_o = −28° 57' 50"—Continued.

No.	Mag.	SECONDS OF TRANSIT. I.	II.	III.	IV.	V.	VI.	VII.	T.	a_1	a_4	MICROMETER.			i	d_1	d_2	Mean Right Ascension, 1850.0.	Mean Declination, 1850.0.
									h. m. s.	s.	s.			r.	' "	"	"	h. m. s.	° ' "
6	9	. .	. .	53.1	. .	. .	. .	. .	12 13 7.47	+31.80	+0.99	.	9	4.25	−41 8.72	− 1.21	− 7.87	12 13 40.26	−29 39 7.80
7	10	50.9	. .	. .	. .	. .	. .	. .	14 33.98	31.80	1.00	II.	7	6.49	32 23.45	1.40	6.86	15 6.78	30 21.71
8	7	. .	55.4	. .	. .	. .	. .	. .	15 24.11	31.79	1.00	II.	7	6.18	32 7.81	1.51	6.82	15 56.90	30 6.15
9	10	. .	. .	. .	38.5	. .	. .	. .	15 38.37	31.79	0.99	.	9	6.43	42 18.31	1.54	8.02	16 11.15	40 17.87
10	10	. .	. .	. .	. .	28.8	. .	. .	16 14.27	31.79	1.00	.	6	4.24	26 10.52	1.61	6.14	16 47.06	24 8.27
11	9	. .	. .	. .	. .	15.8	30.1	. .	17 1.40	31.79	1.01	VI.	3	7.20	13 39.77	1.71	4.71	17 34.20	11 36.19
12	9	20.9	35.2	. .	. .	. .	. .	. .	19 3.85	31.79	1.00	II.	4	4.28	17 12.04	1.97	5.12	19 36.64	15 9.13
13	10	. .	9.2	. .	. .	. .	. .	. .	21 37.93	31.78	0.99	II.	9	10.22	44 8.55	2.29	8.23	22 10.70	42 9.07
14	9	. .	. .	. .	44.3	. .	. .	. .	21 44.14	31.78	0.99	.	9	12.47	45 21.85	2.30	8.39	22 16.91	43 22.54
15	7	39.5	53.8	. .	. .	. .	. .	. .	24 22.44	31.77	1.00	.	3	9.55	14 58.12	2.64	4.86	24 55.21	12 55.62
16	10	. .	30.4	44.6	. .	. .	. .	. .	? 27 59.03	31.77	1.00	.	7	3.46	30 51.36	3.09	6.68	27 31.80	28 51.13
17	9	. .	. .	. .	. .	20.1	34.4	. .	27 5.71	31.77	1.00	.	4	4.53	17 24.84	2.98	5.12	27 38.48	15 22.94
18	11	. .	. .	58.9	. .	. .	. .	. .	30 13.25	31.76	1.00	.	4	8.20	19 9.22	3.36	5.33	30 46.01	17 7.91
19	5.6	. .	. .	. .	32.6	47.2	1.3	. .	30 32.65	31.76	0.99	.	8	7.37	37 46.54	3.40	7.51	31 5.40	35 47.45
20	10	. .	. .	. .	. .	59.3	. .	. .	31 45.04	31.76	1.01	.	2	5.4	7 32.29	3.56	4.01	32 17.81	5 29.86
21	11	. .	44.4	. .	. .	. .	. .	. .	34 13.08	31.75	1.00	.	3	10.28	15 14.76	3.86	4.87	34 45.83	13 13.49
22	10	. .	. .	18.8	. .	. .	. .	. .	34 33.16	31.75	1.00	.	6	6.57	27 27.67	3.90	6.30	35 5.91	25 27.87
23	10	. .	. .	. .	. .	54.0	. .	. .	34 39.66	31.75	1.00	.	6	6.42	27 21.11	3.92	6.28	35 12.41	25 21.31
24	10	. .	17.9	. .	. .	. .	. .	. .	38 46.59	31.74	1.00	.	5	10.48	25 23.05	4.43	6.05	39 19.33	23 23.53
25	6.7	. .	. .	33.8	48.2	. .	. .	. .	41 48.10	31.74	0.99	.	10	7.11	47 33.32	4.83	8.64	42 20.83	45 36.79
26	10	. .	. .	. .	39.9	. .	. .	. .	42 39.78	31.73	0.99	.	9	4.42	41 17.30	4.94	7.91	43 12.50	39 20.15
27	8	. .	. .	25.0	39.5	. .	. .	. .	44 39.44	31.73	1.00	.	3	7.45	13 52.57	5.20	4.73	45 12.17	11 52.50
28	7	. .	. .	38.9	53.2	. .	. .	. .	46 53.25	31.73	1.00	.	4	4.33	17 14.76	5.50	5.09	47 25.98	15 15.35
29	7	. .	48.3	2.5	16.6	. .	. .	. .	50 16.82	31.72	1.00	.	3	6.24	13 11.73	5.94	4.63	50 49.54	11 12.30
30	10	. .	. .	. .	. .	. .	59.3	. .	50 30.59	31.72	1.00	.	6	11.51	29 55.91	5.97	6.57	51 3.31	27 53.45
31	10	. .	58.5	12.6	. .	. .	. .	. .	52 27.07	31.72	1.00	.	5	10.3	25 0.36	6.22	6.00	52 59.79	23 2.58
32	8	. .	35.9	50.3	. .	. .	. .	. .	54 4.63	31.71	1.00	.	7	9.36	33 47.85	6.44	7.03	54 37.34	31 51.37
33	10	. .	. .	30.7	. .	. .	. .	. .	54 45.02	31.71	1.01	.	1	10.0	5 2.55	6.52	3.72	55 17.74	3 2.79
34	10	39.3	. .	. .	. .	. .	. .	. .	56 22.31	31.71	1.00	.	5	10.46	25 22.05	6.74	6.04	56 55.02	23 24.83
35	11	. .	. .	16.8	. .	. .	. .	. .	56 31.15	31.71	1.00	.	4	9.29	19 44.01	6.75	5.39	57 3.86	17 46.15
36	10	. .	. .	1.9	. .	. .	. .	. .	58 16.25	31.71	1.00	.	5	10.38	25 18.01	6.99	6.04	58 48.96	23 21.04
37	8	. .	. .	. .	. .	7.3	22.2	. .	12 58 53.23	31.70	0.99	.	8	11.36	39 47.06	7.03	7.74	12 59 25.92	37 51.83
38	10	57.2	11.6	. .	. .	. .	. .	. .	13 2 40.25	31.70	1.00	.	5	10.56	25 27.09	7.55	6.06	13 3 12.95	23 30.70
39	9	. .	11.8	. .	40.4	. .	. .	. .	3 40.43	31.70	1.00	.	7	6.27	32 12.54	7.68	6.84	4 13.13	30 17.06
40	9	33.1	47.6	. .	. .	. .	. .	. .	6 16.20	31.69	1.00	.	5	10.1	24 59.35	8.01	6.00	6 48.89	23 3.36
41	9	. .	. .	36.9	. .	. .	. .	. .	8 51.27	31.69	1.00	.	8	8.49	38 22.85	8.34	7.58	9 23.96	36 28.77
42	9	9.0	23.4	. .	. .	. .	. .	. .	10 52.10	31.69	1.00	.	7	10.50	34 25.16	8.60	7.10	11 24.79	32 30.86
43	10	. .	26.9	. .	. .	. .	. .	. .	12 55.60	31.68	1.00	.	6	8.7	28 2.97	8.86	6.36	13 28.28	26 8.19
44	10	. .	22.2	. .	. .	. .	. .	. .	13 50.90	31.68	1.00	.	6	5.15	26 36.23	8.97	6.19	14 23.58	24 41.39
45	9	. .	. .	4.9	19.4	. .	. .	. .	16 19.29	31.68	1.00	.	8	7.5	37 30.40	9.28	7.47	16 51.97	35 37.15
46	10	. .	. .	56.2	. .	. .	. .	. .	20 10.56	31.67	1.00	.	6	7.43	27 50.86	9.78	6.34	20 43.23	25 56.98
47	9	. .	. .	. .	. .	31.2	. .	. .	20 16.87	31.67	1.00	.	5	9.40	24 48.76	9.79	5.98	20 49.54	22 54.53
48	9	17.8	32.0	. .	. .	. .	. .	. .	25 0.67	31.67	1.00	.	2	11.56	11 0.04	10.37	4.38	25 33.34	9 4.79
49	10	. .	. .	22.8	. .	. .	. .	. .	28 37.15	31.66	1.00	.	4	11.40	20 50.07	10.82	5.52	29 9.81	18 56.41
50	8	. .	56.2	10.3	. .	. .	. .	. .	29 24.75	31.66	1.00	.	1	12.36	6 21.21	10.91	3.86	29 57.41	4 25.98
51	8	. .	. .	. .	3.6	17.7	. .	. .	30 3.46	31.66	1.00	.	7	6.58	32 28.17	11.00	6.88	30 36.12	30 36.05
52	9	32.2	. .	0.6	. .	. .	. .	. .	32 15.00	31.66	1.00	.	2	11.39	10 51.47	11.26	4.37	32 47.66	8 57.10
53	9	. .	. .	. .	39.3	. .	. .	. .	32 39.38	31.66	1.00	.	4	11.2	20 30.91	11.31	5.48	33 12.04	18 37.70
54	8	17.0	31.4	45.6	. .	. .	. .	. .	13 33 0.03	+31.66	+1.00	.	6	6.35	−27 16.58	−11.35	− 6.27	13 33 32.69	−29 25 24.20

CORRECTIONS.

Date.	Corr. of Clock.	Hourly rate.	m	n	c
1849. h.	s.	s.	s.	s.	s.

INSTRUMENT READINGS.

Date.	Barom.	THERMOM. At.	Ex.
1849. h. m.	in.	°	°

REMARKS.

(241) 16. Minutes assumed as 26 instead of 27.

ZONE 241. MAY 19. K. BELT, −29° 23'. D₀=−28° 57′ 50″—Continued.

No.	Mag.	SECONDS OF TRANSIT. I.	II.	III.	IV.	V.	VI.	VII.	T.	a_1	a_2	MICROMETER.			i	d_1	d_2	Mean Right Ascension, 1850.0.	Mean Declination, 1850.0.
									h. m. s.	s.	s.			r.	′ ″	″	″	h. m. s.	° ′ ″
55	10	..	..	..	26.8	..	..	..	13 36 26.66	+31.66	+0.99	.	9	8.3	−42 58.65	−11.77	−8.13	13 36 59.31	−29 41 8.55
56	9	..	..	..	13.7	..	..	..	37 13.50	31.66	0.99	.	10	12.10	45 3.19	11.86	8.36	37 46.15	43 13.41
57	10	..	5.1	..	..	..	..	..	41 33.81	31.66	1.00	.	7	7.19	32 38.76	12.37	6.90	42 6.47	30 48.03
58	8.7	..	..	..	..	53.9	7.9	..	41 39.36	31.66	1.00	.	2	9.10	9 36.33	12.38	4.23	42 12.02	7 42.94
59	8	..	..	..	..	57.8	12.1	..	42 43.43	31.66	1.00	.	8	6.25	37 10.23	12.52	7.43	43 16.09	35 20.18
60	9	..	..	..	..	..	0.4	..	43 31.76	31.65	1.00	.	10	3.45	45 49.45	12.59	8.46	44 4.41	44 0.50
61	8	..	..	..	43.5	57.9	..	..	44 43.49	31.65	1.00	.	7	10.52	34 26.17	12.72	7.11	45 16.14	32 36.00
62	9	..	27.2	41.6	..	..	..	..	46 55.90	31.65	1.00	.	2	11.25	10 44.40	12.96	4.34	47 28.55	8 51.70
63	5.6	..	..	48.7	..	..	..	..	49 3.02	31.65	1.01	.	1	4.49	2 25.73	13.20	3.41	49 35.68	0 32.34
64	9	..	..	35.8	..	..	..	..	49 50.15	31.65	1.00	.	6	6.36	27 17.09	13.28	6.28	50 22.80	25 26.65
65	8.9	..	44.9	59.1	..	..	..	..	55 13.50	31.65	1.00	.	3	4.51	12 24.83	13.82	4.53	55 46.15	10 33.18
66	10	..	..	35.4	..	..	..	..	58 49.72	31.65	1.01	.	1	7.11	3 37.32	14.15	3.55	13 59 22.38	1 45.02
67	8	..	18.9	33.2	..	..	..	..	13 59 47.57	+31.65	+1.00	.	5	9.55	−24 56.32	−14.25	−5.99	14 0 20.22	−29 23 6.56

ZONE 242. MAY 23. K. BELT, −23° 46′. D₀=−23° 21′ 10″.

No.	Mag.	I.	II.	III.	IV.	V.	VI.	VII.	T.	a_1	a_2	MICROMETER.			i	d_1	d_2	Mean Right Ascension, 1850.0.	Mean Declination, 1850.0.
1	9	..	12.5	26.4	..	..	..	..	16 15 39.95	+39.89	+1.00	.	9	12.59	−45 27.90	−7.68	−7.62	16 16 20.84	−24 6 53.20
2	9	..	..	1.3	..	..	..	..	19 14.97	39.90	1.00	.	10	12.18	50 8.12	8.54	7.99	19 55.87	24 11 34.65
3	8	25.9	39.5	53.2	..	..	..	..	24 6.87	39.91	1.00	.	8	5.20	36 37.45	9.70	6.92	24 47.78	23 58 4.07
4	10	..	..	..	33.6	..	..	..	26 33.67	39.92	1.00	.	4	7.12	18 34.93	10.28	5.51	27 14.59	40 0.72
5	9	8.0	21.8	35.1	..	..	..	..	28 48.95	39.92	1.00	.	6	7.48	28 3.46	10.61	6.23	29 29.87	49 30.50
6	9	52.9	6.4	20.0	..	..	..	..	30 33.71	39.93	1.00	.	3	8.26	14 13.25	11.23	5.16	31 14.64	23 35 39.64
7	9	..	13.2	26.7	..	..	..	..	31 40.45	39.94	1.00	.	9	11.35	44 45.55	11.50	7.56	32 21.39	24 6 14.61
8	5.6	..	..	..	..	3.8	..	..	31 50.11	39.94	1.00	.	10	9.37	48 46.95	11.52	7.88	32 31.05	24 10 16.35
9	10	..	51.8	5.1	..	..	..	..	16 34 18.94	+39.95	+1.00	.	5	4.2	−21 58.33	−12.11	−5.76	16 33 59.89	−23 43 26.20

ZONE 243. JUNE 18. H. BELT, −23° 8′. D₀=−22° 42′ 50″.

No.	Mag.	I.	II.	III.	IV.	V.	VI.	VII.	T.	a_1	a_2	MICROMETER.			i	d_1	d_2	Mean Right Ascension, 1850.0.	Mean Declination, 1850.0.
1	8.9	..	17.	32.5	46.	..	..	..	15 50 46.09	+19.96	+1.00	IV.	1	6.22	−3 12.62	−7.76	−4.43	15 51 7.05	−22 46 14.81
2	9	..	..	..	..	54.4	8.	..	51 40.81	19.96	1.00	V.	6	9.35	28 47.31	7.79	6.28	52 1.77	23 11 51.38
3	9	..	..	..	17.	30.6	..	..	53 16.98	19.96	1.00	V.	7	6.47	32 22.60	7.83	6.53	53 37.94	15 26.96
4	9	..	34.5	48.	2.	..	..	..	55 1.74	19.96	1.00	IV.	8	5.36	36 45.53	7.87	6.85	55 22.70	19 50.25
5	6	..	..	..	1.8	15.	..	..	56 1.58	19.96	1.00	V.	7	6.17	32 7.47	7.89	6.51	56 22.54	15 11.87
6	7	..	..	..	..	2.2	16.	..	56 48.69	19.96	1.00	VI.	6	9.1	28 30.03	7.91	6.25	57 9.65	11 34 19
7	9	..	..	..	..	..	..	15.	57 34.22	19.96	1.00	VII.	7	4.16	31 6.17	7.93	6.44	57 55.18	14 10.54
8	7	..	..	12.	25.5	39.	..	..	15 59 25.48	19.96	1.00	V.	7	9.33	33 46.30	7.98	6.62	15 59 46.44	23 16 50.90
9	9	..	..	..	..	12.5	26.	..	16 1 58.82	19.96	1.00	V.	1	10.45	5 25.21	8.03	4.59	16 2 19.78	22 48 27.83
10	10	..	..	..	38.	51.3	..	..	3 37.91	19.96	1.00	V.	3	6.31	13 15.23	8.07	5.15	3 58.87	22 56 18.45
11	9	..	..	..	..	..	9.3	23.	4 42.16	19.96	1.00	VII.	8	12.11	40 4.37	8.09	7.10	5 3.12	23 23 9.56
12	9	..	..	..	1.	14.5	..	..	7 1.01	19.96	1.00	V.	1	7.39	3 51.42	8.14	4.48	7 21.97	22 46 54.04
13	8	..	43.	56.3	10.	..	..	..	9 10.03	19.96	1.00	IV.	3	1.53	10 55.07	8.19	4.98	9 30.99	22 53 58.24
14	8	..	..	..	59.	12.3	26.	..	11 58.82	19.96	1.00	VI.	8	7.5	37 30.24	8.25	6.91	12 19.78	23 20 35.40
15	8	..	..	..	33.	46.7	..	..	13 33.12	19.96	1.00	V.	1	5.9	2 35.78	8.28	4.38	13 54.08	22 45 38.44
16	9	..	..	..	53.	6.5	..	..	14 52.97	19.96	1.00	V.	6	3.3	25 29.63	8.31	6.03	15 13.93	23 8 33.97
17	8	..	..	50.	4.	..	..	..	16 3.82	19.96	1.00	IV.	5	6.53	23 24.55	8.34	5.87	16 24.78	6 28.76
18	6	..	..	..	..	..	..	55.2	16 14.50	19.96	1.00	VII.	5	5.30	22 42.38	8.34	5.82	16 35.46	5 46.54
19	10	..	24.	37.4	51.2	..	..	..	16 31 51.15	+19.97	+1.00	IV.	4	7.1	−18 29.38	−8.63	−5.51	16 32 12.12	−23 1 33.52

CORRECTIONS.

Date.		Corr. of Clock.	Hourly rate.	m	n	c
1849.	h.	s.	s.	s.	s.	s.

INSTRUMENT READINGS.

Date.		Barom.	THERMOM. At.	Ex.
1849.	h. m.	In.	°	°

REMARKS.

(241) 56. Hor. thread assumed as 9 instead of 10.
(242) 5. Micrometer reading assumed as 7r.68 instead of 7r.48.
(242) 9. Minutes assumed as 33 instead of 34.
(243) 1. Time for T. II assumed as 19s instead of 17s.

ZONE 243. JUNE 18. B. BELT, −23° 8'. D₀=−22° 42' 50"—Continued.

No.	Mag.	SECONDS OF TRANSIT. I.	II.	III.	IV.	V.	VI.	VII.	T.	a_1	a_2	MICROMETER.			i	d_1	d_2	Mean Right Ascension, 1850.0.	Mean Declination, 1850.0.
									h. m. s.	s.	s.			r.	' "	"	"	h. m. s.	° ' "
20	9	..	..	14.	27.8	41.2	..	..	16 34 27.70	+19.98	+1.00	V.	2	4.45	− 7 22.88	− 8.66	− 4.71	16 34 48.68	−22 50 26.25
21	8	..	..	..	46.5	0.2	13.8	27.	35?46.53	19.98	1.00	VII.	2	11.47	10 55.18	8.68	4.96	36 7.51	22 53 58.82
22	10	55.	8.3	22.	..	..	..	..	40 35.61	19.98	1.00	III.	6	7.51	27 54.86	8.74	6.20	40 56.59	23 10 59.80
23	8	..	59.	12.4	26.	..	..	..	45 26.05	19.99	1.00	IV.	7	7.17	32 37.75	8.78	6.55	45 47.04	23 15 43.08
24	8	44.	57.3	11.	24.4	..	..	..	47 24.53	19.99	1.00	IV.	3	2.43	11 20.29	8.80	4.99	47 45.52	22 54 24.08
25	11	..	..	..	..	25.	..	..	48 11.43	19.99	1.00	V.	5	3.9	21 31.57	8.80	5.73	48 32.42	23 4 36.10
26	10	..	..	..	..	40.	53.5	..	49 26.40	19.99	1.00	VI.	10	6.40	47 17.54	8.81	7.65	49 47.39	30 24.00
27	9	..	..	59.	12.5	26.	..	..	51 12.48	19.99	1.00	V.	7	9.58	33 58.90	8.82	6.65	51 33.47	17 4.37
28	9	..	..	30.5	44.	..	..	..	52 44.09	20.00	1.00	IV.	4	11.13	20 36.45	8.82	5.66	53 5.09	23 3 40.93
29	9	..	..	44.	57.5	11.	..	..	53 57.53	20.00	1.00	V.	3	5.32	12 45.48	8.82	5.08	54 18.53	22 55 40.38
30	9	..	..	..	..	16.	29.7	..	55 2.45	20.00	1.00	VI.	6	6.33	27 15.42	8.83	6.16	55 23.45	23 10 20.41
31	11	..	35.4	49.	2.6	..	..	..	57 2.58	20.00	1.00	IV.	7	9.00	33 29.69	8.83	6.61	57 23.58	16 35.13
32	10	..	5.	19.	32.3	..	..	..	58 32.35	20.00	1.00	IV.	7	6.23	32 10.52	8.83	6.52	58 53.35	15 15.87
33	10	..	..	..	..	22.5	36.	..	16 59 8.84	20.01	1.00	VI.	4	6.40	18 18.65	8.83	5.49	16 59 29.85	23 1 22.97
34	11	..	..	..	50.	..	..	..	17 2 50.06	20.01	1.00	IV.	3	10.40	15 20.82	8.83	5.28	17 3 11.07	22 58 24.93
35	11	..	..	..	57.3	11.2	..	..	3 57.50	20.01	1.00	V.	4	9.27	19 42.96	8.83	5.60	4 18.51	23 2 47.39
36	11	..	..	..	..	19.4	33.	..	10 5.77	20.02	1.00	VI.	2	3.14	6 36.67	8.84	4.63	10 26.79	22 49 40.14
37	10	..	..	..	39.	52.2	..	..	11 38.86	20.02	1.00	V.	4	4.33	17 14.73	8.84	5.42	11 59.88	23 0 18.99
38	9.10	..	..	..	..	52.2	6.	..	12 38.73	20.03	1.00	VI.	9	6.16½	42 4.79	8.84	7.27	12 59.76	23 25 10.90
39	8.9	..	..	..	..	..	15.	28.4	13 47.75	20.03	1.00	VII.	2	6.59	8 29.95	8.84	4.75	14 8.78	22 51 33.54
40	10	55.	8.4	22.	35.8	..	..	..	16 35.68	20.03	1.00	IV.	4	7.24	18 40.98	8.85	5.52	16 56.71	23 1 45.35
41	10	..	57.	10.8	24.2	38.	..	..	17 19 24.28	+20.04	+1.00	V.	7	13.17	−35 39.24	− 8.85	− 6.80	17 19 45.32	−23 18 44.89

ZONE 244. JUNE 20. H. BELT, −21° 16'. D₀=−20° 51' 0".

No.	Mag.	I.	II.	III.	IV.	V.	VI.	VII.	T.	a_1	a_2	MICROMETER.			i	d_1	d_2	Mean Right Ascension, 1850.0.	Mean Declination, 1850.0.
1	10	..	..	..	..	14.	27.5	..	15 45 0.61	+24.44	+1.00	VI.	3	11.15	−15 38.33	−13.20	− 5.43	15 45 26.05	−21 6 56.96
2	9	..	..	..	38.2	52.	4.8	..	46 38.26	24.44	1.00	VI.	3	7.8	13 33.78	13.20	5.29	47 3.70	4 52.27
3	7	..	..	..	0.	13.	..	..	?48 59.85	24.44	1.00	IV.	3	3.5	11 31.38	13.20	5.17	49 25.29	21 2 49.75
4	10	..	..	..	..	..	..	56.5	49 16.49	24.44	1.00	VII.	2	3.17	6 38.04	13.20	4.88	49 41.93	20 57 56.12
5	11	..	..	..	..	21.	34.2	..	51 7.45	24.44	1.00	VI.	3	4.41	12 19.66	13.19	5.23	51 32.89	21 3 38.08
6	6.7	..	..	..	..	53.3	6.7	..	52 39.90	24.44	1.00	VI.	9	5.45	41 48.93	13.19	7.05	53 5.34	33 9.17
7	10	..	..	37.	50.2	..	..	..	54 50.26	24.44	1.00	IV.	8	9.48	38 52.59	13.18	6.87	55 15.70	30 12.64
8	8	..	1.	14.4	..	..	..	..	56 27.83	24.44	1.00	III.	7	10.14	34 6.98	13.17	6.56	56 53.27	25 26.71
9	11	..	..	20.	..	..	..	..	57 33.42	24.44	1.00	III.	8	10.28	39 12.73	13.16	6.89	57 58.86	30 32.78
10	7	..	..	..	..	..	..	15.	57 34.72	24.44	1.00	VII.	7	13.46	35 53.58	13.16	6.68	15 58 0.16	27 13.42
11	11	..	31.	44.5	..	..	..	..	15 59 57.88	24.44	1.00	III.	9	13.51	40 55.07	13.15	7.00	16 0 23.32	32 15.22
12	11	..	..	..	24.	..	..	..	16 1 24.04	24.44	1.00	IV.	5	7.21	23 38.66	13.14	5.91	1 49.48	21 14 57.71
13	11	..	..	..	52.	..	..	..	2 52.03	24.44	1.00	IV.	2	6.15½	8 8.34	13.12	4.96	3 17.47	20 59 26.42
14	10	..	..	52.4	..	..	..	..	4 5.83	24.44	1.00	III.	9	8.00	42 57.11	13.11	7.12	4 31.27	21 34 17.34
15	7	..	..	..	..	..	53.	6.2	4 26.10	24.44	1.00	VII.	2	8.50	9 25.93	13.11	5.04	4 51.54	0 44.08
16	9	..	..	59.	12.3	..	..	..	6 12.39	24.44	1.00	IV.	4	11.57	20 58.64	13.09	5.75	6 37.83	21 12 17.48
17	8	..	16.5	30.	..	..	..	..	7 43.36	24.44	1.00	III.	1	8.24½	4 14.36	13.08	4.72	8 8.80	20 55 32.16
18	9	..	..	..	29.3	..	..	..	8 29.14	24.44	1.00	IV.	9	12.10	40 4.19	13.07	6.95	8 54.58	21 31 24.21
19	10	..	..	..	22.2	..	..	..	9 22.26	24.44	1.00	IV.	4	13.29	21 45.02	13.07	5.79	9 47.70	13 3.88
20	10	..	..	..	13.	..	..	..	10 13.04	24.45	1.00	IV.	2	11.15	10 39.36	13.06	5.12	10 38.49	1 57.54
21	9	..	..	..	11.	..	..	..	11 11.06	24.45	1.00	IV.	3	14.7	17 5.18	13.05	5.51	11 36.51	8 23.74
22	7.8	..	..	..	..	..	..	5.2	16 11 24.91	+24.45	+1.00	VII.	5	6.20	−37 7.39	−13.04	− 6.76	16 11 50.36	−21 28 27.19

CORRECTIONS.

Date.		Corr. of Clock.	Hourly rate.	m	n	c
1849.	h.	s.	s.	s.	s.	s.

INSTRUMENT READINGS.

Date.		Barom.	THERMOM. At.	Ex.
1849.	h. m.	in.	°	°

REMARKS.

(244) 11. Hor. thread assumed as 8 instead of 9.
(244) 18. Hor. thread assumed as 8 instead of 9.

ZONE 244. JUNE 20. B. BELT, −21° 16′. D₀=−20° 51′ 0″—Continued.

No.	Mag.	SECONDS OF TRANSIT. I.	II.	III.	IV.	V.	VI.	VII.	T.	a_1	a_2	MICROMETER.			i	d_1	d_2	Mean Right Ascension, 1850.0.	Mean Declination, 1850.0.
									h. m. s.	s.	s.			r.	′ ″	″	″	h. m. s.	° ′ ″
23	10	. .	. .	. .	5.3	. .	. .	. .	16 13 51.90	+24.45	+1.00	IV.	6	4.32½	−26 14.81	−13.02	−6.07	16 13 17.35	−21 17 33.90
24	10	. .	. .	. .	. .	8.3	. .	. .	13 54.98	24.45	1.00	V.	1	8.35	4 19.66	13.01	4.73	14 20.43	20 55 37.40
25	10	. .	39.3	52.3	. .	. .	. .	. .	16 5.92	24.45	1.00	III.	6	9.52	28 55.88	12.99	6.23	16 31.37	21 20 15.10
26	9	. .	35.	48.2	. .	. .	. .	. .	20 1.72	24.45	1.00	III.	5	5.6	22 30.57	12.94	5.84	20 27.17	13 49.35
27	10	. .	. .	. .	. .	. .	55.	. .	20 28.08	24.45	1.00	VI.	3	10.45	15 23.21	12.93	5.40	20 53.53	6 41.54
28	4.5	9.	22.4	36.	49.3	. .	. .	. .	22 49.29	24.45	1.00	IV.	3	14.1	17 2.15	12.90	5.50	23 14.74	8 20.55
29	11	. .	. .	. .	15.2	. .	. .	. .	24 15.24	24.45	1.00	IV.	2	11.1	10 32.30	12.88	5.10	24 40.69	1 50.28
30	11	. .	42.8	56.	. .	. .	. .	. .	26 9.53	24.45	1.00	III.	9	6.10	42 1.64	12.84	7.08	26 34.98	33 21.56
31	11	. .	. .	25.	38.4	. .	. .	. .	31 38.40	24.46	1.00	IV.	7	3.58	30 57.40	12.77	6.36	32 3.66	22 16.53
32	10	. .	. .	29.	42.2	. .	. .	. .	33 42.33	24.46	1.00	IV.	3	3.26	11 41.97	12.73	5.17	34 7.79	2 59.87
33	10	. .	. .	. .	. .	7.2	20.5	. .	37 53.71	24.46	1.00	VI.	4	4.29	17 12.61	12.67	5.51	38 19.17	8 30.79
34	10	. .	. .	. .	. .	. .	. .	19.7	39 39.62	24.46	1.00	VII.	3	10.26	15 13.47	12.65	5.39	40 5.08	6 31.51
35	10	. .	12.2	55.6	. .	. .	. .	. .	42 9.03	24.47	1.00	III.	7	6.32	32 15.04	12.61	6.45	42 34.50	23 34.10
36	11	. .	. .	. .	. .	. .	. .	16.3	42 36.24	24.47	1.00	VII.	3	5.59	12 58.83	12.60	5.25	43 1.71	4 16.68
37	9	. .	. .	2.	15.4	. .	. .	. .	45 15.35	24.47	1.00	IV.	10	4.52	46 23.24	12.56	7.37	45 40.82	37 43.17
38	7.8	. .	. .	. .	11.	24.	. .	. .	46 10.80	24.47	1.00	V.	6	7.57	27 57.89	12.55	6.18	46 36.27	19 16.62
39	9	. .	. .	. .	22.5	36.	. .	. .	47 22.47	24.47	1.00	V.	9	5.50	41 51.55	12.53	7.07	47 47.94	33 11.15
40	9	. .	. .	. .	. .	. .	46.	. .	48 19.23	24.47	1.00	VI.	9	3.25	40 38.33	12.52	6.99	48 44.70	31 57.84
41	9	. .	. .	. .	55.	. .	. .	. .	49 55.03	24.47	1.00	IV.	5	9.17	24 37.16	12.49	5.98	50 20.50	15 55.63
42	8	. .	41.	54.5	. .	. .	. .	. .	51 7.87	24.47	1.00	III.	5	4.48	22 21.49	12.47	5.83	51 33.34	13 39.79
43	10	. .	53.	. .	20.3	. .	. .	. .	54 20.10	24.48	1.00	IV.	4	10.27	20 13.26	12.42	5.70	54 45.58	11 31.38
44	7.8	9.	22.	35.4	. .	. .	. .	. .	56 48.96	24.48	1.00	IV.	6	11.28	29 44.31	12.38	6.28	57 14.44	21 2.97
45	10	. .	. .	. .	. .	. .	. .	32.7	16 56 52.64	24.48	1.00	VII.	2	14.5	12 4.79	12.38	5.18	16 57 18.12	3 22.34
46	11	. .	. .	. .	24.3	. .	. .	. .	17 0 26.15	24.49	1.00	IV.	9	10.52	44 23.87	12.32	7.24	17 0 51.64	35 43.43
47	10	. .	29.	42.5	. .	. .	. .	. .	1 55.87	24.49	1.00	III.	4	8.56	19 27.35	12.30	5.66	2 21.36	10 45.31
48	9	. .	. .	27.	40.3	. .	. .	. .	2 40.29	24.49	1.00	IV.	9	11.14	44 34.96	12.28	7.26	3 6.78	35 54.50
49	7.8	. .	. .	. .	. .	29.	42.	. .	3 15.39	24.49	1.00	VI.	7	9.41	33 50.23	12.27	6.55	3 40.86	25 9.05
50	10	. .	. .	44.	57.3	. .	. .	. .	4 57.39	24.49	1.00	IV.	4	8.43½	19 21.07	12.25	5.64	5 22.88	10 38.93
51	9	. .	. .	. .	43.5	. .	. .	. .	5 43.57	24.49	1.00	IV.	4	10.17	20 8.22	12.23	5.70	6 9.06	11 26.15
52	9	. .	. .	. .	0.0	. .	. .	. .	7 59.85	24.50	1.00	IV.	10	10.37	49 17.21	12.19	7.57	8 25.35	40 36.97
53	10	. .	48.	. .	. .	. .	. .	. .	9 14.83	24.50	1.00	II.	7	9.59	33 59.31	12.17	6.57	9 40.33	21 25 18.05
54	5.6	55.2	8.3	22.	. .	. .	. .	. .	11 35.26	24.50	1.00	III.	1	11.7	5 36.30	12.13	4.78	12 0.76	20 56 53.21
55	11	. .	. .	. .	. .	. .	. .	38.5	11 58.25	24.50	1.00	VII.	7	7.25	32 41.49	12.12	6.48	12 23.75	21 24 0.09
56	9	. .	. .	58.3	11.5	. .	. .	. .	14 11.54	24.51	1.00	IV.	9	7.23	42 38.47	12.07	7.13	14 37.05	33 57.67
57	9	. .	. .	. .	. .	. .	. .	3.5	14 23.20	24.51	1.00	VII.	6	8.39	38 17.52	12.07	6.85	14 48.71	29 36.44
58	7	. .	. .	. .	. .	. .	. .	56.3	16 18.10	24.51	1.00	VII.	6	4.59½	26 28.13	12.03	6.09	15 43.61	17 46.25
59	8	. .	. .	5.	18.3	. .	. .	. .	17 18.33	24.51	1.00	IV.	7	. .	29	12.01	6.25	17 43.84	20
60	10	. .	12.3	25.5	. .	. .	. .	. .	21 39.02	24.52	1.00	III.	7	3.00	30 28.13	11.91	6.34	22 4.54	21 46.38
61	11	. .	. .	. .	. .	15.	. .	. .	22 1.59	24.52	1.00	V.	8	2.40	35 16.75	11.90	6.66	22 27.11	26 35.31
62	11	. .	. .	. .	. .	48.3	2.	. .	24 35.01	24.52	1.00	VI.	3	7.4	13 31.77	11.83	5.28	25 0.53	4 46.88
63	10	. .	. .	. .	. .	. .	. .	16.	25 35.92	24.53	1.00	VII.	3	10.25	15 12.96	11.80	5.38	26 1.45	6 30.14
64	9	. .	. .	. .	. .	. .	. .	5.3	26 25.22	24.53	1.00	VII.	3	9.18	14 39.46	11.78	5.35	26 50.75	5 56.59
65	9	. .	. .	5.	18.3	. .	. .	. .	28 18.39	24.53	1.00	IV.	4	6.22	18 9.72	11.73	5.56	28 43.92	9 27.01
66	5.6	. .	. .	47.4	0.6	14.	. .	. .	34 0.63	24.54	1.00	V.	9	11.52	44 54.09	11.57	7.29	34 26.17	36 12.95
67	11	. .	. .	. .	. .	22.8	. .	. .	35 9.39	24.54	1.00	V.	7	5.33	31 45.28	11.54	6.43	35 34.93	23 3.25
68	9	. .	. .	32.4	. .	. .	. .	. .	36 45.82	24.55	1.00	III.	9	3.2	40 26.84	11.49	7.00	37 11.37	31 45.33
69	10	. .	. .	. .	. .	. .	28.	. .	38 1.09	24.55	1.00	VI.	4	3.47	16 51.43	11.46	5.48	38 26.64	21 8 8.37
70	10	. .	. .	36.2	49.5	. .	. .	. .	41 49.56	24.56	1.00	IV.	1	7.30	3 46.91	11.35	4.66	42 15.12	20 55 2.92
71	9	. .	. .	. .	. .	39.	52.4	. .	17 42 25.55	+24.56	+1.00	VI.	2	9.9	− 9 35.70	−11.33	−5.03	17 42 51.11	−21 0 52.06

CORRECTIONS.

Date.		Corr. of Clock.	Hourly rate.	m	n	c
1849.	h.	s.	s.	s.	s.	s.

INSTRUMENT READINGS.

Date.		Barom.	THERMOM. At.	Ex.
1849.	h. m.	in.	°	°

REMARKS.

(244) 23. Transit over T. V assumed as recorded over T. IV.
(244) 46. Transit over T. III assumed as 26ˢ.3 instead of 24ˢ.3. (See Mural Z., 1849, June 18, and Arg. 213. 71.)
(244) 58. Minutes assumed as 15 instead of 16.

ZONE 244. JUNE 20. B. BELT, −21° 16′. D_o = −20° 51′ 0″—Continued.

No.	Mag.	SECONDS OF TRANSIT. I.	II.	III.	IV.	V.	VI.	VII.	T.	a_1	a_4	MICROMETER.			i	d_1	d_2	Mean Right Ascension, 1850.0.	Mean Declination, 1850.0.
									h. m. s.	s.	s.			r.	′ ″	″	″	h. m. s.	° ′ ″
72	10	..	..	..	..	55.	8.5	..	17 43 41.64	+24.56	+1.00	VI.	7	3.47	−30 51.73	−11.29	−6.37	17 44 7.20	−21 22 9.39
73	9	..	..	3.5	17.	..	..	..	45 16.97	24.57	1.00	IV.	2	5.45	7 52.96	11.24	4.90	45 42.54	20 59 9.10
74	9	..	41.	54.3	7.7	..	..	..	17 50 7.76	+24.58	+1.00	IV.	2	10.45½	−10 24.49	−11.10	−5.06	17 50 33.34	−21 1 40.65

ZONE 245. JUNE 22. B. BELT, −21° 53′. D_o = −21° 28′ 30″.

No.	Mag.	I.	II.	III.	IV.	V.	VI.	VII.	T.	a_1	a_4	MICROMETER.			i	d_1	d_2	Mean Right Ascension, 1850.0.	Mean Declination, 1850.0.
1	9	..	..	..	..	59.5	12.5	..	15 54 45.75	+28.89	+1.00	VI.	1	2.58	− 1 29.62	−11.10	−4.53	15 55 15.62	−21 30 15.25
2	10	..	..	..	..	..	2.5	..	56 35.63	28.88	1.00	VII.	9	6.44	42 18.52	11.12	7.11	57 5.51	22 11 6.75
3	10	..	..	..	..	..	24.3	..	57 57.22	28.88	1.00	VI.	2	10.37	10 20.08	11.14	5.08	58 27.10	21 39 6.30
4	10	..	..	..	..	..	33.	..	15 59 6.06	28.88	1.00	VI.	7	4.43	31 19.97	11.15	6.39	15 59 35.94	22 0 7.51
5	8.9	..	..	22.	35.3	..	..	..	16 0 35.37	28.88	1.00	IV.	7	5.30	31 43.80	11.17	6.42	16 1 5.25	22 0 31.39
6	8	..	..	..	..	..	25.	..	0 57.97	28.88	1.00	VI.	4	3.16	16 35.80	11.17	5.47	1 27.85	21 45 22.44
7	9.10	..	..	..	..	..	..	8.4	1 28.11	28.88	1.00	VII.	4	8.27	19 12.45	11.17	5.64	1 57.99	21 47 59.26
8	9	..	..	..	..	..	..	9.2	2 28.71	28.88	1.00	VII.	9	2.38	40 14.47	11.18	6.96	2 58.59	22 9 2.61
9	9	..	..	..	..	1.5	15.	..	3 48.06	28.88	1.00	VI.	8	5.57½	36 56.23	11.19	6.76	4 17.94	22 5 44.18
10	8.7	..	3.	16.	29.7	..	..	..	5 29.69	28.88	1.00	IV.	7	3.56	30 56.40	11.21	6.37	5 59.57	21 59 43.98
11	9	..	58.4	11.5	..	..	..	..	8 25.15	28.88	1.00	III.	1	5.35	2 48.89	11.21	4.60	8 55.03	21 31 34.73
12	10	..	..	11.5	24.8	38.	..	..	11 24.72	28.88	1.00	V.	9	11.48	44 52.07	11.26	7.28	11 54.60	22 13 40.61
13	10	..	..	14.	27.3	..	..	..	14 27.37	28.88	1.00	IV.	7	6.9	32 3.47	11.29	6.44	14 57.25	22 0 51.20
14	7.8	..	..	15.4	29.	..	..	..	15 28.96	28.88	1.00	IV.	5	8.20	24 8.42	11.29	5.94	15 58.84	21 52 55.65
15	7.8	11.	24.5	..	..	..	..	..	17 51.34	28.88	1.00	II.	4	4.53	17 24.70	11.31	5.52	18 21.22	46 11.53
16	8.9	..	..	51.5	5.	..	..	..	26 4.99	28.88	1.00	IV.	1	9.27	4 45.90	11.36	4.73	25 34.87	33 31.99
17	8	..	..	59.4	13.	..	..	..	28 12.96	28.88	1.00	IV.	3	11.47	15 54.60	11.37	5.43	28 42.84	44 41.40
18	10	..	..	..	34.5	..	..	..	29 34.56	28.88	1.00	IV.	3	11.14	15 37.96	11.38	5.41	30 4.44	21 44 24.75
19	7.8	..	..	35.7	49.1	..	..	..	34 49.06	28.89	1.00	IV.	9	12.41½	45 19.07	11.40	7.31	35 18.95	22 14 7.78
20	10	..	35.2	48.5	..	..	..	..	38 2.05	28.89	1.00	II.	5	9.57	24 57.20	11.41	6.00	38 31.94	21 53 44.61
21	11	..	..	..	32.5	..	..	..	38 32.54	28.89	1.00	IV.	5	9.17	24 37.16	11.41	5.97	39 2.43	53 24.54
22	8	..	..	55.	8.4	22.	..	..	40 8.50	28.89	1.00	V.	2	2.33	6 16.12	11.42	4.82	40 38.39	35 2.36
23	10	..	..	..	25.	..	..	..	41 25.00	28.89	1.00	IV.	6	7.24	27 41.28	11.42	6.17	41 54.89	21 56 28.87
24	10	..	..	19.7	33.1	..	..	..	42 33.11	28.89	1.00	IV.	7	11.15	34 37.76	11.43	6.60	43 3.00	22 3 25.79
25	7.8	..	45.	..	..	..	..	..	45 11.94	28.89	1.00	II.	4	7.15	18 36.31	11.43	5.60	45 41.83	21 47 23.34
26	7.8	..	..	..	..	..	38.	..	45 10.91	28.89	1.00	VI.	2	7.55	8 58.33	11.43	4.98	45 40.80	37 44.79
27	9	..	51.2	4.7	..	..	..	..	47 18.15	28.90	1.00	III.	1	8.59	4 31.76	11.43	4.70	47 48.05	33 17.89
28	8	..	..	..	14.6	..	..	..	48 14.62	28.90	1.00	IV.	1	6.31	3 17.16	11.44	4.62	48 44.52	32 3.22
29	10	..	..	..	32.	..	..	..	50 32.02	28.90	1.00	IV.	6	4.4	26 0.44	11.44	6.06	51 1.92	54 47.94
30	10	..	..	..	..	..	..	33.	50 52.65	28.90	1.00	VII.	6	2.41	25 18.28	11.44	6.02	51 22.55	54 5.74
31	8	21.	34.	47.3	1.	..	..	..	16 56 1.04	28.90	1.00	IV.	5	5.39½	22 47.49	11.43	5.86	16 56 30.94	51 34.78
32	10	..	..	1.	14.	..	..	..	17 1 14.26	28.91	1.00	IV.	5	8.43	24 20.02	11.43	5.95	17 1 44.17	53 7.40
33	10	..	..	..	..	..	29.	..	2 2.05	28.91	1.00	VI.	6	11.19	29 39.65	11.42	6.30	2 31.96	58 27.37
34	9	..	28.5	41.7	..	..	..	..	6 55.30	28.92	1.00	III.	2	13.27	11 45.88	11.41	5.16	7 25.22	40 32.45
35	9	..	..	56.	9.3	..	..	..	23 9.41	28.94	1.00	IV.	3	13.29	16 46.02	11.33	5.48	23 39.35	21 45 32.83
36	9	..	..	..	..	..	29.5	43.	24 2.57	28.94	1.00	VII.	7	11.18	34 38.97	11.32	6.61	24 32.51	22 3 26.90
37	6.7	..	..	..	..	0.	13.5	..	25 46.54	28.95	1.00	VI.	6	6.46	27 22.00	11.31	6.15	26 16.49	21 56 9.46
38	7	..	47.	0.3	..	..	..	..	28 13.85	28.95	1.00	III.	4	10.22	20 10.71	11.29	5.69	29 43.80	48 57.69
39	6	..	..	43.	56.3	9.5	..	..	33 56.30	28.96	1.00	IV.	2	4.58	7 29.26	11.24	4.87	34 26.26	36 15.37
40	10	..	..	..	..	..	17.	..	34 49.92	28.96	1.00	VI.	2	11.28	10 45.78	11.24	5.08	35 19.88	39 32.10
41	10	..	..	..	..	..	..	22.7	35 42.49	28.96	1.00	VII.	2	9.10	9 36.03	11.23	5.00	36 12.45	38 22.26
42	8	..	22.2	36.	19.2	..	..	..	17 46 49.27	+28.99	+1.00	IV.	6	5.20	−26 38.75	−11.13	−6.10	17 47 19.26	−21 55 25.98

CORRECTIONS.

Date.		Corr. of Clock.	Hourly rate.	m	n	c
1849.	h.	s.	s.	s.	s.	s.

INSTRUMENT READINGS.

Date.		Barom.	THERMOM. At.	THERMOM. Ex.
1849.	h. m.	in.	°	°

REMARKS.

(245) 16. Minutes assumed as 26 instead of 25, to agree with Mural, 1849, and Argelander.
(245) 37. An error of 10s in one of the transits, supposed in T. V. If T. VI be in error, T=56s.54.
(245) 38. Minutes assumed as 29 instead of 28.

ZONE 245. JUNE 22. B. BELT, −21° 53′. D_0 = −21° 28′ 30″—Continued.

No.	Mag.	Seconds of Transit. I.	II.	III.	IV.	V.	VI.	VII.	T.	a_1	a_2	Micrometer.			i	d_1	d_2	Mean Right Ascension, 1850.0.	Mean Declination, 1850.0.
									h. m. s.	s.	s.			r.	′ ″	″	″	h. m. s	° ′ ″
43	9	..	..	..	..	..	38.	..	17 48 11.14	+28.99	+1.00	VI.	9	9.25	−43 39.86	−11.12	−7.24	17 48 41.13	−22 12 28.22
44	8	26.	39.2	..	..	..	..	..	52 6.30	29.00	1.00	II.	8	8.51	38 23.72	11.07	6.88	52 36.30	22 7 11.67
45	9	..	..	..	..	35.	..	..	52 50.03	29.00	1.00	V.	6	9.00	28 29.66	11.07	6.22	53 20.03	21 57 16.95
46	9	..	..	..	..	29.5	..	..	54 16.02	29.01	1.00	V.	9	12.28	45 12.23	11.04	7.34	54 46.03	22 14 0.61
47	9	..	..	..	52.5	..	..	..	55 52.54	29.01	1.00	IV.	2	12.57	11 30.79	11.02	5.12	56 22.55	21 40 16.93
48	9	..	..	..	..	..	0.	..	56 33.11	29.01	1.00	VI.	8	9.8	38 32.29	11.01	6.89	57 3.12	22 7 20.19
49	8	..	..	..	..	..	..	57.5	57 17.06	29.01	1.00	VII.	7	10.53	34 26.37	11.01	6.60	57 47.07	22 3 13.98
50	9	..	..	..	..	..	..	51.	58 10.67	29.02	1.00	VII.	5	7.7	23 31.31	10.99	5.90	58 40.69	21 52 18.20
51	9	..	..	..	..	..	39.	52.	17 59 11.81	29.02	1.00	VII.	9	5.39	41 45.74	10.98	7.11	17 59 41.83	22 10 33.83
52	9	..	..	57.	11.	..	..	..	18 1 10.65	29.02	1.00	IV.	10	5.59	46 57.02	10.95	7.47	18 1 40.67	22 15 45.44
53	10	..	..	29.7	43.	..	..	..	3 43.11	29.03	1.00	IV.	4	3.27	16 41.47	10.91	5.46	4 13.14	21 45 27.84
54	6.7	..	..	..	45.	58.2	..	..	4 44.92	29.03	1.00	V.	4	2.14	16 4.64	10.89	5.42	5 14.95	21 44 50.95
55	9	..	..	..	..	..	1.6	15.	6 48.11	29.04	1.00	VII.	8	3.29	35 41.18	10.86	6.70	7 18.15	22 4 28.74
56	8	..	..	..	21.	..	..	..	9 21.03	29.04	1.00	IV.	2	4.2	7 1.03	10.81	4.82	9 51.07	21 35 46.66
57	8	..	..	..	25.	..	..	..	12 24.92	29.05	1.00	IV.	8	4.33	36 13.76	10.76	6.73	12 54.97	22 5 1.25
58	8	..	..	..	4.	17.3	..	..	14 3.97	29.06	1.00	V.	3	10.35	15 18.27	10.73	5.37	14 34.03	21 44 4.37
59	10	..	..	..	49.4	..	..	..	15 49.48	29.06	1.00	IV.	4	10.10	20 4.69	10.70	5.68	16 19.54	48 51.07
60	10	..	..	..	57.3	..	..	..	17 57.38	29.07	1.00	IV.	4	9.59	19 59.14	10.66	5.67	18 27.45	48 45.47
61	10	..	..	..	..	..	24.	..	19 56.96	29.07	1.00	VI.	3	9.46	14 53.46	10.62	5.35	20 27.03	43 39.43
62	9	..	..	..	..	..	31.	..	21 4.00	29.08	1.00	VI.	5	4.14	22 4.25	10.60	5.80	21 34.08	21 50 50.65
63	9	..	..	59.	12.4	..	..	..	23 12.39	29.08	1.00	IV.	9	2.59	45 26.26	10.56	7.36	23 42.47	22 14 14.17
64	11	..	..	..	..	4.	..	..	24 50.61	29.09	1.00	V.	3	1.58	10 57.56	10.52	5.08	25 20.70	21 39 43.16
65	9	..	..	..	..	..	3.	..	25 36.10	29.09	1.00	VI.	8	5.27	36 40.85	10.51	6.77	26 6.19	22 5 28.13
66	9	..	31.	44.3	..	..	..	..	27 57.85	29.10	1.00	III.	2	6.18	8 9.57	10.46	3.88	28 27.95	21 36 53.91
67	6	..	..	..	..	38.5	..	..	28 25.15	29.10	1.00	V.	1	4.32	2 17.13	10.45	4.51	28 55.25	21 31 2.09
68	9	..	..	45.	58.7	..	..	..	29 58.55	29.10	1.00	IV.	8	5.22	36 38.46	10.42	6.76	30 28.65	22 5 25.64
69	9	..	..	56.2	10.	..	..	..	18 31 9.84	+29.11	+1.00	IV.	2	7.36	− 8 48.94	−10.40	−4.93	18 31 39.95	−21 37 34.27

CORRECTIONS.

Date.		Corr. of Clock.	Hourly rate.	m	n	c
1849.	h.	s.	s.	s.	s.	s.

INSTRUMENT READINGS.

Date.		Barom.	Thermom. At.	Thermom. Ex.
1849.	h. m.	in.	°	°

REMARKS.

(245) 45. Transit over T. III assumed as 53^m 3^s.5 instead of 52^m 35^s.
(245) 55. Transits over T.'s V and VI assumed as recorded over T.'s VI and VII.
(245) 63. Hor. thread assumed as 10 instead of 9.

www.ingramcontent.com/pod-product-compliance
Lightning Source LLC
Chambersburg PA
CBHW021513210326
41599CB00012B/1237
9783337030704